T0141948

# Pathways in Mathematics

**Series Editors**

T. Hibi
Toyonaka, Japan

W. König
Berlin, Germany

J. Zimmer
Bath, United Kingdom

Each "Pathways in Mathematics" book offers a roadmap to a currently well developing mathematical research field and is a first-hand information and inspiration for further study, aimed both at students and researchers. It is written in an educational style, i.e., in a way that is accessible for advanced undergraduate and graduate students. It also serves as an introduction to and survey of the field for researchers who want to be quickly informed about the state of the art. The point of departure is typically a bachelor/masters level background, from which the reader is expeditiously guided to the frontiers. This is achieved by focusing on ideas and concepts underlying the development of the subject while keeping technicalities to a minimum. Each volume contains an extensive annotated bibliography as well as a discussion of open problems and future research directions as recommendations for starting new projects

More information about this series at http://www.springer.com/series/15133

Ondřej Došlý • Julia Elyseeva •
Roman Šimon Hilscher

# Symplectic Difference Systems: Oscillation and Spectral Theory

 Birkhäuser

Ondřej Došlý (deceased)
Department of Mathematics and Statistics
Faculty of Science
Masaryk University
Brno, Czech Republic

Julia Elyseeva
Department of Applied Mathematics
Moscow State Technological University
"STANKIN"
Moscow, Russia

Roman Šimon Hilscher
Department of Mathematics and Statistics
Faculty of Science
Masaryk University
Brno, Czech Republic

ISSN 2367-3451      ISSN 2367-346X   (electronic)
Pathways in Mathematics
ISBN 978-3-030-19375-1      ISBN 978-3-030-19373-7   (eBook)
https://doi.org/10.1007/978-3-030-19373-7

Mathematics Subject Classification (2010): 39A21, 39A12, 47B39

This book is published under the imprint Birkhäuser, www.birkhauser-science.com, by the registered company Springer Nature Switzerland AG.
The registered company address is: Gewerbestrasse 11, 6330 Cham, Switzerland

# Preface

The principal concern of the book is the qualitative theory of *symplectic difference systems*. These are the first-order systems

$$y_{k+1} = \mathcal{S}_k y_k, \qquad \text{(SDS)}$$

where $y_k \in \mathbb{R}^{2n}$ and $\mathcal{S}_k \in \mathbb{R}^{2n \times 2n}$ are symplectic matrices, i.e.,

$$\mathcal{S}_k^T \mathcal{J} \mathcal{S}_k = \mathcal{J}, \quad \mathcal{J} = \begin{pmatrix} 0 & I \\ -I & 0 \end{pmatrix}.$$

Symplectic difference systems are natural discrete counterparts of the linear Hamiltonian *differential* systems

$$y' = \mathcal{J} \mathcal{H}(t) \, y, \qquad \text{(LHS)}$$

where the coefficient $\mathcal{H}(t) \in \mathbb{R}^{2n \times 2n}$ is a piecewise continuous and symmetric matrix, i.e., $\mathcal{H}^T(t) = \mathcal{H}(t)$. A common feature of (SDS) and (LHS) is that their fundamental matrix is symplectic whenever it has this property at an initial condition. Therefore, (SDS) and (LHS) are the most general linear difference and differential systems, which have symplectic fundamental matrices.

Referring to the linear Hamiltonian differential system (LHS), the classical qualitative theory of (LHS) is deeply developed; basic as well as advanced results of this theory can be found, for example, in the monographs [70, 205, 248, 250, 328] by Coppel, Kratz, Reid, and Yakubovich and in the recent monograph [203] by Johnson, Obaya, Novo, Núñez, and Fabbri. Regarding symplectic difference systems (SDS), the basic theory of (SDS) is presented in the monograph [16] from 1996 by Ahlbrandt and Peterson. As far as we know, there is no book or a survey paper presenting in a unified way the results of the qualitative theory of (SDS), which were obtained in this rapidly developing area in the last more than 20 years. This book represents an attempt to fill in this gap by providing numerous results, methods, and citations, which are till now scattered only in journal papers. Our aim is to cover both the traditional and the most current topics in the oscillation and spectral theory of symplectic systems, which are in the heart of the present research work.

The book is divided into six chapters. The first one is motivating and contains the results concerning the oscillation properties of the second-order Sturm-Liouville difference equations (being the simplest symplectic difference systems) and linear Hamiltonian differential systems (LHS). It also contains the elements of the discrete calculus of variations and optimal control theory, in which symplectic systems arise in very natural way as Jacobi systems in the second-order optimality conditions. Another motivating factor for studying symplectic difference systems originates in the classical Hamiltonian mechanics. In addition, for an easier reference, we present in this chapter an overview of the main tools from matrix analysis needed in the book, in particular about symplectic matrices, the Moore-Penrose pseudoinverse matrices, and certain orthogonal projectors.

Chapter 2 starts with an introduction to the theory of symplectic difference systems. We present the main notions and basic methods for the investigation of symplectic difference systems. We introduce the concept of a conjoined basis, the notion of the nonexistence of focal points, the Riccati difference equation and inequality, and the Picone formula and relate these concepts to the positivity and nonnegativity of a discrete quadratic functional. These results are easily connected to the necessary and sufficient optimality conditions presented in Chap. 1. We also investigate the existence of extremal solutions of (SDS) at infinity, called the recessive and dominant solutions of (SDS) at infinity, under a certain controllability assumption and present basic transformation theory of symplectic systems. We explain in details the role of these concepts in the oscillation theory of system (SDS).

In Chap. 3, we present one of the main tools of the modern oscillation theory of (SDS), namely, a concept of the comparative index of two matrices (which can be regarded as conjoined bases of (SDS)). It turned out to be very useful tool for solving several open problems in the oscillation theory of (SDS). This chapter summarizes basic properties of the comparative index, which will be needed in the subsequent chapters.

In Chap. 4, we introduce the second main concept used in this book, a concept of the multiplicity of a focal point of a conjoined basis of (SDS). During the past years, it was a central problem of the oscillation theory of (SDS), how to count the numbers of focal point of a conjoined basis between two consecutive integers. This problem was resolved in [208] and opened a new area for the investigation of (SDS). The next parts of this chapter relate the multiplicities of focal points with the comparative index. This can be regarded as an essential contribution to the oscillation theory of (SDS). The main results in this chapter also include, among others, the Sturmian separation and comparison theorems and the transformation theory for symplectic systems, which are presented in the form of explicit relations between the multiplicities of focal points. These results now serve as an inspiration for the progress in the investigation of the oscillation theory for continuous time linear Hamiltonian systems (LHS).

Chapter 5 concentrates on the second main topic of the book, the eigenvalue and spectral theory of (SDS). We investigate the distribution of the (finite) eigenvalues of various boundary value problems associated with system (SDS). The main essence of this chapter is that we consider the nonlinear dependence on the spectral parameter in the coefficient matrix, as well as in the boundary conditions. We prove the oscillation theorems for discrete symplectic eigenvalue problems with various boundary conditions (Dirichlet, separated, jointly varying), which relate the number of finite eigenvalues with the number of focal points of a specific conjoined basis of the system. As a special case, we then obtain traditional results, in which the dependence on the spectral parameter is linear, including the variational characterization of the finite eigenvalues and the Rayleigh principle. We also present extensions of the basic theory to the case, when the coefficients in (SDS) and/or in the boundary conditions may oscillate with respect to the spectral parameter.

In Chap. 6, we collect various additional topics from the oscillation and spectral theory of (SDS). We present the relative oscillation theory for symplectic difference systems, as well as the inequalities and interlacing properties for (finite) eigenvalues of symplectic boundary value problems. We also investigate a nonoscillatory system (SDS) on an unbounded interval without any eventual controllability assumption. This leads to an extensive theory of conjoined bases of (SDS) and, in particular, to a new theory of recessive and dominant solutions of (SDS) at infinity. This yields—together with a suitable application of the comparative index theory—to a new singular Sturmian theory for symplectic systems.

The last section in each chapter contains notes about the presented results and additional references regarding the studied topics. The bibliography includes also several references for further reading about related topics, in particular from the oscillation and spectral theory of linear Hamiltonian differential system (LHS).

The intended readership of the book includes the researchers and graduate students in pure and applied mathematics, who are interested in topics related to discrete symplectic and Hamiltonian systems, discrete oscillation theory, and spectral theory of difference equations and systems, covering also Sturm-Liouville difference equations of higher order. Researchers and students in matrix analysis will also benefit from this book by understanding new applications of the theory of matrices in this field, in particular of the Moore-Penrose pseudoinverse matrices (or the generalized inverses), orthogonal projectors, and (symplectic) matrix factorizations. The educational aim of the book is covered by including a variety of methods from the discrete oscillation and eigenvalue theory, which are particularly useful in the theory of symplectic and Hamiltonian systems. We demonstrate the utility of the presented methods by including different proofs of the same result at several occasions. We also support the understanding of the main results and important notions by illustrating examples. The methods can be followed from a relatively simple introduction to advanced constructions and applications, which are on the edge of the current research in this field. Therefore, this book may be used as a reference literature for a graduate course in the methods of the discrete oscillation and spectral theory.

This monograph is written by three experts in the oscillation and spectral theory of symplectic difference systems, who follow the development of this theory since its first steps more than 20 years ago. The original idea about this book came from O. Došlý and J. Elyseeva in December 2013, who initiated the translation of Chapters 1–4 of her monograph [121], which in revised and extended form became constituent parts of this book. In July 2014, they invited R. Šimon Hilscher to join the author team with the aim to prepare a monograph covering the state of the art of the oscillation and spectral theory of symplectic difference systems. It is sad to know that Professor Došlý suddenly passed away in November 2016. The remaining two authors wish to acknowledge the fundamental contribution of Professor Došlý of this monograph by completing it under his coauthorship (Fig. 1).

The authors wish to thank the Czech Science Foundation for the support of this work provided through the grant 16–00611S. This support allowed J. Elyseeva to enjoy friendly environment and hospitality of the Department of Mathematics and Statistics (Faculty of Science of Masaryk University) during the years 2017–2018 to write substantial parts of Chaps. 5 and 6 and to prepare the final version of this monograph. The second author thanks the management of the Moscow State University of Technology Stankin, in particular the chief of the Department of Applied Mathematics Professor Ludmila Uvarova, for their invaluable assistance in solving organizational issues related to the work on the monograph. We also thank our colleagues from the team of Mathematical Analysis at the Department

**Fig. 1** Professor Došlý lecturing on Sturm-Liouville difference equations and symplectic systems on a seminar on differential equations (Pálava, 2002)

of Mathematics and Statistics for creating a constant positive environment for our work. We thank especially Peter Šepitka for his useful comments about parts of Chaps. 1 and 6. We cordially thank Clemens Heine and Luca Sidler from Birkhäuser Verlag for their editorial help and for promoting the book to Pathways in Mathematics. We also thank Rajeswari Rajkumar from Springer for professional typesetting of the final version of the book. Finally, our very special thanks belong to Zuzana Došlá for her continuous support and encouragement to complete this monograph.

Brno, Czech Republic        Ondřej Došlý (deceased)
Moscow, Russia            Julia Elyseeva
Brno, Czech Republic      Roman Šimon Hilscher
December 2018

# Contents

# List of Figures

# Chapter 1
# Motivation and Preliminaries

In this chapter we describe main motivating factors for the investigation of symplectic difference systems in this book. This motivation comes from several sources, mainly from (i) a generalization of the theory of second-order Sturm-Liouville difference equations, (ii) discrete variational analysis, (iii) (classical and discrete) Hamiltonian mechanics, (iv) discrete analogy of the theory of linear Hamiltonian differential systems, and (v) numerical methods for Hamiltonian differential systems preserving the symplectic structure. Some of these topics are covered in the next sections in order to motivate the subsequent detailed study of symplectic difference systems. Some results (in particular those about Hamiltonian mechanics and linear Hamiltonian differential systems) are presented without proofs, as they serve mainly for comparison with the corresponding discrete-time theory. At the end of this chapter, we also provide an overview of matrix analysis needed for this work, in particular about symplectic matrices, the Moore-Penrose pseudoinverse matrices, and symplectic matrix valued functions.

## 1.1 Short History of Symplectic Difference Systems

Symplectic difference systems (SDS) originate, to our knowledge, in two main branches of mathematics. The first one is the numerical analysis of Hamiltonian differential systems in the works [140, 142, 143, 220] by K. Feng and his collaborators. According to [140, pg. 18], symplectic difference systems "represent a proper way, i.e., the Hamiltonian way, for computing the Hamiltonian dynamics." The research in this area has been recently summarized in the book [141]. The second source of symplectic difference systems can be identified in the representation of continued fractions in the works [8, 10] by C. D. Ahlbrandt; see also the references about continued fractions in [16].

© Springer Nature Switzerland AG 2019
O. Došlý et al., *Symplectic Difference Systems: Oscillation and Spectral Theory*, Pathways in Mathematics, https://doi.org/10.1007/978-3-030-19373-7_1

The concept of a symplectic difference system, as we consider in this work, was introduced in 1996 in the book [16] by C. D. Ahlbrandt and A. Peterson. During that time M. Bohner created in [39] the theory of general linear Hamiltonian difference systems, which then essentially led to the final establishment of the symplectic systems theory as an independent subject in the theory of difference equations. An extension of the concepts from [39], such as disconjugacy and implicit Riccati equations and their relationship to the positivity of discrete quadratic functionals, was given in [45] by M. Bohner and O. Došlý. That paper can be regarded as a starting point of the efforts of many mathematicians to develop the theory of symplectic difference systems in a parallel way to the theory of linear Hamiltonian differential systems (LHS). From this time we mention the results on the positivity and nonnegativity of discrete quadratic functionals with various boundary conditions, including explicit and implicit Riccati matrix equations, by R. Šimon Hilscher and V. Zeidan [100, 178, 180–182, 186, 188], discrete Riccati inequality [173, 174], trigonometric and hyperbolic systems [24, 46–48, 85, 108], and finally the theory of Weyl disks and their limit point and limit circle behavior (the Weyl-Titchmarsh theory) by S. Clark, R. Šimon Hilscher, and P. Zemánek [67, 308–313]. In this context we wish to emphasize the contributions of O. Došlý to the transformation theory of symplectic difference systems and linear Hamiltonian difference systems, including the trigonometric (Bohl), Prüfer, and hyperbolic transformations for symplectic systems [46, 47, 83, 85, 93, 96, 98, 108], which became a basic reference for further studies in the discrete oscillation theory.

As a breaking point in the development of the theory of symplectic systems (SDS) in the last 20 years, we can consider the paper [208] by W. Kratz, where the concept of the *multiplicity* of focal points for conjoined bases of (SDS) was introduced. This paper led to the development of the Sturmian and eigenvalue theory for symplectic systems in [56, 102] by M. Bohner, O. Došlý, and W. Kratz. Since then the discrete-time theory became a strong motivation for the development of the linear Hamiltonian differential systems. Another important contribution, which changed rapidly the study of symplectic difference systems, is represented by the papers [114, 115, 117] by J. Elyseeva, who introduced the concept of a *comparative index* for conjoined bases of (SDS); see also [121]. Due to its close relations with the multiplicities of focal points, this turned out to be a very powerful tool which allowed to derive several deep and perhaps unexpected results not only for discrete symplectic systems—for example, the discrete Sturmian theory, the relative oscillation theory, the spectral theory, and the revisited transformation theory [94–96, 116–120, 122, 123, 125, 126] by J. Elyseeva and partly by O. Došlý. Finally, among recent important contributions to this theory, we consider the introduction of the nonlinear dependence on the spectral parameter in symplectic difference systems in [297] by R. Šimon Hilscher and the theory of recessive and dominant solutions of (SDS) at infinity for possibly uncontrollable symplectic systems in [284, 290] by P. Šepitka and R. Šimon Hilscher. The first mentioned topic initiated the study of advanced topics in the spectral theory of symplectic systems, such as the Weyl-Titchmarsh theory, the concept of weighted focal points, oscillation theory for nonconstant rank, etc. The second mentioned topic leads, among others, to a completely new singular Sturmian theory for symplectic difference systems [292].

The current research in oscillation and spectral theory of symplectic difference systems includes, for example, the topics from the spectral theory of linear relations associated with symplectic systems, relative oscillation theory, recessive and dominant solutions, Riccati equations and inequalities, or the unification of the theory of symplectic systems with linear Hamiltonian differential systems in the theory of time scales.

In the traditional setting, such as in [16], the theory of symplectic difference systems was developed in a parallel way to the known continuous time theory of linear Hamiltonian systems. Starting from the milestone papers [39] by M. Bohner and [208] by W. Kratz, the inspiration moved in favor of the discrete-time theory, i.e., since then the symplectic difference systems motivate the progress in the continuous time theory. A typical example of this process can be found in [127, 129, 130, 207, 289, 295]. In current research the inspiration of continuous and discrete theories is mutual, and, roughly speaking, the results for symplectic difference systems and linear Hamiltonian differential systems are "cooked" in the same pot. This process then gives weight and credit to both of these theories.

## 1.2 Sturm-Liouville Difference Equation

The simplest and most frequently studied special case of symplectic difference systems is the second-order Sturm-Liouville difference equation

$$\Delta(r_k \Delta x_k) + p_k x_{k+1} = 0, \quad r_k \neq 0. \tag{1.1}$$

In this section we present essentials of the oscillation theory of (1.1), for a more extended treatment of this topics, we refer to [4, 110, 204]. The comparison of the results for the case of equation (1.1) with those for symplectic difference systems (SDS) should help the reader better understand the general qualitative theory presented in the subsequent chapters.

According to some historical investigations, Sturm in his famous paper [279] from 1836 originally wanted to formulate the results for a *difference* equation but eventually decided to formulate them for a continuous counterpart of (1.1), i.e., for the differential equation

$$\big(r(t) x'\big)' + p(t) x = 0, \quad r(t) > 0. \tag{1.2}$$

Consider, for a moment, that $r(t) \equiv 1$ in (1.2). We want to "discretize" this equation. Suppose that (1.2) is considered on an interval $[a, b]$ and we want to find an approximate solution of this equation using the Euler discretization scheme. We take the equidistant partition of $[a, b]$ with the discretization stepsize $h = (b-a)/n$, and we denote $t_k = a + kh$, $x_k = x(t_k)$, $p_{k-1} = p(t_k)h^2$ for $k = 0, \ldots, n$. We approximate the second derivative $x''(t_k)$ by the second difference, where the values $x_{k-1}, x_k, x_{k+1}$ occur, i.e., $x''(t_k) \approx \Delta^2 x_{k-1}/h^2 = (x_{k+1} - 2x_k + x_{k-1})/h^2$. Hence, we obtain the difference equation $\Delta^2 x_{k-1} + p_{k-1}x_k = 0$, which after relabeling $k$ to $k+1$ gives the equation

$$\Delta^2 x_k + p_k x_{k+1} = 0.$$

These considerations justify why the shift $k + 1$ appears in the second term of (1.1). Another reason is purely mathematical—without this shift we have no discrete analog of the Wronskian identity, and also some other important formulas would be missing when there is no shift at $x$ in (1.1).

### 1.2.1 Generalized Zeros

Consider the second-order recurrence relation

$$x_{k+2} - x_{k+1} - x_k = 0, \tag{1.3}$$

which defines the Fibonacci sequence. Upon looking for its solutions in the form $x_k = \lambda^k$, we find that (1.3) has a pair of linearly independent solutions

$$x_k^{[1]} = \left(\frac{1 + \sqrt{5}}{2}\right)^k, \quad x_k^{[2]} = \left(\frac{1 - \sqrt{5}}{2}\right)^k. \tag{1.4}$$

Clearly, $x^{[1]}$ is a monotonically increasing sequence, while $x^{[2]}$ is an oscillating sequence. From this point of view, it seems that the standard Sturmian theory, which eliminates the coexistence of oscillatory and nonoscillatory solutions of a second-order linear differential equation, is not valid in the discrete case.

Fortunately, this is not the case, and there exists a deeply developed Sturmian theory for (1.1) when the concept of a *generalized zero* (sometimes also called a *focal point* by analogy with the symplectic systems theory) is properly defined. To motivate this definition, consider the continuous counterpart of (1.1), i.e., the second-order Sturm-Liouville differential equation (1.2). Together with this equation, consider its energy functional

$$\mathcal{F}(y) = \int_a^b [r(t)\, y'^2(t) - p(t)\, y^2(t)]\, dt \tag{1.5}$$

and the associated Riccati equation (which related to (1.2) by the substitution $w = r(t)\, x'/x$)

$$w' + p(t) + \frac{w^2}{r(t)} = 0. \tag{1.6}$$

If $w$ is a solution of (1.6) defined on the whole interval $[a, b]$ and $y \in C^1[a, b]$, then computing $(wy^2)'$ (we add and subtract the term $r(t) y'^2$ in the resulting formula) and then integrating the obtained formula from $a$ to $b$, we obtain the so-called *Picone identity*

$$\int_a^b (ry'^2 - py^2)(t)\, dt = w(t)\, y^2(t)\Big|_a^b + \int_a^b \frac{1}{r(t)} \big[r(t)\, y'(t) - w(t)\, y(t)\big]^2 dt. \tag{1.7}$$

This means that the energy functional is positive for a nontrivial $y$ with $y(a) = 0 = y(b)$. In other words, the functional can be "completed to a square" whenever there exists a solution of the Riccati equation (1.6) defined on the whole interval $[a, b]$.

Now we will follow the previous considerations in the discrete case. The Riccati difference equation, which is related to (1.1) by the substitution $w_k = r_k \Delta x_k / x_k$, has the form

$$\Delta w_k + p_k + \frac{w_k^2}{r_k + w_k} = 0 \tag{1.8}$$

or equivalently,

$$w_{k+1} + p_k - \frac{r_k w_k}{w_k + r_k} = 0. \tag{1.9}$$

Indeed, we have

$$\Delta w_k = \Delta \left( \frac{r_k \Delta x_k}{x_k} \right) = \frac{\Delta(r_k \Delta x_k)\, x_k - r_k (\Delta x_k)^2}{x_k x_{k+1}} = -p_k - \frac{r_k^2 (\Delta x_k)^2}{r_k x_k x_{k+1}}$$

$$= -p_k - \frac{r_k^2 (\Delta x_k)^2}{x_k^2} \frac{x_k}{r_k x_{k+1}} = -p_k - \frac{w_k^2}{r_k (x_k + \Delta x_k)/x_k} = -p_k - \frac{w_k^2}{r_k + w_k}.$$

Now, for $y = \{y_k\}_{k=0}^{N+1}$ with $y_0 = 0 = y_{N+1}$, we have

$$\Delta(w_k y_k^2) = w_{k+1} y_{k+1}^2 - w_k y_k^2$$

$$= \left( -p_k + \frac{w_k r_k}{r_k + w_k} \right) y_{k+1}^2 - w_k y_k^2 + r_k (\Delta y_k)^2 - r_k (\Delta y_k)^2$$

$$= r_k (\Delta y_k)^2 - p_k y_{k+1}^2$$

$$\quad - \frac{1}{r_k + w_k} \Big[ -r_k w_k y_{k+1}^2 + (r_k + w_k) w_k y_k^2 + r_k (r_k + w_k)(\Delta y_k)^2 \Big]$$

$$= r_k(\Delta y_k)^2 - p_k y_{k+1}^2$$

$$- \frac{1}{r_k + w_k} \left[ r_k^2 (\Delta y_k)^2 + w_k^2 y_k^2 + r_k w_k (-y_{k+1}^2 + y_k^2 + (\Delta y_k)^2) \right]$$

$$= r_k(\Delta y_k)^2 - p_k y_{k+1}^2 - \frac{1}{r_k + w_k}(r_k \Delta y_k - w_k y_k)^2.$$

The summation of the last formula and using the endpoints condition $y_0 = 0 = y_{N+1}$ give the discrete Picone identity

$$\sum_{k=0}^{N}[r_k(\Delta y_k)^2 - p_k y_{k+1}^2] = \sum_{k=0}^{N} \frac{1}{w_k + r_k}(r_k \Delta y_k - w_k y_k)^2. \qquad (1.10)$$

Therefore, to get the positivity of the energy functional for a nontrivial $y$ with $y_0 = 0 = y_{N+1}$, we need $r_k + w_k > 0$. Using the Riccati substitution, we then have

$$r_k + w_k = r_k + \frac{r_k \Delta x_k}{x_k} = r_k \frac{x_k + \Delta x_k}{x_k} = \frac{r_k x_{k+1}}{x_k}.$$

Now we can formulate the definition of a *generalized zero* of a nontrivial solution of (1.1).

**Definition 1.1** Let $x$ be a nontrivial solution of (1.1). We say that this solution has a *generalized zero* (equivalently, a *focal point*) in the interval $(k, k + 1]$ if $x_k \neq 0$ and $r_k x_k x_{k+1} \leq 0$. More precisely, we say that the generalized zero is at $k + 1$ if $x_{k+1} = 0$, while it is in the interval $(k, k + 1)$ if $r_k x_k x_{k+1} < 0$.

This means that in contrast with the continuous case, where we normally assume $r(t) > 0$, in the discrete case, we only need the condition $r_k \neq 0$. Returning back to the Fibonacci equation (1.3), it can be written in the Sturm-Liouville form as

$$\Delta\big((-1)^k \Delta x_k\big) + (-1)^k x_{k+1} = 0,$$

i.e., $r_k = (-1)^k$ which changes its sign between each $k$ and $k + 1$. Applying Definition 1.1, where also the sequence $r_k$ is incorporated into the notion of a generalized zero, both solutions $x^{[1]}$ and $x^{[2]}$ of (1.3) are actually oscillating solutions (they have infinitely many generalized zeros).

We finish this subsection with the Wronskian identity and with the so-called *d'Alembert formula* (an alternative terminology is the *reduction of order formula*). Let $x$ and $y$ be solutions of (1.1). Then substituting from (1.1), we have $\Delta(r_k x_k \Delta y_k - r_k y_k \Delta x_k) = 0$, i.e., the quantity $w(x, y)_k := r_k(x_k \Delta y_k - y_k \Delta x_k)$ is constant in $k$. Moreover, if $x_k \neq 0$ in some discrete interval, then

$$\Delta\left(\frac{y_k}{x_k}\right) = \frac{r_k x_k \Delta y_k - r_k y_k \Delta x_k}{x_k x_{k+1}} = \frac{w(x, y)}{r_k x_k x_{k+1}}.$$

The summation of this formula gives the so-called *d'Alembert formula*

$$y_k = x_k \sum_{j=0}^{k-1} \frac{w(x, y)}{r_j x_j x_{j+1}}.$$  (1.11)

This enables to express the second linearly independent solution of (1.1), when a solution (being nonzero in some interval) is known.

### 1.2.2 Reid Roundabout Theorem

The next statement, usually referred to as the *Reid roundabout theorem*, describes a relationship between the basic notions from the discrete oscillation theory of equation (1.1). We present this result including its proof so that it can be compared with that of the Reid roundabout theorem for symplectic difference systems, which we present in Sect. 2.3. For discrete intervals we will use the notation

$$[a, b]_{\mathbb{Z}} := \{a, a + 1, \ldots, b - 1, b\}, \quad a, b \in \mathbb{Z}, \ a < b.$$  (1.12)

**Theorem 1.2** *Assume $r_k \neq 0$ for all $k \in [0, N]_{\mathbb{Z}}$. The following statements are equivalent.*

(i) *Equation (1.1) is disconjugate in the discrete interval $[0, N + 1]_{\mathbb{Z}}$, i.e., the solution $x$ of (1.1) given by the initial conditions $x_0 = 0$ and $x_1 = 1/r_0$ has no generalized zero in the interval $(0, N + 1]$.*
(ii) *There exists a solution $x$ of (1.1) such that $r_k x_k x_{k+1} > 0$ for $k \in [0, N]_{\mathbb{Z}}$.*
(iii) *There exists a solution $w$ of (1.8) defined for $k \in [0, N + 1]_{\mathbb{Z}}$ which satisfies $r_k + w_k > 0$ for all $k \in [0, N]_{\mathbb{Z}}$.*
(iv) *The discrete quadratic functional*

$$\mathcal{F}(y) = \sum_{k=0}^{N} \left[ r_k (\Delta y_k)^2 - p_k y_{k+1}^2 \right] > 0$$  (1.13)

*for every nontrivial $y = \{y_k\}_{k=0}^{N+1}$ with $y_0 = 0 = y_{N+1}$.*

*Proof*

(i) $\implies$ (ii): Consider the solution $x^{[\varepsilon]}$ of (1.1) given by the initial conditions $x_0^{[\varepsilon]} = \varepsilon > 0$ and $x_1^{[\varepsilon]} = 1/r_0$. Then, according to the continuous dependence of solutions on the initial condition, we have $x_k^{[\varepsilon]} \to x_k$ as $\varepsilon \to 0^+$ for $k \in [0, N + 1]_{\mathbb{Z}}$, where $x$ is the solution from (i). Then $r_k x_k^{[\varepsilon]} x_{k+1}^{[\varepsilon]} > 0, k \in [1, N]_{\mathbb{Z}}$, if $\varepsilon$ is sufficiently small and also $r_0 x_0^{[\varepsilon]} x_1^{[\varepsilon]} = \varepsilon > 0$.

(ii) $\implies$ (iii): This is just the Riccati substitution $w_k = r_k \Delta x_k^{[\varepsilon]} / x_k^{[\varepsilon]}$ for $k \in [0, N]_{\mathbb{Z}}$.

(iii) $\implies$ (iv): This implication follows immediately from the Picone identity (1.10).

(iv) $\implies$ (i): By contradiction, suppose that the solution $x$ given by the initial condition $x_0 = 0$, $x_1 = 1/r_0$ has a generalized zero in $(m, m+1]$, i.e., $x_m \neq 0$ and $r_m x_m x_{m+1} \leq 0$ for some $m \in [1, N]_{\mathbb{Z}}$. Define the test sequence $y$ by the formula

$$y_k = \begin{cases} x_k, & k \in [0, m]_{\mathbb{Z}}, \\ 0 & k \in [m+1, N+1]_{\mathbb{Z}}. \end{cases}$$

Denote by $L[x]_k$ the left-hand side of (1.1), i.e., $L[x]_k = \Delta(r_k \Delta x_k) + p_k x_{k+1}$. Then $y_m = x_m \neq 0$ (i.e., $y$ is nontrivial) and $y_0 = x_0 = 0 = y_{N+1}$ and by the summation by parts of the first term in $\mathcal{F}$, we obtain

$$\mathcal{F}(y) = r_k y_k \Delta y_k \Big|_0^{N+1} - \sum_{k=0}^{N} y_{k+1} L[y]_k$$

$$= -\left( \sum_{k=0}^{m-2} x_{k+1} L[x]_k \right) - x_m L[y]_{m-1} - y_{m+1} L[y]_m$$

$$= x_m[(r_m + r_{m-1} - p_{m-1}) x_m - r_{m-1} x_{m-1}] = x_m r_m x_{m+1} \leq 0,$$

which yields a contradiction. In the previous computation, we have used the construction of $y$ as a solution of (1.1) for $k \in [0, m-2]_{\mathbb{Z}}$ and the fact that (1.1) can be written as the three-term symmetric recurrence relation

$$r_{k+1} x_{k+2} + s_k x_{k+1} + r_k x_k = 0, \quad s_k = -r_{k+1} - r_k + p_k, \tag{1.14}$$

where we applied (1.14) at $k = m-1$.                                        $\square$

In the oscillation theory of (1.1), the equivalence (i) $\iff$ (iii) is called the *Riccati technique*, while the equivalence (i) $\iff$ (iv) is called the *variational principle*. As a consequence of the roundabout theorem, we get that (1.1) can be classified as *oscillatory* or *nonoscillatory*, similarly as in the case of its continuous counterpart, the second-order Sturm-Liouville differential equation. This follows from the next statement.

**Theorem 1.3** *Let $x$ be a solution of (1.1) with generalized zeros in the intervals $(m, m+1]$ and $(n, n+1]$, where $m, n \in [0, N]_{\mathbb{Z}}$ with $m < n$. Then any other solution of this equation has a generalized zero in the interval $(m, n+1]$.*

*Proof* The idea of the proof is the following. The existence of a nontrivial solution without any generalized zero in $(m, n + 1]$ implies that the functional

$$\mathcal{F}(y) = \sum_{k=m}^{n} \left[ r_k (\Delta y_k)^2 - p_k y_{k+1}^2 \right] > 0$$

for every nontrivial $y = \{y_k\}_{k=m}^{n+1}$ with $y_m = y_{n+1} = 0$, by Theorem 1.2. On the other hand, a construction similar to the one in the proof of the implication (vi) $\Longrightarrow$ (i) in Theorem 1.2 yields a test sequence $y$, for which $\mathcal{F}(y) \le 0$. This way we obtain a contradiction.                                                                             □

We say that a nontrivial solution $x$ of (1.1) is *oscillatory* (at $\infty$) if there exists a sequence of integers $n_k \to \infty$ as $k \to \infty$ such that interval $(n_k, n_k + 1]$ contains a generalized zero of $x$ for every $k$. In the opposite case, the solution $x$ of (1.1) is said to be *nonoscillatory*. Equation (1.1) is said to be *oscillatory*, if it possesses an oscillatory solution, and it is said to be *nonoscillatory*, when there exists a nonoscillatory solution. The previous theorem eliminates the coexistence of oscillatory and nonoscillatory solutions of (1.1); hence it justifies the classification of (1.1) as being oscillatory or nonoscillatory.

In the literature there exist numerous oscillation and nonoscillation criteria for equation (1.1). As an example, we give here the so-called Leighton-Wintner oscillation criterion.

**Theorem 1.4** *Suppose that* $\sum^{\infty} 1/r_k = \infty = \sum^{\infty} p_k$. *Then equation* (1.1) *is oscillatory.*

*Proof* It can be verified that for any $K \in \mathbb{N}$, there exist integers $K < L < M < N$ such that for the sequence

$$y_k = \begin{cases} 0 & \text{for } k < K, \\ \left( \sum_{j=K}^{k-1} 1/r_j \right) \left( \sum_{j=K}^{L-1} 1/r_j \right)^{-1} & \text{for } k \in [K, L]_{\mathbb{Z}}, \\ 1 & \text{for } k \in [L+1, M-1]_{\mathbb{Z}}, \\ \left( \sum_{j=k}^{N-1} 1/r_j \right) \left( \sum_{j=M}^{N-1} 1/r_j \right)^{-1} & \text{for } k \in [M, N-1]_{\mathbb{Z}}, \\ 0 & \text{for } k \ge N, \end{cases}$$

we have $\mathcal{F}(y) = \sum_{k=K}^{M} [r_k (\Delta y_k)^2 - p_k y_{k+1}^2] \le 0$ if $M$ is sufficiently large. To verify it, one needs to apply the second mean value theorem of summation calculus (see [80, Lemma 3.2]) in computing the sum $\sum_{k=M}^{N-1} p_k y_{k+1}^2$. Other than that the computation is straightforward. Note that the second mean value theorem of summation calculus is not needed under the additional assumption that $p_k \ge 0$ for large $k$.                                                                             □

Let us recall that recurrence (1.14) (considered for $k \in [0, \infty)_{\mathbb{Z}}$) and for $x = \{x_k\}_{k=0}^{\infty}$ with $x_0 = 0$) can be rewritten using the so-called Jacobi matrix. This is defined to be an (infinite) three-diagonal symmetric matrix

$$
J = \begin{pmatrix}
s_0 & r_1 & 0 & 0 & 0 & \ldots 0 \ldots \\
r_1 & s_1 & r_2 & 0 & 0 & \ldots 0 \ldots \\
0 & r_2 & s_2 & r_3 & 0 & \ldots 0 \ldots \\
0 & 0 & r_3 & s_4 & r_4 & \ldots 0 \ldots \\
& & & & & \ddots
\end{pmatrix},
$$

as (1.1) is equivalent with the equality $Jx = 0$. We note that the (mainly spectral) theory of these matrices is deeply developed; see, e.g., [315].

The following transformation formula represents another important tool in the qualitative theory of (1.1).

**Theorem 1.5** *Let $h_k \neq 0$ in some discrete interval. Then the transformation $x_k = h_k y_k$ transforms equation (1.1) into the equation of the same form, i.e., to*

$$
\Delta(\hat{r}_k \Delta y_k) + \hat{p}_k y_{k+1} = 0, \tag{1.15}
$$

*where $\hat{r}_k$ and $\hat{p}_k$ are given by the formulas*

$$
\hat{r}_k = r_k h_k h_{k+1}, \quad \hat{p}_k = \Delta(r_k \Delta h_k) + p_k h_{k+1}. \tag{1.16}
$$

*Proof* We have

$$
h_{k+1}\big[\Delta(r_k \Delta x_k) + p_k x_{k+1}\big]
$$
$$
= h_{k+1}\big[\Delta(r_k h_{k+1} \Delta y_k + r_k \Delta h_k y_k) + p_k h_{k+1} y_{k+1}\big]
$$
$$
= h_{k+1}\Delta(r_k h_{k+1} \Delta y_k) + h_{k+1}\Delta(r_k \Delta h_k y_k) + p_k h_{k+1}^2 y_{k+1}
$$
$$
= \Delta(r_k h_k h_{k+1} \Delta y_k) - \Delta h_k r_k h_{k+1} \Delta y_k + h_{k+1}\Delta(r_k \Delta h_k) y_{k+1}
$$
$$
+ h_{k+1} r_k \Delta h_k \Delta y_k + p_k h_{k+1}^2 y_{k+1}
$$
$$
= \Delta(r_k h_k h_{k+1} \Delta y_k) + h_{k+1}[\Delta(r_k \Delta h_k) + p_k h_{k+1}] y_{k+1}.
$$

Hence, (1.1) is indeed transformed into (1.15) with $\hat{r}_k$ and $\hat{p}_k$ given by (1.16).  □

If the sequence $h$ is a solution of equation (1.1), then $\hat{p}_k = 0$, and (1.15) becomes the equation $\Delta(r_k h_k h_{k+1} \Delta y_k) = 0$. The summation of this equation gives an alternative proof of the d'Alembert formula (1.11) (with $h$ instead of $x$).

We finish this subsection by the so-called *oscillation theorem*, which concerns the discrete Sturm-Liouville eigenvalue problem with the Dirichlet boundary conditions (the statement remains to hold also for general self-adjoint boundary conditions, but for simplicity we formulate the statement only for the Dirichlet conditions). The proof of this statement can be found in [204, Theorem 7.6]. Note that this reference deals with the case of $r_k > 0$ only, but the presented proof essentially works also when $r_k \neq 0$. Alternatively, this result follows from the more general statement for nonlinear dependence on $\lambda$ (Theorem 1.19) or for symplectic difference systems (Corollary 5.79).

**Theorem 1.6** *Consider the eigenvalue problem*

$$-\Delta(r_k \Delta x_k) + q_k x_{k+1} = \lambda w_k y_{k+1}, \quad k \in [0, N-1]_{\mathbb{Z}}, \quad x_0 = 0 = x_{N+1} \quad (1.17)$$

*with $r_k \neq 0$ and $w_k > 0$ for all $k \in [0, N]_{\mathbb{Z}}$. Then the number of eigenvalues of (1.17), which are less than or equal to a given $\lambda_0 \in \mathbb{R}$, is equal to the number of generalized zeros in $(0, N + 1]$ of the solution $x(\lambda_0)$ of the difference equation in (1.17) with $\lambda = \lambda_0$, which satisfies the initial conditions $x_0(\lambda_0) = 0$ and $x_1(\lambda_0) = 1/r_0$.*

### 1.2.3  Recessive and Dominant Solutions

In this subsection we will suppose that equation (1.1) is *nonoscillatory*. This means that there exists $M \in \mathbb{N}$ such that the solution $x^{[M]}$ given by the initial condition $x_M = 0$ and $x_{M+1} = 1/r_M$ has no generalized zero in the interval $(M, \infty)$, i.e., $r_k x_k^{[M]} x_{k+1}^{[M]} > 0$ for all $k \geq M$. We claim that under these conditions, there exists a solution $\tilde{x}$ of (1.1) with the property that

$$\lim_{k \to \infty} \frac{\tilde{x}_k}{x_k} = 0 \quad (1.18)$$

for every solution $x$ of (1.1), which is linearly independent of $\tilde{x}$. The solution $\tilde{x}$ is called a *recessive solution* (at $\infty$). Any solution linearly independent of the recessive solution is called a *dominant solution* (at $\infty$)

In order to show the existence of a recessive solution of (1.1), we consider two linearly independent solutions $x$ and $y$ of (1.1) such that $x$ has no generalized zero in $(N, \infty)$ for some $N \in \mathbb{N}$. Then their Wronskian $c := w(x, y)_k$ is constant on $[N, \infty)$, and hence $\Delta(y_k/x_k) = c/r_k x_k x_{k+1}$ is of constant sign (equal to the sign of $c$); see Sect. 1.2.1. Therefore, the sequence $y_k/x_k$ is monotone, and hence there exists (finite or infinite) $\lim_{k \to \infty} y_k/x_k = L$. If $L = 0$, then we put $\tilde{x} = y$. Then every solution linearly independent of $\tilde{x}$ is of the form $\alpha \tilde{x} + \beta x$ with $\beta \neq 0$ and

$$\lim_{k \to \infty} \frac{\tilde{x}_k}{\alpha \tilde{x}_k + \beta x_k} = \lim_{k \to \infty} \frac{\tilde{x}_k/x_k}{\alpha (\tilde{x}_k/x_k) + \beta} = 0.$$

If $0 \neq L \in \mathbb{R}$, then the principal solution is $\tilde{x} = x - Ly$, and similarly as above, we have $\lim_{k \to \infty} \tilde{x}_k / x_k = 0$ for every solution linearly independent of $\tilde{x}$. Finally, if $L = \pm \infty$, we put $\tilde{x} = x$. Therefore we have just proved the following statement.

**Theorem 1.7** *A recessive solution of (1.1) exists whenever (1.1) is nonoscillatory. Moreover, the recessive solution is unique up to a constant nonzero multiplicative factor. Finally, a solution $x$ of (1.1) is recessive if and only if*

$$\sum_{k=N}^{\infty} \frac{1}{r_k x_k x_{k+1}} = \infty, \tag{1.19}$$

*where $N$ is sufficiently large.*

Another characterization of the recessive solution is closely connected with the concept of *eventually minimal* solution (an equivalent terminology is the *distinguished solution*) of the Riccati equation (1.8).

**Theorem 1.8** *Let $\tilde{x}$ be a solution of (1.1) with $\tilde{x}_k \neq 0$ for $k$ sufficiently large. Let $\tilde{w}_k = (r_k \Delta \tilde{x}_k)/\tilde{x}_k$ be the corresponding solution of the associated Riccati equation (1.8). Then $\tilde{x}$ is the recessive solution of (1.1) if and only if for any other solution $w$ of (1.8) we have $w_k > \tilde{w}_k$ for all $k$ sufficiently large.*

*Proof* Let $\tilde{x}$ be the recessive solution of (1.1) and define $\tilde{w}$ as in the theorem. Let $w_k$ be any other solution of the Riccati equation (1.8). By the Reid roundabout theorem (Theorem 1.2), we may assume that $r_k + w_k > 0$ for large $k$ and that $w_k = (r_k \Delta x_k)/x_k$ for large $k$ for some solution $x$ of (1.1) with $x_k \neq 0$ for large $k$. Then we have

$$w_k - \tilde{w}_k = \frac{r_k \Delta x_k}{x_k} - \frac{r_k \Delta \tilde{x}_k}{\tilde{x}_k} = \frac{r_k (\tilde{x}_k \Delta x_k - x_k \Delta \tilde{x}_k)}{x_k \tilde{x}_k} = \frac{d}{x_k \tilde{x}_k}, \tag{1.20}$$

where $d = w(\tilde{x}, x)_k$ is the constant Wronskian of $\tilde{x}$ and $x$. Since $r_k x_k x_{k+1} > 0$ and $r_k \tilde{x}_k \tilde{x}_{k+1} > 0$ for large $k$, we have

$$\frac{x_k x_{k+1}}{\tilde{x}_k \tilde{x}_{k+1}} > 0 \quad \text{i.e.,} \quad \frac{x_k}{\tilde{x}_k} \cdot \frac{x_{k+1}}{\tilde{x}_{k+1}} > 0.$$

This means that $x_k/\tilde{x}_k$, and hence also $x_k \tilde{x}_k$, does not change its sign for large $k$. Replacing $x$ by $-x$ if necessary, we may suppose without loss of generality that $x_k \tilde{x}_k > 0$, i.e., $\tilde{x}_k/x_k > 0$. Since $\tilde{x}_k/x_k \to 0$ as $k \to \infty$ and $\Delta(\tilde{x}_k/x_k) = c/(r_k x_k x_{k+1})$ with positive denominator of the right-hand side, it follows that $c = w(x, \tilde{x})_k < 0$ for large $k$. Hence, $d = -c > 0$ and from (1.20) we obtain that $w_k - \tilde{w}_k > 0$ for large $k$.

Conversely, suppose that $\tilde{w}$ is the eventually minimal solution of equation (1.8), i.e., we have for large $k$

$$w_k = \frac{r_k \Delta x_k}{x_k} > \tilde{w}_k = \frac{r_k \Delta \tilde{x}_k}{\tilde{x}_k}, \quad \text{i.e.,} \quad \frac{d}{x_k \tilde{x}_k} = w_k - \tilde{w}_k > 0,$$

where $d = r_k(\tilde{x}_k \Delta x_k - x_k \Delta \tilde{x}_k)$. Using the same argument as in the first part of the proof, we have $x_k \tilde{x}_k > 0$ for large $k$, hence $d > 0$. We have

$$\Delta(\tilde{x}_k/x_k) = -\frac{d}{r_k x_k x_{k+1}} < 0.$$

for large $k$. This means that $\tilde{x}_k/x_k$ is a monotonically decreasing sequence of positive numbers; hence it tends to some limit $L \geq 0$ as $k \to \infty$. Suppose that $L > 0$ and put $\hat{x}_k = \tilde{x}_k - Lx_k$. Then we have

$$\frac{r_k \Delta \hat{x}_k}{\hat{x}_k} - \frac{r_k \Delta \tilde{x}_k}{\tilde{x}_k} = \frac{r_k \tilde{x}_k(\Delta \tilde{x}_k - L\Delta x_k) - (\tilde{x}_k - Lx_k) r_k \Delta \tilde{x}_k}{\hat{x}_k x_k}$$

$$= -L \frac{r_k(\tilde{x}_k \Delta x_k - x_k \Delta \tilde{x}_k)}{\hat{x}_k x_k} = -L \frac{c}{\hat{x}_k x_k} < 0$$

and we get a contradiction with the assumption that $\tilde{w}_k < w_k$ for large $k$ for any other solution $w$ of (1.8). Hence $L = 0$, and $\tilde{x}$ is then the recessive solution. □

Finally, the last statement of this subsection states that the largest generalized zero of the recessive solution (if any) has a certain extremal property.

**Theorem 1.9** *Let the interval $(N, N+1]$ contain the largest generalized zero of the recessive solution $\tilde{x}$ of (1.1). Then any linearly independent solution has at least one generalized zero in the interval $(N, \infty)$.*

*Proof* We outline the proof only. Suppose that $N$ is the largest integer such that the interval $(N, N+1]$ contains a generalized zero of $\tilde{x}$, i.e., $\tilde{x}_N \neq 0$ and $r_N \tilde{x}_N \tilde{x}_{N+1} \leq 0$. We will deal with the case $r_N \tilde{x}_N \tilde{x}_{N+1} < 0$ (in case of the equality, the proof needs minor a modification, which we will not present here). Hence we have $r_N + \tilde{w}_N < 0$, where $\tilde{w}_k = (r_k \Delta \tilde{x}_k)/\tilde{x}_k$. Suppose, by contradiction, that there exists a solution $x$ of (1.1) with $r_k x_k x_{k+1} > 0$ for all $k \in [N, \infty)_{\mathbb{Z}}$. Then $r_k + w_k > 0$, where $w_k = (r_k \Delta x_k)/x_k$, in particular, $r_N + w_N > 0$. This implies $\tilde{w}_N > w_N$. We have (we use (1.9))

$$\tilde{w}_{k+1} - w_{k+1} = \frac{r_k^2}{(r_k + w_k)(r_k + \tilde{w}_k)}(\tilde{w}_k - w_k).$$

Substituting $k = N$, we have $\tilde{w}_{N+1} > w_{N+1}$. Since $r_k + w_k > 0$, $r_k + \tilde{w}_k > 0$, $k \in [N+1, \infty)_{\mathbb{Z}}$, we have $\tilde{w}_k > w_k$ for all $k \in [N+1, \infty)_{\mathbb{Z}}$. This is a contradiction with the fact that the recessive solution of (1.1) generates the eventually minimal solution of (1.8), by Theorem 1.8. □

### 1.2.4 Discrete Prüfer Transformation

The Prüfer transformation is an important tool in the investigation of the oscillatory properties of conjoined bases of system (1.103). In this subsection we introduce the Prüfer transformation for the self-adjoint difference equation (1.1) and use it to obtain the so-called reciprocity principle for this equation.

Let $x$ be a nontrivial solution of (1.1). Then $x_k^2 + (r_k \Delta x_k)^2 \neq 0$ for all $k$ (otherwise $x$ would be identically zero) and we can find real numbers $\varrho_k > 0$ and $\varphi_k$ with $0 \leq \Delta \varphi_k < 2\pi$ such that the equations

$$x_k = \varrho_k \sin \varphi_k, \tag{1.21}$$

$$r_k \Delta x_k = \varrho_k \cos \varphi_k \tag{1.22}$$

are satisfied for all $k$. That is, we set

$$\varrho_k := \sqrt{x_k^2 + (r_k \Delta x_k)^2}, \quad \varphi_k := \operatorname{arccot} \frac{r_k \Delta x_k}{x_k}.$$

We call (1.21) and (1.22) the *discrete Prüfer transformation*.

**Theorem 1.10** *Let $x$ be a nontrivial solution of* (1.1), *and let $\varrho$ and $\varphi$ be defined by* (1.21) *and* (1.22). *Then we have the equations*

$$\Delta \varrho_k = \varrho_k \left( \frac{1}{r_k} \cos \varphi_k \sin \varphi_{k+1} - p_k \sin \varphi_k \cos \varphi_{k+1} \right.$$

$$\left. - \frac{p_k}{r_k} \cos \varphi_k \cos \varphi_{k+1} - \frac{1}{2} \left[ (\Delta \sin \varphi_k)^2 + (\Delta \cos \varphi_k)^2 \right] \right), \tag{1.23}$$

$$\sin \Delta \varphi_k = \frac{1}{r_k} \cos \varphi_k \cos \varphi_{k+1} + p_k \sin \varphi_k \sin \varphi_{k+1} + \frac{p_k}{r_k} \cos \varphi_k \sin \varphi_{k+1}. \tag{1.24}$$

*Proof* Use of the discrete product rule for (1.21) yields

$$\sin \varphi_{k+1} \Delta \varrho_k + \varrho_k \Delta \sin \varphi_k = \Delta (\varrho_k \sin \varphi_k) = \Delta x_k$$

$$= \frac{1}{r_k} (r_k \Delta x_k) = \frac{1}{r_k} \varrho_k \cos \varphi_k,$$

while doing the same for (1.22) implies

$$\cos\varphi_{k+1}\,\Delta\varrho_k + \varrho_k\,\Delta\cos\varphi_k = \Delta(\varrho_k\cos\varphi_k) = \Delta(r_k\Delta x_k)$$

$$= -p_k x_{k+1} = -\frac{p_k}{r_k}(r_k\Delta x_k) - p_k x_k$$

$$= -\frac{p_k}{r_k}\varrho_k\cos\varphi_k - p_k\varrho_k\sin\varphi_k,$$

where we have also used that $x$ is a solution of (1.1). Hence we obtain

$$\sin\varphi_{k+1}\,\Delta\varrho_k + \varrho_k\,\Delta\sin\varphi_k = \frac{\varrho_k}{r_k}\cos\varphi_k, \tag{1.25}$$

$$\cos\varphi_{k+1}\,\Delta\varrho_k + \varrho_k\,\Delta\cos\varphi_k = -\frac{p_k}{r_k}\varrho_k\cos\varphi_k - p_k\varrho_k\sin\varphi_k. \tag{1.26}$$

We now multiply (1.25) by $\sin\varphi_{k+1}$ and (1.26) by $\cos\varphi_{k+1}$ and add the resulting equations to obtain (1.23). To verify (1.24), we multiply (1.25) by $\cos\varphi_{k+1}$ and (1.26) by $-\sin\varphi_{k+1}$ and add the resulting equations. Dividing the obtained equation by $\varrho_k > 0$ directly yields (1.24). □

As an application of Theorem 1.10, we derive the reciprocity principle for equation (1.1), which relates the oscillatory properties of a solution $x$ of (1.1) with the oscillatory properties of the quasi-difference $r\Delta x$. For this we assume that

$$r_k > 0 \quad\text{and}\quad p_k > 0 \quad\text{for all } k. \tag{1.27}$$

First we need an auxiliary result.

**Lemma 1.11** *Assume* (1.27), *let $x$ be a solution of* (1.1), *and define $\varphi$ by* (1.21)–(1.22). *If $x_k x_{k+1} > 0$ for some integer $k$, then $0 < \Delta\varphi_k < \pi$.*

*Proof* First of all use equations (1.1), (1.21), and (1.22) to obtain

$$\cos\varphi_{k+1} + p_k\sin\varphi_{k+1} = \frac{r_{k+1}\Delta x_{k+1} + p_k x_{k+1}}{\varrho_{k+1}} = \frac{r_k\Delta x_k}{\varrho_{k+1}} = \frac{\varrho_k\cos\varphi_k}{\varrho_{k+1}}. \tag{1.28}$$

This is implied by (1.21) and (1.24)

$$\sin\Delta\varphi_k = p_k\sin\varphi_k\sin\varphi_{k+1} + \frac{\cos\varphi_k}{r_k}(\cos\varphi_{k+1} + p_k\sin\varphi_{k+1})$$

$$= \frac{p_k x_k x_{k+1}}{\varrho_k\varrho_{k+1}} + \frac{\varrho_k\cos^2\varphi_k}{r_k\varrho_{k+1}} > 0$$

and hence (observe that by definition $0 \le \Delta\varphi_k < 2\pi$), we have $0 < \Delta\varphi_k < \pi$, which is the conclusion. □

**Theorem 1.12 (Reciprocity Principle)** *Assume* (1.27) *and let x be a solution of* (1.1). *Then $x_k$ is eventually of one sign if and only if $r_k \Delta x_k$ is eventually of one sign.*

*Proof* Let $x$ be a solution of (1.1) and define $\varphi$ by the (1.21)–(1.22) as before. Let $N$ be an integer such that $x_k x_{k+1} > 0$ for all $k \geq N$. Then the points $(r_k \Delta x_k, x_k)$ are either in the upper half plane for all $k \geq N$ or in the lower half plane for all $k \geq N$. This and Lemma 1.11 imply that $\varphi_k \leq \varphi_N + \pi$ for all $k \geq N$. By definition, $\varphi_k \leq \varphi_{k+1}$ for all $k \geq N$, so that $\lim_{k \to \infty} \varphi_k$ must exist. But then also $\lim_{k \to \infty} \cos \varphi_k$ exists, and hence (observe also that $\cos \varphi_k \geq \cos \varphi_{k+1}$ for all $k \geq N$ or $\cos \varphi_k \leq \cos \varphi_{k+1}$ for all $k \geq N$), there exists an integer $M$ such that $\cos \varphi_k \cos \varphi_{k+1} > 0$ for all $k \geq M$. This together with (1.22) proves the result. The converse follows from the fact that the sequence $\tilde{x}_k = r_k \Delta x_k$ satisfies the so-called reciprocal equation

$$\Delta \left( \frac{1}{p_k} \Delta \tilde{x}_k \right) + \frac{1}{r_{k+1}} \tilde{x}_{k+1} = 0,$$

as can be easily verified by direct calculations. One can use then the same argumentation as in the first part of the proof.                                        □

## 1.2.5   Sturm-Liouville Eigenvalue Problems

In this section we consider the second-order Sturm-Liouville difference equation

$$\Delta(r_k(\lambda) \, \Delta x_k) + q_k(\lambda) \, x_{k+1} = 0, \qquad k \in [0, N-1]_{\mathbb{Z}}, \tag{SL$_\lambda$}$$

where $r_k : \mathbb{R} \to \mathbb{R}$ for $k \in [0, N]_{\mathbb{Z}}$ and $q_k : \mathbb{R} \to \mathbb{R}$ for $k \in [0, N-1]_{\mathbb{Z}}$ are given differentiable functions of the spectral parameter $\lambda$ such that

$$r_k(\lambda) \neq 0 \text{ and } \dot{r}_k(\lambda) \leq 0, \quad k \in [0, N]_{\mathbb{Z}}, \qquad \dot{q}_k(\lambda) \geq 0, \quad k \in [0, N-1]_{\mathbb{Z}}. \tag{1.29}$$

Here $N \in \mathbb{N}$ is a fixed number with $N \geq 2$ and $[a, b]_{\mathbb{Z}} := [a, b] \cap \mathbb{Z}$, and the dot denotes the differentiation with respect to $\lambda$. With equation (SL$_\lambda$) we consider the Dirichlet boundary conditions, that is, we study the eigenvalue problem

$$\text{(SL}_\lambda), \quad \lambda \in \mathbb{R}, \quad x_0 = 0 = x_{N+1}. \tag{E$_0$}$$

We recall first the classical setting of Sturm-Liouville difference equations; see, e.g., [16, 23, 28, 204], in which the function $r_k(\cdot)$ is constant (nonzero) in $\lambda$ and the function $q_k(\cdot)$ is linear and increasing in $\lambda$. That is, the traditional assumptions for the oscillation and spectral theory of equation (SL$_\lambda$) are the following:

$$\left. \begin{array}{l} r_k(\lambda) \equiv r_k \neq 0, \quad \text{for all } k \in [0, N]_{\mathbb{Z}}, \\ q_k(\lambda) = q_k + \lambda \, w_k, \quad w_k > 0, \quad \text{for all } k \in [0, N-1]_{\mathbb{Z}}. \end{array} \right\} \tag{1.30}$$

In some publications, such as in [28, 204], the authors also impose the sign condition $r_k > 0$ for all $k \in [0, N]_{\mathbb{Z}}$, but it is well-known nowadays that $r_k \neq 0$ is sufficient to develop the oscillation and spectral theory of these equations; see Sect. 1.2.1 and, e.g., [315, pg. 5] or [304]. The explanation of this phenomenon also follows from the analysis of the general equation $(SL_\lambda)$ discussed below.

Assume for a moment that (1.30) holds. Following [204, Chapter 7] or [28, Chapter 4], a number $\lambda_0 \in \mathbb{C}$ is an *eigenvalue* of $(E_0)$ if there exists a nontrivial solution $x = x(\lambda_0)$ of equation $(SL_{\lambda_0})$ satisfying the Dirichlet endpoints $x_0(\lambda_0) = 0 = x_{N+1}(\lambda_0)$. By the uniqueness of solutions of equation $(SL_{\lambda_0})$, it follows that the eigenvalues of $(E_0)$ are characterized by the condition $\hat{x}_{N+1}(\lambda_0) = 0$, where $\hat{x}(\lambda)$ is the *principal solution* of equation $(SL_\lambda)$, i.e., it is the solution starting with the initial values $\hat{x}_0(\lambda) = 0$ and $\hat{x}_1(\lambda) = 1/r_0$. If $x(\lambda)$ is a solution of $(SL_\lambda)$ with (1.30), then the functions $x_k(\lambda)$ are polynomials in $\lambda$ for every $k \in [0, N + 1]_{\mathbb{Z}}$. Therefore, the zeros of $x_k(\lambda)$ are isolated, showing that the eigenvalues of $(E_0)$ are simple (with the multiplicity equal to one) and isolated. Furthermore, by a standard argument from linear algebra, it follows that the eigenvalues of $(E_0)$ with $\lambda \in \mathbb{C}$ are indeed real and that the eigenfunctions corresponding to different eigenvalues are orthogonal with respect to the inner product $\langle x, y \rangle_w := \sum_{k=0}^{N} w_k \, x_{k+1} \, y_{k+1}$. The oscillation theorem for $(E_0)$ then says that the $j$-th eigenfunction has exactly $j$ generalized zeros in the interval $(0, N + 1]$; see Definition 1.1.

In this subsection we show that some of the above properties can be extended to the eigenvalue problem $(E_0)$ in which the coefficients depend on the spectral parameter $\lambda$ in general nonlinearly and they satisfy the monotonicity assumption (1.29). In particular, we discuss the notions of *finite eigenvalues* and *finite eigenfunctions* for such problems which are appropriate generalizations of the corresponding notions for the case of (1.30).

First we show how certain solutions of $(SL_\lambda)$ behave with respect to $\lambda$. Assumption (1.29) implies that the solutions of $(SL_\lambda)$ are differentiable, hence continuous, in $\lambda$ on $\mathbb{R}$. We will consider the solutions whose initial values

$$x_0(\lambda), \quad r_0(\lambda) \, \Delta x_0(\lambda) \quad \text{do not depend on } \lambda. \tag{1.31}$$

This condition is satisfied, for example, by the *principal solution* $\hat{x}(\lambda)$, for which

$$\hat{x}_0(\lambda) = 0, \quad \hat{x}_1(\lambda) = 1/r_0(\lambda) \quad \text{for all } \lambda \in \mathbb{R}. \tag{1.32}$$

The following result shows that under the monotonicity assumption (1.29), the oscillation behavior in $\lambda$ is not allowed for the above type of solutions near any finite value of $\lambda$.

**Lemma 1.13** *Assume that (1.29) holds and let $x(\lambda) = \{x_k(\lambda)\}_{k=0}^{N+1}$ be a nontrivial solution of $(SL_\lambda)$ satisfying (1.31). Then for each $k \in [0, N+1]_{\mathbb{Z}}$ and $\lambda_0 \in \mathbb{R}$, there exists $\delta > 0$ such that $x_k(\lambda)$ is either identically zero or never zero on $(\lambda_0, \lambda_0 + \delta)$, resp. on $(\lambda_0 - \delta, \lambda_0)$.*

*Proof* Let $\lambda_0 \in \mathbb{R}$ and $k \in [0, N + 1]_{\mathbb{Z}}$ be fixed. If $k = 0$, then the result follows trivially. Also, if $x_k(\lambda_0) \neq 0$, then the statement is a consequence of the continuity of $x_k(\lambda)$ in $\lambda$. Therefore, further on we assume that $k \in [1, N + 1]_{\mathbb{Z}}$ and $x_k(\lambda_0) = 0$. First we construct another solution $y(\lambda) = \{y_j(\lambda)\}_{j=0}^{N+1}$ whose initial conditions do not depend on $\lambda$ as in (1.31) such that $y_k(\lambda_0) \neq 0$ and such that the Casorati determinant

$$C[y(\lambda), x(\lambda)]_j := r_j(\lambda) \begin{vmatrix} y_j(\lambda) & x_j(\lambda) \\ \Delta y_j(\lambda) & \Delta x_j(\lambda) \end{vmatrix} = 1 \quad \text{for all} \ j \in [0, N+1]_{\mathbb{Z}}, \lambda \in \mathbb{R}.$$

This means that the solutions $y(\lambda)$ and $x(\lambda)$ form a normalized pair of solutions of (SL$_\lambda$). The solution $y(\lambda)$ can be constructed from the initial conditions

$$y_0(\lambda) = r_0(\lambda)\,\Delta x_0(\lambda)/\omega_0, \qquad r_0(\lambda)\,\Delta y_0(\lambda) = -x_0(\lambda)/\omega_0,$$

where $\omega_0 := x_0^2(\lambda) + r_0^2(\lambda)\,[\Delta x_0(\lambda)]^2$ is independent of $\lambda$. By the continuity of $y_k(\lambda)$ in $\lambda$, there exists $\varepsilon > 0$ such that $y_k(\lambda) \neq 0$ on $(\lambda_0 - \varepsilon, \lambda_0 + \varepsilon)$. For these values of $\lambda$, a direct calculation shows the formula

$$\frac{d}{d\lambda}\left(\frac{x_k(\lambda)}{y_k(\lambda)}\right) = \frac{1}{y_k^2(\lambda)} \sum_{j=0}^{k-1} \left\{ \dot{q}_j(\lambda) \begin{vmatrix} x_{j+1}(\lambda) & x_k(\lambda) \\ y_{j+1}(\lambda) & y_k(\lambda) \end{vmatrix}^2 - \dot{r}_k(\lambda) \begin{vmatrix} \Delta x_j(\lambda) & x_k(\lambda) \\ \Delta y_j(\lambda) & y_k(\lambda) \end{vmatrix}^2 \right\}.$$

Therefore, under the assumption (1.29) the function $z_k(\lambda) := x_k(\lambda)/y_k(\lambda)$ is nondecreasing in $\lambda$ on $(\lambda_0 - \varepsilon, \lambda_0 + \varepsilon)$. This means that once $z_k(\lambda_0) = 0$, then $z_k(\lambda)$ is either identically zero on $(\lambda_0, \lambda_0 + \delta)$ for some $\delta \in (0, \varepsilon)$, or $z_k(\lambda)$ is positive on $(\lambda_0, \lambda_0 + \varepsilon)$. Similar argument applies also on the left side of $\lambda_0$. And since the zeros of $z_k(\lambda)$ in $(\lambda_0 - \varepsilon, \lambda_0 + \varepsilon)$ are exactly those of $x_k(\lambda)$, the result follows.                                                                                    $\square$

*Remark 1.14* The statement of Lemma 1.13 says that for a nontrivial solution $x(\lambda) = \{x_k(\lambda)\}_{k=0}^{N+1}$ of (SL$_\lambda$) satisfying (1.31), the quantity

$$h_k(\lambda) := \operatorname{rank} x_k(\lambda) \tag{1.33}$$

is piecewise constant in $\lambda$ on $\mathbb{R}$ for every given $k \in [0, N + 1]_{\mathbb{Z}}$.

*Remark 1.15* If a solution $x(\lambda)$ of (SL$_\lambda$) satisfies $x_k(\lambda_0) \neq 0$ at some index $k \in [0, N + 1]_{\mathbb{Z}}$ and $\lambda_0 \in \mathbb{R}$, then there exists $\delta > 0$ such that $x_k(\lambda) \neq 0$ on the interval $(\lambda_0 - \delta, \lambda_0 + \delta)$. Moreover, as in the proof of Lemma 1.13, we can derive for all $\lambda \in (\lambda_0 - \delta, \lambda_0 + \delta)$ the formula

$$\dot{p}_k(\lambda) = -\frac{\dot{r}_k(\lambda)\,x_k^2(\lambda)}{r_k^2(\lambda)\,x_{k+1}^2(\lambda)} + \frac{1}{r_k^2(\lambda)} \sum_{j=0}^{k-1} \left\{ \dot{q}_j(\lambda)\,x_{j+1}^2(\lambda) - \dot{r}_j(\lambda)\,[\Delta x_j(\lambda)]^2 \right\},$$

$$\tag{1.34}$$

where

$$p_k(\lambda) := \frac{x_k(\lambda)}{r_k(\lambda)\, x_{k+1}(\lambda)}. \tag{1.35}$$

Identity (1.34) shows that the function $p_k(\lambda)$ is nondecreasing in $\lambda$ whenever it is defined, i.e., whenever $x_{k+1}(\lambda) \neq 0$. This monotonicity of $p_k(\lambda)$ in $\lambda$ is essential for deriving the oscillation theorem below. Note also that according to Definition 1.1, we have $p_k(\lambda) < 0$ if and only if the solution $x(\lambda)$ has a generalized zero in $(k, k + 1)$.

*Remark 1.16* The uniqueness of solutions of $(SL_\lambda)$ implies that a nontrivial solution $x(\lambda)$ of $(SL_\lambda)$ cannot vanish at any two consecutive points $k$ and $k + 1$. Therefore, if $x_k(\lambda) = 0$, then $x_{k+1}(\lambda) \neq 0$, while if $x_{k+1}(\lambda) = 0$, then $x_k(\lambda) \neq 0$.

Let $x(\lambda) = \{x_k(\lambda)\}_{k=0}^{N+1}$ be a nontrivial solution of $(SL_\lambda)$ and denote by $m_k(\lambda)$ the number of its generalized zeros in $(k, k + 1]$. Then $m_k(\lambda) \in \{0, 1\}$. Our aim is to prove the following local oscillation theorem.

**Theorem 1.17 (Local Oscillation Theorem I)** *Assume that* (1.29) *holds. Consider a nontrivial solution* $x(\lambda) = \{x_k(\lambda)\}_{k=0}^{N+1}$ *of* $(SL_\lambda)$ *satisfying* (1.31). *Fix an index* $k \in [0, N]_{\mathbb{Z}}$ *and denote by* $m_k(\lambda)$ *the number of generalized zeros of* $x(\lambda)$ *in* $(k, k + 1]$. *Then* $m_k(\lambda^-)$ *and* $m_k(\lambda^+)$ *exist and for all* $\lambda \in \mathbb{R}$

$$m_k(\lambda^+) = m_k(\lambda) \leq 1, \tag{1.36}$$

$$m_k(\lambda^+) - m_k(\lambda^-) = h_k(\lambda) - h_k(\lambda^-) + h_{k+1}(\lambda^-) - h_{k+1}(\lambda), \tag{1.37}$$

*where* $h_k(\lambda)$ *and* $h_{k+1}(\lambda)$ *are given in* (1.33).

In the above formula, the value of the function $h_j(\lambda)$ is 1 if $x_j(\lambda) \neq 0$, and it is 0 if $x_j(\lambda) = 0$, for $j \in \{k, k + 1\}$. Moreover, the notation $h_j(\lambda^-)$ means the left-hand limit of the function $h_j(\lambda)$ at the given point $\lambda$. Similarly, the notation $m_k(\lambda^-)$ and $m_k(\lambda^+)$ stands, respectively, for the left-hand and right-hand limits of the function $m_k(\lambda)$ at the point $\lambda$.

*Proof of Theorem 1.17* Let $k \in [0, N]_{\mathbb{Z}}$ and $\lambda_0 \in \mathbb{R}$ be given. By Remark 1.14, the limits $h_k(\lambda_0^-)$ and $h_{k+1}(\lambda_0^-)$ exist. We will show that the left-hand and right-hand limits of the function $m_k(\lambda)$ at $\lambda_0$ also exist and equations (1.36) and (1.37) are satisfied. We split the proof into two parts depending on the rank of $x_{k+1}(\lambda_0)$.

Part I. Assume first that $x_{k+1}(\lambda_0) \neq 0$. Then there exists $\varepsilon > 0$ such that $x_{k+1}(\lambda) \neq 0$ for all $\lambda \in (\lambda_0 - \varepsilon, \lambda_0 + \varepsilon)$. This means that for these values of $\lambda$, the point $k+1$ is not a generalized zero of the solution $x(\lambda)$. According to Remark 1.15, the function $p_k(\lambda)$ in (1.35) is nondecreasing on $(\lambda_0 - \varepsilon, \lambda_0 + \varepsilon)$, and we have on this interval either $m_k(\lambda) = 1$ if $p_k(\lambda) < 0$ or $m_k(\lambda) = 0$ if $p_k(\lambda) \geq 0$. We further distinguish the following three subcases:

(I-a)   $p_k(\lambda_0) < 0$,
(I-b)   $p_k(\lambda_0) > 0$, and
(I-c)   $p_k(\lambda_0) = 0$.

In subcase (I-a), in which $p_k(\lambda_0) < 0$, we have $p_k(\lambda) < 0$ and $x_k(\lambda) \neq 0$ for all $\lambda \in (\lambda_0 - \delta, \lambda_0 + \delta)$ for some $\delta \in (0, \varepsilon)$, so that in this case $m_k(\lambda_0) = m_k(\lambda_0^-) = m_k(\lambda_0^+) = 1$, $h_k(\lambda_0) = h_k(\lambda_0^-) = 1$, and $h_{k+1}(\lambda_0) = h_{k+1}(\lambda_0^-) = 1$. Therefore, the equations in (1.36) and (1.37) hold as the identities $1 = 1$ and $0 = 0$, respectively. Similarly in subcase (I-b), in which $p_k(\lambda_0) > 0$, there is $\delta \in (0, \varepsilon)$ such that $p_k(\lambda) > 0$ and $x_k(\lambda) \neq 0$ for all $\lambda \in (\lambda_0 - \delta, \lambda_0 + \delta)$, so that in this case $m_k(\lambda_0) = m_k(\lambda_0^-) = m_k(\lambda_0^+) = 0$, $h_k(\lambda_0) = h_k(\lambda_0^-) = 1$, and $h_{k+1}(\lambda_0) = h_{k+1}(\lambda_0^-) = 1$. Therefore, both equations (1.36) and (1.37) now hold as the identity $0 = 0$. In subcase (I-c), in which $p_k(\lambda_0) = 0$, we have $x_k(\lambda_0) = 0$. By Lemma 1.13, there is $\delta \in (0, \varepsilon)$ such that one of the additional four subcases applies for the behavior of $x_k(\lambda)$ near $\lambda_0$:

(I-c-i)    $x_k(\lambda) \neq 0$ on $(\lambda_0 - \delta, \lambda_0)$ and on $(\lambda_0, \lambda_0 + \delta)$,

(I-c-ii)   $x_k(\lambda) \neq 0$ on $(\lambda_0 - \delta, \lambda_0)$ and $x_k(\lambda) \equiv 0$ on $(\lambda_0, \lambda_0 + \delta)$,

(I-c-iii)  $x_k(\lambda) \equiv 0$ on $(\lambda_0 - \delta, \lambda_0)$ and $x_k(\lambda) \neq 0$ on $(\lambda_0, \lambda_0 + \delta)$, and

(I-c-iv)   $x_k(\lambda) \equiv 0$ both on $(\lambda_0 - \delta, \lambda_0)$ and on $(\lambda_0, \lambda_0 + \delta)$.

In subcase (I-c-i), the function $p_k(\lambda)$ must be nondecreasing on $(\lambda_0 - \delta, \lambda_0 + \delta)$, which implies that $p_k(\lambda) < 0$ on $(\lambda_0 - \delta, \lambda_0)$ and $p_k(\lambda) > 0$ on $(\lambda_0, \lambda_0 + \delta)$. Therefore, in this case $m_k(\lambda_0^-) = 1$, $m_k(\lambda_0^+) = m_k(\lambda_0) = 0$, $h_k(\lambda_0^-) = 1$, $h_k(\lambda_0) = 0$, and $h_{k+1}(\lambda_0^-) = h_{k+1}(\lambda_0) = 1$. This means that the equations in (1.36) and (1.37) now hold as the identities $0 = 0$ and $-1 = -1$, respectively. In subcase (I-c-ii), the function $p_k(\lambda)$ is nondecreasing on $(\lambda_0 - \delta, \lambda_0]$, which implies that $p_k(\lambda) < 0$ on $(\lambda_0 - \delta, \lambda_0)$ and $p_k(\lambda) \equiv 0$ on $(\lambda_0, \lambda_0 + \delta)$. Thus, as in subcase (I-c-i), we now have $m_k(\lambda_0^-) = 1$, $m_k(\lambda_0^+) = m_k(\lambda_0) = 0$, $h_k(\lambda_0^-) = 1$, $h_k(\lambda_0) = 0$, and $h_{k+1}(\lambda_0^-) = h_{k+1}(\lambda_0) = 1$, so that the equations in (1.36) and (1.37) hold as the identities $0 = 0$ and $-1 = -1$, respectively. In subcase (I-c-iii), the situation is similar with the result that $p_k(\lambda)$ is nondecreasing on $[\lambda_0, \lambda_0 + \delta)$, so that $p_k(\lambda) \equiv 0$ in $(\lambda_0 - \delta, \lambda_0]$ and $p_k(\lambda) > 0$ on $(\lambda_0, \lambda_0 + \delta)$. Thus, in this case $m_k(\lambda_0^-) = m_k(\lambda_0^+) = m_k(\lambda_0) = 0$, $h_k(\lambda_0^-) = h_k(\lambda_0) = 0$, and $h_{k+1}(\lambda_0^-) = h_{k+1}(\lambda_0) = 1$, so that both equations (1.36) and (1.37) hold as the identity $0 = 0$. In the last subcase (I-c-iv), we have $p_k(\lambda) \equiv 0$ on $(\lambda_0 - \delta, \lambda_0 + \delta)$ and in this case $m_k(\lambda_0^-) = m_k(\lambda_0^+) = m_k(\lambda_0) = 0$, $h_k(\lambda_0^-) = h_k(\lambda_0) = 0$, and $h_{k+1}(\lambda_0^-) = h_{k+1}(\lambda_0) = 1$, so that (1.36) and (1.37) hold as the identity $0 = 0$.

Part II. Assume that $x_{k+1}(\lambda_0) = 0$. Then by Remark 1.16, we have $x_k(\lambda_0) \neq 0$, and there exists $\varepsilon > 0$ such that $x_k(\lambda) \neq 0$ for all $\lambda \in (\lambda_0 - \varepsilon, \lambda_0 + \varepsilon)$. By Lemma 1.13, there is $\delta \in (0, \varepsilon)$ such that one of the following four subcases applies for the behavior of $x_{k+1}(\lambda)$ near the point $\lambda_0$:

(II-a)   $x_{k+1}(\lambda) \neq 0$ on $(\lambda_0 - \delta, \lambda_0)$ and on $(\lambda_0, \lambda_0 + \delta)$,

(II-b)   $x_{k+1}(\lambda) \neq 0$ on $(\lambda_0 - \delta, \lambda_0)$ and $x_{k+1}(\lambda) \equiv 0$ on $(\lambda_0, \lambda_0 + \delta)$,

(II-c)   $x_{k+1}(\lambda) \equiv 0$ on $(\lambda_0 - \delta, \lambda_0)$ and $x_{k+1}(\lambda) \neq 0$ on $(\lambda_0, \lambda_0 + \delta)$, and

(II-d)   $x_{k+1}(\lambda) \equiv 0$ both on $(\lambda_0 - \delta, \lambda_0)$ and on $(\lambda_0, \lambda_0 + \delta)$.

In subcase (II-a), the function $p_k(\lambda)$ is well defined on $(\lambda_0 - \delta, \lambda_0)$ and $(\lambda_0, \lambda_0 + \delta)$, so that it is nondecreasing on each of these two intervals, by Remark 1.15. Since $x_k(\lambda_0) \neq 0$, it follows that $p_k(\lambda_0^-) = +\infty$ and $p_k(\lambda_0^+) = -\infty$, which shows that $m_k(\lambda_0^-) = 0$ and $m_k(\lambda_0^+) = 1$. Since in this case we also have $m_k(\lambda_0) = 1$ (by the definition of a generalized zero at $k + 1$) and $h_k(\lambda_0^-) = h_k(\lambda_0) = 1$, $h_{k+1}(\lambda_0^-) = 1$, and $h_{k+1}(\lambda_0) = 0$, it follows that the equations in (1.36) and (1.37) hold as the identity $1 = 1$. In subcase (II-b), the function $p_k(\lambda)$ is well defined and nondecreasing on $(\lambda_0 - \delta, \lambda_0)$, so that $p_k(\lambda_0^-) = +\infty$, and hence $m_k(\lambda_0^-) = 0$. Moreover, $h_k(\lambda_0^-) = h_k(\lambda_0) = 1$, $h_{k+1}(\lambda_0^-) = 1$, $h_{k+1}(\lambda_0) = 0$, and $m_k(\lambda_0^+) = m_k(\lambda_0) = 1$, by the definition of a generalized zero at $k + 1$. This shows that in this case, (1.36) and (1.37) hold again as the identity $1 = 1$. In subcase (II-c), we have $m_k(\lambda_0^-) = m_k(\lambda_0) = 1$ (by the definition of a generalized zero at $k + 1$), $h_k(\lambda_0^-) = h_k(\lambda_0) = 1$, and $h_{k+1}(\lambda_0^-) = h_{k+1}(\lambda_0) = 0$. Moreover, the function $p_k(\lambda)$ is well defined and nondecreasing on $(\lambda_0, \lambda_0 + \delta)$, so that $p_k(\lambda_0^+) = -\infty$, and hence $m_k(\lambda_0^+) = 1$. In this case (1.36) and (1.37) hold as the identities $1 = 1$ and $0 = 0$, respectively. Finally, in subcase (II-d), we have $m_k(\lambda_0^-) = m_k(\lambda_0) = m_k(\lambda_0^+)$ (by the definition of a generalized zero at $k + 1$), while $h_k(\lambda_0^-) = h_k(\lambda_0) = 1$ and $h_{k+1}(\lambda_0^-) = h_{k+1}(\lambda_0) = 0$. Thus, both (1.36) and (1.37) now hold as the identity $0 = 0$. This completes the proof. $\qquad\square$

The above result (Theorem 1.17) now leads to further oscillation theorems for the problem (E$_0$). Denote by

$$n_1(\lambda) := \text{ the number of generalized zeros of } x(\lambda) \text{ in } (0, N + 1]. \qquad (1.38)$$

**Theorem 1.18 (Local Oscillation Theorem II)** *Assume that* (1.29) *holds. Consider a nontrivial solution* $x(\lambda) = \{x_k(\lambda)\}_{k=0}^{N+1}$ *of* (SL$_\lambda$) *satisfying* (1.31). *Then* $n_1(\lambda^-)$ *and* $n_1(\lambda^+)$ *exist and for all* $\lambda \in \mathbb{R}$

$$n_1(\lambda^+) = n_1(\lambda) \le N + 1, \qquad (1.39)$$

$$n_1(\lambda^+) - n_1(\lambda^-) = h_{N+1}(\lambda^-) - h_{N+1}(\lambda) \in \{0, 1\}. \qquad (1.40)$$

*Hence, the function* $n_1(\lambda)$ *is nondecreasing in* $\lambda$ *on* $\mathbb{R}$*; the limit*

$$m := \lim_{\lambda \to -\infty} n_1(\lambda) \qquad (1.41)$$

*exists with* $m \in [0, N + 1]_{\mathbb{Z}}$*, so that for a suitable* $\lambda_0 < 0$*, we have*

$$n_1(\lambda) \equiv m \quad \text{and} \quad h_{N+1}(\lambda^-) - h_{N+1}(\lambda) \equiv 0 \quad \text{for all } \lambda \le \lambda_0. \qquad (1.42)$$

*Proof* The number of generalized zeros of the solution $x(\lambda)$ in $(0, N + 1]$ is by definition

$$n_1(\lambda) = \sum_{k=0}^{N} m_k(\lambda), \qquad \lambda \in \mathbb{R},$$

where, as in Theorem 1.17, $m_k(\lambda)$ is the number of generalized zeros of $x(\lambda)$ in $(k, k + 1]$. The statement in (1.39) follows directly from (1.36). The expression in (1.40) is calculated by the telescope sum of the expression in (1.37). This yields that

$$n_1(\lambda^+) - n_1(\lambda^-) = h_{N+1}(\lambda^-) - h_{N+1}(\lambda) - h_0(\lambda^-) + h_0(\lambda), \quad \lambda \in \mathbb{R}.$$

But since by (1.31) the initial conditions of $x(\lambda)$ do not depend on $\lambda$, we have $h_0(\lambda^-) = h_0(\lambda)$ for all $\lambda \in \mathbb{R}$, which shows (1.40). From the two conditions (1.39) and (1.40), we then have that the function $n_1(\lambda)$ is nondecreasing in $\lambda$ on $\mathbb{R}$. Since the values of $n_1(\lambda)$ are nonnegative integers, the limit in (1.41) exists with $m \in \mathbb{N} \cup \{0\}$. Consequently, $n_1(\lambda) \equiv m$ for $\lambda$ sufficiently negative, say for all $\lambda \leq \lambda_0$ for some $\lambda_0 < 0$. Hence, $n_1(\lambda^+) - n_1(\lambda^-) \equiv 0$ for $\lambda \leq \lambda_0$. Applying (1.40) once more then yields the second equation in (1.42). This completes the proof.                                □

Now we relate the above oscillation results with the eigenvalue problem (E$_0$). We say that a number $\lambda_0 \in \mathbb{R}$ is a *finite eigenvalue* of (E$_0$), provided there exists a nontrivial solution $x(\lambda) = \{x_k(\lambda)\}_{k=0}^{N+1}$ of (E$_0$) such that $x_{N+1}(\lambda_0) = 0$ and

$$x_{N+1}(\lambda) \neq 0 \text{ for } \lambda \text{ in some left neighborhood of } \lambda_0. \tag{1.43}$$

Note that such a requirement is justified by Lemma 1.13. We observe that every finite eigenvalue of (E$_0$) is also a traditional eigenvalue, for which the "nondegeneracy condition" (1.43) is dropped. From the uniqueness of solutions of equation (SL$_\lambda$), it then follows that $\lambda_0$ is a finite eigenvalue of (E$_0$) if and only if the principal solution $\hat{x}(\lambda)$ (see (1.32)) satisfies $\hat{x}_{N+1}(\lambda_0) = 0$ and $\hat{x}_{N+1}(\lambda) \neq 0$ for $\lambda$ in some left neighborhood of $\lambda_0$. Or equivalently, the principal solution $\hat{x}(\lambda)$ has $h_{N+1}(\lambda_0^-) = 1$ and $h_{N+1}(\lambda_0) = 0$. This shows that the difference $h_{N+1}(\lambda_0^-) - h_{N+1}(\lambda_0)$, whenever it is positive, indicates a finite eigenvalue of problem (E$_0$).

From Lemma 1.13 we obtain that under the assumption (1.29), the finite eigenvalues of (E$_0$) are isolated. This property was also proven for the classical eigenvalues of (SL$_\lambda$) in [44] under the strict monotonicity of $r_k(\lambda)$ and $q_k(\lambda)$. Such a strict monotonicity assumption is not required in this subsection.

Thus, we finally arrive at the following global oscillation theorem. Denote by

$$n_2(\lambda) := \text{ the number of finite eigenvalues of (E$_0$) in } (-\infty, \lambda]. \tag{1.44}$$

Then from this definition, we have

$$n_2(\lambda^+) = n_2(\lambda), \quad n_2(\lambda) - n_2(\lambda^-) = h_{N+1}(\lambda^-) - h_{N+1}(\lambda) \quad \text{for all } \lambda \in \mathbb{R}, \tag{1.45}$$

i.e., the positivity of the difference $n_2(\lambda) - n_2(\lambda^-)$ indicates a finite eigenvalue at $\lambda$.

**Theorem 1.19 (Global Oscillation Theorem)** *Assume* (1.29). *Then for all* $\lambda \in \mathbb{R}$

$$n_2(\lambda^+) = n_2(\lambda) \leq 1, \tag{1.46}$$

$$n_2(\lambda^+) - n_2(\lambda^-) = n_1(\lambda^+) - n_1(\lambda^-) \in \{0, 1\}, \tag{1.47}$$

*and there exists* $m \in [0, N+1]_{\mathbb{Z}}$ *such that*

$$n_1(\lambda) = n_2(\lambda) + m \qquad \text{for all } \lambda \in \mathbb{R}. \tag{1.48}$$

*Moreover, for a suitable* $\lambda_0 < 0$, *we have*

$$n_2(\lambda) \equiv 0 \quad \text{and} \quad n_1(\lambda) \equiv m \quad \text{for all } \lambda \leq \lambda_0. \tag{1.49}$$

*Proof* The result follows directly from Theorem 1.18. $\qquad\square$

**Corollary 1.20** *Under the assumption* (1.29), *the finite eigenvalues of* (E$_0$) *are isolated and bounded from below.*

*Proof* From Lemma 1.13 we know that the finite eigenvalues of (E$_0$) are isolated. The second statement follows from condition (1.49) of Theorem 1.19, since $n_2(\lambda) \equiv$ 0 for all $\lambda \leq \lambda_0$ means that there are no finite eigenvalues of (E$_0$) in the interval $(-\infty, \lambda_0]$. $\qquad\square$

It remains to connect the above global oscillation theorem with the traditional statement saying that the $j$-th eigenfunction has exactly $j$ generalized zeros in the interval $(0, N+1]$. We will see that under some additional assumption, the statement of this result remains exactly the same when we replace the eigenfunctions of (E$_0$) by its finite eigenfunctions. This additional assumption is formulated in terms of the associated discrete quadratic functional (1.13)

$$F(\eta, \lambda) := \sum_{k-0}^{N} \left\{ r_k(\lambda) (\Delta \eta_k)^2 - q_k(\lambda) \eta_{k+1}^2 \right\},$$

where $\eta = \{\eta_k\}_{k=0}^{N+1}$ is a sequence such that $\eta_0 = 0 = \eta_{N+1}$. The functional $F(\cdot, \lambda)$ is *positive*; we write $F(\cdot, \lambda) > 0$, if $F(\eta, \lambda) > 0$ for every sequence $\eta$ with $\eta_0 = 0 = \eta_{N+1}$ and $\eta \neq 0$.

**Theorem 1.21 (Oscillation Theorem)** *Assume* (1.29). *Then*

$$n_1(\lambda) = n_2(\lambda) \qquad \text{for all } \lambda \in \mathbb{R} \tag{1.50}$$

*if and only if there exists* $\lambda_0 < 0$ *such that* $F(\cdot, \lambda_0) > 0$. *In this case, if* $\lambda_1 < \lambda_2 < \cdots < \lambda_r$ *(where* $r \leq N+1$) *are the finite eigenvalues of* (E$_0$) *with the corresponding finite eigenfunctions* $x^{(1)}, x^{(2)}, \ldots, x^{(r)}$, *then for each* $j \in \{1, \ldots, r\}$ *the finite eigenfunction* $x^{(j)}$ *has exactly* $j$ *generalized zeros in* $(0, N+1]$.

*Remark 1.22*

(i) Note that since the finite eigenfunction $x^{(j)}$ has $x_{N+1}^{(j)} = 0$, it satisfies $x_N^{(j)} \neq 0$, by Remark 1.16. Therefore, the point $N + 1$ is one of the generalized zeros of $x^{(j)}$, and consequently the remaining $j - 1$ generalized zeros of $x^{(j)}$ are in the open interval $(0, N + 1)$. This complies with the traditional continuous time statement.

(ii) Conditions of Theorem 1.21 are automatically satisfied for the classical Sturm-Liouville problem (1.30), i.e., (1.50). Indeed, for the case $q_k(\lambda) = q_k + w_k\lambda$ and $w_k > 0$, we can estimate

$$F(\eta, \lambda) := \sum_{k=0}^{N} \left\{ r_k \, (\Delta\eta_k)^2 - q_k \, \eta_{k+1}^2 \right\} - \lambda \sum_{k=0}^{N} w_k\eta_{k+1}^2 = F_0(\eta, \lambda) - \lambda \sum_{k=0}^{N} w_k\eta_{k+1}^2$$

such that

$$p_k = r_k \, (\Delta\eta_k)^2 - q_k \, \eta_{k+1}^2 = \begin{pmatrix} \eta_k & \eta_{k+1} \end{pmatrix} \begin{pmatrix} r_k & -r_k \\ -r_k & r_k - q_k \end{pmatrix} \begin{pmatrix} \eta_k \\ \eta_{k+1} \end{pmatrix}$$

obeys the inequality $|p_k| \leq c \, (\eta_k^2 + \eta_{k+1}^2)$ for some $c > 0$, and then introducing the notation $d = \min_{k \in [0,N]_{\mathbb{Z}}} w_k > 0$, we see that there exists $\lambda_0 < 0$ such that for all $\lambda < \lambda_0$

$$F(\eta, \lambda) = F_0(\eta, \lambda) - \lambda \sum_{k=0}^{N} w_k\eta_{k+1}^2 \geq -2c\|\eta\|^2 - \lambda d \|\eta\|^2 > 0,$$

where we use that $\|\eta\| \neq 0$ (see also [206, Theorem 3]).

*Proof of Theorem 1.21* If $n_1(\lambda) = n_2(\lambda)$ for all $\lambda \in \mathbb{R}$, then the number $m$ in equation (1.48) of Theorem 1.19 is zero. This implies through condition (1.49) that $n_1(\lambda) \equiv 0$ for all $\lambda \leq \lambda_0$ with some $\lambda_0 < 0$. By Theorem 1.2, the latter condition is equivalent to the positivity of the functional $F(\cdot, \lambda)$ for every $\lambda \leq \lambda_0$, in particular for $\lambda = \lambda_0$. Conversely, assume that $F(\cdot, \lambda_0) > 0$ for some $\lambda_0 < 0$. Then $n_1(\lambda_0) = 0$, by Theorem 1.2, and since the function $n_1(\cdot)$ is nondecreasing in $\lambda$ on $\mathbb{R}$ (see Theorem 1.18), it follows that $n_1(\lambda) \equiv 0$ for all $\lambda \leq \lambda_0$. From this we see that $m = 0$ in (1.49) and hence also in (1.48). Equality (1.50) is therefore established. Finally, assume that (1.50) holds and let $\lambda_j$ (where $j \in \{1, \ldots, r\}$) be the $j$-th finite eigenvalue of $(E_0)$ with the corresponding finite eigenfunction $x^{(j)}$. Then $n_2(\lambda_j) = j$, and from (1.50), we get $n_1(\lambda_j) = j$, i.e., $x^{(j)}$ has exactly $j$ generalized zeros in $(0, N + 1]$. The proof is complete. $\square$

In the last part of this section, we present certain results on the existence of finite eigenvalues of $(E_0)$, in particular a necessary condition and a sufficient condition for the existence of a finite eigenvalue and a characterization of the smallest finite eigenvalue. The proofs of these results are also based on Theorem 1.2 (see also more general Theorems 5.32–5.34 for symplectic systems).

**Theorem 1.23** *Assume* (1.29). *If* ($E_0$) *has a finite eigenvalue, then there exist* $\lambda_0, \lambda_1 \in \mathbb{R}$ *with* $\lambda_0 < \lambda_1$ *and* $m \in \mathbb{N} \cup \{0\}$ *such that* $n_1(\lambda) \equiv m$ *for all* $\lambda \leq \lambda_0$ *and* $F(\cdot, \lambda_1) \not\equiv 0$.

**Theorem 1.24** *Assume* (1.29). *If there exist* $\lambda_0, \lambda_1 \in \mathbb{R}$ *with* $\lambda_0 < \lambda_1$ *such that* $F(\cdot, \lambda_0) > 0$ *and* $F_0(\cdot, \lambda_1) \not\equiv 0$, *then* ($E_0$) *has at least one finite eigenvalue.*

**Theorem 1.25** *Assume* (1.29). *Let there exist* $\lambda_0, \lambda_1 \in \mathbb{R}$ *with* $\lambda_0 < \lambda_1$ *such that* $F(\cdot, \lambda_0) > 0$ *and* $F(\cdot, \lambda_1) \not\equiv 0$. *Then the eigenvalue problem* ($E_0$) *possesses the smallest finite eigenvalue* $\lambda_{\min}$, *which is characterized by any of the conditions:*

$$\lambda_{\min} = \sup\{\lambda \in \mathbb{R}, \ F(\cdot, \lambda) > 0\}, \quad \lambda_{\min} = \min\{\lambda \in \mathbb{R}, \ F(\cdot, \lambda) \not\equiv 0\}.$$

## 1.3 Discrete Variational Theory

In this section we explain the motivation and origin of the symplectic difference systems (SDS) in the discrete variational theory, in particular in the discrete calculus of variations and discrete optimal control theory. We shall see that in both cases symplectic difference systems arise in these variational problems naturally as the second-order systems (i.e., as the Jacobi systems). Moreover, in Theorems 1.29 and 1.34, we justify the fact that the theory of symplectic difference systems, rather than the theory linear Hamiltonian difference systems, is the proper platform for studying the second-order optimality conditions in the discrete variational theory.

### 1.3.1 Discrete Calculus of Variations

Given an index $N \in \mathbb{N}$, we consider the classical nonlinear discrete calculus of variations problem

$$\text{minimize} \quad \mathcal{F}(x) := K(x_0, x_{N+1}) + \sum_{k=0}^{N} L(k, x_{k+1}, \Delta x_k) \tag{1.51}$$

subject to sequences $x = \{x_k\}_{k=0}^{N+1}$ satisfying the endpoints constraint

$$\varphi(x_0, x_{N+1}) = 0. \tag{1.52}$$

Here $x : [0, N+1]_\mathbb{Z} \to \mathbb{R}^n$ is the state variable, $K : \mathbb{R}^{2n} \to \mathbb{R}$ is the endpoints cost, $L : [0, N]_\mathbb{Z} \times \mathbb{R}^n \times \mathbb{R}^n \to \mathbb{R}$ is the Lagrangian, and $\varphi : \mathbb{R}^{2n} \to \mathbb{R}^r$ with $r \leq 2n$ is the constraint function. We assume that the data are sufficiently smooth, i.e., $C^1$ for the first-order optimality conditions and $C^2$ for the second-order optimality conditions.

A sequence $x = \{x_k\}_{k=0}^{N+1}$ is *feasible* if it satisfies the boundary condition (1.52). A feasible sequence $\hat{x}$ is called a *local minimum* for (1.51) if there exists $\varepsilon > 0$ such that $\mathcal{F}(\hat{x}) \leq \mathcal{F}(x)$ for all feasible $x$ satisfying $\|x_k - \hat{x}_k\| < \varepsilon$ for all $k \in [0, N+1]_\mathbb{Z}$, where $\|\cdot\|$ is any norm in $\mathbb{R}^n$. Moreover, a local minimum $\hat{x}$ is *strict* if $\mathcal{F}(\hat{x}) < \mathcal{F}(x)$ for all feasible $x \neq \hat{x}$ with $\|x_k - \hat{x}_k\| < \varepsilon$ for all $k \in [0, N+1]_\mathbb{Z}$. Note that the concepts of a weak and strong local extremum now coincide, since problem (1.51) is finite dimensional.

In the next two theorems, we present the first- and second-order necessary and sufficient optimality conditions for problem (1.51). The proofs of these results can be found, e.g., in [16, 178, 204]; see also the comments in Sect. 1.7. In the sequel, we shall denote by $L_x$ and $L_u$ the gradients of $L$ (i.e., the row vectors of partial derivatives of $L$) with respect to the second and third variables. Similarly, we denote by $L_{xx}$, $L_{xu}$, $L_{uu}$ the matrices containing the corresponding second-order partial derivatives of $L$. We note that in contrast with the exposition in [16, Chapter 4], the gradients $L_x$ and $L_u$ are here the *row* vectors. The gradient of the functions $K$ and $\varphi$ will be denoted by the symbol $\nabla$.

Linearizing the problem (1.51) along a feasible sequence $\hat{x}$, we define the $r \times 2n$ matrix

$$M := \nabla\varphi(\hat{x}_0, \hat{x}_{N+1}). \tag{1.53}$$

A sequence $\eta = \{\eta_k\}_{k=0}^{N+1}$ is called *admissible* if it satisfies the endpoints constraint

$$M \begin{pmatrix} \eta_0 \\ \eta_{N+1} \end{pmatrix} = 0. \tag{1.54}$$

The second variation of the functional $\mathcal{F}$ at a feasible sequence $\hat{x}$ is the discrete quadratic functional

$$\mathcal{F}''(\hat{x}, \eta) := \begin{pmatrix} \eta_0 \\ \eta_{N+1} \end{pmatrix}^T \Gamma \begin{pmatrix} \eta_0 \\ \eta_{N+1} \end{pmatrix} + \sum_{k=0}^{N} \begin{pmatrix} \eta_{k+1} \\ \Delta\eta_k \end{pmatrix}^T \begin{pmatrix} P_k & Q_k \\ Q_k^T & R_k \end{pmatrix} \begin{pmatrix} \eta_{k+1} \\ \Delta\eta_k \end{pmatrix}, \tag{1.55}$$

where the symmetric $2n \times 2n$ matrix $\Gamma$ and the $n \times n$ matrices $P_k$, $Q_k$, $R_k$ are given by

$$P_k := \hat{L}_{xx}(k), \quad Q_k := \hat{L}_{xu}(k), \quad R_k := \hat{L}_{uu}(k), \tag{1.56}$$

$$\Gamma := \nabla^2 K^T(\hat{x}_0, \hat{x}_{N+1}) + \gamma^T \nabla^2 \varphi^T(\hat{x}_0, \hat{x}_{N+1}) \tag{1.57}$$

with some $\gamma \in \mathbb{R}^r$ and with the partial derivatives of $L$ evaluated at $(k, \hat{x}_{k+1}, \Delta\hat{x}_k)$. Note that the matrices $P_k$ and $R_k$ are symmetric, as the Lagrangian $L$ is assumed to be $C^2$ in the second and third variables.

We say that the second variation $\mathcal{F}''$ is *nonnegative* at $\hat{x}$ if $\mathcal{F}''(\hat{x}, \eta) \geq 0$ for all admissible sequences $\eta = \{\eta_k\}_{k=0}^{N+1}$. An alternative terminology is nonnegative definite or positive semidefinite at $\hat{x}$. The quadratic functional is *positive* at $\hat{x}$ if $\mathcal{F}(\hat{x}, \eta) > 0$ for all admissible sequences $\eta = \{\eta_k\}_{k=0}^{N+1}$ satisfying $\eta \not\equiv 0$. An alternative terminology is positive definite at $\hat{x}$.

**Theorem 1.26** *Let $\{\hat{x}_k\}_{k=0}^{N+1}$ be a local minimum for problem (1.51) and assume that the matrix $M$ in (1.53) has full rank $r$. Then there exists a vector $\gamma \in \mathbb{R}^r$ such that the following conditions hold:*

(i) *the Euler-Lagrange difference equation*

$$\Delta \hat{L}_u(k) = \hat{L}_x(k), \quad k \in [0, N-1]_{\mathbb{Z}}, \tag{1.58}$$

*where the partial derivatives of $L$ are evaluated at $(k, \hat{x}_{k+1}, \Delta \hat{x}_k)$,*
(ii) *the transversality condition*

$$\left( \hat{L}_u(0), -\hat{L}_u(N) - \hat{L}_x(N) \right) = \nabla K(\hat{x}_0, \hat{x}_{N+1}) + \gamma^T M, \tag{1.59}$$

(iii) *the second variation $\mathcal{F}''$ with the data from (1.56) and (1.57) evaluated at $(k, \hat{x}_{k+1}, \Delta \hat{x}_k)$ is nonnegative at $\hat{x}$.*

*Remark 1.27* If we *define $\hat{L}_u$ at $k = N + 1$ by $\hat{L}_u(N + 1) := \hat{L}_u(N) + \hat{L}_x(N)$,* then the Euler-Lagrange difference equation (1.58) is satisfied for all $k \in [0, N]_{\mathbb{Z}}$, and the transversality condition in (1.59) can be written in the form known in the continuous time theory, i.e.,

$$\left( \hat{L}_u(0), -\hat{L}_u(N + 1) \right) = \nabla K(\hat{x}_0, \hat{x}_{N+1}) + \gamma^T M.$$

Next we formulate sufficient optimality conditions for problem (1.51).

**Theorem 1.28** *Suppose that a feasible sequence $\hat{x} = \{\hat{x}_k\}_{k=0}^{N+1}$ satisfies, for some $\gamma \in \mathbb{R}^r$, the first-order optimality conditions (i) and (ii) in Theorem 1.26 with the matrix $M$ in (1.53) having full rank $r$. Furthermore, suppose that the second variation $\mathcal{F}''$ is positive at $\hat{x}$. Then $\hat{x}$ is a strict local minimum for problem (1.51).*

If the second variation $\mathcal{F}''$ is nonnegative or even positive at $\hat{x}$, then the zero sequence is its minimum. Applying Theorem 1.26 to this situation, we obtain the *Jacobi equation* for problem (1.51), which is the Euler-Lagrange difference equation for the functional $\mathcal{F}''$. Thus, from (1.58) we get the second-order difference equation

$$\Delta(R_k \Delta \eta_k + Q_k^T \eta_{k+1}) = P_k \eta_{k+1} + Q_k \Delta \eta_k, \quad k \in [0, N-1]_{\mathbb{Z}}. \tag{1.60}$$

Note that when $n = 1$ and $Q_k = 0$ for all $k \in [0, N-1]_{\mathbb{Z}}$, equation (1.60) becomes the Sturm-Liouville difference equation (1.1) studied in Sect. 1.2.

It is well-known in the literature that equation (1.60) can be written as a special symplectic difference system (SDS) when the matrices

$$R_k \text{ and } R_k + Q_k^T \text{ are invertible for all } k \in [0, N]_{\mathbb{Z}}. \tag{1.61}$$

More precisely, in [16, Example 3.17] it is shown that under (1.61) the substitution

$$\left.\begin{array}{ll} x_k := \eta_k, & k \in [0, N+1]_{\mathbb{Z}}, \\ u_k := R_k \Delta \eta_k + Q_k^T \eta_{k+1}, & k \in [0, N]_{\mathbb{Z}}, \\ u_{N+1} := u_N + P_N \eta_{N+1} + Q_N \Delta \eta_N \end{array}\right\}  \tag{1.62}$$

transforms equation (1.60) into the linear Hamiltonian difference system

$$\Delta x_k = A_k x_{k+1} + B_k u_k, \quad \Delta u_k = C_k x_{k+1} - A_k^T u_k, \quad k \in [0, N]_{\mathbb{Z}}, \tag{1.63}$$

with symmetric $B_k$ and $C_k$ and invertible $I - A_k$. In turn, it is shown in [16, Example 3.10] that system (1.63) is a special symplectic system (SDS) with invertible $\mathcal{A}_k = (I - A_k)^{-1}$ for all $k \in [0, N]_{\mathbb{Z}}$; see also Sect. 2.1.2.

In the next result, we present an alternative method, which transforms the Jacobi equation (1.60) directly into the symplectic system (SDS). This method uses the weaker assumption that only the matrix

$$R_k + Q_k^T \text{ is invertible for all } k \in [0, N]_{\mathbb{Z}}. \tag{1.64}$$

Therefore, in comparison with the result in [16, Section 3.6], we allow the matrices $R_k$ to be singular.

**Theorem 1.29** *Assume that* (1.64) *holds and define for* $k \in [0, N]_{\mathbb{Z}}$ *the* $2n \times 2n$ *matrices* $\mathcal{S}_k := \begin{pmatrix} \mathbb{A}_k & \mathbb{B}_k \\ \mathbb{C}_k & \mathbb{D}_k \end{pmatrix}$ *by*

$$\mathbb{A}_k = (R_k + Q_k^T)^{-1} R_k, \quad \mathbb{C}_k = P_k (R_k + Q_k^T)^{-1} R_k - Q_k (R_k + Q_k^T)^{-1} Q_k^T,$$
$$\mathbb{B}_k = (R_k + Q_k^T)^{-1}, \quad \mathbb{D}_k = (R_k + Q_k + Q_k^T + P_k)(R_k + Q_k^T)^{-1}.$$

*Then* $\mathcal{S}_k$ *is a symplectic matrix for all* $k \in [0, N]_{\mathbb{Z}}$. *Consequently, the Jacobi equation* (1.60) *is a special symplectic system* (SDS), *in which* $y_k := (x_k^T, u_k^T)^T$ *is defined by* (1.62). *In addition, the resulting symplectic system* (SDS) *is a Hamiltonian system* (1.63) *if and only if the matrix* $R_k$ *is invertible for all* $k \in [0, N]_{\mathbb{Z}}$.

*Proof* By direct calculations and with the aid of the symmetry of $R_k$ and $P_k$, one can easily verify that $\mathcal{S}_k^T \mathcal{J} \mathcal{S}_k = \mathcal{J}$, which proves the statement. Alternatively, see Theorem 1.34 in combination with Remark 1.35.                                    □

*Remark 1.30* Analogously to problem (1.51), we now consider the discrete calculus of variations problem without the shift in $x_{k+1}$ in the Lagrangian $L$, i.e.,

$$\text{minimize} \quad \underline{\mathcal{F}}(x) := K(x_0, x_{N+1}) + \sum_{k=0}^{N} \underline{L}(k, x_k, \Delta x_k) \tag{1.65}$$

subject to sequences $x = \{x_k\}_{k=0}^{N+1}$ satisfying the endpoints constraint (1.52). Then a similar analysis as in Theorem 1.26 yields the Euler-Lagrange difference equation

$$\Delta \hat{\underline{L}}_u(k) = \hat{\underline{L}}_x(k+1), \quad k \in [0, N-1]_{\mathbb{Z}}, \tag{1.66}$$

or equivalently

$$\Delta[\hat{\underline{L}}_u(k) - \hat{\underline{L}}_x(k)] = \hat{\underline{L}}_x(k), \quad k \in [0, N-1]_{\mathbb{Z}}, \tag{1.67}$$

where the partial derivatives of $\underline{L}$ are evaluated at $(k, \hat{x}_k, \Delta\hat{x}_k)$ and the transversality condition

$$\left( \hat{\underline{L}}_u(0) - \hat{\underline{L}}_x(0), \; -\hat{\underline{L}}_u(N) \right) = \nabla K(\hat{x}_0, \hat{x}_{N+1}) + \gamma^T M. \tag{1.68}$$

In fact, these results for problem (1.65) can also be obtained from problem (1.51) upon writing $x_k = x_{k+1} - \Delta x_k$ and considering the Lagrangian $L(k, x, u) := \underline{L}(k, x - u, u)$ in (1.51). Applying (1.67) to the second variation $\underline{\mathcal{F}}''$, the resulting Jacobi difference equation then has the form

$$\Delta[(\underline{R}_k - \underline{Q}_k)\, \Delta\eta_k + (\underline{Q}_k^T - \underline{P}_k)\, \eta_k] = \underline{P}_k\eta_k + \underline{Q}_k\Delta\eta_k, \quad k \in [0, N-1]_{\mathbb{Z}}. \tag{1.69}$$

Equation (1.69) can be written as the symplectic system (SDS) when the matrix $\underline{R}_k - \underline{Q}_k^T$ (and not necessarily $\underline{R}_k$) is invertible for all $k \in [0, N]_{\mathbb{Z}}$. These systems are studied in more detail in [294, Section 2].

## 1.3.2   Discrete Optimal Control Theory

We now extend the considerations in the previous subsection to discrete optimal control setting. Thus, for a fixed index $N \in \mathbb{N}$, we consider the discrete optimal control problem

$$\text{minimize} \quad \mathcal{G}(x, u) := K(x_0, x_{N+1}) + \sum_{k=0}^{N} L(k, x_{k+1}, u_k) \tag{1.70}$$

subject to the pairs $(x, u)$ of sequences $x = \{x_k\}_{k=0}^{N+1}$ and $u = \{u_k\}_{k=0}^{N}$ satisfying the difference equation

$$\Delta x_k = f(k, x_{k+1}, u_k), \quad k \in [0, N]_{\mathbb{Z}}, \tag{1.71}$$

and the endpoints constraint (1.52). Here $x : [0, N+1]_{\mathbb{Z}} \to \mathbb{R}^n$ is the state variable, $u : [0, N]_{\mathbb{Z}} \to \mathbb{R}^m$ with $m \leq n$ is the control variable, $K : \mathbb{R}^{2n} \to \mathbb{R}$ is the endpoints cost, $L : [0, N]_{\mathbb{Z}} \times \mathbb{R}^n \times \mathbb{R}^m \to \mathbb{R}$ is the Lagrangian, $f : [0, N]_{\mathbb{Z}} \times \mathbb{R}^n \times \mathbb{R}^m \to \mathbb{R}^n$

is the dynamics, and $\varphi : \mathbb{R}^{2n} \to \mathbb{R}^r$ with $r \leq 2n$ is the constraint function. We assume that the functions $K$, $L$, $f$, $\varphi$ are sufficiently smooth, i.e., $C^1$ for the first-order optimality conditions and $C^2$ for the second-order optimality conditions. In order to be able to solve equation (1.71) for $x_{k+1}$ in terms of $x_k$ and $u_k$, we assume that the $n \times n$ matrix

$$I - f_x(k, x, u) \text{ is invertible for every } k \in [0, N]_{\mathbb{Z}}, x \in \mathbb{R}^n, u \in \mathbb{R}^m. \qquad (1.72)$$

A pair $(x, u)$ of sequences $x = \{x_k\}_{k=0}^{N+1}$ and $u = \{u_k\}_{k=0}^{N}$ is *feasible* if it satisfies (1.71) and (1.52). A feasible pair $(\hat{x}, \hat{u})$ is called a *local minimum* for (1.70) if there exists $\varepsilon > 0$ such that $\mathcal{G}(\hat{x}, \hat{u}) \leq \mathcal{G}(x, u)$ for all feasible pairs $(x, u)$ satisfying $\|x_k - \hat{x}_k\| < \varepsilon$ for all $k \in [0, N + 1]_{\mathbb{Z}}$ and $\|u_k - \hat{u}_k\| < \varepsilon$ for all $k \in [0, N]_{\mathbb{Z}}$. Here $\|\cdot\|$ is any norm in $\mathbb{R}^n$, resp., in $\mathbb{R}^m$. Moreover, a local minimum $(\hat{x}, \hat{u})$ is *strict* if $\mathcal{G}(\hat{x}, \hat{u}) < \mathcal{G}(x, u)$ for all such feasible pairs $(x, u) \neq (\hat{x}, \hat{u})$.

The *Hamiltonian* $\mathcal{H} : [0, N]_{\mathbb{Z}} \times \mathbb{R}^n \times \mathbb{R}^m \times \mathbb{R}^n \times \mathbb{R} \to \mathbb{R}$ corresponding to problem (1.70), (1.71) is defined by

$$\mathcal{H}(k, x, u, p, \lambda) := p^T f(k, x, u) + \lambda L(k, x, u). \qquad (1.73)$$

Denote the gradients of the function $f$ along a feasible pair $(\hat{x}, \hat{u})$ by

$$\mathcal{A}_k := f_x(k, \hat{x}_{k+1}, \hat{u}_k), \quad \mathcal{B}_k := f_u(k, \hat{x}_{k+1}, \hat{u}_k) \qquad (1.74)$$

and define the $r \times 2n$ matrix $M$ by (1.53). Then we have $\mathcal{A} : [0, N]_{\mathbb{Z}} \to \mathbb{R}^{n \times n}$ and $\mathcal{B} : [0, N]_{\mathbb{Z}} \to \mathbb{R}^{n \times m}$, and, according to (1.72), the matrix $I - \mathcal{A}_k$ is invertible for all $k \in [0, N]_{\mathbb{Z}}$. We say that the linear system

$$\Delta \eta_k = \mathcal{A}_k \eta_{k+1} + \mathcal{B}_k v_k, \quad k \in [0, N]_{\mathbb{Z}}, \qquad (1.75)$$

is *M-controllable* if for every $d \in \mathbb{R}^r$ there exists a vector $\alpha \in \mathbb{R}^n$ and a sequence $v = \{v_k\}_{k=0}^{N}$ such that the solution $\eta = \{\eta_k\}_{k=0}^{N+1}$ of the initial value problem (1.75) with $\eta_0 = \alpha$ satisfies

$$M \begin{pmatrix} \eta_0 \\ \eta_{N+1} \end{pmatrix} = d.$$

In [192, Proposition 4.5] it is proved that if the matrix $M$ has full rank $r$ and $I - \mathcal{A}_k$ is invertible for all $k \in [0, N]_{\mathbb{Z}}$, then the $M$-controllability of system (1.75) is equivalent with the following normality condition on problem (1.70): the system

$$\Delta p_k = -\mathcal{A}_k^T p_k, \quad \mathcal{B}_k p_k = 0, \quad k \in [0, N]_{\mathbb{Z}}, \quad \begin{pmatrix} -p_0 \\ p_{N+1} \end{pmatrix} = M^T \gamma, \qquad (1.76)$$

where $\gamma \in \mathbb{R}^r$ possesses only the trivial solution $p_k \equiv 0$ on $[0, N + 1]_{\mathbb{Z}}$ (and then also $\gamma = 0$).

A pair $(\eta, v)$ with $\eta = \{\eta_k\}_{k=0}^{N+1}$ and $v = \{v_k\}_{k=0}^{N}$ is said to be *admissible* if it satisfies equation (1.75) and the boundary condition (1.54). For a normal problem, (1.70) we define the second variation of the functional $\mathcal{G}$ at a feasible pair $(\hat{x}, \hat{u})$ as the discrete quadratic functional

$$\mathcal{G}''(\hat{x}, \hat{u}, \eta, v) := \begin{pmatrix} \eta_0 \\ \eta_{N+1} \end{pmatrix}^T \Gamma \begin{pmatrix} \eta_0 \\ \eta_{N+1} \end{pmatrix} + \sum_{k=0}^{N} \begin{pmatrix} \eta_{k+1} \\ v_k \end{pmatrix}^T \begin{pmatrix} P_k & Q_k \\ Q_k^T & R_k \end{pmatrix} \begin{pmatrix} \eta_{k+1} \\ v_k \end{pmatrix}.$$

(1.77)

The $2n \times 2n$ matrix $\Gamma$ is given in (1.57), and the $n \times n$, $n \times m$, $m \times m$ matrices $P_k$, $Q_k$, $R_k$, respectively, are defined by

$$\left. \begin{aligned} P_k &:= p_k^T \hat{f}_{xx}(k) + \hat{L}_{xx}(k), \\ Q_k &:= p_k^T \hat{f}_{xu}(k) + \hat{L}_{xu}(k), \\ R_k &:= p_k^T \hat{f}_{uu}(k) + \hat{L}_{uu}(k), \end{aligned} \right\}$$

(1.78)

for some sequence $p = \{p_k\}_{k=0}^{N+1}$, see Theorem 1.32. The partial derivatives of $f$ and $L$ are evaluated at $(k, \hat{x}_{k+1}, \hat{u}_k)$. Note that $P_k$ and $R_k$ are symmetric.

*Remark 1.31* When the endpoints in problem (1.70) or in problem (1.60) are fixed, i.e., $x_0 = A$ and $x_{N+1} = B$ with some given $A, B \in \mathbb{R}^n$, we have $r = 2n$ and $M = I_{2n}$ in (1.53). In this case the endpoints cost $K(x_0, x_{N+1}) = K(A, B)$ is constant; the variations $\eta = \{\eta_k\}_{k=0}^{N+1}$ satisfy $\eta_0 = 0 = \eta_{N+1}$, and $\Gamma = 0_{2n}$ in (1.57).

We say that the functional $\mathcal{G}''$ is nonnegative (or nonnegative definite or positive semidefinite) at $(\hat{x}, \hat{u})$ if $\mathcal{G}''(\hat{x}, \hat{u}, \eta, v) \geq 0$ for every admissible pair $(\eta, v)$. We say that $\mathcal{G}''$ is positive (or positive definite) at $(\hat{x}, \hat{u})$ if $\mathcal{G}''(\hat{x}, \hat{u}, \eta, v) > 0$ for every admissible pair $(\eta, v) \neq (0, 0)$.

Necessary and sufficient optimality conditions for problem (1.70) are formulated in terms of the weak Pontryagin (maximum) principle and the definiteness of the second variation $\mathcal{G}''$.

**Theorem 1.32** *Assume that $(\hat{x}, \hat{u})$ is a local minimum for problem (1.70) and define the matrices $\mathcal{A}_k$, $\mathcal{B}_k$, $M$ by (1.74) and (1.53) and suppose that $M$ has full rank $r$. Then there exist a constant $\lambda_0 \geq 0$, a vector $\gamma \in \mathbb{R}^r$ and a sequence $p : [0, N + 1]_{\mathbb{Z}} \to \mathbb{R}^n$ such that $\lambda_0 + \sum_{k=0}^{N+1} \|p_k\| \neq 0$ and satisfying:*

(i) *the adjoint equation $-\Delta p_k = \mathcal{H}_x(k, \hat{x}_{k+1}, \hat{u}_k, p_k, \lambda_0)$, i.e.,*

$$-\Delta p_k = \mathcal{A}_k^T p_k + \lambda_0 \hat{L}_x^T(k), \quad k \in [0, N]_{\mathbb{Z}},$$

(1.79)

(ii) *the stationarity condition $\mathcal{H}_u(k, \hat{x}_{k+1}, \hat{u}_k, p_k, \lambda_0) = 0$, i.e.,*

$$\mathcal{B}_k^T p_k + \lambda_0 \hat{L}_u^T(k) = 0, \quad k \in [0, N]_{\mathbb{Z}},$$

(1.80)

(iii) *the transversality condition*

$$\begin{pmatrix} -p_0 \\ p_{N+1} \end{pmatrix} = \lambda_0 \, \nabla K^T(\hat{x}_0, \hat{x}_{N+1}) + M^T \gamma. \qquad (1.81)$$

*If, in addition, system* (1.75) *is M-controllable, then we may take* $\lambda_0 = 1$, *the sequence* $p = \{p_k\}_{k=0}^{N+1}$ *and the vector* $\gamma$ *are unique, and*

(iv) *the second variation* $\mathcal{G}''$ *is nonnegative at* $(\hat{x}, \hat{u})$.

**Theorem 1.33** *Assume that* $(\hat{x}, \hat{u})$ *is a feasible pair for problem* (1.70); *the matrix M in* (1.53) *has full rank r, and assume that there exist a vector* $\gamma \in \mathbb{R}^r$ *and a sequence* $p : [0, N+1]_{\mathbb{Z}} \to \mathbb{R}^n$ *satisfying the conditions* (i)–(iii) *in Theorem 1.32 with* $\lambda_0 := 1$. *Furthermore, suppose that the second variation* $\mathcal{G}''$ *is positive at* $(\hat{x}, \hat{u})$. *Then* $(\hat{x}, \hat{u})$ *is a strict local minimum for problem* (1.70).

We now proceed in deriving the main result of this section. By Theorem 1.32, the second variation $\mathcal{G}''$ is a nonnegative functional at a local minimum $(\hat{x}, \hat{u})$ with the minimum value zero. Thus, applying the weak Pontryagin principle, i.e., equations (1.79) and (1.80), to the second variation $\mathcal{G}''$, we obtain for $k \in [0, N]_{\mathbb{Z}}$ the Jacobi system (here we substitute $q_k := -p_k$)

$$\Delta q_k = -A_k^T q_k + P_k \eta_{k+1} + Q_k v_k, \quad -B_k^T q_k + Q_k^T \eta_{k+1} + R_k v_k = 0. \qquad (1.82)$$

Our aim is to show that system (1.75) and (1.82) has a natural symplectic structure. More precisely, we will show that under a natural invertibility assumption, the pair $(\eta, q)$ solves a symplectic difference system associated with (1.75) and (1.82). For that we need to solve the second equation in (1.82) for $v_k$, which of course can be done if, e.g., $R_k$ is invertible. However, similarly to treatment of the discrete calculus of variations in Theorem 1.29, we wish to avoid the invertibility of $R_k$ and use a more natural invertibility condition.

Following assumption (1.72) and the notation in (1.74), we define the $n \times n$ matrix $\tilde{A}_k$ and the $m \times m$ matrix $S_k$ by

$$\tilde{A}_k := (I - A_k)^{-1}, \quad S_k := R_k + Q_k^T \tilde{A}_k B_k. \qquad (1.83)$$

Note that $A_k$ and $\tilde{A}_k$ commute. We impose the following structural assumption that the matrix

$$S_k \text{ in } (1.83) \text{ is invertible for all } k \in [0, N]_{\mathbb{Z}}. \qquad (1.84)$$

We will discuss the connection of condition (1.84) with the invertibility of $R_k$ in Lemma 1.36 and Corollary 1.37 below. Whenever the matrix $R_k$ is invertible and whenever the matrix $S_k$ is invertible, we define the $n \times n$ matrices $T_k$ and $V_k$ by

$$T_k := I + \tilde{A}_k B_k R_k^{-1} Q_k^T, \quad V_k := I - \tilde{A}_k B_k S_k^{-1} Q_k^T. \qquad (1.85)$$

The next result shows that under (1.84), the Jacobi system (1.82) is a symplectic difference system.

**Theorem 1.34** *Assume that the matrices $\tilde{A}_k$, $S_k$, $V_k$ are defined in* (1.83) *and* (1.85) *and that condition* (1.84) *holds. Define for $k \in [0, N]_{\mathbb{Z}}$ the $2n \times 2n$ matrix $\mathcal{S}_k :=$ $\begin{pmatrix} \mathbb{A}_k & \mathbb{B}_k \\ \mathbb{C}_k & \mathbb{D}_k \end{pmatrix}$ by*

$$
\left.
\begin{aligned}
\mathbb{A}_k &= V_k \tilde{A}_k, & \mathbb{C}_k &= (P_k V_k - Q_k S_k^{-1} Q_k^T)\, \tilde{A}_k, \\
\mathbb{B}_k &= \tilde{A}_k \mathcal{B}_k S_k^{-1} \mathcal{B}_k^T, & \mathbb{D}_k &= (Q_k + P_k \tilde{A}_k \mathcal{B}_k)\, S_k^{-1} \mathcal{B}_k^T + I - \mathcal{A}_k^T.
\end{aligned}
\right\}
\tag{1.86}
$$

*Then $\mathcal{S}_k$ is a symplectic matrix for all $k \in [0, N]_{\mathbb{Z}}$. Consequently, the Jacobi system* (1.75) *and* (1.82) *is a special symplectic system (SDS), in which $y_k := (\eta_k^T, q_k^T)^T$. In addition, the resulting symplectic system (SDS) is a Hamiltonian system* (1.63) *if and only if the matrix $V_k$ (or equivalently $R_k$) is invertible for all $k \in [0, N]_{\mathbb{Z}}$.*

*Remark 1.35* In the discrete calculus of variations setting, we have $\Delta x_k = u_k$ for all $k \in [0, N]_{\mathbb{Z}}$, i.e., $f(k, x, u) = u$. In this case $m = n$ and (1.74) and (1.83) imply that $\mathcal{A}_k = 0$, $\mathcal{B}_k = I$, $\tilde{A}_k = I$, and $S_k = P_k + Q_k^T$. Therefore, we see that the statement of Theorem 1.34 reduces exactly to the statement of Theorem 1.29, and that in both cases, the invertibility of $R_k$ is not needed in order to reveal the symplectic structure of the corresponding Jacobi system.

The proof of Theorem 1.34 is displayed at the end of this section. First we clarify the relationship between the invertibility conditions on $R_k$, $S_k$, $T_k$, and $V_k$.

**Lemma 1.36** *Assume that the matrices $\tilde{A}_k$, $S_k$, $T_k$, $V_k$ are defined in* (1.83) *and* (1.85). *Then the following statements are equivalent:*

  (i)   *the matrices $R_k$ and $T_k$ are invertible,*
 (ii)   *the matrices $R_k$ and $S_k$ are invertible,*
(iii)   *the matrices $S_k$ and $V_k$ are invertible.*

*Proof* For the proof we adopt a known matrix inversion formula. Given any matrices $A, B, C, D$ such that the products below are defined, then the invertibility of $A$, $D$, and $D - CA^{-1}B$ implies the invertibility of $A - BD^{-1}C$ with

$$
(A - BD^{-1}C)^{-1} = A^{-1} + A^{-1}B\,(D - CA^{-1}B)^{-1}CA^{-1}.
\tag{1.87}
$$

Based on this fact, we prove the implications

$$
R_k \text{ and } T_k \text{ invertible} \implies S_k \text{ invertible},
\tag{1.88}
$$

$$
R_k \text{ and } S_k \text{ invertible} \implies V_k \text{ and } T_k \text{ invertible},
\tag{1.89}
$$

$$
S_k \text{ and } V_k \text{ invertible} \implies R_k \text{ invertible}.
\tag{1.90}
$$

Indeed, for (1.88) we set $A := R_k$, $B := -Q_k^T$, $C := B_k$, $D := \tilde{A}_k^{-1}$, which yields by (1.87) that the matrix $S_k$ is invertible. For (1.89) we set $A := I$, $B := \tilde{A}_k B_k$, $C := Q_k^T$, $D := S_k$, and then (1.87) yields the invertibility of $V_k$ with the inverse $V_k^{-1} = T_k$. For (1.90) we set $A := S_k$, $B := Q_k^T$, $C := \tilde{A}_k B_k$, $D := I$, and then by (1.87), we obtain the invertibility of $R_k$.

Suppose now that condition (i) holds. Then (1.88) yields that $S_k$ is invertible, so that condition (ii) holds. Next, assuming (ii), we have from (1.89) that $V_k$ is invertible, i.e., condition (iii) holds. Finally, if we assume (iii), then (1.90) implies $R_k$ invertible, and then in turn (1.89) yields the invertibility of $T_k$. Therefore, condition (i) holds, which completes the proof. □

**Corollary 1.37** *Assume that the matrices $\tilde{A}_k$, $S_k$, $T_k$, $V_k$ are defined in (1.83) and (1.85).*

(i) *Assume that $R_k$ is invertible. Then $T_k$ is invertible if and only if $S_k$ is invertible. In this case the matrix $V_k$ is also invertible and $V_k^{-1} = T_k$.*
(ii) *Assume that $S_k$ is invertible. Then $V_k$ is invertible if and only if $R_k$ is invertible. In this case the matrix $T_k$ is also invertible and $T_k^{-1} = V_k$.*

Note that the invertibility of the matrix $S_k$ alone (without assuming the invertibility of $R_k$) in general does not imply the invertibility of $V_k$ or $R_k$.

*Proof of Theorem 1.34* Let the triple $(\eta, v, q)$ with $\eta = \{\eta_k\}_{k=0}^{N+1}$, $v = \{v_k\}_{k=0}^{N}$, $q = \{q_k\}_{k=0}^{N+1}$ solve the Jacobi system (1.75) and (1.82). We solve equation (1.75) for $\eta_{k+1}$ to get $\eta_{k+1} = \tilde{A}_k A_k \eta_k + \tilde{A}_k B_k v_k$. We insert this expression into the second equation in (1.82), which we solve for $v_k$. Then we get $v_k = S_k^{-1}(B_k^T q_k - Q_k^T \tilde{A}_k \eta_k)$. Substituting this into the first equation in (1.82) and into the above formula for $\eta_{k+1}$, we obtain that the pair $(\eta, q)$ solve the equations $\eta_{k+1} = \mathbb{A}_k \eta_k + \mathbb{B}_k q_k$ and $q_{k+1} = \mathbb{C}_k \eta_k + \mathbb{D}_k q_k$, whose coefficients are given by (1.86). Upon verifying condition (1.145) from Sect. 1.6.1 below, we show that the coefficient matrix $\mathcal{S}_k$ of this system is indeed symplectic. Since $R_k$ is symmetric, from the definition of $S_k$ in (1.83), we have

$$S_k - Q_k^T \tilde{A}_k B_k = R_k = S_k^T - B_k^T \tilde{A}_k^T Q_k. \tag{1.91}$$

By using the definition of $V_k$ in (1.85), it follows that the matrices

$$\mathbb{C}_k^T \mathbb{A}_k = \tilde{A}_k^T (V_k^T P_k - Q_k S_k^{T-1} Q_k^T) V_k \tilde{A}_k$$

$$\stackrel{(1.91)}{=} \tilde{A}_k^T V_k^T P_k V_k \tilde{A}_k - \tilde{A}_k^T Q_k S_k^{T-1} R_k S_k^{-1} Q_k^T \tilde{A}_k,$$

$$\mathbb{D}_k^T \mathbb{B}_k = [B_k S_k^{T-1}(Q_k^T + B_k^T \tilde{A}_k^T P_k) + I - A_k] \tilde{A}_k B_k S_k^{-1} B_k^T$$

$$\stackrel{(1.91)}{=} B_k S_k^{T-1}(S_k + S_k^T - R_k + B_k^T \tilde{A}_k^T P_k \tilde{A}_k B_k) S_k^{-1} B_k^T$$

are symmetric and that

$$
\begin{aligned}
\mathbb{A}_k^T \mathbb{D}_k - \mathbb{C}_k^T \mathbb{B}_k &= \tilde{A}_k^T V_k^T [(Q_k + P_k \tilde{A}_k \mathcal{B}_k) S_k^{-1} \mathcal{B}_k^T + \tilde{A}_k^{T-1}] \\
&\quad - \tilde{A}_k^T (V_k^T P_k - Q_k S_k^{T-1} Q_k^T) \tilde{A}_k \mathcal{B}_k S_k^{-1} \mathcal{B}_k^T \\
&= \tilde{A}_k^T (I - Q_k S_k^{T-1} \mathcal{B}_k^T \tilde{A}_k^T)(Q_k S_k^{-1} \mathcal{B}_k^T + \tilde{A}_k^{T-1}) \\
&\quad + \tilde{A}_k^T Q_k S_k^{T-1} Q_k^T \tilde{A}_k \mathcal{B}_k S_k^{-1} \mathcal{B}_k^T \\
&= I + \tilde{A}_k^T Q_k S_k^{T-1} (S_k^T - \mathcal{B}_k^T \tilde{A}_k^T Q_k + Q_k^T \tilde{A}_k \mathcal{B}_k - S_k) S_k^{-1} \mathcal{B}_k^T \\
&\overset{(1.91)}{=} I.
\end{aligned}
$$

Therefore, by (1.145) the matrix $\mathcal{S}_k$ in Theorem 1.34 with the block entries in (1.86) is indeed a symplectic matrix. This shows that the Jacobi system (1.75) and (1.82) is a symplectic difference system (SDS), in which $y_k := (\eta_k^T, q_k^T)^T$. Moreover, this symplectic difference system is a Hamiltonian system (1.63) if and only if the matrix $\mathbb{A}_k$ is invertible for all $k \in [0, N]_{\mathbb{Z}}$, i.e., if and only if the matrix $V_k$ (or equivalently $R_k$ by Corollary 1.37) is invertible for all $k \in [0, N]_{\mathbb{Z}}$. The proof is complete.  $\square$

*Remark 1.38* Similarly to Remark 1.30, we may consider an alternative discrete optimal control problem, in which the dynamics $f$ and the Lagrangian $L$ have no shift in the state variable, i.e., they are evaluated at $(k, x_k, u_k)$. Such a problem leads to a dual Jacobi system, where there is no shift in $\eta_k$, but the adjoint variable appears with the shift as $p_{k+1}$ (and hence also as $q_{k+1}$). More precisely, the resulting equations in the discrete weak Pontryagin principle have the form

$$
-\Delta p_k = \underline{A}_k^T p_{k+1} + \lambda_0 \hat{\underline{L}}_x^T(k), \quad k \in [0, N]_{\mathbb{Z}}, \tag{1.92}
$$

$$
\underline{\mathcal{B}}_k^T p_{k+1} + \lambda_0 \hat{\underline{L}}_u^T(k) = 0, \quad k \in [0, N]_{\mathbb{Z}}, \tag{1.93}
$$

compare with equations (1.79) and (1.80) in Theorem 1.32. Here the matrices $\underline{A}_k$ and $\underline{\mathcal{B}}_k$ are given by the formulas

$$
\underline{A}_k := \underline{f}_x(k, \hat{x}_k, \hat{u}_k), \quad \underline{\mathcal{B}}_k := \underline{f}_u(k, \hat{x}_k, \hat{u}_k). \tag{1.94}
$$

Similarly to Theorem 1.34, the associated Jacobi system also has the symplectic structure. More details in this direction can be found in the paper [303].

## 1.4 Symplectic Structure of Phase Flow in Hamiltonian Mechanics

The attention concentrated to linear symplectic systems as well as to simple models of discrete Hamiltonian mechanics was initiated in the paper [8] of C. D. Ahlbrandt. This paper was devoted to basic equations of discrete Lagrangian and Hamiltonian mechanics and to first applications of the discrete oscillation theory in the discrete calculus of variations. Ahlbrandt also formulated some open problems in oscillation theory associated with this topic. The fundamental theorem of the classical Hamiltonian mechanics states that the evolution of a Hamiltonian system in time is a symplectic transformation. From this point of view, every Hamiltonian system has the *symplectic structure*; see [27].

Let us consider a system consisting of $l$-mass points with masses $m_i$ and coordinates $r_i = (x_i, y_i, z_i)$, $i = 1, \ldots, l$. Suppose that the potential energy of the system is determined by the function $U(r_1, \ldots, r_l)$, then the evolution of the system in the potential field is described by the Newton equations of motion (see [27, Section 10])

$$\frac{d}{dt}(m_i \dot{r}_i) + \frac{\partial U}{\partial r_i} = 0, \quad i = 1, \ldots, l, \tag{1.95}$$

together with associated initial conditions $r_i(t_0) = r_i^0$ and $\dot{r}_i(t_0) = \dot{r}_i^0$ for $i \in \{1, \ldots, l\}$. According to the principle of least action, equation of motion (1.95) of a given mechanical system complies with extremals of the functional

$$\Phi(r) = \int_{t_0}^{t_1} L(t, q, \dot{q}) \, dt,$$

where $L = T - U$ and $T = \frac{1}{2} \sum_{i=1}^{l} m_i (\dot{x}_i^2 + \dot{y}_i^2 + \dot{z}_i^2)$ is the kinetic energy of the system. Extremals of this functional are given by its Euler-Lagrange equation

$$\frac{d}{dt}\left(\frac{\partial L}{\partial \dot{q}}\right) - \frac{\partial L}{\partial q} = 0, \tag{1.96}$$

where $q = (q_1, \ldots, q_n)$ are the generalized coordinates, $\dot{q} = (\dot{q}_1, \ldots, \dot{q}_n)$ are the generalized velocities, $n := 3l$, and

$$L(t, q, \dot{q}) = T - U = \frac{1}{2} \sum_{i=1}^{n} m_i \dot{q}_i^2 - U(q_1, \ldots, q_n)$$

is the Lagrangian of the system. Under the strict convexity assumption on the Lagrangian with respect to the vector variable $\dot{q}$, i.e., that the matrix of second derivatives $\frac{\partial^2 L}{\partial \dot{q}_i \partial \dot{q}_j}$, $i, j \in \{1, \ldots, n\}$, is positive definite, it is known (see, e.g., [27, pg. 65]) that (1.96) is equivalent with the system of $2n$ first-order equations

$$\dot{q} = \frac{\partial H}{\partial p}, \quad \dot{p} = -\frac{\partial H}{\partial q}, \tag{1.97}$$

where the Hamiltonian $H(t, p, q)$ is the Legendre (sometimes also called Fenchel or conjugate) transformation of the Lagrange function with respect to $\dot{q}$, i.e.,

$$H(t, p, q) = \sup_{\dot{q}}(p\dot{q} - L(t, q, \dot{q})),$$

where $p\dot{q}$ is the scalar product of $p$ and $\dot{q}$. Introducing the variable $y = \binom{p}{q} \in \mathbb{R}^{2n}$, system (1.97) can be written in the form

$$\dot{y} = \mathcal{J} \frac{\partial H}{\partial y}(t, y). \tag{1.98}$$

In particular, if the Hamiltonian is quadratic with

$$H(y, t) = \frac{1}{2} y^T \mathcal{H}(t) y, \quad \mathcal{H}(t) = \mathcal{H}^T(t),$$

then system (1.98) takes the form of the linear Hamiltonian differential system

$$y' = \mathcal{J}\mathcal{H}(t) y, \quad \mathcal{H}(t) = \begin{pmatrix} -C(t) & A^T(t) \\ A(t) & B(t) \end{pmatrix}, \quad \mathcal{H}(t) = \mathcal{H}^T(t). \tag{1.99}$$

Recall (see [147, Chapter 4]) that the transformation $\tilde{y} = R(t, y)$ of solutions of system (1.98) is called *canonical* if the transformation carries Hamiltonian system (1.98) again into a Hamiltonian system

$$\dot{\tilde{y}} = \mathcal{J} \frac{\partial \tilde{H}}{\partial \tilde{y}}(t, \tilde{y}). \tag{1.100}$$

The importance of studying canonical transformations is due to the fact that these transformations permit replacing the Hamiltonian system (1.98) by another Hamiltonian system (1.100), in which the Hamiltonian $\tilde{H}$ is of a simpler structure than $H$. If in a phase space we perform successively two canonical transformations, then the resulting transformation will again be canonical. Furthermore, a transformation that is inverse to a certain canonical transformation will always be canonical, and the identical transformation is again canonical. Therefore, all canonical transformations taken together form a group.

Next we recall two basic statements from the Hamiltonian mechanics (see [27, 147]).

**Theorem 1.39** *For a certain transformation* $\tilde{y} = R(t, y)$ *to be canonical, it is necessary and sufficient that the Jacobian matrix* $S$ *corresponding to this transformation is a generalized-symplectic matrix with constant valence $c$, i.e.,*

$$S^T \mathcal{J} S = c\mathcal{J}, \ c \neq 0.$$

*In particular, in the case of a univalent transformation with* $c = 1$, *the matrix* $S$ *is a symplectic matrix. Then, the condition of the symplectic nature of* (1.98) *must hold identically relative to all the variables* $t$, $q$, $p$.

If in addition $R(t, y)$ is linear, i.e., $R(t, y) = S(t)y$, then in a given basis in $\mathbb{R}^{2n}$ this mapping can be identified with its matrix $S(t)$.

For system (1.100) consider the evolution operator $g_H^t$ mapping each point $y \in \mathbb{R}^{2n}$ regarded as the initial value $y := y(0)$ to the point $y(t) = g_H^t(y(0))$ on the corresponding trajectory through $y$. This operator is called the *Hamiltonian phase flow* of (1.100). We have the following basic property of the Hamiltonian phase flow (see [147, p.139] and [27]).

**Theorem 1.40** *A transformation of phase space (from the variables* $y(0)$ *to the variables* $y(t)$*) carried out by the Hamiltonian phase flow* $\{g_H^t\}$ *of* (1.100) *is canonical and univalent, and its Jacobian matrix is symplectic.*

Basic properties of symplectic matrices will be recalled in Sect. 1.6.1, in particular, symplectic matrices form a group with respect to the matrix multiplication. This group of $2n$-dimensional symplectic matrices is usually denoted by $Sp(2n)$.

In the particular case when $H(y) = \frac{1}{2} y^T \mathcal{H} y$ is autonomous, i.e., the matrix $\mathcal{H}$ is constant in $t$, the phase flow can be expressed in the form $y(t) = g_H^t y(0)$, where $g_H^t = \exp(\mathcal{J}\mathcal{H}t)$, and the phase flow $\exp(\mathcal{J}\mathcal{H}t)$ is a symplectic matrix for every $t$. In other words, the infinitesimal generator of a symplectic flow is the matrix $\mathcal{J}\mathcal{H}$ with symmetric $\mathcal{H}$. The matrix $A := \mathcal{J}\mathcal{H}$ is called a *Hamiltonian matrix* in the literature, as the Hamiltonian matrices of order $2n$ are defined by the property $A^T \mathcal{J} + \mathcal{J} A = 0$.

The study of the symplectic structure of the phase flow of a *discrete* Hamiltonian system

$$\Delta q_k = \frac{\partial H(k, q_{k+1}, p_k)}{\partial p}, \quad \Delta p_k = -\frac{\partial H(k, q_{k+1}, p_k)}{\partial q}, \quad (1.101)$$

where $p_k$, $q_k \in \mathbb{R}^n$, was motivated by the investigation of discrete models of various mechanical systems (see [145, 162] and the comprehensive references given therein) and also by the construction of symplectic algorithms in the Hamiltonian mechanics (see [141, 163, 164, 232] and the references therein). In particular, the central difference scheme

$$\Delta y_k = \mathcal{J} h H_z \left( \frac{y_{k+1} + y_k}{2} \right)$$

for the autonomous system in (1.98) is symplectic (see [140, 141, 163]). The symplectic structure of the phase flow for the discrete linear Hamiltonian system

$$\begin{pmatrix} \Delta x_k \\ \Delta u_k \end{pmatrix} = \mathcal{J} \mathcal{H}_k \begin{pmatrix} x_{k+1} \\ u_k \end{pmatrix}, \quad \mathcal{H}_k = \mathcal{H}_k^T, \quad (1.102)$$

with

$$\mathcal{H}_k = \begin{pmatrix} -C_k & A_k^T \\ A_k & B_k \end{pmatrix}, \quad \det (I - A_k) \neq 0,$$

was investigated under the assumption that $B_k > 0$ (i.e., $B_k$ is positive definite) in [134–137]. The proof of the symplecticity of the discrete flow follows easily from the fact that system (1.102) can be written as symplectic system (see Sect. 2.1.2).

The following theorem concerning symplectic structure of the phase flow of (1.101) was proved in [263, Theorem 2.1]; see also [8, Theorem 3] and [13, Theorem 11].

**Theorem 1.41** *Let the Hamiltonian $H(t, q, p)$ be twice continuously differentiable with respect to $p, q \in \mathbb{R}^n$ and*

$$\det \left( I - \left( \frac{\partial^2 H}{\partial q \, \partial p} \right) \right) \neq 0$$

*in some domain $D \subseteq \mathbb{R}^{2n}$. Then the phase flow $(f^k, g^k) : \mathbb{R}^{2n} \to \mathbb{R}^{2n}$ of (1.101), written in the form*

$$q_{k+1} = f(k, q_k, p_k), \quad p_{k+1} = g(k, q_k, p_k),$$

*where*

$$(f^{k+1}, g^{k+1})(p, q) := (f, g)(k, (f^k, g^k)(p, q)), \quad \left( f^{k_0}, g^{k_0} \right)(p, q) := (p, q)$$

*for some fixed $k_0$, is a discrete one-parametric group of symplectic transformations.*

Observe that one of the important consequences of the existence of the symplectic structure of the phase flow of continuous and discrete Hamiltonian systems (1.98) and (1.101) is the volume-preserving property of a bounded domain of the phase space $\mathbb{R}^{2n}$ (see [27, Liouville theorem, pg. 69], [263, Theorem 2.2]). Concerning autonomous system (1.98), also the energy-preserving law (see [27, pg. 207]) is satisfied, which says that the Hamiltonian $H(p(t), q(t))$ is constant along the solutions $p(t), q(t)$ of (1.98) (i.e., the Hamiltonian $H$ is the first integral of system (1.98)). However, for autonomous *discrete* system (1.101), this energy law is no longer satisfied in general; see [8, Example 4]. Various examples of conservative and nonconservative symplectic difference schemes for (1.98) can be found in [163, Chap. II.16].

## 1.5   Linear Hamiltonian Differential Systems

Linear Hamiltonian differential systems (LHS) represent a natural continuous counterpart of symplectic difference systems (SDS), in the sense that they are the most general first-order linear differential systems, whose fundamental matrix (i.e., the phase flow) is symplectic, whenever it has this property at an initial condition. The purpose of this section is to present some basic results from the oscillation theory of linear Hamiltonian differential systems (LHS), which will be later compared with corresponding results for the symplectic difference systems (SDS). In order to keep the content clear and compact, the results in this section are presented without the proofs. For a comprehensive treatment of linear Hamiltonian differential systems, we refer to the books [70, 203, 205, 248, 250].

### 1.5.1   Basic Properties of Solutions of Hamiltonian Systems

In this section we consider the self-adjoint linear differential system (1.99), i.e.,

$$y' = \mathcal{J}\mathcal{H}(t)\,y, \quad \mathcal{J} = \begin{pmatrix} 0 & I \\ -I & 0 \end{pmatrix}, \; \mathcal{H}(t) = \begin{pmatrix} -C(t) & A^T(t) \\ A(t) & B(t) \end{pmatrix}, \tag{1.103}$$

with $\mathcal{H}(t) = \mathcal{H}^T(t) \in \mathbb{R}^{2n \times 2n}$. Sometimes it is suitable to write (1.103) componentwise, i.e., we split $y = \binom{x}{u}$ with $x, u \in \mathbb{R}^n$, and then (1.103) can be written as

$$x' = A(t)\,x + B(t)\,u, \quad u' = C(t)\,x - A^T(t)\,u. \tag{1.104}$$

When referring to the Hamiltonian system (1.104), we always assume that the real $n \times n$ matrices $A(t)$, $B(t)$, $C(t)$ are piecewise continuous on the interval $\mathcal{I}$ (by $\mathcal{I}$ we always denote a nondegenerate interval of the real line which may be open, closed, or half-open) although integrability on $I$ would suffice for most results. We will start with the notion of a conjoined basis of system (1.104). A $2n \times n$ matrix solution $Y = \binom{X}{U}$ is said to be the *conjoined basis* if

$$Y^T(t)\,\mathcal{J}\,Y(t) = X^T(t)\,U(t) - U^T(t)\,X(t) = 0, \quad \text{rank}\begin{pmatrix} X(t) \\ U(t) \end{pmatrix} = n. \tag{1.105}$$

An alternative terminology for solutions satisfying the first condition in (1.105) is an *isotropic solution* [70] or a *prepared solution* [168].

Bases, which are not conjoined, play in the oscillation theory of (1.103) a similar "destroying" role as the complex solutions in the oscillation theory of the Sturm-Liouville differential equation (being a special case of system (1.103))

$$\big(r(t)x'\big)' + p(t)x = 0. \tag{1.106}$$

More precisely, the function $x(t) = e^{it}$ is a never vanishing solution of the oscillatory equation $x'' + x = 0$, i.e., it "violates" the Sturmian theory of (1.106). In the higher dimension, we consider the Hamiltonian system consisting of two "copies" of the differential equation $x'' + x = 0$, i.e.,

$$\begin{pmatrix} x_1 \\ x_2 \end{pmatrix}'' + \begin{pmatrix} x_1 \\ x_2 \end{pmatrix} = 0,$$

which is "evidently" oscillatory, as the determinant of the $X$-component of the conjoined basis

$$X(t) = \begin{pmatrix} \sin t & 0 \\ 0 & \cos t \end{pmatrix}, \quad U(t) = \begin{pmatrix} \cos t & 0 \\ 0 & -\sin t \end{pmatrix}$$

oscillates; see the definition of the (non)oscillation of (1.103) given below. But at the same time,

$$X(t) = \begin{pmatrix} \cos t & \sin t \\ -\sin t & \cos t \end{pmatrix}, \quad U(t) = \begin{pmatrix} -\sin t & \cos t \\ -\cos t & -\sin t \end{pmatrix} \tag{1.107}$$

is a $2n \times n$ matrix solution, whose first component is everywhere nonsingular. Note that $Y = \begin{pmatrix} X \\ U \end{pmatrix}$ in (1.107) is not a conjoined solution, as can be verified by a direct computation.

Other particular cases of linear Hamiltonian systems (1.103) include the $2n$-order Sturm-Liouville differential equation

$$\sum_{k=0}^{n} (-1)^k \left( r_k(t) x^{(k)} \right)^{(k)} = 0, \quad r_n(t) > 0, \tag{1.108}$$

and the Jacobi matrix differential equation

$$\left( R(t) x' + Q^T(t) x \right)' = Q(t) x' + P(t) x \tag{1.109}$$

with symmetric $R(t)$ and $P(t)$ and positive definite $R(t)$; see, e.g.,[250]. Equation (1.108) can be written as (1.103) with

$$A(t) \equiv (A_{i,j})_{i,j=1}^n = \begin{cases} 1 & \text{if } j = i+1, \\ 0 & \text{otherwise,} \end{cases} \quad B(t) = \text{diag} \left\{ 0, \ldots, 0, \frac{1}{r_n(t)} \right\}$$

and

$$C(t) = \text{diag}\{r_0(t), \ldots, r_{n-1}(t)\}.$$

Concerning equation (1.109), substituting $u = R(t) x' + Q^T(t) x$ we obtain system (1.103) with

$$A(t) = -R^{-1}(t) Q^T(t), \quad B(t) = R^{-1}(t), \quad C(t) = P(t) - Q(t) R^{-1}(t) Q^T(t).$$

If $Y = \binom{X}{U}$ and $\tilde{Y} = \binom{\tilde{X}}{\tilde{U}}$ are $2n \times n$ (not necessarily conjoined) solutions of (1.103), then we have the *Wronskian identity*

$$Y^T(t) \mathcal{J} \tilde{Y}(t) = X^T(t) \tilde{U}(t) - U^T(t) \tilde{X}(t) \equiv M, \tag{1.110}$$

where $M \in R^{n \times n}$ is a constant matrix. Every conjoined basis of (1.103) can be completed to a fundamental symplectic matrix of this system. To show this, let $Y = \binom{X}{U}$ be a conjoined basis (1.103) and let $\tilde{Y} = \binom{\tilde{X}}{\tilde{U}}$ to be a solution of (1.103) given by the initial condition $\tilde{Y}(t_0) = -\mathcal{J} Y(t_0) K^{-1}$, where $K := Y^T(t_0) Y(t_0)$. Then by a direct computation, one can verify that $Z(t) := (Y(t) \, \tilde{Y}(t))$ is a symplectic fundamental matrix of (1.103), because it is symplectic at $t = t_0$ (see Lemma 1.58(v)). From this point of view, any conjoined basis of (1.103) generates a one half of the solution space of (1.103).

Throughout this section we suppose that system (1.103) is *identically normal* on an interval $\mathcal{I}$ (an alternative terminology is *completely controllable*), i.e., if a solution $z = \binom{x}{u}$ of (1.103) satisfies $x(t) \equiv 0$ on a nondegenerate subinterval of $\mathcal{I}$, then $u(t) \equiv 0$ for $t \in \mathcal{I}$ as well. Note that there is a recent theory of focal points for linear Hamiltonian differential systems (1.103) without the assumption of identical normality; see [127, 129, 130, 201, 207, 283, 285–289, 291, 295, 321], but for our motivation, it is suitable to consider identically normal systems. We also suppose throughout this section (without mentioning it later) that the so-called *Legendre condition* is satisfied

$$\text{the matrix } B(t) \text{ is nonnegative definite for } t \in \mathcal{I} \tag{1.111}$$

In the matrix notation introduced in Sect. 1.6 below, the Legendre condition is written as $B(t) \geq 0$. Next we introduce the concept of a focal point of a conjoined basis of (1.103).

**Definition 1.42** Suppose that (1.103) is identically normal and the Legendre condition (1.111) holds on $\mathcal{I}$. We say that a conjoined basis $Y = \binom{X}{U}$ has a *focal point* at $t_0 \in \mathcal{I}$ if $\det X(t_0) = 0$. Its *multiplicity* is then defined as

$$m(t_0) = \text{def } X(t_0) = \dim \text{Ker } X(t_0) = n - \text{rank } X(t_0), \tag{1.112}$$

where Ker is the kernel of the indicated matrix.

Note that this definition implies that the focal points of conjoined bases of (1.103) are *isolated*. Moreover, under assumption (1.111) system (1.103) is completely controllable on $\mathcal{I}$ if and only if the focal points of $Y(t)$ are isolated in $\mathcal{I}$ (see [205, Theorem 4.1.3]). Based on this property, we introduce the notion of an *oscillatory* conjoined basis of (1.103) at $\infty$.

**Definition 1.43**  A conjoined basis $Y(t)$ of (1.103) is called *oscillatory* at $\infty$ if

$$\lim_{b \to \infty} P(Y, a, b) = \lim_{b \to \infty} \sum_{t \in (a,b)} m(t) = \infty, \tag{1.113}$$

where $P(Y, a, b)$ is the total number of focal points in $(a, b)$. In opposite case $Y(t)$ is called *nonoscillatory* at $\infty$.

Let us recall first some results of the Sturmian theory for conjoined bases of a completely controllable linear Hamiltonian differential systems; see [250]. We note that all focal points are counted including their multiplicities.

**Theorem 1.44 (Sturmian Separation Theorem)**  *Let $Y = \binom{X}{U}$ and $\hat{Y} = \binom{\hat{X}}{\hat{U}}$ be conjoined bases of linear Hamiltonian differential system* (1.103). *Then the numbers of their focal points in an interval $\mathcal{I} \subset \mathbb{R}$ differ by at most n. In particular if $Y_a = \binom{X_a}{U_a}$ and $Y_b = \binom{X_b}{U_b}$ are the conjoined bases of* (1.103) *given by the initial conditions $X_a(a) = 0$, $U_a(a) = I$ and $X_b(b) = 0$, $U_b(b) = -I$, then the number of focal points of $Y_a$ in the interval $(a, b]$ is the same as the number of focal point of $Y_b$ in the interval $[a, b)$.*

In particular, Theorem 1.44 implies that the existence of a (non)oscillatory conjoined basis of (1.103) at $\infty$ is equivalent to the property that *all* conjoined bases of this system are (non)oscillatory at $\infty$. In this case we say that system (1.103) is (non)oscillatory at $\infty$.

We have also Sturmian comparison theorem for a completely controllable system (1.103), which reads as follows.

**Theorem 1.45**  *Along with* (1.103) *consider another system of the same form*

$$y' = \mathcal{J}\tilde{\mathcal{H}}(t)\,y \tag{1.114}$$

*and suppose that this system is minorant to* (1.103), *i.e., $\mathcal{H}(t) \geq \tilde{\mathcal{H}}(t)$ for $t \in [a, b]$. Let $Y_a = \binom{X_a}{U_a}$ and $\tilde{Y} = \binom{\tilde{X}_a}{\tilde{X}_b}$ be the solutions of* (1.103) *and* (1.114), *respectively, given by the initial condition $Y_a(a) = \binom{0}{I} = \tilde{Y}_a(a)$. If $\tilde{Y}_a$ has $m \in \mathbb{N}$ focal points in the interval $(a, b]$, then $Y_a$ has at least $m$ focal points in this interval.*

### 1.5.2  Symplectic Transformations

Let $\mathcal{R}(t)$ be a symplectic matrix with differentiable entries in some interval $\mathcal{I} \subseteq \mathbb{R}$. Consider the transformation

$$y = \mathcal{R}(t) \, w. \tag{1.115}$$

Then $w$ is a solution of the system

$$w' = \mathcal{J}\hat{\mathcal{H}}(t) \, w, \quad \hat{\mathcal{H}}(t) := \mathcal{J}\mathcal{R}^{-1}(t) \, [\mathcal{R}'(t) - \mathcal{J}\mathcal{H}(t) \, \mathcal{R}(t)]. \tag{1.116}$$

We claim that system (1.116) is again a linear Hamiltonian system. Indeed, we need to prove that the matrix $\hat{\mathcal{H}}(t)$ is symmetric. By Lemma 1.58(ii) we note that $\mathcal{R}^{-1}(t) = -\mathcal{J}\mathcal{R}^T(t)\,\mathcal{J}$, since $\mathcal{R}(t)$ is symplectic. Moreover, (suppressing the argument $t$ in the calculations below)

$$(\mathcal{R}^T)'\mathcal{J}\mathcal{R} + \mathcal{R}^T \mathcal{J}\mathcal{R}' = 0 \quad \text{i.e.,} \quad \mathcal{J}\mathcal{R}^{-1}\mathcal{R}' = -(\mathcal{R}^T)' \mathcal{R}^{T-1}\mathcal{J}.$$

This implies that the matrix $\mathcal{J}\mathcal{R}^{-1}\mathcal{R}'$ is symmetric. Moreover,

$$\mathcal{J}\mathcal{R}^{-1}\mathcal{J}\mathcal{H}\mathcal{R} = \mathcal{R}^T \mathcal{J}\mathcal{J}\mathcal{H}\mathcal{R} = -\mathcal{R}^T\mathcal{H}\mathcal{R}$$

is also symmetric, which yields that $\hat{\mathcal{H}}$ is symmetric.

In the oscillation theory of linear Hamiltonian differential systems, an important role is played by the lower block-triangular symplectic transformations. Thus, we consider the matrix $\mathcal{R}$ in the form

$$\mathcal{R} = \begin{pmatrix} H & 0 \\ G & H^{T-1} \end{pmatrix} \tag{1.117}$$

with $H^T G = G^T H$. Then the block entries of the matrix $\hat{\mathcal{H}}$ in (1.116) are

$$\begin{aligned} \hat{A} &= H^{-1}(-H' + AH + BG), \\ \hat{B} &= H^{-1}BH^{T-1}, \\ \hat{C} &= -G^T(-H' + AH + BG) + H^T(-G' + CH - A^T G) \end{aligned} \tag{1.118}$$

as can be verified by a direct computation.

*Remark 1.46* Recall that for any Hamiltonian system (1.103), there exist a block diagonal transformation matrix

$$\mathcal{R}(t) = \begin{pmatrix} H(t) & 0 \\ 0 & H^{T-1}(t) \end{pmatrix} \tag{1.119}$$

which transforms (1.103) into (1.116) with $\hat{A}(t) \equiv 0$. Indeed, it is sufficient to assume that $H(t)$ is a fundamental matrix of the differential system

$$H' = A(t) H, \tag{1.120}$$

then according to (1.118), we have $\hat{A}(t) \equiv 0$.

Suppose that $Y = \binom{X}{U}$ is a conjoined basis of (1.103) with $X(t)$ invertible for $t \in \mathcal{I} \subseteq \mathbb{R}$ and consider the (symplectic) transformation matrix

$$\mathcal{R}(t) = \begin{pmatrix} X(t) & 0 \\ U(t) & X^{T-1}(t) \end{pmatrix}.$$

Then $\hat{A}(t) = 0$, $\hat{B}(t) = X^{-1}(t) B(t) X^{T-1}(t)$, and $\hat{C}(t) = 0$, i.e., the resulting system for $w = \binom{w_1}{w_2}$, $w_1, w_2 \in \mathbb{R}^n$, has the form

$$w_1' = X^{-1}(t) B(t) X^{T-1}(t) w_2, \quad w_2' = 0.$$

If we replace vectors $w_1, w_2$ by $n \times n$ matrices $W_1, W_2$, we get a solution

$$W_2(t) = M, \quad W_1(t) = \int_a^t X^{-1}(s) B(s) X^{T-1}(s)\, ds\, M + N, \quad a \in \mathcal{I}, \tag{1.121}$$

where $M$, $N$ are constant $n \times n$ matrices. Substituting $t = a$, we see that this solution is a conjoined basis if and only if $M^T N = N^T M$ and rank $\binom{M}{N} = n$. If we denote $S(t) := \int_a^t X^{-1}(s) B(s) X^{T-1}(s)\, ds$ in (1.121) and we take $M = I$ and $N = 0$, then substituting into the transformation formula $\tilde{Y} = \binom{\tilde{X}}{\tilde{U}} = \mathcal{R}(t) \binom{W_1}{W_2}$ we see by a direct computation that

$$\tilde{X}(t) = X(t) S(t), \quad \tilde{U}(t) = U(t) S(t) + X^{T-1}(t) \tag{1.122}$$

is a conjoined basis of (1.103). Moreover, we have $Y^T \mathcal{J} \tilde{Y} = X^T \tilde{U} - U^T \tilde{X} = I$, so that $Z = (Y, \tilde{Y})$ is a symplectic fundamental matrix of (1.103).

### 1.5.3 Riccati Equation and Reid Roundabout Theorem

Suppose that $Y = \binom{X}{U}$ is a conjoined basis of (1.103) with $X(t)$ invertible in some interval $\mathcal{I} \subseteq \mathbb{R}$. Then $Q = U X^{-1}$ is a symmetric solution of the Riccati matrix differential equation

$$Q' - C(t) + A^T(t) Q + Q A(t) + Q B(t) Q = 0. \tag{1.123}$$

Another important object associated with (1.103) is its quadratic energy functional. We say that a pair of functions $x(t)$, $u(t)$ for $t \in [a, b]$ is *admissible* for the quadratic functional

$$\mathcal{F}(x, u) = \int_a^b \left[ u^T(t) B(t) u(t) + x^T(t) C(t) x(t) \right] dt \qquad (1.124)$$

if $x$ and $u$ satisfy $x'(t) = A(t) x(t) + B(t) u(t)$ on $\mathcal{I}$, i.e., the first equation in the Hamiltonian system (1.103). This equation is again called the *equation of motion* or the *admissibility equation*. For an admissible pair $y = (x, u)$ such that $u$ is differentiable, we then have

$$\mathcal{F}(y) = \int_a^b \left[ u^T(x' - Ax) + x^T Cx \right](t) \, dt$$

$$= \int_a^b \left[ u^T x' + (u^T)' x - x^T u' - u^T Ax + x^T Cx \right](t) \, dt$$

$$= x^T u \Big|_a^b + \int_a^b \left[ x^T(-u' - A^T u + Cx) \right](t) \, dt.$$

Consequently, if $y$ satisfies also second equation in (1.103), called the *Euler-Lagrange equation*, then we have $\mathcal{F}(y) = x^T(t) u(t) \big|_a^b$.

Now we show how (1.124) transforms under lower block-triangular symplectic transformation with transformation matrix given in (1.117). Substituting (1.118) into the admissibility equation for the quadratic functional associated with the transformed system, we see by a short computation that $\binom{w_1}{w_2} = \mathcal{R}^{-1}(t) \binom{x}{u}$ is admissible for this functional, i.e., $w_1'(t) = \hat{A}(t) w_1(t) + \hat{B}(t) w_2(t)$, where $\hat{A}(t)$ and $\hat{B}(t)$ are given by (1.118). Further, again substituting $\hat{A}(t)$, $\hat{B}(t)$, $\hat{C}(t)$ from (1.118) and using that $w_1 = H^{-1}(t) x$ and $w_2 = -G^T(t) x + H^T(t) u$, we derive the identity

$$w_2^T \hat{B}(t) w_2 + w_1^T \hat{C}(t) w_1 - (w_1^T w_2)' = u^T B(t) u + x^T C(t) x - (x^T u)',$$

which upon integration yields the transformation formula

$$\int_a^b [u^T Bu + x^T Cx](t) \, dt = (w_1^T G^T H w_1)(t) \Big|_a^b + \int_a^b [w_2^T \hat{B} w_2 + w_1^T \hat{C} w_1](t) \, dt.$$

If the transformation matrix is $\mathcal{R}(t) = \left( \begin{smallmatrix} I & 0 \\ Q(t) & I \end{smallmatrix} \right)$, where $Q(t)$ is a symmetric solution of Riccati equation (1.123), then $\hat{A} = A + BQ$, $\hat{B} = B$, $\hat{C} = 0$, and we obtain the so-called *Picone identity*

$$\mathcal{F}(x, u) = x^T(t) Q(t) x(t) \Big|_a^b + \int_a^b \left[ (u - Qx)^T B(u - Qx) \right](t) \, dt. \qquad (1.125)$$

In particular, this means that $\mathcal{F}(x, u) > 0$ for admissible $x, u$ with $x(a) = 0 = x(b)$ and $x(t) \not\equiv 0, t \in [a, b]$, whenever there exists a conjoined basis $Y = \binom{X}{U}$ with $X(t)$ invertible for $t \in [a, b]$.

The previous considerations are summarized in the next statement, the so-called *Reid roundabout theorem* for completely controllable linear Hamiltonian differential systems (1.103), see [250, Theorem V.6.3]. We recall that we assume the standing hypothesis (1.111) about the matrix $B(t)$.

**Theorem 1.47** *The following statements are equivalent.*

(i) *System (1.103) is disconjugate on $[a, b]$, i.e., the solution $Y = \binom{X}{U}$ of (1.103) given by the initial condition $X(a) = 0$, $U(a) = I$ has no focal point in $(a, b]$, i.e., it satisfies $\det X(t) \neq 0$ for $t \in (a, b]$.*

(ii) *There exists a conjoined basis $Y = \binom{X}{U}$ of (1.103) with $X(t)$ invertible for $t \in [a, b]$.*

(iii) *Riccati equation (1.123) has a symmetric solution $Q(t)$ defined in the whole interval $[a, b]$.*

(iv) *The quadratic functional $\mathcal{F}$ is positive definite, i.e., $\mathcal{F}(x, u) > 0$ for every admissible $y = (x, u)$ with $x(a) = 0 = x(b)$ and $x(t) \not\equiv 0$ for $t \in [a, b]$.*

(v) *The Riccati inequality*

$$Q' - C(t) + A^T(t) Q + QA(t) + QB(t) Q \leq 0, \quad t \in [a, b], \qquad (1.126)$$

*has a symmetric solution $Q(t)$ defined in the whole interval $[a, b]$.*

The following classical result (see [205, Theorem 5.1.2]) concerning solvability and inequalities for solutions of Riccati equation (1.123) play an important role in the consideration of their discrete analogs in Chap. 4.

**Theorem 1.48 (Riccati Inequality)** *With the linear Hamiltonian system (1.103), consider another linear Hamiltonian system*

$$\hat{y}' = \mathcal{J}\hat{\mathcal{H}}(t)\, \hat{y}, \quad \hat{\mathcal{H}}(t) = \begin{pmatrix} -\hat{C}(t) & \hat{A}^T(t) \\ \hat{A}(t) & \hat{B}(t) \end{pmatrix}, \quad \hat{\mathcal{H}}(t) = \hat{\mathcal{H}}^T(t), \quad t \in \mathcal{I}$$

$$(1.127)$$

*with the piecewise continuous blocks $\hat{A}(t)$, $\hat{B}(t)$, $\hat{C}(t)$ for $t \in \mathcal{I} = [a, b)$. Suppose that system (1.127) is controllable on $\mathcal{I}$, $\hat{B}(t) \geq 0$ for $t \in \mathcal{I}$, and for the Hamiltonians $\mathcal{H}(t)$ and $\hat{\mathcal{H}}(t)$, the majorant condition*

$$\mathcal{H}(t) - \hat{\mathcal{H}}(t) \geq 0, \quad t \in [a, b) \qquad (1.128)$$

*holds, where $A \geq 0$ means that the symmetric matrix $A$ is nonnegative definite. Consider conjoined bases $Y = \binom{X}{U}$ and $\hat{Y} = \binom{\hat{X}}{\hat{U}}$ of systems (1.103) and (1.127) and assume that the initial values $Y(a)$ and $\hat{Y}(a)$ obey the conditions*

$$\hat{X}^T(a)\, \hat{U}(a) \geq D^T X^T(a)\, U(a)\, D$$

*for some matrix $D$ such that $\hat{X}(a) = X(a)\,D$. Assume that $Y$ does not have focal points in $(a, b)$ and $Q(t) = Q^T(t) = U(t)\,X^{-1}(t)$ solves (1.123) on $(a, b)$. Then the solution $\hat{Q}(t) = \hat{Q}^T(t)$ of*

$$\hat{Q}' - \hat{C}(t) + \hat{A}^T(t)\,\hat{Q} + \hat{Q}\hat{A}(t) + \hat{Q}\hat{B}(t)\,\hat{Q} = 0,$$

*where $\hat{Q}(t) = \hat{U}(t)\,\hat{X}^{-1}(t)$ exists on $(a, b)$ and*

$$\hat{Q}(t) \geq Q(t), \quad t \in (a, b).$$

Note that when formulating Theorem 1.48, we do not assume that $\det X(a) \neq 0$ in the initial point $t = a$. One of the principal tools we use in the proofs of our results in oscillation theory of symplectic difference systems is the notion of a comparative index of conjoined bases; see Sect. 3.1. The following statement, proved in [205, Theorem 7.3.1], can be regarded as a continuous analog of the most important results of the comparative index theory.

**Theorem 1.49** *Let $Y = \begin{pmatrix} X \\ U \end{pmatrix}$ and $\hat{Y} = \begin{pmatrix} \hat{X} \\ \hat{U} \end{pmatrix}$ be conjoined bases of systems (1.103) and (1.127) under the majorant condition (1.128), and system (1.127) is controllable. Then for the numbers $P(Y, a, b)$ and $P(\hat{Y}, a, b)$ of focal points in $(a, b)$ of $Y(t)$ and $\hat{Y}(t)$, we have*

$$\begin{aligned}
P(\hat{Y}, a, b) - P(Y, a, b) &\leq \operatorname{ind}[Q(b^-) - \hat{Q}(b^-)] - \operatorname{ind}[Q(a^+) - \hat{Q}(a^+)], \\
P(\hat{Y}, a, b) - P(Y, a, b) &\leq \operatorname{ind}[\hat{Q}(a^+) - Q(a^+)] - \operatorname{ind}[\hat{Q}(b^-) - Q(b^-)],
\end{aligned} \tag{1.129}$$

*where $Q(t) = U(t)\,X^{-1}(t)$ and $\hat{Q}(t) = \hat{U}(t)\,\hat{X}^{-1}(t)$ and* ind *denotes the index, i.e., the number of negative eigenvalues of a symmetric matrix.*

For a symmetric matrix valued function $M(t)$, the notation $\operatorname{ind} M(t_0^+)$ and $\operatorname{ind} M(t_0^-)$ mean the right-hand and left-hand limits of the piecewise constant quantity $\operatorname{ind} M(t)$ for $t \to t_0^+$ and $t \to t_0^-$.

According to [205, Theorem 7.3.1], inequalities (1.129) turn into equalities for $\mathcal{H}(t) \equiv \hat{\mathcal{H}}(t)$, and then we derive from Theorem 1.49 the following separation result (see [205, Theorem 5.2.1]).

**Theorem 1.50** *Let $Y$ and $\hat{Y}$ be conjoined bases of a controllable linear Hamiltonian system (1.103) with condition (1.111). Define $Q(t) := U(t)X(t)^{-1}$ and $\hat{Q}(t) := \hat{U}(t)\hat{X}(t)^{-1}$. Then for all $t \in (a, b)$, we have*

$$\begin{aligned}
m(t) - \hat{m}(t) &= \operatorname{ind}[Q(t^-) - \hat{Q}(t^-)] - \operatorname{ind}[Q(t^+) - \hat{Q}(t^+)], \\
m(t) - \hat{m}(t) &= \operatorname{ind}[\hat{Q}(t^+) - Q(t^+)] - \operatorname{ind}[\hat{Q}(t^-) - Q(t^-)],
\end{aligned} \tag{1.130}$$

*where $m(t)$ and $\hat{m}(t)$ are the multiplicities of focal points of $Y$ and $\hat{Y}$ given by* (1.112). *Moreover, instead of inequalities* (1.129), *we have*

$$P(\hat{Y}, a, b) - P(Y, a, b) = \text{ind}\,[Q(b^-) - \hat{Q}(b^-)] - \text{ind}\,[Q(a^+) - \hat{Q}(a^+)],$$
$$P(\hat{Y}, a, b) - P(Y, a, b) = \text{ind}\,[\hat{Q}(a^+) - Q(a^+)] - \text{ind}\,[\hat{Q}(b^-) - Q(b^-)].$$
$$(1.131)$$

As a direct consequence of Theorem 1.50, we derive the estimate

$$|P(\hat{Y}, a, b) - P(Y, a, b)| \le n, \qquad (1.132)$$

which holds for any conjoined bases $Y$ and $\hat{Y}$ of (1.103). In particular, Theorem 1.50 implies the first statement of Theorem 1.44.

Next we have a look at the difference between positivity and nonnegativity of (1.124). As we have shown, positivity is equivalent to disconjugacy of (1.103) in the *closed* interval $[a, b]$. A slightly modified considerations prior to Theorem 1.47 show that a necessary and sufficient condition for the *nonnegativity* of the functional $\mathcal{F}$ in (1.124) over admissible pairs $y = \binom{x}{u}$ and the Dirichlet boundary conditions at $a$ and $b$ is invertibility of $X(t)$ in the *open* interval $(a, b)$. Here the system (1.103) is again assumed to be identically normal, and $Y = \binom{X}{U}$ is the conjoined basis of (1.103) given by $X(a) = 0$ and $U(a) = I$ as in Theorem 1.47. Hence, the difference between the positivity and nonnegativity of $\mathcal{F}$ is just in the possible noninvertibility of $X(b)$. We have mentioned this topic because later we will show that conditions for the positivity or nonnegativity of quadratic functionals associated with symplectic difference systems considerably differ.

Finally, we note that the concept of disconjugacy of (1.103) can be equivalently defined using the concept of *conjugate points* of vector solutions of (1.103). We say that two points $t_1 < t_2$ are conjugate relative to (1.103) if there exists a solution $y = \binom{x}{u} \in \mathbb{R}^{2n}$ such that $x(t_1) = 0 = x(t_2)$ and $x(t) \not\equiv 0$ for $t \in (t_1, t_2)$. Now, we can define the disconjugacy of system (1.103) in $[a, b]$ (equivalently to item (i) of Theorem 1.47) as follows. System (1.103) is disconjugate on $[a, b]$ if there is no solution of this system with two or more conjugate points in $[a, b]$.

### 1.5.4  Trigonometric and Prüfer Transformations

The trigonometric transformation (an alternative terminology is the *Bohl transformation*) and the Prüfer transformation for classical Sturm-Liouville equation (1.106) present formulas, where either a pair of linearly independent solutions (for the case of the trigonometric transformation) or a solution and its quasiderivative (for the case of the Prüfer transformation) are expressed via the sine and cosine functions.

To formulate a similar result for Hamiltonian systems, we need the concept of *trigonometric Hamiltonian system*. Consider the special Hamiltonian system

$$S' = Q(t)\, C, \quad C' = -Q(t)\, S \tag{1.133}$$

with symmetric and nonnegative definite coefficient matrix $Q(t) \in \mathbb{R}^{n \times n}$. This system is called a *trigonometric system* (see [32, 245]), and its solution are the sine and cosine matrices. This terminology is motivated by the fact that in the scalar case $n = 1$, the functions $S(t) = \sin \int^t Q(s)\, ds$, $C(t) = \cos \int^t Q(s)\, ds$ are solutions of (1.133). Also, if $\binom{S}{C}$ is a conjoined basis of (1.133), then $\binom{C}{-S}$ is a conjoined basis of this system as well. Moreover, if the initial condition (say at $t = a$) is such that the matrix $Z = \left( \begin{smallmatrix} C(a) & -S(a) \\ S(a) & C(a) \end{smallmatrix} \right)$ is orthogonal (i.e., $Z^T Z = I$, e.g., $S(a) = 0$, $C(a) = I$), then this matrix is orthogonal everywhere, and we have the "trigonometric" identities

$$S^T(t)\, S(t) + C^T(t)\, C(t) = I = S(t)\, S^T(t) + C(t)\, C^T(t), \tag{1.134}$$

$$S^T(t)\, C(t) = C^T(t)\, S(t), \quad S(t)\, C^T(t) = C(t)\, S^T(t).$$

In this subsection we will denote the block entries of $\mathcal{H}$ by the calligraphic letters $\mathcal{A}, \mathcal{B}, \mathcal{C}$, since the letter $C$ will be reserved for the cosine matrix function.

**Theorem 1.51 (Matrix Prüfer Transformation)** *Let $Y = \binom{X}{U}$ be a conjoined basis of (1.103). There exist a nonsingular matrix $H(t)$ and a symmetric matrix $Q(t)$ on $\mathcal{I}$ such that*

$$X(t) = S^T(t)\, H(t), \quad U(t) = C^T(t)\, H(t), \tag{1.135}$$

*where $\binom{S}{C}$ is a conjoined basis of (1.133) with*

$$Q(t) = \binom{S^T(t)}{C^T(t)}^T \mathcal{H}(t) \binom{S^T(t)}{C^T(t)}$$

*and $H(t)$ solves the first-order differential system*

$$H' = \binom{S^T(t)}{C^T(t)}^T \mathcal{J}\mathcal{H}(t) \binom{S^T(t)}{C^T(t)} H.$$

Recall that the "angular" equation in the Prüfer transformation for (1.106) $(x = \rho \sin \varphi, \, rx' = \rho \cos \varphi)$ is

$$\varphi' = \frac{1}{r(t)} \cos^2 \varphi + p(t) \sin^2 \varphi = \binom{\sin \varphi}{\cos \varphi}^T \left( \begin{smallmatrix} p(t) & 0 \\ 0 & 1/r(t) \end{smallmatrix} \right) \binom{\sin \varphi}{\cos \varphi}.$$

Therefore, $Q$ plays the role of $\varphi'$ when (1.103) is rewritten Sturm-Liouville equation (1.106). This is in a good agreement with the fact that $S(t) = \sin \int^t Q(s)\,ds$ and $C(t) = \cos \int^t Q(s)\,ds$ is a solution of (1.133) in this scalar case.

The trigonometric (Bohl) transformation expresses a normalized pair of conjoined bases via trigonometric matrices; see [77]. Recall that conjoined bases $Y = \binom{X}{U}$, $\tilde{Y} = \binom{\tilde{X}}{\tilde{U}}$ form a normalized pair of conjoined bases if $Y^T \mathcal{J} \tilde{Y} = X^T \tilde{U} - U^T \tilde{X} = I$.

**Theorem 1.52** *Let $Y = \binom{X}{U}$ and $\tilde{Y} = \binom{\tilde{X}}{\tilde{U}}$ be a pair of normalized conjoined bases of (1.103). There exist continuously differentiable $n \times n$ matrix functions $H(t)$ and $G(t)$ such that $H^T(t)\,G(t) = G^T(t)\,H(t)$ and the transformation (1.115) with $\mathcal{R}(t)$ given by (1.117) transforms system (1.103) into (1.133). In particular, the first components of $Y$, $\tilde{Y}$ can be expressed as $X(t) = H(t)\,S(t)$ and $\tilde{X}(t) = H(t)\,C(t)$. Moreover, the matrices $H(t)$ and $G(t)$ are given by the formulas (suppressing the argument $t$)*

$$HH^T = XX^T + \tilde{X}\tilde{X}^T, \quad G = (UX^T + \tilde{U}\tilde{X}^T)\,H^{T-1},$$

*and the matrix $Q(t)$ in (1.133) is given by the formula*

$$Q(t) = H^{-1}(t)\,B(t)\,H^{T-1}(t).$$

Next we show some applications of this transformation in oscillation theory of differential Hamiltonian systems. It is known (see [250]) that an eventually controllable trigonometric differential system (1.133) is oscillatory if and only if

$$\int^\infty \mathrm{Tr} Q(s)\,ds = \infty, \tag{1.136}$$

where Tr denotes the trace, i.e., the sum of the diagonal elements of a matrix. Combining this result with the transformation given in Theorem 1.52, we have the following oscillation criterion for (1.103)

**Theorem 1.53** *Suppose that $B(t)$ is nonnegative definite for large $t$, system (1.103) is eventually controllable, $\int^\infty \mathrm{Tr}\,B(s)\,ds = \infty$, and there exists a constant $M$ such that eventually $\|X(t)\| \leq M$ for any conjoined basis $(X, U)$ of (1.103). Then system (1.103) is oscillatory.*

### 1.5.5  Principal and Nonprincipal Solutions

Recall that we suppose throughout this section that (1.103) is identically normal. In this subsection we suppose that this assumption is satisfied in an interval of the form $[t_0, \infty)$. We also suppose that system (1.103) is *nonoscillatory*, i.e., this system is disconjugate on $[t_0, \infty)$ if $t_0$ is sufficiently large. These assumptions imply that $X(t)$ is invertible for large $t$ for any conjoined basis $Y = \binom{X}{U}$ of (1.103).

A conjoined basis $\tilde{Y} = \binom{\tilde{X}}{\tilde{U}}$ of (1.103) is said to be a *principal solution* (at $\infty$) if

$$\lim_{t \to \infty} X^{-1}(t)\, \tilde{X}(t) = 0 \tag{1.137}$$

for any conjoined basis $Y = \binom{X}{U}$ such that the (constant) matrix

$$Y^T(t)\, \mathcal{J} \tilde{Y}(t) = X^T(t)\, \tilde{U}(t) - U^T(t)\, \tilde{X}(t) \quad \text{is nonsingular.} \tag{1.138}$$

Any conjoined basis $Y$ satisfying (1.138) is then said to be a *nonprincipal solution* of (1.103) (at $\infty$); an alternative terminology is an *antiprincipal solution* (at $\infty$). Principal and nonprincipal solutions of Hamiltonian systems can be characterized equivalently as conjoined bases whose first component satisfies

$$\lim_{t \to \infty} \left( \int_{t_0}^{t} X^{-1}(s)\, B(s)\, X^{T-1}(s)\, ds \right)^{-1} = 0 \tag{1.139}$$

for the principal solution, while

$$\lim_{t \to \infty} \left( \int_{t_0}^{t} X^{-1}(s)\, B(s)\, X^{T-1}(s)\, ds \right)^{-1} = T,$$

where $T$ is an invertible matrix, for the nonprincipal solution. A principal solution of a nonoscillatory and identically normal system (1.103) is unique up to a constant right nonsingular multiple (see Theorem 1.54 below).

**Theorem 1.54** *The principal solution of* (1.103) *exists whenever this system is nonoscillatory and eventually identically normal. This solution is unique up to a right multiplication by a constant invertible $n \times n$ matrix.*

Another equivalent characterization of the principal solution of (1.123) is via the associated Riccati matrix equation. A conjoined basis $\binom{\tilde{X}}{\tilde{U}}$ is the principal solution if and only if $\tilde{Q} = \tilde{U}\tilde{X}^{-1}$ is the eventually minimal solution of (1.123) in the sense that for any other symmetric solution $Q$ of this equation, we have $Q(t) \geq \tilde{Q}(t)$.

The largest focal point of the principal solution at $\infty$, if any, has the following extremal property; see [105].

**Theorem 1.55** *Let $T$ be the largest focal point of the principal solution $\tilde{Y} = \binom{\tilde{X}}{\tilde{U}}$ of* (1.103) *at $\infty$. Then any other conjoined basis $Y = \binom{X}{U}$ has a focal point in $[T, \infty)$. More precisely, denote by $\tilde{P}(T)$ and $P(T)$ the number of focal points in $[T, \infty)$ of $\tilde{Y}$ and of $Y$ (including multiplicities), respectively. Then*

$$P(T) \geq \tilde{P}(T).$$

## 1.5.6  Nonlinear Dependence on Spectral Parameter

In this final subsection, we consider linear Hamiltonian systems with generally nonlinear dependence on a spectral parameter. The presented results are taken from the book [205]. We consider the linear Hamiltonian system

$$y' = \mathcal{J}\mathcal{H}(t, \lambda)\, y, \quad \mathcal{H}(t, \lambda) = \begin{pmatrix} -C(t, \lambda) & A^T(t, \lambda) \\ A(t, \lambda) & B(t, \lambda) \end{pmatrix}, \tag{1.140}$$

on the interval $[a, b]$ with symmetric $\mathcal{H}(t, \lambda)$ and the Dirichlet boundary conditions

$$x(a) = 0 = x(b). \tag{1.141}$$

We suppose that the matrices $A, B, C$ are piecewise continuous with respect to $t$ on $[a, b]$, that system (1.140) is identically normal, and that the Legendre condition $B(t) \geq 0$ for $t \in [a, b]$ holds. Furthermore, we suppose the following.

(i) For every $t \in [a, b]$ the Hamiltonian $\mathcal{H}(t, \lambda)$ is nondecreasing on $\mathbb{R}$ with respect to the spectral parameter $\lambda$, i.e., $\mathcal{H}(t, \lambda_1) \leq \mathcal{H}(t, \lambda_2)$ for $\lambda_1 < \lambda_2$.

(ii) System (1.140) is *strictly normal*, i.e., if $y(t, \lambda)$ is a solution of (1.140) on $[a, b]$ for two different values $\lambda_1$ and $\lambda_2$ of the spectral parameter, then $y(t, \lambda_i) \equiv 0$ for $t \in [a, b]$ and $i \in \{1, 2\}$.

(iii) There exists $\lambda_0$ such that for all $\lambda \leq \lambda_0$ the associated quadratic functional is positive definite, i.e.,

$$\mathcal{F}(y, \lambda) = \int_a^b [x^T(t)\, C(t, \lambda)\, x(t) + u^T(t)\, B(t, \lambda)\, u(t)]\, dt > 0$$

for all admissible $y = (x, u)$ with $x(a) = 0 = x(b)$ and $x(t) \not\equiv 0$ on $[a, b]$.

Note that these assumptions imply that focal points of any conjoined basis of system (1.140) are isolated (this we have already recalled earlier) and also that the spectrum of (1.140), (1.141) is discrete and bounded below. The following statement, which can be found in [205, Theorem 7.2.1], we will partially "discretize" in the later parts of the book.

**Theorem 1.56 (Global Oscillation Theorem)** *Let the above assumptions be satisfied. Then the number of eigenvalues of* (1.140), (1.141), *which are less than or equal to a given value* $\lambda \in \mathbb{R}$, *is equal to the number of focal points in* $(a, b]$ *of the solution* $Y(t, \lambda)$ *of* (1.140) *satisfying* $Y(a, \lambda) = (0\ I)^T$.

## 1.6 Linear Algebra and Matrix Analysis

In this section we present basic notions and results from linear algebra needed for our treatment. Naturally, we concentrate on symplectic matrices but also on the Moore-Penrose pseudoinverse matrices, which play important role in the discrete oscillation theory. We also discuss some results related to the $LU$ factorization for symplectic matrices.

We use a standard matrix notation. The transpose and inverse of a matrix $M$ are denoted by $M^T$ and $M^{-1}$. Instead of $(M^T)^{-1}$ we will write $M^{T-1}$. By $M^*$ we denote the conjugate transpose of $M$, i.e., $M^* = (\overline{M})^T$, where $\overline{M}$ is the matrix, whose entries are complex conjugate of entries of $M$. Similarly as above, the notation $M^{*-1}$ means $(M^*)^{-1}$. If $M$ is a symmetric matrix, then the inequalities $M > 0$, $M \geq 0$, $M < 0$, $M \leq 0$ mean that $M$ is positive definite, positive semidefinite, negative definite, and negative semidefinite, respectively. If $M$ and $N$ are symmetric matrices, then the inequalities $M > N$, $M \geq N$, $M < N$, $M \leq N$ mean that $M - N > 0$, $M - N \geq 0$, $M - N < 0$, $M - N \leq 0$. By ind $M$ we denote the index of $M$, i.e., the number of negative eigenvalues of $M$ (including the algebraic multiplicities of the eigenvalues).

If $M$ is an $m \times n$ matrix with real entries, we will write $M \in \mathbb{R}^{m \times n}$. By Ker $M$ and Im $M$, we denote the kernel and image of $M$, i.e.,

$$\text{Ker } M = \{c \in \mathbb{R}^n : Mc = 0\}, \quad \text{Im } M = \{d \in \mathbb{R}^m : \exists c \in \mathbb{R}^n \text{ with } d = Ac\}.$$

By def $M$ and rank $M$, we denote the defect and the rank of $M$, i.e., the dimension of Ker $M$ and Im $M$, respectively. If $M \in \mathbb{R}^{n \times n}$, then def $M = n - \text{rank } M$. Moreover, for any matrices $M$ and $N$ of suitable dimensions, we have the formula

$$\text{rank}(MN) = \text{rank } N - \dim (\text{Ker } M \cap \text{Im } N). \tag{1.142}$$

If $M(t)$ is a matrix valued function, then we will use the notation

$$\text{rank } M(t_0^{\pm}), \quad \text{ind } M(t_0^{\pm}), \quad \text{def } M(t_0^{\pm})$$

for the right-hand and left-hand limits at the point $t_0$ of the piecewise constant quantities rank $M(t)$, ind $M(t)$, def $M(t)$. In agreement with the above, the notation $\mu(Y_1(t_0^{\pm}), Y_2(t_0^{\pm}))$ involving the comparative index (see Sect. 3.1.1) means the one-sided limits of the piecewise constant quantity $\mu(Y_1(t), Y_2(t))$. In a similar way, we will use the notation Ker $M(t_0^{\pm})$ for the constant subspace Ker $M(t)$ in the right and left neighborhoods of $t_0$.

If $a_1, \ldots, a_n \in \mathbb{R}$, we denote by diag$\{a_1, \ldots, a_n\}$ the diagonal matrix $A = \{A_{ij}\}_{i,j=1}^n$ with $A_{ij} = 0$ for $i \neq j$ and $A_{ii} = a_i$ for $i = 1, \ldots, n$. A similar notation will be used for block diagonal matrices. If some additional matrix notation will be used only in a particular place in this book, we will explain it at that place where it is used.

In Sect. 1.6.2 we will recall the concept of the Moore-Penrose generalized inverse (or the pseudoinverse) of a matrix $M$. This will be denoted by $M^{\dagger}$.

We will also use a standard notation for matrix groups. By $GL(n)$ we denote the group of invertible $n \times n$ matrices with real entries; $O(n)$ denotes the group of $n \times n$ orthogonal matrices, i.e., square matrices $M$ with $M^T = M^{-1}$. Finally, $Sp(2n)$ denotes the group of $2n \times 2n$ symplectic matrices, whose definition is given below in Sect. 1.6.1.

## 1.6.1 Symplectic Matrices

In this subsection we provide the definition and basic properties of symplectic matrices, which are of course fundamental for the theory of symplectic difference systems.

**Definition 1.57** A real $2n \times 2n$ matrix $S$ is *symplectic*, we write $S \in Sp(2n)$, if

$$S^T \mathcal{J} S = \mathcal{J}, \quad \mathcal{J} = \begin{pmatrix} 0 & I \\ -I & 0 \end{pmatrix}. \tag{1.143}$$

In (1.143), the matrix $I$ is the $n \times n$ identity matrix. The matrix $\mathcal{J}$ is the canonical skew-symmetric $2n \times 2n$ matrix, and of course $\mathcal{J}$ is symplectic, i.e., $\mathcal{J} \in Sp(2n)$ with $\mathcal{J}^{-1} = \mathcal{J}^T = -\mathcal{J}$. If we write the matrix $S$ in the block form

$$S = \begin{pmatrix} \mathcal{A} & \mathcal{B} \\ \mathcal{C} & \mathcal{D} \end{pmatrix} \tag{1.144}$$

with $n \times n$ matrices $\mathcal{A}, \mathcal{B}, \mathcal{C}, \mathcal{D}$, then substituting into (1.143), we get the identities

$$\mathcal{A}^T \mathcal{D} - \mathcal{C}^T \mathcal{B} = I, \quad \mathcal{A}^T \mathcal{C} = \mathcal{C}^T \mathcal{A}, \quad \mathcal{B}^T \mathcal{D} = \mathcal{D}^T \mathcal{B}. \tag{1.145}$$

The following lemma summarizes basic properties of symplectic matrices.

**Lemma 1.58** *Symplectic matrices have the following properties.*

(i) *The product of two symplectic matrices is also a symplectic matrix, i.e., if $S_1, S_2 \in Sp(2n)$, then also the product $S_1 S_2 \in Sp(2n)$.*

(ii) *Every symplectic matrix is invertible with determinant equal to 1, and the inverse of a symplectic matrix is also symplectic, i.e., for $S \in Sp(2n)$ we have $\det S = 1$ and $S^{-1} \in Sp(2n)$. In the block notation (1.144), this means that*

$$S^{-1} = \mathcal{J}^T S^T \mathcal{J} = \begin{pmatrix} \mathcal{D}^T & -\mathcal{B}^T \\ -\mathcal{C}^T & \mathcal{A}^T \end{pmatrix}.$$

(iii) *The transpose of a symplectic matrix is also symplectic, i.e., for $S \in Sp(2n)$, we have $S^T \in Sp(2n)$. This means in terms of the blocks of $S$ that*

$$\mathcal{B}\mathcal{A}^T = \mathcal{A}\mathcal{B}^T, \quad \mathcal{C}\mathcal{D}^T = \mathcal{D}\mathcal{C}^T, \quad \mathcal{A}\mathcal{D}^T - \mathcal{B}\mathcal{C}^T = I. \tag{1.146}$$

(iv) *If $P$ is a matrix such that $P^T \mathcal{J} P = \mathcal{J}^T$, then $S \in Sp(2n)$ if and only if $PSP^{-1} \in Sp(2n)$.*

(v) *If $S = (Y \ \tilde{Y})$ with $2n \times n$ matrices $Y$ and $\tilde{Y}$, then $S \in Sp(2n)$ if and only if $Y$ and $\tilde{Y}$ satisfy*

$$Y^T \mathcal{J} Y = 0, \quad \tilde{Y}^T \mathcal{J} \tilde{Y} = 0, \quad \operatorname{rank} Y = n = \operatorname{rank} \tilde{Y} \tag{1.147}$$

*and*

$$w(Y, \tilde{Y}) := Y^T \mathcal{J} \tilde{Y} = I. \tag{1.148}$$

(vi) *If a $2n \times n$ matrix $Y = \binom{X}{U}$ satisfies conditions in (1.147), then there exist symplectic matrices $\mathcal{Z}$ and $\tilde{\mathcal{Z}}$ such that $\mathcal{Z} = \left( \begin{smallmatrix} \hat{X} & X \\ \hat{U} & U \end{smallmatrix} \right)$ and $\tilde{\mathcal{Z}} = \left( \begin{smallmatrix} X & \tilde{X} \\ U & \tilde{U} \end{smallmatrix} \right)$, i.e., i.e., $\mathcal{Z}\binom{0}{I} = Y$ and $\tilde{\mathcal{Z}}\binom{I}{0} = Y$.*

(vii) *If the $2n \times n$ matrix $Y$ satisfies (1.147), then we have*

$$\operatorname{Ker} Y^T = \operatorname{Im}(\mathcal{J}Y). \tag{1.149}$$

*If $\mathcal{M} \in \mathbb{R}^{n \times n}$ is invertible, and $S \in Sp(2n)$, then the matrix $S Y \mathcal{M}$ also satisfies (1.147).*

(viii) *For any $2n \times n$ matrix $Y = \binom{X}{U}$ satisfying (1.147) there exists $\sigma \in \mathbb{R}$ such that*

$$\det(X + \sigma U) \neq 0. \tag{1.150}$$

*Proof* The properties (i)–(iii) and (vii) are well-known properties of symplectic matrices which are proved, e.g., in [74, 205, 328, 332]. To prove the property (iv), we use the definition of the group $Sp(2n)$ and the arguments

$$S \in Sp(2n) \iff S^T \mathcal{J} S = \mathcal{J} \iff S^T \mathcal{J}^T S = \mathcal{J}^T$$

$$\iff S^T P^T \mathcal{J} P S = P^T \mathcal{J} P \iff P^{T-1} S^T P^T \mathcal{J} P S P^{-1} = \mathcal{J}$$

$$\iff P S P^{-1} \in Sp(2n).$$

The property (v) follows from the definition of $Sp(2n)$ and property (ii). The proof of the property (vi) can be found in [205, Corollary 3.3.9]. To construct the matrix $\mathcal{Z}$, we set $K := (Y^T Y)^{-1} = (X^T X + U^T U)^{-1}$. The matrix $K$ is invertible because of the rank condition in (1.147). Then

$$Z = \begin{pmatrix} UK & X \\ -XK & U \end{pmatrix}, \quad \tilde{Z} = \begin{pmatrix} X & -UK \\ U & XK \end{pmatrix}.$$

The matrices $Z$ and $\tilde{Z}$ are really symplectic in view of property (v).

The proof of (viii) about the existence of $\sigma$ such that (1.150) holds, which we present here, can be found in [1, pg. 41] or [205, Lemma 3.3.1, pg. 86]. Consider the polynomial $q(\sigma) = \det(X + \sigma U)$, $\sigma \in \mathbb{C}$. Then for the choice of $\sigma = i$, the matrix $(X + iU)^*(X + iU) = X^T X + U^T U$ is nonsingular in view of the third condition in (1.147). This implies that $q(\sigma) \not\equiv 0$ and $q(\sigma) = 0$ only for a finite number of its roots. Hence there exists $\sigma \in \mathbb{R}$ such that (1.150) holds. $\qquad \square$

Particular cases of the matrix $P$ from (iv) of the previous lemma which will be used later in this book are

$$P_1 = \begin{pmatrix} 0 & I \\ I & 0 \end{pmatrix}, \quad P_2 = \mathrm{diag}\{I, -I\}, \quad P_3 = \mathrm{diag}\{-I, I\}. \tag{1.151}$$

None of these matrices is symplectic, but they satisfy the assumptions of property (iv), as one can easily verify.

Examples of symplectic matrices, which we will frequently use in this book, are the lower block-triangular matrix

$$L = \begin{pmatrix} H & 0 \\ G & H^{T-1} \end{pmatrix}, \quad H^T G = G^T H, \quad \text{with } H \text{ invertible}, \tag{1.152}$$

and its special case (unit lower block-triangular matrix)

$$L = \begin{pmatrix} I & 0 \\ Q & I \end{pmatrix}, \quad \text{with } Q \text{ symmetric}. \tag{1.153}$$

Analogically, given (1.152) or (1.153), then the matrices

$$R_1 = \begin{pmatrix} H & G \\ 0 & H^{T-1} \end{pmatrix}, \quad R_2 = \begin{pmatrix} I & Q \\ 0 & I \end{pmatrix} \tag{1.154}$$

are also symplectic (unit upper block triangular in the case of $R_2$), as a result of Lemma 1.58(iv) with $P := P_1$ from (1.151).

We will also need matrices which are symplectic and orthogonal at the same time. Recall that an invertible matrix $M$ is orthogonal if $M^{-1} = M^T$. Orthogonal matrices form a group with respect to the matrix multiplication; this group is for matrices with dimension $2n$ denoted by $O(2n)$. If $S \in Sp(2n) \cap O(2n)$, then the matrix $S$ has the block form

$$S = \begin{pmatrix} \mathcal{P} & \mathcal{Q} \\ -\mathcal{Q} & \mathcal{P} \end{pmatrix}, \tag{1.155}$$

where $\mathcal{P}$ and $\mathcal{Q}$ are $n \times n$ matrices satisfying

$$\mathcal{P}^T\mathcal{P} + \mathcal{Q}^T\mathcal{Q} = I = \mathcal{P}\mathcal{P}^T + \mathcal{Q}\mathcal{Q}^T, \quad \mathcal{P}^T\mathcal{Q} = \mathcal{Q}^T\mathcal{P}, \; \mathcal{P}\mathcal{Q}^T = \mathcal{Q}\mathcal{P}^T. \tag{1.156}$$

The structure in (1.155) and the properties in (1.156) follow from the structure in (1.144) and from (1.145) and (1.146).

*Remark 1.59* If $Q \in \mathbb{R}^{n \times n}$ is orthogonal, then the matrix diag$\{Q, Q\}$ is orthogonal and symplectic, i.e., the matrices diag$\{Q, Q\}$ and $\mathcal{J}$ commute.

### 1.6.2  Moore-Penrose Pseudoinverse

Pseudoinverse matrices play an important role in the analysis of symplectic difference systems. A comprehensive treatment of pseudoinverse matrices can be found, e.g., in the book [34]; further references are given in Sect. 1.7. Here we present only the results, which we need in this book.

Let $A \in \mathbb{R}^{m \times n}$. Then there exists a uniquely determined matrix $B \in \mathbb{R}^{n \times m}$ satisfying the following four properties

$$BAB = B, \quad ABA = A, \quad BA \text{ and } AB \text{ are symmetric.} \tag{1.157}$$

The matrix $B$ is called the *pseudoinverse* (or the *Moore-Penrose generalized inverse*) of $A$, and it is denoted by $A^\dagger$. It can be explicitly given by the formula

$$A^\dagger = \lim_{t \to 0+} (A^T A + tI)^{-1} A^T = \lim_{t \to 0+} A^T (AA^T + tI)^{-1}.$$

*Remark 1.60*

(i) For a matrix $A \in \mathbb{R}^{m \times n}$, we have $(A^T)^\dagger = (A^\dagger)^T$, $(A^\dagger)^\dagger = A$, and $\operatorname{Im} A^\dagger = \operatorname{Im} A^T$, $\operatorname{Ker} A^\dagger = \operatorname{Ker} A^T$.

(ii) If $A \in \mathbb{R}^{m \times n}$ and if $P$ and $Q$ are orthogonal matrices of suitable dimensions, then the formula $(QAP)^\dagger = P^T A^\dagger Q^T$ holds.

(iii) If $A \in \mathbb{R}^{m \times n}$, then $\operatorname{Ker} A^T = \operatorname{Ker}(AA^\dagger)$.

(iv) For matrices $A \in \mathbb{R}^{m \times n}$ and $B \in \mathbb{R}^{n \times p}$, we have $\operatorname{Ker}(AB) = \operatorname{Ker}(A^\dagger AB)$.

(v) If $\{A_j\}_{j=1}^\infty$ is a sequence of $m \times n$ matrices such that $A_j \to A$ for $j \to \infty$, then the limit of $A_j^\dagger$ as $j \to \infty$ exists, i.e., $A_j^\dagger \to B$ for $j \to \infty$, if and only if there exists $j_0 \in \mathbb{N}$ such that rank $A_j = $ rank $A$ for all $j \geq j_0$. And in this case $B = A^\dagger$.

(vi) Let $A$ and $B$ be symmetric and positive semidefinite matrices such that $A \leq B$. Then $AB^\dagger A \leq A$ or equivalently $A^\dagger AB^\dagger A^\dagger A \leq A^\dagger$. Furthermore, the inequality $B^\dagger \leq A^\dagger$ is equivalent with $\operatorname{Im} A = \operatorname{Im} B$ or with $\operatorname{rank} A = \operatorname{rank} B$.

(vii) For matrices $A, B \in \mathbb{R}^{m \times n}$ satisfying $A^T B = 0$ and $BA^T = 0$, we have $(A + B)^\dagger = A^\dagger + B^\dagger$ and $\operatorname{rank}(A + B) = \operatorname{rank} A + \operatorname{rank} B$.

The following result will also be useful in Sect. 6.3.5.

**Lemma 1.61** *Let* $\{A_j\}_{j=1}^\infty$ *be a sequence of matrices. Assume that*

(i) $A_j \to A$ *for* $j \to \infty$,

(ii) *there exists an index* $j_0 \in \mathbb{N}$ *such that* $\operatorname{Im} A_j \subseteq \operatorname{Im} A$ *and* $\operatorname{Im} A_j^T \subseteq \operatorname{Im} A^T$ *for all* $j \geq j_0$.

*Then there exists an index* $j_1 \in \mathbb{N}$ *such that* $\operatorname{Im} A_j = \operatorname{Im} A$ *and* $\operatorname{Im} A_j^T = \operatorname{Im} A^T$ *for all* $j \geq j_1$ *and* $A_j^\dagger \to A^\dagger$ *for* $j \to \infty$.

The Moore-Penrose pseudoinverse is closely related with constructing orthogonal projectors. If $V$ is a linear subspace in $\mathbb{R}^n$, then we denote by $P_V$ the $n \times n$ corresponding orthogonal projector onto $V$.

*Remark 1.62* For any $A \in \mathbb{R}^{m \times n}$ the matrix $AA^\dagger$ is the orthogonal projector onto $\operatorname{Im} A$, and the matrix $A^\dagger A$ is the orthogonal projector onto $\operatorname{Im} A^T$. Moreover, $\operatorname{rank} A = \operatorname{rank} AA^\dagger = \operatorname{rank} A^\dagger A$. In addition, for matrices $A$ and $B$, we have

$$(AB)^\dagger = (P_{\operatorname{Im} A^T} B)^\dagger (A P_{\operatorname{Im} B})^\dagger = (A^\dagger AB)^\dagger (ABB^\dagger)^\dagger. \tag{1.158}$$

One of the main reasons why pseudoinverse matrices appear in oscillation theory of discrete symplectic systems is that if $V, W \in \mathbb{R}^{n \times n}$, then we have the equivalences

$$\operatorname{Ker} V \subseteq \operatorname{Ker} W \quad \Longleftrightarrow \quad W = WV^\dagger V, \tag{1.159}$$

$$\operatorname{Im} W \subseteq \operatorname{Im} V \quad \Longleftrightarrow \quad VV^\dagger W = W. \tag{1.160}$$

In our later treatment, we will also need the following properties of pseudoinverse matrices. The next lemma can be found in [34, Theorem 8 and Lemma 3]. This result also follows from the facts that the matrix $XX^\dagger$ is the orthogonal projector onto $\operatorname{Im} X$ and the matrix $X^\dagger X$ is the orthogonal projector onto $\operatorname{Im} X^T$.

**Lemma 1.63** *Let* $X, \tilde{X} \in \mathbb{R}^{m \times n}$. *If* $\operatorname{Im} X = \operatorname{Im} \tilde{X}$, *then* $XX^\dagger = \tilde{X}\tilde{X}^\dagger$. *Analogously, if* $\operatorname{Ker} X = \operatorname{Ker} \tilde{X}$, *then* $X^\dagger X = \tilde{X}^\dagger \tilde{X}$.

As a corollary to Lemma 1.63, we derive the following result which plays an important role in the proofs of Chap. 3.

**Lemma 1.64** *Suppose that the square matrices* $X, \tilde{X}, \mathcal{A}, \mathcal{B}$ *satisfy*

$$\tilde{X} = \mathcal{A} X \mathcal{B}, \quad \det \mathcal{A} \neq 0, \quad \det \mathcal{B} \neq 0.$$

*Then*

$$I - \tilde{X}\tilde{X}^{\dagger} = T(I - XX^{\dagger})\,\mathcal{A}^{-1}, \quad I - \tilde{X}^{\dagger}\tilde{X} = \mathcal{B}^{-1}(I - X^{\dagger}X)\,\tilde{T}, \qquad (1.161)$$

*where $T$ and $\tilde{T}$ are invertible matrices independent of $\mathcal{B}$ and $\mathcal{A}$, respectively, such that $T = I$ for $\mathcal{A} = I$ and $\tilde{T} = I$ for $\mathcal{B} = I$.*

*Proof* Suppose that the singular value decomposition (SVD, see [157, p.70]) of $X$ is of the form $X = V\Sigma U^{T}$, where $\Sigma$ is the diagonal matrix containing the singular values of $X$, and $U$ and $V$ are orthogonal matrices. Then

$$I - XX^{\dagger} = VGV^{T}, \quad I - X^{\dagger}X = UGU^{T}, \quad G := I - F, \quad F := \Sigma\Sigma^{\dagger}.$$
$$(1.162)$$

Observe that according to Lemma 1.63 applied to the particular case $\mathcal{A} = I$ (in this case $\operatorname{Im}\tilde{X} = \operatorname{Im}X$), we have $I - \tilde{X}\tilde{X}^{\dagger} = I - XX^{\dagger}$. Applying this result to the matrices $\tilde{X}$ and $\tilde{X}\mathcal{B}^{-1}U$, we obtain

$$\begin{aligned}
I - \tilde{X}\tilde{X}^{\dagger} &= I - (\mathcal{A}V\Sigma)\,(\mathcal{A}V\Sigma)^{\dagger} \\
&= [I - (\mathcal{A}V\Sigma)\,(\mathcal{A}V\Sigma)^{\dagger}]\,(\mathcal{A}VF + \mathcal{A}VG)\,(\mathcal{A}V)^{-1} \\
&= [I - (\mathcal{A}V\Sigma)\,(\mathcal{A}V\Sigma)^{\dagger}]\,\mathcal{A}VGV^{T}\mathcal{A}^{-1} \\
&= \mathcal{A}V[I - F\Sigma\,(\mathcal{A}V\Sigma)^{\dagger}\mathcal{A}VG]\,GV^{T}\mathcal{A}^{-1} \\
&= TVGV^{T}\mathcal{A}^{-1} = T(I - XX^{\dagger})\,\mathcal{A}^{-1}, \\
T &= \mathcal{A}V[I - F\Sigma\,(\mathcal{A}V\Sigma)^{\dagger}\mathcal{A}VG]\,V^{T},
\end{aligned}$$

where (1.162) has been used. Observe that the matrix $T$ is invertible, since it is a product of the matrices $\mathcal{A}$, $V$, and $I - F\Sigma\,(\mathcal{A}V\Sigma)^{\dagger}\mathcal{A}VG$, the last matrix being invertible by the (easily to verify) implication

$$PR = 0 \quad \Longrightarrow \quad (I - RUP)\,(I + RUP) = I \qquad (1.163)$$

with $P := G$, $R := F$, $U := \Sigma\,(\mathcal{A}V\Sigma)^{\dagger}\mathcal{A}V$. Using Remark 1.60 (ii) it is easy to verify that $T = I$ if $\mathcal{A} = I$. Hence, the first statement in (1.161) is proved. Applying this result in computing $I - \tilde{X}^{T}(\tilde{X}^{T})^{\dagger} = I - \tilde{X}^{\dagger}\tilde{X}$, we obtain also the second identity in (1.161). $\qquad\square$

Next we consider results on solvability of the matrix equations

$$X^{T}QX = X^{T}U. \qquad (1.164)$$

associated with the $n \times n$ blocks $X$, $U$ of the matrix $Y = \begin{pmatrix} X \\ U \end{pmatrix}$ with conditions (1.105).

**Lemma 1.65** *Assume that the matrix* $Y \in \mathbb{R}^{2n \times n}$ *satisfies the first condition in* (1.147), *then equation* (1.164) *is solvable, and the general solution of* (1.164) *is of the form*

$$Q = XX^{\dagger}UX^{\dagger} + E - XX^{\dagger}EXX^{\dagger}, \tag{1.165}$$

*where* $E \in \mathbb{R}^{n \times n}$ *is arbitrary.*

*Proof* According to [215, Theorem 6.11] the matrix equation

$$AQC = B \tag{1.166}$$

has a solution $Q$ if and only if $AA^{\dagger}BC^{\dagger}C = B$. In this case the general solution of (1.166) is of the form

$$Q = A^{\dagger}BC^{\dagger} + E - A^{\dagger}AECC^{\dagger},$$

where $E \in \mathbb{R}^{n \times n}$ is arbitrary. Putting $A := X^T$, $C := X$, $B := X^T U$ and using (1.147), we see that the relation $X^T(X^T)^{\dagger}X^T UX^{\dagger}X = X^T U$ does hold and the general solution of (1.164) has the form (1.165). $\qquad \square$

*Remark 1.66* In the oscillation theory, we use symmetric solutions of (1.164). Applying (1.147) we see that the first addend $XX^{\dagger}UX^{\dagger} = (X^{\dagger})^T X^T UX^{\dagger}$ in (1.165) is symmetric. Then, to construct the general symmetric solution of (1.164), we demand the symmetry of the last addends in (1.165)

$$\mathcal{E} = E - XX^{\dagger}EXX^{\dagger}$$
$$= XX^{\dagger}E(I - XX^{\dagger}) + (I - XX^{\dagger})EXX^{\dagger} + (I - XX^{\dagger})E(I - XX^{\dagger})$$
$$= \mathcal{E}^T. \tag{1.167}$$

Finally, the general symmetric solution of (1.164) is of the form

$$Q = XX^{\dagger}UX^{\dagger} + \mathcal{E}, \quad \mathcal{E} = \mathcal{E}^T, \quad X^T \mathcal{E}X = 0. \tag{1.168}$$

We complete this section by the following important complement of properties (vi) and (viii) in Lemma 1.58.

**Lemma 1.67** *Under assumptions and the notation of Lemma 1.58(vi), there exists* $\sigma \in \mathbb{R}$ *such that the symplectic matrix* $Z\begin{pmatrix} I & 0 \\ \sigma I & I \end{pmatrix} = \begin{pmatrix} \hat{X} + \sigma X & X \\ \hat{U} + \sigma U & U \end{pmatrix}$ *obeys the condition* $\det(\hat{X} + \sigma X) \neq 0$. *Moreover, the number* $\sigma \in \mathbb{R}$ *can be chosen in such a way that*

$$(\hat{X} + \sigma X)^{-1}X \geq 0, \tag{1.169}$$

*or*

$$(\hat{X} + \sigma X)^{-1} X \le 0. \tag{1.170}$$

*The same assertion holds for* $\tilde{Z} \begin{pmatrix} I & \sigma I \\ 0 & I \end{pmatrix} = \begin{pmatrix} X & \tilde{X} + \sigma X \\ U & \tilde{U} + \sigma U \end{pmatrix}.$

*Proof* Putting $Y := (\hat{X} \ X)^T$ in (1.150) of Lemma 1.58(viii), we derive $\det(\hat{X}^T + \sigma X^T) = \det(\hat{X} + \sigma X) \ne 0$. For the proof of (1.169) (or (1.170)), we note that this condition is equivalent to

$$X(\hat{X} + \sigma X)^T = X(Q + \sigma I)X^T \ge 0 \quad (X(Q + \sigma I)X^T \le 0),$$

where $Q = Q^T$ solves the matrix equation $XQX^T = X\hat{X}^T$ (see Remark 1.66). So we see that for $\sigma \ge -\min_{1 \le i \le n} \lambda_i$ (or $\sigma \le -\max_{1 \le i \le n} \lambda_i$), where $\lambda_i$ for $i \in \{1, \ldots, n\}$ are the eigenvalues of $Q$, condition (1.169) (or (1.170)) holds. The proof is complete.                                                                                    □

### 1.6.3 Symplectic Matrices and Generalized LU Factorization

In this subsection we prove some results concerning the representation of a symplectic matrix in the form of a product of lower (upper) block-triangular matrices and orthogonal symplectic matrices. Matrices in $Y \in \mathbb{R}^{2n \times n}$ satisfying (1.147) frequently appear in the proofs of statements in this subsection. These matrices are expressed in the form of a product of a lower block-triangular matrix, an orthogonal matrix, and an invertible matrix. The results of this section play an important role in the proofs of some properties of the comparative index introduced in Chap. 3

In the first auxiliary statement, we present necessary and sufficient conditions for a matrix $Y \in \mathbb{R}^{2n \times n}$ to satisfy (1.147).

**Lemma 1.68** *A $2n \times n$ matrix $Y = \begin{pmatrix} X \\ U \end{pmatrix}$ satisfies (1.147) if and only if*

$$XX^\dagger U(I - X^\dagger X) = 0, \tag{1.171}$$

$$Q = Q^T, \quad Q := XX^\dagger U X^\dagger, \tag{1.172}$$

$$\mathrm{rank}\,[(I - XX^\dagger)\,U(I - X^\dagger X)] = \mathrm{rank}\,(I - X^\dagger X). \tag{1.173}$$

*Moreover, if (1.147) holds, then we have*

$$\det \mathfrak{L} \ne 0, \quad \mathfrak{L} := X - (I - XX^\dagger)\,U(I - X^\dagger X), \tag{1.174}$$

$$\det \mathfrak{M} \ne 0, \quad \mathfrak{M} := X - (I - XX^\dagger)\,U. \tag{1.175}$$

*Proof*

(i) Sufficiency. Let (1.171)–(1.173) hold. We will show that (1.173) implies (1.174), (1.175). In fact, using Remark 1.60(vii) we obtain

$$\text{rank } \mathfrak{L} = \text{rank } X + \text{rank} \left[ (I - XX^{\dagger}) U (I - X^{\dagger}X) \right] = n,$$

which proves the invertibility of $\mathfrak{L}$ in (1.174). As for the matrix $\mathfrak{M}$, we have

$$\mathfrak{M} = X - (I - XX^{\dagger}) U = X - (I - XX^{\dagger}) U (I - X^{\dagger}X) - (I - XX^{\dagger}) U (X^{\dagger}X)$$
$$= [I - (I - XX^{\dagger}) U X^{\dagger}] \mathfrak{L},$$

where the matrix $I - (I - XX^{\dagger}) U X^{\dagger}$ is invertible. This conclusion follows from (1.163) with $P := X^{\dagger}$ and $R := (I - XX^{\dagger})$. Hence, (1.175) holds. Now by (1.171) we have

$$X = XX^{\dagger} \mathfrak{M},$$
$$U = XX^{\dagger}U + (I - XX^{\dagger}) U = XX^{\dagger}UX^{\dagger}X + (I - XX^{\dagger}) U \qquad (1.176)$$
$$= [XX^{\dagger}UX^{\dagger} - (I - XX^{\dagger})] \mathfrak{M}.$$

Hence,

$$X^T U = \mathfrak{M}^T XX^{\dagger}UX^{\dagger} \mathfrak{M} = \mathfrak{M}^T Q \mathfrak{M}. \qquad (1.177)$$

Therefore, (1.172) implies that $X^T U = U^T X$, which is the same as $Y^T \mathcal{J} Y = 0$. Furthermore,

$$Y^T Y = X^T X + U^T U = \mathfrak{M}^T [XX^{\dagger} + (I - XX^{\dagger}) + (XX^{\dagger}UX^{\dagger})^2] \mathfrak{M}$$
$$= \mathfrak{M}^T (I + Q^2) \mathfrak{M}.$$

The last matrix is invertible in view of (1.172). Indeed, $Q^2 = X^{\dagger T} U^T XX^{\dagger}UX^{\dagger} \geq 0$ and hence, rank $Y = n$, which completes the sufficiency part of the proof.

(ii) Necessity. The condition $Y^T \mathcal{J} Y = 0$, i.e., $X^T U = U^T X$, implies

$$(X^{\dagger})^T X^T U (I - X^{\dagger}X) = (X^{\dagger})^T U^T X (I - X^{\dagger}X) = 0,$$

which means that (1.171) holds. Further,

$$Q = XX^{\dagger}UX^{\dagger} = (X^{\dagger})^T X^T UX^{\dagger} = (X^{\dagger})^T U^T XX^{\dagger} = (XX^{\dagger}UX^{\dagger})^T = Q^T,$$

which implies that (1.172) holds. Concerning (1.173), we have

$$\mathrm{rank}(X^T X + U^T U)\,(I - X^\dagger X) = \mathrm{rank}(I - X^\dagger X) = \mathrm{rank}\, U^T U(I - X^\dagger X)$$

$$= \mathrm{rank}\, U^T (I - X X^\dagger)\, U(I - X^\dagger X)$$

$$\le \mathrm{rank}(I - X^\dagger X)\, U(I - X^\dagger X) \le \mathrm{rank}(I - X^\dagger X),$$

where we have used that $\mathrm{rank}\, Y = n = \mathrm{rank}(X^T X + U^T U)$ and that for a product of matrices $\mathrm{rank}\, AB \le \min\{\mathrm{rank}\, A, \mathrm{rank}\, B\}$. Thus, we proved (1.171), and the proof is completed.

□

*Remark 1.69*  Recall (see Remark 1.66) that if $Y \in \mathbb{R}^{2n \times n}$ satisfies (1.147), then the matrix $Q$ given in (1.172) is a symmetric solution of the equation

$$X^T Q X = X^T U. \tag{1.178}$$

In particular, if $X$ is invertible, then we have $Q = U X^{-1}$.

We mentioned at the end of Sect. 1.6.1 the matrices, which belong to $Sp(2n) \cap O(2n)$. In the results of this subsection, we will need special matrices of this class, which are constructed as follows. Let $P \in \mathbb{R}^{n \times n}$. Then $P$ is an orthogonal projector (i.e., $P$ is symmetric and $P^2 = P$) if and only if the $2n \times 2n$ matrix

$$\mathfrak{N}_P := \begin{pmatrix} P & I - P \\ -(I - P) & P \end{pmatrix} \tag{1.179}$$

is orthogonal. In this case the matrix $\mathfrak{N}_P$ is also symplectic, i.e., it belongs to $Sp(2n) \cap O(2n)$.

In the theorem below, we will use the matrix $\mathfrak{N}_{XX^\dagger}$ defined in (1.179) via the projector $P := X X^\dagger$. It concerns a factorization of $Y$ satisfying conditions (1.147) and plays crucial role in proofs of Chap. 3.

**Theorem 1.70**  *Let $Y = \begin{pmatrix} X \\ U \end{pmatrix} \in \mathbb{R}^{2n \times n}$ satisfy (1.147), the matrix $\mathfrak{N}_{XX^\dagger}$ be given by (1.179), and $Q$ be a symmetric matrix satisfying (1.178). Then $Y$ can be expressed in the form*

$$Y = L\, \mathfrak{N}_{XX^\dagger} \begin{pmatrix} M \\ 0 \end{pmatrix}, \quad L := \begin{pmatrix} I & 0 \\ Q & I \end{pmatrix}, \quad \det M \ne 0, \tag{1.180}$$

*where*

$$M := X - (I - X X^\dagger)\,(U - Q X). \tag{1.181}$$

*Proof* Observe that if a matrix $Y$ satisfies (1.147), then in view of property (vii) of Lemma 1.58, the matrix $\tilde{Y} = L^{-1}Y = \begin{pmatrix} X \\ U-QX \end{pmatrix}$ also satisfies (1.147). Then it suffices to apply Lemma 1.68 to $\tilde{Y}$. Using (1.175), the matrix $M$ in (1.181) is invertible, $X = XX^{\dagger}M$, and by (1.178), we get

$$U - QX = (I - XX^{\dagger})(U - QX) = -(I - XX^{\dagger})M.$$

Then $\tilde{Y} = \mathfrak{N}_{XX^{\dagger}}\begin{pmatrix} M \\ 0 \end{pmatrix}$, i.e., $Y = L\,\mathfrak{N}_{XX^{\dagger}}\begin{pmatrix} M \\ 0 \end{pmatrix}$, what we needed to prove.                    □

*Remark 1.71*   Formula (1.180) implies the representation for the blocks $X$ and $U$ in the form

$$X = XX^{\dagger}M \tag{1.182}$$

$$U = [QXX^{\dagger} - (I - XX^{\dagger})]M, \tag{1.183}$$

and

$$X^{T}U = M^{T}XX^{\dagger}UX^{\dagger}M, \tag{1.184}$$

where $\det M \neq 0$. In particular,

$$\operatorname{rank} U = \operatorname{rank} X^{T}U + n - \operatorname{rank} X, \tag{1.185}$$

because for the matrix $Q := XX^{\dagger}UX^{\dagger}$ (see Remark 1.66), we have

$$UM^{-1} = XX^{\dagger}UX^{\dagger} - (I - XX^{\dagger})$$

and then it follows that

$$\operatorname{rank} U = \operatorname{rank}(XX^{\dagger}UX^{\dagger}) + n - \operatorname{rank} X, \quad \operatorname{rank} XX^{\dagger}UX^{\dagger} = \operatorname{rank} X^{T}U.$$

Theorem 1.70 leads also to the following statement concerning a factorization of symplectic matrices.

**Lemma 1.72** *Let $Z = \begin{pmatrix} X & \tilde{X} \\ U & \tilde{U} \end{pmatrix} \in Sp(2n)$ and $Q$ be a symmetric solution of (1.178). Then the matrix $Z$ can be expressed in the form*

$$Z = L\,\mathfrak{N}_{XX^{\dagger}}\,\operatorname{diag}\{M, M^{T-1}\}H, \quad H := \begin{pmatrix} I & \tilde{Q} \\ 0 & I \end{pmatrix}, \tag{1.186}$$

*where the matrices $L, M, \mathfrak{N}_{XX^{\dagger}}$ are given in Theorem 1.70 and*

$$\tilde{Q}^{T} = \tilde{Q}, \quad \tilde{Q} := \tilde{X}^{T}XX^{\dagger}(\tilde{U} - Q\tilde{X}) - (\tilde{U} - Q\tilde{X})^{T}(I - XX^{\dagger})\tilde{X}, \tag{1.187}$$

*with $\tilde{Q}$ solving the equation $X\tilde{Q}X^{T} = X\tilde{X}^{T}$.*

*Proof* Using factorization (1.180) for $Y = \binom{X}{U}$, being the first column of the matrix $Z$, one can verify by a direct computations that

$$\mathfrak{N}_{XX^\dagger}^T L^{-1} Z = \begin{pmatrix} M\, XX^\dagger \tilde{X} - (I - XX^\dagger)\,(\tilde{U} - Q\tilde{X}) \\ 0 \qquad\qquad M^{T-1} \end{pmatrix} = \mathrm{diag}\{M, M^{T-1}\}\, H,$$

where the matrices $M$ and $H$ are determined by (1.181) and (1.186) and

$$M^{T-1} = (I - XX^\dagger)\, \tilde{X} + XX^\dagger (\tilde{U} - Q\tilde{X}). \qquad (1.188)$$

Then, by (1.188), we have $\tilde{Q} = M^{-1}[XX^\dagger \tilde{X} - (I - XX^\dagger)\,(\tilde{U} - Q\tilde{X})]$, i.e., (1.187) holds. Consequently, from the last equality, we get $X\tilde{Q} = XX^\dagger M\tilde{Q} = XX^\dagger \tilde{X}$, so that $X\tilde{Q}X^T = X\tilde{X}^T$ follows by the symmetry of $\tilde{X}X^T$; see Lemma 1.58(iii).  □

Under the additional assumption regarding the invertibility of the matrix $X$, we obtain the following result concerning the $n \times n$ block $LU$-factorization of a symplectic matrix. This result was for the first time proved in [219, Proposition 2.36].

**Corollary 1.73** *Consider the symplectic matrix* $Z = \begin{pmatrix} X & \tilde{X} \\ U & \tilde{U} \end{pmatrix}$ *with invertible matrix* $X$. *Then the factorization in* (1.186) *takes the form*

$$Z = \begin{pmatrix} I & 0 \\ UX^{-1} & I \end{pmatrix} \begin{pmatrix} X & 0 \\ 0 & X^{T-1} \end{pmatrix} \begin{pmatrix} I & X^{-1}\tilde{X} \\ 0 & I \end{pmatrix}. \qquad (1.189)$$

The result in Corollary 1.73 justifies that (1.180) and (1.189) can be called a *generalized block LU-factorization* of $Y$ and $Z$, respectively. Note that the matrices $L$ and $\mathfrak{N}_{XX^\dagger}$ in these formulas generally do not commute (they commute if a solution of (1.178) is taken in the form (1.172)).

As a consequence of Corollary 1.73 and Lemma 1.58(viii), we obtain the next statement concerning the factorization of a symplectic matrix. The proof (different from ours) can be found in [332, Theorem 6.2].

**Theorem 1.74** *Every symplectic matrix* $Z \in Sp(2n)$ *can be expressed as the product*

$$Z = H_1\, L_2\, H_3 = \mathcal{J} L_1\, \mathcal{J} L_2\, \mathcal{J} L_3\, \mathcal{J},$$

*where* $H_1$ *and* $H_3$ *are upper block-triangular symplectic matrices and* $L_1$, $L_2$, *and* $L_3$ *are lower block-triangular symplectic matrices.*

*Proof* Put $H_1 = \begin{pmatrix} I & -\sigma I \\ 0 & I \end{pmatrix}$. Then by Lemma 1.58(viii) for $Y := Z\binom{I}{0} =: \binom{X}{U}$ one can choose $\sigma \in \mathbb{R}$ such hat (1.150) holds. Hence, the matrix $(I\ 0)H_1^{-1} Z\binom{I}{0}$ is invertible, and by Corollary 1.73, there exists a block $LU$-factorization of the matrix $H_1^{-1} Z = L_2 H_3$. The multiplication of this equality by $H_1$ proves the theorem.  □

### 1.6.4 Symplectic Matrices Depending on Parameter

In this section we consider symplectic matrices $W(\lambda)$ depending on a parameter $\lambda \in \mathbb{R}$. We assume that $W(\lambda)$ is piecewise continuously differentiable in $\lambda \in \mathbb{R}$, i.e., it is continuous on $\mathbb{R}$ with the piecewise continuous derivative. Then we can introduce the following matrix

$$\Psi(W(\lambda)) = -\mathcal{J}\frac{d}{d\lambda}(W(\lambda)) W^{-1}(\lambda) = \mathcal{J}\frac{d}{d\lambda}(W(\lambda)) \mathcal{J} W^T(\lambda) \mathcal{J} \qquad (1.190)$$

defined for any piecewise continuously differentiable symplectic matrix $W(\lambda)$. The first result concerning the matrix $\Psi(W(\lambda))$ is the following proposition.

**Proposition 1.75** *A matrix function $W : \mathbb{R} \to \mathbb{R}^{2n \times 2n}$ of the real variable $\lambda$ is symplectic for all $\lambda \in \mathbb{R}$ if and only if $W(0)$ is symplectic and*

$$\dot{W}(\lambda) = \mathcal{J}\mathcal{H}(\lambda) W(\lambda), \quad \lambda \in \mathbb{R}, \qquad (1.191)$$

*with a symmetric matrix $\mathcal{H}(\lambda)$ for all $\lambda \in \mathbb{R}$.*

*Proof* The proof can be found in [216, pg. 229]. First note that (1.191) is the Hamiltonian differential system (1.103), where we replace the variable $t$ by the variable $\lambda$. It is well known (see [328] or [205]) that under the assumption that $W(0)$ is symplectic for the fundamental matrix $W(t)$ of (1.191), it follows that $W(\lambda) \in Sp(2n)$ for $\lambda \in \mathbb{R}$. In the opposite direction, differentiating the identity $W^T(\lambda)\mathcal{J}W(\lambda) = \mathcal{J}$ with respect to $\lambda$, we get

$$\frac{d}{d\lambda}(W^T(\lambda) \mathcal{J} W(\lambda)) = 0 = \dot{W}^T(\lambda) \mathcal{J} W(\lambda) + W^T(\lambda) \mathcal{J} \dot{W}(\lambda),$$

then, multiplying the previous identity by $W^{-1\,T}(\lambda)$ from the left side and by $W^{-1}(\lambda)$ from the right side, we derive the desired identity

$$\mathcal{H}(\lambda) = \mathcal{J}^T \dot{W}(\lambda) W^{-1}(\lambda) = W^{T-1}(\lambda) \dot{W}^T(\lambda) \mathcal{J} = \Psi(W(\lambda)). \qquad (1.192)$$

One can now see the symmetry of $\mathcal{H}(\lambda) = \Psi(W(\lambda))$ directly from (1.192). $\square$

In particular, in terms of the blocks $A(\lambda)$, $B(\lambda)$, $C(\lambda)$, $D(\lambda)$ of the symplectic matrix $W(\lambda) = \begin{pmatrix} A(\lambda) & B(\lambda) \\ C(\lambda) & D(\lambda) \end{pmatrix}$ the operator (1.190) takes the form

$$\Psi(W(\lambda)) = \begin{pmatrix} \dot{D}(\lambda) C^T(\lambda) - \dot{C}(\lambda) D^T(\lambda) & \dot{C}(\lambda) B^T(\lambda) - \dot{D}(\lambda) A^T(\lambda) \\ \dot{A}(\lambda) D^T(\lambda) - \dot{B}(\lambda) C^T(\lambda) & \dot{B}(\lambda) A^T(\lambda) - \dot{A}(\lambda) B^T(\lambda) \end{pmatrix}. \qquad (1.193)$$

The proof is based on direct computations incorporating the identities

$$
\left.
\begin{aligned}
A^T(\lambda)\,C(\lambda) &= C^T(\lambda)\,A(\lambda), & B^T(\lambda)\,D(\lambda) &= D^T(\lambda)\,B(\lambda), \\
A(\lambda)\,B^T(\lambda) &= B(\lambda)\,A^T(\lambda), & D(\lambda)\,C^T(\lambda) &= C(\lambda)\,D^T(\lambda), \\
A^T(\lambda)\,D(\lambda) - C^T(\lambda)\,B(\lambda) &= I, & D(\lambda)\,A^T(\lambda) - C(\lambda)\,B^T(\lambda) &= I.
\end{aligned}
\right\}
$$

which hold due to the symplecticity of $W(\lambda)$. Then upon differentiating the above formulas with respect to $\lambda$, we get

$$
\left.
\begin{aligned}
\dot{A}^T(\lambda)\,C(\lambda) + A^T(\lambda)\,\dot{C}(\lambda) &= \dot{C}^T(\lambda)\,A(\lambda) + C^T(\lambda)\,\dot{A}(\lambda), \\
\dot{B}^T(\lambda)\,D(\lambda) + B^T(\lambda)\,\dot{D}(\lambda) &= \dot{D}^T(\lambda)\,B(\lambda) + D^T(\lambda)\,\dot{B}(\lambda), \\
\dot{A}^T(\lambda)\,D(\lambda) + A^T(\lambda)\,\dot{D}(\lambda) &= \dot{C}^T(\lambda)\,B(\lambda) + C^T(\lambda)\,\dot{B}(\lambda), \\
\dot{D}(\lambda)\,C^T(\lambda) + D(\lambda)\,\dot{C}^T(\lambda) &= \dot{C}(\lambda)\,D^T(\lambda) + C(\lambda)\,\dot{D}^T(\lambda), \\
\dot{A}(\lambda)\,B^T(\lambda) + A(\lambda)\,\dot{B}^T(\lambda) &= \dot{B}(\lambda)\,A^T(\lambda) + B(\lambda)\,\dot{A}^T(\lambda), \\
\dot{D}(\lambda)\,A^T(\lambda) + D(\lambda)\,\dot{A}^T(\lambda) &= \dot{C}(\lambda)\,B^T(\lambda) + C(\lambda)\,\dot{B}^T(\lambda).
\end{aligned}
\right\} \tag{1.194}
$$

Now we are prepared to formulate the main results of this section concerning properties of the symmetric operator (1.190).

**Proposition 1.76** *The following statements hold.*

(i) *For arbitrary piecewise continuously differentiable symplectic matrices $W(\lambda)$ and $V(\lambda)$, we have*

$$
\Psi(V(\lambda)W(\lambda)) = \Psi(V(\lambda)) + V^{T-1}(\lambda)\,\Psi(W(\lambda))\,V^{-1}(\lambda). \tag{1.195}
$$

(ii) *If $W(\lambda)$ is piecewise continuously differentiable, then*

$$
\Psi(W^{-1}(\lambda)) = -W^T(\lambda)\,\Psi(W(\lambda))\,W(\lambda). \tag{1.196}
$$

(iii) *For arbitrary constant symplectic matrices $R$ and $P$, we have*

$$
\Psi(R^{-1}W(\lambda)P) = R^T\,\Psi(W(\lambda))\,R. \tag{1.197}
$$

(iv) *For arbitrary piecewise continuously differentiable symplectic matrix $W(\lambda)$ and constant symplectic matrices $R$ and $P$, we have*

$$
\left.
\begin{aligned}
\Psi(W(\lambda)) \geq 0 &\Leftrightarrow \Psi(R^{-1}W(\lambda)\,P) \geq 0 \Leftrightarrow \Psi(W^{-1}(\lambda)) \leq 0 \\
&\Leftrightarrow \Psi(W^T(\lambda)) \leq 0 \quad\;\; \Leftrightarrow \Psi(W^{T-1}(\lambda)) \geq 0.
\end{aligned}
\right\} \tag{1.198}
$$

*Proof* Applying (1.192) to the product $V(\lambda)W(\lambda)$, we have

$$
\begin{aligned}
\Psi(V(\lambda)\,W(\lambda)) &= \mathcal{J}^T \frac{d}{d\lambda}(V(\lambda)\,W(\lambda))\,(V(\lambda)\,W(\lambda)))^{-1} \\
&= \mathcal{J}^T (\dot{V}(\lambda)\,W(\lambda) + V(\lambda)\,\dot{W}(\lambda))\,W^{-1}(\lambda)\,V^{-1}(\lambda) \\
&= \mathcal{J}^T \dot{V}(\lambda)\,V^{-1}(\lambda) + \mathcal{J}^T V(\lambda)\,\dot{W}(\lambda)\,W^{-1}(\lambda)\,V^{-1}(\lambda) \\
&= \Psi(V(\lambda)) + V^{T-1}(\lambda)\,\Psi(W(\lambda))\,V^{-1}(\lambda),
\end{aligned}
$$

where we used $\mathcal{J}^T V(\lambda) = V^{T-1}(\lambda)\,\mathcal{J}^T$ according to Lemma 1.58(ii). So we have proved (i). For the proof of property (ii), we put $V(\lambda) := W^{-1}(\lambda)$ in property (i) and then use that $\Psi(W^{-1}(\lambda)\,W(\lambda)) = \Psi(I) = 0$. The proof of (iii) is based on the subsequent application of property (i). So we have

$$
\Psi((R^{-1}W(\lambda))\,P) = \Psi(R^{-1}W(\lambda)) = R^T\Psi(W(\lambda))R.
$$

Property (iv) is the direct consequence of (iii), (ii), and Lemma 1.58(ii), when property (iii) is applied to the cases $W^T = \mathcal{J}^T W^{-1} \mathcal{J}$ and $W^{T-1} = \mathcal{J}^T W \mathcal{J}$. $\quad\square$

Another important consequence from Proposition 1.76 are presented by the following lemma.

**Lemma 1.77** *Assume that for a piecewise continuously differentiable symplectic matrix $W(\lambda)$, there exist constant symplectic matrices $R$ and $P$ such that*

$$
W(\lambda) = R\tilde{W}(\lambda)\,P^{-1}, \quad \tilde{W}(\lambda) = \begin{pmatrix} \tilde{A}(\lambda) & \tilde{B}(\lambda) \\ \tilde{C}(\lambda) & \tilde{D}(\lambda) \end{pmatrix}, \quad \det\tilde{B}(\lambda) \neq 0, \; \lambda \in \mathbb{R}. \tag{1.199}
$$

*Then we have (omitting the argument $\lambda$)*

$$
\Psi(\tilde{W}) = -\mathcal{J} \begin{pmatrix} \tilde{B} & 0 \\ \tilde{D} & I \end{pmatrix} \frac{d}{d\lambda} \begin{pmatrix} \tilde{B}^{-1}\tilde{A} & -\tilde{B}^{-1} \\ -\tilde{B}^{T-1} & \tilde{D}\tilde{B}^{-1} \end{pmatrix} \begin{pmatrix} \tilde{B} & 0 \\ \tilde{D} & I \end{pmatrix}^T \mathcal{J}^T, \tag{1.200}
$$

*and then the condition*

$$
\Psi(W(\lambda)) \geq 0, \quad \lambda \in \mathbb{R} \tag{1.201}
$$

*is equivalent to*

$$
\frac{d}{d\lambda}\tilde{Q}(\lambda) \leq 0, \quad \tilde{Q}(\lambda) = \begin{pmatrix} \tilde{B}^{-1}(\lambda)\,\tilde{A}(\lambda) & -\tilde{B}^{-1}(\lambda) \\ -\tilde{B}^{T-1}(\lambda) & \tilde{D}(\lambda)\,\tilde{B}^{-1}(\lambda) \end{pmatrix}, \quad \lambda \in \mathbb{R}. \tag{1.202}
$$

*Proof* By Proposition 1.76(iv) we see that $\Psi(\tilde{W}(\lambda)) \geq 0$, and one can verify by direct computations (applying Proposition 1.76(i)) that representation (1.200) holds. Then, using the nonsingularity of $\tilde{B}(\lambda)$, we prove that (1.201) is equivalent to (1.202). The proof is completed.                                                                            □

In the following results, we will use Lemma 1.77 locally, i.e., in a sufficiently small neighborhood of $\lambda_0 \in \mathbb{R}$. So we have the following property.

**Corollary 1.78** *Assume that a piecewise continuously differentiable symplectic matrix* $W(\lambda) = \begin{pmatrix} A(\lambda) & B(\lambda) \\ C(\lambda) & D(\lambda) \end{pmatrix}$ *obeys* (1.201) *and for some* $\lambda_0 \in \mathbb{R}$, *we have*

$$\det A(\lambda_0) \neq 0. \tag{1.203}$$

*Then there exists* $\varepsilon > 0$ *such that* $\det A(\lambda) \neq 0$ *for* $\lambda \in (\lambda_0 - \varepsilon, \lambda_0 + \varepsilon)$ *and*

$$\frac{d}{d\lambda}(A^{-1}(\lambda) B(\lambda)) \geq 0, \quad \frac{d}{d\lambda}(C(\lambda) A^{-1}(\lambda)) \leq 0, \quad \lambda \in (\lambda_0 - \varepsilon, \lambda_0 + \varepsilon). \tag{1.204}$$

*Similarly, if for some* $\lambda_0 \in \mathbb{R}$

$$\det D(\lambda_0) \neq 0, \tag{1.205}$$

*then there exists* $\varepsilon > 0$ *such that* $\det D(\lambda) \neq 0$ *for* $\lambda \in (\lambda_0 - \varepsilon, \lambda_0 + \varepsilon)$ *and*

$$\frac{d}{d\lambda}(B(\lambda) D^{-1}(\lambda)) \geq 0, \quad \frac{d}{d\lambda}(D^{-1}(\lambda) C(\lambda)) \leq 0, \quad \lambda \in (\lambda_0 - \varepsilon, \lambda_0 + \varepsilon). \tag{1.206}$$

*Proof* Putting $R := I$ and $P := \mathcal{J}$ in Lemma 1.77, we see that $\tilde{W}(\lambda) = W(\lambda)\,\mathcal{J}$ has the form $\tilde{W}(\lambda) = \begin{pmatrix} -B(\lambda) & A(\lambda) \\ -D(\lambda) & C(\lambda) \end{pmatrix}$, and then by (1.202), we derive (1.204). Similarly, if $R := \mathcal{J}^T$ and $P = I$ in Lemma 1.77, then $\tilde{W}(\lambda) = \mathcal{J}\,W(\lambda) = \begin{pmatrix} C(\lambda) & D(\lambda) \\ -A(\lambda) & -B(\lambda) \end{pmatrix}$. Substituting the blocks of $\tilde{W}(\lambda)$ into (1.202), we derive (1.206).                                    □

The following theorem is the most important result of this section.

**Theorem 1.79** *Assume that* $W(\lambda) = \begin{pmatrix} A(\lambda) & B(\lambda) \\ C(\lambda) & D(\lambda) \end{pmatrix}$ *is piecewise continuously differentiable on* $\mathbb{R}$ *and obeys assumption* (1.201). *Then* $\operatorname{Ker} B(\lambda)$ *is piecewise constant in* $\lambda$, *i.e., for any* $\lambda_0 \in \mathbb{R}$ *there exists* $\delta > 0$ *such that*

$$\operatorname{Ker} B(\lambda) \equiv \operatorname{Ker} B(\lambda_0^-) \subseteq \operatorname{Ker} B(\lambda_0) \quad \text{for all } \lambda \in (\lambda_0 - \delta, \lambda_0), \tag{1.207}$$

$$\operatorname{Ker} B(\lambda) \equiv \operatorname{Ker} B(\lambda_0^+) \subseteq \operatorname{Ker} B(\lambda_0) \quad \text{for all } \lambda \in (\lambda_0, \lambda_0 + \delta), \tag{1.208}$$

*Proof* The proof is based on Corollary 1.78 and Lemma 1.67. Putting $Z := W(\lambda_0)$ in Lemma 1.67, we have that there exist $\sigma > 0$ such that for the blocks of

$$\tilde{W} := W(\lambda_0) \begin{pmatrix} I & 0 \\ \sigma I & I \end{pmatrix} = \begin{pmatrix} A(\lambda_0) + \sigma B(\lambda_0) & B(\lambda_0) \\ C(\lambda_0) + \sigma D(\lambda_0) & D(\lambda_0) \end{pmatrix}$$

we have $\det[A(\lambda_0) + \sigma B(\lambda_0)] \neq 0$ and

$$(A(\lambda_0) + \sigma B(\lambda_0))^{-1} B(\lambda_0) \leq 0. \tag{1.209}$$

Then we apply Corollary 1.78 to the matrix $\tilde{W}(\lambda) = W(\lambda) \begin{pmatrix} I & 0 \\ \sigma I & I \end{pmatrix} = \begin{pmatrix} \tilde{A}(\lambda) & B(\lambda) \\ \tilde{C}(\lambda) & D(\lambda) \end{pmatrix}$. Since $W(\lambda)$ obeys (1.201), i.e., $\Psi(W(\lambda)) \geq 0$, then the same condition holds for $\tilde{W}(\lambda)$ by Proposition 1.76(iv). Moreover, the matrix $\tilde{W}(\lambda)$ obeys the assumption $\det \tilde{A}(\lambda_0) \neq 0$. Applying Corollary 1.78 we see that there exist $\varepsilon > 0$ such that $\tilde{A}^{-1}(\lambda) B(\lambda)$ is nondecreasing matrix function for $\lambda \in (\lambda_0 - \varepsilon, \lambda_0 + \varepsilon)$. Choose $c \in \text{Ker } B(\lambda)$ for some $\lambda \in (\lambda_0 - \varepsilon, \lambda_0)$. Then the monotonicity of $\tilde{A}^{-1}(\lambda) B(\lambda)$ and (1.209) imply

$$0 = c^T \tilde{A}^{-1}(\lambda) B(\lambda) c \leq c^T \tilde{A}^{-1}(v) B(v) c \leq c^T \tilde{A}^{-1}(\lambda_0) B(\lambda_0) c \leq 0$$

for all $v \in [\lambda, \lambda_0]$. Hence, $c^T \tilde{A}^{-1}(v) B(v) c = 0$ and so $c \in \text{Ker } B(v)$ for every $v \in [\lambda, \lambda_0]$. Therefore, $\text{Ker } B(\lambda) \subseteq \text{Ker } B(v)$ for all $\lambda, v \in (\lambda_0 - \varepsilon, \lambda_0]$ with $\lambda \leq v$. This means that the set $\text{Ker } B(\lambda)$ is nondecreasing in $\lambda$ on $(\lambda_0 - \varepsilon, \lambda_0]$. This implies that condition (1.207) is satisfied for some sufficiently small $\delta \in (0, \varepsilon)$. For (1.208) we proceed in the same way except that we choose $\sigma$ according to Lemma 1.67 such that

$$(A(\lambda_0) + \sigma B(\lambda_0))^{-1} B(\lambda_0) \geq 0. \tag{1.210}$$

So we have proved that $\text{Ker } B(\lambda)$ is piecewise constant in $\lambda \in \mathbb{R}$. □

*Remark 1.80* The assertions of Theorem 1.79 hold true if we replace (1.201) by the monotonicity assumption

$$\Psi(W(\lambda)) \leq 0, \ \lambda \in \mathbb{R}. \tag{1.211}$$

Indeed, in the proof of Theorem 1.79, we used Proposition 1.76(iv) and Corollary 1.78, where the replacement of (1.201) by (1.211) derives the respective replacements of all signs $\geq$ and $\leq$ by the opposite signs $\leq$ and $\geq$. In particular, in this case we have (see the proof of Theorem 1.79) that $\tilde{A}^{-1}(\lambda) B(\lambda)$ is nonincreasing matrix function for $\lambda \in (\lambda_0 - \varepsilon, \lambda_0 + \varepsilon)$. Then, under assumption (1.211) we prove (1.208) repeating the proof of (1.207) under assumption (1.201). Similarly, we use the proof of (1.208) under assumption (1.201) to prove (1.207) using (1.211).

Based on Remark 1.80, we prove another important fact connected with (1.201).

**Theorem 1.81** *Assume that* $W(\lambda) = \begin{pmatrix} A(\lambda) & B(\lambda) \\ C(\lambda) & D(\lambda) \end{pmatrix}$ *is piecewise continuously differentiable on* $\mathbb{R}$ *and obeys assumption* (1.201). *Then* $\operatorname{Im} B(\lambda)$ *is piecewise constant in* $\lambda$, *i.e., for any* $\lambda_0 \in \mathbb{R}$ *there exists* $\delta > 0$ *such that*

$$\operatorname{Im} B(\lambda_0) \subseteq \operatorname{Im} B(\lambda) \equiv \operatorname{Im} B(\lambda_0^-) \quad \text{for all } \lambda \in (\lambda_0 - \delta, \lambda_0), \tag{1.212}$$

$$\operatorname{Im} B(\lambda_0) \subseteq \operatorname{Im} B(\lambda) \equiv \operatorname{Im} B(\lambda_0^+) \quad \text{for all } \lambda \in (\lambda_0, \lambda_0 + \delta). \tag{1.213}$$

*Proof* To prove the result, we note that (1.201) is equivalent to

$$\Psi(W^{-1}(\lambda)) \leq 0, \quad \lambda \in \mathbb{R}, \tag{1.214}$$

by Proposition 1.76(iv) and by Lemma 1.58(ii)

$$W^{-1}(\lambda) = \begin{pmatrix} D^T(\lambda) & -B^T(\lambda) \\ -C^T(\lambda) & A^T(\lambda) \end{pmatrix}.$$

Then, by Remark 1.80 we can apply Theorem 1.79 to $W^{-1}(\lambda)$ which implies that $\operatorname{Ker} B^T(\lambda)$ is piecewise constant in $\lambda$, i.e., for any $\lambda_0 \in \mathbb{R}$ there exists $\delta > 0$ such that

$$\operatorname{Ker} B^T(\lambda) \equiv \operatorname{Ker} B^T(\lambda_0^-) \subseteq \operatorname{Ker} B^T(\lambda_0) \quad \text{for all } \lambda \in (\lambda_0 - \delta, \lambda_0), \tag{1.215}$$

$$\operatorname{Ker} B^T(\lambda) \equiv \operatorname{Ker} B^T(\lambda_0^+) \subseteq \operatorname{Ker} B^T(\lambda_0) \quad \text{for all } \lambda \in (\lambda_0, \lambda_0 + \delta). \tag{1.216}$$

Conditions (1.215), (1.216) are equivalent to (1.212), (1.213), because $\operatorname{Ker} B^T(\lambda)$ is the orthogonal complement to $\operatorname{Im} B(\lambda)$. □

Now we formulate several corollaries to Theorems 1.79, and 1.81.

**Theorem 1.82** *Assume* (1.201) *(or* (1.211)*). Then the following three conditions are equivalent:*

$$\operatorname{rank} B(\lambda) \text{ is constant for } \lambda \in \mathbb{R}, \tag{1.217}$$

$$\operatorname{Ker} B(\lambda) \text{ is constant for } \lambda \in \mathbb{R}, \tag{1.218}$$

$$\operatorname{Im} B(\lambda) \text{ is constant for } \lambda \in \mathbb{R}. \tag{1.219}$$

*Proof* It is clear that the constancy of $\operatorname{Ker} B(\lambda)$ in $\lambda \in \mathbb{R}$ (or $\operatorname{Im} B(\lambda)$ in $\lambda \in \mathbb{R}$) implies condition (1.217). Conversely, assume (1.217). Then, by Theorem 1.79 we have $\operatorname{Ker} B(\lambda_0^{\pm}) \subseteq \operatorname{Ker} B(\lambda_0)$, which implies $\operatorname{Ker} B(\lambda_0^{\pm}) = \operatorname{Ker} B(\lambda_0)$ for any $\lambda_0 \in \mathbb{R}$. Similarly, by Theorem 1.81, (1.217) and the inclusion $\operatorname{Im} B(\lambda_0) \subseteq \operatorname{Im} B(\lambda_0^{\pm})$ implies the equality $\operatorname{Im} B(\lambda_0) = \operatorname{Im} B(\lambda_0^{\pm})$ for any $\lambda_0 \in \mathbb{R}$. □

Based on Theorems 1.79 and 1.81 and by Lemma 1.63, we also have the following corollary.

**Corollary 1.83** *Assume* (1.201) *(or* (1.211)*). Then, for any* $\lambda_0 \in \mathbb{R}$ *there exists* $\delta > 0$ *such that the matrices* $B^\dagger(\lambda)\, B(\lambda)$ *and* $B(\lambda)\, B^\dagger(\lambda)$ *are constant for* $\lambda \in (\lambda_0 - \delta, \lambda_0)$. *A similar assertion holds also for* $\lambda \in (\lambda_0, \lambda_0 + \delta)$.

*Remark 1.84* We note that by Proposition 1.76(iv), all the monotonicity properties of $\Psi(W(\lambda))$ formulated in Theorems 1.79, 1.81, and 1.82 and in Corollary 1.83 hold also for $\Psi(R^{-1} W(\lambda)\, P)$, where $R$ and $P$ are constant symplectic matrices. These properties are satisfied for all blocks of $W(\lambda)$ and their linear combinations.

### 1.6.5  Monotone Matrix-Valued Functions

In this subsection we present two useful results about the behavior of symmetric monotone matrix-valued functions. The first theorem describes the change in the index (i.e., the number of negative eigenvalues) of a monotone matrix-valued function when its argument crosses a singularity. This result will be utilized in Sect. 5.1 in order to derive oscillation theorems for discrete eigenvalue problems for symplectic difference systems.

We use a standard monotonicity definition for symmetric matrix-valued functions, i.e., a matrix-valued function $A \,:\, (0, \varepsilon) \,\rightarrow\, \mathbb{R}^{n \times n}$ is nonincreasing (nondecreasing) on $(0, \varepsilon)$, if the scalar function $d^T A(t)\, d$ is nonincreasing (nondecreasing) on $(0, \varepsilon)$ for every $d \in \mathbb{R}^n$. Moreover, the notation $f(0^+)$ and $f(0^-)$ stands for the right-hand and left-hand limits of the function $f(t)$ at $t = 0$.

**Theorem 1.85 (Index Theorem)** *Let* $X(t)$, $U(t)$, $R_1(t)$, $R_2(t)$ *be given real* $m \times m$-*matrix-valued functions on* $[0, \varepsilon)$ *such that*

$$\left. \begin{array}{l} R_1(t)\, R_2^T(t) \text{ and } X^T(t)\, U(t) \text{ are symmetric,} \\[4pt] \mathrm{rank}\,(R_1(t),\ R_2(t)) = \mathrm{rank}\,(X^T(t),\ U^T(t)) = m \end{array} \right\} \quad \textit{for } t \in [0, \varepsilon), \qquad (1.220)$$

*and assume that* $X(t)$, $U(t)$, $R_1(t)$, *and* $R_2(t)$ *are continuous at* 0, *i.e.,*

$$\left. \begin{array}{l} \lim_{t \to 0^+} R_1(t) = R_1 := R_1(0), \quad \lim_{t \to 0^+} X(t) = X := X(0), \\[6pt] \lim_{t \to 0^+} R_2(t) = R_2 := R_2(0), \quad \lim_{t \to 0^+} U(t) = U := U(0), \end{array} \right\} \qquad (1.221)$$

*and that* $X(t)$ *is invertible for* $t \in (0, \varepsilon)$. *Moreover, denote*

$$\left. \begin{array}{ll} M(t) := R_1(t)\, R_2^T(t) + R_2(t)\, U(t)\, X^{-1}(t)\, R_2^T(t), & \\[4pt] \Lambda(t) := R_1(t)\, X(t) + R_2(t)\, U(t), & \Lambda := \Lambda(0) \\[4pt] S(t) := X^\dagger R_2^T(t), & S := S(0), \\[4pt] S^*(t) := R_2^T(t) - X S(t) = (I - X X^\dagger)\, R_2^T(t), & S^* := S^*(0), \end{array} \right\} \qquad (1.222)$$

*and suppose that the functions* $U(t) X^{-1}(t)$ *and* $M(t)$ *are either both nonincreasing or both nondecreasing on* $(0, \varepsilon)$ *and that*

$$\operatorname{rank} R_2(t) \equiv \operatorname{rank} R_2 \quad \text{and} \quad \operatorname{rank} S^*(t) \equiv \operatorname{rank} S^* =: m - r \qquad (1.223)$$

*are constant on* $[0, \varepsilon)$. *Finally, let* $T \in \mathbb{R}^{m \times r}$ *be such that*

$$\operatorname{rank} T = r, \ T^T T = I_{r \times r}, \ \operatorname{Im} T = \operatorname{Ker} S^*, \ and \ Q := T^T \Lambda \, ST \in \mathbb{R}^{r \times r}. \tag{1.224}$$

*Then the matrix* $Q$ *is symmetric, and* $\operatorname{ind} M(0^+)$, $\operatorname{ind} M(0^-)$, *and* $\operatorname{def} \Lambda(0^+)$ *exist with*

$$\operatorname{ind} M(0^+) = \operatorname{ind} Q + m - \operatorname{rank} T + \operatorname{def} \Lambda - \operatorname{def} \Lambda(0^+) - \operatorname{def} X \qquad (1.225)$$

*if* $M(t)$ *and* $U(t) X^{-1}(t)$ *are nonincreasing on* $(0, \varepsilon)$ *and*

$$\operatorname{ind} M(0^+) = \operatorname{ind} Q + m - \operatorname{rank} T \qquad (1.226)$$

*if* $M(t)$ *and* $U(t) X^{-1}(t)$ *are nondecreasing on* $(0, \varepsilon)$.

*Proof* We refer to [210, Theorem 2.1]. □

The most significant disadvantage of condition (1.223) is that it depends also on the matrix $X = X(0)$, which makes it "not really" suitable in practical applications. The following special case of Theorem 1.85 removes this disadvantage, since the crucial assumption is formulated only in terms of $R_2(t)$.

**Corollary 1.86 (Index Theorem)** *With the notation* (1.220)–(1.222) *and the assumptions of Theorem 1.85, suppose that*

$$\operatorname{Im} R_2^T(t) \equiv \operatorname{Im} R_2^T \quad \text{is constant on } [0, \varepsilon) \qquad (1.227)$$

*instead of* (1.223). *Then the assertions* (1.225) *and* (1.226) *of Theorem 1.85 hold.*

*Proof* If $\operatorname{Im} R_2^T(t)$ is constant on $[0, \varepsilon)$, then (1.223) is trivially satisfied. The statement then follows from Theorem 1.85. Observe that this constant image assumption does not depend on $X$, while of course (1.223) depends in general on $R_2(t)$ and also on $X$. □

Next we discuss a limit theorem for symmetric monotone matrix-valued functions, which concerns invertible matrices.

**Proposition 1.87 (Limit Theorem)** *Let* $S(t)$ *be a real, symmetric, positive definite, and increasing* $m \times m$ *matrix-valued function on* $[a, \infty)$. *Then the limit* $X := \lim_{t \to \infty} S^{-1}(t)$ *exists;* $X$ *is symmetric and positive semidefinite, and*

$$\lim_{t \to \infty} X S(t) X = X. \qquad (1.228)$$

*Proof* The result follows from a general Limit Theorem in [205, Theorem 3.3.7], in which we take $X(t) := S(1/t)$ and $U(t) \equiv I$ on $(0, \varepsilon]$ with $\varepsilon := 1/a$ (without loss of generality, we assume that $a > 0$). $\qquad\square$

We extend the above result to noninvertible matrices $S(t)$.

**Theorem 1.88 (Limit Theorem)** *Let $S(t)$ be a real symmetric, positive semidefinite, and nondecreasing $m \times m$ matrix-valued function on the interval $[a, \infty)$. Then the limit $X := \lim_{t \to \infty} S^\dagger(t)$ exists; $X$ is symmetric and positive semidefinite, and equality* (1.228) *holds.*

*Proof* The monotonicity of $S(t)$ implies that $\mathrm{Ker}\, S(t)$ is constant on some interval $[\alpha, \infty)$ with $\alpha \geq a$. Let $\ell := \mathrm{rank}\, S(t)$ and $k := \mathrm{def}\, S(t)$ for $t \in [\alpha, \infty)$, so that $\ell + k = m$. Choose $K \in \mathbb{R}^{m \times k}$ and $L \in \mathbb{R}^{m \times \ell}$ such that $\mathrm{Ker}\, S(t) \equiv \mathrm{Im}\, K$ on $[\alpha, \infty)$ and the matrix $V := (K, L) \in \mathbb{R}^{m \times m}$ is orthogonal. Then for all $t \in [\alpha, \infty)$

$$V^T S(t)\, V = \begin{pmatrix} 0 & 0 \\ 0 & \tilde{S}(t) \end{pmatrix}, \qquad S(t) = V \begin{pmatrix} 0 & 0 \\ 0 & \tilde{S}(t) \end{pmatrix} V^T, \tag{1.229}$$

where the $\ell \times \ell$ matrix-valued function $\tilde{S}(t) := L^T S(t)\, L$ is symmetric, positive definite, and increasing on $[\alpha, \infty)$. By Remark 1.60(ii) we get from (1.229) that

$$S^\dagger(t) = V \begin{pmatrix} 0 & 0 \\ 0 & \tilde{S}^{-1}(t) \end{pmatrix} V^T, \qquad X := V \begin{pmatrix} 0 & 0 \\ 0 & \tilde{X} \end{pmatrix} V^T, \tag{1.230}$$

where $X = \lim_{t \to \infty} S^\dagger(t)$ and where the matrix $\tilde{X} := \lim_{t \to \infty} \tilde{S}^{-1}(t)$ exists by Proposition 1.87. Moreover, the matrices $\tilde{X}$ and $X$ are symmetric and positive semidefinite, and $\lim_{t \to \infty} \tilde{X} \tilde{S}(t)\, \tilde{X} = \tilde{X}$ holds. From (1.229) and (1.230), we get

$$\lim_{t \to \infty} X S(t)\, X = \lim_{t \to \infty} V \begin{pmatrix} 0 & 0 \\ 0 & \tilde{X} \tilde{S}(t)\, \tilde{X} \end{pmatrix} V^T = V \begin{pmatrix} 0 & 0 \\ 0 & \tilde{X} \end{pmatrix} V^T = X,$$

which completes the proof. $\qquad\square$

## *1.6.6   Miscellaneous Topics from Matrix Analysis*

In this subsection we collect various result from matrix analysis, which will be needed in the subsequent chapters of this book. The first result is a generalization of the statement that every symmetric matrix $G$ is a limit of a sequence of invertible symmetric matrices $G_\nu$; compare with [70, pg. 40]. In the present context, the matrices $G_\nu$ are no longer invertible, but their image is equal to the image of some fixed orthogonal projector.

**Lemma 1.89** *Let $G \in \mathbb{R}^{n \times n}$ be a symmetric matrix, and let $Q$ be an orthogonal projector with $\operatorname{Im} G \subseteq \operatorname{Im} Q$. Then there exists a sequence $\{G_\nu\}_{\nu=1}^{\infty}$ of symmetric matrices such that $\operatorname{Im} G_\nu = \operatorname{Im} Q$ for all $\nu \in \mathbb{N}$ and $G_\nu \to G$ as $\nu \to \infty$.*

*Proof* Let $g := \operatorname{rank} G$ and $q := \operatorname{rank} Q$, so that $g \leq q$. If $g = q$, then $\operatorname{Im} G = \operatorname{Im} Q$, and we may take the constant sequence $G_\nu := G$ for all $\nu \in \mathbb{N}$. Suppose now that $g < q$. Then there exists an orthogonal matrix $V \in \mathbb{R}^{n \times n}$ such that its first $g$ columns form an orthonormal basis of $\operatorname{Im} G$ and at the same time its first $q$ columns form an orthonormal basis of $\operatorname{Im} Q$. This means that we have the equalities $V^T G V = \operatorname{diag}\{\Gamma_g, 0_{n-g}\}$ and $V^T Q V = \operatorname{diag}\{\Theta_q, 0_{n-q}\}$, where $\Gamma_g \in \mathbb{R}^{g \times g}$ and $\Theta_q \in \mathbb{R}^{q \times q}$ are symmetric and nonsingular. Since $Q$ is an orthogonal projector, $Q^2 = Q$. It follows that $\Theta_q^2 = \Theta_q$, which implies that $\Theta_q = I_q$. Therefore, we have $G = V \operatorname{diag}\{\Gamma_g, 0_{n-g}\} V^T$ and $Q = V \operatorname{diag}\{I_q, 0_{n-q}\} V^T$. Consider the sequence $\{G_\nu\}_{\nu=1}^{\infty}$ of matrices, where

$$G_\nu := V \begin{pmatrix} \Gamma_g & 0 & 0 \\ 0 & \frac{1}{\nu} I_{q-g} & 0 \\ 0 & 0 & 0_{n-q} \end{pmatrix} V^T \quad \text{for all } \nu \in \mathbb{N}. \tag{1.231}$$

It is obvious that for each $\nu \in \mathbb{N}$, the matrix $G_\nu$ is symmetric and $Q G_\nu = G_\nu$. And since $\operatorname{rank} G_\nu = q$, we have $\operatorname{Im} G_\nu = \operatorname{Im} Q$. Moreover, from (1.231) it follows that $\lim_{\nu \to \infty} G_\nu = G$, which completes the proof. $\qquad\qquad\square$

Next we present some special properties of orthogonal projectors. We recall that every orthogonal projector is a diagonalizable matrix (being symmetric) with the spectrum consisting of only two values 0 and 1. More precisely, if $P \in \mathbb{R}^{n \times n}$ is an orthogonal projector and $p := \operatorname{rank} P$, then there exists an $n \times n$ orthogonal matrix $V$ such that

$$P = V \operatorname{diag}\{I_p, 0_{n-p}\} V^T. \tag{1.232}$$

**Lemma 1.90** *Let $P_* \in \mathbb{R}^{n \times n}$ be an orthogonal projector with $p_* := \operatorname{rank} P_*$, and let $V_* \in \mathbb{R}^{n \times n}$ be the corresponding orthogonal matrix from (1.232), i.e.,*

$$P_* = V_* \operatorname{diag}\{I_{p_*}, 0_{n-p_*}\} V_*^T. \tag{1.233}$$

*Let $p \in \mathbb{N}$ satisfy $p_* \leq p \leq n$. Then $P \in \mathbb{R}^{n \times n}$ is an orthogonal projector with*

$$\operatorname{Im} P_* \subseteq \operatorname{Im} P \quad \text{and} \quad \operatorname{rank} P = p \tag{1.234}$$

*if and only if $P$ has the form*

$$P = V_* \operatorname{diag}\{I_{p_*}, R_*\} V_*^T, \tag{1.235}$$

*where $R_* \in \mathbb{R}^{(n-p_*) \times (n-p_*)}$ is an orthogonal projector with rank equal to $p - p_*$.*

*Proof* It is easy to see that every matrix $P$ of the form (1.235) is symmetric and idempotent (i.e., it is an orthogonal projector) and (1.234) holds. Conversely, suppose that $P \in \mathbb{R}^{n \times n}$ is an orthogonal projector satisfying (1.234). Then we may write

$$P = V_* \begin{pmatrix} K_* & L_* \\ L_*^T & R_* \end{pmatrix} V_*^T, \qquad (1.236)$$

where $K_* \in \mathbb{R}^{p_* \times p_*}$ and $R_* \in \mathbb{R}^{(n-p_*) \times (n-p_*)}$ are symmetric and $L_* \in \mathbb{R}^{p_* \times (n-p_*)}$. The first condition in (1.234) is equivalent with the equality $P P_* = P_*$, from which we obtain by using the representations in (1.233) and (1.236) that $K_* = I_{p_*}$ and $L_* = 0_{p_* \times (n-p_*)}$. Thus, $P$ has the form in (1.235), where rank $R_*$ is equal to rank $P - p_* = p - p_*$ according to (1.234). Finally, the idempotence of $P$ now implies the idempotence of $R_*$, showing that $R_*$ is an orthogonal projector.   □

**Theorem 1.91** *Let* $P_*, P, \tilde{P} \in \mathbb{R}^{n \times n}$ *be orthogonal projectors satisfying*

$$\operatorname{Im} P_* \subseteq \operatorname{Im} P, \quad \operatorname{Im} P_* \subseteq \operatorname{Im} \tilde{P}, \quad \operatorname{rank} P = \operatorname{rank} \tilde{P}. \qquad (1.237)$$

*Then there exists an invertible matrix* $E \in \mathbb{R}^{n \times n}$ *such that*

$$E P_* = P_* \quad and \quad \operatorname{Im} E P = \operatorname{Im} \tilde{P}.$$

*Proof* Let $p_* := \operatorname{rank} P_*$ and $p := \operatorname{rank} P = \operatorname{rank} \tilde{P}$. Then obviously $p \geq p_*$. Let $V_* \in \mathbb{R}^{n \times n}$ be the orthogonal matrix in (1.232) associated with projector $P_*$, that is, (1.233) holds. According to Lemma 1.90, there exist orthogonal projectors $R_*, \tilde{R}_* \in \mathbb{R}^{(n-p_*) \times (n-p_*)}$ such that

$$P = V_* \operatorname{diag}\{I_{p_*}, R_*\} V_*^T, \quad \tilde{P} = V_* \operatorname{diag}\{I_{p_*}, \tilde{R}_*\} V_*^T, \qquad (1.238)$$

and rank $R_* = \operatorname{rank} \tilde{R}_* = p - p_*$. Let $Z_*, \tilde{Z}_* \in \mathbb{R}^{(n-p_*) \times (n-p_*)}$ be orthogonal matrices in (1.232) associated with the projectors $R_*$ and $\tilde{R}_*$, that is, we have $R_* = Z_* \operatorname{diag}\{I_{p-p_*}, 0_{n-p}\} Z_*^T$ and $\tilde{R}_* = \tilde{Z}_* \operatorname{diag}\{I_{p-p_*}, 0_{n-p}\} \tilde{Z}_*^T$. It follows that

$$\tilde{Z}_* Z_*^T R_* = \tilde{Z}_* \operatorname{diag}\{I_{p-p_*}, 0_{n-p}\} Z_*^T = \tilde{R}_* \tilde{Z}_* Z_*^T. \qquad (1.239)$$

We set $E := V_* \operatorname{diag}\{I_{p_*}, \tilde{Z}_* Z_*^T\} V_*^T \in \mathbb{R}^{n \times n}$. Then $E$ is nonsingular and from (1.233) it follows that $E P_* = P_*$. Finally, by (1.238) and (1.239), we obtain

$$E P = V_* \operatorname{diag}\{I_{p_*}, \tilde{Z}_* Z_*^T R_*\} V_*^T = V_* \operatorname{diag}\{I_{p_*}, \tilde{R}_* \tilde{Z}_* Z_*^T\} V_*^T = \tilde{P} E,$$

which shows that $\operatorname{Im} E P = \operatorname{Im} \tilde{P} E = \operatorname{Im} \tilde{P}$. The proof is complete.   □

## 1.7   Notes and References

Concerning the oscillatory and spectral properties of Sturm-Liouville difference equations (1.1), this topic is treated in detail in [204]. A comprehensive treatment of oscillation theory of various difference equations and systems, including an introduction to oscillation theory of symplectic difference systems, is presented in [4]. In particular, [4, Section 2.9] contains a relevant list of references for the oscillation theory of the second-order Sturm-Liouville difference equation (1.1). Other monographs devoted to various aspects of linear difference equations are [2, 110] and also the paper [169]. The Leighton-Wintner oscillation criterion (Theorem 1.4) is based on the second mean value theorem of summation calculus (a discrete analog of the second mean value theorem of integral calculus) proven in [80, Lemma 3.2]. Applications of this technique were developed, e.g., in [82, Theorem 2] and [259, Theorem 4]. The discrete Prüfer transformation presented in Sect. 1.2.4 was obtained in [47], and the oscillation theorems in Sect. 1.2.5 were derived in [298].

The first-order optimality conditions in Theorem 1.26 can be proven via the mathematical programming approach as presented in [62, 178, 179, 223] or via the variational approach as in [3, 177] and [16, Chapter 4]; see also the scalar case (i.e., $n = 1$) in [204, Chapter 8]. An overview of the second-order optimality conditions for discrete calculus of variations problem (1.51), including a historical development of these conditions, is presented in the survey paper [186]. Further necessary and sufficient conditions for discrete calculus of variations problems with variable endpoints in terms of coupled intervals are derived in [180]. In the discrete optimal control setting, such conditions are presented in [177–179, 184, 225–227, 330]. The symplectic structure of the Jacobi equations (1.63) and (1.69) discussed in Theorems 1.29 and Remark 1.30 is derived in [294]; see also [303, Corollaries 5.2 and 5.8]. The discrete calculus of variations problem (1.65) without the shift in the state variable is analyzed in [294, Section 2].

The discrete weak Pontryagin (maximum) principle in Theorem 1.32 can be proven via the mathematical programming method (i.e., the Lagrange multipliers rule) as presented in [178, 181, 223] or via a variational method (based on a generalized DuBois-Reymond lemma) as presented in [192]. These references also contain a more general optimal control problem (1.70) involving the pointwise equality control constraints. Problems with state inequality constraints are considered, e.g., in [225–227]. The symplectic structure of the Jacobi system for the discrete optimal control problem (1.70), in which only the matrix $S_k$ in (1.83) is invertible while $R_k$ may be singular, is proven in [303]. The matrix inversion formula (1.87) is from [146]. Optimal control problems with and without a shift in the state variable are studied in [192], where it is also shown that they can transformed one to another by using the implicit function theorem. Symplectic difference systems were recently applied in [331] to study constrained linear-quadratic control problems with and without shift in their data.

The literature related to the symplectic phase flow in Hamiltonian mechanics is given at particular places Sect. 1.4. Let us mention here at least the books [16, 27] and the paper [8].

Classical qualitative theory of linear Hamiltonian differential systems (1.103) is developed in the books [28, 70, 205, 248, 250]. The trigonometric and Prüfer transformations for linear Hamiltonian systems are discussed in [32, 77, 92, 93, 245, 247, 250]. In particular, Theorem 1.51 is the first part of [247, Theorem 4.1], while Theorems 1.52, 1.53 are proved in [77, Theorems 1, 3]. Properties of principal and nonprincipal solutions for completely controllable system (1.103) are proven in [6, 7, 78, 79, 246], as well as in the above general references on linear Hamiltonian systems. An overview of applications of principal solutions of (1.103) at infinity is also presented in [283]. The oscillation and eigenvalue theory of these systems without the complete controllability (or identical normality assumption) was initiated in [207] and further developed in [209, 295, 296, 321]. In particular, the theory of principal and antiprincipal solutions of possibly abnormal linear Hamiltonian systems (1.103) is developed in [283, 285–290, 293]. Uncontrollable systems (1.103) were also considered in [138, 200–203] in the relation with the notion of a weak disconjugacy of (1.103) and dissipative control processes. Applications of the theory of comparative index to linear Hamiltonian systems are derived in [127, 129, 130, 289, 293]. A generalization of the oscillation theorem (Theorem 1.56) for linear Hamiltonian systems (1.140) under no strict normality assumption is proven in [57]. A theory of Riccati matrix differential equations for linear Hamiltonian systems without the controllability assumption was recently developed in [282].

As we mentioned in Sect. 1.6.1, properties of symplectic matrices are discussed, e.g., in [27, 139, 216, 328, 332] and in the papers [35, 74, 75, 219, 222, 224, 326, 327]. We note that there is also a notion of a complex symplectic matrix (symplectic matrix with complex entries), which is sometimes called a conjugate symplectic matrix. In the complex case, some of the properties of symplectic matrices remain the same, but some other are slightly different. We refer to the abovementioned literature for a comparison. The theory of Moore-Penrose pseudoinverses and its properties is presented, e.g., in the books [34, 36, 64, 205]. In particular, the limit result for sequences of Moore-Penrose pseudoinverses in Remark 1.60(v) is from [64, Theorem 10.4.1]. The inequality in Remark 1.60(vi) is proven in [170, Lemma 1]; see also [36, Facts 8.15.7 and 8.15.5]. For the properties in Remark 1.60(vii), we refer to [36, Facts 6.4.32 and 2.10.8]. The statement in Lemma 1.61 is proven in [283, Corollary 10.5 and Lemma 10.4]; Lemma 1.64 is derived in [113, Lemma 2.4 and Remark 2.5]. The Moore-Penrose pseudoinverses are efficiently used in the relation with orthogonal projectors. Applications of this type in symplectic difference systems are contained in the recent papers [284, 290, 292].

The results of the auxiliary Lemma 1.67 also follow from [205, Theorem 3.1.2 and Corollary 3.3.9].

Much of matrix factorization theory comes from numerical linear algebra. Results concerning factorization of symplectic matrices play an important role in

applications; see [35, 74, 75, 139, 219, 222, 224, 234, 239] and the references given
therein. Several special types of symplectic factorizations are highly important for
Sturmian theory of discrete symplectic systems as well. For example, the solvability
of the discrete matrix Riccati equation (2.52) is equivalent to the existence the
symplectic block $LU$ factorization for fundamental solution matrices $Z_k$ of (SDS),
see [219]. The trigonometric transformations considered in Sect. 2.6 are based on
the symplectic block $QR$ factorization from [63] for $Z_k^T$, where $Q$ is a symplectic
orthogonal matrix and $R$ is a symplectic upper block-triangular matrix of the
form (1.154). The symplectic singular value decomposition (SVD), see [239,
Theorem 2.1], is a basic tool of the oscillation theory for discrete trigonometric
systems (2.146) in [96]. The results of Sect. 1.6.3 related to the so-called generalized
$LU$ factorizations of symplectic matrices present another example of applications
of the factorization methods in the discrete oscillation theory.

Several statements of Lemma 1.68 were proven originally in [114, Lemma 2.1].
In particular, it was shown that the conditions in (1.147) are sufficient for (1.171) and
(1.173)–(1.175). The result of Theorem 1.70 is a special case of [114, Theorem 3.1],
which is proved under more general assumptions. Consider symplectic and orthogonal matrices $\mathfrak{N}_P$ given by (1.179). Then [114, Theorem 3.1] states that a $2n \times n$
matrix $Y_k$ is a conjoined basis of (SDS) if and only if there exists an orthogonal
projector $P_k$, a lower block-triangular symplectic matrix $L_k$, and a nonsingular $n \times n$
matrix $M_k$ such that

$$Y_k = L_k \mathfrak{N}_{P_k} \begin{pmatrix} M_k \\ 0 \end{pmatrix}, \qquad L_{k+1}^{-1} S_k L_k = \mathfrak{N}_{P_{k+1}} H_k \mathfrak{N}_{P_k}^T,$$

where $H_k (I\ 0)^T = \begin{pmatrix} M_{k+1} M_k^{-1} \\ 0 \end{pmatrix}$. In particular, the matrix $P_k$ can be chosen in the
form $P_k = I - X_k X_k^\dagger$ assumed in Theorem 1.70. Other special cases of the matrices
$\mathfrak{N}_{P_k}$ are considered in [112, 113]. Also, it should be pointed out that the results of
Lemma 1.68 and Theorem 1.70 imply some special properties of matrices $Y$ with
conditions (1.147) proved in [205]. For instance, formula (1.185) was derived for
the first time in [205, Theorem 3.1.2(iii)].

The results of Sect. 1.6.4 are closely related to the oscillation theory of the
differential Hamiltonian systems (1.99) where $t := \lambda$. This relation is illustrated
by Proposition 1.75. In this connection, a part of the results in Sect. 1.6.4 are well-
known from [205, 207], where the oscillation properties of (1.99) are investigated
under the Legendre condition (1.111). Under the notation of Sect. 1.6.4, condition
(1.111) coincides with

$$(0\ I)\ \Psi(W(\lambda))\ (0\ I)^T \geq 0. \qquad (1.240)$$

Remark that (1.240) and (1.203) imply the first inequality in (1.204) and then the
statements of Theorem 1.79, see the proof of [207, Theorem 3], where $t := \lambda$
and $(X(t)\ U(t)) := W(\lambda)(0\ I)^T$. The operator (1.190) and monotonicity condition
(1.201) for the symplectic coefficient matrix $S_k(\lambda)$ were introduced by the third

author in [297] as the main basic tool for the investigation of discrete symplectic eigenvalue problems with the nonlinear dependence on $\lambda$. This notion is closely related to the so-called *multiplicative derivative* for the matrix functions $X(t)$ defined as $D_t X = X'(t) X^{-1}(t)$ (see [148, Chapter 15] and the references therein). In Sect. 1.6.4 we unify the monotonicity results from [55, 57, 102, 205, 207, 297] using the factorization approach; see [125, Lemma 3.3] and the proof of [125, Lemma 4.3]. In particular, we derive that (1.201) implies also the piecewise constant image of $B(\lambda)$ (see Theorem 1.81) and other results based on the equivalence of (1.201) and (1.214); see the parts of Theorem 1.82 and Corollary 1.83 related to the properties of Im $B(\lambda)$. This connection points out that the "image properties" can be derived from the restricted monotonicity condition (compare also with (1.214))

$$(0 \; I) \, \Psi(W^{-1}(\lambda)) \, (0 \; I)^T \leq 0, \qquad (1.241)$$

while the "kernel properties" follow from (1.240).

Index theorems for monotone matrix-valued functions are often utilized in the oscillation theory of symplectic and Hamiltonian systems; see, e.g., [205, Section 3.4] and [297, Proposition 2.5]. In these references the index theorems are considered with the constant matrix $R_2(t) \equiv R_2$ on $[0, \varepsilon)$. The generalized versions in Theorem 1.85 and Corollary 1.86 with variable $R_2(t)$ are from the paper [210, Theorem 2.1 and Corollary 2.3]. The dependence of $R_2(t)$ on $t$ is crucial for the applications in the oscillation theorems in Sect. 5.1. The extended limit theorem (Theorem 1.88) and its proof were communicated to the authors by Werner Kratz in June 2014.

The statement in Lemma 1.89 about symmetric matrices is proven in [283, Lemma 10.2]. The results in Lemma 1.90 and Theorem 1.91 about orthogonal projectors are from [285, Lemma 9.1 and Theorem 9.2].

We recommend the following additional related references for further reading about the topics presented in this chapter: [5, 31, 76, 83, 199] for the Sturm-Liouville difference equations, [30, 241, 249, 255, 336] for the Sturm-Liouville differential equations, [65, 236, 256] for integration of Hamiltonian systems and variational analysis, and [110, 204, 214] for general theory of difference equations.

# Chapter 2
# Basic Theory of Symplectic Systems

In this chapter we present basic theory of symplectic difference systems. We show that these systems incorporate as special cases many important equations or systems, such as the Sturm-Liouville difference equations, symmetric three-term recurrence equations, Jacobi difference equations, linear Hamiltonian difference systems, or trigonometric and hyperbolic systems. We investigate the definiteness of the associated discrete quadratic functional and its relationship with the nonexistence of focal points of conjoined bases and with the solvability of the Riccati matrix difference equation. We pay special attention to recessive and dominant solutions of a nonoscillatory and eventually controllable symplectic system. We study general and special symplectic transformations, such as the trigonometric and Prüfer transformations.

## 2.1 Symplectic Systems and Their Particular Cases

The central concept of our book is the symplectic difference system

$$y_{k+1} = \mathcal{S}_k y_k \qquad \text{(SDS)}$$

where $y_k \in \mathbb{R}^{2n}$ and the matrices $\mathcal{S}_k \in \mathbb{R}^{2n \times 2n}$ are *symplectic*, i.e.,

$$\mathcal{S}_k^T \mathcal{J} \mathcal{S}_k = \mathcal{J}, \quad \mathcal{J} := \begin{pmatrix} 0 & I \\ -I & 0 \end{pmatrix}.$$

If we write the matrix $\mathcal{S}_k$ in the block form

$$\mathcal{S}_k = \begin{pmatrix} \mathcal{A}_k & \mathcal{B}_k \\ \mathcal{C}_k & \mathcal{D}_k \end{pmatrix}, \qquad (2.1)$$

© Springer Nature Switzerland AG 2019
O. Došlý et al., *Symplectic Difference Systems: Oscillation and Spectral Theory*,
Pathways in Mathematics, https://doi.org/10.1007/978-3-030-19373-7_2

and $y_k$ as $y_k = \binom{x_k}{u_k}$ with $x_k, u_k \in \mathbb{R}^n$, then system (SDS) can be written as

$$x_{k+1} = \mathcal{A}_k x_k + \mathcal{B}_k u_k, \quad u_{k+1} = \mathcal{C}_k x_k + \mathcal{D}_k u_k. \tag{2.2}$$

In analogy with the terminology for linear Hamiltonian differential systems (see Sect. 1.5), we call the first equation in (2.2) as the *equation of motion* while the second equation as the *Euler-Lagrange equation*.

Basic property of (SDS) is that its fundamental matrix (sometimes also called the *discrete phase flow*) is symplectic whenever it is symplectic at an initial condition, say at $k = 0$. This easily follows from the fact that we have for the fundamental matrix $Z$ of (SDS) the expression

$$Z_k = \mathcal{S}_{k-1} \mathcal{S}_{k-2} \cdots \mathcal{S}_1 \mathcal{S}_0 Z_0 = \left( \prod_{i=0}^{k-1} \mathcal{S}_{k-1-i} \right) Z_0, \tag{2.3}$$

since the symplectic matrices form a group with respect to the matrix multiplication (see Lemma 1.58). Here we use the convention that the matrices under the matrix product sign are ordered from the left to the right, i.e., the matrix staying on the left in the product corresponds to the lower index in the product, and the matrix staying in the right corresponds to the upper index in the product operator. Also, if the upper index is less than the lower one, we take the product equal to the identity matrix $I$.

### 2.1.1 Conjoined Bases and Wronskian

Next we define basic concepts of the theory of symplectic systems as presented in [45]. If $Y = \binom{X}{U}$ and $\hat{Y} = \binom{\hat{X}}{\hat{U}}$ are two $2n \times n$ solutions of (SDS), then their *Wronskian*

$$Y_k^T \mathcal{J} \hat{Y}_k = X_k^T \hat{U}_k - U_k^T \hat{X}_k \equiv L \tag{2.4}$$

is constant with respect to $k$, where $L$ is a constant $n \times n$ matrix. We will denote this constant matrix by $w(Y, \hat{Y})$.

**Definition 2.1** A $2n \times n$ solution $Y = \binom{X}{U}$ of (SDS) is a *conjoined basis* if

$$w(Y, Y) = X_k^T U_k - U_k^T X_k = 0 \quad \text{and} \quad \operatorname{rank} Y_k = n. \tag{2.5}$$

The first condition in (2.5) means that the matrix $X_k^T U_k$ is symmetric. A special case of a conjoined basis is the so-called principal solution of (SDS) at the point $k = j$.

**Definition 2.2** For a fixed index $j$, let $Y^{[j]} = \binom{X^{[j]}}{U^{[j]}}$ be the solution of (SDS) given by the initial conditions $X_j^{[j]} = 0$ and $U_j^{[j]} = I$. Then this solution is said to be the *principal solution at* $k = j$.

*Remark 2.3* If condition (2.5) holds at one particular index $k$, then it holds for all indices. Indeed, suppose that (2.5) holds for $k = 0$. Denote by $K := Y_0^T Y_0$ and consider the $2n \times n$ solution $\tilde{Y} = \binom{\tilde{X}}{\tilde{U}}$ of (SDS) given by the initial condition $\tilde{Y}_0 = -\mathcal{J} Y_0 K^{-1}$. Then $Z_k := (Y_k \ \tilde{Y}_k)$ is symplectic for $k = 0$, see the proof of Lemma 1.58(vi). It follows by (2.3) that $Z_k$ is symplectic for all $k$, in particular (2.5) holds for every index $k$.

Throughout the book we will concentrate on the conjoined bases of (SDS) only, since $2n \times n$ solutions of (SDS) which are not conjoined play a "destructive" role in the oscillation theory of (SDS) similarly as their continuous counterparts in the theory of linear Hamiltonian differential systems. This reasoning has been explained in more details in Sect. 1.5

If $w(Y, \bar{Y}) = I$, then we say that $Y$ and $\bar{Y}$ form a pair of *normalized* conjoined bases. In this case, see Lemma 1.58(v),

$$Z_k := (Y_k \ \bar{Y}_k) = \begin{pmatrix} X_k & \bar{X}_k \\ U_k & \bar{U}_k \end{pmatrix}$$

is a symplectic fundamental matrix of (SDS). By equations (1.145) and (1.146) in Lemma 1.58, we then have the properties

$$X_k^T \bar{U}_k - U_k^T \bar{X}_k = I, \quad X_k^T U_k = U_k^T X_k, \quad \bar{X}_k^T \bar{U}_k = \bar{U}_k^T \bar{X}_k, \tag{2.6}$$

$$X_k \bar{U}_k^T - \bar{X}_k U_k^T = I, \quad X_k \bar{X}_k^T = \bar{X}_k X_k^T, \quad U_k \bar{U}_k^T = \bar{U}_k U_k^T. \tag{2.7}$$

We note that given a conjoined basis $Y$ of (SDS), there always exists another conjoined basis $\hat{Y}$, which together with $Y$ forms a pair of normalized conjoined bases. The proof of this claim is essentially contained in Remark 2.3.

**Lemma 2.4** *For any conjoined bases* $Y$, $\bar{Y}$, $\tilde{Y}$ *of (SDS) such that* $w(Y, \bar{Y}) = I$ *the* $n \times n$ *matrix* $w(\tilde{Y}, Y) [w(\tilde{Y}, \bar{Y})]^T$ *is symmetric.*

*Proof* The proof follows from the properties in (2.7) by direct calculation of the product (suppressing the index $k$)

$$w(\tilde{Y}, Y) [w(\tilde{Y}, \bar{Y})]^T$$

$$= -\tilde{Y}^T \mathcal{J} Y \bar{Y}^T \mathcal{J} \tilde{Y} \overset{(2.7)}{=} (\tilde{U}^T \ -\tilde{X}^T) \begin{pmatrix} X\bar{X}^T & X\bar{U}^T \\ \bar{U}X^T - I & U\bar{U}^T \end{pmatrix} \begin{pmatrix} \tilde{U} \\ -\tilde{X} \end{pmatrix}$$

$$= \tilde{U}^T X \bar{X}^T \tilde{U} - (\tilde{U}^T X \bar{U}^T \tilde{X} + \tilde{X}^T \bar{U} X^T \tilde{U}) + \tilde{X}^T \tilde{U} + \tilde{X}^T U \bar{U}^T \tilde{X}. \tag{2.8}$$

By (2.5) and (2.7), we know that the matrices $\tilde{X}^T \tilde{U}$, $X \bar{X}^T$, and $U \bar{U}^T$ are symmetric. Therefore, the sum in (2.8) above is also a symmetric matrix.                                    □

### 2.1.2  Special Symplectic Difference Systems

In this subsection we present several important examples of symplectic difference systems. We start this subsection with a system which is equivalent with (SDS).

*Example 2.5* We consider the *time-reversed* symplectic difference system

$$y_k = \mathcal{S}_k^{-1} y_{k+1}. \tag{2.9}$$

In [45] an alternative terminology—a *reciprocal system*—is used. Using the formula for the inverse of a symplectic matrix from Lemma 1.58(ii), system (2.9) can be written as

$$x_k = \mathcal{D}_k^T x_{k+1} - \mathcal{B}_k^T u_{k+1}, \quad u_k = -\mathcal{C}_k x_{k+1} + \mathcal{A}^T u_{k+1}. \tag{2.10}$$

All results obtained for (SDS) can be "translated" in a natural way to system (2.9) by reversing the direction of the independent variable. Indeed, for $\mathcal{S}_k^* := \mathcal{S}_{N-k}^{-1}$ for $k \in [0, N]_{\mathbb{Z}}$ and $y_k^* := y_{N+1-k}$ for $k \in [0, N + 1]_{\mathbb{Z}}$, we obtain from (2.9) the equivalent system

$$y_{k+1}^* = y_{N-k} = \mathcal{S}_{N-k}^{-1} y_{N-k+1} = \mathcal{S}_k^* y_k^*,$$

where the matrix $\mathcal{S}_k^*$ is symplectic.

*Example 2.6* Next we consider the so-called *trigonometric* symplectic difference systems. The trigonometric system is a system (SDS) where the matrix $\mathcal{S}$ satisfies the additional condition

$$\mathcal{J}^T \mathcal{S}_k \mathcal{J} = \mathcal{S}_k, \tag{2.11}$$

which says that if $y = \binom{x}{u}$ is a solution of (SDS), then $\tilde{y} = -\mathcal{J}y = \binom{-u}{x}$ is a solution of (SDS) as well. Again, substituting into (2.11), we have that (SDS) is a trigonometric symplectic system if and only if

$$\mathcal{D}_k = \mathcal{A}_k, \quad \mathcal{C}_k = -\mathcal{B}_k.$$

Consequently, combining this with (1.145) and (1.146), we see that a trigonometric symplectic system is a system of the form

$$x_{k+1} = \mathcal{A}_k x_k + \mathcal{B}_k u_k, \quad u_{k+1} = -\mathcal{B}_k x_k + \mathcal{A}_k u_k, \tag{2.12}$$

where

$$A_k^T A_k + B_k^T B_k = I, \quad A_k^T B_k = B_k^T A_k, \tag{2.13}$$

which is equivalent to

$$A_k A_k^T + B_k B_k^T = I, \quad A_k B_k^T = B_k A_k^T. \tag{2.14}$$

The terminology "trigonometric system" is justified by the fact that in the scalar case $n = 1$, the equalities in (2.13) imply that $A_k = \cos \varphi_k$ and $B_k = \sin \varphi_k$ for some $\varphi_k$ and then solutions of (2.12) are

$$(x_k, u_k) = \left( \cos \left( \sum_{}^{k-1} \varphi_j \right), \sin \left( \sum_{}^{k-1} \varphi_j \right) \right),$$

$$(\tilde{x}_k, \tilde{u}_k) = \left( - \sin \left( \sum_{}^{k-1} \varphi_j \right), \cos \left( \sum_{}^{k-1} \varphi_j \right) \right).$$

Basic properties of solutions of trigonometric symplectic systems with nonsingular $B_k$ are established in [25].

*Example 2.7*  Next we consider the *linear Hamiltonian difference system*

$$\Delta \begin{pmatrix} x_k \\ u_k \end{pmatrix} = \mathcal{J} \mathcal{H}_k \begin{pmatrix} x_{k+1} \\ u_k \end{pmatrix}, \tag{2.15}$$

where

$$\mathcal{H}_k = \mathcal{H}_k^T, \quad \mathcal{H}_k = \begin{pmatrix} -C_k & A_k^T \\ A_k & B_k \end{pmatrix}, \quad \det(I - A_k) \neq 0. \tag{2.16}$$

System (2.15) can be equivalently written as

$$\Delta x_k = A_k x_{k+1} + B_k u_k, \quad \Delta u_k = C_k x_{k+1} - A_k^T u_k. \tag{2.17}$$

Expanding the forward differences in (2.17), we obtain the system

$$\begin{pmatrix} x_{k+1} \\ u_{k+1} \end{pmatrix} = \mathcal{S}_k^{[H]} \begin{pmatrix} x_k \\ u_k \end{pmatrix},$$

where

$$\mathcal{S}_k^{[H]} = \begin{pmatrix} (I - A_k)^{-1} & (I - A_k)^{-1} B_k \\ C_k (I - A_k)^{-1} & C_k (I - A_k)^{-1} B_k + I - A_k^T \end{pmatrix}. \tag{2.18}$$

By direct substitution into (1.145), we see that the matrix $\mathcal{S}_k^{[H]}$ is symplectic.

The exact relationship between symplectic and Hamiltonian systems is described in the next statement.

**Theorem 2.8** *A symplectic system* (SDS) *with the matrix $S_k$ in* (2.1) *is a rewritten Hamiltonian system* (2.15) *if and only if the matrix $\mathcal{A}_k$ is invertible for all $k$.*

*Proof* It suffices to prove that invertibility of $\mathcal{A}_k$ implies that (SDS) can be written as (2.15). Define

$$A_k := I - \mathcal{A}_k^{-1}, \quad B_k := \mathcal{A}_k^{-1}\mathcal{B}_k, \quad C_k := \mathcal{C}_k \mathcal{A}_k^{-1}. \tag{2.19}$$

Then (1.145) and (1.146) imply that the matrices $B_k$ and $C_k$ are symmetric and that

$$\mathcal{D}_k = \mathcal{A}_k^{T-1}(I + \mathcal{C}_k^T \mathcal{B}_k) = \mathcal{A}_k^{T-1} + \mathcal{C}_k \mathcal{A}_k^{-1} \mathcal{B}_k = I - A_k^T + C_k(I - A_k)^{-1}B_k.$$

Then, in view of (2.18), symplectic system (SDS) can be written as a linear Hamiltonian system (2.15). □

Sometimes, when the matrix $\mathcal{A}_k$ in a symplectic system is invertible, we say that this system has a *Hamiltonian structure*.

*Example 2.9* As an example of discrete symplectic systems with the Hamiltonian structure, consider the so-called *hyperbolic* systems in the form (SDS), where $S_k$ obeys the additional condition

$$P_1 S_k P_1 = S_k, \qquad P_1 = \begin{pmatrix} 0 & I \\ I & 0 \end{pmatrix}. \tag{2.20}$$

Here the matrix $P_1$ introduced in Sect. 1.6.1 (see (1.151)) obeys all assumptions of Lemma 1.58(iv); in particular, we have that $P_1 S_k P_1 \in Sp(2n)$. Relation (2.20) says that if $y = \binom{x}{u}$ is a solution of (SDS), then $\binom{u}{x}$ is a solution as well. By analogy with the trigonometric case, we have that (SDS) is a hyperbolic symplectic system if and only if

$$\mathcal{D}_k = \mathcal{A}_k, \quad \mathcal{C}_k = \mathcal{B}_k.$$

Consequently, combining this with (1.145) and (1.146), we see that a hyperbolic symplectic system is a system of the form

$$x_{k+1} = \mathcal{A}_k x_k + \mathcal{B}_k u_k, \quad u_{k+1} = \mathcal{B}_k x_k + \mathcal{A}_k u_k, \tag{2.21}$$

with $n \times n$ matrices $\mathcal{A}_k$ and $\mathcal{B}_k$ satisfying

$$\mathcal{A}_k^T \mathcal{A}_k - \mathcal{B}_k^T \mathcal{B}_k = I = \mathcal{A}_k \mathcal{A}_k^T - \mathcal{B}_k \mathcal{B}_k^T, \quad \mathcal{A}_k^T \mathcal{B}_k - \mathcal{B}_k^T \mathcal{A}_k = 0 = \mathcal{A}_k \mathcal{B}_k^T - \mathcal{B}_k \mathcal{A}_k^T. \tag{2.22}$$

The first equality in (2.22) implies that matrix $\mathcal{A}_k$ is nonsingular (use $\mathcal{A}_k^T \mathcal{A}_k = I + \mathcal{B}_k^T \mathcal{B}_k$, where $\mathcal{B}_k^T \mathcal{B}_k \geq 0$), and then according to Theorem 2.8 system (2.21) is a special case of the discrete Hamiltonian system (2.15) with the blocks (see (2.19))

$$A_k := I - \mathcal{A}_k^{-1}, \quad B_k := \mathcal{A}_k^{-1} \mathcal{B}_k, \quad C_k := \mathcal{B}_k \mathcal{A}_k^{-1}.$$

Moreover, since

$$(\mathcal{A}_k^T + \mathcal{B}_k^T)(\mathcal{A}_k - \mathcal{B}_k) = \mathcal{A}_k^T \mathcal{A}_k + \mathcal{B}_k^T \mathcal{A}_k - \mathcal{A}_k^T \mathcal{B}_k - \mathcal{B}_k^T \mathcal{B}_k = I,$$

the matrices $\mathcal{A}_k + \mathcal{B}_k$ and $\mathcal{A}_k - \mathcal{B}_k$ are nonsingular, too. Furthermore, we have

$$(\mathcal{A}_k - \mathcal{B}_k)^{-1} = \mathcal{A}_k^T + \mathcal{B}_k^T, \quad (\mathcal{A}_k + \mathcal{B}_k)^{-1} = \mathcal{A}_k^T - \mathcal{B}_k^T. \tag{2.23}$$

The terminology "hyperbolic system" is justified by the fact that in the scalar case $n = 1$, the equalities in (2.22) imply that $\mathcal{A}_k^2 - \mathcal{B}_k^2 = 1$ and the solution of (2.21) defined by the initial condition $x_0 = 0$ and $u_0 = 1$ is

$$x_k = \left( \prod_{i=0}^{k-1} \operatorname{sgn} \mathcal{A}_i \right) \sinh \left( \sum_{i=0}^{k-1} \ln |\mathcal{A}_i + \mathcal{B}_i| \right),$$

$$u_k = \left( \prod_{i=0}^{k-1} \operatorname{sgn} \mathcal{A}_i \right) \cosh \left( \sum_{i=0}^{k-1} \ln |\mathcal{A}_i + \mathcal{B}_i| \right).$$

Basic properties of solutions of hyperbolic symplectic systems are established in [108].

*Example 2.10* Another special case of (SDS) is the symmetric three-term recurrence equation and the equivalent Jacobi matrix difference equation (1.60). Consider the equation

$$\Delta(R_k \Delta x_k + Q_k^T x_{k+1}) - (Q_k \Delta x_k + P_k x_{k+1}) = 0 \tag{2.24}$$

with $P_k, Q_k, R_k \in \mathbb{R}^{n \times n}$, $R_k$ and $P_k$ symmetric and $R_k + Q_k^T$ invertible, and its particular case, the matrix Sturm-Liouville equation

$$\Delta(R_k \Delta x_k) - P_k x_{k+1} = 0. \tag{2.25}$$

Expanding the forward difference in (2.24), we obtain equivalent *three-term matrix recurrence equation*

$$K_{k+1} x_{k+2} - L_{k+1} x_{k+1} + K_k^T x_k = 0, \tag{2.26}$$

where

$$K_k := R_k + Q_k^T, \quad L_k := R_k + R_{k-1} + Q_{k-1}^T + Q_{k-1} + P_{k-1},$$

i.e., $K_k$ is an invertible matrix and $L_k$ is symmetric.

*Example 2.11* Consider now equation (2.26) with $K_k$ invertible, and assume that $R_k$ are any symmetric matrices. Then, following [304, Theorem 3.1, Corollary 3.4] (compare also with Theorem 1.29) system (2.26) can be written as a symplectic difference system (SDS) with the matrices

$$A_k = K_k^{-1} R_k, \quad C_k = (L_{k+1} - R_{k+1}) K_k^{-1} R_k - K_k^T,$$
$$B_k = K_k^{-1}, \quad D_k = (L_{k+1} - R_{k+1}) K_k^{-1}.$$

Indeed, if we define $u_k := K_k x_{k+1} - R_k x_k$ for all $k \in [0, N]_\mathbb{Z}$ together with the additional value $u_{N+1} := (L_{N+1} - R_{N+1}) x_{N+1} - K_N^T x_N$, then

$$x_{k+1} = K_k^{-1} R_k x_k + K_k^{-1} u_k = A_k x_k + B_k u_k,$$
$$u_{k+1} = K_{k+1} x_{k+2} - R_{k+1} x + k + 1 = (L_{k+1} - R_{k+1}) x_{k+1} - K_k^T x_k$$
$$= C_k x_k + D_k u_k,$$

for all $k \in [0, N]_\mathbb{Z}$. Conversely, if the matrices $B_k$ for $k \in [0, N]_\mathbb{Z}$ in (2.2) are invertible, then this system can be written as (2.26) with

$$K_k := B_k^{-1}, \quad L_k := B_k^{-1} A_k + D_{k-1} B_{k-1}.$$

Special choices of the matrices $R_k$ (such as $R_k := 0$ or $R_k := K_k$ when $K_k$ is also symmetric) yield different representations of (2.26) as a symplectic difference system (SDS). For a more detailed treatment of the relationship between (SDS) and (2.24), we refer to [304].

*Example 2.12* Consider the *2n-order Sturm-Liouville difference equation*

$$\sum_{v=0}^{n} (-1)^v \Delta^v \left( r_k^{[v]} \Delta^v y_{k+n-v} \right) = 0, \quad r_k^{[n]} \neq 0. \tag{2.27}$$

We put

$$x_k = \begin{pmatrix} y_{k+n-1} \\ \Delta y_{k+n-2} \\ \vdots \\ \Delta^{n-1} y_k \end{pmatrix}, \quad u_k = \begin{pmatrix} \sum_{j=1}^{n} (-\Delta)^{j-1} \left( r_k^{[j]} \Delta^j y_{k+n-j} \right) \\ \vdots \\ -\Delta \left( r_k^{[n]} \Delta^n y_k \right) + r_k^{[n-1]} \Delta^{n-1} y_{k+1} \\ r_k^{[n]} \Delta^n y_k \end{pmatrix}. \tag{2.28}$$

Then $y = \binom{x}{u}$ is a solution of the Hamiltonian system (2.17) with

$$A_k \equiv A = \{a_{ij}\} = \begin{cases} 1 & \text{if } j = i+1, \\ 0 & \text{otherwise} \end{cases}, \qquad B_k = \text{diag}\left\{0, \ldots, 0, \frac{1}{r_k^{[n]}}\right\}, \qquad (2.29)$$

$$C_k = \text{diag}\left\{r_k^{[0]}, r_k^{[1]}, \ldots, r_k^{[n-1]}\right\}, \qquad (2.30)$$

with $I - A_k$ upper triangular and $\det(I - A_k) = 1$. Hence $I - A_k$ is invertible with

$$(I - A_k)^{-1} = \begin{pmatrix} 1 & 1 & \ldots & 1 & 1 \\ 0 & 1 & \ldots & 1 & 1 \\ \vdots & \vdots & \ddots & \vdots & \vdots \\ 0 & 0 & \ldots & 1 & 1 \\ 0 & 0 & \ldots & 0 & 1 \end{pmatrix}. \qquad (2.31)$$

Therefore, equation (2.27) is also a special case of system (SDS), in which according to (2.18) and (2.31), the coefficients are

$$\mathcal{A}_k = \begin{pmatrix} 1 & 1 & \ldots & 1 & 1 \\ 0 & 1 & \ldots & 1 & 1 \\ \vdots & \vdots & \ddots & \vdots & \vdots \\ 0 & 0 & \ldots & 1 & 1 \\ 0 & 0 & \ldots & 0 & 1 \end{pmatrix}, \qquad \mathcal{B}_k = \frac{1}{r_k^{[n]}} \begin{pmatrix} 0 & \ldots & 0 & 1 \\ 0 & \ldots & 0 & 1 \\ \vdots & \ddots & \vdots & \vdots \\ 0 & \ldots & 0 & 1 \\ 0 & \ldots & 0 & 1 \end{pmatrix}, \qquad (2.32)$$

$$\mathcal{C}_k = \begin{pmatrix} r_k^{[0]} & r_k^{[0]} & \ldots & r_k^{[0]} & r_k^{[0]} \\ 0 & r_k^{[1]} & \ldots & r_k^{[1]} & r_k^{[1]} \\ \vdots & & \ddots & \vdots & \vdots \\ 0 & 0 & \ldots & r_k^{[n-2]} & r_k^{[n-2]} \\ 0 & 0 & \ldots & 0 & r_k^{[n-1]} \end{pmatrix}, \qquad (2.33)$$

$$\mathcal{D}_k = \begin{pmatrix} 1 & 0 & 0 & \ldots & 0 & 0 & r_k^{[0]}/r_k^{[n]} \\ -1 & 1 & 0 & \ldots & 0 & 0 & r_k^{[1]}/r_k^{[n]} \\ 0 & -1 & 1 & \ldots & 0 & 0 & r_k^{[2]}/r_k^{[n]} \\ \vdots & \vdots & \vdots & \ddots & \vdots & \vdots & \vdots \\ 0 & 0 & 0 & \ldots & 1 & 0 & r_k^{[n-3]}/r_k^{[n]} \\ 0 & 0 & 0 & \ldots & -1 & 1 & r_k^{[n-2]}/r_k^{[n]} \\ 0 & 0 & 0 & \ldots & 0 & -1 & 1 + r_k^{[n-1]}/r_k^{[n]} \end{pmatrix}. \qquad (2.34)$$

*Example 2.13* Finally, consider the classical *second-order Sturm-Liouville differ-*
*ence equation*

$$\Delta\left(r_k \Delta x_k\right) + p_k x_{k+1} = 0, \quad r_k \neq 0. \tag{2.35}$$

The symplectic system corresponding to this equation (with $u_k := r_k \Delta x_k$) is

$$\begin{pmatrix} x_{k+1} \\ u_{k+1} \end{pmatrix} = \begin{pmatrix} 1 & 1/r_k \\ -p_k & 1 - p_k/r_k \end{pmatrix} \begin{pmatrix} x_k \\ u_k \end{pmatrix}, \tag{2.36}$$

as can be verified by a direct computation after expanding the forward differences
in the system $\Delta x_k = (1/r_k)\, u_k$ and $\Delta u_k = -p_k x_{k+1}$.

Let us summarize that the above-described particular cases of (SDS) are equiva-
lent to the invertibility of certain blocks in the matrix $\mathcal{S}_k$. If $\mathcal{A}_k$ are invertible, then
(SDS) is equivalent to the linear Hamiltonian difference system (2.15), while if $\mathcal{B}_k$
are invertible, then (SDS) is equivalent to the Jacobi equation (2.24), which is in
turn equivalent to a symmetric three-term recurrence equation (2.26). Finally, if $\mathcal{B}_k$
are invertible and symmetric, then (SDS) is equivalent to a matrix Sturm-Liouville
equation (2.25).

## 2.2   Focal Points

The central concept of the oscillation theory of symplectic difference systems (SDS)
is the concept of a *focal point* of a conjoined basis of this system and the concept
of its *multiplicity*. The concept of a focal point is treated here, while its multiplicity
will be discussed later.

### 2.2.1   *Focal Points of Conjoined Bases*

The following notion is motivated by the work [39] and [45] by Bohner and Došlý.

**Definition 2.14** We say that a conjoined basis $Y = \begin{pmatrix} X \\ U \end{pmatrix}$ of symplectic difference
system (SDS) has *no (forward) focal point* in $(k, k + 1]$ if the following conditions
hold

$$\operatorname{Ker} X_{k+1} \subseteq \operatorname{Ker} X_k, \quad P_k := X_k X_{k+1}^{\dagger} \mathcal{B}_k \geq 0. \tag{2.37}$$

The first condition in (2.37) is usually called the *kernel condition*, while the
second condition in (2.170) is called the *P-condition*. This means that a conjoined
basis $Y$ does have a focal point in the interval $(k, k + 1]$ if one of the conditions
in (2.170) is violated. The term "forward" focal point refers to the direction of the
kernel condition in (2.37). The following lemma provides equivalent formulations
of the kernel condition in (2.37).

**Lemma 2.15** *Let* $Y = \binom{X}{U}$ *be a conjoined basis of* (SDS). *Then*

$$\operatorname{Ker} X_{k+1} \subseteq \operatorname{Ker} X_k \iff \operatorname{Ker} X_{k+1}^T \subseteq \operatorname{Ker} \mathcal{B}_k^T \iff \operatorname{Im} \mathcal{B}_k \subseteq \operatorname{Im} X_{k+1} \quad (2.38)$$

$$\iff X_{k+1} X_{k+1}^\dagger \mathcal{B}_k = \mathcal{B}_k. \quad (2.39)$$

*Proof* The equivalence of the second and third condition in (2.38) with (2.39) follows from (1.159) and (1.160). We will show the equivalence of the first two conditions in (2.38). Suppose that the first inclusion holds. Let $c \in \operatorname{Ker} X_{k+1}^T$ and $\tilde{Y} = \binom{\tilde{X}}{\tilde{U}}$ be the conjoined basis which together with $Y = \binom{X}{U}$ forms a normalized pair of conjoined bases, i.e., $X_k \tilde{U}_k^T - \tilde{X}_k U_k^T = I$. Then $X_k \tilde{X}_k^T = \tilde{X}_k X_k^T$ by (1.146) for the block entries of a symplectic matrix so that $X_{k+1} \tilde{X}_{k+1}^T c = \tilde{X}_{k+1} X_{k+1}^T c = 0$. This means that $\tilde{X}_{k+1}^T c \in \operatorname{Ker} X_{k+1} \subseteq \operatorname{Ker} X_k$. Therefore,

$$\mathcal{B}_k^T c = \tilde{X}_k X_{k+1}^T c + \mathcal{B}_k^T c = \tilde{X}_k X_k^T A_k^T c + (I + \tilde{X}_k U_k^T) \mathcal{B}_k^T c$$
$$= X_k \tilde{X}_k^T A_k^T c + X_k \tilde{U}_k^T \mathcal{B}_k^T c = X_k \tilde{X}_{k+1}^T c = 0.$$

Conversely, let $c \in \operatorname{Ker} X_{k+1}$. By (2.39), we have $\mathcal{B}_k^T = \mathcal{B}_k^T (X_{k+1}^\dagger)^T X_{k+1}^T$. Then using (2.10), we get

$$X_k c = (\mathcal{D}_k^T X_{k+1} - \mathcal{B}_k^T U_{k+1}) c = -\mathcal{B}_k^T (X_{k+1}^\dagger)^T X_{k+1}^T U_{k+1} c$$
$$= -\mathcal{B}_k^T (X_{k+1}^\dagger)^T U_{k+1}^T X_{k+1} c = 0,$$

hence $c \in \operatorname{Ker} X_k$. The proof is complete.  $\square$

*Remark 2.16* Note that if the kernel condition in (2.37) holds, then the matrix $P_k$ is really symmetric. Indeed, let $Q_k$ be a symmetric matrix satisfying the equality $Q_k X_k = U_k X_k^\dagger X_k$. Such a matrix really exists, e.g., it is the matrix

$$Q_k := U_k X_k^\dagger + (U_k X_k^\dagger \tilde{X}_k - \tilde{U}_k)(I - X_k^\dagger X) U_k^T, \quad (2.40)$$

where $\tilde{Y} = \binom{\tilde{X}}{\tilde{U}}$ is a conjoined basis which together with $Y = \binom{X}{U}$ forms a pair of normalized conjoined bases. Substituting for $X_k$ from (2.10), and using formula (2.39) in Lemma 2.15, we have

$$P_k = X_k X_{k+1}^\dagger \mathcal{B}_k = (\mathcal{D}_k^T X_{k+1} - \mathcal{B}_k^T U_{k+1}) X_{k+1}^\dagger \mathcal{B}_k$$
$$= \mathcal{D}_k^T X_{k+1} X_{k+1}^\dagger \mathcal{B}_k - \mathcal{B}_k^T U_{k+1} X_{k+1}^\dagger X_{k+1} X_{k+1}^\dagger \mathcal{B}_k$$
$$= \mathcal{D}_k^T \mathcal{B}_k - \mathcal{B}_k^T Q_{k+1} \mathcal{B}_k,$$

which is symmetric.

Observe that Definition 2.14 is in good agreement with the definition of a focal point of a solution of the second- order equation (2.35), when this equation is written as a symplectic system. Since by (2.36) the matrix $\mathcal{B}_k = 1/r_k \neq 0$ in this case, no focal point of a solution $x$ of (2.35) in the interval $(k, k+1]$ means that $r_k x_k x_{k+1} > 0$. This implies that if $x_k \neq 0$, then $x_{k+1} \neq 0$ as well (the kernel condition), and $r_k x_k x_{k+1} > 0$ is the $P$-condition taking into account that $r_k \neq 0$.

As we will show later in this section, oscillatory properties of (SDS) can be equivalently defined via generalized zeros of vector solutions. We conclude this subsection with the definition of this concept.

**Definition 2.17** We say that a solution $y = \binom{x}{u} \in \mathbb{R}^{2n}$ of (SDS) has a *generalized zero* in the interval $(k, k+1]$ if

$$x_k \neq 0, \quad x_{k+1} \in \operatorname{Im} \mathcal{B}_k, \quad \text{and} \quad x_k^T \mathcal{B}_k^\dagger x_{k+1} \leq 0. \tag{2.41}$$

**Definition 2.18** Symplectic system (SDS) is said to be *disconjugate* on the interval $[0, N+1]$ if no solution of (SDS) has more than one generalized zero in $(0, N+1]$ and the solution $y = \binom{x}{u}$ with $x_0 = 0$ has no generalized zero in $(0, N+1]$.

**Definition 2.19** Symplectic system (SDS) is said to be *nonoscillatory* at $\infty$ if there exists $M$ such that this system is disconjugate on $[M, N+1]$ for any $N \geq M$. In the opposite case, system (SDS) is called *oscillatory* at $\infty$.

*Remark 2.20* Let $y = \binom{x}{u}$ be a vector solution of (SDS), and suppose that $x_{k+1} = \mathcal{B}_k c$ for some $c \in \mathbb{R}^n$, i.e., $x_{k+1} \in \operatorname{Im} \mathcal{B}_k$. Then $x_k^T \mathcal{B}_k^\dagger x_{k+1} = x_k^T c$. To see this, note that $x_{k+1} = \mathcal{B}_k c$ implies

$$x_k = \mathcal{D}_k^T x_{k+1} - \mathcal{B}_k^T u_{k+1} = \mathcal{D}_k^T \mathcal{B}_k c - \mathcal{B}_k^T u_{k+1} = \mathcal{B}_k^T (\mathcal{D}_k c - u_{k+1})$$

and

$$x_k^T \mathcal{B}_k^\dagger x_{k+1} = (\mathcal{D}_k c - u_{k+1})^T \mathcal{B}_k \mathcal{B}_k^\dagger \mathcal{B}_k c = x_k^T c.$$

*Remark 2.21* Consider now that (SDS) a rewritten even order Sturm-Liouville difference equation (2.27), i.e., a rewritten linear Hamiltonian system (2.15) with $x_k$ and $u_k$ given by (2.28) and $\mathcal{A}_k, \mathcal{B}_k, \mathcal{C}_k, \mathcal{D}_k$ given by (2.32)–(2.34). Then by a direct computation (verifying the four properties in (1.157)), we have

$$\mathcal{B}_k^\dagger = \frac{r_k^{[n]}}{n} \begin{pmatrix} 0 & \dots & 0 \\ \vdots & \ddots & \vdots \\ 0 & \dots & 0 \\ 1 & \dots & 1 \end{pmatrix}. \tag{2.42}$$

Taking into account formulas (2.28) and expanding the higher-order difference in the formula for $x$, we see that $x_{k+1} = \mathcal{B}_k c \in \operatorname{Im} \mathcal{B}_k$ with $c = (c_1, \dots, c_n)^T \in \mathbb{R}^n$ if and only if $x_{k+1} = \left(c_n/r_k^{[n]}\right)(1, \ldots, 1)^T$. This means that all the entries of $x_{k+1}$ in (2.28) are equal and hence

$$y_{k+1} = y_{k+2} = \cdots = y_{k+n-1} = 0, \quad c_n = r_k^{[n]} y_{k+n}. \tag{2.43}$$

We then obtain that

$$x_k = \begin{pmatrix} 0 \\ \vdots \\ 0 \\ (-1)^{n-1} y_k \end{pmatrix}, \quad x_{k+1} = y_{k+n} \begin{pmatrix} 1 \\ \vdots \\ 1 \end{pmatrix}, \tag{2.44}$$

and

$$x_k^T \mathcal{B}_k^\dagger x_{k+1} = x_k^T c = (-1)^{n-1} r_k^{[n]} y_k y_{k+n}. \tag{2.45}$$

Therefore, combining (2.43), (2.44), and (2.45), we can see that a solution $\binom{x}{u}$ of (SDS) has a forward focal point in the interval $(k, k+1]$ according to Definition 2.17 if and only if the solution $y$ of (2.27) satisfies

$$y_k \neq 0, \quad y_{k+1} = y_{k+2} = \cdots = y_{k+n-1} = 0, \quad (-1)^{n-1} r_k^{[n]} y_k y_{k+n} \leq 0. \tag{2.46}$$

Condition (2.46) agrees with the notion of a generalized zero in $(k, k+n]$ for a solution $y$ of (2.27) introduced by Hartman in [169, pg. 2] or by Bohner in [41, Remark 3(iii)]. Also, when $n = 1$, then the second condition in (2.46) is vacuous, and the first and third conditions yield the definition of a generalized zero in $(k, k+1]$ for a solution $y$ of (2.35) (or (1.1); see Definition 1.1).

## 2.2.2 Backward Focal Points

The adjective "forward" focal point is used in the previous subsection to distinguish this concept from the concept of a *backward focal point* which is defined below. We will use the convention that "focal point" without any adjective means by default a *forward focal point*.

As we have mentioned in the previous section, symplectic system (SDS) is equivalent to the so-called *time-reversed symplectic system*

$$z_k = S_k^{-1} z_{k+1}, \quad S^{-1} = \begin{pmatrix} \mathcal{D}_k^T & -\mathcal{B}_k^T \\ -\mathcal{C}_k^T & \mathcal{A}_k^T \end{pmatrix}, \tag{2.47}$$

and this system motivates the definition of the *backward focal point* as follows.

**Definition 2.22** Let $Y = \binom{X}{U}$ be a conjoined basis of (SDS). We say that $Y$ has *no backward focal point* in $[k, k + 1)$ if

$$\operatorname{Ker} X_k \subseteq \operatorname{Ker} X_{k+1} \quad \text{and} \quad X_{k+1} X_k^\dagger \mathcal{B}_k^T \geq 0. \tag{2.48}$$

We can also define the concept of a generalized zero of a vector solution in the interval $[k, k + 1)$ which is associated with the concept of backward focal point and it is defined as follows.

**Definition 2.23** Let $y = \binom{x}{u} \in \mathbb{R}^{2n}$ be a solution of (SDS). This solution has a *generalized zero* in the interval $[k, k + 1)$ if

$$x_{k+1} \neq 0, \quad x_k \in \operatorname{Im} \mathcal{B}_k^T, \quad \text{and} \quad x_k^T \mathcal{B}_k^\dagger x_{k+1} \leq 0. \tag{2.49}$$

Then, similarly to generalized zeros in $(k, k + 1]$, we have for $x_k = \mathcal{B}_k^T c$

$$x_{k+1} = A_k x_k + B_k u_k = A_k \mathcal{B}_k^T c + B_k u_k = B_k (A_k c + u_k)$$

and

$$x_k^T \mathcal{B}_k^\dagger x_{k+1} = c^T \mathcal{B}_k \mathcal{B}_k^\dagger B_k (A_k c + u_k) = c^T B_k (A_k^T c + u_k)$$
$$= c^T x_{k+1} = x_{k+1}^T c.$$

*Remark 2.24* Returning to the $2n$-th order Sturm-Liouville difference equation (2.27), the concept of a backward focal point in $[k, k + 1)$ for a solution $y$ of (2.27) in Definition 2.22 translates as follows: according to (2.49) and (2.32), we have $x_k = \mathcal{B}_k^T c \in \operatorname{Im} \mathcal{B}_k^T$ for some $c = (c_1, \ldots, c_n)^T \in \mathbb{R}^n$ if and only if

$$x_k = \frac{1}{r_k^{[n]}} (0, \ldots, 0, c_0)^T, \quad c_0 := \sum_{j=1}^n c_j.$$

This means in view of the definition of $x_k$ in (2.28) that

$$y_{k+1} = \cdots = y_{k+n-1} = 0, \quad x_k = (0, \ldots, 0, (-1)^{n-1} y_k)^T,$$
$$c_0 = (-1)^{n-1} r_k^{[n]} y_k, \quad x_{k+1} = y_{k+n} (1, \ldots, 1)^T.$$

Therefore, $x_{k+1} \neq 0$ if and only if $y_{k+n} \neq 0$, and

$$x_k^T \mathcal{B}_k^\dagger x_{k+1} = x_k^T c = y_{k+n} c_0 = (-1)^{n-1} r_k^{[n]} y_k y_{k+n}.$$

This shows that a solution $\left(\begin{smallmatrix} x \\ u \end{smallmatrix}\right)$ of (SDS) has a backward focal point in the interval $[k, k+1)$ according to Definition 2.23 if and only if the solution $y$ of (2.27) satisfies

$$y_{k+n} \neq 0, \quad y_{k+1} = y_{k+2} = \cdots = y_{k+n-1} = 0, \quad (-1)^{n-1} r_k^{[n]} y_k y_{k+n} \leq 0. \tag{2.50}$$

This is what we define as a generalized zero in the interval $[k, k+n)$ for a solution $y$ of (2.27). For the second-order Sturm-Liouville difference equation (2.35) (or (1.1)), we then obtain from (2.50) the definition of a generalized zero of a solution $y$ of (2.35) in $[k, k+1)$, which reads as $y_{k+1} \neq 0$ and $r_k y_k y_{k+1} \leq 0$.

## 2.3 Riccati Equation and Quadratic Functional

In this section, we treat two basic concepts which are associated with symplectic difference systems, namely, the Riccati matrix difference equation and the associated (energy) quadratic functional. Picone's identity is then an identity which relates these two concepts and shows that, similarly to the scalar case, the existence of a symmetric solution of Riccati equation enables to "complete to the square" the energy functional, and hence it implies its positivity for any nontrivial admissible sequence. The main result of this section is formulated in Theorem 2.36 (the Reid roundabout theorem). We also discuss the nonnegativity of the energy quadratic functional.

### 2.3.1 Riccati Matrix Difference Equation

Let us recall that in Sect. 1.5, we have supposed that the considered linear Hamiltonian differential system is *identically* normal, which implied (together with the Legendre condition (1.111)) that focal points of any conjoined basis are isolated. In the discrete case, all the points in underlaying set $\mathbb{Z}$ (or its subsets) are isolated by themselves, so the assumption of identical normality is of different character.

We start with the simple case. Consider a conjoined basis $Y = \left(\begin{smallmatrix} X \\ U \end{smallmatrix}\right)$ with $X_k$ invertible for all $k \in [0, N+1]_{\mathbb{Z}}$. Then by a direct computation we verify that $Q_k := U_k X_k^{-1}$ is a symmetric solution of the Riccati matrix difference equation

$$Q_{k+1} = (\mathcal{C}_k + \mathcal{D}_k Q_k)(\mathcal{A}_k + \mathcal{B}_k Q_k)^{-1}, \tag{2.51}$$

see also Theorem 2.28 below. In some cases, we will need this equation in a slightly different form with the so-called *Riccati operator*

$$R_k[Q] = 0, \quad R_k[Q] := Q_{k+1}(\mathcal{A}_k + \mathcal{B}_k Q_k) - (\mathcal{C}_k + \mathcal{D}_k Q_k). \tag{2.52}$$

Another important quantity associated with the matrix $Q$ and used frequently in the later parts is the matrix

$$P_k[Q] := (\mathcal{D}_k^T - \mathcal{B}_k^T Q_{k+1})\, \mathcal{B}_k. \tag{2.53}$$

When no identical normality is supposed, it may happen that the first component $X_k$ of a conjoined basis is noninvertible on some discrete interval. A typical example is the symplectic system corresponding to the $2n$-th order difference equation $\Delta^{2n} y_k = 0$. Then for the conjoined basis of this system given by the initial condition $Y_0 = (0\ I)^T$, we have rank $X_k = k$ for all $k = [0, n]_{\mathbb{Z}}$, as can be verified by a direct computation. Nevertheless, also in this case, we can exhibit a kind of Riccati-type difference equation, sometimes called an *implicit* Riccati equation. Another form of the implicit Riccati equation appears later in Theorem 2.36. In this respect equation (2.51) is called the *explicit* Riccati equation.

**Lemma 2.25** *Let* $Y = \binom{X}{U}$ *be a conjoined basis of* (SDS) *with* $\operatorname{Ker} X_{k+1} \subseteq \operatorname{Ker} X_k$ *for* $k \in [0, N]_{\mathbb{Z}}$ *and suppose that* $Q_k$ *is a symmetric matrix with* $Q_k X_k = U_k X_k^{\dagger} X_k$. *Then we have*

$$R_k[Q]\, X_k = 0 \quad \text{and} \quad P_k[Q] = X_k X_{k+1}^{\dagger} \mathcal{B}_k.$$

*Proof* From (1.159) we have $X_k X_{k+1}^{\dagger} X_{k+1} = X_k$ and $X_{k+1} X_{k+1}^{\dagger} \mathcal{B}_k = \mathcal{B}_k$, which yields

$$R_k[Q]\, X_k = \begin{pmatrix} I \\ Q_{k+1} \end{pmatrix}^T \mathcal{J}^T \, \mathcal{S}_k \begin{pmatrix} X_k \\ U_k \end{pmatrix} X_k^{\dagger} X_k = \begin{pmatrix} Q_{k+1} \\ -I \end{pmatrix}^T \begin{pmatrix} X_{k+1} \\ U_{k+1} \end{pmatrix} X_k^{\dagger} X_k = 0.$$

The second claim is shown in Remark 2.16.                                       □

The following identity relates the matrix $\mathcal{A}_k + \mathcal{B}_k Q_k$ from (2.51) with the matrix $\mathcal{D}_k^T - \mathcal{B}_k^T Q_{k+1}$ in (2.53).

**Lemma 2.26** *Assume that the matrices* $Q_k$ *and* $Q_{k+1}$ *are symmetric. Then*

$$(\mathcal{D}_k - Q_{k+1}\mathcal{B}_k)\,(\mathcal{A}_k^T + Q_k \mathcal{B}_k^T) = I - R_k[Q]\, \mathcal{B}_k^T. \tag{2.54}$$

*Proof* By (1.146) we know that $\mathcal{D}_k \mathcal{A}_k^T = I + \mathcal{C}_k \mathcal{B}_k^T$ and $\mathcal{A}_k \mathcal{B}_k^T$ is symmetric. Then

$$\begin{aligned}
(\mathcal{D}_k - Q_{k+1}\mathcal{B}_k)\,(\mathcal{A}_k^T + Q_k \mathcal{B}_k^T) &= \mathcal{D}_k \mathcal{A}_k^T + \mathcal{D}_k Q_k \mathcal{B}_k^T - Q_{k+1}\mathcal{B}_k \mathcal{A}_k^T - Q_{k+1}\mathcal{B}_k Q_k \mathcal{B}_k^T \\
&= I + [(\mathcal{C}_k + \mathcal{D}_k Q_k) - Q_{k+1}(\mathcal{A}_k + \mathcal{B}_k Q_k)]\, \mathcal{B}_k^T \\
&= I - R_k[Q]\, \mathcal{B}_k^T,
\end{aligned}$$

which shows the result.                                                          □

When the matrices $Q_k$ and $Q_{k+1}$ solve the Riccati equation (2.51), we obtain from Lemma 2.26 the following important property.

**Corollary 2.27** *Assume that $Q_k$ and $Q_{k+1}$ are symmetric. Then they satisfy the implicit Riccati equation $R_k[Q]\mathcal{B}_k^T = 0$ if and only if the matrices $\mathcal{A}_k + \mathcal{B}_k Q_k$ and $\mathcal{D}_k^T - \mathcal{B}_k^T Q_{k+1}$ are invertible and they are inverses of each other, i.e.,*

$$(\mathcal{A}_k + \mathcal{B}_k Q_k)^{-1} = \mathcal{D}_k^T - \mathcal{B}_k^T Q_{k+1}. \tag{2.55}$$

*In particular, property (2.55) holds when $Q_k$ and $Q_{k+1}$ satisfy the explicit Riccati equation (2.51).*

*Proof* The result follows from (2.54) with $R_k[Q]\mathcal{B}_k^T = 0$, resp. with $R_k[Q] = 0$. $\qquad\square$

The following results connects the symmetric solutions of the explicit Riccati equation (2.52) with conjoined bases $Y$ of (SDS) with $X_k$ invertible.

**Theorem 2.28** *The Riccati equation (2.52) on $[0, N]_\mathbb{Z}$ has a symmetric solution $Q_k$ defined on $[0, N + 1]_\mathbb{Z}$ if and only if there exists a conjoined basis $Y$ of (SDS) with $X_k$ invertible on $[0, N + 1]_\mathbb{Z}$. In this case $Q_k = U_k X_k^{-1}$ on $[0, N + 1]_\mathbb{Z}$ and $\mathcal{A}_k + \mathcal{B}_k Q_k = X_{k+1} X_k^{-1}$ is invertible on $[0, N]_\mathbb{Z}$.*

*Proof* Assume that $Y$ is a conjoined (SDS) with $X_k$ invertible on $[0, N + 1]_\mathbb{Z}$. We define $Q_k := U_k X_k^{-1}$ on $[0, N + 1]_\mathbb{Z}$. Then it easily follows that $Q_k$ is symmetric on $[0, N + 1]_\mathbb{Z}$ and that it solves equation (2.52) on $[0, N]_\mathbb{Z}$ . Conversely, if $Q_k$ is symmetric solution of (2.52) on $[0, N]_\mathbb{Z}$, then by Corollary 2.27, we know that $\mathcal{A}_k + \mathcal{B}_k Q_k$ is invertible on $[0, N]_\mathbb{Z}$. Consider the solution $X_k$ on $[0, N + 1]_\mathbb{Z}$ of the linear system $X_{k+1} = (\mathcal{A}_k + \mathcal{B}_k Q_k) X_k$ for $k \in [0, N]_\mathbb{Z}$ with the initial condition $X_0 = I$. Then $X_k$ is invertible on $[0, N + 1]_\mathbb{Z}$, and we set $U_k := Q_k X_k^{-1}$ on $[0, N + 1]_\mathbb{Z}$. It then follows from (2.52) that $Y = \binom{X}{U}$ is a conjoined basis of (SDS), which completes the proof. $\qquad\square$

### 2.3.2 Energy Quadratic Functional

In this subsection we will study the quadratic functional, for which the symplectic system (SDS) is the associated Jacobi system in the spirit of Sect. 1.3.2. Denote

$$\mathcal{K} := \begin{pmatrix} 0 & 0 \\ I & 0 \end{pmatrix}$$

and consider the quadratic functional

$$\mathcal{F}(y) = \sum_{k=0}^{N} y_k^T \left( \mathcal{S}_k^T \mathcal{K} \mathcal{S}_k - \mathcal{K} \right) y_k, \quad y_k = \begin{pmatrix} x_k \\ u_k \end{pmatrix} \in \mathbb{R}^{2n}. \tag{2.56}$$

If we write the matrix $\mathcal{S}_k$ in the block form, we have

$$\mathcal{S}_k^T \mathcal{K} \mathcal{S}_k - \mathcal{K} = \begin{pmatrix} \mathcal{C}_k^T \mathcal{A}_k & \mathcal{C}_k^T \mathcal{B}_k \\ \mathcal{B}_k^T \mathcal{C}_k & \mathcal{D}_k^T \mathcal{B}_k \end{pmatrix},$$

and hence we can write the energy functional equivalently as

$$\mathcal{F}(y) = \sum_{k=0}^{N} \{ x_k^T \mathcal{C}_k^T \mathcal{A}_k x_k + 2 x_k^T \mathcal{C}_k^T \mathcal{B}_k u_k + u_k^T \mathcal{B}_k^T \mathcal{D}_k u_k \}. \tag{2.57}$$

**Definition 2.29** We say that the sequence $y = \{y_k\}_{k=0}^{N+1}$ with $y_k = \begin{pmatrix} x_k \\ u_k \end{pmatrix} \in \mathbb{R}^{2n}$ is *admissible* for the quadratic functional $\mathcal{F}$ if

$$x_{k+1} = \mathcal{A}_k x_k + \mathcal{B}_k u_k, \quad k \in [0, N]_{\mathbb{Z}}. \tag{2.58}$$

The admissibility of $y = \begin{pmatrix} x \\ u \end{pmatrix}$ means that the pair $(x, u)$ satisfies the first equation in symplectic system (the equation of motion). The admissibility of $y$ can be also characterized as follows. Define the so-called *controllability matrices* by

$$G_k := \left( \prod_{j=1}^{k-1} \mathcal{A}_{k-j} \mathcal{B}_0 \quad \prod_{j=1}^{k-2} \mathcal{A}_{k-j} \mathcal{B}_1 \quad \cdots \quad \mathcal{A}_{k-1} \mathcal{B}_{k-2} \quad \mathcal{B}_{k-1} \right) \in \mathbb{R}^{n \times kn}. \tag{2.59}$$

It follows by induction that $y = \begin{pmatrix} x \\ u \end{pmatrix}$ with $x_0 = 0$ is admissible if and only if

$$x_k = G_k \begin{pmatrix} u_0 \\ \vdots \\ u_{k-1} \end{pmatrix} \quad \text{for all } k \in [1, N+1]_{\mathbb{Z}}. \tag{2.60}$$

**Lemma 2.30** Let $y = \begin{pmatrix} x \\ u \end{pmatrix}$ be admissible for (2.57). Then this functional can be expressed in the form

$$\mathcal{F}(x, u) = x_k^T u_k \Big|_{k=0}^{N+1} + \sum_{k=0}^{N} x_{k+1}^T (-u_{k+1} + \mathcal{C}_k x_k + \mathcal{D}_k u_k). \tag{2.61}$$

*Proof* Using the admissibility of $y = \begin{pmatrix} x \\ u \end{pmatrix}$, we have

$$\mathcal{F}(x, u) = \sum_{k=0}^{N} \{ x_k^T \mathcal{C}_k^T (\mathcal{A}_k x_k + \mathcal{B}_k u_k) + (x_k^T \mathcal{C}_k^T + u_k^T \mathcal{D}_k^T) \mathcal{B}_k u_k \}$$

$$= \sum_{k=0}^{N} \{ x_k^T \mathcal{C}_k^T (\mathcal{A}_k x_k + \mathcal{B}_k u_k) + x^T (\mathcal{A}_k^T \mathcal{D}_k - I) u_k + u_k^T \mathcal{D}_k^T (x_{k+1} - \mathcal{A}_k x_k) \}$$

$$= \sum_{k=0}^{N} \{x_{k+1}^T C_k x_k - x_k^T u_k + x_{k+1}^T \mathcal{D}_k u_k + x_{k+1}^T u_{k+1} - x_{k+1}^T u_{k+1}\}$$

$$= \sum_{k=0}^{N} \Delta(x_k^T u_k) + \sum_{k=0}^{N} x_{k+1}^T(-u_{k+1} + C_k x_k + \mathcal{D}_k u_k)$$

$$= x_k^T u_k \Big|_{k=0}^{N+1} + \sum_{k=0}^{N} x_{k+1}^T(-u_{k+1} + C_k x_k + \mathcal{D}_k u_k),$$

which shows the result. $\qquad\square$

The quadratic functional (2.56) is given by the symmetric bilinear form

$$\mathcal{F}(\tilde{y}, y) = \sum_{k=1}^{N} \tilde{y}_k^T (\mathcal{S}_k^T \mathcal{K} \mathcal{S}_k - \mathcal{K}) \, y_k$$

$$= \sum_{k=0}^{N} \{\tilde{x}_k^T C_k^T A_k x_k + \tilde{x}_k^T C_k^T B_k u_k + \tilde{u}_k^T B_k^T C_k x_k + \tilde{u}_k^T B_k^T \mathcal{D}_k u_k\}.$$

Similarly as in the previous lemma, we have

$$\mathcal{F}(\tilde{y}, y) = \tilde{x}_k^T u_k \Big|_{k=0}^{N+1} + \sum_{k=0}^{N} \tilde{x}_{k+1}^T(-u_{k+1} + C_k x_k + \mathcal{D}_k u_k). \qquad (2.62)$$

By a direct computation, one can verify that for admissible $y$, $\tilde{y}$

$$\mathcal{F}(y + \tilde{y}) = \mathcal{F}(y) + 2\mathcal{F}(y, \tilde{y}) + \mathcal{F}(\tilde{y}),$$

which has the following consequence frequently used in the discrete calculus of variations and optimal control.

**Theorem 2.31** *Suppose that $\mathcal{F}(y) \geq 0$ for any admissible $y = \binom{x}{u}$ with $x_0 = 0 = x_{N+1}$. Let $\tilde{y} = \binom{\tilde{x}}{\tilde{u}}$ be a solution of (SDS) for which $\tilde{x}_0 = p \in \mathbb{R}^n$ and $\tilde{x}_{N+1} = q \in \mathbb{R}^n$. Then for any admissible $y = \binom{x}{u}$ with $x_0 = p$ and $x_{N+1} = q$, we have $\mathcal{F}(y) \geq \mathcal{F}(\tilde{y})$.*

*Proof* Denote $\delta y_k := \tilde{y}_k - y_k$, i.e., $\tilde{y}_k = y_k + \delta y_k$. Then $\delta y$ is also admissible and $\delta x_0 = 0 = \delta x_{N+1}$, where $\delta y_k = \binom{\delta x_k}{\delta u_k}$, i.e., $\mathcal{F}(\delta y) \geq 0$. We have

$$\mathcal{F}(\tilde{y}) = \mathcal{F}(y + \delta y) = \mathcal{F}(y) + 2\mathcal{F}(\delta y, y) + \mathcal{F}(\delta y) \geq \mathcal{F}(y),$$

since according to (2.62), we have $\mathcal{F}(\delta y, y) = 0$. $\qquad\square$

Finally, note that another way how to write $\mathcal{F}(y)$ for an admissible $y$ is to replace $\mathcal{B}_k u_k$ in the last two terms in (2.57) by $x_{k+1} - \mathcal{A}_k x_k$. Using this approach, one can eliminate the variable $u$ in the functional, and this functional can be written in the form

$$\mathcal{F}(y) = \sum_{k=0}^{N} \begin{pmatrix} x_k \\ x_{k+1} \end{pmatrix}^T \mathcal{G}_k \begin{pmatrix} x_k \\ x_{k+1} \end{pmatrix}, \tag{2.63}$$

where the symmetric $2n \times 2n$ matrix $\mathcal{G}_k$ and the symmetric $n \times n$ matrix $\mathcal{E}_k$ are defined by

$$\mathcal{G}_k := \begin{pmatrix} \mathcal{A}_k^T \mathcal{E}_k \mathcal{A}_k - \mathcal{A}_k^T \mathcal{C}_k \, \mathcal{C}_k^T - \mathcal{A}_k^T \mathcal{E}_k \\ \mathcal{C}_k^T \mathcal{E}_k \mathcal{A}_k \qquad \mathcal{E}_k \end{pmatrix}, \quad \mathcal{E}_k := \mathcal{B}_k \mathcal{B}_k^\dagger \mathcal{D}_k \mathcal{B}_k^\dagger. \tag{2.64}$$

This elimination of $u$ turned to be useful in proving the Sturmian theorems for symplectic difference systems in Chaps. 4 and 5.

### 2.3.3 Picone Identity

The next lemma presents the so-called *Picone identity*, which is used in establishing conditions for positivity of the energy functional $\mathcal{F}$ in the class of admissible pairs $(x, u)$ with $x_0 = 0 = x_{N+1}$.

**Lemma 2.32** *Let $y = \begin{pmatrix} x \\ u \end{pmatrix}$ be an admissible pair for $\mathcal{F}$ and put $z_k := u_k - Q_k x_k$, where $Q_k \in \mathbb{R}^{n \times n}$ is a symmetric matrix. Then we have the identities*

$$\left. \begin{aligned} \Delta(x_k^T Q_k x_k) - x_k^T \mathcal{C}_k^T \mathcal{A}_k x_k - 2x_k^T \mathcal{C}_k^T \mathcal{B}_k u_k - u_k^T \mathcal{B}_k^T \mathcal{D}_k u_k + z_k^T P_k[Q] z_k \\ = 2\, u_k^T \mathcal{B}_k^T R_k[Q] x_k + x_k^T \{R_k^T[Q] \mathcal{A}_k - Q_k \mathcal{B}_k^T R_k[Q]\} x_k, \end{aligned} \right\} \tag{2.65}$$

*and*

$$(\mathcal{D}_k^T - \mathcal{B}_k^T Q_{k+1}) x_{k+1} = x_k + P_k[Q] z_k - \mathcal{B}_k^T R_k[Q] x_k. \tag{2.66}$$

*In particular, if $Q$ is a solution of Riccati equation (2.52), i.e., $R_k[Q] = 0$, and $x_0 = 0 = x_{N+1}$, we have the Picone identity*

$$\mathcal{F}(y) = \sum_{k=0}^{N} z_k^T P_k[Q] z_k. \tag{2.67}$$

*Proof* Formula (2.65), the so-called *Picone identity* in difference form, can be derived by direct calculations, which are however rather technical. We will present a more elegant proof of (2.65) at the end of Sect. 2.6. As for identity (2.66), we prove it as follows. Denote the left-hand side of this identity by $L$. Then we have

$$
\begin{aligned}
L &= (\mathcal{D}_k^T - \mathcal{B}_k^T Q_{k+1}) (\mathcal{A}_k x_k + \mathcal{B}_k u_k) \\
&= \mathcal{D}_k^T \mathcal{A}_k x_k - \mathcal{B}_k^T Q_{k+1} \mathcal{A}_k x_k + \mathcal{D}_k^T \mathcal{B}_k u_k - \mathcal{B}_k^T Q_{k+1} \mathcal{B}_k u_k \\
&= (I + \mathcal{B}_k^T \mathcal{C}_k) x_k - \mathcal{B}_k^T Q_{k+1} \mathcal{A}_k x_k + P_k[Q] (u_k - Q_k x_k) \\
&\qquad\qquad\qquad\qquad + (\mathcal{B}_k^T \mathcal{D}_k - \mathcal{B}_k^T Q_{k+1} \mathcal{B}_k) Q_k x_k \\
&= x_k + P_k[Q] z_k - \mathcal{B}_k^T R_k[Q] x_k,
\end{aligned}
$$

which is the right-hand side of (2.66). Formula (2.67) is obtained by the summation of (2.65) from $k = 0$ to $k = N$ by using $R_k[Q] = 0$ and $x_0 = 0 = x_{N+1}$. $\qquad\square$

### 2.3.4 Reid Roundabout Theorem

Now we are ready to formulate and prove the Reid roundabout theorem for symplectic difference system. This theorem plays the fundamental role in the oscillation theory of these systems. In the proof we will need the following auxiliary result, which shows that the kernel condition (2.37) implies the so-called *image condition* for admissible pairs.

**Lemma 2.33** *Let* $Y = \binom{X}{U}$ *be a conjoined basis of* (SDS) *satisfying the kernel condition in* (2.37). *Then for any admissible sequence* $y = \binom{x}{u}$ *with* $x_0 \in \operatorname{Im} X_0$, *we have* $x_k \in \operatorname{Im} X_k$ *for all* $k \in [0, N+1]_{\mathbb{Z}}$.

*Proof* Suppose that $x_k \in \operatorname{Im} X_k$, i.e., $x_k = X_k c$ for some $c \in \mathbb{R}^n$ and $k \in [0, N]_{\mathbb{Z}}$. The admissibility of $y$ means that

$$
x_{k+1} = \mathcal{A} x_k + \mathcal{B}_k u_k = X_{k+1} c + \mathcal{B}_k (U_k c + u_k) \in \operatorname{Im} X_{k+1},
$$

since $\operatorname{Im} \mathcal{B}_k \subseteq \operatorname{Im} X_{k+1}$ by Lemma 2.15. $\qquad\square$

As we will see later, in the investigation of the positivity or nonnegativity of the quadratic functional $\mathcal{F}$ associated with a symplectic difference system, an important role is played by the principal solution at $k = 0$ given in Definition 2.2.

**Lemma 2.34** *Suppose that for all solutions* $y = \binom{x}{u}$ *of* (SDS) *with* $x_0 = 0$, *we have that* $x_k^T c > 0$ *whenever* $x_k \neq 0$ *and* $x_{k+1} = \mathcal{B}_k c$ *hold for* $k \in [0, N]_{\mathbb{Z}}$, *i.e.,* $y$ *has no generalized zero in* $(0, N+1]$. *Then the principal solution of* (SDS) *at* $k = 0$ *has no focal points in* $(0, N+1]$.

*Proof* Let $Y = \binom{X}{U}$ be the principal solution of (SDS) at $k = 0$. First, let $\alpha \in \operatorname{Ker} X_{m+1}$ for some $m \in [0, N]_{\mathbb{Z}}$ and put $y_k := Y_k \alpha$ on $[0, N+1]_{\mathbb{Z}}$. Then $y$ solves

(SDS) with $x_0 = 0$ and $x_{m+1} = 0 \in \operatorname{Im}\mathcal{B}_m$, so that $x_m = 0$ as well. Otherwise we would have $x_m^T c > 0$ with $c = 0$ by the assumptions of the lemma. Hence $\operatorname{Ker} X_{m+1} \subseteq \operatorname{Ker} X_m$ holds. Now, let $c \in \mathbb{R}^n$, $\alpha := X_{m+1}^\dagger \mathcal{B}_m c$, and define again $y_k := Y_k \alpha$ on $[0, N+1]_\mathbb{Z}$. Then $y$ solves (SDS) with $x_0 = 0$, $x_{m+1} = \mathcal{B}_m c$, and $x_m^T c = c^T X_m X_{m+1}^\dagger \mathcal{B}_m c$. Therefore, $X_m X_{m+1}^\dagger \mathcal{B}_m \geq 0$ holds as well, and $Y = \binom{X}{U}$ has no focal points in $(0, N+1]$.                                                  □

**Lemma 2.35** *If $\mathcal{F}(y) > 0$ for all admissible pairs $y = \binom{x}{u}$ with $x_0 = 0 = x_{N+1}$ and $x \not\equiv 0$, then (SDS) is disconjugate on the interval $[0, N+1]$.*

*Proof* Suppose that (SDS) is not disconjugate. Then by Remark 2.20, there exist a solution $y = \binom{x}{u}$ of (SDS), the indices $m, p \in [0, N]_\mathbb{Z}$ with $m < p$, and vectors $c_m, c_p \in \mathbb{R}^n$ such that

$$x_m \neq 0, \ x_{m+1} = \mathcal{B}_m c_m, \ x_m^T c_m \leq 0, \quad x_p \neq 0, \ x_{p+1} = \mathcal{B}_p c_p, \ x_p^T c_p \leq 0.$$

This means that the solution $y$ has generalized zeros in the intervals $(m, m+1]$ and $(p, p+1]$. Define

$$\tilde{x}_k := \begin{cases} x_k, & m+1 \leq k \leq p, \\ 0 & \text{otherwise}, \end{cases} \qquad \tilde{u}_k := \begin{cases} c_m, & k = m, \\ u_k, & m+1 \leq k \leq p-1, \\ u_p - c_p, & k = p, \\ 0, & \text{otherwise}. \end{cases}$$

Then $\tilde{x}_0 = 0 = \tilde{x}_{N+1}$ and $\tilde{x} \not\equiv 0$ hold. Moreover, $\tilde{y} = \binom{\tilde{x}}{\tilde{u}}$ is easily checked to be admissible on $[0, N+1]_\mathbb{Z}$, and by Lemma 2.30 (with $y := \tilde{y}$), we get

$$\mathcal{F}(\tilde{y}) = \tilde{x}_k^T \tilde{u}_k \big|_0^{N+1} + \sum_{k=0}^{N} \tilde{x}_{k+1}^T \{ \mathcal{C}_k \tilde{x}_k + \mathcal{D}_k \tilde{u}_k - \tilde{u}_{k+1} \}$$

$$= \sum_{k \in \{m, p-1\}} \tilde{x}_{k+1}^T \{ \mathcal{C}_k \tilde{x}_k + \mathcal{D}_k \tilde{u}_k - \tilde{u}_{k+1} \} + \sum_{k=m+1}^{p-2} \tilde{x}_{k+1}^T \{ \mathcal{C}_k \tilde{x}_k + \mathcal{D}_k \tilde{u}_k - \tilde{u}_{k+1} \}$$

$$= x_{m+1}^T (\mathcal{D}_m c_m - u_{m+1}) + x_p^T (\mathcal{C}_{p-1} x_{p-1} + \mathcal{D}_{p-1} u_{p-1} + c_p - u_p)$$

$$\overset{(2.10)}{=} (x_m^T + u_{m+1}^T \mathcal{B}_m) c_m - x_{m+1}^T u_{m+1} + x_p^T c_p$$

$$= x_m^T c_m + x_p^T c_p \leq 0.$$

Thus $\mathcal{F}(y) \not> 0$, which completes the proof.                                          □

We are now ready to prove our main result of this section, the Reid roundabout theorem for symplectic systems (SDS), which states several conditions which are equivalent to the positivity of the functional $\mathcal{F}$ defined in (2.56). We recall that the Riccati operator $R_k[Q]$ is given by (2.52), $P_k[Q]$ is given by (2.53), and the controllability matrix $G_k$ is defined in (2.59).

**Theorem 2.36 (Reid Roundabout Theorem)** *The following statements are equivalent.*

(i) *The quadratic functional $\mathcal{F}(y) > 0$ for all admissible sequences $y = \binom{x}{u}$ with $x_0 = 0 = x_{N+1}$ and $x \not\equiv 0$.*

(ii) *System (SDS) is disconjugate on $[0, N+1]$ according to Definition 2.18.*

(iii) *No solution $y = \binom{x}{u}$ of (SDS) with $x_0 = 0$ has any generalized zero in the interval $(0, N+1]$.*

(iv) *The principal solution $Y^{[0]} = \binom{X^{[0]}}{U^{[0]}}$ of (SDS) has no focal points in the interval $(0, N+1]$, i.e., condition (2.37) holds for all $k \in [0, N]_{\mathbb{Z}}$.*

(v) *The implicit Riccati equation $R_k[Q] G_k = 0$, $k \in [0, N]_{\mathbb{Z}}$, has a symmetric solution $Q$ defined on $[0, N+1]_{\mathbb{Z}}$ such that $P_k[Q] \geq 0$ for all $k \in [0, N]_{\mathbb{Z}}$.*

*Proof* The statement (i) implies (ii) by Lemma 2.35, while (iii) follows from (ii) trivially, and condition (iii) implies (iv) by Lemma 2.34. Now, assume that (iv) holds. Let $Y := Y^{[0]}$ be the principal solution at $k = 0$ and let $\tilde{Y} = \binom{\tilde{X}}{\tilde{U}}$ be the associated solution of (SDS) at $k = 0$, i.e., $\tilde{X}_0 = I$ and $\tilde{U}_0 = 0$. The matrix $Q_k$ defined by (2.40) is symmetric and satisfies the assumptions of Lemma 2.25, so that $P_k[Q] \geq 0$ holds. Moreover, Lemma 2.39 and condition (2.60) imply that $R_k[Q] G_k = 0$, which completes the proof of (v). Suppose now that (v) is true with some symmetric $Q$. Let $y = \binom{x}{u}$ be admissible with $x_0 = 0 = x_{N+1}$. Then $R_k[Q] x_k = 0$ for all $k \in [0, N]_{\mathbb{Z}}$ by (2.60). From (2.67) in Lemma 2.32 we obtain

$$\mathcal{F}(y) = \sum_{k=0}^{N} z_k^T P_k[Q] z_k \geq 0.$$

To show positive definiteness, assume that $\mathcal{F}(y) = 0$ for some admissible $y = \binom{x}{u}$ with $x_0 = 0 = x_{N+1}$. Then $P_k[Q] z_k = 0$ for all $k \in [0, N]_{\mathbb{Z}}$ and (2.66) shows that $x \equiv 0$. Thus $\mathcal{F}(y) > 0$, showing (i). In conclusion, we have proven that the statements (i) through (v) are equivalent. □

*Remark 2.37* The proof of the implications (iv) $\Rightarrow$ (v) $\Rightarrow$ (i) shows that the following more general result is true. The functional $\mathcal{F}$ is positive definite if and only if there exists a conjoined basis $Y = \binom{X}{U}$ of (SDS) with no focal points in the interval $(0, N+1]$. Indeed, for such a conjoined basis $Y = \binom{X}{U}$ of (SDS) and the corresponding symmetric $Q$ in (2.40) we have $R_k[Q] X_k = 0$ for $k \in [0, N]_{\mathbb{Z}}$. Then for any admissible $y = \binom{x}{u}$ with $x_0 = 0 = x_{N+1}$ we have by Lemma 2.33 that $x_k \in \operatorname{Im} X_k$ for all $k \in [0, N+1]_{\mathbb{Z}}$. The positivity of the functional $\mathcal{F}$ now follows from the Picone formula (Lemma 2.32) as in the proof of (v) $\Rightarrow$ (i) in Theorem 2.36. Conversely, the positivity of $\mathcal{F}$ implies condition (iv); hence, there indeed exists a conjoined basis $Y = \binom{X}{U}$ of (SDS) with no focal points in $(0, N+1]$, as we claim in this remark.

Next we present a construction of an admissible pair for (2.57) for which $\mathcal{F} \not\geq 0$, when kernel condition or $P$-condition in (2.37) are violated (compare with the proof of Theorem 2.44 in Sect. 2.3.5). This result also shows a direct proof of the implication (i) $\Rightarrow$ (iv) in Theorem 2.36 above.

**Proposition 2.38** *Let* $Y = \binom{X}{U}$ *be a conjoined basis of* (SDS).

(i) *If there exists* $m \in [0, N]_{\mathbb{Z}}$ *such that* $\operatorname{Ker} X_{m+1} \not\subseteq \operatorname{Ker} X_m$, *i.e., there exists* $\alpha \in \operatorname{Ker} X_{m+1} \setminus \operatorname{Ker} X_m$, *then the pair* $y = \binom{x}{u}$ *defined by*

$$
x_k := \begin{cases} X_k \alpha & 0 \leq k \leq m, \\ 0 & m+1 \leq k \leq N+1, \end{cases}
$$

$$
u_k := \begin{cases} U_k \alpha & 0 \leq k \leq m, \\ 0 & m+1 \leq k \leq N+1, \end{cases}
$$

*is admissible on* $[0, N+1]_{\mathbb{Z}}$, $x \not\equiv 0$, *and we have* $\mathcal{F}(y) = -\alpha^T X_0 U_0 \alpha$.

(ii) *If there exists* $m \in [0, N]_{\mathbb{Z}}$ *such that* $\operatorname{Ker} X_{m+1} \subseteq \operatorname{Ker} X_m$ *and* $P_m \not\geq 0$, *i.e., there exists* $c \in \mathbb{R}^n$ *such that* $c^T P_m c < 0$, *then for* $d := X_{m+1}^\dagger \mathcal{B}_m c$ *the pair* $y = \binom{x}{u}$ *defined by*

$$
x_k := \begin{cases} X_k d & 0 \leq k \leq m, \\ 0 & m+1 \leq k \leq N+1, \end{cases}
$$

$$
u_k := \begin{cases} U_k d & 0 \leq k \leq m-1, \\ U_k d - c & k = m, \\ 0 & m+1 \leq k \leq N+1, \end{cases}
$$

*is admissible on* $[0, N+1]_{\mathbb{Z}}$, $x \not\equiv 0$, *and* $\mathcal{F}(y) = -\alpha^T X_0 U_0 \alpha + c^T P_m c$.

(iii) *In particular, if* $Y = Y^{[0]}$ *is the principal solution of* (SDS) *at* $k = 0$, *then* $\mathcal{F}(y) = 0$ *holds in case* (i), *and* $\mathcal{F}(y) < 0$ *holds in case* (ii).

*Proof* We prove the statement (ii), the proof of (i) is similar. We have $x_0 = x_{N+1} = 0$, $x_{k+1} = \mathcal{A}_k x_k + \mathcal{B}_k u_k$ for $0 \leq k \leq N$ with $k \neq m$ and

$$
\mathcal{A}_m x_m + \mathcal{B}_m u_m = X_{m+1} d - \mathcal{B}_m c = 0 = x_{m+1},
$$

because $X_{m+1} X_{m+1}^\dagger \mathcal{B}_m = \mathcal{B}_m$ by (4.1) and (4.2). Hence $y = \binom{x}{u}$ is admissible. Moreover, $x_m = X_m d = P_m c \neq 0$, which shows that $x \not\equiv 0$ (in case (i) we use $x_m = X_m \alpha \neq 0$, since $\alpha \notin \operatorname{Ker} X_m$). Using Lemma 2.30 and

$$
\mathcal{C}_{m-1} x_{m-1} + \mathcal{D}_{m-1} u_{m-1} = U_m d = u_m + c,
$$

we obtain

$$\mathcal{F}(y) = \sum_{k=0}^{N} x_{k+1}^T \{\mathcal{C}_k x_k + \mathcal{D}_k u_k - u_{k+1}\}$$

$$= x_m^T \{\mathcal{C}_{m-1} x_{m-1} + \mathcal{D}_{m-1} u_{m-1} - u_m\} = c^T X_m d = c^T P_m c < 0.$$

Hence $\mathcal{F} \not\geq 0$ and the proof is complete. $\qquad\square$

The following result is a direct extension of the equivalence of conditions (i) and (iv) in Theorem 2.36.

**Theorem 2.39** *Let $Y$ be any conjoined basis of* (SDS). *Then $Y$ has no forward focal points in* $(0, N+1]$ *if and only if* $\mathcal{F}(y) + d^T X_0^T U_0 d > 0$ *for all admissible* $y = \left(\begin{smallmatrix} x \\ u \end{smallmatrix}\right)$ *with* $x_0 = X_0 d$, $x_{N+1} = 0$, *and* $x \not\equiv 0$.

*Proof* The necessity follows from the Picone identity (Lemma 2.32) similarly as the proof of the corresponding parts in Theorem 2.36. The sufficiency is a direct consequence of Proposition 2.38. $\qquad\square$

We conclude this subsection with a statement completing Theorem 2.36 with further conditions equivalent to the positivity of the functional $\mathcal{F}$ in (2.57). In contrast with conditions (iv) and (v) in the previously mentioned result, we now consider a conjoined basis $Y$ with $X_k$ invertible on the whole interval $[0, N+1]_{\mathbb{Z}}$ and the explicit Riccati equation (2.52).

**Theorem 2.40** *The following statements are equivalent.*

(i) *The quadratic functional $\mathcal{F}$ is positive definite, i.e., condition (i) in Theorem 2.36 holds.*

(ii) *There exists a conjoined basis $Y$ of* (SDS) *with no forward focal points in* $(0, N+1]$ *such that $X_k$ is invertible for all $k \in [0, N+1]_{\mathbb{Z}}$, i.e., $X_k X_{k+1}^{-1} \mathcal{B}_k \geq 0$ for $k \in [0, N]_{\mathbb{Z}}$.*

(iii) *The explicit Riccati equation (2.52), $k \in [0, N]_{\mathbb{Z}}$, has a symmetric solution $Q_k$ on $[0, N+1]_{\mathbb{Z}}$ such that*

$$\mathcal{A}_k + \mathcal{B}_k Q_k \text{ is invertible and } (\mathcal{A}_k + \mathcal{B}_k Q_k)^{-1} \mathcal{B}_k \geq 0, \quad k \in [0, N]_{\mathbb{Z}}.$$
$$(2.68)$$

*Proof* Assume that (i) holds. It follows by a perturbation argument in [258, Theorem 2(iii')] or by [172, Theorem 7] that there exists $\alpha > 0$ such that $\mathcal{F}(y) + \alpha \|x_0\|^2 > 0$ for all admissible $y = \left(\begin{smallmatrix} x \\ u \end{smallmatrix}\right)$ with $x_{N+1} = 0$ and $x \not\equiv 0$. Then, by Theorem 2.39, the conjoined basis $Y$ of (SDS) starting with the initial conditions $X_0 = (1/\alpha) I$ and $U_0 = I$ has no forward focal points in $(0, N+1]$. Since $X_0$ is invertible, the kernel condition in (2.37) implies that $X_k$ is invertible on $[0, N+1]_{\mathbb{Z}}$ and $X_k X_{k+1}^{-1} \mathcal{B}_k \geq 0$ on $[0, N]_{\mathbb{Z}}$, as we claim in part (ii). Next, if (ii) holds then we set $Q_k := U_k X_k^{-1}$ on $[0, N+1]_{\mathbb{Z}}$, which satisfies condition

(iii) by the Riccati equivalence in Theorem 2.28. Finally, if (iii) holds, then $P_k[Q] = (A_k + B_k Q_k)^{-1} B_k \geq 0$ on $[0, N]_{\mathbb{Z}}$ and the Picone formula (2.67) in Lemma 2.32 implies that the functional $\mathcal{F}$ is nonnegative definite, i.e., $\mathcal{F}(y) \geq 0$ for any admissible $y = \binom{x}{u}$ with $x_0 = 0 = x_{N+1}$. If now $\mathcal{F}(y) = 0$ for such an admissible $y$, then $P_k[Q] z_k = 0$ with $z_k := u_k - Q_k x_k$ on $[0, N]_{\mathbb{Z}}$, so that $(\mathcal{D}_k^T - B_k^T Q_{k+1}) x_{k+1} = x_k$ on $[0, N]_{\mathbb{Z}}$ by (2.66). From $x_{N+1} = 0$ we then get $x_k = 0$ for all $k \in [0, N+1]_{\mathbb{Z}}$, showing that the functional $\mathcal{F}$ is actually positive definite, i.e., condition (i) holds.                                                                □

The Reid roundabout theorem has its analogy for the time-reversed symplectic system (2.9).

**Theorem 2.41 (Reid Roundabout Theorem for Time-Reversed System)** *The following statements are equivalent:*

  (i) *The quadratic functional*

$$\mathcal{F}_R(y) := \sum_{k=0}^{N} y_{k+1}^T \{ S_k^{T-1} \mathcal{K} S_k^{-1} - \mathcal{K} \} y_{k+1} \qquad (2.69)$$

$$= \sum_{k=0}^{N} \{ -x_{k+1} C_k \mathcal{D}_k^T x_{k+1} + 2 x_{k+1}^T C_k B_k^T u_{k+1} - u_{k+1}^T A_k B_k^T u_{k+1} \} < 0$$

*for all $y = \binom{x}{u}$ such that $x_k = \mathcal{D}_k^T x_{k+1} - B_k^T u_{k+1}$ for $k \in [0, N]_{\mathbb{Z}}$ and such that $x_0 = 0 = x_{N+1}$ and $x \not\equiv 0$.*
 (ii) *System (2.9) is disconjugate on $[0, N+1]$, i.e., no solution of (2.9) has more than one and no solution $y = \binom{x}{u}$ of (2.9) with $x_{N+1} = 0$ has any generalized zeros in the interval $[0, N+1)$. Here the interval $[m, m+1)$ contains a generalized zero of a solution $y = \binom{x}{u}$ of (2.9) if*

$$x_{m+1} \neq 0, \quad x_m \in \operatorname{Im} B_m^T, \quad and \quad x_m^T B_m^\dagger x_{m+1} \leq 0.$$

(iii) *No solution $y = \binom{x}{u}$ of (2.9) with $x_{N+1} = 0$ has any generalized zero in the interval $[0, N+1)$.*
(iv) *The solution $Y = \binom{X}{U}$ of (2.9) with $Y_{N+1} = \binom{0}{-I}$ has no backward focal points in the interval $[0, N+1)$, i.e., condition (2.48) holds for all $k \in [0, N]_{\mathbb{Z}}$.*
 (v) *The equation $\tilde{R}_k[Q] \tilde{G}_k = 0$, $k \in [0, N]_{\mathbb{Z}}$, has a symmetric solution $Q$ defined on on $[0, N+1]_{\mathbb{Z}}$ such that $\tilde{P}_k[Q] \geq 0$ for all $k \in [0, N]_{\mathbb{Z}}$, where*

$$\tilde{R}_k[Q] := (Q_k B_k^T - A_k^T) Q_{k+1} + (Q_k \mathcal{D}_k^T - C_k^T),$$

$$\tilde{P}_k[Q] := B_k A_k^T - B_k Q_k B_k^T,$$

$$\tilde{G}_k := \left( \prod_{j=k+1}^{N-1} \mathcal{D}_j^T B_N^T \prod_{j=k+1}^{N-2} \mathcal{D}_j^T B_{N-1}^T \cdots \mathcal{D}_{k+1}^T B_{k+2}^T \ B_{k+1}^T \right).$$

*Proof* The proof is based on the transformation of the time-reversed system to a standard symplectic system according to Example 2.5, to which Theorem 2.36 is applied.                                                                                    □

*Remark 2.42* Similarly to Remark 2.37, condition (i) in Theorem 2.41 is equivalent with the existence of a conjoined basis $Y = \binom{X}{U}$ of (2.9) with no backward focal points in the interval $[0, N+1)$.

Actually, all the statements in Theorems 2.36 and 2.41 are mutually equivalent, as we show in the last statement of this subsection.

**Theorem 2.43** *The statements (i) of Theorem 2.36 and (i) of Theorem 2.41 are equivalent.*

*Proof* Assume that (i) of Theorem 2.36 holds and let $y = \binom{x}{u}$ be admissible for the functional $\mathcal{F}_R$ in (2.69) corresponding to (2.9), i.e., $x_k = \mathcal{D}_k^T x_{k+1} - \mathcal{B}_k^T u_{k+1}$ for $k \in [0, N]_{\mathbb{Z}}$, $x_0 = 0 = x_{N+1}$, and $x \not\equiv 0$. Put $y_k^* = \binom{x_k^*}{u_k^*} := \mathcal{S}_k^{-1} y_{k+1}$ for $k \in [0, N]_{\mathbb{Z}}$ and $x_{N+1}^* := 0$. Then $x_k^* = \mathcal{D}_k^T x_{k+1} - \mathcal{B}_k^T u_{k+1} = x_k$ for all $k \in [0, N]_{\mathbb{Z}}$ and $x_{N+1}^* = 0 = x_{N+1}$. It follows that $x_{k+1}^* = x_{k+1} = \mathcal{A}_k x_k^* + \mathcal{B}_k u_k^*$ holds for all $k \in [0, N]_{\mathbb{Z}}$ and (i) of Theorem 2.36 implies $0 < \mathcal{F}(y^*) = -\mathcal{F}_R(y)$. This shows that (i) of Theorem 2.41 is satisfied. Conversely, if (i) of Theorem 2.41 is true, then for an admissible $y = \binom{x}{u}$ for $\mathcal{F}$ such that $x_0 = 0 = x_{N+1}$ and $x \not\equiv 0$, we put $y_{k+1}^* := \mathcal{S}_k y_k$ for $k \in [0, N]_{\mathbb{Z}}$ and $y_0^* := 0$. Similarly as above, it follows by (i) of Theorem 2.41 that $0 > \mathcal{F}_R(y^*) = -\mathcal{F}(y)$, so that (i) of Theorem 2.36 holds, which completes the proof.                                                                     □

### 2.3.5  Nonnegative Quadratic Functional

In Sect. 1.3, we discussed the importance of the nonnegativity and positivity of discrete quadratic functionals in the second-order necessary and sufficient optimality conditions for discrete calculus of variations and optimal control problems (in particular, Theorems 1.26 and 1.32 contain necessary optimality conditions and Theorems 1.28 and 1.33 contain sufficient optimality conditions). Along with the results on the positivity of the discrete quadratic functional $\mathcal{F}$ in (2.57) associated with symplectic system (SDS) in Sect. 2.3.4, we now present conditions, which are equivalent with the nonnegativity of $\mathcal{F}$. We recall the definition of the matrix $P_k := X_k X_{k+1}^\dagger \mathcal{B}_k$ in (2.37), and also define the new matrices

$$M_k := (I - X_{k+1} X_{k+1}^\dagger) \mathcal{B}_k, \quad T_k := I - M_k^\dagger M_k. \tag{2.70}$$

Note that $T_k$ is symmetric. By Lemma 2.15 we know that $M_k = 0$ (and hence $T_k = I$) on $[0, N]_{\mathbb{Z}}$ if and only if the kernel condition

$$\operatorname{Ker} X_{k+1} \subseteq \operatorname{Ker} X_k, \quad k \in [0, N]_{\mathbb{Z}}, \tag{2.71}$$

holds. We note that the matrices $M_k$ and $T_k$ will be used in Sect. 4.1 in the definition of the multiplicities of forward focal points of $Y$ (see Definition 4.1).

**Theorem 2.44** *The quadratic functional $\mathcal{F}$ is nonnegative, i.e., $\mathcal{F}(y) \geq 0$ for every admissible $y = \binom{x}{u}$ with $x_0 = 0 = x_{N+1}$ if and only if the principal solution $Y^{[0]} = \binom{X^{[0]}}{U^{[0]}}$ of (SDS) satisfies the image condition*

$$x_k \in \operatorname{Im} X_k^{[0]}, \quad k \in [0, N+1]_{\mathbb{Z}}, \tag{2.72}$$

*for every admissible $y = \binom{x}{u}$ with $x_0 = 0 = x_{N+1}$ and the P-condition*

$$T_k P_k T_k \geq 0, \quad k \in [0, N]_{\mathbb{Z}}. \tag{2.73}$$

The main difference between the statements of Theorem 2.36 (condition (iv)) and the above theorem lies in the relationship between the kernel condition in (2.71) and the image condition (2.72). Of course, (2.71) implies (2.72), which follows from Lemma 2.33. The converse is not in general true, however. Furthermore, under (2.71) the P-condition (2.73) reduces to the one in (2.37).

Before presenting the proof of Theorem 2.44, we state several additional properties of the matrices $P_k$, $M_k$, and $T_k$ and derive a version of the Picone formula (see Lemma 2.32) under the image condition (2.72). The properties of Moore-Penrose inverses (see Sect. 1.6.2) imply that

$$X_{k+1}^T M_k = 0, \quad M_k^\dagger X_{k+1} = 0, \quad M_k T_k = 0, \quad \mathcal{B}_k T_k = X_{k+1} X_{k+1}^\dagger \mathcal{B}_k T_k. \tag{2.74}$$

With a given conjoined basis $Y = \binom{X}{U}$ we associate on $[0, N+1]_{\mathbb{Z}}$ the matrix

$$Q_k := U_k X_k^\dagger + (U_k X_k^\dagger)^T (I - X_k X_k^\dagger). \tag{2.75}$$

**Lemma 2.45** *Let $Y = \binom{X}{U}$ be a conjoined basis of (SDS), $Q_k$ be defined by (2.75), and let $y = \binom{x}{u}$ be an admissible pair satisfying the image condition (2.72). Then $Y$, $Q$, and $y$ satisfy the following conditions.*

*(i) The matrix $Q_k$ is symmetric and $Q_k X_k = U_k X^\dagger X_k$ on $[0, N+1]_{\mathbb{Z}}$.*
*(ii) The matrix $T_k P_k T_k$ is symmetric for all $k \in [0, N]_{\mathbb{Z}}$.*
*(iii) If $z_k := u_k - Q_k x_k$, then $M_k z_k = 0$ (i.e., $z_k = T_k z_k$) on $[0, N]_{\mathbb{Z}}$.*
*(iv) We have the Picone identity*

$$x_k^T C_k^T A_k x_k + 2 x_k^T C_k^T \mathcal{B}_k u_k + u_k^T D_k^T \mathcal{B}_k u_k = \Delta(x_k^T Q_k x_k) + z_k^T P_k z_k, \tag{2.76}$$

$$\{D_k^T - \mathcal{B}_k^T Q_{k+1}\} x_{k+1} = x_k + P_k z_k, \tag{2.77}$$

*for all $k \in [0, N]_{\mathbb{Z}}$, and consequently*

$$\mathcal{F}(y) = \sum_{k=0}^{N} z_k^T P_k z_k = \sum_{k=0}^{N} z_k^T T_k P_k T_k z_k. \tag{2.78}$$

*Proof* The matrix $Q_k$ in (2.75) is symmetric by the symmetry of $X_k^T U_k$ and $X_k X_k^{\dagger}$. The identity $Q_k X_k = U_k X^{\dagger} X_k$ follows from (2.75) trivially. By using the time-reversed symplectic system (2.10) and the fourth identity in (2.74), we obtain

$$P_k T_k = (\mathcal{D}_k^T - \mathcal{B}_k^T Q_{k+1}) \, \mathcal{B}_k T_k. \tag{2.79}$$

Since $\mathcal{D}_k^T \mathcal{B}_k$ is symmetric, it follows that the matrix $T_k P_k T_k$ is symmetric on $[0, N]_{\mathbb{Z}}$ as well. Since $x_k \in \operatorname{Im} X_k$, we have $x_k = X_k X_k^{\dagger} x_k$ on $[0, N+1]_{\mathbb{Z}}$. Then we have

$$
\begin{aligned}
M_k u_k &= (I - X_{k+1} X_{k+1}^{\dagger}) \, (x_{k+1} - \mathcal{A}_k x_k) \\
&= (I - X_{k+1} X_{k+1}^{\dagger}) \, (X_{k+1} X_{k+1}^{\dagger} x_{k+1} - \mathcal{A}_k X_k X_k^{\dagger} x_k) \\
&= (I - X_{k+1} X_{k+1}^{\dagger}) \, (\mathcal{B}_k U_k - X_{k+1}) \, X_k^{\dagger} x_k \\
&= M_k U_k X_k^{\dagger} x_k = M_k U_k X_k^{\dagger} X_k X_k^{\dagger} x_k = M_k Q_k x_k
\end{aligned}
$$

on $[0, N]_{\mathbb{Z}}$. This shows that $M_k z_k = 0$ on $[0, N]_{\mathbb{Z}}$. Finally, formulas (2.76) and (2.77) follow from the Picone formula in Lemma 2.32; since for the matrix $Q_k$ in (2.75) we have $X_{k+1}^T R_k[Q] X_k = 0$ on $[0, N]_{\mathbb{Z}}$ by Lemma 2.58(ii). $\qquad\square$

*Proof of Theorem 2.44* Let $Y := Y^{[0]}$ be the principal solution of (SDS) at $k = 0$, i.e., $X_0 = 0$ and $U_0 = I$. First we show that conditions (2.72) and (2.73) are necessary for the nonnegativity of the functional $\mathcal{F}$. Thus, assume that $\mathcal{F}(y) \geq 0$ for every admissible $y = \binom{x}{u}$ with $x_0 = 0 = x_{N+1}$. Assume that there exists an admissible $y = \binom{x}{u}$ with $x_0 = 0 = x_{N+1}$ and $m \in [1, N-1]_{\mathbb{Z}}$ such that $x_k \in \operatorname{Im} X_k$ for all $k \in [0, m]_{\mathbb{Z}}$ but $x_{m+1} \notin \operatorname{Im} X_{m+1}$. First we observe that the $n \times n$ matrices

$$K := X_{m+1}^T X_{m+1} + U_{m+1}^T U_{m+1} > 0, \quad S' := K^{-1} U_{m+1}^T M_{m+1}$$

satisfy the equations

$$X_{m+1} S' = 0, \quad U_{m+1} S' = M_m. \tag{2.80}$$

Indeed, for the $2n \times 2n$ matrix, $Z$ defined below has the property

$$Z := \begin{pmatrix} K^{-1} X_{m+1}^T & K^{-1} U_{m+1}^T \\ -U_{m+1}^T & X_{m+1}^T \end{pmatrix}, \quad Z^{-1} = \begin{pmatrix} X_{m+1} & -U_{m+1} K^{-1} \\ U_{m+1} & X_{m+1} K^{-1} \end{pmatrix},$$

so that

$$Z \begin{pmatrix} 0 \\ M_m \end{pmatrix} = \begin{pmatrix} S' \\ 0 \end{pmatrix}, \quad \begin{pmatrix} 0 \\ M_m \end{pmatrix} = Z^{-1} \begin{pmatrix} S' \\ 0 \end{pmatrix} = \begin{pmatrix} X_{m+1}S' \\ U_{m+1}S' \end{pmatrix}.$$

Next we prove that $M_m^T x_{m+1} \neq 0$. Assume that $M_m^T x_{m+1} = 0$. Then

$$\begin{aligned} 0 = M_m^T x_{m+1} &= \mathcal{B}_m^T (I - X_{m+1} X_{m+1}^\dagger) x_{m+1} \\ &= \mathcal{B}_m^T (I - X_{m+1} X_{m+1}^\dagger) (A_m X_m \alpha + \mathcal{B}_m u_m) \\ &= \mathcal{B}_m^T (I - X_{m+1} X_{m+1}^\dagger) [(X_{m+1} - \mathcal{B}_m U_m) \alpha + \mathcal{B}_m u_m] \\ &= \mathcal{B}_m^T (I - X_{m+1} X_{m+1}^\dagger)^2 \mathcal{B}_m (u_m - U_m \alpha) = M_m^T M_m (u_m - U_m \alpha). \end{aligned}$$

This implies that $\| M_m (u_m - U_m \alpha) \|^2 = 0$, i.e., $M_m (u_m - U_m \alpha) = 0$. Then in turn

$$\begin{aligned} 0 = M_m (u_m - U_m \alpha) &= (I - X_{m+1} X_{m+1}^\dagger)(\mathcal{B}_m u_m - \mathcal{B}_m U_m \alpha) \\ &= (I - X_{m+1} X_{m+1}^\dagger) [x_{m+1} - A_m X_m \alpha - (X_{m+1} - A_m X_m) \alpha] \\ &= (I - X_{m+1} X_{m+1}^\dagger)(x_{m+1} - X_{m+1} \alpha) = (I - X_{m+1} X_{m+1}^\dagger) x_{m+1}. \end{aligned}$$

Therefore, $x_{m+1} = X_{m+1} X_{m+1}^\dagger x_{m+1} \in \operatorname{Im} X_{m+1}$, which is a contradiction. Therefore, $M_m^T x_{m+1} \neq 0$ must hold. Now we define the sequence $\tilde{y} = \begin{pmatrix} \tilde{x} \\ \tilde{u} \end{pmatrix}$ as follows:

$$\tilde{x}_k := \begin{cases} X_k (\alpha + \tilde{\alpha}), \, k \in [0, m]_{\mathbb{Z}}, \\ x_k, \qquad\quad k \in [m+1, N+1]_{\mathbb{Z}}, \end{cases}$$

$$\tilde{u}_k := \begin{cases} U_k (\alpha + \tilde{\alpha}), \, k \in [0, m-1]_{\mathbb{Z}}, \\ u_m + U_m \tilde{\alpha}, \, k = m, \\ u_k, \qquad\quad k \in [m+1, N+1]_{\mathbb{Z}}, \end{cases}$$

where $\tilde{\alpha} := t \, S' M_m^T x_{m+1}$ with $t \in \mathbb{R}$, $t \neq 0$. Now $\tilde{x}_0 = X_0 (\alpha + \tilde{\alpha}) = 0$ and $\tilde{x}_{N+1} = x_{N+1} = 0$. For $k \in [0, m-1]_{\mathbb{Z}}$ we have

$$A_k \tilde{x}_k + \mathcal{B}_k \tilde{u}_k = (A_k X_k + \mathcal{B}_k U_k)(\alpha + \tilde{\alpha}) = X_{m+1}(\alpha + \tilde{\alpha}) = \tilde{x}_{k+1},$$

for $k \in [m+1, N]_{\mathbb{Z}}$ we have

$$A_k \tilde{x}_k + \mathcal{B}_k \tilde{u}_k = A_k x_k + \mathcal{B}_k u_k = x_{k+1} = \tilde{x}_{k+1},$$

and for $k = m$ we have

$$A_k \tilde{x}_k + B_k \tilde{u}_k = A_m X_m (\alpha + \tilde{\alpha}) + B_m (u_m + U_m \tilde{\alpha})$$
$$= A_m x_m + B_m u_m + (A_m X_m + B_m U_m) \tilde{\alpha}$$
$$= x_{m+1} + X_{m+1} \tilde{\alpha} = x_{m+1} = \tilde{x}_{m+1},$$

because $X_{m+1} \tilde{\alpha} = t\, X_{m+1} S' M_m^T x_{m+1} = 0$ by (2.80). Therefore, $\tilde{y}$ is admissible with $\tilde{x}_0 = 0 = \tilde{x}_{N+1}$. It follows by Lemma 2.30 and by splitting the total sum over $[0, N]_\mathbb{Z}$ into four sums over the sets $[0, m - 2]_\mathbb{Z}, \{m - 1\}, \{m\}, [m + 1, N]_\mathbb{Z}$ that

$$\mathcal{F}(\tilde{y}) = \left\{ \sum_{k=0}^{m-2} + \sum_{k=m-1}^{m-1} + \sum_{k=m}^{m} + \sum_{k=m+1}^{N} \right\} \tilde{x}_{k+1}^T (\mathcal{C}_k \tilde{x}_k + \mathcal{D}_k \tilde{u}_k - \tilde{u}_{k+1})$$

$$= \tilde{x}_m^T (\mathcal{C}_{m-1} \tilde{x}_{m-1} + \mathcal{D}_{m-1} \tilde{u}_{m-1} - \tilde{u}_m) + \tilde{x}_{m+1}^T (\mathcal{C}_m \tilde{x}_m + \mathcal{D}_m \tilde{u}_m - \tilde{u}_{m+1})$$

$$+ \sum_{k=m+1}^{N} \tilde{x}_{k+1}^T (\mathcal{C}_k \tilde{x}_k + \mathcal{D}_k \tilde{u}_k - \tilde{u}_{k+1})$$

$$= \mathcal{F}_1(\alpha) + \mathcal{F}_2(\alpha, \tilde{\alpha}),$$

where

$$\mathcal{F}_1(\alpha) := x_{m+1}^T (\mathcal{C}_m X_m \alpha + \mathcal{D}_m u_m - u_{m+1}) + \alpha^T X_m^T (U_m \alpha - u_m)$$

$$+ \sum_{k=m+1}^{N} x_{k+1}^T (\mathcal{C}_k x_k + \mathcal{D}_k u_k - u_{k+1}),$$

$$\mathcal{F}_2(\alpha, \tilde{\alpha}) := \tilde{\alpha}^T X_m^T (U_m \alpha - u_m) + x_{m+1}^T U_{m+1} \tilde{\alpha},$$

i.e., the term $\mathcal{F}_1(\alpha)$ does not depend on $\tilde{\alpha}$ (and hence on $t$), while the term $\mathcal{F}_2(\alpha, \tilde{\alpha})$ does depend on $\tilde{\alpha}$ (and hence on $t$). Since $X_{m+1} \tilde{\alpha} = 0$, $X_m - \mathcal{D}_m^T X_{m+1} - \mathcal{B}_m^T U_{m+1}$ by (2.10), and $U_{m+1}^T X_{m+1}$ is symmetric, the first term in $\mathcal{F}_2(\alpha, \tilde{\alpha})$ is equal to

$$\tilde{\alpha}^T X_m^T (U_m \alpha - u_m) = \tilde{\alpha}^T (X_{m+1}^T \mathcal{D}_m - U_{m+1}^T \mathcal{B}_m) (U_m \alpha - u_m)$$

$$= \tilde{\alpha}^T U_{m+1}^T (\mathcal{B}_m u_m - \mathcal{B}_m U_m \alpha)$$

$$= \tilde{\alpha}^T U_{m+1}^T [x_{m+1} - A_m X_m \alpha - (X_{m+1} - A_m X_m) \alpha]$$

$$= \tilde{\alpha}^T U_{m+1}^T (x_{m+1} - X_{m+1} \alpha) = \tilde{\alpha}^T U_{m+1}^T x_{m+1}.$$

Hence, by (2.80) we get the expression

$$\mathcal{F}_2(\alpha, \tilde{\alpha}) = \tilde{\alpha}^T U_{m+1}^T x_{m+1} + x_{m+1}^T U_{m+1} \tilde{\alpha} = 2t \, x_{m+1}^T U_{m+1}^T S' M_m^T x_{m+1}$$

$$= 2t \, x_{m+1}^T M_m M_m^T x_{m+1} = 2t \, \|M_m^T x_{m+1}\|^2,$$

as well as

$$\mathcal{F}(\tilde{y}) = \mathcal{F}_1(\alpha) + \mathcal{F}_2(\alpha, \tilde{\alpha}) = \mathcal{F}_1(\alpha) + 2t \, \|M_m^T x_{m+1}\|^2.$$

But since $M_m^T x_{m+1} \neq 0$, it follows that $\mathcal{F}(\tilde{y}) \to -\infty$ as $t \to -\infty$, which contradicts the assumed nonnegativity of the functional $\mathcal{F}$.

Next we prove that the $P$-condition (2.73) is necessary for the nonnegativity of $\mathcal{F}$. Suppose that there exists $c \in \mathbb{R}^n$, $c \neq 0$, such that $c^T T_m P_m T_m c < 0$ for some $m \in [0, N]_{\mathbb{Z}}$. Set $d := X_{m+1}^\dagger \mathcal{B}_m T_m c$ and define the pair $y = \binom{x}{y}$ by

$$x_k := \begin{cases} X_k d, & k \in [0, m-1]_{\mathbb{Z}}, \\ P_m T_m c, & k = m, \\ 0, & k \in [m+1, N+1]_{\mathbb{Z}}, \end{cases}$$

$$u_k := \begin{cases} U_k d, & k \in [0, m-1]_{\mathbb{Z}}, \\ -\mathcal{A}_m^T (\mathcal{D}_m - Q_{m+1} \mathcal{B}_m) \, T_m c, & k = m, \\ 0, & k \in [m+1, N]_{\mathbb{Z}}, \end{cases}$$

where $Q_{m+1}$ is defined by (2.75). Then $y$ is admissible for $k \in [0, m-2]_{\mathbb{Z}}$ as well as for $k \in [m+1, N]_{\mathbb{Z}}$. For $k = m-1$ we have

$$\mathcal{A}_{m-1} x_{m-1} + \mathcal{B}_{m-1} u_{m-1} = (\mathcal{A}_{m-1} X_{m-1} + \mathcal{B}_{m-1} U_{m-1}) \, d = X_m d$$

$$= X_m X_{m+1}^\dagger \mathcal{B}_m T_m c = P_m T_m c = x_m,$$

while for $k = m$ we use (1.145) and (2.79) to obtain

$$\mathcal{A}_m x_m + \mathcal{B}_m u_m = \mathcal{A}_m P_m T_m c - \mathcal{B}_m \mathcal{A}_m^T (\mathcal{D}_m - Q_{m+1} \mathcal{B}_m) \, T_m c$$

$$= \mathcal{A}_m (P_m T_m - P_m T_m) \, c = 0 = x_{m+1}.$$

Thus, $y$ is admissible for $\mathcal{F}$. Moreover, $x_0 = X_0 d = 0$ and $x_{N+1} = 0$. To compute the value of $\mathcal{F}(y)$ we use Lemma 2.30, (2.79), and (2.74) to obtain

$$\mathcal{F}(y) = \sum_{k=0}^{N} x_{k+1}^T (\mathcal{C}_k x_k + \mathcal{D}_k u_k - u_{k+1})$$

$$= \sum_{k=0}^{m-1} x_{k+1}^T (\mathcal{C}_k x_k + \mathcal{D}_k u_k - u_{k+1}) = x_m^T (\mathcal{C}_{m-1} x_{m-1} + \mathcal{D}_{m-1} u_{m-1} - u_m)$$

$$= d^T X_m^T \left\{ (\mathcal{C}_{m-1} X_{m-1} + \mathcal{D}_{m-1} U_{m-1}) d + \mathcal{A}_m^T (\mathcal{D}_m - \mathcal{Q}_{m+1} \mathcal{B}_m) T_m c \right\}$$

$$= d^T X_m^T U_m d + c^T T_m P_m^T \mathcal{A}_m^T (\mathcal{D}_m - \mathcal{Q}_{m+1} \mathcal{B}_m) T_m c$$

$$= d^T X_m^T U_m d + c^T T_m \mathcal{B}_m^T (X_{m+1}^\dagger)^T (X_{m+1}^T - U_m^T \mathcal{B}_m^T) (\mathcal{D}_m - \mathcal{Q}_{m+1} \mathcal{B}_m) T_m c$$

$$= d^T X_m^T U_m d + c^T T_m (\mathcal{B}_m^T \mathcal{D}_m - \mathcal{B}_m^T \mathcal{Q}_{m+1} \mathcal{B}_m) T_m c - d^T U_m^T P_m T_m c$$

$$= c^T T_m P_m T_m c < 0.$$

This contradicts the assumed nonnegativity of the functional $\mathcal{F}$.

The remaining part of the proof is devoted to showing that (2.72) and (2.73) are sufficient conditions for the nonnegativity of $\mathcal{F}$. Thus, let $y = \binom{x}{u}$ be admissible with $x_0 = 0 = x_{N+1}$. Then by (2.72) the sequence $y$ satisfies $x_k \in \operatorname{Im} X_k$ for all $k \in [0, N+1]_{\mathbb{Z}}$. In turn, the Picone formula (2.78) in Lemma 2.45 together with condition (2.73) imply that

$$\mathcal{F}(y) = \sum_{k=0}^{N} z_k^T P_k z_k \geq 0.$$

The proof of Theorem 2.44 is complete. $\qquad\square$

*Remark 2.46* In view of the notion of multiplicities of focal points introduced in Definition 4.1 in Sect. 4.1, the statement in Theorem 2.44 says that the functional $\mathcal{F}$ is nonnegative if and only if the principal solution $Y^{[0]}$ of (SDS) at $k = 0$ has no forward focal points in the intervals $(k, k+1)$ for every $k \in [0, N]_{\mathbb{Z}}$ and the image condition (2.72) holds.

We conclude this subsection with the analysis of the image condition (2.72). It will be further utilized in Sects. 5.4.1 and 5.4.2. We recall the definition of the matrix $M_k$ in (2.70) and its properties in (2.74).

**Lemma 2.47** *Suppose that $Y$ is a conjoined basis of system* (SDS) *and fix an index $k \in [0, N]_{\mathbb{Z}}$. Then we have*

(i) *the condition $x_{k+1} \in \operatorname{Im} X_{k+1}$ implies that $M_k^T x_{k+1} = 0$,*
(ii) *the equalities $x_{k+1} = \mathcal{A}_k x_k + \mathcal{B}_k u_k$, $M_k^T x_{k+1} = 0$, and $x_k \in \operatorname{Im} X_k$ imply that $x_{k+1} \in \operatorname{Im} X_{k+1}$.*

*Proof* First, the condition $x_{k+1} = X_{k+1} c \in \operatorname{Im} X_{k+1}$ implies by (2.74) that $M_k^T x_{k+1} = M_k^T X_{k+1} c = 0$. Hence, (i) is true. Next, if $x_{k+1} = \mathcal{A}_k x_k + \mathcal{B}_k u_k$ holds together with conditions $x_k = X_k c \in \operatorname{Im} X_k$ and $M_k^T x_{k+1} = 0$, then

$$0 = M_k^T x_{k+1} = M_k^T (\mathcal{A}_k X_k c + \mathcal{B}_k u_k) = M_k^T [X_{k+1} c + \mathcal{B}_k (u_k - U_k c)]$$

$$= M_k^T \mathcal{B}_k (u_k - U_k c) = \mathcal{B}_k^T (I - X_{k+1} X_{k+1}^\dagger)^T (I - X_{k+1} X_{k+1}^\dagger) \mathcal{B}_k (u_k - U_k c)$$

$$= M_k^T M_k (u_k - U_k c),$$

so that

$$0 = M_k(u_k - U_k c) = (I - X_{k+1} X_{k+1}^\dagger) \mathcal{B}_k(u_k - U_k c).$$

Therefore, the equality

$$x_{k+1} = X_{k+1} c + \mathcal{B}_k(u_k - U_k c) = X_{k+1} c + X_{k+1} X_{k+1}^\dagger \mathcal{B}_k(u_k - U_k c) \in \operatorname{Im} X_{k+1}$$

holds, which proves part (ii). □

The following result provides a characterization of the image condition in terms of the matrices $M_k$. The importance of this condition is highlighted in Theorem 2.44, where we discuss the nonnegativity of the quadratic functional $\mathcal{F}$. Note that for $x_0 = 0 = x_{N+1}$, the two inclusions $x_0 \in \operatorname{Im} X_0$ and $x_{N+1} \in \operatorname{Im} X_{N+1}$ are satisfied trivially.

**Lemma 2.48** *Suppose that $Y$ is a conjoined basis of the symplectic system* (SDS) *and let $y = (x, u)$ be admissible with $x_0 = 0 = x_{N+1}$. Then*

$$x_k \in \operatorname{Im} X_k \quad \text{for all } k \in [1, N]_{\mathbb{Z}} \tag{2.81}$$

*holds if and only if*

$$M_k^T x_{k+1} = 0 \quad \text{for all } k \in [0, N-1]_{\mathbb{Z}}. \tag{2.82}$$

*Proof* First, (2.81) implies (2.82) by Lemma 2.47(i). Next, suppose that $y = (x, u)$ is admissible, $x_0 = 0 = x_{N+1}$, and (2.82) holds. Then $x_0 = 0 \in \operatorname{Im} X_0$, and inductively we get by Lemma 2.47(ii) that $x_{k+1} \in \operatorname{Im} X_{k+1}$ for all $k \in [0, N-1]_{\mathbb{Z}}$, i.e., (2.81) is satisfied. □

### 2.3.6 Quadratic Functionals with General Endpoints

In this subsection we present extensions of the equivalence (i) and (iv) in Theorem 2.36 (positivity of $\mathcal{F}$) and of Theorem 2.44 (nonnegativity of $\mathcal{F}$) to quadratic functionals with separated and jointly varying (or coupled) boundary conditions. Thus, we consider the matrices $R_0, R_0^*, R_{N+1}, R_{N+1}^*, \Gamma_0, \Gamma_{N+1} \in \mathbb{R}^{n \times n}$ satisfying

$$\left.\begin{aligned}
R_0^* R_0^T &= R_0 (R_0^*)^T, \quad \operatorname{rank}(R_0^*, R_0) = n, \\
R_{N+1}^* R_{N+1}^T &= R_{N+1}(R_{N+1}^*)^T, \quad \operatorname{rank}(R_{N+1}^*, R_{N+1}) = n, \\
\Gamma_0 &:= -R_0^\dagger R_0^* R_0^\dagger R_0, \quad \Gamma_{N+1} := R_{N+1}^\dagger R_{N+1}^* R_{N+1}^\dagger R_{N+1},
\end{aligned}\right\} \tag{2.83}$$

where the matrices $\Gamma_0$ and $\Gamma_{N+1}$ are symmetric. We will investigate the definiteness of the quadratic functional

$$\mathcal{G}(y) := x_0^T \, \Gamma_0 \, x_0 + x_{N+1}^T \, \Gamma_{N+1} \, x_{N+1} + \mathcal{F}(y), \qquad (2.84)$$

over admissible sequences $y = (x, u)$ satisfying the separated boundary conditions

$$x_0 \in \operatorname{Im} R_0^T, \quad x_{N+1} \in \operatorname{Im} R_{N+1}^T, \qquad (2.85)$$

where $\mathcal{F}(y)$ is the quadratic functional defined in (2.57). We will also utilize a special conjoined basis $\bar{Y} = \begin{pmatrix} \bar{X} \\ \bar{U} \end{pmatrix}$ of (SDS), called the *natural conjoined basis*, which is given by the initial conditions $\bar{Y}_0 = (-R_0, \ R_0^*)^T$, i.e.,

$$\bar{X}_0 = -R_0^T, \quad \bar{U}_0 = (R_0^*)^T. \qquad (2.86)$$

Note that for $R_0 = 0 = R_{N+1}$ and $R_0^* = I = R_{N+1}^*$ the separated boundary conditions (2.85) reduce to the Dirichlet boundary conditions $x_0 = 0 = x_{N+1}$, and the natural conjoined basis $\bar{Y}$ reduces to the principal solution $Y^{[0]}$ at $k = 0$.

**Theorem 2.49 (Nonnegativity for Separated Endpoints)** *Assume that the conditions in (2.83) hold. Then the following statement are equivalent.*

(i) *The quadratic functional $\mathcal{G}$ in (2.84) is nonnegative over the separated endpoints (2.85), i.e., $\mathcal{G}(y) \geq 0$ for all admissible sequences $y = (x, u)$ satisfying (2.85).*

(ii) *The natural conjoined basis $\bar{Y}$ of (SDS), given by the initial conditions (2.86), satisfies the following three conditions: the image condition*

$$x_k \in \operatorname{Im} \bar{X}_k, \quad k \in [0, N + 1]_{\mathbb{Z}}, \qquad (2.87)$$

*for every admissible $y = (x, u)$ satisfying (2.85), the P-condition (2.73), and the final endpoint inequality*

$$\Gamma_{N+1} + \bar{U}_{N+1} \bar{X}_{N+1}^{\dagger} \geq 0 \quad \text{on } \operatorname{Im} R_{N+1}^T \cap \operatorname{Im} \bar{X}_{N+1}. \qquad (2.88)$$

*Proof* The result is proven by a matrix diagonalization argument in [50, Theorem 2]. Alternatively, we can use the Picone identity (Lemma 2.45) or the transformation of this problem to the Dirichlet endpoints described in Sect. 5.2.2 and apply Theorem 2.44. $\qquad \square$

For the positivity of the functional $\mathcal{G}$, we use the appropriate strengthening of the conditions in Theorem 2.49.

**Theorem 2.50 (Positivity for Separated Endpoints)** *Assume that the conditions in (2.83) hold. Then the following statement are equivalent.*

(i) *The quadratic functional $\mathcal{G}$ in (2.84) is positive over the separated endpoints (2.85), i.e., $\mathcal{G}(y) > 0$ for all admissible sequences $y = (x, u)$ satisfying (2.85) and $x \not\equiv 0$.*

(ii) *The natural conjoined basis $\bar{Y}$ of (SDS), given by the initial conditions (2.86), has no forward focal points in the interval $(0, N + 1]$ and satisfies the final endpoint inequality*

$$\Gamma_{N+1} + \bar{U}_{N+1}\bar{X}_{N+1}^\dagger > 0 \quad \text{on } \operatorname{Im} R_{N+1}^T \cap \operatorname{Im} \bar{X}_{N+1}. \tag{2.89}$$

*Proof* The result is proven by using the Picone formula (Lemma 2.32) in [50, Theorem 2]. Alternatively, we can use the transformation of this problem to the Dirichlet endpoints described in Sect. 5.2.2 and apply the equivalence of conditions (i) and (iv) in Theorem 2.36.                                                                 □

*Remark 2.51* The final endpoint inequality in (2.88) is equivalent to

$$\bar{X}_{N+1}^T(\Gamma_{N+1}\bar{X}_{N+1} + \bar{U}_{N+1}) \geq 0 \quad \text{on } \operatorname{Ker}[(I - R_{N+1}^\dagger R_{N+1})\, \bar{X}_{N+1}]. \tag{2.90}$$

Similarly, the final endpoint inequality in (2.89) is equivalent to (2.90) together with the condition

$$\left. \begin{aligned} &\operatorname{Ker}[R_{N+1}^\dagger R_{N+1}(\Gamma_{N+1}\bar{X}_{N+1} + \bar{U}_{N+1})] \\ &\qquad \cap \operatorname{Ker}[(I - R_{N+1}^\dagger R_{N+1})\,\bar{X}_{N+1}] \subseteq \operatorname{Ker}\bar{X}_{N+1}. \end{aligned} \right\} \tag{2.91}$$

These conditions are used in [174, Theorem 2] and [181, Theorem 5] (with the matrix $\mathcal{M} := I - R_{N+1}^\dagger R_{N+1}$).

For the case of joint (or coupled) boundary conditions, we consider the matrices $R_1, R_2, \Gamma \in \mathbb{R}^{2n \times 2n}$ satisfying

$$\left. \begin{aligned} R_1 R_2^T = R_2 R_1^T, \quad \operatorname{rank}(R_1, R_2) = 2n, \\ \Gamma := R_2^\dagger R_1 R_2^\dagger R_2, \end{aligned} \right\} \tag{2.92}$$

where the matrix $\Gamma$ is symmetric. We will investigate the definiteness of the quadratic functional

$$\mathcal{G}(y) := \begin{pmatrix} x_0 \\ x_{N+1} \end{pmatrix}^T \Gamma \begin{pmatrix} x_0 \\ x_{N+1} \end{pmatrix} + \mathcal{F}(y), \tag{2.93}$$

over admissible sequences $y = (x, u)$ satisfying the joint boundary conditions

$$\begin{pmatrix} x_0 \\ x_{N+1} \end{pmatrix} \in \operatorname{Im} R_2^T, \tag{2.94}$$

where $\mathcal{F}(y)$ is the quadratic functional defined in (2.57). We note that for the choice of $R_1 = \mathrm{diag}\{R_0^*, R_{N+1}^*\}$ and $R_2 = \mathrm{diag}\{R_0, R_{N+1}\}$ the joint boundary conditions (2.94) reduce to the separated boundary conditions (2.85). Also, for the choice of $R_1 = I_{2n}$ and $R_2 = 0_{2n}$, the joint boundary conditions (2.94) reduce to the Dirichlet boundary conditions $x_0 = 0 = x_{N+1}$.

The results below are formulated in terms of the principal solution $Y^{[0]}$ (SDS) at $k = 0$ and the associated conjoined basis $\tilde{Y}$, which are given by the initial conditions

$$Y_0^{[0]} = (0, \ I)^T, \quad \tilde{Y}_0 = (I, \ 0)^T, \tag{2.95}$$

and which form the symplectic fundamental matrix $Z = (\tilde{Y}, \ Y^{[0]})$ of (SDS) with $Z_0 = I_{2n}$. Then we define the $2n \times 2n$ matrices

$$\mathcal{X}_k := \begin{pmatrix} 0 & I \\ X_k^{[0]} & \tilde{X}_k \end{pmatrix}, \quad \mathcal{U}_k := \begin{pmatrix} -I & 0 \\ U_k^{[0]} & \tilde{U}_k \end{pmatrix}, \quad k \in [0, N+1]_{\mathbb{Z}}. \tag{2.96}$$

**Theorem 2.52 (Nonnegativity for Joint Endpoints)**  *Assume that the conditions in (2.92) hold. Then the following statement are equivalent.*

(i) *The quadratic functional $\mathcal{G}$ in (2.93) is nonnegative over the joint endpoints (2.94), i.e., $\mathcal{G}(y) \geq 0$ for all admissible sequences $y = (x, u)$ satisfying (2.94).*

(ii) *The principal solution $Y^{[0]}$ of (SDS) together with the associated conjoined basis $\tilde{Y}$, given by the initial conditions (2.95), satisfies the following three conditions: the image condition*

$$x_k - \tilde{X}_k x_0 \in \mathrm{Im}\, X_k^{[0]}, \quad k \in [0, N+1]_{\mathbb{Z}}, \tag{2.97}$$

*for every admissible $y = (x, u)$ satisfying (2.94), the P-condition (2.73), and the final endpoint inequality*

$$\Gamma + \mathcal{U}_{N+1} \mathcal{X}_{N+1}^\dagger \geq 0 \quad \text{on } \mathrm{Im}\, R_2^T \cap \mathrm{Im}\, \mathcal{X}_{N+1}, \tag{2.98}$$

*where $\mathcal{X}_{N+1}$ and $\mathcal{U}_{N+1}$ are defined in (2.96).*

*Proof*  The result is proven in [174, Theorem 2] by transforming this problem to the separated endpoints (see Sect. 5.2.2) and applying Theorem 2.49. □

**Theorem 2.53 (Positivity for Joint Endpoints)**  *Assume that the conditions in (2.83) hold. Then the following statement are equivalent.*

(i) *The quadratic functional $\mathcal{G}$ in (2.93) is positive over the joint endpoints (2.94), i.e., $\mathcal{G}(y) > 0$ for all admissible sequences $y = (x, u)$ satisfying (2.94) and $x \not\equiv 0$.*

(ii) *The principal solution $Y^{[0]}$ of (SDS) has no forward focal points in the interval $(0, N+1]$ and satisfies together with the associated conjoined basis $\tilde{Y}$, given* *by the initial conditions* (2.95), *the final endpoint inequality*

$$\Gamma + \mathcal{U}_{N+1}\mathcal{X}_{N+1}^{\dagger} > 0 \quad on \ \operatorname{Im} R_2^T \cap \operatorname{Im} \mathcal{X}_{N+1}, \tag{2.99}$$

*where* $\mathcal{X}_{N+1}$ *and* $\mathcal{U}_{N+1}$ *are defined in* (2.96).

*Proof* The result is proven in [181, Theorem 10] by transforming this problem to the separated endpoints (see Sect. 5.2.2) and applying Theorem 2.50. $\qquad\square$

Further conditions, which are equivalent to the positivity and nonnegativity of the quadratic functionals $\mathcal{G}$ over the separated or joint endpoints (2.85) or (2.94), are discussed in Sect. 2.7.

## 2.4  Riccati-Type Inequalities

In this section we present the results concerning inequalities involving solutions of the Riccati difference equation (2.52) or solutions of a discrete time inequality involving the Riccati operator $R_k[Q]$.

### 2.4.1  Inequalities for Riccati-Type Quotients

This subsection contains a comparison result for solutions of two discrete symplectic systems and the corresponding Riccati-type quotients. It is a generalization of the corresponding result for linear Hamiltonian difference systems (2.15) known in [51, Theorem 1]. Therefore, with system (2.15) we consider another system of the same form

$$\Delta \begin{pmatrix} x_k \\ u_k \end{pmatrix} = \mathcal{J}\underline{\mathcal{H}}_k \begin{pmatrix} x_{k+1} \\ u_k \end{pmatrix}, \tag{2.100}$$

where

$$\underline{\mathcal{H}}_k = \underline{\mathcal{H}}_k^T, \quad \underline{\mathcal{H}}_k = \begin{pmatrix} -\underline{C}_k & \underline{A}_k^T \\ \underline{A}_k & \underline{B}_k \end{pmatrix}, \quad \det(I - \underline{A}_k) \neq 0. \tag{2.101}$$

**Theorem 2.54 (Riccati Inequality for Two Hamiltonian Systems)** *Let us assume that the coefficients of systems* (2.15) *and* (2.100) *satisfy* (2.16) *and* (2.101) *and*

$$\mathcal{H}_k \geq \underline{\mathcal{H}}_k, \quad B_k \geq \underline{B}_k \underline{B}_k^{\dagger} B_k, \quad \operatorname{Ker} B_k \subseteq \operatorname{Ker} \underline{B}_k, \quad k \in [0, N]_{\mathbb{Z}}. \tag{2.102}$$

*Let* $Y$ *and* $\underline{Y}$ *be conjoined bases of systems* (2.15) *and* (2.100), *respectively, and that* $Q_k$ *and* $\underline{Q}_k$ *are symmetric with* $X_k^T Q_k X_k = U_k^T X_k$ *and* $\underline{X}_k^T \underline{Q}_k \underline{X}_k = \underline{U}_k^T X_k$

*for* $k \in [0, N+1]_{\mathbb{Z}}$. *If* $\operatorname{Im} \underline{X}_0 \subseteq \operatorname{Im} X_0$ *and* $\underline{X}_0^T (\underline{Q}_0 - Q_0) \underline{X}_0^T \geq 0$, *and if the conjoined basis* $Y$ *has no forward focal points in* $(0, N+1]$, *then* $\underline{Y}$ *has no forward focal points in* $(0, N+1]$ *either and*

$$\operatorname{Im} \underline{X}_k \subseteq \operatorname{Im} X_k, \quad \underline{X}_k^T (\underline{Q}_k - Q_k) \underline{X}_k^T \geq 0, \quad k \in [0, N+1]_{\mathbb{Z}}.$$

Below we present an extension of Theorem 2.54 to symplectic systems. Hence, with system (SDS) we consider another discrete symplectic system

$$\underline{y}_{k+1} = \underline{S}_k \, \underline{y}_k, \tag{2.103}$$

where

$$\underline{S}_k^T \mathcal{J} \underline{S}_k = \mathcal{J}, \quad \underline{S}_k = \begin{pmatrix} \underline{A}_k & \underline{B}_k \\ \underline{C}_k & \underline{D}_k \end{pmatrix}.$$

Let $\underline{\mathcal{F}}$ be the discrete quadratic functional, as in (2.57) and (2.63), corresponding to symplectic system (2.103) with the associated symmetric matrices $\underline{\mathcal{E}}_k$ and $\underline{\mathcal{G}}_k$, which are defined in a parallel way as the matrices $\mathcal{E}_k$ and $\mathcal{G}_k$ in (2.64). Admissible pairs $(\underline{x}, \underline{u})$ are now defined by the equation $\underline{x}_{k+1} = \underline{A}_k \underline{x}_k + \underline{B}_k \underline{u}_k$ for $k \in [0, N]_s \mathbb{Z}$, and we shall emphasize this fact by saying that $(\underline{x}, \underline{u})$ is *admissible with respect to* $(\underline{A}, \underline{B})$.

**Theorem 2.55 (Riccati Inequality for Two Symplectic Systems)** *Let* $Y$ *and* $\underline{Y}$ *be any conjoined bases of* (SDS) *and* (2.103), *respectively. Furthermore, let* $Q_k$ *and* $\underline{Q}_k$ *be symmetric matrices such that* $X_k^T Q_k X_k = X_k^T U_k$ *and* $\underline{X}_k^T \underline{Q}_k \underline{X}_k = \underline{X}_k^T \underline{U}_k$ *on* $[0, N+1]_{\mathbb{Z}}$, *and conditions*

$$\operatorname{Im} \left( \underline{A}_k - A_k \, \underline{B}_k \right) \subseteq \operatorname{Im} \mathcal{B}_k \quad k \in [0, N]_{\mathbb{Z}}, \tag{2.104}$$

$$\operatorname{Im} \underline{X}_0 \subseteq \operatorname{Im} X_0, \quad \underline{X}_0^T (\underline{Q}_0 - Q_0) \underline{X}_0 \geq 0, \tag{2.105}$$

$$\mathcal{G}_k \leq \underline{\mathcal{G}}_k \quad k \in [0, N]_{\mathbb{Z}}, \tag{2.106}$$

*hold. If* $Y$ *has no forward focal points in* $(0, N+1]$, *then* $\underline{Y}$ *has no forward focal points in* $(0, N+1]$ *either, and*

$$\operatorname{Im} \underline{X}_k \subseteq \operatorname{Im} X_k, \quad \underline{X}_k^T (\underline{Q}_k - Q_k) \underline{X}_k \geq 0, \quad k \in [0, N+1]_{\mathbb{Z}}. \tag{2.107}$$

Before presenting the proof of Theorem 2.55, we need some preparatory considerations. If $Y$ is a conjoined basis of (SDS) satisfying kernel condition in (2.37), then for any admissible $(x, u)$ with $x_0 \in \operatorname{Im} X_0$, we have $x_k \in \operatorname{Im} X_k$ for all $k \in [0, N+1]_{\mathbb{Z}}$ and, by Lemma 2.32,

$$x_k^T C_k^T A_k x_k + 2 x_k^T C_k^T \mathcal{B}_k u_k + u_k^T \mathcal{D}_k^T \mathcal{B}_k u_k = \Delta(x_k^T \bar{Q}_k x_k) + \bar{w}_k^T \mathcal{P}_k \bar{w}_k, \tag{2.108}$$

where the symmetric matrix $\bar{Q}_k := X_k X_k^\dagger U_k X_k^\dagger$ on $[0, N+1]_{\mathbb{Z}}$ and the vector $\bar{w}_k := u_k - \bar{Q}_k x_k$. It is interesting to observe that once formula (2.108) is established, then it is satisfied with *any* symmetric $Q_k$ such that $X_k^T Q_k X_k = X_k^T \bar{Q}_k X_k$. In order to see this, we let $x_k = X_k c_k$ for $k \in [0, N+1]_{\mathbb{Z}}$ and some $c_k \in \mathbb{R}^n$, and then

$$x_k^T \bar{Q}_k x_k = c_k^T X_k^T \bar{Q}_k X_k c_k = c_k^T X_k^T Q_k X_k c_k = x_k^T Q_k x_k. \qquad (2.109)$$

*Proof of Theorem 2.55* Let $Y$ be a conjoined basis of (SDS) with no focal points in $(0, N + 1]$. We proceed in the proof by showing the following steps.

*Claim 1* For all $k \in [0, N]_{\mathbb{Z}}$, we have the inclusions

$$\mathrm{Im}(\underline{X}_{k+1} - \mathcal{A}_k \underline{X}_k) \subseteq \mathrm{Im}\, \mathcal{B}_k \subseteq \mathrm{Im}\, X_{k+1}.$$

The first inclusion follows from the identity

$$\underline{X}_{k+1} - \mathcal{A}_k \underline{X}_k = (\underline{\mathcal{A}}_k - \mathcal{A}_k)\, \underline{X}_k + \underline{\mathcal{B}}_k \underline{U}_k$$

and from assumption (2.104). The second inclusion is equivalent to the kernel condition $\mathrm{Ker}\, X_{k+1} \subseteq \mathrm{Ker}\, X_k$, by Lemma 2.15, hence it is satisfied.

*Claim 2* For all $k \in [0, N + 1]_{\mathbb{Z}}$ we have the inclusion

$$\mathrm{Im}\, \underline{X}_k \subseteq \mathrm{Im}\, X_k. \qquad (2.110)$$

We shall prove it by induction. By assumption (2.105), the statement holds for $k = 0$. Suppose that condition (2.110) holds for some $k \in [0, N]$. In this case $\underline{X}_k = X_k X_k^\dagger \underline{X}_k$ by (1.160), and then we have

$$\begin{aligned}
\underline{X}_{k+1} &= \underline{X}_{k+1} - \mathcal{A}_k \underline{X}_k + \mathcal{A}_k X_k X_k^\dagger \underline{X}_k \\
&= X_{k+1} X_k^\dagger \underline{X}_k + (\underline{X}_{k+1} - \mathcal{A}_k \underline{X}_k) - (X_{k+1} - \mathcal{A}_k X_k) X_k^\dagger \underline{X}_k \\
&= X_{k+1} X_k^\dagger \underline{X}_k + (\underline{X}_{k+1} - \mathcal{A}_k \underline{X}_k) - \mathcal{B}_k U_k X_k^\dagger \underline{X}_k. \qquad (2.111)
\end{aligned}$$

The inclusion $\mathrm{Im}\, \underline{X}_{k+1} \subseteq \mathrm{Im}\, X_{k+1}$ then follows from Claim 1, since each of the three terms in (2.111) above is contained in $\mathrm{Im}\, X_{k+1}$.

*Claim 3* If $(x, \underline{u})$ is admissible with respect to $(\underline{\mathcal{A}}, \underline{\mathcal{B}})$, then $x_{k+1} - \mathcal{A}_k x_k \in \mathrm{Im}\, \mathcal{B}_k$ for all $k \in [0, N]_{\mathbb{Z}}$, i.e., there exist vectors $\{u_k\}_{k=0}^N$ such that $(x, u)$ is admissible with respect to $(\mathcal{A}, \mathcal{B})$.

This is a consequence of assumption (2.104), since

$$x_{k+1} - A_k x_k = (\underline{A}_k - A_k) x_k + \underline{B}_k \underline{u}_k \in \operatorname{Im} B_k.$$

*Claim 4*  If $\underline{Y}$ is a solution of (2.103), then

$$\underline{X}_k^T \underline{C}_k^T A_k \underline{X}_k + 2 \underline{X}_k^T \underline{C}_k^T B_k \underline{U}_k + \underline{U}_k^T \underline{D}_k^T B_k \underline{U}_k = \Delta(\underline{X}_k^T \underline{U}_k) = \Delta(\underline{X}_k^T \underline{Q}_k \underline{X}_k).$$
$$(2.112)$$

This follows by a direct calculation.

*Claim 5*  For all $k \in [0, N]$ we have $\underline{X}_k^T (\underline{Q}_k - Q_k) \underline{X}_k \geq 0$, i.e., condition (2.107) holds.

We will show that $\Delta[\underline{X}_k^T (\underline{Q}_k - Q_k) \underline{X}_k] \geq 0$ for $k \in [0, N]_{\mathbb{Z}}$, which together with initial condition (2.105) imply the statement. Let $c \in \mathbb{R}^n$ be arbitrary and put $x_k := \underline{X}_k c$ and $\underline{u}_k := \underline{U}_k c$ on $[0, N+1]_{\mathbb{Z}}$. Then $(x, \underline{u})$ is admissible with respect to $(\underline{A}, \underline{B})$ and

$$c^T \Delta[\underline{X}_k^T (\underline{Q}_k - Q_k) \underline{X}_k] c = c^T \Delta(\underline{X}_k^T \underline{U}_k) c - c^T \Delta(\underline{X}_k^T Q_k \underline{X}_k) c$$

$$\overset{(2.112),\ (2.63)}{=} \begin{pmatrix} x_k \\ x_{k+1} \end{pmatrix}^T \underline{\mathcal{G}}_k \begin{pmatrix} x_k \\ x_{k+1} \end{pmatrix} - \Delta(x_k^T Q_k x_k). \qquad (2.113)$$

By Claim 3, there exists $u = \{u_k\}_{k=0}^N$ such that $(x, u)$ is admissible with respect to $(A, B)$, while Claim 2 yields that $x_k \in \operatorname{Im} X_k$ for $k \in [0, N+1]_{\mathbb{Z}}$. Hence, we get

$$\Delta(x_k^T Q_k x_k) \overset{(2.109)}{=} \Delta(x_k^T \bar{Q}_k x_k) \overset{(2.108)}{=} \begin{pmatrix} x_k \\ x_{k+1} \end{pmatrix}^T \mathcal{G}_k \begin{pmatrix} x_k \\ x_{k+1} \end{pmatrix} - \bar{w}_k^T \mathcal{P}_k \bar{w}_k.$$

Using this identity in formula (2.113) and using assumption (2.106), we obtain

$$c^T \Delta[\underline{X}_k^T (\underline{Q}_k - Q_k) \underline{X}_k] c = \begin{pmatrix} x_k \\ x_{k+1} \end{pmatrix}^T (\underline{\mathcal{G}}_k - \mathcal{G}_k) \begin{pmatrix} x_k \\ x_{k+1} \end{pmatrix} + \bar{w}_k^T \mathcal{P}_k \bar{w}_k \geq 0,$$

where we used the fact that $\mathcal{P}_k = P_k = X_k X_{k+1}^\dagger B_k \geq 0$, since $Y$ is assumed to have no focal points in $(0, N+1]$.

*Claim 6*  The conjoined basis $\underline{Y}$ has no focal points in $(0, N+1]$.

We shall prove this via Theorem 2.39. Let $(x, \underline{u})$ be admissible with respect to $(\underline{A}, \underline{B})$ with $x_0 = \underline{X}_0 \underline{d}$ for some $\underline{d} \in \mathbb{R}^n$, $x_{N+1} = 0$, and $x \neq 0$. Then, by Claim 3, there is $u = \{u_k\}_{k=0}^N$ such that $(x, u)$ is admissible with respect to $(A, B)$ and, by assumption (2.105), $x_0 = X_0 d$ for some $d \in \mathbb{R}^n$. Hence, we have

$$\mathcal{F}(x, \underline{u}) + \underline{d}^T \underline{X}_0^T \underline{U}_0 \underline{d} \stackrel{(2.63)}{=} \sum_{k=0}^{N} \begin{pmatrix} x_k \\ x_{k+1} \end{pmatrix}^T \underline{\mathcal{G}}_k \begin{pmatrix} x_k \\ x_{k+1} \end{pmatrix} + \underline{d}^T \underline{X}_0^T \underline{Q}_0 \underline{X}_0 \underline{d}$$

$$\stackrel{(2.106),\ (2.105)}{\geq} \sum_{k=0}^{N} \begin{pmatrix} x_k \\ x_{k+1} \end{pmatrix}^T \mathcal{G}_k \begin{pmatrix} x_k \\ x_{k+1} \end{pmatrix} + \underline{d}^T \underline{X}_0^T Q_0 \underline{X}_0 \underline{d}$$

$$\stackrel{(2.63)}{=} \mathcal{F}(x, u) + d^T X_0^T U_0 d > 0,$$

since $Y$ is assumed to have no focal points in $(0, N + 1]$. Thus, by Theorem 2.39, the conjoined basis $\underline{Y}$ has no focal points in $(0, N + 1]$. This theorem is now proven.

$\square$

When the two systems (SDS) and (2.103) are the same, we obtain from Theorem 2.55 the following Riccati inequality.

**Corollary 2.56 (Inequality for One Symplectic System)** *Let $Y$ and $\underline{Y}$ be conjoined bases of* (SDS), *and let $Q_k$ and $\underline{Q}_k$ be symmetric $n \times n$ matrices such that $X_k^T Q_k X_k = X_k^T U_k$ and $\underline{X}_k^T \underline{Q}_k \underline{X}_k = \underline{X}_k^T \underline{U}_k$ on $[0, N + 1]_{\mathbb{Z}}$, and conditions (2.105) hold. If $Y$ has no forward focal points in $(0, N + 1]$, then $\underline{Y}$ has no forward focal points in $(0, N + 1]$ either, and inequality (2.107) is satisfied.*

## 2.4.2  Discrete Riccati Inequality

In Theorems 2.36 and 2.40, we presented several conditions which are equivalent to the positivity of the discrete quadratic functional $\mathcal{F}$ in (2.57), which are analogous to the continuous time conditions (i)–(iv) in Theorem 1.47. However, in contrast with (1.123), the discrete Riccati operator $R_k[Q]$ defined in (2.52) is not symmetric even when $Q_k$ is symmetric. This creates a difficulty in understanding what should be the discrete version of the Riccati inequality in condition (v) of Theorem 1.47. In this subsection, we answer this question and derive a complete discrete time analog of the latter result.

**Theorem 2.57 (Discrete Riccati Inequality)** *The functional $\mathcal{F}$ is positive definite if and only if either of the following equivalent conditions is satisfied.*

*(i) The system*

$$\left. \begin{array}{l} X_{k+1} = \mathcal{A}_k X_k + \mathcal{B}_k U_k, \\ N_k := X_{k+1}^T (U_{k+1} - \mathcal{C}_k X_k - \mathcal{D}_k U_k) \leq 0, \end{array} \right\} \tag{2.114}$$

$k \in [0, N]_{\mathbb{Z}}$, *has a solution $Y = \begin{pmatrix} X \\ U \end{pmatrix}$ such that $X_k^T U_k$ is symmetric and $X_k$ is invertible for all $k \in [0, N + 1]_{\mathbb{Z}}$ and $X_k X_{k+1}^{-1} \mathcal{B}_k \geq 0$ on $[0, N]_{\mathbb{Z}}$.*

*(ii)  The discrete Riccati inequality*

$$R_k[Q](\mathcal{A}_k + \mathcal{B}_k Q_k)^{-1} \le 0, \qquad (2.115)$$

$k \in [0, N]_\mathbb{Z}$, *has a symmetric solution $Q_k$ defined on $[0, N+1]_\mathbb{Z}$ satisfying the conditions in (2.68), i.e., $\mathcal{A}_k + \mathcal{B}_k Q_k$ is invertible and $(\mathcal{A}_k + \mathcal{B}_k Q_k)^{-1} \mathcal{B}_k \ge 0$ for all $k \in [0, N]_\mathbb{Z}$.*

The proof of Theorem 2.57 is shown below after the following auxiliary result.

**Lemma 2.58** *Let $k \in [0, N]_\mathbb{Z}$ be fixed and assume that $X_j$ and $U_j$ are $n \times n$ matrices for $j \in [k, k+1]_\mathbb{Z}$ such that $X_{k+1} = \mathcal{A}_k X_k + \mathcal{B}_k U_k$. Then the following conditions hold.*

(i) *If $X_j^T U_j$ is symmetric for $j \in [k, k+1]_\mathbb{Z}$, then the matrix*

$$X_{k+1}^T(U_{k+1} - \mathcal{C}_k X_k - \mathcal{D}_k U_k)$$
$$= \Delta(X_k^T U_k) - (X_k^T \mathcal{C}_k^T \mathcal{A}_k X_k + 2\, X_k^T \mathcal{C}_k^T \mathcal{B}_k U_k + U_k^T \mathcal{D}_k^T \mathcal{B}_k U_k)$$

*is symmetric as well.*

(ii) *If $X_{k+1}^T U_{k+1}$ is symmetric and if $Q_j$ is symmetric with $Q_j X_j = U_j X_j^\dagger X_j$ for $j \in [k, k+1]_\mathbb{Z}$, then*

$$X_{k+1}^T R_k[Q] X_k = X_{k+1}^T (U_{k+1} - \mathcal{C}_k X_k - \mathcal{D}_k U_k)\, X_k^\dagger X_k. \qquad (2.116)$$

(iii) *If $X_j^T U_j$ and $Q_j$ are symmetric with $Q_j X_j = U_j X_j^\dagger X_j$ for $j \in [k, k+1]_\mathbb{Z}$ and if $X_k$ is invertible, then the matrix*

$$X_{k+1}^T R_k[Q] X_k = X_{k+1}^T (U_{k+1} - \mathcal{C}_k X_k - \mathcal{D}_k U_k) \qquad (2.117)$$

*is symmetric.*

*Proof*  Part (i) is a simple calculation. For part (ii), we first derive

$$R_k[Q] X_k = [Q_{k+1} X_{k+1} - (\mathcal{C}_k X_k + \mathcal{D}_k U_k)] X_k^\dagger X_k$$
$$= [U_{k+1} - \mathcal{C}_k X_k - \mathcal{D}_k U_k - U_{k+1}(I - X_{k+1}^\dagger X_{k+1})] X_k^\dagger X_k.$$

Then, after multiplying by $X_{k+1}^T$ from the left and by using the symmetry of $X_{k+1}^T U_{k+1}$, we obtain (2.116). Identity (2.117) in part (iii) then follows directly from (i) and (ii).  □

*Proof of Theorem 2.57*  If not specified otherwise, conditions (i) and (ii) refer to Theorem 2.57. By Theorem 2.40, the positivity of $\mathcal{F}$ implies condition (i) trivially with $N_k \equiv 0$. Condition (i) implies (ii) by the Riccati substitution $Q_k := U_k X_k^{-1}$ on

$[0, N + 1]_{\mathbb{Z}}$. Next, we show that condition (ii) implies (i). Let

$$F_k := R_k[Q] (\mathcal{A}_k + \mathcal{B}_k Q_k)^{-1} \leq 0$$

be the matrix defining the inequality (2.115), where $Q_k$ satisfies condition (ii). Let $X$ be the solution of the equation $X_{k+1} = (\mathcal{A}_k + \mathcal{B}_k Q_k) X_k$, $k \in [0, N]_{\mathbb{Z}}$, given by the initial condition $X_0 = I$. Then $X_k$ is invertible on $[0, N + 1]_k$. If we set $U_k := Q_k X_k$ on $[0, N + 1]_{\mathbb{Z}}$, then $(X, U)$ satisfies $X_{k+1} = \mathcal{A}_k X_k + \mathcal{B}_k U_k$ and

$$N_k = X_{k+1}^T [Q_{k+1} - (C_k + D_k Q_k) X_k X_{k+1}^{-1}] X_{k+1} = X_{k+1}^T F_k X_{k+1} \leq 0$$

for all $k \in [0, N]_s Z$, that is, $(X, U)$ solves system (2.114). Note that the matrices $N_k$ and $F_k = X_{k+1}^{T-1} N_k X_{k+1}^{-1}$ are symmetric, by Lemma 2.58(iii).

The rest of the proof is about showing that condition (i) implies the positivity of the functional $\mathcal{F}$. With $N_k$ as in (2.114) we put $F_k := X_{k+1}^{T-1} N_k X_{k+1}^{-1} \leq 0$. Define $\underline{\mathcal{A}}_k := \mathcal{A}_k$, $\underline{\mathcal{B}}_k := \mathcal{B}_k$, $\underline{\mathcal{C}}_k := \mathcal{C}_k + F_k \mathcal{A}_k$, $\underline{\mathcal{D}}_k := \mathcal{D}_k + F_k \mathcal{B}_k$, and

$$\underline{\mathcal{S}}_k := \begin{pmatrix} \underline{\mathcal{A}}_k & \underline{\mathcal{B}}_k \\ \underline{\mathcal{C}}_k & \underline{\mathcal{D}}_k \end{pmatrix} = \mathcal{S}_k + \mathcal{R}_k \quad \text{with } \mathcal{R}_k := \begin{pmatrix} 0 & 0 \\ F_k \mathcal{A}_k & F_k \mathcal{B}_k \end{pmatrix}. \tag{2.118}$$

The proof will be finished by showing the following claims.

*Claim 1*  The matrix $\underline{\mathcal{S}}_k$ is symplectic.

This follows from the observation that $\mathcal{S}_k^T \mathcal{J} \mathcal{R}_k = (\mathcal{A}_k \ \mathcal{B}_k)^T F_k (\mathcal{A}_k \ \mathcal{B}_k)$ is symmetric, $\mathcal{R}_k^T \mathcal{J} \mathcal{R}_k = 0$, and from the calculation

$$\underline{\mathcal{S}}_k^T \mathcal{J} \underline{\mathcal{S}}_k = (\mathcal{S}_k + \mathcal{R}_k)^T \mathcal{J} (\mathcal{S}_k + \mathcal{R}_k) = \mathcal{S}_k^T \mathcal{J} \mathcal{S}_k + \mathcal{S}_k^T \mathcal{J} \mathcal{R}_k + \mathcal{R}_k^T \mathcal{J} \mathcal{S}_k + \mathcal{R}_k^T \mathcal{J} \mathcal{R}_k$$

$$= \mathcal{J} + \mathcal{R}_k^T \mathcal{J}^T \mathcal{S}_k + \mathcal{R}_k^T \mathcal{J} \mathcal{S}_k = \mathcal{J}.$$

*Claim 2*  The pair $(X, U)$ solves the system (2.103); hence it is a conjoined basis of (2.103) with no focal points in $(0, N + 1]$. This follows from the invertibility of $X_k$ on $[0, N + 1]_{\mathbb{Z}}$, the calculations

$$\underline{\mathcal{A}}_k X_k + \underline{\mathcal{B}}_k U_k = \mathcal{A}_k X_k + \mathcal{B}_k U_k = X_{k+1}$$

$$\underline{\mathcal{C}}_k X_k + \underline{\mathcal{D}}_k U_k = C_k X_k + D_k U_k + F_k (\mathcal{A}_k X_k + \mathcal{B}_k U_k)$$

$$= C_k X_k + D_k U_k + X_{k+1}^{T-1} N_k = U_{k+1},$$

and from $\underline{P}_k := X_k X_{k+1}^{-1} \underline{\mathcal{B}}_k = X_k X_{k+1}^{-1} \mathcal{B}_k \geq 0$.

*Claim 3*  The functional $\mathcal{F}$ is positive definite. We have by Theorem 2.40 applied to (2.103) that the functional $\underline{\mathcal{F}}$ is positive definite. Next, the definition of $\underline{\mathcal{A}}_k$ and $\underline{\mathcal{B}}_k$ implies that $\mathrm{Im}\,(\mathcal{A}_k - \underline{\mathcal{A}}_k \ \mathcal{B}_k) = \mathrm{Im}\,\mathcal{B}_k = \mathrm{Im}\,\underline{\mathcal{B}}_k$. Furthermore, the symmetric matrix $\underline{\mathcal{E}}_k := \mathcal{E}_k + F_k$ satisfies

$$\underline{\mathcal{D}}_k^T \underline{\mathcal{B}}_k = \mathcal{D}_k^T \mathcal{B}_k + \mathcal{B}_k^T F_k \mathcal{B}_k = \mathcal{B}_k^T (\mathcal{E}_k + F_k) \mathcal{B}_k = \underline{\mathcal{B}}_k^T \underline{\mathcal{E}}_k \underline{\mathcal{B}}_k,$$

$$\mathcal{G}_k - \underline{\mathcal{G}}_k = \begin{pmatrix} \mathcal{A}_k^T \mathcal{E}_k \mathcal{A}_k - \mathcal{A}_k^T \mathcal{C}_k \, \mathcal{C}_k^T - \mathcal{A}_k^T \mathcal{E}_k \\ \mathcal{C}_k - \mathcal{E}_k \mathcal{A}_k \qquad \mathcal{E}_k \end{pmatrix}$$

$$- \begin{pmatrix} \mathcal{A}_k^T (\mathcal{E}_k + F_k) \mathcal{A}_k - \mathcal{A}_k^T (\mathcal{C}_k + F_k \mathcal{A}_k) \, \mathcal{C}_k^T + \mathcal{A}_k^T F_k - \mathcal{A}_k^T (\mathcal{E}_k + F_k) \\ \mathcal{C}_k + F_k \mathcal{A}_k - (\mathcal{E}_k + F_k) \mathcal{A}_k \qquad \mathcal{E}_k + F_k \end{pmatrix}$$

$$= \begin{pmatrix} 0 & 0 \\ 0 & -F_k \end{pmatrix} \geq 0.$$

Consequently, the conditions

$$\mathrm{Im} \left( \mathcal{A}_k - \underline{\mathcal{A}}_k \, \mathcal{B}_k \right) \subseteq \mathrm{Im} \, \underline{\mathcal{B}}_k, \quad \underline{\mathcal{G}}_k \leq \mathcal{G}_k, \quad k \in [0, N]_{\mathbb{Z}},$$

are satisfied, compare with conditions (2.104)–(2.106) in which the roles of systems (SDS) and (2.103) are interchanged. Therefore, by Theorem 2.55 in this latter context with $\underline{Y}_k := Y_k$ and $\underline{Q}_k = Q_k = U_k X_k^{-1}$ on $[0, N+1]_{\mathbb{Z}}$, we obtain that $Y$ is a conjoined basis of (SDS) with no forward focal points in $(0, N+1]$. Consequently, by Theorem 2.40, the positivity of $\mathcal{F}$ follows, and the proof of this theorem is now complete.

$\square$

*Remark 2.59* In the proof above, we used a matrix $\underline{\mathcal{S}}_k$ of the form $\underline{\mathcal{S}}_k = \mathcal{S}_k + \mathcal{R}_k$ with $\mathcal{R}_k = \begin{pmatrix} 0 & 0 \\ G_k & H_k \end{pmatrix}$. This matrix $\underline{\mathcal{S}}_k$ is symplectic if and only if $G_k^T A_k$ and $H_k^T B_k$ are symmetric, and the identity $H_k^T A_k = B_k^T G_k$ holds. The choice $G_k := F_k A_k$ and $H_k := F_k B_k$ with symmetric $F_k$ is then natural, which was first observed in [55] in connection with an eigenvalue problem associated with system (SDS), see Sect. 5.3.

Riccati inequality is often used in nonoscillation criteria for differential and difference equations, since it is easier to find a solution of the *inequality* (which corresponds to a solution of *some* majorant equation) than a solution to the *equality*. In the following examples, we show a situation when a symmetric $Q_k$ solves the Riccati inequality (2.115) and satisfies condition (ii) in Theorem 2.57, but it does not solve the Riccati equation (2.52) so that condition (iii) in Theorem 2.40 is not satisfied with this $Q_k$.

*Example 2.60*

(i) Let $A_k \equiv 0$, $B_k \equiv -C^{T-1}$, $C_k \equiv C$, $D_k \equiv -C^{T-1}-C-K$, where $C$ is a constant nonsingular matrix, $K \neq 0$, and $CK^T = KC^T \geq 0$. Then $Q_k \equiv I$ satisfies condition (ii) in Theorem 2.57, since $A_k + B_k Q_k = -C^{T-1}$ is invertible, $(A_k + B_k Q_k)^{-1} B_k \equiv I > 0$, and $R_k[Q](A_k + B_k Q_k)^{-1} = -KC^T \leq 0$, while the Riccati equation is $R_k[Q] = K \neq 0$. Another (more specific) example can be obtained when we take, e.g., $C = K = I$. Note also that since $\mathcal{A}_k$ is not invertible, the Hamiltonian Riccati inequality in [176, Corollary 4.1] cannot be applied to this setting.

(ii) Let $\mathcal{A}_k$ and $\mathcal{C}_k$ be invertible, $\mathcal{B}_k \equiv 0$, and $\mathcal{D}_k = \mathcal{A}_k^{T-1}$, with $\mathcal{C}_k^T \mathcal{A}_k > 0$. Then $Q_k \equiv 0$ satisfies condition (ii) in Theorem 2.57, since $\mathcal{A}_k + \mathcal{B}_k Q_k = \mathcal{A}_k$ is invertible, $(\mathcal{A}_k + \mathcal{B}_k Q_k)^{-1} \mathcal{B}_k \equiv 0$, and $R_k[Q](\mathcal{A}_k + \mathcal{B}_k Q_k)^{-1} = -\mathcal{C}_k \mathcal{A}_k^{-1} < 0$, while the Riccati equation is $R_k[Q] = -\mathcal{C}_k \neq 0$.

## 2.5  Recessive and Dominant Solutions

In this section we develop the notions of a recessive and dominant solution for a nonoscillatory symplectic system (SDS) under the eventual controllability condition. These concepts extend to the matrix case the corresponding notions for the second-order equation (1.1) in Sect. 1.2. We note that the exposition in this subsection differs from the presentation of recessive and dominant solutions in [16, Sections 3.11–3.16]. For controllable systems, these two approaches are however equivalent (see Remark 2.68).

### 2.5.1  Definitions and Basic Properties

First we note that the nonoscillation of (SDS) at $\infty$ (Definition 2.19) means that every conjoined basis $Y$ of (SDS) does not have eventually any focal points, i.e., there exists $N \in [0, \infty)_{\mathbb{Z}}$ such that condition (2.37) holds for every $k \in [N, \infty)_{\mathbb{Z}}$. This implies that for every conjoined basis $Y$ of a nonoscillatory system (SDS),

$$\text{the kernel of } X_k \text{ is eventually constant.} \qquad (2.119)$$

System (SDS) is said to be *(completely) controllable* on some discrete interval $[N, \infty)_{\mathbb{Z}}$ if for any subinterval $[N_1, N_2]_{\mathbb{Z}}$ of $[N, \infty)_{\mathbb{Z}}$ with at least two points the trivial solution $(x, u) \equiv (0, 0)$ is the only solution of (SDS) for which $x_k \equiv 0$ on $[N_1, N_2]_{\mathbb{Z}}$. System (SDS) is *eventually (completely) controllable* if there exists $N \in [0, \infty)_{\mathbb{Z}}$ such that it is (completely) controllable on $[N, \infty)_{\mathbb{Z}}$.

The eventual controllability of (SDS) in combination with property (2.119) means the matrix $X_k$ is eventually invertible. That is, if the system (SDS) is nonoscillatory at $\infty$ and eventually controllable, then for every conjoined basis $Y$, there exists $N \in \mathbb{Z}$ such that

$$X_k \text{ is invertible and } X_k X_{k+1}^{-1} \mathcal{B}_k \geq 0 \text{ for all } k \in [N, \infty)_{\mathbb{Z}}. \qquad (2.120)$$

*Remark 2.61*

(i) The above condition of eventual controllability is satisfied, for example, by systems (SDS) with $\mathcal{B}_k$ invertible for large $k$, such as equation (1.1). Other types of such systems are discussed in Sect. 2.1.2. Equation (2.27) is an example of a controllable system with singular $\mathcal{B}_k$.

(ii) On the other hand, system (SDS) with $\mathcal{B}_k \equiv 0$ is not eventually controllable. In this case we have by (1.146) that $\mathcal{A}_k$ and $\mathcal{D}_k = \mathcal{A}_k^{T-1}$ are invertible and the first equation of (SDS) has the form $x_{k+1} = \mathcal{A}_k x_k$. This means that if $Y$ is a conjoined basis with singular $X_N$, then $X_k$ is singular for all $k \in [N, \infty)_{\mathbb{Z}}$.

Under (2.120) we may associate with the conjoined basis $Y$ the symmetric matrix

$$S_k := \sum_{j=N}^{k-1} X_{j+1}^{-1} \mathcal{B}_j X_j^{T-1}, \quad k \in [N+1, \infty)_{\mathbb{Z}}, \quad S_N := 0. \tag{2.121}$$

The matrix $S_k$ is symmetric, positive semidefinite, and nondecreasing in $k$, since

$$\Delta S_k = X_{k+1}^{-1} \mathcal{B}_k X_k^{T-1} = X_k^{-1} (X_k X_{k+1}^{-1} \mathcal{B}_k) X_k^{T-1} \geq 0, \quad k \in [N, \infty)_{\mathbb{Z}}. \tag{2.122}$$

In fact, we shall prove a stronger result under the complete controllability assumption on $[N, \infty)_{\mathbb{Z}}$.

**Lemma 2.62** *Assume that system (SDS) is nonoscillatory at $\infty$ and eventually completely controllable. Then for any conjoined basis $Y$ of (SDS), there exists $N \in [0, \infty)_{\mathbb{Z}}$ such that the matrix $S_k$ defined in (2.121) is positive definite for all $k \in [N+2, \infty)_{\mathbb{Z}}$.*

*Proof* The assumptions imply that for some $N \in [0, \infty)_{\mathbb{Z}}$, condition (2.120) holds. Assume that $S_k d = 0$ for some $k \in [N+2, \infty)_{\mathbb{Z}}$. Then $X_{j+1}^{-1} \mathcal{B}_j X_j^{T-1} d \equiv 0$ for all $j \in [N, k-1]_{\mathbb{Z}}$, i.e., $\mathcal{B}_j X_j^{T-1} d \equiv 0$ for all $j \in [N, k-1]_{\mathbb{Z}}$. We now define $x_j := 0$, $u_j := X_j^{T-1} d$, and $Q_j := U_j X_j^{-1}$ for $j \in [N, k]_{\mathbb{Z}}$. It follows that for every $j \in [N, k-1]_{\mathbb{Z}}$ we have $\mathcal{B}_j u_j = 0$ and, by (2.55) in Corollary 2.27,

$$u_{j+1} = X_{j+1}^{T-1} d = (\mathcal{A}_j X_j + \mathcal{B}_j U_j)^{T-1} d = (\mathcal{A}_j^T + Q_j \mathcal{B}_j^T) X_j^{T-1} d$$

$$\overset{(2.55)}{=} (\mathcal{D}_j - Q_{j+1} \mathcal{B}_j) u_k = \mathcal{D}_j u_j.$$

This implies that the pair $(x \equiv 0, u)$ solves system (SDS) on the interval $[N, k-1]_{\mathbb{Z}}$. By the complete controllability assumption, we obtain that $u_j \equiv 0$ on $[N, k]_{\mathbb{Z}}$, i.e., $d = 0$. This shows that the matrix $S_k$ is invertible, hence positive definite. $\qquad \square$

The result in Lemma 2.62 yields that for any conjoined basis $Y$ of a nonoscillatory and eventually controllable system (SDS), the matrix $S_k^{-1}$ eventually exists as a positive definite matrix, which is nonincreasing in $k$. Therefore, there exists the limit

$$T := \lim_{k \to \infty} S_k^{-1}, \quad T \geq 0. \tag{2.123}$$

Depending on the form of the limiting matrix $T$, we make the following definition:

**Definition 2.63** A conjoined basis $\tilde{Y}$ of a nonoscillatory and eventually controllable system (SDS) is called a *recessive solution* at $\infty$ if the matrix $T$ defined in (2.123) satisfies $T = 0$, while it is called a *dominant solution* at $\infty$ if $T > 0$.

*Remark 2.64* One can easily see from the limit properties of invertible matrices that $Y$ is a dominant solution at $\infty$ if and only if the limit of $S_k$ itself exists as $k \to \infty$, and in this case

$$\lim_{k \to \infty} S_k = T^{-1}.$$

The following reduction of order theorem will be useful for the construction and properties of recessive and dominant solutions of (SDS).

**Theorem 2.65** *Assume that $Y$ is a conjoined basis of (SDS) such that $X_k$ is invertible on some interval $[N, M]_{\mathbb{Z}}$. Then $\tilde{Y}$ is a solution of (SDS) if and only if for some constant matrices $K$ and $L$, we have*

$$\tilde{X}_k = X_k (K + S_k L), \quad \tilde{U}_k = U_k (K + S_k L) + X_k^{T-1} L, \quad k \in [N, M]_{\mathbb{Z}}, \tag{2.124}$$

*where the matrix $S_k$ is defined in (2.121). In this case $K = X_N^{-1} \tilde{X}_N$ and $L = Y^T \mathcal{J} \tilde{Y}$. Moreover, $\tilde{Y}$ is also a conjoined basis of (SDS) if and only if $K^T L$ is symmetric.*

*Proof* The proof is rather straightforward in both implications. The details can be found in [16, Theorem 3.32, pp. 105–109]. □

The following two results justify the existence of recessive and dominant solutions at $\infty$ for a nonoscillatory system (SDS), as well as the uniqueness of the recessive solution at $\infty$. They provide an extension of Theorem 1.7 to symplectic systems.

**Theorem 2.66** *Assume that system (SDS) is nonoscillatory at $\infty$ and eventually controllable. Then there exists a recessive solution of (SDS) at $\infty$. In addition, the recessive solution is unique up to a right constant nonsingular multiple.*

*Proof* Consider the principal solution $\hat{Y}$ at $k = 0$, i.e., $\hat{X}_0 = 0$ and $\hat{U}_0 = I$. Then for some $N \in [1, \infty)_{\mathbb{Z}}$, the matrix $\hat{X}_k$ satisfies (2.120), and we may define the matrices $\hat{S}_k$ and $\hat{T}$ by (2.121) and (2.123) through $\hat{X}_k$. Let $\tilde{Y}$ be the conjoined basis of (SDS) given by the initial conditions $\bar{X}_N = 0$ and $\bar{U}_N = \hat{X}_N^{T-1}$. Then $\hat{Y}_N^T \mathcal{J} \bar{Y}_N = I$, so that the conjoined bases $\hat{Y}$ and $\bar{Y}$ are normalized. We will show that the sequence

$$\begin{pmatrix} \tilde{X}_k \\ \tilde{U}_k \end{pmatrix} := \begin{pmatrix} \hat{X}_k & \bar{X}_k \\ \hat{U}_k & \bar{U}_k \end{pmatrix} \begin{pmatrix} I \\ -\hat{T} \end{pmatrix}, \quad k \in [0, \infty)_{\mathbb{Z}}, \tag{2.125}$$

is a conjoined basis of (SDS) and that $\tilde{Y}$ is a recessive solution of (SDS) at $\infty$. By Theorem 2.65 (with $K = 0$ and $L = I$) we know that

$$\bar{X}_k = \hat{X}_k \hat{S}_k, \quad \bar{U}_k = \hat{U}_k \hat{S}_k + \hat{X}_k^{T-1}, \quad k \in [N, \infty)_{\mathbb{Z}}. \tag{2.126}$$

Then, substituting (2.126) into (2.125), we obtain that

$$\tilde{X}_k = \hat{X}_k (I - \hat{S}_k \hat{T}), \quad \tilde{U}_k = \hat{U}_k (I - \hat{S}_k \hat{T}) - \hat{X}_k^{T-1} \hat{T}, \quad k \in [N, \infty)_{\mathbb{Z}}, \tag{2.127}$$

so that by Theorem 2.65 (with $Y := \hat{Y}$, $K = I$, and $L = -\hat{T}$), we conclude that $\tilde{Y}$ is a conjoined basis of (SDS). The assumptions of the theorem imply that $\tilde{X}_k$ is invertible for all $k \in [L, \infty)_{\mathbb{Z}}$ for some index $L \in [N, \infty)_{\mathbb{Z}}$. Then by (2.127) we know that also $I - \hat{S}_k \hat{T}$ is invertible for $k \in [L, \infty)_{\mathbb{Z}}$. Therefore, for the symmetric matrices $G_k := \hat{T}(I - \hat{S}_k \hat{T})$, $k \in [N, \infty)_{\mathbb{Z}}$, we have

$$G_k \geq 0, \quad k \in [N, \infty)_{\mathbb{Z}}, \quad \{G_k\}_{k=N}^{\infty} \text{ is nonincreasing on} [N, \infty)_{\mathbb{Z}}, \tag{2.128}$$

$$\operatorname{Im} G_k = \operatorname{Im} [\hat{T}(I - \hat{S}_k \hat{T})] = \operatorname{Im} \hat{T}, \quad k \in [L, \infty)_{\mathbb{Z}}. \tag{2.129}$$

But from the definition of $G_k$, we also have

$$\operatorname{Im} G_k = \operatorname{Im} [\hat{T}(I - \hat{S}_k \hat{T})] \subseteq \operatorname{Im} \hat{T}, \quad k \in [N, L-1]_{\mathbb{Z}}. \tag{2.130}$$

Hence, from (2.128)–(2.130) we obtain that $\operatorname{Im} G_k = \operatorname{Im} \hat{T}$ for all $k \in [N, \infty)_{\mathbb{Z}}$. By taking the orthogonal complements (using the symmetry of $G_k$ and $\hat{T}$), we get

$$\operatorname{Ker} [\hat{T}(I - \hat{S}_k \hat{T})] = \operatorname{Ker} G_k = \operatorname{Ker} \hat{T}, \quad k \in [N, \infty)_{\mathbb{Z}}. \tag{2.131}$$

Assume now that $\tilde{X}_k d = 0$ for some $k \in [N, \infty)_{\mathbb{Z}}$ and $d \in \mathbb{R}^n$. It follows by (2.127) and by the invertibility of $\hat{X}_k$ that $(I - \hat{S}_k \hat{T}_k) d = 0$, i.e., $d = \hat{S}_k \hat{T}_k d$. At the same time, $G_k d = \hat{T}(I - \hat{S}_k \hat{T}_k) d = 0$, so that $d \in \operatorname{Ker} G_k = \operatorname{Ker} \hat{T}$ by (2.131). Therefore, $d = \hat{S}_k \hat{T}_k d = 0$. This shows that the matrix $\tilde{X}_k$ is invertible for all $k \in [N, \infty)_{\mathbb{Z}}$.

Define the associated matrix $\tilde{S}_k$ as in (2.121) through $\tilde{X}_k$ in $[N, \infty)_{\mathbb{Z}}$. Since $\tilde{Y}_N^T \mathcal{J} \tilde{Y}_N = I$, we get by Theorem 2.65 (with $K = 0$ and $L = I$ again) that

$$\bar{X}_k = \tilde{X}_k \tilde{S}_k, \quad \bar{U}_k = \tilde{U}_k \tilde{S}_k + \tilde{X}_k^{T-1}, \quad k \in [N, \infty)_{\mathbb{Z}}. \tag{2.132}$$

Combining the first equations in (2.126) and (2.132) yields the equality $\hat{X}_k \hat{S}_k = \tilde{X}_k \tilde{S}_k$ on $[N, \infty)_{\mathbb{Z}}$. Therefore, together with the expression for $\tilde{X}_k$ in (2.127), we obtain that (note that $\hat{S}_k$ and $\tilde{S}_k$ are invertible for $k \geq N + 2$ by Lemma 2.62)

$$\tilde{S}_k^{-1} = \hat{S}_k^{-1} \hat{X}_k^{-1} \tilde{X}_k = \hat{S}_k^{-1}(I - \hat{S}_k \hat{T}) = \hat{S}_k^{-1} - \hat{T}, \quad k \in [N + 2, \infty)_{\mathbb{Z}}.$$

Thus, the matrix $\tilde{T}$ defined in (2.121) through $\tilde{S}_k$ satisfies

$$\tilde{T} = \lim_{k \to \infty} \tilde{S}_k^{-1} = \lim_{k \to \infty} (\hat{S}_k^{-1} - \hat{T}) = 0,$$

which shows that $\tilde{Y}$ is a recessive solution of (SDS) at $\infty$ according to Definition 2.63.

Now we prove the essential uniqueness of the recessive solution $\tilde{Y}$. Let $\tilde{Y}^{(0)}$ be another recessive solution of (SDS) at $\infty$. Let $N \in [0, \infty)_{\mathbb{Z}}$ be an index such that (2.120) holds for $\tilde{X}_k$ and $\tilde{X}_k^{(0)}$. Let $\tilde{S}_k$ and $\tilde{S}_k^{(0)}$ be the associated matrices defined in (2.121) through $\tilde{X}_k$ and $\tilde{X}_k^{(0)}$, respectively, and let $\tilde{T}$ and $\tilde{T}^{(0)}$ be the matrices in (2.123), which according to Definition 2.63 satisfy $\tilde{T} = 0 = \tilde{T}^{(0)}$. By Theorem 2.65, we know that

$$\tilde{X}_k^{(0)} = \tilde{X}_k(K + \tilde{S}_k L), \quad \tilde{U}_k^{(0)} = \tilde{U}_k(K + \tilde{S}_k L) + \tilde{X}_k^{T-1} L, \quad k \in [N, \infty)_{\mathbb{Z}},$$
$$(2.133)$$

where $K = \tilde{X}_N^{-1} \tilde{X}_N^{(0)}$ is regular and $L = \tilde{Y}^T \mathcal{J} \tilde{Y}^{(0)}$ is the Wronskian of $\tilde{Y}$ and $\tilde{Y}^{(0)}$. By [284, Proposition 4.20], the matrices $\tilde{S}_k$ and $\tilde{S}_k^{(0)}$ are related by the formula

$$\left(\tilde{S}_k^{(0)}\right)^{-1} = K^T \tilde{S}_k^{-1} K + K^T L,$$

which means upon taking the limit as $k \to \infty$ that

$$0 = \tilde{T}^{(0)} = K^T \tilde{T} K + K^T L = K^T L.$$

Since $K$ is regular, it follows that $L = 0$, and hence by (2.133) we obtain $\tilde{Y}_k^{(0)} = \tilde{Y}_k K$ on $[N, \infty)_{\mathbb{Z}}$. Thus, the recessive solution $\tilde{Y}^{(0)}$ is a regular constant multiple of the recessive solution $\tilde{Y}$. Conversely, if $\tilde{Y}$ is a recessive solution of (SDS) at $\infty$, satisfying (2.120), and if $K$ is a regular $n \times n$ matrix, then we define the conjoined basis $\tilde{Y}^{(0)}$ by $\tilde{Y}_k^{(0)} := \tilde{Y}_k K$ on $[N, \infty)_{\mathbb{Z}}$. Then $\tilde{X}_k^{(0)}$ is invertible on $[N, \infty)_{\mathbb{Z}}$ and

$$\left.\begin{array}{c}\left(\tilde{X}_{k+1}^{(0)}\right)^{-1} \mathcal{B}_k \left(\tilde{X}_k^{(0)}\right)^{T-1} = K^{-1} \tilde{X}_{k+1}^{-1} \mathcal{B} \tilde{X}_k^{T-1} K^{T-1} \geq 0 \\ \tilde{S}_k^{(0)} = K^{-1} \tilde{S}_k K^{T-1},\end{array}\right\} \quad k \in [N, \infty)_{\mathbb{Z}},$$

so that

$$\tilde{T}^{(0)} = \lim_{k \to \infty} \left( \tilde{S}_k^{(0)} \right)^{-1} = \lim_{k \to \infty} K^T \tilde{S}_k^{-1} K = K^T \tilde{T} K = 0.$$

Hence, $\tilde{Y}^{(0)}$ is also a recessive solution of (SDS) according to Definition 2.63. The proof is complete.  □

The next result shows that the recessive solution at $\infty$ is the smallest solution among all conjoined bases $Y$ of (SDS) which are linearly independent with $\tilde{Y}$, measured in terms of their first components $\tilde{X}_k$ and $X_k$. This property extends the limit (1.18) to symplectic systems.

**Theorem 2.67** *Assume that system* (SDS) *is nonoscillatory at* $\infty$ *and eventually controllable. Let* $\tilde{Y}$ *and* $Y$ *be two conjoined bases of* (SDS) *such that the Wronskian* $W := w(\tilde{Y}, Y) = \tilde{Y}^T \mathcal{J} Y$ *is invertible. Then* $\tilde{Y}$ *is the recessive solution of* (SDS) *at* $\infty$ *if and only if*

$$\lim_{k \to \infty} X_k^{-1} \tilde{X}_k = 0. \tag{2.134}$$

*In this case* $Y$ *is a dominant solution of* (SDS) *at* $\infty$.

*Proof* Let $N \in [0, \infty)_{\mathbb{Z}}$ be an index such that $\tilde{X}_k$ and $X_k$ satisfy (2.120). Let $\tilde{S}_k$, $\tilde{T}$ and $S_k$, $T$ be the matrices in (2.121) and (2.123) associated with $\tilde{X}_k$ and $X_k$, respectively. Then by Theorem 2.65 (with the invertible matrix $K := X_N^{-1} \tilde{X}_N$ and with $L := -W^T$), we have

$$\tilde{X}_k = X_k (K - S_k W^T), \quad X_k = \tilde{X}_k (K^{-1} + \tilde{S}_k W), \quad k \in [N, \infty)_{\mathbb{Z}}. \tag{2.135}$$

Assume that $\tilde{Y}$ is the recessive solution of (SDS) at $\infty$, i.e., $\tilde{T} = 0$. Then (2.135) implies that $K^{-1} + \tilde{S}_k W = \tilde{X}_k^{-1} X_k$ is invertible on $[N, \infty)_{\mathbb{Z}}$, as well as $\tilde{S}_k$ is invertible on $[N + 2, \infty)_{\mathbb{Z}}$ by Lemma 2.62. Hence,

$$\lim_{k \to \infty} X_k^{-1} \tilde{X}_k = \lim_{k \to \infty} (K^{-1} + \tilde{S}_k W)^{-1} = \lim_{k \to \infty} (\tilde{S}_k^{-1} K^{-1} + W)^{-1} \tilde{S}_k^{-1}$$

$$= (\tilde{T} K^{-1} + W)^{-1} \tilde{T} = 0.$$

Conversely, assume that (2.134) holds. The second equation in (2.135) implies that $\tilde{S}_k = (\tilde{X}_k^{-1} X_k - K^{-1}) W^{-1}$ on $[N, \infty)_{\mathbb{Z}}$, so that

$$\tilde{T} = \lim_{k \to \infty} \tilde{S}_k^{-1} = \lim_{k \to \infty} W (\tilde{X}_k^{-1} X_k - K^{-1})^{-1}$$

$$= \lim_{k \to \infty} W (I - X_k^{-1} \tilde{X}_k K^{-1})^{-1} X_k^{-1} \tilde{X}_k \stackrel{(2.134)}{=} 0.$$

This proves that $\tilde{Y}$ is the recessive solution of (SDS) at $\infty$. Finally, knowing that $\tilde{Y}$ is the recessive solution of (SDS) at $\infty$, then from (2.135), we obtain

$$K - S_k W^T = X_k^{-1} \tilde{X}_k = (K^{-1} + \tilde{S}_k W)^{-1}, \quad k \in [N, \infty)_{\mathbb{Z}},$$

so that

$$S_k = [K - (\tilde{S}_k^{-1} K^{-1} + W)^{-1} \tilde{S}_k^{-1}] W^{T-1}, \quad k \in [N+2, \infty)_{\mathbb{Z}}.$$

Upon taking the inverse and using that $\tilde{T} = 0$, we get

$$T = \lim_{k \to \infty} S_k^{-1} = \lim_{k \to \infty} W^T [K - (\tilde{S}_k^{-1} K^{-1} + W)^{-1} \tilde{S}_k^{-1}]^{-1}$$

$$= W^T [K - (\tilde{T} K^{-1} + W)^{-1} \tilde{T}]^{-1} = W^T K^{-1}.$$

Therefore, the matrix $T$ is invertible. At the same time, we know that $T \geq 0$ by (2.123), so that $T > 0$ follows. Hence, $Y$ is a dominant solution of (SDS) at $\infty$ according to Definition 2.63. The proof is complete.                                    □

*Remark 2.68* The result in Theorem 2.67 shows that the notions of a recessive solution at $\infty$ for a nonoscillatory and completely controllable system (SDS) in Definition 2.63 and in [16, pg. 115] are equivalent.

*Remark 2.69* The results in Theorems 2.66 and 2.67 show that, under the eventual controllability assumption, the nonoscillation of (SDS) implies the existence of the recessive solution of (SDS) at $\infty$, as well as the existence of a dominant solution of (SDS) at $\infty$. However, from the quantitative results on focal points (i.e., Sturmian separation theorems) presented later in Sects. 4.2.4 and 5.5.1, it follows that also the converse to Theorem 2.66 holds. Namely, the existence of a nonoscillatory conjoined basis of (SDS) at $\infty$ implies that every conjoined basis of (SDS) is also nonoscillatory at $\infty$, i.e., system (SDS) is nonoscillatory at $\infty$.

## 2.5.2 Minimal Solution of Riccati Equation

The recessive solution of (SDS) gives rise to a minimal solution of the associated Riccati difference equation (2.51). This statement extends Theorem 1.8 to symplectic systems.

**Theorem 2.70** *Assume that (SDS) is nonoscillatory and eventually controllable, and let $\tilde{Y}$ be its recessive solution at $\infty$. Then the solution $\tilde{Q}_k := \tilde{U}_k \tilde{X}_k^{-1}$ of the Riccati equation (2.51) is eventually minimal, i.e., if $Q_k$ is any symmetric solution of (2.51) such that $(\mathcal{A}_k + \mathcal{B}_k Q_k)^{-1} \mathcal{B}_k \geq 0$ for all $k$ in some interval $[N, \infty)_{\mathbb{Z}}$, then*

$$Q_k \geq \tilde{Q}_k \quad \text{for all } k \in [N, \infty)_{\mathbb{Z}}. \tag{2.136}$$

*Proof* Let $\tilde{Q}_k$ and $Q_k$ be as in the theorem. Then the matrices $\mathcal{A}_k + \tilde{Q}_k \mathcal{B}_k$ and $\mathcal{A}_k + Q_k \mathcal{B}_k$ are invertible on $[N, \infty)_{\mathbb{Z}}$ by Corollary 2.27. By Theorem 2.28 we associate with $Q_k$ a conjoined basis $Y$ of (SDS) such that $X_k$ is invertible and $Q_k = U_k X_k^{-1}$ on $[N, \infty)_{\mathbb{Z}}$. Then $Y$ satisfies condition (2.120) by the assumptions in the theorem. Denote by $W := w(\tilde{Y}, Y) = \tilde{Y}^T \mathcal{J} Y$ the (constant) Wronskian of $\tilde{Y}$ and $Y$. Then for $k \in [N, \infty)_{\mathbb{Z}}$ we have

$$Q_k - \tilde{Q}_k = U_k X_k^{-1} - \tilde{U}_k \tilde{X}_k^{-1} = \tilde{X}_k^{T-1} W X_k^{-1} = \tilde{X}_k^{T-1} (W X_k^{-1} \tilde{X}_k) \tilde{X}_k^{-1}. \tag{2.137}$$

Consider now the symmetric matrix function $W X_k^{-1} \tilde{X}_k$ on $[N, \infty)_{\mathbb{Z}}$. Then

$$\begin{aligned}
\Delta(W X_k^{-1} \tilde{X}_k) &= W(X_{k+1}^{-1} \tilde{X}_{k+1} - X_k^{-1} \tilde{X}_k) \\
&= W[X_{k+1}^{-1}(\mathcal{A}_k \tilde{X}_k + \mathcal{B}_k \tilde{U}_k) - X_{k+1}^{-1}(\mathcal{A}_k X_k + \mathcal{B}_k U_k) X_k^{-1} \tilde{X}_k] \\
&= W X_{k+1}^{-1} \mathcal{B}_k(\tilde{U}_k - U_k X_k^{-1} \tilde{X}_k) = -W X_{k+1}^{-1} \mathcal{B}_k X_k^{T-1} W^T \leq 0,
\end{aligned}$$

so that the function $W X_k^{-1} \tilde{X}_k$ is nonincreasing on $[N, \infty)_{\mathbb{Z}}$. At the same time, we know from Theorem 2.67 that $W X_k^{-1} \tilde{X}_k \to 0$ as $k \to \infty$. This implies that $W X_k^{-1} \tilde{X}_k \geq 0$ on $[N, \infty)_{\mathbb{Z}}$ and hence, equation (2.137) yields that $Q_k - \tilde{Q}_k \geq 0$ for all $k \in [N, \infty)_{\mathbb{Z}}$. Therefore, (2.136) is satisfied. $\square$

## 2.6 Transformations of Symplectic Systems

Transformation approach to various differential and difference equations is historically a very effective method. The basic idea is simple. An equation is transformed into an "easier" equation, this equation is then investigated, and finally the obtained results are transformed back to the original equation. Of course, important are invariants of the applied transformations. In the investigation of symplectic systems, the situation is similar, and naturally the main role will be played by symplectic transformations.

### 2.6.1 General Symplectic Transformation

Our first result shows that symplectic transformations preserve the type (i.e., the symplectic property) of the underlying system.

**Lemma 2.71** *Let $\mathcal{R}_k \in \mathbb{R}^{2n \times 2n}$ for $k \in [0, N + 1]_{\mathbb{Z}}$ be a sequence of symplectic matrices, and consider the transformation*

$$y_k = \mathcal{R}_k w_k. \tag{2.138}$$

*Then (2.138) transforms system (SDS) into another symplectic system*

$$w_{k+1} = \tilde{S}_k w_k, \quad \tilde{S}_k := \mathcal{R}_{k+1}^{-1} S_k \mathcal{R}_k = -\mathcal{J} \mathcal{R}_{k+1}^T \mathcal{J} S_k \mathcal{R}_k, \qquad (2.139)$$

*where the matrix $\tilde{S}_k$ is symplectic for $k \in [0, N]_{\mathbb{Z}}$. Moreover, $Y_k$ is a conjoined basis of (SDS) if and only if $\tilde{Y}_k = \mathcal{R}_k^{-1} Y_k$ is a conjoined basis of (2.139).*

*Proof* By (2.138) we have

$$y_{k+1} = \mathcal{R}_{k+1} w_{k+1} = S_k y_k = S_k \mathcal{R}_k w_k.$$

Hence, the new variable $w$ is a solution of the system (2.139). Moreover, the matrix $\tilde{S}_k$ is symplectic, being a product of symplectic matrices (see Lemma 1.58(i)). Recall that the multiplication by a symplectic (hence invertible) matrix preserves both conditions, which define a conjoined basis (i.e., the rank condition and the symmetry of $\tilde{X}_k^T \tilde{U}_k$), see Lemma 1.58(vii). This property proofs the last statement of the lemma.                                                                                    □

Sometimes we will need to know the block structure of the matrix $\tilde{S}_k$ in the transformed system (2.139). Hence, consider the transformation matrix $\mathcal{R}_k$ in the form

$$\mathcal{R}_k = \begin{pmatrix} H_k & M_k \\ G_k & N_k \end{pmatrix}, \qquad (2.140)$$

and denote by $\tilde{A}_k, \tilde{B}_k, \tilde{C}_k, \tilde{D}_k$ the block in the matrix $\tilde{S}_k$. Then by a direct computation, we have the formulas

$$\left.\begin{aligned}
\tilde{A}_k &= N_{k+1}^T (A_k H_k + B_k G_k) - M_{k+1}^T (C_k H_k + D_k G_k), \\
\tilde{B}_k &= N_{k+1}^T (A_k M_k + B_k N_k) - M_{k+1}^T (C_k M_k + D_k N_k), \\
\tilde{C}_k &= -G_{k+1}^T (A_k H_k + B_k G_k) + H_{k+1}^T (C_k H_k + D_k G_k), \\
\tilde{D}_k &= -G_{k+1}^T (A_k M_k + B_k N_k) + H_{k+1}^T (C_k M_k + D_k N_k).
\end{aligned}\right\} \qquad (2.141)$$

Summarizing, a linear transformation with a symplectic transformation matrix transforms a symplectic system again into a symplectic systems. In our treatment, we will mainly concentrate on oscillatory properties of symplectic systems, and the above-described transformation generally does not preserve an oscillatory nature of the transformed systems. The oscillatory properties are preserved, when the transformation matrix is lower block triangular, i.e., when $M_k = 0$ (see Lemma 2.72 below). In this case, the transformation matrix has the form

$$\mathcal{R}_k = \begin{pmatrix} H_k & 0 \\ G_k & H_k^{T-1} \end{pmatrix}, \qquad (2.142)$$

and the formulas for $\tilde{A}_k, \tilde{B}_k, \tilde{C}_k, \tilde{D}_k$ in (2.141) simplify to

$$\left.\begin{aligned}
\tilde{A}_k &= H_{k+1}^{-1}(A_k H_k + B_k G_k), \\
\tilde{B}_k &= H_{k+1}^{-1} B_k H_k^{T-1}, \\
\tilde{C}_k &= -G_{k+1}^T(A_k H_k + B_k G_k) + H_{k+1}^T(C_k H_k + D_k G_k), \\
\tilde{D}_k &= -G_{k+1}^T B_k H_k^{T-1} + H_{k+1}^T D_k H_k^{T-1}.
\end{aligned}\right\} \tag{2.143}$$

**Lemma 2.72** *Suppose that $Y$ is a conjoined basis of* (SDS) *and $\tilde{Y}_k = \binom{\tilde{X}_k}{\tilde{U}_k} = \mathcal{R}_k^{-1} Y_k$, where $\mathcal{R}_k$ is a lower block-triangular symplectic matrix. Then $Y$ has no focal point in $(k, k+1]$ if and only if $\tilde{Y}$ has no focal point in this interval.*

*Proof* Obviously, we have $\operatorname{Ker} X_{k+1} \subseteq \operatorname{Ker} X_k$ if and only if $\operatorname{Ker} \tilde{X}_{k+1} \subseteq \operatorname{Ker} \tilde{X}_k$ and, by (2.39), $\mathcal{B}_k = X_{k+1} X_{k+1}^\dagger \mathcal{B}_k$ or $\tilde{\mathcal{B}}_k = H_{k+1}^{-1} X_{k+1} X_{k+1}^\dagger \mathcal{B}_k H_k^{T-1}$ according to (2.143). Substituting the last representation for $\tilde{\mathcal{B}}_k$ into $\tilde{P}_k := \tilde{X}_k \tilde{X}_{k+1}^\dagger \tilde{\mathcal{B}}_k$ and using Lemma 1.63, we have

$$\begin{aligned}
\tilde{P}_k = \tilde{X}_k \tilde{X}_{k+1}^\dagger \tilde{\mathcal{B}}_k &= H_k^{-1} X_k (H_{k+1}^{-1} X_{k+1})^\dagger H_{k+1}^{-1} X_{k+1} X_{k+1}^\dagger \mathcal{B}_k H_k^{T-1} \\
&= H_k^{-1} X_k X_{k+1}^\dagger \mathcal{B}_k H_k^{T-1} = H_k^{-1} P_k H_k^{T-1},
\end{aligned}$$

i.e., $P_k \geq 0$ if and only if $\tilde{P}_k \geq 0$. This completes the proof.  □

Note that a symplectic transformation with a lower block-triangular transformation matrix preserves also the so-called *multiplicity* of a focal point. We will present and prove this statement later in this book (see Sect. 4.4.2).

In the next statement, we prove that the quadratic functional associated with (SDS) transforms under a transformation with lower block-triangular symplectic transformation matrix (2.142) essentially in the same way as the corresponding symplectic system.

**Lemma 2.73** *Let $y = \binom{x}{u}$ be an admissible pair for functional (2.57) and let $\tilde{y}_k = \mathcal{R}_k^{-1} y_k$ with $\tilde{y} = \binom{\tilde{x}}{\tilde{u}}$, where $\mathcal{R}_k$ is a lower block-triangular transformation matrix given in (2.142). Denote by $\tilde{\mathcal{F}}(\tilde{y})$ the quadratic functional corresponding to transformed system (2.139), i.e., to the symplectic system with block entries given in (2.143). Then we have*

$$\mathcal{F}(y) = \tilde{x}_k^T G_k^T H_k \tilde{x}_k \big|_{k=0}^{N+1} + \tilde{\mathcal{F}}(\tilde{y}). \tag{2.144}$$

*Proof* Following (2.57), denote by $\Delta \mathcal{F}_k$ and $\Delta \tilde{\mathcal{F}}_k$ the expressions

$$\begin{aligned}
\Delta \mathcal{F}_k(y) &:= x_k^T C_k^T A_k x_k + 2 x_k^T C_k^T B_k u_k + u_k^T B_k^T D_k u_k, \\
\Delta \tilde{\mathcal{F}}_k(\tilde{y}) &:= \tilde{x}_k^T \tilde{C}_k^T \tilde{A}_k \tilde{x}_k + 2 \tilde{x}_k^T \tilde{C}_k^T \tilde{B}_k \tilde{u}_k + \tilde{u}_k^T \tilde{B}_k^T \tilde{D}_k \tilde{u}_k.
\end{aligned}$$

By using Lemma 2.30, we then prove that

$$\Delta \tilde{\mathcal{F}}_k(\tilde{y}) - \Delta(\tilde{x}_k^T \tilde{u}_k) = \Delta \mathcal{F}_k(y) - \Delta(x_k^T u_k)$$

or

$$\Delta \mathcal{F}_k(y) = \Delta(\tilde{x}_k^T G_k^T H_k \tilde{x}_k) + \Delta \tilde{\mathcal{F}}_k(\tilde{y}). \tag{2.145}$$

Indeed

$$\Delta \tilde{\mathcal{F}}_k(\tilde{y}) - \Delta(\tilde{x}_k^T \tilde{u}_k)$$

$$= \tilde{x}_{k+1}^T(-\tilde{u}_{k+1} + \tilde{C}_k \tilde{x}_k + \tilde{D}_k \tilde{u}_k)$$

$$= \tilde{x}_{k+1}^T(-\tilde{u}_{k+1} + (0 \ I)\, \tilde{S}_k \tilde{y}_k)$$

$$= x_{k+1}^T H_{k+1}^{T-1}\big(G_{k+1}^T x_{k+1} - H_{k+1}^T u_{k+1} + (-G_{k+1}^T, \ H_{k+1}^T)\mathcal{S}_k y_k\big)$$

$$= x_{k+1}^T H_{k+1}^{T-1}\big(G_{k+1}^T x_{k+1} - H_{k+1}^T u_{k+1} - G_{k+1}^T(I \ 0)\, \mathcal{S}_k y_k + H_{k+1}^T(0 \ I)\, \mathcal{S}_k y_k\big)$$

$$= x_{k+1}^T\big(-u_{k+1} + (0 \ I)\, \mathcal{S}_k y_k\big)$$

$$= x_{k+1}^T(-u_{k+1} + \mathcal{C}_k x_k + \mathcal{D}_k u_k)$$

$$= \Delta \mathcal{F}_k(y) - \Delta(x_k^T u_k).$$

Finally, using the connection $\tilde{y}_k = \mathcal{R}_k^{-1} y_k$, we derive (2.145). Summing this identity for $k \in [0, N]_{\mathbb{Z}}$, we prove (2.144). $\qquad\square$

We finish this section with the proof of formula (2.65) in Lemma 2.32.

*Alternative proof of Lemma 2.32* Consider the transformation

$$\begin{pmatrix} x_k \\ u_k \end{pmatrix} = \begin{pmatrix} I & 0 \\ Q_k & I \end{pmatrix} \begin{pmatrix} \tilde{x}_k \\ z_k \end{pmatrix}, \quad \text{i.e.,} \quad \tilde{x}_k = x_k, \ z_k = u_k - Q_k x_k.$$

Then by (2.143) we have

$$\tilde{\mathcal{A}}_k = \mathcal{A}_k + \mathcal{B}_k Q_k, \quad \tilde{\mathcal{B}}_k = \mathcal{B}_k, \quad \tilde{\mathcal{C}}_k = -R_k[Q], \quad \tilde{\mathcal{D}}_k = -Q_{k+1}\mathcal{B}_k + \mathcal{D}_k.$$

Then the summand in the transformed quadratic functional is

$$\Delta \tilde{\mathcal{F}}_k(\tilde{x}, z) = -\tilde{x}_k^T R_k^T[Q]\, (\mathcal{A}_k + \mathcal{B}_k Q_k)\, \tilde{x}_k - 2\, \tilde{x}_k^T R_k^T[Q]\, \mathcal{B}_k z_k$$

$$\qquad\qquad + z_k^T \mathcal{B}_k^T(-Q_{k+1}\mathcal{B}_k + \mathcal{D}_k)\, z_k$$

$$\qquad = -x_k^T R_k^T[Q]\, (\mathcal{A}_k + \mathcal{B}_k Q_k)\, x_k - 2\, x_k^T R_k^T[Q]\, \mathcal{B}_k(u_k - Q_k x_k)$$

$$+ z_k^T P_k[Q] z_k$$
$$= x_k^T R_k^T [Q] (\mathcal{B}_k Q_k - A_k) x_k - 2 x_k^T R_k^T [Q] \mathcal{B}_k u_k + z_k^T P_k[Q] z_k.$$

Substituting the last expression into (2.145), we derive (2.65).                     □

## 2.6.2   Trigonometric or Bohl Transformation

Now, consider the trigonometric symplectic system (2.12) in the matrix form

$$S_{k+1} = \mathcal{P}_k S_k + \mathcal{Q}_k C_k, \quad C_{k+1} = -\mathcal{Q}_k S_k + \mathcal{P}_k C_k \tag{2.146}$$

with $\mathcal{P}_k$ and $\mathcal{Q}_k$ satisfying (see (2.13))

$$\mathcal{P}_k^T \mathcal{P}_k + \mathcal{Q}_k^T \mathcal{Q}_k = I, \quad \mathcal{P}_k^T \mathcal{Q}_k = \mathcal{Q}_k^T \mathcal{P}_k.$$

The fact that trigonometric systems are self-reciprocal, i.e., transformation (2.138) with $\mathcal{R}_k = \mathcal{J}$ transforms this system into itself, implies that if $(S, C)$ is a solution of (2.146), then $(C, -S)$ solves this system as well. Consequently, if the matrix $\begin{pmatrix} S_k & C_k \\ C_k & -S_k \end{pmatrix}$ is symplectic for some $k$, then it is symplectic everywhere and the following identities hold:

$$C_k^T C_k + S_k^T S_k = I, \quad C_k^T S_k = S_k^T C_k, \tag{2.147}$$
$$C_k C_k^T + S_k S_k^T = I, \quad S_k C_k^T = C_k S_k^T. \tag{2.148}$$

The following theorem presents the discrete analog of Theorem 1.52.

**Theorem 2.74** *There exist $n \times n$-matrices $H_k$ and $K_k$ such that $H_k$ is nonsingular and $H_k^T K_k = K_k^T H_k$ for $k \in [0, N + 1]_{\mathbb{Z}}$, and the transformation*

$$\begin{pmatrix} s_k \\ c_k \end{pmatrix} = \begin{pmatrix} H_k^{-1} & 0 \\ -K_k^T & H_k^T \end{pmatrix} \begin{pmatrix} x_k \\ u_k \end{pmatrix} \tag{2.149}$$

*transforms the symplectic system (SDS) into the trigonometric system (2.146) without changing the oscillatory behavior. Moreover, the matrices $\mathcal{P}_k$ and $\mathcal{Q}_k$ from (2.146) may be explicitly given by*

$$\mathcal{P}_k = H_{k+1}^{-1} A_k H_k + H_{k+1}^{-1} \mathcal{B}_k K_k, \quad \mathcal{Q}_k = H_{k+1}^{-1} \mathcal{B}_k H_k^{T-1}. \tag{2.150}$$

*Proof* We let $\tilde{Y} = (\tilde{X}, \tilde{U})$ and $Y = (X, U)$ be normalized conjoined bases of (SDS). Then they form the symplectic fundamental solution matrix $Z = (\tilde{Y}, Y)$, and then

$Z_k Z_k^T > 0$ because $Z_k$ is nonsingular. In particular, the matrix $X_k X_k^T + \tilde{X}_k \tilde{X}_k^T$ is positive definite, and there exists a nonsingular $n \times n$-matrix $H_k$ with

$$H_k H_k^T = X_k X_k^T + \tilde{X}_k \tilde{X}_k^T, \quad k \in [0, N+1]_{\mathbb{Z}}. \tag{2.151}$$

We put

$$K_k := \left( U_k X_k^T + \tilde{U}_k \tilde{X}_k^T \right) H_k^{T-1}, \tag{2.152}$$

so that the following identity hold

$$Z_k Z_k^T = \mathcal{R}_k \mathcal{R}_k^T, \quad \mathcal{R}_k := \begin{pmatrix} H_k & 0 \\ K_k & H_k^{T-1} \end{pmatrix}. \tag{2.153}$$

Indeed, since $Z_k Z_k^T$ is symplectic, we have that the matrix $H_k H_k^T K_k H_k^T$ is symmetric, and then, due to nonsingularity of $H_k$, so is $H_k^T K_k$, i.e., $H_k^T K_k = K_k^T H_k$. Moreover, $U U^T + \tilde{U} \tilde{U}^T = K_k K_k^T + (H_k H_k^T)^{-1}$.

So we have proved that the block lower triangular matrix $\mathcal{R}_k$ in the right-hand side of (2.153) is symplectic, and then the product of two symplectic matrices $\tilde{Z}_k = \begin{pmatrix} H_k^{-1} & 0 \\ -K_k^T & H_k^T \end{pmatrix} Z_k$ satisfies the condition $\tilde{Z}_k \tilde{Z}_k^T = I$, i.e.,

$$\tilde{Z}_k = \begin{pmatrix} C_k & S_k \\ -S_k & C_k \end{pmatrix} = \mathcal{R}_k^{-1} Z_k \tag{2.154}$$

is symplectic and orthogonal matrix. Moreover since $\tilde{S}_k = \tilde{Z}_{k+1} \tilde{Z}_k^{-1}$ is also symplectic and orthogonal, we have that $\tilde{Z}_k$ solves trigonometric system (2.146) with the blocks given by (2.150) according to formulas (2.143) for the transformation matrix $\mathcal{R}_k$ in (2.153). The proof is completed. □

*Remark 2.75* Observe that every trigonometric system (2.146) can be transformed into another trigonometric system with symmetric and positive semidefinite matrices $\tilde{\mathcal{Q}}_k$ at the position of the matrices $\mathcal{Q}_k$, using the transformation

$$\begin{pmatrix} \tilde{s}_k \\ \tilde{c}_k \end{pmatrix} = \begin{pmatrix} H_k^{-1} & 0 \\ 0 & H_k^T \end{pmatrix} \begin{pmatrix} s_k \\ c_k \end{pmatrix},$$

where the matrices $H_k$ are recursively defined by $H_0 = I$ and $H_{k+1} = G_k^{-1} H_k$ with orthogonal matrices $G_k$, i.e., $G_k^T G_k = I$, such that $G_k \mathcal{Q}_k$ are symmetric and positive semidefinite. Such matrices $G_k$ exist according to the well-known principle of polar decomposition (see, e.g., [195, Theorem 3.1.9(c)]). This setting implies that all matrices $H_k$ are orthogonal and hence that the transformation matrices

$\mathrm{diag}\{H_k^{-1}, H_k^T\}$ are symplectic and orthogonal. The transformed system then reads

$$\tilde{S}_{k+1} = \tilde{\mathcal{P}}_k \tilde{S}_k + \tilde{\mathcal{Q}}_k \tilde{C}_k, \qquad \tilde{C}_{k+1} = -\tilde{\mathcal{Q}}_k \tilde{S}_k + \tilde{\mathcal{P}}_k \tilde{C}_k,$$

where

$$\tilde{\mathcal{P}}_k = H_{k+1}^{-1} \mathcal{P}_k H_k, \quad \tilde{\mathcal{Q}}_k = H_{k+1}^{-1} \mathcal{Q}_k H_k = H_k^{-1} G_k Q_k H_k^{T-1}$$

so that indeed all matrices $\tilde{\mathcal{Q}}_k$ are symmetric and positive semidefinite.

By analogy with the continuous case (see Sect. 1.5.4), we formulate a necessary and sufficient condition of oscillation of (2.146) (see formula (1.136)).

**Theorem 2.76** *Assume that the matrices $\mathcal{Q}_k$ in the trigonometric system (2.146) are symmetric and positive definite. Then this system is nonoscillatory at $\infty$ if and only if*

$$\sum_{k=0}^{\infty} \operatorname{arccot} \lambda^{(1)}(\mathcal{Q}_k^{-1} \mathcal{P}_k) < \infty, \tag{2.155}$$

*where $\lambda^{(1)}(\cdot)$ denotes the minimal eigenvalue of the matrix indicated.*

*Proof* The proof of this result can be found in [46]. □

*Remark 2.77* The previous theorem requires the matrices $\mathcal{Q}_k$ in (2.146) to be positive definite. By the trigonometric transformation of (SDS) given in Theorem 2.74, the matrices $\mathcal{Q}_k$ are given by $\mathcal{Q}_k = H_{k+1}^{-1} \mathcal{B}_k H_k^{T-1}$, hence a necessary condition for positive definiteness of $\mathcal{Q}_k$ is nonsingularity of $\mathcal{B}_k$. However, symplectic systems (SDS) with $\mathcal{B}_k$ nonsingular do not cover a relatively large class of equations and systems, e.g., the higher-order Sturm-Liouville difference equations (2.27), which can be written in the form (SDS) with (see Example 2.21)

$$\mathcal{B}_k = \frac{1}{r_k^{(n)}} \begin{pmatrix} 0 \dots 0\, 1 \\ \vdots \quad \vdots\, \vdots \\ 0 \dots 0\, 1 \end{pmatrix}.$$

The extension of Theorem 2.74 to trigonometric systems with arbitrary $\mathcal{Q}_k$ is given in [96].

## 2.6.3 Hyperbolic Transformation

In this section we prove that any symplectic difference systems (SDS) satisfying certain additional condition can be transformed into a hyperbolic system (see

Sect. 2.1.2). Consider the hyperbolic system (2.21) in the matrix form

$$S_{k+1} = \mathcal{P}_k S_k + \mathcal{Q}_k C_k, \quad C_{k+1} = \mathcal{Q}_k S_k + \mathcal{P}_k C_k, \tag{2.156}$$

with $\mathcal{P}_k$ and $\mathcal{Q}_k$ satisfying (see (2.22))

$$\mathcal{P}_k^T \mathcal{P}_k - \mathcal{Q}_k^T \mathcal{Q}_k = I = \mathcal{P}_k \mathcal{P}_k^T - \mathcal{Q}_k \mathcal{Q}_k^T, \quad \mathcal{P}_k^T \mathcal{Q}_k - \mathcal{Q}_k^T \mathcal{P}_k = 0 = \mathcal{P}_k \mathcal{Q}_k^T - \mathcal{Q}_k \mathcal{P}_k^T.$$

The fact that transformation (2.138) with $\mathcal{R}_k = P_1$ (see (2.20)) transforms this system into itself implies that if $(S, C)$ is a solution of (2.156), then $(C, S)$ solves this system as well. Consequently, if the matrix $\begin{pmatrix} S_k & C_k \\ C_k & S_k \end{pmatrix}$ is symplectic for some $k$, then it is symplectic everywhere and the following identities hold:

$$S_k^T C_k - C_k^T S_k = 0 = S_k C_k^T - C_k S^T, \tag{2.157}$$

$$C_k^T C_k - S_k^T S_k = I = C_k C_k^T - S_k S_k^T. \tag{2.158}$$

**Theorem 2.78** *Suppose that symplectic system (SDS) possesses normalized conjoined bases* $Y = \begin{pmatrix} X \\ U \end{pmatrix}$, $\tilde{Y} = \begin{pmatrix} \tilde{X} \\ \tilde{U} \end{pmatrix}$ *such that* $X_k \tilde{X}_k^T$ *is positive definite in a given discrete interval. Then, in this interval, there exist* $n \times n$-*matrices* $H_k$ *and* $K_k$ *such that* $H_k$ *is nonsingular,* $H_k^T K_k = K_k^T H_k$, *and the transformation*

$$\begin{pmatrix} s_k \\ c_k \end{pmatrix} = \begin{pmatrix} H_k^{-1} & 0 \\ -K_k^T & H_k^T \end{pmatrix} \begin{pmatrix} x_k \\ u_k \end{pmatrix} \tag{2.159}$$

*transforms symplectic system (SDS) into the hyperbolic system (2.156) without changing the oscillatory behavior, i.e., a conjoined basis* $Y_k$ *of (SDS) has a focal point in* $(k, k+1]$ *if and only if* $\begin{pmatrix} H_k^{-1} X_k \\ -K_k^T X_k + H_k^T U_k \end{pmatrix}$ *has a focal point there. Moreover, the matrices* $\mathcal{P}_k$ *and* $\mathcal{Q}_k$ *are given by the formulas*

$$\mathcal{P}_k = H_{k+1}^{-1} A_k H_k + H_{k+1}^{-1} B_k K_k, \quad \mathcal{Q}_k = H_{k+1}^{-1} B_k H_k^{T-1}. \tag{2.160}$$

*Proof* Let $Y = \begin{pmatrix} X \\ U \end{pmatrix}$ and $\tilde{Y} = \begin{pmatrix} \tilde{X} \\ \tilde{U} \end{pmatrix}$ be normalized conjoined bases of (SDS) such that $X_k \tilde{X}_k^T$ is positive definite, $H$ be any $n \times n$ matrix satisfying

$$H_k H_k^T = 2 X_k \tilde{X}_k^T \tag{2.161}$$

and let

$$K_k := (U_k \tilde{X}_k^T + \tilde{U}_k X_k^T) H_k^{T-1}. \tag{2.162}$$

Then $H_k$ is nonsingular by the assumption in Theorem 2.78. Consider the symplectic fundamental solution matrix of system (SDS) of the form

$$Z_k = \frac{1}{\sqrt{2}} (\tilde{Y}_k, \ Y_k) \begin{pmatrix} I & -I \\ I & I \end{pmatrix}, \quad \text{and}$$

then, by analogy with the proof of Theorem 2.74, we prove the identity

$$Z_k P_2 Z_k^T P_2 = \mathcal{R}_k P_2 \mathcal{R}_k^T P_2, \quad \mathcal{R}_k := \begin{pmatrix} H_k & 0 \\ K_k & H_k^{T-1} \end{pmatrix}, \quad P_2 = \text{diag}\{I, -I\},$$

$$(2.163)$$

where $H_k$ and $K_k$ are given by (2.161) and (2.162). Indeed, using that the matrix in the left-hand side of this relation is symplectic (see Lemma 1.58 (iv)), we have by (2.161) and (2.162) that $H_k H_k^T K_k H_k^T$ is symmetric together with $H_k^T K_k$. Moreover, $U_k \tilde{U}_k^T + \tilde{U}_k U_k^T = 2 U_k \tilde{U}_k^T = K_k K_k^T - (H_k H_k^T)^{-1}$. Then we have proved (2.163) and the symplecticity of the transformation matrix $\mathcal{R}_k$. Identity (2.163) implies the relation

$$\tilde{Z}_k P_2 \tilde{Z}_k^T P_2 = I, \ \tilde{Z}_k = \mathcal{R}_k^{-1} Z_k$$

for the symplectic matrix $\tilde{Z}$, then

$$\tilde{Z}_k = P_2 \tilde{Z}_k^{T-1} P_2 = P_2 \mathcal{J} \tilde{Z}_k \mathcal{J}^T P_2 = P_1 \tilde{Z}_k P_1, \quad \text{where } P_1 := \begin{pmatrix} 0 & I \\ I & 0 \end{pmatrix},$$

that is,

$$\tilde{Z}_k = \begin{pmatrix} C_k & S_k \\ S_k & C_k \end{pmatrix} = \mathcal{R}_k^{-1} Z_k \tag{2.164}$$

is a symplectic and hyperbolic matrix. Moreover, since $\tilde{S}_k = \tilde{Z}_{k+1} \tilde{Z}_k^{-1}$ is also symplectic and hyperbolic, we have that $\tilde{Z}_k$ solves hyperbolic system (2.156) with the blocks given by (2.160) according to formulas (2.143) for the transformation matrix $\mathcal{R}_k$ in (2.153). The proof is completed. □

Remark 2.79 By the same arguments as in Remark 2.75, one can transform any hyperbolic system (2.156) into another hyperbolic system with symmetric and positive semidefinite matrices $\tilde{Q}_k$ at the position of the matrices $Q_k$, using the transformation

$$\begin{pmatrix} \tilde{x}_k \\ \tilde{u}_k \end{pmatrix} = \begin{pmatrix} H_k^{-1} & 0 \\ 0 & H_k^T \end{pmatrix} \begin{pmatrix} x_k \\ u_k \end{pmatrix},$$

where the matrices $H_k$ are recursively defined by $H_0 = I$ and $H_{k+1} = G_k^{-1} H_k$ with orthogonal matrices $G_k$, i.e., $G_k^T G_k = I$, such that $G_k Q_k$ are symmetric and positive semidefinite. The transformed system then reads

$$\tilde{x}_{k+1} = \tilde{P}_k \tilde{x}_k + \tilde{Q}_k \tilde{u}_k, \qquad \tilde{u}_{k+1} = \tilde{Q}_k \tilde{x}_k + \tilde{P}_k \tilde{u}_k,$$

where

$$\tilde{P}_k = H_{k+1}^{-1} P_k H_k \quad \text{and} \quad \tilde{Q}_k = H_{k+1}^{-1} Q_k H_k = H_k^{-1} G_k Q_k H_k^{T-1},$$

so that indeed all matrices $\tilde{Q}_k$ are symmetric and positive semidefinite.

*Remark 2.80* This remark concerns the assumption of the existence of a pair of normalized conjoined bases $Y = \begin{pmatrix} X \\ U \end{pmatrix}$ and $\tilde{Y} = \begin{pmatrix} \tilde{X} \\ \tilde{U} \end{pmatrix}$ such that $X_k \tilde{X}_k^T$ is positive definite. If (SDS) is nonoscillatory at $\infty$ (and eventually controllable), then this system possesses the recessive and dominant solutions $\bar{Y} = \begin{pmatrix} \bar{X} \\ \bar{U} \end{pmatrix}$ and $\hat{Y} = \begin{pmatrix} \hat{X} \\ \hat{U} \end{pmatrix}$ at $\infty$, respectively, such that $w(\bar{Y}, \hat{Y}) = I$. Then

$$Y_k = \begin{pmatrix} X_k \\ U_k \end{pmatrix} = \frac{1}{\sqrt{2}} (\hat{Y}_k + \bar{Y}_k), \quad \tilde{Y}_k = \begin{pmatrix} \tilde{X}_k \\ \tilde{U}_k \end{pmatrix} = \frac{1}{\sqrt{2}} (\bar{Y}_k - \hat{Y}_k),$$

is a normalized pair of conjoined bases for which $X_k \tilde{X}_k^T = \frac{1}{2} (\hat{X}_k \hat{X}_k^T - \bar{X}_k \bar{X}_k^T)$ is positive definite eventually, since

$$\lim_{k \to \infty} \hat{X}_k^{-1} \bar{X}_k = 0.$$

Consequently, any nonoscillatory symplectic difference system (SDS) at $\infty$ can be transformed into a hyperbolic symplectic system.

## 2.6.4   Prüfer Transformation

In this section, we present a discrete analog of Theorem 1.51 for systems (SDS) which in turn generalizes results of Sect. 1.2.4 for the scalar Sturm-Liouville equations.

**Theorem 2.81** *Let* $Y = \begin{pmatrix} X \\ U \end{pmatrix}$ *be a conjoined basis of* (SDS). *There exist nonsingular matrices* $H_k$ *and* $n \times n$ *matrices* $S_k$ *and* $C_k$ *such that*

$$X_k = S_k^T H_k, \quad U_k = C_k^T H_k, \tag{2.165}$$

*where $\binom{S}{C}$ is a conjoined basis of (2.146) with*

$$\mathcal{P}_k = H_{k+1}^{T-1} Y_k^T S_k^T Y_k H_k^{-1}, \quad \mathcal{Q}_k = H_{k+1}^{T-1} Y_k^T S_k^T \mathcal{J} Y_k H_k^{-1}, \tag{2.166}$$

*and $H_k$ solve the first order difference system*

$$H_{k+1} = \tilde{Y}_{k+1}^T S_k \tilde{Y}_k H_k, \quad \tilde{Y}_k = \begin{pmatrix} S_k^T \\ C_k^T \end{pmatrix}. \tag{2.167}$$

*Proof* Let $Y = \binom{X}{U}$ be a conjoined basis of (SDS), and let $H_k$ be nonsingular matrices satisfying $H_k^T H_k = X_k^T X_k + U_k^T U_k > 0$. Introduce the $2n \times 2n$ matrix

$$\mathcal{Z}_k = P_1 \, (\mathcal{J} Y_k H_k^{-1}, \; Y_k H_k^{-1}) \, P_1 = \begin{pmatrix} C_k^T & -S_k^T \\ S_k^T & C_k^T \end{pmatrix},$$

where $P_1 = \begin{pmatrix} 0 & I \\ I & 0 \end{pmatrix}$. Then, according to properties (iv), (v), and (vii) in Lemma 1.58, we have that $\mathcal{Z}_k$ is symplectic and orthogonal (see Sect. 1.6.1), where we used the fact

$$w(\mathcal{J} Y_k H_k^{-1}, \; Y_k H_k^{-1}) = H_k^{T-1} Y_k^T Y_k H_k^{-1} = I.$$

Moreover, it is easy to verify directly that

$$\mathcal{Z}_{k+1}^T \mathcal{Z}_k = \tilde{S}_k = \begin{pmatrix} \mathcal{P}_k & \mathcal{Q}_k \\ -\mathcal{Q}_k & \mathcal{P}_k \end{pmatrix}, \tag{2.168}$$

where $\mathcal{P}_k$ and $\mathcal{Q}_k$ are given by (2.166). Indeed, from (2.168) we have

$$(C_{k+1} \; S_{k+1}) \begin{pmatrix} C_k^T \\ S_k^T \end{pmatrix} = (S_{k+1} \; C_{k+1}) \begin{pmatrix} S_k^T \\ C_k^T \end{pmatrix} = H_{k+1}^{T-1} Y_{k+1}^T Y_k H_k^{-1}$$

$$= H_{k+1}^{T-1} Y_k^T S_k^T Y_k H_k^{-1} = \mathcal{P}_k.$$

Similarly,

$$(C_{k+1} \; S_{k+1}) \begin{pmatrix} -S_k^T \\ C_k^T \end{pmatrix} = (S_{k+1} \; C_{k+1}) \begin{pmatrix} C_k^T \\ -S_k^T \end{pmatrix}$$

$$= H_{k+1}^{T-1} Y_k^T S_k^T \mathcal{J} Y_k H_k^{-1} = \mathcal{Q}_k.$$

Certainly the matrix in the right-hand side of (2.168) is also symplectic and orthogonal. Equation (2.168) implies $\binom{S_{k+1}}{C_{k+1}} = \tilde{S}_k \binom{S_k}{C_k}$, i.e., the matrix $\binom{S_k}{C_k}$ is a conjoined basis of (2.146). Finally, from (SDS), we have

$$\binom{S_{k+1}^T}{C_{k+1}^T} H_{k+1} = S_k \binom{S_k^T}{C_k^T} H_k, \text{ and} \tag{2.169}$$

then, using the property $S_k S_k^T + C_k C_k^T = I$, we derive (2.167) by multiplying (2.169) from the left by $(S_{k+1} \ C_{k+1})$.                                                  □

## 2.7  Notes and References

Main references for various special symplectic systems are the books [4, 16]. In addition, Jacobi matrix difference equations and symmetric three-term recurrence equations are studied from the point of view of symplectic systems in [304]. For properties of discrete linear Hamiltonian systems, we refer to the survey paper [3] and to the book [4, Sections 2.3–2.7]. Basic properties of solutions of trigonometric symplectic systems with nonsingular $\mathcal{B}_k$ are established in [25] and of hyperbolic symplectic systems in [108].

To define properly the concept of a focal point in a given interval (more precisely, the concept of "no focal points" in this interval) of a conjoined basis of linear Hamiltonian difference systems was a difficult problem, which resisted its solution for rather long time. The paper [39] of M. Bohner from 1996 constitutes a breaking point in this direction, in which he presented the definition of this notion for the linear Hamiltonian systems (2.15). More precisely, a conjoined basis $Y = \binom{X}{U}$ of (2.15) has *no focal point* in the interval $(k, k+1]$ if

$$\operatorname{Ker} X_{k+1} \subseteq \operatorname{Ker} X_k, \quad P_k := X_k X_{k+1}^\dagger (I - A_k)^{-1} B_k \geq 0. \tag{2.170}$$

In earlier works, oscillatory properties of linear Hamiltonian difference systems were studied only under the assumption that the matrix $B_k$ is positive definite (see the papers of Erbe and Yan [134–137]). This situation corresponds, in some sense, to a controllable system (2.15). The notion of "no (forward) focal points" in the interval $(k, k+1]$ for conjoined bases of symplectic system (SDS) in Definition 2.14 was introduced in [45] by M. Bohner and O. Došlý, and it was motivated by the corresponding notion in (2.170) from [39]. In addition, the notion of "no backward focal points" in $[k, k+1)$ is also from [45]. A transformation between the notions of no forward and backward focal points is presented in [300].

Identities in Lemma 2.26 and Corollary 2.27 were established in [189]. In this respect it is surprising that the solvability of the explicit Riccati equation (2.52) implies the invertibility of the matrices $A_k + B_k Q_k$ and $\mathcal{D}_k^T - B_k^T Q_{k+1}$. This fact is not very very well known, and it brings the result in Theorem 2.28 to the same form as in the continuous case—parts (ii) and (iii) of Theorem 1.47.

The importance of the study of equivalent conditions for the positivity and nonnegativity of discrete quadratic functionals is justified by the results in Theorems 1.26, 1.28, 1.32, and 1.33. The results on the nonnegativity of the functional $\mathcal{F}$ (Theorem 2.44 in Sect. 2.3.5) is essentially from [54]. However, the latter reference uses the necessity of the $P$-condition (2.73) for the nonnegativity of $\mathcal{F}$ from [100, Theorem 1]. The results on discrete Reid roundabout theorems (Theorems 2.36 and 2.41) are from [45]. A different proof of the equivalence of parts (i), (ii), and (iv) in Theorem 2.36, based on a matrix diagonalization technique, is presented in [171]. Extension of this technique to separated endpoints and to the nonnegativity of a quadratic functional (Theorem 2.49) is presented in [50, Theorem 2]. The extension of this nonnegativity result to the jointly varying endpoints (Theorem 2.52) is derived in [174, Theorem 2]. The corresponding results regarding the positivity of a quadratic functional (Theorems 2.50 and 2.53) are from [181, Theorems 5 and 10]. We note that the latter paper contains also further equivalent conditions for the positivity of the functional $\mathcal{G}$ in (2.84) in terms of a conjoined basis $Y$ of (SDS) with no forward focal points in $(0, N + 1]$ and with invertible $X_k$ for all $k \in [0, N + 1]_{\mathbb{Z}}$. This then yields an additional initial equality or inequality at $k = 0$ (see [181, Theorems 6, 7, and 11]). First results on the positivity of $\mathcal{G}$ with joint endpoints are proven in [42, Theorem 1]. A survey of these conditions for continuous and discrete case is presented in [306]. In the literature one can find additional conditions, which are equivalent to the positivity of discrete quadratic functional $\mathcal{F}$ and $\mathcal{G}$, such as the nonexistence of conjugate intervals [181], the nonexistence of coupled intervals [180, 184, 188], the solvability of implicit Riccati equations [173, 174, 181], or the positivity and nonnegativity of perturbed quadratic functionals [175, 258]. Various Picone-type identities for symplectic systems are derived in [191, 302]. Transformations which reduce the problem with separated endpoints into a problem with zero endpoints on an extended interval, and a problem with general jointly varying endpoints into a problem with separated endpoints in the double dimension $2n$ (see Sect. 5.2.1 for more details) are known in [39, 42, 53, 103] due to Bohner, Došlý, and Kratz, or earlier in [109] due to Dwyer and Zettl. These techniques are also used in [173, 174, 181, 185]. Recently, another transformation of this type was developed in [196, 308]. In this respect the statement in Theorem 2.40 is a special case of [181, Theorem 7].

The inequalities for the Riccati-type quotients for symplectic systems presented in Sect. 2.4.1 are from [173]. The results on discrete Riccati inequality (2.115) in Sect. 2.4.2 are from [174]. For the discrete linear Hamiltonian system (1.102), the Riccati inequality was derived in [176]. Further inequalities between symmetric solutions of two Riccati equations (2.52) or Riccati inequalities (2.115) in the spirit of Theorem 2.55 are proven in [300].

The results on recessive and dominant solutions of (SDS) at infinity are essentially from [81], with some small adjustments from the recent work [284]. In particular, the recessive and dominant solutions are introduced in Sect. 2.5.1 through the summation property (2.123) with (2.121), which was not the case in [16, Section 3.11] and [25, 81]. In the latter references, the limit property in Theorem 2.67 was used for this purpose as the definition. We refer to Remark 6.117

for more discussion on this subject. Recently, a theory of recessive and dominant solutions of (SDS) was initiated in [284, 290] without the controllability assumption. We will discuss these more general concepts in Sect. 6.3. Observe that the arguments in the construction of the recessive solution at $\infty$ of an eventually controllable symplectic system (in the proof of Theorem 2.66) follow the more general considerations presented in Sect. 6.3.5 (see the proof of Theorem 6.103). This explicit construction of the recessive solution of (SDS) at $\infty$ is also new in the controllable case.

Transformation theory of symplectic systems was initiated by Bohner and Došlý in [45]. The results on trigonometric, Prüfer, and hyperbolic transformations in Sects. 2.6.2–2.6.4 were obtained in [46, 47, 108]. Applications of trigonometric and hyperbolic transformations in the oscillation theory of symplectic system are derived in [48, 85, 106, 107]. Discrete matrix trigonometric and hyperbolic functions, as solutions of discrete trigonometric and hyperbolic symplectic systems, are defined in [24, 194, 333].

We recommend the following additional related references for further reading about the topics presented in this chapter: [9, 14, 40, 111, 237] for discrete Riccati equations, [11, 15] for recessive and dominant solutions of special symplectic systems, and [17, 84, 86, 88, 183, 266, 323, 324] for general theory of symplectic difference systems. Moreover, results about more general theory of dynamic equations on time scales, in particular on Hamiltonian or symplectic dynamic systems on time scales, are presented in [12, 49, 58, 59, 97, 99, 187, 190, 299, 307, 334].

# Chapter 3
# Comparative Index Theory

In this chapter we introduce the comparative index as a main mathematical tool for the results in the subsequent chapters of this book. Recall from [27] that a subspace $\text{span}\{u_1, \ldots, u_n\} \subseteq \mathbb{R}^{2n}$ is a *Lagrangian subspace* if it has dimension $n$ and $u_i^T \mathcal{J} u_j = 0$ for all $i$ and $j$. We introduce a notion of the comparative index $\mu(Y, \hat{Y})$ for a pair of Lagrangian subspaces represented by $2n \times n$ matrices $Y$ and $\hat{Y}$ with conditions

$$Y^T \mathcal{J} Y = 0, \quad \text{rank } Y = n, \tag{3.1}$$

(and similarly for $\hat{Y}$). Depending on the formulation of the problem (see Chaps. 4 and 5), the matrices $Y$ and $\hat{Y}$ can represent Lagrangian subspaces associated with conjoined bases of system (SDS) or with symplectic coefficient matrices $\mathcal{S}_k$ and $\hat{\mathcal{S}}_k$ of two symplectic systems (SDS) or with their fundamental solution matrices. For the special case of $Y := Y_{k+1}$ and $\hat{Y} := \mathcal{S}_k(0 \; I)^T$, the comparative index $\mu(Y, \hat{Y})$ represents the main concept of the discrete oscillation theory—the multiplicity of a focal point (see Chap. 4). Because of this connection, algebraic properties of the comparative index turned out to be an essential effective tool for solving several important problems in the discrete oscillation theory.

## 3.1 Comparative Index and Its Properties

In this section we define one of the fundamental concepts of the oscillation theory of symplectic difference systems, the concept of a *comparative index* of two matrices satisfying (3.1).

© Springer Nature Switzerland AG 2019
O. Došlý et al., *Symplectic Difference Systems: Oscillation and Spectral Theory*,
Pathways in Mathematics, https://doi.org/10.1007/978-3-030-19373-7_3

### 3.1.1  Definition of Comparative Index

Consider two $2n \times n$ matrices $Y = \left(\begin{smallmatrix} X \\ U \end{smallmatrix}\right)$ and $\hat{Y} = \left(\begin{smallmatrix} \hat{X} \\ \hat{U} \end{smallmatrix}\right)$ satisfying (3.1) and let

$$w := w(Y, \hat{Y}) = Y^T \mathcal{J} \hat{Y} \tag{3.2}$$

be their Wronskian.

**Definition 3.1 (Comparative Index)**  Let

$$\mu_1(Y, \hat{Y}) := \operatorname{rank} \mathcal{M}, \quad \mathcal{M} := (I - X^\dagger X) \, w, \tag{3.3}$$

and

$$\mu_2(Y, \hat{Y}) := \operatorname{ind} \mathcal{P}, \quad \mathcal{P} := \mathcal{T}^T (w^T X^\dagger \hat{X}) \mathcal{T}, \quad \mathcal{T} := I - \mathcal{M}^\dagger \mathcal{M}, \tag{3.4}$$

where $\mathcal{M}$ is defined in (3.3). The quantity

$$\mu(Y, \hat{Y}) = \mu_1(Y, \hat{Y}) + \mu_2(Y, \hat{Y})$$

is called the *comparative index* of the matrices $Y$ and $\hat{Y}$.

The next result will be important in proving the properties of the comparative index.

**Theorem 3.2**  *The following statements hold.*

  (i) *The matrix $\mathcal{M}$ in (3.3), (3.4) can be replaced by the matrix*

$$\tilde{\mathcal{M}} := (I - XX^\dagger) \, \hat{X}, \tag{3.5}$$

  *i.e., $\operatorname{rank} \mathcal{M} = \operatorname{rank} \tilde{\mathcal{M}}$ and*

$$\mathcal{T} = I - \mathcal{M}^\dagger \mathcal{M} = I - \tilde{\mathcal{M}}^\dagger \tilde{\mathcal{M}}. \tag{3.6}$$

  (ii) *The condition $\mu_1(Y, \hat{Y}) = 0$ is equivalent to the condition $\operatorname{Im} \hat{X} \subseteq \operatorname{Im} X$. In this case $\mathcal{T} = I$ and*

$$\mu(Y, \hat{Y}) = \mu_2(Y, \hat{Y}) = \operatorname{ind} [\hat{X}^T (\hat{Q} - Q) \hat{X}], \tag{3.7}$$

  *where $Q$ and $\hat{Q}$ are symmetric matrices such that*

$$X^T Q X = X^T U, \quad \hat{X}^T \hat{Q} \hat{X} = \hat{X}^T \hat{U}. \tag{3.8}$$

*In particular, if $X$ and $\hat{X}$ are invertible, then*

$$\mu(Y, \hat{Y}) = \mu_2(Y, \hat{Y}) = \text{ind}(\hat{Q} - Q), \quad Q := U X^{-1}, \quad \hat{Q} := \hat{U}\hat{X}^{-1}. \tag{3.9}$$

*(iii) The matrix $\mathcal{P}$ in (3.4) is symmetric and can be presented in the form*

$$\mathcal{P} = \mathcal{T}^T [\hat{X}^T (\hat{Q} - Q)\, \hat{X}]\, \mathcal{T}, \tag{3.10}$$

*where $Q$ and $\hat{Q}$ are any symmetric matrices satisfying (3.8) and $\mathcal{T}$ is defined by (3.6) with $\mathcal{M}$ and $\tilde{\mathcal{M}}$ given by (3.3) and (3.5).*
*(iv) The equality $\mu(Y, \hat{Y}) = 0$ holds if and only if*

$$\text{Im}\,\hat{X} \subseteq \text{Im}\,X \quad \text{and} \quad \hat{X}^T (\hat{Q} - Q)\,\hat{X} \geq 0, \tag{3.11}$$

*where $Q = Q^T$ and $\hat{Q} = \hat{Q}^T$ are again given by (3.8).*

*Proof*

(i) By (3.3) we have

$$\begin{aligned}
\mathcal{M} &= (I - X^\dagger X)\, w = (I - X^\dagger X)\, (X^T \hat{U} - U^T \hat{X}) \\
&= -(I - X^\dagger X)\, U^T \hat{X} = -(I - X^\dagger X)\, U^T (I - X X^\dagger)\, \hat{X} \\
&= \mathcal{L}^T (I - X X^\dagger)\, \hat{X} = \mathcal{L}^T \tilde{\mathcal{M}},
\end{aligned}$$

where we have used (1.171) and where the (invertible) matrix $\mathcal{L}$ is defined in (1.174). Then rank $\mathcal{M} = $ rank $\tilde{\mathcal{M}}$. Furthermore, the equality $\mathcal{M}^\dagger \mathcal{M} = \tilde{\mathcal{M}}^\dagger \tilde{\mathcal{M}}$ holds by Lemma 1.63. Hence, we proved (3.6), and the matrix $\mathcal{M}$ in (3.3) and (3.4) can be replaced by $\tilde{\mathcal{M}}$.
(ii) By the previous part of the proof, we have $\mu_1(Y, \hat{Y}) = 0$ if and only if $\mathcal{M} = 0 = \tilde{\mathcal{M}}$. Then $\hat{X} = X X^\dagger \hat{X}$, which is equivalent by (1.160) to the condition $\text{Im}\,\hat{X} \subseteq \text{Im}\,X$. In this case $\mathcal{T} - I - \mathcal{M}^\dagger \mathcal{M} = I$ and the matrix

$$\mathcal{P} = w^T X^\dagger \hat{X} = (\hat{U}^T X - \hat{X}^T U)\, X^\dagger \hat{X} = \hat{U}^T \hat{X} - \hat{X}^T (X^T)^\dagger X^T U X^\dagger \hat{X}$$

are symmetric, since we suppose (3.1). In addition, if symmetric matrices $Q$ and $\hat{Q}$ satisfy (3.8), then

$$\mathcal{P} = \hat{X}^T \hat{Q}\hat{X} - \hat{X}^T (X^T)^\dagger X^T Q X X^\dagger \hat{X} = \hat{X}^T (\hat{Q} - Q)\,\hat{X},$$

where we have again used that $\tilde{\mathcal{M}} = 0$, i.e., $\hat{X} = X X^\dagger \hat{X}$. In this case we have $\mu(Y, \hat{Y}) = \mu_1(Y, \hat{Y}) + \mu_2(Y, \hat{Y}) = \mu_2(Y, \hat{Y}) = \text{ind}\,\mathcal{P}$. In particular, for invertible $X$ and $\hat{X}$, obviously $\mu_1(Y, \hat{Y}) = 0$ holds, and, hence, $Q = U X^{-1}$, $\hat{Q} = \hat{U}\hat{X}^{-1}$, and (3.9) follows from (3.7).

(iii) In the general case, we have $\mathcal{M}\mathcal{T} = 0$, i.e., $\text{Im}(\hat{X}\mathcal{T}) \subseteq \text{Im}\, X$. Then the matrix $w^T X^\dagger \hat{X}$ in (3.4) is symmetric, when restricted to the subspace $\text{Im}\,\mathcal{T}$. Indeed, repeating the arguments from part (ii) of this proof, we obtain

$$\mathcal{P} = \mathcal{T}^T[\hat{U}^T\hat{X} - \hat{X}^T(X^T)^\dagger X^T U X^\dagger \hat{X}]\mathcal{T} = \mathcal{T}X^T(\hat{Q} - Q)\hat{X}\mathcal{T} = \mathcal{P}^T,$$

so that $\mathcal{P}$ is symmetric. Recall that we proved (3.6) in part (i).

(iv) This statement is a direct consequence of part (ii).

$\square$

**Remark 3.3** Observe that the comparative index is not a commutative function of $Y$ and $\hat{Y}$. Later, in Theorem 3.5(v) we present the connection between $\mu(Y, \hat{Y})$ and $\mu(\hat{Y}, Y)$.

**Remark 3.4** We now evaluate the comparative index for some special cases.

(i) As it was already mentioned in Theorem 3.2(ii), when $\det X \neq 0$ and $\det \hat{X} \neq 0$, then $\mathcal{M} = \tilde{\mathcal{M}} = 0$, and $Q = UX^{-1}$, $\hat{Q} = \hat{U}\hat{X}^{-1}$, so that we get $\mu(Y, \hat{Y}) = \text{ind}(\hat{Q} - Q)$.

(ii) If $\hat{X} = 0$, then $\mu(Y, \hat{Y}) = 0$ for any $Y$ according to (3.5) and (3.10).

(iii) If $X = 0$, then $\mu_2(Y, \hat{Y}) = 0$ by (3.4) and $\mu(Y, \hat{Y}) = \mu_1(Y, \hat{Y}) = \text{rank}\,\hat{X}$.

(iv) If $U = 0$, then $\det X \neq 0$, and $\mu_1(Y, \hat{Y}) = 0$, $\mu_2(Y, \hat{Y}) = \text{ind}\,\hat{X}^T\hat{U}$.

### 3.1.2 Dual Comparative Index

In this subsection we introduce the so-called *dual comparative index* $\mu^*(Y, \hat{Y})$, whose definition is very similar to that of the standard comparative index (Definition 3.1), only the matrix $\mathcal{J}$ in (3.3) and (3.4), and in particular in the definition of the Wronskian (3.2), is replaced by the matrix $\mathcal{J}^T = -\mathcal{J}$. We define

$$\mu_1^*(Y, \hat{Y}) := \mu_1(Y, \hat{Y}), \quad \mu_2^*(Y, \hat{Y}) := \text{ind}(-\mathcal{P}) \tag{3.12}$$

with $\mathcal{P}$ given in (3.4). The *dual comparative index* $\mu^*(Y, \hat{Y})$ is defined as

$$\mu^*(Y, \hat{Y}) := \mu_1^*(Y, \hat{Y}) + \mu_2^*(Y, \hat{Y}).$$

Observe that if the $2n \times n$ matrices $Y$ and $\hat{Y}$ satisfy (3.1), then the matrices $P_2 Y$ and $P_2 \hat{Y}$ with $P_2 = \text{diag}\{I, -I\}$ given in (1.151) satisfy the same condition, because the matrix $P_2$ is invertible and

$$P_2^T \mathcal{J} P_2 = -\mathcal{J}, \quad (P_2 Y)^T \mathcal{J}(P_2 Y) = Y^T P_2^T \mathcal{J} P_2 Y = -Y^T \mathcal{J} Y = 0.$$

Consequently, the dual comparative index can be determined by the formula

$$\mu_i^*(Y, \hat{Y}) = \mu_i(P_2 Y, P_2 \hat{Y}), \quad i \in \{1, 2\}, \tag{3.13}$$

with $P_2$ given by (1.151). Observe also that the dual comparative index as introduced in (3.12) makes it possible to compute the signature and the rank of $\mathcal{P}$, which are the difference and the sum of the numbers of positive and negative eigenvalues of $\mathcal{P}$, see [148, Chapter 10]. Indeed, we have the formulas

$$\operatorname{sgn} \mathcal{P} = \mu_2^*(Y, \hat{Y}) - \mu_2(Y, \hat{Y}) = \mu^*(Y, \hat{Y}) - \mu(Y, \hat{Y}),$$

$$\operatorname{rank} \mathcal{P} = \mu_2^*(Y, \hat{Y}) + \mu_2(Y, \hat{Y}).$$

The terminology "dual" comparative index will be explained later in this section.

### 3.1.3 Basic Properties of Comparative Index

Let $Z$ and $\hat{Z}$ be symplectic matrices, and let $Y$ and $\hat{Y}$ be $2n \times n$ matrices given by the formulas

$$Y = \begin{pmatrix} X \\ U \end{pmatrix} = Z \begin{pmatrix} 0 \\ I \end{pmatrix}, \quad \hat{Y} = \begin{pmatrix} \hat{X} \\ \hat{U} \end{pmatrix} = \hat{Z} \begin{pmatrix} 0 \\ I \end{pmatrix}. \tag{3.14}$$

Then, by Lemma 1.58(v), the matrices $Y$ and $\hat{Y}$ obey condition (3.1). Conversely, as we showed in Lemma 1.58(vi), given a $2n \times n$ matrix $Y$ satisfying (3.1), there exists another $2n \times n$ matrix $\tilde{Y}$ satisfying (3.1) such that $Z = (\tilde{Y}\ Y)$ is a symplectic matrix, for which (3.14) holds. Moreover, in this case the matrices $\tilde{Y}$ and $Y$ are normalized in the sense that $w(\tilde{Y}, Y) = I$, by Lemma 1.58(v).

In the next theorem, we present basic properties of the comparative index.

**Theorem 3.5** *Let* $Y = \begin{pmatrix} X \\ U \end{pmatrix}$ *and* $\hat{Y} = \begin{pmatrix} \hat{X} \\ \hat{U} \end{pmatrix}$ *be* $2n \times n$ *matrices satisfying* (3.1), *and let* $w(Y, \hat{Y})$ *be their Wronskian defined in* (3.2). *The comparative index has the following properties.*

(i) $\mu_i(Y C_1, \hat{Y} C_2) = \mu_i(Y, \hat{Y})$, $i \in \{1, 2\}$, *where* $C_1, C_2$ *are invertible* $n \times n$ *matrices. This property can be reformulated for* $Z$ *and* $\hat{Z}$ *given by* (3.14) *as*

$$\mu_i \left( Z(0\ I)^T, \hat{Z}(0\ I)^T \right) = \mu_i \left( ZL_1(0\ I)^T, \hat{Z}L_2(0\ I)^T \right), \quad i \in \{1, 2\},$$

*where* $L_1, L_2$ *are arbitrary* $2n \times 2n$ *lower block-triangular symplectic matrices.*

(ii) $\mu_i(LY, L\hat{Y}) = \mu_i(Y, \hat{Y})$, $i \in \{1, 2\}$, and hence $\mu(LY, L\hat{Y}) = \mu(Y, \hat{Y})$, where $L$ is any $2n \times 2n$ lower block-triangular symplectic matrix.

(iii) For any symplectic matrices $Z$ and $\hat{Z}$, we have

$$\mu_i \left( Z(0 \ I)^T, \ \hat{Z}(0 \ I)^T \right) = \mu_i^* \left( Z^{-1}(0 \ I)^T, Z^{-1}\hat{Z}(0 \ I)^T \right), \quad i \in \{1, 2\}.$$

(iv) $\mu_2(Y, \hat{Y}) = \mu_2^*(\hat{Y}, Y)$.

(v) $\mu(Y, \hat{Y}) + \mu(\hat{Y}, Y) = \operatorname{rank} w(Y, \hat{Y})$. This property can be reformulated for the matrices $Z$ and $\hat{Z}$ given by (3.14) as

$$\mu \left( Z(0 \ I)^T, \ \hat{Z}(0 \ I)^T \right) + \mu \left( \hat{Z}(0 \ I)^T, \ Z(0 \ I)^T \right) = \mu \left( (0 \ I)^T, \ Z^{-1}\hat{Z}(0 \ I)^T \right).$$

(vi) $\mu_1(Y, \hat{Y}) = \operatorname{rank} \hat{X} - \operatorname{rank} X + \mu_1(\hat{Y}, Y)$. In view of (iv), this is equivalent to

$$\mu(Y, \hat{Y}) = \operatorname{rank} \hat{X} - \operatorname{rank} X + \mu^*(\hat{Y}, Y).$$

(vii) $0 \le \mu(Y, \hat{Y}) \le p$ and $0 \le \mu^*(Y, \hat{Y}) \le p$, where $p = p(Y, \hat{Y})$ is defined by

$$p := \min \{ \operatorname{rank} w(Y, \hat{Y}), \ \operatorname{rank} \hat{X}, \ \operatorname{rank} \hat{X} - \operatorname{rank} X + \operatorname{rank} w(Y, \hat{Y}) \}.$$
$$\tag{3.15}$$

In particular,

$$\mu(Y, \hat{Y}) = \operatorname{rank} \hat{X}$$

$$\Leftrightarrow \begin{cases} \mu_1(Y, \hat{Y}) = \operatorname{rank} w(Y, \hat{Y}) - \operatorname{rank} X, \\ \mu_2(Y, \hat{Y}) = \operatorname{rank} \hat{X} + \operatorname{rank} X - \operatorname{rank} w(Y, \hat{Y}), \end{cases}$$

$$\Leftrightarrow \mu(\hat{Y}, Y) = \operatorname{rank} w(Y, \hat{Y}) - \operatorname{rank} \hat{X},$$

$$\Leftrightarrow \begin{cases} \mu_1(\hat{Y}, Y) = \operatorname{rank} w(Y, \hat{Y}) - \operatorname{rank} \hat{X}, \\ \mu_2(\hat{Y}, Y) = 0, \end{cases}$$

$$\Leftrightarrow \mu^*(Y, \hat{Y}) = \operatorname{rank} w(Y, \hat{Y}) - \operatorname{rank} X,$$

$$\Leftrightarrow \begin{cases} \mu_1^*(Y, \hat{Y}) = \operatorname{rank} w(Y, \hat{Y}) - \operatorname{rank} X, \\ \mu_2^*(Y, \hat{Y}) = 0, \end{cases}$$

$$\Leftrightarrow \mu^*(\hat{Y}, Y) = \operatorname{rank} X,$$

$$\Leftrightarrow \begin{cases} \mu_1^*(\hat{Y}, Y) = \operatorname{rank} w(Y, \hat{Y}) - \operatorname{rank} \hat{X}, \\ \mu_2^*(\hat{Y}, Y) = \operatorname{rank} X + \operatorname{rank} \hat{X} - \operatorname{rank} w(Y, \hat{Y}), \end{cases}$$

*and*

$$\mu(Y, \hat{Y}) = \operatorname{rank} w(Y, \hat{Y})$$

$$\Leftrightarrow \begin{cases} \mu_1(Y, \hat{Y}) = \operatorname{rank} \hat{X} - \operatorname{rank} X, \\ \mu_2(Y, \hat{Y}) = \operatorname{rank} w(Y, \hat{Y}) - \operatorname{rank} \hat{X} + \operatorname{rank} X, \end{cases}$$

$$\Leftrightarrow \mu(\hat{Y}, Y) = 0,$$

$$\Leftrightarrow \mu^*(\hat{Y}, Y) = \operatorname{rank} X - \operatorname{rank} \hat{X} + \operatorname{rank} w(Y, \hat{Y}),$$

$$\Leftrightarrow \begin{cases} \mu_1^*(\hat{Y}, Y) = 0, \\ \mu_2^*(\hat{Y}, Y) = \operatorname{rank} X - \operatorname{rank} \hat{X} + \operatorname{rank} w(Y, \hat{Y}), \end{cases}$$

$$\Leftrightarrow \mu^*(Y, \hat{Y}) = \operatorname{rank} \hat{X} - \operatorname{rank} X,$$

$$\Leftrightarrow \begin{cases} \mu_1^*(Y, \hat{Y}) = \operatorname{rank} \hat{X} - \operatorname{rank} X, \\ \mu_2^*(Y, \hat{Y}) = 0, \end{cases}$$

*and*

$$\mu(Y, \hat{Y}) = \operatorname{rank} \hat{X} - \operatorname{rank} X + \operatorname{rank} w(Y, \hat{Y})$$

$$\Leftrightarrow \begin{cases} \mu_1(Y, \hat{Y}) = 0, \\ \mu_2(Y, \hat{Y}) = \operatorname{rank} \hat{X} - \operatorname{rank} X + \operatorname{rank} w(Y, \hat{Y}), \end{cases}$$

$$\Leftrightarrow \mu^*(Y, \hat{Y}) = 0,$$

$$\Leftrightarrow \mu^*(\hat{Y}, Y) = \operatorname{rank} w(Y, \hat{Y}),$$

$$\Leftrightarrow \begin{cases} \mu_1^*(\hat{Y}, Y) = \operatorname{rank} X - \operatorname{rank} \hat{X}, \\ \mu_2^*(\hat{Y}, Y) = \operatorname{rank} \hat{X} - \operatorname{rank} X + \operatorname{rank} w(Y, \hat{Y}), \end{cases}$$

$$\Leftrightarrow \mu(\hat{Y}, Y) = \operatorname{rank} X - \operatorname{rank} \hat{X},$$

$$\Leftrightarrow \begin{cases} \mu_1(\hat{Y}, Y) = \operatorname{rank} X - \operatorname{rank} \hat{X}, \\ \mu_2(\hat{Y}, Y) = 0. \end{cases}$$

*(viii) It holds*

$$\mu\left(\mathcal{J}Y, \mathcal{J}\hat{Y}\right) = \mu(Y, \hat{Y}) + \mu\left(\mathcal{J}Y, (I\ 0)^T\right) - \mu\left(\mathcal{J}\hat{Y}, (I\ 0)^T\right).$$

*(ix) For arbitrary symplectic matrices $W$, $Z$, $\hat{Z}$, we have*

$$\mu\left(WZ(0\ I)^T, W(0\ I)^T\right) - \mu\left(W\hat{Z}(0\ I)^T, W(0\ I)^T\right)$$
$$= \mu\left(\hat{Z}(0\ I)^T, W^{-1}(0\ I)^T\right) - \mu\left(Z(0\ I)^T, W^{-1}(0\ I)^T\right).$$

The proof of the previous theorem is postponed for later (see Sects. 3.1.5 and 3.2.2).

At this moment we present a result which plays a key role in the comparative index theory. In particular it shows how the comparative index behaves under a symplectic transformation.

**Theorem 3.6 (Main Theorem on the Comparative Index)** *Let* $Z$, $\hat{Z}$, $W$ *be arbitrary* $2n \times 2n$ *symplectic matrices. Then*

$$
\mu \left( W Z(0\ I)^T, W \hat{Z}(0\ I)^T \right) - \mu \left( Z(0\ I)^T, \hat{Z}(0\ I)^T \right)
$$
$$
= \mu \left( W Z(0\ I)^T, W(0\ I)^T \right) - \mu \left( W \hat{Z}(0\ I)^T, W(0\ I)^T \right).
$$
(3.16)

*Proof* First of all, observe that properties (ii) and (viii) of the the previous theorem are particular cases of (3.16) with $W = L$ or $W = \mathcal{J}$, since it is not difficult to verify that

$$
\mu \left( L Z(0\ I)^T, L(0\ I)^T \right) = \mu \left( L \hat{Z}(0\ I)^T, L(0\ I)^T \right) = 0
$$

for a lower block-triangular matrix $L$ (see (ii) of Remark 3.4). Then, to prove the statement of theorem, it suffices to apply Theorem 1.74 concerning the representation of a symplectic matrix $W$ as a product $W = H_1 L_2 H_3 = \mathcal{J} L_1 \mathcal{J} L_2 \mathcal{J} L_3 \mathcal{J}$, where $H_1$, $H_2$ are upper block-triangular symplectic matrices and $L_1, L_2, L_3$ are lower block-triangular symplectic matrices. We prove that if (3.16) holds for symplectic matrices $W_1$, $W_2$ (instead of $W$) for arbitrary symplectic matrices $Z$, $\hat{Z}$, then this formula holds also for their product $W = W_1 W_2$. To show this, we use first (3.16) for $W = W_2$ and then for $W = W_1$. So we have

$$
\mu \left( W_2 Z(0\ I)^T, W_2 \hat{Z}(0\ I)^T \right) - \mu \left( Z(0\ I)^T, \hat{Z}(0\ I)^T \right)
$$
$$
= \mu \left( W_2 Z(0\ I)^T, W_2(0\ I)^T \right) - \mu \left( W_2 \hat{Z}(0\ I)^T, W_2(0\ I)^T \right)
$$
$$
= \mu \left( W_1^{-1} W_2 Z(0\ I)^T, W_1^{-1} W_2(0\ I)^T \right) + \mu \left( W_2 Z(0\ I)^T, W_1(0\ I)^T \right)
$$
$$
- \mu \left( W_1^{-1} W_2 \hat{Z}(0\ I)^T, W_1^{-1} W_2(0\ I)^T \right) - \mu \left( W_2 \hat{Z}(0\ I)^T, W_1(0\ I)^T \right).
$$

On the other hand, by using the assumption that (3.16) holds for $W = W_1$ and arbitrary symplectic $Z$ and $\hat{Z}$, we obtain

$$
\mu \left( W_2 Z(0\ I)^T, W_2 \hat{Z}(0\ I)^T \right) = \mu \left( W_1(W_1^{-1} W_2 Z)(0\ I)^T, W_1(W_1^{-1} W_2 \hat{Z})(0\ I)^T \right)
$$
$$
= \mu \left( W_1^{-1} W_2 Z(0\ I)^T, W_1^{-1} W_2 \hat{Z}(0\ I)^T \right) + \mu \left( W_2 Z(0\ I)^T, W_1(0\ I)^T \right)
$$
$$
- \mu \left( W_2 \hat{Z}(0\ I)^T, W_1(0\ I)^T \right).
$$

If we compare the obtained expressions and cancel the same terms, then we get

$$\mu\left(W_1^{-1}W_2\,Z(0\ I)^T,\,W_1^{-1}W_2\,\hat{Z}(0\ I)^T\right)-\mu\left(Z(0\ I)^T,\,\hat{Z}(0\ I)^T\right)$$

$$=\mu\left(W_1^{-1}W_2\,Z(0\ I)^T,\,W_1^{-1}W_2(0\ I)^T\right)-\mu\left(W_1^{-1}W_2\,\hat{Z}(0\ I)^T,\,W_1^{-1}W_2(0\ I)^T\right).$$

Consequently, formula (3.16) holds for $W = W_1^{-1}W_2$. In particular, if (3.16) holds for $W_1$, then it holds also for $W_1^{-1}$. Now it suffices to apply the already proved statement to the pair of matrices $W_1^{-1}$ and $W_2$. Hence, we obtain validity of (3.16) for $W = W_1W_2$. Applying the proved "multiplicative" rule to the product $W = \mathcal{J}L_1\mathcal{J}L_2\mathcal{J}L_3\mathcal{J}$, we obtain that (3.16) holds for any symplectic matrix $W$.  □

*Remark 3.7* Given that formula (3.16) holds with $W = \mathcal{J}$ and $W = L$ (where $L$ is a lower block-triangular symplectic matrix), we can use the factorization of a symplectic matrix $W$ into the product $W = \mathcal{J}L_1\mathcal{J}L_2\mathcal{J}L_3\mathcal{J}$ (where $L_1, L_2, L_3$ are lower block-triangular symplectic matrices) to show that (3.16) holds for any symplectic matrix $W$. This yields another proof of Theorem 3.6. Indeed, it suffices to prove that if (3.16) holds for a symplectic $W$, then it holds also for $\mathcal{J}W$ and $LW$. However, the validity of (3.16) for $\mathcal{J}W$ and $LW$ follows by a double application of properties (viii) and (ii) of Theorem 3.5, respectively.

By using (ix) of Theorem 3.5, formula (3.16) can be restated as

$$\mu\left(WZ(0\ I)^T,\,W\hat{Z}(0\ I)^T\right)-\mu\left(Z(0\ I)^T,\,\hat{Z}(0\ I)^T\right)$$

$$=\mu\left(\hat{Z}(0\ I)^T,\,W^{-1}(0\ I)^T\right)-\mu\left(Z(0\ I)^T,\,W^{-1}(0\ I)^T\right).$$

$$(3.17)$$

From (3.17) we derive the following property, which we call the "triangle inequality" for the comparative index.

**Theorem 3.8 (Triangle Inequality)** *For arbitrary symplectic matrices $W_1$, $W_2$, and $W_3$ of dimension $2n$, we have*

$$\mu\left(W_1(0\ I)^T,\,W_3(0\ I)^T\right)$$

$$\leq \mu\left(W_1(0\ I)^T,\,W_2(0\ I)^T\right)+\mu\left(W_2(0\ I)^T,\,W_3(0\ I)^T\right). \qquad (3.18)$$

*Moreover, the equality in (3.18) holds if and only if*

$$\mu\left(W_3^{-1}W_1(0\ I)^T,\,W_3^{-1}W_2(0\ I)^T\right)$$

$$=\mu^*\left(W_1^{-1}W_3(0\ I)^T,\,W_1^{-1}W_2(0\ I)^T\right)=0. \qquad (3.19)$$

*Proof* From (3.17) we have

$$\mu\big(Z(0\ I)^T, \hat{Z}(0\ I)^T\big) + \mu\big(\hat{Z}(0\ I)^T, W^{-1}(0\ I)^T\big)$$
$$= \mu\left(WZ(0\ I)^T, W\hat{Z}(0\ I)^T\right) + \mu\left(Z(0\ I)^T, W^{-1}(0\ I)^T\right).$$

Putting $Z := W_1$, $\hat{Z} := W_2$ and $W^{-1} := W_3$, we derive

$$\mu\left(W_1(0\ I)^T, W_2(0\ I)^T\right) + \mu\left(W_2(0\ I)^T, W_3(0\ I)^T\right)$$
$$= \mu\left(W_3^{-1}W_1(0\ I)^T, W_3^{-1}W_2(0\ I)^T\right) + \mu\left(W_1(0\ I)^T, W_3(0\ I)^T\right).$$

Using that $\mu\left(W_3^{-1}W_1(0\ I)^T, W_3^{-1}W_2(0\ I)^T\right) \geq 0$, we derive (3.18). From the last identity, we have (3.19), where we also used Theorem 3.5(iii). □

The following lower bounds of the comparative index complement the upper bounds in Theorem 3.5(vii).

**Lemma 3.9** *Let $Y$ and $\tilde{Y}$ be $2n \times n$ matrices satisfying (3.1). Then*

$$\min\{\mu(Y, \hat{Y}), \mu^*(Y, \hat{Y})\}$$
$$\geq \max\{0,\ \text{rank}\,\hat{X} - \text{rank}\,X,\ \text{rank}\,w(Y, \hat{Y}) - \text{rank}\,X\}. \qquad (3.20)$$

*Proof* We know from property (vi) in Theorem 3.5 and from $\mu^*(\hat{Y}, Y) \geq 0$ that the inequality $\mu(Y, \hat{Y}) \geq \text{rank}\,\hat{X} - \text{rank}\,X$ holds. Similarly from property (vi) in Theorem 3.5 applied to $\mu(\hat{Y}, Y) \geq 0$, we get $\mu^*(Y, \hat{Y}) \geq \text{rank}\,\hat{X} - \text{rank}\,X$. Next, the properties (iii) and (vi) in Theorem 3.5 imply that

$$\mu(Y, \hat{Y}) = \mu^*(Z^{-1}(0\ I)^T, Z^{-1}\hat{Y})$$
$$= \text{rank}\,w(Y, \hat{Y}) - \text{rank}\,X + \mu(Z^{-1}\hat{Y}, Z^{-1}(0\ I)^T) \qquad (3.21)$$
$$= \text{rank}\,w(Y, \hat{Y}) - \text{rank}\,X + \mu^*(\hat{Z}^{-1}Y, \hat{Z}^{-1}(0\ I)^T) \qquad (3.22)$$

and then $\mu(Y, \hat{Y}) \geq \text{rank}\,w(Y, \hat{Y}) - \text{rank}\,X$. In a similar way, we obtain from the properties (iii) and (vi) in Theorem 3.5 for the dual comparative index (see also Theorem 3.11 below) that $\mu^*(Y, \hat{Y}) \geq \text{rank}\,w(Y, \hat{Y}) - \text{rank}\,X$. Therefore, the estimates in (3.20) are proved. □

*Remark 3.10*

(i) From Lemma 3.9 and property (vii) in Theorem 3.5, we have the estimates

$$
\left.
\begin{aligned}
r \le \mu(Y, \hat{Y}) \le p, \quad r \le \mu^*(Y, \hat{Y}) \le p, \\
r := \max\{0, \operatorname{rank} \hat{X} - \operatorname{rank} X, \operatorname{rank} w(Y, \hat{Y}) - \operatorname{rank} X\}, \\
p := \min\{\operatorname{rank} w(Y, \hat{Y}), \operatorname{rank} \hat{X}, \operatorname{rank} \hat{X} - \operatorname{rank} X + \operatorname{rank} w(Y, \hat{Y})\},
\end{aligned}
\right\}
$$
(3.23)

where the lower and upper bounds $r$ and $p$ are such that

$$
r + p = \operatorname{rank} \hat{X} - \operatorname{rank} X + \operatorname{rank} w(Y, \hat{Y}).
$$
(3.24)

Identity (3.24) can be easy verified by direct computations.

(ii) It follows from property (vii) of Theorem 3.5 that $\mu(Y, \hat{Y})$ achieves the upper bound $p$ if and only if the dual index $\mu^*(Y, \hat{Y})$ coincides with the lower bound $r$ in (3.23). Replacing the roles of $Y$ and $\hat{Y}$ in the formulation of Theorem 3.5 (vii), we see that the condition $\mu^*(Y, \hat{Y}) = p$ is equivalent to $\mu(Y, \hat{Y}) = r$.

(iii) Putting $a := \operatorname{rank} X$, $b := \operatorname{rank} \hat{X}$, and $c := \operatorname{rank} w(Y, \hat{Y})$, we see from (3.23) that the quantities $a \ge 0$, $b \ge 0$, $c \ge 0$ obey the triangle inequalities $a \le b+c$, $c \le a + b$, $b \le a + c$. This fact can be proved independently using the well-known inequalities $\operatorname{rank}(AB) \le \min\{\operatorname{rank} A, \operatorname{rank} B\}$ and $\operatorname{rank}(A+B) \le \operatorname{rank} A + \operatorname{rank} B$.

### 3.1.4 Duality Principle for Comparative Index

Observe that the properties of $\mu(Y, \hat{Y})$ displayed in the previous subsection can be reformulated for the dual comparative $\mu^*(Y, \hat{Y})$. This reformulation is given in the next theorem.

**Theorem 3.11** *Concerning the dual comparative index the following holds.*

(i) *The properties (i)–(ix) of Theorem 3.5 hold also for the dual comparative index $\mu^*$.*

(ii) *Any formula, proved using properties (i)–(ix) of Theorem 3.5, holds also for the dual comparative index in the "dual form"; the comparative index $\mu$ is replaced by the dual comparative index $\mu^*$ and vice versa.*

*Proof* We will use the matrices $P_2$ given in (1.151). In view of (3.12) and (3.13), we have

$$
\mu_i^{**}(Y, \hat{Y}) = (\mu_i^*(Y, \hat{Y}))^* = (\mu_i(P_2Y, P_2\hat{Y}))^* = \mu_i(Y, \hat{Y}), \quad i \in \{1, 2\}.
$$

Since property (i) of Theorem 3.5 holds for any matrices satisfying (3.1), we have
$\mu_i(P_2YC_1, P_2\hat{Y}C_2) = \mu_i(P_2Y, P_2\hat{Y})$, $i \in \{1, 2\}$. Consequently, this property holds
also for the dual comparative index. Then obviously

$$
\mu\left(\prod_{i=1}^{k} W_i(0 \ I)^T, \prod_{j=1}^{l} V_j(0 \ I)^T\right) = \mu\left(\prod_{i=1}^{k} W_i(0 \ -I)^T, \prod_{j=1}^{l} V_j(0 \ -I)^T\right)
$$

$$
= \mu\left(P_2\prod_{i=1}^{k}(P_2 W_i P_2)(0 \ I)^T, \ P_2\prod_{j=1}^{l}(P_2 V_j P_2)(0 \ I)^T\right)
$$

$$
= \mu^*\left(\prod_{i=1}^{k}(P_2 W_i P_2)(0 \ I)^T, \prod_{j=1}^{l}(P_2 V_j P_2)(0 \ I)^T\right) \tag{3.25}
$$

for the products $\prod_{i=1}^{k} W_i$ and $\prod_{j=1}^{l} V_j$ of arbitrary symplectic matrices $W_i$ and $V_j$.
Since $P_2\mathcal{J}P_2 = -\mathcal{J}$ in view of property (iv) of Lemma 1.58, we have

$$
W_i, \ V_j \in Sp(2n) \ \Leftrightarrow \ P_2 W_i P_2, \ P_2 V_j P_2 \in Sp(2n), \quad i \in \{1, \ldots, k\}, \ j \in \{1, \ldots, l\}.
$$

Consequently, since each of the properties (i)–(vii), (ix) holds for *any* symplectic
matrices $Z$, $\hat{Z}$, $W$, it holds also for the matrices $P_2 Z P_2$, $P_2\hat{Z}P_2$, $P_2 W P_2$. Therefore,
because of (3.25), these properties hold for $\mu^*$. Further, a matrix $L \in Sp(2n)$
is lower block-triangular if and only if $P_2 L P_2 \in Sp(2n)$, where $P_2 L P_2$ is also
lower block-triangular. This means in view of (3.25) that property (ii) for the dual
comparative index is proved. Now we prove property (viii) for the dual comparative
index. It follows from property (i) and $\mathcal{J}^T = -\mathcal{J}$ that the matrix $\mathcal{J}$ in (viii) can be
replaced by $\mathcal{J}^T$. Then we have

$$
\mu\left(\mathcal{J}^T Z(0 \ I)^T, \mathcal{J}^T\hat{Z}(0 \ I)^T\right) = \mu\left(Z(0 \ I)^T, \hat{Z}(0 \ I)^T\right)
$$

$$
+ \mu\left(\mathcal{J}^T Z(0 \ I)^T, \mathcal{J}^T(0 \ I)^T\right) - \mu\left(\mathcal{J}^T\hat{Z}(0 \ I)^T, \mathcal{J}^T(0 \ I)^T\right)
$$

for any symplectic matrices $Z$ and $\hat{Z}$. Upon taking $P_2 Z P_2$ and $P_2\hat{Z}P_2$ instead of $Z$
and $\hat{Z}$ and using that $\mathcal{J}^T = P_2\mathcal{J}P_2$, we derive

$$
\mu\left(P_2\mathcal{J}Z(0 \ -I)^T, P_2\mathcal{J}\hat{Z}(0 \ -I)^T\right) = \mu\left(P_2 Z(0 \ -I)^T, P_2\hat{Z}(0 \ -I)^T\right)
$$

$$
+ \mu\left(P_2\mathcal{J}Z(0 \ -I)^T, P_2\mathcal{J}(0 \ -I)^T\right) - \mu\left(P_2\mathcal{J}\hat{Z}(0 \ -I)^T, P_2\mathcal{J}(0 \ -I)^T\right),
$$

or the same formula in the dual form

$$\mu^* \left( \mathcal{J} Z(0 \ -I)^T, \mathcal{J} \hat{Z}(0 \ -I)^T \right) = \mu^* \left( Z(0 \ -I)^T, \hat{Z}(0 \ I)^T \right)$$
$$+ \mu^* \left( \mathcal{J} Z(0 \ -I)^T, \mathcal{J}(0 \ -I)^T \right) - \mu^* \left( \mathcal{J} \hat{Z}(0 \ -I)^T, \mathcal{J}(0 \ -I)^T \right).$$

Since property (i) holds for the dual comparative index, we derive property (viii) for the dual index as well. This completes the proof of the first part of theorem. The proof of the second part is obvious. □

In particular, the following "dual" reformulation of (3.17) holds:

$$\mu^* \left( Z(0 \ I)^T, \hat{Z}(0 \ I)^T \right) - \mu^* \left( W Z(0 \ I)^T, W \hat{Z}(0 \ I)^T \right)$$
$$= \mu^* \left( Z(0 \ I)^T, W^{-1}(0 \ I)^T \right) - \mu^* \left( \hat{Z}(0 \ I)^T, W^{-1}(0 \ I)^T \right).$$
$$(3.26)$$

And for further reference, we also present the dual version of the main theorem for the comparative index (Theorem 3.6).

**Corollary 3.12** *Let* $Z$, $\hat{Z}$, $W$ *be arbitrary* $2n \times 2n$ *symplectic matrices. Then*

$$\mu^* \left( W Z(0 \ I)^T, W \hat{Z}(0 \ I)^T \right) - \mu^* \left( Z(0 \ I)^T, \hat{Z}(0 \ I)^T \right)$$
$$= \mu^* \left( W Z(0 \ I)^T, W(0 \ I)^T \right) - \mu^* \left( W \hat{Z}(0 \ I)^T, W(0 \ I)^T \right).$$
$$(3.27)$$

### 3.1.5 Proof of Properties (i)–(vii), (ix) of Comparative Index

As we have already mentioned before, the proof of Theorem 3.5 substantially uses the results of Sect. 1.6.3. The proof of the first two properties uses Lemmas 1.63 and 1.64.

*Proof of Property (i)* According to Theorem 3.2,

$$\mu_1(Y C_1, \hat{Y} C_2) = \text{rank} \left[ (I - (X C_1)(X C_1)^\dagger] \hat{X} C_2 \right.$$
$$= \text{rank} \, (I - X X^\dagger) \, \hat{X} C_2 = \mu_1(Y, \hat{Y}),$$

where we have used Lemma 1.63. Applying Lemma 1.64 in computing the comparative index $\mu_2(Y C_1, \hat{Y} C_2)$ by (3.10), we have

$$\mathcal{T} = I - [(I - X X^\dagger) \hat{X} C_2]^\dagger (I - X X^\dagger) \, \hat{X} C_2$$
$$= C_2^{-1} \left( I - [(I - X X^\dagger) \hat{X}]^\dagger (I - X X^\dagger) \hat{X} \right) \tilde{T}, \quad \det \tilde{T} \neq 0.$$

Since $\det C_1 \neq 0$, then

$$C_1^T X^T Q X C_1 = C_1^T X^T U C_1 \iff X^T Q X = X^T U,$$

so that in the computation of $\mu_2(YC_1, \hat{Y}C_2)$ by using (3.10), one can take any symmetric matrices $Q$ and $\hat{Q}$ satisfying (3.8). Hence, we obtain

$$\mu_2(YC_1, \hat{Y}C_2) = \operatorname{ind}\left(\mathcal{T}^T C_2^T \hat{X}^T (\hat{Q} - Q) \hat{X} C_2 \mathcal{T}\right) = \mu_2(Y, \hat{Y}),$$

where we have used the above formula for $\mathcal{T}$ and the fact that the matrix $\tilde{\mathcal{T}}$ is invertible. This proves property (i).                                                          □

*Proof of Property (ii)* Recall that any symplectic lower block-triangular matrix $L$ is of the form $L = \begin{pmatrix} K & 0 \\ P & K^{T-1} \end{pmatrix}$, where $PK^{-1} = (PK^{-1})^T$. Then obviously

$$w(LY, L\hat{Y}) = Y^T L^T \mathcal{J} L \hat{Y} = Y^T \mathcal{J} \hat{Y} = w(Y, \hat{Y}),$$

and $I - (KX)^\dagger(KX) = I - X^\dagger X$ because of Lemma 1.63. According to Definition 3.1, $\mu_1(LY, L\hat{Y}) = \mu_1(Y, \hat{Y})$ and the multiplication of $Y$ and $\hat{Y}$ by $L$ do not change the matrix $\mathcal{T}$. Consequently, it is sufficient to prove that

$$\mathcal{P} = \mathcal{T}^T \left(w^T (KX)^\dagger K \hat{X}\right) \mathcal{T} = \mathcal{T}^T \left(w^T X^\dagger \hat{X}\right) \mathcal{T}.$$

Observe that by Theorem 3.2, the matrix $\mathcal{M}$ in (3.3) can be replaced by the matrix $\tilde{\mathcal{M}} = (I - XX^\dagger)\hat{X}$ and that $\tilde{\mathcal{M}}\mathcal{T} = 0$, which is equivalent to $\hat{X}\mathcal{T} = XX^\dagger \hat{X}\mathcal{T}$ or $K\hat{X}\mathcal{T} = KXX^\dagger \hat{X}\mathcal{T}$. Consequently,

$$(KX)^\dagger K \hat{X} \mathcal{T} = (KX)^\dagger KXX^\dagger \hat{X} \mathcal{T} = X^\dagger XX^\dagger \hat{X} \mathcal{T} = X^\dagger \hat{X} \mathcal{T},$$

and hence $\mu_2(LY, L\hat{Y}) = \mu_2(Y, \hat{Y})$. The property (ii) is now proved.        □

*Proof of Property (iii)* The proof is based on Theorem 3.2. Indeed,

$$-w\left(Z^{-1}(0 \ I)^T, Z^{-1}\hat{Z}(0 \ I)^T\right) = (0 \ I) Z^{T-1} \mathcal{J}^T Z^{-1} \hat{Z}(0 \ I)^T$$

$$= (0 \ I) \mathcal{J}^T \hat{Z}(0 \ I)^T = \hat{X}, \qquad (3.28)$$

where we have used that $Z^{T-1} \mathcal{J}^T Z^{-1} = \mathcal{J}^T$. Further, $(I \ 0) Z^{-1}(0 \ I)^T = -X^T$, and hence the number $\mu_1^*$ in the right-hand side of property (iii) computed according to (3.12), (3.3) complies with $\mu_1(Y, \hat{Y})$, computed using the first statement of Theorem 3.2. It is not difficult to verify that $(I \ 0) Z^{-1}\hat{Z}(0 \ I)^T = -w(Y, \hat{Y})$, and, hence, the index $\mu_2^*$, computed in the right-hand side of property (iii), is $\mu_2(Y, \hat{Y})$, where we also used the symmetry of the matrix $\mathcal{P}$ in (3.4).                              □

The proof of properties (iv) and (v) is based on the results of Lemma 1.68 and Theorem 1.70.

*Proof of Property (iv)* We will use Theorem 1.70 applying factorizations (1.180) to $Y$ and $\hat{Y}$. Using the already proved properties (i) and (ii),

$$\mu(Y, \hat{Y}) = \mu\big(\mathfrak{N}(I\ 0)^T,\ L^{-1}\hat{L}\,\hat{\mathfrak{N}}\,(I\ 0)^T\big), \tag{3.29}$$

where the symplectic matrices $L$ and $\mathfrak{N} := \mathfrak{N}_{XX^\dagger}$ are determined by (1.180) and (1.179) with $Y$, and the symplectic matrices $\hat{L}$ and $\hat{\mathfrak{N}} := \mathfrak{N}_{\hat{X}\hat{X}^\dagger}$ are determined in a similar way with $\hat{Y}$. In order to shorten the notation, we define the matrices

$$F := XX^\dagger, \quad G := I - XX^\dagger, \quad \hat{F} := \hat{X}\hat{X}^\dagger, \quad \hat{G} := I - \hat{X}\hat{X}^\dagger.$$

Then, evaluating (3.29) according to formulas (3.5) and (3.10), we have $\mathcal{M} = G\hat{F}$, $\mathcal{P} = T\hat{F}(\hat{Q} - Q)\,\hat{F}T$, and

$$\hat{F}T = \hat{F}(I - \mathcal{M}^\dagger\mathcal{M}) = (F + G)\,\hat{F}(I - \mathcal{M}^\dagger\mathcal{M}) = F\hat{F}(I - \mathcal{M}^\dagger\mathcal{M}). \tag{3.30}$$

The matrix $\mathfrak{N}^T\hat{\mathfrak{N}}\,(0\ I)^T$ satisfies (3.1), and then using (1.171) for $\mathfrak{N}^T\hat{\mathfrak{N}}\,(0\ I)^T$ and Remark 1.60(vii), we obtain

$$\begin{aligned}
F\hat{F}(I - \mathcal{M}^\dagger\mathcal{M}) &= F(F\hat{F} + G\hat{G})\,[I - \mathcal{M}^\dagger\mathcal{M} - (\mathcal{M}^{*T})^\dagger\mathcal{M}^{*T}] \\
&= F\,[I - \mathcal{M}\mathcal{M}^\dagger - \mathcal{M}^{*T}(\mathcal{M}^{*T})^\dagger]\,(F\hat{F} + G\hat{G}) \times \\
&\qquad \times [I - \mathcal{M}^\dagger\mathcal{M} - (\mathcal{M}^{*T})^\dagger\mathcal{M}^{*T}] \\
&= -F(I - \mathcal{M}^{*\dagger}\mathcal{M}^*)\,\mathfrak{L},
\end{aligned}$$

where $\mathcal{M}^* = \hat{G}F$ and the invertible matrix $\mathfrak{L}$ are determined for $\mathfrak{N}^T\hat{\mathfrak{N}}\,(0\ I)^T$ by formula (1.174). Therefore, we have proved that

$$\left.\begin{aligned}
\hat{F}T &= -FT^*\mathfrak{L}, \quad T = I - \mathcal{M}^\dagger\mathcal{M}, \quad T^* = I - \mathcal{M}^{*\dagger}\mathcal{M}^*, \\
\mathcal{M} &= G\hat{F}, \quad \mathcal{M}^* = \hat{G}F, \quad \det\mathfrak{L} \neq 0,
\end{aligned}\right\} \tag{3.31}$$

or $T\hat{F}(\hat{Q} - Q)\,\hat{F}T = \mathfrak{L}^T T^* F(\hat{Q} - Q)\,FT^*\mathfrak{L}$. Therefore, we have

$$\mu_2(Y, \hat{Y}) = \operatorname{ind} T\hat{F}(\hat{Q} - Q)\,\hat{F}T = \operatorname{ind} T^* F(\hat{Q} - Q)\,FT^* = \mu_2^*(\hat{Y}, Y), \tag{3.32}$$

which shows that the property (iv) holds. $\qquad\square$

*Proof of Property (v)* We first substitute factorization (1.180) for $Y$ and $\hat{Y}$ into $w(Y, \hat{Y})$. Using the notation of Theorem 1.70, we obtain

$$w(Y, \hat{Y}) = M^T[G\hat{F} - F\hat{G} + F(\hat{Q} - Q)\,\hat{F}]\,\hat{M}$$
$$= M^T M_1[\mathcal{M} - \mathcal{M}^{*T} + \mathcal{T}^*F(\hat{Q} - Q)\,\hat{F}\mathcal{T}]\,M_2\,\hat{M},$$

where the invertible matrices $M$ and $\hat{M}$ are determined by (1.180) via $Y$ and $\hat{Y}$, the matrices $\mathcal{M}$ and $\mathcal{M}^*$ are given by (3.31), and the matrices

$$M_1 := I + F\mathcal{T}^*(\hat{Q} - Q)\,\mathcal{M}^\dagger G, \ M_2 = I - \hat{G}\mathcal{M}^{*T\dagger}(\hat{Q} - Q)\,\hat{F}$$

are obviously also invertible in view of (1.163) and the orthogonality of the projectors $F$, $G$, $\hat{F}$, $\hat{G}$. Using Lemma 1.68 again for the matrix $\mathfrak{N}^T\hat{\mathfrak{N}}(0 \ I)^T$ (see the proof of the property (iv)) and using (3.31), we have

$$w(Y, \hat{Y}) = -M^T M_1[\mathcal{M}\,\mathcal{M}^\dagger + \mathcal{M}^{*\dagger}\mathcal{M}^* + \mathcal{T}^*F(\hat{Q} - Q)\,F\mathcal{T}^*]\,\mathfrak{L}\,M_2\,\hat{M}.$$

Consequently,

$$\operatorname{rank} w(Y, \hat{Y}) = \operatorname{rank}[\mathcal{M}\mathcal{M}^\dagger + \mathcal{M}^{*\dagger}\mathcal{M}^* + \mathcal{T}^*F(\hat{Q} - Q)\,F\mathcal{T}^*]$$
$$= \operatorname{rank}\mathcal{M} + \operatorname{rank}\mathcal{M}^* + \operatorname{rank}\mathcal{T}^*F(\hat{Q} - Q)\,F\mathcal{T}^*$$
$$= \mu_1(Y, \hat{Y}) + \mu_1(\hat{Y}, Y) + \operatorname{ind}\mathcal{T}^*F(\hat{Q} - Q)\,F\mathcal{T}^*$$
$$+ \operatorname{ind}\mathcal{T}^*F(-\hat{Q} + Q)\,F\mathcal{T}^* = \mu(Y, \hat{Y}) + \mu(\hat{Y}, Y),$$

where we have used (3.32), and in computing the rank of a sum of two matrices, we have used that $\operatorname{rank}(A + B) = \operatorname{rank} A + \operatorname{rank} B$ if $A^T B = AB^T = 0$ (see Remark 1.60(vii)). This completes the proof of property (v). □

*Proof of Property (vi)* This property is a consequence of formula (1.142) for the rank of a product of two matrices. But it can also be obtained from the proofs of the previous properties. In fact, using properties (iii) and (v), we obtain

$$\mu(Y, \hat{Y}) = \mu^*\left(Z^{-1}(0 \ I)^T, Z^{-1}\hat{Z}(0 \ I)^T\right)$$
$$= \operatorname{rank}\hat{X} - \mu^*\left(Z^{-1}\hat{Z}(0 \ I)^T, Z^{-1}(0 \ I)^T\right), \qquad (3.33)$$

where, in addition, it is used in formula (3.28). Further

$$\mu^*\left(Z^{-1}\hat{Z}(0 \ I)^T, Z^{-1}(0 \ I)^T\right) = \mu\left(\hat{Z}^{-1}Z(0 \ I)^T, \hat{Z}^{-1}(0 \ I)^T\right)$$
$$= \operatorname{rank} X - \mu\left(\hat{Z}^{-1}(0 \ I)^T, \hat{Z}^{-1}Z(0 \ I)^T\right)$$
$$= \operatorname{rank} X - \mu^*\left(\hat{Z}(0 \ I)^T, Z(0 \ I)^T\right),$$

where we have again used properties (iii) and (v). Finally, using property (iv), we conclude the proof of property (vi). □

*Proof of Property (vii)* Observe that $\mu(Y, \hat{Y}) \leq \min\{\text{rank } w, \text{rank } \hat{X}\}$ follows from property (v) and (3.33). Combining properties (v) and (vi), we also have the identity

$$\mu(Y, \hat{Y}) + \mu^*(Y, \hat{Y}) = \text{rank } \hat{X} - \text{rank } X + \text{rank } w(Y, \hat{Y}) \geq 0, \tag{3.34}$$

which means that $\mu(Y, \hat{Y}) \leq p(Y, \hat{Y})$, where $p(Y, \hat{Y})$, is given by (3.15). A similar estimate holds also for $\mu^*(Y, \hat{Y})$. In view of the property (v) and the dual form of the properties (vi) and (v), it is also obvious that

$$\mu(Y, \hat{Y}) = \text{rank } \hat{X} \overset{(v)}{\Leftrightarrow} \mu(\hat{Y}, Y) = \text{rank } w(Y, \hat{Y}) - \text{rank } \hat{X}$$

$$\overset{(vi)}{\Leftrightarrow} \mu^*(Y, \hat{Y}) = \text{rank } w(Y, \hat{Y}) - \text{rank } X$$

$$\overset{(v)}{\Leftrightarrow} \mu^*(\hat{Y}, Y) = \text{rank } X. \tag{3.35}$$

Further, by (3.33) we have

$$\mu(Y, \hat{Y}) = \text{rank } \hat{X} \quad \Leftrightarrow \quad \mu^* \left( Z^{-1} \hat{Z}(0 \ I)^T, Z^{-1}(0 \ I)^T \right) = 0,$$

so that

$$\mu_1^* \left( Z^{-1} \hat{Z}(0 \ I)^T, Z^{-1}(0 \ I)^T \right) = \text{rank } [I - w(Y, \hat{Y}) \, w(Y, \hat{Y})^\dagger] \, X^T = 0.$$

Using the properties (iii) and (vi) with $Z^{-1}(0 \ I)^T$ and $Z^{-1} \hat{Z}(0 \ I)^T$, we obtain

$$\mu_1(Y, \hat{Y}) = \mu_1^* \left( Z^{-1}(0 \ I)^T, Z^{-1} \hat{Z}(0 \ I)^T \right) = \text{rank } w(Y, \hat{Y}) - \text{rank } X. \tag{3.36}$$

Hence, we showed that the equality $\mu(Y, \hat{Y}) = \text{rank } \hat{X}$ is necessary and sufficient for the identities

$$\left. \begin{array}{l} \mu_1(Y, \hat{Y}) = \text{rank } w(Y, \hat{Y}) - \text{rank } X, \\ \mu_2(Y, \hat{Y}) = \text{rank } \hat{X} + \text{rank } X - \text{rank } w(Y, \hat{Y}) \end{array} \right\} \tag{3.37}$$

Applying the property (vi), we see that (3.36) is equivalent with

$$\mu_1(\hat{Y}, Y) = \text{rank } w(Y, \hat{Y}) - \text{rank } \hat{X}. \tag{3.38}$$

Combining conditions (3.36), (3.38) with (3.35), we derive all equalities referred to the case $\mu(Y, \hat{Y}) = \text{rank } \hat{X}$. Similar equivalences hold also for the case $\mu^*(Y, \hat{Y}) = \text{rank } \hat{X}$ because of the duality principle (see Sect. 3.1.4). The remaining cases of

$\mu(Y, \hat{Y}) = \operatorname{rank} w(Y, \hat{Y})$ and $\mu(Y, \hat{Y}) = \operatorname{rank} \hat{X} - \operatorname{rank} X + \operatorname{rank} w(Y, \hat{Y})$ can be treated analogously. $\qquad\square$

*Proof of Property (ix)* Applying properties (v), (iii), and finally (vi), we obtain

$$\mu\left(WZ(0\ I)^T, W(0\ I)^T\right) - \mu\left(W\hat{Z}(0\ I)^T, W(0\ I)^T\right)$$

$$= \mu\left((0\ I)^T, Z(0\ I)^T\right) - \mu\left((0\ I)^T, \hat{Z}(0\ I)^T\right)$$

$$\quad - \mu\left(W(0\ I)^T, WZ(0\ I)^T\right) + \mu\left(W(0\ I)^T, W\hat{Z}(0\ I)^T\right)$$

$$= \mu\left((0\ I)^T, Z(0\ I)^T\right) - \mu\left((0\ I)^T, \hat{Z}(0\ I)^T\right)$$

$$\quad - \mu^*\left(W^{-1}(0\ I)^T, Z(0\ I)^T\right) + \mu^*\left(W^{-1}(0\ I)^T, \hat{Z}(0\ I)^T\right)$$

$$= \mu\left(\hat{Z}(0\ I)^T, W^{-1}(0\ I)^T\right) - \mu\left(Z(0\ I)^T, W^{-1}(0\ I)^T\right).$$

This completes the proofs of properties (i)–(vii) and (ix) in Theorem 3.5. $\qquad\square$

Concerning the proof of the property (viii), this proof requires results about index of a block symmetric operator. The connection of the comparative index and the index of some symmetric operators is the subject of the next section.

## 3.2   Comparative Index and Symmetric Operators

This section is devoted to establishing a connection between the comparative index and the negative inertia index of some symmetric operator. In the next chapter, we will show that the multiplicity of a focal point in the interval $(k, k + 1]$ is fully determined by a suitable comparative index, and, hence, there is a deep connection between the (non)oscillation and the (non)existence of negative eigenvalues of a symmetric operator associated with (SDS).

### 3.2.1   Index of Block Symmetric Matrices

In this subsection we prove results concerning the index of symmetric $2n \times 2n$ matrices consisting of four $n \times n$ matrices.

Let $B \overset{A}{\sim} C$ mean that the matrices $B$ and $C$ are congruent, i.e., $B = A^T C A$ with $\det A \neq 0$. An important role in our treatment is played by the next lemma.

**Lemma 3.13** *Let $D, X \in \mathbb{R}^{n \times n}$, where $D$ is symmetric. Then*

$$\text{ind} \begin{pmatrix} 0 & X \\ X^T & D \end{pmatrix} = \text{rank } X + \text{ind} \, (I - X^{\dagger}X) \, D \, (I - X^{\dagger}X).$$

*Proof* Suppose that the singular value decomposition of $X$ is of the form $V \Sigma U^T$, where $\Sigma$ is the diagonal matrix having the singular values of $X$ on the diagonal, and $U$ and $V$ are orthogonal matrices. Then

$$\mathfrak{K} := \begin{pmatrix} 0 & X \\ X^T & D \end{pmatrix} \overset{A}{\sim} \begin{pmatrix} 0 & F \\ F & GU^TDUG \end{pmatrix}, \quad A := \begin{pmatrix} (G + \Sigma) V^T & S \\ 0 & U^T \end{pmatrix},$$

where $F := \Sigma \Sigma^{\dagger}$, $G := I - F$ and $S := \frac{1}{2} U^T DU(I + G) U^T$. Then

$$\text{ind} \, \mathfrak{K} = \text{ind} \left( \begin{pmatrix} 0 & F \\ F & 0 \end{pmatrix} + \text{diag}\{0, \, GU^TDUG\} \right)$$

$$= \text{ind} \begin{pmatrix} 0 & F \\ F & 0 \end{pmatrix} + \text{ind} \, (GU^TDUG)$$

and

$$\text{ind} \begin{pmatrix} 0 & F \\ F & 0 \end{pmatrix} = \text{rank } F = \text{rank } X,$$

and, in view of Remark 1.60 (ii),

$$\text{ind} \, (GU^TDUG) = \text{ind} \, (I - X^{\dagger}X)^T D \, (I - X^{\dagger}X).$$

The proof is complete. $\qquad \square$

An important consequence of Lemma 3.13 and of Theorem 3.2 is formulated in the next statement.

**Lemma 3.14** *Let $Y$ and $\hat{Y}$ satisfy (3.1). Then, using the notation of Theorem 3.2, we have*

$$\mu(Y, \hat{Y}) = i_-(\Phi), \quad \mu^*(Y, \hat{Y}) = i_+(\Phi),$$
$$\left. \Phi = \begin{pmatrix} 0 & (I - XX^{\dagger}) \, \hat{X} \\ \hat{X}^T (I - XX^{\dagger}) & \hat{X}^T (\hat{Q} - Q) \, \hat{X} \end{pmatrix}, \quad Y = \begin{pmatrix} X \\ U \end{pmatrix}, \quad \hat{Y} = \begin{pmatrix} \hat{X} \\ \hat{U} \end{pmatrix}, \right\}$$

$$(3.39)$$

*where $i_-(\Phi) = \text{ind}(\Phi)$ is the number of negative eigenvalues, and $i_+(\Phi)$ is the number of positive eigenvalues of the (symmetric) matrix $\Phi$.*

*Proof* The result follows directly from Lemma 3.13, Theorem 3.2, and (3.12).   □

In particular, for the cases mentioned in Remark 3.4, the corresponding matrices have the following form.

(i) If $\det X \neq 0$ and $\det \hat{X} \neq 0$, then $\Phi = \mathrm{diag}\{0,\ \hat{X}^T (\hat{Q} - Q)\, \hat{X}\}$ and

$$\mu(Y, \hat{Y}) = i_-(\hat{Q} - Q), \quad \mu^*(Y, \hat{Y}) = i_+(\hat{Q} - Q).$$

(ii) If $X = 0$, it follows that

$$\Phi = \begin{pmatrix} 0 & \hat{X} \\ \hat{X}^T & \hat{X}^T (\hat{Q} - Q)\, \hat{X} \end{pmatrix}, \quad \mu(Y, \hat{Y}) = \mu^*(Y, \hat{Y}) = \mathrm{rank}\, \hat{X}.$$

(iii) If $U = 0$, then $\det X \neq 0$ and hence

$$\Phi = \mathrm{diag}\{0,\ \hat{X}^T \hat{Q} \hat{X}\}, \quad \mu(Y, \hat{Y}) = i_-(\hat{X}^T \hat{U}), \quad \mu^*(Y, \hat{Y}) = i_+(\hat{X}^T \hat{U}).$$

The next statement is a generalization of Lemma 3.13.

**Lemma 3.15** *Let* $A^T = A$, $D^T = D$, *and* $X$ *be square matrices. Then*

$$\mathrm{ind} \begin{pmatrix} A & X \\ X^T & D \end{pmatrix} = \mathrm{ind}\, A + \mathrm{ind} \begin{pmatrix} 0 & (I - AA^\dagger)\, X \\ [(I - AA^\dagger)X]^T & D - X^T A^\dagger X \end{pmatrix}$$

$$= \mathrm{ind}\, A + \mathrm{rank}\, M$$

$$+ \mathrm{ind}\, (I - M^\dagger M)\, (D - X^T A^\dagger X)\, (I - M^\dagger M)$$

$$= \mathrm{ind}\, D + \mathrm{ind} \begin{pmatrix} A - X D^\dagger X^T & X^T (I - DD^\dagger) \\ (I - DD^\dagger)\, X & 0 \end{pmatrix}$$

$$= \mathrm{ind}\, D + \mathrm{rank}\, \tilde{M}$$

$$+ \mathrm{ind}\, (I - \tilde{M}^\dagger \tilde{M})\, (A - X D^\dagger X^T)\, (I - \tilde{M}^\dagger \tilde{M}),$$

*where* $\tilde{M} := (I - DD^\dagger)\, X^T$ *and* $M := (I - AA^\dagger)\, X$.

*Proof* Using the congruent transformation, we obtain

$$\mathfrak{K} = \begin{pmatrix} A & X \\ X^T & D \end{pmatrix} \overset{P}{\sim} \begin{pmatrix} A & (I - AA^\dagger)\, X \\ [(I - AA^\dagger)\, X]^T & D - X^T A^\dagger X \end{pmatrix}, \quad P = \begin{pmatrix} I & A^\dagger X \\ 0 & I \end{pmatrix}.$$

Consequently, in view of Lemma 3.13,

$$\operatorname{ind} \mathfrak{K} = \operatorname{ind}\left( \operatorname{diag}\{A, 0\} + \begin{pmatrix} 0 & (I - AA^\dagger)\,X \\ [(I - AA^\dagger)\,X]^T & D - X^T A^\dagger X \end{pmatrix} \right)$$

$$= \operatorname{ind} A + \operatorname{ind}\begin{pmatrix} 0 & (I - AA^\dagger)\,X \\ [(I - AA^\dagger)\,X]^T & D - X^T A^\dagger X \end{pmatrix}$$

$$= \operatorname{ind} A + \operatorname{rank} M + \operatorname{ind}\left[(I - M^\dagger M)\,(D - X^T A^\dagger X)\,(I - M^\dagger M)\right],$$

where $M = (I - AA^\dagger)\,X$. Observe that

$$\mathfrak{K} \overset{P_1}{\sim} \begin{pmatrix} D & X^T \\ X & A \end{pmatrix} = \tilde{\mathfrak{K}}, \quad P_1 = \begin{pmatrix} 0 & I \\ I & 0 \end{pmatrix}.$$

Consequently, applying the just proved result to the congruent matrix $\tilde{\mathfrak{K}}$, we obtain

$$\operatorname{ind} \mathfrak{K} = \operatorname{ind} \tilde{\mathfrak{K}} = \operatorname{ind} D + \operatorname{rank} \tilde{M} + \operatorname{ind}\left[(I - \tilde{M}^\dagger \tilde{M})\,(A - X D^\dagger X^T)\,(I - \tilde{M}^\dagger \tilde{M})\right],$$

where $\tilde{M} = (I - DD^\dagger)\,X^T$. The statement is proved.                                      $\square$

The following corollary to Lemma 3.15 was used in the formulation of a comparison theorem for differential and difference Hamiltonian systems; see [43, 51, 205].

**Corollary 3.16** *The symmetric matrix $\begin{pmatrix} A & X \\ X^T & D \end{pmatrix}$ is nonnegative definite if and only if*

$$A \geq 0, \quad D - X^T A^\dagger X \geq 0, \quad \operatorname{Im} X \subseteq \operatorname{Im} A, \tag{3.40}$$

*which is equivalent to*

$$D \geq 0, \quad A - X D^\dagger X^T \geq 0, \quad \operatorname{Ker} D \subseteq \operatorname{Ker} X. \tag{3.41}$$

## 3.2.2  Symmetric Operator $\Lambda[V]$ and Proof of Property (viii)

In the definition of the quadratic functional (2.56), we have used the symmetry of the operator

$$\Lambda[\mathcal{S}_k] := \mathcal{S}_k^T \begin{pmatrix} 0 & I \\ 0 & 0 \end{pmatrix} \mathcal{S}_k - \begin{pmatrix} 0 & I \\ 0 & 0 \end{pmatrix}$$

acting on the group of symplectic matrices determining (SDS). Recall that according to property (v) of Lemma 1.58, the $2n \times n$ blocks of $\mathcal{S}_k$ satisfy (1.147) and (1.148).

A generalization of $\Lambda[\mathcal{S}_k]$ is the operator, acting on pairs of matrices $Y, \hat{Y}$ satisfying only (3.1). Let a $2n \times 2n$ matrix $V$ consist of $2n \times n$ matrices

$$V = (Y \ \hat{Y}), \quad Y = \begin{pmatrix} X \\ U \end{pmatrix}, \quad \hat{Y} = \begin{pmatrix} \hat{X} \\ \hat{U} \end{pmatrix}. \tag{3.42}$$

Then under the assumption

$$X^T U = U^T X, \quad \hat{X}^T \hat{U} = \hat{U}^T \hat{X}, \tag{3.43}$$

the operator

$$\Lambda[V] := \mathrm{diag}\{U^T, \hat{X}^T\} V = \begin{pmatrix} U^T X & U^T \hat{X} \\ \hat{X}^T U & \hat{X}^T \hat{U} \end{pmatrix}$$

$$= V^T \begin{pmatrix} 0 & I \\ 0 & 0 \end{pmatrix} V - \begin{pmatrix} 0 & w(Y, \hat{Y}) \\ 0 & 0 \end{pmatrix} \tag{3.44}$$

is symmetric (here $w(Y, \hat{Y}) = Y^T \mathcal{J} \hat{Y}$). Next we prove some properties of this operator.

**Lemma 3.17**

*(i) Let $W$ be a symplectic matrix, and $V$ be determined by (3.42) and (3.43). Then*

$$\Lambda[WV] = V^T \Lambda[W] V + \Lambda[V], \tag{3.45}$$

*(ii) It holds*

$$\Lambda[\mathcal{J} V \mathcal{J}^T] = -\mathcal{J} \Lambda[V] \mathcal{J}^T. \tag{3.46}$$

*Proof*

(i) By the definition of the operator $\Lambda$, we have

$$\Lambda[WV] = V^T W^T \begin{pmatrix} 0 & I \\ 0 & 0 \end{pmatrix} WV - \begin{pmatrix} 0 & w(WY, W\hat{Y}) \\ 0 & 0 \end{pmatrix}$$

$$= V^T \left\{ W^T \begin{pmatrix} 0 & I \\ 0 & 0 \end{pmatrix} W - \begin{pmatrix} 0 & I \\ 0 & 0 \end{pmatrix} \right\} V + V^T \begin{pmatrix} 0 & I \\ 0 & 0 \end{pmatrix} V - \begin{pmatrix} 0 & w(Y, \hat{Y}) \\ 0 & 0 \end{pmatrix}$$

$$= V^T \Lambda[W] V + \Lambda[V].$$

where we have used (1.148) for $W$ and

$$w(WY, W\hat{Y}) = Y^T W^T \mathcal{J} W \hat{W} = Y^T \mathcal{J} \hat{Y} = w(Y, \hat{Y}).$$

This proves (3.45).

(ii) Again, by the definition of $\Lambda$, we have

$$-\mathcal{J}\Lambda[V]\mathcal{J}^T = -\mathcal{J}\begin{pmatrix} U^T X & U^T \hat{X} \\ \hat{X}^T U & \hat{X}^T \hat{U} \end{pmatrix} \mathcal{J}^T$$

$$= \begin{pmatrix} -\hat{X}^T \hat{U} & \hat{X}^T U \\ U^T \hat{X} & -U^T X \end{pmatrix} = \Lambda[\mathcal{J}\hat{Y}, -\mathcal{J}Y] = \Lambda[\mathcal{J}V\mathcal{J}^T],$$

which proves (3.46).

$\square$

An important role in our treatment is played by the following results describing the relationship between $\Lambda[V]$ and the comparative index; see [115, Lemma 4.4].

**Proposition 3.18** *If $V = (Y, \hat{Y})$ and $Y, \hat{Y}$ satisfy (3.1), then*

$$\text{ind } \Lambda[V] = \text{ind } X^T U + \mu(Y, \hat{Y}) = \text{ind } \hat{X}^T \hat{U} + \mu^*(\mathcal{J}\hat{Y}, \mathcal{J}Y), \qquad (3.47)$$

*hence, the comparative index $\mu(Y, \hat{Y})$ can be expressed as*

$$\mu(Y, \hat{Y}) = \text{ind } \Lambda[V] - \text{ind } X^T U. \qquad (3.48)$$

*Proof* To compute ind $\Lambda[V]$, we use (1.180) for $Y$ and $\hat{Y}$. Using the same notation as in the proof of property (iv) in Theorem 3.5, it is possible to verify that

$$\Lambda[V] = \mathcal{A}^T \begin{pmatrix} FQF & -G\hat{F} \\ -\hat{F}G & 2\hat{F}G\hat{F} + \hat{F}\hat{Q}\hat{F} - \hat{F}FQF\hat{F} \end{pmatrix} \mathcal{A},$$

$$\mathcal{A} = \begin{pmatrix} (I - GQF)M & \hat{F}\hat{M} \\ 0 & \hat{M} \end{pmatrix},$$

where det $M \neq 0$, det $\hat{M} \neq 0$, det$(I - GQF) \neq 0$ and, hence, the matrix $\mathcal{A}$ is invertible. Consequently, by Lemma 3.15, we have

$$\text{ind } \Lambda[V] = \text{ind } FQF + \text{ind } \begin{pmatrix} 0 & -G\hat{F} \\ -\hat{F}G & 2\hat{F}G\hat{F} + \hat{F}\hat{Q}\hat{F} - \hat{F}FQF\hat{F} \end{pmatrix}$$

$$= \text{ind } M^T FQFM + \text{rank } M + \text{ind } \mathcal{T}\hat{F}(\hat{Q} - Q)\hat{F}\mathcal{T}$$

$$= \text{ind } X^T QX + \mu(Y, \hat{Y}) = \text{ind } X^T U + \mu(Y, \hat{Y}),$$

where we have used (3.31), (3.32), and (3.30). This proves the first part of formula (3.47). To prove the second part, we use property (ii) of Lemma 3.17. Then

$$\text{ind } \Lambda[V] = \text{ind } (-\Lambda[\mathcal{J}V\mathcal{J}^T]) = \text{ind } (-\Lambda[\mathcal{J}\hat{Y}, -\mathcal{J}Y])$$

$$= \text{ind } \hat{X}^T \hat{U} + \mu^*(\mathcal{J}\hat{Y}, \mathcal{J}Y),$$

where ind $(-\Lambda[\mathcal{J}\hat{Y}, -\mathcal{J}Y])$ is computed analogously as ind $\Lambda[V]$ in the proof of the first equality in (3.47). The proof is now complete.                                      □

As a corollary to Proposition 3.18, we have a similar result for the dual comparative index $\mu^*(Y, \hat{Y})$.

**Corollary 3.19** *Under the assumptions of Proposition 3.18, we have*

$$\text{ind} (-\Lambda[V]) = \text{ind} (-X^T U) + \mu^*(Y, \hat{Y}) = \text{ind} (-\hat{X}^T \hat{U}) + \mu(\mathcal{J}\hat{Y}, \mathcal{J}Y).$$
$$(3.49)$$

*Hence, the comparative index $\mu^*(Y, \hat{Y})$ can be expressed as*

$$\mu^*(Y, \hat{Y}) = \text{ind} (-\Lambda[V]) - \text{ind} (-X^T U),$$   (3.50)

*and then*

$$\mu(Y, \hat{Y}) + \mu^*(Y, \hat{Y}) = \text{rank} \Lambda[V] - \text{rank} X^T U$$

$$= \text{rank} \hat{X} - \text{rank} X + \text{rank} w(Y, \hat{Y}).$$   (3.51)

*Proof* We apply Proposition 3.18 to the case $P_2Y$, $P_2\hat{Y}$, where $P_2 = \text{diag}\{I, -I\}$ (see Lemma 1.58). It is easy to verify that $\Lambda[\tilde{V}] = -\Lambda[V]$, where $\tilde{V} := (P_2Y \ P_2\hat{Y})$ and $V := (Y \ \hat{Y})$. Then the proof of (3.49) and (3.50) is completed. Formula (3.51) follows from (3.34). Note that for the case of $w(Y, \hat{Y}) = I$, we have rank$(\Lambda[V]) = \text{rank } U + \text{rank } \hat{X}$, where the nonsingularity of $V$ is used. For this special case, formula (3.51) takes the form

$$\mu(Y, \hat{Y}) + \mu^*(Y, \hat{Y}) = \text{rank } U + \text{rank } \hat{X} - \text{rank}(X^T U) = \text{rank } \hat{X} - \text{rank } X + n,$$

which can also be derived from (1.185).                                              □

We conclude this section with a statement, which yields the proof of property (viii) of Theorem 3.5.

**Proposition 3.20** *Formula (3.47) and property (viii) of Theorem 3.5 are equivalent.*

*Proof* By property (vi) of Theorem 3.5, we obtain the equality

$$\mu^*(\mathcal{J}\hat{Y}, \mathcal{J}Y) = \text{rank } U - \text{rank } \hat{U} + \mu(\mathcal{J}Y, \mathcal{J}\hat{Y}).$$

Consequently, (3.47) equivalent to

$$\mu^*(\mathcal{J}\hat{Y}, \mathcal{J}Y) = \mu(Y, \hat{Y}) + \text{ind } X^T U - \text{ind } \hat{X}^T \hat{U},$$

and this is equivalent to

$$\mu(\mathcal{J}Y, \mathcal{J}\hat{Y}) = \mu(Y, \hat{Y}) + p, \quad p := \text{ind } X^T U - \text{rank } U + \text{rank } \hat{U} - \text{ind } \hat{X}^T \hat{U}.$$

On the other hand, it is possible to prove that $p = \mu(\mathcal{J}Y, (I\ 0)^T) - \mu(\mathcal{J}\hat{Y}, (I\ 0)^T)$. Indeed, if we compute the above comparative indices by using Theorem 3.2, we obtain

$$\mu(\mathcal{J}Y, (I\ 0)^T) = \text{rank}\,(I - UU^\dagger) + \text{ind}\,X^T U = n - \text{rank}\,U + \text{ind}\,X^T U,$$

$$\mu(\mathcal{J}\hat{Y}, (I\ 0)^T) = \text{rank}\,(I - \hat{U}\hat{U}^\dagger) + \text{ind}\,\hat{X}^T \hat{U} = n - \text{rank}\,\hat{U} + \text{ind}\,\hat{X}^T \hat{U}.$$

The proof is complete. □

## 3.3 Comparative Index for Symplectic Matrices

In this section we introduce the comparative index for a pair of matrices determined by the blocks of symplectic matrices $W, \hat{W} \in \mathbb{R}^{2n \times 2n}$. The comparative index for symplectic matrices plays a fundamental role in comparison theorems for symplectic systems. It enables to formulate an exact relationship between the number of focal points of conjoined bases of two symplectic systems. Another problem, where we apply the results of this section, is the theory of boundary value problems associated with (SDS).

### 3.3.1 Basic Properties

We introduce the following notation. For any matrix $W = \left(\begin{smallmatrix} A & B \\ C & D \end{smallmatrix}\right) \in \mathbb{R}^{2n \times 2n}$, we define the $4n \times 2n$ matrix

$$\langle W \rangle := \begin{pmatrix} I & 0 \\ A & B \\ 0 & -I \\ C & D \end{pmatrix}. \tag{3.52}$$

For easier application of the identities below, we note that when $W$ is a symplectic matrix, then its inverse satisfies $W^{-1} = \left(\begin{smallmatrix} D^T & -B^T \\ -C^T & A^T \end{smallmatrix}\right)$; see Lemma 1.58(ii).

**Lemma 3.21** *Let* $W, \hat{W} \in \mathbb{R}^{2n \times 2n}$ *be symplectic matrices. Then the matrices* $\langle W \rangle$ *and* $\langle \hat{W} \rangle$ *defined in (3.52) have the following properties.*

(i) *The matrix* $\langle W \rangle$ *satisfies (3.1), i.e.,* $\text{rank}\langle W \rangle = 2n$, $\langle W \rangle^T \mathcal{J} \langle W \rangle = 0$.
(ii) *For the Wronskian of* $\langle W \rangle$ *and* $\langle \hat{W} \rangle$, *we have the relations*

$$w(\langle W \rangle, \langle \hat{W} \rangle) = \langle W \rangle^T \mathcal{J} \langle \hat{W} \rangle = -\mathcal{J}(I - W^{-1}\hat{W}),$$

$$\text{rank}\,w(\langle W \rangle, \langle \hat{W} \rangle) = \text{rank}(W - \hat{W}),$$

*and the comparative index for $\langle W \rangle$ and $\langle \hat{W} \rangle$ obeys the estimate*

$$\mu(\langle W \rangle, \langle \hat{W} \rangle)$$
$$\leq \min \left\{ \operatorname{rank}(W - \hat{W}), \; n + \operatorname{rank} \hat{\mathcal{B}}, \; \operatorname{rank}(W - \hat{W}) + \operatorname{rank} \hat{\mathcal{B}} - \operatorname{rank} \mathcal{B} \right\}.$$
$$(3.53)$$

*(iii)  The comparative index of $\langle W \rangle$ and $\langle \hat{W} \rangle$ satisfies*

$$\mu_i(\langle W \rangle, \langle \hat{W} \rangle) = \mu_i^*(\langle W^{-1} \rangle, \langle \hat{W}^{-1} \rangle), \; i \in \{1, 2\}.$$

*(iv)  The comparative index of $\langle W \rangle$ and $\langle \hat{W} \rangle$ satisfies*

$$\mu(\langle W \rangle, \langle \hat{W} \rangle) = \operatorname{rank} \hat{\mathcal{B}} - \operatorname{rank} \mathcal{B} + \mu^*(\langle \hat{W} \rangle, \langle W \rangle).$$

*(v)  The comparative index of $\langle W \rangle$ and $\langle \hat{W} \rangle$ satisfies*

$$\mu(\langle W \rangle, \langle \hat{W} \rangle) = \mu \left( W(0\ I)^T, \hat{W}(0\ I)^T \right) + \mu \left( \langle \hat{W}^{-1} W \rangle, \langle I \rangle \right) \qquad (3.54)$$
$$= \mu^* \left( W^{-1}(0\ I)^T, \hat{W}^{-1}(0\ I)^T \right) + \mu \left( \langle W \hat{W}^{-1} \rangle, \langle I \rangle \right). \qquad (3.55)$$

*(vi)  We have the identity*

$$\mu \left( W(0\ I)^T, \hat{W}(0\ I)^T \right) - \mu^* \left( W^{-1}(0\ I)^T, \hat{W}^{-1}(0\ I)^T \right)$$
$$= \mu \left( \langle W \hat{W}^{-1} \rangle, \langle I \rangle \right) - \mu \left( \langle \hat{W}^{-1} W \rangle, \langle I \rangle \right).$$

*Proof of Parts (i)–(iv)*  For part (i) we have that $\operatorname{rank}\langle W \rangle = \operatorname{rank} W = 2n$ and

$$\begin{pmatrix} I & 0 \\ A & B \end{pmatrix}^T \begin{pmatrix} 0 & -I \\ C & D \end{pmatrix} = \begin{pmatrix} C^T A & C^T B \\ B^T C & B^T D \end{pmatrix} = \begin{pmatrix} A^T C & C^T B \\ B^T C & D^T B \end{pmatrix} = \begin{pmatrix} 0 & -I \\ C & D \end{pmatrix}^T \begin{pmatrix} I & 0 \\ A & B \end{pmatrix},$$

where we have used property (1.145) of a symplectic matrix.

For part (ii) we have

$$\langle W \rangle^T \mathcal{J} \langle \hat{W} \rangle = \begin{pmatrix} I & 0 \\ A & B \end{pmatrix}^T \begin{pmatrix} 0 & -I \\ \hat{C} & \hat{D} \end{pmatrix} - \begin{pmatrix} 0 & -I \\ C & D \end{pmatrix}^T \begin{pmatrix} I & 0 \\ \hat{A} & \hat{B} \end{pmatrix} = \mathcal{J}^T (I - W^{-1} \hat{W}).$$

Consequently, we obtain the identity

$$\operatorname{rank} \left( \langle W \rangle^T \mathcal{J} \langle \hat{W} \rangle \right) = \operatorname{rank} (I - W^{-1} \hat{W}) = \operatorname{rank} [W^{-1}(W - \hat{W})]$$
$$= \operatorname{rank} (W - \hat{W}).$$

Estimate (3.53) follows from the upper bound for the comparative index in property (vii) of Theorem 3.5, where we evaluate

$$\text{rank} \begin{pmatrix} I & 0 \\ A & B \end{pmatrix} = \text{rank} \begin{pmatrix} I & 0 \\ A & I \end{pmatrix} \text{diag}\{I, B\}$$

$$= \text{rank diag}\{I, B\} = n + \text{rank } B.$$

By a similar way, we evaluate the rank of the $2n \times 2n$ upper block of $\langle \hat{W} \rangle$.

For the proof of property (iii), we note that by Theorem 3.5(i)–(ii), we obtain

$$\mu_i(\langle W \rangle, \langle \hat{W} \rangle) = \mu_i(L\langle W \rangle W^{-1}, L\langle \hat{W} \rangle \hat{W}^{-1}) = \mu_i(P_2\langle W^{-1} \rangle, P_2\langle \hat{W}^{-1} \rangle)$$

for $i \in \{1, 2\}$, where

$$L := \text{diag} \left\{ \begin{pmatrix} 0 & I \\ I & 0 \end{pmatrix}, \begin{pmatrix} 0 & I \\ I & 0 \end{pmatrix} \right\}, \quad P_2 = \text{diag}\{I_{2n}, -I_{2n}\},$$

and, in view of (3.13), $\mu_i(P_2 Y, P_2 \hat{Y}) = \mu_i^*(Y, \hat{Y})$ for any $Y = \begin{pmatrix} X \\ U \end{pmatrix}$ and $\hat{Y} = \begin{pmatrix} \hat{X} \\ \hat{U} \end{pmatrix}$ satisfying (3.1).

Property (iv) is a consequence of Theorem 3.5(vi). Obviously, for our particular case, we evaluate the ranks of the upper blocks of $\langle W \rangle$ and $\langle \hat{W} \rangle$ by analogy with the proof of (ii) and substitute them into Theorem 3.5(vi). This then yields the proof of part (iv) of this lemma. □

The proof of properties (v) and (vi) will be given in Sect. 3.3.6. Here we point out that this rather technical proof is based on a generalization of Theorem 3.6 (see Lemma 3.23 below) and on additional algebraic properties of the comparative indices for (3.52) (see Proposition 3.37).

The main result of this section is the generalization of Theorem 3.6 to the case where instead of $Z, \hat{Z}, W$, four matrices $Z, \hat{Z}, W, \hat{W}$ are used. For this purpose introduce the notation

$$\left. \begin{aligned} \mathcal{L}(Y, \hat{Y}, W, \hat{W}) := \mu(\hat{W}\hat{Y}, \hat{W}(0 \ I)^T) - \mu(WY, W(0 \ I)^T) \\ + \mu(WY, \hat{W}\hat{Y}) - \mu(Y, \hat{Y}), \end{aligned} \right\} \quad (3.56)$$

where $Y = Z(0 \ I)^T$ and $\hat{Y} = \hat{Z}(0 \ I)^T$ are $2n \times n$ matrices satisfying conditions (3.1). We prove the following properties of operator (3.56).

**Lemma 3.22** Let $Y$ and $\hat{Y}$ be $2n \times n$ matrices satisfying (3.1). Then the following properties of operator in (3.56) hold.

(i) For any symplectic matrix $W$, we have $\mathcal{L}(Y, \hat{Y}, W, W) = 0$.
(ii) For $p \in \mathbb{N}$ and symplectic matrices $W_1, \ldots, W_p$ and $\hat{W}_1, \ldots, \hat{W}_p$, we define

$$Z(r) := \prod_{k=1}^{r} W_{r-k+1} = W_r W_{r-1} \ldots W_1, \quad \hat{Z}(r) := \prod_{k=1}^{r} \hat{W}_{r-k+1}, \quad r = 1, \ldots, p,$$

*and* $Z(0) := I$, $\hat{Z}(0) := I$. *Then we have the following "multiplicative property" of operator* (3.56):

$$
\left.\begin{aligned}
\mathcal{L}(Y, \hat{Y}, Z(p), \hat{Z}(p)) &= \sum_{r=1}^{p} \mathcal{L}\big(Z(r-1)\,Y, \hat{Z}(r-1)\,\hat{Y}, W_r, \hat{W}_r\big) \\
&+ \sum_{r=1}^{p} \left[ \mu\big(Z(r)\,(0\ I)^T, W_r(0\ I)^T\big) - \mu\big(\hat{Z}(r)\,(0\ I)^T, \hat{W}_r(0\ I)^T\big) \right].
\end{aligned}\right\}
$$
(3.57)

*Proof* For the proof of (i), we use Theorem 3.6, which is equivalent to property (i). For the proof of (ii), we note that by the definition in (3.56), we have on the left-hand side of (3.57) that

$$
\left.\begin{aligned}
\mathcal{L}(Y, \hat{Y}&, Z(p), \hat{Z}(p)) \\
&= \mu\big(\hat{Z}(p)\,\hat{Y}, \hat{Z}(p)\,(0\ I)^T\big) - \mu\big(Z(p)\,Y, Z(p)\,(0\ I)^T\big) \\
&+ \mu\big(Z(p)\,Y, \hat{Z}(p)\,\hat{Y}\big) - \mu(Y, \hat{Y}).
\end{aligned}\right\}
$$
(3.58)

Similarly, for the operator $\mathcal{L}\big(Z(r-1)\,Y, \hat{Z}(r-1)\,\hat{Y}, W_r, \hat{W}_r\big)$ on the right-hand side of (3.57), we derive

$$
\begin{aligned}
\mathcal{L}\big(Z(r-1)&\,Y, \hat{Z}(r-1)\,\hat{Y}, W_r, \hat{W}_r\big) \\
&= \mu\big(\hat{Z}(r)\,\hat{Y}, \hat{W}_r(0\ I)^T\big) - \mu(Z(r)\,Y, W_r(0\ I)^T) + \Delta\mu(Z(r-1)\,Y, \hat{Z}(r-1)\,\hat{Y}),
\end{aligned}
$$

and then

$$
\begin{aligned}
\mathcal{L}\big(Z(r-1)&\,Y, \hat{Z}(r-1)\,\hat{Y}, W_r, \hat{W}_r\big) - \mu(\hat{Z}(r)\,(0\ I)^T, \hat{W}_r(0\ I)^T) \\
&\quad + \mu(Z(r)\,(0\ I)^T, W_r(0\ I)^T) \\
&= -\mu\big(\hat{Z}(r)\,(0\ I)^T, \hat{W}_r(0\ I)^T\big) + \mu\big(\hat{Z}(r)\,\hat{Y}, \hat{W}_r(0\ I)^T\big) \\
&\quad + \mu\big(Z(r)\,(0\ I)^T, W_r(0\ I)^T\big) - \mu\big(Z(r)\,Y, W_r(0\ I)^T\big) \\
&\quad + \Delta\mu\big(Z(r-1)\,Y, \hat{Z}(r-1)\,\hat{Y}\big) \\
&= -\mathcal{L}\big(\hat{Z}(r-1)\,\hat{Y}, \hat{Z}(r-1)\,(0\ I)^T, \hat{W}_r, \hat{W}_r\big) + \Delta\mu\big(\hat{Z}(r-1)\,\hat{Y}, \hat{Z}(r-1)\,(0\ I)^T\big) \\
&\quad + \mathcal{L}\big(Z(r-1)\,Y, Z(r-1)\,(0\ I)^T, W_r, W_r\big) - \Delta\mu\big(Z(r-1)\,Y, Z(r-1)\,(0\ I)^T\big) \\
&\quad + \Delta\mu\big(Z(r-1)\,Y, \hat{Z}(r-1)\,\hat{Y}\big).
\end{aligned}
$$
(3.59)

By property (i) we have

$$
\mathcal{L}(\hat{Z}(r-1)\,\hat{Y}, \hat{Z}(r-1)\,(0\ I)^T, \hat{W}_r, \hat{W}_r) = 0,
$$
$$
\mathcal{L}(Z(r-1)\,Y, Z(r-1)\,(0\ I)^T, W_r, W_r) = 0.
$$

Then summing (3.59) from $r = 1$ to $r = p$, we derive equality (3.57). The proof is complete.                                                                    □

Using notation (3.56) we begin with the following auxiliary result, which turns out to be highly important for future applications.

**Lemma 3.23** *Let* $W$, $Z$, $\hat{W}$, $\hat{Z}$ *be symplectic matrices. Then for the operator* $\mathcal{L}(Y, \hat{Y}, W, \hat{W})$ *given by* (3.56) *with* $Y = Z(0\ I)^T$ *and* $\hat{Y} = \hat{Z}(0\ I)^T$, *we have*

$$
\mathcal{L}(Y, \hat{Y}, W, \hat{W})
$$

$$
= \mu\left(W(0\ I)^T, \hat{W}(0\ I)^T\right) + \mu\left(\hat{W}^{-1}WZ(0\ I)^T, Z(0\ I)^T\right)
$$

$$
- \mu\left(\hat{Z}^{-1}\hat{W}^{-1}WZ(0\ I)^T, \hat{Z}^{-1}Z(0\ I)^T\right)
$$

$$
- \mu\left(\hat{W}^{-1}WZ(0\ I)^T, \hat{W}^{-1}W(0\ I)^T\right), \tag{3.60}
$$

*and*

$$
\mathcal{L}(Y, \hat{Y}, W, \hat{W})
$$

$$
= \mu\left(W(0\ I)^T, W\hat{W}^{-1}(0\ I)^T\right) + \mu\left(\hat{Z}^{-1}Z(0\ I)^T, \hat{Z}^{-1}\hat{W}^{-1}WZ(0\ I)^T\right)
$$

$$
- \mu\left(\hat{W}Z(0\ I)^T, WZ(0\ I)^T\right) - \mu\left(WZ(0\ I)^T, W\hat{W}^{-1}(0\ I)^T\right). \tag{3.61}
$$

*Proof* For the proof of equality (3.60), we apply Lemma 3.22(ii) to the case $p = 2$, $W_1 := \hat{W}^{-1}W$, $W_2 := \hat{W}$, $\hat{W}_1 := I$, $\hat{W}_2 := \hat{W}$, and then by (3.57), we have

$$
\mathcal{L}(Y, \hat{Y}, W, \hat{W}) = \mathcal{L}(Y, \hat{Y}, \hat{W}^{-1}W, I) + \mu(W(0\ I)^T, \hat{W}(0\ I)^T), \tag{3.62}
$$

where we have used that $\mathcal{L}(\hat{W}^{-1}WY, \hat{Y}, \hat{W}, \hat{W}) = 0$ by Lemma 3.22(i). By (3.56),

$$
\mathcal{L}(Y, \hat{Y}, \hat{W}^{-1}W, I) = -\mu(\hat{W}^{-1}WZ(0\ I)^T, \hat{W}^{-1}W(0\ I)^T)
$$

$$
+ \mu(\hat{W}^{-1}WZ(0\ I)^T, \hat{Z}(0\ I)^T) - \mu(Z(0\ I)^T, \hat{Z}(0\ I)^T),
$$

and by Theorem 3.6 for the last two addends in the previous formula, we have

$$
\mu(\hat{W}^{-1}WZ(0\ I)^T, \hat{Z}(0\ I)^T) - \mu(Z(0\ I)^T, \hat{Z}(0\ I)^T)
$$

$$
= \mu(\hat{W}^{-1}WZ(0\ I)^T, Z(0\ I)^T) - \mu(\hat{Z}^{-1}\hat{W}^{-1}WZ(0\ I)^T, \hat{Z}^{-1}Z(0\ I)^T).
$$

Finally, for the operator $\mathcal{L}(Y, \hat{Y}, \hat{W}^{-1}W, I)$ in (3.62), we have the representation

$$\mathcal{L}(Y, \hat{Y}, \hat{W}^{-1}W, I) = -\mu(\hat{W}^{-1}WZ(0\ I)^T, \hat{W}^{-1}W(0\ I)^T)$$
$$+ \mu(\hat{W}^{-1}WZ(0\ I)^T, Z(0\ I)^T) - \mu(\hat{Z}^{-1}\hat{W}^{-1}WZ(0\ I)^T, \hat{Z}^{-1}Z(0\ I)^T)$$

which completes the proof of (3.60).

Similarly, the proof of (3.61) is derived by using Lemma 3.22(ii) with $p = 2$, $W_1 := \hat{W}$, $W_2 := W\hat{W}^{-1}$, $\hat{W}_1 := \hat{W}$, $\hat{W}_2 := I$, and then by (3.57) and Lemma 3.22(i), we have

$$\mathcal{L}(Y, \hat{Y}, W, \hat{W}) = \mathcal{L}(\hat{W}Y, \hat{W}\hat{Y}, W\hat{W}^{-1}, I) + \mu(W(0\ I)^T, W\hat{W}(0\ I)^T). \qquad (3.63)$$

By (3.56), we get

$$\mathcal{L}(\hat{W}Y, \hat{W}\hat{Y}, W\hat{W}^{-1}, I) = -\mu(WZ(0\ I)^T, W\hat{W}^{-1}(0\ I)^T)$$
$$+ \mu(WZ(0\ I)^T, \hat{W}\hat{Z}(0\ I)^T) - \mu(\hat{W}Z(0\ I)^T, \hat{W}\hat{Z}(0\ I)^T),$$

and by Theorem 3.6 for the last two addends in the previous formula we have

$$\mu(\hat{W}Z(0\ I)^T, \hat{W}\hat{Z}(0\ I)^T) - \mu(WZ(0\ I)^T, \hat{W}\hat{Z}(0\ I)^T)$$
$$= \mu(\hat{W}Z(0\ I)^T, WZ(0\ I)^T) - \mu(\hat{Z}^{-1}Z(0\ I)^T, \hat{Z}^{-1}\hat{W}^{-1}WZ(0\ I)^T).$$

Finally, the operator $\mathcal{L}(\hat{W}Y, \hat{W}\hat{Y}, W\hat{W}^{-1}, I)$ takes the form

$$\mathcal{L}(\hat{W}Y, \hat{W}\hat{Y}, W\hat{W}^{-1}, I) = -\mu(WZ(0\ I)^T, W\hat{W}^{-1}(0\ I)^T)$$
$$- \mu(\hat{W}Z(0\ I)^T, WZ(0\ I)^T) + \mu(\hat{Z}^{-1}Z(0\ I)^T, \hat{Z}^{-1}\hat{W}^{-1}WZ(0\ I)^T).$$

Substituting the last formula into (3.63), we complete the proof of (3.61).  □

Using Lemma 3.23 and the comparative index for a pair of symplectic matrices, we can now approach the main result of this section, which generalizes Theorem 3.6 to the case of two different matrices $W$ and $\hat{W}$.

**Theorem 3.24** *Let* $W$, $Z$, $\hat{W}$, $\hat{Z}$ *be symplectic matrices. Then for the operator* $\mathcal{L}(Y, \hat{Y}, W, \hat{W})$ *given by (3.56) with* $Y = Z(0\ I)^T$ *and* $\hat{Y} = \hat{Z}(0\ I)^T$, *we have*

$$\mathcal{L}(Y, \hat{Y}, W, \hat{W}) = \mu(\langle W \rangle, \langle \hat{W} \rangle) - \mu\left(\langle \hat{Z}^{-1}\hat{W}^{-1}WZ \rangle, \langle \hat{Z}^{-1}Z \rangle\right), \qquad (3.64)$$

*and*

$$\mu(\langle W \rangle, \langle \hat{W} \rangle) - \mu\left(\langle \hat{Z}^{-1}\hat{W}^{-1}WZ \rangle, \langle \hat{Z}^{-1}Z \rangle\right)$$
$$= \mu\left(\langle \hat{Z}^{-1}Z \rangle, \langle \hat{Z}^{-1}\hat{W}^{-1}WZ \rangle\right) - \mu(\langle \hat{W} \rangle, \langle W \rangle). \qquad (3.65)$$

*Proof* To prove (3.64) we show that this identity is equivalent to (3.60). Using successively formulas (3.54) and (3.55), we obtain the following representation for the comparative indices on the right-hand side of (3.60):

$$\mu\left(W(0\ I)^T, \hat{W}(0\ I)^T\right) = \mu(\langle W\rangle, \langle\hat{W}\rangle) - \mu(\langle\hat{W}^{-1}W\rangle, \langle I\rangle),$$

$$\mu\left(\hat{W}^{-1}WZ(0\ I)^T, Z(0\ I)^T\right) = \mu\left(\langle\hat{W}^{-1}WZ\rangle, \langle Z\rangle\right) - \mu\left(\langle Z^{-1}\hat{W}^{-1}WZ\rangle, \langle I\rangle\right)$$

$$= \mu\left(\hat{W}^{-1}WZ(0\ I)^T, \hat{W}^{-1}W(0\ I)^T\right)$$

$$+ \mu(\langle\hat{W}^{-1}W\rangle, \langle I\rangle) - \mu\left(\langle Z^{-1}\hat{W}^{-1}WZ\rangle, \langle I\rangle\right),$$

and

$$\mu\left(\hat{Z}^{-1}\hat{W}^{-1}WZ(0\ I)^T, \hat{Z}^{-1}Z(0\ I)^T\right)$$

$$= \mu\left(\langle\hat{Z}^{-1}\hat{W}^{-1}WZ\rangle, \langle\hat{Z}^{-1}Z\rangle\right) - \mu\left(\langle Z^{-1}\hat{W}^{-1}WZ\rangle, \langle I\rangle\right).$$

Substituting the obtained equalities into (3.60) and canceling the same summands, we get (3.64). To prove (3.65) it is sufficient to observe that by Lemma 3.21(ii)

$$\text{rank}(\langle W\rangle^T \mathcal{J}\langle\hat{W}\rangle) = \text{rank}(W - \hat{W}), \tag{3.66}$$

and similarly

$$\text{rank}\,\langle\hat{Z}^{-1}\hat{W}^{-1}WZ\rangle^T \mathcal{J}\langle\hat{Z}^{-1}Z\rangle = \text{rank}(\hat{Z}^{-1}\hat{W}^{-1}WZ - \hat{Z}^{-1}Z)$$

$$= \text{rank}\,\hat{Z}^{-1}\hat{W}^{-1}(W - \hat{W})\,Z$$

$$= \text{rank}(W - \hat{W}). \tag{3.67}$$

Consequently, using property (v) of Theorem 3.5, we obtain (3.65). □

## 3.3.2 General Case

Next we apply the properties of the comparative index established in Sects. 3.1.2–3.1.4 to the comparative index of a pair of symplectic matrices. We also present particular cases of the comparative index for matrices appearing in discrete linear Hamiltonian systems and Sturm-Liouville difference equations.

Applying Theorem 3.2 to matrices (3.52), we obtain the following result.

**Lemma 3.25** *Let $W$ and $\hat{W}$ be symplectic matrices. The comparative index $\mu(\langle W \rangle, \langle \hat{W} \rangle)$ can be computed by the following formulas*

$$\mu_1(\langle W \rangle, \langle \hat{W} \rangle) = \operatorname{rank} \mathcal{M}, \tag{3.68}$$

$$\mu_2(\langle W \rangle, \langle \hat{W} \rangle) = \operatorname{ind} \mathcal{T} \begin{pmatrix} I & \hat{A}^T \\ 0 & \hat{B}^T \end{pmatrix} (Q_{\langle \hat{W} \rangle} - Q_{\langle W \rangle}) \begin{pmatrix} I & 0 \\ \hat{A} & \hat{B} \end{pmatrix} \mathcal{T}, \tag{3.69}$$

$$\mathcal{M} := (I - \mathcal{B}\mathcal{B}^\dagger)(\hat{A} - A, \hat{B}), \quad \mathcal{T} := I_{2n} - \mathcal{M}^\dagger \mathcal{M}, \tag{3.70}$$

*where the symmetric matrices $Q_{\langle \hat{W} \rangle}$ and $Q_{\langle W \rangle}$ satisfy*

$$\begin{pmatrix} I & A^T \\ 0 & B^T \end{pmatrix} Q_{\langle W \rangle} \begin{pmatrix} I & 0 \\ A & B \end{pmatrix} = \Lambda[W], \quad \Lambda[W] := \begin{pmatrix} C^T A & C^T B \\ B^T C & B^T D \end{pmatrix}$$

*and are determined by the formulas (analogously for $Q_{\langle \hat{W} \rangle}$)*

$$Q_{\langle W \rangle} := \begin{pmatrix} I & -A^T \\ 0 & I \end{pmatrix} \left[ \begin{pmatrix} C^T A & C^T B \mathcal{B}^\dagger \\ \mathcal{B}^\dagger C & \mathcal{B}^\dagger D \mathcal{B}^\dagger \end{pmatrix} + \mathfrak{C} \right] \begin{pmatrix} I & 0 \\ -A & I \end{pmatrix}, \tag{3.71}$$

$$\mathfrak{C} = \mathfrak{C}^T, \quad \operatorname{diag}\{I, B^T\} \, \mathfrak{C} \, \operatorname{diag}\{I, B\} = 0. \tag{3.72}$$

*Proof* By a direct computation, one can verify that

$$\langle W \rangle = \operatorname{diag} \left\{ \begin{pmatrix} I & 0 \\ A & I \end{pmatrix}, \begin{pmatrix} I & -A^T \\ 0 & I \end{pmatrix} \right\} \operatorname{diag}\{I, B, C^T, I\} \begin{pmatrix} I_{2n} \\ W \end{pmatrix}. \tag{3.73}$$

The first matrix on the right-hand side of (3.73) is a symplectic block diagonal matrix; hence by (ii) of Theorem 3.5, we obtain

$$\mu_1(\langle W \rangle, \langle \hat{W} \rangle) = \mu_1(\mathfrak{D}\langle W \rangle, \mathfrak{D}\langle \hat{W} \rangle) = \operatorname{rank} \mathcal{M},$$

$$\mathfrak{D} := \operatorname{diag} \left\{ \begin{pmatrix} I & 0 \\ -A & I \end{pmatrix}, \begin{pmatrix} I & A^T \\ 0 & I \end{pmatrix} \right\},$$

where $\mathcal{M}$ is given in (3.70). Hence, formula (3.68) is proved. Further, using (3.73), it is not difficult to obtain general solution of Riccati-type relation (3.8) according to (1.168). A general solution of relation (3.8) for

$$\mathcal{Y} = \operatorname{diag}\{I, B, C^T, I\} \begin{pmatrix} I_{2n} \\ W \end{pmatrix} = \begin{pmatrix} I & 0 \\ 0 & B \\ C^T A & C^T B \\ C & D \end{pmatrix}$$

has the form, in view of (1.168),

$$\tilde{Q} = \text{diag}\,\{I,\ \mathcal{B}\mathcal{B}^\dagger\} \begin{pmatrix} \mathcal{C}^T\!\mathcal{A} & \mathcal{C}^T\!\mathcal{B} \\ \mathcal{C} & \mathcal{D} \end{pmatrix} \text{diag}\,\{I,\ \mathcal{B}^\dagger\} + \mathfrak{C},$$

where the matrix $\mathfrak{C}$ satisfies (3.72). Then the general solution of (3.8) for $\langle W \rangle$ is of the form (3.71). One particular solution of (3.8) for $\langle W \rangle$ which coincides with the matrix $\mathcal{G}_k$ given by (2.64) is

$$
\begin{aligned}
Q_{\langle W \rangle} &= \begin{pmatrix} I & -\mathcal{A}^T \\ 0 & I \end{pmatrix} \begin{pmatrix} \mathcal{C}^T\!\mathcal{A} & \mathcal{C}^T \\ \mathcal{C} & \mathcal{B}\mathcal{B}^\dagger\mathcal{D}\mathcal{B}^\dagger \end{pmatrix} \begin{pmatrix} I & 0 \\ -\mathcal{A} & I \end{pmatrix} \\
&= \begin{pmatrix} \mathcal{A}^T\mathcal{B}\mathcal{B}^\dagger\mathcal{D}\mathcal{B}^\dagger\mathcal{A} - \mathcal{A}^T\mathcal{C}\,\mathcal{C}^T - \mathcal{A}^T\mathcal{B}\mathcal{B}^\dagger\mathcal{D}\mathcal{B}^\dagger & \\ \mathcal{C} - (\mathcal{A}^T\mathcal{B}\mathcal{B}^\dagger\mathcal{D}\mathcal{B}^\dagger)^T & \mathcal{B}\mathcal{B}^\dagger\mathcal{D}\mathcal{B}^\dagger \end{pmatrix},
\end{aligned}
\tag{3.74}
$$

which we obtain from (3.71) with

$$\mathfrak{C} := \begin{pmatrix} 0 & \mathcal{C}^T(I - \mathcal{B}\mathcal{B}^\dagger) \\ (I - \mathcal{B}\mathcal{B}^\dagger)\mathcal{C} & 0 \end{pmatrix}.$$

This proves the statement of this lemma.                                                          $\square$

**Corollary 3.26** *Let $V$ be a symplectic matrix, and then for $W := V$ and $\hat{W} := I$, the formulas in Lemma 3.25 take the form*

$$\mu(\langle V \rangle, \langle I \rangle) = \text{rank}\ \mathcal{M} + \text{ind}\ \mathcal{T}\,[(\mathcal{D} - I)\,\mathcal{B}^\dagger(\mathcal{A} - I) - \mathcal{C}]\,\mathcal{T}, \tag{3.75}$$

$$V = \begin{pmatrix} \mathcal{A} & \mathcal{B} \\ \mathcal{C} & \mathcal{D} \end{pmatrix}, \quad \mathcal{M} := (I - \mathcal{B}\mathcal{B}^\dagger)\,(\mathcal{A} - I), \quad \mathcal{T} := I - \mathcal{M}^\dagger\mathcal{M}.$$

*Proof* Observe that by (3.53), we have the estimate

$$\mu(\langle V \rangle, \langle I \rangle) \leq n,$$

because $\hat{\mathcal{B}} = 0$. The formulas of Lemma 3.25 for this case reduce to determining the rank and the index of $n \times n$ matrices. The solution $Q_{\hat{W}}$ can be taken to be equal to zero. Using $Q_W$ in the form (3.71) with $\mathfrak{C} = 0$ leads to the formula

$$\mu_2(\langle V \rangle, \langle I \rangle) = \text{ind}\left[ -\mathcal{T}\,(I,\ I - \mathcal{A}^T) \begin{pmatrix} \mathcal{C}^T\!\mathcal{A} & \mathcal{C}^T\mathcal{B}\mathcal{B}^\dagger \\ \mathcal{B}\mathcal{B}^\dagger\mathcal{C} & \mathcal{B}\mathcal{B}^\dagger\mathcal{D}\mathcal{B}^\dagger \end{pmatrix} \begin{pmatrix} I \\ I - \mathcal{A} \end{pmatrix} \mathcal{T} \right],$$

from which we derive the second addend in (3.75) because of

$$\mathcal{M}\mathcal{T} = 0 \quad \Leftrightarrow \quad \mathcal{B}\mathcal{B}^\dagger(\mathcal{A} - I)\mathcal{T} = (\mathcal{A} - I)\mathcal{T}$$

and (1.145). The proof is complete.                                                              $\square$

From the previous computations and Lemma 3.21(v), we obtain another way of determining the comparative index $\mu(\langle W \rangle, \langle \hat{W} \rangle)$.

**Lemma 3.27** *Let $W$ and $\hat{W}$ be symplectic matrices. We have the formula*

$$\mu\left(\langle W \rangle, \langle \hat{W} \rangle\right) = \mu\left(\begin{pmatrix} B \\ D \end{pmatrix}, \begin{pmatrix} \hat{B} \\ \hat{D} \end{pmatrix}\right) + \mu(\langle \hat{W}^{-1} W \rangle, \langle I \rangle), \tag{3.76}$$

*where $\mu(\langle \hat{W}^{-1} W \rangle, \langle I \rangle)$ is determined by (3.75) with $V := \hat{W}^{-1} W$. Analogously,*

$$\mu(\langle W \rangle, \langle \hat{W} \rangle) = \mu^*\left(\begin{pmatrix} -B^T \\ A^T \end{pmatrix}, \begin{pmatrix} -\hat{B}^T \\ \hat{A}^T \end{pmatrix}\right) + \mu(\langle W \hat{W}^{-1} \rangle, \langle I \rangle), \tag{3.77}$$

*where $\mu(\langle W \hat{W}^{-1} \rangle, \langle I \rangle)$ is given by (3.75) with $V := W \hat{W}^{-1}$.*

In the particular cases, when the matrices $\hat{W}^{-1} W$ and $W \hat{W}^{-1}$ are unit lower (or unit upper) block triangular (see Sect. 1.6.1), the formulas in Corollary 3.26 and Lemma 3.27 reduce as follows.

**Corollary 3.28** *Let $W$ and $\hat{W}$ be symplectic matrices. If $\hat{W}^{-1} W$ or $W \hat{W}^{-1}$ is a unit lower block-triangular matrix of the form*

$$\begin{pmatrix} I & 0 \\ Q & I \end{pmatrix}, \quad Q = Q^T, \tag{3.78}$$

*then*

$$\mu(\langle W \rangle, \langle \hat{W} \rangle) = \mathrm{ind}\,(-Q).$$

*If $\hat{W}^{-1} W$ is unit upper block-triangular matrix of the form*

$$\begin{pmatrix} I & Q \\ 0 & I \end{pmatrix}, \quad Q = Q^T, \tag{3.79}$$

*then*

$$\mu(\langle W \rangle, \langle \hat{W} \rangle) = \mu\left(\begin{pmatrix} B \\ D \end{pmatrix}, \begin{pmatrix} \hat{B} \\ \hat{D} \end{pmatrix}\right).$$

*If the analogical assumption (3.79) holds for $W \hat{W}^{-1}$, then*

$$\mu(\langle W \rangle, \langle \hat{W} \rangle) = \mu^*\left(\begin{pmatrix} -B^T \\ A^T \end{pmatrix}, \begin{pmatrix} -\hat{B}^T \\ \hat{A}^T \end{pmatrix}\right).$$

*Proof* The first terms on the right-hand side of (3.76) and (3.77) in Lemma 3.27 equal to zero for the matrices of the form (3.78) in view of Remark 3.4(ii). For the second terms in (3.76) and (3.77), we obtain according to (3.75) the first statement of the corollary. Concerning the second statement for the upper block-triangular matrices, $\mu(\langle \hat{W}^{-1}W \rangle, \langle I \rangle) = 0$ and $\mu(\langle W\hat{W}^{-1} \rangle, \langle I \rangle) = 0$, by (3.75). Consequently, computing $\mu(\langle W \rangle, \langle \hat{W} \rangle)$ by Lemma 3.27, we obtain the proof of the second statement of this corollary. $\qquad\square$

### 3.3.3 Invertible Block $\mathcal{B}$

Suppose that the blocks $\mathcal{B}$ and $\hat{\mathcal{B}}$ in the matrices $W = \left(\begin{smallmatrix} A & B \\ C & D \end{smallmatrix}\right)$ and $\hat{W} = \left(\begin{smallmatrix} \hat{A} & \hat{B} \\ \hat{C} & \hat{D} \end{smallmatrix}\right)$ are invertible. Since

$$\begin{pmatrix} I & 0 \\ A & B \end{pmatrix} = \begin{pmatrix} I & 0 \\ A & I \end{pmatrix}\begin{pmatrix} I & 0 \\ 0 & B \end{pmatrix}, \tag{3.80}$$

then the upper blocks of the matrices $\langle W \rangle$ and $\langle \hat{W} \rangle$ are invertible as well. Then obviously $\mu_1(\langle W \rangle, \langle \hat{W} \rangle) = 0$, and by Remark 3.4(i), the comparative index is given by $\mu(\langle W \rangle, \langle \hat{W} \rangle) = \mu_2(\langle W \rangle, \langle \hat{W} \rangle)$. Consequently,

$$
\begin{aligned}
\mu(\langle W \rangle, \langle \hat{W} \rangle) &= \mathrm{ind}\left[\begin{pmatrix} 0 & -I \\ \hat{C} & \hat{D} \end{pmatrix}\begin{pmatrix} I & 0 \\ \hat{A} & \hat{B} \end{pmatrix}^{-1} - \begin{pmatrix} 0 & -I \\ C & D \end{pmatrix}\begin{pmatrix} I & 0 \\ A & B \end{pmatrix}^{-1}\right] \\
&= \mathrm{ind}\begin{pmatrix} \hat{B}^{-1}\hat{A} - B^{-1}A & B^{-1} - \hat{B}^{-1} \\ (B^{-1} - \hat{B}^{-1})^T & \hat{D}\hat{B}^{-1} - DB^{-1} \end{pmatrix}.
\end{aligned} \tag{3.81}
$$

Observe that that the above formula can also be obtained from Lemma 3.25. Now we consider simple examples of computations of the comparative index for a pair of symplectic matrices

*Example 3.29* Consider a pair of Sturm-Liouville difference equations (compare with Sect. 1.2)

$$\Delta\left(r_k^{[1]}\Delta x_k\right) - r_k^{[0]}x_{k+1} = 0, \quad \Delta\left(\hat{r}_k^{[1]}\Delta x_k\right) - \hat{r}_k^{[0]}x_{k+1} = 0.$$

According to Example 2.13, these equations can be written as $2 \times 2$ symplectic systems with the matrices $\mathcal{S}_k$ and $\hat{\mathcal{S}}_k$ of the form

$$\mathcal{S}_k = \begin{pmatrix} 1 & 1/r_k^{[1]} \\ r_k^{[0]} & 1 + r_k^{[0]}/r_k^{[1]} \end{pmatrix}, \quad \hat{\mathcal{S}}_k = \begin{pmatrix} 1 & 1/\hat{r}_k^{[1]} \\ \hat{r}_k^{[0]} & 1 + \hat{r}_k^{[0]}/\hat{r}_k^{[1]} \end{pmatrix}.$$

Computing $\mu(\langle \mathcal{S}_k \rangle, \langle \hat{\mathcal{S}}_k \rangle)$ according to (3.81), we obtain

$$\mu(\langle \mathcal{S}_k \rangle, \langle \hat{\mathcal{S}}_k \rangle) = \text{ind}\,(\hat{\mathcal{U}}_k \hat{\mathcal{X}}_k^{-1} - \mathcal{U}_k \mathcal{X}_k^{-1})$$

$$= \text{ind}\begin{pmatrix} 1 & 0 \\ -1 & 1 \end{pmatrix} \begin{pmatrix} \hat{r}_k^{[1]} - r_k^{[1]} & 0 \\ 0 & \hat{r}_k^{[0]} - r_k^{[0]} \end{pmatrix} \begin{pmatrix} 1 & -1 \\ 0 & 1 \end{pmatrix}$$

$$= \text{ind}\,(\hat{r}_k^{[1]} - r_k^{[1]}) + \text{ind}\,(\hat{r}_k^{[0]} - r_k^{[0]}).$$

In particular, $\mu(\langle \mathcal{S}_k \rangle, \langle \hat{\mathcal{S}}_k \rangle) = 0$ if and only if $\hat{r}_k^{[i]} \geq r_k^{[i]}$ for $i \in \{0, 1\}$.

Next we present a generalization of the previous example to the comparative index for a pair of matrices of the symplectic system corresponding to the vector Sturm-Liouville equation (2.25), i.e.,

$$\Delta(\mathcal{R}_k \Delta x_k) - \mathcal{P}_k x_{k+1} = 0.$$

*Example 3.30*  Consider a pair of equations (2.25) with the corresponding symplectic matrices (see Sect. 2.1)

$$\mathcal{S}_k = \begin{pmatrix} I & \mathcal{R}_k^{-1} \\ \mathcal{P}_k & I + \mathcal{P}_k \mathcal{R}_k^{-1} \end{pmatrix}, \quad \hat{\mathcal{S}}_k = \begin{pmatrix} I & \hat{\mathcal{R}}_k^{-1} \\ \hat{\mathcal{P}}_k & I + \hat{\mathcal{P}}_k \hat{\mathcal{R}}_k^{-1} \end{pmatrix}.$$

The invertibility of the blocks $\mathcal{B}_k := \mathcal{R}_k^{-1}$ and $\hat{\mathcal{B}}_k := \hat{\mathcal{R}}_k^{-1}$ enables to compute $\mu(\langle \mathcal{W}_k \rangle, \langle \hat{\mathcal{W}}_k \rangle)$ by (3.81)

$$\mu(\langle \mathcal{W}_k \rangle, \langle \hat{\mathcal{W}}_k \rangle) = \text{ind}\begin{pmatrix} I & 0 \\ -I & I \end{pmatrix} \begin{pmatrix} \hat{\mathcal{R}}_k - \mathcal{R}_k & 0 \\ 0 & \hat{\mathcal{P}}_k - \mathcal{P}_k \end{pmatrix} \begin{pmatrix} I & -I \\ 0 & I \end{pmatrix}$$

$$= \text{ind}\,(\hat{\mathcal{R}}_k - \mathcal{R}_k) + \text{ind}\,(\hat{\mathcal{P}}_k - \mathcal{P}_k).$$

In particular, $\mu(\langle \mathcal{S}_k \rangle, \langle \hat{\mathcal{S}}_k \rangle) = 0$ if and only if $\hat{\mathcal{R}}_k \geq \mathcal{R}_k$ and $\hat{\mathcal{P}}_k \geq \mathcal{P}_k$.

### 3.3.4  Invertible Block $\mathcal{A}$

In this subsection we consider the situation when the matrices $W = \begin{pmatrix} \mathcal{A} & \mathcal{B} \\ \mathcal{C} & \mathcal{D} \end{pmatrix}$ and $\hat{W} = \begin{pmatrix} \hat{\mathcal{A}} & \hat{\mathcal{B}} \\ \hat{\mathcal{C}} & \hat{\mathcal{D}} \end{pmatrix}$ have $\mathcal{A}$ and $\hat{\mathcal{A}}$ invertible. In this case we can recover a Hamiltonian structure from the symplectic matrices $W$ and $\hat{W}$; see Example 2.7 in Sect. 2.1.2.

**Lemma 3.31** *Suppose that symplectic matrices $W$ and $\hat{W}$ satisfy*

$$\det \mathcal{A} \neq 0, \quad \det \hat{\mathcal{A}} \neq 0. \tag{3.82}$$

*Then*

$$\mu(\langle W \rangle, \langle \hat{W} \rangle) = \mu\left(\langle W \rangle_H, \langle \hat{W} \rangle_H\right), \tag{3.83}$$

$$\langle W \rangle_H := \begin{pmatrix} I & 0 \\ I - \mathcal{A}^{-1} & \mathcal{A}^{-1}\mathcal{B} \\ \mathcal{C}\mathcal{A}^{-1} & -(I - \mathcal{A}^{T-1}) \\ 0 & I \end{pmatrix}, \tag{3.84}$$

*where*

$$\langle W \rangle_H^T \, \mathcal{J} \, \langle \hat{W} \rangle_H = H - \hat{H}, \quad H := \begin{pmatrix} -\mathcal{C}\mathcal{A}^{-1} & I - \mathcal{A}^{T-1} \\ I - \mathcal{A}^{-1} & \mathcal{A}^{-1}\mathcal{B} \end{pmatrix} \tag{3.85}$$

*and where $\hat{H}$ is defined similarly through $\hat{W}$.*

*Proof* Assumption (3.82) implies the block $LU$-factorization of $W$ and $\hat{W}$ (compare with Corollary 1.73)

$$W = \begin{pmatrix} I & 0 \\ \mathcal{C}\mathcal{A}^{-1} & I \end{pmatrix} \operatorname{diag}\left\{\mathcal{A}, \quad \mathcal{A}^{T-1}\right\} \begin{pmatrix} I & \mathcal{A}^{-1}\mathcal{B} \\ 0 & I \end{pmatrix} \tag{3.86}$$

and an analogical factorization holds for $\hat{W}$. Using the factorization (3.86), properties (i) and (ii) in Theorem 3.5, we obtain

$$\mu(\langle W \rangle, \langle \hat{W} \rangle) = \mu\left(L\langle W \rangle \, C_1, L\langle \hat{W} \rangle \, C_2\right) = \mu\left(\langle W \rangle_H, \langle \hat{W} \rangle_H\right),$$

where

$$L := \operatorname{diag}\left\{\begin{pmatrix} 0 & I \\ -I & I \end{pmatrix}, \begin{pmatrix} I & I \\ -I & 0 \end{pmatrix}\right\},$$

$$C_1 = \begin{pmatrix} \mathcal{A}^{-1} & -\mathcal{A}^{-1}\mathcal{B} \\ 0 & I \end{pmatrix}, \quad C_2 = \begin{pmatrix} \hat{\mathcal{A}}^{-1} & -\hat{\mathcal{A}}^{-1}\hat{\mathcal{B}} \\ 0 & I \end{pmatrix}.$$

Hence, formula (3.83) is proved. Identity (3.85) follows by direct computations. □

Now we present how to compute the comparative index $\mu\left(\langle W \rangle_H, \langle \hat{W} \rangle_H\right)$.

**Lemma 3.32** *Let W and Ŵ be symplectic matrices satisfying (3.82). Then*

$$\mu\left(\langle W\rangle_H, \langle \hat{W}\rangle_H\right) = \text{ind}\,(H - \hat{H}) + \text{ind}\,(\hat{A}^{-1}\hat{B}) - \text{ind}\,(A^{-1}B),  \quad (3.87)$$

*where $H = H^T$ and $\hat{H} = \hat{H}^T$ are given by (3.85).*

*Proof* Observe that $\langle W\rangle_H = \mathfrak{N}^T\left(\begin{smallmatrix} I_{2n} \\ -H \end{smallmatrix}\right)$ holds, where

$$\mathfrak{N} := \begin{pmatrix} I & 0 & 0 & 0 \\ 0 & 0 & 0 & I \\ 0 & 0 & I & 0 \\ 0 & -I & 0 & 0 \end{pmatrix}.$$

Applying (3.17) to the matrices $Z(0\ I)^T = \left(\begin{smallmatrix} I_{2n} \\ -H \end{smallmatrix}\right)$, $\hat{Z}(0\ I)^T = \left(\begin{smallmatrix} I_{2n} \\ -\hat{H} \end{smallmatrix}\right)$, and $W = \mathfrak{N}$, we obtain

$$\mu\left(\langle W\rangle_H, \langle \hat{W}\rangle_H\right) = \mu\left(\begin{pmatrix} I_{2n} \\ -H \end{pmatrix}, \begin{pmatrix} I_{2n} \\ -\hat{H} \end{pmatrix}\right) + \mu\left(\begin{pmatrix} I_{2n} \\ -\hat{H} \end{pmatrix}, \mathfrak{N}^T\,(0_{2n}\ I_{2n})^T\right)$$

$$- \mu\left(\begin{pmatrix} I_{2n} \\ -H \end{pmatrix}, \mathfrak{N}^T\,(0_{2n}\ I_{2n})^T\right).$$

Computing the comparative indices on the right-hand side of the above formula, we obtain (3.87). □

Another method of computing $\mu(\langle W\rangle_H, \langle \hat{W}\rangle_H)$ is based on Theorem 3.2 and Lemma 3.25.

**Lemma 3.33** *Let W and Ŵ be symplectic matrices satisfying (3.82). Then the comparative index (3.83) can be computed in the following way*

$$\mu_1\left(\langle W\rangle_H, \langle \hat{W}\rangle_H\right) = \text{rank}\,\mathcal{M},$$

$$\mathcal{M} := [I - (A^{-1}B)(A^{-1}B)^\dagger]\,(A^{-1} - \hat{A}^{-1},\ \hat{A}^{-1}\hat{B}),$$

$$\mu_2\left(\langle W\rangle_H, \langle \hat{W}\rangle_H\right) = \text{ind}\left[T\begin{pmatrix} I\ I - \hat{A}^{T\,-1} \\ 0\ \ \hat{A}^{-1}\hat{B} \end{pmatrix}(\hat{G} - G)\begin{pmatrix} I & 0 \\ I - \hat{A}^{-1}\ \hat{A}^{-1}\hat{B} \end{pmatrix}T\right],$$

*where $T := I - \mathcal{M}^\dagger\mathcal{M}$. The matrix G is a symmetric solution of the matrix equation*

$$\begin{pmatrix} I\ I - A^{-T} \\ 0\ \ A^{-1}B \end{pmatrix}G\begin{pmatrix} I & 0 \\ I - A^{-1}\ A^{-1}B \end{pmatrix} = \begin{pmatrix} \mathcal{C}A^{-1} & 0 \\ 0 & A^{-1}B \end{pmatrix},  \quad (3.88)$$

*and it is determined by the formula*

$$G = \begin{pmatrix} I & -I + A^{T-1} \\ 0 & I \end{pmatrix} \left[ \begin{pmatrix} \mathcal{C}A^{-1} & 0 \\ 0 & (A^{-1}B)^{\dagger} \end{pmatrix} + \mathfrak{C} \right] \begin{pmatrix} I & 0 \\ -I + A^{-1} & I \end{pmatrix}, \qquad (3.89)$$

$$\mathfrak{C} = \mathfrak{C}^T, \quad \text{diag}\{I,\ A^{-1}B\}\ \mathfrak{C}\ \text{diag}\{I,\ A^{-1}B\} = 0.$$

*Similar formulas to (3.88) and (3.89) hold also for the matrix* $\hat{G}$.

*Proof* We use the representation

$$\langle W \rangle_H = \text{diag} \left\{ \begin{pmatrix} I & 0 \\ I - A^{-1} & I \end{pmatrix}, \begin{pmatrix} I & -(I - A^{T-1}) \\ 0 & I \end{pmatrix} \right\} \begin{pmatrix} I & 0 \\ 0 & A^{-1}B \\ \mathcal{C}A^{-1} & 0 \\ 0 & I \end{pmatrix}$$

and follow the idea of the proof of Lemma 3.25.                                      □

*Example 3.34* We analyze the comparative index for matrices associated with discrete linear Hamiltonian systems. Consider the system (2.15), i.e.,

$$\Delta \begin{pmatrix} x_k \\ u_k \end{pmatrix} = \mathcal{J}\mathcal{H}_k \begin{pmatrix} x_{k+1} \\ u_k \end{pmatrix}, \quad \mathcal{H}_k = \begin{pmatrix} -C_k & A_k^T \\ A_k & B_k \end{pmatrix}, \quad \mathcal{J} = \begin{pmatrix} 0 & I \\ -I & 0 \end{pmatrix}$$

and

$$\Delta \begin{pmatrix} x_k \\ u_k \end{pmatrix} = \mathcal{J}\hat{\mathcal{H}}_k \begin{pmatrix} x_{k+1} \\ u_k \end{pmatrix}, \quad \hat{\mathcal{H}}_k = \begin{pmatrix} -\hat{C}_k & \hat{A}_k^T \\ \hat{A}_k & \hat{B}_k \end{pmatrix}. \qquad (3.90)$$

Using (3.87), it is easy to see that

$$\mu\left( \langle S_k^{[H]} \rangle_H, \langle \hat{S}_k^{[H]} \rangle_H \right) = \text{ind}\,(\mathcal{H}_k - \hat{\mathcal{H}}_k) + \text{ind}\,\hat{B}_k - \text{ind}\,B_k, \qquad (3.91)$$

where $S_k^{[H]}$ is the symplectic matrix given in (2.18), with $\hat{S}_k^{[H]}$ being defined analogously. From (3.47) it follows that

$$\text{ind}\,B_k + \mu\left( \begin{pmatrix} B_k \\ I \end{pmatrix}, \begin{pmatrix} \hat{B}_k \\ I \end{pmatrix} \right) = \text{ind}\,\hat{B}_k + \mu^*\left( \mathcal{J}\begin{pmatrix} \hat{B}_k \\ I \end{pmatrix}, \mathcal{J}\begin{pmatrix} B_k \\ I \end{pmatrix} \right),$$

or in other terms,

$$\text{ind}\,\hat{B}_k - \text{ind}\,B_k = \mu\left( \begin{pmatrix} B_k \\ I \end{pmatrix}, \begin{pmatrix} \hat{B}_k \\ I \end{pmatrix} \right) - \text{ind}\,(B_k - \hat{B}_k). \qquad (3.92)$$

Substituting the last formula into (3.91), we obtain

$$
\mu\left(\langle S_k^{[H]}\rangle_H, \langle \hat{S}_k^{[H]}\rangle_H\right)
$$

$$
= \mu\left(\binom{B_k}{I}, \binom{\hat{B}_k}{I}\right) + \left\{\operatorname{ind}(\mathcal{H}_k - \hat{\mathcal{H}}_k) - \operatorname{ind}(B_k - \hat{B}_k)\right\}. \tag{3.93}
$$

Observe that by Lemma 3.15, the expression in braces in the last computation is nonnegative and represents the index of a symmetric operator with the zero block on the main diagonal. Simplifying formula (3.77) for the "Hamiltonian" case, it is possible to show that the expression in braces is equal to the comparative index $\mu\left(\langle S_k^{[H]}(\hat{S}_k^{[H]})^{-1}\rangle, \langle I\rangle\right)$.

Analogously, the formulas in Lemma 3.33 for the pair of systems (2.15) and (3.90) have the form

$$
\mu_1\left(\langle S_k^{[H]}\rangle_H, \langle \hat{S}_k^{[H]}\rangle_H\right) = \operatorname{rank}\mathcal{M}_k, \tag{3.94}
$$

$$
\mu_2\left(\langle S_k^{[H]}\rangle_H, \langle \hat{S}_k^{[H]}\rangle_H\right) = \operatorname{ind}\left[\mathcal{T}_k\begin{pmatrix} I & \hat{A}_k^T \\ 0 & \hat{B}_k \end{pmatrix}(G_k - \hat{G}_k)\begin{pmatrix} I & 0 \\ \hat{A}_k & \hat{B}_k \end{pmatrix}\mathcal{T}_k\right], \tag{3.95}
$$

where

$$
\mathcal{M}_k := (I - B_k B_k^\dagger)\,(\hat{A}_k - A_k, \ \hat{B}_k), \quad \mathcal{T}_k := I - \mathcal{M}_k^\dagger \mathcal{M}_k,
$$

and where $G_k$ is a symmetric solution of the equation

$$
\begin{pmatrix} I & A_k^T \\ 0 & B_k \end{pmatrix} G_k \begin{pmatrix} I & 0 \\ A_k & B_k \end{pmatrix} = \begin{pmatrix} C_k & 0 \\ 0 & B_k \end{pmatrix}, \tag{3.96}
$$

i.e., $G_k$ is given by the formula

$$
G_k = \begin{pmatrix} I & -A_k^T \\ 0 & I \end{pmatrix}\left[\begin{pmatrix} C_k & 0 \\ 0 & B_k^\dagger \end{pmatrix} + \mathfrak{C}_k\right]\begin{pmatrix} I & 0 \\ -A_k & I \end{pmatrix}, \tag{3.97}
$$

$$
\mathfrak{C}_k = \mathfrak{C}_k^T, \quad \operatorname{diag}\{I, B_k\}\,\mathfrak{C}_k\,\operatorname{diag}\{I, B_k\} = 0.
$$

Similar formulas to (3.96) and (3.97) hold also for the matrix $\hat{G}_k$.

*Remark 3.35* Note that formula (3.92) generalizes Remark 1.60(vi) to the case when $A = A^T$ and $B = B^T$ are indefinite by sign. Indeed, under the assumptions $A \geq 0$ and $B \geq 0$, we have from (3.92) that

$$
\mu\left(\binom{B}{I}, \binom{A}{I}\right) = \operatorname{ind}(B - A). \tag{3.98}
$$

Then ind $(B - A) = 0$ is equivalent to $\mu\left(\binom{B}{I}, \binom{A}{I}\right) = 0$, and the last condition coincides with

$$\operatorname{Im} A \subseteq \operatorname{Im} B, \quad A - AB^\dagger A \geq 0,$$

by Theorem 3.2(iv). By a similar way, the condition $B^\dagger \leq A^\dagger$ is equivalent to $\mu\left(\binom{A^\dagger}{I}, \binom{B^\dagger}{I}\right) = 0$ or to

$$\operatorname{Im} B^\dagger \subseteq \operatorname{Im} A^\dagger, \quad B^\dagger - B^\dagger A B^\dagger \geq 0.$$

Using these equivalences it is possible to prove easily the last assertion in Remark 1.60(vi).

*Example 3.36* In this example we turn our attention to the $2n$-th order Sturm-Liouville difference equation (2.27), i.e.,

$$\sum_{\nu=0}^{n} (-1)^\nu \Delta^\nu \left(r_k^{[\nu]} \Delta^\nu y_{k+n-\nu}\right) = 0,$$

where $r_k^{[n]} \neq 0$ for all $k$. This equation can be written as the Hamiltonian system (2.15) with the coefficient matrices $A_k$, $B_k$, and $C_k$ given in (2.29) and (2.30). Together with (2.27) we consider the equation of the same form with the coefficients $\hat{r}_k^{[j]}$ for $j = 0, \ldots, n$, i.e.,

$$\sum_{\nu=0}^{n} (-1)^\nu \Delta^\nu \left(\hat{r}_k^{[\nu]} \Delta^\nu y_{k+n-\nu}\right) = 0 \tag{3.99}$$

with $\hat{r}_k^{[n]} \neq 0$ for all $k$. In this particular case, $A_k = \hat{A}_k$ and $\mathcal{M}_k = 0$, so that $\mu_1\left(\langle \mathcal{S}_k^{[H]}\rangle_H, \langle \hat{\mathcal{S}}_k^{[H]}\rangle_H\right) = 0$. Simplifying equation (3.95), we obtain

$$\mu\left(\langle \mathcal{S}_k^{[H]}\rangle_H, \langle \hat{\mathcal{S}}_k^{[H]}\rangle_H\right) = \operatorname{ind} \operatorname{diag}\left\{\hat{r}_k^{[0]} - r_k^{[0]}, \ldots, \hat{r}_k^{[n]} - r_k^{[n]}\right\}$$

$$= \sum_{\nu=0}^{n} \operatorname{ind}\left(\hat{r}_k^{[\nu]} - r_k^{[\nu]}\right). \tag{3.100}$$

The same result follows directly from (3.93) and (2.29), (2.30). In particular, $\mu\left(\langle \mathcal{S}_k^{[H]}\rangle_H, \langle \hat{\mathcal{S}}_k^{[H]}\rangle_H\right) = 0$ if and only if $\hat{r}_k^{[j]} \geq r_k^{[j]}$ for all $j = 0, \ldots, n$.

## 3.3.5   Additional Properties of Comparative Index

In this section we derive additional properties of the comparative index, which we will use in subsequent chapters when dealing with boundary value problems for symplectic system with general (nonseparated) boundary conditions.

We introduce the following notation. Let $W$ and $V$ be matrices separated into $n \times n$ blocks

$$
W = \begin{pmatrix} A & B \\ C & D \end{pmatrix}, \quad V = \begin{pmatrix} \hat{A} & \hat{B} \\ \hat{C} & \hat{D} \end{pmatrix}.
$$

Introduce the notation

$$
\{W, V\} := \begin{pmatrix} A & 0 & -B & 0 \\ 0 & \hat{A} & 0 & \hat{B} \\ -C & 0 & D & 0 \\ 0 & \hat{C} & 0 & \hat{D} \end{pmatrix}, \quad \{W\} := \{I, W\} = \begin{pmatrix} I & 0 & 0 & 0 \\ 0 & A & 0 & B \\ 0 & 0 & I & 0 \\ 0 & C & 0 & D \end{pmatrix}, \tag{3.101}
$$

and

$$
S_0 := \begin{pmatrix} 0 & 0 & I & 0 \\ 0 & I & I & 0 \\ -I & 0 & 0 & -I \\ 0 & 0 & 0 & I \end{pmatrix}. \tag{3.102}
$$

Then one can directly verify that $S_0 \in Sp(4n)$ and under the additional assumption $W, V \in Sp(2n)$, the $4n \times 4n$ matrices $\{W, V\}$ and $\{W\}$ are symplectic as well, i.e., $\{W, V\}, \{W\} \in Sp(4n)$.

We have the multiplicative property

$$
\{WV, \hat{W}\hat{V}\} = \{W, \hat{W}\}\{V, \hat{V}\}, \tag{3.103}
$$

in particular, $\{W, \hat{W}\}^{-1} = \{W^{-1}, \hat{W}^{-1}\}$. We also derive the following relations

$$
\{W, \hat{W}\} S_0 = \begin{pmatrix} B & 0 & A & B \\ 0 & \hat{A} & \hat{A} & \hat{B} \\ -D & 0 & -C & -D \\ 0 & \hat{C} & \hat{C} & \hat{D} \end{pmatrix}, \quad \{W\} S_0 (0_{2n}, I_{2n})^T = \langle W \rangle, \tag{3.104}
$$

which can be easily verified by direct computations. Moreover, by (3.104) for any symplectic $2n \times 2n$ matrices $P$ and $R$, we have

$$
\langle R^{-1} W P \rangle = \{P^{-1}, R^{-1}\} \langle W \rangle P, \tag{3.105}
$$

where the matrix $\langle W \rangle$ is defined in (3.52). Using the properties of the comparative index, we can prove the following result.

**Proposition 3.37** Let $W, V, R, P \in Sp(2n)$ and $S_0$ be given by (3.102). Then, using the notation in (3.101) and (3.52), we have

$$\mu \left( \langle W \rangle, \{W\} \, (0_{2n} \; I_{2n})^T \right) = 0, \tag{3.106}$$

$$\mu \left( \langle W \rangle, \{V\} \, (0_{2n} \; I_{2n})^T \right) = \mu \left( \{W\} \, (0_{2n} \; I_{2n})^T, \{V\} \, (0_{2n} \; I_{2n})^T \right) \tag{3.107}$$

$$= \mu \left( W(0 \; I)^T, V(0 \; I)^T \right), \tag{3.108}$$

$$\mu \left( \langle \{W\} \rangle, \langle \{V\} \rangle \right) = \mu \left( \langle W \rangle, \langle V \rangle \right). \tag{3.109}$$

*Proof* Using (3.104), properties (i), (iii) of Theorem 3.5 and (3.13), we obtain

$$\mu \left( \langle W \rangle, \{W\} \, (0_{2n} \; I_{2n})^T \right) = \mu \left( \{W\} \, S_0 \, (0_{2n} \; I_{2n})^T, \{W\} \, (0_{2n} \; I_{2n})^T \right)$$

$$= \mu^* \left( (\{W\} S_0)^{-1} (0_{2n} \; I_{2n})^T, S_0^{-1} (0_{2n} \; I_{2n})^T \right) = \mu \left( \begin{pmatrix} I & A^T \\ 0 & B^T \\ 0 & 0 \\ 0 & A^T \end{pmatrix}, \begin{pmatrix} I & I \\ 0 & 0 \\ 0 & 0 \\ 0 & I \end{pmatrix} \right)$$

$$= \mu \left( \begin{pmatrix} I & 0 \\ 0 & B^T \\ 0 & 0 \\ 0 & A^T \end{pmatrix} \begin{pmatrix} I & A^T \\ 0 & I \end{pmatrix}, \begin{pmatrix} I & 0 \\ 0 & 0 \\ 0 & 0 \\ 0 & I \end{pmatrix} \begin{pmatrix} I & I \\ 0 & I \end{pmatrix} \right) = \mu \left( \begin{pmatrix} I & 0 \\ 0 & B^T \\ 0 & 0 \\ 0 & A^T \end{pmatrix}, \begin{pmatrix} I & 0 \\ 0 & 0 \\ 0 & 0 \\ 0 & I \end{pmatrix} \right).$$

The last comparative index can be easily computed as being equal to zero. Thus, we have proved (3.106). Now we proceed with the proof of formula (3.107). By Theorem 3.6,

$$\mu \left( \langle W \rangle, \{V\} \, (0_{2n} \; I_{2n})^T \right) - \mu \left( \{W\} \, (0_{2n} \; I_{2n})^T, \{V\} \, (0_{2n} \; I_{2n})^T \right)$$

$$= \mu \left( \langle W \rangle, \{W\} \, (0_{2n} \; I_{2n})^T \right) - \mu \left( \langle V^{-1} W \rangle, \{V^{-1} W\} \, (0_{2n} \; I_{2n})^T \right).$$

From (3.106) it follows that the right-hand side of the last formula equals to zero, and then (3.107) follows. Since the $2n \times 2n$ blocks of the matrices $\{W\} \, (0_{2n} \; I_{2n})^T$ and $\{V\} \, (0_{2n} \; I_{2n})^T$ have a block diagonal structure, then by the definition of the comparative index

$$\mu \left( \{W\} \, (0_{2n} \; I_{2n})^T, \{V\} \, (0_{2n} \; I_{2n})^T \right) = \mu \left( W(0 \; I)^T, V(0 \; I)^T \right),$$

which proves formula (3.108). Finally, formula (3.109) can be verified by direct computations by using the block diagonal structure of the $2n \times 2n$ blocks of the matrices $\{W\}$ and $\{V\}$ in (3.101).                                                     □

*Remark 3.38* Note that we have the identity

$$\text{diag}\{P_1, P_1\}\{W, I\} \text{diag}\{P_1, P_1\} = \{I, P_2 W P_2\} := \{P_2 W P_2\}, \qquad (3.110)$$

where $P_1$ and $P_2$ are given by (1.151), and the block diagonal matrix $\text{diag}\{P_1, P_1\}$ is symplectic. Using (3.110) and Theorem 3.5(i), (ii), one can prove that the comparative indices associated with $\{W, I\}$ have properties similar to (3.106), (3.107), (3.108). In particular, we obtain

$$\mu\left(\{W, I\}\langle I\rangle, \{W, I\}\,(0_{2n}\ I_{2n})^T\right) = 0, \qquad (3.111)$$

$$\mu\left(\{W, I\}\langle I\rangle, \{V, I\}\,(0_{2n}\ I_{2n})^T\right) = \mu\left(\{W, I\}\,(0_{2n}\ I_{2n})^T, \{V, I\}\,(0_{2n}\ I_{2n})^T\right)$$

$$= \mu\left(P_2 W P_2 (0\ I)^T, P_2 V P_2 (0\ I)^T\right)$$

$$= \mu^*\left(W(0\ I)^T, V(0\ I)^T\right). \qquad (3.112)$$

The statements in Proposition 3.37 and Remark 3.38 are used in the proof of the following result, which will be further applied in the subsequent chapters.

**Lemma 3.39** *Let $W, V, R, P$ be symplectic matrices and define $\tilde{W} := R^{-1} W P$ and $\tilde{V} := R^{-1} V P.$. Then*

$$\mu(\langle \tilde{W}\rangle, \langle \tilde{V}\rangle) - \mu(\langle W\rangle, \langle V\rangle)$$

$$= \mu\left(\langle \tilde{W}\rangle, \{P^{-1}, R^{-1}\}\,(0_{2n}\ I_{2n})^T\right) - \mu\left(\langle \tilde{V}\rangle, \{P^{-1}, R^{-1}\}\,(0_{2n}\ I_{2n})^T\right)$$

$$\hspace{10cm} (3.113)$$

$$= \mu\left(\langle V\rangle, \{P, R\}\,(0_{2n}\ I_{2n})^T\right) - \mu\left(\langle W\rangle, \{P, R\}\,(0_{2n}\ I_{2n})^T\right). \qquad (3.114)$$

*Moreover, each term in (3.114) can be computed by using the formulas*

$$\mu\left(\langle W\rangle, \{P, R\}\,(0_{2n}\ I_{2n})^T\right)$$

$$= \mu\left(W(0\ I)^T, WP(0\ I)^T\right) + \mu\left(WP(0\ I)^T, R(0\ I)^T\right) \qquad (3.115)$$

$$= \mu^*\left(W^{-1} R(0\ I)^T, P(0\ I)^T\right) + \mu\left(W(0\ I)^T, R(0\ I)^T\right). \qquad (3.116)$$

*In particular, we have*

$$\mu(\langle \tilde{W} \rangle, \langle \tilde{V} \rangle) - \mu(\langle W \rangle, \langle V \rangle)$$
$$= \mu \left( \tilde{W} (0 \ I)^T, \ R^{-1} (0 \ I)^T \right) - \mu \left( \tilde{V} (0 \ I)^T, \ R^{-1} (0 \ I)^T \right)$$
$$+ \mu^* \left( V^{-1} (0 \ I)^T, \ P (0 \ I)^T \right) - \mu^* \left( W^{-1} (0 \ I)^T, \ P (0 \ I)^T \right),$$
$$\tag{3.117}$$

*or*

$$\mu(\langle \tilde{W} \rangle, \langle \tilde{V} \rangle) - \mu(\langle W \rangle, \langle V \rangle)$$
$$= \mu \left( R^{-1} (0 \ I)^T, \ \tilde{V} (0 \ I)^T \right) - \mu \left( R^{-1} (0 \ I)^T, \ \tilde{W} (0 \ I)^T \right)$$
$$+ \mu^* \left( P (0 \ I)^T, \ W^{-1} (0 \ I)^T \right) - \mu^* \left( P (0 \ I)^T, \ V^{-1} (0 \ I)^T \right).$$
$$\tag{3.118}$$

The proof of this lemma is postponed to Sect. 3.3.6.

Based on the results in Lemma 3.39, we derive the next corollary, which will be used in the subsequent chapters.

**Corollary 3.40** *We have the following statements.*

*(i) Let $R, P \in Sp(2n)$ be lower block-triangular. Then*

$$\mu \left( \langle R^{-1} W P \rangle, \langle R^{-1} V P \rangle \right) = \mu \left( \langle W \rangle, \langle V \rangle \right).$$

*(ii) Let $R = I$ and $P = \mathcal{J}$. Then*

$$\mu \left( \langle W \rangle, \langle V \rangle \right) = \mu^* \left( \langle V \mathcal{J} \rangle, \langle W \mathcal{J} \rangle \right) + \mu^* \left( (I \ 0)^T, \ V^{-1} (0 \ I)^T \right)$$
$$- \mu^* \left( (I \ 0)^T, \ W^{-1} (0 \ I)^T \right).$$

*(iii) Let $\tilde{V} := R^{-1} V P = I$. Then*

$$\mu \left( \langle \tilde{W} \rangle, \langle I \rangle \right) - \mu \left( \langle W \rangle, \langle R P^{-1} \rangle \right)$$
$$= \mu \left( \tilde{W} (0 \ I)^T, \ R^{-1} (0 \ I)^T \right) - \mu^* \left( W^{-1} (0 \ I)^T, \ P (0 \ I)^T \right)$$
$$- \mu \left( P^{-1} (0 \ I)^T, \ R^{-1} (0 \ I)^T \right)$$
$$\tag{3.119}$$

$$= \mu^* \left( P(0\ I)^T,\ W^{-1}(0\ I)^T \right) - \mu \left( R^{-1}(0\ I)^T,\ \tilde{W}(0\ I)^T \right)$$

$$- \mu \left( P^{-1}(0\ I)^T,\ R^{-1}(0\ I)^T \right). \tag{3.120}$$

(iv) *If* $W = \left( \begin{smallmatrix} I & 0 \\ -Q & I \end{smallmatrix} \right)$, *where* $Q = Q^T$ *and* $Q \geq 0$, *then for arbitrary* $R \in Sp(2n)$, *we have*

$$\mu \left( \langle R^{-1} W R \rangle,\ \langle I \rangle \right) = 0. \tag{3.121}$$

(v) *For any* $R \in Sp(4n)$ *we have*

$$\mu \left( \langle R^{-1}\ \{W\}\ S_0 \rangle,\ \langle R^{-1}\ \{V\}\ S_0 \rangle \right) = \mu \left( R^{-1}\ \langle W \rangle,\ R^{-1}\ \langle V \rangle \right). \tag{3.122}$$

*Proof* Particular cases (i) and (iii), formula (3.120) follows immediately from (3.117) and (3.118), respectively, where by Theorem 3.5(iii), we have

$$\mu^*(P(0\ I)^T,\ PR^{-1}(0\ I)^T) = \mu(P^{-1}(0\ I)^T,\ R^{-1}(0\ I)^T). \tag{3.123}$$

For the proof of formula (3.119) in (iii), we apply (3.117), where

$$\mu \left( (0\ I)^T,\ R^{-1}(0\ I)^T \right) - \mu^* \left( PR^{-1}(0\ I)^T,\ P(0\ I)^T \right) = \mu^* \left( P(0\ I)^T,\ PR^{-1}(0\ I)^T \right)$$

according to Theorem 3.5(v). Then, using (3.123), we derive (3.119).

We turn our attention to the proof of (ii). From (3.118) we obtain

$$\mu \left( \langle W\mathcal{J} \rangle,\ \langle V\mathcal{J} \rangle \right) - \mu \left( \langle W \rangle,\ \langle V \rangle \right)$$

$$= \mu \left( (0\ I)^T,\ V\mathcal{J}(0\ I)^T \right) - \mu \left( (0\ I)^T,\ W\mathcal{J}(0\ I)^T \right)$$

$$+ \mu^* \left( (I\ 0)^T,\ W^{-1}(0\ I)^T \right) - \mu^* \left( (I\ 0)^T,\ V^{-1}(0\ I)^T \right).$$

Putting the first two summands on the right-hand side of the last computation to the left and then using Lemma 3.21(iv), we obtain the proof of (ii).

Property (iv) follows from (3.120) with $R = P$, since on the right-hand side of (3.120) we have in our case $\mu^* \left( P(0\ I)^T,\ W^{-1}(0\ I)^T \right) = 0$, as $W^{-1}$ is lower block-triangular, and

$$\mu \left( R^{-1}(0\ I)^T,\ \tilde{W}(0\ I)^T \right) = \mu^*(R(0\ I)^T,\ WR(0\ I)^T) = \text{ind}\ R_{12}^T Q R_{12} = 0,$$

where $R_{12}$ is the right upper block of $R$, and $\mu\left(R^{-1}(0\ I)^T, R^{-1}(0\ I)^T\right) = 0$. Further, $\mu(\langle W\rangle, \langle RR^{-1}\rangle) = \mu(\langle W\rangle, \langle I\rangle) = \text{ind}\,Q = 0$ by Corollary 3.28. The property (iv) is proved.

Finally, we prove (v). By (3.114) with $P := S_0$, we have

$$\mu\left(\langle R^{-1}\{W\}S_0\rangle, \langle R^{-1}\{V\}S_0\rangle\right)$$
$$= \mu\left(\langle\{W\}\rangle, \langle\{V\}\rangle\right) + \mu\left(\langle\{V\}\rangle, \{S_0, R\}\,(0_{2n}\ I_{2n})^T\right)$$
$$- \mu\left(\langle\{W\}\rangle, \{S_0, R\}\,(0_{2n}\ I_{2n})^T\right)$$
$$= \mu\left(\langle W\rangle, \langle V\rangle\right) + \mu\left(\langle\{V\}\rangle, \{S_0, R\}\,(0_{2n}\ I_{2n})^T\right)$$
$$- \mu\left(\langle\{W\}\rangle, \{S_0, R\}\,(0_{2n}\ I_{2n})^T\right),$$

where we have used (3.109). In addition, applying formula (3.115), we obtain

$$\mu\left(\langle\{V\}\rangle, \{S_0, R\}\,(0_{2n}\ I_{2n})^T\right) = \mu\left(\{V\}\,(0_{2n}\ I_{2n})^T, \{V\}S_0\,(0_{2n}\ I_{2n})^T\right)$$
$$+ \mu\left(\{V\}\,S_0\,(0_{2n}\ I_{2n})^T, R\,(0_{2n}\ I_{2n})^T\right).$$

Using the notation $\langle V\rangle = \{V\}\,S_0\,(0_{2n}\ I_{2n})^T$, we obtain

$$\mu\left(\{V\}\,(0_{2n}\ I_{2n})^T, \langle V\rangle\right) = n - \mu\left(\langle V\rangle, \{V\}\,(0_{2n}\ I_{2n})^T\right) = n,$$

where we have used property (v) of Theorem 3.5 and (3.106). Applying similar computations for $\mu\left(\langle\{W\}\rangle, \{S_0, R\}\,(0_{2n}\ I_{2n})^T\right)$, we obtain

$$\mu\left(\langle\{V\}\rangle, \{S_0, R\}\,(0_{2n}\ I_{2n})^T\right) - \mu\left(\langle\{W\}\rangle, \{S_0, R\}\,(0_{2n}\ I_{2n})^T\right)$$
$$= \mu\left(\langle V\rangle, R\,(0_{2n}\ I_{2n})^T\right) - \mu\left(\langle W\rangle, R\,(0_{2n}\ I_{2n})^T\right),$$

and, hence,

$$\mu\left(\langle R^{-1}\{W\}\,S_0\rangle, \langle R^{-1}\{V\}\,S_0\rangle\right)$$
$$= \mu\left(\langle W\rangle, \langle V\rangle\right) + \mu\left(\langle V\rangle, R\,(0_{2n}\ I_{2n})^T\right) - \mu\left(\langle W\rangle, R\,(0_{2n}\ I_{2n})^T\right)$$
$$= \mu\left(R^{-1}\langle W\rangle, R^{-1}\langle V\rangle\right),$$

where we have used Theorem 3.6. This proves (3.122). $\qquad\square$

*Remark 3.41* Note that for the comparative indices associated with $\{W, \hat{W}\}$ and $\{W, \hat{W}\}S_0$ and their special cases with $W := I$ or $\hat{W} := I$, the duality principle holds (see Sect. 3.1.4). Introduce the matrix $\mathcal{P}_2 := \mathrm{diag}\{I_{2n}, -I_{2n}\}$, which is an analog of $P_2 = \mathrm{diag}\{I, -I\}$ defined by (1.151). According to (3.13), the dual comparative index can be defined using $\mathcal{P}_2$. Similarly, we have

$$\mu^*(\langle W\rangle, \langle \hat{W}\rangle) = \mu(\mathcal{P}_2\langle W\rangle, \mathcal{P}_2\langle \hat{W}\rangle)$$

$$= \mu(\mathcal{P}_2\langle W\rangle P_2, \mathcal{P}_2\langle \hat{W}\rangle P_2)$$

$$= \mu(\langle P_2 W P_2\rangle, \langle P_2 \hat{W} P_2\rangle),$$

where we have applied Theorem 3.5(i). By a similar way, using the relations

$$\mathcal{P}_2\{W, \hat{W}\}\mathcal{P}_2 = \{P_2 W P_2, P_2 \hat{W} P_2\},$$

$$\mathcal{P}_2 S_0 \,\mathrm{diag}\{-P_2, P_2\} = S_0,$$

Theorem 3.5(i), and representation (3.13) for the dual comparative index associated with $\mathcal{P}_2 \in \mathbb{R}^{4n}$, one can prove that all results of Sect. 3.3 hold for the dual indices as well.

### 3.3.6  Comparative Index for Symplectic Matrices: Proofs

In this section we present the proofs of properties of the comparative index, which we displayed earlier in Sect. 3.3. In particular, we present the proofs of properties (v) and (vi) of Lemma 3.21 and the proof of Lemma 3.39.

*Proof of Properties (v) and (vi) of Lemma 3.21* The proof of the property (v) in Lemma 3.21 is based on Proposition 3.37; see formulas (3.106) and (3.108). Applying identity (3.60) from Lemma 3.23 to the case $W := \{W\}$, $\hat{W} := \{\hat{W}\}$, $Z = \hat{Z} = S_0$ with $S_0$ given by (3.102), we have $Y = \hat{Y} := \langle I\rangle$, and using (3.106) for the matrices $W$, $\hat{W}$, $\hat{W}^{-1}W$, we obtain

$$\mathcal{L}(\langle I\rangle, \langle I\rangle, \{W\}, \{\hat{W}\}) = \mu\left(\{W\}\langle I\rangle, \{\hat{W}\}\langle I\rangle\right)$$

$$= \mu\left(\{W\}(0\ I)^T, \{\hat{W}\}(0\ I)^T\right) + \mu\left(\{\hat{W}^{-1}W\}\langle I\rangle, \langle I\rangle\right)$$

$$= \mu\left(W(0\ I)^T, \hat{W}(0\ I)^T\right) + \mu\left(\langle \hat{W}^{-1}W\rangle, \langle I\rangle\right),$$

$$(3.124)$$

where we also incorporate the last equality in (3.108). Hence, equality (3.54) is proved. Using property (iii) of Lemma 3.21 and repeating the proof of (3.124) for

$\mu^*(\langle W^{-1}\rangle, \langle \hat{W}^{-1}\rangle)$, we obtain

$$\mu\left(\langle W\rangle, \langle \hat{W}\rangle\right) = \mu^*\left(\langle W^{-1}\rangle, \langle \hat{W}^{-1}\rangle\right)$$

$$= \mu^*\left(W^{-1}\,(0\ I)^T,\ \hat{W}^{-1}\,(0\ I)^T\right) + \mu^*\left(\langle \hat{W}W^{-1}\rangle, \langle I\rangle\right),$$

i.e., equality (3.55) holds as well. Using part (iii) of Lemma 3.21 again, we get $\mu^*(\langle \hat{W}W^{-1}\rangle, \langle I\rangle) = \mu(\langle W\hat{W}^{-1}\rangle, \langle I\rangle)$. Finally, part (iii) of Theorem 3.5 yields the identity $\mu^*(W^{-1}\,(0\ I)^T,\ \hat{W}^{-1}\,(0\ I)^T) = \mu(W\,(0\ I)^T,\ W\hat{W}^{-1}\,(0\ I)^T)$, which finishes the proof of (3.55). The proof of property (vi) follows directly from (3.54) and (3.55). □

*Proof of Lemma 3.39* Remark that by (3.105) and part (i) of Theorem 3.5, the invertible matrices $P$ in the comparative index

$$\mu(\langle \tilde{W}\rangle, \langle \tilde{V}\rangle) = \mu\left(\{P^{-1}, R^{-1}\}\langle W\rangle\, P,\ \{P^{-1}, R^{-1}\}\langle V\rangle P\right)$$

can be neglected. Now, using Theorem 3.6, we obtain the proof of (3.113). Formula (3.114) follows from (3.17) or from property (ix) of Theorem 3.5.

Now we prove (3.116). By (i) of Theorem 3.5, we have

$$\mu\left(\langle W\rangle,\ \{P, R\}\,(0_{2n}\ I_{2n})^T\right) = \mu\left(\langle W\rangle P,\ \{P, R\}\,(0_{2n}\ I_{2n})^T\right),$$

and further, using (3.105) for the case $R := I$ and (3.103), we derive

$$\mu\left(\langle W\rangle P,\ \{P, R\}\,(0_{2n}\ I_{2n})^T\right) = \mu\left(\{P, I\}\langle WP\rangle,\ \{P, R\}\,(0_{2n}\ I_{2n})^T\right)$$

$$= \mu\left(\{P, I\}\langle WP\rangle,\ \{P, I\}\{I, R\}\,(0_{2n}\ I_{2n})^T\right)$$

$$= \mu\left(\langle WP\rangle,\ \{I, R\}\,(0_{2n}\ I_{2n})^T\right) + \mu\left(\{I, R\}\,(0_{2n}\ I_{2n})^T,\ \{P^{-1}, I\}\,(0_{2n}\ I_{2n})^T\right)$$

$$- \mu\left(\langle WP\rangle,\ \{P^{-1}, I\}\,(0_{2n}\ I_{2n})^T\right),$$

where the last sum is obtained via (3.17). By (3.108),

$$\mu\left(\langle WP\rangle, \{I, R\}\,(0\ I)^T\right) = \mu\left(\langle WP\rangle, \{R\}\,(0\ I)^T\right) = \mu\left(WP\,(0\ I)^T,\ R\,(0\ I)^T\right),$$

which yields the second summand on the right-hand side of (3.115). Further, by (3.101) and Definition 3.1 of the comparative index, we have

$$\mu\left(\{I, R\}\,(0_{2n}\ I_{2n})^T,\ \{P^{-1}, I\}\,(0_{2n}\ I_{2n})^T\right) = \mu\left((0\ I)^T,\ P^{-1}\,(0\ I)^T\right).$$

To determine the last comparative index, we use again Theorem 3.5(i) and (3.105). We obtain

$$\mu\left(\langle WP\rangle, \{P^{-1}, I\}\, (0_{2n}\ I_{2n})^T\right) = \mu\left(\langle WP\rangle(WP)^{-1}, \ \{P^{-1}, I\}\, (0_{2n}\ I_{2n})^T\right)$$

$$= \mu\left(\{P^{-1}W^{-1}, I\}\, \langle I\rangle, \ \{P^{-1}, I\}\, (0_{2n}\ I_{2n})^T\right).$$

Using Remark 3.38 we have by (3.112) that

$$\mu\left(\{P^{-1}W^{-1}, I\}\, \langle I\rangle, \{P^{-1}, I\}(0_{2n}\ I_{2n})^T\right) = \mu^*\left(P^{-1}W^{-1}(0\ I)^T, \ P^{-1}(0\ I)^T\right).$$

Finally, we derive

$$\mu\left((0\ I)^T, \ P^{-1}(0\ I)^T\right) - \mu^*\left(P^{-1}W^{-1}(0\ I)^T, \ P^{-1}(0\ I)^T\right)$$

$$= \mu^*\left(W^{-1}(0\ I)^T, \ P(0\ I)^T\right) = \mu\left(W\,(0\ I)^T, \ WP(0\ I)^T\right),$$

where we have used properties (ix) and (iii) of Theorem 3.5. This proves formula (3.115). Applying Theorem 3.6 to just proved formula, we have

$$\mu\left(W\,(0\ I)^T, \ WP\,(0\ I)^T\right) + \mu\left(WP(0\ I)^T, \ R\,(0\ I)^T\right)$$

$$= \mu\left(R^{-1}W\,(0\ I)^T, \ R^{-1}WP\,(0\ I)^T\right) + \mu\left(W\,(0\ I)^T, \ R\,(0\ I)^T\right)$$

$$= \mu^*\left(W^{-1}R\,(0\ I)^T, \ P\,(0\ I)^T\right) + \mu\left(W\,(0\ I)^T, \ R\,(0\ I)^T\right).$$

This proves formula (3.116).

The proof of (3.117) we get from (3.113) in the following way:

$$\mu\left(\langle\tilde{W}\rangle, \langle\tilde{V}\rangle\right) - \mu\left(\langle W\rangle, \langle V\rangle\right)$$

$$= \mu\left(\langle\tilde{W}\rangle, \{P^{-1}, R^{-1}\}\, (0_{2n}\ I_{2n})^T\right) - \mu\left(\langle\tilde{V}\rangle, \{P^{-1}, R^{-1}\}\, (0_{2n}\ I_{2n})^T\right),$$

where, on the right-hand side, we twice apply formula (3.116) obtaining

$$\mu\left(\langle\tilde{W}\rangle, \{P^{-1}, R^{-1}\}\, (0_{2n}\ I_{2n})^T\right) - \mu\left(\langle\tilde{V}\rangle, \{P^{-1}, R^{-1}\}\, (0_{2n}\ I_{2n})^T\right)$$

$$= \mu\left(\tilde{W}\,(0\ I)^T, \ R^{-1}\,(0\ I)^T\right) - \mu\left(\tilde{V}(0\ I)^T, \ R^{-1}\,(0\ I)^T\right)$$

$$+ \mu^*\left(P^{-1}W^{-1}\,(0\ I)^T, \ P^{-1}(0\ I)^T\right) - \mu^*\left(P^{-1}V^{-1}\,(0\ I)^T, \ P^{-1}\,(0\ I)^T\right).$$

Applying property (ix) of Theorem 3.5 to the difference of last summands completes the proof of (3.117).

To prove (3.118), we apply twice property (v) of Theorem 3.5 on the right-hand side of the just proved formula (3.117) obtaining the identities

$$\mu\left(\tilde{W}\,(0\ I)^T,\ R^{-1}\,(0\ I)^T\right)$$
$$= \mu\left((0\ I)^T,\ P^{-1}W^{-1}\,(0\ I)^T\right) - \mu\left(R^{-1}\,(0\ I)^T,\ \tilde{W}\,(0\ I)^T\right),$$
$$\mu^*\left(W^{-1}\,(0\ I)^T,\ P(0\ I)^T\right)$$
$$= \mu\left((0\ I)^T,\ WP\,(0\ I)^T\right) - \mu^*\left(P\,(0\ I)^T,\ W^{-1}\,(0\ I)^T\right).$$

Consequently, we have

$$\mu\left(\tilde{W}\,(0\ I)^T,\ R^{-1}\,(0\ I)^T\right) - \mu^*\left(W^{-1}\,(0\ I)^T,\ P\,(0\ I)^T\right)$$
$$= \mu^*\left(P\,(0\ I)^T,\ W^{-1}\,(0\ I)^T\right) - \mu\left(R^{-1}\,(0\ I)^T,\ \tilde{W}\,(0\ I)^T\right).$$

An analogous identity holds also for the matrix $V$. Substituting the obtained differences into (3.117), we obtain the proof of (3.118). The proof of this lemma is completed. □

## 3.4 Notes and References

The concept of a comparative index was introduced in [114]. The origin of this notion is twofold. On one hand, $\mu(Y, \hat{Y})$ gives the possibility to compare matrices $Y$ and $\hat{Y}$ with (3.1) presenting a quantitative measure of violation of conditions (3.11). On the other hand, according to the results of Sect. 3.2, the quantity $\mu(Y, \hat{Y})$ is equal to the index, i.e., the number of negative eigenvalues of a symmetric matrix associated with $Y$ and $\hat{Y}$.

The results in Theorem 3.2 were proven in [114], while Theorem 3.6 and the properties of the comparative index in the form presented in the book (see Theorem 3.5) are from the paper [115]. The lower bounds for the comparative index in Lemma 3.9 are improved versions of the lower bounds derived in [289].

The statement in Lemma 3.13 was proven in [113, Lemma 2.7] and then used together with Lemma 3.14 in the proofs of the properties of the comparative index in [115, Lemma 4.4]. The proof of Corollary 3.16 is presented in [205, Lemma 3.1.10]. Finally, let us note that generalizations of Lemmas 3.13 and 3.14 for blocks of different dimensions can be found in [316, Theorem 2.3] and [317, Lemma 1.4].

In Sect. 3.2.2 we constructed a symmetric operator $\Lambda[V]$ associated with $V = (Y, \hat{Y})$, where $Y, \hat{Y}$ satisfy (3.1). Computing the index of $\Lambda[V]$, we prove the key property (viii) of Theorem 3.5 which is basic for the proof of the main theorem on the comparative index (see Theorem 3.6). For the special case $w(Y, \hat{Y}) = Y^T \mathcal{J} \hat{Y} = I$, the matrix $V$ is symplectic. Note that one can construct other symmetric operators associated with $V \in Sp(2n)$. For example, in [154, 155] the symmetric operator $\mathcal{J}(V - V^{-1})$ and its generalizations are used for the classification of eigenvalues of symplectic matrices; other examples can be found in [163, 164], where symplectic difference schemes are constructed using the symmetry of the Hamiltonian matrix in (1.103). In [132] we offered an algorithm for the computation of focal points based on Proposition 3.18.

The main results of Sect. 3.3 can be found in [117, 121], while Lemma 3.22(ii) is proven in [124, Lemma 2.3]. We point out that $4n \times 2n$ matrices in the form of (3.52) are well known and often used for different purposes in the discrete and continuous Sturmian and spectral theory; see, for example, [39, 42, 54, 205, 248, 250].

Symmetric matrices (3.74) were originally derived in [44, 171, 172]. They were used in comparison results for (SDS), e.g., in [56, 173, 258]. In the present work, we derive (3.74) as a particular solution of matrix equation (3.8) associated with the $4n \times 2n$ matrix $\langle S \rangle$. The result in Lemma 3.32 and formula (3.91) was proven for the first time in [128, Lemma 1]. Majority of the result in Sect. 3.3.5 can be found in [121, Chapter 2].

Note also that the comparative index theory itself as based on the notion of the Moore-Penrose generalized inversion and index results for block symmetric matrices can be considered as a new perspective tool in the matrix linear algebra. In this connection we mention the index results in Proposition 3.18, Lemma 3.22, and Remark 3.35 and other results of this chapter which deserve future investigations.

# Chapter 4
# Oscillation Theory of Symplectic Systems

In this chapter we present the oscillation theory of symplectic difference system (SDS). In particular, we show that it is possible to develop this theory in a similar way to the oscillation theory of linear Hamiltonian systems or their special cases, the Sturm-Liouville differential and difference equations. A crucial role in these results plays the notion of a multiplicity of a focal point for conjoined bases of (SDS), which allows to describe in quantitative way the behavior of the solutions. We define this notion in Sect. 4.1 for the case of forward focal points in the interval $(k, k + 1]$ and backward focal points in the interval $[k, k + 1)$ and connect these notions with the comparative index from Chap. 3. The algebraic properties of the comparative index, in particular Theorems 3.6 and 3.24, make it possible to present a completely new approach to the oscillation theory of (SDS) by deriving classical separation and comparison results in the form of explicit relations between the multiplicities of focal points. In Sects. 4.2 and 4.3, we provide the Sturmian separation and comparison theory for symplectic difference systems (SDS). In Sect. 4.4 we study symplectic transformations and their effect on focal points on conjoined bases of (SDS) by presenting explicit relations between their multiplicities.

## 4.1 Multiplicity of Focal Points

In this section we introduce the multiplicity of focal points of conjoined bases of (SDS) as the main notion of the modern oscillation theory. We also present results, which relate two basic concepts of the oscillation theory of discrete symplectic systems, namely, the concepts of the *multiplicity* of a *focal point* and the *comparative index*. Among others, we also establish basic relationship between forward and backward focal points.

© Springer Nature Switzerland AG 2019
O. Došlý et al., *Symplectic Difference Systems: Oscillation and Spectral Theory*,
Pathways in Mathematics, https://doi.org/10.1007/978-3-030-19373-7_4

### 4.1.1   Definition and Main Properties

The notion of the nonexistence of a focal point for a conjoined basis of (SDS) solved the problem of positivity of the associated quadratic functional (see Sect. 2.3). But the problem of the Sturmian separation and comparison theory for symplectic difference systems remained untouched. In the definition of the multiplicity of a focal point in the discrete case, the problem was how to "measure" the violation of the kernel condition and the $P$-condition in (2.37). In particular, the problem was that if the kernel condition is violated, then the matrix $P$ is no longer symmetric, hence to count something like its negative eigenvalues or so had no meaning. This problem was finally resolved in the fundamental paper of W. Kratz [208] in 2003, where the multiplicity of a focal point of a conjoined basis of (SDS) was defined as follows.

**Definition 4.1** Let $Y = \binom{X}{U}$ be a conjoined basis of (SDS), and consider the following matrices

$$M_k := (I - X_{k+1}X_{k+1}^\dagger)\,\mathcal{B}_k, \quad T_k := I - M_k^\dagger M_k, \tag{4.1}$$

$$P_k := T_k^T X_k X_{k+1}^\dagger \mathcal{B}_k T_k, \tag{4.2}$$

The *multiplicity* of a *forward focal point* (or a *left focal point*) of the conjoined basis $Y$ in the interval $(k, k+1]$ is defined as the number

$$m(Y_k) = m(k) := \operatorname{rank} M_k + \operatorname{ind} P_k. \tag{4.3}$$

Recall that ind denotes the index of the matrix indicated, i.e., the number of its negative eigenvalues. This definition is correct, since the matrix $P$ in (4.2) is really symmetric (see Proposition 4.4 or Lemma 2.45(ii) with a slightly different notation of $P_k$). In the next treatment, we will denote

$$m_1(Y_k) = m_1(k) := \operatorname{rank} M_k, \quad m_2(Y_k) = m_2(k) := \operatorname{ind} P_k.$$

The quantity $m_1(k)$ characterizes "how much" the kernel condition is violated (it is also called the multiplicity of a focal point at $k+1$), and the quantity $m_2(k)$ "measures" violation of $P$-condition; it is also called the multiplicity of a focal point in the interval $(k, k+1)$; compare also with Remark 2.46.

*Remark 4.2* There are equivalent definitions of the multiplicity of a focal point. For example, the multiplicity of a focal point at $k+1$ can be defined as

$$m_1(k) = \operatorname{rank} \mathcal{M}_k, \quad \mathcal{M}_k := (I - X_{k+1}^\dagger X_{k+1})\,X_k^T,$$

and the multiplicity of a focal point in $(k, k+1)$ can be defined as

$$m_2(k) = \operatorname{ind} P_k, \quad P_k := \mathcal{T}_k\,(\mathcal{B}_k^T \mathcal{D}_k - \mathcal{B}_k^T \mathcal{Q}_{k+1}\mathcal{B}_k)\,\mathcal{T}_k, \quad \mathcal{T}_k := I - \mathcal{M}_k^\dagger \mathcal{M}_k,$$

where $Q$ is a symmetric matrix satisfying $X_k^T Q_k X_k = X_k^T U_k$. The equivalence of the above given formula with those given in Definition 4.1 can be proved by a direct computation; see [193]. However, we prefer to postpone the proof to the next subsection, where we obtain the equivalence of the two definitions of the multiplicity of a focal point from the equivalence of various definitions of the comparative index and from a formula connecting the comparative index with the multiplicity of a focal point.

The adjective "forward" focal point is used here to distinguish this concept from the concept of a *backward focal point* which is given below. We will use the convention that when no adjective by "focal point" is used, we actually mean a *forward focal point*.

Consider the time-reversed symplectic system (2.47), i.e., the system

$$Y_k = \mathcal{S}_k^{-1} Y_{k+1}, \quad \mathcal{S}^{-1} = \begin{pmatrix} \mathcal{D}^T & -\mathcal{B}^T \\ -\mathcal{C}^T & \mathcal{A}^T \end{pmatrix}.$$

Recall that this system motivates Definition 2.22 of a backward focal point, which can be generalized as follows.

**Definition 4.3** Let $Y = \begin{pmatrix} X \\ U \end{pmatrix}$ be a conjoined basis of (SDS), and consider the following matrices:

$$\tilde{M}_k := (I - X_k X_k^\dagger) \mathcal{B}_k^T, \quad \tilde{T}_k := I - \tilde{M}_k^\dagger \tilde{M}_k, \tag{4.4}$$

$$\tilde{P}_k := \tilde{T}_k^T X_{k+1} X_k^\dagger \mathcal{B}_k^T \tilde{T}_k. \tag{4.5}$$

The *multiplicity* of a *backward focal point* (or a *right focal point*) of the conjoined basis $Y = \begin{pmatrix} X \\ U \end{pmatrix}$ in the interval $[k, k+1)$ is defined as the number

$$m^*(Y_k) = m^*(k) = \operatorname{rank} \tilde{M}_k + \operatorname{ind} \tilde{P}_k. \tag{4.6}$$

In the next treatment, we will denote

$$m_1^*(Y_k) - m_1^*(k) := \operatorname{rank} \tilde{M}_k, \quad m_2^*(Y_k) = m_2^*(k) := \operatorname{ind} \tilde{P}_k.$$

The quantity $m_1^*(k)$ is called the multiplicity in the point $k$, while $m_2^*(k)$ is called the multiplicity in the interval $(k, k+1)$. The main properties of $m(k)$ and $m^*(k)$ are the following.

**Proposition 4.4** *Let $m(k)$ and $m^*(k)$ be the multiplicities of forward and backward focal points of a conjoined basis $Y = \begin{pmatrix} X \\ U \end{pmatrix}$ of (SDS) in $(k, k+1]$ and $[k, k+1)$, respectively. Then we have the following properties :*

*(i) The matrices $P_k$ and $\tilde{P}_k$ in (4.2) and (4.5) are always symmetric.*

*(ii) We have $m_1(k) = 0$ if and only if the kernel condition in (2.37) holds, i.e.,*

$$\operatorname{Ker} X_{k+1} \subseteq \operatorname{Ker} X_k.$$

*Similarly, $m_1(k) = 0$ if and only if the image condition*

$$\operatorname{Im} \mathcal{B}_k \subseteq \operatorname{Im} X_{k+1}$$

*holds. In this case $T_k = I$, $P_k = X_k X_{k+1}^\dagger \mathcal{B}_k$, and $m(k) = m_2(k) = \operatorname{ind} P_k$.*

(iii) *We have $m_1^*(k) = 0$ if and only if the reversed kernel condition in (2.48) holds, i.e.,*

$$\operatorname{Ker} X_k \subseteq \operatorname{Ker} X_{k+1}.$$

*Similarly, $m_1^*(k) = 0$ if and only if the image condition*

$$\operatorname{Im} \mathcal{B}_k^T \subseteq \operatorname{Im} X_k$$

*holds. In this case $\tilde{T}_k = I$, $\tilde{P}_k = X_{k+1} X_k^\dagger \mathcal{B}_k^T$, and $m^*(k) = m_2^*(k) = \operatorname{ind} \tilde{P}_k$.*

(iv) *We have $m(k) = 0$ if and only if the conjoined basis $Y$ has no forward focal points in $(k, k+1]$, i.e., the conditions in (2.37) hold. Similarly, $m^*(k) = 0$ if and only if $Y$ has no backward focal points in $[k, k+1)$, i.e., the conditions in (2.48) are satisfied.*

(v) *The following inequalities are satisfied:*

$$0 \le r_k \le m(k) \le R_k \le n, \quad r_k + R_k = \operatorname{rank} \mathcal{B}_k - \Delta \operatorname{rank} X_k,$$

$$r_k := \max\{0, -\Delta \operatorname{rank} X_k, \operatorname{rank} \mathcal{B}_k - \operatorname{rank} X_{k+1}\},$$

$$R_k := \min\{\operatorname{rank} \mathcal{B}_k, \operatorname{rank} X_k, \operatorname{rank} \mathcal{B}_k - \Delta \operatorname{rank} X_k\},$$

*and similarly*

$$0 \le r_k^* \le m^*(k) \le R_k^* \le n, \quad r_k^* + R_k^* = \operatorname{rank} \mathcal{B}_k + \Delta \operatorname{rank} X_k,$$

$$r_k^* := \max\{0, \Delta \operatorname{rank} X_k, \operatorname{rank} \mathcal{B}_k - \operatorname{rank} X_k\},$$

$$R_k^* := \min\{\operatorname{rank} \mathcal{B}_k, \operatorname{rank} X_{k+1}, \operatorname{rank} \mathcal{B}_k + \Delta \operatorname{rank} X_k\}.$$

(vi) *The multiplicities $m(k)$ and $m^*(k)$ are connected by the formulas*

$$m_1^*(k) - m_1(k) = \operatorname{rank} X_{k+1} - \operatorname{rank} X_k, \tag{4.7}$$

$$m_2^*(k) = m_2(k), \tag{4.8}$$

$$m^*(k) - m(k) = \operatorname{rank} X_{k+1} - \operatorname{rank} X_k. \tag{4.9}$$

The proof of Proposition 4.4 follows from the comparative index properties and will be postponed later in Sect. 4.1.2.

*Remark 4.5* Recall that the second summands $m_2(k)$ and $m_2^*(k)$ in Definitions 4.1 and 4.3 of $m(k)$ and $m^*(k)$ are called the multiplicity of a focal point in the open interval $(k, k + 1)$ (or between $k$ and $k + 1$). Formula (4.8) says that these quantities are really the same justifying the terminology used in Definitions 4.1 and 4.3.

Define the number of (forward) focal points of a conjoined basis $Y = \binom{X}{U}$ in the interval $(M, N + 1]$ by the formula

$$l(Y, M, N + 1) := \sum_{k=M}^{N} m(k). \qquad (4.10)$$

Analogously, the number of backward focal points in $[M, N + 1)$ is defined by

$$l^*(Y, M, N + 1) := \sum_{k=M}^{N} m^*(k). \qquad (4.11)$$

From Proposition 4.4(vi), it immediately follows

**Corollary 4.6** *We have for the conjoined basis $Y = \binom{X}{U}$*

$$l^*(Y, M, N + 1) - l(Y, M, N + 1) = \operatorname{rank} X_{N+1} - \operatorname{rank} X_M, \qquad (4.12)$$

*in particular*

$$\left| l^*(Y, M, N + 1) - l(Y, M, N + 1) \right| \leq \max\{\operatorname{rank} X_{N+1}, \operatorname{rank} X_M\} \leq n. \qquad (4.13)$$

### 4.1.2 Multiplicity of a Focal Point and Comparative Index

The first statement of this subsection presents a basic formula relating the multiplicity of a focal point in $(k, k + 1]$ and the comparative index.

**Lemma 4.7** *Let $Z$ be a symplectic fundamental matrix of* (SDS)*, and let $Y = Z(0 \ I)^T$. Then*

$$m_i(k) = \mu_i\left(Y_{k+1}, S_k(0 \ I)^T\right) = \mu_i^*\left(Z_{k+1}^{-1}(0 \ I)^T, Z_k^{-1}(0 \ I)^T\right), \quad i \in \{1, 2\}, \qquad (4.14)$$

*where $m(k) = m_1(k) + m_2(k)$ is the number of focal points of $Y$ in $(k, k + 1]$, $\mu = \mu_1 + \mu_2$ is the comparative index of the matrices indicated, and $\mu^* = \mu_1^* + \mu_2^*$ is the dual comparative index.*

*Proof* In view of property (iii) of Theorem 3.5 and that $Z_k^{-1} = Z_{k+1}^{-1} S_k$, the second equality in (4.14) is obvious. Now we prove the first equality. To this end, it is sufficient to apply the statement of Theorem 3.2(i) as follows

$$\mu_1\left(Y_{k+1}, S_k(0 \ I)^T\right) = \operatorname{rank}(I - X_{k+1} X_{k+1}^\dagger) \mathcal{B}_k = m_1(k),$$

and also the Wronskian identity (2.4)

$$w\left(Y_{k+1}, S_k(0\ I)^T\right) = w\left(Y_k, (0\ I)^T\right) = X_k^T \tag{4.15}$$

when computing

$$\mu_2\left(Y_{k+1}, S_k(0\ I)^T\right) = \text{ind}\ \mathcal{T}_k(X_k X_{k+1}^\dagger \mathcal{B}_k)\ \mathcal{T}_k = m_2(k),$$

$$\mathcal{T}_k = I - \tilde{\mathcal{M}}_k^\dagger \tilde{\mathcal{M}}_k, \quad \tilde{\mathcal{M}}_k = (I - X_{k+1} X_{k+1}^\dagger)\mathcal{B}_k.$$

This proves the lemma.                                                                        □

Similarly as in Lemma 4.7, we can prove the following statement.

**Lemma 4.8** *Let Z be a symplectic fundamental matrix of* (SDS) *(which is also a fundamental matrix of* (2.9)*), and* $Y = Z(0\ I)^T$. *Then*

$$m_i^*(k) = \mu_i^*\left(Y_k, S_k^{-1}(0\ I)^T\right) = \mu_k\left(Z_k^{-1}(0\ I)^T, Z_{k+1}^{-1}(0\ I)^T\right), \quad i \in \{1, 2\}, \tag{4.16}$$

*where* $m^*(k) = m_1^*(k) + m_2^*(k)$ *is the number of backward focal points of Y in* $[k, k+1)$.

*Proof* The second equality in (4.16) follows from property (iii) of Theorem 3.5 and the duality principle (Theorem 3.11), where the equality $Z_k^{-1} S_k^{-1} = Z_{k+1}^{-1}$ has been used. As for the first equality in (4.16), using the definition of the dual comparative index (3.12), we have by (i) of Theorem 3.5

$$\mu_1^*\left(Y_k, S_k^{-1}(0\ I)^T\right) = \mu_1\left(Y_k, S_k^{-1}(0\ I)^T\right)$$

$$= \mu_1\left(\begin{pmatrix}X_k\\U_k\end{pmatrix}, \begin{pmatrix}-\mathcal{B}_k^T\\\mathcal{A}_k^T\end{pmatrix}\right) = \text{rank}\ (I - X_k X_k^\dagger)\ \mathcal{B}_k^T = m_1^*(k).$$

Using the Wronskian identity (2.4)

$$w\left(Y_k, S_k^{-1}(0\ I)^T\right) = w\left(Y_{k+1}, (0\ I)^T\right) = X_{k+1}^T \tag{4.17}$$

and the definition of $\mu_2^*$, we obtain

$$\mu_2^*\left(Y_k, S_k^{-1}(0\ I)^T\right) = \text{ind}\ [-\tilde{\mathcal{T}}_k X_{k+1} X_k^\dagger (-\mathcal{B}_k^T)\tilde{\mathcal{T}}_k] = m_2^*(k).$$

This lemma is proved.                                                                        □

*Proof of Proposition 4.4*

(i): By Lemma 4.7,

$$m_2(k) = \mu_2(Y_{k+1}, \mathcal{S}_k(0\ I)^T) = \operatorname{ind} \mathcal{P},$$

where the matrix $\mathcal{P}$, evaluated by the definition of the comparative index (see Definition 3.1), coincides with $P_k$ given by (4.2). Then, $P_k$ is symmetric by Theorem 3.2(iii). Similarly, by Lemma 4.8, Theorem 3.2(iii), and the definition of the dual comparative index (3.12), we prove the symmetry of $\tilde{P}_k$.

(ii): By Lemma 4.7, we have

$$m_1(k) = \mu_1(Y_{k+1}, \mathcal{S}_k(0\ I)^T) = \mu_1^*(Z_{k+1}^{-1}(0\ I)^T, Z_k^{-1}(0\ I)^T)$$
$$= \mu_1(Z_{k+1}^{-1}(0\ I)^T, Z_k^{-1}(0\ I)^T),$$

and by Theorem 3.2(ii) the equality $\mu_1(Z_{k+1}^{-1}(0\ I)^T, Z_k^{-1}(0\ I)^T) = 0$ holds if and only if $\operatorname{Im} X_k^T \subseteq \operatorname{Im} X_{k+1}^T$ or if and only if $\operatorname{Ker} X_{k+1} \subseteq \operatorname{Ker} X_k$. Similarly, $\mu_1(Y_{k+1}, \mathcal{S}_k(0\ I)^T) = 0$ is equivalent to $\operatorname{Im} \mathcal{B}_k \subseteq \operatorname{Im} X_{k+1}$ by Theorem 3.2(ii). In both cases $T_k = I$ and $m(k) = m_2(k) = \mu_2(Y_{k+1}, \mathcal{S}_k(0\ I)^T) = \operatorname{ind} P_k$, again by Theorem 3.2(ii).

(iii): Applying Lemma 4.8 we have

$$m_1^*(k) = \mu_1^*(Y_k, \mathcal{S}_k^{-1}(0\ I)^T) = \mu_1(Y_k, \mathcal{S}_k^{-1}(0\ I)^T)$$
$$= \mu_1(Z_k^{-1}(0\ I)^T, Z_{k+1}^{-1}(0\ I)^T).$$

By Theorem 3.2(ii), the equality $\mu_1(Z_k^{-1}(0\ I)^T, Z_{k+1}^{-1}(0\ I)^T) = 0$ holds if and only if $\operatorname{Im} X_{k+1}^T \subseteq \operatorname{Im} X_k^T$ or if and only if $\operatorname{Ker} X_k \subseteq \operatorname{Ker} X_{k+1}$. Similarly, condition $\mu_1(Y_k, \mathcal{S}_k^{-1}(0\ I)^T) = 0$ is equivalent to $\operatorname{Im} \mathcal{B}_k^T \subseteq \operatorname{Im} X_k$. In both vases, by Theorem 3.2(ii), we have $T_k = I$ and $m^*(k) = m_2^*(k) - \mu_2^*(Y_k, \mathcal{S}_k^{-1}(0\ I)^T) = \operatorname{ind} \tilde{P}_k$.

(iv): The proof follows from (ii) and (iii).

(v): According to Lemma 4.7, $m(k) = \mu(Y, \hat{Y})$ holds, where $Y := Y_{k+1}$ and $\hat{Y} := \mathcal{S}_k(0\ I)^T$, and then $w(Y, \hat{Y}) = X_k^T$ (see (4.15)) and $\hat{X} := \mathcal{B}_k$. We see that the first estimate follows from Theorem 3.5(vii), Lemma 3.9, and Remark 3.10. In a similar way, the second estimate follows from Lemma 4.8 (4.17) and again from Theorem 3.5(vii), Lemma 3.9, and Remark 3.10.

(vi): By Lemma 4.8, we have $m_i^*(k) = \mu_i\left(Z_k^{-1}(0\ I)^T, Z_{k+1}^{-1}(0\ I)^T\right)$ for $i \in \{1, 2\}$. Using property (vi) of Theorem 3.5 in the right-hand side of the last formula, we obtain

$$\mu_1\left(Z_k^{-1}(0\ I)^T, Z_{k+1}^{-1}(0\ I)^T\right) = \operatorname{rank} X_{k+1} - \operatorname{rank} X_k + \mu_1^*\left(Z_{k+1}^{-1}(0\ I)^T, Z_k^{-1}(0\ I)^T\right).$$

Using the identity $\mu_1^* \left( Z_{k+1}^{-1}(0\ I)^T, Z_k^{-1}(0\ I)^T \right) = m_1(k)$ from Lemma 4.7, we obtain (4.7). Analogously, from (iv) of Theorem 3.5, it follows that

$$m_2^*(k) = \mu_2 \left( Z_k^{-1}(0\ I)^T, Z_{k+1}^{-1}(0\ I)^T \right) = \mu_2^* \left( Z_{k+1}^{-1}(0\ I)^T, Z_k^{-1}(0\ I)^T \right) = m_2(k),$$

which proves (4.8). Formula (4.9) now follows from the definition of $m^*(k)$ as the sum of $m_1^*(k)$ and $m_2^*(k)$ (analogously for $m(k)$) and from (4.7) and (4.8).

$\square$

Note that by Theorems 3.2 and 3.5(iii), it is possible to formulate six equivalent definitions of the comparative index $\mu(Y, \hat{Y})$; as for any choice of $\mathcal{M}$ or $\tilde{\mathcal{M}}$ given by (3.3) or (3.5), there exist three different representations of $\mathcal{P}$ defined by (3.4), (3.10), and the similar representation for the second component of the comparative index $\mu(Z^{-1}(0\ I)^T, Z^{-1}\hat{Z}(0\ I)^T)$. By Lemmas 4.7 and 4.8, these definitions of the comparative index imply equivalent definitions of the multiplicity of a focal point in $(k, k+1]$ and $[k, k+1)$, as we have already mentioned in Remark 4.2. Now we formulate some of them and leave it to the reader to show how other definitions follow from the properties of the comparative index.

**Definition 4.9** The number of forward focal points $m(k) = m_1(k) + m_2(k)$ in $(k, k+1]$ of a conjoined basis $Y = \binom{X}{U}$ of (SDS) is given by the formula

$$m_1(k) = \operatorname{rank} \mathcal{M}_k, \quad \mathcal{M}_k := (I - X_{k+1}^\dagger X_{k+1})\, X_k^T, \quad \mathcal{T}_k := I - \mathcal{M}_k^\dagger \mathcal{M}_k, \tag{4.18}$$

$$m_2(k) = \operatorname{ind} P_k, \quad P_k = \mathcal{T}_k X_k X_{k+1}^\dagger B_k \mathcal{T}_k, \tag{4.19}$$

where the matrix $P_k$ in (4.19) can be expressed as

$$P_k = \mathcal{T}_k \left( B_k^T \mathcal{D}_k - B_k^T Q_{k+1} B_k \right) \mathcal{T}_k, \tag{4.20}$$

and $Q_k$ is any symmetric matrix satisfying

$$X_k^T Q_k X_k = X_k^T U_k. \tag{4.21}$$

**Definition 4.10** Let $Z = (\tilde{Y}\ Y)$ be the fundamental symplectic matrix of (SDS), where $\tilde{Y} = \binom{\tilde{X}}{\tilde{U}}$ and $Y = \binom{X}{U}$ are conjoined bases of (SDS). Then the number of focal points $m(k) = m_1(k) + m_2(k)$ in $(k, k+1]$ of the conjoined basis $Y$ is given by the formula (4.18) and

$$m_2(k) = \operatorname{ind} \mathcal{T}_k X_k (\Delta \tilde{Q}_k)\, X_k^T \mathcal{T}_k, \tag{4.22}$$

where the symmetric matrix $\tilde{Q}_k$ satisfies

$$X_k \tilde{Q}_k X_k^T = -\tilde{X}_k X_k^T. \tag{4.23}$$

In a similar way, for the multiplicity of backward focal points, one can formulate the following definitions.

**Definition 4.11** The number of backward focal points $m^*(k) = m_1^*(k) + m_2^*(k)$ in $[k, k+1)$ of a conjoined basis $Y = \binom{X}{U}$ of (SDS) is given by the formula

$$m_1^*(k) = \operatorname{rank} \tilde{\mathcal{M}}_k, \quad \tilde{\mathcal{M}}_k := (I - X_k^\dagger X_k) X_{k+1}^T, \quad \tilde{\mathcal{T}}_k := I - \tilde{\mathcal{M}}_k^\dagger \tilde{\mathcal{M}}_k, \tag{4.24}$$

$$m_2^*(k) = \operatorname{ind} \tilde{P}_k, \quad \tilde{P}_k = \tilde{\mathcal{T}}_k^T X_{k+1} X_k^\dagger \mathcal{B}_k^T \tilde{\mathcal{T}}_k, \tag{4.25}$$

where the matrix $\tilde{P}_k$ in (4.25) can be expressed as

$$\tilde{P}_k = \tilde{\mathcal{T}}_k (\mathcal{B}_k Q_k \mathcal{B}_k^T + \mathcal{B}_k \mathcal{A}_k^T) \tilde{\mathcal{T}}_k, \tag{4.26}$$

and $Q_k$ is any symmetric matrix satisfying

$$X_k^T Q_k X_k = X_k^T U_k. \tag{4.27}$$

**Definition 4.12** Let $Z = (\tilde{Y} \; Y)$ be the fundamental symplectic matrix of (SDS), where $\tilde{Y} = \binom{\tilde{X}}{\tilde{U}}$ and $Y = \binom{X}{U}$ are conjoined bases of (SDS). Then the number of backward focal points $m^*(k) = m_1^*(k) + m_2^*(k)$ in $[k, k+1)$ of the conjoined basis $Y$ is given by the formula (4.24) for $m_1^*(k)$ and $m_2^*(k) = m_2(k)$ with $m_2(k)$ given by (4.22), (4.23).

**Proposition 4.13** *Definitions 4.1, 4.9, and 4.10 are equivalent. Similarly, Definitions 4.3, 4.11, and 4.12 are equivalent.*

*Proof* The equivalence of Definitions 4.1 and 4.9 follows from Lemma 4.7. Indeed, the computation of $\mu \left( Y_{k+1}, S_k(0 \; I)^T \right)$ by Definition 3.1, using (4.15), gives (4.18) and (4.19). Formulas (4.20) and (4.21) follow from (iii) of Theorem 3.2. The proof of the equivalence of Definitions 4.1 and 4.10 follows from the second formula in Lemma 4.7. Indeed, we compute the quantity $\mu^* \left( Z_{k+1}^{-1}(0 \; I)^T, Z_k^{-1}(0 \; I)^T \right)$ by (i) and (iii) of Theorem 3.2, using the formula $Z_k^{-1}(0 \; I)^T = \binom{-X_k^T}{\tilde{X}_k^T}$ and the definition of the dual comparative index (see (3.12)). This proves (4.18) and (4.22). In a similar way, based on Lemma 4.8 and the properties of the comparative index, we prove the equivalence of Definitions 4.3, 4.11, and 4.12.                                                      □

*Remark 4.14*

(i) A consequence of the just proved Proposition 4.13 was already mentioned
in Lemma 2.15, where we have proved the equivalence of the inclusions
$\operatorname{Ker} X_{k+1} \subseteq \operatorname{Ker} X_k$ and $\operatorname{Im} \mathcal{B}_k \subseteq \operatorname{Im} X_{k+1}$, which is the consequence of
the fact that $m_1(k) = 0$. Proposition 4.13 also implies the equivalence of
the inequalities $X_k X_{k+1}^\dagger \mathcal{B}_k \geq 0$, $\mathcal{B}_k^T \mathcal{D}_k - \mathcal{B}_k^T \mathcal{Q}_{k+1} \mathcal{B}_k \geq 0$, and $X_k(\tilde{\mathcal{Q}}_{k+1} -$
$\tilde{\mathcal{Q}}_k) X_k^T \geq 0$ when $m_2(k) = 0$.

(ii) The properties of the matrix $\tilde{\mathcal{Q}}$ in Definition 4.10 satisfying (4.23) have
an interesting relationship with the analogously defined matrix for linear
Hamiltonian differential system (1.103). Under certain assumptions (see
Sect. 1.5), the invertibility of the upper entry $X$ of a conjoined basis $Y = \binom{X}{U}$
of (1.103) implies the strict monotonicity of the matrix $\tilde{Q}(t) = -X^{-1}(t)\tilde{X}(t)$,
i.e., $\tilde{Q}_1(t_1) < \tilde{Q}(t_2)$ for $t_1 < t_2$, where $\tilde{Y} = \binom{\tilde{X}}{\tilde{U}}$ and $Y$ are normalized
conjoined bases of (1.103); see [205, pg. 125]. Note that the assumption of
invertibility of the matrix $X(t)$ in the continuous case corresponds to the
absence of focal points of $Y = \binom{X}{U}$ in the discrete case, i.e., to $m(k) = 0$ for all
$k \in [0, N]_\mathbb{Z}$. For this case and under the additional assumption $\det X_k \neq 0$, we
have $\det X_l \neq 0$ and $\Delta \tilde{Q}_{l-1} \geq 0$ for $l \in [k+1, N+1]_\mathbb{Z}$ with $\tilde{Q}_l := -X_l^{-1}\tilde{X}_l$
for $l \in [k, N+1]_\mathbb{Z}$ (see Remark 4.14(i)), and the strict monotonicity $\Delta \tilde{Q}_k > 0$
holds if and only if $\det \mathcal{B}_k \neq 0$. This fact follows from (iii) of Theorem 3.2 and
from the equivalence of Definitions 4.1 and 4.10. It is interesting to note that
this strict monotonicity does not require $\mathcal{B}_k > 0$.

(iii) Consider symplectic system (SDS) for the scalar case ($n = 1$)

$$\begin{pmatrix} x_{k+1} \\ u_{k+1} \end{pmatrix} = \begin{pmatrix} a_k & b_k \\ c_k & d_k \end{pmatrix} \begin{pmatrix} x_k \\ u_k \end{pmatrix}, \quad a_k d_k - c_k b_k = 1. \tag{4.28}$$

We will show that the number $m(y_k)$, resp., $m^*(y_k)$, evaluated for a solution
$y_k = (x_k, u_k)^T$ with $x_k^2 + u_k^2 \neq 0$, takes the maximal value 1 if and
only if $y_k$ has a generalized zero in $(k, k+1]$, resp., in $[k, k+1)$ (see
Definitions 2.17, 2.23). First of all let us note that the case $b_k = 0$ is excluded
from the consideration. Indeed, condition $b_k = 0$ implies that $m(y_k) = 0$, resp.,
$m^*(y_k) = 0$, by the estimates in Proposition 4.4(v). In a similar way, if $b_k = 0$,
then $y_k$ has no generalized zero in $(k, k+1]$, resp., in $[k, k+1)$, indeed, for
this case the conditions $x_k \neq 0$ and $x_{k+1} = 0$ in (2.41), resp., $x_{k+1} \neq 0$ and
$x_k = 0$ in (2.49), are impossible by (4.28), where $a_k d_k - c_k b_k = a_k d_k = 1$.

Next assume that $m(y_k) = 1$. Then, by Definition 4.9 we have

$$m_1(y_k) = 1 \iff \operatorname{rank}[(1 - x_{k+1}x_{k+1}^\dagger) x_k] = 1 \iff x_k \neq 0, \ x_{k+1} = 0,$$

$$m_2(y_k) = 1 \iff x_k x_{k+1}^{-1} b_k < 0.$$

The last conditions constitute Definition 2.17 for $n = 1$ (compare also with Definition 1.1). Similarly, assuming that $m^*(y_k) = 1$, we have by Definition 4.11 that

$$m_1^*(y_k) = 1 \iff \text{rank}\,[(1 - x_k x_k^\dagger)\, x_{k+1}] = 1 \iff x_{k+1} \neq 0, \ x_k = 0,$$

$$m_2^*(y_k) = 1 \iff x_{k+1} x_k^{-1} b_k < 0.$$

The last conditions constitute Definition 2.23 for $n = 1$.

*Remark 4.15* In Remark 6.60 we will present additional formulas for the multiplicities of forward and backward focal points $m(k)$ and $m^*(k)$ by using the comparative index.

## 4.2 Sturmian Separation Theorems and Its Corollaries

In this section we present discrete analogs of classical Sturmian separation theorems (see Theorem 1.44) and its generalizations (see Theorem 1.50). The consideration is based on the fundamental notion of the multiplicity of a focal point.

### 4.2.1 Separation Theorem: First Step

In this subsection we present the results of the paper [101], where the first step toward separation theorem which counts focal points *including multiplicities* has been made. The construction in the proof of the main result of this subsection shows that if the principal solution $Y^{[0]}$ of (SDS) at $k = 0$ has $m \in \mathbb{N}$ (forward) focal points in the interval $(0, N+1]$, then there exist $m$ admissible pairs $y = (x, u)$ with linearly independent components $x = \{x_k\}_{k=1}^N \in \mathbb{R}^{Nn}$ with $x_0 = 0 = x_{N+1}$, such that the associated quadratic functional $\mathcal{F}(y) \leq 0$.

**Theorem 4.16 (Sturmian Separation Theorem)** *Suppose that the principal solution $Y^{[0]}$ of (SDS) at $0$ has no forward focal points in $(0, N + 1]$. Then any other conjoined basis of (SDS) has at most n focal points in $(0, N + 1]$.*

The general idea of the proof is simple. The assumptions of theorem imply that $\mathcal{F}(y) > 0$ for every nontrivial admissible $y$ with $x_0 = 0 = x_{N+1}$ (see Theorem 2.36). We show that the existence of a conjoined basis with more than $n$ focal points enables the construction of an admissible $y$ for which $\mathcal{F}(y) < 0$.

In Proposition 2.38 we presented a construction of an admissible pair for (2.57) for which $\mathcal{F} \not\geq 0$, when kernel condition or $P$-condition in (2.37) are violated. In the following result we discuss this construction in the framework of the multiplicities of forward focal points. The next lemma is a consequence of [205, Lemmas 3.1.5 and 3.1.6]; see also [208, pg. 142].

**Lemma 4.17** *Let* $Y = \begin{pmatrix} X \\ U \end{pmatrix}$ *be a conjoined basis of* (SDS), $M_k$ *be given by* (4.1) *and* $k \in [0, N]_{\mathbb{Z}}$. *Then there exists an* $n \times n$ *matrix* $S_k$ *such that*

$$\text{rank } S_k = \text{rank } M_k, \quad X_{k+1} S_k = 0, \quad \text{Ker } X_k \cap \text{Im } S_k = \{0\}. \tag{4.29}$$

In the next two lemmas, the matrices $M_k$, $P_k$, $T_k$ are defined by (4.1) and (4.2).

**Lemma 4.18** *Let* rank $M_k = p$. *Then there exist linearly independent vectors* $\alpha_1, \ldots, \alpha_p \in \mathbb{R}^n$ *such that*

$$X_{k+1} \alpha_j = 0, \quad X_k \alpha_j \neq 0, \quad j = 1, \ldots, p.$$

*Proof* Let $S_k$ be the $n \times n$ matrix for which (4.29) holds, and let $\alpha_1, \ldots, \alpha_p$ be a basis of Im $S_k$. Then Im $S_k \subseteq \text{Ker } X_{k+1}$ implies $X_{j+1} \alpha_j = 0$, and in addition Ker $X_k \cap \text{Im } S_k = \{0\}$ implies $X_j \alpha_j \neq 0$ for $j \in \{1, \ldots, p\}$. $\qquad\square$

**Lemma 4.19** *Let* $(k, k + 1]$ *contain a focal point of multiplicity* $p + q \leq n$ *of a conjoined basis* $Y = \begin{pmatrix} X \\ U \end{pmatrix}$ *of* (SDS), *where* $p = \text{rank } M_k$ *and* $q = \text{ind } P_k$. *Further, let* $\alpha_1, \ldots, \alpha_p$ *be the same as in the Lemma 4.18 and* $\beta_1, \ldots, \beta_q$ *be orthogonal vectors corresponding to the negative eigenvalues of* $P_k$, *i.e.,* $\beta_j^T P_k \beta_j < 0$ *for* $j \in \{1, \ldots, q\}$. *Denote* $\gamma_j := X_{k+1}^\dagger \mathcal{B}_k T_k \beta_j$. *Then the vectors* $\alpha_1, \ldots \alpha_p, \gamma_1, \ldots, \gamma_q$ *are linearly independent.*

*Proof* First we prove that $\gamma_1, \ldots, \gamma_q$ are linearly independent. Suppose that this is not the case, i.e., there exists a nontrivial linear combination $\sum_{j=1}^{q} \mu_j \gamma_j = 0$, and let $\beta = \sum_{j=1}^{q} \mu_j \beta_j$. Then

$$0 > \beta^T P_k \beta = \beta^T T_k^T X_k \left( \sum_{j=1}^{q} \mu_j X_{k+1}^\dagger \mathcal{B}_k T_k \beta_j \right) = \beta^T T_k^T X_k \left( \sum_{j=1}^{q} \mu_j \gamma_j \right) = 0,$$

which is a contradiction. Now suppose that $\gamma = \sum_{j=1}^{q} \mu_j \gamma_j = \sum_{j=1}^{p} \lambda_j \alpha_j \neq 0$, and let $\beta = \sum_{j=1}^{q} \mu_j \beta_j$ be as before. Then

$$0 > \beta^T P_k \beta = \beta^T T_k^T X_k X_{k+1}^\dagger \mathcal{B}_k T_k \beta = \beta^T T_k^T X_k X_{k+1}^\dagger X_{k+1} \gamma = 0,$$

which is a contradiction. $\qquad\square$

Recall that if $(k, k + 1]$ contains a focal point of multiplicity $p + q$, where $p = \text{rank } M_k$, $q = \text{ind } P_k$, we say that $p$ focal points are at $k + 1$, and $q$ focal points are in the open interval $(k, k + 1)$.

*Proof of Theorem 4.16* Let $Y = \begin{pmatrix} X \\ U \end{pmatrix}$ be a conjoined basis of (SDS), and let the intervals

$$(k_i, k_i + 1] \subseteq (0, N + 1], \quad i \in \{1, \ldots, l\}, \ 0 \le k_1 < k_2 < \cdots < k_l \le N,$$

contain focal points of $Y$ of multiplicities $m_i$, $i \in \{1, \ldots, l\}$. Let $m_i = p_i + q_i$, where $p_i = \operatorname{rank} M_{k_i}$, $q_i = \operatorname{ind} P_{k_i}$. For each interval $(k_i, k_i + 1]$ define the admissible pairs $y^{[i,j]}$ as follows. For $j \in \{1, \ldots, p_i\}$ we set

$$
x_k^{[i,j]} = \begin{cases} X_k \alpha_j^{[i]}, & 0 \le k \le k_i, \\ 0, & k_i + 1 \le k \le N + 1, \end{cases}
$$
$$
u_k^{[i,j]} = \begin{cases} U_k \alpha_j^{[i]}, & 0 \le k \le k_i, \\ 0, & k_i + 1 \le k \le N + 1, \end{cases}
$$

(4.30)

where $\alpha_j^{[i]} \in \operatorname{Ker} X_{k_i+1} \setminus \operatorname{Ker} X_{k_i}$ are linearly independent $n$-dimensional vectors (see Lemma 4.19). For $j \in \{p_i + 1, \ldots, p_i + q_i\}$ we define

$$
x_k^{[i,j]} = \begin{cases} X_k \gamma_j^{[i]}, & 0 \le k \le k_i, \\ 0, & k_i + 1 \le k \le N + 1, \end{cases}
$$
$$
u_k^{[i,j]} = \begin{cases} U_k \gamma_j^{[i]}, & 0 \le k \le k_i - 1, \\ U_k \gamma_j^{[i]} - T_k \beta_j^{[i]}, & k = k_i, \\ 0, & k_i + 1 \le k \le N + 1, \end{cases}
$$

(4.31)

where $\beta_j^{[i]}$, $j \in \{1, \ldots, q_i\}$ are orthogonal eigenvectors corresponding to the negative eigenvalues of the matrix $P_{k_i}$ and $\gamma_j^{[i]} = X_{k_i+1}^\dagger \mathcal{B}_{k_i} \beta_j^{[i]}$. By Proposition 2.38, we have for any $i \in \{1, \ldots, l\}$

$$\mathcal{F}(y^{[i,j]}) = (x_0^{[i,j]})^T u_0^{[i,j]}, \quad j \in \{1, \ldots, p_i\},$$
$$\mathcal{F}(y^{[i,j]}) = (x_0^{[i,j]})^T u_0^{[i,j]} + (\beta_j^{[i]})^T P_{k_1} \beta_j^{[i]}, \ j \in \{p_i + 1, \ldots, p_i + q_i\}.$$

To simplify some of the next computations, we relabel occasionally the quantities $x^{[i,j]}, u^{[i,j]}, \alpha_j^{[i]}, \ldots$ as follows. We introduce the index $\ell \in \{1, \ldots, \sum_{i=1}^l m_i\}$ by

$$[i, j] \longmapsto \ell = \sum_{s=0}^{i-1} m_s + j, \ m_0 := 0.$$

Now suppose, by contradiction, that the number of focal points of $Y$ in $(0, N + 1]$ exceeds $n$, i.e., $m := \sum_{i=1}^l m_i > n$. In order to make the idea of the proof more understandable, we will first suppose that $q_i = 0$ for $i \in \{1, \ldots, l\}$, i.e., all focal points are at $k_i + 1$ (the kernel condition is violated but all $P_{k_i} \ge 0$). Since

$\sum_{i=1}^{l} m_i = \sum_{i=1}^{l} p_i = m > n$, there exists a nontrivial linear combination

$$\sum_{\ell=1}^{m} \mu_\ell x_0^{[\ell]} = 0, \tag{4.32}$$

i.e., the admissible pair $y = \binom{x}{u}$ given by

$$x_k := \sum_{\ell=1}^{m} \mu_\ell x_k^{[\ell]}, \quad u_k := \sum_{\ell=1}^{m} \mu_\ell u_k^{[\ell]}, \quad k \in [1, N+1]_{\mathbb{Z}}, \tag{4.33}$$

satisfies $x_0 = 0 = x_{N+1}$. Moreover, the $Nn$-dimensional vector $x = \{x_k\}_{k=1}^{N}$ is *nonzero*. Indeed, consider first the largest focal point $k_l + 1$ in $(0, N+1]$. According to the construction of $x^{[i,j]}$ (returning to the original labeling), we have

$$x_{k_l}^{[i,j]} = 0, \quad i \in \{1, \ldots, l-1\}, \ j \in \{1, \ldots, p_i\},$$

so if $x = 0$, i.e., in particular, $x_{k_l} = 0$, we have

$$\sum_{j=1}^{p_l} \mu_{l,j} x_{k_l}^{[l,j]} = \sum_{j=1}^{p_l} \mu_{l,j} X_{k_l} \alpha_j^{[l]} = X_{k_l} \left( \sum_{j=1}^{p_l} \mu_{l,j} \alpha_j^{[l]} \right) = 0. \tag{4.34}$$

Since the vectors $\alpha_j^{[l]}$, $j \in \{1, \ldots, p_l\}$ form the basis of the space Im $S_{k_l}$, where $S_{k_l}$ is the same as $S_k$ in the proof of Lemma 4.18 (here with $k = k_l$) and at the same time by (4.34)

$$\sum_{j=1}^{p_l} \mu_{l,j} \alpha_j^{[l]} \in \text{Ker } X_{k_l},$$

we have $\sum_{j=1}^{p_l} \mu_{l,j} \alpha_j^{[l]} = 0$ because of Lemma 4.17, which means that the numbers $\mu_{l,j} = 0$, $j \in \{1, \ldots, p_l\}$, since the vectors $\alpha_j^{[l]}$, are linearly independent. Repeating the previous argument for $k \in \{k_{l-1}, \ldots, k = k_1\}$, we find that $\mu_{i,j} = 0$ for $i \in \{1, \ldots, l\}$ and $j \in \{1, \ldots, p_i\}$, which contradicts our assumption that the linear combination (4.32) is nontrivial. Therefore, $x \neq 0$ in the admissible pair given by (4.33).

Now, let $y^{[\kappa]} = (x^{[\kappa]}, u^{[\kappa]})$ and $y^{[\ell]} = (x^{[\ell]}, u^{[\ell]})$ for $\kappa, \ell \in \{1, \ldots, m\}$ be two admissible pairs constructed by (4.30). Then using (2.62) we obtain

$$\mathcal{F}(y^{[\kappa]}; y^{[\ell]}) = \begin{cases} 0, & \kappa \neq \ell, \\ (x_0^{[\ell]})^T u_0^{[\ell]}, & \kappa = \ell. \end{cases} \tag{4.35}$$

Consequently, for $y = \binom{x}{u}$ given by (4.33), we have

$$\mathcal{F}(y) = \mathcal{F}\left(\sum_{\ell=1}^{m} \mu_l y^{[\ell]}\right) = \sum_{\kappa,\ell=1}^{m} \mu_\kappa \mu_\ell\, \mathcal{F}(y^{[\kappa]}; y^{[\ell]})$$

$$= \left(\sum_{\ell=1}^{m} \mu_\ell x_0^{[\ell]}\right)^T \left(\sum_{\ell=1}^{m} \mu_\ell u_0^{[\ell]}\right) = x_0^T u_0 = 0,$$

since $x_0 = 0$ by (4.32). This contradicts the positivity of $\mathcal{F}$.

Suppose now that at least one of the $q_i$, $i \in \{1, \dots, l\}$ is positive. Then we have for this index and for $j \in \{1, \dots, p_i\}$

$$\mathcal{F}(y^{[i,j]}) = (x_0^{[i,j]})^T u_0^{[i,j]} + (\beta_j^{[i]})^T P_{k_i} \beta_j^{[i]},$$

and we have admissible pairs defined both by (4.30) and (4.31). In the previous part of the proof, we have already computed $\mathcal{F}(z^{[\kappa]}; z^{[\ell]})$ for admissible pairs given by (4.30). It remains to compute this bilinear form if one or both admissible pairs are of the form (4.31). We will perform the computation in the latter case. In the former case (i.e., one of the admissible pairs is given by (4.30) and the second one by (4.31)), substituting into the formula into (2.62), we get again (4.35). So, let $y^{[\kappa]}$ and $y^{[\ell]}$ be two admissible pairs given (4.31). If they are associated with the different focal intervals (i.e., the integers $k_i$ in (4.31) are different for $y^{[\kappa]}$ and $y^{[\ell]}$), using (2.62) we find again that (4.35) holds. Therefore, suppose finally that $y^{[\kappa]}$ and $y^{[\ell]}$ correspond to the same focal interval $(k_i, k_i + 1)$. Then

$$\mathcal{F}(y^{[\kappa]}; y^{[\ell]}) = (x_0^{[\kappa]})^T u_0^{[\ell]} + (x_{k_i}^{[\kappa]})^T \{\mathcal{C}_{k_i-1} x_{k_i-1}^{[\ell]} + \mathcal{D}_{k_i-1} u_{k_i-1}^{[\ell]} - u_{k_i}^{[\ell]}\}$$

$$= (x_0^{[\kappa]})^T u_0^{[\ell]} + (\gamma^{[\kappa]})^T X_{k_i} T_{k_i} \beta^{[\ell]}$$

$$= (x_0^{[\kappa]})^T u_0^{[\ell]} + (\beta^{[\kappa]})^T P_{k_i} \beta^{[\ell]}.$$

If $\kappa \neq \ell$, then the vectors $\beta^{[\kappa]}$ and $\beta^{[\ell]}$ are orthogonal eigenvectors of the matrix $P_{k_i}$, and thus $(\beta^{[\kappa]})^T P_{k_i} \beta^{[\ell]} = 0$.

Summarizing our previous computations, for $z = (x, u)$ given by (4.33) (i.e., $x_0 = 0$ by (4.32)), we have (again with the two-indices labeling)

$$\mathcal{F}(y) = \sum_{i=1}^{l} \sum_{j=1}^{q_i} (\beta_j^{[i]})^T P_{k_i} \beta_j^{[i]} < 0,$$

which again contradicts the positivity of $\mathcal{F}$. Note that $x = \{x_k\}_{k=1}^{N}$ is again nontrivial, since for each $i \in \{1, \dots, l\}$ the vectors $\alpha_j^{[i]}$ and $\gamma_s^{[i]}$ for $j \in \{1, \dots, p_i\}$ and $s \in \{1, \dots, q_i\}$ are linearly independent (by Lemma 4.19) and one can repeat the same argument as used in that part of the proof where we supposed that $q_i = 0$ for $i \in \{1, \ldots, l\}$. □

## 4.2.2   Separation Theorems and Comparative Index

In this section we present a discrete version of Theorem 1.50 and derive corollaries of this result, in particular we prove a discrete version of Theorem 1.44. We prove these results using the relationship between the concept of a focal point and the comparative index elaborated in Sect. 4.1.2. This relationship gives the possibility to apply algebraic properties of the comparative index, in particular Theorem 3.6, which is crucial for the results of this section. We start with a discrete version of Theorem 1.50.

**Theorem 4.20** *Let $Y$ and $\hat{Y}$ be conjoined bases of* (SDS). *Then*

$$\Delta\mu(Y_k, \hat{Y}_k) = m(k) - \hat{m}(k), \tag{4.36}$$

*where $\mu(Y_k, \hat{Y}_k)$ is the comparative index of $Y_k$ and $\hat{Y}_k$ and where $m(k)$ and $\hat{m}(k)$ are the multiplicities of forward focal points of $Y$ and $\hat{Y}$, respectively, in the interval $(k, k+1]$.*

*Proof* The proof easily follows from Theorem 3.6 and Lemma 4.7. We put $Z := Z_k$, $\hat{Z} := \hat{Z}_k$, $W := \mathcal{S}_k$ in (3.16), where $Z$ and $\hat{Z}$ are symplectic fundamental matrices of (SDS) such that $Y = Z(0\ I)^T$ and $\hat{Y} = \hat{Z}(0\ I)^T$. Then

$$WZ = Z_{k+1} = \mathcal{S}_k Z_k, \quad W\hat{Z} = \hat{Z}_{k+1} = \mathcal{S}_k \hat{Z}_k,$$

and using Lemma 4.7 and (3.16), we have

$$\mu(Y_{k+1}, \hat{Y}_{k+1}) - \mu(Y_k, \hat{Y}_k) = m(k) - \hat{m}(k)$$

which is equality (4.36). □

Similarly, using (3.26) and Lemma 4.8, we obtain the "dual" identity

$$\mu^*(Y_k, \hat{Y}_k) - \mu^*(Y_{k+1}, \hat{Y}_{k+1}) = m^*(k) - \hat{m}^*(k),$$

which is essentially a formula (4.36) for reversed symplectic system (2.47). This result is formulated in the next theorem.

**Theorem 4.21** *Let $Y$ and $\hat{Y}$ be conjoined bases of* (SDS) *(which are also conjoined bases of* (2.47)*). Then*

$$-\Delta\mu^*(Y_k, \hat{Y}_k) = m^*(k) - \hat{m}^*(k), \tag{4.37}$$

where $\mu^*(Y_k, \hat{Y}_k)$ is the dual comparative index of $Y$ and $\hat{Y}$ and where $m^*(k)$ and $\hat{m}^*(k)$ are the numbers of backward focal points of $Y$ and $\hat{Y}$ in the interval $[k, k+1)$.

**Remark 4.22** In Theorems 4.20 and 4.21, the conjoined bases $Y$ and $\hat{Y}$ play the same role. It means that interchanging them in (4.36), (4.37), we obtain

$$-\Delta\mu(\hat{Y}_k, Y_k) = m(k) - \hat{m}(k), \tag{4.38}$$

and

$$\Delta\mu^*(\hat{Y}_k, Y_k) = m^*(k) - \hat{m}^*(k). \tag{4.39}$$

Note that comparing left-hand sides of (4.36) and (4.38) (analogously for (4.37) and (4.39)), we obtain

$$-\Delta\mu(\hat{Y}_k, Y_k) = \Delta\mu(Y_k, \hat{Y}_k), \quad \text{resp.} \quad -\Delta\mu^*(\hat{Y}_k, Y_k) = \Delta\mu^*(Y_k, \hat{Y}_k).$$

The last equalities, taking into account property (v) of Theorem 3.5, are consequences of the Wronskian identity (2.4) for conjoined bases $Y_k$, $\hat{Y}_k$ of (SDS).

The first important consequence of (4.36) is a discrete version of Theorem 1.44 for linear Hamiltonian differential systems. It says, among others, that the numbers of focal points of any two conjoined bases of (SDS) in a given interval differ by at most $n$.

**Theorem 4.23** *Let the number of (forward) focal points of $Y$ and $\hat{Y}$ in the interval $(M, N + 1]$ be given by (4.10). Then*

$$l(Y, M, N + 1) - l(\hat{Y}, M, N + 1) = \mu(Y_{N+1}, \hat{Y}_{N+1}) - \mu(Y_M, \hat{Y}_M), \tag{4.40}$$

*and*

$$\mu(Y_{N+1}, \hat{Y}_{N+1}) - \mu(Y_M, \hat{Y}_M)$$
$$= \mu(Y_{N+1}, Y_{N+1}^{[M]}) - \mu(\hat{Y}_{N+1}, Y_{N+1}^{[M]}) \tag{4.41}$$
$$= \mu(\hat{Y}_M, Y_M^{[N+1]}) - \mu(Y_M, Y_M^{[N+1]}), \tag{4.42}$$

*where $Y^{[M]}$ and $Y^{[N+1]}$ are the principal solutions of (SDS) at $M$ and $N + 1$, respectively.*

*Proof* The summation of both sides of (4.36) from $k = M$ to $k = N$ yields equality (4.40). Formula (4.41) follows from Theorem 3.6 with $W := Z_{N+1}Z_M^{-1}$, $Y := Y_M$, and $\hat{Y} := \hat{Y}_M$, where $Z_k$ is a fundamental solution matrix of (SDS). Formula (4.42) is derived from (4.41), applying property (ix) in Theorem 3.5, where we also use the obvious relations $Y_{N+1}^{[M]} = Z_{N+1}Z_M^{-1}(0\ I)^T$ and $Y_M^{[N+1]} = Z_M Z_{N+1}^{-1}(0\ I)^T$. $\quad\square$

Quite analogously we derive similar results for the backward focal points.

**Theorem 4.24** *Let the number of backward focal points of conjoined bases $Y$ and $\hat{Y}$ of* (SDS) *in* $[M, N + 1)$ *be given by* (4.11). *Then*

$$l^*(Y, M, N + 1) - l^*(\hat{Y}, M, N + 1) = \mu^*(Y_M, \hat{Y}_M) - \mu^*(Y_{N+1}, \hat{Y}_{N+1}), \quad (4.43)$$

*and*

$$\mu^*(Y_M, \hat{Y}_M) - \mu^*(Y_{N+1}, \hat{Y}_{N+1})$$
$$= \mu^*(\hat{Y}_{N+1}, Y_{N+1}^{[M]}) - \mu^*(Y_{N+1}, Y_{N+1}^{[M]}) \quad (4.44)$$
$$= \mu^*(Y_M, Y_M^{[N+1]}) - \mu^*(\hat{Y}_M, Y_M^{[N+1]}), \quad (4.45)$$

*where $Y^{[M]}$ and $Y^{[N+1]}$ are the principal solutions of* (SDS) *at $M$ and $N + 1$, respectively.*

*Proof* By the summation of both sides of (4.37) from $k = M$ to $k = N$, we obtain (4.43). Formula (4.45) follows from the dual version of Theorem 3.6 (see formula (3.26)) for the case $W := Z_{N+1} Z_M^{-1}$, $Y := Y_M$, $\hat{Y} := \hat{Y}_M$, where $Z_k$ is a fundamental solution matrix of (SDS). Formula (4.44) is derived from (4.45) applying the dual version of property (ix) in Theorem 3.5. $\qquad\square$

In the remaining part of this subsection, we present important corollaries to Theorems 4.23 and 4.24. In particular, applying property (vii) of Theorem 3.5, it is possible to derive a new set of inequalities for the difference $l(Y, M, N + 1) - l(\hat{Y}, M, N + 1)$ of the multiplicities of focal points. The first result in this direction, based on (4.40), is presented below.

**Corollary 4.25** *Let $Y$ and $\hat{Y}$ be two conjoined bases of* (SDS). *Then the following estimate holds*

$$\left| l(Y, M, N + 1) - l(\hat{Y}, M, N + 1) \right|$$
$$\leq \min \left\{ \operatorname{rank} w(Y, \hat{Y}), \; r_{M,N+1}, \; \hat{r}_{M,N+1} \right\} \leq n, \quad (4.46)$$

*where $n$ is the dimension of block entries of the matrix $S_k$ in* (SDS), *$w(Y, \hat{Y})$ is the (constant) Wronskian of $Y$ and $\hat{Y}$, and*

$$\left.\begin{array}{l} r_{M,N+1} := \max\{\operatorname{rank} X_M, \operatorname{rank} X_{N+1}\}, \\ \hat{r}_{M,N+1} := \max\{\operatorname{rank} \hat{X}_M, \operatorname{rank} \hat{X}_{N+1}\}. \end{array}\right\} \quad (4.47)$$

*Proof* Estimate (4.46) follows from (vii) of Theorem 3.5 and from the Wronskian identity $w(Y_M, \hat{Y}_M) = w(Y_{N+1}, \hat{Y}_{N+1})$ in (2.4). More specifically, the left-hand side of (4.46) is equal to $|\mu_{N+1} - \mu_M|$, where $\mu_k$ is the abbreviation for $\mu(Y_k, \hat{Y}_k)$. From property (vii) in Theorem 3.5, we know that $\mu_M \leq \min\{\operatorname{rank} w(Y, \hat{Y}), \operatorname{rank} \hat{X}_M\}$ and $\mu_{N+1} \leq \min\{\operatorname{rank} w(Y, \hat{Y}), \operatorname{rank} \hat{X}_{N+1}\}$, which yields that

$$|\mu_{N+1} - \mu_M| \leq \min\{\operatorname{rank} w(Y, \hat{Y}), \hat{r}_{M,N+1}\}, \tag{4.48}$$

where $\hat{r}_{M,N+1}$ is given in (4.47). If we now switch the roles of the conjoined bases $Y$ and $\hat{Y}$, abbreviate $\mu(\hat{Y}_k, Y_k)$ as $\hat{\mu}_k$, and use that the Wronskian of $\hat{Y}$ and $Y$ is equal to $-w^T(Y, \hat{Y})$, then the formula $\mu_k + \hat{\mu}_k = \operatorname{rank} w(Y, \hat{Y})$ in property (v) of Theorem 3.5 yields

$$|\mu_{N+1} - \mu_M| = |\hat{\mu}_{N+1} - \hat{\mu}_M| \overset{(4.48)}{\leq} \min\{\operatorname{rank} w(Y, \hat{Y}), r_{M,N+1}\} \tag{4.49}$$

with $r_{M,N+1}$ given again in (4.47). By combining (4.48) and (4.49), we obtain the estimate in (4.46). $\qquad\square$

By a similar way, we derive estimates for the backward focal points.

**Corollary 4.26** *Let $Y$ and $\hat{Y}$ be two conjoined bases of (SDS). Then we have*

$$\left|l^*(Y, M, N+1) - l^*(\hat{Y}, M, N+1)\right|$$
$$\leq \min\left\{\operatorname{rank} w(Y, \hat{Y}), r_{M,N+1}, \hat{r}_{M,N+1}\right\} \leq n, \tag{4.50}$$

*where $r_{M,N+1}$ and $\hat{r}_{M,N+1}$ are given in (4.47).*

Quite analogously we obtain the next statement, based on (4.41), (4.42), (4.44), and (4.45). In contrast with Corollaries 4.25 and 4.26, we obtain the estimates, which deal with conjoined bases of (SDS) taken in the same point $M$ or $N+1$.

**Corollary 4.27** *For any conjoined bases $Y$ and $\hat{Y}$ of (SDS), we have the estimates*

$$\left|l(Y, M, N+1) - l(\hat{Y}, M, N+1)\right| \leq \min\left\{q_M, q_{N+1}, \operatorname{rank} X_M^{[N+1]}\right\} \leq n, \tag{4.51}$$

$$\left|l^*(Y, M, N+1) - l^*(\hat{Y}, M, N+1)\right| \leq \min\left\{q_M, q_{N+1}, \operatorname{rank} X_M^{[N+1]}\right\} \leq n, \tag{4.52}$$

*where*

$$q_k := \max\{\operatorname{rank} X_k, \operatorname{rank} \hat{X}_k\}. \tag{4.53}$$

*Proof* From formula (4.41) we get the estimate

$$\left|l(Y, M, N+1) - l(\hat{Y}, M, N+1)\right| \leq \max\left\{\mu(Y_{N+1}, Y_{N+1}^{[M]}), \mu(\hat{Y}_{N+1}, Y_{N+1}^{[M]})\right\}$$

and by a similar way, from (4.42) we have

$$\left| l(Y, M, N + 1) - l(\hat{Y}, M, N + 1) \right| \leq \max \left\{ \mu(\hat{Y}_M, Y_M^{[N+1]}), \mu(Y_M, Y_M^{[N+1]}) \right\}.$$

Then, applying property (vii) in Theorem 3.5 to the comparative indices in these inequalities and using the Wronskian identity (2.4) when computing

$$\text{rank } w(Y_k, Y_k^{[i]}) = \text{rank } X_i, \quad \text{rank } w(\hat{Y}_k, Y_k^{[i]}) = \text{rank } \hat{X}_i, \ i \in \{M, N + 1\},$$

we derive

$$\left| l(Y, M, N + 1) - l(\hat{Y}, M, N + 1) \right|$$
$$\leq \max \left\{ \min\{\text{rank } X_M, \ \text{rank } X_{N+1}^{[M]}\}, \ \min\{\text{rank } \hat{X}_M, \ \text{rank } X_{N+1}^{[M]}\} \right\},$$

and

$$\left| l(Y, M, N + 1) - l(\hat{Y}, M, N + 1) \right|$$
$$\leq \max \left\{ \min\{\text{rank } X_{N+1}, \ \text{rank } X_M^{[N+1]}\}, \ \min\{\text{rank } \hat{X}_{N+1}, \ \text{rank } X_M^{[N+1]}\} \right\}.$$

Formula (4.51) now follows from

$$\text{rank } w(Y_k^{[N+1]}, Y_k^{[M]}) = \text{rank } X_M^{[N+1]} = \text{rank } X_{N+1}^{[M]}$$

and from the fact that for any real numbers $x$, $y$, $z$ we have the equality

$$\max\{\min\{x, z\}, \ \min\{y, z\}\} = \min\{\max\{x, y\}, \ z\}.$$

In a similar way we obtain (4.52) from (4.44), (4.45), and from property (vii) in Theorem 3.5. □

*Remark 4.28*

(i) The results in Corollary 4.27 have interesting consequences. In particular, it is possible to obtain an estimate for the difference of the forward focal points of two conjoined bases $Y$ and $\hat{Y}$ in the interval $(M, N + 1]$, which does not depend on the values at the right (or left) endpoint $k = N + 1$ (or $k = M$). More precisely, inequality (4.51) implies that

$$\left| l(Y, M, N + 1) - l(\hat{Y}, M, N + 1) \right| \leq \max\{\text{rank } X_M, \text{rank } \hat{X}_M\} \leq n,$$
$$(4.54)$$

$$\left| l(Y, M, N + 1) - l(\hat{Y}, M, N + 1) \right| \leq \max\{\text{rank } X_{N+1}, \text{rank } \hat{X}_{N+1}\} \leq n.$$
$$(4.55)$$

(ii) Analogously, inequality (4.52) yields similar estimates for the quantity $|l^*(Y, M, N + 1) - l^*(\hat{Y}, M, N + 1)|$. The results in (4.54)–(4.55) allow to compare the numbers of focal points of $Y$ and $\hat{Y}$ in unbounded intervals, when the system (SDS) is nonoscillatory.

(iii) It also follows from Corollary 4.27 that we can compare the numbers of focal points of $Y$ and $\hat{Y}$ using estimates which do not depend on $Y$ and $\hat{Y}$. In more details we have from (4.51) and (4.52)

$$\left| l(Y, M, N + 1) - l(\hat{Y}, M, N + 1) \right| \le \operatorname{rank} X_{N+1}^{[M]} = \operatorname{rank} X_M^{[N+1]} \le n, \tag{4.56}$$

$$\left| l^*(Y, M, N + 1) - l^*(\hat{Y}, M, N + 1) \right| \le \operatorname{rank} X_M^{[N+1]} = \operatorname{rank} X_{N+1}^{[M]} \le n. \tag{4.57}$$

The next group of corollaries to Theorems 4.20 and 4.21 is connected with inequalities for the Riccati type quotients, which are partly considered in Sect. 2.4.1. We begin with the shortest proof of Corollary 2.56. Recall that we assumed in this corollary (we use the notation of this subsection) that the two conjoined bases $Y$ and $\hat{Y}$ of (SDS) satisfy

$$\operatorname{Im} \hat{X}_k \subseteq \operatorname{Im} X_k, \quad \hat{X}_k^T (\hat{Q}_k - Q_k) \hat{X}_k \ge 0 \tag{4.58}$$

for $k = 0$, i.e., $\mu(Y_0, \hat{Y}_0) = 0$ according to Theorem 3.2(iv), and $Y$ has no focal point in $(0, N + 1]$, i.e., $m(k) = 0$ for $k \in [0, N]_{\mathbb{Z}}$. Then $\hat{Y}$ has no focal point in $(0, N + 1]$ either, and (4.58) holds for all $k \in [0, N + 1]_{\mathbb{Z}}$. The symmetric matrices $Q$ and $\hat{Q}$ solve the equations $X_k^T Q_k X_k = X_k^T U_k$ and $\hat{X}_k^T \hat{Q}_k \hat{X}_k = \hat{X}_k^T \hat{U}_k$ for $k \in [0, N + 1]_{\mathbb{Z}}$.

*Proof of Corollary 2.56* We use formula (4.36) and the fact that all four numbers involved in this identity are nonnegative. Then the equalities $m(k) = 0$ and $\mu(Y_k, \hat{Y}_k) = 0$ are equivalent to $\hat{m}(k) = 0$ and $\mu(Y_{k+1}, \hat{Y}_{k+1}) = 0$. Consequently, the proof follows from the fact that $m(k) = 0$ and $\mu(Y_k, \hat{Y}_k) = 0$ imply $\hat{m}(k) = 0$ and $\mu(Y_{k+1}, \hat{Y}_{k+1}) = 0$. □

The following result concerning the condition $\mu(Y_{N+1}, \hat{Y}_{N+1}) = 0$ can be proved quite analogically.

**Corollary 4.29** *Suppose that two conjoined bases $Y$ and $\hat{Y}$ of (SDS) satisfy (4.58) for $k = N + 1$ and $\hat{Y}$ does not have any forward focal point in $(M, N + 1]$. Then $Y$ has no forward focal point in $(M, N + 1]$ either, and (4.58) holds for all $k \in [M, N + 1]_{\mathbb{Z}}$.*

*Proof* See the proof of Corollary 2.56 above, where we showed that the conditions $\hat{m}(k) = 0$, $\mu(Y_{k+1}, \hat{Y}_{k+1}) = 0$ are equivalent to $m(k) = 0$, $\mu(Y_k, \hat{Y}_k) = 0$ by formula (4.36). □

Note that analogical statements can be obtained from (4.39) for the backward focal points.

**Corollary 4.30** *Suppose that*

$$\text{Im } X_k \subseteq \text{Im } \hat{X}_k, \quad X_k^T (\hat{Q}_k - Q_k) X_k \geq 0 \tag{4.59}$$

*holds for* $k = N + 1$. *If* $\hat{Y}$ *has no backward focal point in* $[M, N + 1)$, *then* $Y$ *has no backward focal in this interval either, and* (4.59) *holds for all* $k \in [M, N + 1]_{\mathbb{Z}}$.

*Proof* We use formula (4.39), i.e.,

$$m^*(k) - \hat{m}^*(k) = \Delta \mu^*(\hat{Y}_k, Y_k).$$

Then conditions $\hat{m}^*(k) = 0$ and $\mu^*(\hat{Y}_{k+1}, Y_{k+1}) = 0$ are equivalent to the conditions $m^*(k) = 0$ and $\mu^*(\hat{Y}_k, Y_k) = 0$. The proof now follows from the fact that $\hat{m}^*(k) = 0$ and $\mu^*(\hat{Y}_{k+1}, Y_{k+1}) = 0$ imply $m^*(k) = 0$ and $\mu^*(\hat{Y}_k, Y_k) = 0$. $\quad \square$

Certainly the proof the the previous result also implies a similar statement connected with the condition (4.59) given at the initial point $k = 0$. We leave the formulation of this result to the reader.

## 4.2.3 Number of Focal Points and Principal Solutions

In this subsection we derive optimal bounds for the numbers of forward and backward focal points of any conjoined basis $Y$ in $(M, N + 1]$ and $[M, N + 1)$, respectively. These bounds are optimal in a sense that they are formulated in terms of quantities, which do not depend on the chosen conjoined basis $Y$. More precisely, they are formulated in terms of the numbers of forward and backward focal points of principal solutions $Y^{[M]}$ and $Y^{[N+1]}$.

Important consequences of formulas (4.36) and (4.37) are related to properties of the principal solutions of symplectic system (SDS) and its reversed system (2.9). Recall that the principal solution of (SDS) at $k$ is the solution satisfying the condition $Y_k^{[k]} = (0 \; I)^T$. Note that $\mu(Y_k, Y_k^{[k]}) = 0$ for any conjoined basis $Y$. In particular, from (4.40) it follows

$$l(Y, M, N + 1) - l(Y^{[M]}, M, N + 1) = \mu(Y_{N+1}, Y_{N+1}^{[M]}), \tag{4.60}$$

where $Y^{[M]}$ is the principal solution at $k = M$. Analogously,

$$l(Y, M, N + 1) - l(Y^{[N+1]}, M, N + 1) = -\mu(Y_M, Y_M^{[N+1]}), \tag{4.61}$$

where $Y^{[N+1]}$ is the principal solution at $k = N + 1$. As a consequence of (4.60) and (4.61), we have the estimate

$$l(Y^{[M]}, M, N + 1) \leq l(Y, M, N + 1) \leq l(Y^{[N+1]}, M, N + 1), \tag{4.62}$$

$$l(Y^{[N+1]}, M, N + 1) = l(Y^{[M]}, M, N + 1) + \operatorname{rank} X_{N+1}^{[M]} \tag{4.63}$$

for the number of forward focal points of any conjoined basis $Y$ of (SDS) in $(0, N + 1]$. Similarly, for the number of backward focal points, we have

$$l^*(Y, M, N + 1) - l^*(Y^{[M]}, M, N + 1) = -\mu^*(Y_{N+1}, Y_{N+1}^{[M]}), \tag{4.64}$$

$$l^*(Y, M, N + 1) - l^*(Y^{[N+1]}, M, N + 1) = \mu^*(Y_M, Y_M^{[N+1]}), \tag{4.65}$$

where $\mu^*(Y, \hat{Y})$ is the dual comparative index. Then we also have

$$l^*(Y^{[N+1]}, M, N + 1) \leq l^*(Y, M, N + 1) \leq l^*(Y^{[M]}, M, N + 1), \tag{4.66}$$

$$l^*(Y^{[M]}, M, N + 1) = l^*(Y^{[N+1]}, M, N + 1) + \operatorname{rank} X_M^{[N+1]} \tag{4.67}$$

for the number of backward focal points of any conjoined basis $Y$ of (SDS) in the interval $[M, N + 1)$.

*Remark 4.31* In Theorems 4.34 and 4.35, we will show that the lower bounds in (4.62) and (4.66) are the same, as well as the upper bounds in (4.62) and (4.66). Moreover, these lower and upper bounds are independent on the conjoined basis $Y$. Since these bounds are attained for the specific choices of $Y := Y^{[M]}$ and $Y := Y^{[N+1]}$, the inequalities in (4.62) and (4.66) are universal and cannot be improved— in the sense that the estimates (4.62) and (4.66) are satisfied *for all* conjoined bases $Y$ of (SDS).

In (4.60) and (4.65), we derived the exact formulas

$$l(Y, M, N + 1) = l(Y^{[M]}, M, N + 1) + \mu(Y_{N+1}, Y_{N+1}^{[M]}), \tag{4.68}$$

$$l^*(Y, M, N + 1) = l^*(Y^{[N+1]}, M, N + 1) + \mu(Y_M, Y_M^{[N+1]}), \tag{4.69}$$

which show how to calculate the number of forward or backward focal points of an arbitrary conjoined basis $Y$ of (SDS) as a sum of a quantity which does not depend on $Y$ and the comparative index of $Y$ with $Y^{[M]}$ at $k = N + 1$ or the dual comparative index of $Y$ with $Y^{[N+1]}$ at $k = M$. For practical purposes, e.g., in the oscillation theory on unbounded intervals, it is convenient to have estimates for the numbers $l(Y, M, N + 1)$ and $l^*(Y, M, N + 1)$, which do not explicitly involve the possible complicated evaluation of the comparative index. In Theorem 4.32 below, we present such estimates of $l(Y, M, N + 1)$ and $l^*(Y, M, N + 1)$. At the same time, we show that the universal lower and upper bounds for $l(Y, M, N + 1)$ and

$l^*(Y, M, N + 1)$ in (4.62) and (4.66) can be improved for some particular choice of the conjoined basis $Y$.

**Theorem 4.32** *For any conjoined basis $Y$ of (SDS), we have the inequalities*

$$r_N \leq l(Y, M, N + 1) - l(Y^{[M]}, M, N + 1) \leq R_N,$$

$$r_N^* \leq l^*(Y, M, N + 1) - l^*(Y^{[N+1]}, M, N + 1) \leq R_N^*,$$

*where*

$$r_N = \max\{0, \text{rank } X_{N+1}^{[M]} - \text{rank } X_{N+1}, \text{rank } X_M - \text{rank } X_{N+1}\}, \tag{4.70}$$

$$R_N = \min\{\text{rank } X_M, \text{rank } X_{N+1}^{[M]}, \text{rank } X_{N+1}^{[M]} - \text{rank } X_{N+1} + \text{rank } X_M\}, \tag{4.71}$$

$$r_N^* = \max\{0, \text{rank } X_M^{[N+1]} - \text{rank } X_M, \text{rank } X_{N+1} - \text{rank } X_M\}, \tag{4.72}$$

$$R_N^* = \min\{\text{rank } X_{N+1}, \text{rank } X_M^{[N+1]}, \text{rank } X_{N+1} - \text{rank } X_M + \text{rank } X_M^{[N+1]}\}. \tag{4.73}$$

*Moreover, we have*

$$r_N + R_N = \text{rank } X_{N+1}^{[M]} - \text{rank } X_{N+1} + \text{rank } X_M, \tag{4.74}$$

$$r_N^* + R_N^* = \text{rank } X_M^{[N+1]} + \text{rank } X_{N+1} - \text{rank } X_M. \tag{4.75}$$

*Proof* The values of the lower and upper bounds in (4.70), (4.71), and equality (4.74) follow from (4.60) and Remark 3.10(i) (see (3.23), (3.24)), in which we take $Y := Y_{N+1}, \hat{Y} := Y_{N+1}^{[M]}$, and $w(Y, \hat{Y}) = X_M^T$ being the Wronskian of $Y$ and $Y^{[M]}$. In a similar way we obtain (4.72), (4.73), and (4.75) from (4.65) and Remark 3.10(i) for the dual comparative index, in which we take $Y := Y_M, \hat{Y} := Y_M^{[N+1]}$, and $w(Y, \hat{Y}) = X_{N+1}^T$ being the Wronskian of $Y$ and $Y^{[N+1]}$.                    □

Now we present a discrete version of the second part of Theorem 1.44. This result is concerned with the numbers of forward focal points of the principal solution $Y^{[M]}$ in $(M, N + 1]$ and the number of backward focal points of the principal solution $Y^{[N+1]}$ in $[M, N + 1)$. We begin with the result in a more general form.

**Lemma 4.33** *Let $Y$ and $\hat{Y}$ be conjoined bases of (SDS) and $Z$ and $\hat{Z}$ be symplectic fundamental matrices of (SDS), such that $Y = Z(0 \, I)^T$ and $\hat{Y} = \hat{Z}(0 \, I)^T$. Then the number $m(k)$ of forward focal points of $Y$ and the number $\hat{m}^*(k)$ of backward focal points of $\hat{Y}$ are connected by the formula*

$$\hat{m}^*(k) - m(k) = \Delta \mu(\hat{Z}_k^{-1} Y_k, \hat{Z}_k^{-1}(0 \, I)^T), \tag{4.76}$$

*where*

$$\Delta\mu(\hat{Z}_k^{-1} Y_k, \hat{Z}_k^{-1}(0\ I)^T) = \Delta\mu^*(Z_k^{-1}\hat{Y}_k, Z_k^{-1}(0\ I)^T). \qquad (4.77)$$

*Moreover,*

$$l^*(\hat{Y}, M, N+1) - l(Y, M, N+1)$$
$$= \mu(\hat{Z}_{N+1}^{-1} Y_{N+1}, \hat{Z}_{N+1}^{-1}(0\ I)^T) - \mu(\hat{Z}_M^{-1} Y_M, \hat{Z}_M^{-1}(0\ I)^T), \qquad (4.78)$$

*and*

$$\mu(\hat{Z}_{N+1}^{-1} Y_{N+1}, \hat{Z}_{N+1}^{-1}(0\ I)^T) - \mu(\hat{Z}_M^{-1} Y_M, \hat{Z}_M^{-1}(0\ I)^T)$$
$$= \mu(Y_M, Y_M^{[N+1]}) - \mu^*(\hat{Y}_{N+1}, Y_{N+1}^{[M]}) \qquad (4.79)$$
$$= \mu^*(\hat{Y}_M, Y_M^{[N+1]}) - \mu(Y_{N+1}, Y_{N+1}^{[M]}), \qquad (4.80)$$

*where $Y^{[M]}$ and $Y^{[N+1]}$ are the principal solutions of* (SDS) *at $M$ and $N+1$, respectively.*

*Proof* The proof is based on Theorem 4.20 and formula (4.9). Substituting the value $\hat{m}(k) = \hat{m}^*(k) - \Delta \operatorname{rank} \hat{X}_k$ into (4.36), we have

$$m(k) - \hat{m}^*(k) = \Delta\mu(Y_k, \hat{Y}_k) - \Delta \operatorname{rank} \hat{X}_k. \qquad (4.81)$$

Then we replace $\mu(Y_k, \hat{Y}_k) - \operatorname{rank} \hat{X}_k$ by $\mu(Y_k, \hat{Z}_k(0\ I)^T) - \mu((0\ I)^T, \hat{Z}_k(0\ I)^T)$ on the right-hand side of the last identity and derive (4.76) by application of property (ix) of Theorem 3.5. Formula (4.77) is based on the application of property (iii) of Theorem 3.5; see also (3.33). Summing (4.76) from $k = M$ to $k = N$, we derive (4.78). Formulas (4.80) and (4.79) follow from (4.41) and (4.42) by subtracting the substitution rank $\hat{X}_k|_M^{N+1}$ from both sides according to (4.81), and then using formula (3.22) applied to $\mu(\hat{Y}_{N+1}, Y_{N+1}^{[M]})$ in (4.41) and $\mu(\hat{Y}_M, Y_M^{[N+1]})$ in (4.42), respectively.                                                                      □

From Lemma 4.33 we derive the main results of this subsection.

**Theorem 4.34** *Let $l(Y^{[M]}, M, N+1)$ be the number of forward focal points of $Y^{[M]}$ in $(M, N+1]$ and $l^*(Y^{[N+1]}, M, N+1)$ be the number of backward focal points of $Y^{[N+1]}$ in $[M, N+1)$. Then*

$$l(Y^{[M]}, M, N+1) = l^*(Y^{[N+1]}, M, N+1). \qquad (4.82)$$

*Proof* Putting $Y := Y^{[M]}$ and $\hat{Y} := Y^{[N+1]}$ in (4.78), we see that the substitution on the right-hand side of (4.78) equals to zero.                                          □

Similarly we prove the next statement, which illustrates the duality principle in the comparative index theory.

**Theorem 4.35** *Let $l^*(Y^{[M]}, M, N+1)$ be the number of backward focal points of $Y^{[M]}$ in $[M, N+1)$, and $l(Y^{[N+1]}, M, N+1)$ be the number of forward focal points of $Y^{[N+1]}$ in $(0, N+1]$. Then*

$$l^*(Y^{[M]}, M, N+1) = l(Y^{[N+1]}, M, N+1). \tag{4.83}$$

*Proof* Putting $\hat{Y} := Y^{[M]}$ and $Y := Y^{[N+1]}$ in (4.78), we see that the substitution on the right-hand side of (4.78) equals to zero. □

The next group of corollaries to Lemma 4.33 is connected with estimates for the difference $l^*(\hat{Y}, M, N+1) - l(Y, M, N+1)$, which we derive by analogy with a similar results in Sect. 4.2.2.

**Lemma 4.36** *We have the following estimates*

$$\left| l(Y, M, N+1) - l^*(\hat{Y}, M, N+1) \right|$$
$$\leq \max\left\{ \min\{\operatorname{rank} X_M, \operatorname{rank} \hat{X}_M\}, \min\{\operatorname{rank} X_{N+1}, \operatorname{rank} \hat{X}_{N+1}\} \right\} \leq n. \tag{4.84}$$

*and*

$$\left| l(Y, M, N+1) - l^*(\hat{Y}, M, N+1) \right| \leq \operatorname{rank} X_M^{[N+1]} = \operatorname{rank} X_{N+1}^{[M]} \leq n \tag{4.85}$$

*In particular, for one conjoined basis $Y$ of (SDS), we have*

$$\left| l(Y, M, N+1) - l^*(Y, M, N+1) \right|$$
$$\leq \min\left\{ \max\{\operatorname{rank} X_M, \operatorname{rank} X_{N+1}\}, \operatorname{rank} X_{N+1}^{[M]} \right\} \leq n. \tag{4.86}$$

*Proof* Estimate (4.84) follows from (4.78) and property (vii) in Theorem 3.5. Indeed, by (4.78) we have

$$\left| l^*(\hat{Y}, M, N+1) - l(Y, M, N+1) \right|$$
$$\leq \max\{\mu(\hat{Z}_{N+1}^{-1} Y_{N+1}, \hat{Z}_{N+1}^{-1}(0\ I)^T), \mu(\hat{Z}_M^{-1} Y_M, \hat{Z}_M^{-1}(0\ I)^T)\},$$

then it is sufficient to apply Theorem 3.5(vii), taking into account that

$$\operatorname{rank} w(\hat{Z}_k^{-1} Y_k, \hat{Z}_k^{-1}(0\ I)^T) = \operatorname{rank} X_k.$$

Note also that according to distributivity of the operation max with respect to min we have that the right-hand side in (4.84) less than or equal to each of the numbers $\max\{\operatorname{rank} X_l, \operatorname{rank} X_p\}$, $\max\{\operatorname{rank} X_l, \operatorname{rank} \hat{X}_p\}$, and $\max\{\operatorname{rank} \hat{X}_l, \operatorname{rank} \hat{X}_p\}$ for $p, l \in \{M, N+1\}$, $p \neq l$.

Estimate (4.85) follows from (4.79) using the inequality

$$\left| l^*(\hat{Y}, M, N+1) - l(Y, M, N+1) \right|$$
$$\leq \max\{\mu(Y_M, Y_M^{[N+1]}), \mu^*(\hat{Y}_{N+1}, Y_{N+1}^{[M]})\} \leq \operatorname{rank} X_M^{[N+1]}$$
$$= \operatorname{rank} X_{N+1}^{[M]}$$

derived by Theorem 3.5 (vii). □

### 4.2.4 Separation Results on Singular Intervals

In this subsection we apply the separation results of the previous sections to the case of symplectic systems (SDS), which are nonoscillatory near $\pm\infty$.

**Definition 4.37** System (SDS) is said to be *nonoscillatory* at $\infty$ if there exists $M \in \mathbb{N}$ such that the principal solution of (SDS) at $M$, i.e., the conjoined basis $Y^{[M]}$ given by the initial condition $Y_M^{[M]} = (0\ I)^T$, has no forward focal points in the interval $(M, \infty)$. Similarly, (SDS) is nonoscillatory at $-\infty$ if there exists $M$ such the principal solution at $M$ has no backward focal points in $(-\infty, M)$. In the opposite case system, (SDS) is said to be *oscillatory* at $\infty$, resp., at $-\infty$.

An equivalent definition of the (non)oscillation of system (SDS) at $\pm\infty$ in terms of arbitrary conjoined bases of (SDS) is presented in Sect. 6.3.2 (see Proposition 6.61).

**Lemma 4.38** *Assume that the system* (SDS) *is nonoscillatory at* $\infty$. *Then for any conjoined basis $Y$ of* (SDS) *and for arbitrary $M \in \mathbb{Z}$, there exist the finite limits*

$$\left. \begin{array}{c} l(Y, M, \infty) = \displaystyle\sum_{k=M}^{\infty} m(k), \quad l^*(Y, M, \infty) = \displaystyle\sum_{k=M}^{\infty} m^*(k), \\[2mm] \displaystyle\lim_{k\to\infty} \operatorname{rank} X_k \end{array} \right\} \tag{4.87}$$

*for the numbers of forward focal points in $(M, \infty)$ and backward focal points in $[M, \infty)$ connected by the equality*

$$l^*(Y, M, \infty) - l(Y, M, \infty) = \lim_{k\to\infty} \operatorname{rank} X_k - \operatorname{rank} X_M. \tag{4.88}$$

*Similarly, if the system* (SDS) *is nonoscillatory at* $-\infty$, *then for any conjoined basis*
$Y$ *of* (SDS) *and for any* $N \in \mathbb{Z}$, *there exist the finite limits*

$$
\left.
\begin{aligned}
l(Y, -\infty, N+1) = \sum_{k=-\infty}^{N} m(k), \quad l^*(Y, -\infty, N+1) = \sum_{k=-\infty}^{N} m^*(k), \\
\lim_{k \to -\infty} \operatorname{rank} X_k
\end{aligned}
\right\}
$$

$$(4.89)$$

*for the numbers of forward focal points of* $Y$ *in* $(-\infty, N+1]$ *and backward focal
points in* $(-\infty, N+1)$, *and*

$$l^*(Y, -\infty, N+1) - l(Y, -\infty, N+1) = \operatorname{rank} X_{N+1} - \lim_{k \to -\infty} \operatorname{rank} X_k. \quad (4.90)$$

*Proof* Note that by (4.60)

$$0 \leq l(Y, M, N+1) - l(Y^{[M]}, M, N+1) = \mu(Y_{N+1}, Y_{N+1}^{[M]}) \leq n$$

for arbitrary $M \in \mathbb{Z}$, then the first finite limit in (4.87) does exist for any conjoined
basis $Y$ of nonoscillatory system (SDS). Next, according to (4.13)

$$|l^*(Y, M, N+1) - l(Y, M, N+1)| \leq \max(\operatorname{rank} X_{N+1}, \operatorname{rank} X_M) \leq n,$$

then the second limit in (4.87) exists as well. The existence of the limit of rank $X_k$
as $k \to \infty$ can be proved directly using the kernel condition $\operatorname{Ker} X_{k+1} \subseteq \operatorname{Ker} X_k$
in (2.37), which implies $0 \leq \operatorname{rank} X_k \leq \operatorname{rank} X_{k+1} \leq n$ and then rank $X_k$ is
bounded nondecreasing function of the index $k$. However the existence of the limit
$\lim_{k \to \infty} \operatorname{rank} X_k$ also follows from identity (4.12), which says that the existence of
the first two limits in (4.87) implies that the third one does exist as well, and then
taking the limit in the above identity as $N \to \infty$ we derive (4.88). The proof of the
second claim of Lemma 4.38 is similar.                                                        □

Next we present a singular version of Theorems 4.23 and 4.24.

**Theorem 4.39 (Singular Separation Theorem)** *Assume that* (SDS) *is nonoscillatory at* $\infty$. *Then for any two conjoined bases* $Y$ *and* $\hat{Y}$ *of this system, there exists
the finite limit of the comparative index*

$$\mu_\infty(Y, \hat{Y}) := \lim_{k \to \infty} \mu(Y_k, \hat{Y}_k) \quad (4.91)$$

*connecting the numbers of forward focal points of these conjoined bases in* $(M, \infty)$

$$l(Y, M, \infty) - l(\hat{Y}, M, \infty) = \mu_\infty(Y, \hat{Y}) - \mu(Y_M, \hat{Y}_M). \quad (4.92)$$

*Similarly, there exists the finite limit of the dual comparative index*

$$\mu_\infty^*(Y, \hat{Y}) := \lim_{k \to \infty} \mu^*(Y_k, \hat{Y}_k) \tag{4.93}$$

*such that the numbers of backward focal points in* $[M, \infty)$ *are connected by the identity*

$$l^*(Y, M, \infty) - l^*(\hat{Y}, M, \infty) = \mu^*(Y_M, \hat{Y}_M) - \mu_\infty^*(Y, \hat{Y}). \tag{4.94}$$

*Proof* The proof follows immediately from Lemma 4.38 and from Theorems 4.23 and 4.24, where we evaluate the limits of the summands in (4.40) and (4.43) as $N$ tends to $\infty$.                                                                    □

We note that a similar theorem can be formulated for the case when system (SDS) is nonoscillatory at $-\infty$.

*Remark 4.40* Assume that system (SDS) is nonoscillatory at $\infty$, then for arbitrary conjoined bases of this system, we have the following connections for the finite limits (4.91) and (4.93):

$$\mu_\infty(Y, \hat{Y}) + \mu_\infty(\hat{Y}, Y) = \operatorname{rank} w(Y_k, \hat{Y}_k), \tag{4.95}$$

$$\mu_\infty^*(Y, \hat{Y}) + \mu_\infty^*(\hat{Y}, Y) = \operatorname{rank} w(Y_k, \hat{Y}_k), \tag{4.96}$$

$$\mu_\infty(Y, \hat{Y}) = \lim_{k \to \infty} \operatorname{rank} \hat{X}_k - \lim_{k \to \infty} \operatorname{rank} X_k + \mu_\infty^*(\hat{Y}, Y). \tag{4.97}$$

Indeed, all limits in (4.95)–(4.97) exist by Lemma 4.38 and Theorem 4.39, and then we derive identities (4.95)–(4.97) using the properties of the comparative index according to Theorem 3.5(v), (vi) by taking limits as $k \to \infty$.

*Remark 4.41* Further singular separation theorems on unbounded intervals, e.g., of the form $[M, \infty)_{\mathbb{Z}}$ as in this subsection, will be presented in Sect. 6.4. They are based on comparison with the numbers of focal points of the principal solution $Y^{[M]}$ and the (minimal) recessive solution $Y^{[\infty]}$ of (SDS) at $\infty$ from Sect. 6.3. As we shall see, the latter approach also allows to evaluate the limits of the comparative indices $\mu_\infty(Y, \hat{Y})$ and $\mu_\infty^*(Y, \hat{Y})$ from Remark 4.40 explicitly; see Corollaries 6.174 and 6.186.

## 4.3  Comparison Theorems and Its Corollaries

In this section we consider together with (SDS) another system of the same form (which we write now in the $2n \times n$ matrix form)

$$\hat{Y}_{k+1} = \hat{\mathcal{S}}_k \hat{Y}_k, \quad \hat{\mathcal{S}}_k^T \mathcal{J} \hat{\mathcal{S}}_k = \mathcal{J}. \tag{4.98}$$

If $Y$ and $\hat{Y}$ are conjoined bases of (SDS) and (4.98), respectively, then we can introduce the Wronskian of $Y$ and $\hat{Y}$ using notation (3.2) as

$$w(Y_k, \hat{Y}_k) := Y_k^T \mathcal{J} \hat{Y}_k,$$

but, generally speaking, the Wronskian identity (2.4) is lost due to the formula

$$\Delta w(\hat{Y}_k, Y_k) = \hat{Y}_k^T (\hat{\mathcal{S}}_k^T \mathcal{J} \mathcal{S}_k - \mathcal{J}) Y_k = \hat{Y}_{k+1}^T (\mathcal{J} - \hat{\mathcal{S}}_k^{T-1} \mathcal{J} \mathcal{S}_k^{-1}) Y_{k+1}. \qquad (4.99)$$

Note also that by the properties of symplectic matrices (see Lemma 1.58(ii)), for any symplectic fundamental matrices $Z$ and $\hat{Z}$ of (SDS) and (4.98) such that $Y = Z(0\ I)^T$, $\hat{Y} = \hat{Z}(0\ I)^T$, we have the important connection

$$w(\hat{Y}_k, Y_k) = \hat{Y}_k^T \mathcal{J} Y_k = -(I\ 0)\, \hat{Z}_k^{-1} Z_k(0\ I)^T = -(I\ 0)\, \hat{Z}_k^{-1} Y_k \qquad (4.100)$$

between the matrix $\hat{Z}_k^{-1} Y_k$ and the Wronskian (3.2) of two conjoined bases $\hat{Y}$ and $Y$ of systems (4.98) and (SDS). This fact will be used in the subsequent results. For example, identity (4.99) is a direct consequence of the relation

$$\Delta(\hat{Z}_k^{-1} Z_k) = \hat{Z}_{k+1}^{-1} Z_{k+1} - \hat{Z}_k^{-1} Z_k = \hat{Z}_k^{-1} (\hat{\mathcal{S}}_k^{-1} \mathcal{S}_k - I)\, Z_k$$
$$= \hat{Z}_{k+1}^{-1} (I - \hat{\mathcal{S}}_k \mathcal{S}_k^{-1})\, Z_{k+1}. \qquad (4.101)$$

The results of this section show that there exists a deep connection between numbers of focal points of conjoined bases of (SDS), (4.98), and $\tilde{Y}_k = \hat{Z}_k^{-1} Y_k$.

### 4.3.1  Sturmian Comparison Theorems

We start with some auxiliary statements. The first one is based on Lemma 3.23.

**Lemma 4.42** *Let $Z$ and $\hat{Z}$ be symplectic fundamental matrices of* (SDS) *and* (4.98), *respectively, $Y = Z(0\ I)^T$, $\hat{Y} = \hat{Z}(0\ I)^T$, and*

$$\mathcal{L}(Y_k, \hat{Y}_k, \mathcal{S}_k, \hat{\mathcal{S}}_k) := \hat{m}(k) - m(k) + \Delta \mu(Y_k, \hat{Y}_k). \qquad (4.102)$$

*Then*

$$\mathcal{L}(Y_k, \hat{Y}_k, \mathcal{S}_k, \hat{\mathcal{S}}_k)$$
$$= \mu(\mathcal{S}_k(0\ I)^T, \hat{\mathcal{S}}_k(0\ I)^T) + \mu(\hat{\mathcal{S}}_k^{-1} \mathcal{S}_k Y_k, Y_k)$$
$$- \mu(\hat{Z}_{k+1}^{-1} Z_{k+1}(0\ I)^T, \hat{Z}_k^{-1} Z_k(0\ I)^T) - \mu(\hat{\mathcal{S}}_k^{-1} \mathcal{S}_k Y_k, \hat{\mathcal{S}}_k^{-1} \mathcal{S}_k(0\ I)^T),$$
$$(4.103)$$

*or*

$$\mathcal{L}(Y_k, \hat{Y}_k, S_k, \hat{S}_k)$$
$$= \mu\big(S_k(0\ I)^T, S_k\hat{S}_k^{-1}(0\ I)^T\big) + \mu\big(\hat{Z}_k^{-1}Z_k(0\ I)^T, \hat{Z}_{k+1}^{-1}Z_{k+1}(0\ I)^T\big)$$
$$- \mu\big(\hat{S}_k S_k^{-1}Y_{k+1}, Y_{k+1}\big) - \mu\big(Y_{k+1}, S_k\hat{S}_k^{-1}(0\ I)^T\big). \tag{4.104}$$

*Proof* We put $Z := Z_k$, $\hat{Z} := \hat{Z}_k$, $W := S_k$, $\hat{W} := \hat{S}_k$ in Lemma 3.23. Since $Z_{k+1} = S_k Z_k$, $\hat{Z}_{k+1} = \hat{S}_k \hat{Z}_k$, using the formula for the relationship between the number of focal points and the comparative index in Lemma 4.7, we have the required statement. □

Now we turn our attention to two particular cases of (4.103) and (4.104), which will be important in applications.

**Lemma 4.43** *Suppose that*

$$\hat{S}_k^{-1}S_k = \begin{pmatrix} I & 0 \\ \mathcal{Q}_k & I \end{pmatrix} \tag{4.105}$$

*holds. Then*

$$\mathcal{L}(Y_k, \hat{Y}_k, S_k, \hat{S}_k)$$
$$= \text{ind}(-X_k^T \mathcal{Q}_k X_k) - \mu\big(\hat{Z}_{k+1}^{-1}Z_{k+1}(0\ I)^T, \hat{Z}_k^{-1}Z_k(0\ I)^T\big)$$
$$= \mu\big(\hat{Z}_k^{-1}Z_k(0\ I)^T, \hat{Z}_{k+1}^{-1}Z_{k+1}(0\ I)^T\big) - \text{ind}(X_k^T \mathcal{Q}_k X_k). \tag{4.106}$$

*Similarly, if*

$$S_k\hat{S}_k^{-1} = \begin{pmatrix} I & 0 \\ \mathcal{R}_k & I \end{pmatrix}, \tag{4.107}$$

*then*

$$\mathcal{L}(Y_k, \hat{Y}_k, S_k, \hat{S}_k)$$
$$= \mu\big(\hat{Z}_k^{-1}Z_k(0\ I)^T, \hat{Z}_{k+1}^{-1}Z_{k+1}(0\ I)^T\big) - \text{ind}(X_{k+1}^T \mathcal{R}_k X_{k+1}) \tag{4.108}$$
$$= \text{ind}(-X_{k+1}^T \mathcal{R}_k X_{k+1}) - \mu\big(\hat{Z}_{k+1}^{-1}Z_{k+1}(0\ I)^T, \hat{Z}_k^{-1}Z_k(0\ I)^T\big).$$

*Proof* Recall that the matrices $\mathcal{Q}_k$ and $\mathcal{R}_k$ in (4.105) and (4.107) are symmetric by (1.153). If (4.105) holds, then obviously by (iii) of Theorem 3.5, we have

$$\mu\big(\mathcal{S}_k(0\ I)^T,\ \hat{\mathcal{S}}_k(0\ I)^T\big) = \mu^*\big(\mathcal{S}_k^{-1}(0\ I)^T,\ \mathcal{S}_k^{-1}\hat{\mathcal{S}}_k(0\ I)^T\big)$$

$$= \mu^*\big(\mathcal{S}_k^{-1}(0\ I)^T,\ (0\ I)^T\big) = 0,$$

$$\mu\big(\hat{\mathcal{S}}_k^{-1}\mathcal{S}_k Y_k,\ \hat{\mathcal{S}}_k^{-1}\mathcal{S}_k(0\ I)^T\big) = \mu\big(\hat{\mathcal{S}}_k^{-1}\mathcal{S}_k Y_k,\ (0\ I)^T\big) = 0,$$

$$\mu(\hat{\mathcal{S}}_k^{-1}\mathcal{S}_k Y_k,\ Y_k) = \mu\left(\begin{pmatrix} X_k \\ \mathcal{Q}_k X_k + U_k \end{pmatrix},\ \begin{pmatrix} X_k \\ U_k \end{pmatrix}\right) = \mathrm{ind}\,(-X_k^T \mathcal{Q}_k X_k).$$

Substituting these formulas for comparative indices into the right-hand side of (4.103), we prove the first part of (4.106). Note that by Theorem 3.5(v), we obtain

$$\mu\big(\hat{\mathcal{S}}_k^{-1}\mathcal{S}_k Y_k,\ Y_k\big) = \mu\big((0\ I)^T,\ Z_k^{-1}\mathcal{S}_k^{-1}\hat{\mathcal{S}}_k Y_k\big) - \mu\big(Y_k,\ \hat{\mathcal{S}}_k^{-1}\mathcal{S}_k Y_k\big),$$

and analogously

$$\mu\big(\hat{Z}_{k+1}^{-1}Z_{k+1}(0\ I)^T,\ \hat{Z}_k^{-1}Z_k(0\ I)^T\big)$$

$$= \mu\big((0\ I)^T,\ Z_k^{-1}\mathcal{S}_k^{-1}\hat{\mathcal{S}}_k Y_k\big) - \mu\big(\hat{Z}_k^{-1}Z_k(0\ I)^T,\ \hat{Z}_{k+1}^{-1}Z_{k+1}(0\ I)^T\big).$$

Consequently,

$$\mu\big(\hat{\mathcal{S}}_k^{-1}\mathcal{S}_k Y_k,\ Y_k\big) - \mu\big(\hat{Z}_{k+1}^{-1}Z_{k+1}(0\ I)^T,\ \hat{Z}_k^{-1}Z_k(0\ I)^T\big)$$

$$= \mu\big(\hat{Z}_k^{-1}Z_k(0\ I)^T,\ \hat{Z}_{k+1}^{-1}Z_{k+1}(0\ I)^T\big) - \mu\big(Y_k,\ \hat{\mathcal{S}}_k^{-1}\mathcal{S}_k Y_k\big),$$

where $\mu\big(Y_k,\ \hat{\mathcal{S}}_k^{-1}\mathcal{S}_k Y_k\big) = \mathrm{ind}(X_k^T \mathcal{Q}_k X_k)$. This proves (4.106). Formula (4.108) can be proved analogously using (4.104). $\qquad\square$

The representation for the operator $\mathcal{L}$ in (4.102) of Lemma 4.42 can be simplified if we apply Theorem 3.24 to this operator. Recall that in the previous chapter, we have introduced notation (3.52) for a symplectic matrix $W$, i.e.,

$$\langle W \rangle = \begin{pmatrix} I & 0 \\ A & B \\ 0 & -I \\ C & D \end{pmatrix}, \qquad W = \begin{pmatrix} A & B \\ C & D \end{pmatrix}.$$

Using notation (3.52) we introduce the notion of the relative oscillation numbers.

**Definition 4.44** For the pair of symplectic difference systems (SDS) and (4.98) and their symplectic fundamental matrices $Z$ and $\hat{Z}$, respectively, we define the *relative oscillation number* $\#_k(\hat{Z}, Z)$ at index $k$ of these fundamental matrices by the formula

$$\left.\begin{aligned}\#_k(\hat{Z}, Z) &:= \mu\big(\langle S_k \rangle, \langle \hat{S}_k \rangle\big) - \mu\big(\langle \hat{Z}_{k+1}^{-1} Z_{k+1} \rangle, \langle \hat{Z}_k^{-1} Z_k \rangle\big) \\ &= \mu\big(\langle \hat{Z}_k^{-1} Z_k \rangle, \langle \hat{Z}_{k+1}^{-1} Z_{k+1} \rangle\big) - \mu\big(\langle \hat{S}_k \rangle, \langle S_k \rangle\big)\end{aligned}\right\} \tag{4.109}$$

and the relative oscillation number of $Z$ and $\hat{Z}$ in the interval $[M, N]$ by the formula

$$\#(\hat{Z}, Z, M, N) := \sum_{k=M}^{N} \#_k(\hat{Z}, Z). \tag{4.110}$$

Now we formulate the Sturmian comparison theorem for the pair of symplectic difference systems (SDS), (4.98).

**Theorem 4.45 (Sturmian Comparison Theorem)** *Let $Z$ and $\hat{Z}$ be symplectic fundamental matrices of (SDS) and (4.98), respectively, and $Y = Z(0\ I)^T$ and $\hat{Y} = \hat{Z}(0\ I)^T$. Then the operator $\mathcal{L}$ from (4.102) can be expressed as*

$$\mathcal{L}(Y_k, \hat{Y}_k, S_k, \hat{S}_k) = \#_k(\hat{Z}, Z), \tag{4.111}$$

*where $\#_k(\hat{Z}, Z)$ is the relative oscillation number given by (4.109).*

*Proof* The proof is similar to that of Lemma 4.42. We put $Z = Z_k$, $\hat{Z} = \hat{Z}_k$, $W = S_k$, $\hat{W} = \hat{S}_k$ in Theorem 3.24. Then $Z_{k+1} = S_k Z_k$, $\hat{Z}_{k+1} = \hat{S}_k \hat{Z}_k$, and using Lemma 4.7, we obtain the proof of (4.111) with (4.109). $\qquad\square$

*Remark 4.46* Consider the meaning of the addends in the definition of the relative oscillation numbers in details.

(i) The comparative index

$$\mu(\langle \hat{Z}_{k+1}^{-1} Z_{k+1} \rangle, \langle \hat{Z}_k^{-1} Z_k \rangle) = \mu^*(\langle Z_{k+1}^{-1} \hat{Z}_{k+1} \rangle, \langle Z_k^{-1} \hat{Z}_k \rangle)$$

in the first equality of (4.109) describes the multiplicity of a forward focal point of a conjoined basis of a symplectic $4n \times 4n$ system associated with (SDS) and (4.98). Indeed, by (3.101)–(3.104), the $4n \times 2n$ matrix $\mathcal{Y}_k = S_0^{-1}\{\hat{Z}_k^{-1} Z_k\}(0_{2n}\ I_{2n})^T = \mathcal{Z}_k(0_{2n}\ I_{2n})^T$ is a conjoined basis of the symplectic system

$$\mathcal{Y}_{k+1} = S_0^{-1}\{\hat{Z}_k^{-1} \hat{S}_k^{-1} S_k \hat{Z}_k\} S_0\, \mathcal{Y}_k. \tag{4.112}$$

Then, according to (4.14) in Lemma 4.7 and by (3.104), the multiplicity $m(\mathcal{Y}_k)$ of a forward focal point of this basis in $(k, k+1]$ is given by the comparative index

$$\mu^*(\mathcal{Z}_{k+1}^{-1}(0_{2n}\ I_{2n})^T, \mathcal{Z}_k^{-1}(0_{2n}\ I_{2n})^T) = \mu^*(\langle Z_{k+1}^{-1}\hat{Z}_{k+1}\rangle, \langle Z_k^{-1}\hat{Z}_k\rangle).$$

(ii) Applying Lemma 3.21(v) to the comparative index $\mu(\langle\hat{Z}_{k+1}^{-1}Z_{k+1}\rangle, \langle\hat{Z}_k^{-1}Z_k\rangle)$ in the first equation of (4.109), we have the representation

$$\mu(\langle\hat{Z}_{k+1}^{-1}Z_{k+1}\rangle, \langle\hat{Z}_k^{-1}Z_k\rangle)$$
$$= \mu^*(Z_{k+1}^{-1}\hat{Z}_{k+1}(0\ I)^T, Z_k^{-1}\hat{Z}_k(0\ I)^T) + \mu(\langle\hat{Z}_k^{-1}\hat{S}_k^{-1}S_k\hat{Z}_k\rangle, \langle I\rangle),$$
(4.113)

where the first addend is the multiplicity of a forward focal point of the conjoined basis $\tilde{Y}_k = \hat{Z}_k^{-1}Y_k$ of the symplectic system

$$\tilde{Y}_{k+1} = \hat{Z}_k^{-1}\hat{S}_k^{-1}S_k\hat{Z}_k\ \tilde{Y}_k \tag{4.114}$$

by (4.14) in Lemma 4.7. Recall that the upper block of $\tilde{Y}_k$ is associated with the Wronskian of $Y_k$ and $\hat{Y}_k$; see (4.100).

(iii) Similarly, the comparative index $\mu(\langle\hat{Z}_k^{-1}Z_k\rangle, \langle\hat{Z}_{k+1}^{-1}Z_{k+1}\rangle)$ in the second formula of (4.109) represents the multiplicity $m^*(\mathcal{Y}_k)$ of a backward focal point in the interval $[k, k+1)$ of the conjoined basis $\mathcal{Y}_k = S_0^{-1}\{Z_k^{-1}\hat{Z}_k\}(0_{2n}\ I_{2n})^T = \mathcal{Z}_k(0_{2n}\ I_{2n})^T$ of the symplectic system

$$\mathcal{Y}_{k+1} = S_0^{-1}\{Z_k^{-1}S_k^{-1}\hat{S}_k Z_k\}S_0\mathcal{Y}_k, \tag{4.115}$$

according to (4.16) in Lemma 4.8. The first addend in the formula

$$\mu(\langle\hat{Z}_k^{-1}Z_k\rangle, \langle\hat{Z}_{k+1}^{-1}Z_{k+1}\rangle)$$
$$= \mu(\hat{Z}_k^{-1}Z_k(0\ I)^T, \hat{Z}_{k+1}^{-1}Z_{k+1}(0\ I)^T) + \mu(\langle Z_k^{-1}S_k^{-1}\hat{S}_k Z_k\rangle, \langle I\rangle),$$
(4.116)

derived by Lemma 3.21(v) is the multiplicity of a backward focal point of the conjoined basis $\tilde{Y}_k = Z_k^{-1}\hat{Y}_k$ of the symplectic system

$$\tilde{Y}_{k+1} = Z_k^{-1}S_k^{-1}\hat{S}_k Z_k\tilde{Y}_k. \tag{4.117}$$

Using the results of Sects. 3.3 and 3.3.5, we can prove the following properties of the relative oscillation numbers.

**Lemma 4.47** *Let $Z$ and $\hat{Z}$ be symplectic fundamental matrices of* (SDS) *and* (4.98), *respectively. Then we have the following properties for their relative oscillation numbers:*

*(i) If $Y = Z(0\ I)^T$ and $\hat{Y}_i = \hat{Z}(0\ I)^T$, then*

$$\#_k(\hat{Z}, Z) + \#_k(Z, \hat{Z}) = \Delta\mathrm{rank}\, w(Y_k, \hat{Y}_k), \quad w(Y_k, \hat{Y}_k) = Y_k^T \mathcal{J} \hat{Y}_k,$$
(4.118)

*(ii) We have the estimate*

$$|\#_k(\hat{Z}, Z)| \leq \mathrm{rank}\,(\mathcal{S}_k - \hat{\mathcal{S}}_k).$$
(4.119)

*(iii) Let $Z$ and $\Phi$ be symplectic fundamental matrices of* (SDS), *and let $\hat{Z}$ and $\hat{\Phi}$ be symplectic fundamental matrices of* (4.98). *Then we have the formula*

$$\#_k(\hat{Z}, Z) - \#_k(\hat{\Phi}, \Phi) = \Delta\mu\big(\langle\hat{\Phi}_k^{-1}\Phi_k\rangle, \{W, V\}(0_{2n}, I_{2n})^T\big),$$
$$W = \Phi_k^{-1}Z_k, \quad V = \hat{\Phi}_k^{-1}\hat{Z}_k.$$
(4.120)

*Consequently*

$$\#(\hat{Z}, Z, M, N) - \#(\hat{\Phi}, \Phi, M, N)$$
$$= \mu\big(\langle\hat{\Phi}_k^{-1}\Phi_k\rangle, \{W, V\}(0_{2n}, I_{2n})^T\big)\Big|_M^{N+1},$$
(4.121)

*where $\{W, V\}$ is defined by* (3.101).
*(iv) If $W$ and $V$ are lower block-triangular matrices, then*

$$\#_k(\hat{Z}, Z, M, N) = \#_k(\Phi, \hat{\Phi}, M, N).$$

*Proof*

(i) Using (4.109) and Lemma 3.21 (iii), (iv) we obtain

$$\#_k(Z, \hat{Z}) = \mu(\langle Z_k^{-1}\hat{Z}_k\rangle, \langle Z_{k+1}^{-1}\hat{Z}_{k+1}\rangle) - \mu(\langle\mathcal{S}_k\rangle, \langle\hat{\mathcal{S}}_k\rangle)$$
$$= \mu(\langle\hat{\mathcal{S}}_k\rangle, \langle\mathcal{S}_k\rangle) - \mu(\langle Z_{k+1}^{-1}\hat{Z}_{k+1}\rangle, \langle Z_k^{-1}\hat{Z}_k\rangle)$$
$$= \mu(\langle\hat{\mathcal{S}}_k\rangle, \langle\mathcal{S}_k\rangle) - \mu^*(\langle\hat{Z}_{k+1}^{-1}Z_{k+1}\rangle, \langle\hat{Z}_k^{-1}Z_k\rangle)$$
$$= \mu(\langle\hat{\mathcal{S}}_k\rangle, \langle\mathcal{S}_k\rangle) - \mu(\langle\hat{Z}_k^{-1}Z_k\rangle, \langle\hat{Z}_{k+1}^{-1}Z_{k+1}\rangle) + \Delta\mathrm{rank}\, w(Y_k, \hat{Y}_k),$$

where in the last equality, we have also incorporated equation (4.100) when applying Lemma 3.21(iv). Consequently, $\#_k(Z, \hat{Z}) = -\#_k(\hat{Z}, Z) + \Delta\, \mathrm{rank}\, w(Y_k, \hat{Y}_k)$ holds, which is the same as (4.118).

(ii) The proof follows from Lemma 3.21 (ii) (see (3.53)) and equation (4.101), which implies

$$\mu(\langle \hat{Z}_k^{-1} Z_k \rangle, \langle \hat{Z}_{k+1}^{-1} Z_{k+1} \rangle) \le \operatorname{rank} \Delta(\hat{Z}_k^{-1} Z_k) = \operatorname{rank}(\mathcal{S}_k - \hat{\mathcal{S}}_k)$$

and

$$\mu(\langle \hat{\mathcal{S}}_k \rangle, \langle \mathcal{S}_k \rangle) \le \operatorname{rank}(\mathcal{S}_k - \hat{\mathcal{S}}_k).$$

(iii) Note that the matrix $W$ in (4.120) is constant for two fundamental matrices $Z$, $\Phi$ of the same system (SDS) (similarly for $V$ and $\hat{Z}$ and $\hat{\Phi}$), then $\Phi_l^{-1}\Phi_l = V\hat{Z}_l^{-1} Z_l W^{-1}$, $l = k$, $k+1$. Using Lemma 3.39 we obtain

$$\mu\left(\langle \hat{\Phi}_k^{-1} \Phi_k \rangle, \langle \hat{\Phi}_{k+1}^{-1} \Phi_{k+1} \rangle\right)$$
$$= \mu\left(\langle \hat{Z}_k^{-1} Z_k \rangle, \langle \hat{Z}_{k+1}^{-1} Z_{k+1} \rangle\right) - \Delta\mu\left(\langle \hat{\Phi}_k^{-1} \Phi_k \rangle, \{W, V\}(0_{2n}, \ I_{2n})^T\right).$$

Consequently,

$$\mu\left(\langle \hat{Z}_k^{-1} Z_k \rangle, \langle \hat{Z}_{k+1}^{-1} Z_{k+1} \rangle\right) - \mu\left(\langle \hat{\Phi}_k^{-1} \Phi_k \rangle, \langle \hat{\Phi}_{k+1}^{-1} \Phi_{k+1} \rangle\right)$$
$$= \Delta\mu\left(\langle \hat{\Phi}_k^{-1} \Phi_k \rangle, \{W, V\}(0_{2n}, \ I_{2n})^T\right).$$

Hence the last formula implies (4.120) and its summation from $k = M$ to $k = N$ yields (4.121).

(iv) The proof follows from the elementary fact that if $W$ and $V$ are symplectic lower block-triangular matrices, then $\{W, V\}(0 \ I)^T$ defined by (3.101) has upper block equal to zero, and hence, we have $\mu(\langle \hat{\Phi}_i^{-1} \Phi_i \rangle, \{W, V\}(0_{2n}, \ I_{2n})^T) = 0$.

□

Note that part (iv) of the previous lemma shows that the results based on Theorem 4.45 depend really only on the second block column of the considered fundamental matrices, while the first matrix column (which is not determined by the second matrix column uniquely) is disregarded.

*Remark 4.48* Note also that the operator $\mathcal{L}$ in (4.102), in view of properties (v) and (vi) of Theorem 3.5, can be expressed in various forms as

$$\mathcal{L}(Y_k, \hat{Y}_k, \mathcal{S}_k, \hat{\mathcal{S}}_k) = \hat{m}(k) - m(k) + \Delta \operatorname{rank} w(Y_k, \hat{Y}_k) - \Delta\mu(\hat{Y}_k, Y_k), \tag{4.122}$$

$$\mathcal{L}(Y_k, \hat{Y}_k, \mathcal{S}_k, \hat{\mathcal{S}}_k) = \hat{m}^*(k) - m^*(k) + \Delta\mu^*(\hat{Y}_k, Y_k), \tag{4.123}$$

$$\mathcal{L}(Y_k, \hat{Y}_k, W_k, \hat{W}_k) = \hat{m}^*(k) - m^*(k) + \Delta \operatorname{rank} w(Y_k, \hat{Y}_k) - \Delta\mu^*(Y_k, \hat{Y}_k). \tag{4.124}$$

Using property (4.118) of the numbers $\#_k(\hat{Z}, Z)$, we obtain from (4.122) a generalization of formula (4.38) for the case of conjoined bases of two different systems

$$\hat{m}(k) - m(k) - \Delta\mu(\hat{Y}_k, Y_k) = -\#_k(Z, \hat{Z}), \tag{4.125}$$

Analogously, from (4.123) it follows a generalization of (4.39):

$$\hat{m}^*(k) - m^*(k) + \Delta\mu^*(\hat{Y}_k, Y_k) = \#_k(\hat{Z}, Z). \tag{4.126}$$

Finally, from (4.124) it follows the generalization of (4.37)

$$\hat{m}^*(k) - m^*(k) - \Delta\mu^*(Y_k, \hat{Y}_k) = -\#_k(Z, \hat{Z}). \tag{4.127}$$

Now we formulate a consequence of Theorem 4.45. We use there the concept of the number of (forward) focal points of the conjoined bases of (SDS) in $(M, N + 1]$ defined by (4.10) and

$$l(\hat{Y}, M, N + 1) = \sum_{i=M}^{N} \hat{m}(i). \tag{4.128}$$

The next theorem extends formula (4.40) in Theorem 4.23 and Corollary 4.25 to the case when $Y$ and $\hat{Y}$ are conjoined bases of two different symplectic systems.

**Theorem 4.49** *Suppose that $Y$ and $\hat{Y}$ are conjoined bases of (SDS) and (4.98), respectively, their numbers of focal points in $(M, N + 1]$ are defined by (4.10), (4.128). Further suppose that the relative oscillation number in $[M, N]$ of symplectic fundamental matrices $Z$ and $\hat{Z}$ is given by (4.110) and these matrices are related to $Y$ and $\hat{Y}$ by the relations $Y = Z(0\ I)^T$ and $\hat{Y} = \hat{Z}(0\ I)^T$. Then*

$$l(Y, M, N + 1) - l(\hat{Y}, M, N + 1) + \#(\hat{Z}, Z, M, N)$$
$$= \mu(Y_{N|1}, \hat{Y}_{N|1}) - \mu(Y_M, \hat{Y}_M), \tag{4.129}$$

*in particular,*

$$\left| l(Y, M, N + 1) - l(\hat{Y}, M, N + 1) + \#(\hat{Z}, Z, M, N) \right|$$
$$\leq \min\{\hat{r}_{M,N+1}, \bar{r}_{M,N+1}\} \leq n, \tag{4.130}$$

*where $\hat{r}_{M,N+1}$ is given by (4.47),*

$$\bar{r}_{M,N+1} = \max\{\operatorname{rank} w(Y_M, \hat{Y}_M), \operatorname{rank} w(Y_{N+1}, \hat{Y}_{N+1})\},$$

*and $n$ is dimension of blocks of $S_k$ in (SDS).*

*Proof* The summation of (4.111) from $k = M$ to $k = N$ gives (4.129). The estimate (4.130) is derived by analogy with the proof of (4.46) in Corollary 4.25, where we cannot use that the Wronskian $w(Y, \hat{Y})$ is constant.                                          $\square$

The next theorems extend Corollaries 2.56 and 4.30 to the case of two symplectic systems (SDS) and (4.98). The first result is a discrete counterpart of Theorem 1.48 (Riccati inequality) and generalizes Theorem 2.54 to two discrete Hamiltonian systems and Theorem 2.55 to two symplectic systems.

**Theorem 4.50** *Suppose that conjoined bases* $Y = \begin{pmatrix} X \\ U \end{pmatrix}$ *and* $\hat{Y} = \begin{pmatrix} \hat{X} \\ \hat{U} \end{pmatrix}$ *of* (SDS) *and* (4.98) *satisfy* (4.58) *for* $k = M$, *the relative oscillation number* (4.109) *for symplectic fundamental matrices* $Z$ *and* $\hat{Z}$ *of these systems associated with* $Y$ *and* $\hat{Y}$ *such that* $Y = Z\,(0\ I)^T$ *and* $\hat{Y} = \hat{Z}\,(0\ I)^T$ *satisfies*

$$\#_k(\hat{Z}, Z) \leq 0, \quad k \in [M, N]_{\mathbb{Z}}, \tag{4.131}$$

*and that* $Y$ *has no forward focal points in* $(M, N + 1]$. *Then* $\hat{Y}$ *has no forward focal points in this interval either,*

$$\#(\hat{Z}, Z, M, N) = 0, \tag{4.132}$$

*and condition* (4.58) *holds for all* $k \in [M, N + 1]_{\mathbb{Z}}$.

*Proof* Rewrite formula (4.111) in the form

$$\hat{m}(k) + \mu(Y_{k+1}, \hat{Y}_{k+1}) + (-\#_k(\hat{Z}, Z)) = \mu(Y_k, \hat{Y}_k) + m(k).$$

Because all summands in the left-hand side are nonnegative, the fact that the right-hand side equals zero by assumptions implies that all three summands also equal zero for all indices $k \in [M, N]_{\mathbb{Z}}$.                              $\square$

*Remark 4.51*

(i) The comparative indices for the symplectic coefficient matrices of (SDS) and (4.98) in (4.109) are traditionally assumed to be zero in the classical oscillation theory. In this case we will call the conditions

$$\mu\big(\langle S_k \rangle, \langle \hat{S}_k \rangle\big) = 0 \tag{4.133}$$

and

$$\mu\big(\langle \hat{S}_k \rangle, \langle S_k \rangle\big) = 0 \tag{4.134}$$

as the (Sturmian) *majorant conditions* for the pair of symplectic systems (SDS) and (4.98). These conditions can be rewritten in terms of the blocks of $S_k$ and $\hat{S}_k$ according to the results in Sect. 3.3.2. For example, the majorant condition

(4.134) is equivalent to the conditions

$$\left.\begin{array}{r} \mathrm{Im}\,(A_k - \hat{A}_k, \quad B_k) \subseteq \mathrm{Im}\,\hat{B}_k, \\[2mm] \begin{pmatrix} I & A_k^T \\ 0 & B_k^T \end{pmatrix}(Q_{\langle S_k\rangle} - Q_{\langle \hat{S}_k\rangle})\begin{pmatrix} I & 0 \\ A_k & B_k \end{pmatrix}, \end{array}\right\} \tag{4.135}$$

where the symmetric matrix $Q_{\langle S_k\rangle}$ is defined by (3.71) with $W := S_k$ and $Q_{\langle \hat{S}_k\rangle}$ is defined similarly for system (4.98) (see Lemma 3.25). In particular, these matrices can be taken in form (3.74), which is equivalent to (2.64), and one can replace the second condition in (4.135) by the more strict majorant condition

$$\mathcal{G}_k - \hat{\mathcal{G}}_k \geq 0, \tag{4.136}$$

where $\mathcal{G}_k$ is defined by (2.64) and $\hat{\mathcal{G}}_k$ is defined similarly for system (4.98). Note that (4.136) is sufficient for the second condition in (4.135).

(ii) Consider linear Hamiltonian difference system (2.15) and another system of the same form with the Hamiltonian $\hat{\mathcal{H}} = \begin{pmatrix} -\hat{C} & \hat{A}^T \\ \hat{A} & \hat{B} \end{pmatrix}$. Then using Example 3.34 and formula (3.93), we see that (4.133) is equivalent to the conditions

$$\mathrm{Ker}\,B_k \subseteq \mathrm{Ker}\,\hat{B}_k, \quad \hat{B}_k \geq \hat{B}_k B_k^\dagger \hat{B}_k, \tag{4.137}$$

$$\left.\begin{array}{r} \mathrm{Im}(A_k - \hat{A}_k) \subseteq \mathrm{Im}(B_k - \hat{B}_k), \\[2mm] \hat{C}_k - C_k - (A_k - \hat{A}_k)^T (B_k - \hat{B}_k)^\dagger (A_k - \hat{A}_k) \geq 0. \end{array}\right\} \tag{4.138}$$

Then, (4.137) is equivalent to the first summand in (3.93) being zero, while (4.138) is equivalent to the fact that the second summand in (3.93) takes the value zero. Let us now examine the relationship of (4.137) and (4.138) with conditions (2.102) in Theorem 2.54, which read as follows:

$$\mathcal{H}_k - \hat{\mathcal{H}}_k \geq 0, \quad \mathrm{Ker}\,B_k \subseteq \mathrm{Ker}\,\hat{B}_k, \quad \hat{B}_k \geq \hat{B}_k B_k^\dagger \hat{B}_k. \tag{4.139}$$

The conditions in (4.139) represent a discrete counterpart of conditions (1.128) for linear Hamiltonian differential systems (1.99), i.e., of the conditions

$$\mathcal{H}(t) - \hat{\mathcal{H}}(t) \geq 0, \quad \hat{B}(t) \geq 0.$$

Note that by Corollary 3.16, the condition $\mathcal{H}_k - \hat{\mathcal{H}}_k \geq 0$ in (4.139) is equivalent to three conditions, namely, to (4.138) and the additional condition $B_k - \hat{B}_k \geq 0$. In this way, all three conditions in (4.139) are sufficient for (4.137) and (4.138) and, in turn, also for (4.133). The last fact confirms that Corollary 4.50 is a generalization of Theorem 2.54. Note, however, that (4.137) and (4.138) do not imply the inequality $B_k - \hat{B}_k \geq 0$. Moreover, if (4.137) and $B_k \geq \hat{B}_k$

simultaneously hold, then ind $B_k = $ ind $\hat{B}_k$, which follows from (3.92) (see also Remark 3.35). In particular, for a pair of $2n$-th order Sturm-Liouville difference equations (2.27), the leading coefficients $r_k^{[n]}$ and $\hat{r}_k^{[n]}$ must have in this case the same sign; see also Example 3.36. For the continuous time case, the conditions in (1.128) imply the Legendre condition (1.111) for system (1.103), and then in this case ind $\hat{B}(t) = $ ind $B(t) = 0$. In the discrete case, we need neither the Legendre condition ind $\hat{B}_k = $ ind $B_k = 0$ nor the more general assumption ind $\hat{B}_k = $ ind $B_k$. Hence, the sufficient conditions of Theorem 2.54 are slightly overdefined by the condition $B_k - \hat{B}_k \geq 0$, which is included in $\mathcal{H}_k - \hat{\mathcal{H}}_k \geq 0$. On the other hand, the last condition is convenient for applications of this theorem in some particular cases.

(iii) Concerning special cases of the majorant condition in (4.133) for the scalar and matrix Sturm-Liouville difference equations and for the $2n$-th order Sturm-Liouville difference equations, see, respectively, Examples 3.29, 3.30, and 3.36.

Note the majorant condition (4.133) represents a simple assumption, which implies (4.131). This follows from (4.109). In this case the results of Theorem 4.50 can be strengthened in the following direction.

**Corollary 4.52** *Suppose that all assumptions of Theorem 4.50 hold and (4.131) is replaced by majorant condition (4.133) for $k \in [M, N]_{\mathbb{Z}}$. Then all statements of Theorem 4.50 are true, and we have instead of (4.132) the equality*

$$\mu\big(\langle \hat{Z}_{k+1}^{-1} Z_{k+1} \rangle, \langle \hat{Z}_k^{-1} Z_k \rangle\big) = 0, \quad k \in [M, N]_{\mathbb{Z}}, \tag{4.140}$$

*i.e., the conjoined bases $\mathcal{Y}$ and $\tilde{Y}$ of systems (4.112) and (4.114) do not have any forward focal points in the interval $(M, N + 1]$.*

In a similar way, in order to present an extension of Corollary 4.30, we use formula (4.126), which connects the multiplicities of backward focal points of conjoined bases of (SDS) and (4.98).

**Theorem 4.53** *Suppose that for conjoined bases $Y$ and $\hat{Y}$ of systems (SDS) and (4.98) condition (4.59) holds at $k = N + 1$, $\hat{Y}$ has no backward focal points in the interval $[M, N+1)$, and the relative oscillation number for symplectic fundamental matrices $Z$ and $\hat{Z}$ associated with $Y$ and $\hat{Y}$ obeys the condition*

$$\#_k(\hat{Z}, Z) \geq 0, \quad k \in [M, N]_{\mathbb{Z}}. \tag{4.141}$$

*Then the conjoined basis $Y$ has no backward focal points in this interval either*

$$\#_k(\hat{Z}, Z, M, N) = 0, \tag{4.142}$$

*and (4.59) holds for all $k \in [M, N + 1]$.*

*Proof* We rewrite (4.126) in the form

$$\hat{m}^*(k) + \mu^*(\hat{Y}_{k+1}, Y_{k+1}) = \#_k(\hat{Z}, Z) + m^*(k) + \mu^*(\hat{Y}_k, Y_k).$$

Then taking into account the assumptions of the theorem, we have that the condition $\hat{m}^*(k) = \mu^*(\hat{Y}_{k+1}, Y_{k+1}) = 0$ implies $\#_k(\hat{Z}, Z) = m^*(k) = \mu^*(\hat{Y}_k, Y_k) = 0$. □

In view of formula (4.109), the majorant condition (4.134) is a simplest sufficient condition for (4.141).

**Corollary 4.54** *Suppose that instead of* (4.141) *the majorant condition* (4.134) *holds for* $k \in [M, N]_{\mathbb{Z}}$, *while all the other assumptions of Theorem 4.53 are satisfied. Then all statements of this theorem are true, and, additionally, we have that the conjoined bases* $\mathcal{Y}$ *and* $\tilde{Y}$ *of systems* (4.115) *and* (4.117) *do not have any backward focal points in the interval* $[M, N + 1)$.

### 4.3.2   Comparison Theorems for Principal Solutions

We have proved in Theorem 4.34 the equality for the numbers of forward and backward focal points of principal solutions at $k = M$ and $k = N + 1$, i.e.,

$$l(Y^{[M]}, M, N + 1) = l^*(Y^{[N+1]}, M, N + 1).$$

Now we will turn our attention to the situation, when the conjoined bases in the previous formula are solutions of different symplectic systems. For the proof we need the following auxiliary result on a representation of the relative oscillation numbers for the special case of the principal solutions.

**Lemma 4.55** *Let* $Z^{[M]}$ *and* $\hat{Z}^{[N+1]}$ *be fundamental symplectic matrices of* (SDS) *and* (4.98) *such that* $Y^{[M]} = Z^{[M]} (0 \ I)^T$ *and* $\hat{Y}^{[N+1]} = \hat{Z}^{[N+1]} (0 \ I)^T$ *are the principal solutions of these systems at* $k = M$ *and* $k = N + 1$, *respectively. We define the relative oscillation number of these matrices by formula* (4.109), *i.e.,*

$$\#_k(\hat{Z}^{[N+1]}, Z^{[M]}) := \mu(\langle \mathcal{G}_k \rangle, \langle \mathcal{G}_{k+1} \rangle) - \mu(\langle \hat{\mathcal{S}}_k \rangle, \langle \mathcal{S}_k \rangle), \quad k \in [M, N]_{\mathbb{Z}},$$
$$(4.143)$$

*where*

$$\mathcal{G}_k := (\hat{Z}_k^{[N+1]})^{-1} Z_k^{[M]}.$$

*Then, for the endpoints $k = M$ and $k = N$ of the interval, the formula in* (4.143) *is specified as*

$$\#_M(\hat{Z}^{[N+1]}, Z^{[M]}) = \mu^*(\hat{Y}_M^{[N+1]}, \hat{S}_M^{-1} S_M (0\ I)^T)$$
$$- \mu(\hat{S}_M(0\ I)^T, S_M(0\ I)^T), \quad (4.144)$$

$$\#_N(\hat{Z}^{[N+1]}, Z^{[M]}) = \mu(\hat{S}_N Y_N^{[M]}, \hat{S}_N S_N^{-1}(0\ I)^T)$$
$$- \mu^*(\hat{S}_N^{-1}(0\ I)^T, S_N^{-1}(0\ I)^T). \quad (4.145)$$

*Proof* The specification of the values $\#_k(\hat{Z}^{[N+1]}, Z^{[M]})$ for $k = M$ and $k = N$ follows from Lemma 3.21. In fact, using the condition $Z_M^{[M]}(0\ I)^T = (0\ I)^T$, we obtain that

$$\mu(\langle \mathcal{G}_M \rangle, \langle \mathcal{G}_{M+1} \rangle) = \mu(\mathcal{G}_M(0\ I)^T, \mathcal{G}_{M+1}(0\ I)^T) + \mu(\langle \mathcal{G}_{M+1}^{-1} \mathcal{G}_M \rangle, \langle I \rangle)$$
$$= \mu(\mathcal{G}_M(0\ I)^T, \mathcal{G}_{M+1}(0\ I)^T) + \mu(\langle S_M^{-1} \hat{S}_M \rangle, \langle I \rangle),$$

where we have used Lemma 3.21(v). Analogously

$$\mu(\langle \hat{S}_M \rangle, \langle S_M \rangle) = \mu(\hat{S}_M(0\ I)^T, S_M(0\ I)^T) + \mu(\langle S_M^{-1} \hat{S}_M \rangle, \langle I \rangle).$$

Consequently, by Theorem 3.5(iii),

$$\mu(\langle \mathcal{G}_M \rangle, \langle \mathcal{G}_{M+1} \rangle) - \mu(\langle \hat{S}_M \rangle, \langle S_M \rangle)$$
$$= \mu(\mathcal{G}_M(0\ I)^T, \mathcal{G}_{M+1}(0\ I)^T) - \mu(\hat{S}_M(0\ I)^T, S_M(0\ I)^T)$$
$$= \mu((\hat{Z}_M^{[N+1]})^{-1}(0\ I)^T, (\hat{Z}_{M+1}^{[N+1]})^{-1} S_M(0\ I)^T) - \mu(\hat{S}_M(0\ I)^T, S_M(0\ I)^T)$$
$$= \mu^*(\hat{Z}_M^{[N+1]}(0\ I)^T, \hat{S}_M^{-1} S_M(0\ I)^T) - \mu(\hat{S}_M(0\ I)^T, S_M(0\ I)^T),$$

which proves (4.144). Analogously, by Lemma 3.21(v), we have

$$\mu(\langle \mathcal{G}_N \rangle, \langle \mathcal{G}_{N+1} \rangle) = \mu^*(\mathcal{G}_N^{-1}(0\ I)^T, \mathcal{G}_{N+1}^{-1}(0\ I)^T) + \mu(\langle \mathcal{G}_N \mathcal{G}_{N+1}^{-1} \rangle, \langle I \rangle)$$
$$= \mu^*(\mathcal{G}_N^{-1}(0\ I)^T, \mathcal{G}_{N+1}^{-1}(0\ I)^T) + \mu(\langle \hat{S}_N S_N^{-1} \rangle, \langle I \rangle),$$

and similarly

$$\mu(\langle \hat{S}_N \rangle, \langle S_N \rangle) = \mu^*(\hat{S}_N^{-1}(0\ I)^T, S_N^{-1}(0\ I)^T) + \mu(\langle \hat{S}_N S_N^{-1} \rangle, \langle I \rangle).$$

Consequently, by Theorem 3.5(iii) we have proved that

$$
\begin{aligned}
\mu\big(\langle \mathcal{G}_N \rangle, \langle \mathcal{G}_{N+1} \rangle\big) &- \mu\big(\langle \hat{\mathcal{S}}_N \rangle, \langle \mathcal{S}_N \rangle\big) \\
&= \mu^*\big(\mathcal{G}_N^{-1}(0\ I)^T, \mathcal{G}_{N+1}^{-1}(0\ I)^T\big) - \mu^*\big(\hat{\mathcal{S}}_N^{-1}(0\ I)^T, \mathcal{S}_N^{-1}(0\ I)^T\big) \\
&= \mu^*\big((Z_N^{[M]})^{-1}\hat{\mathcal{S}}_N^{-1}(0\ I)^T, (Z_{N+1}^{[M]})^{-1}(0\ I)^T\big) - \mu^*\big(\hat{\mathcal{S}}_N^{-1}(0\ I)^T, \mathcal{S}_N^{-1}(0\ I)^T\big) \\
&= \mu\big(\hat{\mathcal{S}}_N Z_N^{[M]}(0\ I)^T, \hat{\mathcal{S}}_N \mathcal{S}_N^{-1}(0\ I)^T\big) - \mu^*\big(\hat{\mathcal{S}}_N^{-1}(0\ I)^T, \mathcal{S}_N^{-1}(0\ I)^T\big).
\end{aligned}
$$

The proof is complete. $\qquad\square$

Now we present a generalization of Theorems 4.34 and 4.35 to two symplectic systems.

**Theorem 4.56 (Sturmian Comparison Theorem)** *Consider the symplectic fundamental matrices* $Z^{[M]}$, $Z^{[N+1]}$ *and* $\hat{Z}^{[M]}$, $\hat{Z}^{[N+1]}$ *of systems* (SDS) *and* (4.98), *which are associated with the principal solutions* $Y^{[M]}$, $Y^{[N+1]}$ *and* $\hat{Y}^{[M]}$, $\hat{Y}^{[N+1]}$ *of these systems via* (3.14). *Then*

$$
\begin{aligned}
l^*(\hat{Y}^{[N+1]}, M, N+1) &- l(Y^{[M]}, M, N+1) \\
&= l(\hat{Y}^{[M]}, M, N+1) - l(Y^{[M]}, M, N+1) = \#(\hat{Z}^{[N+1]}, Z^{[M]}, M, N),
\end{aligned}
\tag{4.146}
$$

*and*

$$
\begin{aligned}
l(\hat{Y}^{[N+1]}, M, N+1) &- l^*(Y^{[M]}, M, N+1) \\
&= l^*(\hat{Y}^{[M]}, M, N+1) - l^*(Y^{[M]}, M, N+1) = \#(\hat{Z}^{[M]}, Z^{[N+1]}, M, N),
\end{aligned}
\tag{4.147}
$$

*where the quantities* $l^*(Y, M, N+1)$, $l(Y, M, N+1)$, *and* $\#(\hat{Z}, Z, M, N)$ *are given by formulas* (4.10), (4.11), *and* (4.109), (4.110). *Moreover, for the relative oscillation numbers in the right-hand sides of* (4.146), (4.147) *we have the relations*

$$
\#(\hat{Z}^{[N+1]}, Z^{[M]}, M, N) = -\#(Z^{[N+1]}, \hat{Z}^{[M]}, M, N),
\tag{4.148}
$$

$$
\#(\hat{Z}^{[M]}, Z^{[N+1]}, M, N) = -\#(Z^{[M]}, \hat{Z}^{[N+1]}, M, N).
\tag{4.149}
$$

*Proof* For the proof of (4.146), we use Theorem 4.49 for the particular case when $\hat{Y} := \hat{Y}^{[N+1]}$ is the principal solution of (4.98) at $k = N+1$ and $Y := Y^{[M]}$ is the principal solution of (SDS) at $k = M$. We rewrite (4.129) into the form

$$
\begin{aligned}
l(\hat{Y}^{[N+1]}, M, N+1) &- l(Y^{[M]}, M, N+1) - \operatorname{rank} \hat{X}_M^{[N+1]} \\
&= \#(\hat{Z}^{[N+1]}, Z^{[M]}, M, N),
\end{aligned}
\tag{4.150}
$$

where we have used that $\mu\big(Y_M^{[M]}, \hat{Y}_M^{[N+1]}\big) = \text{rank } \hat{X}_M^{[N+1]}$ and $\mu\big(Y_{N+1}^{[M]}, \hat{Y}_{N+1}^{[N+1]}\big) = 0$, according to the definition of the comparative index. We modify the summands in the left-hand side of (4.150) as

$$l(\hat{Y}^{[N+1]}, M, N+1) - \text{rank } \hat{X}_M^{[N+1]} = l^*(\hat{Y}^{[N+1]}, M, N+1) = l(\hat{Y}^{[M]}, M, N+1),$$

where we have used (4.12) for $\hat{Y}_k^{[N+1]}$ and (4.82) for $l^*(\hat{Y}^{[N+1]}, M, N+1)$ and $l(\hat{Y}^{[M]}, M, N+1)$. This proves formula (4.146).

In a similar way, by putting $\hat{Y} := \hat{Y}^{[M]}$ and $Y := Y^{[N+1]}$ in Theorem 4.49, we have (4.129) in the form

$$l(\hat{Y}^{[M]}, M, N+1) - l(Y^{[N+1]}, M, N+1) + \text{rank } \hat{X}_{N+1}^{[M]}$$
$$= \#(\hat{Z}^{[M]}, Z^{[N+1]}, M, N). \tag{4.151}$$

Applying (4.12) we have $l(\hat{Y}^{[M]}, M, N+1) + \text{rank } \hat{X}_{N+1}^{[M]} = l^*(\hat{Y}^{[M]}, M, N+1)$, and by (4.83) we get $l(Y^{[N+1]}, M, N+1) = l^*(Y^{[M]}, M, N+1)$. Substituting the last representations into (4.151), we complete the proof of (4.147). Relations (4.148) and (4.149) follow from interchanging the roles of $Y^{[M]}$ and $\hat{Y}^{[M]}$ in the proof of (4.146) and (4.147). Note that they can also be derived independently by using parts (i) and (iii) of Lemma 4.47.                                                                         □

The result in Theorem 4.56 plays a fundamental role in the relative oscillation theory of eigenvalue problems in Chap. 6. As a consequence of Theorem 4.56, we have the following corollary.

**Corollary 4.57** *The condition*

$$\#(\hat{Z}^{[N+1]}, Z^{[M]}, M, N) \geq 0 \tag{4.152}$$

*for the relative oscillation number at the right-hand side of* (4.146) *is necessary and sufficient for the inequality*

$$l(\hat{Y}^{[M]}, M, N+1) \geq l(Y^{[M]}, M, N+1) \tag{4.153}$$

*concerning the number of (forward) focal points of the principal solutions at $k = M$ of* (SDS) *and* (4.98), *respectively.*

*Remark 4.58* The above corollary implies the statements of Theorems 5.89 and 5.90 in the next chapter, which are there proved by using a relationship of the investigated eigenvalue problems to their quadratic energy functionals. Formula (4.153) together with (4.62) implies the statements of Theorems 5.89 and 5.90 as follows:

(i) The difference between the number of focal points of any conjoined basis of (SDS) and the number of focal points of the principal solution at $k = M$ of (4.98) in $(M, N + 1]$ is at most $n$, proven in [56, Theorem 1.2].

(ii) The number of focal points in $(M, N + 1]$ of any conjoined basis of (4.98) is greater or equal to the number of focal points of the principal solution of (SDS) at $k = M$, proven in [56, Theorem 1.3].

Note that statements (i) and (ii) are simple consequences of (4.153) and (4.60), resp. of (4.62). Indeed, using (4.60) and

$$l(Y^{[M]}, M, N + 1) = l(Y, M, N + 1) - \mu\left(Y_{N+1}, Y_{N+1}^{[M]}\right)$$

we obtain from (4.153) that

$$l(\hat{Y}^{[M]}, M, N + 1) - l(Y, M, N + 1) + \mu(Y_{N+1}, Y_{N+1}^{[M]}) \geq 0,$$

or

$$l(Y, M, N + 1) - l(\hat{Y}^{[M]}, M, N + 1) \leq \mu(Y_{N+1}, Y_{N+1}^{[M]}) \leq n.$$

The last equality implies the statement (i). As for (ii), from (4.60) it follows that $l(\hat{Y}, M, N + 1) \geq l(\hat{Y}^{[M]}, M, N + 1)$ and then inequality (4.153) proves that $l(\hat{Y}, M, N + 1) \geq l(Y^{[M]}, M, N + 1)$ for any conjoined basis $\hat{Y}$ of (4.98). Note also that according to (4.143) the majorant condition (4.134) is the simplest sufficient condition for (4.152).

### 4.3.3 Singular Comparison Theorems

Now we turn our attention to Theorem 4.45 and its corollaries for the singular case presenting a generalization of Theorem 4.39 for two symplectic systems (SDS) and (4.98) We introduce the following notation

$$\#(\hat{Z}, Z, M, +\infty) := \sum_{k=M}^{\infty} \#_k(\hat{Z}, Z) \tag{4.154}$$

for the infinite series of the relative oscillation numbers $\#_k(\hat{Z}, Z)$ given by (4.109) and introduce a similar notation for the case of $-\infty$, i.e.,

$$\#(\hat{Z}, Z, -\infty, N) = \sum_{k=-\infty}^{N} \#_k(\hat{Z}, Z). \tag{4.155}$$

The singular version of Theorem 4.39 is the following.

**Theorem 4.59 (Singular Sturmian Comparison Theorem)** *Assume that the majorant condition (4.133) holds for all $k \geq M_0$ for some $M_0 \in \mathbb{Z}$ and system* (SDS) *is nonoscillatory at $\infty$. Then system (4.98) is nonoscillatory at $\infty$ as well, for arbitrary symplectic fundamental matrices $Z$ and $\hat{Z}$ of* (SDS) *and (4.98) such that $Y = Z (0 \ I)^T$ and $\hat{Y} = \hat{Z} (0 \ I)^T$ and for arbitrary $M \in \mathbb{Z}$ there exist finite limits (4.154), (4.87), (4.91), (4.93), which are connected by the identities*

$$
\left.
\begin{aligned}
\#(\hat{Z}, Z, M, +\infty) &= l(\hat{Y}, M, +\infty) - l(Y, M, +\infty) \\
&\quad + \mu_{+\infty}(Y, \hat{Y}) - \mu(Y_M, \hat{Y}_M), \\
\#(Z, \hat{Z}, M, +\infty) &= l^*(Y, M, +\infty) - l^*(\hat{Y}, M, +\infty) \\
&\quad + \mu^*_{+\infty}(Y, \hat{Y}) - \mu^*(Y_M, \hat{Y}_M).
\end{aligned}
\right\}
\tag{4.156}
$$

*Proof* If system (SDS) is nonoscillatory at $\infty$, then applying Corollary 4.52 for the pair of the principal solutions $Y^{[M]}$ and $\hat{Y}^{[M]}$ of (SDS) and (4.98) at $k = M$ such that $l(Y^{[M]}, M, +\infty) = 0$, we see that $l(\hat{Y}^{[M]}, M, +\infty) = 0$, i.e., system (4.98) is nonoscillatory as well. By Lemma 4.38, there exist finite limits (4.87) for arbitrary conjoined bases $Y$ and $\hat{Y}$ of (SDS) and (4.98), and then, by identity (4.129) and by a similar identity

$$
l^*(Y, M, N + 1) - l^*(\hat{Y}, M, N + 1) + \mu^*(Y_{N+1}, \hat{Y}_{N+1}) - \mu^*(Y_M, \hat{Y}_M)
$$
$$
= \#(Z, \hat{Z}, M, N)
\tag{4.157}
$$

derived from (4.127), we see that the relative oscillation numbers are bounded, i.e., for some positive $C_1$ and $C_2$, we have

$$
|\#(\hat{Z}, Z, M, N)| \leq C_1, \quad |\#(Z, \hat{Z}, M, N)| \leq C_2, \quad i \in \{1, 2\}.
\tag{4.158}
$$

Moreover, the relative oscillation numbers are monotonic with respect to $N \geq M_0$, by the majorant condition (4.133), which holds for all $k \geq M_0$. Then the sum in (4.154) and the similar defined sum $\#(Z, \hat{Z}, M, +\infty)$ are finite, and, by taking the limits as $N \to +\infty$ in (4.129) and (4.157), we prove that there exist finite limits of the comparative index (4.91) and (4.93) for conjoined bases $Y$ and $\hat{Y}$ of (SDS) and (4.98). In this way we derive (4.156), which completes the proof.                        □

*Remark 4.60* Note that according to Remark 4.46 and Corollary 4.52, systems (4.112) and (4.114) are nonoscillatory at $+\infty$, i.e., condition (4.140) for the multiplicities of forward focal points of their conjoined bases holds for all sufficiently large $k$. Then, by Lemma 4.38 and by Theorem 4.59, the multiplicities of backward focal points of these conjoined bases also equal zero for all sufficiently large $k$, i.e.,

$$
\mu(\langle Z_k^{-1} \hat{Z}_k^{-1} \rangle, \langle Z_{k+1}^{-1} \hat{Z}_{k+1}^{-1} \rangle) = 0, \quad k \geq M_1.
$$

Moreover, by (4.118) in Lemma 4.47, we have the relation

$$\#(\hat{Z}, Z, M, +\infty) + \#(Z, \hat{Z}, M, +\infty) = \lim_{k \to \infty} w(Y_k, \hat{Y}_k) - w(Y_M, \hat{Y}_M) \quad (4.159)$$

between the finite limits in (4.156), where $\lim_{k \to \infty} w(Y_k, \hat{Y}_k)$ is the finite limit of the Wronskian. Note also that under the assumptions of Theorem 4.59 for arbitrary conjoined bases of systems (SDS) and (4.98), we have the following generalization of relations (4.95)–(4.97):

$$\mu_{+\infty}(Y, \hat{Y}) + \mu_{+\infty}(\hat{Y}, Y) = \lim_{k \to \infty} \text{rank } w(Y_k, \hat{Y}_k), \quad (4.160)$$

$$\mu^*_{+\infty}(Y, \hat{Y}) + \mu^*_{+\infty}(\hat{Y}, Y) = \lim_{k \to \infty} \text{rank } w(Y_k, \hat{Y}_k), \quad (4.161)$$

$$\mu_{+\infty}(Y, \hat{Y}) = \lim_{k \to \infty} \text{rank } \hat{X}_k - \lim_{k \to \infty} \text{rank } X_k + \mu^*_{+\infty}(\hat{Y}, Y). \quad (4.162)$$

Recall that all quantities $\mu(Y_k, \hat{Y}_k)$, $\mu^*(Y_k, \hat{Y}_k)$, rank $X_k$, and rank $w(Y_k, \hat{Y}_k)$ take their values from the set $\{0, 1, \ldots, n\}$ and then the existence of the limits of these quantities for $k \to \infty$ means that they are constant for sufficiently large $k$.

Obviously, Theorem 4.59 holds when $Y$ and $\hat{Y}$ are conjoined bases of the same system (SDS). In this case Theorem 4.59 turns into Theorem 4.39. For the case of the nonoscillation at $-\infty$, we have the following analog of Theorem 4.59.

**Theorem 4.61 (Singular Sturmian Comparison Theorem)** *Suppose that the majorant condition (4.134) holds for all $k \leq M_0$ for some $M_0 \in \mathbb{Z}$ and system (4.98) is nonoscillatory at $-\infty$. Then system (SDS) is nonoscillatory at $-\infty$ as well, for arbitrary symplectic fundamental matrices $Z$ and $\hat{Z}$ of (SDS) and (4.98) such that $Y = Z (0 \; I)^T$ and $\hat{Y} = \hat{Z} (0 \; I)^T$ and for arbitrary $N \in \mathbb{Z}$ there exist finite limits (4.155), (4.89), and*

$$\mu_{-\infty}(\hat{Y}, Y) := \lim_{k \to -\infty} \mu(\hat{Y}_k, Y_k), \quad \mu^*_{-\infty}(\hat{Y}, Y) := \lim_{k \to -\infty} \mu^*(\hat{Y}_k, Y_k),$$

*which are connected by the identities*

$$\left.\begin{aligned}
\#(Z, \hat{Z}, -\infty, N) &= l(Y, -\infty, N+1) - l(\hat{Y}, -\infty, N+1) \\
&\quad + \mu(\hat{Y}_{N+1}, Y_{N+1}) - \mu_{-\infty}(\hat{Y}, Y) \\
\#(\hat{Z}, Z, -\infty, N) &= l^*(\hat{Y}, -\infty, N+1) - l^*(Y, -\infty, N+1) \\
&\quad + \mu^*(\hat{Y}_{N+1}, Y_{N+1}) - \mu^*_{-\infty}(\hat{Y}, Y).
\end{aligned}\right\} \quad (4.163)$$

*Proof* Assuming that system (4.98) is nonoscillatory at $-\infty$, we see that system (SDS) is nonoscillatory at $-\infty$ as well by Corollary 4.54. Then applying Lemma 4.38 to the case $-\infty$, we obtain that the limits (4.89) exist for arbitrary conjoined bases $Y$ and $\hat{Y}$ of (SDS) and (4.98). By (4.125) and (4.126), the relative oscillation numbers are then bounded, i.e., (4.158) holds as $M \to -\infty$. Using majorant condition (4.134) for all sufficiently negative $k$, we see that these numbers are also monotonic with respect to $M \leq M_0$. Then there exist finite limits (4.155) and #$(Z, \hat{Z}, -\infty, N)$, and we derive (4.163) from (4.125) and (4.126) by analogy with the proof of Theorem 4.59.                                                                 □

## 4.4   Focal Points and Symplectic Transformations

In Sect. 2.6 we presented basic elements of the theory of symplectic transformations of symplectic difference system. Recall that if $\mathcal{R}_k$ are symplectic $2n \times 2n$ matrices, then transformation (2.138), i.e., $y_k = \mathcal{R}_k w_k$, transforms (SDS) into another symplectic system (2.139), i.e., $w_{k+1} = \tilde{\mathcal{S}}_k w_k$. We rewrite the latter system in the matrix form as

$$\tilde{Y}_{k+1} = \tilde{\mathcal{S}}_k \tilde{Y}_k, \quad \tilde{\mathcal{S}}_k = \mathcal{R}_{k+1}^{-1} \mathcal{S}_k \mathcal{R}_k \tag{4.164}$$

for the transformed conjoined basis $\tilde{Y}$ given by

$$Y_k = \mathcal{R}_k \tilde{Y}_k. \tag{4.165}$$

An important question is what are the invariants of transformation (4.165). In particular, what additional assumptions on $\mathcal{R}_k$ imply that (4.165) preserves oscillatory nature of the transformed systems (e.g., the number of focal points, definiteness of the associated quadratic functionals, etc.).

### 4.4.1   Focal Points and General Symplectic Transformation

We start with the formula relating the number of focal points of a conjoined basis $Y$ of (SDS) and the conjoined basis $\tilde{Y} := \mathcal{R}^{-1} Y$ of (4.164).

**Theorem 4.62** *Suppose that the conjoined bases $Y$ and $\tilde{Y}$ of (SDS) and of (4.164), respectively, are related by (4.165). Then we have*

$$m(\tilde{Y}_k) - m(Y_k) - \Delta\mu\big(\tilde{Y}_k, \mathcal{R}_k^{-1}(0 \ I)^T\big) = u_k, \tag{4.166}$$

*where*

$$u_k = \mu\big(\mathcal{R}_{k+1}^{-1}(0 \ I)^T, \tilde{\mathcal{S}}_k(0 \ I)^T\big) - \mu^*\big(\mathcal{R}_k(0 \ I)^T, \mathcal{S}_k^{-1}(0 \ I)^T\big) \tag{4.167}$$

$$= \mu^*\big(\mathcal{S}_k^{-1}(0 \ I)^T, \mathcal{R}_k(0 \ I)^T\big) - \mu\big(\tilde{\mathcal{S}}_k(0 \ I)^T, \mathcal{R}_{k+1}^{-1}(0 \ I)^T\big), \tag{4.168}$$

and where $m(Y_k)$ and $m(\tilde{Y}_k)$ are the numbers of focal points of the indicated conjoined bases in $(k, k+1]$.

*Proof* Let $Z$ and $\tilde{Z} = \mathcal{R}^{-1} Z$ be symplectic fundamental matrices of (SDS) and (4.164), respectively, such that $Y = Z (0 \ I)^T$ and $\tilde{Y} = \tilde{Z} (0 \ I)^T$. We introduce the symplectic difference system for the transformation matrix $\mathcal{R}_k$ as

$$\mathcal{R}_{k+1} = \hat{\mathcal{S}}_k \mathcal{R}_k, \quad \hat{\mathcal{S}}_k := \mathcal{R}_{k+1} \mathcal{R}_k^{-1}. \tag{4.169}$$

Now we apply Theorem 4.45 to systems (SDS) and (4.169) by putting $\hat{Z}_k := \mathcal{R}_k$ and $\hat{Y} := \mathcal{R}_k (0 \ I)^T$ and then derive the formula

$$m(\hat{Y}_k) - m(Y_k) + \Delta\mu(Y_k, \mathcal{R}_k(0 \ I)^T) = \#(\mathcal{R}_k, Z_k), \tag{4.170}$$

where

$$\#(\mathcal{R}_k, Z_k) = \mu(\langle \hat{\mathcal{S}}_k \rangle, \langle \mathcal{R}_{k+1} \mathcal{R}_k^{-1} \rangle) - \mu(\langle \tilde{Z}_{k+1} \rangle, \langle \tilde{Z}_k \rangle).$$

Evaluating the relative oscillation number $\#(\mathcal{R}_k, Z_k)$ according to Remark 4.46(ii), we have by (4.113) the formula

$$\#(\mathcal{R}_k, Z_k) = \mu(\langle \hat{\mathcal{S}}_k \rangle, \langle \mathcal{R}_{k+1} \mathcal{R}_k^{-1} \rangle) - \mu(\langle \tilde{\mathcal{S}}_k \rangle, \langle I \rangle) - m(\tilde{Y}_k).$$

Applying Corollary 3.40(iii) to the difference $\mu(\langle \hat{\mathcal{S}}_k \rangle, \langle \mathcal{R}_{k+1} \mathcal{R}_k^{-1} \rangle) - \mu(\langle \tilde{\mathcal{S}}_k \rangle, \langle I \rangle)$ on the right-hand side of the last identity, we derive the representation

$$\#(\mathcal{R}_k, Z_k) = u_k + m^*(\hat{Y}_k) - m(\tilde{Y}_k), \tag{4.171}$$

where $u_k$ is given by (4.167), (4.168) and $m^*(\hat{Y}_k) = \mu(\mathcal{R}_k^{-1}(0 \ I)^T, \mathcal{R}_{k+1}^{-1}(0 \ I)^T)$ is the multiplicity of a backward focal point of $\hat{Y}_k = \mathcal{R}_k (0 \ I)^T$ in $[k, k+1)$. Substituting (4.171) into (4.170) and using the connection $m^*(\hat{Y}_k) - m(\hat{Y}_k) = \Delta\mu((0 \ I)^T, \mathcal{R}_k (0 \ I)^T)$ between the multiplicities of backward and forward focal points, we derive (4.166) in the form

$$m(\tilde{Y}_k) - m(Y_k) + \Delta\mu(Y_k, \mathcal{R}_k (0 \ I)^T) = u_k + \Delta\mu((0 \ I)^T, \mathcal{R}_k(0 \ I)^T). \tag{4.172}$$

The final representation (4.166) now follows from the formula

$$\mu(Y_k, \mathcal{R}_k (0 \ I)^T) = \mu((0 \ I)^T, \mathcal{R}_k (0 \ I)^T) - \mu(\tilde{Y}_k, \mathcal{R}_k^{-1}(0 \ I)^T), \tag{4.173}$$

which is derived according to Theorem 3.5(ix). The proof is complete.  □

Note that we can interchange systems (SDS) and (4.164). Then $\mathcal{R}_k$ and $\mathcal{R}_k^{-1}$ also change their role, and as a result of this approach, we obtain another formulas expressing the difference $\tilde{m}(k) - m(k)$. The results are summarized in the next corollary.

**Corollary 4.63** *Suppose that the assumptions of Theorem 4.62 hold. Then*

$$m(\tilde{Y}_k) - m(Y_k) + \Delta\mu\big(Y_k, \mathcal{R}_k\, (0\ I)^T\big) = \tilde{u}_k, \qquad (4.174)$$

*where*

$$\tilde{u}_k = \mu^*\big(\mathcal{R}_k^{-1}(0\ I)^T_k,\ \tilde{\mathcal{S}}_k^{-1}(0\ I)^T\big) - \mu\big(\mathcal{R}_{k+1}(0\ I)^T_k,\ \mathcal{S}_k(0\ I)^T\big) \qquad (4.175)$$

$$= \mu\big(\mathcal{S}_k\,(0\ I)^T_k,\ \mathcal{R}_{k+1}\,(0\ I)^T\big) - \mu^*\big(\tilde{\mathcal{S}}_k^{-1}(0\ I)^T_k,\ \mathcal{R}_k^{-1}(0\ I)^T\big) \qquad (4.176)$$

*and where the sequences $\tilde{u}_k$ and $u_k$ given by (4.167) are connected by the formula*

$$\tilde{u}_k - u_k = \Delta\big(\operatorname{rank}(I\ 0)\,\mathcal{R}_k\,(0\ I)^T\big). \qquad (4.177)$$

*Proof* As it was mentioned above, we derive (4.174) and (4.175) just by interchanging the roles of (SDS) and (4.164) in (4.166). Formula (4.177) follows from (4.172) and (4.173), where we evaluate the comparative index $\mu\big((0\ I)^T_k,\ \mathcal{R}_k\,(0\ I)^T\big)$ according to Remark 3.4(iii).                                                                     □

Summation of formulas (4.166) and (4.174) from $k = M$ to $k = N$ gives the fundamental result relating the number of (forward) focal points in $(M, N + 1]$ by a general symplectic transformation which is formulated in the next theorem.

**Theorem 4.64** *The numbers $l(Y, M, N + 1)$ and $l(\tilde{Y}, M, N + 1)$ of forward focal points of $Y$ and $\tilde{Y}$ related by (4.165) in $(M, N + 1]$ satisfy the formulas*

$$\left.\begin{aligned} &l(\tilde{Y}, M, N + 1) - l(Y, M, N + 1) \\ &\qquad - \mu\big(\tilde{Y}_k, \mathcal{R}_k^{-1}(0\ I)^T\big)\Big|_M^{N+1} = S(M, N), \\ &S(M, N) = \tilde{S}(M, N) - \operatorname{rank}\big((I\ 0)\,\mathcal{R}_k\,(0\ I)^T\big)\Big|_M^{N+1}, \\ &S(M, N) := \sum_{k=M}^{N} u_k, \quad \tilde{S}(M, N) := \sum_{k=M}^{N} \tilde{u}_k, \end{aligned}\right\} \qquad (4.178)$$

*where $u_k$ and $\tilde{u}_k$ are the sequences given in (4.167) and (4.175).*

*Remark 4.65*

(i) Note that for the partial sums $S(M, N)$ and $\tilde{S}(M, N)$ in (4.178), we have by (4.177) the estimate

$$|S(M, N) - \tilde{S}(M, N)|$$
$$\leq \max\big\{\operatorname{rank}(I\ 0)^T\mathcal{R}_{N+1}\,(0\ I)^T_k,\ \operatorname{rank}(I\ 0)^T\mathcal{R}_M\,(0\ I)^T\big\} \leq n. \qquad (4.179)$$

In particular, $S(M, N) = \tilde{S}(M, N)$ for the case when the transformation matrix $\mathcal{R}_k$ is constant, i.e., $\mathcal{R}_k \equiv \mathcal{R}$. It follows from (4.179) that either the partial sums $S(M, N)$ and $\tilde{S}(M, N)$ are simultaneously bounded for a fixed $M \in \mathbb{Z}$ as $N \to \infty$, i.e., the inequalities

$$|S(M, N)| \leq C(M), \quad |\tilde{S}(M, N)| \leq \tilde{C}(M), \quad N \geq M, \tag{4.180}$$

hold for some positive constants $C(M)$ and $\tilde{C}(M)$, or these sums are simultaneously unbounded.

(ii) The left-hand side of (4.166) can be written in the equivalent form

$$m(\tilde{Y}_k) - m(Y_k) - \Delta\mu\big(\tilde{Y}_k, \mathcal{R}_k^{-1}\,(0\ I)^T\big)$$
$$= m^*(\tilde{Y}_k) - m^*(Y_k) - \Delta\mu^*\big(Y_k, \mathcal{R}_k\,(0\ I)^T\big), \tag{4.181}$$

where $m^*(Y_k)$ and $m^*(\tilde{Y}_k)$ are the multiplicities of backward focal points of $Y$ and $\tilde{Y}$ in $[k, k + 1)$. The proof of (4.181) is based on Proposition 4.4(vi) and formulas (4.173), (3.34). Identity (4.181) implies that the numbers of focal points $l^*(Y, M, N + 1)$ and $l^*(\tilde{Y}, M, N + 1)$ can be expressed by a formula similar to (4.178), i.e., by

$$l^*(\tilde{Y}, M, N + 1) - l^*(Y, M, N + 1) - \mu^*\big(Y_k, \mathcal{R}_k\,(0\ I)^T\big)\Big|_M^{N+1} = S(M, N). \tag{4.182}$$

### 4.4.2   Focal Points and Special Symplectic Transformations

We consider now formula (4.166) in the particular case when the transformation matrix is $\mathcal{R}_k = \mathcal{J}$.

**Corollary 4.66** *When $\mathcal{R}_k \equiv \mathcal{J}$, formulas (4.166) and (4.167) read as*

$$\left.\begin{aligned} m(\mathcal{J}^T Y_k) - m(Y_k) - \Delta\mu\big(\mathcal{J}^T Y_k, \mathcal{J}^T(0\ I)^T\big) \\ = \operatorname{ind}(-A_k^T C_k) - \operatorname{ind}(A_k B_k^T), \\ \mu\big(\mathcal{J}^T Y_k, \mathcal{J}^T(0\ I)^T\big) = \operatorname{rank}(I - U_k U_k^\dagger) + \operatorname{ind}(X_k^T U_k), \end{aligned}\right\} \tag{4.183}$$

*and formulas (4.174) and (4.175) can be written as*

$$\left.\begin{aligned} m(\mathcal{J}^T Y_k) - m(k) + \Delta\mu\big(Y_k, \mathcal{J}(0\ I)^T\big) = \operatorname{ind}(-C_k^T D_k) - \operatorname{ind}(B_k^T D_k), \\ \mu\big(Y_k, \mathcal{J}(0\ I)^T\big) = \operatorname{rank}(I - X_k X_k^\dagger) + \operatorname{ind}(-X_k^T U_k). \end{aligned}\right\} \tag{4.184}$$

*and*

$$u_k = \text{ind}\,(-\mathcal{A}_k^T \mathcal{C}_k) - \text{ind}\,(\mathcal{A}_k \mathcal{B}_k^T) = \text{ind}\,(-\mathcal{C}_k^T \mathcal{D}_k) - \text{ind}\,(\mathcal{B}_k^T \mathcal{D}_k) = \tilde{u}_k. \quad (4.185)$$

*Proof* In the particular case $\mathcal{R}_k = \mathcal{J}$, we obtain from (4.167) the formulas

$$\mu\big(\mathcal{R}_{k+1}^{-1}(0\ I)_k^T\,\tilde{S}_k(0\ I)^T\big) = \mu\big((-I\ 0)^T,\ \mathcal{J}^T S_k \mathcal{J}(0\ I)^T\big)$$

$$= \mu_2\big((-I\ 0)^T,\ (-\mathcal{C}_k^T,\ \mathcal{A}_k^T)^T\big) = \text{ind}\,(-\mathcal{A}_k^T \mathcal{C}_k),$$

$$\mu^*\big(\mathcal{R}_k(0\ I)_k^T\,S_k^{-1}(0\ I)^T\big) = \mu^*\big((I\ 0)^T,\ (-\mathcal{B}_k,\ \mathcal{A}_k)^T\big)$$

$$= \mu_2^*\big((I\ 0)^T,\ (-\mathcal{B}_k,\ \mathcal{A}_k)^T\big) = \text{ind}\,(\mathcal{A}_k \mathcal{B}_k^T).$$

Further, (4.175) implies

$$\mu\big(\mathcal{R}_{k+1}(0\ I)_k^T\,S_k(0\ I)^T\big) = \mu\big((I\ 0)^T,\ S_k(0\ I)^T\big)$$

$$= \mu_2\big((I\ 0)^T,\ (\mathcal{B}_k^T,\ \mathcal{D}_k^T)^T\big) = \text{ind}\,(\mathcal{B}_k^T \mathcal{D}_k),$$

$$\mu^*\big(\mathcal{R}_k^{-1}(0\ I)_k^T\,\tilde{S}_k^{-1}(0\ I)^T\big) = \mu^*\big((-I\ 0)^T,\ (\mathcal{C}_k,\ \mathcal{D}_k)^T\big) = \text{ind}\,(-\mathcal{C}_k \mathcal{D}_k^T),$$

*and*

$$\mu\big(\mathcal{R}_k^{-1} Y_k,\ \mathcal{R}_k^{-1}(0\ I)^T\big) = \mu\big(\mathcal{J}^T Y_k,\ (-I\ 0)^T\big) = \text{rank}\,(I - U_k U_k^\dagger) + \text{ind}\,(X_k^T U_k),$$

$$\mu\big(Y_k,\ \mathcal{R}_k(0\ I)^T\big) = \mu\big(Y_k,\ (0\ I)^T\big) = \text{rank}\,(I - X_k X_k^\dagger) + \text{ind}\,(-X_k^T U_k),$$

where in all previous computations we have used Remark 3.4(iv) and the definitions of the comparative index and the dual comparative index. Relation (4.185) follows from formula (4.177) for the case when the transformation matrix does not depend on $k$.                                                                         □

Formula (4.166) also implies that multiplicity of a focal point is preserved under transformation with lower block-triangular symplectic matrix, as it is formulated in the next corollary. This is the problem that we have already mentioned in Sect. 2.6.

**Corollary 4.67** *Let* $\mathcal{R}_k = L_k$, *where* $L_k$ *is a symplectic lower block-triangular matrix. Then for any conjoined basis* $Y$ *of* (SDS), *we have* $m(Y_k) = m(L_k^{-1} Y_k)$.

*Proof* In our particular case $\Delta\mu\big(\tilde{Y}_k,\ L_k^{-1}(0\ I)^T\big) = 0$, consequently, the left-hand side of (4.166) takes the form $m(\tilde{Y}_k) - m(Y_k)$. Further, using definition of $u_k$ in Theorem 4.62 by (4.168), we obtain

$$\mu^*\big(S_k^{-1}(0\ I)_k^T\,L_k(0\ I)^T\big) = \mu\big(\tilde{S}_k(0\ I)_k^T\,L_{k+1}^{-1}(0\ I)^T\big) = 0,$$

which follows directly from Remark 3.4(ii). Consequently, the right-hand side of (4.166) equals zero.                                                                         □

### 4.4.3   Generalized Reciprocity Principle

Recall from Definition 4.37 that symplectic system (SDS) is nonoscillatory (at $\infty$) if there exists $M \in \mathbb{N}$ such that the principal solution $Y^{[M]}$ of (SDS) at $k = M$ has no forward focal point in $(M, \infty)$. In the opposite case, (SDS) is oscillatory (at $\infty$). The inequality

$$\left| l(Y, M, N + 1) - l(Y^{[M]}, M, N + 1) \right| \leq n, \quad N \geq M,$$

from Corollary 4.25 implies that one can take any conjoined basis $Y$ in the definition of (non)oscillation of (SDS) at $\infty$ instead of the principal solution $Y^{[M]}$. In the following result, we present the most general statement in this context. It is a consequence of Theorem 4.64.

**Theorem 4.68 (Generalized Reciprocity Principle for Symplectic Systems)** *Let us define the sequences* $S(M, N)$ *and* $\tilde{S}(M, N)$ *by* (4.178).

(i) *Assume that at least one of the sequences* $S(M, N)$ *or* $\tilde{S}(M, N)$ *is bounded as* $N \to \infty$, *i.e., there exist constants* $C(M)$ *or* $\tilde{C}(M)$ *such that*

$$|S(M, N)| \leq C(M) \quad or \quad |\tilde{S}(M, N)| \leq \tilde{C}(M), \quad N \geq M. \qquad (4.186)$$

*Then systems* (SDS) *and* (4.164) *have the same oscillatory nature at* $\infty$, *i.e., they are oscillatory or nonoscillatory at* $\infty$ *at the same time.*

(ii) *If* (SDS) *and* (4.164) *are simultaneously nonoscillatory at* $\infty$, *then the sequences* $S(M, N)$ *and* $\tilde{S}(M, N)$ *are bounded as* $N \to \infty$.

(iii) *If at least one of the sequences* $S(M, N)$ *and* $\tilde{S}(M, N)$ *is unbounded, then at least one of systems* (SDS) *and* (4.164) *is oscillatory at* $\infty$.

*Proof* Recall that the sequences $S(M, N)$ and $\tilde{S}(M, N)$ are both bounded or both unbounded (see Remark 4.65).

(i) Condition (4.186), formula (4.178), and property (vii) of Theorem 3.5 imply that

$$-C(M) \leq l(\tilde{Y}, M, N + 1) - l(Y, M, N + 1) - \mu(\tilde{Y}_k, \mathcal{R}_k^{-1}) \big|_M^{N+1}$$
$$= S(M, N) \leq C(M),$$

$$-C(M) - n \leq -C(M) - \mu(\tilde{Y}_M, \mathcal{R}_M^{-1}) \leq l(\tilde{Y}, M, N + 1) - l(Y, M, N + 1)$$
$$\leq C(M) + \mu(\tilde{Y}_{N+1}, \mathcal{R}_{N+1}^{-1}) \leq C(M) + n.$$

Consequently,

$$\left| l(\tilde{Y}, M, N + 1) - l(Y, M, N + 1) \right| \leq C(M) + n \quad \text{for all } N \geq M. \qquad (4.187)$$

Obviously, if we replace the integer $M$ by $M_1$, the last estimate remains to hold with some other constant $C(M_1)$. Suppose that (SDS) is nonoscillatory at $\infty$, i.e., for any conjoined basis $Y$, there exists $M_1$ (depending on the conjoined basis $Y$) such that $l(Y, M_1, N) = 0$ for every $N > M_1$. Then from (4.187) we obtain that $l(\mathcal{R}^{-1}Y, M, N + 1)$ is bounded as $N \to \infty$. Since $l(\mathcal{R}^{-1}Y, M_1, N + 1)$ is the partial sum of a series formed by integers or zeros, then its boundedness is possible only if $l(Y, M_2, N+1) = 0$ for some $M_2 > M_1$ and every $N \geq M_2$. Hence, (4.164) is nonoscillatory at $\infty$ as well. Quite similarly we prove that the nonoscillation of (4.164) at $\infty$ implies the nonoscillation of (SDS) at $\infty$.

(ii) If both systems (SDS) and (4.164) are nonoscillatory at $\infty$, then there exists $M_1$ (sufficiently large) such that $l(Y, M_1, N+1) = l(\tilde{Y}, M_1, N+1) = 0$. Then by (4.178) we have $|S(M_1, N)| \leq n$, because of property (vii) of Theorem 3.5. Hence, by (4.179) the sequence $\tilde{S}(M_1, N)$ is also bounded.

(iii) This statement follows immediately from part (ii).

$\qquad\qquad\qquad\qquad\qquad\qquad\qquad\qquad\qquad\qquad\qquad\qquad\qquad\qquad\qquad\quad$ $\square$

The result in Theorem 4.68 implies that only the case of unboundedness of sequences $S(M, N)$ and $\tilde{S}(M, N)$ (case (iii) in the previous theorem) and the oscillation of one of systems (SDS) and (4.164) at $\infty$ needs an additional investigation to answer the question about the (non)oscillation of the other system at $\infty$. In all the remaining cases and under the additional assumption on the oscillatory nature of one of systems (SDS) or (4.164) at $\infty$, Theorem 4.68 provides the answer about the (non)oscillation of the other one at $\infty$.

A simple sufficient condition for the boundedness of $S(M, N)$ is given in the next theorem.

**Theorem 4.69** *Systems* (SDS) *and* (4.164) *oscillate or do not oscillate simultaneously at* $\infty$, *if at least one of the sequences* $u_k$ *or* $\tilde{u}_k$ *given by* (4.167), (4.168), *and* (4.175), (4.176) *tends to zero as* $k \to \infty$, *i.e., there exists* $M > 0$ *such that for all* $k \geq M$ *we have*

$$u_k = 0 \quad \Leftrightarrow \quad \mu\big(\mathcal{R}_{k+1}^{-1}(0\ I)^T\!,\ \tilde{S}_k(0\ I)^T\big) = \mu^*\big(\mathcal{R}_k(0\ I)^T\!,\ S_k^{-1}(0\ I)^T\big)$$

$$(4.188)$$

*or*

$$\tilde{u}_k = 0 \quad \Leftrightarrow \quad \mu^*(\mathcal{R}_k^{-1}(0\ I)^T\!,\ \tilde{S}_k^{-1}(0\ I)^T) = \mu(\mathcal{R}_{k+1}(0\ I)^T\!,\ S_k(0\ I)^T).$$

$$(4.189)$$

*Proof* Under assumption (4.188) we have $S(M, N) = 0$ for all $M \geq N$, and then the first statement follows directly from Theorem 4.68(i). Similarly, (4.189) implies $\tilde{S}(M, N) = 0$ for all $M \geq N$, and then, again by Theorem 4.68(i) both systems oscillate or do not oscillate simultaneously at $\infty$. $\qquad\qquad\qquad\qquad\qquad\qquad\quad$ $\square$

In particular, for $\mathcal{R}_k = \mathcal{J}^T$, we have the following corollary to Theorem 4.69.

**Corollary 4.70** *Systems* (SDS) *and* (4.164) *oscillate or do not oscillate simultaneously at $\infty$ if there exists $M > 0$ such that*

$$\text{ind}\,(-\mathcal{A}_k^T \mathcal{C}_k) = \text{ind}\,(\mathcal{A}_k \mathcal{B}_k^T), \quad k \geq M, \tag{4.190}$$

*and* (4.190) *is equivalent to*

$$\text{ind}\,(-\mathcal{C}_k \mathcal{D}_k^T) = \text{ind}\,(\mathcal{B}_k^T \mathcal{D}_k), \quad k \geq M. \tag{4.191}$$

*Remark 4.71*

(i) Note that for the case when $\text{rank}\,[(I\ 0)\,\mathcal{R}_k(0\ I)^T]$ is constant for $k \geq M$, conditions (4.188) and (4.189) are equivalent according to (4.177). In particular, $\text{rank}\,[(I\ 0)\,\mathcal{R}_k(0\ I)^T] = n$ for the case $\mathcal{R}_k = \mathcal{J}^T$ (see Corollary 4.70).

(ii) Conditions (4.188), (4.189) will be satisfied if we assume for all $k \geq M$

$$\mu(\mathcal{R}_{k+1}^{-1}(0\ I)^T, \tilde{\mathcal{S}}_k(0\ I)^T) = \mu^*(\mathcal{R}_k(0\ I)^T, \mathcal{S}_k^{-1}(0\ I)^T) = 0, \tag{4.192}$$

or

$$\mu^*(\mathcal{R}_k^{-1}(0\ I)^T, \tilde{\mathcal{S}}_k^{-1}(0\ I)^T) = \mu(\mathcal{R}_{k+1}(0\ I)^T, \mathcal{S}_k(0\ I)^T) = 0. \tag{4.193}$$

In particular, for the case $\mathcal{R}_k = \mathcal{J}^T$ by (4.190), (4.191), we have that (4.192) are equivalent to the conditions

$$\mathcal{A}_k^T \mathcal{C}_k \leq 0, \quad \mathcal{A}_k \mathcal{B}_k^T \geq 0, \quad k \geq M, \tag{4.194}$$

while (4.193) implies

$$\mathcal{C}_k \mathcal{D}_k^T \leq 0, \quad \mathcal{B}_k^T \mathcal{D}_k \geq 0. \tag{4.195}$$

Note that in some special cases, to verify (4.193) (or (4.195)) may be easier than to verify (4.192) (or (4.194)). An illustrating example supporting this idea is given in Sect. 4.4.4 (see Example 4.73).

### 4.4.4 Applications and Examples

In this subsection we present several examples, which illustrate the applicability of the above results.

*Example 4.72* Let $\mathcal{S}_k = \left(\begin{smallmatrix} 1 & 0 \\ 3 & 1 \end{smallmatrix}\right)$ be the coefficient matrix of a (nonoscillatory) symplectic system (SDS) at $\infty$. The transformation (4.165) with the matrix $\mathcal{R}_k = \mathcal{J}$ satisfies condition (iii) of Theorem 4.68, since $S(M, N) = \sum_{k=M}^{N} 1 = N - M + 1$

is unbounded. Then, by (iii) of Theorem 4.68, the transformed system (4.164) with the matrix

$$\tilde{S} = \mathcal{J}^T S \mathcal{J} = \begin{pmatrix} 1 & -3 \\ 0 & 1 \end{pmatrix}$$

is oscillatory at $\infty$. Indeed, the conjoined basis $\tilde{Y}_k = (1 \ 0)^T$ of this system has exactly

$$l(\tilde{Y}, M, N+1) = \sum_{k=M}^{N} \mu\big((1 \ 0)^T, (-3 \ 1)^T\big) = N - M + 1$$

forward focal points in the interval $(M, N+1]$.

*Example 4.73* The symplectic difference system corresponding to the equation determining the Fibonacci numbers $x_{k+1} = x_{k+1} + x_k$, which can be written in the self-adjoint form as $\Delta\big((-1)^k \Delta x_k\big) + (-1)^{k+1} = 0$, has the coefficient matrix

$$S_k = \begin{pmatrix} 1 & (-1)^k \\ (-1)^{k+1} & 0 \end{pmatrix}.$$

This system is oscillatory at $\infty$, since the first component $x$ of the principal solution at $k = 0$ is $x_0 = 0$, $x_1 = 1$, and $x_{k+2} = x_{k+1} + x_k$ for $k \geq 0$. Consequently, we have $m(k) = m_2(k) = \text{ind}(-1)^k$ for all $k \geq 1$. Obviously, condition (4.194) is not satisfied, but condition (4.190) is satisfied for all $k$. Consequently, system (4.164) with the transformation matrix $\mathcal{R}_k = \mathcal{J}$ is also oscillatory at $\infty$. Remark also that this system satisfies (4.195), since $\mathcal{D}_k = 0$. Then, for the given example to verify, (4.195) is easier than to verify (4.190).

*Example 4.74* Here we present an example, in which condition (4.190) is not satisfied, but condition (4.186) holds. Consider the nonoscillatory system with the coefficient matrix

$$S_k = \begin{pmatrix} 1 & 0 \\ -(-2)^{k+1} & 1 \end{pmatrix} \begin{pmatrix} 1 & 1 \\ 0 & 1 \end{pmatrix} \begin{pmatrix} 1 & 0 \\ (-2)^k & 1 \end{pmatrix}$$

$$= \begin{pmatrix} 1 + (-2)^k & 1 \\ (-2)^k(3 - (-2)^{k+1}) & 1 - (-2)^{k+1} \end{pmatrix}. \tag{4.196}$$

This system is nonoscillatory at $\infty$, since it is constructed from the nonoscillatory symplectic system $Y_{k+1} = \begin{pmatrix} 1 & 1 \\ 0 & 1 \end{pmatrix} Y_k$ by a symplectic transformation with the lower triangular transformation matrix $\mathcal{R}_k := \begin{pmatrix} 1 & 0 \\ -(-2)^k & 1 \end{pmatrix}$. Note that the latter system is a rewritten Sturm-Liouville difference equation $\Delta^2 x_k = 0$. Concerning the matrix

$\mathcal{S}_k$ defined in (4.196), we have

$$\operatorname{ind}(\mathcal{B}_k\mathcal{A}_k^T) = \operatorname{ind}[1 + (-2)^k] = \begin{cases} 0, & k = 2m, \\ 1, & k = 2m + 1, \end{cases}$$

and

$$\operatorname{ind}(-\mathcal{A}_k^T\mathcal{C}_k) = \operatorname{ind}[(1 + (-2)^k)(-2)^k(-3 + (-2)^{k+1})] = \begin{cases} 1, & k = 2m, \\ 0, & k = 2m + 1. \end{cases}$$

Consequently, the sequence $S(M, N) = \sum_{k=M}^{N}(-1)^k$ is bounded, and by Theorem 4.178 the system with the matrix $\mathcal{J}^T\mathcal{S}_k\mathcal{J}$ is also nonoscillatory at $\infty$.

*Example 4.75* Consider the coefficient matrix

$$\begin{aligned} \mathcal{S}_k &= \begin{pmatrix} 1 & 0 \\ -(-1)^{k+1} & 1 \end{pmatrix}\begin{pmatrix} 1 & 1 \\ 0 & 1 \end{pmatrix}\begin{pmatrix} 1 & 0 \\ (-1)^k & 1 \end{pmatrix} \\ &= \begin{pmatrix} 1 + (-1)^k & 1 \\ (-1)^k(2 + (-1)^k) & 1 + (-1)^k \end{pmatrix}. \end{aligned} \tag{4.197}$$

The system with the matrix (4.197) is again nonoscillatory at $\infty$, by using the same arguments as in Example 4.74. It is not difficult to verify that

$$\operatorname{ind}(-\mathcal{A}_k^T\mathcal{C}_k) = \operatorname{ind}[(1 + (-1)^k)(-1)^{k+1}(2 + (-1)^k)] = \begin{cases} 1, & k = 2m, \\ 0, & k = 2k + 1, \end{cases}$$

and $\operatorname{ind}(\mathcal{B}_k\mathcal{A}_k^T) = \operatorname{ind}[1 + (-1)^k] = 0$. Hence, $S(M, N)$ is unbounded and therefore the system with the matrix $\mathcal{J}^T\mathcal{S}_k\mathcal{J}$ is oscillatory at $\infty$.

## 4.5  Notes and References

As it was mentioned above, the results of this chapter present discrete analogs of well-known classical oscillation results for linear Hamiltonian differential systems (1.103). The basic concept in both of the theories is the multiplicity of focal points of conjoined bases of (1.103) and (SDS). Point out that this notion for the continuous case is based on two assumptions. The first one is the Legendre condition (1.111), while the second one (the identical normality assumption) is completely omitted in the modern consideration of oscillation theory of (1.103); see, for example, [127, 207, 283, 289, 321]). The notion of the multiplicities of focal points of conjoined bases of (SDS) and (1.103) (without the controllability assumption) was for the first time introduced by W. Kratz in his two outstanding papers [208] and [207]. In Sect. 4.1.1 we used the definition of the multiplicity (see Definition 4.3) and the main terminology concerning the numbers $m_1$ and $m_2$ from [208].

The notion of a backward focal point, or more precisely the notion of "no backward focal points" in $[k, k + 1)$, for conjoined bases of (SDS) was introduced in [45]. The definition of the multiplicities of backward focal points (see Definition 4.3) was introduced in [87] and [115], [119]. Properties (i), (ii), and (iv) of Proposition 4.4 concerning the multiplicities of forward focal points and the estimate $m(k) \leq \operatorname{rank} \mathcal{B}_k$ from Proposition 4.4(v) were for the first time proved in [208, Lemma 1]. Property (vi) in Proposition 4.4, which connects the multiplicities of forward and backward focal points, was derived in and [115] and [119]. Point out that a similar relation for the multiplicities of right (backward) and left (forward) proper focal points holds for the continuous case as well (see the resent result [289, Theorem 5.1]). The main difference in the proofs of these results is based on the absence of the Legendre condition (1.111) in the discrete case. Because of this, the leading role in the proof of (4.9) is played by relation (4.8). On the other hand, in the continuous case, we have that (4.8) is trivially satisfied because of the Legendre condition; see [127, Lemma 3.2], where it is proven that the second component of the comparative index associated with the multiplicity of proper focal points is zero. Finally, we note that the main result in Corollary 4.6 coincides with [115, formula (3.12)].

The connection of the comparative index with the multiplicities of focal points in Lemma 4.7 was established in [114, Lemmas 2.2 and 2.3] and then presented together with Lemma 4.8 in [115, Lemmas 3.1 and 3.2]. Among other important consequences, this lemma states the equivalences of different definitions of the multiplicity of forward focal points, in particular, the equivalences of Definitions 4.1 and 4.9; see also [193]. The equivalence with Definition 4.9 stated in Proposition 4.13 is from [121].

Regarding Sect. 4.2.1, the result in Theorem 4.16 is proven in [101, Theorem 1]. An analog of Theorem 4.16 concerning the multiplicities of backward focal points was proven in [87, Theorem 1]. The main results of Sect. 4.2.2 were derived in [114] and [115]. The refinement of the discrete oscillation theory offered by the comparative index theory is based on the possibility to present the relations between focal points in the form of explicit equalities instead of inequalities. Such a possibility is based on the algebraic properties of the comparative index, in particular, on the connection between the comparative index and the negative inertia of some symmetric matrix (see Sect. 3.2.1). Sect. 4.2.2 presents the first results in this direction. The result in Theorem 4.20 (see [115, Theorem 1.1]) can be viewed as a discrete version of Theorem 1.50 (see [205, Theorem 5.2.1]), which was recently generalized to the abnormal differential Hamiltonian systems in [127, Theorem 2.3] and [289, Theorem 4.1] via the comparative index approach. Similarly, Theorem 4.21 for the multiplicity of backward focal points (see [115, formula (3.4)]) is a discrete counterpart of the same result for the right proper focal points in [289, Theorem 4.1]. The result in Theorem 4.23 and estimate (4.46) in Corollary 4.25 concerning $\operatorname{rank} w(Y, \hat{Y})$ are from [115, Corollary 3.1]. Formulas (4.41) and (4.45) were derived in the continuous time settings in [289, Proposition 4.2] via the comparative index approach. The new estimates technique for the numbers of focal points based on the upper bounds for the comparative index in part (vii) of Theorem 3.5 was for the first time applied to discrete eigenvalue problems in [123, Corollary 7]. This new technique is illustrated by the results of Corollaries 4.25, 4.26, and 4.27. Some parts of these results are formulated by analogy with [289, Theorem 5.2 and Corollaries 5.8 and 5.10]. Recall from Sect. 2.7 that Corollary 2.56 was proven in [173] but then again in [115, Theorem 1.2] by the comparative index approach. The result in Corollary 4.30 follows from [94, Corollary 3.6] applied to the case $\mathcal{S}_k \equiv \hat{\mathcal{S}}_k$.

The question about a possible coincidence of the numbers of forward and backward focal points of the principal solutions of (SDS) was first posed as an open problem in [101, Section 4]. The main result of Sect. 4.2.3 (see Theorem 4.34) solves this problem, and it is proven in [115, Lemma 3.3]. Here we present another proof of this result, which follows from Lemma 4.33. Based on the latter lemma, on property (vii) of Theorem 3.5 and Remark 3.10, we present some estimates (see Lemma 4.36 and Theorem 4.32) from [289] in the more complete and improved form.

The consideration in Sect. 4.2.4 is based on results of [94] applied to the case $\mathcal{S}_k \equiv \hat{\mathcal{S}}_k$. Further singular Sturmian separation theorems involving the (minimal) recessive solution of (SDS) at $\infty$ were derived in [292]; see Sect. 6.4. The results of Sect. 4.3.1 are from the paper [117] and from the monograph [121]. The main result (Theorem 4.45) was proven for the first time in [117, Theorem 2.1]; the notion of the relative oscillation numbers (in slightly different notation) was introduced in [124, Theorem 2.1] and then in [94, Definition 3.2]. The comparison results in Theorem 4.45 in terms of the relative oscillation numbers can be viewed as a discrete generalization of Theorem 1.49 (proved in [205, Theorem 7.3.1]), because we now deal with explicit equalities for the multiplicities of focal points instead of inequalities. Point out that a generalization of [205, Theorem 7.3.1] to abnormal Hamiltonian differential systems (1.103) was recently proven in [127, Theorem 2.2], which also deals with equalities for the multiplicities of proper focal points. In both cases (discrete and continuous), the difference between inequalities and equalities is based on incorporating the focal points of conjoined bases of some transformed "Wronskian" system associated with two discrete symplectic (or two differential Hamiltonian) systems. From this point of view, the main results of this section may also belong to the relative oscillation theory for discrete symplectic systems. For controllable linear Hamiltonian differential systems, the relative oscillation theory is developed in the recent paper [92]. For the second-order Sturm-Liouville difference equations (which are a special case of (SDS)), the renormalized and more general relative oscillation theory is established in [22, 314]. The results in Theorems 4.50 and 4.53 were proven in [121]; a special case of these theorems (Corollaries 4.52 and 4.54) is presented in [94, Corollaries 3.4 and 3.6]; see also [117, Corollaries 2.1 and 2.2].

The main result of Sect. 4.3.2 (Theorem 4.56) was proven in [117, Corollary 2.4]. The results of Sect. 4.3.3 are from [94, Section 4]. The statement in Lemma 4.55 is from [121]. Together with Theorem 4.56, it opened the door to the relative oscillation theory for symplectic eigenvalue problems developed in [118, 120, 123, 124]; see also Sect. 6.1. The considerations in Sect. 4.4 about the effect of symplectic transformations on multiplicities of focal points are based on the papers [116, 117, 126]. The results in Theorem 4.62 and Corollary 4.66 (in a slightly different notation) and a part of Theorem 4.64 associated with $u_k$ were proven in [116, Lemma 3.1, Corollary 3.2, Theorem 3.3]. The reciprocity principle in the restricted form based on assumptions (4.192) and (4.194) was proven in [116, Theorem 3.5 and Corollary 3.6]. Then the same result was proven under the more general assumptions (4.188) and (4.190) in [117, Theorem 3.2] for the case of constant transformation matrices. The generalized reciprocity principle in the form presented in this book was proven in [121, 126]. Let us note that condition (4.194) covers as a particular case the reciprocity principle for linear Hamiltonian difference systems in [45, Theorem 3], where this principle is formulated for Hamiltonian systems (2.15) with $A_k = 0$, $C_k \leq 0$, $B_k \geq 0$ and under the identical normality assumption. The proof of this special statement is based on properties of the recessive solution of (2.15) at $\infty$. Similarly, conditions (4.192) cover the results of [98, Theorem 3.4]. In particular, there is an interesting interpretation of the numbers $u_k$ given by (4.167) under the assumptions of [98, Theorem 3.4] (see also [116, Remark (ii)]). It is necessary to point out that the recently proven continuous analogs of Theorems 4.62 and 4.64 were derived for abnormal linear differential Hamiltonian systems in [129],[130], [131] based on the comparative index approach.

Important applications of Theorem 4.62 to the special trigonometric transformations

$$
\mathcal{R}_k = \begin{pmatrix} \cos(\alpha_k)\, I & \sin(\alpha_k)\, I \\ -\sin(\alpha_k)\, I & \cos(\alpha_k)\, I \end{pmatrix}
$$

can be found in [96, 122].

# Chapter 5
# Discrete Symplectic Eigenvalue Problems

In this chapter we investigate eigenvalue problems associated with symplectic system (SDS), where the coefficient matrix depends on a spectral parameter, i.e.,

$$y_{k+1}(\lambda) = \mathcal{S}_k(\lambda)\, y_k(\lambda), \quad k \in [0, N]_{\mathbb{Z}}. \tag{5.1}$$

Here $\lambda \in \mathbb{R}$ is the eigenvalue parameter, and the $2n \times 2n$ matrix $\mathcal{S}_k(\lambda)$ is symplectic for every $\lambda \in \mathbb{R}$. First we study in Sect. 5.1 problem (5.1) with a general nonlinear dependence on $\lambda$ and with the coefficient matrix $\mathcal{B}_k(\lambda)$ having constant rank. Here we assume a natural monotonicity assumption on the behavior of the coefficient matrix $\mathcal{S}_k(\lambda)$ in $\lambda$. This type of monotonicity condition was discussed in details in Sect. 1.6.4. In Sect. 5.2, we present transformations between various boundary conditions for system (5.1), in particular a transformation of separated endpoints into Dirichlet boundary conditions and a transformation of general joint boundary conditions into separated ones. These boundary conditions may depend nonlinearly on the spectral parameter $\lambda$ (under a certain monotonicity condition). In Sects. 5.3–5.5, we proceed with the study of problem (5.1) with a special linear dependence on $\lambda$, as these systems have important applications in the Sturmian theory for system (SDS). Finally, in Sect. 5.6, we also present some extensions of the oscillation theorems from Sect. 5.1 to symplectic systems, whose coefficient $\mathcal{B}_k(\lambda)$ has nonconstant rank.

## 5.1 Nonlinear Dependence on Spectral Parameter

In this section we consider a general eigenvalue problem with symplectic difference system depending nonlinearly on the spectral parameter $\lambda$. We develop the notions of (finite) eigenvalues and (finite) eigenfunctions and their multiplicities and prove the corresponding oscillation theorem for Dirichlet boundary conditions. Consider

© Springer Nature Switzerland AG 2019

O. Došlý et al., *Symplectic Difference Systems: Oscillation and Spectral Theory*,
Pathways in Mathematics, https://doi.org/10.1007/978-3-030-19373-7_5

the system (5.1) in the form

$$\left.\begin{array}{l} x_{k+1}(\lambda) = \mathcal{A}_k(\lambda)\, x_k(\lambda) + \mathcal{B}_k(\lambda)\, u_k(\lambda), \\[2mm] u_{k+1}(\lambda) = \mathcal{C}_k(\lambda)\, x_k(\lambda) + \mathcal{D}_k(\lambda)\, u_k(\lambda), \end{array}\right\} \quad k \in [0, N]_{\mathbb{Z}}, \qquad \text{(SDS}_\lambda\text{)}$$

and the corresponding eigenvalue problem with the Dirichlet boundary conditions

$$y_{k+1}(\lambda) = \mathcal{S}_k(\lambda)\, y_k(\lambda), \quad k \in [0, N]_{\mathbb{Z}}, \quad \lambda \in \mathbb{R}, \quad x_0(\lambda) = 0 = x_{N+1}(\lambda), \tag{E}$$

where $y = (x, u)$. The coefficient matrix $\mathcal{S}_k(\lambda)$ of system (SDS$_\lambda$) is assumed to be symplectic, i.e., for all $k \in [0, N]_{\mathbb{Z}}$ and $\lambda \in \mathbb{R}$

$$\mathcal{S}_k^T(\lambda)\, \mathcal{J} \mathcal{S}_k(\lambda) = \mathcal{J}, \quad \mathcal{S}_k(\lambda) := \begin{pmatrix} \mathcal{A}_k(\lambda) & \mathcal{B}_k(\lambda) \\ \mathcal{C}_k(\lambda) & \mathcal{D}_k(\lambda) \end{pmatrix}, \quad \mathcal{J} := \begin{pmatrix} 0 & I \\ -I & 0 \end{pmatrix}. \tag{5.2}$$

The same property is then satisfied by the fundamental matrix of system (SDS$_\lambda$). In addition, we assume that the matrix $\mathcal{S}_k(\lambda)$ piecewise continuously differentiable, i.e., it is continuous on $\mathbb{R}$ and the derivative $\dot{\mathcal{S}}_k(\lambda) := \frac{d}{d\lambda}\mathcal{S}_k(\lambda)$ is piecewise continuous in the parameter $\lambda \in \mathbb{R}$ for all $k \in [0, N]_{\mathbb{Z}}$. Given the above symplectic matrix $\mathcal{S}_k(\lambda)$, we consider the monotonicity assumption

$$\Psi(\mathcal{S}_k(\lambda)) = \Psi_k(\lambda) := \mathcal{J}\dot{\mathcal{S}}_k(\lambda)\, \mathcal{J}\mathcal{S}_k^T(\lambda)\, \mathcal{J} \geq 0, \quad k \in [0, N]_{\mathbb{Z}}, \quad \lambda \in \mathbb{R}. \tag{5.3}$$

Recall that the matrix $\Psi_k(\lambda) = \Psi(\mathcal{S}_k(\lambda))$ is symmetric for any $k \in [0, N]_{\mathbb{Z}}$ and $\lambda \in \mathbb{R}$ (see Proposition 1.75).

### 5.1.1  Finite Eigenvalues

In this subsection, we derive some monotonicity results based on the assumption (5.3), which lead to the definition of finite eigenvalues for problem (E). Recall that in Sect. 1.6.4 we proved important properties of $\Psi(\mathcal{S}_k(\lambda))$ and corollaries to (5.3) applied to arbitrary piecewise continuously differentiable symplectic matrix $W(\lambda)$. Putting $W(\lambda) := \mathcal{S}_k(\lambda)$ in Theorems 1.79, 1.81, 1.82 and in Corollary 1.83, we derive the following important properties of the symplectic coefficient matrix $\mathcal{S}_k(\lambda)$ under assumption (5.3).

**Theorem 5.1**  *Assume (5.3) for $\mathcal{S}_k(\lambda)$ given by (5.2). Then for any $k \in [0, N]_{\mathbb{Z}}$, the following assertions hold:*

(i) *The set* $\operatorname{Ker} \mathcal{B}_k(\lambda)$ *is piecewise constant in $\lambda$, i.e., for any $\lambda_0 \in \mathbb{R}$, there exists $\delta > 0$ such that*

$$\operatorname{Ker} \mathcal{B}_k(\lambda) \equiv \operatorname{Ker} \mathcal{B}_k(\lambda_0^-) \subseteq \operatorname{Ker} \mathcal{B}_k(\lambda_0) \quad \textit{for all } \lambda \in (\lambda_0 - \delta, \lambda_0), \qquad (5.4)$$

$$\operatorname{Ker} \mathcal{B}_k(\lambda) \equiv \operatorname{Ker} \mathcal{B}_k(\lambda_0^+) \subseteq \operatorname{Ker} \mathcal{B}_k(\lambda_0) \quad \textit{for all } \lambda \in (\lambda_0, \lambda_0 + \delta). \qquad (5.5)$$

(ii) *The set* $\operatorname{Im} \mathcal{B}_k(\lambda)$ *is piecewise constant in $\lambda$, i.e., for any $\lambda_0 \in \mathbb{R}$, there exists $\delta > 0$ such that*

$$\operatorname{Im} \mathcal{B}_k(\lambda_0) \subseteq \operatorname{Im} \mathcal{B}_k(\lambda) \equiv \operatorname{Im} \mathcal{B}_k(\lambda_0^-) \quad \textit{for all } \lambda \in (\lambda_0 - \delta, \lambda_0), \qquad (5.6)$$

$$\operatorname{Im} \mathcal{B}_k(\lambda_0) \subseteq \operatorname{Im} \mathcal{B}_k(\lambda) \equiv \operatorname{Im} \mathcal{B}_k(\lambda_0^+) \quad \textit{for all } \lambda \in (\lambda_0, \lambda_0 + \delta). \qquad (5.7)$$

(iii) *The following three conditions are equivalent:*

$$\operatorname{rank} \mathcal{B}_k(\lambda) \textit{ is constant for } \lambda \in \mathbb{R}, \qquad (5.8)$$

$$\textit{the set } \operatorname{Ker} \mathcal{B}_k(\lambda) \textit{ is constant for } \lambda \in \mathbb{R}, \qquad (5.9)$$

$$\textit{the set } \operatorname{Im} \mathcal{B}_k(\lambda) \textit{ is constant for } \lambda \in \mathbb{R}. \qquad (5.10)$$

(iv) *The matrices $\mathcal{B}_k(\lambda)\, \mathcal{B}_k^\dagger(\lambda)$ and $\mathcal{B}_k^\dagger(\lambda)\, \mathcal{B}_k(\lambda)$ are piecewise constant in $\lambda$.*

Next we show that similar properties hold also for arbitrary symplectic fundamental matrix $\mathcal{Z}_k(\lambda)$ of $(\mathrm{SDS}_\lambda)$ such that $\mathcal{Z}_0(\lambda)$ is piecewise continuously differentiable and $\Psi(\mathcal{Z}_0(\lambda) \geq 0$ for $\lambda \in \mathbb{R}$. In this case, since the matrix $\mathcal{S}_k(\lambda)$ is piecewise continuously differentiable with respect to $\lambda$, it follows that the fundamental matrix $\mathcal{Z}_k(\lambda)$ is piecewise continuously differentiable for all $k \in [0, N+1]_{\mathbb{Z}}$. As a corollary to the multiplicative property (1.195) in Proposition 1.76(i), we have the following important result.

**Proposition 5.2**  *Assume that $\mathcal{Z}_k(\lambda)$ is a symplectic fundamental matrix of (5.3) such that $\mathcal{Z}_0(\lambda)$ is piecewise continuously differentiable. Then under the assumption*

$$\Psi(\mathcal{Z}_k(\lambda)) \geq 0, \quad \lambda \in \mathbb{R} \qquad (5.11)$$

*for the index $k = 0$, we have that (5.11) holds for any $k \in [0, N+1]_{\mathbb{Z}}$.*

*Proof*  Applying Proposition 1.76(i), we have

$$\Psi(\mathcal{Z}_{k+1}(\lambda)) = \Psi(\mathcal{S}_k(\lambda)\, \mathcal{Z}_k(\lambda)) = \mathcal{S}_k^{T-1}(\lambda)\, \Psi(\mathcal{Z}_k(\lambda))\, \mathcal{S}_k^{-1}(\lambda) + \Psi(\mathcal{S}_k(\lambda)),$$

or by Proposition 1.76(ii)

$$\Delta\Psi(\mathcal{Z}_k^{-1}(\lambda)) = -\mathcal{Z}_{k+1}^T(\lambda)\, \Psi(\mathcal{S}_k(\lambda)\, \mathcal{Z}_{k+1}(\lambda)).$$

Then we derive

$$\mathcal{Z}_{k+1}^T(\lambda)\,\Psi(\mathcal{Z}_{k+1}(\lambda))\,\mathcal{Z}_{k+1}(\lambda)$$

$$= \mathcal{Z}_0^T(\lambda)\,\Psi(\mathcal{Z}_0(\lambda))\,\mathcal{Z}_0(\lambda) + \sum_{i=0}^{k}\mathcal{Z}_{i+1}^T(\lambda)\,\Psi(\mathcal{S}_i(\lambda))\,\mathcal{Z}_{i+1}(\lambda), \qquad (5.12)$$

and therefore (5.11) holds for any $k \in [0, N+1]_{\mathbb{Z}}$. $\qquad\qquad\qquad\qquad\qquad\square$

Putting $W(\lambda) := \mathcal{Z}_k(\lambda)$ for $k \in [0, N+1]_{\mathbb{Z}}$ in Theorems 1.79, 1.81, 1.82 and in Corollary 1.83, we now formulate the most important result of this section.

**Theorem 5.3** *Assume* (5.3) *and* $\Psi(\mathcal{Z}_0(\lambda)) \geq 0$ *for* $\lambda \in \mathbb{R}$ *for a symplectic fundamental matrix* $\mathcal{Z}_k(\lambda)$ *of* $(\mathrm{SDS}_\lambda)$ *in the form*

$$\mathcal{Z}_k(\lambda) = \big(\hat{Y}_k(\lambda)\ Y_k(\lambda)\big) = \begin{pmatrix}\hat{X}_k(\lambda)\ X_k(\lambda)\\ \hat{U}_k(\lambda)\ U_k(\lambda)\end{pmatrix}.$$

*Then for every* $k \in [0, N+1]_{\mathbb{Z}}$, *we have*

(i) *The set* $\mathrm{Ker}\,X_k(\lambda)$ *is piecewise constant in* $\lambda$, *i.e., for any* $\lambda_0 \in \mathbb{R}$, *there exists* $\delta > 0$ *such that*

$$\mathrm{Ker}\,X_k(\lambda) \equiv \mathrm{Ker}\,X_k(\lambda_0^-) \subseteq \mathrm{Ker}\,X_k(\lambda_0) \quad \text{for all } \lambda \in (\lambda_0 - \delta, \lambda_0), \qquad (5.13)$$

$$\mathrm{Ker}\,X_k(\lambda) \equiv \mathrm{Ker}\,X_k(\lambda_0^+) \subseteq \mathrm{Ker}\,X_k(\lambda_0) \quad \text{for all } \lambda \in (\lambda_0, \lambda_0 + \delta). \qquad (5.14)$$

(ii) *The set* $\mathrm{Im}\,X_k(\lambda)$ *is piecewise constant in* $\lambda$, *i.e., for any* $\lambda_0 \in \mathbb{R}$, *there exists* $\delta > 0$ *such that*

$$\mathrm{Im}\,X_k(\lambda_0) \subseteq \mathrm{Im}\,X_k(\lambda) \equiv \mathrm{Im}\,X_k(\lambda_0^-) \quad \text{for all } \lambda \in (\lambda_0 - \delta, \lambda_0), \qquad (5.15)$$

$$\mathrm{Im}\,X_k(\lambda_0) \subseteq \mathrm{Im}\,X_k(\lambda) \equiv \mathrm{Im}\,X_k(\lambda_0^+) \quad \text{for all } \lambda \in (\lambda_0, \lambda_0 + \delta). \qquad (5.16)$$

(iii) *The following three conditions are equivalent:*

$$\mathrm{rank}\,X_k(\lambda) \text{ is constant for } \lambda \in \mathbb{R}, \qquad (5.17)$$

$$\text{the set } \mathrm{Ker}\,X_k(\lambda) \text{ is constant for } \lambda \in \mathbb{R}, \qquad (5.18)$$

$$\text{the set } \mathrm{Im}\,X_k(\lambda) \text{ is constant for } \lambda \in \mathbb{R}. \qquad (5.19)$$

(iv) *The matrices* $X_k(\lambda)\,X_k^\dagger(\lambda)$ *and* $X_k^\dagger(\lambda)\,X_k(\lambda)$ *are piecewise constant in* $\lambda$.

*Remark 5.4* We remark that one can associate $\mathcal{Z}_k(\lambda)$ with a given conjoined basis $Y(\lambda) = \left( \begin{smallmatrix} X(\lambda) \\ U(\lambda) \end{smallmatrix} \right)$ of $(\text{SDS}_\lambda)$; see Lemma 1.58(iv). In particular cases, we will assume in addition that the initial conditions of the conjoined basis $Y(\lambda)$ do not depend on $\lambda$, i.e.,

$$Y_0(\lambda) \equiv Y_0 \quad \text{for all } \lambda \in \mathbb{R}. \tag{5.20}$$

It then follows from the proof of Lemma 1.58(iv) that the symplectic fundamental matrix $\mathcal{Z}_k(\lambda)$ such that $Y(\lambda) = \mathcal{Z}(\lambda)(0 \ I)^T$ also does not depend on $\lambda$ for $k = 0$, i.e., $\mathcal{Z}_0(\lambda) \equiv \mathcal{Z}_0$. In this special case, the condition $\Psi(\mathcal{Z}_0(\lambda)) \geq 0$ for $\lambda \in \mathbb{R}$ is trivially satisfied, and we derive all assertions of Theorem 5.3 for the block $X_k(\lambda)$ of a conjoined basis $Y_k(\lambda) = \left( \begin{smallmatrix} X_k(\lambda) \\ U_k(\lambda) \end{smallmatrix} \right)$ of $(\text{SDS}_\lambda)$ with initial conditions (5.20).

In Theorem 5.3, we showed that for every fixed $\lambda_0 \in \mathbb{R}$, the quantity rank $X_k(\lambda)$ is constant on some left and right neighborhoods of $\lambda_0$. This allows to define correctly the notion of a finite eigenvalue of problem (E). Let $Y^{[0]}(\lambda) = \left( \begin{smallmatrix} X^{[0]}(\lambda) \\ U^{[0]}(\lambda) \end{smallmatrix} \right)$ be the *principal solution* of $(\text{SDS}_\lambda)$ at $k = 0$, that is, the solution starting with the initial values

$$X_0^{[0]}(\lambda) \equiv 0, \quad U_0^{[0]}(\lambda) \equiv I \quad \text{for all } \lambda \in \mathbb{R},$$

so that these initial conditions are independent of $\lambda$, and Theorem 5.3 works for this special case (see Remark 5.4).

The result in Theorem 5.3 justifies the introduction of the following notion.

**Definition 5.5 (Finite Eigenvalue)** Under (5.3), a number $\lambda_0 \in \mathbb{R}$ is a *finite eigenvalue* of problem (E) if

$$\theta(\lambda_0) := \text{rank } X_{N+1}^{[0]}(\lambda_0^-) - \text{rank } X_{N+1}^{[0]}(\lambda_0) \geq 1. \tag{5.21}$$

In this case the number $\theta(\lambda_0)$ is called the *algebraic multiplicity* of $\lambda_0$.

*Remark 5.6*

(i) The definition of a finite eigenvalue of (E) is one-sided, that is, it only depends on the behavior of $X_{N+1}^{[0]}(\lambda)$ in a left neighborhood of $\lambda_0$.

(ii) By (5.13), the finite eigenvalues are well defined, since the number $\theta(\lambda_0)$ is always nonnegative. Moreover,

$$\theta(\lambda_0) = \text{def } X_{N+1}^{[0]}(\lambda_0) - \text{def } X_{N+1}^{[0]}(\lambda_0^-)$$

$$= \dim\left([\text{Ker } X_{N+1}^{[0]}(\lambda_0^-)]^\perp \cap \text{Ker } X_{N+1}^{[0]}(\lambda_0)\right). \tag{5.22}$$

(iii) When $X_{N+1}^{[0]}(\lambda)$ is invertible except at isolated values of $\lambda$ (which is the case of "controllable" systems), then def $X_{N+1}^{[0]}(\lambda_0^-) = 0$ for every $\lambda_0 \in \mathbb{R}$. Therefore, in this case a finite eigenvalue reduces to the classical eigenvalue, which is determined by the condition def $X_{N+1}^{[0]}(\lambda_0) \geq 1$, i.e., by the singularity of $X_{N+1}^{[0]}(\lambda_0)$. The algebraic multiplicity is then equal to def $X_{N+1}^{[0]}(\lambda_0)$; see, e.g., [44, Corollary 1].

(iv) When the dependence in $\lambda$ is linear as in (5.238) or (5.241), we will discuss the special features of the finite eigenvalues in Sect. 5.3.3.

The following is a simple consequence of Theorem 5.3 and Definition 5.5.

**Corollary 5.7** *Under* (5.3), *the finite eigenvalues of* (E) *are isolated.*

### 5.1.2  Finite Eigenfunctions

In this subsection, we develop a geometric notion corresponding to the finite eigenvalues from Definition 5.5. First observe that if $\lambda_0$ is a finite eigenvalue of (E) and $c \in \mathrm{Ker}\, X_{N+1}^{[0]}(\lambda_0)$, then $y := Y^{[0]}(\lambda_0)\, c$ is a vector solution satisfying both $(S_{\lambda_0})$ and $x_0 = 0 = x_{N+1}$, i.e., $y$ solves the problem (E) with $\lambda = \lambda_0$. It remains to describe which of these solutions are in a sense "degenerate," that is, which of them do not correspond to a finite eigenvalue $\lambda_0$. This procedure has a parallel strategy in the classical eigenvalue theory, where only the nontrivial solutions of (E) count as the eigenfunctions for the eigenvalue $\lambda_0$.

**Definition 5.8 (Degenerate Solution)** Let $\lambda_0 \in \mathbb{R}$ be given. A solution $y$ of system $(S_{\lambda_0})$ is said to be *degenerate at* $\lambda_0$ (or it is a *degenerate solution*), if there exists $\delta > 0$ such that for all $\lambda \in (\lambda_0 - \delta, \lambda_0]$, the solution $y(\lambda)$ of $(SDS_\lambda)$ given by the initial conditions $y_0(\lambda) = y_0$ satisfies

$$\Psi_k(\lambda)\, y_{k+1}(\lambda) = 0 \quad \text{for all } k \in [0, N]_{\mathbb{Z}}. \tag{5.23}$$

In the opposite case, we say that the solution $y$ is *nondegenerate at* $\lambda_0$.

A degenerate solution at $\lambda_0$ represents in fact a family of solutions $y(\lambda)$ for $\lambda \in (\lambda_0 - \delta, \lambda_0]$, which includes the solution $y$ itself for $\lambda = \lambda_0$. Moreover, this family of solutions is independent of $\lambda$ with respect to the semi-norm induced by the positive semidefinite matrix $\Psi_k(\lambda)$, i.e.,

$$\| y(\lambda) \|_\lambda^2 := \sum_{k=0}^{N} y_{k+1}^T(\lambda)\, \Psi_k(\lambda)\, y_{k+1}(\lambda) \quad \text{for } \lambda \in (\lambda_0 - \delta, \lambda_0].$$

*Remark 5.9*

(i) Since the matrix $\Psi_k(\lambda)$ is symmetric and since $\mathcal{S}_k(\lambda)$ and $\mathcal{J}$ are invertible, it follows from (5.3) that condition (5.23) can be written in the equivalent form

$$\dot{\mathcal{S}}_k^T(\lambda)\, \mathcal{J}\, y_{k+1}(\lambda) = 0 \quad \text{for all } k \in [0, N]_{\mathbb{Z}}.$$

(ii) When the dependence on $\lambda$ is linear as in (5.238) and (5.241), a degenerate solution $y = (x, u)$ at $\lambda_0$ is a solution of $(S_{\lambda_0})$ satisfying

$$\mathcal{W}_k \, x_{k+1} = 0 \quad \text{for all } k \in [0, N]_{\mathbb{Z}}. \tag{5.24}$$

Condition (5.24) is indeed equivalent to (5.23) where $\lambda \in (\lambda_0 - \delta, \lambda_0]$, since under (5.241) any solution of $(S_{\lambda_0})$ satisfying (5.24) is at the same time a solution of $(SDS_\lambda)$ for every $\lambda \in \mathbb{R}$. Hence, degeneracy condition (5.23) is a local property when $(SDS_\lambda)$ depends on $\lambda$ nonlinearly, but it is a global property for the linear dependence on $\lambda$.

Consider the following spaces of solutions of system $(S_{\lambda_0})$:

$$\mathcal{E}(\lambda_0) := \big\{ y = (x, u) \text{ solves system } (S_{\lambda_0}) \text{ with } x_0 = 0 = x_{N+1} \big\},$$
$$\mathcal{W}(\lambda_0) := \big\{ y = (x, u) \in \mathcal{E}(\lambda_0), \; y \text{ is degenerate at } \lambda_0 \big\}.$$

Then it follows that

$$\mathcal{E}(\lambda_0) = \big\{ Y^{[0]}(\lambda_0)\, c, \;\; c \in \operatorname{Ker} X^{[0]}_{N+1}(\lambda_0) \big\}. \tag{5.25}$$

Indeed, the inclusion $\supseteq$ in (5.25) follows from the considerations at the beginning of this section, while the inclusion $\subseteq$ in (5.25) is obtained from the uniqueness of solutions of system $(S_{\lambda_0})$—every solution $y \in \mathcal{E}(\lambda_0)$ is of the form $y = Y^{[0]}(\lambda_0)\, c$, where $c = u_0$ and where $X^{[0]}_{N+1}(\lambda_0)\, c = 0$. Our aim is to prove that the degenerate solutions at $\lambda_0$ correspond to those vectors $c \in \operatorname{Ker} X^{[0]}_{N+1}(\lambda_0)$ which are in $\operatorname{Ker} X^{[0]}_{N+1}(\lambda_0^-)$, i.e., we will prove in Theorem 5.11 below that

$$\mathcal{W}(\lambda_0) = \big\{ Y^{[0]}(\lambda_0)\, c, \;\; c \in \operatorname{Ker} X^{[0]}_{N+1}(\lambda_0^-) \big\}. \tag{5.26}$$

In turn, the finite eigenfunctions are exactly the nondegenerate solutions at $\lambda_0$.

**Definition 5.10 (Finite Eigenfunction)** Under (5.3), every nondegenerate solution $y$ at $\lambda_0$ of (E) with $\lambda = \lambda_0$ is called a *finite eigenfunction* corresponding to the finite eigenvalue $\lambda_0$, and the number

$$\omega(\lambda_0) := \dim \mathcal{E}(\lambda_0) - \dim \mathcal{W}(\lambda_0) \tag{5.27}$$

is called the *geometric multiplicity* of $\lambda_0$.

The following result is a characterization of the finite eigenvalues of (E) with the nonlinear dependence on $\lambda$.

**Theorem 5.11 (Geometric Characterization of Finite Eigenvalues)** *Let assumption (5.3) be satisfied. A number $\lambda_0$ is a finite eigenvalue of (E) with algebraic multiplicity $\theta(\lambda_0) \geq 1$ defined in (5.21) if and only if there exists* *a corresponding finite eigenfunction y. In this case, the geometric multiplicity of $\lambda_0$ defined in (5.27) is equal to its algebraic multiplicity, i.e., $\omega(\lambda_0) = \theta(\lambda_0)$.*

For the proof of Theorem 5.11, we need the following auxiliary result.

**Lemma 5.12** *Let $Y(\lambda) = (X(\lambda), U(\lambda))$ and $\tilde{Y}(\lambda) = (\tilde{X}(\lambda), \tilde{U}(\lambda))$ be normalized conjoined bases of $(\mathrm{SDS}_\lambda)$ such that they form the symplectic fundamental matrix*

$$\tilde{Z}_k(\lambda) = \begin{pmatrix} X_k(\lambda) & \tilde{X}_k(\lambda) \\ U_k(\lambda) & \tilde{U}_k(\lambda) \end{pmatrix} \tag{5.28}$$

*of $(\mathrm{SDS}_\lambda)$. Assume that $\tilde{Z}_0(\lambda) \equiv \tilde{Z}_0$, i.e., it does not depend on $\lambda$, and that $X_{k+1}(\lambda_0)$ is invertible for some $k \in [0, N]_{\mathbb{Z}}$ and $\lambda_0 \in \mathbb{R}$. Then there exists $\varepsilon > 0$ such that*

$$\frac{d}{d\lambda} [X_{k+1}^{-1}(\lambda)\, \tilde{X}_{k+1}(\lambda)] = \sum_{j=0}^{k} \zeta_{k+1,j+1}^T(\lambda)\, \Psi_j(\lambda)\, \zeta_{k+1,j+1}(\lambda), \tag{5.29}$$

*for all $\lambda \in (\lambda_0 - \varepsilon, \lambda_0 + \varepsilon)$, where*

$$\zeta_{k,j}(\lambda) := \tilde{Z}_j(\lambda) \begin{pmatrix} -X_k^{-1}(\lambda)\, \tilde{X}_k(\lambda) \\ I \end{pmatrix}. \tag{5.30}$$

*Proof* Using the assumption $\tilde{Z}_0(\lambda) \equiv \tilde{Z}_0$, Proposition 1.76(ii), and the definition of $\Psi(\tilde{Z}_{k+1}(\lambda))$, we derive from (5.12) that

$$\Psi(\tilde{Z}_{k+1}^{-1}(\lambda)) = \tilde{Z}_{k+1}^T(\lambda)\, \mathcal{J}\, \dot{\tilde{Z}}_{k+1}(\lambda) = -\Omega_k(\lambda), \quad k \in [0, N]_{\mathbb{Z}}, \quad \lambda \in \mathbb{R}, \tag{5.31}$$

where

$$\Omega_k(\lambda) := \sum_{j=0}^{k} \tilde{Z}_{j+1}^T(\lambda)\, \Psi_j(\lambda)\, \tilde{Z}_{j+1}(\lambda). \tag{5.32}$$

Since $X_{k+1}(\lambda_0)$ is invertible, it follows that $X_{k+1}(\lambda)$ is invertible on $(\lambda_0 - \varepsilon, \lambda_0 + \varepsilon)$ for some $\varepsilon > 0$, and then one can apply Lemma 1.77 with $W(\lambda) := \tilde{Z}_{k+1}^{-1}(\lambda)$, $R := -\mathcal{J}$, $P := I$, and $\tilde{W}(\lambda) := \mathcal{J}\tilde{Z}_{k+1}^{-1}(\lambda)$. By (1.200) and Proposition 1.76(i), we derive (suppressing the argument $\lambda$ and index $k + 1$)

$$\mathcal{J}^T \Psi(\tilde{W}(\lambda))\, \mathcal{J} = \Psi(\tilde{Z}_{k+1}^{-1}(\lambda)) = -\begin{pmatrix} X^T & 0 \\ \tilde{X}^T & I \end{pmatrix} \frac{d}{d\lambda} \begin{pmatrix} -UX^{-1} & -X^{T-1} \\ -X^{-1} & X^{-1}\tilde{X} \end{pmatrix} \begin{pmatrix} X & \tilde{X} \\ 0 & I \end{pmatrix}.$$

Substituting the above representation for $\Psi(\tilde{Z}_{k+1}^{-1}(\lambda))$ into formula (5.31), we derive identity (5.29).                                                                                 □

*Proof of Theorem 5.11* If we prove that equality (5.26) holds, then the result will follow since the finite eigenfunctions $y$ for $\lambda_0$ will be of the form $y = Y^{[0]}(\lambda_0)\,c$ with $c \in \operatorname{Ker} X^{[0]}_{N+1}(\lambda_0) \setminus \operatorname{Ker} X^{[0]}_{N+1}(\lambda_0^-)$.

Let $y = (x, u)$ be a solution of (E) with $\lambda = \lambda_0$, and assume that $y$ is degenerate at $\lambda_0$. Let $\delta > 0$ and $y(\lambda) = (x(\lambda), u(\lambda))$ for $\lambda \in (\lambda_0 - \delta, \lambda_0]$ be the constant and the corresponding family of solutions from Definition 5.8. By the uniqueness of solutions of $(\mathrm{SDS}_\lambda)$, we get

$$y_k(\lambda) = Y^{[0]}_k(\lambda)\,c \quad \text{for all } k \in [0, N+1]_{\mathbb{Z}} \text{ and } \lambda \in (\lambda_0 - \delta, \lambda_0] \tag{5.33}$$

for some $c \in \mathbb{R}^n$, in fact for $c = u_0$. From $x_{N+1} = 0$, we must necessarily have $c \in \operatorname{Ker} X^{[0]}_{N+1}(\lambda_0)$. By Theorem 5.3, we may assume that $\operatorname{Ker} X^{[0]}_{N+1}(\lambda) \equiv \operatorname{Ker} X^{[0]}_{N+1}(\lambda_0^-)$ is constant on $(\lambda_0 - \delta, \lambda_0)$. Then for every $k \in [0, N]_{\mathbb{Z}}$ and every $\lambda \in (\lambda_0 - \delta, \lambda_0]$, we have

$$y_{k+1}(\lambda) = \mathcal{S}_k(\lambda)\,y_k(\lambda), \qquad y_k(\lambda) = -\mathcal{J}\mathcal{S}_k^T(\lambda)\,\mathcal{J}\,y_{k+1}(\lambda). \tag{5.34}$$

By taking the derivative of this equation at $\lambda \in (\lambda_0 - \delta, \lambda_0)$, resp., the left derivative of this equation at $\lambda = \lambda_0$, we obtain

$$\dot{y}_{k+1}(\lambda) = \mathcal{S}_k(\lambda)\,\dot{y}_k(\lambda) + \dot{\mathcal{S}}_k(\lambda)\,y_k(\lambda) \overset{(5.34)}{=} \mathcal{S}_k(\lambda)\,\dot{y}_k(\lambda) + \mathcal{J}\,\Psi_k(\lambda)\,y_{k+1}(\lambda)$$

$$\overset{(5.23)}{=} \mathcal{S}_k(\lambda)\,\dot{y}_k(\lambda), \qquad k \in [0, N]_{\mathbb{Z}}, \ \lambda \in (\lambda_0 - \delta, \lambda_0]. \tag{5.35}$$

In addition, since $y_0(\lambda) = y_0$ for every $\lambda \in (\lambda_0 - \delta, \lambda_0]$, the initial conditions of $y(\lambda)$ do not depend on $\lambda$. Hence, $\dot{y}_0(\lambda) = 0$ for all $\lambda \in (\lambda_0 - \delta, \lambda_0]$. By the uniqueness of solutions of system $(\mathrm{SDS}_\lambda)$, it follows from (5.35) that $\dot{y}_k(\lambda) = 0$ for all $k \in [0, N+1]_{\mathbb{Z}}$ and $\lambda \in (\lambda_0 - \delta, \lambda_0]$. This means that the functions $y(\lambda)$ do not depend on $\lambda$ on $(\lambda_0 - \delta, \lambda_0]$, i.e.,

$$y_k(\lambda) \equiv y_k(\lambda_0) \quad \text{for all } \lambda \in (\lambda_0 - \delta, \lambda_0] \text{ and all } k \in [0, N+1]_{\mathbb{Z}}. \tag{5.36}$$

Therefore, for every $\lambda \in (\lambda_0 - \delta, \lambda_0]$ and $k \in [0, N+1]_{\mathbb{Z}}$, we have

$$Y^{[0]}_k(\lambda)\,c \overset{(5.33)}{=} y_k(\lambda) \overset{(5.36)}{=} y_k(\lambda_0) = y_k = (x_k, u_k).$$

The endpoint condition $x_{N+1} = 0$ then yields $X^{[0]}_{N+1}(\lambda)\,c = 0$ for all $\lambda \in (\lambda_0 - \delta, \lambda_0]$, i.e., $c \in \operatorname{Ker} X^{[0]}_{N+1}(\lambda)$ for all $\lambda \in (\lambda_0 - \delta, \lambda_0]$. And since $\operatorname{Ker} X^{[0]}_{N+1}(\lambda)$ is constant on $(\lambda_0 - \delta, \lambda_0)$, it follows that $c \in \operatorname{Ker} X^{[0]}_{N+1}(\lambda_0^-)$.

Conversely, let $c \in \operatorname{Ker} X^{[0]}_{N+1}(\lambda_0^-)$. Then $c \in \operatorname{Ker} X^{[0]}_{N+1}(\lambda)$ for all values $\lambda \in (\lambda_0 - \varepsilon, \lambda_0]$ for some $\varepsilon > 0$. Let $\tilde{Y}(\lambda_0)$ be a conjoined basis of $(\mathrm{S}_{\lambda_0})$ such that $\tilde{Y}(\lambda_0)$ and $Y(\lambda_0)$ are normalized and $\tilde{X}_{N+1}(\lambda_0)$ is invertible (similar to the proof of Theorem 5.3). For each $\lambda \in \mathbb{R}$, let $\tilde{Y}(\lambda)$ be the conjoined basis of $(\mathrm{SDS}_\lambda)$ starting with the initial conditions $\tilde{Y}_0(\lambda) = \tilde{Y}_0(\lambda_0)$, so that these initial conditions are independent of $\lambda$. Then $\tilde{X}_{N+1}(\lambda)$ is invertible for all $\lambda \in (\lambda_0 - \delta, \lambda_0]$ for some $\delta \in (0, \varepsilon)$. For $j \in [0, N+1]_\mathbb{Z}$ and $\lambda \in (\lambda_0 - \delta, \lambda_0]$, we now define the functions

$$y_j(\lambda) := Y_j^{[0]}(\lambda)\, c - \tilde{Y}_j(\lambda)\, \tilde{X}_{N+1}^{-1}(\lambda)\, X_{N+1}^{[0]}(\lambda)\, c = Y_j^{[0]}(\lambda)\, c,$$

compared with $\zeta_{N+1,j}(\lambda)$ in (5.30). Then for every $\lambda \in (\lambda_0 - \delta, \lambda_0]$, the function $z(\lambda)$ solves the system $(\mathrm{SDS}_\lambda)$ with $z_0(\lambda) = (0, c)$. Since

$$c^T \tilde{X}_{N+1}^{-1}(\lambda)\, X_{N+1}^{[0]}(\lambda)\, c = 0 \quad \text{for all } \lambda \in (\lambda_0 - \delta, \lambda_0], \tag{5.37}$$

differentiating equation (5.37) at any $\lambda \in (\lambda_0 - \delta, \lambda_0]$, we get from Lemma 5.12, in which $Y(\lambda) := \tilde{Y}(\lambda)$ and $\tilde{Y}(\lambda) := Y^{[0]}(\lambda)$, that

$$0 \overset{(5.37)}{=} \frac{\mathrm{d}}{\mathrm{d}\lambda}\, c^T \tilde{X}_{N+1}^{-1}(\lambda)\, X_{N+1}^{[0]}(\lambda)\, c \overset{(5.29)}{=} \sum_{j=0}^{N} y_{j+1}^T(\lambda)\, \Psi_j(\lambda)\, y_{j+1}(\lambda)$$

for $\lambda \in (\lambda_0 - \delta, \lambda_0]$. Therefore, by (5.3), $\Psi_j(\lambda)\, y_{j+1}(\lambda) = 0$ for all $j \in [0, N]_\mathbb{Z}$ and $\lambda \in (\lambda_0 - \delta, \lambda_0]$, showing that $y(\lambda_0) = Y^{[0]}(\lambda_0)\, c$ is a degenerate solution at $\lambda_0$. The proof is complete.                                                                                   $\square$

*Remark 5.13* The proof of Theorem 5.11 shows that, under (5.3), for any fixed $k \in [0, N]_\mathbb{Z}$ and any $c \in \mathrm{Ker}\, X_{k+1}(\lambda_0^-)$, there exists $\delta > 0$ such that the function $Y_j(\lambda)\, c$ is independent of $\lambda \in (\lambda_0 - \delta, \lambda_0]$ for all $j \in [0, k+1]_\mathbb{Z}$, where $Y(\lambda)$ is a conjoined basis of $(\mathrm{SDS}_\lambda)$ satisfying (5.20).

*Remark 5.14* Of course, similar statements as above can be proven for functions defined in the right neighborhood of $\lambda_0$, say for $\lambda \in [\lambda_0, \lambda_0 + \delta)$. For example, when $c \in \mathrm{Ker}\, X_{k+1}(\lambda_0^+)$, there exists $\delta > 0$ such that $Y_j(\lambda)\, c$ is independent of $\lambda \in [\lambda_0, \lambda_0 + \delta)$ for all $j \in [0, k+1]_\mathbb{Z}$.

### 5.1.3 Oscillation Theorems for Constant Rank of $\mathcal{B}_k(\lambda)$

In this section we establish the main results on the oscillation properties of system $(\mathrm{SDS}_\lambda)$ under the additional restriction (5.8), i.e.,

$$\mathrm{rank}\, \mathcal{B}_k(\lambda) \text{ is constant for } \lambda \in \mathbb{R}. \tag{5.38}$$

Later, in Sect. 5.6 we completely omit this assumption using the comparative index tools presented in Chaps. 3 and 4.

Recall the definition of focal points and their multiplicities for conjoined bases of $(\mathrm{SDS}_\lambda)$. According to Definition 4.1, a conjoined basis $Y(\lambda) = (X(\lambda), U(\lambda))$

of (SDS$_\lambda$) has a *focal point* in the real interval $(k, k+1]$ provided

$$m_k(\lambda) := \operatorname{rank} M_k(\lambda) + \operatorname{ind} P_k(\lambda) \geq 1 \qquad (5.39)$$

and then the number $m_k(\lambda)$ is its *multiplicity*, where

$$\left.\begin{aligned}
M_k(\lambda) &:= [I - X_{k+1}(\lambda)\, X_{k+1}^\dagger(\lambda)]\, \mathcal{B}_k(\lambda), \\
T_k(\lambda) &:= I - M_k^\dagger(\lambda)\, M_k(\lambda), \\
P_k(\lambda) &:= T_k(\lambda)\, X_k(\lambda)\, X_{k+1}^\dagger(\lambda)\, \mathcal{B}_k(\lambda)\, T_k(\lambda),
\end{aligned}\right\} \qquad (5.40)$$

and the matrix $P_k(\lambda)$ is symmetric. By this definition, all algebraic properties of the multiplicities of focal points $m_k$ formulated in Sect. 4.1.1 remain true for $m_k(\lambda)$ and $\lambda \in \mathbb{R}$. For example, in the subsequent proofs, we will use that the matrix $M_k(\lambda)$ defined in (5.40) satisfies

$$\operatorname{rank} M_k(\lambda) = \operatorname{rank} N_k(\lambda), \quad N_k(\lambda) := [I - X_{k+1}^\dagger(\lambda)\, X_{k+1}(\lambda)]\, X_k^T(\lambda), \qquad (5.41)$$

which gives the number of focal points of the conjoined basis $Y(\lambda)$ located at $k+1$ in terms of $X_k(\lambda)$ and $X_{k+1}(\lambda)$ only, i.e., without explicitly appearing $\mathcal{B}_k(\lambda)$ (see Definition 4.9 and Proposition 4.13).

**Theorem 5.15 (Local Oscillation Theorem I)** *Assume* (5.3). *Let* $Y(\lambda)$ *be a conjoined basis of* (SDS$_\lambda$) *with* (5.20). *Fix* $k \in [0, N]_\mathbb{Z}$ *and suppose* (5.38). *As in* (5.39), *let* $m_k(\lambda)$ *denote the number of focal points of* $Y(\lambda)$ *in* $(k, k+1]$. *Then* $m_k(\lambda^-)$ *and* $m_k(\lambda^+)$ *exist and for all* $\lambda \in \mathbb{R}$

$$m_k(\lambda^+) = m_k(\lambda) \leq n, \qquad (5.42)$$

$$m_k(\lambda^+) - m_k(\lambda^-) = \Delta\left[\operatorname{rank} X_k(\lambda^-) - \operatorname{rank} X_k(\lambda)\right]. \qquad (5.43)$$

The proof of Theorem 5.15 will be presented in Sect. 5.1.6. Next we establish further results based on Theorem 5.15. Denote by, including the multiplicities,

$$n_1(\lambda) := \text{ the number of focal points of } Y(\lambda) \text{ in } (0, N+1], \qquad (5.44)$$

i.e., $n_1(\lambda) = l(Y(\lambda), 0, N+1)$ according to the notation in (4.10).

**Theorem 5.16 (Local Oscillation Theorem II)** *Assume that conditions* (5.3) *and* (5.38) *hold for all* $k \in [0, N]_\mathbb{Z}$. *Let* $Y(\lambda)$ *be a conjoined basis of* (SDS$_\lambda$) *such that* (5.20) *holds. Then* $n_1(\lambda^-)$ *and* $n_1(\lambda^+)$ *exist and for all* $\lambda \in \mathbb{R}$

$$n_1(\lambda^+) = n_1(\lambda) \leq (N+1)\, n < \infty, \qquad (5.45)$$

$$n_1(\lambda^+) - n_1(\lambda^-) = \operatorname{rank} X_{N+1}(\lambda^-) - \operatorname{rank} X_{N+1}(\lambda) \geq 0. \qquad (5.46)$$

*Hence, the function $n_1(\cdot)$ is nondecreasing on $\mathbb{R}$, the limit*

$$m := \lim_{\lambda \to -\infty} n_1(\lambda) \tag{5.47}$$

*exists with $m \in [0, (N+1)n]_{\mathbb{Z}}$, so that for a suitable $\lambda_0 < 0$ we have*

$$n_1(\lambda) \equiv m, \quad \operatorname{rank} X_{N+1}(\lambda^-) - \operatorname{rank} X_{N+1}(\lambda) \equiv 0, \quad \lambda \leq \lambda_0. \tag{5.48}$$

*Proof* Since $n_1(\lambda) = \sum_{k=0}^{N} m_k(\lambda)$ for all $\lambda \in \mathbb{R}$ with $m_k(\lambda)$ given in (5.39), the statement in (5.45) follows directly from (5.42). The expression in (5.46) is then a telescope sum of the expression in (5.43). This yields for all $\lambda \in \mathbb{R}$

$$n_1(\lambda^+) - n_1(\lambda^-) = \operatorname{rank} X_{N+1}(\lambda^-) - \operatorname{rank} X_{N+1}(\lambda) - \operatorname{rank} X_0(\lambda^-) + \operatorname{rank} X_0(\lambda).$$

But since by (5.20) the initial conditions of $Y(\lambda)$ do not depend on $\lambda$, we have $\operatorname{rank} X_0(\lambda^-) = \operatorname{rank} X_0(\lambda)$ for all $\lambda \in \mathbb{R}$, so that the statement in (5.46) follows. From the two conditions (5.45) and (5.46), we then have that the function $n_1(\cdot)$ is nondecreasing on $\mathbb{R}$. Since the values of $n_1(\lambda)$ are nonnegative integers, the limit in (5.47) exists and $m \in \mathbb{N} \cup \{0\}$. Consequently, $n_1(\lambda) \equiv m$ for all $\lambda$ sufficiently negative, say for all $\lambda \leq \lambda_0$ for some $\lambda_0 < 0$, so that $n_1(\lambda^+) - n_1(\lambda^-) \equiv 0$ for $\lambda \leq \lambda_0$. Applying (5.46) once more then yields the second equation in (5.48). □

Now we apply the above local oscillation theorem to the principal solution $Y^{[0]}(\lambda) = (X^{[0]}(\lambda), U^{[0]}(\lambda))$ of (SDS$_\lambda$) at $k = 0$. In this case we have from system (SDS$_\lambda$) and (5.40) that

$$Y_0^{[0]}(\lambda) = \begin{pmatrix} 0 \\ I \end{pmatrix}, \quad Y_1^{[0]}(\lambda) = \begin{pmatrix} \mathcal{B}_0(\lambda) \\ \mathcal{D}_0(\lambda) \end{pmatrix}, \quad \mathcal{M}_0(\lambda) = 0, \quad \mathcal{T}_0(\lambda) = I, \quad P_0(\lambda) = 0.$$

This means that the principal solution $Y^{[0]}(\lambda)$ has no forward focal points in the interval $(0, 1]$ for all $\lambda \in \mathbb{R}$, and hence, according to (5.44), we have

$$n_1(\lambda) = \text{ the number of focal points of } Y^{[0]}(\lambda) \text{ in } (1, N+1], \tag{5.49}$$

i.e., $n_1(\lambda) = l(Y^{[0]}(\lambda), 0, N+1)$ according to (4.10). We also denote by, including the multiplicities,

$$n_2(\lambda) := \text{ the number of finite eigenvalues of (E) in } (-\infty, \lambda]. \tag{5.50}$$

Then from this definition, we have

$$n_2(\lambda^+) = n_2(\lambda), \quad n_2(\lambda) - n_2(\lambda^-) = \theta(\lambda) \quad \text{for all } \lambda \in \mathbb{R}, \tag{5.51}$$

i.e., the difference $n_2(\lambda) - n_2(\lambda^-)$ gives the number of finite eigenvalues at $\lambda$.

**Theorem 5.17 (Global Oscillation Theorem)** *Assume that (5.3) and (5.38) hold for all $k \in [0, N]_{\mathbb{Z}}$. Then with the notation (5.44) and (5.50), we have for all $\lambda \in \mathbb{R}$*

$$n_1(\lambda^+) = n_1(\lambda) \leq Nn, \tag{5.52}$$

$$n_2(\lambda^+) = n_2(\lambda) < \infty, \tag{5.53}$$

$$n_2(\lambda^+) - n_2(\lambda^-) = n_1(\lambda^+) - n_1(\lambda^-) \geq 0, \tag{5.54}$$

*and there exists $m \in [0, Nn]_{\mathbb{Z}}$ such that*

$$n_1(\lambda) = n_2(\lambda) + m \qquad \text{for all } \lambda \in \mathbb{R}. \tag{5.55}$$

*Moreover, for a suitable $\lambda_0 < 0$, we have*

$$n_2(\lambda) \equiv 0 \quad \text{and} \quad n_1(\lambda) \equiv m \quad \text{for all } \lambda \leq \lambda_0. \tag{5.56}$$

*Proof* Conditions (5.53) and (5.54) follow directly from (5.51) and (5.46). Since both functions $n_1(\cdot)$ and $n_2(\cdot)$ are right-continuous and (5.54) holds, then they must differ on $\mathbb{R}$ by a constant $l \in \mathbb{R}$. But by (5.48), we have $n_1(\lambda) \equiv m$ for all $\lambda \leq \lambda_0$ with $m$ given in (5.47). Taking into account (5.49), we see that $m \leq Nn$ and the statement in (5.55) follows. From (5.55) we obtain in turn that $n_2(\lambda) \equiv 0$ for all $\lambda \leq \lambda_0$. $\qquad\square$

**Corollary 5.18** *Under the assumptions of Theorem 5.17, the finite eigenvalues of (E) are bounded from below and from above.*

*Proof* This result follows from (5.56), since $n_2(\lambda) \equiv 0$ for all $\lambda \leq \lambda_0$ means that there are no finite eigenvalues of (E) in the interval $(-\infty, \lambda_0]$. $\qquad\square$

We note that the boundedness of the finite eigenvalues of (E) from above follows from equation (5.55) and estimate (5.52), which is a consequence of the discreteness of the problem. In the continuous time case, the number of focal points (and hence the number of finite eigenvalues) can be unbounded from above; see, e.g., [205, Theorems 7.6.3 and 7.7.1].

The next three subsections are devoted to the proof of Theorem 5.15. This proof is based on a construction of suitable partitioned matrices and an auxiliary symplectic system between the indices $k$ and $k+1$. Note that the entire construction is valid for an arbitrary coefficient $\mathcal{B}_k(\lambda)$. Only at the end assumption (5.38) is invoked in order to apply the index theorem in Corollary 1.86.

## 5.1.4 Construction of Auxiliary Conjoined Basis

In the construction below, we assume that $Y(\lambda)$ is a given conjoined basis of $(\mathrm{SDS}_\lambda)$ whose initial conditions do not depend on $\lambda$, i.e., satisfying (5.20), and condition (5.3) holds. We also fix an index $k \in [0, N]_{\mathbb{Z}}$ and a number $\lambda_0 \in \mathbb{R}$. By Theorem 5.3, we know that (5.13) holds, so that

$$r := \operatorname{rank} X_{k+1}(\lambda_0^-) \equiv \operatorname{rank} X_{k+1}(\lambda) \quad \text{for all } \lambda \in (\lambda_0 - \delta, \lambda_0) \tag{5.57}$$

exists for some $\delta > 0$. The number $\delta$ will also appear in the construction below.

**Lemma 5.19** *There exist orthogonal matrices* $\mathcal{P}, \mathcal{Q} \in \mathbb{R}^{n \times n}$ *such that*

$$\mathcal{Q} X_{k+1}(\lambda) \mathcal{P} = \begin{pmatrix} X_{11}(\lambda) & 0_{r \times (n-r)} \\ 0 & 0_{n-r} \end{pmatrix}, \quad \mathcal{Q} U_{k+1}(\lambda) \mathcal{P} = \begin{pmatrix} U_{11}(\lambda) & 0 \\ U_{21}(\lambda) & U_{22} \end{pmatrix}, \tag{5.58}$$

*for all* $\lambda \in (\lambda_0 - \delta, \lambda_0]$, *and the matrix*

$$X_{11}^T(\lambda) U_{11}(\lambda) \quad \text{is symmetric for all } \lambda \in (\lambda_0 - \delta, \lambda_0]. \tag{5.59}$$

*Here* $X_{11}(\lambda) \in \mathbb{R}^{r \times r}$ *is invertible for all* $\lambda \in (\lambda_0 - \delta, \lambda_0)$, *and* $U_{22} \in \mathbb{R}^{(n-r) \times (n-r)}$ *is invertible.*

*Proof* Let $\mathcal{P}_1 \in \mathbb{R}^{n \times r}$ and $\mathcal{P}_2 \in \mathbb{R}^{n \times (n-r)}$ be matrices whose columns form orthonormal bases for $\operatorname{Im} X_{k+1}^T(\lambda_0^-)$ and $\operatorname{Ker} X_{k+1}(\lambda_0^-)$, respectively. Then since $[\operatorname{Ker} X_{k+1}(\lambda_0^-)]^{\perp} = \operatorname{Im} X_{k+1}^T(\lambda_0^-)$, the matrix $\mathcal{P} := (\mathcal{P}_1, \mathcal{P}_2) \in \mathbb{R}^{n \times n}$ is orthogonal,

$$\operatorname{Im} \mathcal{P}_2 = \operatorname{Ker} X_{k+1}(\lambda_0^-) \equiv \operatorname{Ker} X_{k+1}(\lambda) \quad \text{for all } \lambda \in (\lambda_0 - \delta, \lambda_0). \tag{5.60}$$

$$X_{k+1}(\lambda) \mathcal{P} = \left( X_{k+1}(\lambda) \mathcal{P}_1, \ 0_{n \times (n-r)} \right) \quad \text{for all } \lambda \in (\lambda_0 - \delta, \lambda_0]. \tag{5.61}$$

Note that (5.61) indeed holds also at $\lambda = \lambda_0$ by the continuity of $X_{k+1}$ in $\lambda$. Let $\mathcal{Q}_1 \in \mathbb{R}^{r \times n}$ and $\mathcal{Q}_2 \in \mathbb{R}^{(n-r) \times n}$ be matrices whose columns form orthonormal bases for $\operatorname{Im} X_{k+1}(\lambda_1)$ and $\operatorname{Ker} X_{k+1}^T(\lambda_1)$, respectively, for some fixed $\lambda_1 \in (\lambda_0 - \delta, \lambda_0)$. Then the matrix $\mathcal{Q} := \begin{pmatrix} \mathcal{Q}_1 \\ \mathcal{Q}_2 \end{pmatrix} \in \mathbb{R}^{n \times n}$ is orthogonal, $\operatorname{Im} \mathcal{Q}_2^T = \operatorname{Ker} X_{k+1}^T(\lambda_1)$, and

$$\mathcal{Q} X_{k+1}(\lambda_1) \mathcal{P} = \begin{pmatrix} \mathcal{Q}_1 \\ \mathcal{Q}_2 \end{pmatrix} X_{k+1}(\lambda_1) \left( \mathcal{P}_1, \mathcal{P}_2 \right) = \begin{pmatrix} X_{11} & 0_{r \times (n-r)} \\ 0 & 0_{n-r} \end{pmatrix}, \tag{5.62}$$

where $X_{11} := \mathcal{Q}_1 X_{k+1}(\lambda_1) \mathcal{P}_1 \in \mathbb{R}^{r \times r}$ has rank $X_{11} = r$, i.e., $X_{11}$ is invertible. Define now

$$\begin{pmatrix} X_{11}(\lambda) & X_{12}(\lambda) \\ X_{21}(\lambda) & X_{22}(\lambda) \end{pmatrix} := \mathcal{Q} X_{k+1}(\lambda) \mathcal{P}, \qquad \lambda \in (\lambda_0 - \delta, \lambda_0], \tag{5.63}$$

$$\begin{pmatrix} U_{11}(\lambda) & U_{12}(\lambda) \\ U_{21}(\lambda) & U_{22}(\lambda) \end{pmatrix} := \mathcal{Q} U_{k+1}(\lambda) \mathcal{P}, \qquad \lambda \in (\lambda_0 - \delta, \lambda_0], \tag{5.64}$$

so that by (5.61) and (5.62)

$$X_{12}(\lambda) \equiv 0, \quad X_{22}(\lambda) \equiv 0 \qquad \text{for all } \lambda \in (\lambda_0 - \delta, \lambda_0],$$

$$X_{11}(\lambda_1) = \mathcal{Q}_1 X_{k+1}(\lambda_1) \mathcal{P}_1 = X_{11} \text{ is invertible}, \quad X_{21}(\lambda_1) = \mathcal{Q}_2 X_{k+1}(\lambda_1) \mathcal{P}_1 = 0.$$

Moreover, since the matrices $\mathcal{P}$ and $\mathcal{Q}$ are orthogonal, we have for all $\lambda \in (\lambda_0 - \delta, \lambda_0]$

$$X_{k+1}(\lambda) = \mathcal{Q}^T \begin{pmatrix} X_{11}(\lambda) & 0 \\ X_{21}(\lambda) & 0 \end{pmatrix} \mathcal{P}^T, \quad U_{k+1}(\lambda) = \mathcal{Q}^T \begin{pmatrix} U_{11}(\lambda) & U_{12}(\lambda) \\ U_{21}(\lambda) & U_{22}(\lambda) \end{pmatrix} \mathcal{P}^T. \quad (5.65)$$

Since $Y(\lambda)$ is a conjoined basis of $(\text{SDS}_\lambda)$, the matrix $X_{k+1}^T(\lambda) U_{k+1}(\lambda)$ symmetric and rank $Y_{k+1}(\lambda) = n$ for all $\lambda \in \mathbb{R}$. Hence, with $\lambda = \lambda_1$, we have

$$X_{k+1}^T(\lambda_1) U_{k+1}(\lambda_1) = \mathcal{P} \begin{pmatrix} X_{11}^T U_{11}(\lambda_1) & X_{11}^T U_{12}(\lambda_1) \\ 0 & 0 \end{pmatrix} \mathcal{P}^T.$$

This implies that $X_{11}^T U_{12}(\lambda_1) = 0$ and since $X_{11}$ is invertible, $U_{12}(\lambda_1) = 0$. Now from (5.64), we have $U_{12}(\lambda) = \mathcal{Q}_1 U_{k+1}(\lambda) \mathcal{P}_2$ and $U_{22}(\lambda) = \mathcal{Q}_2 U_{k+1}(\lambda) \mathcal{P}_2$ for all $\lambda \in (\lambda_0 - \delta, \lambda_0]$, which implies by Remark 5.13 that the functions $U_{12}(\lambda)$ and $U_{22}(\lambda)$ do not depend on $\lambda \in (\lambda_0 - \delta, \lambda_0]$. Thus, since $U_{12}(\lambda_1) = 0$, we get

$$U_{12}(\lambda) \equiv 0, \quad U_{22}(\lambda) \equiv U_{22}(\lambda_0^-) =: U_{22} \in \mathbb{R}^{(n-r) \times (n-r)} \quad \text{for all } \lambda \in (\lambda_0 - \delta, \lambda_0].$$

This proves the second formula in (5.58). Moreover, since

$$n = \text{rank} \left( X_{k+1}^T(\lambda_1), \ U_{k+1}^T(\lambda_1) \right) = \text{rank} \begin{pmatrix} X_{11}^T & 0_{r \times (n-r)} & U_{11}^T(\lambda_1) & U_{21}^T(\lambda_1) \\ 0 & 0_{n-r} & 0_{(n-r) \times r} & U_{22}^T \end{pmatrix},$$

it follows that rank $U_{22} = n - r$, that is, $U_{22}$ is invertible. Next, for $\lambda \in (\lambda_0 - \delta, \lambda_0]$, the matrix

$$X_{k+1}^T(\lambda) U_{k+1}(\lambda) = \mathcal{P} \begin{pmatrix} X_{11}^T(\lambda) U_{11}(\lambda) + X_{21}^T(\lambda) U_{21}(\lambda) & X_{21}^T(\lambda) U_{22} \\ 0 & 0 \end{pmatrix} \mathcal{P}^T$$
$$(5.66)$$

is symmetric, so that $X_{21}^T(\lambda) U_{22} \equiv 0$ on $(\lambda_0 - \delta, \lambda_0]$. But since $U_{22}$ is invertible, we get $X_{21}(\lambda) \equiv 0$ on $(\lambda_0 - \delta, \lambda_0]$, showing through (5.65) that also the first formula in (5.58) holds. In addition, since rank $\mathcal{Q} X_{k+1}(\lambda) \mathcal{P} = \text{rank} X_{k+1}(\lambda) \equiv r$ for all $\lambda \in (\lambda_0 - \delta, \lambda_0)$ and rank $X_{11}(\lambda_1) = \text{rank} X_{11} = r$, we get from (5.58) that rank $X_{11}(\lambda) \equiv r$ on $(\lambda_0 - \delta, \lambda_0)$, and so $X_{11}(\lambda)$ is invertible for all $\lambda \in (\lambda_0 - \delta, \lambda_0)$. Finally, from (5.66) we obtain that

$$X_{k+1}^T(\lambda) U_{k+1}(\lambda) = \mathcal{P} \begin{pmatrix} X_{11}^T(\lambda) U_{11}(\lambda) & 0 \\ 0 & 0 \end{pmatrix} \mathcal{P}^T$$

is symmetric for all $\lambda \in (\lambda_0 - \delta, \lambda_0]$. From this we conclude that (5.59) holds, which completes the proof.                                                                                    □

**Corollary 5.20** *In addition to Theorem 5.3(iv), we have*

$$X_{k+1}(\lambda)\, X_{k+1}^{\dagger}(\lambda) \equiv Q^T \begin{pmatrix} I_r & 0 \\ 0 & 0_{n-r} \end{pmatrix} Q, \quad X_{k+1}^{\dagger}(\lambda)\, X_{k+1}(\lambda) \equiv P \begin{pmatrix} I_r & 0 \\ 0 & 0_{n-r} \end{pmatrix} P^T$$

$$(5.67)$$

*for all* $\lambda \in (\lambda_0 - \delta, \lambda_0)$

*Proof* By Lemma 5.19 and Remark 1.60(ii), we have for all $\lambda \in (\lambda_0 - \delta, \lambda_0]$

$$X_{k+1}(\lambda) = Q^T \begin{pmatrix} X_{11}(\lambda) & 0 \\ 0 & 0 \end{pmatrix} P^T, \quad X_{k+1}^{\dagger}(\lambda) = P \begin{pmatrix} X_{11}^{\dagger}(\lambda) & 0 \\ 0 & 0 \end{pmatrix} Q.$$

And since by Lemma 5.19 the matrix $X_{11}(\lambda)$ is invertible on $(\lambda_0 - \delta, \lambda_0)$, we get (5.67). The proof is complete.                                                                   □

Based on the result of Lemma 5.19, we define the matrices

$$\tilde{X}_{k+1}(\lambda) := \begin{pmatrix} X_{11}(\lambda) & 0_{r\times(n-r)} \\ 0 & 0_r \end{pmatrix}, \quad \tilde{U}_{k+1}(\lambda) := \begin{pmatrix} U_{11}(\lambda) & 0 \\ U_{21}(\lambda) & U_{22} \end{pmatrix}, \quad \lambda \in \mathbb{R}.$$

$$(5.68)$$

Then by (5.59) the matrix $\tilde{X}_{k+1}^T(\lambda)\, \tilde{U}_{k+1}(\lambda)$ is symmetric for all $\lambda \in (\lambda_0 - \delta, \lambda_0]$. In addition, by (5.58) we have

$$\tilde{X}_{k+1}(\lambda) = Q\, X_{k+1}(\lambda)\, P, \quad \tilde{U}_{k+1}(\lambda) = Q\, U_{k+1}(\lambda)\, P, \quad \lambda \in (\lambda_0 - \delta, \lambda_0],$$

$$(5.69)$$

which yields that

$$\operatorname{rank}\left(\tilde{X}_{k+1}^T(\lambda),\ \tilde{U}_{k+1}^T(\lambda)\right) \overset{(5.69)}{=} \operatorname{rank}\left(X_{k+1}^T(\lambda),\ U_{k+1}^T(\lambda)\right) = n, \quad \lambda \in (\lambda_0 - \delta, \lambda_0].$$

Next we construct for $\lambda \in (\lambda_0 - \delta, \lambda_0]$ suitable matrices $\tilde{X}_k(\lambda)$ and $\tilde{U}_k(\lambda)$. First we observe that by Remark 5.13 with $j = k$, the functions

$$X_k(\lambda)\, P_2 \text{ and } U_k(\lambda)\, P_2 \text{ do not depend on } \lambda \in (\lambda_0 - \delta, \lambda_0],$$

because $\operatorname{Im} P_2 = \operatorname{Ker} X_{k+1}(\lambda_0^-)$. This implies that the number

$$\rho := \operatorname{rank} X_k(\lambda_0^-)\, P_2 \equiv \operatorname{rank} X_k(\lambda)\, P_2 \quad \text{for all } \lambda \in (\lambda_0 - \delta, \lambda_0) \qquad (5.70)$$

is well defined. Note that by (1.142) and (5.41), the definition of $\rho$ yields

$$
\begin{aligned}
\rho &= \operatorname{rank} \mathcal{P}_2 - \dim \left( \operatorname{Ker} X_k(\lambda) \cap \operatorname{Im} \mathcal{P}_2 \right) \\
&= \operatorname{def} X_{k+1}(\lambda) - \dim \left( \operatorname{Ker} X_k(\lambda) \cap \operatorname{Ker} X_{k+1}(\lambda) \right) \\
&= \operatorname{rank} N_k(\lambda) \overset{(5.41)}{=} \operatorname{rank} M_k(\lambda)
\end{aligned}
$$

for all $\lambda \in (\lambda_0 - \delta, \lambda_0)$. Later in this section, we will prove directly that the number $\rho = \operatorname{rank} M_k(\lambda_0^-)$. In addition, the definition of $M_k(\lambda)$, formula (5.67)(i), and (1.142) yield that

$$
\begin{aligned}
\rho &= \operatorname{rank} M_k^T(\lambda) = \operatorname{rank} \mathcal{B}_k^T(\lambda) \left[ I - X_{k+1}(\lambda) X_{k+1}^\dagger(\lambda) \right] \\
&= \operatorname{rank} \left[ I - X_{k+1}(\lambda) X_{k+1}^\dagger(\lambda) \right] - \dim \left( \operatorname{Ker} \mathcal{B}_k^T(\lambda) \cap \operatorname{Im} \left[ I - X_{k+1}(\lambda) X_{k+1}^\dagger(\lambda) \right] \right) \\
&= n - r - \dim \left( \operatorname{Ker} \mathcal{B}_k^T(\lambda) \cap \operatorname{Ker} X_{k+1}^T(\lambda) \right), \qquad \lambda \in (\lambda_0 - \delta, \lambda_0). \tag{5.71}
\end{aligned}
$$

*Remark 5.21* From equation (5.71), we can see that $\rho \leq n - r$ and that $\rho = n - r$ if and only if $\operatorname{Ker} \mathcal{B}_k^T(\lambda) \cap \operatorname{Ker} X_{k+1}^T(\lambda) = \{0\}$ on $(\lambda_0 - \delta, \lambda_0)$. This latter condition is satisfied, e.g., when $\mathcal{B}_k(\lambda)$ or $X_{k+1}(\lambda)$ is invertible on $(\lambda_0 - \delta, \lambda_0)$.

We now refine the structure of the above matrices to partition $X_k(\lambda)$ and $U_k(\lambda)$. Since $\rho \leq \operatorname{rank} X_k(\lambda_0^-)$, we may put

$$
\tilde{r} := r_k - \rho, \quad \text{where } r_k := \operatorname{rank} X_k(\lambda_0^-). \tag{5.72}
$$

Then since $\operatorname{rank} X_k(\lambda_0^-) \mathcal{P}_2 = \rho$, we must have $\operatorname{rank} X_k(\lambda_0^-) \mathcal{P}_1 = r_k - \rho = \tilde{r}$. But the matrix $X_k(\lambda_0^-) \mathcal{P}_1 \in \mathbb{R}^{n \times r}$, which implies $\tilde{r} \leq r$. The block structure of the matrices below is such that their total dimension $n$ is partitioned as

$$
n = \tilde{r} + (r - \tilde{r}) + (n - r - \rho) + \rho.
$$

**Lemma 5.22** *There are orthogonal matrices $\mathcal{P}$, $\mathcal{Q}$, $\tilde{\mathcal{Q}}$ satisfying Lemma 5.19 and*

$$
\tilde{\mathcal{Q}} X_k(\lambda) \mathcal{P} = \begin{pmatrix} \tilde{X}_{11}(\lambda) & 0 & 0 & 0_{\tilde{r} \times \rho} \\ 0 & 0_{r - \tilde{r}} & 0 & 0 \\ 0 & 0 & 0_{n - r - \rho} & 0 \\ \tilde{X}_{41}(\lambda) & \tilde{X}_{42} & 0 & \tilde{X}_{44} \end{pmatrix}, \quad \lambda \in (\lambda_0 - \delta, \lambda_0], \tag{5.73}
$$

*where $\tilde{X}_{11}(\lambda) \in \mathbb{R}^{\tilde{r} \times \tilde{r}}$ is invertible for all $\lambda \in (\lambda_0 - \delta, \lambda_0)$ and $\tilde{X}_{44} \in \mathbb{R}^{\rho \times \rho}$ is invertible.*

*Proof* Let the matrices $\mathcal{P}$ and $\mathcal{Q}$ be from Lemma 5.19. Since $X_k(\lambda)\,\mathcal{P}_2 \in \mathbb{R}^{n\times(n-r)}$ and rank $X_k(\lambda)\,\mathcal{P}_2 = \rho$ for $\lambda \in (\lambda_0 - \delta, \lambda_0)$, there are orthogonal matrices $\tilde{\mathcal{Q}} \in \mathbb{R}^{n\times n}$ and $\bar{P}_2 \in \mathbb{R}^{(n-r)\times(n-r)}$ such that

$$\tilde{\mathcal{Q}}\, X_k(\lambda)\,\mathcal{P}_2\,\bar{P}_2 = \begin{pmatrix} 0_{\tilde{r}} & 0_{\tilde{r}\times\rho} \\ 0 & 0_{(r-\tilde{r})\times\rho} \\ 0 & 0_{(n-r-\rho)\times\rho} \\ 0 & \tilde{X}_{44} \end{pmatrix}, \quad \lambda \in (\lambda_0 - \delta, \lambda_0), \tag{5.74}$$

where $\tilde{X}_{44} \in \mathbb{R}^{\rho\times\rho}$ is invertible. Let $\lambda_1 \in (\lambda_0 - \delta, \lambda_0)$ be fixed. Since $\tilde{\mathcal{Q}}\, X_k(\lambda_1)\,\mathcal{P}_1 \in \mathbb{R}^{n\times r}$, there are orthogonal matrices $\bar{Q}_1 \in \mathbb{R}^{(n-\rho)\times(n-\rho)}$ and $\bar{P}_1 \in \mathbb{R}^{r\times r}$ such that

$$\bar{Q}\,\tilde{\mathcal{Q}}\, X_k(\lambda_1)\,\mathcal{P}_1\,\bar{P}_1 = \begin{pmatrix} \tilde{X}_{11}(\lambda_1) & 0_{\tilde{r}\times(r-\tilde{r})} \\ 0 & 0_{r-\tilde{r}} \\ 0 & 0_{(n-r-\rho)\times(r-\tilde{r})} \\ \tilde{X}_{41}(\lambda_1) & \tilde{X}_{42}(\lambda_1) \end{pmatrix}, \quad \bar{Q} := \begin{pmatrix} \bar{Q}_1 & 0 \\ 0 & I_\rho \end{pmatrix} \in \mathbb{R}^{n\times n},$$

where $\tilde{X}_{11}(\lambda_1) \in \mathbb{R}^{\tilde{r}\times\tilde{r}}$ is invertible, $\tilde{X}_{41}(\lambda_1) \in \mathbb{R}^{\rho\times\tilde{r}}$, and $\tilde{X}_{42}(\lambda_1) \in \mathbb{R}^{\rho\times(r-\tilde{r})}$. Note that the multiplication of equation (5.74) by the orthogonal matrix $\bar{Q}$ from the left does not change the structure of (5.74). Therefore, since the product $\bar{Q}\tilde{\mathcal{Q}}$ is an orthogonal matrix, we may assume without loss of generality that the matrix $\tilde{\mathcal{Q}}$ in (5.74) is such that

$$\tilde{\mathcal{Q}}\, X_k(\lambda_1)\,\mathcal{P}_1\,\bar{P}_1 = \begin{pmatrix} \tilde{X}_{11}(\lambda_1) & 0_{\tilde{r}\times(r-\tilde{r})} \\ 0 & 0_{r-\tilde{r}} \\ 0 & 0_{(n-r-\rho)\times(r-\tilde{r})} \\ \tilde{X}_{41}(\lambda_1) & \tilde{X}_{42}(\lambda_1) \end{pmatrix},$$

$$\tilde{\mathcal{Q}}\, X_k(\lambda)\,\mathcal{P}_2\,\bar{P}_2 = \begin{pmatrix} 0_{\tilde{r}} & 0_{\tilde{r}\times\rho} \\ 0 & 0_{(r-\tilde{r})\times\rho} \\ 0 & 0_{(n-r-\rho)\times\rho} \\ 0 & \tilde{X}_{44} \end{pmatrix}$$

for $\lambda \in (\lambda_0 - \delta, \lambda_0)$ with both $\tilde{X}_{11}(\lambda_1) \in \mathbb{R}^{\tilde{r}\times\tilde{r}}$ and $\tilde{X}_{44} \in \mathbb{R}^{\rho\times\rho}$ invertible, $\tilde{X}_{41}(\lambda_1) \in \mathbb{R}^{\rho\times\tilde{r}}$, and $\tilde{X}_{42}(\lambda_1) \in \mathbb{R}^{\rho\times(r-\tilde{r})}$. Next we observe that the multiplication of the equations in (5.58) by the orthogonal matrix

$$\bar{P} := \begin{pmatrix} \bar{P}_1 & 0_{r\times(n-r)} \\ 0_{(n-r)\times r} & \bar{P}_2 \end{pmatrix} \in \mathbb{R}^{n\times n}$$

from the right does not change the structure of the formulas in (5.58). Therefore, since the matrix $\mathcal{P}\bar{P}$ is orthogonal, we may assume without loss of generality that the matrix $\mathcal{P}$ is such that

$$\tilde{\mathcal{Q}}\,X_k(\lambda_1)\,\mathcal{P}_1 = \begin{pmatrix} \tilde{X}_{11}(\lambda_1) & 0_{\tilde{r}\times(r-\tilde{r})} \\ 0 & 0_{r-\tilde{r}} \\ 0 & 0_{(n-r-\rho)\times(r-\tilde{r})} \\ \tilde{X}_{41}(\lambda_1) & \tilde{X}_{42}(\lambda_1) \end{pmatrix}, \quad \tilde{\mathcal{Q}}\,X_k(\lambda)\,\mathcal{P}_2 = \begin{pmatrix} 0_{\tilde{r}} & 0_{\tilde{r}\times\rho} \\ 0 & 0_{(r-\tilde{r})\times\rho} \\ 0 & 0_{(n-r-\rho)\times\rho} \\ 0 & \tilde{X}_{44} \end{pmatrix}$$

$$(5.75)$$

for all $\lambda \in (\lambda_0 - \delta, \lambda_0)$ with $\tilde{X}_{11}(\lambda_1) \in \mathbb{R}^{\tilde{r}\times\tilde{r}}$ and $\tilde{X}_{44} \in \mathbb{R}^{\rho\times\rho}$ invertible, $\tilde{X}_{41}(\lambda_1) \in \mathbb{R}^{\rho\times\tilde{r}}$, and $\tilde{X}_{42}(\lambda_1) \in \mathbb{R}^{\rho\times(r-\tilde{r})}$. Therefore, for $\lambda = \lambda_1$ we have from (5.75)

$$\tilde{\mathcal{Q}}\,X_k(\lambda_1)\,\mathcal{P} = \begin{pmatrix} \tilde{X}_{11}(\lambda_1) & 0_{\tilde{r}\times(r-\tilde{r})} & 0_{\tilde{r}} & 0_{\tilde{r}\times\rho} \\ 0 & 0_{r-\tilde{r}} & 0 & 0_{(r-\tilde{r})\times\rho} \\ 0 & 0_{(n-r-\rho)\times(r-\tilde{r})} & 0_{n-r-\rho} & 0_{(n-r-\rho)\times\rho} \\ \tilde{X}_{41}(\lambda_1) & \tilde{X}_{42}(\lambda_1) & 0 & \tilde{X}_{44} \end{pmatrix}$$

$$(5.76)$$

with $\tilde{X}_{11}(\lambda_1) \in \mathbb{R}^{\tilde{r}\times\tilde{r}}$ and $\tilde{X}_{44} \in \mathbb{R}^{\rho\times\rho}$ invertible. We now calculate the sets

$$\operatorname{Im}\mathcal{K} := \operatorname{Ker}\tilde{\mathcal{Q}}\,X_k(\lambda_1)\,\mathcal{P}, \quad \operatorname{Im}\tilde{\mathcal{K}} := \operatorname{Ker}\mathcal{P}^T X_k^T(\lambda_1)\,\tilde{\mathcal{Q}}^T \qquad (5.77)$$

for some matrices $\mathcal{K}$ and $\tilde{\mathcal{K}}$. Since $\operatorname{rank}\tilde{\mathcal{Q}}\,X_k(\lambda_1)\,\mathcal{P} = \operatorname{rank}X(\lambda_1) = r_k = \tilde{r} + \rho$, we have $\operatorname{def}\tilde{\mathcal{Q}}\,X_k(\lambda_1)\,\mathcal{P} = n - \tilde{r} - \rho$. And since $\operatorname{rank}\tilde{\mathcal{Q}}\,X_k(\lambda_1)\,\mathcal{P} = \operatorname{rank}\mathcal{P}^T X_k^T(\lambda_1)\,\tilde{\mathcal{Q}}^T$, we have $\operatorname{def}\mathcal{P}^T X_k^T(\lambda_1)\,\tilde{\mathcal{Q}}^T = n - \tilde{r} - \rho$ as well. Therefore, the matrices $\mathcal{K}, \tilde{\mathcal{K}} \in \mathbb{R}^{n\times(n-\tilde{r}-\rho)}$. Denote within the refined block structure

$$\mathcal{K} = (K_{ij}), \qquad \tilde{\mathcal{K}} = (\tilde{K}_{ij}), \quad \text{where } i \in \{1, 2\}, \ j \in \{1, 2, 3, 4\}.$$

Then the first equality in (5.77) yields through (5.76) that $K_{11} = 0$ and $K_{12} = 0$. Then with the choice $K_{21} := I$, we get $K_{41} = -\tilde{X}_{44}^{-1}\tilde{X}_{42}(\lambda_1)$. With the additional choice $K_{22} := 0$, $K_{31} := 0$, $K_{32} := I$, and $K_{44} := 0$, we obtain

$$\mathcal{K} = \begin{pmatrix} 0_{\tilde{r}\times(r-\tilde{r})} & 0_{\tilde{r}\times(n-r-\rho)} \\ I_{r-\tilde{r}} & 0_{(r-\tilde{r})\times(n-r-\rho)} \\ 0_{(n-r-\rho)\times(r-\tilde{r})} & I_{n-r-\rho} \\ -\tilde{X}_{44}^{-1}\tilde{X}_{42}(\lambda_1) & 0_{\rho\times(n-r-\rho)} \end{pmatrix}.$$

But since $\operatorname{Im} \mathcal{K} = \operatorname{Ker} \tilde{\mathcal{Q}} X_k(\lambda_1) \mathcal{P}$ is independent on $\lambda \in (\lambda_0 - \delta, \lambda_0)$, it follows that the matrix $\tilde{X}_{42}(\lambda) \equiv \tilde{X}_{42}$ is constant on $(\lambda_0 - \delta, \lambda_0)$. Similarly, the second equality in (5.77) yields $\tilde{K}_{41} = 0$, $\tilde{K}_{42} = 0$, and then $\tilde{K}_{11} = 0$ and $\tilde{K}_{12} = 0$. We then choose $\tilde{K}_{21} := I$, $\tilde{K}_{22} := 0$, $\tilde{K}_{31} := 0$, and $\tilde{K}_{32} := I$. Therefore we proved

$$\mathcal{K} = \begin{pmatrix} 0_{\tilde{r} \times (r - \tilde{r})} & 0_{\tilde{r} \times (n - r - \rho)} \\ I_{r - \tilde{r}} & 0_{(r - \tilde{r}) \times (n - r - \rho)} \\ 0_{(n - r - \rho) \times (r - \tilde{r})} & I_{n - r - \rho} \\ -\tilde{X}_{44}^{-1} \tilde{X}_{42} & 0_{\rho \times (n - r - \rho)} \end{pmatrix}, \quad \tilde{\mathcal{K}} = \begin{pmatrix} 0_{\tilde{r} \times (r - \tilde{r})} & 0_{\tilde{r} \times (n - r - \rho)} \\ I_{r - \tilde{r}} & 0_{(r - \tilde{r}) \times (n - r - \rho)} \\ 0_{(n - r - \rho) \times (r - \tilde{r})} & I_{n - r - \rho} \\ 0_{\rho \times (r - \tilde{r})} & 0_{\rho \times (n - r - \rho)} \end{pmatrix}.$$

Now from Theorem 5.3, the sets $\operatorname{Ker} X_k(\lambda)$ and $\operatorname{Ker} X_k^T(\lambda)$ are constant on the interval $(\lambda_0 - \delta, \lambda_0)$, then also the sets $\operatorname{Ker} \tilde{\mathcal{Q}} X_k(\lambda) \mathcal{P}$ and $\operatorname{Ker} \mathcal{P}^T X_k^T(\lambda) \tilde{\mathcal{Q}}^T$ are constant on $(\lambda_0 - \delta, \lambda_0)$, which together with (5.76) implies the equality in (5.73) for $\lambda \in (\lambda_0 - \delta, \lambda_0)$. And by the continuity of $X_k(\cdot)$, we also have the formula in (5.73) at $\lambda = \lambda_0$. Finally, from $r_k = \operatorname{rank} X_k(\lambda) = \operatorname{rank} \tilde{\mathcal{Q}} X_k(\lambda) \mathcal{P}$, equation (5.76), and the invertibility of $\tilde{X}_{44}$, we get $\operatorname{rank} \tilde{X}_{11}(\lambda) = \tilde{r}$ for all $\lambda \in (\lambda_0 - \delta, \lambda_0)$, i.e., $\tilde{X}_{11}(\lambda)$ is invertible on $(\lambda_0 - \delta, \lambda_0)$. $\qquad\square$

Within the refined block structure, we define for any $\lambda \in \mathbb{R}$ the matrices

$$\tilde{X}_k(\lambda) := \tilde{\mathcal{Q}} X_k(\lambda) \mathcal{P} = \big(\tilde{X}_{ij}(\lambda)\big), \quad \tilde{U}_k(\lambda) := \tilde{\mathcal{Q}} U_k(\lambda) \mathcal{P} = \big(\tilde{U}_{ij}(\lambda)\big), \quad (5.78)$$

where $i, j \in \{1, 2, 3, 4\}$ and $\tilde{\mathcal{Q}}, \mathcal{P} \in \mathbb{R}^{n \times n}$ are from Lemma 5.22. Then for $\lambda \in (\lambda_0 - \delta, \lambda_0]$, the matrix $\tilde{X}_k(\lambda)$ is given by formula (5.73).

**Lemma 5.23** *There are orthogonal matrices $\mathcal{P}$, $\mathcal{Q}$, $\tilde{\mathcal{Q}}$ satisfying Lemmas 5.19 and 5.22 and*

$$\tilde{\mathcal{Q}} U_k(\lambda) \mathcal{P} = \begin{pmatrix} \tilde{U}_{11}(\lambda) & \tilde{U}_{12}(\lambda) & 0 & \tilde{U}_{14} \\ \tilde{U}_{21}(\lambda) & \tilde{U}_{22}(\lambda) & 0 & \tilde{U}_{24} \\ \tilde{U}_{31}(\lambda) & \tilde{U}_{32}(\lambda) & \tilde{U}_{33} & \tilde{U}_{34} \\ \tilde{U}_{41}(\lambda) & \tilde{U}_{42} & 0 & \tilde{U}_{44} \end{pmatrix}, \quad \lambda \in (\lambda_0 - \delta, \lambda_0], \quad (5.79)$$

*where $\tilde{U}_{33} \in \mathbb{R}^{(n - r - \rho) \times (n - r - \rho)}$ is invertible.*

*Proof* Since $U_k(\lambda) \mathcal{P}_2$ is independent of $\lambda \in (\lambda_0 - \delta, \lambda_0]$ by Remark 5.13 with $j = k$, the third and fourth block columns of $\tilde{U}_k(\lambda)$ are constant in $\lambda \in (\lambda_0 - \delta, \lambda_0]$, i.e., $\tilde{U}_{ij}(\lambda) \equiv \tilde{U}_{ij}$ on $(\lambda_0 - \delta, \lambda_0]$ for $i \in \{3, 4\}$ and $j \in \{1, 2, 3, 4\}$. Since

$$\tilde{X}_k^T(\lambda) \tilde{U}_k(\lambda) \overset{(5.78)}{=} \mathcal{P}^T X_k^T(\lambda) U_k(\lambda) \mathcal{P}, \quad \lambda \in \mathbb{R},$$

the matrix $\tilde{X}_k^T(\lambda)\,\tilde{U}_k(\lambda)$ is symmetric for all $\lambda \in \mathbb{R}$. From this and (5.73), we conclude that with $\tilde{F}_{1j}(\lambda) := \tilde{X}_{11}^T(\lambda)\,\tilde{U}_{1j}(\lambda) + \tilde{X}_{41}^T(\lambda)\,\tilde{U}_{4j}(\lambda)$, $j \in \{1, 2, 3, 4\}$, the matrix

$$
\begin{pmatrix}
\tilde{F}_{11}(\lambda) & \tilde{F}_{12}(\lambda) & \tilde{F}_{13}(\lambda) & \tilde{F}_{14}(\lambda) \\
\tilde{X}_{42}^T\,\tilde{U}_{41}(\lambda) & \tilde{X}_{42}^T\,\tilde{U}_{42}(\lambda) & \tilde{X}_{42}^T\,\tilde{U}_{43} & \tilde{X}_{42}^T\,\tilde{U}_{44} \\
0 & 0 & 0 & 0 \\
\tilde{X}_{44}^T\,\tilde{U}_{41}(\lambda) & \tilde{X}_{44}^T\,\tilde{U}_{42}(\lambda) & \tilde{X}_{44}^T\,\tilde{U}_{43} & \tilde{X}_{44}^T\,\tilde{U}_{44}
\end{pmatrix}
\quad \text{is symmetric for } \lambda \in (\lambda_0 - \delta, \lambda_0].
$$

Then since $\tilde{X}_{44}$ is invertible, we obtain $\tilde{U}_{43} = 0$, and in turn since $\tilde{X}_{11}(\lambda)$ is invertible for $\lambda \in (\lambda_0 - \delta, \lambda_0)$, we have $\tilde{U}_{13} = 0$. Next, the equality $\tilde{X}_{44}^T\,\tilde{U}_{42}(\lambda) = (\tilde{X}_{42}^T\,\tilde{U}_{44})^T$ yields that $\tilde{U}_{42}(\lambda) \equiv \tilde{X}_{44}^{T-1}\tilde{U}_{44}^T\tilde{X}_{42} =: \bar{U}_{42}$ is constant on $(\lambda_0 - \delta, \lambda_0]$. Next, for every $\lambda \in (\lambda_0 - \delta, \lambda_0]$

$$
n = \operatorname{rank}\left(X_k^T(\lambda),\, U_k^T(\lambda)\right) = \operatorname{rank}\left(\tilde{X}_k^T(\lambda),\, \tilde{U}_k^T(\lambda)\right)
$$

$$
= \operatorname{rank}
\begin{pmatrix}
\tilde{X}_{11}^T(\lambda) & 0 & 0 & \tilde{X}_{41}^T(\lambda) & \tilde{U}_{11}^T(\lambda) & \tilde{U}_{21}^T(\lambda) & \tilde{U}_{31}^T(\lambda) & \tilde{U}_{41}^T(\lambda) \\
0 & 0_{r-\tilde{r}} & 0 & \tilde{X}_{42}^T & \tilde{U}_{12}^T(\lambda) & \tilde{U}_{22}^T(\lambda) & \tilde{U}_{32}^T(\lambda) & \bar{U}_{42}^T \\
0 & 0 & 0_{n-r-\rho} & 0 & 0 & \tilde{U}_{23}^T & \tilde{U}_{33}^T & 0 \\
0 & 0 & 0 & \tilde{X}_{44}^T & \tilde{U}_{14}^T & \tilde{U}_{24}^T & \tilde{U}_{34}^T & \tilde{U}_{44}^T
\end{pmatrix},
$$

where $\tilde{U}_{23} \in \mathbb{R}^{(r-\tilde{r})\times(n-r-\rho)}$, $\tilde{U}_{33} \in \mathbb{R}^{(n-r-\rho)\times(n-r-\rho)}$, and $\operatorname{rank}(\tilde{U}_{23}^T,\, \tilde{U}_{33}^T) = n - r - \rho$. Hence, there exists an orthogonal matrix $\bar{R}_2 \in \mathbb{R}^{(n-\tilde{r}-\rho)\times(n-\tilde{r}-\rho)}$ with

$$
\bar{R}_2 \begin{pmatrix} \tilde{U}_{23} \\ \tilde{U}_{33} \end{pmatrix} = \begin{pmatrix} 0_{(r-\tilde{r})\times(n-r-\rho)} \\ \bar{U}_{33} \end{pmatrix},
$$

where $\bar{U}_{33} \in \mathbb{R}^{(n-r-\rho)\times(n-r-\rho)}$ is invertible. Now the multiplication of $\tilde{X}_k(\lambda)$ and $\tilde{U}_k(\lambda)$ by the orthogonal matrix

$$
\bar{R} := \begin{pmatrix} I_{\tilde{r}} & 0 & 0 \\ 0 & \bar{R}_2 & 0 \\ 0 & 0 & I_\rho \end{pmatrix} \in \mathbb{R}^{n\times n}
$$

from the left does not change the structure of $\tilde{X}_k(\lambda)$ and $\tilde{U}_k(\lambda)$ in (5.78). Therefore, since the matrix $\bar{R}\,\tilde{Q}$ is orthogonal, we may assume without loss of generality that the matrix $\tilde{Q}$ in (5.78) is such that equation (5.79) holds. The proof of this lemma is complete. $\qquad\square$

Finally, we finish the construction of $\tilde{U}_{k+1}(\lambda)$ within the refined matrix block structure.

**Lemma 5.24** *There are orthogonal matrices* $\mathcal{P}$, $\mathcal{Q}$, $\tilde{\mathcal{Q}}$ *satisfying Lemmas 5.19 and 5.22 and 5.23, and such that the matrix* $U_{22} \in \mathbb{R}^{(n-r)\times(n-r)}$ *in* (5.58) *satisfies*

$$U_{22} = \begin{pmatrix} \bar{U}_{33} & \bar{U}_{34} \\ 0 & \bar{U}_{44} \end{pmatrix}, \quad U_{22}^{-1} = \begin{pmatrix} \bar{U}_{33}^{-1} & -\bar{U}_{33}^{-1}\,\bar{U}_{34}\,\bar{U}_{44}^{-1} \\ 0 & \bar{U}_{44}^{-1} \end{pmatrix}, \tag{5.80}$$

*where the matrices* $\bar{U}_{33} \in \mathbb{R}^{(n-r-\rho)\times(n-r-\rho)}$ *and* $\bar{U}_{44} \in \mathbb{R}^{\rho\times\rho}$ *are invertible.*

*Proof* There exists an orthogonal matrix $\bar{S}_2 \in \mathbb{R}^{(n-r)\times(n-r)}$ such that the matrix $\bar{S}_2\,U_{22}$ is upper block-triangular, i.e., we have

$$\bar{S}_2\,U_{22} = \begin{pmatrix} \bar{U}_{33} & \bar{U}_{34} \\ 0 & \bar{U}_{44} \end{pmatrix}$$

with $\bar{U}_{33} \in \mathbb{R}^{(n-r-\rho)\times(n-r-\rho)}$ and $\bar{U}_{44} \in \mathbb{R}^{\rho\times\rho}$ invertible, because $U_{22}$ is invertible. The multiplication of the formulas in (5.68) by the orthogonal matrix

$$\bar{S} := \begin{pmatrix} I_r & 0 \\ 0 & \bar{S}_2 \end{pmatrix} \in \mathbb{R}^{n\times n}$$

from the left does not change the structure of (5.68), and the matrix $\bar{S}\mathcal{Q}$ is orthogonal. Hence, we may assume without loss of generality that the matrix $\mathcal{Q}$ in Lemmas 5.19, 5.22, and 5.23 satisfies the first equation in (5.80). The formula for the inverse of $U_{22}$ is then verified by a direct calculation.                     $\square$

### 5.1.5 Construction of Auxiliary Symplectic System

Recall that the fact that $\mathcal{S}_k(\lambda)$ is symplectic implies by Lemma 1.58(iii) that $\mathcal{S}_k^T(\lambda)$ is symplectic as well. Therefore, the coefficients of system (SDS$_\lambda$) satisfy for every $k \in [0, N]_{\mathbb{Z}}$ and $\lambda \in \mathbb{R}$ the identities

$$\left.\begin{array}{ll} \mathcal{A}_k^T(\lambda)\,\mathcal{C}_k(\lambda) = \mathcal{C}_k^T(\lambda)\,\mathcal{A}_k(\lambda), & \mathcal{B}_k^T(\lambda)\,\mathcal{D}_k(\lambda) = \mathcal{D}_k^T(\lambda)\,\mathcal{B}_k(\lambda), \\ \mathcal{A}_k(\lambda)\,\mathcal{B}_k^T(\lambda) = \mathcal{B}_k(\lambda)\,\mathcal{A}_k^T(\lambda), & \mathcal{D}_k(\lambda)\,\mathcal{C}_k^T(\lambda) = \mathcal{C}_k(\lambda)\,\mathcal{D}_k^T(\lambda), \\ \mathcal{A}_k^T(\lambda)\,\mathcal{D}_k(\lambda) - \mathcal{C}_k^T(\lambda)\,\mathcal{B}_k(\lambda) = I, & \mathcal{D}_k(\lambda)\,\mathcal{A}_k^T(\lambda) - \mathcal{C}_k(\lambda)\,\mathcal{B}_k^T(\lambda) = I. \end{array}\right\}$$
$$\tag{5.81}$$

Upon differentiating the above formulas with respect to $\lambda$, we get

$$
\left.\begin{aligned}
\dot{A}_k^T(\lambda)\, C_k(\lambda) + A_k^T(\lambda)\, \dot{C}_k(\lambda) &= \dot{C}_k^T(\lambda)\, A_k(\lambda) + C_k^T(\lambda)\, \dot{A}_k(\lambda), \\
\dot{B}_k^T(\lambda)\, D_k(\lambda) + B_k^T(\lambda)\, \dot{D}_k(\lambda) &= \dot{D}_k^T(\lambda)\, B_k(\lambda) + D_k^T(\lambda)\, \dot{B}_k(\lambda), \\
\dot{A}_k^T(\lambda)\, D_k(\lambda) + A_k^T(\lambda)\, \dot{D}_k(\lambda) &= \dot{C}_k^T(\lambda)\, B_k(\lambda) + C_k^T(\lambda)\, \dot{B}_k(\lambda), \\
\dot{D}_k(\lambda)\, C_k^T(\lambda) + D_k(\lambda)\, \dot{C}_k^T(\lambda) &= \dot{C}_k(\lambda)\, D_k^T(\lambda) + C_k(\lambda)\, \dot{D}_k^T(\lambda), \\
\dot{A}_k(\lambda)\, B_k^T(\lambda) + A_k(\lambda)\, \dot{B}_k^T(\lambda) &= \dot{B}_k(\lambda)\, A_k^T(\lambda) + B_k(\lambda)\, \dot{A}_k^T(\lambda), \\
\dot{D}_k(\lambda)\, A_k^T(\lambda) + D_k(\lambda)\, \dot{A}_k^T(\lambda) &= \dot{C}_k(\lambda)\, B_k^T(\lambda) + C_k(\lambda)\, \dot{B}_k^T(\lambda).
\end{aligned}\right\}
\tag{5.82}
$$

For the matrix $\Psi_k(\lambda)$ in (5.3), we have

$$
\Psi_k(\lambda) = \begin{pmatrix} \dot{D}_k(\lambda)\, C_k^T(\lambda) - \dot{C}_k(\lambda)\, D_k^T(\lambda) & \dot{C}_k(\lambda)\, B_k^T(\lambda) - \dot{D}_k(\lambda)\, A_k^T(\lambda) \\ \dot{A}_k(\lambda)\, D_k^T(\lambda) - \dot{B}_k(\lambda)\, C_k^T(\lambda) & \dot{B}_k(\lambda)\, A_k^T(\lambda) - \dot{A}_k(\lambda)\, B_k^T(\lambda) \end{pmatrix}.
\tag{5.83}
$$

One can now see the symmetry of $\Psi_k(\lambda)$ directly from (5.82).

With the aid of the orthogonal matrices $\mathcal{P}$, $\mathcal{Q}$, and $\tilde{\mathcal{Q}}$ from Lemma 5.24, we now construct an auxiliary symplectic system between the indices $k$ and $k+1$. For every $\lambda \in \mathbb{R}$, we define the $n \times n$ matrices

$$
\tilde{A}_k(\lambda) := \mathcal{Q}\, A_k(\lambda)\, \tilde{\mathcal{Q}}^T = \begin{pmatrix} A_{11}(\lambda) & A_{12}(\lambda) \\ A_{21}(\lambda) & A_{22}(\lambda) \end{pmatrix}
$$

$$
= \begin{pmatrix} \bar{A}_{11}(\lambda) & \bar{A}_{12}(\lambda) & \bar{A}_{13}(\lambda) & \bar{A}_{14}(\lambda) \\ \bar{A}_{21}(\lambda) & \bar{A}_{22}(\lambda) & \bar{A}_{23}(\lambda) & \bar{A}_{24}(\lambda) \\ \bar{A}_{31}(\lambda) & \bar{A}_{32}(\lambda) & \bar{A}_{33}(\lambda) & \bar{A}_{34}(\lambda) \\ \bar{A}_{41}(\lambda) & \bar{A}_{42}(\lambda) & \bar{A}_{43}(\lambda) & \bar{A}_{44}(\lambda) \end{pmatrix}
$$

and similarly

$$
\tilde{B}_k(\lambda) := \mathcal{Q}\, B_k(\lambda)\, \tilde{\mathcal{Q}}^T = \big(B_{ij}(\lambda)\big)_{i,j \in \{1,2\}} = \big(\bar{B}_{ij}(\lambda)\big)_{i,j \in \{1,2,3,4\}},
$$

$$
\tilde{C}_k(\lambda) := \mathcal{Q}\, C_k(\lambda)\, \tilde{\mathcal{Q}}^T = \big(C_{ij}(\lambda)\big)_{i,j \in \{1,2\}} = \big(\bar{C}_{ij}(\lambda)\big)_{i,j \in \{1,2,3,4\}},
$$

$$
\tilde{D}_k(\lambda) := \mathcal{Q}\, D_k(\lambda)\, \tilde{\mathcal{Q}}^T = \big(D_{ij}(\lambda)\big)_{i,j \in \{1,2\}} = \big(\bar{D}_{ij}(\lambda)\big)_{i,j \in \{1,2,3,4\}},
$$

where the matrices $A_{ij}(\lambda)$, $B_{ij}(\lambda)$, $C_{ij}(\lambda)$, $D_{ij}(\lambda)$ are formed within the block structure with $n = r + (n - r)$ as in Lemma 5.19 and the matrices $\bar{A}_{ij}(\lambda)$, $\bar{B}_{ij}(\lambda)$, $\bar{C}_{ij}(\lambda)$, $\bar{D}_{ij}(\lambda)$ are formed within the refined block structure with $n = \tilde{r} + (r - \tilde{r}) + (n - r - \rho) + \rho$ as in Lemmas 5.22 and 5.23. Then the $2n \times 2n$ matrix

$$
\tilde{S}_k(\lambda) := \begin{pmatrix} \tilde{A}_k(\lambda) & \tilde{B}_k(\lambda) \\ \tilde{C}_k(\lambda) & \tilde{D}_k(\lambda) \end{pmatrix} = \begin{pmatrix} \mathcal{Q} & 0 \\ 0 & \mathcal{Q} \end{pmatrix} S_k(\lambda) \begin{pmatrix} \tilde{\mathcal{Q}}^T & 0 \\ 0 & \tilde{\mathcal{Q}}^T \end{pmatrix}
\tag{5.84}
$$

is symplectic as a product of three symplectic matrices; see Lemma 1.58(i) and Remark 1.59. Consequently, the matrices $\tilde{X}_k(\lambda)$, $\tilde{U}_k(\lambda)$, $\tilde{X}_{k+1}(\lambda)$, $\tilde{U}_{k+1}(\lambda)$ given in (5.78) and (5.68) satisfy for $\lambda \in (\lambda_0 - \delta, \lambda_0]$ the symplectic system

$$\left.\begin{aligned} \tilde{X}_{k+1}(\lambda) &= \tilde{\mathcal{A}}_k(\lambda)\,\tilde{X}_k(\lambda) + \tilde{\mathcal{B}}_k(\lambda)\,\tilde{U}_k(\lambda), \\ \tilde{U}_{k+1}(\lambda) &= \tilde{\mathcal{C}}_k(\lambda)\,\tilde{X}_k(\lambda) + \tilde{\mathcal{D}}_k(\lambda)\,\tilde{U}_k(\lambda), \end{aligned}\right\} \tag{5.85}$$

as well as, by (2.10), the time-reversed equations

$$\left.\begin{aligned} \tilde{X}_k(\lambda) &= \tilde{\mathcal{D}}_k^T(\lambda)\,\tilde{X}_{k+1}(\lambda) - \tilde{\mathcal{B}}_k^T(\lambda)\,\tilde{U}_{k+1}(\lambda), \\ \tilde{U}_k(\lambda) &= -\tilde{\mathcal{C}}_k^T(\lambda)\,\tilde{X}_{k+1}(\lambda) + \tilde{\mathcal{A}}_k^T(\lambda)\,\tilde{U}_{k+1}(\lambda). \end{aligned}\right\} \tag{5.86}$$

It follows that $\big(\tilde{Y}_j(\lambda)\big)_{j\in\{k,k+1\}} = (\tilde{X}_j(\lambda), \tilde{U}_j(\lambda))_{j\in\{k,k+1\}}$ is a conjoined basis of system (5.85). Next, we define for $\lambda \in \mathbb{R}$ the symmetric $2n \times 2n$ matrix

$$\tilde{\Psi}_k(\lambda) := \mathcal{J}\,\tfrac{\mathrm{d}}{\mathrm{d}\lambda}\,[\tilde{\mathcal{S}}_k(\lambda)]\,\mathcal{J}\tilde{\mathcal{S}}_k^T(\lambda)\,\mathcal{J} \overset{(5.84)}{=} \begin{pmatrix} \mathcal{Q} & 0 \\ 0 & \mathcal{Q} \end{pmatrix} \Psi_k(\lambda) \begin{pmatrix} \mathcal{Q}^T & 0 \\ 0 & \mathcal{Q}^T \end{pmatrix} \overset{(5.3)}{\ge} 0. \tag{5.87}$$

Now we analyze the structure of the coefficients of system (5.85).

**Lemma 5.25** *Given the above coefficients $\tilde{\mathcal{A}}_k(\lambda)$, $\tilde{\mathcal{B}}_k(\lambda)$, $\tilde{\mathcal{C}}_k(\lambda)$, $\tilde{\mathcal{D}}_k(\lambda)$ and the matrices $\tilde{X}_k(\lambda)$, $\tilde{U}_k(\lambda)$ in (5.78) and $\tilde{X}_{k+1}(\lambda)$, $\tilde{U}_{k+1}(\lambda)$ in (5.68), we have for all $\lambda \in (\lambda_0 - \delta, \lambda_0]$*

$$\tilde{\mathcal{B}}_k(\lambda) = \begin{pmatrix} B_{11}(\lambda) & B_{12}(\lambda) \\ 0_{(n-r)\times r} & B_{22} \end{pmatrix}, \quad B_{12}(\lambda) = \begin{pmatrix} 0 & \bar{B}_{14}(\lambda) \\ 0 & \bar{B}_{24}(\lambda) \end{pmatrix}, \quad B_{22} = \begin{pmatrix} 0_{n-r-\rho} & 0 \\ 0 & \bar{B}_{44} \end{pmatrix}, \tag{5.88}$$

*where $\bar{B}_{44} \in \mathbb{R}^{\rho\times\rho}$ is invertible,*

$$\tilde{\mathcal{D}}_k(\lambda) = \begin{pmatrix} D_{11}(\lambda) & D_{12}(\lambda) \\ D_{21}(\lambda) & D_{22}(\lambda) \end{pmatrix}, \quad D_{12}(\lambda) = \begin{pmatrix} 0 & \bar{D}_{14}(\lambda) \\ 0 & \bar{D}_{24}(\lambda) \end{pmatrix}, \quad D_{22}(\lambda) = \begin{pmatrix} \bar{D}_{33} & \bar{D}_{34}(\lambda) \\ 0 & \bar{D}_{44}(\lambda) \end{pmatrix}, \tag{5.89}$$

*where $\bar{D}_{33} \in \mathbb{R}^{(n-r-\rho)\times(n-r-\rho)}$ is invertible,*

$$\tilde{\mathcal{A}}_k(\lambda) = \begin{pmatrix} A_{11}(\lambda) & A_{12}(\lambda) \\ A_{21} & A_{22} \end{pmatrix}, \quad A_{21} = \begin{pmatrix} 0 & 0 \\ \bar{A}_{41} & \bar{A}_{42} \end{pmatrix}, \quad A_{22}(\lambda) = \begin{pmatrix} \bar{A}_{33} & 0 \\ \bar{A}_{43} & \bar{A}_{44} \end{pmatrix}, \tag{5.90}$$

*and where $\bar{A}_{33} \in \mathbb{R}^{(n-r-\rho)\times(n-r-\rho)}$ and $\bar{A}_{44} \in \mathbb{R}^{\rho\times\rho}$ are invertible. Moreover,*

$$Y_{11}(\lambda) := \begin{pmatrix} \tilde{X}_{11}(\lambda) & 0 \\ 0 & 0_{r-\tilde{r}} \end{pmatrix} = D_{11}^T(\lambda)\,X_{11}(\lambda) - B_{11}^T(\lambda)\,U_{11}(\lambda), \quad \lambda \in (\lambda_0 - \delta, \lambda_0], \tag{5.91}$$

*and the matrices* $B_{11}(\lambda)$, $D_{11}(\lambda) \in \mathbb{R}^{r \times r}$ *satisfy for all* $\lambda \in (\lambda_0 - \delta, \lambda_0]$

$$B_{11}^T(\lambda)\, D_{11}(\lambda) \text{ is symmetric and } \operatorname{rank}\left(B_{11}^T(\lambda),\, D_{11}^T(\lambda)\right) = r. \qquad (5.92)$$

*Proof* The first equation in (5.86) for $\lambda \in (\lambda_0 - \delta, \lambda_0]$ yields in its diagonal blocks and in the right upper block, respectively, the equations

$$\begin{pmatrix} \tilde{X}_{11}(\lambda) & 0 \\ 0 & 0 \end{pmatrix} = D_{11}^T(\lambda)\, X_{11}(\lambda) - \left[ B_{11}^T(\lambda)\, U_{11}(\lambda) + B_{21}^T(\lambda)\, U_{21}(\lambda) \right],$$

$$\begin{pmatrix} 0 & 0 \\ 0 & \tilde{X}_{44} \end{pmatrix} = -B_{22}^T(\lambda)\, U_{22}, \quad 0 = -B_{21}^T(\lambda)\, U_{22}.$$

Since $U_{22}$ is invertible, we obtain $B_{21}(\lambda) \equiv 0$ on $(\lambda_0 - \delta, \lambda_0]$, and then equation (5.91) is satisfied. Next, by (5.80), for every $\lambda \in (\lambda_0 - \delta, \lambda_0]$, we have

$$\begin{pmatrix} 0 & 0 \\ 0 & \tilde{X}_{44} \end{pmatrix} = - \begin{pmatrix} \bar{B}_{33}^T(\lambda) & \bar{B}_{43}^T(\lambda) \\ \bar{B}_{34}^T(\lambda) & \bar{B}_{33}^T(\lambda) \end{pmatrix} \begin{pmatrix} \bar{U}_{33} & \bar{U}_{34} \\ 0 & \bar{U}_{44} \end{pmatrix}.$$

Since $\bar{U}_{33}$ is invertible, this yields that $\bar{B}_{33}(\lambda) \equiv 0$ and $\bar{B}_{34}(\lambda) \equiv 0$ on $(\lambda_0 - \delta, \lambda_0]$, and then since $\bar{U}_{44}$ is invertible, $\bar{B}_{43}(\lambda) \equiv 0$ on $(\lambda_0 - \delta, \lambda_0]$. Finally, the equation $\tilde{X}_{44} = -\bar{B}_{44}^T(\lambda)\, \bar{U}_{44}$ on $(\lambda_0 - \delta, \lambda_0]$ implies that $\bar{B}_{44}(\lambda) \equiv -\bar{U}_{44}^{T-1} \tilde{X}_{44}^T =: \bar{B}_{44} \in \mathbb{R}^{\rho \times \rho}$ is constant and invertible on $(\lambda_0 - \delta, \lambda_0]$ and $B_{22}(\lambda) \equiv B_{22}$ is constant on $(\lambda_0 - \delta, \lambda_0]$. Next, the first equation in (5.85), in particular its third column in the refined block structure, yields through (5.68), (5.73), and (5.79) that $\bar{B}_{13}(\lambda)\, \tilde{U}_{33} = 0$ and $\bar{B}_{23}(\lambda)\, \tilde{U}_{33} = 0$ hold on $(\lambda_0 - \delta, \lambda_0]$. Since $\tilde{U}_{33}$ is invertible, we get $\bar{B}_{13}(\lambda) \equiv 0$ and $\bar{B}_{23}(\lambda) \equiv 0$ on $(\lambda_0 - \delta, \lambda_0]$. Therefore, we showed that for all $\lambda \in (\lambda_0 - \delta, \lambda_0]$ equation (5.88) holds with the matrix $\bar{B}_{44} \in \mathbb{R}^{\rho \times \rho}$ invertible.

Next, from the second equation in (5.85), in particular from its third column in the refined block structure, we get via (5.68), (5.73), (5.79), and (5.80) that $\bar{D}_{13}(\lambda)\, \tilde{U}_{33} = 0$, $\bar{D}_{23}(\lambda)\, \tilde{U}_{33} = 0$, $\bar{D}_{33}(\lambda)\, \tilde{U}_{33} = \tilde{U}_{33}$, and $\bar{D}_{43}(\lambda)\, \tilde{U}_{33} = 0$ on $(\lambda_0 - \delta, \lambda_0]$. And since $\tilde{U}_{33}$ is invertible, it follows that $\bar{D}_{13}(\lambda) \equiv 0$, $\bar{D}_{23}(\lambda) \equiv 0$, $\bar{D}_{33}(\lambda) \equiv \tilde{U}_{33}\tilde{U}_{33}^{-1} =: \bar{D}_{33}$, and $\bar{D}_{43}(\lambda) \equiv 0$ on $(\lambda_0 - \delta, \lambda_0]$. This shows that equality (5.89) holds.

Since the matrix $\tilde{\mathcal{S}}_k(\lambda)$ is symplectic, we have from (1.145) that

$$\tilde{\mathcal{B}}_k^T(\lambda)\, \tilde{\mathcal{D}}_k(\lambda) = \begin{pmatrix} B_{11}^T(\lambda)\, D_{11}(\lambda) & B_{11}^T(\lambda)\, D_{12}(\lambda) \\ B_{12}^T(\lambda)\, D_{11}(\lambda) + B_{22}^T\, D_{21}(\lambda) & B_{12}^T(\lambda)\, D_{12}(\lambda) + B_{22}^T\, D_{22}(\lambda) \end{pmatrix}$$

is symmetric for all $\lambda \in (\lambda_0 - \delta, \lambda_0]$. This implies that $B_{11}^T(\lambda)\, D_{11}(\lambda)$ is symmetric for $\lambda \in (\lambda_0 - \delta, \lambda_0]$ and that

$$B_{11}^T(\lambda)\, D_{12}(\lambda) = \left[ B_{12}^T(\lambda)\, D_{11}(\lambda) + B_{22}^T\, D_{21}(\lambda) \right]^T, \quad \lambda \in (\lambda_0 - \delta, \lambda_0]. \qquad (5.93)$$

By extracting the second column of (5.93) in the refined block structure, we get

$$B_{11}^T(\lambda) \begin{pmatrix} \bar{D}_{14}(\lambda) \\ \bar{D}_{24}(\lambda) \end{pmatrix} = D_{11}^T(\lambda) \begin{pmatrix} \bar{B}_{14}(\lambda) \\ \bar{B}_{24}(\lambda) \end{pmatrix} + \begin{pmatrix} \bar{D}_{41}^T(\lambda) \\ \bar{D}_{24}^T(\lambda) \end{pmatrix} \bar{B}_{44}, \quad \lambda \in (\lambda_0 - \delta, \lambda_0].$$

And since $\bar{B}_{44}$ is invertible, it follows that

$$\begin{pmatrix} \bar{D}_{41}^T(\lambda) \\ \bar{D}_{24}^T(\lambda) \end{pmatrix} \in \mathrm{Im}\left(B_{11}^T(\lambda), D_{11}^T(\lambda)\right), \quad \lambda \in (\lambda_0 - \delta, \lambda_0]. \tag{5.94}$$

From the identity $\tilde{\mathcal{A}}_k^T(\lambda)\,\tilde{\mathcal{D}}_k(\lambda) - \tilde{\mathcal{C}}_k^T(\lambda)\,\tilde{\mathcal{B}}_k(\lambda) = I$ for all $\lambda \in \mathbb{R}$, compare with (5.81), we have $\mathrm{rank}\left(\tilde{\mathcal{B}}_k^T(\lambda),\ \tilde{\mathcal{D}}_k^T(\lambda)\right) = n$, so that by (5.88) and (5.89) and the invertibility of $\bar{B}_{44}$ and $\bar{D}_{33}$

$$n = \mathrm{rank} \begin{pmatrix} \bar{B}_{11}^T(\lambda) & \bar{B}_{21}^T(\lambda) & 0 & 0 & \bar{D}_{11}^T(\lambda) & \bar{D}_{21}^T(\lambda) & \bar{D}_{31}^T(\lambda) & \bar{D}_{41}^T(\lambda) \\ \bar{B}_{12}^T(\lambda) & \bar{B}_{22}^T(\lambda) & 0 & 0 & \bar{D}_{12}^T(\lambda) & \bar{D}_{22}^T(\lambda) & \bar{D}_{32}^T(\lambda) & \bar{D}_{42}^T(\lambda) \\ 0 & 0 & 0 & 0 & 0 & 0 & \bar{D}_{33}^T & 0 \\ \bar{B}_{14}^T(\lambda) & \bar{B}_{24}^T(\lambda) & 0 & \bar{B}_{44}^T & \bar{D}_{14}^T(\lambda) & \bar{D}_{24}^T(\lambda) & \bar{D}_{34}^T(\lambda) & \bar{D}_{441}^T(\lambda) \end{pmatrix}$$

$$= \mathrm{rank} \begin{pmatrix} \bar{B}_{11}^T(\lambda) & \bar{B}_{21}^T(\lambda) & 0 & 0 & \bar{D}_{11}^T(\lambda) & \bar{D}_{21}^T(\lambda) & 0 & \bar{D}_{41}^T(\lambda) \\ \bar{B}_{12}^T(\lambda) & \bar{B}_{22}^T(\lambda) & 0 & 0 & \bar{D}_{12}^T(\lambda) & \bar{D}_{22}^T(\lambda) & 0 & \bar{D}_{42}^T(\lambda) \\ 0 & 0 & 0 & 0 & 0 & 0 & I_{n-r-\rho} & 0 \\ 0 & 0 & 0 & I_\rho & 0 & 0 & 0 & 0 \end{pmatrix}. \tag{5.95}$$

for $\lambda \in (\lambda_0 - \delta, \lambda_0]$. If we now interchange in (5.95) the third and seventh columns and use condition (5.94), we obtain from (5.95) that for $\lambda \in (\lambda_0 - \delta, \lambda_0]$

$$n = \mathrm{rank} \begin{pmatrix} B_{11}^T(\lambda) & 0_{r\times(n-r)} & D_{11}^T(\lambda) & 0_{r\times(n-r)} \\ 0_{(n-r)\times r} & I_{n-r} & 0_{(n-r)\times r} & 0_{n-r} \end{pmatrix}$$

$$= \mathrm{rank}\left(B_{11}^T(\lambda),\ D_{11}^T(\lambda)\right) + n - r.$$

Therefore, $\mathrm{rank}\left(B_{11}^T(\lambda),\ D_{11}^T(\lambda)\right) = r$ and condition (5.92) is established.

It remains to prove (5.90). From the second equation in (5.86), in particular from its third and fourth columns in the refined block structure, we have

$$A_{21}^T(\lambda)\,U_{22} = \begin{pmatrix} 0 & \tilde{U}_{14} \\ 0 & \tilde{U}_{24} \end{pmatrix}, \quad A_{22}^T(\lambda)\,U_{22} = \begin{pmatrix} \tilde{U}_{33} & \tilde{U}_{34} \\ 0 & \tilde{U}_{44} \end{pmatrix}, \quad \lambda \in (\lambda_0 - \delta, \lambda_0].$$

Since by Lemma 5.24 the matrix $U_{22}$ is invertible, it follows from (5.80) that $A_{21}(\lambda) \equiv A_{21}$ and $A_{22}(\lambda) \equiv A_{22}$ are constant on $(\lambda_0 - \delta, \lambda_0]$ and that on this interval $\bar{A}_{31}(\lambda) \equiv 0$, $\bar{A}_{32}(\lambda) \equiv 0$, $\bar{A}_{41}(\lambda) \equiv \bar{U}_{44}^{T-1} \tilde{U}_{14}^{T} =: \bar{A}_{41}$, $\bar{A}_{42}(\lambda) \equiv \bar{U}_{44}^{T-1} \tilde{U}_{24}^{T} =: \bar{A}_{42}$, $\bar{A}_{33}(\lambda) \equiv \bar{U}_{33}^{T-1} \tilde{U}_{33}^{T} =: \bar{A}_{33}$ is invertible, $\bar{A}_{34}(\lambda) \equiv 0$, $\bar{A}_{43}(\lambda) \equiv \bar{A}_{43}$ is constant, and $\bar{A}_{44}(\lambda) \equiv \bar{U}_{44}^{T-1} \tilde{U}_{44}^{T} =: \bar{A}_{44}$ is invertible. The proof is complete. $\qquad\square$

Following Lemma 5.25, let us now analyze the structure of the matrix $\tilde{\Psi}_k(\lambda)$ in (5.87). By (5.83) we have for any $\lambda \in \mathbb{R}$

$$
\tilde{\Psi}_k(\lambda) = \begin{pmatrix} \dot{\tilde{D}}_k(\lambda)\,\tilde{C}_k^T(\lambda) - \dot{\tilde{C}}_k(\lambda)\,\tilde{D}_k^T(\lambda) & \dot{\tilde{C}}_k(\lambda)\,\tilde{B}_k^T(\lambda) - \dot{\tilde{D}}_k(\lambda)\,\tilde{A}_k^T(\lambda) \\ \dot{\tilde{A}}_k(\lambda)\,\tilde{D}_k^T(\lambda) - \dot{\tilde{B}}_k(\lambda)\,\tilde{C}_k^T(\lambda) & \dot{\tilde{B}}_k(\lambda)\,\tilde{A}_k^T(\lambda) - \dot{\tilde{A}}_k(\lambda)\,\tilde{B}_k^T(\lambda) \end{pmatrix}
$$
$$
=: \begin{pmatrix} \tilde{\mathcal{H}}_k(\lambda) & \tilde{\mathcal{G}}_k(\lambda) \\ \tilde{\mathcal{F}}_k(\lambda) & \tilde{\mathcal{E}}_k(\lambda) \end{pmatrix}, \tag{5.96}
$$

and $\tilde{\Psi}_k(\lambda)$ is symmetric. Then the matrices $\tilde{\mathcal{H}}_k(\lambda)$ and $\tilde{\mathcal{E}}_k(\lambda)$ are symmetric as well. For brevity, we introduce the following notation for the row block columns of the matrices $\tilde{\mathcal{A}}_k(\lambda)$, $\tilde{\mathcal{B}}_k(\lambda)$, $\tilde{\mathcal{C}}_k(\lambda)$, $\tilde{\mathcal{D}}_k(\lambda)$ as

$$
\left.\begin{aligned}
A_1(\lambda) &:= \big(A_{11}(\lambda),\, A_{12}(\lambda)\big), & A_2(\lambda) &:= \big(A_{21}(\lambda),\, A_{22}(\lambda)\big), \\
B_1(\lambda) &:= \big(B_{11}(\lambda),\, B_{12}(\lambda)\big), & B_2(\lambda) &:= \big(B_{21}(\lambda),\, B_{22}(\lambda)\big), \\
C_1(\lambda) &:= \big(C_{11}(\lambda),\, C_{12}(\lambda)\big), & C_2(\lambda) &:= \big(C_{21}(\lambda),\, C_{22}(\lambda)\big), \\
D_1(\lambda) &:= \big(D_{11}(\lambda),\, D_{12}(\lambda)\big), & D_2(\lambda) &:= \big(D_{21}(\lambda),\, D_{22}(\lambda)\big).
\end{aligned}\right\} \tag{5.97}
$$

Then $A_1(\lambda) \in \mathbb{R}^{r \times n}$ and $A_2(\lambda) \in \mathbb{R}^{(n-r) \times n}$ and similarly for the other matrices above. Then, by the form of the coefficients $\tilde{\mathcal{A}}_k(\lambda)$, $\tilde{\mathcal{B}}_k(\lambda)$, $\tilde{\mathcal{D}}_k(\lambda)$ in (5.88)–(5.90), we obtain

$$
\left.\begin{aligned}
\tilde{\mathcal{H}}_k(\lambda) &= \begin{pmatrix} H_{11}(\lambda) & H_{12}(\lambda) \\ H_{21}(\lambda) & H_{22}(\lambda) \end{pmatrix}, & \tilde{\mathcal{G}}_k(\lambda) &= \begin{pmatrix} G_{11}(\lambda) & 0 \\ G_{21}(\lambda) & 0_{n-r} \end{pmatrix}, \\
\tilde{\mathcal{F}}_k(\lambda) &= \begin{pmatrix} F_{11}(\lambda) & F_{12}(\lambda) \\ 0 & 0_{n-r} \end{pmatrix}, & \tilde{\mathcal{E}}_k(\lambda) &= \begin{pmatrix} E_{11}(\lambda) & 0 \\ 0 & 0_{n-r} \end{pmatrix},
\end{aligned}\right\} \tag{5.98}
$$

for all $\lambda \in (\lambda_0 - \delta, \lambda_0]$, where

$$
\left.\begin{aligned}
H_{ij}(\lambda) &:= \dot{D}_i(\lambda)\,C_j^T(\lambda) - \dot{C}_i(\lambda)\,D_j^T(\lambda), \\
G_{ij}(\lambda) &:= \dot{C}_i(\lambda)\,B_j^T(\lambda) - \dot{D}_i(\lambda)\,A_j^T(\lambda), \\
F_{ij}(\lambda) &:= \dot{A}_i(\lambda)\,D_j^T(\lambda) - \dot{B}_i(\lambda)\,C_j^T(\lambda), \\
E_{ij}(\lambda) &:= \dot{B}_i(\lambda)\,A_j^T(\lambda) - \dot{A}_i(\lambda)\,B_j^T(\lambda),
\end{aligned}\right\} \quad i, j \in \{1, 2\}, \tag{5.99}
$$

and where $H_{ii}(\lambda)$ and $E_{ii}(\lambda)$ are symmetric and $G_{ij}^T(\lambda) = F_{ji}(\lambda)$ for $i, j \in \{1, 2\}$.

Our next aim is to analyze the behavior of the $r \times r$ matrix $P_{11}(\lambda)$ defined by

$$P_{11}(\lambda) := B_{11}^T(\lambda)\, D_{11}(\lambda) - B_{11}^T(\lambda)\, Q_{11}(\lambda)\, B_{11}(\lambda), \quad \lambda \in (\lambda_0 - \delta, \lambda_0), \tag{5.100}$$

$$Q_{11}(\lambda) := U_{11}(\lambda)\, X_{11}^{-1}(\lambda), \quad \lambda \in (\lambda_0 - \delta, \lambda_0), \tag{5.101}$$

where $X_{11}(\lambda)$ and $U_{11}(\lambda)$ are from (5.58) and $B_{11}(\lambda)$, $D_{11}(\lambda)$ are from (5.88), (5.89). For this we need to know first the behavior of the function $Q_{11}(\lambda)$ on $(\lambda_0 - \delta, \lambda_0)$. We define the $n \times n$ matrix

$$\tilde{L}_k(\lambda) := \mathcal{P}^T L_k(\lambda)\, \mathcal{P}, \quad L_k(\lambda) := \left(I_n,\, 0_n\right) \Omega_k(\lambda) \left(I_n,\, 0_n\right)^T, \tag{5.102}$$

where $\Omega_k(\lambda)$ is defined in (5.32).

**Lemma 5.26** *The symmetric matrix $Q_{11}(\lambda)$ defined in (5.101) satisfies*

$$\dot{Q}_{11}(\lambda) = -\left(X_{11}^{T-1}(\lambda),\, 0_{r \times (n-r)}\right) \tilde{L}_k(\lambda) \begin{pmatrix} X_{11}^{-1}(\lambda) \\ 0_{(n-r) \times r} \end{pmatrix}, \quad \lambda \in (\lambda_0 - \delta, \lambda_0). \tag{5.103}$$

*Consequently, under (5.3), the function $Q_{11}(\cdot)$ is nonincreasing on $(\lambda_0 - \delta, \lambda_0)$.*

*Proof* We suppress the argument $\lambda$. First note that by (5.31)

$$X_{k+1}^T \dot{U}_{k+1} - U_{k+1}^T \dot{X}_{k+1} = \left(I\ 0\right) \tilde{Z}_{k+1}^T \mathcal{J} \dot{\tilde{Z}}_{k+1} \left(I\ 0\right)^T = -L_k(\lambda) \tag{5.104}$$

on $\mathbb{R}$. From this and from $\frac{d}{d\lambda} X_{11}^{-1} = -X_{11}^{-1} \dot{X}_{11} X_{11}^{-1}$, we conclude that on $(\lambda_0 - \delta, \lambda_0)$

$$
\begin{aligned}
\dot{Q}_{11} &\overset{(5.59)}{=} X_{11}^{T-1}\, (X_{11}^T \dot{U}_{11} - U_{11}^T \dot{X}_{11})\, X_{11}^{-1} \\
&\overset{(5.68)}{=} \left(X_{11}^{T-1},\, 0_{r \times (n-r)}\right) \left[\tilde{X}_{k+1}^T \tfrac{d}{d\lambda} \tilde{U}_{k+1} - \tilde{U}_{k+1}^T \tfrac{d}{d\lambda} \tilde{X}_{k+1}\right] \left(X_{11}^{T-1},\, 0_{r \times (n-r)}\right)^T \\
&\overset{(5.69)}{=} \left(X_{11}^{T-1},\, 0\right) \mathcal{P}^T\, [X_{k+1}^T \dot{U}_{k+1} - U_{k+1}^T \dot{X}_{k+1}]\, \mathcal{P} \left(X_{11}^{T-1},\, 0\right)^T \\
&\overset{(5.104)}{=} -\left(X_{11}^{T-1},\, 0\right) \tilde{L}_k \left(X_{11}^{T-1},\, 0\right)^T,
\end{aligned}
$$

which shows the result in (5.103). Under (5.3) we then get $\dot{Q}_{11}(\lambda) \leq 0$ for all $\lambda \in (\lambda_0 - \delta, \lambda_0)$. Hence, the function $Q_{11}(\cdot)$ is nonincreasing on $(\lambda_0 - \delta, \lambda_0)$.  $\square$

*Remark 5.27* If $X_{k+1}(\lambda)$ is invertible on $(\lambda_0 - \delta, \lambda_0)$, then $r = n$, and so no construction of the matrices $\mathcal{P}$ and $\mathcal{Q}$ in Lemma 5.19 is needed. In this case we get from (5.103), (5.102), and (5.31) the following identity for $\lambda \in (\lambda_0 - \delta, \lambda_0)$

$$\dot{Q}_{k+1}(\lambda) = -X_{k+1}^{T-1}(\lambda) \left[ \sum_{j=0}^{k} Y_{j+1}^T(\lambda) \, \Psi_j(\lambda) \, Y_{j+1}(\lambda) \right] X_{k+1}^{-1}(\lambda),$$

where $Q_{k+1}(\lambda) := U_{k+1}(\lambda) \, X_{k+1}^{-1}(\lambda)$.

We will see in Lemma 5.28 that the behavior of the matrix $P_{11}(\lambda)$ is determined by the behavior of $Q_{11}(\lambda)$. Namely, since $Q_{11}(\lambda)$ is nonincreasing, and hence the function $-B_{11}^T(\lambda) \, Q_{11}(\lambda) \, B_{11}(\lambda)$ appearing in $P_{11}(\lambda)$ is nondecreasing, then this behavior is not destroyed by the additional term $B_{11}^T(\lambda) \, D_{11}(\lambda)$. This is similar as in the linear case in (5.238) with (5.241), in which $Q_{11}(\lambda)$ is (of course) nonincreasing and

$$P_{11}(\lambda) = B_{11}^T D_{11} - \lambda \, B_{11}^T \, W_{11} B_{11} - B_{11}^T \, Q_{11}(\lambda) \, B_{11}$$

is nondecreasing, even though $W_{11} \geq 0$, see [102, pg. 601] and [211, Proposition 4.1].

**Lemma 5.28** *The symmetric matrix $P_{11}(\lambda)$ defined in (5.100) satisfies*

$$\dot{P}_{11}(\lambda) = B_{11}^T(\lambda) \, \Upsilon_{11}(\lambda) \, B_{11}(\lambda) + R_{11}^T(\lambda) \, E_{11}(\lambda) \, R_{11}(\lambda), \quad \lambda \in (\lambda_0 - \delta, \lambda_0), \tag{5.105}$$

*where the matrix $E_{11}(\lambda) \in \mathbb{R}^{r \times r}$ is defined in (5.99) and*

$$\Upsilon_{11}(\lambda) := \left( X_{11}^{T-1}(\lambda), \, 0 \right) \tilde{L}_{k-1}(\lambda) \left( X_{11}^{T-1}(\lambda), \, 0 \right)^T \in \mathbb{R}^{r \times r}, \tag{5.106}$$

$$R_{11}(\lambda) := Q_{11}(\lambda) \, B_{11}(\lambda) - D_{11}(\lambda) \in \mathbb{R}^{r \times r}, \tag{5.107}$$

*and where the matrix $\tilde{L}_{k-1}(\lambda)$ is given by (5.102) with the index $k-1$. Consequently, under (5.3) the function $P_{11}(\cdot)$ is nondecreasing on $(\lambda_0 - \delta, \lambda_0)$.*

*Proof* We suppress the argument $\lambda$ in this proof. First note that by (5.96), (5.98), and the form of $\tilde{X}_{k+1}$ and $\tilde{U}_{k+1}$ in (5.68)

$$\Lambda_{11} := \left( X_{11}^{T-1}, \, 0 \right) \tilde{Y}_{k+1}^T \tilde{\Psi}_k \tilde{Y}_{k+1} \left( X_{11}^{T-1}, \, 0 \right)^T$$

$$\stackrel{(5.68)}{=} \left( I_r, \, 0_{(n-r) \times r}, \, Q_{11}, \, Q_{21}^T \right) \tilde{\Psi}_k \left( I_r, \, 0_{(n-r) \times r}, \, Q_{11}, \, Q_{21}^T \right)^T$$

$$\stackrel{(5.96), \, (5.98)}{=} \left( I, \, Q_{11} \right) \begin{pmatrix} H_{11} & G_{11} \\ F_{11} & E_{11} \end{pmatrix} \begin{pmatrix} I \\ Q_{11} \end{pmatrix}, \tag{5.108}$$

where $Q_{11} = U_{11} X_{11}^{-1}$ as above, $Q_{21} := U_{21} X_{11}^{-1}$, and $H_{11}, G_{11}, F_{11}, E_{11}$ are given in (5.99), implying that $H_{11}$ and $E_{11}$ are symmetric and $F_{11} = G_{11}^T$. Now we use Lemma 5.26 to get

$$
-\dot{Q}_{11} \overset{(5.103)}{=} \left( X_{11}^{T,-1}, 0 \right) \left( \tilde{L}_{k-1} + \tilde{Y}_{k+1}^T \tilde{\Psi}_k \tilde{Y}_{k+1} \right) \left( X_{11}^{T,-1}, 0 \right)^T
$$

$$
\overset{(5.106),\ (5.108)}{=} \Upsilon_{11} + \Lambda_{11} \quad \text{on } (\lambda_0 - \delta, \lambda_0). \tag{5.109}
$$

After these preparatory calculations, we have from (5.100)

$$
\dot{P}_{11} = \dot{B}_{11}^T D_{11} + B_{11}^T \dot{D}_{11} - \dot{B}_{11}^T Q_{11} B_{11} - B_{11}^T \dot{Q}_{11} B_{11} - B_{11}^T Q_{11} \dot{B}_{11}
$$

$$
\overset{(5.109)}{=} B_{11}^T \Upsilon_{11} B_{11} + S_{11}, \tag{5.110}
$$

where the matrix $S_{11}$ is defined by

$$
S_{11} := \dot{B}_{11}^T D_{11} + B_{11}^T \dot{D}_{11} - \dot{B}_{11}^T Q_{11} B_{11} - B_{11}^T Q_{11} \dot{B}_{11} + B_{11}^T \Lambda_{11} B_{11}. \tag{5.111}
$$

If we show that $S_{11} = R_{11}^T E_{11} R_{11}$ with $R_{11}$ given in (5.107), then the required identity in (5.105) will follow from equation (5.110). Consider now the identities in (5.81) written for the auxiliary symplectic system (5.85) with the coefficients $\tilde{A}_k$, $\tilde{B}_k, \tilde{C}_k, \tilde{D}_k$ and with their block structure (5.97). From the symmetry of the matrices $\tilde{D}_k \tilde{C}_k^T$ and $\tilde{A}_k \tilde{B}_k^T$ and from the identity $\tilde{D}_k \tilde{A}_k^T - \tilde{C}_k \tilde{B}_k^T = I_n$, we obtain

$$
D_1 C_1^T = C_1 D_1^T, \quad A_1 B_1^T = B_1 A_1^T, \quad D_1 A_1^T - C_1 B_1^T = I_r, \quad A_2 B_1^T = B_2 A_1^T, \tag{5.112}
$$

and from the symmetry of the matrix $\tilde{B}_k^T \tilde{D}_k$ and from $\tilde{A}_k^T \tilde{D}_k - \tilde{C}_k^T \tilde{B}_k = I_n$, we get

$$
B_{11}^T D_{11} = D_{11}^T B_{11}, \quad B_{12}^T D_{11} + B_{22}^T D_{21} - D_{12}^T B_{11} = 0_{(n-r) \times r}, \tag{5.113}
$$

$$
A_{11}^T D_{11} + A_{21}^T D_{21} - C_{11}^T B_{11} = I_r, \quad A_{12}^T D_{11} + A_{22}^T D_{21} - C_{12}^T B_{11} = 0_{(n-r) \times r}. \tag{5.114}
$$

Consider the identities in (5.82) written in terms of $\tilde{A}_k, \tilde{B}_k, \tilde{C}_k, \tilde{D}_k$. By (5.82)(vi),

$$
\dot{D}_1 A_2^T = \dot{C}_1 B_2^T, \tag{5.115}
$$

while from (5.82)(ii), we have

$$
\dot{B}_{11}^T D_1 + B_{11}^T \dot{D}_1 = \dot{D}_{11}^T B_1 + D_{11}^T \dot{B}_1 + \dot{D}_{21}^T B_2, \tag{5.116}
$$

and finally from (5.82)(iii), we obtain

$$D_{11}^T \dot{A}_1 + \dot{D}_{11}^T A_1 + \dot{D}_{21}^T A_2 = B_{11}^T \dot{C}_1 + \dot{B}_{11}^T C_1. \tag{5.117}$$

Then we calculate by using the definitions of $G_{11}$ and $E_{11}$ in (5.99)

$$B_{11}^T G_{11} \overset{(5.116), (5.117)}{=} D_{11}^T (\dot{A}_1 B_1^T - \dot{B}_1 A_1^T) + \dot{D}_{11}^T (A_1 B_1^T - B_1 A_1^T)$$

$$+ \dot{B}_{11}^T (D_1 A_1^T - C_1 B_1^T) + \dot{D}_{21}^T (A_2 B_1^T - B_2 A_1^T)$$

$$\overset{(5.99), (5.112)}{=} \dot{B}_{11}^T - D_{11}^T E_{11}. \tag{5.118}$$

Therefore, by using the identity

$$\dot{A}_1 D_1^T - \dot{B}_1 C_1^T \overset{(5.99)}{=} F_{11} = G_{11}^T = B_1 \dot{C}_1^T - A_1 \dot{D}_1^T$$

and formulas (5.108) and (5.118), we obtain from expression (5.111) that

$$S_{11} \overset{(5.118)}{=} B_{11}^T H_{11} B_{11} + (\dot{B}_{11}^T - D_{11}^T E_{11}) Q_{11} B_{11} + B_{11}^T Q_{11} (\dot{B}_{11} - E_{11} D_{11})$$

$$+ B_{11}^T Q_{11} E_{11} Q_{11} B_{11} - \dot{B}_{11}^T Q_{11} B_{11} - B_{11}^T Q_{11} \dot{B}_{11}$$

$$+ \dot{B}_{11}^T D_{11} + B_{11}^T \dot{D}_{11}$$

$$= (B_{11}^T Q_{11} - D_{11}^T) E_{11} (Q_{11} B_{11} - D_{11}) + Z_{11}, \tag{5.119}$$

where

$$Z_{11} := B_{11}^T H_{11} B_{11} - D_{11}^T E_{11} D_{11} + \dot{B}_{11}^T D_{11} + B_{11}^T \dot{D}_{11}.$$

We evaluate the matrix $Z_{11}$ by using identities (5.113), (5.114), (5.115) as follows

$$Z_{11} \overset{(5.99)}{=} B_{11}^T (\dot{D}_1 C_1^T - \dot{C}_1 D_1^T) B_{11} - D_{11}^T (\dot{B}_1 A_1^I - \dot{A}_1 B_1^T) D_{11}$$

$$+ \dot{B}_{11}^T D_{11} + B_{11}^T \dot{D}_{11}$$

$$\overset{(5.116), (5.117)}{=} B_{11}^T (\dot{D}_1 C_1^T - \dot{C}_1 D_1^T) B_{11}$$

$$- (\dot{B}_{11}^T D_1 + B_{11}^T \dot{D}_1 - \dot{D}_{11}^T B_1 - \dot{D}_{21}^T B_2) A_1^T D_{11}$$

$$+ (B_{11}^T \dot{C}_1 + \dot{B}_{11}^T C_1 - \dot{D}_{11}^T A_1 - \dot{D}_{21}^T A_2) B_1^T D_{11}$$

$$+ \dot{B}_{11}^T D_{11} + B_{11}^T \dot{D}_{11}$$

$$\overset{(5.113), (5.114)}{=} B_{11}^T (\dot{D}_1 C_1^T - \dot{C}_1 D_1^T) B_{11}$$

$$+ B_{11}^T \left[ \dot{C}_1 \begin{pmatrix} D_{11}^T B_{11} \\ D_{12}^T B_{11} - B_{22}^T D_{21} \end{pmatrix} - \dot{D}_1 \begin{pmatrix} C_{11}^T B_{11} - A_{21}^T D_{21} + I_r \\ C_{12}^T B_{11} - A_{22}^T D_{21} \end{pmatrix} \right]$$

$$+ \dot{B}_{11}^T (C_1 B_1^T - D_1 A_1^T) D_{11} + \dot{D}_{11}^T (B_1 A_1^T - A_1 B_1^T) D_{11}$$

$$+ \dot{D}_{21}^T (B_2 A_1^T - A_2 B_1^T) D_{11} + \dot{B}_{11}^T D_{11} + B_{11}^T \dot{D}_{11}$$

$$\overset{(5.112)}{=} B_{11}^T (\dot{D}_1 C_1^T - \dot{C}_1 D_1^T) B_{11} + B_{11}^T \dot{D}_{11} + B_{11}^T [\dot{C}_1 D_1^T B_{11} - \dot{C}_1 B_2^T D_{21}$$

$$- \dot{D}_1 C_1^T B_{11} + \dot{D}_1 A_2^T D_{21} - \dot{D}_1 \begin{pmatrix} I, & 0 \end{pmatrix}^T ]$$

$$= B_{11}^T \dot{D}_{11} + B_{11}^T (\dot{D}_1 A_2^T - \dot{C}_1 B_2^T) D_{21} - B_{11}^T \begin{pmatrix} \dot{D}_{11}, & \dot{D}_{12} \end{pmatrix} \begin{pmatrix} I, & 0 \end{pmatrix}^T$$

$$\overset{(5.115)}{=} B_{11}^T \dot{D}_{11} - B_{11}^T \dot{D}_{11} = 0.$$

Therefore, upon inserting $Z_{11} = 0$ into equation (5.119) and then expression (5.119) into formula (5.110), we see that the matrix $\dot{P}_{11}$ has the form displayed in (5.105). If now assumption (5.3) holds, then $\tilde{\Psi}_k \geq 0$ by (5.87), as well as $E_{11} \geq 0$ by (5.96) with (5.98). And since in this case $\tilde{L}_{k-1} \geq 0$ by (5.102) and (5.31), formula (5.105) implies that $\dot{P}_{11} \geq 0$ on $(\lambda_0 - \delta, \lambda_0)$. This shows that the function $P_{11}(\cdot)$ is nondecreasing on the interval $(\lambda_0 - \delta, \lambda_0)$. The proof is complete.  □

*Remark 5.29* If $X_{k+1}(\lambda)$ is invertible on $(\lambda_0 - \delta, \lambda_0)$, then $r = n$, and similar to Remark 5.27, we get from (5.105), (5.102), and (5.31) the following identity for $\lambda \in (\lambda_0 - \delta, \lambda_0)$

$$\dot{P}_k(\lambda) = \mathcal{B}_k^T(\lambda) \left[ \sum_{j=0}^{k-1} Y_{j+1}^T(\lambda) \Psi_j(\lambda) Y_{j+1}(\lambda) \right] \mathcal{B}_k(\lambda),$$

$$+ [Q_{k+1}(\lambda) \mathcal{B}_k(\lambda) - \mathcal{D}_k(\lambda)]^T \mathcal{E}_k(\lambda) [Q_{k+1}(\lambda) \mathcal{B}_k(\lambda) - \mathcal{D}_k(\lambda)],$$

where the matrix $Q_{k+1}(\lambda)$ is from Remark 5.27, and

$$P_k(\lambda) := \mathcal{B}_k^T(\lambda) \mathcal{D}_k(\lambda) - \mathcal{B}_k^T(\lambda) Q_{k+1}(\lambda) \mathcal{B}_k(\lambda),$$

$$\mathcal{E}_k(\lambda) := \dot{\mathcal{B}}_k(\lambda) \mathcal{A}_k^T(\lambda) - \dot{\mathcal{A}}_k(\lambda) \mathcal{B}_k^T(\lambda).$$

The above formula is an extension of the discrete version of [211, Lemma 4.3] to the nonlinear dependence on $\lambda$. Note that when the dependence on $\lambda$ is linear as in (5.241), the matrices $\mathcal{A}_k(\lambda) \equiv \mathcal{A}_k$ and $\mathcal{B}_k(\lambda) \equiv \mathcal{B}_k$ are constant on $\mathbb{R}$, so that in this case $\mathcal{E}_k(\lambda) \equiv 0$.

Define now for $\lambda \in \mathbb{R}$ the auxiliary $n \times n$ matrices

$$\tilde{M}_k(\lambda) := \tilde{Q} M_k(\lambda) \tilde{Q}^T, \quad \tilde{T}_k(\lambda) := \tilde{Q} T_k(\lambda) \tilde{Q}^T, \quad \tilde{P}_k(\lambda) := \tilde{Q} P_k(\lambda) \tilde{Q}^T,$$

$$(5.120)$$

where $M_k(\lambda)$, $T_k(\lambda)$, $P_k(\lambda)$ are defined in (5.40). Then

$$\text{rank } \tilde{M}_k(\lambda) = \text{rank } M_k(\lambda), \quad \text{ind } \tilde{P}_k(\lambda) = \text{ind } P_k(\lambda), \qquad \lambda \in \mathbb{R}. \tag{5.121}$$

Since by (5.69) and Remark 1.60(ii), we have for $\lambda \in (\lambda_0 - \delta, \lambda_0]$

$$\tilde{X}_{k+1}^\dagger(\lambda) = \mathcal{P}^T X_{k+1}^\dagger(\lambda)\, \mathcal{Q}^T, \quad \tilde{X}_{k+1}(\lambda)\, \tilde{X}_{k+1}^\dagger(\lambda) = \mathcal{Q}\, X_{k+1}(\lambda)\, X_{k+1}^\dagger(\lambda)\, \mathcal{Q}^T,$$

it follows that

$$\tilde{M}_k(\lambda) = \left[ I - \tilde{X}_{k+1}(\lambda)\, \tilde{X}_{k+1}^\dagger(\lambda) \right] \tilde{B}_k(\lambda), \quad \lambda \in (\lambda_0 - \delta, \lambda_0]. \tag{5.122}$$

Again by Remark 1.60(ii), we have $\tilde{M}_k^\dagger(\lambda) = \tilde{\mathcal{Q}}\, M_k^\dagger(\lambda)\, \mathcal{Q}^T$, so that for $\lambda \in (\lambda_0 - \delta, \lambda_0]$

$$\tilde{T}_k(\lambda) = I - \tilde{M}_k^\dagger(\lambda)\, \tilde{M}_k(\lambda), \quad \tilde{P}_k(\lambda) = \tilde{T}_k(\lambda)\, \tilde{X}_k(\lambda)\, \tilde{X}_{k+1}^\dagger(\lambda)\, \tilde{B}_k(\lambda)\, \tilde{T}_k(\lambda). \tag{5.123}$$

From the form of $\tilde{X}_{k+1}(\lambda)$, $\tilde{B}_k(\lambda)$, $\tilde{M}_k(\lambda)$ in (5.68), (5.88), (5.122), we get

$$\tilde{M}_k(\lambda) = \begin{pmatrix} I - X_{11}(\lambda)\, X_{11}^\dagger(\lambda) & 0 \\ 0 & I_{n-r} \end{pmatrix} \begin{pmatrix} B_{11}(\lambda) & B_{12}(\lambda) \\ 0 & B_{22} \end{pmatrix}, \quad \lambda \in (\lambda_0 - \delta, \lambda_0], \tag{5.124}$$

with the matrices $B_{12}(\lambda)$ and $B_{22}$ as in (5.88). This implies that

$$M_k(\lambda) \equiv \mathcal{Q}^T \begin{pmatrix} 0_r & 0 \\ 0 & B_{22} \end{pmatrix} \tilde{\mathcal{Q}}, \quad B_{22} = \begin{pmatrix} 0_{n-r-\rho} & 0 \\ 0 & \bar{B}_{44} \end{pmatrix}, \quad \lambda \in (\lambda_0 - \delta, \lambda_0),$$

with invertible $\bar{B}_{44} \in \mathbb{R}^{\rho \times \rho}$. This shows that rank $M_k(\lambda) \equiv \rho$ on $(\lambda_0 - \delta, \lambda_0)$, compared with the paragraph preceding Remark 5.21. Now since the first block column of $B_{12}(\lambda)$ in (5.88) is zero and $\bar{B}_{44}$ is invertible, (5.124) yields that

$$\text{Ker } \tilde{M}_k(\lambda) = \text{Ker} \begin{pmatrix} M_{11}(\lambda) & 0 \\ 0 & B_{22} \end{pmatrix} \tag{5.125}$$

for all $\lambda \in (\lambda_0 - \delta, \lambda_0]$, where the matrix $M_{11}(\lambda) \in \mathbb{R}^{r \times r}$ is defined by

$$M_{11}(\lambda) := \left[ I - X_{11}(\lambda)\, X_{11}^\dagger(\lambda) \right] B_{11}(\lambda), \quad \lambda \in \mathbb{R}. \tag{5.126}$$

Consequently, by Lemma 1.63, for any $\lambda \in (\lambda_0 - \delta, \lambda_0]$, we have

$$\tilde{M}_k^\dagger(\lambda)\, \tilde{M}_k(\lambda) = \begin{pmatrix} M_{11}^\dagger(\lambda) & 0 \\ 0 & B_{22}^\dagger \end{pmatrix} \begin{pmatrix} M_{11}(\lambda) & 0 \\ 0 & B_{22} \end{pmatrix} = \begin{pmatrix} M_{11}^\dagger(\lambda)\, M_{11}(\lambda) & 0 \\ 0 & B_{22}^\dagger\, B_{22} \end{pmatrix}.$$

Therefore, by the form of $\tilde{T}_k(\lambda)$ in (5.123),

$$\tilde{T}_k(\lambda) = \begin{pmatrix} T_{11}(\lambda) & 0 \\ 0 & I - B_{22}^{\dagger} B_{22} \end{pmatrix}, \qquad \lambda \in (\lambda_0 - \delta, \lambda_0], \tag{5.127}$$

where the matrix $T_{11}(\lambda) \in \mathbb{R}^{r \times r}$ is defined by

$$T_{11}(\lambda) := I - M_{11}^{\dagger}(\lambda) M_{11}(\lambda), \quad \lambda \in \mathbb{R}, \quad I - B_{22}^{\dagger} B_{22} = \begin{pmatrix} I_{n-r-\rho} & 0 \\ 0 & 0_{\rho} \end{pmatrix}. \tag{5.128}$$

Finally, from the form of the matrices $\tilde{X}_k(\lambda)$, $Y_{11}(\lambda)$, $\tilde{P}_k(\lambda)$, and $\tilde{T}_k(\lambda)$ in (5.73), (5.91), (5.123), and (5.127), we get

$$\tilde{P}_k(\lambda) = \begin{pmatrix} P_{11}(\lambda) & 0 \\ 0 & 0_{n-r} \end{pmatrix}, \qquad \lambda \in (\lambda_0 - \delta, \lambda_0], \tag{5.129}$$

where the matrix $P_{11}(\lambda)$ is given in (5.100)–(5.101) for $\lambda \in (\lambda_0 - \delta, \lambda_0)$ and

$$P_{11}(\lambda_0) := T_{11}(\lambda_0) Y_{11}(\lambda_0) X_{11}^{\dagger}(\lambda_0) B_{11}(\lambda_0) T_{11}(\lambda_0). \tag{5.130}$$

In addition, since $X_{11}(\lambda)$ is invertible for $\lambda \in (\lambda_0 - \delta, \lambda_0)$, we have

$$M_{11}(\lambda) \equiv 0, \quad T_{11}(\lambda) \equiv I, \quad P_{11}(\lambda) = Y_{11}(\lambda) X_{11}^{-1}(\lambda) B_{11}(\lambda), \quad \lambda \in (\lambda_0 - \delta, \lambda_0), \tag{5.131}$$

and the matrix $P_{11}(\lambda)$ is nondecreasing on $(\lambda_0 - \delta, \lambda_0)$, by Lemma 5.28. Thus, we are now ready for the final step in the proof of the first local oscillation theorem (Theorem 5.15).

## 5.1.6  Application of the Index Theorem

In this subsection we utilize the above construction in order to apply the index theorem (Theorem 1.85 and Corollary 1.86) to the reduced quantities in the dimension $r$ on the left neighborhood of $\lambda_0$ and to similarly reduced quantities in the appropriate dimension on the right neighborhood of $\lambda_0$. Here we need to assume that the matrix $B_{11}(\lambda)$ has constant image in $\lambda$, as the matrix $R_2^T(t)$ in Corollary 1.86 is assumed to have constant image in $t$. But under monotonicity assumption (5.3), conditions (5.38) and (5.10) are equivalent (see Theorem 5.1 (iii)), and then condition (5.38) can be used instead of (5.10).

*Proof of Theorem 5.15* Assume that (5.38) holds, then the matrix $\mathcal{B}_k(\lambda)$ has constant image in $\lambda$ on $\mathbb{R}$ (see Theorem 5.1 (iii)); hence the matrix $\tilde{\mathcal{B}}_k(\lambda)$ has constant image in $\lambda$ with $B_{11}(\lambda) \in \mathbb{R}^{r \times r}$ having constant image in $\lambda$, too. For $t \in (0, \delta)$, we define the matrices

$$
\left.
\begin{aligned}
X(t) &:= X_{11}(\lambda_0 - t), & \Lambda(t) &:= Y_{11}(\lambda_0 - t), \\
U(t) &:= -U_{11}(\lambda_0 - t), & M(t) &:= P_{11}(\lambda_0 - t), \\
R_1(t) &:= D_{11}^T(\lambda_0 - t), & R_2(t) &:= B_{11}^T(\lambda_0 - t), \\
S_1(t) &:= \tfrac{1}{2}\left[D_{11}(\lambda_0 - t)\, B_{11}^{\dagger}(\lambda_0 - t) + B_{11}^{\dagger T}(\lambda_0 - t)\, D_{11}^T(\lambda_0 - t)\right], \\
S_2(t) &:= \tfrac{1}{2}\big[D_{11}^T(\lambda_0 - t)\,[I - B_{11}(\lambda_0 - t)\, B_{11}^{\dagger}(\lambda_0 - t)] \\
&\quad + [I - B_{11}^{\dagger}(\lambda_0 - t)\, B_{11}(\lambda_0 - t)]\, D_{11}^T(\lambda_0 - t)\big], \\
S_1 &:= \tfrac{1}{2}\left[D_{11}(\lambda_0)\, B_{11}^{\dagger}(\lambda_0) + B_{11}^{\dagger T}(\lambda_0)\, D_{11}^T(\lambda_0)\right], \\
S_2 &:= \tfrac{1}{2}\big[D_{11}^T(\lambda_0)\,[I - B_{11}(\lambda_0)\, B_{11}^{\dagger}(\lambda_0)] \\
&\quad + [I - B_{11}^{\dagger}(\lambda_0)\, B_{11}(\lambda_0)]\, D_{11}^T(\lambda_0)\big],
\end{aligned}
\right\}
\tag{5.132}
$$

and the matrices

$$
\left.
\begin{aligned}
X &:= X_{11}(\lambda_0), & \Lambda &:= Y_{11}(\lambda_0), & S^* &:= M_{11}(\lambda_0), \\
U &:= -U_{11}(\lambda_0), & R_2 &:= B_{11}^T(\lambda_0) & T &:= T_{11}(\lambda_0), \\
R_1 &:= D_{11}^T(\lambda_0), & S &:= X_{11}^{\dagger}(\lambda_0)\, B_{11}, & Q &:= P_{11}(\lambda_0),
\end{aligned}
\right\}
\tag{5.133}
$$

where the matrices $Y_{11}(\lambda)$, $P_{11}(\lambda)$, $M_{11}(\lambda)$, $T_{11}(\lambda)$ are given in (5.91), (5.100), (5.130), (5.126), and (5.128). We now verify the assumptions of the index theorem (Corollary 1.86).

By Lemma 5.19 and (5.92), the matrices $X^T(t)\, U(t)$, $R_1(t)\, R_2^T(t)$, and $S_1(t)$ are symmetric, $\operatorname{rank}(X^T(t),\ U^T(t)) = r$, as well as $\operatorname{rank}(R_1(t),\ R_2(t)) = r$ for all $t \in [0, \delta)$. Moreover, the symmetry of $B_{11}^{\dagger}(\lambda)\, B_{11}(\lambda)$ and $D_{11}^T(\lambda)\, B_{11}(\lambda)$ in (5.92) yields that $R_1(t) = R_2(t)\, S_1(t) + S_2(t)$ and $S_2(t)\, R_2^T(t) = 0$ for $t \in [0, \delta)$, i.e., $\operatorname{Im} R_2^T(t) \subseteq \operatorname{Ker} S_2(t)$ for $t \in [0, \delta)$. The opposite inclusion $\operatorname{Ker} S_2(t) \subseteq \operatorname{Im} R_2^T(t)$ for $t \in [0, \delta)$ is a consequence of the relation $\operatorname{Ker}(D_{11}^T(\lambda),\ B_{11}^T(\lambda)) = \operatorname{Im}(B_{11}^T(\lambda),\ -D_{11}^T(\lambda))^T$ obtained from (5.92). Hence, $\operatorname{Ker} S_2(t) = \operatorname{Im} R_2^T(t)$ for $t \in [0, \delta)$. Furthermore, by Lemma 5.19, the matrix $X(t)$ is invertible for $t \in (0, \delta)$. Next, from (5.132)–(5.133), we have $X(t) \to X$, $U(t) \to U$, $S_1(t) \to S_1$, $S_2(t) \to S_2$, $R_1(t) \to R_1$, $R_2(t) \to R_2$ for $t \to 0^+$. Equations (5.100), (5.101), (5.91), (5.126), (5.128), and (5.130) imply that the matrices $M(t)$, $\Lambda(t)$, $\Lambda$, $S$, $S^*$, $T$, $Q$ defined in (5.132)–(5.133) are equal to their corresponding defining expressions in Theorem 1.85. Finally, $\operatorname{Im} R_2(t)$ is constant for $t \in [0, \delta)$ by assumption (5.38). By Lemma 5.26, the matrix $U(t)\, X^{-1}(t)$ is nonincreasing on $(0, \delta)$, as $X(t)$ and $U(t)$ are defined in (5.132) through $X_{11}(\lambda)$ and $U_{11}(\lambda)$ with the argument $\lambda = \lambda_0 - t$ and $U(t)$ has the opposite sign to $U_{11}(\lambda_0 - t)$, while by Lemma 5.28 the matrix $M(t)$ is nonincreasing on $(0, \delta)$, as $M(t)$ is defined in (5.132) through $P_{11}(\lambda)$ with the argument $\lambda = \lambda_0 - t$.

This means that equality (1.225) of Theorem 1.85 (Corollary 1.86) holds, in which we take $m = r$. From the construction in Sects. 5.1.4 and 5.1.5, we calculate

$$\text{ind } M(0^+) \stackrel{(5.132)}{=} \text{ind } P_{11}(\lambda_0^-) \stackrel{(5.129)}{=} \text{ind } \tilde{P}_k(\lambda_0^-) \stackrel{(5.120)}{=} \text{ind } P_k(\lambda_0^-),$$

$$\text{def } \Lambda(0^+) \stackrel{(5.132)}{=} \text{def } Y_{11}(\lambda_0^-) \stackrel{(5.91)}{=} r - \tilde{r} \stackrel{(5.72)}{=} r + \rho - \text{rank } X_k(\lambda_0^-),$$

$$\text{def } \Lambda \stackrel{(5.132)}{=} \text{def } Y_{11}(\lambda_0) \stackrel{(5.91)}{=} r - \text{rank } \tilde{X}_{11}(\lambda_0) \stackrel{(5.73)}{=} r + \rho - \text{rank } \tilde{X}_k(\lambda_0)$$

$$\stackrel{(5.78)}{=} r + \rho - \text{rank } X_k(\lambda_0),$$

$$\text{def } X \stackrel{(5.133)}{=} \text{def } X_{11}(\lambda_0) = r - \text{rank } X_{11}(\lambda_0) \stackrel{(5.58)}{=} r - \text{rank } \tilde{X}_{k+1}(\lambda_0)$$

$$\stackrel{(5.69)}{=} r - \text{rank } X_{k+1}(\lambda_0),$$

$$\text{ind } Q \stackrel{(5.133)}{=} \text{ind } P_{11}(\lambda_0) \stackrel{(5.129)}{=} \text{ind } \tilde{P}_k(\lambda_0) \stackrel{(5.120)}{=} \text{ind } P_k(\lambda_0),$$

$$\text{rank } T \stackrel{(5.133)}{=} \text{rank } T_{11}(\lambda_0) \stackrel{(5.128)}{=} r - \text{rank } M_{11}(\lambda_0)$$

$$\stackrel{(5.125)}{=} r - [\text{rank } \tilde{M}_k(\lambda_0) - \rho] \stackrel{(5.121)}{=} r + \rho - \text{rank } M(\lambda_0).$$

When we insert the above data into equation (1.225) and if we recall the definitions of $r = \text{rank } X_{k+1}(\lambda_0^-)$ and $\rho = \text{rank } M_k(\lambda_0^-)$, we get

$$\text{ind } P_k(\lambda_0^-) = \text{ind } M(0^+) \stackrel{(1.225)}{=} \text{ind } Q + m - \text{rank } T + \text{def } \Lambda - \text{def } \Lambda(0^+) - \text{def } X$$

$$= \text{ind } P_k(\lambda_0) + r - [r + \rho - \text{rank } M_k(\lambda_0)] + [r + \rho - \text{rank } X_k(\lambda_0)]$$

$$- [r + \rho - \text{rank } X_k(\lambda_0^-)] - [r - \text{rank } X_{k+1}(\lambda_0)]$$

$$= \text{ind } P_k(\lambda_0) + \text{rank } M_k(\lambda_0) + \text{rank } X_k(\lambda_0^-) - \text{rank } X_k(\lambda_0)$$

$$+ \text{rank } X_{k+1}(\lambda_0) - \text{rank } X_{k+1}(\lambda_0^-) - \text{rank } M_k(\lambda_0^-). \qquad (5.134)$$

In view of the definition of $m_k(\lambda)$ in (5.39) as the number of focal points in $(k, k+1]$, we have from (5.134) the identity

$$m_k(\lambda_0^-) = m_k(\lambda_0) + \text{rank } X_k(\lambda_0^-) - \text{rank } X_k(\lambda_0)$$

$$+ \text{rank } X_{k+1}(\lambda_0) - \text{rank } X_{k+1}(\lambda_0^-). \qquad (5.135)$$

Now we need to consider the construction in Sects. 5.1.4 and 5.1.5 in the right neighborhood of $\lambda_0$, i.e., for $\lambda \in [\lambda_0, \lambda_0 + \delta)$. The analysis of this construction shows that exactly the same principles in obtaining the partitioned auxiliary matrices $\tilde{X}_{k+1}(\lambda)$, $\tilde{U}_{k+1}(\lambda)$, the refined partitioned matrices $\tilde{X}_k(\lambda)$, $\tilde{U}_k(\lambda)$, and the coefficients $\tilde{A}_k(\lambda), \tilde{B}_k(\lambda), \tilde{C}_k(\lambda), \tilde{D}_k(\lambda)$ hold when the dimensions $r$ and $\rho$ in (5.57) and (5.70) are replaced by the dimensions

$$s := \text{rank } X_{k+1}(\lambda_0^+) \quad \text{and} \quad \sigma := \text{rank } X_k(\lambda_0^+) \mathcal{P}_2 = \text{rank } M_k(\lambda_0^+),$$

respectively, and at the same time the intervals $(\lambda_0 - \delta, \lambda_0]$ and $(\lambda_0 - \delta, \lambda_0)$ are replaced by the intervals $[\lambda_0, \lambda_0 + \delta)$ and $(\lambda_0, \lambda_0 + \delta)$. Therefore, we now define for $t \in (0, \delta)$ the $s \times s$ matrices

$$
\left.
\begin{aligned}
X(t) &:= X_{11}(\lambda_0 + t), & \Lambda(t) &:= Y_{11}(\lambda_0 + t), \\
U(t) &:= -U_{11}(\lambda_0 + t), & M(t) &:= P_{11}(\lambda_0 + t), \\
R_1(t) &:= D_{11}^T(\lambda_0 + t), & R_2(t) &:= B_{11}^T(\lambda_0 + t), \\
S_1(t) &:= \tfrac{1}{2}\left[D_{11}(\lambda_0 + t)\, B_{11}^\dagger(\lambda_0 + t) + B_{11}^{\dagger T}(\lambda_0 + t)\, D_{11}^T(\lambda_0 + t)\right], \\
S_2(t) &:= \tfrac{1}{2}\left[D_{11}^T(\lambda_0 + t)\left[I - B_{11}(\lambda_0 + t)\, B_{11}^\dagger(\lambda_0 + t)\right]\right. \\
&\qquad \left. + [I - B_{11}^\dagger(\lambda_0 + t)\, B_{11}(\lambda_0 + t)]\, D_{11}^T(\lambda_0 + t)\right],
\end{aligned}
\right\}
\tag{5.136}
$$

and the matrices $S_1$, $S_2$, $X$, $U$, $R_1$, $\Lambda$, $R_2$, $S$, $S^*$, $T$, $Q$ by the equations in (5.132) and (5.133). It follows by Lemmas 5.26 and 5.28 that the matrices $U(t) X^{-1}(t)$ and $M(t)$ are now nondecreasing on $(0, \delta)$, as they are defined in (5.136) through $X_{11}(\lambda)$, $U_{11}(\lambda)$, $P_{11}(\lambda)$ with the argument $\lambda = \lambda_0 + t$ and $U(t)$ has the opposite sign to $U_{11}(\lambda_0 + t)$. This means that equality (1.226) of Theorem 1.85 (Corollary 1.86) now holds, in which we take $m = s$. We now calculate

$$\text{ind } M(0^+) \overset{(5.136)}{=} \text{ind } P_{11}(\lambda_0^+) = \text{ind } \tilde{P}_k(\lambda_0^+) = \text{ind } P_k(\lambda_0^+),$$

$$\text{ind } Q \overset{(5.133)}{=} \text{ind } P_{11}(\lambda_0) = \text{ind } \tilde{P}_k(\lambda_0) = \text{ind } P_k(\lambda_0),$$

$$\text{rank } T \overset{(5.133)}{=} \text{rank } T_{11}(\lambda_0) = s - \text{rank } M_{11}(\lambda_0) = s - [\text{rank } \tilde{M}_k(\lambda_0) - \sigma]$$

$$= s + \sigma - \text{rank } M(\lambda_0).$$

When we insert the above data into equation (1.226), we obtain

$$\text{ind } P_k(\lambda_0^+) = \text{ind } M(0^+) \overset{(1.226)}{=} \text{ind } Q + m - \text{rank } T$$

$$= \text{ind } P_k(\lambda_0) + s - [s + \sigma - \text{rank } M_k(\lambda_0)]$$

$$= \text{ind } P_k(\lambda_0) + \text{rank } M_k(\lambda_0) - \text{rank } M_k(\lambda_0^+). \tag{5.137}$$

In view of the definition of $m_k(\lambda)$ in (5.39), we get from (5.137) that $m_k(\lambda_0^+) = m_k(\lambda_0)$, which proves the equality in (5.42). By combining (5.42) with equation (5.135), we can see that the equality in (5.43) is also satisfied. The proof of Theorem 5.15 is now complete.                                                                   $\square$

### 5.1.7  Applications and Examples

In this subsection we show some applications of the oscillation theorems from Sect. 5.1.3. For this purpose we define the associated quadratic functional, compare with (2.56),

$$\mathcal{F}(y, \lambda) := \sum_{k=0}^{N} y_k^T \left( \mathcal{S}_k^T(\lambda) \, \mathcal{K} \, \mathcal{S}_k(\lambda) - \mathcal{K} \right) y_k, \quad \mathcal{K} := \begin{pmatrix} 0_n & 0_n \\ I_n & 0_n \end{pmatrix}, \tag{5.138}$$

where $y = (x, u)$ is an *admissible* pair, i.e., $x_{k+1} = \mathcal{A}_k(\lambda) \, x_k + \mathcal{B}_k(\lambda) \, u_k$ for all $k \in [0, N]_{\mathbb{Z}}$, satisfying the Dirichlet endpoints $x_0 = 0 = x_{N+1}$. We say that the functional $\mathcal{F}(\cdot, \lambda)$ is positive, if $\mathcal{F}(z, \lambda) > 0$ for every admissible $z = (x, u)$ with $x_0 = 0 = x_{N+1}$ and $x \not\equiv 0$. Conversely, the functional $\mathcal{F}(\cdot, \lambda)$ is not positive, if $\mathcal{F}(z, \lambda) \leq 0$ for some admissible $z = (x, u)$ with $x_0 = 0 = x_{N+1}$ and $x \not\equiv 0$. These two properties will be denoted by $\mathcal{F}(\cdot, \lambda) > 0$ and $\mathcal{F}(\cdot, \lambda) \not> 0$, respectively. The following result is from Theorem 2.36.

**Proposition 5.30** *Let $\lambda_0 \in \mathbb{R}$ be fixed. Then the functional $\mathcal{F}(\cdot, \lambda_0) > 0$ if and only if $n_1(\lambda_0) = 0$, i.e., the principal solution $Y^{[0]}(\lambda_0)$ of system $(S_{\lambda_0})$ has no focal points in $(0, N + 1]$.*

Additional conditions which are equivalent to the positivity of $\mathcal{F}(\cdot, \lambda_0)$ can be found in the literature, such as the solvability of the explicit and implicit Riccati equations and inequalities, conjugate and coupled intervals, perturbed quadratic functionals, etc.; see, e.g., Theorem 2.36.

From (5.55) and (5.56), it follows that the constant $m \in \mathbb{N} \cup \{0\}$ in Theorem 5.17 is actually zero, if and only if $n_2(\lambda) = n_1(\lambda) \equiv m = 0$ for all or for some $\lambda \in \mathbb{R}$. Combining this observation with Proposition 5.30 yields the following.

**Theorem 5.31** *Assume (5.3) and (5.38) for all $k \in [0, N]_{\mathbb{Z}}$. Then*

$$n_1(\lambda) = n_2(\lambda) \quad \text{for all } \lambda \in \mathbb{R} \tag{5.139}$$

*if and only if there exists $\lambda_0 < 0$ such that $\mathcal{F}(\cdot, \lambda_0) > 0$.*

*Proof* If $\mathcal{F}(\cdot, \lambda_0) > 0$, then $n_1(\lambda_0) = 0$, by Proposition 5.30. Since the function $n_1(\cdot)$ is nondecreasing on $\mathbb{R}$ by Theorem 5.16, it follows that $n_1(\lambda) \equiv 0$ for all $\lambda \leq \lambda_0$, i.e., $m = 0$ in (5.55), showing (5.139). Conversely, if (5.139) holds, then $m = 0$, and so, by (5.56), we have $n_1(\lambda) \equiv m = 0$ for $\lambda \leq \lambda_0$ with $\lambda_0$ from Theorem 5.17. From Proposition 5.30 we then get $\mathcal{F}(\cdot, \lambda_0) > 0$.                              $\square$

Next we present conditions in terms of $\mathcal{F}(\cdot, \lambda) > 0$ which are closely related to the existence of finite eigenvalues of (E).

**Theorem 5.32 (Existence of Finite Eigenvalues: Necessary Condition)** *Let the assumptions (5.3) and (5.38) be satisfied for all $k \in [0, N]_{\mathbb{Z}}$. If (E) has a finite* *eigenvalue, then there exist* $\lambda_0, \lambda_1 \in \mathbb{R}$ *with* $\lambda_0 < \lambda_1$ *and* $m \in \mathbb{N} \cup \{0\}$ *such that* $n_1(\lambda) \equiv m$ *for all* $\lambda \leq \lambda_0$ *and* $\mathcal{F}(\cdot, \lambda_1) \not\equiv 0$.

*Proof* The fact that (E) has a finite eigenvalue of (E) means that $n_2(\lambda_1) \geq 1$ for some $\lambda_1 \in \mathbb{R}$. By Theorem 5.17, we know that equality (5.55) is satisfied for some $m \in \mathbb{N} \cup \{0\}$ and $n_1(\lambda) \equiv m$ for all $\lambda \leq \lambda_0$ for some $\lambda_0 < 0$. Without loss of generality, we may take $\lambda_0 < \lambda_1$, so that the first part of this theorem is proven. Next, from (5.55) with $\lambda = \lambda_1$, we obtain $n_1(\lambda_1) = n_2(\lambda_1) + m \geq n_2(\lambda_1) \geq 1$, showing that the principal solution of $(S_{\lambda_1})$ has at least one focal point in $(0, N+1]$. In turn, Proposition 5.30 shows that $\mathcal{F}(\cdot, \lambda_1) \not\equiv 0$.                    □

**Theorem 5.33 (Existence of Finite Eigenvalues: Sufficient Condition)** *Let the assumptions* (5.3) *and* (5.38) *be satisfied for all* $k \in [0, N]_{\mathbb{Z}}$. *If there exist* $\lambda_0, \lambda_1 \in \mathbb{R}$ *with* $\lambda_0 < \lambda_1$ *such that* $\mathcal{F}(\cdot, \lambda_0) > 0$ *and* $\mathcal{F}(\cdot, \lambda_1) \not\equiv 0$, *then* (E) *has at least one finite eigenvalue.*

*Proof* The positivity of $\mathcal{F}(\cdot, \lambda_0)$ implies by Theorem 5.31 that equality (5.139) holds. If we assume that there is no finite eigenvalue of (E) at all, i.e., if $n_2(\lambda) \equiv 0$ for every $\lambda \in \mathbb{R}$, then $n_1(\lambda) \equiv 0$ for all $\lambda \in \mathbb{R}$ as well. In particular, $n_1(\lambda_1) = 0$. This means by Proposition 5.30 that $\mathcal{F}(\cdot, \lambda_1) > 0$, which contradicts our assumption. Therefore, under the given conditions, the eigenvalue problem (E) must have at least one finite eigenvalue.                    □

The following result characterizes the smallest finite eigenvalue of (E).

**Theorem 5.34** *Assume that* (5.3) *and* (5.38) *hold for all* $k \in [0, N]_{\mathbb{Z}}$. *If there exist* $\lambda_0, \lambda_1 \in \mathbb{R}$ *with* $\lambda_0 < \lambda_1$ *such that* $\mathcal{F}(\cdot, \lambda_0) > 0$ *and* $\mathcal{F}(\cdot, \lambda_1) \not\equiv 0$, *then the eigenvalue problem* (E) *possesses a smallest finite eigenvalue* $\lambda_{\min}$, *which is characterized by any of the following conditions:*

$$\lambda_{\min} = \sup \mathcal{P}, \qquad \mathcal{P} := \{\lambda \in \mathbb{R}, \ \mathcal{F}(\cdot, \lambda) > 0\}, \qquad (5.140)$$

$$\lambda_{\min} = \min \mathcal{N}, \qquad \mathcal{N} := \{\lambda \in \mathbb{R}, \ \mathcal{F}(\cdot, \lambda) \not\equiv 0\}. \qquad (5.141)$$

*Moreover, the algebraic multiplicity of* $\lambda_{\min}$ *is then equal to* $n_1(\lambda_{\min})$, *i.e., to the number of focal points of the principal solution of* $(S_{\lambda_{\min}})$ *in* $(0, N+1]$.

*Proof* From Theorem 5.33, we know that the eigenvalue problem (E) has at least one finite eigenvalue. Since $\mathcal{F}(\cdot, \lambda_0) > 0$ is assumed, then $\lambda_0 \in \mathcal{P}$ and the set $\mathcal{P}$ is nonempty. Moreover, by Proposition 5.30 we have $n_1(\lambda_0) = 0$. Since the function $n_1(\cdot)$ is nondecreasing on $\mathbb{R}$, it follows that $n_1(\lambda) \equiv 0$ for $\lambda \leq \lambda_0$, i.e., $\mathcal{F}(\cdot, \lambda) > 0$ for $\lambda \leq \lambda_0$. This implies that $(-\infty, \lambda_0] \subseteq \mathcal{P}$. In addition, $\lambda_1 \notin \mathcal{P}$, so that $\mathcal{P}$ is bounded from above and therefore $\mathcal{P} = (-\infty, \omega)$, where $\omega = \sup \mathcal{P}$ exists. It follows that $n_1(\omega) \geq 1$, because by Theorem 5.16, the function $n_1$ is right-continuous on $\mathbb{R}$. We will show that $\lambda_{\min} = \omega$ is the smallest finite eigenvalue of (E). From Theorem 5.31, we know that $n_1(\lambda) = n_2(\lambda)$ for all $\lambda \in \mathbb{R}$. Hence, $n_2(\lambda) \equiv 0$ for all $\lambda < \omega$ and $n_2(\omega) = n_1(\omega) \geq 1$, proving that $\omega$ is the smallest finite eigenvalue of (E) with the algebraic multiplicity $n_1(\omega)$. As for (5.141), we note that the set $\mathcal{N}$ is nonempty, because $\lambda_1 \in \mathcal{N}$, and the interval $(-\infty, \lambda_0]$ is not contained in $\mathcal{N}$. Therefore, $\mathcal{N}$ is bounded from below. Let $v \in \mathcal{N}$, i.e., $n_1(v) \geq 1$. Then $n_2(v) = n_1(v) \geq 1$. Since we know from Theorem 5.17 and Corollary 5.7 that the function $n_2$ is right-continuous on $\mathbb{R}$ and the finite eigenvalues are isolated and bounded from below, it follows that $\kappa := \min\{v \in \mathbb{R}, \ n_2(v) \geq 1\} = \min \mathcal{N}$ exists and satisfies $\lambda_0 < \kappa$. Furthermore, by the definition of $\kappa$, we have $n_2(\lambda) \equiv 0$ for all $\lambda < \kappa$ and $n_2(\kappa) \geq 1$. This yields that $\lambda_{\min} = \kappa$ is the smallest finite eigenvalue of (E) with multiplicity $n_2(\kappa) = n_1(\kappa)$.                                    □

Finally, we discuss the applicability of our results to some special discrete symplectic systems (SDS$_\lambda$).

*Example 5.35* Consider the second-order Sturm-Liouville difference equation; see Sect. 1.2, which can also be viewed as the Jacobi matrix,

$$\Delta\big(r_k(\lambda)\, \Delta x_k(\lambda)\big) + q_k(\lambda)\, x_{k+1}(\lambda) = 0, \quad k \in [0, N-1]_{\mathbb{Z}}, \tag{5.142}$$

where the coefficients $r_k, q_k : \mathbb{R} \to \mathbb{R}$ satisfy

$$r_k(\lambda) \neq 0, \quad \dot{r}_k(\lambda) \leq 0, \quad \dot{q}_k(\lambda) \geq 0, \quad k \in [0, N]_{\mathbb{Z}}, \ \lambda \in \mathbb{R}.$$

In this case the matrices $\mathcal{S}_k(\lambda)$ and $\Psi_k(\lambda)$ have the form (see Example 2.13)

$$\mathcal{S}_k(\lambda) = \begin{pmatrix} 1 & 1/r_k(\lambda) \\ -q_k(\lambda) & 1 - q_k(\lambda)/r_k(\lambda) \end{pmatrix},$$

$$\Psi_k(\lambda) = \frac{1}{r_k^2(\lambda)} \begin{pmatrix} r_k(\lambda) & q_k(\lambda) \\ 0 & 1 \end{pmatrix} \begin{pmatrix} \dot{q}_k(\lambda) & 0 \\ 0 & -\dot{r}_k(\lambda) \end{pmatrix} \begin{pmatrix} r_k(\lambda) & 0 \\ q_k(\lambda) & 1 \end{pmatrix}.$$

The square of $r_k(\lambda)$ in the matrix $\Psi_k(\lambda)$ above shows that the oscillation and spectral theory of the second-order Sturm-Liouville difference equations, even in the classical setting where $r_k(\lambda) \equiv r_k \neq 0$ and $q_k(\lambda) = q_k + \lambda\, w_k$ with $w_k > 0$, does not depend on the sign of the coefficient $r_k(\lambda)$ but rather on its monotonicity. The principal solution $\hat{x}(\lambda)$ of (5.142) is the solution starting with the initial conditions $\hat{x}_0(\lambda) \equiv 0$ and $\hat{x}_1(\lambda) = 1/r_0(\lambda)$ for all $\lambda \in \mathbb{R}$. The result in Theorem 5.3 yields that if $x(\lambda)$ is a nontrivial solution of (5.142) and $x_k(\lambda_0) = 0$ for some $k \in [0, N+1]_{\mathbb{Z}}$ and $\lambda_0 \in \mathbb{R}$, then either $x_k(\lambda)$ is identically zero or never zero in some left and right neighborhoods of $\lambda_0$. This implies, according to Definition 5.5, that a number $\lambda_0 \in \mathbb{R}$ is a finite eigenvalue of (E) provided there exists $\delta > 0$ such that $\hat{x}_{N+1}(\lambda_0) = 0$ and $\hat{x}_{N+1}(\lambda) \neq 0$ for all $\lambda \in (\lambda_0 - \delta, \lambda_0)$. The degeneracy condition in (5.23) of Definition 5.8 at $\lambda_0$ then means that the solutions $x(\lambda)$ of (5.142) with $\lambda \in (\lambda_0 - \delta, \lambda_0]$ starting with the initial conditions $x_0(\lambda) = x_0(\lambda_0)$ and $x_1(\lambda) = x_0(\lambda) + r_0(\lambda_0)\, \Delta x_0(\lambda_0)/r_0(\lambda)$ satisfy

$$\dot{r}_k(\lambda)\, \Delta x_k(\lambda) \equiv 0, \quad \dot{q}_k(\lambda)\, x_{k+1}(\lambda) \equiv 0 \quad \text{for all } k \in [0, N]_{\mathbb{Z}} \text{ and } \lambda \in (\lambda_0 - \delta, \lambda_0].$$

Since rank $\mathcal{B}_k(\lambda) = \mathrm{rank}\, 1/r_k(\lambda) \equiv 1$ is constant in $\lambda$, the results in Theorems 5.15 and 5.16 and 5.17 and in Corollary 5.18 hold without any additional assumption. Similar analysis can also be done for the higher-order Sturm-Liouville difference equations; see Example 5.38.

*Example 5.36* Similar to Example 5.35, we consider the second-order matrix Sturm-Liouville equation (see Example 2.10)

$$\Delta\big(R_k(\lambda)\,\Delta x_k(\lambda)\big) + Q_k(\lambda)\, x_{k+1}(\lambda) = 0, \quad k \in [0, N-1]_{\mathbb{Z}}, \tag{5.143}$$

where $R_k, Q_k : \mathbb{R} \to \mathbb{R}^{n \times n}$ are symmetric matrix functions such that $R_k(\lambda)$ is invertible, $\dot{R}_k(\lambda) \leq 0$, and $\dot{Q}_k(\lambda) \geq 0$ for all $k \in [0, N]_{\mathbb{Z}}$ and $\lambda \in \mathbb{R}$. In this case

$$\mathcal{S}_k(\lambda) = \begin{pmatrix} I & R_k^{-1}(\lambda) \\ Q_k(\lambda) & I - Q_k(\lambda)\, R_k^{-1}(\lambda) \end{pmatrix},$$

$$\Psi_k(\lambda) = \begin{pmatrix} I & Q_k(\lambda)\, R_k^{-1}(\lambda) \\ 0 & R_k^{-1}(\lambda) \end{pmatrix} \begin{pmatrix} \dot{Q}_k(\lambda) & 0 \\ 0 & -\dot{R}_k(\lambda) \end{pmatrix} \begin{pmatrix} I & 0 \\ R_k^{-1}(\lambda)\, Q_k(\lambda) & R_k^{-1}(\lambda) \end{pmatrix}.$$

A number $\lambda_0 \in \mathbb{R}$ is a finite eigenvalue of (E) if the principal solution $\hat{X}(\lambda)$ of (5.143), i.e., the solution starting with $\hat{X}_0(\lambda) \equiv 0$ and $\hat{X}_1(\lambda) = R_0^{-1}(\lambda)$ for all $\lambda \in \mathbb{R}$, has $\hat{X}_{N+1}(\lambda_0)$ singular and $\hat{X}_{N+1}(\lambda)$ is invertible for $\lambda \in (\lambda_0 - \delta, \lambda_0)$ for some $\delta > 0$. Since in this case rank $\mathcal{B}_k(\lambda) = \mathrm{rank}\, R_k^{-1}(\lambda) \equiv n$ is constant in $\lambda$, the results in Theorems 5.15 and 5.16 and 5.17 and in Corollary 5.18 then hold without any additional assumption.

*Example 5.37* Consider the linear Hamiltonian system (see Example 2.7)

$$\left.\begin{array}{l} \Delta x_k(\lambda) = A_k(\lambda)\, x_{k+1}(\lambda) + B_k(\lambda)\, u_k(\lambda), \\ \Delta u_k(\lambda) = C_k(\lambda)\, x_{k+1}(\lambda) - A_k^T(\lambda)\, u_k(\lambda), \end{array}\right\} \quad k \in [0, N]_{\mathbb{Z}}, \tag{5.144}$$

where the coefficients $A_k, B_k, C_k : \mathbb{R} \to \mathbb{R}^{n \times n}$ are such that $B_k(\lambda)$ and $C_k(\lambda)$ are symmetric, $I - A_k(\lambda)$ is invertible with $\tilde{A}_k(\lambda) := [I - A_k(\lambda)]^{-1}$, and

$$\dot{\mathcal{H}}_k(\lambda) \geq 0, \quad \mathcal{H}_k(\lambda) := \begin{pmatrix} -C_k(\lambda) & A_k^T(\lambda) \\ A_k(\lambda) & B_k(\lambda) \end{pmatrix}, \quad k \in [0, N]_{\mathbb{Z}}, \ \lambda \in \mathbb{R}.$$

The matrices $\mathcal{S}_k(\lambda)$ and $\Psi_k(\lambda)$ have the form

$$\mathcal{S}_k(\lambda) = \begin{pmatrix} \tilde{A}_k(\lambda) & \tilde{A}_k(\lambda)\, B_k(\lambda) \\ C_k(\lambda)\, \tilde{A}_k(\lambda) & C_k(\lambda)\, \tilde{A}_k(\lambda)\, B_k(\lambda) + I - A_k^T(\lambda) \end{pmatrix}, \tag{5.145}$$

$$\Psi_k(\lambda) = \begin{pmatrix} I & -C_k(\lambda)\, \tilde{A}_k(\lambda) \\ 0 & \tilde{A}_k(\lambda) \end{pmatrix} \dot{\mathcal{H}}_k(\lambda) \begin{pmatrix} I & 0 \\ -\tilde{A}_k^T(\lambda)\, C_k(\lambda) & \tilde{A}_k^T(\lambda) \end{pmatrix}. \tag{5.146}$$

The results in Theorems 5.15 and 5.16 and 5.17 and in Corollary 5.18 then hold under the assumption that the matrix $B_k(\lambda)$ has constant rank in $\lambda$ for every $k \in [0, N]_{\mathbb{Z}}$. In Sect. 5.6 we completely omit this restriction.

*Example 5.38* For a fixed $n \in \mathbb{N}$, we consider the $2n$-th order Sturm-Liouville difference equation (see Example 2.12)

$$\sum_{j=0}^{n} (-1)^j \Delta^j \left( r_k^{[j]}(\lambda) \, \Delta^j \, y_{k+n-j}(\lambda) \right) = 0, \quad k \in [0, N - n]_{\mathbb{Z}}, \tag{5.147}$$

where $r_k^{[i]} : \mathbb{R} \to \mathbb{R}$ are piecewise continuously differentiable for all $i \in \{0, 1 \ldots, n\}$ and $k \in [0, N]_{\mathbb{Z}}$ such that

$$r_k^{[n]}(\lambda) \neq 0, \quad \dot{r}_k^{[i]}(\lambda) \leq 0 \quad \text{for all } i \in \{0, 1, \ldots, n\}.$$

Equation (5.147) can be written as a special linear Hamiltonian system (5.144), whose coefficients are given in (2.29) and (2.30). In particular $A_k(\lambda) \equiv A$ is constant in $\lambda$. In turn by Example 5.37, system (5.144) is a special symplectic difference system (SDS$_\lambda$) with

$$\mathcal{A}_k(\lambda) \equiv \begin{pmatrix} 1 & 1 & \ldots & 1 \\ 0 & 1 & \ldots & 1 \\ \vdots & \vdots & \ddots & \vdots \\ 0 & 0 & \ldots & 1 \end{pmatrix}, \quad \mathcal{B}_k(\lambda) = \frac{1}{r_k^{[n]}(\lambda)} \begin{pmatrix} 0 & \ldots & 0 & 1 \\ 0 & \ldots & 0 & 1 \\ \vdots & \ddots & \vdots & \vdots \\ 0 & \ldots & 0 & 1 \end{pmatrix}, \tag{5.148}$$

$$\mathcal{C}_k(\lambda) = \begin{pmatrix} r_k^{[0]}(\lambda) & r_k^{[0]}(\lambda) & \ldots & r_k^{[0]}(\lambda) \\ 0 & r_k^{[1]}(\lambda) & \ldots & r_k^{[1]}(\lambda) \\ \vdots & \vdots & \ddots & \vdots \\ 0 & 0 & \ldots & r_k^{[n-1]}(\lambda) \end{pmatrix}, \tag{5.149}$$

$$\mathcal{D}_k(\lambda) = \begin{pmatrix} 1 & 0 & 0 & \ldots & 0 & 0 & r_k^{[0]}(\lambda)/r_k^{[n]}(\lambda) \\ -1 & 1 & 0 & \ldots & 0 & 0 & r_k^{[1]}(\lambda)/r_k^{[n]}(\lambda) \\ 0 & -1 & 1 & \ldots & 0 & 0 & r_k^{[2]}(\lambda)/r_k^{[n]}(\lambda) \\ \vdots & \vdots & \vdots & \ddots & \vdots & \vdots & \vdots \\ 0 & 0 & 0 & \ldots & 1 & 0 & r_k^{[n-3]}(\lambda)/r_k^{[n]}(\lambda) \\ 0 & 0 & 0 & \ldots & -1 & 1 & r_k^{[n-2]}(\lambda)/r_k^{[n]}(\lambda) \\ 0 & 0 & 0 & \ldots & 0 & -1 & 1 + r_k^{[n-1]}(\lambda)/r_k^{[n]}(\lambda) \end{pmatrix}. \tag{5.150}$$

The matrix $\Psi_k(\lambda)$ is now given in (5.146), where

$$\dot{\mathcal{H}}_k(\lambda) = \mathrm{diag}\{-\dot{\mathcal{C}}_k(\lambda), \ \dot{\mathcal{B}}_k(\lambda)\}$$

$$= -\mathrm{diag}\left\{\dot{r}_k^{[0]}(\lambda), \ \ldots, \dot{r}_k^{[n-1]}(\lambda), \ 0, \ \ldots, \ 0, \ \frac{\dot{r}_k^{[n]}(\lambda)}{\left(r_k^{[n]}(\lambda)\right)^2}\right\} \geq 0.$$

Since the rank of the matrix $\mathcal{B}_k(\lambda)$ in (5.148) is constant in $\lambda$, it follows that the results in Theorems 5.15 and 5.16 and 5.17 and in Corollary 5.18 then hold without any additional assumption.

*Example 5.39* Assume that the symplectic matrix $\mathcal{S}_k(\lambda)$ in (5.2) is linear in $\lambda$, that is, we have $\mathcal{S}_k(\lambda) := \mathcal{S}_k + \lambda \mathcal{V}_k$ for all $k \in [0, N]_{\mathbb{Z}}$ and $\lambda \in \mathbb{R}$. Then Lemma 1.58(iii) yields that $\mathcal{S}_k^T(\lambda) = \mathcal{S}_k^T + \lambda \mathcal{V}_k^T \in Sp(2n)$, i.e.,

$$(\mathcal{S}_k + \lambda \mathcal{V}_k)\,\mathcal{J}(\mathcal{S}_k + \lambda \mathcal{V}_k)^T = \mathcal{S}_k\,\mathcal{J}\mathcal{S}_k^T + \lambda\,(\mathcal{V}_k\mathcal{J}\mathcal{S}_k^T + \mathcal{S}_k\mathcal{J}\mathcal{V}_k^T) + \lambda^2 \mathcal{V}_k\mathcal{J}\mathcal{V}_k^T = \mathcal{J}$$

for all $\lambda \in \mathbb{R}$. It follows that the matrix $\mathcal{S}_k$ is symplectic, $\mathcal{V}_k\mathcal{J}\mathcal{V}_k^T = 0$, and $\mathcal{V}_k\mathcal{J}\mathcal{S}_k^T = \mathcal{S}_k\mathcal{J}^T\mathcal{V}_k^T$. Using this fact we prove that the symmetric matrix $\Psi_k(\lambda)$ in (5.3) is constant in $\lambda$ and it has the form

$$\Psi_k(\lambda) \equiv \Psi_k := \mathcal{J}\,\mathcal{V}_k\,\mathcal{J}\mathcal{S}_k^T\,\mathcal{J} \geq 0 \quad \text{for all } k \in [0, N]_{\mathbb{Z}}.$$

In this case we can actually *prove* that the finite eigenvalues of (E) are real and that the finite eigenfunctions corresponding to different finite eigenvalues are orthogonal with respect to the semi-inner product

$$\langle z, \tilde{z} \rangle_\Psi := \sum_{k=0}^{N} z_{k+1}^T\,\Psi_k\,\tilde{z}_{k+1}.$$

The proof follows standard arguments from linear algebra. The results in Theorems 5.15, 5.16, and 5.17 and in Corollary 5.18 then hold under the assumption that the matrix $\mathcal{B}_k(\lambda) = \mathcal{B}_k + \lambda\tilde{\mathcal{B}}_k$ has constant rank in $\lambda$ for every $k \in [0, N]_{\mathbb{Z}}$. We note that in Sect. 5.6 we completely omit this restriction.

## 5.2 Eigenvalue Problems with General Boundary Conditions

In this section we present the theory of eigenvalues for system (5.1) with general boundary conditions, i.e., with separated and jointly varying endpoints.

### 5.2.1   Transformations of Boundary Conditions

In this subsection we consider system (SDS$_\lambda$) together with the so-called *self-adjoint boundary conditions*

$$R_1(\lambda) \begin{pmatrix} x_0(\lambda) \\ x_{N+1}(\lambda) \end{pmatrix} + R_2(\lambda) \begin{pmatrix} -u_0(\lambda) \\ u_{N+1}(\lambda) \end{pmatrix} = 0, \tag{5.151}$$

where $R_1(\lambda)$ and $R_2(\lambda)$ are piecewise continuously differentiable matrix-valued functions such that (compare with (2.92))

$$\left. \begin{array}{c} R_1(\lambda), \ R_2(\lambda) \in \mathbb{R}^{2n \times 2n}, \quad \operatorname{rank}(R_1(\lambda) \ R_2(\lambda)) = 2n, \\ R_1(\lambda) \, R_2^T(\lambda) = R_2(\lambda) \, R_1^T(\lambda), \quad \lambda \in \mathbb{R}. \end{array} \right\} \tag{5.152}$$

Moreover, with the notation $\mathcal{R}(\lambda) := (R_1(\lambda) \ R_2(\lambda)) \in \mathbb{R}^{2n \times 4n}$, we impose the following monotonicity restriction

$$\dot{R}_1(\lambda) \, R_2^T(\lambda) - \dot{R}_2(\lambda) \, R_1^T(\lambda) = \dot{\mathcal{R}}(\lambda) \, \mathcal{J}_{4n} \mathcal{R}^T(\lambda) \leq 0. \tag{5.153}$$

As a particular case of matrices $R_1(\lambda)$ and $R_2(\lambda)$ in (5.151), we consider their block diagonal form

$$R_1(\lambda) = \operatorname{diag}\{R_0^*(\lambda), \ R_{N+1}^*(\lambda)\}, \quad R_2(\lambda) = \operatorname{diag}\{-R_0(\lambda), \ R_{N+1}(\lambda)\}. \tag{5.154}$$

Here $R_i(\lambda)$, $R_i^*(\lambda) \in \mathbb{R}^{n \times n}$, $i \in \{0, N+1\}$ and $*$ is just a notation without any connection to the conjugate transpose of a matrix. In this case the boundary conditions are *separated*, i.e.,

$$\left. \begin{array}{c} R_0^*(\lambda) \, x_0(\lambda) + R_0(\lambda) \, u_0(\lambda) = 0, \\ R_{N+1}^*(\lambda) \, x_{N+1}(\lambda) + R_{N+1}(\lambda) \, u_{N+1}(\lambda) = 0, \end{array} \right\} \tag{5.155}$$

and conditions (5.152) translate as (compared with (2.83))

$$\left. \begin{array}{c} R_0^*(\lambda) \, R_0^T(\lambda) = R_0(\lambda) \, R_0^{*T}(\lambda), \quad \operatorname{rank}(R_0^*(\lambda), \ R_0(\lambda)) = n, \\ R_{N+1}^*(\lambda) \, R_{N+1}^T(\lambda) = R_{N+1}(\lambda) \, R_{N+1}^{*T}(\lambda), \quad \operatorname{rank}(R_{N+1}^*(\lambda), \ R_{N+1}(\lambda)) = n, \end{array} \right\} \tag{5.156}$$

while the monotonicity condition (5.153) is rewritten in the form

$$\dot{R}_0^*(\lambda) \, R_0^T(\lambda) - \dot{R}_0(\lambda) \, R_0^{*T}(\lambda) = \dot{\tilde{R}}(\lambda) \, \mathcal{J} \, \tilde{R}^T(\lambda) \geq 0, \tag{5.157}$$

$$\dot{R}_{N+1}^*(\lambda) \, R_{N+1}^T(\lambda) - \dot{R}_{N+1}(\lambda) \, R_{N+1}^{*T}(\lambda) = \dot{R}(\lambda) \, \mathcal{J} R^T(\lambda) \leq 0, \tag{5.158}$$

where

$$\tilde{R}(\lambda) := (R_0^*(\lambda) \ R_0(\lambda)), \quad R(\lambda) := (R_{N+1}^*(\lambda) \ R_{N+1}(\lambda)). \tag{5.159}$$

In particular, if $R_1(\lambda) \equiv R_1$ and $R_2(\lambda) \equiv R_2$ are independent on $\lambda$, then the monotonicity conditions in (5.153), (5.157), (5.158) are automatically satisfied. Moreover, for $R_0^* = I$, $R_0 = 0$, $R_{N+1}^* = I$, and $R_{N+1} = 0$, we obtain the *Dirichlet boundary conditions* used in problem (E).

The discreteness of the underlaying set $\mathbb{N}$ or $(\mathbb{Z})$ enables a construction, which transforms general boundary condition (5.151) to the Dirichlet condition.

**Lemma 5.40** *The eigenvalue problem* (SDS$_\lambda$) *with separated boundary conditions* (5.155) *on* $[0, N + 1]_\mathbb{Z}$ *can be extended to an eigenvalue problem of the same form (the so-called extended system) on the interval* $[-1, N + 2]_\mathbb{Z}$ *with the Dirichlet boundary conditions*

$$\tilde{x}_{-1}(\lambda) = 0 = \tilde{x}_{N+2}(\lambda). \tag{5.160}$$

*Proof* The reduction of separated boundary conditions on $[0, N + 1]_\mathbb{Z}$ to the Dirichlet conditions on $[-1, N + 2]_\mathbb{Z}$ is realized by the following construction. Consider separated boundary conditions (5.155). We construct the *extended system*, where we extend the original eigenvalue problem (5.1), (5.155) considered for $k \in [0, N]_\mathbb{Z}$ to a system for $k \in [-1, N+1]_\mathbb{Z}$, where we transform general separated boundary conditions (5.155) at $k = 0$ and $k = N + 1$ to the *Dirichlet* boundary condition at $k = -1$ and $k = N + 2$.

*Step 1* Firstly, we define the symplectic matrices $\mathcal{S}_{-1}(\lambda)$ and $\mathcal{S}_{N+1}(\lambda)$ according to Lemma 1.58(vi) such that

$$\left. \begin{array}{l} \mathcal{S}_{-1}(\lambda) \, (0 \ I)^T = \mathcal{J}^T \, \tilde{R}^T(\lambda) \, Q_0(\lambda), \quad \det Q_0(\lambda) \neq 0, \\ \mathcal{S}_{N+1}^{-1}(\lambda) \, (0 \ I)^T = \mathcal{J}^T \, R^T(\lambda) \, Q_{N+1}(\lambda), \quad \det Q_{N+1}(\lambda) \neq 0, \end{array} \right\} \quad \lambda \in \mathbb{R}, \tag{5.161}$$

where the $n \times 2n$ matrices $\tilde{R}(\lambda)$ and $R(\lambda)$ are given by (5.159). Note that the matrices $\mathcal{S}_{-1}(\lambda)$ and $\mathcal{S}_{N+1}(\lambda)$ are not defined uniquely by conditions (5.161). For $\mathcal{S}_{-1}(\lambda)$ and $\mathcal{S}_{N+1}(\lambda)$ given by (5.161), we introduce the extended problem

$$\tilde{y}_{k+1}(\lambda) = \mathcal{S}_k(\lambda) \, \tilde{y}_k(\lambda), \quad k \in [-1, N + 1]_\mathbb{Z}, \quad \tilde{x}_{-1}(\lambda) = 0 = \tilde{x}_{N+2}(\lambda). \tag{5.162}$$

One can show that for any choice of piecewise continuously differentiable $Q_0(\lambda)$ and $Q_{N+1}(\lambda)$, if a vector function $y_k(\lambda) = (x_k(\lambda), u_k(\lambda))$ for $k \in [0, N + 1]_\mathbb{Z}$ obeys (SDS$_\lambda$) and (5.155) under assumption (5.156), then $\tilde{y}_k = (\tilde{x}_k(\lambda), \tilde{u}_k(\lambda))$ for

$k \in [-1, N + 2]_{\mathbb{Z}}$ solves the extended problem (5.162), where $\tilde{y}_k(\lambda) \equiv y_k(\lambda)$ for $k \in [0, N + 1]_{\mathbb{Z}}$. The proof is based on Lemma 1.58(vii); see formula (1.149). Indeed, assume that $y_k(\lambda)$ solves (SDS$_\lambda$) and (5.155). Then $y_0(\lambda) \in \mathrm{Ker}\, \tilde{R}(\lambda)$ and $y_{N+1}(\lambda) \in \mathrm{Ker}\, R(\lambda)$, where $\tilde{R}(\lambda)$ and $R(\lambda)$ are defined in (5.159). Note that in this step we do not impose any monotonicity conditions (5.3), (5.157), (5.158). Then, by (1.149), $y_0(\lambda) \in \mathrm{Im}\, \mathcal{J}\tilde{R}^T(\lambda)$ and $y_{N+1}(\lambda) \in \mathrm{Im}\, \mathcal{J}R^T(\lambda)$, compared with the definition of $\mathcal{S}_{-1}(\lambda)$ and $\mathcal{S}_{N+1}(\lambda)$ in (5.161). The last conditions mean that $\tilde{y}_k = (\tilde{x}_k(\lambda), \tilde{u}_k(\lambda))$ for $k \in [-1, N + 2]_{\mathbb{Z}}$ obeys (5.160), i.e., it solves the extended problem (5.162), where $\tilde{y}_k(\lambda) \equiv y_k(\lambda)$ for $k \in [0, N + 1]_{\mathbb{Z}}$. Conversely, if a vector function $\tilde{y}_k(\lambda) = (\tilde{x}_k(\lambda), \tilde{u}_k(\lambda))$ for $k \in [-1, N + 2]_{\mathbb{Z}}$ solves the extended problem (5.162), then conditions (5.160) imply $\tilde{y}_0(\lambda) \in \mathrm{Im}\, \mathcal{J}\tilde{R}^T(\lambda)$ and $\tilde{y}_{N+1}(\lambda) \in \mathrm{Im}\, \mathcal{J}R^T(\lambda)$ or, again by (1.149), $\tilde{y}_0(\lambda) \in \mathrm{Ker}\, \tilde{R}(\lambda)$ and $\tilde{y}_{N+1}(\lambda) \in \mathrm{Ker}\, R(\lambda)$. This shows that $\tilde{y}_0(\lambda)$ and $\tilde{y}_{N+1}(\lambda)$ obey (5.155).

*Step 2* Secondly, we present the construction of $\mathcal{S}_{-1}(\lambda)$ and $\mathcal{S}_{N+1}(\lambda)$ in such a way that conditions (5.157), (5.158) imply

$$\Psi(\mathcal{S}_{-1}(\lambda)) \geq 0, \quad \Psi(\mathcal{S}_{N+1}(\lambda)) \geq 0. \tag{5.163}$$

According to (5.161) and the proof of Lemma 1.58(vi), we introduce the $n \times n$ matrices

$$K_0(\lambda) = [\tilde{R}(\lambda)\, \tilde{R}^T(\lambda)]^{-1}, \quad K_{N+1}(\lambda) = [R(\lambda)\, R^T(\lambda)]^{-1} \tag{5.164}$$

and define the first $2n \times n$ blocks of $\mathcal{S}_{-1}(\lambda)$ and $\mathcal{S}_{N+1}^{-1}(\lambda)$ in (5.161) according to the proof of Lemma 1.58(vii) as

$$\left. \begin{aligned} \mathcal{S}_{-1}(\lambda)\,(I\ 0)^T &= \tilde{R}^T(\lambda)\, K_0(\lambda)\, Q_0^{-1\,T}(\lambda), \\ \mathcal{S}_{N+1}^{-1}(\lambda)\,(I\ 0)^T &= R^T(\lambda)\, K_{N+1}(\lambda)\, Q_0^{-1\,T}(\lambda), \end{aligned} \right\} \tag{5.165}$$

where the matrices $\tilde{R}(\lambda)$ and $R(\lambda)$ are defined in (5.159). Then, by (5.161), (5.165), and (1.193), we have

$$\left. \begin{aligned} \Psi(\mathcal{S}_{-1}^{-1}(\lambda)) &= L_0^T(\lambda) \begin{pmatrix} -\dot{\tilde{R}}(\lambda)\, \mathcal{J}\tilde{R}^T(\lambda) & \tilde{P}(\lambda) \\ \tilde{P}^T(\lambda) & -\dot{\tilde{R}}(\lambda)\, \mathcal{J}\tilde{R}^T(\lambda) \end{pmatrix} L_0(\lambda), \\ \tilde{P}(\lambda) &= K_0^{-1}(\lambda)\, \dot{Q}_0(\lambda)\, Q_0^{-1}(\lambda) + R_0^*(\lambda)\, \dot{R}_0^{*T}(\lambda) + R_0(\lambda)\, \dot{R}_0^T(\lambda), \\ L_0(\lambda) &= \mathrm{diag}\{K_0(\lambda)\, Q_0^{T-1}(\lambda),\ Q_0(\lambda)\} \end{aligned} \right\} \tag{5.166}$$

and in a similar way

$$
\begin{aligned}
\Psi(\mathcal{S}_{N+1}(\lambda)) &= L_{N+1}^T(\lambda) \begin{pmatrix} -\dot{R}(\lambda)\,\mathcal{J}\,R^T(\lambda) & P(\lambda) \\ P^T(\lambda) & -\dot{R}(\lambda)\,\mathcal{J}\,R^T(\lambda) \end{pmatrix} L_{N+1}(\lambda), \\
P(\lambda) &= K_{N+1}^{-1}(\lambda)\,\dot{Q}_{N+1}(\lambda)\,Q_{N+1}^{-1}(\lambda) \\
&\quad + R_{N+1}^*(\lambda)\,\dot{R}_{N+1}^{*T}(\lambda) + R_{N+1}(\lambda)\,\dot{R}_{N+1}^T(\lambda), \\
L_{N+1}(\lambda) &= \operatorname{diag}\{K_{N+1}(\lambda)\,Q_{N+1}^{T-1}(\lambda),\ Q_{N+1}(\lambda)\}.
\end{aligned}
\tag{5.167}
$$

Then, under the assumption that $Q_0(\lambda)$ and $Q_{N+1}(\lambda)$ are the fundamental matrices of the differential equations (compared with Remark 1.46)

$$
\begin{aligned}
\dot{Q}_0(\lambda) &= -K_0(\lambda)\,[R_0^*(\lambda)\,\dot{R}_0^{*T}(\lambda) + R_0(\lambda)\,\dot{R}_0^T(\lambda)]\,Q_0(\lambda), \\
\dot{Q}_{N+1}(\lambda) &= -K_{N+1}(\lambda)\,[R_{N+1}^*(\lambda)\,\dot{R}_{N+1}^{*T}(\lambda) + R_{N+1}(\lambda)\,\dot{R}_{N+1}^T(\lambda)]\,Q_{N+1}(\lambda),
\end{aligned}
\tag{5.168}
$$

we have in (5.166) and (5.167) that $\tilde{P}(\lambda) = P(\lambda) = 0$. Then by (5.157) and (5.158), we obtain

$$
\Psi(\mathcal{S}_{-1}^{-1}(\lambda)) = L_0^T(\lambda) \begin{pmatrix} -\dot{\tilde{R}}(\lambda)\,\mathcal{J}\,\tilde{R}^T(\lambda) & 0 \\ 0 & -\dot{\tilde{R}}(\lambda)\,\mathcal{J}\,\tilde{R}^T(\lambda) \end{pmatrix} L_0(\lambda) \le 0, \tag{5.169}
$$

which yields by using Proposition 1.76(iv) that $\Psi(\mathcal{S}_{-1}(\lambda)) \ge 0$. Similarly,

$$
\Psi(\mathcal{S}_{N+1}(\lambda)) = L_{N+1}^T(\lambda) \begin{pmatrix} -\dot{R}(\lambda)\,\mathcal{J}\,R^T(\lambda) & 0 \\ 0 & -\dot{R}(\lambda)\,\mathcal{J}\,R^T(\lambda) \end{pmatrix} L_{N+1}(\lambda) \ge 0. \tag{5.170}
$$

Thus, we have proved that monotonicity conditions (5.157) and (5.158) really imply (5.169) and (5.170) with the matrices $\mathcal{S}_{-1}(\lambda)$ and $\mathcal{S}_{N+1}(\lambda)$ given by (5.161), (5.165), and (5.168).

*Step 3* Lastly, We remark that according to the above construction, the block diagonal matrices $L_0(\lambda)$ and $L_{N+1}(\lambda)$ are nonsingular, so that conditions (5.169), (5.170), (5.168) imply (5.157), (5.158). $\qquad\square$

As a direct consequence of the proof of Lemma 5.40, we formulate the main result of this section.

**Theorem 5.41** *Under assumptions* (5.2), (5.3), (5.156), (5.157), *and* (5.158) *with all coefficient matrices being piecewise continuously differentiable (with respect to* $\lambda \in \mathbb{R}$*), the problem* (SDS$_\lambda$), (5.155) *is equivalent to the extended problem* (5.162), *where the matrices* $S_{-1}(\lambda)$ *and* $S_{N+1}(\lambda)$ *obey monotonicity conditions* (5.163) *and are defined as*

$$S_{-1}(\lambda) = \begin{pmatrix} \mathcal{A}_{-1}(\lambda) & \mathcal{B}_{-1}(\lambda) \\ \mathcal{C}_{-1}(\lambda) & \mathcal{D}_{-1}(\lambda) \end{pmatrix} := V_{-1}(\lambda) \operatorname{diag}\{Q_0^{T-1}(\lambda), \ Q_0(\lambda)\},$$

$$V_{-1}(\lambda) := \begin{pmatrix} R_0^{*T}(\lambda) K_0(\lambda) & -R_0^{T}(\lambda) \\ R_0^{T}(\lambda) K_0(\lambda) & R_0^{*T}(\lambda) \end{pmatrix}, \tag{5.171}$$

$$K_0(\lambda) := [R_0^*(\lambda) R_0^{*T}(\lambda) + R_0(\lambda) R_0^{T}(\lambda)]^{-1} \tag{5.172}$$

*and*

$$S_{N+1}(\lambda) = \begin{pmatrix} \mathcal{A}_{N+1}(\lambda) & \mathcal{B}_{N+1}(\lambda) \\ \mathcal{C}_{N+1}(\lambda) & \mathcal{D}_{N+1}(\lambda) \end{pmatrix} := \operatorname{diag}\{Q_{N+1}^{T}(\lambda), \ Q_{N+1}^{-1}(\lambda)\} V_{N+1}(\lambda),$$

$$V_{N+1}(\lambda) := \begin{pmatrix} R_{N+1}^*(\lambda) & R_{N+1}(\lambda) \\ -K_{N+1}(\lambda) R_{N+1}(\lambda) & K_{N+1}(\lambda) R_{N+1}^*(\lambda) \end{pmatrix}, \tag{5.173}$$

$$K_{N+1}(\lambda) := [R_{N+1}^*(\lambda) R_{N+1}^{*T}(\lambda) + R_{N+1}(\lambda) R_{N+1}^{T}(\lambda)]^{-1}. \tag{5.174}$$

*Here* $Q_0(\lambda)$ *and* $Q_{N+1}(\lambda)$ *are the fundamental matrices of the differential equations in* (5.168). *In particular, if* $R_i^*(\lambda)$ *and* $R_i(\lambda)$ *do not depend on* $\lambda$, *then one can put* $Q_i(\lambda) \equiv I$ *for* $i \in \{0, N+1\}$.

*Remark 5.42* Formulas (5.171) and (5.173) yield that the coefficients $\mathcal{B}_{-1}(\lambda)$ and $\mathcal{B}_{N+1}(\lambda)$ of the extended system are

$$\mathcal{B}_{-1}(\lambda) = -R_0^{T}(\lambda) Q_0(\lambda), \quad \mathcal{B}_{N+1}(\lambda) = Q_{N+1}^{T}(\lambda) R_{N+1}(\lambda), \quad \lambda \in \mathbb{R}. \tag{5.175}$$

*Remark 5.43* Remark that the principal solution of the extended system in (5.162) at $k = -1$ takes the value

$$Y_0^{[-1]}(\lambda) = S_{-1}(\lambda)\, (0 \ I)^T = \mathcal{J}^T \tilde{R}^T(\lambda)\, Q_0(\lambda), \quad \det Q_0(\lambda) \neq 0, \tag{5.176}$$

where $Q_0(\lambda)$ is the fundamental matrix of (5.168). Similarly, the principal solution of the extended system in (5.162) at $k = N + 2$ takes the value

$$Y_{N+1}^{[N+2]}(\lambda) = S_{N+1}^{-1}(\lambda)\, (0 \ I)^T = \mathcal{J}^T R^T(\lambda)\, Q_{N+1}(\lambda), \quad \det Q_{N+1}(\lambda) \neq 0, \tag{5.177}$$

where $Q_{N+1}(\lambda)$ is the fundamental matrix of the second equation in (5.168). We point out that in the subsequent formulations of the oscillation theorems for (5.162), the nonsingular matrices $Q_0(\lambda)$ and $Q_{N+1}(\lambda)$ can be omitted because of Theorem 3.5(i), which says that the comparative index is invariant with respect to multiplication by nonsingular matrices, i.e., $\mu(YC, \hat{Y}\hat{C}) = \mu(Y, \hat{Y})$ with $\det C \neq 0$ and $\det \hat{C} \neq 0$.

In the theorem below, we present a similar result on the equivalence of problem $(SDS_\lambda)$, (5.151) with the general (nonseparated) boundary conditions and some $4n \times 4n$ augmented problem with the separated boundary conditions. The consideration is based on the notation in Sect. 3.3. Recall that in Sect. 3.3.5, we introduced the notation

$$\{W, V\} := \begin{pmatrix} A & 0 & -B & 0 \\ 0 & \hat{A} & 0 & \hat{B} \\ -C & 0 & D & 0 \\ 0 & \hat{C} & 0 & \hat{D} \end{pmatrix}, \quad \{W\} := \{I, W\} = \begin{pmatrix} I & 0 & 0 & 0 \\ 0 & A & 0 & B \\ 0 & 0 & I & 0 \\ 0 & C & 0 & D \end{pmatrix},$$

where

$$W = \begin{pmatrix} A & B \\ C & D \end{pmatrix}, \quad V = \begin{pmatrix} \hat{A} & \hat{B} \\ \hat{C} & \hat{D} \end{pmatrix}$$

are $2n \times 2n$ matrices; see formula (3.101). We also use the $4n \times n$ matrices; see formula (3.52),

$$\langle W \rangle = \begin{pmatrix} I & 0 \\ A & B \\ 0 & -I \\ C & D \end{pmatrix}.$$

**Theorem 5.44** *Under the assumptions* (5.2), (5.3), (5.152), (5.153), *the problem* $(SDS_\lambda)$ *with* (5.151) *is equivalent to the augmented eigenvalue problem with the separated boundary conditions*

$$\left. \begin{array}{l} \tilde{y}_{k+1}(\lambda) = \{S_k(\lambda)\} \, \tilde{y}_k(\lambda), \quad k \in [0, N]_{\mathbb{Z}}, \\[2mm] \begin{pmatrix} 0 & 0 \\ -I & I \end{pmatrix} \tilde{x}_0(\lambda) + \begin{pmatrix} -I & -I \\ 0 & 0 \end{pmatrix} \tilde{u}_0(\lambda) = 0, \\[2mm] R_1(\lambda) \, \tilde{x}_{N+1}(\lambda) + R_2(\lambda) \, \tilde{u}_{N+1}(\lambda) = 0 \end{array} \right\} \qquad (5.178)$$

*where*

$$\Psi_{4n}(\{S_k(\lambda)\}) := \mathcal{J}_{4n}\{\dot{S}_k(\lambda)\} \, \mathcal{J}_{4n}\{S_k^T(\lambda)\} \, \mathcal{J}_{4n} = \{0, \, \Psi(S_k(\lambda))\} \geq 0. \qquad (5.179)$$

*Proof* Let $Z^{[0]}(\lambda) = (\tilde{Y}(\lambda), \, Y(\lambda))$ be the fundamental symplectic matrix of (5.1) given by the initial condition $Z_0^{[0]}(\lambda) = I$. Then the general solution of (5.1) is of the form $y_k(\lambda) = Z_k^{[0]}(\lambda) \, c$ with $c \in \mathbb{R}^{2n}$. Substituting this solution into (5.151), we obtain

$$R_1(\lambda) \begin{pmatrix} I & 0 \\ \tilde{X}_{N+1}(\lambda) & X_{N+1}(\lambda) \end{pmatrix} c + R_2(\lambda) \begin{pmatrix} 0 & -I \\ \tilde{U}_{N+1}(\lambda) & U_{N+1}(\lambda) \end{pmatrix} c$$

$$= \mathcal{R}(\lambda) \langle Z_{N+1}^{[0]}(\lambda) \rangle c = 0,$$

where $\mathcal{R}(\lambda) = (R_1(\lambda) \ R_2(\lambda))$ as in the beginning of this subsection and where $\langle Z_{N+1}^{[0]}(\lambda) \rangle = \{ Z_{N+1}^{[0]}(\lambda) \} S_0(0 \ I)^T$ according to (3.104) and the notation introduced in (3.101) and (3.102). Observe that matrices $\{W\}$ and $\{V\}$ defined by (3.101) satisfy $\{W\}\{V\} = \{WV\}$ (see (3.103)). Consequently, the $4n \times 2n$ matrix $\langle Z_k^{[0]}(\lambda) \rangle$ is a conjoined basis (see Lemma 3.21(i)) of a symplectic system with the coefficient matrix $\{S_k(\lambda)\}$, and it obviously satisfies (5.178) for $k = 0$. Then $y_k(\lambda) = Z_k^{[0]}(\lambda) c$ is a solution of (5.1) with (5.151) if and only if $\tilde{y}_k(\lambda) = \langle Z_k^{[0]}(\lambda) \rangle c$ is a solution of problem (5.178).                                                                                   □

At the end of this section, we formulate an analog of Theorem 5.44 for the case when boundary conditions (5.151) are transformed into the left endpoint $k = 0$.

**Theorem 5.45** *Under the assumptions* (5.2), (5.3), (5.152), (5.153), *the problem* (SDS$_\lambda$) *with* (5.151) *is equivalent to the eigenvalue problem with the separated boundary conditions*

$$\left. \begin{aligned} \tilde{y}_{k+1}(\lambda) &= \{S_k(\lambda)\} \, \tilde{y}_k(\lambda), \quad k \in [0, N]_{\mathbb{Z}}, \\ \tilde{R}_1(\lambda) \, \tilde{x}_0(\lambda) + \tilde{R}_2(\lambda) \, \tilde{u}_0(\lambda) &= 0, \\ \begin{pmatrix} 0 & 0 \\ -I & I \end{pmatrix} \tilde{x}_{N+1}(\lambda) + \begin{pmatrix} -I & -I \\ 0 & 0 \end{pmatrix} \tilde{u}_{N+1}(\lambda) &= 0, \end{aligned} \right\} \tag{5.180}$$

*where we have monotonicity condition* (5.179) *for* $\{S_k(\lambda)\}$ *and where*

$$\tilde{R}_1(\lambda) = R_1(\lambda) \, P_1, \quad \tilde{R}_2(\lambda) = -R_2(\lambda) \, P_1, \quad P_1 := \begin{pmatrix} 0 & I \\ I & 0 \end{pmatrix}. \tag{5.181}$$

*Here* $\tilde{\mathcal{R}}(\lambda) \, \mathcal{J}_{4n} \tilde{\mathcal{R}}^T(\lambda) = 0$ *and* $\operatorname{rank} \tilde{\mathcal{R}}(\lambda) = 2n$ *with* $\tilde{\mathcal{R}}(\lambda) := (\tilde{R}_1(\lambda) \ \tilde{R}_2(\lambda))$ *and, moreover, instead of* (5.153), *we have (compare with* (5.157))

$$\dot{\tilde{R}}_1(\lambda) \, \tilde{R}_2^T(\lambda) - \dot{\tilde{R}}_2(\lambda) \, \tilde{R}_1^T(\lambda) = \dot{\tilde{\mathcal{R}}}(\lambda) \, \mathcal{J}_{4n} \tilde{\mathcal{R}}^T(\lambda) \geq 0. \tag{5.182}$$

*Proof* Let $Z^{[N+1]}(\lambda) = (\tilde{Y}(\lambda), \ Y(\lambda))$ be the fundamental symplectic matrix of (5.1) given by the initial condition $Z_{N+1}^{[N+1]}(\lambda) = I$. Then the general solution of (5.1) is of the form $y_k = Z_k^{[N+1]}(\lambda) c$ with $c \in \mathbb{R}^{2n}$. Substituting this solution into (5.151), we obtain

$$
R_1(\lambda) \begin{pmatrix} \tilde{X}_0(\lambda) & X_0(\lambda) \\ I & 0 \end{pmatrix} c + R_2(\lambda) \begin{pmatrix} -\tilde{U}_0(\lambda) & -U_0(\lambda) \\ 0 & I \end{pmatrix} c
$$

$$
= \big( R_1(\lambda)\, P_1 - R_2(\lambda)\, P_1 \big) \big\langle Z_0^{[N+1]}(\lambda) \big\rangle c = 0,
$$

where $\big\langle Z_k^{[N+1]}(\lambda) \big\rangle$ is a conjoined basis of the symplectic system in (5.180), and it obviously satisfies (5.180) for $k = N + 1$. Then $y_k(\lambda) = Z_k^{[N+1]}(\lambda)\, c$ is a solution of (5.1) with (5.151) if and only if $\tilde{y}_k(\lambda) = \big\langle Z_k^{[N+1]}(\lambda) \big\rangle c$ is a solution of (5.180). Recall also that the transformation of $\mathcal{R}(\lambda) = (R_1(\lambda)\ R_2(\lambda)) = \tilde{\mathcal{R}}(\lambda)\, \text{diag}\{P_1, -P_1\}$ with the matrix $P_1$ defined in (1.151) (see also Lemma 1.58(iv)) completely preserves the symplectic structure of the boundary conditions, i.e., $\tilde{\mathcal{R}}^T(\lambda)$ again obeys (1.147) and in the monotonicity condition for $\tilde{\mathcal{R}}(\lambda)$ the sign is opposite according to the demands for the separated boundary conditions for $k = 0$; see (5.157).                                    $\square$

*Remark 5.46*  Assume for the moment that we construct extended problem (5.162) for (5.178), and then the principal solution $\mathcal{Y}^{[-1]}(\lambda)$ of this problem obeys the condition

$$
\mathcal{Y}_0^{[-1]}(\lambda) = \langle I \rangle, \tag{5.183}
$$

i.e., instead of this principal solution, one can consider the conjoined basis $\big\langle Z^{[0]}(\lambda) \big\rangle$ of the augmented system in (5.178). Here $Z^{[0]}(\lambda)$ is the fundamental matrix of (5.1) such that $Z_0^{[0]}(\lambda) = I$; see the proof of Theorem 5.44. In a similar way, the principal solution $\mathcal{Y}^{[N+2]}(\lambda)$ at $k = N+2$ of extended problem (5.162) constructed for (5.180) satisfies the condition

$$
\mathcal{Y}_{N+1}^{[N+2]}(\lambda) = \langle I \rangle, \tag{5.184}
$$

and then the conjoined basis $\big\langle Z^{[N+1]}(\lambda) \big\rangle$ of the augmented system in (5.180) can be used instead of $\mathcal{Y}^{[N+2]}(\lambda)$. Here $Z^{[N+1]}(\lambda)$ is the fundamental matrix of (5.1) such that $Z_{N+1}^{[N+1]}(\lambda) = I$; see the proof of Theorem 5.45.

## 5.2.2   Transformation of Quadratic Functionals

In this subsection we extend the transformations from the previous subsection to include the associated quadratic functionals. Following (2.56) and (5.138), we consider the associated quadratic functional for admissible functions $y = (x, u)$, i.e., $x_{k+1} = \mathcal{A}_k(\lambda)\, x_k + \mathcal{B}_k(\lambda)\, u_k$ for $k \in [0, N]_{\mathbb{Z}}$. Given the $2n \times 2n$ matrices $R_1(\lambda)$

and $R_2(\lambda)$ satisfying (5.152), we define the symmetric $2n \times 2n$ matrix (see (2.92))

$$\Gamma(\lambda) := R_2^\dagger(\lambda)\, R_1(\lambda)\, R_2^\dagger(\lambda)\, R_2(\lambda) \tag{5.185}$$

and the quadratic functional

$$\mathcal{G}(y, \lambda) := \mathcal{F}(y, \lambda) + \begin{pmatrix} x_0 \\ x_{N+1} \end{pmatrix}^T \Gamma(\lambda) \begin{pmatrix} x_0 \\ x_{N+1} \end{pmatrix}, \tag{5.186}$$

where $\mathcal{F}(y, \lambda)$ is defined in (5.138) and where $y$ is admissible and satisfies the general boundary conditions

$$\begin{pmatrix} x_0 \\ x_{N+1} \end{pmatrix} \in \operatorname{Im} R_2^T(\lambda). \tag{5.187}$$

The functional $\mathcal{G}(\cdot, \lambda)$ is said to be *positive* if $\mathcal{G}(y, \lambda) > 0$ for every admissible $y = (x, u)$ satisfying (5.187) and $x \not\equiv 0$. In this case we write $\mathcal{G}(\cdot, \lambda) > 0$. Similarly, the functional $\mathcal{G}(\cdot, \lambda)$ is said to be *nonnegative* if $\mathcal{G}(y, \lambda) \geq 0$ for every admissible $y = (x, u)$ satisfying (5.187), we write $\mathcal{G}(\cdot, \lambda) \geq 0$.

When the boundary conditions are separated, i.e., when (5.154) holds with $n \times n$ matrices $R_0(\lambda)$, $R_0^*(\lambda)$, $R_{N+1}(\lambda)$, $R_{N+1}^*(\lambda)$ satisfying (5.156), the functional $\mathcal{G}(y, \lambda)$ in (5.186) takes the form

$$\mathcal{G}(y, \lambda) = \mathcal{F}(y, \lambda) + x_0^T\, \Gamma_0(\lambda)\, x_0 + x_{N+1}^T\, \Gamma_{N+1}(\lambda)\, x_{N+1}, \tag{5.188}$$

where the symmetric $n \times n$ matrices $\Gamma_0(\lambda)$ and $\Gamma_{N+1}(\lambda)$ are defined by (see (2.83))

$$\left.\begin{aligned} \Gamma_0(\lambda) &:= -R_0^\dagger(\lambda)\, R_0^*(\lambda)\, R_0^\dagger(\lambda)\, R_0(\lambda), \\ \Gamma_{N+1}(\lambda) &:= R_{N+1}^\dagger(\lambda)\, R_{N+1}^*(\lambda)\, R_{N+1}^\dagger(\lambda)\, R_{N+1}(\lambda), \end{aligned}\right\} \tag{5.189}$$

so that $\Gamma(\lambda) = \operatorname{diag}\{\Gamma_0(\lambda),\, \Gamma_{N+1}(\lambda)\}$. The boundary conditions in (5.187) are then also separated and have the form

$$x_0 \in \operatorname{Im} R_0^T(\lambda), \quad x_{N+1} \in \operatorname{Im} R_{N+1}^T(\lambda). \tag{5.190}$$

First we consider the quadratic functional $\mathcal{G}(y, \lambda)$ in (5.188) with separated endpoints (5.190). We define the extended quadratic functional $\mathcal{F}_{ext}(y, \lambda)$ in (5.191), compared with (5.138), by

$$\mathcal{F}_{ext}(y, \lambda) := \sum_{k=-1}^{N+1} y_k^T \left( S_k^T(\lambda)\, \mathcal{K}\, S_k(\lambda) - \mathcal{K} \right) y_k, \tag{5.191}$$

$$= \mathcal{F}(y, \lambda) + \gamma_{-1}(\lambda) + \gamma_{N+1}(\lambda), \tag{5.192}$$

for an admissible $y$ on $[-1, N+2]_\mathbb{Z}$ satisfying $x_{-1} = 0 = x_{N+2}$. Here the matrices $\mathcal{S}_{-1}(\lambda)$ and $\mathcal{S}_{N+1}(\lambda)$ are defined by (5.171) and (5.173) and

$$\gamma_k(\lambda) := y_k^T \left( \mathcal{S}_k^T(\lambda) \, \mathcal{K} \, \mathcal{S}_k(\lambda) - \mathcal{K} \right) y_k, \quad k \in \{-1, N+1\}.$$

**Lemma 5.47** *Assume that the matrices $R_0(\lambda)$, $R_0^*(\lambda)$, $R_{N+1}(\lambda)$, $R_{N+1}^*(\lambda)$ satisfy (5.156) and $\Gamma_0(\lambda)$, $\Gamma_{N+1}(\lambda)$ are defined by (5.189). If $y$ is admissible on $[-1, N+2]_\mathbb{Z}$ with $x_{-1} = 0 = x_{N+2}$, then $y$ is admissible on $[0, N+1]_\mathbb{Z}$ with (5.190) and $\mathcal{G}(y, \lambda) = \mathcal{F}_{ext}(y, \lambda)$. Conversely, if $y$ is admissible on $[0, N+1]_\mathbb{Z}$ with (5.190), then it can be extended to be admissible on $[-1, N+2]_\mathbb{Z}$ with $x_{-1} = 0 = x_{N+2}$ and $\mathcal{F}_{ext}(y, \lambda) = \mathcal{G}(y, \lambda)$. In particular, we have $\mathcal{G}(\cdot, \lambda) > 0$ over (5.190) if and only if $\mathcal{F}_{ext}(\cdot, \lambda) > 0$ over $x_{-1} = 0 = x_{N+2}$ and $\mathcal{G}(\cdot, \lambda) \geq 0$ over (5.190) if and only if $\mathcal{F}_{ext}(\cdot, \lambda) \geq 0$ over $x_{-1} = 0 = x_{N+2}$.*

*Proof* First we assume that $y$ is admissible on $[-1, N+2]_\mathbb{Z}$ with $x_{-1} = 0 = x_{N+2}$. Then $y$ is admissible on $[0, N+1]_\mathbb{Z}$ as well and

$$x_0 = \mathcal{A}_{-1}(\lambda)\, x_{-1} + \mathcal{B}_{-1}(\lambda)\, u_{-1} = -R_0^T(\lambda)\, Q_0(\lambda)\, u_{-1} \in \mathrm{Im}\, R_0^T(\lambda),$$

see (5.175). Moreover, we have

$$0 = x_{N+2} = \mathcal{A}_{N+1}(\lambda)\, x_{N+1} + \mathcal{B}_{N+1}(\lambda)\, u_{N+1}$$
$$= Q_{N+1}^T(\lambda)\, [R_{N+1}^*(\lambda)\, x_{N+1} + R_{N+1}(\lambda)\, u_{N+1}],$$

which implies that, see Lemma 1.58(vii) and (5.156),

$$y_{N+1} = \begin{pmatrix} x_{N+1} \\ u_{N+1} \end{pmatrix} \in \mathrm{Ker}\left( R_{N+1}^*(\lambda), \ R_{N+1}(\lambda) \right) = \mathrm{Im} \begin{pmatrix} -R_{N+1}^T(\lambda) \\ R_{N+1}^{*T}(\lambda) \end{pmatrix}.$$

Therefore, for some $d \in \mathbb{R}^n$, we get

$$x_{N+1} = -R_{N+1}^T(\lambda)\, d \in \mathrm{Im}\, R_{N+1}^T(\lambda), \quad u_{N+1} = R_{N+1}^{*T}(\lambda)\, d. \tag{5.193}$$

Then by using $x_{-1} = 0$, $\mathcal{B}_{-1}(\lambda)\, \mathcal{B}_{-1}^\dagger(\lambda) = R_0^\dagger(\lambda)\, R_0(\lambda)$, and the symmetry of the matrix $\mathcal{B}_{-1}^T(\lambda)\, \mathcal{D}_{-1}(\lambda)$, we obtain

$$\gamma_{-1}(\lambda) = u_{-1}^T \mathcal{B}_{-1}^T(\lambda)\, \mathcal{D}_{-1}(\lambda)\, u_{-1} = x_0^T\, \Gamma_0(\lambda)\, x_0. \tag{5.194}$$

Moreover, following the proof of Lemma 2.30 and using (5.193), we have (suppressing the argument $\lambda$ by the coefficients of $\mathcal{S}_{N+1}(\lambda)$)

$$\gamma_{N+1}(\lambda) = x_{N+1}^T \mathcal{C}_{N+1}^T (\mathcal{A}_{N+1} x_{N+1} + \mathcal{B}_{N+1} u_{N+1})$$
$$+ u_{N+1}^T (\mathcal{D}_{N+1}^T \mathcal{A}_{N+1} - I)\, x_{N+1} + u_{N+1}^T \mathcal{D}_{N+1}^T \mathcal{B}_{N+1} u_{N+1}$$

$$= (x_{N+1}^T C_{N+1}^T + u_{N+1}^T D_{N+1}^T) x_{N+2} - u_{N+1}^T x_{N+1}$$

$$= x_{N+1}^T \Gamma_{N+1}(\lambda) x_{N+1}. \tag{5.195}$$

This shows that $\mathcal{G}(y, \lambda) = \mathcal{F}_{ext}(y, \lambda)$. Conversely, if $y$ is admissible on $[0, N+1]_{\mathbb{Z}}$ with (5.190), i.e., $x_0 = R_0^T(\lambda) c$ and $x_{N+1} = R_{N+1}^T(\lambda) d$ for some $c, d \in \mathbb{R}^n$, then we define

$$x_{-1} := 0, \quad u_{-1} := B_{-1}^\dagger(\lambda) x_0, \quad u_{N+1} := R_{N+1}^{*T}(\lambda) d, \quad x_{N+2} := 0.$$

Then we get

$$\mathcal{A}_{-1}(\lambda) x_{-1} + \mathcal{B}_{-1}(\lambda) u_{-1}$$

$$= \mathcal{B}_{-1}(\lambda) B_{-1}^\dagger(\lambda) x_0 = R_0^\dagger(\lambda) R_0(\lambda) R_0^T(\lambda) c = R_0^T(\lambda) c = x_0,$$

$$\mathcal{A}_{N+1}(\lambda) x_{N+1} + \mathcal{B}_{N+1}(\lambda) u_{N+1}$$

$$= Q_{N+1}^T(\lambda) [R_{N+1}^*(\lambda) R_{N+1}^T(\lambda) - R_{N+1}(\lambda) R_{N+1}^{*T}(\lambda)] d = 0 = x_{N+2}.$$

Therefore, $y$ is admissible on $[-1, N+2]_{\mathbb{Z}}$ with $x_{-1} = 0 = x_{N+2}$. The same calculations as in (5.194) and (5.195) then show that also in this case $\mathcal{F}_{ext}(y, \lambda) = \mathcal{G}(y, \lambda)$. Finally, since we extend the $x$ component of an admissible $y$ at $k = -1$ and $k = N+2$ by the zero values $x_{-1} = 0$ and $x_{N+2} = 0$, it follows that $\mathcal{G}(\cdot, \lambda) > 0$ over (5.190) if and only if $\mathcal{F}_{ext}(\cdot, \lambda) > 0$ over $x_{-1} = 0 = x_{N+2}$, and $\mathcal{G}(\cdot, \lambda) \geq 0$ over (5.190) if and only if $\mathcal{F}_{ext}(\cdot, \lambda) \geq 0$ over $x_{-1} = 0 = x_{N+2}$. The proof is complete. □

Next we consider the general quadratic functional $\mathcal{G}(y, \lambda)$ in (5.186) over the jointly varying endpoints (5.187). Given the $2n \times 2n$ matrices $R_1(\lambda)$, $R_2(\lambda)$, and $\Gamma(\lambda)$ satisfying (5.152) and (5.185), we define the auxiliary $2n \times 2n$ matrices

$$\tilde{R}_0^*(\lambda) := \begin{pmatrix} 0 & 0 \\ -I & I \end{pmatrix}, \quad \tilde{R}_0(\lambda) := \begin{pmatrix} -I & -I \\ 0 & 0 \end{pmatrix}, \quad \tilde{R}_{N+1}^*(\lambda) := R_1(\lambda), \quad \left.\begin{matrix} \\ \\ \end{matrix}\right\}$$

$$\tilde{R}_{N+1}(\lambda) := R_2(\lambda), \quad \tilde{\Gamma}_0(\lambda) := 0, \quad \tilde{\Gamma}_{N+1}(\lambda) := \Gamma(\lambda). \tag{5.196}$$

In addition, we define the the augmented quadratic functional

$$\tilde{\mathcal{F}}_{aug}(\tilde{y}, \lambda) := \tilde{\mathcal{F}}(\tilde{y}, \lambda) + \tilde{x}_0^T \tilde{\Gamma}_0 \tilde{x}_0 + \tilde{x}_{N+1}^T \tilde{\Gamma}_{N+1} \tilde{x}_{N+1} \tag{5.197}$$

over the separated endpoints

$$\tilde{x}_0 \in \mathrm{Im}\, \tilde{R}_0^T(\lambda), \quad \tilde{x}_{N+1} \in \mathrm{Im}\, \tilde{R}_{N+1}^T(\lambda), \tag{5.198}$$

where the functional $\tilde{\mathcal{F}}(\tilde{y}, \lambda)$ is defined by, compared with (5.138),

$$\tilde{\mathcal{F}}(\tilde{y}, \lambda) := \sum_{k=0}^{N} \tilde{y}_k^T \left( \tilde{\mathcal{S}}_k^T(\lambda) \, \tilde{\mathcal{K}} \, \tilde{\mathcal{S}}_k(\lambda) - \tilde{\mathcal{K}} \right) \tilde{y}_k, \quad \tilde{\mathcal{K}} := \begin{pmatrix} 0_{2n} & 0_{2n} \\ I_{2n} & 0_{2n} \end{pmatrix}, \tag{5.199}$$

and where the $4n \times 4n$ matrix $\tilde{\mathcal{S}}_k(\lambda)$ is the coefficient matrix of the symplectic system in (5.178), i.e., for $k \in [0, N]_{\mathbb{Z}}$, we set

$$\tilde{\mathcal{S}}_k(\lambda) = \begin{pmatrix} \tilde{\mathcal{A}}_k(\lambda) & \tilde{\mathcal{B}}_k(\lambda) \\ \tilde{\mathcal{C}}_k(\lambda) & \tilde{\mathcal{D}}_k(\lambda) \end{pmatrix} := \{\mathcal{S}_k(\lambda)\} = \begin{pmatrix} I & 0 & 0 & 0 \\ 0 & A_k(\lambda) & 0 & B_k(\lambda) \\ 0 & 0 & I & 0 \\ 0 & C_k(\lambda) & 0 & D_k(\lambda) \end{pmatrix}. \tag{5.200}$$

It is easy to verify by a direct calculation that

$$\tilde{\mathcal{S}}_k^T(\lambda) \, \tilde{\mathcal{K}} \, \tilde{\mathcal{S}}_k(\lambda) - \tilde{\mathcal{K}} = \{0, \mathcal{S}_k^T(\lambda) \, \mathcal{K} \, \mathcal{S}_k(\lambda) - \mathcal{K}\}, \tag{5.201}$$

where we use the notation given by (3.101).

The following result describes the connection between the functionals $\mathcal{G}(y, \lambda)$ in (5.186) and $\tilde{\mathcal{F}}_{aug}(\tilde{y}, \lambda)$ in (5.197).

**Lemma 5.48** *Assume that the matrices $R_1(\lambda)$, $R_2(\lambda)$ satisfy (5.152) and $\Gamma(\lambda)$ is defined by (5.185). If $y$ is $(\mathcal{A}(\lambda), \mathcal{B}(\lambda))$-admissible with (5.187), then $\tilde{y} = (\tilde{x}, \tilde{u})$ defined by*

$$\tilde{x}_k := \begin{pmatrix} x_0 \\ x_k \end{pmatrix}, \quad k \in [0, N+1]_{\mathbb{Z}}, \quad \tilde{u}_k := \begin{pmatrix} -u_0 \\ u_k \end{pmatrix}, \quad k \in [0, N]_{\mathbb{Z}}, \tag{5.202}$$

*is $(\tilde{\mathcal{A}}(\lambda), \tilde{\mathcal{B}}(\lambda))$-admissible with (5.198) and $\tilde{\mathcal{F}}_{aug}(\tilde{y}, \lambda) = \mathcal{G}(y, \lambda)$. Conversely, if $\tilde{y}$ is $(\tilde{\mathcal{A}}(\lambda), \tilde{\mathcal{B}}(\lambda))$-admissible with (5.198), then $y = (x, u)$ given by*

$$x_k := (0 \; I) \, \tilde{x}_k, \quad k \in [0, N+1]_{\mathbb{Z}}, \quad u_k := (0 \; I) \, \tilde{u}_k, \quad k \in [0, N]_{\mathbb{Z}}, \tag{5.203}$$

*is $(\mathcal{A}(\lambda), \mathcal{B}(\lambda))$-admissible with (5.187) and $\mathcal{G}(y, \lambda) = \tilde{\mathcal{F}}_{aug}(\tilde{y}, \lambda)$. In particular, we have $\mathcal{G}(\cdot, \lambda) > 0$ over (5.187) if and only if $\tilde{\mathcal{F}}_{aug}(\cdot, \lambda) > 0$ over (5.198) and $\mathcal{G}(\cdot, \lambda) \geq 0$ over (5.187) if and only if $\tilde{\mathcal{F}}_{aug}(\cdot, \lambda) \geq 0$ over (5.198)*

*Proof* Assume that $y$ is $(\mathcal{A}(\lambda), \mathcal{B}(\lambda))$-admissible with (5.187) and define $\tilde{y} = (\tilde{x}, \tilde{u})$ by (5.202). Then $\tilde{y}$ is $(\tilde{\mathcal{A}}(\lambda), \tilde{\mathcal{B}}(\lambda))$-admissible, (5.198) holds, and the equality $\tilde{\mathcal{F}}_{aug}(\tilde{y}, \lambda) = \mathcal{G}(y, \lambda)$ follows by (5.201). Conversely, assume that $\tilde{y} = (\tilde{x}, \tilde{u})$ is $(\tilde{\mathcal{A}}(\lambda), \tilde{\mathcal{B}}(\lambda))$-admissible with (5.198). Then $\tilde{x}_k$ has the form in (5.202) for $k \in [0, N+1]_{\mathbb{Z}}$, while $\tilde{u}_k = (\beta_k^T, u_k^T)^T$ for some $\beta_k \in \mathbb{R}^n$ for $k \in [0, N]_{\mathbb{Z}}$, where $y := (x, u)$ is $(\mathcal{A}(\lambda), \mathcal{B}(\lambda))$-admissible and satisfies (5.187). The equality $\mathcal{G}(y, \lambda) = \tilde{\mathcal{F}}_{aug}(\tilde{y}, \lambda)$ again follows by (5.201). □

### 5.2.3   Oscillation Theorems for General Endpoints

In this subsection we derive the oscillation theorem for the eigenvalue problem (5.1) with general boundary conditions. First we consider the separated endpoints (5.155). Let $\bar{Y}(\lambda) = (\bar{X}(\lambda), \bar{U}(\lambda))$ be the so-called *natural conjoined basis* of system (5.1), i.e., it the conjoined basis satisfying the initial conditions

$$\bar{X}_0(\lambda) = -R_0^T(\lambda), \quad \bar{U}_0(\lambda) = R_0^{*T}(\lambda), \quad \lambda \in \mathbb{R}. \tag{5.204}$$

Then we define for $\lambda \in \mathbb{R}$ the auxiliary matrices

$$\left.\begin{aligned}
\Lambda(\lambda) &:= R_{N+1}^*(\lambda)\,\bar{X}_{N+1}(\lambda) + R_{N+1}(\lambda)\,\bar{U}_{N+1}(\lambda), \\
M(\lambda) &:= [I - \Lambda(\lambda)\,\Lambda^\dagger(\lambda)]\,R_{N+1}(\lambda), \\
T(\lambda) &:= I - M^\dagger(\lambda)\,M(\lambda), \\
P(\lambda) &:= T(\lambda)\,\bar{X}_{N+1}(\lambda)\,\Lambda^\dagger(\lambda)\,R_{N+1}(\lambda)\,T(\lambda).
\end{aligned}\right\} \tag{5.205}$$

We will see in the proof of the main result below (Theorem 5.50) that under assumptions (5.156)–(5.158) the quantity $\operatorname{rank}\Lambda(\lambda)$ is piecewise constant on $\mathbb{R}$. This fact allows to make the following definition.

**Definition 5.49** We say that $\lambda_0 \in \mathbb{R}$ is a *finite eigenvalue* of problem (5.1) with (5.155) if

$$\theta(\lambda_0) := \operatorname{rank}\Lambda(\lambda_0^-) - \operatorname{rank}\Lambda(\lambda_0) \geq 1. \tag{5.206}$$

In this case the number $\theta(\lambda_0)$ is called an *algebraic multiplicity* of $\lambda_0$ as a finite eigenvalue of (5.1), (5.155).

As in Sect. 5.1.3, we will count the finite eigenvalues including their multiplicities. We will also use the notation:

$$n_1(\lambda) := \text{ number of forward focal points of } \bar{Y}(\lambda) \text{ in } (0, N+1], \tag{5.207}$$

$$n_2(\lambda) := \text{ number of finite eigenvalues of (5.1), (5.155) in } (-\infty, \lambda], \tag{5.208}$$

$$p(\lambda) := \operatorname{rank} M(\lambda) + \operatorname{ind} P(\lambda), \tag{5.209}$$

that is, $n_1(\lambda) = l(\bar{Y}(\lambda), 0, N+1)$ according to notation (4.10). The next result is a generalization of Theorem 5.17 from Dirichlet boundary conditions $x_0(\lambda) = 0 = x_{N+1}(\lambda)$ to general separated boundary conditions (5.155). Namely, the choice of $R_0^*(\lambda) = I = R_{N+1}^*(\lambda)$ and $R_0(\lambda) = 0 = R_{N+1}(\lambda)$ yields Theorem 5.17 from Theorem 5.50.

**Theorem 5.50 (Global Oscillation Theorem for Separated Endpoints)** *Let us assume that the matrices* $R_0(\lambda)$, $R_0^*(\lambda)$, $R_{N+1}(\lambda)$, $R_{N+1}^*(\lambda)$ *are piecewise continuously differentiable on* $\mathbb{R}$ *and satisfy conditions* (5.156)–(5.158). *Furthermore, suppose that*

$$\left. \begin{array}{l} \text{rank } \mathcal{B}_k(\lambda) \text{ is constant for } \lambda \in \mathbb{R} \text{ for all } k \in [0, N]_{\mathbb{Z}} \\ \text{rank } R_0(\lambda) \text{ and rank } R_{N+1}(\lambda) \text{ are constant for } \lambda \in \mathbb{R}. \end{array} \right\} \qquad (5.210)$$

*Then with the notation* (5.207)–(5.209), *we have for every* $\lambda \in \mathbb{R}$ *the identities*

$$n_1(\lambda^+) = n_1(\lambda) \le (N+1)\, n, \quad p(\lambda^+) = p(\lambda), \qquad (5.211)$$

$$n_2(\lambda^+) = n_2(\lambda), \qquad (5.212)$$

$$n_2(\lambda^+) - n_2(\lambda^-) = n_1(\lambda^+) - n_1(\lambda^-) + p(\lambda^+) - p(\lambda^-), \qquad (5.213)$$

*and there exists* $\ell \in [0, (N+2)\, n]_{\mathbb{Z}}$ *such that*

$$n_1(\lambda) + p(\lambda) = n_2(\lambda) + \ell, \quad \lambda \in \mathbb{R}. \qquad (5.214)$$

*Moreover, for a suitable* $\lambda_0 < 0$, *we have*

$$n_2(\lambda) \equiv 0, \quad n_1(\lambda) + p(\lambda) \equiv \ell, \quad \lambda \le \lambda_0. \qquad (5.215)$$

*Remark 5.51* Under the assumptions of Theorem 5.50, it follows by Theorem 5.3 that the finite eigenvalues of (5.1), (5.155) are isolated and bounded from below and from above. The first condition in (5.215) then means that these finite eigenvalues are bounded from below, while conditions (5.214) and (5.211) imply that the finite eigenvalues are bounded from above.

*Proof of Theorem 5.50* The method is to transform (by Lemma 5.40) the problem (5.1), (5.155) into the extended problem (5.162) on the interval $[-1, N+2]_{\mathbb{Z}}$, to which the result in Theorem 5.17 can be applied. Consider the principal solution $Y^{[-1]}(\lambda)$ at $k = -1$ of the extended symplectic system in (5.162). Then by Remark 5.43 (see (5.176)), we have $Y_0^{[-1]}(\lambda) = \bar{Y}_0(\lambda)\, Q_0(\lambda)$, so that

$$\left. \begin{array}{l} Y_k^{[-1]}(\lambda) = \bar{Y}_k(\lambda)\, Q_0(\lambda), \quad k \in [0, N+1]_{\mathbb{Z}}, \\ \hat{Y}_{N+2}^{[-1]}(\lambda) = \mathcal{Q}(\lambda)\, V_{N+1}(\lambda)\, \bar{Y}_{N+1}(\lambda)\, Q_0(\lambda), \\ \mathcal{Q}(\lambda) := \text{diag}\{Q_{N+1}^T(\lambda), Q_{N+1}^{-1}(\lambda)\}, \end{array} \right\} \qquad (5.216)$$

where the symplectic matrix $V_{N+1}(\lambda)$ is given by (5.173). Then we can write $X_{N+2}^{[-1]}(\lambda) = Q_{N+1}^T(\lambda)\, \Lambda(\lambda)\, Q_0(\lambda)$ by the definition of $\Lambda(\lambda)$ in (5.205). This implies by Theorems 1.81 and 1.82 that the image of $X_{N+2}^{[-1]}(\lambda)$ and the rank of $\Lambda(\lambda)$ are piecewise constant on $\mathbb{R}$. Moreover, by (5.216) the multiplicities of focal points of

$Y_k^{[-1]}(\lambda)$ and $\bar{Y}_k(\lambda)$ are connected as follows:

$$m(Y_{-1}^{[-1]}(\lambda)) = 0,$$

$$m(Y_k^{[-1]}(\lambda)) = \mu\big(Y_{k+1}^{[-1]}(\lambda), \mathcal{S}_k(\lambda)\,(0\ I)^T\big) = \mu\big(\bar{Y}_{k+1}(\lambda)\,\mathcal{Q}_0(\lambda), \mathcal{S}_k(\lambda)\,(0\ I)^T\big)$$

$$= \mu\big(\bar{Y}_{k+1}(\lambda), \mathcal{S}_k(\lambda)\,(0\ I)^T\big) = m(\bar{Y}_k(\lambda)), \quad k \in [0, N]_{\mathbb{Z}},$$

where we used Lemma 4.7 (see (4.14)) and Theorem 3.5(i). Moreover, for $k = N + 1$, we have (see (5.216))

$$m(Y_{N+1}^{[-1]}(\lambda)) = \mu\big(Y_{N+2}^{[-1]}(\lambda), \mathcal{S}_{N+1}(\lambda)\,(0\ I)^T\big)$$

$$= \mu\big(\mathcal{Q}(\lambda)\,V_{N+1}(\lambda)\,\bar{Y}_{N+1}(\lambda)\,\mathcal{Q}_0(\lambda), \mathcal{Q}(\lambda)\,V_{N+1}(\lambda)\,(0\ I)^T\big)$$

$$= \mu\big(V_{N+1}(\lambda)\,\bar{Y}_{N+1}(\lambda), V_{N+1}(\lambda)\,(0\ I)^T\big) = p(\lambda),$$

where we again used Lemma 4.7 (see (4.14)), Theorem 3.5(i)–(ii), the definition of the comparative index $\mu\big(V_{N+1}(\lambda)\,\bar{Y}_{N+1}(\lambda), V_{N+1}(\lambda)\,(0\ I)^T\big)$ according to Theorem 3.2(i), and (5.205) for $p(\lambda)$. Note that by Theorem 5.15, we have that the multiplicities $m(Y_k^{[-1]}(\lambda))$ of focal points of the principal solution $Y_k^{[-1]}(\lambda)$ of the extended problem (5.162) on the interval $[-1, N+2]_{\mathbb{Z}}$ are right-continuous in $\lambda$ for $k \in [0, N+1]_{\mathbb{Z}}$, and then the same property holds for $m(\bar{Y}_k(\lambda))$ for $k \in [0, N]_{\mathbb{Z}}$ and $p(\lambda)$. Therefore, we have proved (5.211).

Moreover, the finite eigenvalues of the extended problem (5.162) according to Definition 5.5 coincide with the finite eigenvalues of (5.1), (5.155) according to Definition 5.49. By (5.175) and (5.210), we guarantee that assumption (5.38) is satisfied for all $k \in [-1, N+1]_{\mathbb{Z}}$, and hence, the results in (5.213)–(5.215) follow from the corresponding statements in Theorems 5.16 and 5.17. We remark that the number $\ell$ in (5.214) denotes the number of forward focal points of $Y^{[-1]}(\lambda)$ in $(-1, N+2]$ for $\lambda$ sufficiently negative, and it is estimated, according to (5.47) and (5.56), as

$$\ell = \lim_{\lambda \to -\infty} [n_1(\lambda) + p(\lambda)] \le (N+1)\,n + n = (N+2)\,n. \tag{5.217}$$

The proof is complete.                                                          □

*Remark 5.52* In the proof of Theorem 5.50, we have shown that the function $p(\lambda)$ given by (5.205) can be presented in terms of the comparative index as follows

$$p(\lambda) = \mu(V_{N+1}(\lambda)\,\bar{Y}_{N+1}(\lambda), V_{N+1}(\lambda)\,(0\ I)^T), \tag{5.218}$$

where $V_{N+1}(\lambda)$ is given by (5.173).

The following result is a generalization of Theorem 5.31 from the Dirichlet endpoints to the case of separated endpoints. We utilize the quadratic functional $\mathcal{G}(y, \lambda)$ defined in (5.188), which is considered over the separated endpoints (5.190).

**Theorem 5.53** *Assume that the matrices $R_0(\lambda)$, $R_0^*(\lambda)$, $R_{N+1}(\lambda)$, $R_{N+1}^*(\lambda)$ are piecewise continuously differentiable on $\mathbb{R}$ and satisfy conditions (5.156)–(5.158) and (5.210). Then with the notation (5.207)–(5.209), we have*

$$n_1(\lambda) + p(\lambda) = n_2(\lambda), \quad \lambda \in \mathbb{R}, \tag{5.219}$$

*if and only if there exists $\lambda_0 < 0$ such that the quadratic functional $\mathcal{G}(y, \lambda)$ in (5.188) obeys the condition $\mathcal{G}(\cdot, \lambda_0) > 0$.*

*Proof* The result follows from the application of Theorem 5.31 to the extended problem (5.162) on the interval $[-1, N + 2]_{\mathbb{Z}}$, once we take into account the connection between the quadratic functional $\mathcal{G}(y, \lambda)$ in (5.188) over (5.190) and the extended quadratic functional $\mathcal{F}_{ext}(y, \lambda)$ in (5.191) over $x_{-1} = 0 = x_{N+2}$ in Lemma 5.47. □

Next we consider the eigenvalue problem (5.1) with general jointly varying endpoints (5.151). Under the notation of Theorem 5.44, consider the symplectic fundamental matrix $Z^{[0]}(\lambda)$ of (5.1) such that $Z_0^{[0]}(\lambda) = I$ and for $k \in [0, N + 1]_{\mathbb{Z}}$ define the augmented $4n \times 2n$ matrix $\mathcal{Y}(\lambda) = \langle Z^{[0]}(\lambda) \rangle$ with the $2n \times 2n$ blocks

$$\mathcal{X}_k(\lambda) := \begin{pmatrix} I & 0 \\ \tilde{X}_k(\lambda) & X_k^{[0]}(\lambda) \end{pmatrix}, \quad \mathcal{U}_k(\lambda) := \begin{pmatrix} 0 & -I \\ \tilde{U}_k(\lambda) & U_k^{[0]}(\lambda) \end{pmatrix}, \tag{5.220}$$

where $Y^{[0]}(\lambda) = \begin{pmatrix} X^{[0]}(\lambda) \\ U^{[0]}(\lambda) \end{pmatrix}$ is the principal solution of (5.1) at $k = 0$. We note that the pair $\mathcal{Y}(\lambda) := (\mathcal{X}(\lambda), \mathcal{U}(\lambda))$ constitutes a conjoined basis of the augmented system in (5.178). Finally, following (5.205), we define the $2n \times 2n$ matrices

$$\left. \begin{array}{l} \mathcal{L}(\lambda) := R_1(\lambda)\, \mathcal{X}_{N+1}(\lambda) + R_2(\lambda)\, \mathcal{U}_{N+1}(\lambda), \\ \mathcal{M}(\lambda) := [I - \mathcal{L}(\lambda)\, \mathcal{L}^\dagger(\lambda)]\, R_2(\lambda), \\ \mathcal{T}(\lambda) := I - \mathcal{M}^\dagger(\lambda)\, \mathcal{M}(\lambda), \\ \mathcal{P}(\lambda) := \mathcal{T}(\lambda)\, \mathcal{X}_{N+1}(\lambda)\, \mathcal{L}^\dagger(\lambda)\, R_2(\lambda)\, \mathcal{T}(\lambda). \end{array} \right\} \tag{5.221}$$

We will see as before that under assumptions (5.152) and (5.153), the quantity rank $\mathcal{L}(\lambda)$ is piecewise constant on $\mathbb{R}$. Therefore, we make the following definition.

**Definition 5.54** We say that $\lambda_0 \in \mathbb{R}$ is a *finite eigenvalue* of problem (5.1) with (5.151) if

$$\zeta(\lambda_0) := \operatorname{rank} \mathcal{L}(\lambda_0^-) - \operatorname{rank} \mathcal{L}(\lambda_0) \geq 1. \tag{5.222}$$

In this case the number $\zeta(\lambda_0)$ is called an *algebraic multiplicity* of $\lambda_0$ as a finite eigenvalue of (5.1), (5.151).

We will again count the finite eigenvalues including their multiplicities. We will use the notation:

$$n_1(\lambda) := \text{ number of forward focal points of } Y^{[0]}(\lambda) \text{ in } (0, N+1], \qquad (5.223)$$

$$n_2(\lambda) := \text{ number of finite eigenvalues of (5.1), (5.151) in } (-\infty, \lambda], \qquad (5.224)$$

$$q(\lambda) := \text{rank}\,\mathcal{M}(\lambda) + \text{ind}\,\mathcal{P}(\lambda), \qquad (5.225)$$

that is, $n_1(\lambda) = l(Y^{[0]}(\lambda), 0, N+1)$ according to (4.10). The next result is a generalization of Theorem 5.17 from Dirichlet boundary conditions $x_0(\lambda) = 0 = x_{N+1}(\lambda)$ to general jointly varying boundary conditions (5.151). Namely, the choice of $R_1(\lambda) = I$ and $R_1(\lambda) = 0$ yields Theorem 5.17 from Theorem 5.55.

**Theorem 5.55 (Global Oscillation Theorem for Joint Endpoints)** *Suppose that the matrices $R_1(\lambda)$ and $R_2(\lambda)$ are piecewise continuously differentiable on $\mathbb{R}$ and satisfy conditions (5.152)–(5.153). Furthermore, suppose that*

$$\left.\begin{array}{l} \text{rank}\,\mathcal{B}_k(\lambda) \text{ is constant for } \lambda \in \mathbb{R} \text{ for all } k \in [0, N]_{\mathbb{Z}} \\[4pt] \text{rank}\,R_2(\lambda) \text{ is constant for } \lambda \in \mathbb{R}. \end{array}\right\} \qquad (5.226)$$

*Then with the notation (5.223)–(5.225), we have for every $\lambda \in \mathbb{R}$ the identities*

$$n_1(\lambda^+) = n_1(\lambda) \leq Nn, \quad q(\lambda^+) = q(\lambda), \qquad (5.227)$$

$$n_2(\lambda^+) = n_2(\lambda), \qquad (5.228)$$

$$n_2(\lambda^+) - n_2(\lambda^-) = n_1(\lambda^+) - n_1(\lambda^-) + q(\lambda^+) - q(\lambda^-), \qquad (5.229)$$

*and there exists $\ell \in [0, (N+2)n]_{\mathbb{Z}}$ such that*

$$n_1(\lambda) + q(\lambda) = n_2(\lambda) + \ell, \quad \lambda \in \mathbb{R}. \qquad (5.230)$$

*Moreover, for a suitable $\lambda_0 < 0$, we have*

$$n_2(\lambda) \equiv 0, \quad n_1(\lambda) + q(\lambda) \equiv \ell, \quad \lambda \leq \lambda_0. \qquad (5.231)$$

*Proof* We make the transformation of problem (5.1) with the joint boundary conditions (5.151) to the augmented problem (5.178) with separated boundary conditions, which is described in Theorem 5.44. We define the $2n \times 2n$ matrices $\tilde{R}_0^*(\lambda)$, $\tilde{R}_0(\lambda)$, $\tilde{R}_{N+1}^*(\lambda)$, $\tilde{R}_{N+1}(\lambda)$ by (5.196) and the augmented problem

$$\left.\begin{array}{l} \tilde{y}_{k+1}(\lambda) = \tilde{\mathcal{S}}_k(\lambda)\,\tilde{y}_k(\lambda), \quad k \in [0, N]_{\mathbb{Z}}, \\[4pt] \tilde{R}_0^*(\lambda)\,\tilde{x}_0(\lambda) + \tilde{R}_0(\lambda)\,\tilde{u}_0(\lambda) = 0, \\[4pt] \tilde{R}_{N+1}^*(\lambda)\,\tilde{x}_{N+1}(\lambda) + \tilde{R}_{N+1}(\lambda)\,\tilde{u}_{N+1}(\lambda) = 0. \end{array}\right\} \qquad (5.232)$$

with the $4n \times 4n$ symplectic coefficient matrix $\tilde{\mathcal{S}}_k(\lambda) = \{\mathcal{S}_k(\lambda)\}$ in (5.200). Then the monotonicity assumption (5.157) for the augmented problem (5.232) is trivially satisfied, since the matrices $\tilde{R}_0^*(\lambda)$ and $\tilde{R}_0(\lambda)$ do not depend on $\lambda$, as well as rank $\tilde{R}_0(\lambda) \equiv n$ is constant for $\lambda \in \mathbb{R}$. Assumption (5.158) for the augmented problem (5.232) is also satisfied by condition (5.153). Note the correct inequality "$\leq 0$" in the assumptions (5.158) and (5.153). Finally,

$$\operatorname{rank} \tilde{\mathcal{B}}_k(\lambda) = \operatorname{rank} \operatorname{diag}\{0, \ \mathcal{B}_k(\lambda)\} = \operatorname{rank} \mathcal{B}_k(\lambda) \quad \text{is constant for } \lambda \in \mathbb{R}$$

for all $k \in [0, N]_{\mathbb{Z}}$. Regarding the natural conjoined basis of the augmented problem, by the uniqueness of solutions, we obtain that the augmented conjoined basis $\tilde{\mathcal{Y}}(\lambda) = (\tilde{\mathcal{X}}(\lambda), \tilde{\mathcal{U}}(\lambda))$ of the system in (5.232) satisfying the initial conditions, compared with (5.204),

$$\tilde{\mathcal{X}}_0(\lambda) = -\tilde{R}_0^T(\lambda) = \begin{pmatrix} I & 0 \\ I & 0 \end{pmatrix}, \quad \tilde{\mathcal{U}}_0(\lambda) = \tilde{R}_0^{*T}(\lambda) = \begin{pmatrix} 0 & -I \\ 0 & I \end{pmatrix},$$

is equal to $\mathcal{Y}(\lambda) = (\mathcal{X}(\lambda), \mathcal{U}(\lambda))$ given in (5.220). Thus, the finite eigenvalues of problem (5.232) according to Definition 5.49 translate as the finite eigenvalues of (5.1) with (5.151) according to Definition 5.54. The result in the theorem then follows by the application of Theorem 5.50 to the augmented problem (5.232). In more details, we prove that

$$m(\tilde{\mathcal{Y}}_k(\lambda)) = m(Y_k^{[0]}(\lambda)), \quad k \in [0, N]_{\mathbb{Z}}, \tag{5.233}$$

where $Y^{[0]}(\lambda)$ is the principal solution of (5.1) at $k = 0$. Indeed, we have by Lemma 4.7 and Proposition 3.37 (see (3.107), (3.108))

$$m(\tilde{\mathcal{Y}}_k(\lambda)) = \mu\big(\langle Z_{k+1}^{[0]}\rangle, \{\mathcal{S}_k(\lambda)\}(0_{2n} \ I_{2n})^T\big)$$
$$= \mu\big(\{Z_{k+1}^{[0]}\}(0_{2n} \ I_{2n})^T, \{\mathcal{S}_k(\lambda)\}(0_{2n} \ I_{2n})^T\big)$$
$$= \mu\big(Y_{k+1}^{[0]}(\lambda), S_k(\lambda) (0 \ I)^T\big) = m(Y_k^{[0]}(\lambda)).$$

We note that the number $\ell$ in (5.230) denotes the number of forward focal points of $\tilde{\mathcal{Y}}(\lambda)$ or equivalently of $Y^{[0]}(\lambda)$ in $(0, N + 1]$ for $\lambda$ sufficiently negative, and it satisfies the estimate

$$\ell = \lim_{\lambda \to -\infty} [n_1(\lambda) + q(\lambda)] \leq Nn + 2n = (N + 2)n, \tag{5.234}$$

compared with (5.217). This estimate is better than the one, which we could obtain from a direct application of Theorem 5.50 to the augmented problem in dimension $2n$, which yields that $\ell \leq (N + 2) 2n$. The proof is complete. $\qquad \square$

*Remark 5.56* As in Remark 5.51, we point out that, under the assumptions of Theorem 5.55, the finite eigenvalues of (5.1), (5.151) are isolated and bounded from below and from above.

The following result is a generalization of Theorem 5.31 from the Dirichlet endpoints to jointly varying endpoints. We utilize the quadratic functional $\mathcal{G}(y, \lambda)$ defined in (5.186), which is considered over the jointly varying endpoints (5.187).

**Theorem 5.57** *Assume that $R_1(\lambda)$ and $R_2(\lambda)$ are piecewise continuously differentiable on $\mathbb{R}$ and satisfy conditions (5.152)–(5.153) and (5.226). Then with the notation (5.223)–(5.225), we have*

$$n_1(\lambda) + q(\lambda) = n_2(\lambda), \quad \lambda \in \mathbb{R}, \tag{5.235}$$

*if and only if there exists $\lambda_0 < 0$ such that the quadratic functional $\mathcal{G}(y, \lambda)$ in (5.186) over (5.187) obeys the condition $\mathcal{G}(\cdot, \lambda_0) > 0$.*

*Proof* The result follows from the application of Theorem 5.53 to the augmented problem (5.232), when we take into account the connection between the quadratic functional $\mathcal{G}(y, \lambda)$ in (5.186) over (5.187) and the augmented quadratic functional $\tilde{\mathcal{F}}_{aug}(\tilde{y}, \lambda)$ in (5.197) over (5.198) in Lemma 5.48.  □

*Remark 5.58* The periodic and antiperiodic boundary conditions are included in (5.155) for the special choice of constant matrices $R_1(\lambda)$ and $R_2(\lambda)$. Indeed, for the periodic endpoints, we have

$$\left.\begin{array}{l} y_0(\lambda) = y_{N+1}(\lambda), \quad R_1(\lambda) = \begin{pmatrix} 0 & 0 \\ -I & I \end{pmatrix}, \quad R_2(\lambda) = -\begin{pmatrix} I & I \\ 0 & 0 \end{pmatrix}, \\[2mm] \mathcal{L}(\lambda) = \mathcal{L}_1(\lambda) := \mathcal{J} - \mathcal{J}\left(\tilde{Y}_{N+1}(\lambda)\, Y^{[0]}_{N+1}(\lambda)\right) = \mathcal{J}\left(I - Z^{[0]}(\lambda)\right), \\[2mm] R_2^\dagger(\lambda) = \tfrac{1}{2}\, R_2^T(\lambda), \quad \Gamma = 0, \end{array}\right\} \tag{5.236}$$

where $\Gamma$ is given by (5.185), while for the antiperiodic endpoints, we have

$$\left.\begin{array}{l} y_0(\lambda) = -y_{N+1}(\lambda), \quad R_1(\lambda) = -\begin{pmatrix} 0 & 0 \\ I & I \end{pmatrix}, \quad R_2(\lambda) = \begin{pmatrix} -I & I \\ 0 & 0 \end{pmatrix}, \\[2mm] \mathcal{L}(\lambda) = \mathcal{L}_2(\lambda) := \mathcal{J} + \mathcal{J}\left(\tilde{Y}_{N+1}(\lambda)\, Y^{[0]}_{N+1}(\lambda)\right) = \mathcal{J}\left(I + Z^{[0]}(\lambda)\right), \\[2mm] R_2^\dagger(\lambda) = \tfrac{1}{2}\, R_2^T(\lambda), \quad \Gamma = 0. \end{array}\right\} \tag{5.237}$$

## 5.3  Linear Dependence on Spectral Parameter

In this section we consider an eigenvalue problem with a linear dependence on the spectral parameter. In particular, we will investigate the problem with the special linear dependence on $\lambda$, in which the first equation of (SDS$_\lambda$) does not depend on the spectral parameter $\lambda$, i.e.,

$$\left.\begin{array}{l} x_{k+1}(\lambda) = \mathcal{A}_k x_k(\lambda) + \mathcal{B}_k u_k(\lambda), \\ u_{k+1}(\lambda) = \mathcal{C}_k x_k(\lambda) + \mathcal{D}_k u_k(\lambda) - \lambda \mathcal{W}_k x_{k+1}(\lambda), \end{array}\right\} \quad k \in [0, N]_{\mathbb{Z}}. \tag{5.238}$$

This will allow to use the theory of discrete quadratic functionals (see Sect. 2.3.2), for which the set of admissible pairs $y = (x, u)$ does not depend on $\lambda$. For this special case, the matrix $\mathcal{S}_k(\lambda)$ in (E) has the form

$$\mathcal{S}_k(\lambda) = \begin{pmatrix} I & 0 \\ -\lambda \mathcal{W}_k & I \end{pmatrix} \mathcal{S}_k, \quad \mathcal{S}_k = \begin{pmatrix} \mathcal{A}_k & \mathcal{B}_k \\ \mathcal{C}_k & \mathcal{D}_k \end{pmatrix}, \tag{5.239}$$

where $\mathcal{S}_k$ is symplectic and

$$\mathcal{W}_k = \mathcal{W}_k^T, \quad \mathcal{W}_k \geq 0, \quad k \in [0, N]_{\mathbb{Z}}. \tag{5.240}$$

Then, applying Proposition 1.76(i) to the product $\begin{pmatrix} I & 0 \\ -\lambda \mathcal{W}_k & I \end{pmatrix} \mathcal{S}_k$, we see that the symmetric matrix $\Psi(\mathcal{S}_k(\lambda))$ for this special case has the form

$$\Psi_k(\lambda) \equiv \Psi_k := \begin{pmatrix} \mathcal{W}_k & 0 \\ 0 & 0 \end{pmatrix} \geq 0, \quad k \in [0, N]_{\mathbb{Z}}, \quad \lambda \in \mathbb{R}. \tag{5.241}$$

In this section we apply the results of Sect. 5.1 to problem (E) with this special linear dependence on $\lambda$; in particular, we concentrate on the specific properties of this problem which do not hold for the general nonlinear case. Further results about system (5.238) will be presented in Sects. 5.4 and 5.5.

## 5.3.1 Transformation of Boundary Conditions

In this subsection we consider system (5.238) together with the self-adjoint boundary conditions (5.151), which are in the special form

$$\left.\begin{array}{l} \mathbf{R}_1(\lambda) \begin{pmatrix} x_0(\lambda) \\ x_{N+1}(\lambda) \end{pmatrix} + \mathbf{R}_2(\lambda) \begin{pmatrix} -u_0(\lambda) \\ u_{N+1}(\lambda) \end{pmatrix} = 0, \\ \mathbf{R}_1(\lambda) := R_1 - \lambda R_2 \mathfrak{W}, \quad \mathbf{R}_2(\lambda) := R_2. \end{array}\right\} \tag{5.242}$$

Here the constant matrices $R_1$, $R_2$, $\mathfrak{W}$ obey the conditions (compared with (5.152))

$$\left.\begin{array}{l} R_1, R_2 \in \mathbb{R}^{2n \times 2n}, \quad \operatorname{rank}(R_1 \ R_2) = 2n, \\ R_1 R_2^T = R_2 R_1^T, \quad \mathfrak{W} = \mathfrak{W}^T. \end{array}\right\} \tag{5.243}$$

Note that monotonicity assumption (5.153) for this special case is equivalent to

$$\dot{\mathbf{R}}_1(\lambda) R_2^T = -R_2 \mathfrak{W} R_2^T \leq 0,$$

which is satisfied under the assumption

$$\mathfrak{W} \geq 0. \tag{5.244}$$

For the separated boundary conditions in the block diagonal form

$$\mathbf{R}_1(\lambda) = \operatorname{diag}\{\mathbf{R}_0^*(\lambda), \mathbf{R}_{N+1}^*(\lambda)\}, \quad R_2 = \operatorname{diag}\{-R_0, R_{N+1}\},$$
$$\mathfrak{W} = \operatorname{diag}\{\mathcal{W}_{-1}, \mathcal{W}_{N+1}\}$$

we have from (5.242) the conditions

$$\left.\begin{array}{l} \mathbf{R}_0^*(\lambda) x_0(\lambda) + R_0 u_0(\lambda) = 0, \\ \mathbf{R}_{N+1}^*(\lambda) x_{N+1}(\lambda) + R_{N+1} u_{N+1}(\lambda) = 0, \\ \mathbf{R}_0^*(\lambda) = R_0^* + \lambda R_0 \mathcal{W}_{-1}, \quad \mathbf{R}_{N+1}^*(\lambda) = R_{N+1}^* - \lambda R_{N+1} \mathcal{W}_{N+1}, \end{array}\right\} \tag{5.245}$$

while conditions (5.243) translate as

$$\left.\begin{array}{l} R_0^* R_0^T = R_0 (R_0^*)^T, \quad \operatorname{rank}(R_0^*, \ R_0) = n, \\ R_{N+1}^* R_{N+1}^T = R_{N+1}(R_{N+1}^*)^T, \quad \operatorname{rank}(R_{N+1}^*, \ R_{N+1}) = n, \\ \mathcal{W}_{-1} = \mathcal{W}_{-1}^T, \quad \mathcal{W}_{N+1} = \mathcal{W}_{N+1}^T. \end{array}\right\} \tag{5.246}$$

The monotonicity conditions (5.157), (5.158) are rewritten in the form

$$\dot{\mathbf{R}}_0^*(\lambda) R_0^T = R_0 \mathcal{W}_0 R_0^T \geq 0,$$
$$\dot{\mathbf{R}}_{N+1}^*(\lambda) R_{N+1}^T = -R_{N+1} \mathcal{W}_{N+1} R_{N+1}^T \leq 0.$$

Therefore, according to (5.244), it is sufficient to require that

$$\mathcal{W}_{-1} \geq 0, \tag{5.247}$$
$$\mathcal{W}_{N+1} \geq 0. \tag{5.248}$$

In particular, if $\mathbf{R}_1(\lambda)$ is independent on $\lambda$, then monotonicity conditions (5.244), (5.247), (5.248) are automatically satisfied. Moreover, for the choice of $R_0^* = I$, $R_0 = 0$, $R_{N+1}^* = I$, and $R_{N+1} = 0$, we obtain the Dirichlet boundary conditions in (E).

The main purpose of this subsection is to transform the mentioned above boundary conditions in such a way that new extended boundary value problem (5.162) preserves the special linear dependence on $\lambda$ given by (5.238).

Now we prove the following results for the case of the separated boundary conditions (5.242).

**Lemma 5.59** *Consider the low block triangular transformation* $\tilde{y}_k = L_k(\lambda)\, y_k$ *with the matrices*

$$L_k(\lambda) := I, \quad k \in [0, N]_{\mathbb{Z}}, \quad L_{N+1}(\lambda) := \begin{pmatrix} I & 0 \\ -\lambda \mathcal{W}_{N+1} & I \end{pmatrix}. \tag{5.249}$$

*Then the eigenvalue problem for system* (5.238) *with separated boundary conditions* (5.245) *on* $[0, N+1]_{\mathbb{Z}}$ *takes the form*

$$\left.\begin{array}{l} \tilde{y}_{k+1}(\lambda) = \begin{pmatrix} I & 0 \\ -\lambda \tilde{\mathcal{W}}_k & I \end{pmatrix} S_k \tilde{y}_k(\lambda), \quad k \in [0, N]_{\mathbb{Z}}, \\[2mm] \tilde{\mathcal{W}}_k = \mathcal{W}_k, \quad k \in [0, N-1]_{\mathbb{Z}}, \quad \tilde{\mathcal{W}}_N = \mathcal{W}_N + \mathcal{W}_{N+1} \geq 0, \end{array}\right\} \tag{5.250}$$

*with the boundary conditions*

$$\left.\begin{array}{l} \mathbf{R}_0^*(\lambda)\, \tilde{x}_0(\lambda) + R_0 \tilde{u}_0(\lambda) = 0, \quad \mathbf{R}_0^*(\lambda) = R_0^* + \lambda R_0 \mathcal{W}_{-1} \\[2mm] R_{N+1}^* \tilde{x}_{N+1}(\lambda) + R_{N+1} \tilde{u}_{N+1}(\lambda) = 0. \end{array}\right\} \tag{5.251}$$

*That is, for the transformed problem* (5.250) *with* (5.251), *the boundary conditions for* $k = N + 1$ *are independent on* $\lambda$.

*Proof* The proof follows from the representation of the boundary conditions in the right point $k = N + 1$ as

$$\left( R_{N+1}^* - \lambda R_{N+1} \mathcal{W}_{N+1}\ \ R_{N+1} \right) = \left( R_{N+1}^*\ \ R_{N+1} \right) L_{N+1}(\lambda),$$

where the matrix $L_{N+1}(\lambda)$ is defined in (5.249).                                 □

*Remark 5.60* We note that the symplectic transformation with matrix (5.249) preserves the multiplicities of focal points of any conjoined basis of (SDS$_\lambda$) according to Corollary 4.67.

In the next lemma, we consider the situation when the (separated) boundary conditions (5.245) are independent on $\lambda$ for $k = N + 1$.

**Lemma 5.61** *Under the additional assumption*

$$\mathcal{W}_{N+1} = 0, \tag{5.252}$$

*the eigenvalue problem for system* (5.238) *with separated boundary conditions* (5.245) *on* $[0, N+1]_{\mathbb{Z}}$ *can be extended to an eigenvalue problem of the special linear form* (5.239) *on the interval* $[-1, N+2]_{\mathbb{Z}}$ *with the Dirichlet boundary conditions* (5.160).

*Proof* In Lemma 5.40 for the nonlinear case, we preserve the construction for $\mathcal{S}_{N+1}$, because under assumption (5.252) the matrix $\mathcal{S}_{N+1}$ is independent on $\lambda$ (see the proof of Lemma 5.40). For the left point $k = 0$, we introduce the notation

$$\tilde{R} := \begin{pmatrix} R_0^* & R_0 \end{pmatrix}$$

and construct the constant matrix $\mathcal{S}_{-1}$ such that

$$\mathcal{S}_{-1} = \begin{pmatrix} \tilde{R}^T K & \mathcal{J}^T \tilde{R}^T \end{pmatrix}, \quad K := (\tilde{R}\tilde{R}^T)^{-1}.$$

Then we complete the construction of the matrix $\mathcal{S}_{-1}(\lambda)$ by

$$\mathcal{S}_{-1}(\lambda) = \begin{pmatrix} I & 0 \\ -\lambda W_{-1} & I \end{pmatrix} \begin{pmatrix} \tilde{R}^T K & \mathcal{J}^T \tilde{R}^T \end{pmatrix}. \tag{5.253}$$

The result then follows from Lemma 5.40.                                                      □

Based on the results of Lemmas 5.59 and 5.61, one can reformulate Theorem 5.41 for the case of the special linear dependence on $\lambda$.

**Theorem 5.62** *Under assumptions* (5.240), (5.246), (5.247), *and* (5.248), *the eigenvalue problem for system* (5.238) *with separated boundary conditions* (5.245) *on* $[0, N + 1]_{\mathbb{Z}}$ *is equivalent to the extended problem*

$$\tilde{y}_{k+1}(\lambda) = \begin{pmatrix} I & 0 \\ -\lambda \tilde{W}_k & I \end{pmatrix} S_k \tilde{y}_k(\lambda), \quad k \in [-1, N + 1]_{\mathbb{Z}}, \quad \tilde{x}_{-1}(\lambda) = 0 = \tilde{x}_{N+2}(\lambda),$$

*where the symmetric matrix*

$$\tilde{W}_k := \begin{cases} W_k, & k \in [-1, N - 1]_{\mathbb{Z}}, \\ W_N + W_{N+1}, & k = N, \\ 0, & k = N + 1. \end{cases} \tag{5.254}$$

*is nonnegative definite for all* $k \in [-1, N + 1]_{\mathbb{Z}}$,

$$\left. \begin{aligned} \mathcal{S}_{-1} = \begin{pmatrix} \mathcal{A}_{-1} & \mathcal{B}_{-1} \\ \mathcal{C}_{-1} & \mathcal{D}_{-1} \end{pmatrix} := \begin{pmatrix} R_0^{*T} K_0 & -R_0^T \\ R_0^T K_0 & R_0^{*T} \end{pmatrix}, \\ K_0 := (R_0^* R_0^{*T} + R_0 R_0^T)^{-1} \end{aligned} \right\} \tag{5.255}$$

*and*

$$\left. \begin{aligned} \mathcal{S}_{N+1} = \begin{pmatrix} \mathcal{A}_{N+1} & \mathcal{B}_{N+1} \\ \mathcal{C}_{N+1} & \mathcal{D}_{N+1} \end{pmatrix} := \begin{pmatrix} R_{N+1}^* & R_{N+1} \\ -K_{N+1} R_{N+1} & K_{N+1} R_{N+1}^* \end{pmatrix}, \\ K_{N+1} := (R_{N+1}^* R_{N+1}^{*T} + R_{N+1} R_{N+1}^T)^{-1}. \end{aligned} \right\} \tag{5.256}$$

*Proof* The proof is based on the subsequent application of Lemmas 5.59 and 5.61. At the first step, applying Lemma 5.59, we derive the new boundary problem with the additional assumption (5.252). Applying Lemma 5.61 to this problem and incorporating (5.250), we complete the proof of Theorem 5.62. □

Next we apply Theorem 5.45 for the nonlinear case to the linear problem (5.238) with (5.242).

**Theorem 5.63** *Under assumptions* (5.243) *and* (5.244), *problem* (5.238) *with* (5.242) *is equivalent to the linear eigenvalue problem with the separated boundary conditions*

$$
\left.
\begin{aligned}
\tilde{y}_{k+1}(\lambda) &= \left\{ \begin{pmatrix} I & 0 \\ -\lambda W_k & I \end{pmatrix} \right\} \{S_k\}\, \tilde{y}_k(\lambda), \quad k \in [0, N]_{\mathbb{Z}}, \\
&\tilde{\mathbf{R}}_1(\lambda)\, \tilde{x}_0(\lambda) + \tilde{\mathbf{R}}_2(\lambda)\, \tilde{u}_0(\lambda) = 0, \\
&\tilde{\mathbf{R}}_1(\lambda) := \tilde{R}_1 + \lambda \tilde{R}_2 \tilde{\mathfrak{W}}, \quad \tilde{\mathbf{R}}_2(\lambda) := \tilde{R}_2 \\
&\begin{pmatrix} 0 & 0 \\ -I & I \end{pmatrix} \tilde{x}_{N+1}(\lambda) + \begin{pmatrix} -I & -I \\ 0 & 0 \end{pmatrix} \tilde{u}_{N+1}(\lambda) = 0,
\end{aligned}
\right\}
\tag{5.257}
$$

*where*

$$
\tilde{R}_1 = R_1\, P_1, \quad \tilde{R}_2 = -R_2\, P_1, \quad \tilde{\mathfrak{W}} = P_1 \mathfrak{W} P_1, \quad P_1 = \begin{pmatrix} 0 & I \\ I & 0 \end{pmatrix},
\tag{5.258}
$$

*and*

$$
\dot{\tilde{\mathbf{R}}}_1(\lambda)\, \tilde{R}_2^T = \tilde{R}_2 \tilde{\mathfrak{W}} \tilde{R}_2^T \geq 0.
\tag{5.259}
$$

*Proof* According to Theorem 5.45, the coefficients in (5.242) are transformed as

$$
\tilde{\mathcal{R}}(\lambda) = \left( \tilde{\mathbf{R}}_1(\lambda)\ \tilde{R}_2 \right) = \left( R_1\ R_2 \right) \begin{pmatrix} I & 0 \\ -\lambda \mathfrak{M} & I \end{pmatrix} \operatorname{diag}\{P_1,\ -P_1\}
$$

$$
= \left( R_1\ R_2 \right) \operatorname{diag}\{P_1,\ -P_1\} \operatorname{diag}\{P_1,\ -P_1\} \begin{pmatrix} I & 0 \\ -\lambda \mathfrak{W} & I \end{pmatrix} \operatorname{diag}\{P_1,\ -P_1\}
$$

$$
= \left( \tilde{R}_1\ \tilde{R}_2 \right) \begin{pmatrix} I & 0 \\ \lambda P_1 \mathfrak{W} P_1 & I \end{pmatrix} = \left( \tilde{R}_1\ \tilde{R}_2 \right) \begin{pmatrix} I & 0 \\ \lambda \tilde{\mathfrak{W}} & I \end{pmatrix}
$$

$$
= \left( \tilde{R}_1 + \lambda \tilde{R}_2 \tilde{\mathfrak{W}}\ \ \tilde{R}_2 \right),
$$

where the matrices above are defined by (5.257) and (5.258). Applying Theorem 5.45 to the case under the consideration, we complete the proof. □

*Remark 5.64*

(i) Observe that according to our construction, the matrix of the symplectic system
   in (5.257) is of the same form as the matrix of system (5.239); since by
   definition (3.101), we have

$$\left\{ \begin{pmatrix} I & 0 \\ -\lambda W & I \end{pmatrix} \right\} = \begin{pmatrix} I & 0 & 0 & 0 \\ 0 & I & 0 & 0 \\ 0 & 0 & I & 0 \\ 0 & -\lambda W & 0 & I \end{pmatrix}.$$

   Consequently, the weight matrix by $-\lambda$ is equal to $\mathrm{diag}\{0, W\} \geq 0$.

(ii) Note that in problem (5.257) the boundary condition for $k = N + 1$ is
   independent on $\lambda$. Applying Lemma 5.61, one can extend (5.257) to the
   problem with the Dirichlet boundary conditions preserving the special linear
   dependence on $\lambda$. By a similar way, one can modify Theorem 5.44 for the
   nonlinear case to derive a problem with the separated boundary conditions,
   which are independent on $\lambda$ for $k = 0$, and then apply Theorem 5.62 for the
   construction of the extended problem with the Dirichlet boundary conditions,
   again with the preservation of the special linear dependence on $\lambda$.

### 5.3.2 Quadratic Functionals

In this subsection we analyze the quadratic functional $\mathcal{G}(y, \lambda)$ defined in (5.186)
over the joint endpoints (5.187), respectively, the quadratic functional $\mathcal{G}(y, \lambda)$
defined in (5.188) over the separated endpoints (5.190), when the boundary
conditions depend linearly on $\lambda$ as in Sect. 5.3.1.

   Given system (SDS$_\lambda$) and the $2n \times 2n$ matrices $R_1$, $R_2$, $\mathfrak{W}$ satisfying (5.243)
and (5.244), the functional $\mathcal{G}(y, \lambda)$ in (5.186) takes the form

$$\mathcal{G}(y, \lambda) = \mathcal{F}_0(y) + \begin{pmatrix} x_0 \\ x_{N+1} \end{pmatrix}^T (\Gamma - \lambda \mathfrak{W}) \begin{pmatrix} x_0 \\ x_{N+1} \end{pmatrix} - \lambda \sum_{k=0}^{N} x_{k+1}^T W_k x_{k+1}, \qquad (5.260)$$

where $\mathcal{F}_0(y)$ is the basic functional defined in (2.56) and $\Gamma := R_2^\dagger R_1 R_2^\dagger R_2$ as
in (5.185). (Note that in Sect. 2.3 the functional $\mathcal{F}_0(y)$ is denoted by $\mathcal{F}(y)$.) The
functional $\mathcal{G}(y, \lambda)$ in (5.260) acts on admissible functions $y = (x, u)$ satisfying the
boundary conditions

$$\begin{pmatrix} x_0 \\ x_{N+1} \end{pmatrix} \in \mathrm{Im}\, R_2^T. \qquad (5.261)$$

Observe that the admissibility condition, i.e., the first equation of system (5.238), and the boundary conditions (5.261) do not depend on $\lambda$. The form of the functional $\mathcal{G}(y, \lambda)$ in (5.260) yields the following simple comparison result.

**Proposition 5.65** *Assume that (5.240) and (5.244) hold and consider the functional $\mathcal{G}(y, \lambda)$ in (5.260). Then for every $\lambda, \lambda_0 \in \mathbb{R}$, $\lambda < \lambda_0$, we have the inequality $\mathcal{G}(y, \lambda) \geq \mathcal{G}(y, \lambda_0)$ for every admissible $y$ satisfying (5.261). Consequently, if $\mathcal{G}(\cdot, \lambda_0) > 0$ for some $\lambda_0 \in \mathbb{R}$, then $\mathcal{G}(y, \lambda) > 0$ for all $\lambda < \lambda_0$ as well.*

Similarly, given $n \times n$ matrices $R_0^*$, $R_0$, $R_{N+1}^*$, $R_{N+1}$, and $\mathcal{W}_{-1}$, $\mathcal{W}_{N+1}$ satisfying (5.246), (5.247), and (5.248), the functional $\mathcal{G}(y, \lambda)$ in (5.188) takes the form

$$
\left.
\begin{aligned}
\mathcal{G}(y, \lambda) = \mathcal{F}_0(y) + x_0^T (\Gamma_0 - \lambda \mathcal{W}_{-1}) x_0 \\
+ x_{N+1}^T (\Gamma_{N+1} - \lambda \mathcal{W}_{N+1}) x_{N+1} - \lambda \sum_{k=0}^{N} x_{k+1}^T \mathcal{W}_k x_{k+1},
\end{aligned}
\right\} \quad (5.262)
$$

where $\Gamma_0 := -R_0^\dagger R_0^* R_0^\dagger R_0$ and $\Gamma_{N+1} := R_{N+1}^\dagger R_{N+1}^* R_{N+1}^\dagger R_{N+1}$ as in (5.189). The functional $\mathcal{G}(y, \lambda)$ in (5.262) acts on admissible functions $y = (x, u)$ satisfying the separated boundary conditions

$$
x_0 \in \operatorname{Im} R_0^T, \quad x_{N+1} \in \operatorname{Im} R_{N+1}^T. \quad (5.263)
$$

Observe that the boundary conditions (5.263) do not depend on $\lambda$. Motivated by the statements of the global oscillation theorems (Theorems 5.31, 5.53, and 5.57), our aim is to find conditions, which will guarantee the existence of $\lambda_0 < 0$ such that the functional $\mathcal{G}(\cdot, \lambda_0)$ is positive definite. The statement in Proposition 5.65 then holds for the functional $\mathcal{G}(y, \lambda)$ in (5.262) under the assumptions (5.240), (5.247), and (5.248).

For the next results, we recall the definition in (2.64) of the symmetric $n \times n$ matrix $\mathcal{E}_k = \mathcal{B}_k \mathcal{B}_k^\dagger \mathcal{D}_k \mathcal{B}_k^\dagger$, which satisfies $\mathcal{B}_k^T \mathcal{E}_k \mathcal{B}_k = \mathcal{B}_k^T \mathcal{D}_k$ for $k \in [0, N]_{\mathbb{Z}}$.

**Proposition 5.66** *Assume*

$$
\mathcal{W}_{-1} \geq 0, \quad \mathcal{W}_k > 0, \quad k \in [0, N-1]_{\mathbb{Z}}, \quad \mathcal{W}_N \geq 0, \quad \mathcal{W}_{N+1} \geq 0, \quad (5.264)
$$

*and for some $\lambda < 0$, we have*

$$
\Gamma_0 - \mathcal{C}_0^T \mathcal{A}_0 + \mathcal{A}_0^T \mathcal{E}_0 \mathcal{A}_0 - \lambda \mathcal{W}_{-1} > 0 \quad \text{on } \operatorname{Im} R_0^T \quad (5.265)
$$

$$
\Gamma_{N+1} + \mathcal{E}_N - \lambda (\mathcal{W}_N + \mathcal{W}_{N+1}) > 0 \quad \text{on } \operatorname{Im} R_{N+1}^T. \quad (5.266)
$$

*Then there exists $\lambda_0 < 0$ such that $\mathcal{G}(\cdot, \lambda_0) > 0$ over (5.263).*

*Proof* For an admissible $y$, we rewrite $\mathcal{G}(y, \lambda)$ by using the matrix $\mathcal{E}_k$ as

$$\mathcal{G}(y, \lambda) = x_0^T (\Gamma_0 - C_0^T A_0 + A_0^T \mathcal{E}_0 A_0 - \lambda W_{-1}) x_0$$

$$+ \sum_{k=0}^{N-1} \left\{ 2x_k^T (C_k^T - A_k^T \mathcal{E}_k) x_{k+1} + x_{k+1}^T (\mathcal{E}_k - C_{k+1}^T A_{k+1} - \lambda W_k) x_{k+1} \right\}$$

$$+ 2x_N^T (C_N^T - A_N^T \mathcal{E}_N) x_{N+1} + x_{N+1}^T [\Gamma_{N+1} + \mathcal{E}_N - \lambda(W_N + W_{N+1})] x_{N+1}.$$

Therefore, the positivity assumption on $W_k$ for $k \in [0, N-1]_{\mathbb{Z}}$ in (5.264) and the positivity assumptions (5.265) and (5.266) imply that for some $\lambda_0 < 0$ the functional $\mathcal{G}(y, \lambda_0) > 0$ for all admissible $y$ with (5.263) and $x = \{x_k\}_{k=0}^{N+1} \neq 0$. □

Consider now a special case of the functional $\mathcal{G}(y, \lambda)$ in (5.262) in which the matrices $W_{-1} = 0 = W_{N+1}$, i.e.,

$$\mathcal{G}(y, \lambda) = \mathcal{F}_0(y) + x_0^T \Gamma_0 x_0 + x_{N+1}^T \Gamma_{N+1} x_{N+1} - \lambda \sum_{k=0}^{N} x_{k+1}^T W_k x_{k+1}. \tag{5.267}$$

**Proposition 5.67** *Suppose that*

$$W_k > 0, \quad k \in [0, N]_{\mathbb{Z}}, \tag{5.268}$$

$$\Gamma_0 - A_0^T C_0 + A_0^T \mathcal{E}_0 A_0 > 0 \quad \text{on Im } R_0^T. \tag{5.269}$$

*Then there exists $\lambda_0 < 0$ such that the functional $\mathcal{G}(y, \lambda_0)$ in (5.267) is positive definite over (5.263).*

*Proof* The result follows from Proposition 5.66, since assumptions (5.268) and (5.269) imply the validity of (5.264)–(5.266) with $W_{-1} := 0$ and $W_{N+1} := 0$. □

*Remark 5.68* One may certainly formulate numerous other sufficient conditions for (5.264)–(5.266) to be satisfied involving nonzero weight matrices $W_{-1}$ and/or $W_{N+1}$. For example, one such condition could be

$$W_k > 0, \quad k \in [-1, N-1]_{\mathbb{Z}}, \quad W_N + W_{N+1} > 0, \quad W_N \geq 0, \quad W_{N+1} \geq 0, \tag{5.270}$$

or the following ones, implying (5.270),

$$W_k > 0, \quad k \in [-1, N]_{\mathbb{Z}}, \quad W_{N+1} \geq 0, \tag{5.271}$$

$$W_k > 0, \quad k \in [-1, N+1]_{\mathbb{Z}}. \tag{5.272}$$

Condition (5.269) is trivially satisfied for the Dirichlet boundary conditions $x_0 = 0 = x_{N+1}$, i.e., for $R_0 = 0 = R_{N+1}$. In this case, in agreement with the notation in (5.138), the functional $\mathcal{G}(y, \lambda)$ in (5.267) has the form

$$\mathcal{F}(y, \lambda) = \mathcal{F}_0(y) - \lambda \sum_{k=0}^{N} x_{k+1}^T \mathcal{W}_k\, x_{k+1}. \tag{5.273}$$

Thus, we obtain from Proposition 5.67 the following.

**Corollary 5.69** *Suppose that (5.268) holds. Then there exists $\lambda_0 < 0$ such that the functional $\mathcal{F}(y, \lambda)$ in (5.273) is positive definite over $x_0 = 0 = x_{N+1}$.*

*Remark 5.70* The question of finding some simple and easy to verify sufficient conditions for the positivity of the functional $\mathcal{G}(\cdot, \lambda_0)$ for some $\lambda_0 < 0$ in (5.260) over joint endpoints (5.261) in terms of the coefficients of system (5.1), e.g., a "positivity condition" on the weight matrices $\mathcal{W}_k$ and $\mathfrak{W}$ in the spirit of Proposition 5.67 and Corollary 5.69, remains an open problem.

### 5.3.3  Finite Eigenvalues and Finite Eigenfunctions

In this subsection we analyze special properties of the finite eigenvalues of problem (SDS$_\lambda$) with (5.239) and with the boundary conditions depending linearly on $\lambda$ as in Sect. 5.3.1. In particular, we will discuss the results of the corresponding global oscillation theorems, including special properties of finite eigenvalues and finite eigenfunctions, which can be derived for this case.

  First we consider the general boundary conditions (5.242) determined by the $2n \times 2n$ matrices $R_1$, $R_2$, and $\mathfrak{W}$ satisfying (5.243) and (5.244). According to Definition 5.54, the finite eigenvalues are determined by the behavior of the function $\mathcal{L}(\lambda)$ in (5.221) defined through the matrices $\mathcal{X}_k(\lambda)$ and $\mathcal{U}_k(\lambda)$ in (5.220) and through the conjoined bases $Y^{[0]}(\lambda)$ and $\tilde{Y}(\lambda)$ of (5.238) such that $\tilde{Y}_0(\lambda) = (I\ 0)^T$ and $Y^{[0]}(\lambda)$ is the principal solution at $k = 0$. Since the dependence on $\lambda$ in system (SDS$_\lambda$) and in the boundary conditions (5.242) is now linear, and since the initial conditions of $\mathcal{X}(\lambda)$ and $\mathcal{U}(\lambda)$ at $k = 0$ do not depend on $\lambda$, it follows that $\mathcal{L}(\lambda)$ is a matrix polynomial. This means that for each $\lambda_0 \in \mathbb{R}$, the one-sided limits of the rank of $\mathcal{L}(\lambda)$ satisfy

$$\operatorname{rank} \mathcal{L}(\lambda_0^+) = r = \operatorname{rank} \mathcal{L}(\lambda_0^-), \quad r := \max_{\lambda \in \mathbb{R}} \operatorname{rank} \mathcal{L}(\lambda). \tag{5.274}$$

This means that the algebraic multiplicity $\zeta(\lambda_0)$ in (5.222) is equal to

$$\zeta(\lambda_0) = r - \operatorname{rank} \mathcal{L}(\lambda_0) \tag{5.275}$$

with $r$ given in (5.274). In order to understand the definition of a finite eigenfunction for the problem (SDS$_\lambda$) with (5.242), we combine the results in Sects. 5.1.2 and 5.3.1.

**Definition 5.71** Assume (5.239), (5.240), (5.243), and (5.244). A solution $y = (x, u)$ of $(SDS_\lambda)$ and (5.242) with $\lambda = \lambda_0$ is called a *finite eigenfunction* corresponding to a finite eigenvalue $\lambda_0$ if it satisfies the nondegeneracy condition

$$\{\mathcal{W}_k x_{k+1}\}_{k=0}^N \neq 0 \quad \text{or} \quad \mathfrak{W} \begin{pmatrix} x_0 \\ x_{N+1} \end{pmatrix} \neq 0. \tag{5.276}$$

The dimension $\omega(\lambda_0)$ of the space of all finite eigenfunctions $y$ corresponding to the finite eigenvalue $\lambda_0$ is called a *geometric multiplicity* of $\lambda_0$.

In the following result, we provide a geometric characterization of the finite eigenvalues of $(SDS_\lambda)$ with (5.239). We thus extend Theorem 5.11 to general jointly varying endpoints.

**Theorem 5.72** *Let assumptions (5.239), (5.240), (5.243), and (5.244) be satisfied. A number $\lambda_0$ is a finite eigenvalue of $(SDS_\lambda)$ with (5.239) with algebraic multiplicity $\zeta(\lambda_0) \geq 1$ in (5.275) if and only if there exists a corresponding finite eigenfunction $y$ for $\lambda_0$. In this case, the geometric multiplicity of $\lambda_0$ is equal to its algebraic multiplicity, i.e., $\omega(\lambda_0) = \zeta(\lambda_0)$.*

*Proof* We proceed by adopting the transformations to the augmented eigenvalue problem in Theorem 5.44 and then to the extended eigenvalue problem in Theorem 5.62 (see Remark 5.64(ii)). In this way we obtain an extended augmented problem over the interval $[-1, N + 2]_{\mathbb{Z}}$ with the augmented functions $\tilde{x}_k = \begin{pmatrix} x_0 \\ x_k \end{pmatrix}$ and $\tilde{u}_k = \begin{pmatrix} -u_0 \\ u_k \end{pmatrix}$ and the augmented matrices $\tilde{\mathcal{W}}_k$, which have according to (5.254) the form

$$\begin{rcases} \tilde{\mathcal{W}}_{-1} = 0, \quad \tilde{\mathcal{W}}_k = \text{diag}\{0, \mathcal{W}_k\}, \quad k \in [0, N - 1]_{\mathbb{Z}}, \\ \tilde{\mathcal{W}}_N = \text{diag}\{0, \mathcal{W}_N\} + \mathfrak{W}, \quad \tilde{\mathcal{W}}_{N+1} = 0. \end{rcases} \tag{5.277}$$

The weight matrix $\tilde{\Psi}_k(\lambda) \equiv \tilde{\Psi}_k \geq 0$ on $[-1, N + 1]_{\mathbb{Z}}$ corresponding to (5.241) then has the form $\tilde{\Psi}_k = \text{diag}\{\tilde{\mathcal{W}}_k, 0\}$ for $k \in [-1, N + 1]_{\mathbb{Z}}$. The degeneracy condition in Definition 5.8 for this extended augmented problem reads as $\tilde{\mathcal{W}}_k \tilde{x}_{k+1} \equiv 0$ on $[-1, N + 1]_{\mathbb{Z}}$, i.e.,

$$\mathcal{W}_k x_{k+1} \equiv 0 \quad \text{on } [0, N - 1]_{\mathbb{Z}}, \quad \mathcal{W}_N x_{N+1} + \mathfrak{W} \begin{pmatrix} x_0 \\ x_{N+1} \end{pmatrix} = 0. \tag{5.278}$$

Since $\mathcal{W}_N \geq 0$ and $\mathfrak{W} \geq 0$ is assumed, condition (5.278) and hence the degeneracy condition is equivalent to $\mathcal{W}_k x_{k+1} \equiv 0$ on $[0, N]_{\mathbb{Z}}$ and $\mathfrak{W} \begin{pmatrix} x_0 \\ x_{N+1} \end{pmatrix} = 0$. Therefore, our nondegeneracy condition in (5.276) means that the corresponding solution of the extended augmented problem is also nondegenerate on $[-1, N + 2]_{\mathbb{Z}}$. The statement in the theorem then follows from the application of Theorem 5.11 to the extended augmented problem. $\square$

In the last part of this subsection, we will discuss the properties of the finite eigenvalues of $(SDS_\lambda)$ with (5.242) from the point of view of self-adjoint boundary value problems. Based on Theorem 5.72, we may define a finite eigenvalue of $(SDS_\lambda)$ with (5.239) in the context of the linear dependence on the spectral parameter in the symplectic system and in the boundary conditions as a number $\lambda_0 \in \mathbb{C}$, for which there exists corresponding a solution $y = (x, u)$ of $(SDS_\lambda)$ with (5.242) satisfying (5.276). This means, that we now allow complex finite eigenvalues. Then we can actually prove that these finite eigenvalues are real (Proposition 5.74) and the finite eigenfunctions corresponding to different finite eigenvalues are orthogonal (Proposition 5.73). Here we consider the semi-inner product

$$\langle y, \tilde{y} \rangle := \sum_{k=0}^{N} x_{k+1}^T W_k \, \tilde{x}_{k+1} + \begin{pmatrix} x_0 \\ x_{N+1} \end{pmatrix}^T \mathfrak{W} \begin{pmatrix} \tilde{x}_0 \\ \tilde{x}_{N+1} \end{pmatrix}. \tag{5.279}$$

**Proposition 5.73** *Assume* (5.239), (5.240), (5.243), *and* (5.244). *Let* $y$ *and* $\tilde{y}$ *be finite eigenfunctions corresponding to the finite eigenvalues* $\lambda$ *and* $\tilde{\lambda}$ *of* $(SDS_\lambda)$ *with* (5.242), *and let* $\lambda \neq \tilde{\lambda}$. *Then* $\langle y, \tilde{y} \rangle = 0$, *i.e., the finite eigenfunctions* $y$ *and* $\tilde{y}$ *are orthogonal with respect to the inner product* (5.279).

*Proof* Let $y$ satisfy $(SDS_\lambda)$ with (5.239), and let $\tilde{y}$ satisfy $(SDS_\lambda)$ with (5.239) for the spectral parameter $\tilde{\lambda}$. Then

$$(\lambda - \tilde{\lambda}) \langle y, \tilde{y} \rangle = (\lambda - \tilde{\lambda}) \sum_{k=0}^{N} x_{k+1}^T W_k \, \tilde{x}_{k+1} + (\lambda - \tilde{\lambda}) \begin{pmatrix} x_0 \\ x_{N+1} \end{pmatrix}^T \mathfrak{W} \begin{pmatrix} \tilde{x}_0 \\ \tilde{x}_{N+1} \end{pmatrix}$$

$$= \sum_{k=0}^{N} \left\{ (u_{k+1}^T - x_k^T C_k^T - u_k^T D_k^T) \, \tilde{x}_{k+1} - x_{k+1}^T (\tilde{u}_{k+1} - C_k \tilde{x}_k - D_k \tilde{u}_k) \right\}$$

$$+ (\lambda - \tilde{\lambda}) \begin{pmatrix} x_0 \\ x_{N+1} \end{pmatrix}^T \mathfrak{W} \begin{pmatrix} \tilde{x}_0 \\ \tilde{x}_{N+1} \end{pmatrix}$$

$$= \sum_{k=0}^{N} \left\{ x_k^T (A_k^T D_k - C_k^T B_k - I) \, \tilde{u}_k - u_k^T (B_k^T C_k - D_k^T A_k + I) \, \tilde{x}_k \right\}$$

$$+ (u_k^T \tilde{x}_k - x_k^T \tilde{u}_k) \Big|_0^{N+1} + (\lambda - \tilde{\lambda}) \begin{pmatrix} x_0 \\ x_{N+1} \end{pmatrix}^T \mathfrak{W} \begin{pmatrix} \tilde{x}_0 \\ \tilde{x}_{N+1} \end{pmatrix}$$

$$= \begin{pmatrix} -u_0 \\ u_{N+1} \end{pmatrix}^T \begin{pmatrix} \tilde{x}_0 \\ \tilde{x}_{N+1} \end{pmatrix} - \begin{pmatrix} x_0 \\ x_{N+1} \end{pmatrix}^T \begin{pmatrix} -\tilde{u}_0 \\ \tilde{u}_{N+1} \end{pmatrix}$$

$$+ (\lambda - \tilde{\lambda}) \begin{pmatrix} x_0 \\ x_{N+1} \end{pmatrix}^T \mathfrak{W} \begin{pmatrix} \tilde{x}_0 \\ \tilde{x}_{N+1} \end{pmatrix}. \tag{5.280}$$

The assumptions on $R_1$, $R_2$, $\mathfrak{W}$ in (5.243) and (5.244) imply that

$$\mathrm{Ker}\left(R_1 - \lambda R_2 \mathfrak{W},\ R_2\right) = \mathrm{Im}\begin{pmatrix} -R_2^T \\ (R_1 - \lambda R_2 \mathfrak{W})^T \end{pmatrix} \tag{5.281}$$

and similarly for $\tilde{\lambda}$. Therefore, since $y$ satisfies the boundary conditions (5.239) and $\tilde{y}$ satisfies (5.239) with $\lambda := \tilde{\lambda}$, it follows that there exist $d, \tilde{d} \in \mathbb{R}^{2n}$ such that

$$\begin{pmatrix} x_0 \\ x_{N+1} \end{pmatrix} = -R_2^T d, \quad \begin{pmatrix} -u_0 \\ u_{N+1} \end{pmatrix} = (R_1 - \lambda R_2 \mathfrak{W})^T d,$$

$$\begin{pmatrix} \tilde{x}_0 \\ \tilde{x}_{N+1} \end{pmatrix} = -R_2^T \tilde{d}, \quad \begin{pmatrix} -\tilde{u}_0 \\ \tilde{u}_{N+1} \end{pmatrix} = (R_1 - \tilde{\lambda} R_2 \mathfrak{W})^T \tilde{d}.$$

Substituting this into (5.280), we obtain

$$(\lambda - \tilde{\lambda})\, \langle y, \tilde{y} \rangle = -d^T (R_1 - \lambda R_2 \mathfrak{W})\, R_2^T \tilde{d} + d^T R_2 (R_1 - \tilde{\lambda} R_2 \mathfrak{W})^T \tilde{d}$$

$$+ (\lambda - \tilde{\lambda})\, d^T R_2 \mathfrak{W} R_2^T \tilde{d} = 0.$$

And since $\lambda \neq \tilde{\lambda}$, it follows that $\langle y, \tilde{y} \rangle = 0$, i.e., the finite eigenfunctions $y$ and $\tilde{y}$ are orthogonal with respect to (5.279). $\qquad\square$

**Proposition 5.74** *Assume* (5.239), (5.240), (5.243), *and* (5.244). *Then the finite eigenvalues of* (SDS$_\lambda$) *with* (5.242) *are real and bounded from below and from above. Moreover, the total number of finite eigenvalues of* (SDS$_\lambda$) *with* (5.242) *can be estimated by*

$$\sum_{k=0}^{N-1} \mathrm{rank}\, \mathcal{W}_k + \mathrm{rank}\left(\mathrm{diag}\{0, \mathcal{W}_N\} + \mathfrak{W}\right) \le (N+2)\, n. \tag{5.282}$$

*Proof* Since the coefficients of system (SDS$_\lambda$) and the boundary conditions (5.242) are real, then with $\lambda$ being its finite eigenvalue with the finite eigenfunction $y$, the complex conjugate number $\bar{\lambda}$ is also a finite eigenvalue with the finite eigenfunction $\bar{y}$. In this case $\langle \bar{y}, \bar{y} \rangle = \overline{\langle y, y \rangle} = \langle y, y \rangle > 0$, by (5.276). Moreover, by Lemma 2.61 applied to system (SDS$_\lambda$) and the functional $\mathcal{G}(y, \lambda)$, we have

$$\mathcal{G}(y, \lambda) = \begin{pmatrix} -u_0 \\ u_{N+1} \end{pmatrix}^T \begin{pmatrix} x_0 \\ x_{N+1} \end{pmatrix} + \begin{pmatrix} x_0 \\ x_{N+1} \end{pmatrix}^T (\Gamma - \lambda \mathfrak{W}) \begin{pmatrix} x_0 \\ x_{N+1} \end{pmatrix} = 0,$$

where we used (5.281) and the fact that $y$ satisfies the boundary conditions (5.242). Similarly, $\mathcal{G}(\bar{y}, \bar{\lambda}) = 0$. Denote $\mathcal{F}(y) := \mathcal{G}(y, 0)$ and $\mathcal{F}(\bar{y}) := \mathcal{G}(\bar{y}, 0)$. It follows that $\mathcal{F}(y) = \overline{\mathcal{F}(y)} = \mathcal{F}(\bar{y})$, and thus by (5.260), we obtain

$$\lambda = \frac{\mathcal{F}(y)}{\langle y, y \rangle} = \frac{\mathcal{F}(\bar{y})}{\langle \bar{y}, \bar{y} \rangle} = \bar{\lambda}.$$

This shows that $\lambda \in \mathbb{R}$. The boundedness from below of the finite eigenvalues follows from Remark 5.56. The estimate in (5.282) follows from the estimate for the sum $\sum_{k=0}^{N} \operatorname{rank} \tilde{\mathcal{W}}_k$, where $\tilde{\mathcal{W}}_k$ are given by (5.277). □

*Remark 5.75* For the Dirichlet boundary conditions

$$x_0(\lambda) = 0 = x_{N+1}(\lambda), \tag{5.283}$$

i.e., for $R_1 = I$ and $R_2 = \Gamma = \mathfrak{W} = 0$, the finite eigenvalues of (SDS$_\lambda$) with (5.283) are determined by the condition

$$\theta(\lambda_0) = r - \operatorname{rank} X_{N+1}^{[0]}(\lambda_0), \quad 1 \leq \theta(\lambda_0) \leq n, \tag{5.284}$$

where according to (5.284), the value $r$ is given by

$$r = \max_{\lambda \in \mathbb{R}} \operatorname{rank} X_{N+1}^{[0]}(\lambda) = \operatorname{rank} X_{N+1}^{[0]}(\lambda_0^+) = \operatorname{rank} X_{N+1}^{[0]}(\lambda_0^-) \tag{5.285}$$

and the function $X_{N+1}^{[0]}(\lambda)$ comes from the principal solution $Y^{[0]}(\lambda)$ of (SDS$_\lambda$) at $k = 0$. In this case $X_{N+1}^{[0]}(\lambda)$ is a matrix polynomial. The nondegeneracy condition (5.276) for $y$ being a finite eigenfunction has the form

$$\{\mathcal{W}_k x_{k+1}\}_{k=0}^{N-1} \neq 0. \tag{5.286}$$

By Propositions 5.73 and 5.74, the finite eigenvalues of (SDS$_\lambda$) with (5.283) are real, and the finite eigenfunctions corresponding to different finite eigenvalues are orthogonal with respect to the inner product

$$\langle y, \tilde{y} \rangle = \sum_{k=0}^{N-1} x_{k+1}^T \mathcal{W}_k \tilde{x}_{k+1}. \tag{5.287}$$

The finite eigenvalues are bounded from below, by Corollary 5.18. Finally, by (5.286) the total number of finite eigenvalues of (SDS$_\lambda$) with (5.283) can be estimated by

$$\sum_{k=0}^{N-1} \operatorname{rank} \mathcal{W}_k \leq nN. \tag{5.288}$$

*Remark 5.76* For the separated boundary conditions (5.245), i.e., when the matrices $R_1 = \operatorname{diag}\{R_0^*, R_{N+1}^*\}$ and $R_2 = \operatorname{diag}\{-R_0, R_{N+1}\}$, as well as $\Gamma = \operatorname{diag}\{\Gamma_0, \Gamma_{N+1}\}$ and $\mathfrak{W} = \operatorname{diag}\{\mathcal{W}_{-1}, \mathcal{W}_{N+1}\}$, are block diagonal, the finite eigenvalues of (SDS$_\lambda$) with (5.245) are determined by the condition; see (5.206),

$$\theta(\lambda_0) = r - \operatorname{rank} \Lambda(\lambda_0), \quad 1 \leq \theta(\lambda_0) \leq n, \tag{5.289}$$

where the value $r$ is given by

$$r = \max_{\lambda \in \mathbb{R}} \text{rank } \Lambda(\lambda) = \text{rank } \Lambda(\lambda_0^+) = \text{rank } \Lambda(\lambda_0^-). \tag{5.290}$$

The function $\Lambda(\lambda)$ is defined in (5.205) and comes from the natural conjoined basis $\bar{Y}(\lambda)$ of (SDS$_\lambda$) at $k = 0$. In this case $\Lambda(\lambda)$ is again a matrix polynomial. The nondegeneracy condition (5.276) for $y$ being a finite eigenfunction is

$$\{\mathcal{W}_k x_{k+1}\}_{k=-1}^{N-1} \neq 0 \quad \text{or} \quad (\mathcal{W}_N + \mathcal{W}_{N+1}) x_{N+1} \neq 0. \tag{5.291}$$

The finite eigenvalues of (SDS$_\lambda$) with (5.245) are real, and the finite eigenfunctions corresponding to different finite eigenvalues are orthogonal with respect to the inner product

$$\langle y, \tilde{y} \rangle = \sum_{k=-1}^{N-1} x_{k+1}^T \mathcal{W}_k \tilde{x}_{k+1} + x_{N+1}^T (\mathcal{W}_N + \mathcal{W}_{N+1}) x_{N+1}. \tag{5.292}$$

By Remark 5.51, the finite eigenvalues of (SDS$_\lambda$) with (5.245) are bounded from below. By (5.291), the total number of finite eigenvalues of (SDS$_\lambda$) with (5.245) can be estimated by

$$\sum_{k=-1}^{N-1} \text{rank } \mathcal{W}_k + \text{rank } (\mathcal{W}_N + \mathcal{W}_{N+1}) \leq (N+2)n. \tag{5.293}$$

*Example 5.77* The following example shows that the conclusions in Remark 5.75 do not hold in general without the assumption $\mathcal{W}_k \geq 0$. Put $N = 3$,

$$\mathcal{A}_k = \text{diag}\{1, 0\}, \quad \mathcal{B}_k = \text{diag}\{0, 1\}, \quad \mathcal{C}_k = \text{diag}\{0, -1\}, \quad \mathcal{D}_k = \text{diag}\{1, 0\}$$

for $k \in [0, 3]_\mathbb{Z}$, and

$$\mathcal{W}_0 = \text{diag}\{0, 1\}, \quad \mathcal{W}_1 = 0, \quad \mathcal{W}_2 = \text{diag}\{0, -1\}.$$

Then the system is symplectic, $\mathcal{W}_0 \geq 0$, $\mathcal{W}_1 \geq 0$, but $\mathcal{W}_2 \not\geq 0$. The calculation of the principal solution $Y^{[0]}(\lambda)$ by the difference system yields that

$$X_0^{[0]}(\lambda) = 0_2, \quad X_1^{[0]}(\lambda) = \text{diag}\{0, 1\}, \quad X_2^{[0]}(\lambda) = \text{diag}\{0, \lambda\},$$

$$X_3^{[0]}(\lambda) = \text{diag}\{0, -1\}, \quad X_4^{[0]}(\lambda) = X_{N+1}^{[0]}(\lambda) \equiv 0_2.$$

Hence $r_{N+1} = 0$, so that there are no finite eigenvalues by (5.284) in Remark 5.75. But every $\lambda$ possesses a finite eigenfunction, namely, the function $y = \{y_k\}_{k=0}^4$ with $y_k = Y_k^{[0]}(\lambda) \binom{0}{1}$, so that $x_0(\lambda) = \binom{0}{0}$, $x_1(\lambda) = \binom{0}{1}$, $x_2(\lambda) = \binom{0}{-\lambda}$, and $x_4(\lambda) = \binom{0}{0}$.

### 5.3.4 Global Oscillation Theorem

In this subsection we present the global oscillation theorem for problem (5.238) with (5.283) and comment on its special features resulting from the linear dependence on $\lambda$. The statement of this special global oscillation theorem will be of particular interest in several applications later in this section as well as in Sect. 5.5. First we recall the notation (5.49) for $n_1(\lambda)$ and the notation (5.50) for $n_2(\lambda)$, being the number of focal points of $Y^{[0]}(\lambda)$ in $(1, N+1]$ and the number of finite eigenvalues of (5.238) with (5.283) in $(-\infty, \lambda]$, respectively.

**Theorem 5.78 (Global Oscillation Theorem for Linear Dependence on $\lambda$)**
*Consider system* (5.238), *i.e., system* (5.1) *with* (5.239) *and with Dirichlet boundary conditions* (5.283). *Suppose that* (5.240) *holds, i.e.,* $\mathcal{W}_k \geq 0$ *for all* $k \in [0, N]_{\mathbb{Z}}$. *Then we have for all* $\lambda \in \mathbb{R}$

$$n_1(\lambda^+) = n_1(\lambda) \leq L, \tag{5.294}$$

$$n_2(\lambda^+) = n_2(\lambda) < \infty, \tag{5.295}$$

$$n_2(\lambda^+) - n_2(\lambda^-) = n_1(\lambda^+) - n_1(\lambda^-) \geq 0, \tag{5.296}$$

*and there exists* $m \in [0, L]_{\mathbb{Z}}$, *where* $L := \sum_{k=1}^{N} \operatorname{rank} \mathcal{B}_k \leq Nn$, *such that*

$$n_1(\lambda) = n_2(\lambda) + m \quad \text{for all } \lambda \in \mathbb{R}. \tag{5.297}$$

*Moreover, for a suitable* $\lambda_0 < 0$, *we have*

$$n_2(\lambda) \equiv 0 \quad \text{and} \quad n_1(\lambda) \equiv m \quad \text{for all } \lambda \leq \lambda_0. \tag{5.298}$$

*Proof* The statement follows directly from Theorem 5.17. Here we realize that assumption (5.38) is automatically satisfied for system (5.238), since the coefficient matrix $\mathcal{B}_k(\lambda) \equiv \mathcal{B}_k$ does not depend on $\lambda$. Also, in view of Proposition 4.4, the number $n_1(\lambda)$ is in this case bounded from above by the number $L$ defined in the theorem. $\square$

When the weight matrices $\mathcal{W}_k$ are positive definite, we obtain the following more specific oscillation theorem.

**Corollary 5.79 (Global Oscillation Theorem for Linear Dependence on $\lambda$)**
*Consider system* (5.238), *i.e., system* (5.1) *with* (5.239), *with Dirichlet boundary conditions* (5.283). *Suppose that* (5.268) *holds, i.e.,* $\mathcal{W}_k > 0$ *for all* $k \in [0, N]_{\mathbb{Z}}$. *Then we have*

$$n_1(\lambda) = n_2(\lambda) \quad \text{for all } \lambda \in \mathbb{R}, \tag{5.299}$$

*and for a suitable* $\lambda_0 < 0$, *we have* $n_1(\lambda) = n_2(\lambda) \equiv 0$ *for all* $\lambda \leq \lambda_0$.

*Proof* By Corollary 5.69 we know that the positivity assumption (5.268) implies the existence of $\lambda_0 < 0$ such that the functional $\mathcal{F}_0(y) - \lambda \langle y, y \rangle_W > 0$ for all admissible $y = (x, u)$ with (5.283) $x \not\equiv 0$. This means by Theorem 5.31 that the equality in (5.299) holds (i.e., the number $m = 0$ in Theorem 5.78).  □

We make the following remarks about the above oscillation theorems for linear dependence on $\lambda$.

*Remark 5.80*

(i) The second equality in (5.298) measures the "index" of the quadratic functional $\mathcal{F}_0(y, \lambda) := \mathcal{F}_0(y) - \lambda \langle y, y \rangle_W$. More precisely, the number $m = 0$ in Theorem 5.78 if and only if $\mathcal{F}_0(\cdot, \lambda) > 0$ for $\lambda < \lambda_1$, where $\lambda_1$ is the smallest finite eigenvalue of (5.238) with (5.283), compared with Theorem 5.31. Moreover, in general by the methods of Sect. 5.4, we have that for $\lambda < \lambda_1$

$$m = \dim\{x = \{x_k\}_{k=0}^{N+1},\ y = (x, u) \text{ is admissible such that } \mathcal{F}_0(y; \lambda) \leq 0\}.$$

(ii) The eigenvalue problem (5.238) with (5.283) is *self-adjoint* in the following sense. All finite eigenvalues are real, and the finite eigenfunctions corresponding to different finite eigenvalues are orthogonal, as we discuss in Remark 5.75.

(iii) The eigenvalue problem (5.238) with (5.283) is equivalent with the corresponding eigenvalue problem for a $2n(N+1) \times 2n(N+1)$ matrix pencil $\mathbf{A} + \lambda \mathbf{B}$ with the eigenvectors $(u_0, x_1, u_1, \ldots, x_N, u_N, u_{N+1})$ and with the block diagonal matrix

$$\mathbf{B} = \text{diag}\{0, W_0, 0, W_1, \ldots, 0, W_{N-1}, 0, 0\}.$$

We omit here to write down $\mathbf{A}$ explicitly, because it is not used here. Then it follows quite easily from the difference system and the definition of the principal solution $Y^{[0]}(\lambda)$ of (5.238) that

$$\det (\mathbf{A} + \lambda \mathbf{B}) = \det X_{N+1}^{[0]}(\lambda).$$

The property in (5.284) and the references on matrix pencils (cf. [148, Ch. XII] or [61, 318, 319]) imply that $r_{N+1}$ is the *normal rank* of the pencil, that our notion of a *finite eigenvalue* (or zeros) coincides with the corresponding notion for pencils, and, particularly, that the assumption (A2) of [55], i.e., $\det X_{N+1}^{[0]} \not\equiv 0$, means that the pencil is *regular*. Hence, by omitting this assumption, we consider the *singular* case. We also want to mention here that all the *minimal indices* of our special matrix pencil (occurring in the Kronecker canonical form, cf. [148, Ch. XII] or [71, 157, 318]), equal to zero; see also [148, Section XII.6]. This fact does simplify the Kronecker canonical form of the pencil considerably, but it is not used directly further on.

(iv) The concept of eigenvalues and eigenvectors for (5.1) of the main reference [55], i.e., $\lambda$ is an eigenvalue if and only if $\det X_{N+1}^{[0]}(\lambda) = 0$, stems from the continuous eigenvalue problems for linear Hamiltonian differential systems (1.103). If $\det X_{N+1}^{[0]}(\lambda) \not\equiv 0$, i.e., if (A2) of [55] holds, then as mentioned above, the corresponding matrix pencil is regular, and the definitions in (5.284) and in [55] coincide. But if the pencil is singular, then the definition of eigenvalues in [55] is not appropriate anymore. Instead, the concept of finite eigenvalues from (5.284) is the right one for the singular case as one can see, and this concept stems from the theory of matrix pencils. Actually, there exists also the notion of *infinite eigenvalues* (or zeros) in the theory of matrix pencils (cf. [148] or [319]), but it does not play any role here. In particular, the geometric meaning of the concept of finite eigenvalues as formulated in Remark 5.75 depends on the special structure (e.g., self-adjointness) of the corresponding matrix pencil, where the assumption $\mathcal{W}_k \geq 0$ plays an important role.

(v)  We conclude this remark by pointing out some possible applications of formula (5.299); see also [55, Remark 3(iv)]. Let $\lambda_0 \in \mathbb{R}$ be given. If we want to know how many finite eigenvalues of (5.1) are less than or equal to $\lambda_0$, we can calculate recursively the principal solution $Y^{[0]}(\lambda)$ at $k = 0$ of (5.238) and determine the number of finite eigenvalues (zeros) of (5.238), (5.283), resp., of $X_{N+1}^{[0]}(\lambda)$, that are less than or equal to $\lambda_0$. However, $X_{N+1}^{[0]}(\lambda)$ is a polynomial, and it might be difficult to calculate this number. Alternatively, if the number $m$ as discussed above is known, then by Theorem 5.78 we could calculate the principal solution at $k = 0$ of (5.238) for the particular $\lambda_0$ in question and count the number of its focal points in the interval $(0, N + 1]$. This procedure could possibly lead to a numerical algorithm to treat the algebraic eigenvalue problem (5.238) with (5.283) also in this singular case, although it is well known that singular matrix pencils have in general ill-posed eigenstructure (cf. [318, 319] or [71, pg. 180]). For the numerical treatment of the algebraic eigenvalue problem for symmetric, banded matrices via Sturm-Liouville difference equations (note that this is a very special case of (5.238)), the theory shows that $\det X_k^{[0]}(\lambda)$ for a "Sturmian chain," which may be used similarly as for treating symmetric tridiagonal matrices; see [206] and [212].

## 5.4  Variational Description of Finite Eigenvalues

In this section we present further results related to the eigenvalue problems for system (5.1) with linear dependence on the spectral parameter.

In linear algebra, it is well known that the eigenvalues of a real symmetric matrix can be calculated as the minimum of the associated quadratic form over an appropriate space of vectors. More precisely, let $A$ be a real symmetric $n \times n$ matrix and $\lambda_1 \leq \cdots \leq \lambda_n$ be its eigenvalues, each eigenvalue repeated according to its multiplicity. Denote by $y_1, \ldots, y_n$ the corresponding system of eigenvectors.

Then the eigenvalues of $A$ satisfy the formula

$$\lambda_m = \min\left\{\frac{\langle Ay, y\rangle}{\langle y, y\rangle},\ y \perp y_k,\ k = 1, \ldots, m-1,\ y \neq 0\right\}. \tag{5.300}$$

The main result of this section presents a similar formula for the finite eigenvalues of (5.1) with (5.283), resp., with (5.245) or with (5.242). First we need some preliminary computations, which may be interesting regardless their later application.

### 5.4.1   Extended Picone Identity

In formula (2.67) in Lemma 2.32, we derived the Picone identity, which presents the quadratic functional $\mathcal{F}(y)$ as a sum of "squares." In the following statement, we extend this result to the quadratic functional $\mathcal{F}_0(y, \lambda)$ by incorporating the finite eigenfunctions of (5.1) with (5.283).

**Theorem 5.81 (Extended Picone's Identity)** *Suppose that $Y$ is a conjoined basis of the symplectic system (5.1) for a fixed $\lambda \in \mathbb{R}$. Let $Q$ be symmetric with $QX = UX^\dagger X$, and define $M$, $T$, $P$ by (4.1) and (4.2). Let $\lambda_1, \ldots, \lambda_m$ be the finite eigenvalues of (5.1) and (5.283) with the corresponding orthonormal finite eigenfunctions $y^{(j)} = (x^{(j)}, u^{(j)})$, $1 \leq j \leq m$. Let $\beta_1, \ldots, \beta_m \in \mathbb{R}$ and define*

$$\hat{y} := \sum_{j=1}^{m} \beta_j y^{(j)}.$$

*Finally, suppose that $y = (x, u)$ is admissible, i.e., $x_{k+1} = \mathcal{A}_k x_k + \mathcal{B}_k u_k$ for $k \in [0, N]_{\mathbb{Z}}$, put $\tilde{y} := y + \hat{y}$ and $\tilde{z} := \tilde{u} - Q\tilde{x}$, and assume that*

$$y \perp \beta_j \lambda_j y^{(j)}, \quad i.e., \quad \left\langle y,\ \beta_j \lambda_j y^{(j)} \right\rangle = 0, \quad \text{for all } 1 \leq j \leq m \tag{5.301}$$

*and that*

$$\tilde{x}_k \in \operatorname{Im} X_k \quad \text{for all } k \in [1, N]_{\mathbb{Z}}. \tag{5.302}$$

*Then we have that*

$$\mathcal{F}_0(y, \lambda) - x_k^T u_k \Big|_{k=0}^{N+1} = \sum_{k=0}^{N} \tilde{z}_k^T P_k \tilde{z}_k + \sum_{j=1}^{m} (\lambda - \lambda_j) \beta_j^2$$

$$+ \tilde{x}_k^T Q_k \tilde{x}_k \Big|_{k=0}^{N+1} - \tilde{x}_k^T \tilde{u}_k \Big|_{k=0}^{N+1}. \tag{5.303}$$

*Proof* First we note that since the finite eigenfunctions $y^{(j)}$ satisfy the boundary conditions (5.283), it follows that $\hat{x}_0 = 0 = \hat{x}_{N+1}$, $\tilde{x}_0 = x_0$, and $\tilde{x}_{N+1} = x_{N+1}$. From (2.67) in Lemma 2.32, we obtain

$$\mathcal{F}_0(\tilde{y}, \lambda) = \sum_{k=0}^{N} \left\{ \tilde{x}_{k+1}^T \tilde{Q}_k \tilde{x}_{k+1} - \tilde{x}_k^T \tilde{Q}_k \tilde{x}_k + \tilde{z}_k^T P_k \tilde{z}_k \right\}$$

$$= \tilde{x}_k^T Q_k \tilde{x}_k \Big|_{k=0}^{N+1} + \sum_{k=0}^{N} \tilde{z}_k^T P_k \tilde{z}_k,$$

because $\tilde{y}$ is admissible and (5.302) holds. Next, by using the recursion in system (5.1) with $\lambda = \lambda_j$ for $1 \le j \le m$ and by (2.61) in Lemma 2.30, we conclude that

$$\mathcal{F}_0(\hat{y}) = \sum_{k=0}^{N} \hat{x}_{k+1}^T \left\{ \mathcal{C}_k \hat{x}_k + \mathcal{D}_k \hat{u}_k - \hat{u}_{k+1} \right\}$$

$$= \sum_{k=0}^{N} \sum_{j=1}^{m} \beta_j \lambda_j \hat{x}_{k+1}^T \mathcal{W}_k x_{k+1}^{(j)} = \sum_{j=1}^{m} \lambda_j \beta_j^2.$$

Moreover, by using $y \perp y^{(j)}$ from (5.301), we have

$$\mathcal{F}_0(y, \hat{y}) = x_k^T \hat{u}_k \Big|_{k=0}^{N+1} + \sum_{k=0}^{N} \sum_{j=1}^{m} \beta_j \lambda_j x_{k+1}^T \mathcal{W}_k x_{k+1}^{(j)} = x_k^T \hat{u}_k \Big|_{k=0}^{N+1}$$

and similarly

$$\mathcal{F}_0(\hat{y}, y) = \hat{x}_k^T u_k \Big|_{k=0}^{N+1} = 0.$$

Altogether, by using $x_0 = \tilde{x}_0$ and $x_{N+1} = \tilde{x}_{N+1}$ and the above calculations, we get

$$\mathcal{F}_0(y, \lambda) - x_k^T u_k \Big|_{k=0}^{N+1} = \mathcal{F}(y) - \lambda \langle y, y \rangle - x_k^T u_k \Big|_{k=0}^{N+1}$$

$$= \mathcal{F}(y) + \mathcal{F}(\hat{y}) + \mathcal{F}(y, \hat{y}) + \mathcal{F}(\hat{y}, y) - \sum_{j=1}^{m} \lambda_j \beta_j^2 - x_k^T \hat{u}_k \Big|_{k=0}^{N+1}$$

$$- \lambda \big( \langle y, y \rangle + \langle \hat{y}, \hat{y} \rangle \big) + \lambda \langle \hat{y}, \hat{y} \rangle - x_k^T u_k \Big|_{k=0}^{N+1}$$

$$= \mathcal{F}(\tilde{y}) - \sum_{j=1}^{m} \lambda_j \beta_j^2 - x_k^T \hat{u}_k \Big|_{k=0}^{N+1} - \lambda \langle \tilde{y}, \tilde{y} \rangle + \lambda \sum_{j=1}^{m} \beta_j^2 - x_k^T u_k \Big|_{k=0}^{N+1}$$

$$= \mathcal{F}(\tilde{y}, \lambda) + \sum_{j=1}^{m} (\lambda - \lambda_j) \beta_j^2 - x_k^T (\hat{u}_k + u_k) \Big|_{k=0}^{N+1}$$

$$= \tilde{x}_k^T Q_k \tilde{x}_k \Big|_{k=0}^{N+1} + \sum_{k=0}^{N} \tilde{z}_k^T P_k \tilde{z}_k + \sum_{j=1}^{m} (\lambda - \lambda_j) \beta_j^2 - \tilde{x}_k^T \tilde{u}_k \Big|_{k=0}^{N+1}.$$

Therefore, the proof of (5.303) is complete.                                                □

We note that the role of assumption (5.302) in Theorem 5.81 is characterized by Lemma 2.47 in Sect. 2.3.5.

*Remark 5.82* Condition (5.301) in Theorem 5.81 is slightly more general than the corresponding condition used in [56, Theorem 4.2]. Here we incorporate the coefficients $\beta_j$ and the finite eigenvalues $\lambda_j$ into the orthogonality condition (5.301), which shows that in the case of $\beta_j = 0$ or $\lambda_j = 0$, this condition is not needed.

### 5.4.2   Rayleigh Principle

In this subsection we extend formula (5.300) to the finite eigenvalues of (5.1) with (5.283), resp., with (5.245) or with (5.242). The main results are essentially based on the validity of the oscillation theorem (Corollary 5.79) saying that

$$n_1(\lambda) = n_2(\lambda), \quad n_1(\lambda^+) = n_1(\lambda), \quad n_2(\lambda^+) = n_2(\lambda) \quad \text{for all } \lambda \in \mathbb{R}.$$
$$(5.304)$$

The latter reference states that the conditions in (5.304) are equivalent to the positivity of the quadratic functional $\mathcal{F}(\cdot, \lambda_0)$ in (5.273) for some $\lambda_0 < 0$. For practical applications, it will be convenient to guarantee this positivity assumption by conditions imposed on the coefficients of system (5.1).

If $\lambda_1 \leq \ldots \leq \lambda_m$ are the finite eigenvalues of (5.1) with (5.283), including their multiplicities, then we put for convenience $\lambda_0 := -\infty$ and $\lambda_{m+1} := \infty$. Note that $m \leq nN < \infty$ by (5.288) in Remark 5.75.

**Theorem 5.83 (Rayleigh Principle for Dirichlet Boundary Conditions)** *Assume that (5.239), (5.240), and (5.273) hold and*

$$\text{there exists } \lambda_0 < 0 \text{ such that } \mathcal{F}(\cdot, \lambda_0) > 0. \quad (5.305)$$

*Let $\lambda_1 \leq \lambda_2 \leq \cdots \leq \lambda_m$ denote the finite eigenvalues of (5.1) with (5.283), with the corresponding orthonormal finite eigenfunctions $y^{(j)}$, $1 \leq j \leq m$, with respect to* the inner product (5.287). Then for every $0 \leq j \leq m$, we have

$$\lambda_{j+1} = \min \left\{ \frac{\mathcal{F}(y, 0)}{\langle y, y \rangle}, \; y = (x, u) \text{ is admissible with } x_0 = 0 = x_{N+1}, \right.$$
$$\left. y \perp y^{(1)}, \dots, y^{(j)}, \text{ and } \{W_k x_{k+1}\}_{k=0}^{N-1} \neq 0 \right\}.$$

(5.306)

Note that we include the case of $j = 0$, where the orthogonality condition on $y$ becomes empty, as well as the case of $j = m$, where $\lambda_{m+1} = \infty$.

*Proof of Theorem 5.83* Let $0 \leq j \leq m$ be fixed, and choose $\lambda \in (\lambda_j, \lambda_{j+1})$. Then, by (5.304), we have $n_2(\lambda) = n_1(\lambda) = j$, so that the principal solution $Y(\lambda) := Y^{[0]}(\lambda)$ at $k = 0$ possesses exactly $j$ focal points in the interval $(0, N + 1]$. The fact that $N + 1$ is actually not a focal point of $Y(\lambda)$ follows from assumption (5.305), Proposition 5.65, and Theorem 2.36, which imply that rank $M_k(\tilde{\lambda}) = 0$ for all $\tilde{\lambda} \leq \lambda_0$. Hence,

$$\text{rank } M_k(\lambda^+) = \text{rank } M_k(\lambda) = 0 \quad \text{for all } k \in [0, N]_{\mathbb{Z}}, \quad \lambda \in (\lambda_m, \lambda_{m+1}).$$

In particular, we have rank $M_N(\lambda) = 0$. First, we apply the extended Picone identity (Theorem 5.81) to $y = 0$, which yields

$$\tilde{y} = \hat{y} = \sum_{i=1}^{m} \beta_i y^{(i)} \quad \text{with } \tilde{x}_0 = 0 = \tilde{x}_{N+1}, \; x_0 = 0 = x_{N+1}.$$

Suppose that $\beta_1, \dots, \beta_j$ satisfy the $j$ linear homogeneous equations

$$\left.\begin{array}{l} M_k^T \tilde{x}_{k+1} = 0, \quad k \in [0, N-1]_{\mathbb{Z}}, \\ \tilde{z}_k \perp \{\alpha \in \mathbb{R}^n, \; \alpha \text{ is an eigenvector corresponding} \\ \quad \text{to a negative eigenvalue of } P_k\}, \quad k \in [0, N]_{\mathbb{Z}}, \end{array}\right\} \quad (5.307)$$

where $\tilde{z}_k = \tilde{u}_k - Q_k \tilde{x}_k$. Note that the number of these equations is equal to the number of of focal points of $Y$ in the open interval $(0, N+1)$. Then, by Lemma 2.48, assumption (5.302) holds, and we obtain from Theorem 5.81 that

$$0 = \mathcal{F}(y, \lambda) = \sum_{k=0}^{N} \tilde{z}_k^T P_k \tilde{z}_k + \sum_{i=1}^{j} (\lambda - \lambda_i) \beta_i^2, \quad (5.308)$$

where $\tilde{z}_k^T P_k \tilde{z}_k \geq 0$ for $k \in [0, N]_{\mathbb{Z}}$ by (5.307) and $\lambda - \lambda_i \geq \lambda - \lambda_j > 0$ for $1 \leq i \leq j$. Hence, (5.308) implies that $\beta_1 = \cdots = \beta_j = 0$, so that (5.307) possesses only the trivial solution. Thus we have shown that

the coefficient matrix corresponding to system (5.307) is nonsingular. (5.309)

Now suppose that $y = (x, u)$ is admissible with $x_0 = 0 = x_{N+1}$ and $y \perp y^{(i)}$ for $1 \leq i \leq j$. According to (5.309), there exists a unique set of constants $\beta_1, \dots, \beta_j$ such that the function $\tilde{y} := y + \hat{y} = y + \sum_{i=1}^{j} \beta_i y^{(i)}$ satisfies the $j$ linear inhomogeneous equations defined by (5.307). Consequently, Theorem 5.81 implies that

$$\mathcal{F}(y, \lambda) = \mathcal{F}(y, \lambda) - x_k^T u_k \Big|_{k=0}^{N+1}$$

$$= \sum_{k=0}^{N} \tilde{z}_k^T P_k \tilde{z}_k + \sum_{i=1}^{j} (\lambda - \lambda_i) \beta_i^2 \geq \sum_{i=1}^{j} (\lambda - \lambda_i) \beta_i^2 \geq 0.$$

This shows that

$$\mathcal{F}(y) \geq \lambda \langle y, y \rangle \quad \text{for all } \lambda \in (\lambda_j, \lambda_{j+1}). \tag{5.310}$$

Hence, $\mathcal{F}(y) \geq \lambda_{j+1} \langle y, y \rangle$ by the continuity and taking $\lambda \to \lambda_{j+1}^-$ in (5.310). Moreover, we have $\mathcal{F}(y) = \lambda_{j+1} \langle y, y \rangle$ for $y = y^{(j+1)}$. For multiple finite eigenvalues, use the fact that $\mathcal{F}(y, \lambda) = \mathcal{F}(y + \hat{y}, \lambda)$ if $y$ is admissible with $x_0 = 0 = x_{N+1}$ and if $\hat{y}$ is a finite eigenvector corresponding to $\lambda$. Hence the assertion follows. □

In later applications in Sect. 5.5, we will use the statement of Theorem 5.83 under stronger positivity assumption (5.268) instead of (5.305), i.e., under $\mathcal{W}_k > 0$ for all $k \in [0, N]_{\mathbb{Z}}$.

**Corollary 5.84** *Assume that* (5.239), (5.240), (5.246), (5.268), *and* (5.273) *hold. Then the finite eigenvalues* $\lambda_1 \leq \cdots \leq \lambda_m$ *of* (5.1) *with* (5.283) *satisfy for every* $0 \leq j \leq m$ *the equality in* (5.306), *in which we consider admissible* $y$ *with* $x_0 = 0 = x_{N+1}$, $y \perp y^{(1)}, \dots, y^{(j)}$, *and* $\{x_k\}_{k=1}^{N} \neq 0$.

*Proof* The result follows from Theorem 5.83 and Corollary 5.69. □

Now we consider the Rayleigh principle for the eigenvalue problem (5.1) with separated boundary conditions (5.245). Here we use the quadratic functional $\mathcal{G}(\cdot, \lambda)$ given in (5.262). We note that by (5.293) the total number $m$ of finite eigenvalues of (5.1) with (5.245) satisfies $m \leq (N + 2) n$.

**Theorem 5.85 (Rayleigh Principle for Separated Boundary Conditions)** *Assume that* (5.239), (5.240), (5.246), (5.247), (5.248), *and* (5.262) *hold and*

$$\text{there exists } \lambda_0 < 0 \text{ such that } \mathcal{G}(\cdot, \lambda_0) > 0. \tag{5.311}$$

*Let* $\lambda_1 \leq \lambda_2 \leq \cdots \leq \lambda_m$ *denote the finite eigenvalues of* (5.1) *with* (5.245), *with the corresponding orthonormal finite eigenfunctions* $y^{(j)}$, $1 \leq j \leq m$, *with respect to* *the inner product (5.292). Then for every $0 \leq j \leq m$, we have*

$$\lambda_{j+1} = \min \left\{ \frac{\mathcal{G}(y, 0)}{\langle y, y \rangle}, \; y = (x, u) \text{ is admissible with (5.263),} \right.$$
$$y \perp y^{(1)}, \ldots, y^{(j)}, \text{ and } \{\mathcal{W}_k \, x_{k+1}\}_{k=-1}^{N-1} \neq 0$$
$$\left. \text{or } (\mathcal{W}_N + \mathcal{W}_{N+1}) \, x_{N+1} \neq 0 \right\}. \tag{5.312}$$

*Proof* This statement follows from Theorem 5.83 upon applying it to the equivalent extended eigenvalue problem on the interval $[-1, N + 2]_{\mathbb{Z}}$ with the Dirichlet boundary conditions $x_{-1}(\lambda) = 0 = x_{N+2}(\lambda)$. This equivalence is discussed in details in Theorem 5.41 for the two eigenvalue problems and in Lemma 5.47 for the associated quadratic functionals. □

Assumption (5.311) can be guaranteed by several conditions described in Propositions 5.66 and 5.67 and in Remark 5.68. Condition (5.311) can also be tested by the properties of the natural conjoined basis $\bar{Y}(\lambda_0)$ of system (5.238) with $\lambda := \lambda_0$, which are presented in Theorem 2.50.

**Corollary 5.86** *Assume that (5.239), (5.246), (5.270), and (5.262) hold. Then the finite eigenvalues $\lambda_1 \leq \cdots \leq \lambda_m$ of (5.1) with (5.245) satisfy for every index $0 \leq j \leq m$ the equality in (5.312), in which we consider admissible $y$ with (5.263), $y \perp y^{(1)}, \ldots, y^{(j)}$, and $\{x_k\}_{k=0}^{N+1} \neq 0$.*

*Proof* The result follows from Theorem 5.85 and Remark 5.68. □

When $\mathcal{W}_{-1} = 0$, $\mathcal{W}_{N+1} = 0$, and $\mathcal{W}_k > 0$ for all $k \in [0, N]_{\mathbb{Z}}$, we obtain another generalization of Corollary 5.84 to the separated boundary conditions

$$R_0^* \, x_0(\lambda) + R_0 \, u_0(\lambda) = 0, \quad R_{N+1}^* \, x_{N+1}(\lambda) + R_{N+1} \, u_{N+1}(\lambda) = 0, \tag{5.313}$$

which in this case do not depend on the spectral parameter $\lambda$. In this case the inner product in (5.292) reduces to the traditional expression

$$\langle y, \tilde{y} \rangle = \sum_{k=0}^{N} x_{k+1}^T \mathcal{W}_k \, \tilde{x}_{k+1}. \tag{5.314}$$

**Corollary 5.87** *Assume that (5.239), (5.246), (5.268), (5.269), and (5.267) hold. Then the finite eigenvalues $\lambda_1 \leq \cdots \leq \lambda_m$ of (5.1) with (5.313) satisfy for every index $0 \leq j \leq m$ the equality in (5.312), in which we consider inner product (5.314) and admissible $y$ with (5.263), $y \perp y^{(1)}, \ldots, y^{(j)}$, and $\{x_k\}_{k=0}^{N+1} \neq 0$.*

*Proof* The result follows from Theorem 5.85 and Proposition 5.67. □

As a final result in this section, we consider the Rayleigh principle for the eigenvalue problem consisting of system (5.1) with the joint endpoints (5.242).

**Theorem 5.88 (Rayleigh Principle for Joint Boundary Conditions)** *Assume that* (5.239), (5.240), (5.243), (5.244), *and* (5.260) *hold and*

$$\text{there exists } \lambda_0 < 0 \text{ such that } \mathcal{G}(\cdot, \lambda_0) > 0. \tag{5.315}$$

*Let* $\lambda_1 \leq \lambda_2 \leq \cdots \leq \lambda_m$ *denote the finite eigenvalues of* (5.1) *with* (5.242), *with the corresponding orthonormal finite eigenfunctions* $y^{(j)}$, $1 \leq j \leq m$, *with respect to the inner product* (5.279). *Then for every* $0 \leq j \leq m$, *we have*

$$\lambda_{j+1} = \min \left\{ \frac{\mathcal{G}(y, 0)}{\langle y, y \rangle}, \ y = (x, u) \text{ is admissible with (5.261)}, \atop y \perp y^{(1)}, \dots, y^{(j)}, \text{ and } \{\mathcal{W}_k \, x_{k+1}\}_{k=0}^{N-1} \neq 0 \atop \text{or } \mathfrak{W} \begin{pmatrix} x_0 \\ x_{N+1} \end{pmatrix} \neq 0 \right\}. \tag{5.316}$$

*Proof* This statement follows from Theorem 5.85 upon applying it to the equivalent augmented eigenvalue problem on the interval with the separated. This equivalence is discussed in details in Theorem 5.44 for the two eigenvalue problems.               □

Assumption (5.315) can also be tested by the properties of the principal solution $Y^{[0]}(\lambda_0)$ of system (5.238) with $\lambda := \lambda_0$, which are presented in Theorem 2.53.

## 5.5  Applications of Oscillation Theorems

In this section we present results on the numbers the focal points of conjoined bases of one or two symplectic systems. In the derivation of these results, we utilize the eigenvalue theory and the oscillation theorems for boundary value problems for symplectic systems depending linearly on the spectral parameter, i.e., essentially based on the results in Sects. 5.3–5.4.

### 5.5.1  Sturmian Comparison and Separation Theorems

In this subsection we present an alternative approach to the Sturmian comparison and separation theorems based on the variational techniques from Sect. 5.4. Comparison theory for symplectic difference systems has been already treated in Sect. 4.3 (see Corollary 4.57). There, the main tool was the relationship between the comparative index and the multiplicity of a focal point. Here we present a proof based on the eigenvalue problem (5.1) and the associated discrete quadratic functional.

We recall from Sects. 2.3.2 and 5.3.4 the quadratic functionals

$$\mathcal{F}(y) = \sum_{k=0}^{N} \left\{ x_k^T \mathcal{A}_k^T \mathcal{C}_k x_k + 2 x_k^T \mathcal{C}_k^T \mathcal{B}_k u_k + u_k^T \mathcal{B}_k^T \mathcal{D}_k u_k \right\},$$

$$\mathcal{F}(y, \lambda) = \mathcal{F}(y) - \lambda \langle y, y \rangle,$$

where $y = (x, u)$ is admissible, i.e., $x_{k+1} = \mathcal{A}_k x_k + \mathcal{B}_k u_k$ for $k \in [0, N]_{\mathbb{Z}}$, and $x_0 = 0 = x_{N+1}$. We will utilize the symmetric $2n \times 2n$ matrices $\mathcal{G}_k$ and the $n \times n$ matrices $\mathcal{E}_k$ defined in (2.64), which are used to write the functional $\mathcal{F}(y)$ in terms of the $x$ component of $y$ only; see (2.63). With systems (SDS) and (5.1), we consider another symplectic systems

$$\hat{y}_{k+1} = \hat{\mathcal{S}}_k \hat{y}_k, \quad k \in [0, N]_{\mathbb{Z}}, \tag{5.317}$$

and

$$\hat{y}_{k+1}(\lambda) = \hat{\mathcal{S}}_k(\lambda) \hat{y}_k(\lambda), \quad k \in [0, N]_{\mathbb{Z}}, \tag{5.318}$$

with the block structure as in (5.239) for the coefficient matrices $\mathcal{S}_k(\lambda)$ and $\hat{\mathcal{S}}_k(\lambda)$. In this case we define the corresponding matrices $\hat{\mathcal{G}}$ and $\hat{\mathcal{E}}_k$ and the quadratic functionals $\hat{\mathcal{F}}(\hat{y})$ and $\hat{\mathcal{F}}(\hat{y}, \lambda)$ analogously to (2.64) and (5.273). Finally, in this section we suppose that the weight matrices satisfy

$$\mathcal{W}_k = I = \hat{\mathcal{W}}_k, \quad k \in [0, N]_{\mathbb{Z}}, \tag{5.319}$$

so that the number $m = 0$ in formula (5.297), i.e.,

$$n_1(\lambda) = n_2(\lambda) \quad \text{for all } \lambda \in \mathbb{R} \tag{5.320}$$

for system (SDS), as well as for system (5.317). In particular, the Rayleigh principle in Corollary 5.84 holds for both systems (5.1) and (5.318) with the Dirichlet boundary conditions.

Two following two results represent Sturmian-type comparison theorems for symplectic systems (SDS) and (5.317).

**Theorem 5.89 (Sturmian Comparison Theorem)** *Suppose that*

$$\mathcal{G}_k \geq \hat{\mathcal{G}}_k \text{ and } \mathrm{Im}\,(\mathcal{A}_k - \hat{\mathcal{A}}_k, \quad \mathcal{B}_k) \subseteq \mathrm{Im}\,\hat{\mathcal{B}}_k \tag{5.321}$$

*for all $k \in [0, N]_{\mathbb{Z}}$. If the principal solution of (5.317) at $k = 0$ has $q$ focal points in $(0, N + 1]$, then any conjoined basis of (SDS) has at most $q + n$ focal points in $(0, N + 1]$.*

*Proof* Let $Y$ be a conjoined basis of (SDS) and suppose that it has $p$ focal points in $(0, N + 1]$. Then, as in Proposition 2.38 and the proof of Theorem 4.16, we can construct for each focal point of $Y$ the corresponding admissible $y^{[j]} = (x^{[j]}, u^{[j]})$ such that $x_{N+1}^{[j]} = 0$ for all $1 \leq j \leq p$. Furthermore, since the principal solution $Y^{[0]}$ of (5.317) at $k = 0$ is assumed to have $q$ focal points in $(0, N+1]$, we know by Corollary 5.79 with $\lambda = 0$ applied to the eigenvalue problem (5.318) with $\mathcal{W} = I$ that this eigenvalue problem has $q$ nonpositive finite eigenvalues $\lambda_i$, $1 \leq i \leq q$, with the corresponding orthonormal finite eigenvectors $\hat{y}^{(i)} = (\hat{x}^{(i)}, \hat{u}^{(i)})$, $1 \leq i \leq q$. Moreover, by Corollary 5.84, we have $\mathcal{F}(y) \geq \hat{\mathcal{F}}(y) > 0 \cdot \langle y, y \rangle = 0$ for $(\hat{\mathcal{A}}, \hat{\mathcal{B}})$-admissible $y = (x, u)$ satisfying

$$y \perp \hat{y}^{(i)}, \quad \text{i.e.,} \quad \left\langle y, \hat{y}^{(i)} \right\rangle = \sum_{k=0}^{N} x_{k+1}^{T} \hat{x}_{k+1}^{(i)} = 0, \quad 1 \leq i \leq q, \quad x \neq 0.$$

Now suppose that $p > q + n$. Then there exists a nontrivial linear combination

$$\sum_{j=1}^{p} c_j \begin{pmatrix} \left\langle y^{[j]}, \hat{y}^{(1)} \right\rangle \\ \left\langle y^{[j]}, \hat{y}^{(2)} \right\rangle \\ \vdots \\ \left\langle y^{[j]}, \hat{y}^{(q)} \right\rangle \\ x_0^{[j]} \end{pmatrix} = 0.$$

Define

$$y = (x, u) = \sum_{j=1}^{p} c_j y^{[j]}.$$

By construction, $x_{N+1} = 0$ and $y$ is admissible. Moreover, $\sum_{j=1}^{p} c_j x_0^{[j]} = 0$ implies that $x_0 = 0$, as well as

$$0 = \sum_{j=1}^{p} c_j \left\langle y^{[j]}, \hat{y}^{(i)} \right\rangle = \left\langle y, \hat{y}^{(i)} \right\rangle \quad \text{for all } 1 \leq i \leq q,$$

so that $y \perp \hat{y}^{(i)}$ for all $1 \leq i \leq q$. As in the proof of Theorem 4.16 we have that $x \neq 0$ (since not all $c_j = 0$) and $\mathcal{F}(z) \leq 0$. From the second condition in (5.321), it follows that there exists $\hat{u} = \{\hat{u}_k\}_{k=0}^{N}$ such that $x_{k+1} = \mathcal{A}_k x_k + \mathcal{B}_k \hat{u}_k$ for all $k \in [0, N]_{\mathbb{Z}}$, and hence $\hat{y} := (x, \hat{u})$ is admissible for $\hat{\mathcal{F}}$. By using the representation of quadratic functional $\hat{\mathcal{F}}$ via the matrix $\hat{\mathcal{G}}$, as in (2.63), coupled with the first condition in (5.321), we find that

$$\hat{\mathcal{F}}(\hat{y}) = \sum_{k=0}^{N} \begin{pmatrix} x_k \\ x_{k+1} \end{pmatrix}^{T} \hat{\mathcal{G}}_k \begin{pmatrix} x_k \\ x_{k+1} \end{pmatrix} \leq \sum_{k=0}^{N} \begin{pmatrix} x_k \\ x_{k+1} \end{pmatrix}^{T} \mathcal{G}_k \begin{pmatrix} x_k \\ x_{k+1} \end{pmatrix} = \mathcal{F}(y) \leq 0.$$

Hence we have found an $(\hat{A}, \hat{B})$-admissible $y = (x, u)$ with $x \not\equiv 0$, $x_0 = 0 = x_{N+1}$, $y \perp \hat{y}^{(i)}$ for all $1 \le i \le q$, and $\hat{\mathcal{F}}(y) \le 0$. This contradicts the Rayleigh principle (Corollary 5.84), by which $\hat{\mathcal{F}}(y) > 0$ for all admissible $y$ with $x_0 = 0 = x_{N+1}$, $x \not\equiv 0$, and $y \perp \hat{y}^{(i)}$ for all $1 \le i \le q$. Therefore, we must have $p \le q + n$, which proves the result. $\qquad\square$

**Theorem 5.90 (Sturmian Comparison Theorem)** *Suppose that* (5.321) *holds. If the principal solution of* (SDS) *at* $k = 0$ *has* $q$ *focal points in* $(0, N + 1]$, *then any conjoined basis of* (5.317) *has at least* $q$ *focal points in* $(0, N + 1]$.

*Proof* Let $Y^{[0]}$ be the principal solution of the (SDS) at $k = 0$, and let $\hat{Y}$ be a conjoined basis of (5.317). For any $\lambda \in \mathbb{R}$, let $Y^{[0]}(\lambda)$ be the principal solution of (5.1) at $k = 0$, and let $\hat{Y}(\lambda)$ be the conjoined basis of (5.318) such that $\hat{Y}_0(\lambda) \equiv \hat{Y}_0$, so that these initial conditions are constant in $\lambda \in \mathbb{R}$. We denote by $n_1(\lambda)$ and $p(\lambda)$ the numbers of focal points of $Y^{[0]}(\lambda)$ and $\hat{Y}(\lambda)$ in the interval $(0, N + 1]$. The assumptions then imply that $q = n_1(0)$, and we need to prove that $n_1(0) \le p(0)$. We will show that

$$n_1(\lambda) \le p(\lambda) \quad \text{for all } \lambda \in \mathbb{R}. \tag{5.322}$$

By the oscillation theorem (Corollary 5.79), there are exactly $n_2(0) = n_1(0) = q$ nonpositive finite eigenvalues of (5.1), (5.283). We denote by $\lambda_i$ for $1 \le i \le r$ all the finite eigenvalues of (5.1), (5.283) with the corresponding orthonormal eigenfunctions $y^{(i)}$, $1 \le i \le r$ (so that $q \le r$). Fix now $\lambda \in \mathbb{R}$. Then we have

$$\lambda \in [\lambda_m, \lambda_{m+1}) \quad \text{for some } m \in \{0, \dots, r\},$$

where we put $\lambda_0 := -\infty$ and $\lambda_{r+1} := \infty$. By (5.320), this means that $n_1(\lambda) = n_2(\lambda) = m$. First suppose that $\lambda$ is not a finite eigenvalue of (5.1), (5.283) so that $\lambda \in (\lambda_m, \lambda_{m+1})$. Let $\tilde{p}$ be the number of focal points of $\hat{Y}(\lambda)$ in the open interval $(0, N + 1)$ so that $\tilde{p} \le p(\lambda)$. Put $\tilde{y} = (\tilde{x}, \tilde{u}) := \sum_{j=1}^{m} \beta_j y^{(j)}$, where the constants $\beta_1, \dots, \beta_j$ are chosen in such a way that $\tilde{y}$ satisfies $\tilde{p}$ linear homogeneous conditions

$$\left.\begin{aligned}
&\hat{M}_k^T(\lambda)\,\tilde{x}_{k+1} = 0, \quad k \in [0, N-1]_{\mathbb{Z}}, \\
&\tilde{z}_k \perp \{\alpha \in \mathbb{R}^n, \ \alpha \text{ is an eigenvector corresponding} \\
&\qquad\qquad \text{to a negative eigenvalue of } \hat{P}_k(\lambda)\}, \quad k \in [0, N]_{\mathbb{Z}}
\end{aligned}\right\} \tag{5.323}$$

where

$$\tilde{z}_k = \tilde{u}_k - \hat{U}_k(\lambda)\,\hat{X}_k^\dagger(\lambda)\,\tilde{x}_k = \sum_{j=1}^{m} \beta_j \big[\tilde{u}_k^{(j)} - \hat{U}_k(\lambda)\,\hat{X}_k^\dagger(\lambda)\,\tilde{x}_k^{(j)}\big]$$

and where the matrices $\hat{M}_k(\lambda)$ and $\hat{P}_k(\lambda)$ are given by (5.40) with $\hat{Y}(\lambda)$ instead of $Y(\lambda)$. The sequence $\tilde{y}$ is admissible for $\mathcal{F}$, and by the second condition in (5.321), there exists $\hat{u}$ such that the pair $\hat{y} := (\tilde{x}, \hat{u})$ is admissible for $\hat{\mathcal{F}}$. Since the value of the quadratic functional does not depend on the second component of an admissible sequence $y = (x, u)$, see (2.63), we can write also $\tilde{y} = (\tilde{x}, \hat{u})$. Then by the extended Picone identity (Theorem 5.81), we have

$$\hat{\mathcal{F}}(\tilde{y}, \lambda) := \hat{\mathcal{F}}(\tilde{y}) - \lambda \langle \tilde{y}, \tilde{y} \rangle = \sum_{k=0}^{N} \tilde{z}_k^T \hat{P}_k \tilde{z}_k \geq 0.$$

We note that the first condition in (5.323) implies that $\tilde{x}_k \in \operatorname{Im} \hat{X}_k$ for all $k \in [0, N+1]_{\mathbb{z}}$ by Lemma 2.48, and hence Theorem 5.81 can be applied. At the same time, by a direct computation using the orthonormality of $y^{(1)}, \ldots, y^{(m)}$, we have

$$\mathcal{F}_\lambda(\tilde{y}) := \mathcal{F}(\tilde{y}) - \lambda \langle \tilde{y}, \tilde{y} \rangle = \sum_{j=1}^{m} (\lambda_j - \lambda) \beta_j^2.$$

Now again by (2.63) and the first condition in (5.321), we have

$$\sum_{j=1}^{m} (\lambda_j - \lambda) \beta_j^2 \geq \sum_{k=0}^{N} \tilde{z}_k^T \hat{P}_k(\lambda) \tilde{z}_k \geq 0,$$

since $\lambda > \lambda_j$ for all $1 \leq j \leq m$. This is however possible only if $\beta_j = 0$ for all $1 \leq j \leq m$. This means that the system (5.323) of $\tilde{p}$ linear homogeneous conditions has only the trivial solution, and hence the number of conditions $\tilde{p}$ is greater than or equal to the number of parameters $\beta_j$ (which is equal to $q$). This proves the required statement when $\lambda$ is not a finite eigenvalue of (5.1), (5.283). Finally, since the functions $n_1(\lambda)$ and $p(\lambda)$ are right-continuous, by letting $\lambda \to \lambda_m^+$, we obtain the statement also in the case when $\lambda = \lambda_m$ is a finite eigenvalue of (5.1), (5.283). The proof is complete.    □

When the two systems (SDS) and (5.317) are the same, then the assumptions in (5.321) are trivially satisfied, and we obtain from Theorems 5.89 and 5.90 the following, compare with Sect. 4.2.3 and in particular formula (4.68), where a more precise estimate for the number of focal points was derived by using the comparative index.

**Corollary 5.91 (Sturmian Separation Theorem)** *Suppose that the principal solution of* (SDS) *at* $k = 0$ *has* $q$ *focal points in* $(0, N+1]$. *Then any conjoined basis of* (SDS) *has at at least* $q$ *and at most* $q + n$ *focal points in* $(0, N+1]$.

## 5.5.2  Modifications of Global Oscillation Theorem

In this subsection we present some modifications of the global oscillation theorem (Theorem 5.78). These extensions are obtained using the number $l^*(Y, M, N)$ of the backward focal points in the interval $[M, N)$ of the conjoined basis $Y$ (see Definition 4.3 and Sect. 4.2.3). The first application of the concept of backward focal points (which is the same as "focal points" of the reversed symplectic system (2.47)) consists in replacing the quantity

$$n_1(\lambda) = n_1(Y^{[0]}(\lambda)) := l(Y^{[0]}(\lambda), 0, N + 1)$$

by the quantity

$$n_1^*(Y^{[N+1]}(\lambda)) := l^*(Y^{[N+1]}(\lambda), 0, N + 1), \tag{5.324}$$

where $l^*(Y^{[N+1]}(\lambda), 0, N + 1)$ is number of backward focal points of the principal solution of (5.238) at $N + 1$ in $[0, N + 1)$; see Theorem 4.34. Hence, as a corollary of Theorem 4.34, we obtain the next statement.

**Corollary 5.92** *Under the assumptions of Theorem 5.78, the number $n_1(\lambda)$ in (5.294), (5.296), (5.297), (5.298) can be replaced by $n_1^*(Y^{[N+1]}(\lambda))$ defined in (5.324), where $l^*(Y^{[N+1]}(\lambda), 0, N + 1)$ is the number of backward focal points of the principal solution $Y^{[N+1]}(\lambda)$ of (5.238) at $N + 1$ in $[0, N + 1)$.*

Another possibility how to apply the number of backward focal points in the interval $[M, N)$ is the following modification of the global oscillation theorem to eigenvalues which are *strictly* less than a given value $\lambda \in \mathbb{R}$.

**Theorem 5.93** *Under the assumptions of Theorem 5.78, there exists an index $m^* \in \mathbb{N} \cup \{0\}$ such that*

$$n_1^*(Y^{[0]}(\lambda)) = n_2^o(\lambda) + m^*, \quad n_2^o(\lambda) = \sum_{\mu < \lambda} \theta(\mu), \tag{5.325}$$

*where $\sum_{\mu < \lambda} \theta(\mu)$ is the number of finite eigenvalues of (5.238), (5.283) which are less than $\lambda \in \mathbb{R}$, $Y^{[0]}(\lambda)$ is the principal solution of (5.238) at $k = 0$, and $n_1^*(Y^{[0]}(\lambda)) := l^*(Y^{[0]}(\lambda), 0, N + 1)$ is the number of backward focal points of $Y^{[0]}(\lambda)$ in $[0, N + 1)$.*

*Proof* Let $\lambda_0 < \lambda_{\min} = \min \sigma$, where $\sigma$ is the finite spectrum of (5.238), (5.283). Then

$$n_2^o(\lambda) = n_2(\lambda) - \theta(\lambda) = n_2(\lambda) + \text{rank } X_{N+1}^{[0]}(\lambda)\big|_{\lambda_0}^{\lambda},$$

where $\theta(\lambda)$ is the (algebraic) multiplicity of $\lambda \in \sigma$ and $\theta(\lambda) = 0$ for $\lambda \notin \sigma$. Hence, using Corollary 4.6, we obtain

$$
\begin{aligned}
n_2^o(\lambda) &= l(Y^{[0]}(\lambda), 0, N+1) - l(Y^{[0]}(\lambda_0), 0, N+1) + \operatorname{rank} X_{N+1}^{[0]}(\lambda)\big|_{\lambda_0}^{\lambda} \\
&= l^*(Y^{[0]}(\lambda), 0, N+1) - l^*(Y^{[0]}(\lambda_0), 0, N+1) \\
&= n_1^*(Y^{[0]}(\lambda)) - n_1^*(Y^{[0]}(\lambda_0)),
\end{aligned}
$$

where $l^*(Y, 0, N+1)$ is the number of backward focal points of a given conjoined basis in $[0, N+1)$. If we denote $m^* := n_1^*(Y^{[0]}(\lambda_0))$, then the theorem is proved.
$\square$

Using the equality $l^*(Y^{[0]}(b), 0, N+1) = l(Y^{[N+1]}(b), 0, N+1)$ (see Theorem 4.35), we derive the "dual version" of Theorem 5.93.

**Corollary 5.94 (Global Oscillation Theorem for Linear Dependence on $\lambda$)**
*Under the assumptions of Theorem 5.78, there exists an index $m^* \in \mathbb{N} \cup \{0\}$ such that*

$$
n_1(Y^{[N+1]}(\lambda)) = n_2^o(\lambda) + m^*, \quad n_2^o(\lambda) = \sum_{\mu < \lambda} \theta(\mu), \tag{5.326}
$$

*where $\sum_{\mu < \lambda} \theta(\mu)$ is the number of finite eigenvalues of (5.238), (5.283) which are less than $\lambda \in \mathbb{R}$, $Y^{[N+1]}(\lambda)$ is the principal solution of (5.238) at $k = N+1$, and $n_1(Y^{[N+1]}(\lambda))$ is the number of forward focal points of $Y^{[N+1]}(\lambda)$ in $(0, N+1]$.*

The advantage of the just proved statements for the multiplicities of backward focal points is that, in view of Lemma 4.8, the previous formulas involve the blocks of $S_k^{-1}(0 \ I)^T$, which do not contain the spectral parameter $\lambda$.

## 5.6   Oscillation Theorems for Variable Rank of $\mathcal{B}_k(\lambda)$

In this section we extend the results of Sect. 5.1 for the nonlinear eigenvalue problems (E), i.e.,

$$
y_{k+1} = \mathcal{S}_k(\lambda)\, y_k, \quad y_k = \begin{pmatrix} x_k \\ u_k \end{pmatrix}, \quad \lambda \in \mathbb{R}, \quad x_0 = 0 = x_{N+1},
$$

by applying the comparative index theory (see Chap. 3). The comparative index approach gives us the possibility to remove the restrictive assumption (5.38), i.e.,

$$
\operatorname{rank} \mathcal{B}_k(\lambda) \ \text{is constant in } \lambda \text{ on } \mathbb{R}
$$

which was used in Sect. 5.1.3 in the local and global oscillation theorems (see Theorems 5.15, 5.16, 5.17 and Corollary 5.18). Recall that condition (5.38) is automatically satisfied for nonlinear eigenvalue problems for the scalar and matrix Sturm-Liouville equations (see Examples 5.35 and 5.36), because the symplectic coefficient matrix $S_k(\lambda)$ of the associated symplectic eigenvalue problem (E) obeys the condition $\det B_k(\lambda) \neq 0$ for all $\lambda \in \mathbb{R}$. It was shown in Example 5.38 that the higher- order Sturm-Liouville difference equations written as a discrete symplectic system also obey (5.38). However there are important classes of spectral problems (E), for which condition (5.38) imposes serious restrictions on the applicability of the local and global oscillation theorems. In Example 5.39, for spectral problems (E) for the real symplectic difference systems with the general linear dependence, $S_k(\lambda) = S_k + \lambda V_k$ on $\lambda$ condition (5.38) implies the absence of real finite eigenvalues of the linear matrix pencil $\mathbb{B}_k(\lambda) = B_k + \lambda \tilde{B}_k$. For the most important special case of (E), for the linear Hamiltonian difference systems (see Example 5.37) condition (5.38) means that rank $[I - A_k(\lambda)]^{-1} B_k(\lambda) = \operatorname{rank} B_k(\lambda)$ is constant for $\lambda \in \mathbb{R}$.

Recall that the proofs of Theorems 5.15, 5.16,  5.17 are based on two main grounds. The first one is monotonicity condition (5.3) and its corollaries, which imply almost all basic properties of the spectral problem (E), including the main notion of finite eigenvalues (Definition 5.5) and their isolated character (see Corollary 5.7). The second foundation is the index theorem (see Theorem 1.85 and Corollary 1.86) imposing the restriction in (5.38). The lower boundedness of the spectrum of (E) (see Corollary 5.18) follows from Theorem 5.17 based on Corollary 1.86, and then (5.3) together with (5.38) implies that all eigenvalues of (E) are isolated and bounded from below.

In this section we retain the first ground assuming that (5.3) holds but change the second one replacing Corollary 1.86 by the main theorem of the comparative index theory (see Theorem 3.24). By applying Theorem 3.24, we avoid (5.38) in the extended version of the local oscillation theorem (see Theorem 5.95) incorporating the jump discontinuities of the piecewise constant function rank $B_k(\lambda)$ for $k \in [0, N]_{\mathbb{Z}}$. As a corollary we show that the finite spectrum of the symplectic eigenvalue problem (E) is bounded from below if and only if so is the set of discontinuity points of rank $B_k(\lambda)$ (see Corollary 5.96).

For the proof of the extended version of the global oscillation theorem (see Theorem 5.98), we assume additionally to (5.3) that rank $B_k(\lambda)$ is constant for all sufficiently negative $\lambda$. Note that this assumption is naturally satisfied for problems (E) with a polynomial dependence on $\lambda$, in particular for a general linear dependence on $\lambda$ (see Example 5.39), as well as for the Hamiltonian eigenvalue problems with the nonlinear dependence in $\lambda$ considered in Example 5.37.

### 5.6.1  Statement of Main Results

The consideration in this section is based on Theorem 5.1. In the previous results, under assumption (5.38), we did not need assertions (i) and (ii) of this theorem, because we have that (5.9) and (5.10) hold. In this section, assertions (i) and (ii) of Theorem 5.1 play a leading role.

By Theorem 5.1(i), we conclude that the set $\operatorname{Ker} \mathcal{B}_k(\lambda)$ is piecewise constant in $\lambda$ on $\mathbb{R}$. That is, for every $\lambda_0 \in \mathbb{R}$, there exists $\delta > 0$ such that

$$\operatorname{Ker} \mathcal{B}_k(\lambda) \equiv \operatorname{Ker} \mathcal{B}_k(\lambda_0^-) \subseteq \operatorname{Ker} \mathcal{B}_k(\lambda_0) \quad \text{for all } \lambda \in (\lambda_0 - \delta, \lambda_0), \tag{5.327}$$

$$\operatorname{Ker} \mathcal{B}_k(\lambda) \equiv \operatorname{Ker} \mathcal{B}_k(\lambda_0^+) \subseteq \operatorname{Ker} \mathcal{B}_k(\lambda_0) \quad \text{for all } \lambda \in (\lambda_0, \lambda_0 + \delta), \tag{5.328}$$

and the quantity $\operatorname{rank} \mathcal{B}_k(\lambda)$ is constant on some left and right neighborhoods of $\lambda_0$. By analogy with Definition 5.5, we introduce the numbers

$$\vartheta_k(\lambda_0) := \operatorname{rank} \mathcal{B}_k(\lambda_0^-) - \operatorname{rank} \mathcal{B}_k(\lambda_0), \quad k \in [0, N]_{\mathbb{Z}}, \tag{5.329}$$

which count the jumps of $\operatorname{rank} \mathcal{B}_k(\lambda)$ in the left neighborhood of $\lambda_0$. Moreover, Theorem 5.1(i) implies that all break points of $\operatorname{rank} \mathcal{B}_k(\lambda)$ are isolated. Now we present an extended version of Theorems 5.15 and 5.16 without condition (5.38). Recall that according to definition (5.39), we denote by $m_k(\lambda)$ the multiplicity of focal points for conjoined bases of $(\mathrm{SDS}_\lambda)$ in the interval $(k, k+1]$ (see (5.39)). The number $n_1(\lambda) = l(Y(\lambda), 0, N+1)$ is then the number of forward focal points of a conjoined basis $Y(\lambda) = (X(\lambda), U(\lambda))$ in $(0, N+1]$, including the multiplicities.

**Theorem 5.95 (Local Oscillation Theorem for $\operatorname{rank} \mathcal{B}(\lambda) \neq \mathrm{const}$)** *Let assumption (5.3) be satisfied. Let $Y(\lambda)$ be a conjoined basis of $(\mathrm{SDS}_\lambda)$ with (5.20). Then $m_k(\lambda^-)$ and $m_k(\lambda^+)$ exist for all $\lambda \in \mathbb{R}$ and*

$$\left. \begin{aligned} m_k(\lambda^+) &= m_k(\lambda) \leq n, \\ m_k(\lambda^+) - m_k(\lambda^-) &+ \operatorname{rank} \mathcal{B}_k(\lambda^-) - \operatorname{rank} \mathcal{B}_k(\lambda) \\ &= \operatorname{rank} X_k(\lambda) - \operatorname{rank} X_k(\lambda^-) + \operatorname{rank} X_{k+1}(\lambda^-) - \operatorname{rank} X_{k+1}(\lambda). \end{aligned} \right\} \tag{5.330}$$

*Moreover,*

$$\left. \begin{aligned} n_1(\lambda^+) &= n_1(\lambda) \leq n(N+1), \\ n_1(\lambda^+) - n_1(\lambda^-) &+ \sum_{k=0}^{N} \vartheta_k(\lambda) \\ &= \operatorname{rank} X_{N+1}(\lambda^-) - \operatorname{rank} X_{N+1}(\lambda) = \theta(\lambda). \end{aligned} \right\} \tag{5.331}$$

The proof of Theorem 5.95 is presented in Sect. 5.6.4.

Recall again that by Corollary 5.7 all break points of rank $X_{N+1}(\lambda)$ and of rank $\mathcal{B}_k(\lambda)$ are isolated, so that by using that $n_1(\lambda)$ is right-continuous we derive from (5.331) for any $a, b \in \mathbb{R}$ with $a < b$

$$n_1(b) - n_1(a) + \sum_{\mu \in (a,b]} \sum_{k=0}^{N} \vartheta_k(\mu) = \sum_{\mu \in (a,b]} \theta(\mu), \tag{5.332}$$

where the sums $\sum_{\mu \in (a,b]} \sum_{k=0}^{N} \vartheta_k(\mu)$ and $\sum_{\mu \in (a,b]} \theta(\mu)$ are finite. If we assume additionally that the function rank $\mathcal{B}_k(\lambda)$ is left-continuous for all sufficiently small $\lambda$, i.e., if there exists $\lambda_{00} \in \mathbb{R}$ such that

$$\text{rank } \mathcal{B}_k(\lambda) = \text{rank } \mathcal{B}_k(\lambda^-) \quad \text{for all } \lambda < \lambda_{00}, \quad k \in [0, N]_{\mathbb{Z}}, \tag{5.333}$$

then by applying Theorem 5.95, we have

$$\left. \begin{array}{r} n_1(\lambda^+) = n_1(\lambda), \\ n_1(\lambda^+) - n_1(\lambda^-) = \text{rank } X_{N+1}(\lambda^-) - \text{rank } X_{N+1}(\lambda) \geq 0, \end{array} \right\} \quad \lambda < \lambda_{00}. \tag{5.334}$$

Repeating the proof of Theorem 5.16 we conclude by (5.334) that $n_1(\lambda)$ is bounded and nondecreasing for all $\lambda < \lambda_{00}$, then there exists $\lambda_0$ such that

$$n_1(\lambda) \equiv m \quad \text{for all } \lambda < \lambda_0.$$

So we see by (5.334) that

$$\text{rank } X_{N+1}(\lambda) = \text{rank } X_{N+1}(\lambda^-) \quad \text{for all } \lambda < \lambda_0, \tag{5.335}$$

i.e., the finite spectrum of (E) is bounded from below. Moreover, using the boundedness of $n_1(\lambda)$, we can prove the following corollary to Theorem 5.95.

**Corollary 5.96** *Assume* (5.3). *Then, condition* (5.333) *holds if and only if the finite spectrum of* (E) *is bounded from below.*

*Proof* It was already proved above that (5.333) implies (5.335). Conversely, assume (5.335). It follows from (5.332) that

$$\left| \sum_{\mu \in (a,b]} \sum_{k=0}^{N} \vartheta_k(\mu) - \sum_{\mu \in (a,b]} \theta(\mu) \right| = |n_1(b) - n_1(a)| \leq n(N+1).$$

Moreover, under assumption (5.335), we have $\sum_{\mu \in (a,b]} \theta(\mu) = 0$ for $b < \lambda_0$. It follows that $\sum_{\mu \in (a,b]} \sum_{k=0}^{N} \vartheta_k(\mu)$ is bounded as $a \to -\infty$ and $\vartheta_k(\mu) \geq 0$. Finally, there exists $\lambda_{00}$ such that (5.333) holds, i.e., $\vartheta_k(\lambda) \equiv 0$. $\qquad \square$

*Remark 5.97*

(i) Note that conditions (5.3) and (5.333) imply that there exists $\tilde{\lambda}$ such that

$$\operatorname{rank} \mathcal{B}_k(\lambda) = \text{const}, \quad \operatorname{rank} X_{N+1}(\lambda) = \text{const} \quad \text{for all } \lambda < \tilde{\lambda}. \tag{5.336}$$

Indeed, by (5.327), (5.328) $\operatorname{rank} \mathcal{B}_k(\lambda_0) \leq \operatorname{rank} \mathcal{B}_k(\lambda)$ for $\lambda \in (\lambda_0 - \delta, \lambda_0 + \delta)$ and the last inequality coupled with (5.333) implies that $0 \leq \operatorname{rank} \mathcal{B}_k(\lambda) \leq n$ is a nondecreasing function with respect to $\lambda$ for $\lambda < \lambda_{00}$. Then there exists the finite limit of $\operatorname{rank} \mathcal{B}_k(\lambda)$ as $\lambda \to -\infty$, i.e., (5.336) holds. By a similar way, condition (5.335) implies that $\operatorname{rank} X_{N+1}(\lambda) = \text{const}$ for all sufficiently small $\lambda$.

(ii) For the linear Hamiltonian difference system (5.144) condition (5.333) is automatically satisfied under (5.3). Indeed, according to Example 5.37, condition (5.3) is equivalent to

$$\dot{\mathcal{H}}_k(\lambda) \geq 0, \quad \mathcal{H}_k(\lambda) = \begin{pmatrix} -C_k(\lambda) & A_k^T(\lambda) \\ A_k(\lambda) & B_k(\lambda), \end{pmatrix}, \tag{5.337}$$

and then the symmetric matrix $B_k(\lambda)$ is nondecreasing matrix function with respect to $\lambda$. By Weyl's inequality, see [270, Corollary 4.9], all eigenvalues of $B_k(\lambda)$ are nondecreasing with respect to $\lambda$. So we have that $\operatorname{ind} B_k(\lambda)$ and $\operatorname{ind}[-B_k(\lambda)]$ are nonincreasing and nondecreasing functions of $\lambda$, so that there exist finite limits

$$\lim_{\lambda \to -\infty} \operatorname{ind} B_k(\lambda) < \infty, \quad \lim_{\lambda \to -\infty} \operatorname{ind}[-B_k(\lambda)] < \infty.$$

This means that the number of positive and negative eigenvalues of $B_k(\lambda)$ is constant for all $\lambda < \tilde{\lambda}$. Consequently, we obtain that $\operatorname{rank} B_k(\lambda) = \operatorname{rank} \mathcal{B}_k(\lambda) = \text{const}$ for all $\lambda < \tilde{\lambda}$.

Recall now the notation from Sect. 5.1.3

$$n_1(\lambda) \leq Nn, \quad n_2(\lambda) := \sum_{\mu \leq \lambda} \theta(\mu)$$

for the number of forward focal points of the principal solution $Y^{[0]}(\lambda)$ in $(1, N+1]$ and for the number of finite eigenvalues of (E) less than or equal to $\lambda$, and introduce a similar notation for the step function

$$n_B(\lambda) := \sum_{\mu \leq \lambda} \sum_{k=0}^{N} \vartheta_k(\mu), \tag{5.338}$$

where $\vartheta_k(\mu)$ is defined by (5.329). Note that conditions (5.3) and (5.333) imply that there exists $\tilde{\lambda} \in \mathbb{R}$ such that

$$n_2(\lambda) \equiv 0, \quad n_B(\lambda) \equiv 0, \quad n_1(\lambda) \equiv m \quad \text{for all } \lambda < \tilde{\lambda}, \tag{5.339}$$

by formula (5.331) and Corollary 5.96. Under assumptions (5.3) and (5.333), the functions $n_2(\lambda)$ and $n_B(\lambda)$ are also right-continuous, i.e.,

$$n_2(\lambda) = n_2(\lambda^+), \quad n_B(\lambda) = n_B(\lambda^+),$$

by Theorem 5.3 and Corollary 5.7. From Corollary 5.96 and Theorem 5.95, we then derive the following generalization of the global oscillation theorem (Theorem 5.17) for the case when rank $\mathcal{B}_k(\lambda)$ is not constant.

**Theorem 5.98 (Global Oscillation Theorem for rank $\mathcal{B}_k(\lambda) \neq$ const)** *Assume that (5.3) and (5.333) hold. Then there exists $m \in [0, nN]_{\mathbb{Z}}$ such that for any $\lambda \in \mathbb{R}$, we have*

$$n_2(\lambda) + m = n_1(\lambda) + n_B(\lambda), \tag{5.340}$$

*where $m$ is given by (5.339).*

*Proof* The proof follows from (5.332) for $b := \lambda$, where we evaluate the finite limits of the quantities $n_1(a)$, $\sum_{a < \mu \leq \lambda} \sum_{k=0}^{N} \vartheta_k(\mu)$, and $\sum_{a < \mu \leq \lambda} \theta(\mu)$ as $a \to -\infty$ according to (5.339), (5.335), and (5.333). $\qquad\square$

Note that under the assumption (5.38), we have in (5.340) that $n_B(\lambda) = 0$ for $\lambda \in \mathbb{R}$. In this case Theorems 5.95 and 5.98 present the results of Theorems 5.15, 5.16, and 5.17. However, the same results follow from Theorem 5.95 under the weakened assumption rank $\mathcal{B}_k(\lambda) = $ rank $\mathcal{B}_k(\lambda^-)$ for $\lambda \in \mathbb{R}$, i.e., the following corollary holds.

**Corollary 5.99** *Assume (5.3) and suppose that (5.333) holds for all $\lambda \in \mathbb{R}$. Then under the notation of Theorem 5.98, there exists $m \in [0, nN]_{\mathbb{Z}}$ such that*

$$n_2(\lambda) + m = n_1(\lambda) \quad \text{for all } \lambda \in \mathbb{R}, \tag{5.341}$$

*where $m$ is given by (5.339).*

The previous version of the global oscillation theorem (Theorem 5.17) implies that the number $n_1(\lambda)$ of focal points of $Y^{[0]}(\lambda)$ in $(0, N + 1]$ is monotonic in $\lambda$ on $\mathbb{R}$. The present version (Theorem 5.98) says that this function in general is not monotonic, but there exists a representation of the quantity $n_1(\lambda) - m$ as the difference $n_2(\lambda) - n_B(\lambda)$ of two nondecreasing functions in accordance with the *Jordan decomposition* of a function of bounded variation. In the last part of this subsection, we provide examples illustrating the results stated above.

*Example 5.100* This example is devoted to applications of formula (5.332). Consider problem (E) for the trigonometric difference system with

$$\mathcal{S}_k(\lambda) = \begin{pmatrix} \cos \lambda & \sin \lambda \\ -\sin \lambda & \cos \lambda \end{pmatrix}, \quad k \in [0, N]_{\mathbb{Z}}. \tag{5.342}$$

According to (5.3), we have $\Psi(\mathcal{S}_k(\lambda)) = I > 0$, so that the the monotonicity condition (5.3) holds for $\lambda \in \mathbb{R}$. The principal solution $Y^{[0]}(\lambda)$ of (E) with (5.342) has the form $Y_k^{[0]}(\lambda) = \big(\sin(k\lambda),\ \cos(k\lambda)\big)^T$, and then the finite eigenvalues of this problem are $\lambda_p = \pi p / (N + 1)$, $p \in \mathbb{Z}$. The multiplicities of focal points of $Y^{[0]}(\lambda)$ in $(k, k + 1]$ are given by

$$m_k(\lambda) = \begin{cases} 1, \ \lambda = \pi p/(k + 1), \ \lambda \neq \pi l, \ p, l \in \mathbb{Z}, \\ 1, \ \sin(\lambda) \sin(k\lambda) \sin((k + 1)\lambda) < 0, \\ 0, \ \text{otherwise.} \end{cases}$$

So we see that $m_k(\lambda) = m_k(\lambda + \pi l)$ for $l \in \mathbb{Z}$, and then $n_1(\lambda)$ is periodic with the minimal period $T = \pi$ and nondecreasing in any interval $[a, b] \subseteq [\pi l, \pi(l + 1))$, $l \in \mathbb{Z}$. For example, for $N = 3$, we have

$$n_1(\lambda) = \begin{cases} 0, \ \lambda \in [0, \pi/4), \\ 1, \ \lambda \in [\pi/4, \pi/2), \\ 2, \ \lambda \in [\pi/2, 3\pi/4), \\ 3, \ \lambda \in [3\pi/4, \pi), \end{cases} \qquad n_1(\lambda) = n_1(\lambda + \pi l), \quad l \in \mathbb{Z},$$

and by (5.332) the quantity $n_1(b) - n_1(a)$ presents the number of eigenvalues of (E), (5.342) in $(a, b]$ for $\pi l \leq a < b < \pi(l + 1)$, $l \in \mathbb{Z}$. However, if $b = \pi(l + 1)$, then the function $n_1(\lambda)$ loses the monotonicity, and in accordance with (5.332), we restore it by adding the total number of zeros of $\mathcal{B}_k(\lambda) = \sin(\lambda)$, $k \in [0, 3]_{\mathbb{Z}}$, in $(a, b]$. So we have that $n_1(\pi(l + 1)) - n_1(a) + 4 = 4 - n_1(a)$ presents the number of eigenvalues of (E), (5.342) in $(a, \pi(l + 1)]$ for $\pi l \leq a \leq \pi(l + 1)$, $l \in \mathbb{Z}$. Similarly, for the case $a = -\pi/2$, $b = 2\pi$, and $N = 3$, the number of focal points of $Y^{[0]}(\lambda)$ equals to $n_1(2\pi) = n_1(0) = 0$, $n_1(-\pi/2) = n_1(\pi/2) = 2$, the number of eigenvalues of (E) in $(-\pi/2, 2\pi]$ equals to 10, the sum $\sum_{-\pi/2 < \mu \leq 2\pi} \sum_{k=0}^{3} \vartheta_k(\mu) = 4 \cdot 3 = 12$, and (5.332) says that $n_1(2\pi) - n_1(-\pi/2) + \sum_{-\pi/2 < \mu \leq 2\pi} \sum_{k=0}^{3} \vartheta_k(\mu) = -2 + 12 = 10$.

Consider an example illustrating Theorem 5.98.

*Example 5.101*  Introduce symplectic difference system (E) with

$$\mathcal{S}_k(\lambda) = \begin{pmatrix} \lambda - k + 1 & \lambda - k \\ -\lambda + k & 1 - \lambda + k \end{pmatrix}.$$

Then we have $\Psi_k(\mathcal{S}_k(\lambda)) = \begin{pmatrix} 1 & 1 \\ 1 & 1 \end{pmatrix} \geq 0$ and condition (5.336) for the coefficient $\mathcal{B}_k(\lambda) = \lambda - k$ for $k \in [0, N]_{\mathbb{Z}}$ holds for all $\lambda < \tilde{\lambda} = 0$. For this case, problem (E) has only one finite eigenvalue for any $N$. For $N = 3$, the principal solution at 0 is

$$Y_0^{[0]}(\lambda) = \begin{pmatrix} 0 \\ 1 \end{pmatrix}, \quad Y_1^{[0]}(\lambda) = \begin{pmatrix} \lambda \\ 1 - \lambda \end{pmatrix}, \quad Y_2^{[0]}(\lambda) = \begin{pmatrix} 2\lambda - 1 \\ 2 - 2\lambda \end{pmatrix},$$

$$Y_3^{[0]}(\lambda) = \begin{pmatrix} 3\lambda - 3 \\ 4 - 3\lambda \end{pmatrix}, \quad Y_4^{[0]}(\lambda) = \begin{pmatrix} 4\lambda - 6 \\ 7 - 4\lambda \end{pmatrix}.$$

Then the problem has the unique eigenvalue $\lambda_1 = \frac{3}{2}$. The multiplicities of focal points for this solution are

$$m_0(\lambda) = 0, \quad m_1(\lambda) = \begin{cases} 1, \ \lambda \in (-\infty, 0) \cup [\frac{1}{2}, 1), \\ 0, \ \text{otherwise}, \end{cases}$$

$$m_2(\lambda) = \begin{cases} 1, \ \lambda \in (-\infty, \frac{1}{2}) \cup [1, 2), \\ 0, \ \text{otherwise}, \end{cases}$$

$$m_3(\lambda) = \begin{cases} 1, \ \lambda \in (-\infty, 1) \cup [\frac{3}{2}, 3), \\ 0, \ \text{otherwise}. \end{cases}$$

Then the numbers $n_1(\lambda)$ of focal points in $(0, 4]$ and the function $n_B(\lambda)$ are

$$n_1(\lambda) = \begin{cases} 3, \ \lambda \in (-\infty, 0), \\ 2, \ \lambda \in [0, 1) \cup [\frac{3}{2}, 2), \\ 1, \ \lambda \in [1, \frac{3}{2}) \cup [2, 3) \\ 0, \ \lambda \in [3, \infty). \end{cases} \quad n_B(\lambda) = \begin{cases} 0, \ \lambda \in (-\infty, 0), \\ 1, \ \lambda \in [0, 1), \\ 2, \ \lambda \in [1, 2), \\ 3, \ \lambda \in [2, 3), \\ 4, \ \lambda \in [3, \infty). \end{cases}$$

So we see that the constant $m$ in Theorem 5.98 equals to 3. Finally, we have

$$n_1(\lambda) - m + n_B(\lambda) = n_2(\lambda) = \begin{cases} 0, \ \lambda \in (-\infty, \frac{3}{2}), \\ 1, \ \lambda \in [\frac{3}{2}, \infty). \end{cases}$$

*Example 5.102* Consider $4 \times 4$ symplectic system $(SDS_\lambda)$ for $k \in [0, 2]_Z$ with

$$S_0(\lambda) = S_2(\lambda) = \begin{pmatrix} 1 & 0 & 0 & 0 \\ 0 & 0 & 0 & 1 \\ -\lambda & 0 & 1 & 0 \\ 0 & -1 & 0 & -\lambda \end{pmatrix},$$

$$\left. \vphantom{\begin{pmatrix} 1 \\ 0 \\ -\lambda \\ 0 \end{pmatrix}} \right\}$$

$$S_1(\lambda) = \begin{pmatrix} 1 & 0 & \lambda^3 & 0 \\ 0 & 0 & 0 & 1 \\ -\lambda & 0 & 1-\lambda^4 & 0 \\ 0 & -1 & 0 & -\lambda-\lambda^3 \end{pmatrix}. \tag{5.343}$$

Condition (5.3) takes the form $\Psi(S_k(\lambda)) = \mathrm{diag}\{0,\ I\} \geq 0$ for $k \in \{0, 2\}$ and

$$\Psi(S_1(\lambda)) = \begin{pmatrix} 3\lambda^4 & 0 & 3\lambda^3 & 0 \\ 0 & 1+3\lambda^2 & 0 & 0 \\ 3\lambda^3 & 0 & 3\lambda^2 & 0 \\ 0 & 0 & 0 & 0 \end{pmatrix} \geq 0.$$

The upper blocks of the principal solution $Y^{[0]}(\lambda)$ are

$$X_0^{[0]}(\lambda) = 0, \quad X_1^{[0]}(\lambda) = \begin{pmatrix} 0 & 0 \\ 0 & 1 \end{pmatrix}, \quad X_2^{[0]}(\lambda) = \begin{pmatrix} \lambda^3 & 0 \\ 0 & -\lambda \end{pmatrix},$$

$$X_3^{[0]}(\lambda) = \begin{pmatrix} \lambda^3 & 0 \\ 0 & \lambda^2+\lambda^4-1 \end{pmatrix}.$$

Then the finite eigenvalues of (E), (5.343) are

$$\lambda_1 = -\sqrt{\frac{-1+\sqrt{5}}{2}}, \quad \lambda_2 = 0, \quad \lambda_3 = \sqrt{\frac{-1+\sqrt{5}}{2}},$$

and the multiplicities of focal points of the principal solution $Y^{[0]}(\lambda)$ are

$$m_0(\lambda) = 0, \quad m_1(\lambda) = \begin{cases} 1, & \lambda \geq 0, \\ 0, & \lambda < 0, \end{cases}$$

$$m_2(\lambda) = \begin{cases} 1, & \lambda\,(\lambda^2 + \frac{1-\sqrt{5}}{2}) \geq 0 \text{ and } \lambda \neq 0, \\ 0, & \text{otherwise.} \end{cases}$$

Then the function $n_1(\lambda) = m_0(\lambda) + m_1(\lambda) + m_2(\lambda)$ takes the form

$$n_1(\lambda) = \begin{cases} 0, \ \lambda < \lambda_1, \\ 1, \ \lambda_1 \le \lambda < \lambda_3, \\ 2, \ \lambda_3 \le \lambda. \end{cases}$$

So we see that $n_1(\lambda)$ does not count the eigenvalue $\lambda_2 = 0$. By (5.340), we need to add the function $n_B(\lambda) = 0$ for $\lambda < 0$ and $n_B(\lambda) = 1$ for $\lambda \ge 0$ to derive $n_2(\lambda) = n_1(\lambda) + n_B(\lambda)$ counting all (real) eigenvalues of the problem.

In the last example, we illustrate Corollary 5.99.

*Example 5.103* For the illustration of Corollary 5.99, we change the statement of the problem in Example 5.102 replacing $\mathcal{S}_1(\lambda)$ in (5.343) by the matrix

$$\mathcal{S}_1(\lambda) = \begin{pmatrix} 1 & 0 & 0 & 0 \\ 0 & 0 & 0 & 1 \\ -\lambda & 0 & 1 & 0 \\ 0 & -1 & 0 & -\lambda \end{pmatrix}, \quad \text{for } \lambda \le 0,$$

$$\mathcal{S}_1(\lambda) = \begin{pmatrix} 1 & 0 & \lambda^3 & 0 \\ 0 & 0 & 0 & 1 \\ -\lambda & 0 & 1 - \lambda^4 & 0 \\ 0 & -1 & 0 & -\lambda - \lambda^3 \end{pmatrix}, \quad \text{for } \lambda > 0,$$

which is continuously differentiable with respect to $\lambda \in \mathbb{R}$. Then, the matrices $\mathcal{S}_k(\lambda)$ for $k \in [0, 2]_{\mathbb{Z}}$ obey condition (5.333) for all $\lambda \in \mathbb{R}$. The modified principal solution $Y^{[0]}(\lambda)$ has the upper blocks

$$X_0^{[0]}(\lambda) = 0, \quad X_1^{[0]}(\lambda) = \begin{pmatrix} 0 & 0 \\ 0 & 1 \end{pmatrix},$$

$$X_2^{[0]}(\lambda) = \begin{pmatrix} 0 & 0 \\ 0 & -\lambda \end{pmatrix} \text{ for } \lambda \le 0, \quad X_2^{[0]}(\lambda) = \begin{pmatrix} \lambda^3 & 0 \\ 0 & -\lambda \end{pmatrix} \text{ for } \lambda > 0,$$

$$X_3^{[0]}(\lambda) = \begin{pmatrix} 0 & 0 \\ 0 & \lambda^2 - 1 \end{pmatrix} \text{ for } \lambda \le 0, \quad X_3^{[0]}(\lambda) = \begin{pmatrix} \lambda^3 & 0 \\ 0 & \lambda^2 + \lambda^4 - 1 \end{pmatrix} \text{ for } \lambda > 0.$$

By Definition 5.5, this modified problem has the finite eigenvalues

$$\tilde{\lambda}_1 = -1, \quad \tilde{\lambda}_2 = \sqrt{\frac{-1 + \sqrt{5}}{2}},$$

and $\lambda = 0$ is not a finite eigenvalue anymore. The multiplicities $m_0(\lambda)$ and $m_1(\lambda)$ of focal points for this principal solution are the same as in Example 5.102, while

$m_2(\lambda)$ is given by

$$m_2(\lambda) = \begin{cases} 1, & \lambda \in [-1, 0) \cup [\tilde{\lambda}_2, \infty), \\ 0, & \text{otherwise.} \end{cases}$$

Then it follows that $m_0(\lambda) + m_1(\lambda) + m_2(\lambda) = n_1(\lambda) = n_2(\lambda)$ in accordance with Corollary 5.99 (with $m = 0$).

## 5.6.2  Monotonicity and the Comparative Index

Since monotonicity condition (5.3) is formulated for the symplectic coefficient matrix $\mathcal{S}_k(\lambda)$, the main purpose of this subsection is to relate (5.3) with the limit behavior of the comparative index for symplectic matrices introduced in Sect. 3.3.1. Recall the notation

$$\langle W \rangle = \begin{pmatrix} \mathcal{X} \\ \mathcal{U} \end{pmatrix}, \quad \mathcal{X} = \begin{pmatrix} I & 0 \\ A & B \end{pmatrix}, \quad \mathcal{U} = \begin{pmatrix} 0 & -I \\ C & D \end{pmatrix}, \quad W = \begin{pmatrix} A & B \\ C & D \end{pmatrix} \quad (5.344)$$

for arbitrary symplectic matrix $W$ separated into the $n \times n$ blocks $A$, $B$, $C$, $D$. Basic properties of the comparative index $\mu(\langle W \rangle, \langle \hat{W} \rangle)$ are given in Lemma 3.21. Recall that according to the duality principle (see Theorem 3.11 and Remark 3.41), all identities in Lemma 3.21 hold also for the dual index $\mu^*(\langle W \rangle, \langle \hat{W} \rangle)$. For example, instead of (v) in Lemma 3.21, we have

$$\left. \begin{aligned} \mu^*(\langle W \rangle, \langle \hat{W} \rangle) &= \mu^*(W(0\ I)^T, \hat{W}(0\ I)^T) + \mu^*(\langle \hat{W}^{-1} W \rangle, \langle I \rangle) \\ &= \mu(W^{-1}(0\ I)^T, \hat{W}^{-1}(0\ I)^T) + \mu^*(\langle W\hat{W}^{-1} \rangle, \langle I \rangle). \end{aligned} \right\} \quad (5.345)$$

For the nonsingular case $\det B \neq 0$ and $\det \hat{B} \neq 0$, it follows that $\det \mathcal{X} \neq 0$ and $\det \hat{\mathcal{X}} \neq 0$ for $\mathcal{X}$ and $\hat{\mathcal{X}}$ defined by (5.344), and then according to Sect. 3.3.3 (see (3.81))

$$\left. \begin{aligned} \mu(\langle W \rangle, \langle \hat{W} \rangle) &= \mu^*(\langle \hat{W} \rangle, \langle W \rangle) = \text{ind}(\hat{Q} - Q), \\ Q = \mathcal{U}\mathcal{X}^{-1} &= \begin{pmatrix} B^{-1}A & -B^{-1} \\ -B^{T-1} & DB^{-1} \end{pmatrix}, \end{aligned} \right\} \quad (5.346)$$

with $\hat{Q}$ defined analogously to $Q$.

Connections between monotonicity condition (5.3) and the comparative index theory are based on the following lemma.

**Lemma 5.104** *Assume that a symplectic matrix $W(\lambda)$ is piecewise continuously differentiable with respect to $\lambda \in \mathbb{R}$ and obeys (1.201), i.e., $\Psi(W(\lambda)) \geq 0$ for* all $\lambda \in \mathbb{R}$. Then for any $\lambda_0 \in \mathbb{R}$, there exists $\delta = \delta(\lambda_0) > 0$ such that for all $a, b \in (\lambda_0 - \delta, \lambda_0 + \delta)$ with $a \leq b$

$$\left.\begin{array}{l} \mu(\langle W(b)\rangle, \langle W(a)\rangle) = \mu(W(b)\,(0\ I)^T, W(a)\,(0\ I)^T), \\ \mu^*(\langle W(a)\rangle, \langle W(b)\rangle) = \mu^*(W(a)\,(0\ I)^T, W(b)\,(0\ I)^T), \end{array}\right\} \tag{5.347}$$

and

$$\left.\begin{array}{ll} \mu(\langle W(b)\rangle, \langle W(a)\rangle) = 0 & \text{for all } a, b \in [\lambda_0, \lambda_0 + \delta),\ a \leq b, \\ \mu^*(\langle W(a)\rangle, \langle W(b)\rangle) = 0 & \text{for all } a, b \in (\lambda_0 - \delta, \lambda_0],\ a \leq b. \end{array}\right\} \tag{5.348}$$

*Proof* For (5.347) we fix $\lambda_0 \in \mathbb{R}$ and introduce the symplectic matrix

$$R = \begin{pmatrix} I & G \\ 0 & I \end{pmatrix}, \quad G := I - B(\lambda_0)B^\dagger(\lambda_0). \tag{5.349}$$

Note that the $2n \times n$ submatrix $[B^T(\lambda)\ D^T(\lambda)]^T$ of $W(\lambda)$ obeys condition (3.1), and then $\tilde{W}(\lambda) = R^{-1}W(\lambda) = \begin{pmatrix} \tilde{A}(\lambda) & \tilde{B}(\lambda) \\ \tilde{C}(\lambda) & \tilde{D}(\lambda) \end{pmatrix}$ has the nonsingular block

$$\tilde{B}(\lambda) := B(\lambda) - GD(\lambda)$$

for $\lambda = \lambda_0$; see formula (1.175) in Lemma 1.68. Since $\det \tilde{B}(\lambda)$ is continuous with respect to $\lambda$ and $\det \tilde{B}(\lambda_0) \neq 0$, there exists $\delta = \delta(\lambda_0) > 0$ such that $\det \tilde{B}(\lambda) \neq 0$ for all $\lambda \in (\lambda_0 - \delta, \lambda_0 + \delta)$. By Lemma 1.77 applied to the case under the consideration (with $P = I$), condition (1.202) holds, then we have for all $a, b \in (\lambda_0 - \delta, \lambda_0 + \delta)$ with $a \leq b$

$$\left.\begin{array}{l} \mu(\langle R^{-1}W(b)\rangle, \langle R^{-1}W(a)\rangle) = \mu^*(\langle R^{-1}W(a)\rangle, \langle R^{-1}W(b)\rangle) \\ \qquad\qquad\qquad\qquad\qquad = \text{ind}\,[\tilde{Q}(a) - \tilde{Q}(b)] = 0, \end{array}\right\} , \tag{5.350}$$

where $\overset{\*}{Q}(\lambda)$ is defined via the blocks of $\tilde{W}(\lambda) := R^{-1}W(\lambda)$ according to (5.346). By Lemma 3.21(v),(iii) condition (5.350) implies

$$\mu(\langle W^{-1}(a)W(b)\rangle, \langle I\rangle) = \mu^*(\langle W^{-1}(b)W(a)\rangle, \langle I\rangle) = 0$$

for all $a, b \in (\lambda_0 - \delta, \lambda_0 + \delta)$ with $a \leq b$. Then by Lemma 3.21(v) and (5.345), we derive (5.347). Next we shall prove (5.348). By Lemma 3.21(v),(iv), it follows from (5.350) that

$$\mu(R^{-1}W(b)\,(0\ I)^T, R^{-1}W(a)\,(0\ I)^T) = \mu^*(R^{-1}W(a)\,(0\ I)^T, R^{-1}W(b)\,(0\ I)^T) = 0,$$

where $R$ is given by (5.349). Then, according to Theorem 3.6 and Theorem 3.5(ix),

$$\mu(W(b)\,(0\ I)^T, W(a)\,(0\ I)^T) = \mu(R(R^{-1}W(b))\,(0\ I)^T, R(R^{-1}W(a))\,(0\ I)^T)$$

$$= \mu(R^{-1}W(b)\,(0\ I)^T, R^{-1}W(a)\,(0\ I)^T) + \mu(W(b)\,(0\ I)^T, R(0\ I)^T)$$

$$- \mu(W(a)\,(0\ I)^T, R(0\ I)^T)$$

$$= \mu(R^{-1}W(a)\,(0\ I)^T, R^{-1}(0\ I)^T) - \mu(R^{-1}W(b)\,(0\ I)^T, R^{-1}(0\ I)^T).$$

So we have derived that for all $a, b \in (\lambda_0 - \delta, \lambda_0 + \delta)$ with $a \leq b$

$$\left.\begin{aligned}\mu(\langle W(b)\rangle, \langle W(a)\rangle) &= \mu(W(b)\,(0\ I)^T, W(a)\,(0\ I)^T)\\ &= \mu(\tilde{W}(a)\,(0\ I)^T, R^{-1}(0\ I)^T) - \mu(\tilde{W}(b)\,(0\ I)^T, R^{-1}(0\ I)^T),\end{aligned}\right\} \tag{5.351}$$

where the matrix $\tilde{W}(\lambda) = R^{-1}W(\lambda)$ has the nonsingular block $\tilde{B}(\lambda)$. Evaluating the first comparative index on the right-hand side above according to Definition 3.1, we obtain

$$\mu(\tilde{W}(\lambda)\,(0\ I)^T, R^{-1}(0\ I)^T) = \mu_2(\tilde{W}(\lambda)\,(0\ I)^T, R^{-1}(0\ I)^T)$$

$$= \operatorname{ind}[-GB(\lambda)\,\tilde{B}^{-1}(\lambda)\,G].$$

Since $GB(\lambda_0) = 0$ by (5.349), it follows that $\operatorname{ind}[-GB(\lambda_0)\,\tilde{B}(\lambda_0)^{-1}G] = 0$. Putting $a := \lambda_0$ and $b \in [\lambda_0, \lambda_0 + \delta)$ in (5.351), we derive

$$0 \leq \mu(\langle W(b)\rangle, \langle W(\lambda_0)\rangle) = \mu(W(b)\,(0\ I)^T, W(\lambda_0)\,(0\ I)^T)$$

$$= \mu(\tilde{W}(\lambda_0)\,(0\ I)^T, R^{-1}(0\ I)^T) - \mu(\tilde{W}(b)\,(0\ I)^T, R^{-1}(0\ I)^T)$$

$$= -\mu(\tilde{W}(b)\,(0\ I)^T, R^{-1}(0\ I)^T) \leq 0.$$

Thus, we have proved

$$\mu(\tilde{W}(\lambda)\,(0\ I)^T, R^{-1}(0\ I)^T) = 0 \quad \text{for all } \lambda \in [\lambda_0, \lambda_0 + \delta). \tag{5.352}$$

Finally, we complete the proof of the first equality in (5.348) by using (5.352) and (5.351). The proof of the second equality in (5.348) is similar. Instead of (5.351), we have for all $a, b \in (\lambda_0 - \delta, \lambda_0 + \delta)$ with $a \leq b$

$$\left.\begin{aligned}\mu^*(\langle W(a)\rangle, \langle W(b)\rangle) &= \mu^*(W(a)\,(0\ I)^T, W(b)\,(0\ I)^T)\\ &= \mu^*(\tilde{W}(b)\,(0\ I)^T, R^{-1}(0\ I)^T) - \mu^*(\tilde{W}(a)\,(0\ I)^T, R^{-1}(0\ I)^T),\end{aligned}\right\} \tag{5.353}$$

where $\mu^*(\tilde{W}(\lambda)\,(0\ I)^T, R^{-1}(0\ I)^T) = \text{ind}\,[G\,B(\lambda)\,\tilde{B}^{-1}(\lambda)\,G]$. Then (5.352) is replaced by

$$\mu^*(\tilde{W}(\lambda)\,(0\ I)^T, R^{-1}(0\ I)^T) = 0 \quad \text{for all } \lambda \in (\lambda_0 - \delta, \lambda_0] \tag{5.354}$$

and we complete the proof of (5.348) combining (5.354) and (5.353). $\qquad\square$

Using Lemma 3.21(iv), we see that conditions (5.348) are equivalent to

$$\left.\begin{aligned} \mu^*(\langle W(a)\rangle, \langle W(b)\rangle) &= \text{rank}\,B(b) - \text{rank}\,B(a) \\ &\text{for all } a, b \in [\lambda_0, \lambda_0 + \delta),\ a \leq b, \\[4pt] \mu(\langle W(b)\rangle, \langle W(a)\rangle) &= \text{rank}\,B(a) - \text{rank}\,B(b) \\ &\text{for all } a, b \in (\lambda_0 - \delta, \lambda_0],\ a \leq b, \end{aligned}\right\} \tag{5.355}$$

where $B(\lambda)$ is the block of $W(\lambda)$ given by (5.344).

We proved Lemma 5.104 in the local sense, i.e., all results of this lemma hold in a neighborhood of $\lambda_0$. If we add the assumption

$$\text{rank}\,B(\lambda) \text{ is constant for } \lambda \in \mathbb{R}, \tag{5.356}$$

then we can reformulate Lemma 5.104 in the global sense.

**Lemma 5.105** *Under the assumptions of Lemma 5.104 suppose additionally that (5.356) holds. Then for all $a, b \in \mathbb{R}$ with $a \leq b$*

$$\mu(\langle W(b)\rangle, \langle W(a)\rangle) = \mu^*(\langle W(a)\rangle, \langle W(b)\rangle) = 0. \tag{5.357}$$

*Proof* First of all, let us recall that under assumption (1.201), condition (5.356) is equivalent to

the set $\text{Im}\,B(\lambda)$ is constant for $\lambda \in \mathbb{R}$

or

$$B(\lambda)\,B^\dagger(\lambda) \text{ is constant in } \lambda \in \mathbb{R}; \tag{5.358}$$

see Theorem 1.82 and Corollary 1.83. Then one can prove (5.357) repeating the main steps of the proof of Lemma 5.104 applied in the global sense under assumption (5.356). We note that the same result follows also from Lemma 5.104 by the "triangle inequality" (see Theorem 3.8) applied to the matrices $\langle W(a)\rangle$, $\langle W(b)\rangle$, $\langle W(c)\rangle$ for $a \leq b \leq c$. Indeed, by (5.348), we have

$$\mu(\langle W(b)\rangle, \langle W(a)\rangle) = 0, \quad a, b \in [\lambda_0, \lambda_0 + \delta),\ a \leq b,$$

while from the second equality in (5.355) and (5.356), we get

$$\mu(\langle W(b)\rangle, \langle W(a)\rangle) = 0, \quad a, b \in (\lambda_0 + \delta, \lambda_0], \ a \le b.$$

Applying inequality (3.18), we derive that $\mu(\langle W(b)\rangle, \langle W(a)\rangle) = 0$ holds for all $a, b \in (\lambda_0 + \delta, \lambda_0 + \delta)$ with $a \le b$. By a similar way, we obtain that $\mu(\langle W(c)\rangle, \langle W(b)\rangle) = 0$ and $\mu(\langle W(b)\rangle, \langle W(a)\rangle) = 0$ imply $\mu(\langle W(c)\rangle, \langle W(a)\rangle) = 0$ for any $a \le b \le c$.                                                                  $\square$

Equalities (5.348) and (5.355) imply the main theorem of this section.

**Theorem 5.106** *Assume* (5.3) *and suppose that* $Z(\lambda)$ *is a symplectic fundamental matrix of* $(SDS_\lambda)$ *satisfying* (5.11) *for* $k = 0$. *Then for any* $\lambda_0 \in \mathbb{R}$, *there exist the finite limits*

$$\left.\begin{aligned}
\mu\big(\langle \mathcal{S}_k(\lambda_0^+)\rangle, \langle \mathcal{S}_k(\lambda_0)\rangle\big) &= 0, \\
\mu\big(\langle \mathcal{S}_k(\lambda_0)\rangle, \langle \mathcal{S}_k(\lambda_0^-)\rangle\big) &= \vartheta_k(\lambda_0),
\end{aligned}\right\} k \in [0, N]_{\mathbb{Z}}, \tag{5.359}$$

$$\left.\begin{aligned}
\mu\big(\langle Z_k(\lambda_0^+)\rangle, \langle Z_k(\lambda_0)\rangle\big) &= 0, \\
\mu\big(\langle Z_k(\lambda_0)\rangle, \langle Z_k(\lambda_0^-)\rangle\big) &= \operatorname{rank} X_k(\lambda_0^-) - \operatorname{rank} X_k(\lambda_0),
\end{aligned}\right\} k \in [0, N+1]_{\mathbb{Z}}, \tag{5.360}$$

*where* $Y_k^T(\lambda) = (X_k^T(\lambda), U_k^T(\lambda))^T = Z_k(\lambda)\,(0\ I)^T$, *and* $\vartheta_k(\lambda_0)$ *are given by* (5.329). *In particular, for* $Z_0(\lambda) = I$, *we have*

$$\left.\begin{aligned}
\mu\big(\langle Z_{N+1}(\lambda_0^+)\rangle, \langle Z_{N+1}(\lambda_0)\rangle\big) &= 0, \\
\mu\big(\langle Z_{N+1}(\lambda_0)\rangle, \langle Z_{N+1}(\lambda_0^-)\rangle\big) &= \theta(\lambda_0),
\end{aligned}\right\} \tag{5.361}$$

*where* $\theta(\lambda_0)$ *is the (algebraic) multiplicity of the finite eigenvalue* $\lambda_0$ *of* (E).

*Proof* Recall that (5.3) and (5.11) for $k = 0$ imply (5.11) for $k \in [0, N]_{\mathbb{Z}}$ (see Sect. 5.1.1). Then $\mathcal{S}_k(\lambda)$ and $Z_k(\lambda)$ meet all requirements of Lemma 5.104. The existence of the limits $\mu(\langle \mathcal{S}_k(\lambda_0^+)\rangle, \langle \mathcal{S}_k(\lambda_0)\rangle) = 0$ and $\mu(\langle Z_k(\lambda_0^+)\rangle, \langle Z_k(\lambda_0)\rangle) = 0$ follows directly from (5.348) with $a := \lambda_0$, where we put $W(\lambda) := \mathcal{S}_k(\lambda)$ and $W(\lambda) := Z_k(\lambda)$. The existence of the limits $\operatorname{rank} \mathcal{B}_k(\lambda_0^-)$ and $\operatorname{rank} X_k(\lambda_0^-)$ follows from Theorems 5.1 and 5.3, but we can also prove this fact independently by using Lemma 5.104. Indeed, for the case $a \le b < \lambda_0$, we have by the second equality in (5.355) that $\operatorname{rank} \mathcal{B}_k(\lambda)$ and $\operatorname{rank} X_k(\lambda)$ are nonincreasing with respect to $\lambda \in (\lambda_0 - \delta, \lambda_0)$, and then there exist the finite limits $\operatorname{rank} \mathcal{B}_k(\lambda_0^-)$ and $\operatorname{rank} X_k(\lambda_0^-)$, i.e., there exists $\tilde{\delta} \in (0, \delta]$ such that

$$\left.\begin{aligned}
\operatorname{rank} \mathcal{B}_k(\lambda) &\equiv \operatorname{rank} \mathcal{B}_k(\lambda_0^-), \\
\operatorname{rank} X_k(\lambda) &\equiv \operatorname{rank} X_k(\lambda_0^-),
\end{aligned}\right\} \text{ for all } \lambda \in (\lambda_0 - \tilde{\delta}, \lambda_0). \tag{5.362}$$

By using the second equality in (5.355), putting $W(\lambda) := S_k(\lambda)$ with $b := \lambda_0$, we derive the second equality in (5.359). By a similar way, putting $W(\lambda) := Z_k(\lambda)$, we then complete the proof of (5.360).                                                                    □

*Remark 5.107*

(i) We remark that the comparative index theory provides another proof of Theorem 1.81 (or Theorems 5.1(ii) and 5.3(ii)) under assumption (1.201). For example, applying (5.362) and the second equalities in (5.355) and (5.348), we see that there exists $\tilde{\delta} > 0$ such that

$$\left.\begin{aligned}
\mu\big(\langle S_k(b)\rangle, \langle S_k(a)\rangle\big) = 0 \quad \text{for all } a, b \in (\lambda_0 - \tilde{\delta}, \lambda_0), \ a \le b, \\
\mu^*\big(\langle S_k(a)\rangle, \langle S_k(b)\rangle\big) = 0 \quad \text{for all } a, b \in (\lambda_0 - \tilde{\delta}, \lambda_0], \ a \le b.
\end{aligned}\right\} \quad (5.363)$$

Then by Lemma 3.21(v) and (5.345), we obtain

$$\mu_1(S_k(b) (0\ I)^T, S_k(a) (0\ I)^T) = 0 \quad \text{for all } a, b \in (\lambda_0 - \tilde{\delta}, \lambda_0), \ a \le b,$$
$$\mu_1^*(S_k(a) (0\ I)^T, S_k(b) (0\ I)^T) = \mu_1(S_k(a) (0\ I)^T, S_k(b) (0\ I)^T) = 0$$
$$\text{for all } a, b \in (\lambda_0 - \tilde{\delta}, \lambda_0], \ a \le b.$$

By Theorem 3.2(iv), the last conditions are equivalent to (5.6) in Theorem 5.1(ii). The case (5.7) can be derived similarly.

(ii) Analogously, Lemma 5.104 provides an alternative proof of Theorem 1.79 (or Theorems 5.1(i) 5.3(i)). In more details, by Lemma 3.21(v) and (5.345), conditions (5.363) imply

$$\mu_1^*(S_k^{-1}(b) (0\ I)^T, S_k^{-1}(a) (0\ I)^T) = \mu_1(S_k^{-1}(b) (0\ I)^T, S_k^{-1}(a) (0\ I)^T) = 0$$
$$\text{for all } a, b \in (\lambda_0 - \tilde{\delta}, \lambda_0), \ a \le b,$$
$$\mu_1(S_k^{-1}(a) (0\ I)^T, S_k^{-1}(b) (0\ I)^T) = 0 \quad \text{for all } a, b \in (\lambda_0 - \tilde{\delta}, \lambda_0], \ a \le b.$$

Then by Theorem 3.2(iv), we derive (5.4) in Theorem 5.1(i) (i.e., condition (5.327)). By a similar way, conditions (5.13) can be derived. By analogy with the previous proof, we can also show that (5.328) and (5.14) follow from the first equalities in (5.355) and (5.348).

(iii) Note that Lemma 5.104 allows to count the jumps of rank $B_k(\lambda)$ and rank $X_k(\lambda)$ in the right neighborhood of $\lambda_0$. By analogy with the proof of Theorem 5.106, we can show that there exist the finite limits

$$\left.\begin{aligned}
\mu^*\big(\langle S_k(\lambda_0)\rangle, \langle S_k(\lambda_0^+)\rangle\big) = \operatorname{rank} B_k(\lambda_0^+) - \operatorname{rank} B_k(\lambda_0), \\
\mu^*\big(\langle S_k(\lambda_0^-)\rangle, \langle S_k(\lambda_0)\rangle\big) = 0,
\end{aligned}\right\} \quad k \in [0, N]_{\mathbb{Z}},$$

$$\left.\begin{aligned}
\mu^*\big(\langle Z_k(\lambda_0)\rangle, \langle Z_k(\lambda_0^+)\rangle\big) = \operatorname{rank} X_k(\lambda_0^+) - \operatorname{rank} X_k(\lambda_0), \\
\mu^*\big(\langle Z_k(\lambda_0^-)\rangle, \langle Z_k(\lambda_0)\rangle\big) = 0,
\end{aligned}\right\} \quad k \in [0, N+1]_{\mathbb{Z}}.$$

### 5.6.3  Monotonicity and the Cayley Transform

In this section we investigate oscillatory properties of the following $4n \times 4n$ symplectic system

$$\left. \begin{array}{c} \mathcal{Y}_{k+1} = W_k^c(\lambda)\,\mathcal{Y}_k, \quad k \in [0, N]_{\mathbb{z}}, \\[2mm] W_k^c(\lambda) = \dfrac{1}{2} \begin{pmatrix} I + W_k^{T-1}(\lambda) & \mathcal{J}\,[I - W_k(\lambda)] \\[2mm] -\mathcal{J}\,[I - W_k^{T-1}(\lambda)] & I + W_k(\lambda) \end{pmatrix} \end{array} \right\} \tag{5.364}$$

associated with the $2n \times 2n$ symplectic difference system

$$Y_{k+1} = W_k(\lambda)\,Y_k, \quad k \in [0, N]_{\mathbb{z}}, \tag{5.365}$$

with the symplectic matrix $W_k(\lambda)$ satisfying the monotonicity condition (1.201). We will call (5.364) the *Cayley system* associated with (5.365). Motivation for this terminology is the following. Assume additionally that

$$\det\,[I + W_k(\lambda)] \neq 0. \tag{5.366}$$

Then the Cayley transform of $W_k(\lambda)$ is defined as

$$\mathbf{C}(W_k(\lambda)) := [I - W_k(\lambda)]\,[I + W_k(\lambda)]^{-1}, \tag{5.367}$$

and the matrix $\mathcal{J}\mathbf{C}(W_k(\lambda))$ is symmetric (see [326, Lemma 1.1]). Then, if we introduce the notation

$$W_k^c(\lambda) = \begin{pmatrix} \mathfrak{A}_k(\lambda) & \mathfrak{B}_k(\lambda) \\ \mathfrak{C}_k(\lambda) & \mathfrak{D}_k(\lambda) \end{pmatrix}, \tag{5.368}$$

for the blocks of the coefficient matrix in (5.364), then the symmetric matrices

$$\mathfrak{A}_k^{-1}(\lambda)\,\mathfrak{B}_k(\lambda), \quad \mathfrak{C}_k(\lambda)\,\mathfrak{A}_k^{-1}(\lambda), \quad \mathfrak{B}_k(\lambda)\,\mathfrak{D}_k^{-1}(\lambda), \quad \mathfrak{D}_k^{-1}(\lambda)\,\mathfrak{C}_k(\lambda)$$

are connected with the Cayley transforms of the matrices $W_k^{-1}(\lambda)$, $W_k^{T-1}(\lambda)$, $W_k(\lambda)$, $W_k^T(\lambda)$ by the formulas

$$\left. \begin{array}{l} \mathfrak{A}_k^{-1}(\lambda)\,\mathfrak{B}_k(\lambda) = -\mathcal{J}\mathbf{C}(W_k^{-1}(\lambda)), \quad \mathfrak{C}_k(\lambda)\,\mathfrak{A}_k^{-1}(\lambda) = -\mathcal{J}\mathbf{C}(W_k^{-1\,T}(\lambda)), \\[2mm] \mathfrak{B}_k(\lambda)\,\mathfrak{D}_k^{-1}(\lambda) = \mathcal{J}\mathbf{C}(W_k(\lambda)), \quad\quad \mathfrak{D}_k^{-1}(\lambda)\,\mathfrak{C}_k(\lambda) = \mathcal{J}\mathbf{C}(W_k^T(\lambda)). \end{array} \right\} \tag{5.369}$$

System (5.364) is taken into consideration by a modified version of Theorem 4.45, which is the basic tool for the proof of Theorem 5.95. For that proof, we will need the following equivalent form of formulas (4.111) and (4.109) in Theorem 4.45 regarding the multiplicities of focal points of conjoined bases of (4.98) and (SDS).

**Proposition 5.108 (Sturmian Comparison Theorem)** *Equality* (4.111) *in Theorem 4.45 can be written as*

$$
\left.
\begin{aligned}
& m(\hat{Y}_k) - m(Y_k) + \mu(\langle \hat{S}_k \rangle, \langle S_k \rangle) \\
& = \Delta\mu(\langle \hat{Z}_k \rangle, \langle Z_k \rangle) + \mu(R^{-1}\langle Z_{k+1}^{-1} \hat{Z}_{k+1} \rangle, R^{-1}\langle Z_k^{-1} \hat{Z}_k \rangle),
\end{aligned}
\right\}
$$
(5.370)

*where the symplectic orthogonal $4n \times 4n$ matrix $R$ is given by*

$$
R = \frac{1}{\sqrt{2}}
\begin{pmatrix}
0 & -I & I & 0 \\
0 & I & I & 0 \\
-I & 0 & 0 & -I \\
-I & 0 & 0 & I
\end{pmatrix}
$$
(5.371)

*and the comparative index*

$$
\mu\big(R^{-1}\langle Z_{k+1}^{-1} \hat{Z}_{k+1} \rangle, R^{-1}\langle Z_k^{-1} \hat{Z}_k \rangle\big) = \mu^*\big(R^{-1}\langle \hat{Z}_{k+1}^{-1} Z_{k+1} \rangle, R^{-1}\langle \hat{Z}_k^{-1} Z_k \rangle\big)
$$
(5.372)

*is equal to the number of forward focal points in $(k, k+1]$ of the conjoined basis $\mathcal{Y}_k = R^{-1}\langle Z_k^{-1} \hat{Z}_k \rangle$ of the Cayley system*

$$
\mathcal{Y}_{k+1} = W_k^c \mathcal{Y}_k, \quad k \in [0, N]_{\mathbb{Z}}, \quad \text{with } W_k := Z_{k+1}^{-1} \hat{S}_k Z_k.
$$

The proof of Proposition 5.108 is given in Sect. 5.6.4. Point out that we will apply (5.370) to the case $S_k := S_k(\lambda_0)$ and $\hat{S}_k := S_k(\lambda)$, where $\lambda$ belongs to a neighborhood of $\lambda_0$. Then we have to investigate the oscillatory properties of (5.364) under the monotonicity condition (1.201).

**Lemma 5.109** *Let $W(\lambda) \in \mathbb{R}^{2n \times 2n}$ be a continuously differentiable symplectic matrix. Then condition* (1.201) *is equivalent to*

$$
\Psi(W^c(\lambda)) = \mathcal{J}_{4n} \tfrac{d}{d\lambda} [W^c(\lambda)] \, \mathcal{J}_{4n} \, [W^c(\lambda)]^T \mathcal{J}_{4n} \geq 0,
$$

*where $\mathcal{J}_{4n} \in \mathbb{R}^{4n \times 4n}$ and $W^c(\lambda)$ is given by* (5.364). *Moreover, if we assume that $W_k(\lambda)$ obeys* (5.366) *for $k \in [0, N]_{\mathbb{Z}}$, then each of the following conditions for the blocks in* (5.368)

$$
\tfrac{d}{d\lambda} [\mathfrak{A}_k^{-1}(\lambda) \, \mathfrak{B}_k(\lambda)] \geq 0,
$$
(5.373)

$$
\tfrac{d}{d\lambda} [\mathfrak{C}_k(\lambda) \, \mathfrak{A}_k^{-1}(\lambda)] \leq 0,
$$
(5.374)

$$\frac{d}{d\lambda} [\mathfrak{B}_k(\lambda) \mathfrak{D}_k^{-1}(\lambda)] \geq 0, \tag{5.375}$$

$$\frac{d}{d\lambda} [\mathfrak{D}_k^{-1}(\lambda) \mathfrak{C}_k(\lambda)] \leq 0 \tag{5.376}$$

*is equivalent to* (1.201) *for* $W(\lambda) := W_k(\lambda)$.

*Proof* For the proof of the first claim, we use the following representation:

$$W^c = R^{-1}\{I, W\} R, \quad \{I, W\} = \begin{pmatrix} I & 0 & 0 & 0 \\ 0 & A & 0 & B \\ 0 & 0 & I & 0 \\ 0 & C & 0 & D \end{pmatrix}, \quad W = \begin{pmatrix} A & B \\ C & D \end{pmatrix}, \tag{5.377}$$

where $R$ is given by (5.371). Moreover, the matrix $\{I, W\}$ is symplectic if and only if so is $W$ (see Sect. 3.3.5 for more details). Formula (5.377) justifies the symplectic structure of the matrices $W_k^c(\lambda)$ in (5.364) and the construction of conjoined bases of (5.364) in the subsequent proofs (see Lemma 5.110). Applying (5.377), we have

$$\Psi(W^c(\lambda)) = R^T \Psi(\{I, W(\lambda)\}) R = R^T \{0, \Psi(W(\lambda))\} R,$$

where the $4n \times 4n$ matrix $\{0, \Psi(W(\lambda))\}$ is symmetric and nonnegative if and only if so is $\Psi(W(\lambda))$. The proof of the first claim is completed. Next, it is easy to verify that

$$\frac{d}{d\lambda} \left( \frac{1}{2} \mathcal{J} C(W(\lambda)) \right) = [I + W(\lambda)]^{T-1} W^T(\lambda) \Psi(W(\lambda)) W(\lambda) [I + W(\lambda)]^{-1},$$

and then condition (1.201) for $W_k(\lambda)$ is equivalent to

$$\frac{d}{d\lambda} (\mathcal{J} C(W_k(\lambda))) \geq 0. \tag{5.378}$$

Moreover, using the representations (see Proposition 1.76(ii)), we get

$$\Psi(W^{-1}(\lambda)) = -W^T(\lambda) \Psi(W(\lambda)) W(\lambda),$$

$$\Psi(W^{T-1}(\lambda)) = \Psi(\mathcal{J} W(\lambda) \mathcal{J}^T) = \mathcal{J} \Psi(W(\lambda)) \mathcal{J}^T,$$

and $\Psi(W^T(\lambda)) = \Psi(\mathcal{J} W^{-1}(\lambda) \mathcal{J}^T) = \mathcal{J} \Psi(W^{-1}(\lambda)) \mathcal{J}^T$. Thus, we see that equation (1.201) holds if and only if (see Proposition 1.76 (iv))

$$\Psi(W^{-1}(\lambda)) \leq 0, \quad \Psi(W^{-1T}(\lambda)) \geq 0, \quad \Psi(W^T(\lambda)) \leq 0.$$

So we also have that (1.201) for $W_k(\lambda)$ is equivalent to each of the inequalities

$$\frac{d}{d\lambda}\,[\mathcal{J}\,\mathbf{C}(W_k^{-1}(\lambda))] \le 0, \quad \frac{d}{d\lambda}\,[\mathcal{J}\,\mathbf{C}(W_k^{T-1}(\lambda))] \ge 0, \left.\begin{array}{c}\\ \\ \end{array}\right\} \tag{5.379}$$
$$\frac{d}{d\lambda}\,[\mathcal{J}\,\mathbf{C}(W_k^{T}(\lambda))] \le 0.$$

Using (5.369), (5.378), and (5.379), we then complete the proof of the second claim.

□

Recall that we will apply the results of this section to the special case of system (5.364) with $W_k(\lambda) := Z_{k+1}^{-1}(\lambda_0)\,S_k(\lambda)\,Z_k(\lambda_0)$, where $Z^{-1}(\lambda)$ is a fundamental matrix of (SDS$_\lambda$). In this case obviously $W_k(\lambda_0) = I$. In this connection, we prove the following lemma.

**Lemma 5.110** *Assume that a symplectic matrix $W_k(\lambda)$ is continuously differentiable and (1.201) holds for $W(\lambda) := W_k(\lambda)$ for $k \in [0, N]_\mathbb{Z}$. Suppose that for some $\lambda_0 \in \mathbb{R}$*

$$W_k(\lambda_0) = I, \quad k \in [0, N]_\mathbb{Z}. \tag{5.380}$$

*Then there exists $\delta > 0$ such that the principal solution $\mathcal{Y}_k(\lambda)$ at 0 of the Cayley system (5.364) does not have any forward focal points in the interval $(0, N + 1]$ for all $\lambda \in [\lambda_0, \lambda_0 + \delta)$, i.e.,*

$$m(\mathcal{Y}_k(\lambda)) = \mu\big(\mathcal{Y}_{k+1}(\lambda), W_k^c(\lambda)\,(0\ I)^T\big) = 0, \quad k \in [0, N]_\mathbb{Z}. \tag{5.381}$$

*Similarly, there exists $\delta > 0$ such that for all $\lambda \in (\lambda_0 - \delta, \lambda_0]$*

$$\mu^*\big(\mathcal{Y}_{k+1}(\lambda), W_k^c(\lambda)\,(0\ I)^T\big) = 0, \quad k \in [0, N]_\mathbb{Z}. \tag{5.382}$$

*Proof* Note first that (5.380) implies that there exists $\hat{\delta} > 0$ such that (5.366) holds for all $\lambda \in (\lambda_0 - \hat{\delta}, \lambda_0 + \hat{\delta})$, i.e., the continuous matrices $I + W_k(\lambda) = 2I + [W_k(\lambda) - I]$ are nonsingular for $k \in [0, N]$. Then condition (1.201) implies (5.373) and (5.374) for the blocks of $W_k^c(\lambda)$. Moreover, by (5.380), we have $\mathfrak{A}_k^{-1}(\lambda_0)\,\mathfrak{B}_k(\lambda_0) = \mathfrak{C}_k(\lambda_0)\,\mathfrak{A}_k^{-1}(\lambda_0) = 0$, and then we get

$$\mathfrak{A}_k^{-1}(\lambda)\,\mathfrak{B}_k(\lambda) \ge 0, \quad \mathfrak{C}_k(\lambda)\,\mathfrak{A}_k^{-1}(\lambda) \le 0, \quad \lambda \in [\lambda_0, \lambda_0 + \hat{\delta}), \tag{5.383}$$
$$\mathfrak{A}_k^{-1}(\lambda)\,\mathfrak{B}_k(\lambda) \le 0, \quad \mathfrak{C}_k(\lambda)\,\mathfrak{A}_k^{-1}(\lambda) \ge 0, \quad \lambda \in (\lambda_0 - \hat{\delta}, \lambda_0]. \tag{5.384}$$

Secondly, point out that the principal solution at 0 of system (5.364) has the form

$$\mathcal{Y}_k(\lambda) = \begin{pmatrix} \mathcal{X}_k(\lambda) \\ \mathcal{U}_k(\lambda) \end{pmatrix} = Z_k^c(\lambda)\,(0\ I)^T = \frac{1}{2}\begin{pmatrix} \mathcal{J}\,[I - Z_k(\lambda)] \\ I + Z_k(\lambda) \end{pmatrix}, \quad Z_0(\lambda) = I, \tag{5.385}$$

where the symplectic $2n \times 2n$ matrix $Z_k(\lambda)$ solves (5.365). Then (5.11) holds for all $k \in [0, N + 1]_{\mathbb{Z}}$ (see Proposition 5.2). Condition (5.380) implies that $Z_k(\lambda_0) = I$ for $k \in [0, N + 1]_{\mathbb{Z}}$, and then the continuous matrices $I + Z_k(\lambda) = 2I + [Z_k(\lambda) - I]$ for $k \in [0, N + 1]_{\mathbb{Z}}$ are nonsingular in a sufficiently small neighborhood of $\lambda_0$. Then, applying Lemma 5.109 for the case $W(\lambda) := Z_k(\lambda)$, we see by (5.375) that $\frac{\mathrm{d}}{\mathrm{d}\lambda}[\mathcal{X}_k(\lambda)\mathcal{U}_k^{-1}(\lambda)] \geq 0$ and $\mathcal{X}_k(\lambda_0)\mathcal{U}_k^{-1}(\lambda_0) = 0$. Then we can conclude that there exists $\tilde{\delta} > 0$ such that

$$\mathcal{X}_k(\lambda)\mathcal{U}_k^{-1}(\lambda) \geq 0 \quad \text{for all } \lambda \in [\lambda_0, \lambda_0 + \tilde{\delta}), \tag{5.386}$$

$$\mathcal{X}_k(\lambda)\mathcal{U}_k^{-1}(\lambda) \leq 0 \quad \text{for all } \lambda \in (\lambda_0 - \tilde{\delta}, \lambda_0]. \tag{5.387}$$

Finally, we prove (5.381) and (5.382) by using relations which connect the number of focal points of $\mathcal{Y}_k(\lambda)$ and the transformed conjoined basis $\mathcal{J}_{4n}^T \mathcal{Y}_k(\lambda)$, which has the nonsingular upper block $-\mathcal{U}_k(\lambda)$, $\lambda \in (\lambda_0 - \tilde{\delta}, \lambda_0 + \tilde{\delta})$, given by (5.385). By Corollary 4.66 (see (4.183)), we have

$$\left.\begin{aligned}
m(\mathcal{J}_{4n}^T \mathcal{Y}_k(\lambda)) - m(\mathcal{Y}_k(\lambda)) - \Delta\mu\big(\mathcal{J}_{4n}^T \mathcal{Y}_k(\lambda), \mathcal{J}_{4n}^T (0\ I)^T\big) \\
= \operatorname{ind}[-\mathfrak{A}_k^T(\lambda)\,\mathfrak{C}_k(\lambda)] - \operatorname{ind}[\mathfrak{A}_k(\lambda)\,\mathfrak{B}_k^T(\lambda)], \\
\mu(\mathcal{J}_{4n}^T \mathcal{Y}_k(\lambda), \mathcal{J}_{4n}^T (0\ I)^T) = \operatorname{ind}[\mathcal{X}_k^T(\lambda)\,\mathcal{U}_k(\lambda)].
\end{aligned}\right\} \tag{5.388}$$

By (5.386) and (5.383), we have $m(\mathcal{J}_{4n}^T \mathcal{Y}_k(\lambda)) = m(\mathcal{Y}_k(\lambda))$ for all $\lambda \in [\lambda_0, \lambda_0 + \delta)$, where $\delta := \min(\hat{\delta}, \tilde{\delta})$. Moreover, $m(\mathcal{J}_{4n}^T \mathcal{Y}_k(\lambda)) = \operatorname{ind} \mathcal{P}_k(\lambda)$, where

$$\mathcal{P}_k(\lambda) = -\mathcal{U}_k(\lambda)\mathcal{U}_{k+1}^{-1}(\lambda)\,\mathfrak{C}_k(\lambda) = -\mathfrak{A}_k^T(\lambda)\,\mathfrak{C}_k(\lambda) + \mathfrak{C}_k^T(\lambda)\,\mathcal{X}_{k+1}(\lambda)\mathcal{U}_{k+1}^{-1}(\lambda)\,\mathfrak{C}_k(\lambda)$$

and $\mathcal{P}_k(\lambda_0) = 0$. Then, by (5.386) and (5.383), we obtain

$$\mathcal{P}_k(\lambda) = -\mathcal{U}_k(\lambda)\mathcal{U}_{k+1}^{-1}(\lambda)\,\mathfrak{C}_k(\lambda) \geq 0 \quad \text{for all } \lambda \in [\lambda_0, \lambda_0 + \delta), \tag{5.389}$$

$$\mathcal{P}_k(\lambda) = -\mathcal{U}_k(\lambda)\mathcal{U}_{k+1}^{-1}(\lambda)\,\mathfrak{C}_k(\lambda) \leq 0 \quad \text{for all } \lambda \in (\lambda_0 - \delta, \lambda_0]. \tag{5.390}$$

So we complete the proof of (5.381) using (5.389), which implies that $m(\mathcal{Y}_k) = m(\mathcal{J}_{4n}^T \mathcal{Y}_k) = \operatorname{ind} \mathcal{P}_k(\lambda) = 0$ for $\lambda \in [\lambda_0, \lambda_0 + \delta)$. For the proof of (5.382), we use instead of (5.388) the dual identity (see Theorem 3.11)

$$\left.\begin{aligned}
\mu^*\big(\mathcal{Y}_{k+1}(\lambda), W_k^c(\lambda)(0\ I)^T\big) - \operatorname{ind}[-\mathcal{P}_k(\lambda)] - \Delta\mu^*\big(\mathcal{J}_{4n}^T \mathcal{Y}_k(\lambda), \mathcal{J}_{4n}^T (0\ I)^T\big) \\
= \operatorname{ind}[\mathfrak{A}_k^T(\lambda)\,\mathfrak{C}_k(\lambda)] - \operatorname{ind}[-\mathfrak{A}_k(\lambda)\,\mathfrak{B}_k^T(\lambda)], \\
\mu^*(\mathcal{J}_{4n}^T \mathcal{Y}_k(\lambda), \mathcal{J}_{4n}^T (0\ I)^T) = \operatorname{ind}[-\mathcal{X}_k^T(\lambda)\,\mathcal{U}_k(\lambda)],
\end{aligned}\right\}$$

which implies that $\mu^*(\mathcal{Y}_{k+1}(\lambda), W_k^c(\lambda)(0\ I)^T) = 0$ for all $\lambda \in (\lambda_0 - \delta, \lambda_0]$ due to (5.390), (5.387), and (5.384). The proof is complete.  $\qquad\square$

### 5.6.4  Proofs of the Main Results

In this section we provide the rather technical proof of Proposition 5.108 (see Sect. 5.6.3) and the proof of the local oscillation theorem (Theorem 5.95).

*Proof of Proposition 5.108* For the proof of (5.370), we apply the main properties of the comparative index presented in Chap. 3. Note first that with $R$ defined by (5.371), we have $R\,(0_{2n}\ I_{2n})^T = (1/\sqrt{2})\,\langle I_{2n}\rangle$ (see notation (5.344)), where the constant $\sqrt{2}$ can be omitted in computations using Theorem 3.5(i). Applying the properties of the comparative index, we have by Theorem 3.5(v)

$$\Delta\mu(Y_k, \hat{Y}_k) = \Delta \operatorname{rank} w(Y_k, \hat{Y}_k) - \Delta\mu(\hat{Y}_k, Y_k),$$

and by Lemma 3.21(iv)

$$\mu\big(\langle \hat{Z}_k^{-1} Z_k\rangle, \langle \hat{Z}_{k+1}^{-1} Z_{k+1}\rangle\big) = \mu^*\big(\langle \hat{Z}_{k+1}^{-1} Z_{k+1}\rangle, \langle \hat{Z}_k^{-1} Z_k\rangle\big) + \Delta \operatorname{rank} w(Y_k, \hat{Y}_k),$$

where $w(Y_k, \hat{Y}_k)$ is the Wronskian given by (3.2). Here we also use the calculation $(I\ 0)\,\hat{Z}_k^{-1} Z_k (0\ I)^T = w^T(Y_k, \hat{Y}_k)$ (see (4.100)). After substituting the last two formulas into (4.111) and (4.109) in Theorem 4.45, we derive

$$\left. \begin{aligned} m(\hat{Y}_k) - m(Y_k) + \mu\big(\langle \hat{S}_k\rangle, \langle S_k\rangle\big) \\ = \Delta\mu(\hat{Y}_k, Y_k) + \mu^*\big(\langle \hat{Z}_k^{-1} Z_k\rangle, \langle \hat{Z}_{k+1}^{-1} Z_{k+1}\rangle\big). \end{aligned} \right\} \tag{5.391}$$

Now we apply the transformations in Lemma 3.21(v) to get

$$\Delta\mu(\hat{Y}_k, Y_k) = \Delta\mu\big(\langle \hat{Z}_k\rangle, \langle Z_k\rangle\big) - \Delta\mu\big(\langle Z_k^{-1}\hat{Z}_k\rangle, \langle I\rangle\big),$$

in Lemma 3.21(iii) to get

$$\mu^*\big(\langle \hat{Z}_{k+1}^{-1} Z_{k+1}\rangle, \langle \hat{Z}_k^{-1} Z_k\rangle\big) = \mu\big(\langle Z_{k+1}^{-1}\hat{Z}_{k+1}\rangle, \langle Z_k^{-1}\hat{Z}_k\rangle\big),$$

and in Theorem 3.6 to get

$$\mu\big(\langle Z_{k+1}^{-1}\hat{Z}_{k+1}\rangle, \langle Z_k^{-1}\hat{Z}_k\rangle\big) = \mu\big(R(R^{-1}\langle Z_{k+1}^{-1}\hat{Z}_{k+1}\rangle), R(R^{-1}\langle Z_k^{-1}\hat{Z}_k\rangle)\big)$$

$$= \mu\big(R^{-1}\langle Z_{k+1}^{-1}\hat{Z}_{k+1}\rangle, R^{-1}\langle Z_k^{-1}\hat{Z}_k\rangle\big) + \Delta\mu\big(\langle Z_k^{-1}\hat{Z}_k\rangle, \langle I\rangle\big).$$

A subsequent substitution all the formulas derived above into (5.391) leads to the proof of (5.370). Next, identity (5.372) follows from the relation

$$\mu\big(R^{-1}\langle W\rangle, R^{-1}\langle \hat{W}\rangle\big) = \mu^*\big(R^{-1}\langle W^{-1}\rangle, R^{-1}\langle \hat{W}^{-1}\rangle\big), \tag{5.392}$$

which is similar to identity (iii) in Lemma 3.21. Consider the proof of (5.392). Applying Theorem 3.6, we have

$$\mu\big(R^{-1}\langle W\rangle, R^{-1}\langle \hat{W}\rangle\big) = \mu\big(\langle W\rangle, \langle \hat{W}\rangle\big) + \mu\big(\langle \hat{W}\rangle, \langle I\rangle\big) - \mu\big(\langle W\rangle, \langle I\rangle\big),$$

where by Lemma 3.21(iii)

$$\mu\big(\langle W\rangle, \langle \hat{W}\rangle\big) = \mu^*\big(\langle W^{-1}\rangle, \langle \hat{W}^{-1}\rangle\big), \quad \mu\big(\langle \hat{W}\rangle, \langle I\rangle\big) = \mu^*\big(\langle \hat{W}^{-1}\rangle, \langle I\rangle\big),$$
$$\mu\big(\langle W\rangle, \langle I\rangle\big) = \mu^*\big(\langle \hat{W}^{-1}\rangle, \langle I\rangle\big).$$

Then

$$\mu\big(R^{-1}\langle W\rangle, R^{-1}\langle \hat{W}\rangle\big)$$
$$= \mu^*\big(\langle W^{-1}\rangle, \langle \hat{W}^{-1}\rangle\big) + \mu^*\big(\langle \hat{W}^{-1}\rangle, \langle I\rangle\big) - \mu^*\big(\langle W^{-1}\rangle, \langle I\rangle\big)$$
$$= \mu^*\big(R^{-1}\langle W^{-1}\rangle, R^{-1}\langle \hat{W}^{-1}\rangle\big)$$

by the dual version of Theorem 3.6 (see Corollary 3.12). The proof of (5.392) and (5.372) is completed. Finally, note that by Lemma 4.7, the comparative index

$$\mu^*\big(R^{-1}\langle \hat{Z}_{k+1}^{-1} Z_{k+1}\rangle, R^{-1}\langle \hat{Z}_k^{-1} Z_k\rangle\big)$$

is equal to the number of forward focal points in $(k, k+1]$ of the conjoined basis $\mathcal{Y}_k = R^{-1}\langle Z_k^{-1}\hat{Z}_k\rangle = \mathcal{Z}_k(0\ I)^T$ associated with the fundamental matrix

$$\mathcal{Z}_k = \sqrt{2}\, R^{-1}\{I, Z_k^{-1}\hat{Z}_k\}\, R$$

(see notation (5.377)). It is easy to verify directly that $\mathcal{Z}_k$ solves the Cayley system

$$\mathcal{Y}_{k+1} = W_k^c\, \mathcal{Y}_k, \quad k \in [0, N]_{\mathbb{Z}}, \quad \text{with } W_k := Z_{k+1}^{-1}\hat{S}_k Z_k.$$

The proof of Proposition 5.108 is complete.                                    $\square$

*Proof of Theorem 5.95* Fix any $\lambda_0 \in \mathbb{R}$, and apply (5.370) to the case

$$\hat{S}_k := S_k(\lambda), \quad S_k := S_k(\lambda_0), \quad \hat{Z}_k := Z_k(\lambda), \quad Z_k := Z_k(\lambda_0), \quad \lambda \ge \lambda_0,$$

where we assume that symplectic fundamental matrix $Z_k(\lambda)$ does not depend on $\lambda$ for $k = 0$, i.e., $\hat{Z}_0 = Z_0$. Using the notation $m(Y_k(\lambda)) = m_k(\lambda)$, where $Y_k(\lambda) = Z_k(\lambda)\,(0\ I)^T$, we have

$$\left.\begin{aligned}
m_k(\lambda) &- m_k(\lambda_0) + \mu\big(\langle S_k(\lambda)\rangle, \langle S_k(\lambda_0)\rangle\big) \\
&= \Delta\mu\big(\langle Z_k(\lambda)\rangle, \langle Z_k(\lambda_0)\rangle\big) \\
&\quad + \mu^*\big(R^{-1}\langle Z_{k+1}^{-1}(\lambda)Z_{k+1}(\lambda_0)\rangle, R^{-1}\langle Z_k^{-1}(\lambda)Z_k(\lambda_0)\rangle\big),
\end{aligned}\right\} \tag{5.393}$$

where we rewrite the last term according to (5.372). By (5.359) and (5.360) in Theorem 5.106, we evaluate the right-hand limits

$$\mu\big(\langle \mathcal{S}_k(\lambda_0^+)\rangle, \langle \mathcal{S}_k(\lambda_0)\rangle\big) = \Delta\mu\big(\langle Z_k(\lambda_0^+)\rangle, \langle Z_k(\lambda_0)\rangle\big) = 0.$$

According to Lemma 4.7, the last term in (5.393) is equal to the number of forward focal points in $(k, k+1]$ of the conjoined basis

$$\mathcal{Y}_k(\lambda) = R^{-1}\langle Z_k^{-1}(\lambda_0)Z_k(\lambda)\rangle \tag{5.394}$$

of the Cayley system (5.364) with the matrix

$$W_k(\lambda) := Z_{k+1}^{-1}(\lambda_0)\,\mathcal{S}_k(\lambda)\,Z_k(\lambda_0). \tag{5.395}$$

Note that the matrices $Z_k(\lambda_0)$ and $Z_{k+1}(\lambda_0)$ in (5.395) are constant for the fixed $\lambda_0$, and then $\Psi(W_k(\lambda)) = Z_{k+1}^T(\lambda_0)\,\Psi(\mathcal{S}_k(\lambda))\,Z_{k+1}(\lambda_0) \geq 0$ and $W_k(\lambda_0) = I$. Applying (5.381) in Lemma 5.110 and Theorem 3.5(i), we then obtain

$$\mu^*\big(R^{-1}\langle Z_{k+1}^{-1}(\lambda_0^+)Z_{k+1}(\lambda_0)\rangle, R^{-1}\langle Z_k^{-1}(\lambda_0^+)Z_k(\lambda_0)\rangle\big)$$
$$= m(\mathcal{Y}_k(\lambda_0^+)) = \mu\big(\mathcal{Y}_k(\lambda_0^+), W_k^c(\lambda_0^+)\,(0\ I)^T\big) = 0,$$

where $\mathcal{Y}_k(\lambda)$ and $W_k(\lambda)$ are given by (5.394) and (5.395). Substituting all the limits derived above into (5.393), we conclude that there exists the finite limit

$$m_k(\lambda_0^+) = m_k(\lambda_0).$$

Next, applying (5.370) to the case

$$\hat{\mathcal{S}}_k := \mathcal{S}_k(\lambda_0),\ \ \mathcal{S}_k := \mathcal{S}_k(\lambda),\ \ \hat{Z}_k := Z_k(\lambda_0),\ \ Z_k := Z_k(\lambda),\ \ \hat{Z}_0 = Z_0,\ \ \lambda < \lambda_0,$$

we derive

$$\left.\begin{aligned}
m_k(\lambda_0)\, &-m_k(\lambda) + \mu\big(\langle \mathcal{S}_k(\lambda_0)\rangle, \langle \mathcal{S}_k(\lambda)\rangle\big)\\
&= \Delta\mu\big(\langle Z_k(\lambda_0)\rangle, \langle Z_k(\lambda)\rangle\big)\\
&\quad + \mu\big(R^{-1}\langle Z_{k+1}^{-1}(\lambda)Z_{k+1}(\lambda_0)\rangle, R^{-1}\langle Z_k^{-1}(\lambda)Z_k(\lambda_0)\rangle\big).
\end{aligned}\right\} \tag{5.396}$$

Then we evaluate the left-hand limits

$$\mu\big(\langle \mathcal{S}_k(\lambda_0)\rangle, \langle \mathcal{S}_k(\lambda_0^-)\rangle\big) = \vartheta_k(\lambda_0),$$
$$\Delta\mu\big(\langle Z_k(\lambda_0)\rangle, \langle Z_k(\lambda_0^-)\rangle\big) = \Delta\,[\mathrm{rank}\,X_k(\lambda_0^-) - \mathrm{rank}\,X_k(\lambda_0)],$$

according to (5.359) and (5.360) in Theorem 5.106. Note that the last term in (5.396) is dual to the last term in (5.393), i.e.,

$$\mu\big(R^{-1}\langle Z_{k+1}^{-1}(\lambda)Z_{k+1}(\lambda_0)\rangle, R^{-1}\langle Z_k^{-1}(\lambda)Z_k(\lambda_0)\rangle\big) = \mu^*\big(\mathcal{Y}_{k+1}(\lambda), W_k^c(\lambda)\,(0\ I)^T\big),$$

where $\mathcal{Y}_k(\lambda)$ and $W_k(\lambda)$ are given by (5.394) and (5.395). Then, by (5.382) the left-hand limit of this quantity equals to zero, and we derive from (5.396) that

$$m_k(\lambda_0) - m_k(\lambda_0^-) + \vartheta_k(\lambda_0) = \Delta\,[\mathrm{rank}\,X_k(\lambda_0^-) - \mathrm{rank}\,X_k(\lambda_0)].$$

The proof of (5.330) is completed. Upon summing equality (5.330) for $k \in [0, N]_{\mathbb{Z}}$, we get (5.331). The proof of Theorem 5.95 is complete.                              □

## 5.6.5   Weighted Focal Points

In this section we consider the discrete Hamiltonian eigenvalue problems (5.144) presented in Example 5.37, i.e.,

$$\left.\begin{array}{l} \Delta x_k(\lambda) = A_k(\lambda)\,x_{k+1}(\lambda) + B_k(\lambda)\,u_k(\lambda), \\ \Delta u_k(\lambda) = C_k(\lambda)\,x_{k+1}(\lambda) - A_k^T(\lambda)\,u_k(\lambda), \end{array}\right\} \quad k \in [0, N]_{\mathbb{Z}},$$

with the Dirichlet boundary conditions

$$x_0(\lambda) = 0 = x_{N+1}(\lambda), \tag{5.397}$$

where the coefficients $A_k, B_k, C_k : \mathbb{R} \to \mathbb{R}^{n \times n}$ are piecewise continuously differentiable in $\lambda$, $B_k(\lambda)$ and $C_k(\lambda)$ are symmetric, and $I - A_k(\lambda)$ is invertible with $\tilde{A}_k(\lambda) := [I - A_k(\lambda)]^{-1}$. The symmetric Hamiltonian $\mathcal{H}_k(\lambda)$ is defined by

$$\mathcal{H}_k(\lambda) := \begin{pmatrix} -C_k(\lambda) & A_k^T(\lambda) \\ A_k(\lambda) & B_k(\lambda) \end{pmatrix}$$

and it obeys the monotonicity condition (5.337), i.e.,

$$\dot{\mathcal{H}}_k(\lambda) \geq 0, \quad \lambda \in \mathbb{R}, \quad k \in [0, N]_{\mathbb{Z}}.$$

Recall that problem (5.144), (5.397) is the most important special case of the symplectic eigenvalue problem (E), where $\mathcal{S}_k(\lambda)$ has the form (5.145), i.e.,

$$\mathcal{S}_k(\lambda) = \begin{pmatrix} [I - A_k(\lambda)]^{-1} & [I - A_k(\lambda)]^{-1}B_k(\lambda) \\ C_k(\lambda)\,[I - A_k(\lambda)]^{-1} & C_k(\lambda)\,[I - A_k(\lambda)]^{-1}B_k(\lambda) + I - A_k^T(\lambda) \end{pmatrix}.$$

In the previous sections, we have proved generalized versions of the discrete oscillation theorems for (E) with the nonconstant rank of $\mathcal{B}_k(\lambda)$. This approach gives a possibility to compute the number of finite eigenvalues of (E) between arbitrary points $a, b \in \mathbb{R}$ with $a \leq b$ by using the information on the distribution of jumps of the piecewise constant function rank $\mathcal{B}_k(\lambda)$ for $k \in [0, N]_{\mathbb{Z}}$ in $(a, b]$. It is clear that obtaining such an information for problems (E) of big dimension $n \gg 1$ imposes considerable computational difficulties.

In this section we offer an approach how to avoid these difficulties for the special case (5.144), (5.397). For this purpose we introduce a new notion of *weighted focal points* for a conjoined basis of (5.144). As it is shown in Sect. 5.6.1, the classical function $n_1(\lambda)$ of the number of forward focal points in $(0, N+1]$ can lose the monotonic character with respect to $\lambda$ if (5.38) is not satisfied. We offer to introduce a monotonic, but in general indefinite by sign, function $\#(\lambda)$ of the number of weighted focal points in $(0, N+1]$ such that the quantity $\#(b) - \#(a)$ represents the number of finite eigenvalues of (5.144), (5.397) in $(a, b]$ for arbitrary $a, b \in \mathbb{R}$ with $a \leq b$. We prove modified versions of Theorems 5.95 and 5.98, which refer to this new notion of weighted focal points.

Based on the definition of forward focal points and its multiplicities (see Definition 4.1), we introduce the number of weighted focal points as follows.

**Definition 5.111** A conjoined basis $Y(\lambda)$ of the linear Hamiltonian difference system (5.144) has a *weighted (forward) focal point in* $(k, k+1]$ if

$$m(Y_k(\lambda)) - \text{ind } B_k(\lambda) \neq 0.$$

In this case the number of weighted (forward) focal points in $(k, k+1]$ is defined as the quantity

$$\#_k(\lambda) = \#(Y_k(\lambda)) := m(Y_k(\lambda)) - \text{ind } B_k(\lambda), \tag{5.398}$$

where $m(Y_k(\lambda))$ is given by (4.3).

Note that we have the estimate for $m(Y_k(\lambda))$ (see Proposition 4.4(v))

$$0 \leq m(Y_k(\lambda)) \leq \text{rank } B_k(\lambda),$$

and then by the inequality ind $B_k(\lambda) \leq \text{rank } B_k(\lambda)$, we derive that

$$|\#(Y_k(\lambda))| \leq \text{rank } B_k(\lambda) \leq n. \tag{5.399}$$

In the following example, we illustrate Definition 5.111 for the scalar case $n = 1$. This example also shows that the quantity $\#(Y_k(\lambda))$ can be negative.

*Example 5.112* Consider system (5.144) for the case $n = 1$ with $A_k(\lambda) \equiv 0$. Then the number of weighted focal points takes values from the set $\{0, \pm 1\}$ and

$$\#(Y_k(\lambda)) = \begin{cases} 1, & \text{if } B_k(\lambda) > 0, \ x_k(\lambda) \neq 0, \ x_k(\lambda)\, x_{k+1}(\lambda) \leq 0, \\ -1, & \text{if } B_k(\lambda) < 0, \ x_{k+1}(\lambda) \neq 0, \ x_k(\lambda)\, x_{k+1}(\lambda) \leq 0, \\ 0, & \text{otherwise.} \end{cases}$$

Let $Y(\lambda) = (X(\lambda), U(\lambda))$ be a conjoined basis of (5.144) such that (5.20) holds. We denote the number of weighted focal points of $Y(\lambda)$ in $(0, N + 1]$ by

$$\#(\lambda) := \sum_{k=0}^{N} \#_k(\lambda). \tag{5.400}$$

The main result of this section is the following modification of the local oscillation theorem (see Theorem 5.95) for problem (5.144), (5.397) dealing with the number of weighted focal points of a conjoined basis $Y(\lambda)$ of (5.144).

**Theorem 5.113 (Generalized Local Oscillation Theorem)**  *Suppose that (5.337) holds and $Y(\lambda) = (X(\lambda), U(\lambda))$ is a conjoined basis of (5.144) satisfying (5.20). Then the left- and right-hand limits $\#_k(\lambda^-)$ and $\#_k(\lambda^+)$ exist for all $k \in [0, N]_{\mathbb{Z}}$ and $\lambda \in \mathbb{R}$, and*

$$\left. \begin{aligned} \#_k(\lambda^+) &= \#_k(\lambda), \\ \#_k(\lambda^+) - \#_k(\lambda^-) &= \operatorname{rank} X_k(\lambda) - \operatorname{rank} X_k(\lambda^-) \\ &\quad + \operatorname{rank} X_{k+1}(\lambda^-) - \operatorname{rank} X_{k+1}(\lambda). \end{aligned} \right\} \tag{5.401}$$

*Moreover, with the notation in (5.400), we have*

$$\left. \begin{aligned} \#(\lambda^+) &= \#(\lambda), \quad |\#(\lambda)| \leq n(N + 1), \\ \#(\lambda^+) - \#(\lambda^-) &= \operatorname{rank} X_{N+1}(\lambda^-) - \operatorname{rank} X_{N+1}(\lambda). \end{aligned} \right\} \tag{5.402}$$

*Proof* Note that monotonicity condition (5.337) implies $\dot{B}_k(\lambda) \geq 0$, and then all eigenvalues $\mu_i(\lambda)$ for $i \in \{1, 2, \dots, n\}$ of $B_k(\lambda) = B_k^T(\lambda)$ are real nondecreasing functions of $\lambda$. The function $\operatorname{ind} \mu_i(\lambda)$ of the scalar argument $\mu_i(\lambda)$ is piecewise constant, nonincreasing, and right-continuous, then so is the function $\operatorname{ind} B_k(\lambda)$ for $k \in [0, N]_{\mathbb{Z}}$. Similarly, the function $\operatorname{ind}[-\mu_i(\lambda)]$ of the scalar argument $-\mu_i(\lambda)$ is piecewise constant, nondecreasing, and left-continuous, so is the function $\operatorname{ind}[-B_k(\lambda)]$ for $k \in [0, N]_{\mathbb{Z}}$. Hence, for any $\lambda \in \mathbb{R}$, there exist the following left-hand and right-hand limits

$$\operatorname{ind} B_k(\lambda^+) = \operatorname{ind} B_k(\lambda), \quad \operatorname{ind}[-B_k(\lambda^-)] = \operatorname{ind}[-B_k(\lambda)], \tag{5.403}$$

$$\operatorname{ind} B_k(\lambda^-) = \operatorname{rank} B_k(\lambda^-) - \operatorname{rank} B_k(\lambda) + \operatorname{ind} B_k(\lambda), \tag{5.404}$$

$$\operatorname{ind}[-B_k(\lambda^+)] = \operatorname{rank} B_k(\lambda^+) - \operatorname{rank} B_k(\lambda) + \operatorname{ind}[-B_k(\lambda)]. \tag{5.405}$$

By Theorem 5.95 proved for (E) with the nonconstant rank of $\mathcal{B}_k(\lambda)$, we have that (see (5.330))

$$m_k(\lambda^+) = m_k(\lambda) \le n,$$

$$m_k(\lambda^+) - m_k(\lambda^-) + \operatorname{rank} \mathcal{B}_k(\lambda^-) - \operatorname{rank} \mathcal{B}_k(\lambda)$$

$$= \operatorname{rank} X_k(\lambda) - \operatorname{rank} X_k(\lambda^-) + \operatorname{rank} X_{k+1}(\lambda^-) - \operatorname{rank} X_{k+1}(\lambda),$$

where $\operatorname{rank} \mathcal{B}_k(\lambda) = \operatorname{rank} B_k(\lambda)$ according to (5.145). Then, by (5.403) and (5.330), the function $\#_k(\lambda)$ is right-continuous, and the left-hand side of the second equality in (5.330) can be replaced by $\#_k(\lambda^+) - \#_k(\lambda^-)$ according to (5.404). The proof of (5.401) is completed. Note that the inequality in (5.402) follows from (5.399) and the definition of $\#(\lambda)$. Summing (5.401) for $k \in [0, N]_{\mathbb{Z}}$ and using that $Y_k(\lambda)$ does not depend on $\lambda$ for $k = 0$, we complete the proof of (5.402).                    $\square$

Note that (5.337) also implies that there exist the finite limits

$$\lim_{\lambda \to -\infty} \operatorname{ind} \mathcal{B}_k(\lambda) = c_k < \infty, \quad k \in [0, N]_{\mathbb{Z}},$$

i.e., there exists $\lambda_{00}$ such that

$$\operatorname{ind} \mathcal{B}_k(\lambda) = c_k, \quad \sum_{k=0}^{N} \operatorname{ind} \mathcal{B}_k(\lambda) = \sum_{k=0}^{N} c_k =: C, \quad \text{for } \lambda < \lambda_{00} \tag{5.406}$$

(we use again that $\operatorname{ind} \mathcal{B}_k(\lambda)$ is monotonic and bounded). Repeating the proof of Theorem 5.16, we conclude by (5.402) that $\#(\lambda)$ is bounded and nondecreasing for all $\lambda \in \mathbb{R}$, so that there exists $\lambda_0$ such that

$$\#(\lambda) \equiv p, \quad p = m - C, \quad n_1(\lambda) \equiv m, \quad \text{for } \lambda < \lambda_0. \tag{5.407}$$

Finally, we see by (5.407) and (5.402) that

$$\operatorname{rank} X_{N+1}(\lambda) = \operatorname{rank} X_{N+1}(\lambda^-), \quad \text{for } \lambda < \lambda_0, \tag{5.408}$$

i.e., the finite spectrum of (5.144), (5.397) is bounded from below (see also Remark 5.97(ii)).

Recall the notation

$$n_2(\lambda) := \sum_{\mu \le \lambda} \theta(\mu)$$

for the number of finite eigenvalues of (5.144), (5.397) less than or equal to $\lambda$. As a corollary to Theorem 5.113, we derive the following modification of Theorem 5.98 for the special case (5.144), (5.397).

**Theorem 5.114 (Generalized Global Oscillation Theorem)**   *Assume that* (5.337) *holds. Then the finite spectrum of* (5.144), (5.397) *is bounded from below, and there exists* $p \in \{0, \pm 1, \ldots, \pm n(N + 1)\}$ *such that*

$$n_2(\lambda) + p = \#(\lambda) \quad \text{for all } \lambda \in \mathbb{R},   \tag{5.409}$$

*where p is given by* (5.407).

*Proof* Recall again that all break points of rank $X_{N+1}(\lambda)$ are isolated, then using that $\#(\lambda)$ is right-continuous we derive for $a, b \in \mathbb{R}$ with $a < b$ that

$$\#(b) - \#(a) = \sum_{\mu \in (a,b]} \theta(\mu),   \tag{5.410}$$

where the sum $\sum_{\mu \in (a,b]} \theta(\mu)$ is finite. The proof of (5.409) follows from (5.410) for $b := \lambda$, where we evaluate the finite limits of the quantities $\#(a)$ and $\sum_{\mu \in (a,\lambda]} \theta(\mu)$ as $a \to -\infty$ according to (5.407) and (5.408).                                                                  □

**Corollary 5.115**   *Assume* (5.337) *and suppose that the functions* ind $B_k(\lambda)$ *for* $k \in [0, N]_{\mathbb{Z}}$ *are constant in* $\lambda \in \mathbb{R}$ *(depending on k), i.e.,* (5.406) *holds for all* $\lambda \in \mathbb{R}$. *Then under the notation of Theorem 5.114 we have that there exists* $m \in [0, nN]_{\mathbb{Z}}$ *such that*

$$n_2(\lambda) + m = n_1(\lambda) \quad \text{for all } \lambda \in \mathbb{R},   \tag{5.411}$$

*where m is given by* (5.407).

The next example shows that $n_1(\lambda)$ is generally nonmonotonic in the setting of this section. However, the addition of the "correction term" $-$ ind $B_k$, according to (5.398), implies that under (5.337) the function $\#(\lambda)$ is monotonic in $\lambda$.

*Example 5.116*   Consider problem (5.144), (5.397) for $n = 1$ and $N = 3$ with the Hamiltonian

$$\mathcal{H}_k(\lambda) = \text{diag}\{\lambda, \ \lambda - k - 1\},$$

for which condition (5.337) is obviously satisfied. By (5.145), the matrix of the associated symplectic system is

$$S_k(\lambda) = \begin{pmatrix} 1 & \lambda - k - 1 \\ -\lambda & -\lambda^2 + (k+1)\lambda + 1 \end{pmatrix}.$$

The upper components of the principal solution at $k = 0$ are

$$x_0^{[0]}(\lambda) = 0, \quad x_1^{[0]}(\lambda) = \lambda - 1, \quad x_2^{[0]}(\lambda) = -\lambda^3 + 3\lambda^2 - 3,$$

$$x_3^{[0]}(\lambda) = \lambda^5 - 6\lambda^4 + 7\lambda^3 + 10\lambda^2 - 11\lambda - 6,$$

$$x_4^{[0]}(\lambda) = -\lambda^7 + 10\lambda^6 - 29\lambda^5 + 5\lambda^4 + 69\lambda^3 - 20\lambda^2 - 50\lambda - 10.$$

The finite eigenvalues are the roots of the equation $x_4^{[0]}(\lambda) = 0$, i.e.,

$$\lambda_1 \approx -1.0710, \quad \lambda_2 \approx -0.6565, \quad \lambda_3 \approx -0.2422, \quad \lambda_4 \approx 1.4278,$$

$$\lambda_5 \approx 2.7005, \quad \lambda_6 \approx 3.5417, \quad \lambda_7 \approx 4.2998.$$

The first graph of Fig. 5.1 illustrates nonmonotonic character of the number $n_1(\lambda)$ of forward focal points in $(0, 4]$ evaluated according to (5.49) in the interval $\lambda \in [-1.3, 4.5]$. The second graph of Fig. 5.1 shows the number of weighted focal points in $(0, 4]$ computed according to Example 5.112. We can see that the function $\#(\lambda)$ is nondecreasing with respect to $\lambda$ with the jump discontinuities located in the zeros $\lambda_i$ for $i \in \{1, \ldots, 7\}$ of the function $x_4^{[0]}(\lambda)$ (see the last graph of Fig. 5.1).

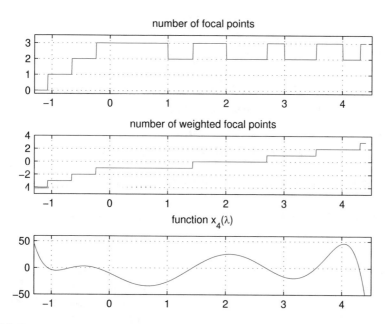

**Fig. 5.1** The graphs of Example 5.116

### 5.6.6  General Endpoints and Nonconstant Rank

In this subsection we present generalizations of the results of Sect. 5.2.3 without the assumptions (5.210), i.e., without

$$\left.\begin{array}{l} \text{rank } \mathcal{B}_k(\lambda) \text{ is constant for } \lambda \in \mathbb{R} \text{ for all } k \in [0, N]_\mathbb{z} \\ \text{rank } R_0(\lambda) \text{ and } \text{rank } R_{N+1}(\lambda) \text{ are constant for } \lambda \in \mathbb{R} \end{array}\right\}$$

in the global oscillation theorem for separated endpoints (Theorem 5.50). Or, for the more general case, we omit assumptions (5.226), i.e.,

$$\left.\begin{array}{l} \text{rank } \mathcal{B}_k(\lambda) \text{ is constant for } \lambda \in \mathbb{R} \text{ for all } k \in [0, N]_\mathbb{z} \\ \text{rank } R_2(\lambda) \text{ is constant for } \lambda \in \mathbb{R} \end{array}\right\}$$

used in Theorem 5.55 for the joint endpoints.

Under the notation of Sect. 5.2.3, we consider the eigenvalue problem (5.1) with general boundary conditions. First we consider the separated endpoints (5.155).

Recall that we use the notation $\bar{Y}(\lambda) = (\bar{X}(\lambda), \bar{U}(\lambda))$ for the natural conjoined basis of system (5.1) with the initial conditions (5.204)

$$\bar{X}_0(\lambda) = -R_0^T(\lambda), \quad \bar{U}_0(\lambda) = R_0^{*T}(\lambda), \quad \lambda \in \mathbb{R}.$$

and the auxiliary matrices (5.205)

$$\left.\begin{array}{l} \Lambda(\lambda) := R_{N+1}^*(\lambda)\, \bar{X}_{N+1}(\lambda) + R_{N+1}(\lambda)\, \bar{U}_{N+1}(\lambda), \\ M(\lambda) := [I - \Lambda(\lambda)\, \Lambda^\dagger(\lambda)]\, R_{N+1}(\lambda), \\ T(\lambda) := I - M^\dagger(\lambda)\, M(\lambda), \\ P(\lambda) := T(\lambda)\, \bar{X}_{N+1}(\lambda)\, \Lambda^\dagger(\lambda)\, R_{N+1}(\lambda)\, T(\lambda). \end{array}\right\}$$

Then, according to (5.207), (5.208), and (5.209), we introduce the numbers

$$n_1(\lambda) := \text{ number of forward focal points of } \bar{Y}(\lambda) \text{ in } (0, N+1], \tag{5.412}$$

$$p(\lambda) := \text{rank } M(\lambda) + \text{ind } P(\lambda). \tag{5.413}$$

The consideration is based on Theorem 5.41 about the equivalence between problem (5.1), (5.155) and the extended problem (5.162). According to Theorem 5.41, the matrices $S_{-1}(\lambda)$ and $S_{N+1}(\lambda)$ given by (5.171) and (5.173) obey the monotonicity assumption (5.163). Then we consider the numbers

$$\left.\begin{array}{l} \vartheta_{-1}(\lambda_0) := \text{rank } R_0(\lambda_0^-) - \text{rank } R_0(\lambda_0), \\ \vartheta_{N+1}(\lambda_0) := \text{rank } R_{N+1}(\lambda_0^-) - \text{rank } R_{N+1}(\lambda_0), \end{array}\right\} \tag{5.414}$$

which count the jumps of the rank of the matrices $R_0(\lambda)$ and $R_{N+1}(\lambda)$ in the boundary conditions (5.155). Moreover, one can apply the local oscillation theorem with nonconstant rank (see Theorem 5.95) to the principal solution $Y_k^{[-1]}(\lambda)$ for $k \in [-1, N+1]_\mathbb{Z}$ of problem (5.162) and then incorporate the connection between $Y_k^{[-1]}(\lambda)$ for $k \in [-1, N+1]_\mathbb{Z}$ and $\bar{Y}(\lambda) = (\bar{X}(\lambda), \bar{U}(\lambda))$ given by (5.216). By analogy with the proof of Theorem 5.50, we derive the following result.

**Theorem 5.117 (Local Oscillation Theorem for Separated Endpoints)** *Assume that* (5.3), (5.156), (5.157), *and* (5.158) *hold. Let* $\bar{Y}(\lambda)$ *be the natural conjoined basis of* $(SDS_\lambda)$ *with* (5.204) *and* $m_k(\lambda)$ *for* $k \in [0, N]_\mathbb{Z}$ *are the multiplicities of forward focal points of* $\bar{Y}(\lambda)$. *Then* $m_k(\lambda^-)$ *and* $m_k(\lambda^+)$ *exist for all* $\lambda \in \mathbb{R}$ *and satisfy* (5.330) *for* $k \in [0, N]_\mathbb{Z}$. *Moreover, for* $k \in \{-1, N+1\}$, *we have*

$$\vartheta_{-1}(\lambda) = \operatorname{rank} X_0(\lambda^-) - \operatorname{rank} X_0(\lambda) \tag{5.415}$$

*and*

$$\left. \begin{aligned} p(\lambda^+) &= p(\lambda), \\ p(\lambda^+) - p(\lambda^-) + \vartheta_{N+1}(\lambda) &= \operatorname{rank} X_{N+1}(\lambda) - \operatorname{rank} X_{N+1}(\lambda^-) \\ &\quad + \operatorname{rank} \Lambda(\lambda^-) - \operatorname{rank} \Lambda(\lambda), \end{aligned} \right\} \tag{5.416}$$

*where* $p(\lambda)$ *is given by* (5.209). *Instead of* (5.331), *we have*

$$\left. \begin{aligned} n_1(\lambda^+) &= n_1(\lambda) \leq n(N+1), \\ n_1(\lambda^+) - n_1(\lambda^-) + p(\lambda^+) - p(\lambda^-) + \sum_{k=-1}^{N+1} \vartheta_k(\lambda) \\ &= \operatorname{rank} \Lambda(\lambda^-) - \Lambda(\lambda) = \theta(\lambda), \end{aligned} \right\} \tag{5.417}$$

*where* $\theta(\lambda)$ *is given by* (5.206).

*Proof* For the proof it is sufficient to repeat the proof of Theorem 5.50 replacing Theorems 5.15 and 5.16 by Theorem 5.95 formulated for the nonconstant rank of $\mathcal{B}_k(\lambda)$. In particular, we incorporate that rank $X_{N+2}^{[-1]}(\lambda) = \operatorname{rank} \Lambda(\lambda)$,

$$m(Y_{-1}^{[-1]}(\lambda^-)) = m(Y_{-1}^{[-1]}(\lambda)) = m(Y_{-1}^{[-1]}(\lambda^+)) = 0,$$

$$m(Y_k^{[-1]}(\lambda)) = m(\bar{Y}_k(\lambda)), \quad k \in [0, N]_\mathbb{Z},$$

$$m(Y_{N+1}^{[-1]}(\lambda)) = p(\lambda)$$

which completes the proof. $\qquad\qquad\square$

To formulate the global oscillation theorem, we assume as in the previous sections that there exists $\lambda_{00} < 0$ such that

$$\operatorname{rank} \mathcal{B}_k(\lambda) = \operatorname{rank} \mathcal{B}_k(\lambda^-) \quad \text{for all } \lambda < \lambda_{00}, \quad k \in [0, N]_{\mathbb{Z}},$$

for the blocks of the symplectic matrix $\mathcal{S}_k(\lambda)$. In addition, we introduce a similar assumption for the blocks of the separated boundary conditions (5.155)

$$\left.\begin{aligned}
\operatorname{rank} R_0(\lambda) &= \operatorname{rank} R_0(\lambda^-) \quad \text{for all } \lambda < \lambda_{00}, \\
\operatorname{rank} R_{N+1}(\lambda) &= \operatorname{rank} R_{N+1}(\lambda^-) \quad \text{for all } \lambda < \lambda_{00}.
\end{aligned}\right\} \tag{5.418}$$

Based on Corollary 5.96, we prove a necessary and sufficient condition for the boundedness from below the spectrum of problem (5.1), (5.155) with the nonconstant rank of $R_0(\lambda)$ and $R_{N+1}(\lambda)$.

**Corollary 5.118** *Assume that* (5.3), (5.156), (5.157), *and* (5.158) *hold. Then conditions* (5.333) *and* (5.418) *are satisfied if and only if the finite spectrum of* (5.1), (5.155) *is bounded from below.*

*Proof* The proof is based on Theorem 5.41 about the equivalence between problem (5.1), (5.155) and the extended problem (5.162). Then it is sufficient to apply Corollary 5.96 to the extended problem, where the blocks of $\mathcal{S}_{-1}(\lambda)$ and $\mathcal{S}_{N+1}(\lambda)$ are defined by the matrices in (5.155) according to (5.171) and (5.173).  □

Under assumptions (5.333), (5.418), we introduce the notation

$$n_2(\lambda) := \text{ number of finite eigenvalues of (5.1), (5.155)}$$

$$\text{in the interval } (-\infty, \lambda],$$

$$n_{\tilde{\mathcal{B}}}(\lambda) := \sum_{\mu \leq \lambda} \sum_{k=-1}^{N+1} \vartheta_k(\mu), \tag{5.419}$$

where $\vartheta_k(\mu)$ for $k \in [0, N+1]_{\mathbb{Z}}$ are defined by (5.329) and (5.414). Note that conditions (5.3), (5.333), (5.418), and formula (5.417) imply the existence of $\tilde{\lambda} \in \mathbb{R}$ such that

$$n_2(\lambda) \equiv 0, \quad n_{\tilde{\mathcal{B}}}(\lambda) \equiv 0, \quad n_1(\lambda) + p(\lambda) \equiv \ell \quad \text{for all } \lambda < \tilde{\lambda}. \tag{5.420}$$

Moreover, the functions $n_2(\lambda)$ and $n_{\tilde{\mathcal{B}}}(\lambda)$ are right-continuous by Theorems 5.3, 5.1 and Corollary 5.7. From Corollary 5.118 and Theorem 5.117, we derive the following generalization of the global oscillation theorem (Theorem 5.50).

**Theorem 5.119 (Global Oscillation Theorem for Separated Endpoints)**
*Assume* (5.3), (5.156), (5.157), (5.158), (5.333), *and* (5.418). *Then there exists a number* $\ell \in [0, (N+2)\,n]_{\mathbb{Z}}$ *such that*

$$n_1(\lambda) + p(\lambda) + n_{\tilde{\mathcal{B}}}(\lambda) = n_2(\lambda) + \ell \quad \text{for all } \lambda \in \mathbb{R}, \tag{5.421}$$

*where* $\ell$ *is given by* (5.420).

*Proof* The proof is based on Theorem 5.117, and it repeats the proof of Theorem 5.98. The details are therefore omitted.      □

*Remark 5.120* Note that the number $n_{\tilde{\mathcal{B}}}(\lambda)$ given by (5.419) can be presented in the form

$$n_{\tilde{\mathcal{B}}}(\lambda) = n_{\mathcal{B}}(\lambda) + n_L(\lambda) + n_R(\lambda), \tag{5.422}$$

where

$$n_L(\lambda) := \sum_{\mu \le \lambda} \vartheta_{-1}(\mu), \tag{5.423}$$

$$n_R(\lambda) := \sum_{\mu \le \lambda} \vartheta_{N+1}(\mu), \tag{5.424}$$

where $n_{\mathcal{B}}(\lambda)$ is defined by (5.338). These three numbers characterize the contribution of $\mathcal{S}_k(\lambda)$ for $k \in [0, N]_{\mathbb{Z}}$ and the left and the right boundary condition, respectively, to the spectrum of problem (5.1), (5.155). Point out that the matrices $\mathcal{S}_k(\lambda)$ for $k \in [0, N]_{\mathbb{Z}}$ and $R_0(\lambda)$ and $R_{N+1}(\lambda)$ (and then also the quantities $n_{\mathcal{B}}(\lambda)$, $n_L(\lambda)$, and $n_R(\lambda)$) are known, while the numbers $n_1(\lambda)$ and $p(\lambda)$ have to be computed using the conjoined basis $\bar{Y}(\lambda)$.

In the following example, we analyze the main results of this subsection for a discrete Sturm-Liouville eigenvalue problem with linear and quadratic dependence on $\lambda$ in the boundary conditions (5.155). This analysis is motivated by the results in [150, 166, 167].

*Example 5.121* Consider the Sturm-Liouville eigenvalue problem

$$\left. \begin{array}{l} -\Delta(r_k \Delta x_k) + q_k x_{k+1} = \lambda w_k x_{k+1}, \quad r_k \ne 0, \quad w_k > 0, \quad k \in [0, N-1]_{\mathbb{Z}}, \\ R_0^*(\lambda)\, x_0 + R_0(\lambda)\,(r_0 \Delta x_0) = 0, \quad R_{N+1}^*(\lambda)\, x_{N+1} + R_{N+1}(\lambda)\,(r_N \Delta x_N) = 0. \end{array} \right\} \tag{5.425}$$

Recall that according to Example 5.35, problem (5.425) can be rewritten in form (5.1) (or (SDS$_\lambda$)), (5.155) after the replacement $u_k := r_k \Delta x_k$, $k \in [0, N]_{\mathbb{Z}}$, $u_{N+1} := u_N$, with the same matrices in the boundary conditions. For the case $n = 1$, conditions (5.157)–(5.158) mean that the function in the left-hand side is nonnegative, resp., nonpositive, for all $\lambda \in \mathbb{R}$, while (5.156) mean that the pair of functions $R_j^*(\lambda)$ and $R_j(\lambda)$ for $j \in \{0, N + 1\}$ do not have common zeros. In particular, if $R_j^*(\lambda)$ and $R_j(\lambda)$ are polynomials, then one can use the resultant res $(R_j^*(\lambda), R_j(\lambda))$ (or the eliminant) of two polynomials (see [152]), which is a polynomial expression of their coefficients, to formulate a necessary and sufficient condition for the validity of (5.156). Note that we exclude the situation when $R_j^*(\lambda) \equiv 0$ (or $R_j(\lambda) \equiv 0$) from the consideration, because in this case the boundary conditions in (5.425) can be reduced to be independent on $\lambda$ (after dividing by $R_j(\lambda) \neq 0$ for $\lambda \in \mathbb{R}$ or, respectively, after dividing by $R_j^*(\lambda) \neq 0$ for $\lambda \in \mathbb{R}$).

For example, for the general quadratic dependence on $\lambda$

$$R_j^*(\lambda) = a_j^* \lambda^2 + b_j^* \lambda + c_j^*, \quad R_j(\lambda) = a_j \lambda^2 + b_j \lambda + c_j, \quad j \in \{0, N+1\} \quad (5.426)$$

(including the general linear dependence for $a_j^* = 0 = a_j$) conditions (5.157)–(5.158) take the form

$$\left.\begin{array}{c} (-1)^{j/(N+1)} \left(-m_j \lambda^2 + 2 l_j \lambda - h_j\right) \geq 0, \\ m_j := b_j^* a_j - a_j^* b_j, \quad l_j := a_j^* c_j - c_j^* a_j, \quad h_j := c_j^* b_j - b_j^* c_j, \\ j \in \{0, N + 1\}, \end{array}\right\} \quad (5.427)$$

while condition (5.156) holds if and only if the resultant res $(R_j^*(\lambda), R_j(\lambda)) \neq 0$, $\lambda \in \mathbb{R}$ (see [152]). In particular, for the linear dependence $(a_j^* = 0 = a_j)$, we have res $(R_j^*(\lambda), R_j(\lambda)) = -h_j$ for the case $b_j \neq 0$ and $b_j^* \neq 0$, and similarly res $(R_j^*(\lambda), R_j(\lambda)) = c_j$ (or res $(R_j^*(\lambda), R_j(\lambda)) = c_j^*$) for the case $b_j = 0$ and $b_j^* \neq 0$ (or $b_j^* = 0$ and $b_j \neq 0$). Therefore, the condition $(-1)^{j/(N+1)} h_j < 0$ for $j \in \{0, N + 1\}$ is necessary and sufficient for $(b_j^*)^2 + b_j^2 > 0$, (5.427), and (5.156) (compare with the main assumptions in [150, 166]). For the quadratic case $a_j \neq 0$ and $a_j^* \neq 0$, the resultant takes the form res $(R_j^*(\lambda), R_j(\lambda)) = m_j h_j - l_j^2$. Therefore, we can see that in this case, the condition $(-1)^{j/(N+1)} M_j < 0$ for $j \in \{0, N + 1\}$, where $M_j := \begin{pmatrix} h_j & l_j \\ l_j & m_j \end{pmatrix}$, is necessary and sufficient for (5.427), (5.156). The situation when $a_j = 0$ or $a_j^* = 0$ can be investigated separately by using the general definition of res $(R_j^*(\lambda), R_j(\lambda))$ (see [152] for more details). In any case one can use $(-1)^{j/(N+1)} M_j < 0$ for $j \in \{0, N + 1\}$ as a sufficient condition for (5.427) and (5.156); see [167, Lemmas 4.6 and 4.7]. Regarding the eigenvalues of problem (5.425), it was proven in [150, 166] that for the linear case $(a_j^* = 0 = a_j)$ and under the assumptions $r_k > 0$, (5.156), and (5.157)–(5.158), the problem (5.425) has a finite number of real simple eigenvalues. The same property is proven in [167] for the case (5.426) with the boundary conditions $x_0 = 0$ and $R_{N+1}^*(\lambda) x_{N+1} + R_{N+1}(\lambda) (r_N \Delta x_N) = 0$, where $M_{N+1} > 0$.

Consider now the eigenvalue problem (5.425) with the quadratic dependence in $\lambda$ in the boundary conditions

$$\left.\begin{array}{c} -\Delta^2 x_k - 2 x_{k+1} = \lambda x_{k+1}, \quad k \in [0, N - 1]_{\mathbb{Z}}, \\ x_0 = 0, \quad (\lambda^2 + \lambda - 1) x_{N+1} - \lambda \Delta x_N = 0, \end{array}\right\} \quad (5.428)$$

where the parameters $l_{N+1} = 0$, $m_{N+1} = h_{N+1} = 1$ given by (5.427) obey the condition $M_{N+1} > 0$. Then by [167] the problem (5.428) has a finite number of real simple eigenvalues. By putting $u_k := x_{k+1}$ and $y_k = (x_k, u_k)^T$, we rewrite this problem in the form of (5.1), (5.155), i.e.,

$$\left. \begin{array}{c} y_{k+1} = \begin{pmatrix} 0 & 1 \\ -1 & -\lambda \end{pmatrix} y_k, \quad k \in [0, N-1]_{\mathbb{Z}}, \\ x_0 = 0, \quad \lambda x_N + (\lambda^2 - 1) u_N = 0. \end{array} \right\} \tag{5.429}$$

Note that assumptions (5.155), (5.156) hold for (5.429), and then consider the application of Theorem 5.119 to this problem for the case $N = 3$. The principal solution $y^{[0]}(\lambda)$ of (5.429) at $k = 0$ and the function $\Lambda(\lambda)$ in (5.205) have the form

$$y_0^{[0]}(\lambda) = \begin{pmatrix} 0 \\ 1 \end{pmatrix}, \quad y_1^{[0]}(\lambda) = \begin{pmatrix} 1 \\ -\lambda \end{pmatrix}, \quad y_2^{[0]}(\lambda) = \begin{pmatrix} -\lambda \\ -1 + \lambda^2 \end{pmatrix},$$

$$y_3^{[0]}(\lambda) = \begin{pmatrix} -1 + \lambda^2 \\ 2\lambda - \lambda^3 \end{pmatrix}, \quad \Lambda(\lambda) = \lambda(\lambda^2 - 1)(3 - \lambda^2).$$

In this case we have the real simple eigenvalues

$$\lambda_1 = -\sqrt{3}, \quad \lambda_2 = -1, \quad \lambda_3 = 0, \quad \lambda_4 = 1, \quad \lambda_5 = \sqrt{3}.$$

We compute the multiplicities of focal points of the principal solution $y^{[0]}(\lambda)$ by

$$m_0(\lambda) = 0, \quad m_1(\lambda) = \begin{cases} 1, & \lambda \in [0, \infty), \\ 0, & \lambda \in (-\infty, 0), \end{cases}$$

$$m_2(\lambda) = \begin{cases} 1, & \lambda \in [-1, 0) \cup [1, \infty), \\ 0, & \text{otherwise,} \end{cases}$$

and the number $p(\lambda)$ given by (5.413) as follows

$$p(\lambda) = \begin{cases} 1, & \lambda \in [-\sqrt{3}, -1) \cup [0, 1) \cup [\sqrt{3}, \infty), \\ 0, & \text{otherwise.} \end{cases}$$

Then we have

$$n_1(\lambda) + p(\lambda) = \begin{cases} 0, & \lambda \in (-\infty, -\sqrt{3}), \\ 1, & \lambda \in [-\sqrt{3}, 0), \\ 2, & \lambda \in [0, \sqrt{3}), \\ 3, & \lambda \in [\sqrt{3}, \infty), \end{cases}$$

and we can see that the sum $n_1(\lambda) + p(\lambda)$ does not calculate the eigenvalues $\lambda_2 = -1$ and $\lambda_4 = 1$. However, according to Theorem 5.119, we need to incorporate the influence of the nonconstant rank of the coefficient $\lambda^2 - 1$ in the right endpoint $N + 1$. For this case we have by (5.424)

$$n_B(\lambda) \equiv 0, \quad n_L(\lambda) \equiv 0, \quad n_{\tilde{B}}(\lambda) = n_R(\lambda) = \begin{cases} 0, \ \lambda \in (-\infty, -1), \\ 1, \ \lambda \in [-1, 1), \\ 2, \ \lambda \in [1, \infty). \end{cases}$$

Then, by Theorem 5.119 with $\ell := 0$, the sum $n_1(\lambda) + p(\lambda) + n_{\tilde{B}}(\lambda)$ determines all the eigenvalues of (5.429), i.e.,

$$n_2(\lambda) = n_1(\lambda) + p(\lambda) + n_{\tilde{B}}(\lambda) = \begin{cases} 0, \ \lambda \in (-\infty, -\sqrt{3}), \\ 1, \ \lambda \in [-\sqrt{3}, -1), \\ 2, \ \lambda \in [-1, 0), \\ 3, \ \lambda \in [0, 1), \\ 4, \ \lambda \in [1, \sqrt{3}), \\ 5, \ \lambda \in [\sqrt{3}, \infty). \end{cases}$$

Now we consider the generalization of Theorem 5.55 for joint endpoints to the case when assumptions (5.226) do not hold. Recall the notation (see (5.220))

$$\mathcal{X}_k(\lambda) := \begin{pmatrix} I & 0 \\ \tilde{X}_k(\lambda) & X_k^{[0]}(\lambda) \end{pmatrix}, \quad \mathcal{U}_k(\lambda) := \begin{pmatrix} 0 & -I \\ \tilde{U}_k(\lambda) & U_k^{[0]}(\lambda) \end{pmatrix},$$

where $Y^{[0]}(\lambda)$ is the principal solution of (5.1) at $k = 0$ and, by (5.221),

$$\left. \begin{aligned} \mathcal{L}(\lambda) &:= R_1(\lambda)\,\mathcal{X}_{N+1}(\lambda) + R_2(\lambda)\,\mathcal{U}_{N+1}(\lambda), \\ \mathcal{M}(\lambda) &:= [I - \mathcal{L}(\lambda)\,\mathcal{L}^\dagger(\lambda)]\,R_2(\lambda), \\ \mathcal{T}(\lambda) &:= I - \mathcal{M}^\dagger(\lambda)\,\mathcal{M}(\lambda), \\ \mathcal{P}(\lambda) &:= \mathcal{T}(\lambda)\,\mathcal{X}_{N+1}(\lambda)\,\mathcal{L}^\dagger(\lambda)\,R_2(\lambda)\,\mathcal{T}(\lambda). \end{aligned} \right\}$$

The multiplicity $\zeta(\lambda_0)$ of finite eigenvalue of problem (SDS$_\lambda$), (5.151) is presented by Definition 5.54.

The consideration is based on Theorem 5.44, which guarantees that under assumptions (5.2), (5.3), (5.152), and (5.153), problem (SDS$_\lambda$), (5.151) is equivalent to the augmented problem (5.178) or (5.232) (see the proof of Theorem 5.55). Note that problem (5.232) obeys all assumptions of the local oscillation theorem for the separated endpoints (see Theorem 5.117). Moreover, we again use that

$$\text{rank } \tilde{\mathcal{B}}_k(\lambda) = \text{rank diag}\{0, \ \mathcal{B}_k(\lambda)\} = \text{rank } \mathcal{B}_k(\lambda), \quad k \in [0, N]_{\mathbb{Z}},$$

where $\tilde{\mathcal{B}}_k(\lambda)$ is the block of $\{\mathcal{S}_k(\lambda)\}$ (see (5.200)). Then the definition of the numbers $\tilde{\vartheta}_k(\lambda_0)$ for $k \in [0, N]_{\mathbb{Z}}$ for the augmented problem (5.232) stays the same as before (see (5.329)), since

$$
\left.\begin{aligned}
\tilde{\vartheta}_k(\lambda_0) &= \operatorname{rank} \operatorname{diag}\{0,\ \mathcal{B}_k(\lambda_0^-)\} - \operatorname{rank} \operatorname{diag}\{0,\ \mathcal{B}_k(\lambda_0)\} \\
&= \operatorname{rank} \mathcal{B}_k(\lambda_0^-) - \operatorname{rank} \mathcal{B}_k(\lambda_0) = \vartheta_k(\lambda_0), \quad k \in [0, N]_{\mathbb{Z}}.
\end{aligned}\right\}
\tag{5.430}
$$

As the boundary condition for $k = 0$ does not depend on $\lambda$, we have according to (5.414) applied to (5.232) that

$$
\left.\begin{aligned}
\tilde{\vartheta}_{-1}(\lambda_0) &= 0, \\
\tilde{\vartheta}_{N+1}(\lambda_0) &= \operatorname{rank} R_2(\lambda_0^-) - \operatorname{rank} R_2(\lambda_0).
\end{aligned}\right\}
\tag{5.431}
$$

Introduce the notation (see (5.223) and (5.225))

$$
n_1(\lambda) := \text{ number of forward focal points of } Y^{[0]}(\lambda) \text{ in } (0, N+1],
$$

$$
q(\lambda) := \operatorname{rank} \mathcal{M}(\lambda) + \operatorname{ind} \mathcal{P}(\lambda).
$$

Then Theorem 5.117 applied to augmented problem (5.232) reads as follows.

**Theorem 5.122 (Local Oscillation Theorem for Joint Endpoints)** *Assume* (5.2), (5.3), (5.152), *and* (5.153). *Let* $Y^{[0]}(\lambda)$ *be the principal solution at* $k = 0$ *of* (SDS$_\lambda$), *and* $m_k(\lambda)$ *for* $k \in [0, N]_{\mathbb{Z}}$ *are the multiplicities of focal points of* $Y^{[0]}(\lambda)$. *Then* $m_k(\lambda^-)$ *and* $m_k(\lambda^+)$ *exist for all* $\lambda \in \mathbb{R}$ *and satisfy* (5.330) *for all* $k \in [0, N]_{\mathbb{Z}}$. *Moreover, for* $k = N + 1$, *we have*

$$
\left.\begin{aligned}
q(\lambda^+) &= q(\lambda), \\
q(\lambda^+) - q(\lambda^-) + \tilde{\vartheta}_{N+1}(\lambda) &= \operatorname{rank} X_{N+1}(\lambda) - \operatorname{rank} X_{N+1}(\lambda^-) \\
&\quad + \operatorname{rank} \mathcal{L}(\lambda^-) - \operatorname{rank} \mathcal{L}(\lambda),
\end{aligned}\right\}
\tag{5.432}
$$

*where* $q(\lambda)$ *is given by* (5.225). *Instead of* (5.417), *we have*

$$
\left.\begin{aligned}
n_1(\lambda^+) &= n_1(\lambda) \le n(N+1), \\
n_1(\lambda^+) - n_1(\lambda^-) + q(\lambda^+) - q(\lambda^-) + \sum_{k=0}^{N+1} \tilde{\vartheta}_k(\lambda) & \\
&= \operatorname{rank} \mathcal{L}(\lambda^-) - \mathcal{L}(\lambda) = \zeta(\lambda),
\end{aligned}\right\}
\tag{5.433}
$$

*where* $\zeta(\lambda)$ *is given by* (5.222).

To formulate the global oscillation theorem, we replace assumptions (5.418) by the more general assumption for joint endpoints

$$\operatorname{rank} R_2(\lambda) = \operatorname{rank} R_2(\lambda^-) \quad \text{for all } \lambda < \lambda_{00}. \tag{5.434}$$

Using the notion of finite eigenvalue of problem (SDS$_\lambda$), (5.151) and conditions (5.333) and (5.434), we prove an analog of Corollary 5.118 for joint endpoints.

**Corollary 5.123** *Under assumptions* (5.2), (5.3), (5.152), *and* (5.153), *conditions* (5.333) *and* (5.434) *hold if and only if the finite spectrum of problem* (SDS$_\lambda$), (5.151) *is bounded from below.*

*Proof* We apply Corollary 5.118 to augmented problem (5.178) or (5.232) replacing (5.418) by (5.434).                                                                        □

Introduce the notation

$$n_2(\lambda) := \text{ number of finite eigenvalues of (5.1), (5.151) in } (-\infty, \lambda],$$

$$\tilde{n}_{\tilde{\mathcal{B}}}(\lambda) := \sum_{\mu \leq \lambda} \sum_{k=0}^{N+1} \tilde{\vartheta}_k(\mu), \tag{5.435}$$

where $\tilde{\vartheta}_k(\lambda)$ are given by (5.430) and (5.431). Note that under assumptions (5.3), (5.333), and (5.434) instead of (5.420) for the separated endpoints, we have for some $\tilde{\lambda} \in \mathbb{R}$

$$n_2(\lambda) \equiv 0, \quad \tilde{n}_{\tilde{\mathcal{B}}}(\lambda) \equiv 0, \quad n_1(\lambda) + q(\lambda) \equiv \ell \quad \text{for all } \lambda < \tilde{\lambda}, \tag{5.436}$$

where we apply (5.433). Based on Corollary 5.123 and Theorem 5.122, we derive a generalization of Theorem 5.55 without assumptions (5.226).

**Theorem 5.124 (Global Oscillation Theorem for Joint Endpoints)** *Assume that* (5.2), (5.3), (5.152), (5.153), (5.333), *and* (5.434) *hold. Then there exists a number* $\ell \in [0, (N+2)\,n]_\mathbb{Z}$ *such that*

$$n_1(\lambda) + q(\lambda) + \tilde{n}_{\tilde{\mathcal{B}}}(\lambda) = n_2(\lambda) + \ell \quad \text{for all } \lambda \in \mathbb{R}, \tag{5.437}$$

*where $\ell$ is given by* (5.436).

*Proof* The proof repeats the proof of Theorem 5.55, where instead of Theorem 5.50, we apply Theorem 5.119.                                                                        □

*Remark 5.125* For the case when the endpoints are separated, formula (5.437) turns into (5.421). Indeed, in this case $\tilde{n}_{\tilde{\mathcal{B}}}(\lambda) = n_{\tilde{\mathcal{B}}}(\lambda)$ and by (5.218)

$$q(\lambda) = \mu\big(\tilde{V}_{N+1}(\lambda)\langle Z^{[0]}_{N+1}(\lambda)\rangle,\ \tilde{V}_{N+1}(\lambda)\,(0_{2n}\ I_{2n})^T\big), \tag{5.438}$$

where the symplectic matrix $\tilde{V}_{N+1}(\lambda)$ is associated with the boundary condition in (5.232) at $N+1$ by (5.173). For the case of separated endpoints, it is easy to verify by direct calculations that

$$\tilde{V}_{N+1}(\lambda) = \{V_{-1}^{-1}(\lambda), V_{N+1}(\lambda)\}, \tag{5.439}$$

where $V_{-1}(\lambda)$, $V_{N+1}(\lambda)$ are given by (5.171) and (5.173) for separated boundary conditions (5.155) (remark that we use notation (3.101)). Then it is possible to prove the relation

$$q(\lambda) = \mu\big(\bar{Y}_{N+1}(\lambda), Y_{N+1}^{[0]}(\lambda)\big) + p(\lambda), \quad \lambda \in \mathbb{R}, \tag{5.440}$$

which holds for $q(\lambda)$ and $p(\lambda)$ given by (5.209) and (5.225) for the case of the separated boundary conditions. Here $\bar{Y}_{N+1}(\lambda)$ is the conjoined basis of (5.1) given by (5.204). Here we present the outline of the rather technical (direct) proof of this result based on algebraic properties of the comparative index derived in Sect. 3.3.5. Applying (5.438) and (5.439), we have by (3.105)

$$q(\lambda) = \mu\big(\langle V_{N+1}(\lambda)Z_{N+1}^{[0]}V_{-1}(\lambda)\rangle, \{V_{-1}^{-1}(\lambda), V_{N+1}(\lambda)\}(0_{2n}\ I_{2n})^T\big),$$

where we applied Theorem 3.5(i). Next, by (3.116)

$$\mu\big(\langle V_{N+1}(\lambda)\ Z_{N+1}^{[0]}(\lambda)\ V_{-1}(\lambda)\rangle, \{V_{-1}^{-1}(\lambda), V_{N+1}(\lambda)\}(0_{2n}\ I_{2n})^T\big)$$

$$= \mu^*\big(V_{-1}^{-1}(\lambda)\ [Z_{N+1}^{[0]}(\lambda)]^{-1}(0\ I)^T, V_{-1}^{-1}(\lambda)\ (0\ I)^T\big)$$

$$+ \mu\big(V_{N+1}(\lambda)\ \bar{Y}_{N+1}(\lambda), V_{N+1}(\lambda)\ (0\ I)^T\big)$$

$$= \mu\big(\bar{Y}_{N+1}(\lambda), Y_{N+1}^{[0]}(\lambda)\big) + p(\lambda),$$

where we use Theorem 3.5(iii), (i) and the definition of $p(\lambda)$ according to (5.218). Applying the Sturmian separation result (see formula (4.60))

$$l(\bar{Y}(\lambda), 0, N+1) - l(Y^{[0]}(\lambda), 0, N+1) = \mu\big(\bar{Y}_{N+1}(\lambda), Y_{N+1}^{[0]}(\lambda)\big),$$

it is easy to see that (5.437) indeed turns into (5.421) for the separated endpoints.

*Remark 5.126* The periodic and antiperiodic boundary conditions are included in (5.155) for the special choice of constant matrices $R_1(\lambda)$ and $R_2(\lambda)$, as we discussed in Remark 5.58 in Sect. 5.2.3. In both cases (5.236) and (5.237), the matrices $R_2(\lambda)$ are constant in $\lambda$, so that by (5.430) and (5.431) the function $\tilde{n}_{\bar{B}}(\lambda)$ in (5.436) coincides with the function $n_B(\lambda)$ in (5.338). Recall now from (5.371) the symplectic and orthogonal matrix

$$R := \frac{1}{\sqrt{2}} \begin{pmatrix} 0 & -I & I & 0 \\ 0 & I & I & 0 \\ -I & 0 & 0 & -I \\ -I & 0 & 0 & I \end{pmatrix}.$$

which was used in Sect. 5.6.3. Then the matrix $\tilde{S}_{N+1}(\lambda) \in Sp(4n)$ associated with the boundary condition (5.236) at $N+1$ of the augmented problem (5.178) can be taken in the form $\tilde{S}_{N+1}(\lambda) \equiv \tilde{S}_{N+1} := R^T$. For this choice of $\tilde{S}_{N+1}$, the number $q(\lambda)$ given by (5.438) for the periodic boundary conditions (5.236) takes the form (see the first equality in formula (5.377) and (3.104))

$$\left. \begin{aligned} q(\lambda) = q_1(\lambda) &:= \mu\left(R^T \langle Z_{N+1}^{[0]}(\lambda)\rangle, \ R^T (0 \ I)^T\right) \\ &= \mu\left(\begin{pmatrix} \mathcal{L}_1(\lambda) \\ J^T \mathcal{L}_2(\lambda) \end{pmatrix}, \ R^T (0 \ I)^T\right) \leq n. \end{aligned} \right\} \tag{5.441}$$

Moreover, consider a generalized version of the periodic and antiperiodic boundary conditions (5.236) and (5.237) in the form

$$y_0(\lambda) = \hat{S}(\lambda) \, y_{N+1}(\lambda), \tag{5.442}$$

where $\hat{S}(\lambda)$ is a given symplectic piecewise continuously differentiable matrix for $\lambda \in \mathbb{R}$ satisfying $\Psi(\hat{S}(\lambda)) \geq 0$, where $\Psi(\hat{S}(\lambda))$ is given by (5.3). The problem (5.1), (5.442) is then equivalent to the following extended problem with the periodic endpoints

$$\left. \begin{aligned} y_{k+1}(\lambda) = S_k(\lambda) \, y_k(\lambda), \quad k \in [0, N+1]_{\mathbb{z}}, \quad S_{N+1}(\lambda) &:= \hat{S}(\lambda), \\ y_0(\lambda) = y_{N+2}(\lambda), \end{aligned} \right\} \tag{5.443}$$

where the matrix $S_k(\lambda)$ obeys (5.3) for $k \in [0, N+1]_{\mathbb{z}}$. Applying the above results for (5.236) to problem (5.443), we get

$$\mathcal{L}(\lambda) = \mathcal{J}\left(I - \hat{S}(\lambda) \, Z_{N+1}^{[0]}(\lambda)\right).$$

Here we used that $Z_{N+2}^{[0]}(\lambda) = \hat{S}(\lambda) \, Z_{N+1}^{[0]}(\lambda)$, while $q(\lambda)$ given by (5.438) for conditions (5.442) takes the form

$$\begin{aligned} q(\lambda) &= \mu\left(R^T \langle Z_{N+2}^{[0]}(\lambda)\rangle, \ R^T (0 \ I)^T\right) + \mu\left(Y_{N+2}^{[0]}(\lambda), \ \hat{S}(\lambda) (0 \ I)^T\right) \\ &= \mu\left(\begin{pmatrix} \mathcal{L}(\lambda) \\ I + \hat{S}(\lambda) \, Z_{N+1}^{[0]}(\lambda) \end{pmatrix}, \ R^T (0 \ I)^T\right) + \mu\left(\hat{S}(\lambda) \, Y_{N+1}^{[0]}(\lambda), \ \hat{S}(\lambda) (0 \ I)^T\right), \end{aligned}$$

where we incorporated the multiplicity $\mu\left(Y^{[0]}_{N+2}(\lambda),\ \hat{S}(\lambda)\,(0\ I)^T\right)$ of the focal point of the principal solution of (5.443) at $N + 1$ according to (4.14) in Lemma 4.7. In particular, for the antiperiodic boundary conditions (5.237), we have $\hat{S}(\lambda) := -I$ in (5.443), and then for this case

$$
\begin{aligned}
q(\lambda) = q_2(\lambda) &:= \mu\left(R^T\,\langle-Z^{[0]}_{N+1}(\lambda)\rangle,\ R^T\,(0\ I)^T\right) \\
&= \mu\left(\begin{pmatrix}\mathcal{L}_2(\lambda)\\ J^T\mathcal{L}_1(\lambda)\end{pmatrix},\ R^T\,(0\ I)^T\right) \le n,
\end{aligned}
\right\}
\tag{5.444}
$$

where $\mathcal{L}_1(\lambda)$ and $\mathcal{L}_2(\lambda)$ are given in (5.236) and (5.237) and where we used Theorem 3.5(i) in the calculations.

*Example 5.127* Consider problem (5.1), (5.151) in the form

$$
y_{k+1}(\lambda) = \begin{pmatrix}I - P_k\lambda^2\,\lambda I\\ -\lambda P_k \quad I\end{pmatrix} y_k(\lambda), \quad k \in [0,\,N]_{\mathbb{Z}}, \quad y_{N+1}(\lambda) = G\,y_0(\lambda),
\tag{5.445}
$$

where $G$ is a constant symplectic matrix. It was proven in [165, Section 3] that under the assumptions $P_k$ symmetric, $P_k \ge 0$, and $\sum_{k=0}^{N} P_k > 0$, problem (5.445) is self-adjoint, and the width of the central zone of stability of a discrete linear Hamiltonian system associated with (5.445) can be estimated by using the properties of the eigenvalues of this problem. Note that problem (5.445) satisfies all the assumptions of Theorem 5.124, and then this theorem together with Remarks 5.58 and 5.126 can be applied for the calculation of the eigenvalues of (5.445).

For example, consider the case $n = 1$, $P_k = 1$, $G = I_2$, and $N = 1$. Then the symplectic fundamental matrix of the symplectic system in (5.445) and $\mathcal{L}_1(\lambda)$ in (5.236) have the form

$$
Z^{[0]}_0(\lambda) \equiv I, \quad Z^{[0]}_1(\lambda) = \begin{pmatrix}I - \lambda^2\,\lambda I\\ -\lambda I \quad I\end{pmatrix}, \quad Z^{[0]}_2(\lambda) = \begin{pmatrix}\lambda^4 - 3\lambda^2 + 1 & 2\lambda - \lambda^3\\ -2\lambda + \lambda^3 & 1 - \lambda^2\end{pmatrix},
$$

$$
\mathcal{L}_1(\lambda) = \begin{pmatrix}-2\lambda + \lambda^3 & -\lambda^2\\ -\lambda^4 + 3\lambda^2 & -2\lambda + \lambda^3\end{pmatrix},
$$

with $\det \mathcal{L}_1(\lambda) = \lambda^2(2 - \lambda)(2 + \lambda)$. Then $\lambda_2 = 0$ has the multiplicity $\theta(\lambda_2) = 2$, while the eigenvalues $\lambda_1 = -2$ and $\lambda_3 = 2$ are simple. The multiplicities $m_k(\lambda) = m_k(Y^{[0]}(\lambda))$, $k \in [0,\,1]_{\mathbb{Z}}$ of focal points of $Y^{[0]}(\lambda)$ and the function $\tilde{n}_{\tilde{\mathcal{B}}}(\lambda) = n_{\mathcal{B}}(\lambda)$ given by (5.435) and (5.338) calculating the zeros of $\mathcal{B}_k(\lambda) = \lambda$ for $k \in [0,\,1]_{\mathbb{Z}}$ are

$$
m_0(\lambda) = 0, \quad m_1(\lambda) = \begin{cases}0, & \lambda \in (-\infty,\,-\sqrt{2}),\\ 1, & \lambda \in [-\sqrt{2},\,0),\\ 0, & \lambda \in [0,\,\sqrt{2}),\\ 1, & \lambda \in [\sqrt{2},\,\infty),\end{cases} \quad n_{\mathcal{B}}(\lambda) = \begin{cases}0, & \lambda \in (-\infty,\,0),\\ 2, & \lambda \in [0,\,\infty).\end{cases}
$$

It follows that

$$m_1(\lambda) + n_B(\lambda) = \begin{cases} 0, \ \lambda \in (-\infty, -\sqrt{2}), \\ 1, \ \lambda \in [-\sqrt{2}, 0), \\ 2, \ \lambda \in [0, \sqrt{2}), \\ 3, \ \lambda \in [\sqrt{2}, \infty), \end{cases} \qquad q(\lambda) = \begin{cases} 0, \ \lambda \in (-\infty, -2), \\ 1, \ \lambda \in [-2, -\sqrt{2}), \\ 0, \ \lambda \in [-\sqrt{2}, 0), \\ 1, \ \lambda \in [0, \sqrt{2}), \\ 0, \ \lambda \in [\sqrt{2}, 2), \\ 1, \ \lambda \in [2, \infty), \end{cases}$$

where by Theorem 5.98 the sum $m_1(\lambda) + n_B(\lambda)$ calculates the eigenvalues of problem (5.445) with the Dirichlet boundary conditions (i.e., the zeros of the polynomial $X_2^{[0]}(\lambda) = 2\lambda - \lambda^3$). On the other hand, for the calculation of the number of eigenvalues of (5.445) according to Theorem 5.124, we have to add the number $q(\lambda)$ evaluated according to (5.441) and the definition of the comparative index (see Definition 3.1). Finally, we have

$$n_2(\lambda) = m_1(\lambda) + n_B(\lambda) + q(\lambda) = \begin{cases} 0, \ \lambda \in (-\infty, -2), \\ 1, \ \lambda \in [-2, 0), \\ 3, \ \lambda \in [0, 2), \\ 4, \ \lambda \in [2, \infty), \end{cases}$$

where $n_2(\lambda)$ is the number of the eigenvalues (including multiplicities) of (5.445) less than or equal to the given $\lambda$ in the accordance with Theorem 5.124.

## 5.7  Notes and References

The oscillation theorems for symplectic eigenvalue problems were first considered in [55, 102] for systems with linear dependence on $\lambda$ and the Dirichlet boundary conditions. More precisely, in [55], a certain finite exceptional set of values $\lambda$ was excluded from the statement of the main result, while in [102] all values $\lambda \in \mathbb{R}$ were covered. This issue was also closely related to the introduction of the multiplicities of (forward) focal points in [208], as shown in Definition 4.1. Extension of the oscillation theorems in [102] to separated boundary conditions via a transformation to the Dirichlet boundary conditions was derived in [55, 103].

Nonlinear dependence on the spectral parameter in symplectic difference systems, as it is presented in Sect. 5.1, was introduced in [297]. There the oscillation theorems were proven under the condition that the coefficient $\mathcal{B}_k(\lambda) \equiv \mathcal{B}_k$ is constant in $\lambda$ on $\mathbb{R}$, which was also the case in [102]. This assumption was weakened to the constant $\mathrm{Im}\,\mathcal{B}_k(\lambda) \equiv \mathbb{R}$ in $\lambda$ in [298] for the scalar second-order Sturm-Liouville difference equations $(\mathrm{SL}_\lambda)$ with (1.30) and to an arbitrary constant image of $\mathcal{B}_k(\lambda)$ in $\lambda$ in [210] for general symplectic difference systems with Dirichlet boundary conditions. Discrete oscillation theorems for eigenvalue problems with separated or general boundary conditions (under $\text{Im}\,\mathcal{B}_k(\lambda)$ constant in $\lambda \in \mathbb{R}$) were obtained in [305] for systems with linear dependence on $\lambda$ and in [301] for systems with nonlinear dependence on $\lambda$.

Finally, oscillation theorems without condition $\text{Im}\,\mathcal{B}_k(\lambda)$ constant in $\lambda$ for problem (E) under monotonicity assumption (5.3) were proven in [125, Theorem 2.4]. It particular, it was shown there that assumption (5.3) implies that $\text{rank}\,\mathcal{B}_k(\lambda)$, $\text{Im}\,\mathcal{B}_k(\lambda)$, and $\text{Ker}\,\mathcal{B}_k(\lambda)$ are piecewise constant in $\lambda$ (see [125, Remark 3.6]). Generalizations of these oscillation theorems to symplectic difference systems with general boundary conditions, which also incorporate possible oscillations in the rank of the coefficients in the system and in the boundary conditions, were obtained recently in [133].

The treatment of discrete symplectic eigenvalue problems with linear dependence on $\lambda$ both in the system and in the boundary conditions as presented in Sect. 5.3 is motivated by the results for linear Hamiltonian differential systems (1.103) in [205, Section 2.2]. The positivity result in Proposition 5.67 was proven in [103, Proposition 2]. The Rayleigh principle for symplectic eigenvalue problems with Dirichlet boundary conditions (Theorem 5.83 and Corollary 5.84) was proven in [56, Theorem 4.6]. Extensions of this result to separated or joint boundary conditions (constant in $\lambda$) were obtained in [103, Theorem 2] and [305, Theorems 3.2], respectively. In this respect the results with Rayleigh principle for symplectic eigenvalue problems with separated and joint boundary conditions depending (linearly) on $\lambda$ (in Theorems 5.85 and 5.88 as well as in Corollaries 5.86 and 5.87) are also new in the literature. The Sturmian comparison and separation theorems obtained in Sect. 5.5.1 via the Rayleigh principle are from [56].

It should be noted that linear Hamiltonian system (5.144) with nonlinear dependence on $\lambda$ was first studied in [44, 264] under a strict normality assumption, which corresponds to the strict monotonicity of the matrix $H_k(\lambda)$, and further in [210, 297] under (5.337) and the restriction that $\text{Im}\,[\tilde{A}_k(\lambda)\,B_k(\lambda)]$ is constant in $\lambda \in \mathbb{R}$ (see Example 5.37). This condition is equivalent to $\text{rank}\,B_k(\lambda)$ constant in $\lambda \in \mathbb{R}$ by Theorem 5.1(iii). Later, in [95], the last condition was completely omitted.

There are several other topics in the spectral theory of discrete systems, which were studied within symplectic difference systems (SDS) or linear Hamiltonian difference systems (2.15). In this context we wish to mention first the literature regarding the discrete Weyl-Titchmarsh theory, i.e., the theory of square summable solutions of symplectic difference systems depending on $\lambda \in \mathbb{C}$. The study in this direction was initiated in [60, 67] for symplectic difference systems (5.238) with special linear dependence on the spectral parameter $\lambda$, i.e., for the matrix $\mathcal{S}_k(\lambda)$ in (5.239) with (5.241). Weyl-Titchmarsh theory for symplectic difference systems with general linear dependence on $\lambda$ was developed in [308–310, 313] and with polynomial and analytic dependence on $\lambda$ in [89, 90, 311, 312]. The results in [310, 313] show an interesting phenomenon in the discrete time theory, where the limit circle invariance (and the Weyl alternative) holds without any additional assumption in comparison with the corresponding continuous time result. Extensions of several classical concepts from the theory of square summable solutions for symplectic difference systems to linear relations were presented in

[68, 335]. Properties of eigenvalues and the spectrum of symplectic difference systems were studied in [120, 122].

In the literature, there are many studies about the spectral properties of linear Hamiltonian difference systems (5.144). Those are special symplectic difference systems as we showed in Example 5.37. Below we review the relevant literature regarding selected spectral properties of (5.144) with various dependence on $\lambda$ (often linear in $\lambda$). Stability zones were studied in [242–244]. Spectral properties of system (5.144) on bounded interval was studied in [264] and on unbounded interval in [66, 221, 252, 253, 265, 271, 272, 275, 277]. Spectral properties of linear subspaces associated with system (5.144) were studied in [251, 269]. Further results from spectral theory of (5.144) were obtained in [33, 44, 144, 235, 261, 337].

We recommend the following additional related references for further reading about the topics presented in this chapter: [31, 199, 273, 274, 276] for discrete Sturm-Liouville eigenvalue problems and [18–20, 29, 72, 73, 158–161, 197, 198, 217, 218, 228–231, 233, 240, 260, 278, 320, 325, 329] for applications of symplectic systems, symplectic algorithms, and related numerical analysis.

# Chapter 6
# Miscellaneous Topics on Symplectic Systems

In this chapter we will present some additional topics from the theory of symplectic difference systems (SDS), which are closely related to their oscillation or spectral theory. These topics cover the relative and renormalized oscillation theorems for symplectic eigenvalue problems with nonlinear dependence on spectral parameter and the theory of symplectic difference systems, which do not impose any controllability assumption. The latter one includes, in particular, a general theory of recessive and dominant solutions at $\infty$ (as a generalization of Sect. 2.5) and their applications in the singular Sturmian theory (as an extension and completion of Sect. 4.2.4).

## 6.1 Relative Oscillation Theory

Relative oscillation theory—rather than measuring the spectrum of one single problem—measures the difference between the spectra of two different problems. This is done by replacing focal points of conjoined bases of one problem by matrix analogs of weighted zeros of Wronskians of conjoined bases of two different problems. In this section we develop the relative oscillation theory for symplectic boundary value problems.

We consider two symplectic eigenvalue problems which may differ both in the coefficient matrix of the system and also in the boundary conditions. Together with problem (5.1), (5.151), we consider the problem in the same form for system (5.318), i.e.,

$$\hat{y}_{k+1}(\lambda) = \hat{\mathcal{S}}_k(\lambda)\, \hat{y}_k(\lambda), \quad k \in [0, N]_{\mathbb{Z}}, \tag{6.1}$$

© Springer Nature Switzerland AG 2019
O. Došlý et al., *Symplectic Difference Systems: Oscillation and Spectral Theory*,
Pathways in Mathematics, https://doi.org/10.1007/978-3-030-19373-7_6

where matrix $\hat{\mathcal{S}}_k(\lambda)$ is symplectic for every $\lambda \in \mathbb{R}$ and

$$\hat{\mathcal{S}}_k(\lambda) = \begin{pmatrix} \hat{\mathcal{A}}_k(\lambda) & \hat{\mathcal{B}}_k(\lambda) \\ \hat{\mathcal{C}}_k(\lambda) & \hat{\mathcal{D}}_k(\lambda) \end{pmatrix}, \tag{6.2}$$

With (6.1) we consider boundary conditions of the same form as (5.151), that is,

$$\hat{R}_1(\lambda) \begin{pmatrix} \hat{x}_0(\lambda) \\ \hat{x}_{N+1}(\lambda) \end{pmatrix} + \hat{R}_2(\lambda) \begin{pmatrix} -\hat{u}_0(\lambda) \\ \hat{u}_{N+1}(\lambda) \end{pmatrix} = 0, \tag{6.3}$$

where $\hat{R}_1(\lambda)$ and $\hat{R}_2(\lambda)$ are piecewise continuously differentiable matrix-valued functions such that

$$\left. \begin{array}{l} \hat{R}_1(\lambda),\ \hat{R}_2(\lambda) \in \mathbb{R}^{2n \times 2n}, \quad \operatorname{rank}(\hat{R}_1(\lambda)\ \hat{R}_2(\lambda)) = 2n, \\[4pt] \hat{R}_1(\lambda)\ \hat{R}_2^T(\lambda) = \hat{R}_2(\lambda)\ \hat{R}_1^T(\lambda), \quad \lambda \in \mathbb{R}. \end{array} \right\} \tag{6.4}$$

By analogy with (5.3) and (5.153), we impose the monotonicity assumptions

$$\Psi(\hat{\mathcal{S}}_k(\lambda)) := \mathcal{J}\dot{\hat{\mathcal{S}}}_k(\lambda)\, \mathcal{J}\hat{\mathcal{S}}_k^T(\lambda)\, \mathcal{J} \geq 0, \quad k \in [0, N]_{\mathbb{Z}}, \quad \lambda \in \mathbb{R}. \tag{6.5}$$

and, using the notation $\hat{\mathcal{R}}(\lambda) := (\hat{R}_1(\lambda)\ \hat{R}_2(\lambda)) \in \mathbb{R}^{2n \times 4n}$,

$$\dot{\hat{R}}_1(\lambda)\, \hat{R}_2^T(\lambda) - \dot{\hat{R}}_2(\lambda)\, \hat{R}_1^T(\lambda) = \dot{\hat{\mathcal{R}}}(\lambda)\, \mathcal{J}_{4n} \hat{\mathcal{R}}^T(\lambda) \leq 0. \tag{6.6}$$

We will investigate the pair of eigenvalue problems (5.1), (5.151) and (6.1), (6.3) and their particular cases which deal with separated and the Dirichlet boundary conditions.

The fundamental role in our treatment is played by the relative oscillation numbers introduced in Chap. 4 (see Definition 4.44) for systems (SDS) and (4.98) and by Theorem 4.56 for the principal solutions of these systems. For this reason we return in this chapter back to notation (4.10) and (4.11) from Chap. 4. In particular, we use

$$l(Y, M, N+1) := \sum_{k=M}^{N} m(Y_k), \quad l^*(Y, M, N+1) := \sum_{k=M}^{N} m^*(Y_k) \tag{6.7}$$

for the numbers of forward and backward focal points of $Y$ in $(M, N+1]$ and $[M, N+1)$, instead of using $n_1(Y(\lambda))$ and $n_1^*(Y(\lambda))$ as in Chap. 5. Moreover, we introduce the notation

$$\#\{\nu \in \sigma \mid \nu \in \mathcal{I}\}, \quad \#\{\nu \in \hat{\sigma} \mid \nu \in \mathcal{I}\} \tag{6.8}$$

for the number of finite eigenvalues of problems (5.1), (5.151) and (6.1), (6.3), where $\sigma$ and $\hat{\sigma}$ are the finite spectra of these problems, respectively, and $\mathcal{I} \subseteq \mathbb{R}$. In particular, we have

$$n_2(\lambda) = \#\{\nu \in \sigma \mid \nu \leq \lambda\},$$

where $n_2(\lambda)$ denotes the number of finite eigenvalues of (5.1), (5.151) (including special cases) less than or equal to a given $\lambda$ under the notation in Chap. 5.

## 6.1.1   Sturm-Liouville Difference Equations

In this subsection we briefly recall the results of G. Teschl and his collaborators in [22, 314, 315] concerning the relative oscillation and spectral theory for discrete Sturm-Liouville eigenvalue problems.

Consider a pair of discrete Sturm-Liouville eigenvalue problems

$$- \Delta(r_k \Delta x_k) + p_k x_{k+1} = \lambda x_{k+1}, \quad x_0 = 0 = x_{N+1} \tag{6.9}$$

and

$$- \Delta(r_k \Delta \hat{x}_k) + \hat{p}_k x_{k+1} = \lambda \hat{x}_{k+1}, \quad \hat{x}_0 = 0 = \hat{x}_{N+1} \tag{6.10}$$

with $r_k > 0$. Introduce the Wronskian for solutions $x$, $\hat{x}$ of (6.9), (6.10)

$$w_k(x, \hat{x}) = -r_k(x_k \hat{x}_{k+1} - x_{k+1} \hat{x}_k) = -r_k[x_k \Delta \hat{x}_k - (\Delta x_k) \hat{x}_k]. \tag{6.11}$$

In accordance with [314], Wronskian (6.11) has a generalized zero in the interval $[k, k+1)$ if either $w_k w_{k+1} < 0$ or $w_k = 0$ and $w_{k+1} \neq 0$. Then, by [314, Theorem 4.3], in case $p_k = \hat{p}_k$ for any $a < b$ it holds

$$\#(x^{[0]}(a), \hat{x}^{[N+1]}(b)) = \#\{\lambda \in \sigma \mid a < \lambda < b\}, \tag{6.12}$$

where $\#(x, \hat{x})$ denotes the number of generalized zeros of $w_k(x, \hat{x})$ in $(0, N+1)$ and $\#\{\lambda \in \sigma \mid a < \lambda < b\}$ is the number of eigenvalues of (6.9) between $a$ and $b$. Here $x^{[0]}(\lambda)$ and $\hat{x}^{[N+1]}(\lambda)$ are the principal solutions of (6.9) (or (6.10)) at $k = 0$ and $k = N + 1$, respectively, i.e., the nontrivial solutions satisfying $x_0^{[0]} = 0$ and $\hat{x}_{N+1}^{[N+1]} = 0$.

The main result of [22] extends (6.12) to the case $p_k \neq \hat{p}_k$. Assume that $\hat{x}^{[N+1]}(\lambda)$ is the principal solution of (6.10) at $k = N + 1$. According to [22, Theorem 1.2], the number of *weighted* generalized zeros of the Wronskian on $(0, N+1)$ equals the difference of the number of eigenvalues of (6.10) less than

$b$ and the number of eigenvalues (6.9) less than or equal to $a$,

$$\#(x^{[0]}(a),\, \hat{x}^{[N+1]}(b)) = \#(x^{[N+1]}(a),\, \hat{x}^{[0]}(b))$$

$$= \#\{v \in \hat{\sigma} \mid v < b\} - \#\{v \in \sigma \mid v \leq a\}. \tag{6.13}$$

The concept of weighted zero of the Wronskian is defined as follows.

**Definition 6.1** Denote

$$q_k(a, b) := p_k - \hat{p}_k + b - a. \tag{6.14}$$

The Wronskian $w_k(x, \hat{x})$ has a *weighted zero at $k$* if $\#_k(x, \hat{x}) = \pm 1$, where

$$\#_k(x, \hat{x}) := \begin{cases} 1, & q_k(a, b) > 0, \ w_k w_{k+1} \leq 0, \ w_{k+1} \neq 0, \\ -1, & q_k(a, b) < 0, \ w_{k+1} w_k \leq 0, \ w_k \neq 0, \end{cases} \tag{6.15}$$

and $\#_k(x, \hat{x}) := 0$ otherwise. The number of weighted zeros in $(0, N + 1)$ is given by the formula

$$\#(x, \hat{x}) := \sum_{k=0}^{N} \#_k(x, \hat{x}) - \begin{cases} 0, & \text{if } w_0(x, \hat{x}) \neq 0, \\ 1, & \text{if } w_0(x, \hat{x}) = 0. \end{cases} \tag{6.16}$$

Note that in case $p_k = \hat{p}_k$ and $b > a$ the above definition reduces to the definition of the generalized zero of the Wronskian in $[k, k+1)$ and the number of generalized zeros in $(0, N + 1)$.

## 6.1.2 Dirichlet Boundary Value Problems

In this subsection we consider systems (5.1) and (6.1) together with the Dirichlet boundary conditions

$$\left. \begin{array}{l} y_{k+1}(\lambda) = \mathcal{S}_k(\lambda)\, y_k(\lambda), \ k \in [0, N]_{\mathbb{Z}}, \ \lambda \in \mathbb{R}, \ x_0(\lambda) = 0 = x_{N+1}(\lambda), \\ \hat{y}_{k+1}(\lambda) = \hat{\mathcal{S}}_k(\lambda)\, \hat{y}_k(\lambda), \ k \in [0, N]_{\mathbb{Z}}, \ \lambda \in \mathbb{R}, \ \hat{x}_0(\lambda) = 0 = \hat{x}_{N+1}(\lambda), \end{array} \right\} \tag{6.17}$$

We assume the monotonicity conditions (5.3) and (6.5) and suppose that (5.38) hold for both matrices $\mathcal{S}_k(\lambda)$ and $\hat{\mathcal{S}}_k(\lambda)$ for $k \in [0, N]_{\mathbb{Z}}$, i.e.,

$$\text{rank } \mathcal{B}_k(\lambda) \text{ and } \text{rank } \hat{\mathcal{B}}_k(\lambda) \text{ are constant for } \lambda \in \mathbb{R} \text{ and } k \in [0, N]_{\mathbb{Z}}. \tag{6.18}$$

The global oscillation theorems established for problem (E) (see Theorem 5.17) relates the number of finite eigenvalues of (E) less than or equal to a given number

$\lambda := \lambda_1$ to the number of focal points (counting multiplicity) of the principal solution of the difference system in (E) with $\lambda = \lambda_1$. Our aim is to add a new aspect to this classical result by showing that matrix analogs of *weighted zeros* of the Wronskian for two suitable matrix solutions of the difference systems in (6.17) can be used to count the difference between the number of finite eigenvalues of problems (6.17).

Note that problems (6.17) under conditions (5.3), (6.5), and (6.18) obey all assumptions of Theorem 5.17, and then the finite spectra $\sigma$ and $\hat{\sigma}$ are bounded, and we can order the finite eigenvalues of (6.17) into nondecreasing sequences

$$\left.\begin{aligned} -\infty < \lambda_1 \leq \lambda_2 \leq \cdots \leq \lambda_m < +\infty, \\ -\infty < \hat{\lambda}_1 \leq \hat{\lambda}_2 \leq \cdots \leq \hat{\lambda}_l < +\infty. \end{aligned}\right\} \tag{6.19}$$

We put $\lambda_1 = \infty$ ($\hat{\lambda}_1 = \infty$) if $\sigma = \emptyset$ ($\hat{\sigma} = \emptyset$). We also define

$$\lambda_0 = \min\{\lambda_1, \hat{\lambda}_1\}. \tag{6.20}$$

Recall that according to Theorem 5.17

$$\left.\begin{aligned} l(Y^{[0]}(a), 0, N+1) = \#\{v \in \sigma \mid v \leq a\} + m, \\ l(\hat{Y}^{[0]}(b), 0, N+1) = \#\{v \in \hat{\sigma} \mid v \leq b\} + \hat{m}, \end{aligned}\right\} \tag{6.21}$$

where $Y^{[0]}(\lambda)$ and $\hat{Y}^{[0]}(\lambda)$ are the principal solutions of the symplectic systems in (6.17) at 0, and where

$$m := l(Y^{[0]}(\lambda), 0, N+1), \quad \hat{m} = l(\hat{Y}^{[0]}(\lambda), 0, N+1), \quad \lambda < \lambda_0. \tag{6.22}$$

Then we have from (6.21) and (6.22) that

$$\left.\begin{aligned} \#\{v \in \hat{\sigma} \mid v \leq b\} - \#\{v \in \sigma \mid v \leq a\} \\ = l(\hat{Y}^{[0]}(b), 0, N+1) - l(Y^{[0]}(a), 0, N+1) - (\hat{m} - m), \end{aligned}\right\} \tag{6.23}$$

where

$$\hat{m} - m = l(\hat{Y}^{[0]}(\lambda), 0, N+1) - l(Y^{[0]}(\lambda), 0, N+1), \quad \lambda < \lambda_0. \tag{6.24}$$

Applying Theorem 4.56 connecting the differences of the multiplicities of focal points on the right-hand sides of (6.23) and (6.24) with the relative oscillation numbers (see Definition 4.44), we prove the following central results.

**Theorem 6.2 (Relative Oscillation Theorem)**     *Assume that (5.3), (6.5), (6.18) hold for systems (5.1) and (6.1). Let $Z^{[0]}(\lambda)$ and $\hat{Z}^{[N+1]}(\lambda)$ be symplectic fundamental matrices of these systems associated with the principal solutions $Y^{[0]}(\lambda) = Z^{[0]}(\lambda)\,(0\ I)^T$ and $\hat{Y}^{[N+1]}(\lambda) = \hat{Z}^{[N+1]}(\lambda)\,(0\ I)^T$ at $k = 0$ and $k = N+1$,* *respectively. Define the relative oscillation numbers according to* (4.143) *(with* $\hat{Z}^{[N+1]} := \hat{Z}^{[N+1]}(b)$, $Z^{[M]} := Z^{[M]}(a)$, $M := 0$), *i.e.,*

$$\#_k(\hat{Z}^{[N+1]}(b), Z^{[0]}(a)) := \mu(\langle \mathcal{G}_k(a,b)\rangle, \langle \mathcal{G}_{k+1}(a,b)\rangle) - \mu(\langle \hat{S}_k(b)\rangle, \langle S_k(a)\rangle) \tag{6.25}$$

*for* $k \in [0, N]_{\mathbb{Z}}$, *where*

$$\mathcal{G}_k(a, b) := [\hat{Z}_k^{[N+1]}(b)]^{-1} Z_k^{[0]}(a).$$

*Then there exists a constant* $\mathfrak{P} \in [-nN, nN]_{\mathbb{Z}}$ *such that for any* $a, b \in \mathbb{R}$ *we have for the spectra* $\sigma$ *and* $\hat{\sigma}$ *of problems* (6.17)

$$\#\{\nu \in \hat{\sigma} \mid \nu \le b\} - \#\{\nu \in \sigma \mid \nu \le a\} = \#(\hat{Z}^{[N+1]}(b), Z^{[0]}(a), 0, N) - \mathfrak{P}, \tag{6.26}$$

*where*

$$\#(\hat{Z}^{[N+1]}(b), Z^{[0]}(a), 0, N) = \sum_{k=0}^{N} \#_k(\hat{Z}^{[N+1]}(b), Z^{[0]}(a)) \tag{6.27}$$

*and, with* $m$, $\hat{m}$, *and* $\lambda_0$ *given by* (6.22) *and* (6.20),

$$\mathfrak{P} = \hat{m} - m = \#(\hat{Z}^{[N+1]}(\lambda), Z^{[0]}(\lambda), 0, N), \quad \lambda < \lambda_0. \tag{6.28}$$

*Proof* For the proof we use Theorem 4.56 (see (4.107)) and connections (6.23) and (6.24) derived above.                                                            □

*Remark 6.3* Recall that according to Lemma 4.55 the relative oscillation numbers (6.25) can be specified at the endpoints $k = 0$ and $k = N$ as follows:

$$\#_0(\hat{Z}^{[N+1]}(b), Z^{[0]}(a)) = \mu^*(\hat{Y}_0^{[N+1]}(b), \hat{S}_0^{-1}(b)\, S_0(a)\, (0\ I)^T)$$
$$- \mu(\hat{S}_0(b)\, (0\ I)^T,\, S_0(a)\, (0\ I)^T), \tag{6.29}$$
$$\#_N(\hat{Z}^{[N+1]}(b), Z^{[0]}(a)) = \mu(\hat{S}_N(b)\, Y_N^{[0]}(a),\, \hat{S}_N(b)\, S_N^{-1}(a)\, (0\ I)^T)$$
$$- \mu^*(\hat{S}_N^{-1}(b)\, (0\ I)^T,\, S_N^{-1}(a)\, (0\ I)^T). \tag{6.30}$$

In particular, under assumptions (5.3), (6.5) for the case $S_0(\lambda) \equiv \hat{S}_0(\lambda)$ (respectively, for $S_N(\lambda) \equiv \hat{S}_N(\lambda)$) we have that $\mu(\hat{S}_0(b)\,(0\ I)_,^T\, S_0(a)\,(0\ I)^T) = 0$ (respectively, $\mu^*(\hat{S}_N^{-1}(b)\,(0\ I)_,^T\, S_N^{-1}(a)\,(0\ I)^T) = 0$); see the proof of Theorem 6.4 below.

For the case when $S_k(\lambda) \equiv \hat{S}_k(\lambda)$ for $k \in [0, N]_{\mathbb{Z}}$, we have the so-called *renormalized* oscillation theorem for (E), which is a corollary to Theorem 6.2.

**Theorem 6.4 (Renormalized Oscillation Theorem)** *Assume that conditions* (5.3) *and* (6.18) *hold for system* (5.1). *Let* $Z^{[0]}(\lambda)$ *and* $Z^{[N+1]}(\lambda)$ *be symplectic fundamental matrices of* (5.1) *such that* $Y^{[0]}(\lambda) = Z^{[0]}(\lambda)(0 \; I)^T$ *and* $Y^{[N+1]}(\lambda) = Z^{[N+1]}(\lambda)(0 \; I)^T$ *are the principal solutions of this system at* $k = 0$ *and* $k = N + 1$, *respectively. Then for any* $a, b \in \mathbb{R}$ *with* $a < b$ *the number of finite eigenvalues of problem* (E) *in* $(a, b]$ *is given by the formula*

$$
\left.
\begin{aligned}
\#\{\nu \in \sigma \mid a < \nu \le b\} &= \sum_{k=0}^{N} \mu\big(\langle \mathcal{G}_k(a, b)\rangle, \langle \mathcal{G}_{k+1}(a, b)\rangle\big), \\
\mathcal{G}_k(a, b) &:= [Z_k^{[N+1]}(b)]^{-1} Z_k^{[0]}(a).
\end{aligned}
\right\}
\tag{6.31}
$$

*Moreover, with* $\lambda_1 = \min \sigma$ *we have*

$$
\left.
\begin{aligned}
\#\{\nu \in \sigma \mid \nu \le b\} &= \sum_{k=0}^{N} \mu\big(\langle \mathcal{G}_k(\lambda, b)\rangle, \langle \mathcal{G}_{k+1}(\lambda, b)\rangle\big), \\
\mathcal{G}_k(\lambda, b) &:= [Z_k^{[N+1]}(b)]^{-1} Z_k^{[0]}(\lambda), \quad \lambda < \lambda_1.
\end{aligned}
\right\}
\tag{6.32}
$$

*Proof* Remark that for $S_k(\lambda) \equiv \hat{S}_k(\lambda)$ for $k \in [0, N]_{\mathbb{Z}}$, we have according to Lemma 5.105 that $\mu\big(\langle S_k(b)\rangle, \langle S_k(a)\rangle\big) = 0$ for $k \in [0, N]_{\mathbb{Z}}$, where assumption (6.18) plays a key role. Moreover, for the case $S_k(\lambda) \equiv \hat{S}_k(\lambda)$ for $k \in [0, N]_{\mathbb{Z}}$, the left-hand side of (6.26) takes the form of the left-hand side of (6.31), while the constant $\mathfrak{P} = 0$ by (6.28). Then the proof of (6.31) follows from (6.25) and (6.26). Formula (6.32) is derived from (6.31) by taking the limit $a \to -\infty$ and using that the spectrum of (E) under conditions (5.3) and (6.18) is bounded. $\qquad\square$

*Remark 6.5*

(i) Note that by (4.148) in Theorem 4.56, we have

$$
\#(\hat{Z}^{[N+1]}(b), Z^{[0]}(a), 0, N) = -\#(Z^{[N+1]}(a), \hat{Z}^{[0]}(b), 0, N).
$$

Then one can replace the relative oscillation numbers in (6.26) by

$$
-\#(Z^{[N+1]}(a), \hat{Z}^{[0]}(b), 0, N) = -\sum_{k=0}^{N} \#_k(Z^{[N+1]}(a), \hat{Z}^{[0]}(b)),
$$

where, according to Definition 4.44, instead of (6.25) we have

$$
\left.
\begin{aligned}
\#_k(&Z^{[N+1]}(a), \hat{Z}^{[0]}(b)) \\
&= \mu\big(\langle \tilde{\mathcal{G}}_k(a, b)\rangle, \langle \tilde{\mathcal{G}}_{k+1}(a, b)\rangle\big) - \mu\big(\langle S_k(a)\rangle, \langle \hat{S}_k(b)\rangle\big) \\
&= \mu\big(\langle \hat{S}_k(b)\rangle, \langle S_k(a)\rangle\big) - \mu\big(\langle \tilde{\mathcal{G}}_{k+1}(a, b)\rangle, \langle \tilde{\mathcal{G}}_k(a, b)\rangle\big),
\end{aligned}
\right\}
\tag{6.33}
$$

with

$$\tilde{\mathcal{G}}_k(a, b) := [Z_k^{[N+1]}(a)]^{-1} \hat{Z}_k^{[0]}(b).$$

In particular, this reformulation leads to the representation

$$\left. \begin{aligned} \#\{\nu \in \sigma \mid a < \nu \le b\} &= \sum_{k=0}^{N} \mu\big(\langle \tilde{\mathcal{G}}_{k+1}(a, b)\rangle, \langle \tilde{\mathcal{G}}_k(a, b)\rangle\big), \\ \tilde{\mathcal{G}}_k(a, b) &:= [Z_k^{[N+1]}(a)]^{-1} Z_k^{[0]}(b), \end{aligned} \right\} \tag{6.34}$$

which is equivalent to (6.31).

(ii) We recall that according to Remark 4.46(i) the value of the comparative index $\mu\big(\langle \tilde{\mathcal{G}}_{k+1}(a, b)\rangle, \langle \tilde{\mathcal{G}}_k(a, b)\rangle\big)$ in (6.33) and (6.34) presents the multiplicities of forward focal points of the transformed $4n$-dimensional (Wronskian) system (4.112) associated with (5.1) and (6.1). Similarly, the comparative index $\mu\big(\langle \mathcal{G}_k(a, b)\rangle, \langle \mathcal{G}_{k+1}(a, b)\rangle\big)$ in (6.26) and (6.31) is equal to the multiplicity of backward focal points of system (4.115) (see Remark 4.46(iii)). Recall also that by Remark 4.46(ii), (iv) both comparative indices are presented by (4.113) and (4.116), where the first addends are associated with transformed $2n$-dimensional (Wronskian) systems (4.114) and (4.117), i.e., with the oscillations of the Wronskian of $Y(\lambda)$ and $\hat{Y}(\lambda)$. In subsequent sections we will present some special cases, when formulas (6.25) and (6.33) can be simplified by using the special structure of the symplectic coefficient matrices $\mathcal{S}_k(\lambda)$ and $\hat{\mathcal{S}}_k(\lambda)$. In particular, in these cases they are associated with the oscillation of conjoined bases of the $2n$-dimensional Wronskian systems (4.114) and (4.117).

### 6.1.3  Lower Block-Triangular Perturbation

In this subsection we investigate the most closely related matrix analog of the results for Sturm-Liouville problems (6.9), (6.10). Consider the pair of eigenvalue problems (6.17) with the Dirichlet boundary conditions, whose coefficient matrices obey the conditions

$$\mathcal{A}_k(\lambda) = \hat{\mathcal{A}}_k(\lambda) \equiv \mathcal{A}_k, \quad \mathcal{B}_k(\lambda) = \hat{\mathcal{B}}_k(\lambda) \equiv \mathcal{B}_k, \quad k \in [0, N]_{\mathbb{Z}}. \tag{6.35}$$

Then we have that conditions (6.18) are satisfied and under assumptions (5.3) and (6.5) one can apply Theorems 6.2 and 6.4 to this special case. It follows from (6.35) that for arbitrary fixed $\lambda := \beta$ we have

$$\mathcal{S}_k(\lambda)\,\mathcal{S}_k^{-1}(\beta) = \begin{pmatrix} I & 0 \\ Q_k(\lambda, \beta) & I \end{pmatrix}, \quad \hat{\mathcal{S}}_k(\lambda)\,\hat{\mathcal{S}}_k^{-1}(\beta) = \begin{pmatrix} I & 0 \\ \hat{Q}_k(\lambda, \beta) & I \end{pmatrix},$$

or (compare with (5.239) where $\beta = 0$)

$$
\mathcal{S}_k(\lambda) = \begin{pmatrix} I & 0 \\ Q_k(\lambda, \beta) & I \end{pmatrix} \mathcal{S}_k(\beta), \quad \hat{\mathcal{S}}_k(\lambda) = \begin{pmatrix} I & 0 \\ \hat{Q}_k(\lambda, \beta) & I \end{pmatrix} \hat{\mathcal{S}}_k(\beta), \tag{6.36}
$$

where the symmetric matrices $Q_k(\lambda, \beta)$ and $\hat{Q}_k(\lambda, \beta)$ (see Sect. 1.6.1) can be expressed in terms of the blocks of $\mathcal{S}_k(a)$ and $\hat{\mathcal{S}}_k(b)$ as

$$
\left.\begin{aligned}
Q_k(\lambda, \beta) &= C_k(\lambda)\, D_k^T(\beta) - D_k(\lambda)\, C_k^T(\beta) = P_k(\lambda)\, \mathcal{J} P_k^T(\beta), \\
P_k(\lambda) &:= \left( C_k(\lambda)\ D_k(\lambda) \right), \\
\hat{Q}_k(\lambda, \beta) &= \hat{C}_k(\lambda)\, \hat{D}_k^T(\beta) - \hat{D}_k(\lambda)\, \hat{C}_k^T(\beta) = \hat{P}_k(\lambda)\, \mathcal{J} \hat{P}_k^T(\beta), \\
\hat{P}_k(\lambda) &:= \left( \hat{C}_k(\lambda)\ \hat{D}_k(\lambda) \right).
\end{aligned}\right\} \tag{6.37}
$$

Moreover, for two arbitrary values of $\lambda = a$ and $\lambda = b$, we have

$$
\mathcal{S}_k(a)\, \hat{\mathcal{S}}_k^{-1}(b) = \begin{pmatrix} I & 0 \\ Q_k(a, b) & I \end{pmatrix}, \tag{6.38}
$$

where the symmetric matrix $Q_k(a, b) = Q_k^T(a, b)$ is given by

$$
\left.\begin{aligned}
Q_k(a, b) &= C_k(a)\, \hat{D}_k^T(b) - D_k(a)\, \hat{C}_k^T(b) = P_k(a)\, \mathcal{J} \hat{P}_k^T(b) \\
&= Q_k(a, \beta) - \hat{Q}_k(b, \beta) + P_k(\beta)\, \mathcal{J} \hat{P}_k^T(\beta),
\end{aligned}\right\} \tag{6.39}
$$

with $Q_k(\lambda, \beta)$ and $\hat{Q}_k(\lambda, \beta)$ defined by (6.37). In particular,

$$
\left.\begin{aligned}
Q_k(a, b) &= P_k(a)\, \mathcal{J} P_k^T(b) = Q_k(a, \beta) - Q_k(b, \beta), \\
&\text{for the case } \mathcal{S}_k(\lambda) \equiv \hat{\mathcal{S}}_k(\lambda), \quad k \in [0, N]_{\mathbb{Z}}.
\end{aligned}\right\} \tag{6.40}
$$

The matrices $\Psi(\mathcal{S}_k(\lambda))$ and $\Psi(\hat{\mathcal{S}}_k(\lambda))$ in (5.3) and (6.5) for the case (6.35) take the form (see (1.193))

$$
\Psi(\mathcal{S}_k(\lambda)) = \mathrm{diag}\{\dot{P}_k(\lambda)\, \mathcal{J}^T P_k^T(\lambda),\ 0\}, \quad \Psi(\hat{\mathcal{S}}_k(\lambda)) = \mathrm{diag}\{\dot{\hat{P}}_k(\lambda)\, \mathcal{J}^T \hat{P}_k^T(\lambda),\ 0\}.
$$

Then the monotonicity conditions will be satisfied under the assumption

$$
\left.\begin{aligned}
\dot{P}_k(\lambda)\, \mathcal{J} P_k^T(\lambda) &= \dot{Q}_k(\lambda, \beta) \leq 0, \\
\dot{\hat{P}}_k(\lambda)\, \mathcal{J} \hat{P}_k^T(\lambda) &= \dot{\hat{Q}}_k(\lambda, \beta) \leq 0,
\end{aligned}\right\} \quad k \in [0, N]_{\mathbb{Z}}, \quad \lambda \in \mathbb{R}. \tag{6.41}
$$

For the special linear dependence on $\lambda$ (see (5.239) and (5.240)), we then have

$$\mathcal{S}_k(\lambda) = \begin{pmatrix} I & 0 \\ -\lambda \mathcal{W}_k & I \end{pmatrix} \begin{pmatrix} \mathcal{A}_k & \mathcal{B}_k \\ \mathcal{C}_k & \mathcal{D}_k \end{pmatrix}, \quad \hat{\mathcal{S}}_k(\lambda) = \begin{pmatrix} I & 0 \\ -\lambda \hat{\mathcal{W}}_k & I \end{pmatrix} \begin{pmatrix} \mathcal{A}_k & \mathcal{B}_k \\ \hat{\mathcal{C}}_k & \hat{\mathcal{D}}_k \end{pmatrix}, \tag{6.42}$$

so that the matrices $\mathcal{S}_k(\lambda)$ and $\hat{\mathcal{S}}_k(\lambda)$ obey (6.35), (6.38) for any $a, b \in \mathbb{R}$, and

$$\mathcal{Q}_k(a, b) = b \hat{\mathcal{W}}_k - a \mathcal{W}_k + \mathcal{C}_k \hat{\mathcal{D}}_k^T - \mathcal{D}_k \hat{\mathcal{C}}_k^T. \tag{6.43}$$

In particular,

$$\mathcal{Q}_k(a, b) = (b - a) \mathcal{W}_k \quad \text{for } \mathcal{S}_k(\lambda) \equiv \hat{\mathcal{S}}_k(\lambda), \quad k \in [0, N]_{\mathbb{Z}}. \tag{6.44}$$

For the scalar case of two Sturm-Liouville difference equations (6.9) and (6.10), which are rewritten in the matrix form (2.36), we then have

$$\mathcal{S}_k(\lambda) = \begin{pmatrix} 1 & 1/r_k \\ p_k - \lambda & 1 + (p_k - \lambda)/r_k \end{pmatrix}, \quad \hat{\mathcal{S}}_k(\lambda) = \begin{pmatrix} 1 & 1/r_k \\ \hat{p}_k - \lambda & 1 + (\hat{p}_k - \lambda)/r_k \end{pmatrix}. \tag{6.45}$$

Then, with the number $q_k(a, b)$ from Definition 6.1,

$$\mathcal{Q}_k(a, b) = b - a + p_k - \hat{p}_k = q_k(a, b).$$

Recall that, as we mentioned above, the relative oscillation theory for problems (6.17) under assumption (6.35) is the most closely related analog of the results for Sturm-Liouville problems (6.9) and (6.10). Assume that $Z(a)$ and $\hat{Z}(b)$ are symplectic fundamental matrices of systems (5.1) and (6.1) associated with conjoined bases $Y(\lambda)$ and $\hat{Y}(\lambda)$ of these systems according to (3.14). Then we have the connections (see (4.100))

$$\left. \begin{aligned} w(\hat{Y}_k(b), Y_k(a)) &= -(I \ 0) \hat{Z}_k^{-1}(b) Y_k(a), \\ w(Y_k(a), \hat{Y}_k(b)) &= -(I \ 0) Z_k^{-1}(a) \hat{Y}_k(b), \\ w(\hat{Y}_k(b), Y_k(a)) &= -w^T(Y_k(a), \hat{Y}_k(b)). \end{aligned} \right\} \tag{6.46}$$

We will show that for the case (6.35) the role of generalized zeros of the Wronskian is played by focal points of the transformed conjoined bases

$$\hat{Z}_k^{-1}(b) Y_k(a), \quad Z_k^{-1}(a) \hat{Y}_k(b)$$

of the discrete symplectic systems (4.114) and (4.117) rewritten in the form

$$\tilde{Y}_{k+1} = \hat{Z}_{k+1}^{-1}(b) \mathcal{S}_k(a) \hat{\mathcal{S}}_k^{-1}(b) \hat{Z}_{k+1}(b) \tilde{Y}_k, \quad \tilde{Y}_k = \hat{Z}_k^{-1}(b) Y_k(a), \tag{6.47}$$

$$\bar{Y}_{k+1} = Z_{k+1}^{-1}(a) \hat{\mathcal{S}}_k(b) \mathcal{S}_k^{-1}(a) Z_{k+1}(a) \bar{Y}_k, \quad \bar{Y}_k = Z_k^{-1}(a) \hat{Y}_k(b). \tag{6.48}$$

Assume (6.35), then by (6.38) the blocks $\mathfrak{B}_k(a, b)$ and $\tilde{\mathfrak{B}}_k(a, b)$ in the right upper corner of the symplectic matrices in (6.47) and (6.48) have the form

$$\left.\begin{aligned}
\mathfrak{B}_k(a, b) &= -\hat{X}_{k+1}^T(b) \, \mathcal{Q}_k(a, b) \, \hat{X}_{k+1}(b), \\
\tilde{\mathfrak{B}}_k(a, b) &= X_{k+1}^T(a) \, \mathcal{Q}_k(a, b) \, X_{k+1}(a),
\end{aligned}\right\} \tag{6.49}$$

while systems (6.47) and (6.48) take the form

$$-\mathcal{J}\Delta\tilde{Y}_k = \hat{Z}_{k+1}^T(b) \, \operatorname{diag}\{-\mathcal{Q}_k(a, b), \, 0\} \, \hat{Z}_{k+1}(b) \, \tilde{Y}_k, \quad \tilde{Y}_k = \hat{Z}_k^{-1}(b) \, Y_k(a), \tag{6.50}$$

$$-\mathcal{J}\Delta\bar{Y}_k = Z_{k+1}^T(a) \, \operatorname{diag}\{\mathcal{Q}_k(a, b), \, 0\} \, Z_{k+1}(a) \, \bar{Y}_k, \quad \bar{Y}_k = Z_k^{-1}(a) \, \hat{Y}_k(b). \tag{6.51}$$

For systems (5.1) and (6.1) under restriction (6.35), the formulas for the relative oscillation numbers in (6.26) and (6.31) are simplified as follows.

**Lemma 6.6** *Suppose that the matrices $S(\lambda)$ and $\hat{S}(\lambda)$ in (5.1) and (6.1) satisfy (6.35) and that $Z(\lambda)$ and $\hat{Z}(\lambda)$ are symplectic fundamental matrices of (5.1) and (6.1) associated with conjoined bases $Y(\lambda)$ and $\hat{Y}(\lambda)$ of (5.1) and (6.1) such that*

$$Z(\lambda) \, (0 \ I)^T = Y(\lambda) = \begin{pmatrix} X(\lambda) \\ U(\lambda) \end{pmatrix}, \quad \hat{Z}(\lambda) \, (0 \ I)^T = \hat{Y}(\lambda) = \begin{pmatrix} \hat{X}(\lambda) \\ \hat{U}(\lambda) \end{pmatrix}.$$

*Then the relative oscillation numbers (4.109) are given by the formulas*

$$\left.\begin{aligned}
\#_k(\hat{Z}(b), Z(a)) &= m^*(Z_k^{-1}(a) \, \hat{Y}_k(b)) - \operatorname{ind} \tilde{\mathfrak{B}}_k(a, b) \\
&= \operatorname{ind} \mathfrak{B}_k(a, b) - m(\hat{Z}_k^{-1}(b) \, Y_k(a)),
\end{aligned}\right\} \tag{6.52}$$

*where the matrices $\mathfrak{B}_k(a, b)$ and $\tilde{\mathfrak{B}}_k(a, b)$ are given by (6.49), and where the quantities $m^*(Z_k^{-1}(a) \, \hat{Y}_k(b))$ and $m(\hat{Z}_k^{-1}(a) \, Y_k(b))$ represent the number of backward focal points in $[k, k + 1)$ and the number of forward focal points in $(k, k + 1]$ of the conjoined bases $Z^{-1}(a) \, \hat{Y}(b)$ and $\hat{Z}^{-1}(a) \, Y(b)$ of systems (6.51) and (6.50), respectively. In addition, we have the estimate*

$$|\#_k(\hat{Z}(b), Z(a))| \leq \min \left\{\operatorname{rank} \mathfrak{B}_k(a, b), \operatorname{rank} \tilde{\mathfrak{B}}_k(a, b)\right\} \leq n. \tag{6.53}$$

*Proof* The matrices $S_k(\lambda)$ and $\hat{S}_k(\lambda)$ in (5.1) and (6.1) satisfy (6.38), and then assumption (4.107) of Lemma 4.43 is also satisfied. Using (4.108) and incorporating (4.16) in Lemma 4.8, we see that

$$\mu\big(\hat{Z}_k^{-1}(b) \, Z_k(a) \, (0 \ I)^T, \, \hat{Z}_{k+1}^{-1}(b) \, Z_{k+1}(a) \, (0 \ I)^T\big) = m^*(Z_k^{-1}(a) \, \hat{Y}_k(b))$$

(compare with Remark 4.46(iii)). So we obtain the proof of the first identity in (6.52).

To prove the second identity, we use (4.118). Indeed, replacing the roles $Y$ and $\hat{Y}$ in the first (already proved) identity (6.52), we obtain by (4.118) that

$$\#_k(\hat{Z}(b), Z(a)) = \Delta \operatorname{rank} w(Y_k(a), \hat{Y}_k(b)) - \#_k(Z(a), \hat{Z}(b))$$

$$= \Delta \operatorname{rank} w(Y_k(a), \hat{Y}_k(b)) - m^*(\hat{Z}_k^{-1}(b) Y_k(a))$$

$$+ \operatorname{ind} [-\hat{X}_{k+1}^T(b) Q_k(a, b) \hat{X}_{k+1}(b)]$$

$$= \operatorname{ind} [-\hat{X}_{k+1}^T(b) Q_k(a, b) \hat{X}_{k+1}(b)] - m(\hat{Z}_k^{-1}(b) Y_k(a)),$$

where, in addition, Proposition 4.4(vi) (see (4.9)) is used.

For the proof of estimate (6.53), we incorporate that the matrices $\mathcal{B}_k(a, b)$ and $\tilde{\mathcal{B}}_k(a, b)$ given by (6.49) are the blocks of the symplectic matrices in (6.47) and (6.48) in the right upper corner. Then by Proposition 4.4(v)

$$m^*(Z_k^{-1}(a) \hat{Y}_k(b)) \leq \operatorname{rank} \tilde{\mathcal{B}}_k(a, b), \quad m(\hat{Z}_k^{-1}(b) Y_k(a)) \leq \operatorname{rank} \mathcal{B}_k(a, b).$$

Taking in mind that $\operatorname{ind} \tilde{\mathcal{B}}_k(a, b) \leq \operatorname{rank} \tilde{\mathcal{B}}_k(a, b)$ and $\operatorname{ind} \mathcal{B}_k(a, b) \leq \operatorname{rank} \mathcal{B}_k(a, b)$, we derive estimate (6.53). The proof is complete.  □

As it was mentioned above, it is possible to relate the number (6.52) and the notion of a weighted generalized zero of the Wronskian for solutions of second-order Sturm-Liouville difference equations (6.9) and (6.10) presented in Sect. 6.1.1.

**Proposition 6.7** *The Wronskian of two nontrivial solutions $x_k(a)$ and $\hat{x}_k(b)$ of (6.9) and (6.10) has a weighted generalized zero according to Definition 6.1 if and only if the relative oscillation number defined in Lemma 6.6 for $n = 1$ with $Q_k(a, b) := q_k(a, b)$ takes the value $\pm 1$.*

*Proof* Note that by (6.52) $\#_k(\hat{Z}(b), Z(a)) = 1$ if and only if $m^*(Z_k^{-1}(a) \hat{Y}_k(b)) = 1$ and $\operatorname{ind} \tilde{\mathcal{B}}_k(a, b) = 0$. According to Remark 4.14(iii), the case $m^*(Z_k^{-1}(a) \hat{Y}_k(b)) = 1$ is equivalent to the existence of a generalized zero in $[k, k + 1)$ of the solution $Z_k^{-1}(a) \hat{Y}_k(b)$ for $n = 1$. Using connection (6.46) we have that the Wronskian $w_k := w(Y_k, \hat{Y}_k)$ is associated with the upper block of the solution $Z_k^{-1}(a) \hat{Y}_k(b)$ (up to the sign, which is not important). Then by Remark 4.14(iii), we have that $m^*(Z_k^{-1}(a) Y_k(b)) = 1$ if and only if $w_k = 0$, $w_{k+1} \neq 0$, or $w_k w_{k+1} \tilde{\mathcal{B}}_k(a, b) < 0$, where $\tilde{\mathcal{B}}_k(a, b) = x_{k+1}^2 q_k(a, b)$ according to (6.49). Then, under the assumption $q_k(a, b) > 0$, the last condition is equivalent with $w_k w_{k+1} < 0$ (note that we use $\operatorname{ind} \tilde{\mathcal{B}}_k(a, b) = 0$, and the case $\tilde{\mathcal{B}}_k(a, b) = 0$ is excluded by Remark 4.14(iii)). Thus, we have proved that the conditions $m^*(Z_k^{-1}(a) \hat{Y}_k(b)) = 1$ and $\operatorname{ind} \tilde{\mathcal{B}}_k(a, b) = 0$ are equivalent to the first case in Definition 6.1.

In a similar way, we have that $\#_k(\hat{Z}(b), Z(a)) = -1$ if and only if the two conditions $m(\hat{Z}_k^{-1}(b)Y_k(a)) = 1$ and ind $\mathfrak{B}_k(a, b) = 0$ hold. By Remark 4.14(iii), we have $m(\hat{Z}_k^{-1}(b)Y_k(a)) = 1$ if and only if $w_{k+1} = 0$, $w_k \neq 0$, or $w_k w_{k+1} \mathfrak{B}_k(a, b) < 0$, where $\mathfrak{B}_k(a, b) = -\hat{x}_{k+1}^2 q_k(a, b)$ according to (6.49). Then, under the assumption $q_k(a, b) < 0$, the last condition is equivalent with $w_k w_{k+1} < 0$. Thus, we have proved that $\#_k(\hat{Z}(b), Z(a)) = -1$ or the two conditions $m(\hat{Z}_k^{-1}(b)Y_k(a)) = 1$, ind $\mathfrak{B}_k(a, b) = 0$ are equivalent to the second case in Definition 6.1. □

Based on Lemma 6.6, one can specialize the results of Theorem 6.2 as follows.

**Theorem 6.8** *Assume that conditions (6.35) and (6.41) hold for systems (5.1) and (6.1). Let $Z^{[0]}(\lambda)$ and $\hat{Z}^{[N+1]}(\lambda)$ be symplectic fundamental matrices of (5.1) and (6.1) associated with the principal solutions $Y^{[0]}(\lambda) = Z^{[0]}(\lambda)(0 \ I)^T$ and $\hat{Y}^{[N+1]}(\lambda) = \hat{Z}^{[N+1]}(\lambda)(0 \ I)^T$ of these systems at $k = 0$ and $k = N + 1$, respectively. Then we have that all the formulas in Theorem 6.2 hold with the relative oscillation numbers defined by (6.52), (6.49), (6.39), where $\hat{Z}(\lambda) := \hat{Z}^{[N+1]}(\lambda)$ and $Z(\lambda) := Z^{[0]}(\lambda)$.*

*Proof* For the proof we use Theorem 6.2 and Lemma 6.6. □

In a similar way, we derive the special case of Theorem 6.4.

**Theorem 6.9** *Assume that conditions (6.35) and (6.41) hold for system (5.1). Let $Z^{[0]}(\lambda)$ and $Z^{[N+1]}(\lambda)$ be symplectic fundamental matrices of (5.1) such that $Y^{[0]}(\lambda) = Z^{[0]}(\lambda)(0 \ I)^T$ and $Y^{[N+1]}(\lambda) = Z^{[N+1]}(\lambda)(0 \ I)^T$ are the principal solutions of this system at $k = 0$ and $k = N + 1$, respectively. Then for any $a, b \in \mathbb{R}$ with $a < b$ we have that the number of finite eigenvalues of problem (E) in $(a, b]$ is counted by the number of backward focal points of the conjoined basis $[Z^{[0]}(a)]^{-1}Y^{[N+1]}(b)$ of system (6.51) in the interval $[0, N + 1)$, i.e.,*

$$\#\{v \in \sigma \mid a < v \leq b\} = l^*([Z^{[0]}(a)]^{-1}Y^{[N+1]}(b), 0, N + 1), \quad (6.54)$$

*where system (6.51) is considered for the case $\mathcal{S}_k(\lambda) = \hat{\mathcal{S}}_k(\lambda)$ with $Z_k(u) := Z_k^{[0]}(u)$ and $\mathcal{Q}_k(a, b) \geq 0$ is given by (6.40). Moreover*

$$\#\{v \in \sigma \mid v \leq b\} = l^*([Z^{[0]}(\lambda)]^{-1}Y^{[N+1]}(b), 0, N + 1), \quad \lambda < \lambda_1, \quad (6.55)$$

*where $\lambda_1 = \min \sigma$.*

*Proof* We use Theorem 6.4 for the case (6.35). Remark that for $\mathcal{S}_k(\lambda) \equiv \hat{\mathcal{S}}_k(\lambda)$, $k \in [0, N]_{\mathbb{Z}}$, according to (6.40) and (6.41)

$$\mathcal{Q}_k(a, b) \geq 0, \quad a, b \in \mathbb{R}, \ a \leq b, \quad (6.56)$$

Then the relative oscillation number in (6.52) takes the form (here we use that $\tilde{\mathfrak{B}}_k(a, b) \geq 0$ and $\mathfrak{B}_k(a, b) \leq 0$)

$$\begin{aligned}
\#_k(\hat{Z}(b), Z(a)) &= m^*(Z_k^{[0]\,-1}(a)\, Y_k^{[N+1]}(b)) \\
&= \operatorname{rank} \mathfrak{B}_k(a, b) - m(Z_k^{[N+1]\,-1}(b)\, Y_k^{[0]}(a)), \\
\mathfrak{B}_k(a, b) &= -X_{k+1}^{[N+1]\,T}(b)\, \mathcal{Q}_k(a, b)\, X_{k+1}^{[N+1]}(b),
\end{aligned} \right\} \qquad (6.57)$$

where $Z_k^{[0]}(\lambda)$ and $Z^{[N+1]}(\lambda)$ are symplectic fundamental matrices of system (5.1) associated with the principal solutions $Y_k^{[0]}(\lambda)$ and $Y_k^{[N+1]}(\lambda)$. Substituting the first formula in (6.57) into (6.31), we derive (6.55).                                               □

*Remark 6.10*

(i) Recall that according to Remark 6.5(i) one can modify the representations of the relative oscillation numbers using (6.33). For the special case under the consideration, we have instead of (6.33) the formula

$$\begin{aligned}
\#_k(Z^{[N+1]}(a), \hat{Z}^{[0]}(b)) &= m^*([\hat{Z}_k^{[0]}(b)]^{-1}\, Y_k^{[N+1]}(a)) - \operatorname{ind}[\mathfrak{B}_k(a, b)] \\
&= \operatorname{ind}[\tilde{\mathfrak{B}}_k(a, b)] - m([Z_k^{[N+1]}(a)]^{-1}\, \hat{Y}_k^{[0]}(b)),
\end{aligned} \right\} \qquad (6.58)$$

where the matrices $\mathfrak{B}_k(a, b)$ and $\tilde{\mathfrak{B}}_k(a, b)$ are given by (6.49) with $\hat{X}(b) := \hat{X}^{[0]}(b)$ and $X(a) := X^{[N+1]}(a)$. In particular, this reformulation leads to the representation

$$\#\{\nu \in \sigma \mid a < \nu \leq b\} = l([Z^{[N+1]}(a)]^{-1} Y^{[0]}(b), 0, N+1), \qquad (6.59)$$

which is equivalent to (6.54) and presents $\#\{\nu \in \sigma \mid a < \nu \leq b\}$ in terms of the multiplicities of forward focal points of the conjoined basis $[Z^{[N+1]}(a)]^{-1} Y^{[0]}(b)$ of the transformed system (6.51).

(ii) For the special linear dependence on parameter $\lambda$ according to (6.42), (6.43), (6.44), one can use Theorem 5.93 and property (4.147) in Theorem 4.56 to derive the formulas

$$\#\{\nu \in \hat{\sigma} \mid \nu < b\} - \#\{\nu \in \sigma \mid \nu < a\} = \#(\hat{Z}^{[0]}(b), Z^{[N+1]}(a), 0, N) - \mathfrak{P}^*, \qquad (6.60)$$

where

$$\mathfrak{P}^* = \hat{m}^* - m^* = \#(\hat{Z}^{[0]}(\lambda), Z^{[N+1]}(\lambda), 0, N), \qquad \lambda < \lambda_0. \qquad (6.61)$$

Similarly, instead of (6.54) we have

$$\#\{\nu \in \sigma \mid a \leq \nu < b\} = l^*([Z^{[N+1]}(a)]^{-1} Y^{[0]}(b), 0, N+1). \qquad (6.62)$$

Formulas (6.60), (6.61), and (6.62) can be also derived directly from (6.26), (6.28), and (6.31) by using Lemma 4.47(iii).

### 6.1.4 Matrix Sturm-Liouville Eigenvalue Problems

In this subsection we consider the discrete matrix Sturm-Liouville spectral problems (see Example 5.36)

$$\left.\begin{array}{ll} \Delta(R_k(\lambda)\,\Delta x_k(\lambda)) - Q_k(\lambda)\,x_{k+1}(\lambda) = 0, & k \in [0, N-1]_{\mathbb{Z}}, \\ x_0(\lambda) = 0 = x_{N+1}(\lambda), & \det R_k(\lambda) \neq 0, \quad k \in [0, N]_{\mathbb{Z}}, \end{array}\right\} \tag{6.63}$$

and

$$\left.\begin{array}{ll} \Delta(\hat{R}_k(\lambda)\,\Delta \hat{x}_k(\lambda)) - \hat{Q}_k(\lambda)\,\hat{x}_{k+1}(\lambda) = 0, & k \in [0, N-1]_{\mathbb{Z}}, \\ \hat{x}_0(\lambda) = 0 = \hat{x}_{N+1}(\lambda), & \det \hat{R}_k(\lambda) \neq 0, \quad k \in [0, N]_{\mathbb{Z}}, \end{array}\right\} \tag{6.64}$$

where $x_k(\lambda) \in \mathbb{R}^n$ for $n \geq 1$ and $\lambda \in \mathbb{R}$ is the spectral parameter and the real symmetric $n \times n$ matrix-valued functions $R_k(\lambda)$, $\hat{R}_k(\lambda)$, $Q_k(\lambda)$, $\hat{Q}_k(\lambda)$ for $k \in [0, N]_{\mathbb{Z}}$ are piecewise continuously differentiable in the variable $\lambda$ and obey the conditions

$$\dot{R}_k(\lambda) \leq 0, \quad \dot{Q}_k(\lambda) \leq 0, \quad k \in [0, N]_{\mathbb{Z}}, \tag{6.65}$$

$$\dot{\hat{R}}_k(\lambda) \leq 0, \quad \dot{\hat{Q}}_k(\lambda) \leq 0, \quad k \in [0, N]_{\mathbb{Z}}. \tag{6.66}$$

According to Example 5.36, conditions (6.65) and (6.65) together with the nonsingularity assumptions $\det R_k(\lambda) \neq 0$ and $\det \hat{R}_k(\lambda) \neq 0$ imply (5.3), (5.5), (6.18). Then one can apply Theorems 6.2 and 6.4 to this special case. As in the previous subsection, we refine formulas for the relative oscillation numbers using the special structure of the symplectic matrices $\mathcal{S}_k(\lambda)$ and $\hat{\mathcal{S}}_k(\lambda)$ associated with (6.63) and (6.64) (see Sect. 2.1.2), i.e.,

$$\left.\begin{array}{l} \mathcal{S}_k(\lambda) = \begin{pmatrix} I & R_k^{-1}(\lambda) \\ Q_k(\lambda)\,I + Q_k(\lambda)\,R_k^{-1}(\lambda) \end{pmatrix} = L_k(\lambda)\,H_k(\lambda), \\[2mm] L_k(\lambda) := \begin{pmatrix} I & 0 \\ Q_k(\lambda) & I \end{pmatrix}, \quad H_k(\lambda) := \begin{pmatrix} I & R_k^{-1}(\lambda) \\ 0 & I \end{pmatrix}, \end{array}\right\} \tag{6.67}$$

and

$$\left.\begin{array}{l} \hat{\mathcal{S}}_k(\lambda) = \begin{pmatrix} I & \hat{R}_k^{-1}(\lambda) \\ \hat{Q}_k(\lambda)\,I + \hat{Q}_k(\lambda)\,\hat{R}_k^{-1}(\lambda) \end{pmatrix} = \hat{L}_k(\lambda)\,\hat{H}_k(\lambda), \\[2mm] \hat{L}_k(\lambda) := \begin{pmatrix} I & 0 \\ \hat{Q}_k(\lambda) & I \end{pmatrix}, \quad \hat{H}_k(\lambda) := \begin{pmatrix} I & \hat{R}_k^{-1}(\lambda) \\ 0 & I \end{pmatrix}. \end{array}\right\} \tag{6.68}$$

Recall that spectral problems (6.63) and (6.64) are the matrix analogs of the Sturm-Liouville problems (6.9) and (6.10) with different coefficients $r_k(\lambda)$ and $\hat{r}_k(\lambda)$ and with the Wronskian $w_k(x, \hat{x}) = -r_k x_{k+1} \hat{x}_k + \hat{r}_k x_k \hat{x}_{k+1}$ (compare with (6.11)). The relative oscillation theory for (6.9) and (6.10) with $r_k \neq \hat{r}_k$ and $p_k \neq \hat{p}_k$ is presented in [21].

By (6.67) and (6.68), we see that problems (6.63) and (6.64) under the additional assumption $R_k(\lambda) \equiv \hat{R}_k(\lambda) \equiv R_k$ obey conditions (6.35). However, in the general case of nonconstant $R_k(\lambda)$ and/or $\hat{R}_k(\lambda)$, the theory developed in Sect. 6.1.3 needs to be generalized to the case when the matrices $R_k(\lambda)$ and $\hat{R}_k(\lambda)$ are different.

Consider two conjoined bases $Y(\lambda) = \begin{pmatrix} X(\lambda) \\ U(\lambda) \end{pmatrix}$ and $\hat{Y}(\lambda) = \begin{pmatrix} \hat{X}(\lambda) \\ \hat{U}(\lambda) \end{pmatrix}$ of the symplectic systems in (6.17) associated with (6.63) and (6.64), where we have the connections (see Sect. 2.1.2)

$$\left. \begin{aligned} U_k(\lambda) &= R_k(\lambda) \, \Delta X_k(\lambda), \quad \hat{U}_k(\lambda) = \hat{R}_k(\lambda) \, \Delta \hat{X}_k(\lambda), \quad k \in [0, N]_{\mathbb{Z}}, \\ U_{N+1}(\lambda) &= U_N(\lambda) + Q_N(\lambda) \, X_{N+1}(\lambda), \\ \hat{U}_{N+1}(\lambda) &= \hat{U}_N(\lambda) + \hat{Q}_N(\lambda) \, \hat{X}_{N+1}(\lambda). \end{aligned} \right\} \tag{6.69}$$

Note that the coefficients $Q_N(\lambda)$ and $\hat{Q}_N(\lambda)$ are not needed in equations (6.63) and (6.64), but for convenience we define them at $k = N$ such that (6.65) and (6.66) hold. In Remark 6.12 we will show that the results of this section do not depend on the definition of $Q_N(\lambda)$ and $\hat{Q}_N(\lambda)$.

Recall that for two fixed values $\lambda = a$ and $\lambda = b$ $(a, b \in \mathbb{R})$ and $k \in [0, N+1]_{\mathbb{Z}}$, the Wronskian (3.2) of $Y_k(a)$ and $\hat{Y}_k(b)$ has the form

$$w_k(Y(a), \hat{Y}(b)) := w_k(a, b) = X_k^T(a) \, \hat{U}_k(b) - U_k^T(a) \, \hat{X}_k(b). \tag{6.70}$$

Moreover, it satisfies for $k \in [0, N]_{\mathbb{Z}}$ the equation (see (4.99))

$$\left. \begin{aligned} -\Delta w_k(a, b) &= (\Delta X_k^T(a)) \, [R_k(a) - \hat{R}_k(b)] \, \Delta \hat{X}_k(b) \\ &\quad + X_{k+1}^T(a) \, [Q_k(a) - \hat{Q}_k(b)] \, \hat{X}_{k+1}(b). \end{aligned} \right\} \tag{6.71}$$

For the convenience of the presentation of the subsequent results, we introduce the notation for the number of backward focal points of the conjoined basis $Z^{-1}\hat{Y}$ associated with the Wronskian by (6.46) using the notation

$$m^*(w_k, w_{k+1}) := m^*(Z_k^{-1} \hat{Y}_k) = \mu\big(\hat{Z}_k^{-1} Y_k, \hat{Z}_{k+1}^{-1} Y_{k+1}\big).$$

Then we have using Definition 4.11

$$\left. \begin{aligned} m^*(w_k, w_{k+1}) &= \operatorname{rank} \mathcal{M}_k + \operatorname{ind} \mathcal{P}_k, \\ \mathcal{M}_k := (I - w_k^\dagger w_k) \, w_{k+1}^T, \quad \mathcal{T}_k &:= I - \mathcal{M}_k^\dagger \mathcal{M}_k, \quad \mathcal{P}_k := \mathcal{T}_k w_{k+1} w_k^\dagger \, \mathfrak{C}_k \mathcal{T}_k, \\ \mathfrak{C}_k := (\hat{Z}_k^{-1} Y_k)^T \mathcal{J} \hat{Z}_{k+1}^{-1} Y_{k+1} &= Y_k^T \mathcal{J} \hat{S}_k^{-1} S_k Y_k = Y_{k+1}^T \mathcal{J} S_k \hat{S}_k^{-1} Y_{k+1}. \end{aligned} \right\} \tag{6.72}$$

Note that according to Proposition 4.4(v), we have the estimate

$$m^*(w_k, w_{k+1}) \leq \operatorname{rank} \mathfrak{C}_k \leq n. \tag{6.73}$$

As we already mentioned above, for the case $R_k(\lambda) \equiv \hat{R}_k(\lambda) \equiv R_k$, one can apply Lemma 6.6 to evaluate the relative oscillation numbers for problems (6.63) and (6.64). Here (for convenience) we reformulate this lemma in terms of the coefficients $Q_k(\lambda)$ and $\hat{Q}_k(\lambda)$ in (6.63) and (6.64).

**Lemma 6.11 (Case I)** *Assume that for the two spectral problems* (6.63), (6.65) *and* (6.64), (6.66), *the matrices* $R_k(\lambda)$ *and* $\hat{R}_k(\lambda)$ *obey the conditions*

$$\dot{R}_k(\lambda) = \hat{\dot{R}}_k(\lambda) = 0, \quad R_k \equiv \hat{R}_k, \quad k \in [0, N]_{\mathbb{Z}}. \tag{6.74}$$

*Then the relative oscillation numbers* (4.109) *have the form*

$$\left. \begin{aligned} \#_k^I(\hat{Z}(b), Z(a)) &= m^*(w_k(a, b), w_{k+1}(a, b)) - \operatorname{ind} \mathfrak{C}_k(a, b), \\ \mathfrak{C}_k(a, b) &= X_{k+1}^T(a)\,[Q_k(a) - \hat{Q}_k(b)]\,X_{k+1}(a), \end{aligned} \right\} \tag{6.75}$$

*where* $m^*(w_k(a, b), w_{k+1}(a, b)) := m^*(Z_k^{-1}(a)\,\hat{Y}_k(b))$ *is the number of backward focal points of the conjoined basis* $Z_k^{-1}(a)\,\hat{Y}_k(b)$ *according to* (6.72) *with the matrix* $\mathfrak{C}_k := \mathfrak{C}_k(a, b)$.

*Proof* We use formula (6.52) putting $\mathfrak{C}_k(a, b) := \tilde{\mathfrak{B}}_k(a, b)$. □

*Remark 6.12* Note that in the definition of symplectic systems (6.67) and (6.68) one can put $Q_N(\lambda) = \hat{Q}_N(\lambda) = 0$ and then $\#_N^I(\hat{Z}(b), Z(a)) = 0$. However for the case when $\hat{Z}_k(b) := \hat{Z}_k^{[N+1]}(b)$, we have $\#_N^I(\hat{Z}(b), Z(a)) = 0$ for any choice of $Q_N(\lambda)$ and $\hat{Q}_N(\lambda)$. Indeed, for this case by (6.49), the matrix $\mathfrak{B}_N(a, b) = 0$, and then $\#_N^I(\hat{Z}(b), Z(a)) = 0$ by estimate (6.53).

**Lemma 6.13 (Case II)** *Assume that for the two spectral problems* (6.63), (6.65) *and* (6.64), (6.66), *the matrices* $Q_k(\lambda)$ *and* $\hat{Q}_k(\lambda)$ *obey the conditions*

$$\dot{Q}_k(\lambda) = \hat{\dot{Q}}_k(\lambda) = 0, \quad Q_k \equiv \hat{Q}_k, \quad k \in [0, N]_{\mathbb{Z}}. \tag{6.76}$$

*Then the relative oscillation numbers* (4.109) *take the form*

$$\left. \begin{aligned} \#_k^{II}(\hat{Z}(b), Z(a)) &= m^*(w_k(a, b), w_{k+1}(a, b)) - \operatorname{ind} \tilde{\mathfrak{C}}_k(a, b) + P_k, \\ \tilde{\mathfrak{C}}_k(a, b) &= U_k^T(a)\,[\hat{R}_k^{-1}(b) - R_k^{-1}(a)]\,U_k(a), \quad U_k(\lambda) := R_k(\lambda)\,\Delta X_k(\lambda). \end{aligned} \right\} \tag{6.77}$$

Here $m^*(w_k(a, b), w_{k+1}(a, b)) := m^*(Z_k^{-1}(a)\,\hat{Y}_k(b))$ is the number of backward focal points of $Z_k^{-1}(a)\,\hat{Y}_k(b)$ given by (6.72) with $\mathfrak{C}_k := \tilde{\mathfrak{C}}_k(a, b)$, and $P_k$ is the *constant (with respect to* $\lambda$*) defined by*

$$P_k := \operatorname{ind} \hat{R}_k(\lambda_0) - \operatorname{ind} R_k(\lambda_0), \quad \lambda_0 \in \mathbb{R}. \tag{6.78}$$

*Proof* Note that for case (6.76) matrices (6.67), (6.68) obey the condition

$$\hat{S}_k^{-1}(b)\, S_k(a) = \hat{H}_k^{-1}(b)\, H_k(a) = \begin{pmatrix} I & R_k^{-1}(a) - \hat{R}_k^{-1}(b) \\ 0 & I \end{pmatrix}. \tag{6.79}$$

The symplectic upper block-triangular factors $H_k(a)$ and $\hat{H}_k(b)$ in (6.79) can be represented in the form

$$H_k(a) = -\mathcal{J}\, K_k(a)\, \mathcal{J}, \quad \hat{H}_k(b) = -\mathcal{J}\, \hat{K}_k(b)\, \mathcal{J},$$

where $K_k(a)$ and $\hat{K}_k(b)$ are the symplectic lower block-triangular matrices. Assumption (6.76) then implies that $L_k(a) \equiv \hat{L}_k(b) = L_k$ in (6.67) and (6.68).

Consider operator (3.56) introduced in Sect. 3.3.1 (see also (4.102)), i.e.,

$$\mathcal{L}(Y, \hat{Y}, \mathcal{S}, \hat{\mathcal{S}}) := \mu(\hat{S}\hat{Y}, \hat{S}\,(0\ I)^T) - \mu(SY, \mathcal{S}\,(0\ I)^T) + \mu(SY, \hat{S}\hat{Y}) - \mu(Y, \hat{Y}). \tag{6.80}$$

Applying the multiplicative property of this operator in Lemma 3.22(ii) (with the data $p := 4$, $W_1 = \hat{W}_1 := \mathcal{J}$, $W_2 := K_k(a)$, $\hat{W}_2 := \hat{K}_k(b)$, $W_3 = \hat{W}_3 := -\mathcal{J}$, $W_4 = \hat{W}_4 := L_k$, $Y := Y_k$, and $\hat{Y} := \hat{Y}_k$), we obtain

$$\mathcal{L}(Y_k, \hat{Y}_k, \mathcal{S}_k(a), \hat{\mathcal{S}}_k(b))$$
$$= \mathcal{L}\big(Y_k, \hat{Y}_k, -L_k\mathcal{J}K_k(a)\,\mathcal{J}, -L_k\mathcal{J}\hat{K}_k(b)\,\mathcal{J}\big)$$
$$= \mathcal{L}(Y_k, \hat{Y}_k, \mathcal{J}, \mathcal{J}) + \mathcal{L}\big(\mathcal{J}Y_k, \mathcal{J}\hat{Y}_k, K_k(a), \hat{K}_k(b)\big)$$
$$+ \mathcal{L}\big(K_k(a)\,\mathcal{J}Y_k, \hat{K}_k(b)\,\mathcal{J}\hat{Y}_k, -\mathcal{J}, -\mathcal{J}\big) + \mathcal{L}\big(H_k(a)\,Y_k(a), \hat{H}_k(b)\,\hat{Y}_k(b), L_k, L_k\big)$$
$$+ \Big\{\mu\big(H_k(a)\,(0\ I)_,^T -\mathcal{J}(0\ I)^T\big) - \mu\big(\hat{H}_k(b)\,(0\ I)_,^T -\mathcal{J}(0\ I)^T\big)\Big\},$$

where the addends in the braces correspond to the last sum in (3.57). Taking into account that the right-hand side of operator (6.80) equals zero for $\mathcal{S}_k = \hat{\mathcal{S}}_k$ (see Lemma 3.22(i)) and evaluating the difference

$$\Big\{\mu\big(H_k(a)\,(0\ I)_,^T -\mathcal{J}(0\ I)^T\big) - \mu\big(\hat{H}_k(b)\,(0\ I)_,^T -\mathcal{J}(0\ I)^T\big)\Big\}$$
$$= \mu\big(-\mathcal{J}(0\ I)_,^T \hat{H}_k(b)\,(0\ I)^T\big) - \mu\big(-\mathcal{J}(0\ I)_,^T H_k(a)\,(0\ I)^T\big)$$

using Theorem 3.5(v), we have

$$\mathcal{L}\big(Y_k(a), \hat{Y}_k(b), \mathcal{S}_k(a), \hat{\mathcal{S}}_k(b)\big) = \mathcal{L}\big(\mathcal{J}Y_k(a), \mathcal{J}\hat{Y}_k(b), K_k(a), \hat{K}_k(b)\big)$$
$$+ \text{ind } \hat{R}_k(b) - \text{ind } R_k(a).$$

Recall that the symmetric nonsingular matrices $R_k(\lambda)$ and $\hat{R}_k(\lambda)$ are continuous functions in $\lambda$ and then their eigenvalues have the constant sign for $\lambda \in \mathbb{R}$. So we have ind $\hat{R}_k(a) = \text{ind } \hat{R}_k(\lambda_0)$ and ind $R_k(a) = \text{ind } R_k(\lambda_0)$ for any $\lambda_0 \in \mathbb{R}$.

Note that in the operator $\mathcal{L}\big(\mathcal{J}Y_k, \mathcal{J}\hat{Y}_k, K_k(a), \hat{K}_k(b)\big)$, the symplectic matrices $K_k(a)$ and $\hat{K}_k(b)$ are unit lower block-triangular (see Sect. 1.6.1), and then they obey condition (6.38), i.e.,

$$K_k(a)\, \hat{K}_k^{-1}(b) = \mathcal{J} H_k(a)\, \hat{H}_k^{-1}(b)\, \mathcal{J}^T = \begin{pmatrix} I & 0 \\ \hat{R}_k^{-1}(b) - R_k^{-1}(a) & I \end{pmatrix}.$$

Evaluating $\mathcal{L}\big(\mathcal{J}Y_k(a), \mathcal{J}\hat{Y}_k(b), K_k(a), \hat{K}_k(b)\big)$ according to Lemma 6.6, where the quantity $\tilde{\mathfrak{B}}_k(a, b)$ is replaced by $\tilde{\mathfrak{C}}_k(a, b) = U_k^T(a)\,[\hat{R}_k^{-1}(b) - R_k^{-1}(a)]\, U_k(a)$, we derive (6.77) with $P_k$ given by (6.78) (note that the Wronskian $w(\mathcal{J}Y_k(a), \mathcal{J}\hat{Y}_k(b))$ is equal to the Wronskian $w(Y_k(a), \hat{Y}_k(b))$). The proof is completed. $\qquad\square$

Consider the evaluation of the relative oscillation numbers for the general case. Introduce the following Wronskian

$$w_{k^*}(a, b) := X_{k+1}^T(a)\, \hat{U}_k(b) - U_k^T(a)\, \hat{X}_{k+1}(b), \quad k^* \in (k, k+1), \qquad (6.81)$$

for $U_k(\lambda)$ and $\hat{U}_k(\lambda)$ defined as in (6.69). Here we use the intermediate point $k^* \in (k, k+1)$ for a convenient interpretation of the subsequent results. Note that

$$w_{k^*}(a, b) = w_k(a, b) - U_k^T(a)\, [\hat{R}_k^{-1}(b) - R_k^{-1}(a)]\, \hat{U}_k(b), \qquad (6.82)$$

$$w_{k+1}(a, b) = w_{k^*}(a, b) - X_{k+1}^T(a)\, [Q_k(a) - \hat{Q}_k(b)]\, \hat{X}_{k+1}(b), \qquad (6.83)$$

and then summing (6.82) and (6.83) we derive (6.71). In particular, if case I takes place (i.e., conditions (6.74) hold), then we have $w_{k^*}(a, b) = w_k(a, b)$ by (6.82), and similarly $w_{k+1}(a, b) = w_{k^*}(a, b)$ by (6.83) in case II.

**Theorem 6.14 (General Case)** *For spectral problems* (6.63), (6.65) *and* (6.64), (6.66) *associated with symplectic matrices* (6.67) *and* (6.68), *the relative oscillation numbers* (4.109) *have the form*

$$\#_k(\hat{Z}(b), Z(a)) = \#_{II}(k, k^*) + \#_I(k^*, k+1), \qquad (6.84)$$

*where the numbers*

$$\#_{II}(k, k^*) := m^*(w_k(a, b), w_{k^*}(a, b)) - \text{ind } \tilde{\mathfrak{C}}_k(a, b) + P_k, \tag{6.85}$$

$$\#_I(k^*, k + 1) := m^*(w_{k^*}(a, b), w_{k+1}(a, b)) - \text{ind } \mathfrak{C}_k(a, b) \tag{6.86}$$

*are evaluated according to (6.77), (6.78), and (6.75), respectively, with the quantities $w_{k+1}(a, b) := w_{k^*}(a, b)$ for case II and $w_k(a, b) := w_{k^*}(a, b)$ for case I.*

*Proof* For the proof we use factorizations (6.67), (6.68) and Lemma 3.22. Using Lemma 3.22(ii) (with $p = 2$, $W_1 := H_k(a)$ $\hat{W}_1 := \hat{H}_k(b)$, $W_2 := L_k(a)$, $\hat{W}_2 := \hat{L}_k(b)$, $Y := Y_k(a)$, $\hat{Y} := \hat{Y}_k(b)$), we derive

$$\mathcal{L}\big(Y_k(a), \hat{Y}_k(b), S_k(a), \hat{S}_k(b)\big) = \mathcal{L}\big(Y_k(a), \hat{Y}_k(b), L_k(a) H_k(a), \hat{L}_k(b) \hat{H}_k(b)\big)$$

$$= \mathcal{L}\big(Y_k(a), \hat{Y}_k(b), H_k(a), \hat{H}_k(b)\big) + \mathcal{L}\big(H_k(a) Y_k(a), \hat{H}_k(b) \hat{Y}_k(b), L_k(a), \hat{L}_k(b)\big),$$

where $\mathcal{L}\big(Y_k(a), \hat{Y}_k(b), H_k(a), \hat{H}_k(b)\big)$ and $\mathcal{L}\big(H_k(a) Y_k(a), \hat{H}_k(b) \hat{Y}_k(b), L_k(a), \hat{L}_k(b)\big)$ can be evaluated according to cases II and I, respectively. For case II we have that the conjoined bases $Y_k(a)$ and $\hat{Y}_k(b)$ obey the symplectic systems $Y_{k^*}(a) = H_k(a) Y_k(a)$ and $\hat{Y}_{k^*}(b) = \hat{H}_k(b) \hat{Y}_k(b)$ for $k^* \in (k, k + 1)$, and then we have to use the Wronskian $Y_{k^*}(a)^T J \hat{Y}_{k^*}(b) = w_{k^*}(a, b)$ given by (6.81) instead of $w_{k+1}(a, b)$. Similarly, in case I we use that $Y_{k^*}(a)$ and $\hat{Y}_{k^*}(b)$ obey the symplectic systems $Y_{k+1}(a) = L_k(a) Y_{k^*}(a)$ and $\hat{Y}_{k+1}(b) = \hat{L}_k(b) \hat{Y}_{k^*}(b)$, and then we apply (6.75) replacing $w_k(a, b)$ by $w_{k^*}(a, b)$. Finally, point out that such modifications of (6.77) and (6.75) do not touch the matrices $\tilde{\mathfrak{C}}_k(a, b)$ and $\mathfrak{C}_k(a, b)$ according to their definitions in (6.77) and (6.75). The proof is completed. □

Now we formulate some properties of the relative oscillation numbers in Theorem 6.14.

**Proposition 6.15** *The relative oscillation numbers in (6.84), (6.85), (6.86) satisfy the following properties.*

(i) *If case I takes place (i.e., the conditions in (6.74) hold), then in (6.84) we have $\#_k(\hat{Z}(b), Z(a)) = \#_k^I(\hat{Z}(b), Z(a))$ for $\#_k^I(\hat{Z}(b), Z(a))$ given by (6.75). Similarly, for case II we have $\#_k(\hat{Z}(b), Z(a)) = \#_k^{II}(\hat{Z}(b), Z(a))$ with $\#_k^{II}(\hat{Z}(b), Z(a))$ given by (6.77).*

(ii) *If the conditions*

$$Q_k(a) \geq \hat{Q}_k(b), \quad R_k(a) \geq \hat{R}_k(b) \tag{6.87}$$

*hold, then the relative oscillation numbers given by (6.84), (6.85), ad (6.86) are nonnegative. If additionally to (6.87) we have $\hat{R}_k(\lambda) > 0$ or if*

$$Q_k(\lambda) \equiv \hat{Q}_k(\lambda), \quad R_k(\lambda) \equiv \hat{R}_k(\lambda), \quad a \leq b,$$

*then for $a \leq b$ the relative oscillation numbers are presented in the form*

$$\#_k(\hat{Z}(b), Z(a)) = m^*(w_k(a, b), w_{i*}(a, b)) + m^*(w_{i*}(a, b), w_{k+1}(a, b)) \geq 0,$$
(6.88)

*where $m^*(w_k, w_{i*})$ and $m^*(w_{i*}, w_{k+1})$ are defined by (6.72) with the quantities $\mathfrak{C}_k := \tilde{\mathfrak{C}}_k(a, b) \geq 0$ and $\mathfrak{C}_k := \mathfrak{C}_k(a, b) \geq 0$ given by (6.77) and (6.75), respectively.*

(iii) *For the relative oscillation numbers in (6.84), we have the estimate*

$$\left|\#_k(\hat{Z}(b), Z(a))\right| \leq \text{rank}\,[R_k(a) - \hat{R}_k(b)] + \text{rank}\,[Q_k(a) - \hat{Q}_k(b)] \leq 2n.$$
(6.89)

*Proof* For the proof of (i), we use (6.82) and (6.83). Case I implies (see (6.82)) that $w_k(a, b) = w_{i*}(a, b)$ and the matrices $\tilde{\mathfrak{C}}_k(a, b)$ and $P_k$ in (6.77) and (6.78) are equal to the zero matrix. Then $\#_{II}(i, i^*) = 0$ and $\#_k(\hat{Z}(b), Z(a)) = \#_k^I(\hat{Z}(b), Z(a))$. In a similar way, for case II we have $\mathfrak{C}_k(a, b) = 0$, and we get from (6.83) that $w_{k+1}(a, b) = w_{i*}(a, b)$. Finally, it follows that $\#_I(i^*, i + 1) = 0$ and $\#_k(\hat{Z}(b), Z(a)) = \#_k^{II}(\hat{Z}(b), Z(a))$.

For the proof of (ii), we note that under assumptions (6.87) we have by Example 3.30 that $\mu(\langle \hat{S}_k(b)\rangle, \langle S_k(a)\rangle) = 0$ and then the relative oscillation numbers are nonnegative because of their definition in (6.25). If, additionally, we have $\hat{R}_k(\lambda) > 0$, then by (6.87) we have $R_k(\lambda) > 0$, and moreover (6.87) is equivalent to $\hat{R}_k^{-1}(b) - R_k^{-1}(a) \geq 0$ (see Remark 3.35) and $P_k = 0$, $\text{ind}\,\mathfrak{C}_k(a, b) = \text{ind}\,\tilde{\mathfrak{C}}_k(a, b) = 0$. A similar situation occurs for $Q_k(\lambda) \equiv \hat{Q}_k(\lambda)$, $R_k(\lambda) \equiv \hat{R}_k(\lambda)$, and $a \leq b$, because of the monotonicity assumptions (6.65). Then the proof of (ii) is completed.

By (4.119) (see also the proof of Theorem 6.14), the relative oscillation numbers (6.85) and (6.86) obey the inequalities

$$|\#_{II}(i, i^*)| \leq \text{rank}\,[H_k(a) - \hat{H}_k(b)] = \text{rank}\,[R_k(a) - \hat{R}_k(b)],$$

$$|\#_I(l^*, i + 1)| \leq \text{rank}\,[L_k(a) - \hat{L}_k(b)] = \text{rank}\,[Q_k(a) - \hat{Q}_k(b)],$$

and then for the relative oscillation numbers in (6.84), we have estimate (6.89). The proof is completed. □

Finally, we reformulate Theorems 6.2 and 6.4 for the special case (6.63) and (6.64), i.e., for the matrix Sturm-Liouville difference equations.

**Theorem 6.16 (Relative Oscillation Theorem)** *Let $\sigma$ and $\hat{\sigma}$ be the finite spectra, and let $Y^{[0]}(\lambda)$ and $\hat{Y}^{[N+1]}(\lambda)$ be the principal solutions of problems (6.63) and (6.64) with conditions (6.65) and (6.66). Then there exists a constant $\mathfrak{P} \in [-nN, nN]_\mathbb{Z}$ such that for all $a, b \in \mathbb{R}$ the statements of Theorem 6.2 hold with the relative oscillation numbers defined by (6.84), (6.85), and (6.86), where $\hat{Z}(\lambda) := \hat{Z}^{[N+1]}(\lambda)$ and $Z(\lambda) := Z^{[0]}(\lambda)$.*

For the case $R_k(\lambda) \equiv \hat{R}_k(\lambda)$ and $Q_k(\lambda) \equiv \hat{Q}_k(\lambda)$ for $b > a$, Theorem 6.16 presents the number of finite eigenvalues of (6.63) in $(a, b]$.

**Theorem 6.17 (Renormalized Oscillation Theorem)**  *For problem* (6.63), (6.65), *all statements of Theorem 6.4 (including the case $a := \lambda$, $\lambda < \lambda_1$) hold with*

$$\mu\big(\langle G_k(a, b)\rangle, \langle G_{k+1}(a, b)\rangle\big) = m^*(w_k, w_{k*}) + m^*(w_{k*}, w_{k+1}) \geq 0, \qquad (6.90)$$

*where the Wronskians $w_l := w_l(a, b)$, $l = k, k + 1$ and $w_{k*} := w_{k*}(a, b)$ are defined by (6.70) and (6.81) with $Y(a) := Y^{[0]}(a)$, $\hat{Y}(b) := Y^{[N+1]}(b)$ and where $m^*(w_k, w_{k*})$ and $m^*(w_{k*}, w_{k+1})$ are given by (6.72) with the quantities $\tilde{\mathfrak{C}}_k := \tilde{\mathfrak{C}}_k(a, b) \geq 0$ and $\mathfrak{C}_k := \mathfrak{C}_k(a, b) \geq 0$ given by (6.77) and (6.75), respectively.*

*Proof*  Applying Proposition 6.15(ii) for the case $R_k(\lambda) \equiv \hat{R}_k(\lambda)$, $Q_k(\lambda) \equiv \hat{Q}_k(\lambda)$ and Theorem 6.4, we complete the proof of Theorem 6.17.                          $\square$

*Remark 6.18*  In the definition of (6.85), we use the number $P_k$ given by (6.78), which does not depend on $a$ and $b$. Then it makes sense to introduce the new constant

$$\tilde{\mathfrak{P}} := \mathfrak{P} - \sum_{i=0}^{N} P_k \qquad (6.91)$$

and use identity (6.26) in the form

$$\left.\begin{array}{c} \#\{\nu \in \hat{\sigma} \mid \nu \leq b\} - \#\{\nu \in \sigma \mid \nu \leq a\} \\ = \displaystyle\sum_{k=0}^{N} \left\{ \#_k(\hat{Z}^{[N+1]}(b), Z^{[0]}(a)) - P_k \right\} - \tilde{\mathfrak{P}}. \end{array}\right\} \qquad (6.92)$$

For the numbers $\#_k(\hat{Z}^{[N+1]}(b), Z^{[0]}(a)) - P_k$, one can also improve the estimate (6.89) as follows:

$$\left| \#_k(\hat{Z}(b), Z(a)) - P_k \right| \leq \operatorname{rank} \tilde{\mathfrak{C}}_k(a, b) + \operatorname{rank} \mathfrak{C}_k(a, b),$$

where $\tilde{\mathfrak{C}}_k(a, b)$ and $\mathfrak{C}_k(a, b)$ are given by (6.77) and (6.75). The proof follows from inequality (6.53).

### 6.1.5  Examples

This section is devoted to examples, which illustrate the applications of Theorems 6.16 and 6.17 to the scalar spectral problems (6.63) and (6.64).

*Example 6.19* Consider problem (6.63) for the scalar Sturm-Liouville difference equation

$$\Delta(r_k(\lambda)\,\Delta x_k(\lambda)) - q_k(\lambda)\,x_{k+1}(\lambda) = 0, \quad k \in [0, 3]_{\mathbb{Z}}, \left.\begin{array}{r}\\ \\ \\ \end{array}\right\}$$
$$x_0(\lambda) = 0 = x_4(\lambda), \qquad\qquad (6.93)$$
$$r_k(\lambda) := (-1)^{k+1}\exp((-1)^k\lambda), \quad q_k(\lambda) = -\lambda.$$

Then for the principal solution of (6.93) at 0 defined by the initial conditions $x_0^{[0]}(\lambda) = 0$ and $x_1^{[0]}(\lambda) = 1/r_0(\lambda)$, we have

$$x_4^{[0]}(\lambda) = -\lambda^3 - 6\lambda^2\sinh(\lambda) - 8\lambda\sinh^2(\lambda) + 2\lambda + 4\sinh(\lambda),$$

and the finite eigenvalues of (6.93) are the zeros of $x_4(\lambda)$, i.e., $\lambda_1 \approx -0.6167186$, $\lambda_2 = 0$, $\lambda_3 \approx 0.6167186$. According to Theorem 6.17, we have

$$\#_k(\hat{Z}(b), Z(a)) = m^*(w_k, w_{k^*}) + m^*(w_{k^*}, w_{k+1}),$$

where for the scalar case

$$m^*(w_k, w_{k^*}) = \begin{cases} 1, & w_{k^*} \neq 0,\ w_k w_{k^*} \leq 0, \\ 0, & \text{otherwise,} \end{cases}$$
$$m^*(w_{k^*}, w_{k+1}) = \begin{cases} 1, & w_{k+1} \neq 0,\ w_{k+1}w_{k^*} \leq 0, \\ 0, & \text{otherwise.} \end{cases} \qquad (6.94)$$

Then, according to (6.94) and Theorem 6.17, the number of finite eigenvalues of problem (6.93) in the interval $(a, b]$ is equal to the total number of generalized zeros of the Wronskian in all intervals $[k, k^*)$ and $[k^*, k+1)$ for $k \in [0, N]_{\mathbb{Z}}$ and $k^* \in (k, k+1)$. For example, if $a = -0.8$ and $b = 1.8$, then we have the three sign changes of the Wronskian $w_k(a, b)$ (see Fig. 6.1), and then according to Theorem 6.17, the three finite eigenvalues of problem (6.93), i.e., the numbers $\lambda_1 \approx -0.6167186$, $\lambda_2 = 0$, $\lambda_3 \approx 0.6167186$, are located in the interval $(-0.8, 1.8]$. Note that the relative oscillation number $0 \leq \#_k(\hat{Z}^{[N+1]}(b), Z^{[0]}(a)) \leq 2$ achieves its maximal value 2 at the point $k = 2$.

*Example 6.20* Consider spectral problem (6.93) and the following spectral problem (referred to as "problem 1" and "problem 2" in Fig. 6.2):

$$\Delta(\hat{r}_k(\lambda)\,\Delta\hat{x}_k(\lambda)) - \hat{q}_k(\lambda)\,\hat{x}_{k+1}(\lambda) = 0, \quad k \in [0, 3]_{\mathbb{Z}}, \left.\begin{array}{r}\\ \\ \\ \end{array}\right\}$$
$$\hat{x}_0(\lambda) = 0 = \hat{x}_4(\lambda), \qquad\qquad (6.95)$$
$$\hat{r}_k(\lambda) = (-1)^k, \quad \hat{q}_k(\lambda) = -(\lambda + 2).$$

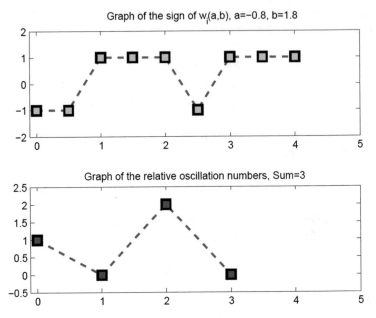

**Fig. 6.1** The graphs of the sign of the Wronskian and the relative oscillation numbers in Example 6.19 for the values $a = -0.8$ and $b = 1.8$

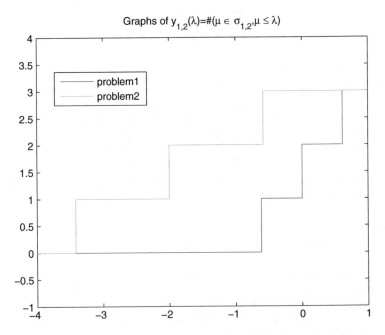

**Fig. 6.2** The graphs of the functions $y_{1,2}(\lambda)$ of the number of finite eigenvalues below or equal to $\lambda$ for $\lambda \in \mathbb{R}$ for problems 1 and 2 in Example 6.20

The finite eigenvalues of (6.95) are the zeros of the equation

$$\hat{x}_4^{[0]}(\lambda) = (\lambda + 2)^3 - 2\lambda - 4 = 0,$$

$$\hat{\lambda}_1 = -\sqrt{2} - 2 \approx -3.4142136, \quad \hat{\lambda}_2 = -2, \quad \hat{\lambda}_3 = \sqrt{2} - 2 \approx -0.5857864.$$

The localization of the finite eigenvalues of (6.93) and (6.95) is shown in Fig. 6.2, where we present the functions $y_{1,2}(\lambda)$ of the number of finite eigenvalues below or equal to $\lambda$ for $\lambda \in \mathbb{R}$.

According to Theorem 6.16, we can calculate the difference between numbers of finite eigenvalues of (6.95) and (6.93) using the relative oscillation numbers $\#_k(\hat{Z}^{[N+1]}(b), Z^{[0]}(a)) = \#_{II}(k, k^*) + \#_I(k^*, k+1)$. For the scalar case, we have

$$\#_{II}(k, k^*) - P_k = \begin{cases} 1, & \hat{r}_k^{-1}(b) - r_k^{-1}(a) > 0 \text{ and } w_{k*} \neq 0, \ w_k w_{k*} \leq 0, \\ -1, & \hat{r}_k^{-1}(b) - r_k^{-1}(a) < 0 \text{ and } w_k \neq 0, \ w_k w_{k*} \leq 0, \\ 0, & \text{otherwise}, \end{cases}$$

$$(6.96)$$

where the number $P_k$ given by (6.78) is defined as

$$P_k := \text{ind}\,\hat{r}_k(\lambda_0) - \text{ind}\,r_k(\lambda_0) = (-1)^{k+1}.$$

Then we can say that the Wronskian has a weighted zero at the point $k$ if the quantity $\#_{II}(k, k^*) - P_k = \pm 1$. According to Remark 6.18, we can consider the sum $\sum_{k=0}^{3} P_k = 0$ as the parameter of problem (6.95).

Similarly, for the scalar case, we derive

$$\#_I(k^*, k+1) = \begin{cases} 1, & q_k(a) - \hat{q}_k(b) > 0 \text{ and } w_{k+1} \neq 0, \ w_{k+1} w_{k*} \leq 0, \\ -1, & q_k(a) - \hat{q}_k(b) < 0 \text{ and } w_{k*} \neq 0, \ w_{k+1} w_{k*} \leq 0, \\ 0, & \text{otherwise}, \end{cases}$$

$$(6.97)$$

and say that the Wronskian has a weighted node at $k^*$ if $\#_I(k^*, k+1) = \pm 1$.

Denote $C_k(a, b) = q_k(a) - \hat{q}_k(b)$ and $B_k(a, b) = \hat{r}_k^{-1}(b) - r_k^{-1}(a)$, and observe that we can calculate the constant $\tilde{\mathfrak{P}}$ given by (6.91) evaluating the sum of (6.96) and (6.97) for $a = b = \lambda_0 < \min\{\lambda_1, \hat{\lambda}_1\}$. By Fig. 6.3 we conclude that $\tilde{\mathfrak{P}} = 0$, where we take $a = b = \lambda_0 = -4$. Then, according to Theorem 6.16, we can evaluate the difference $\#\{v \in \sigma_2 \mid v \leq b\} - \#\{v \in \sigma_1 \mid v \leq a\}$ between the numbers of finite eigenvalues of (6.95) and (6.93) calculating the total number of weighted zeros of the Wronskian for all $k$ and $k^*$, $k \in [0, N]_{\mathbb{Z}}$, $k^* \in (k, k+1)$.

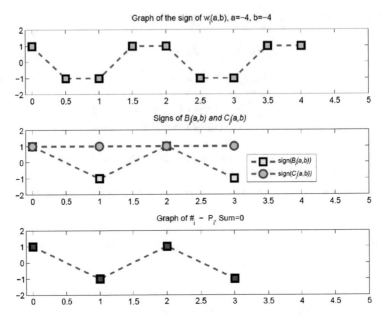

**Fig. 6.3** The graphs of the signs of the Wronskian, $C_k(a, b) = q_k(a) - \hat{q}_k(b)$, $B_k(a, b) = \hat{r}_k^{-1}(b) - r_k^{-1}(a)$, and the relative oscillation numbers in Example 6.20 for the values $a = b = -4$

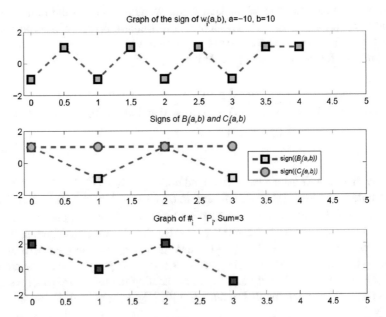

**Fig. 6.4** The graphs of the signs of the Wronskian, $C_k(a, b) = q_k(a) - \hat{q}_k(b)$, $B_k(a, b) = \hat{r}_k^{-1}(b) - r_k^{-1}(a)$, and the relative oscillation numbers in Example 6.20 for the values $a = -10$ and $b = 10$

Next, in Fig. 6.4 we present the graphs of the signs of the Wronskian, $C_k(a, b) = q_k(a) - \hat{q}_k(b)$ and $B_k(a, b) = \hat{r}_k^{-1}(b) - r_k^{-1}(a)$ and the relative oscillation numbers $\#_k(\hat{Z}^{[N+1]}(b), Z^{[0]}(a)) - P_k$ for $k \in [0, 3]_\mathbb{Z}$. For example, we have

$$\#_0(\hat{Z}^{[N+1]}(b), Z^{[0]}(a)) - P_0 = \#_{II}(0, k^*) - P_0 + \#_I(k^*, 1) = 2, \quad k^* = 1/2,$$

because $B_0(a, b) > 0$, $w_0 w_{k^*} < 0$, $C_0(a, b) > 0$, $w_1 w_{k^*} < 0$. Similarly, in the next point $k = 1$, we have $B_1(a, b) < 0$, $w_1 w_{k^*} < 0$, $C_1(a, b) > 0$, $w_2 w_{k^*} < 0$, and $k^* = 3/2$, and then $\#_1(\hat{Z}^{[N+1]}(b), Z^{[0]}(a)) - P_1 = -1 + 1 = 0$. Finally, according to Fig. 6.2, we have

$$\#\{\nu \in \sigma_2 \mid \nu \leq 10\} - \#\{\nu \in \sigma_1 \mid \nu \leq -10\} = 3 - 0 = 3,$$

and the sum of the relative oscillation numbers by Fig. 6.4 equals to 3.

### 6.1.6 Separated Boundary Conditions

We consider symplectic systems (5.1) and (6.1) with different separated boundary conditions (5.155), i.e., the problems

$$\left.\begin{aligned}
y_{k+1}(\lambda) &= \mathcal{S}_k(\lambda)\, y_k(\lambda), \quad k \in [0, N]_\mathbb{Z}, \\
R_0^*(\lambda)\, x_0(\lambda) &+ R_0(\lambda)\, u_0(\lambda) = 0, \\
R_{N+1}^*(\lambda)\, x_{N+1}(\lambda) &+ R_{N+1}(\lambda)\, u_{N+1}(\lambda) = 0, \\
\hat{y}_{k+1}(\lambda) &= \hat{\mathcal{S}}_k(\lambda)\, \hat{y}_k(\lambda), \quad k \in [0, N]_\mathbb{Z}, \\
\hat{R}_0^*(\lambda)\, \hat{x}_0(\lambda) &+ \hat{R}_0(\lambda)\, \hat{u}_0(\lambda) = 0, \\
\hat{R}_{N+1}^*(\lambda)\, \hat{x}_{N+1}(\lambda) &+ \hat{R}_{N+1}(\lambda)\, \hat{u}_{N+1}(\lambda) = 0,
\end{aligned}\right\} \tag{6.98}$$

where we assume that the matrices $R_k^*(\lambda)$, $R_k(\lambda)$, $\hat{R}_k^*(\lambda)$, $\hat{R}_k(\lambda)$ for $k \in \{0, N + 1\}$ obey assumptions (5.156), (5.157), and (5.158). Moreover, we derive all results under restriction (5.210)

$$\left.\begin{aligned}
\text{rank } \mathcal{B}_k(\lambda),\ \text{rank } \hat{\mathcal{B}}_k(\lambda) \text{ are constant for } \lambda \in \mathbb{R} \text{ for all } k \in [0, N]_\mathbb{Z} \\
\text{rank } R_0(\lambda) \text{ and rank } R_{N+1}(\lambda) \text{ are constant for } \lambda \in \mathbb{R}, \\
\text{rank } \hat{R}_0(\lambda) \text{ and rank } \hat{R}_{N+1}(\lambda) \text{ are constant for } \lambda \in \mathbb{R}.
\end{aligned}\right\} \tag{6.99}$$

Under the assumptions of Theorem 5.50 for problems (6.98), the finite spectra $\sigma$ and $\hat{\sigma}$ of (6.98) are bounded, and one can order the finite eigenvalues of (6.98) according to (6.19) using notation (6.20).

The consideration in this subsection is based on Theorem 5.41. Introduce the following matrices associated with the separated boundary conditions in (6.98) (see (5.171) and (5.173))

$$
\left.
\begin{aligned}
V_{-1}(\lambda) &:= \begin{pmatrix} R_0^{*T}(\lambda)\, K_0(\lambda) & -R_0^T(\lambda) \\ R_0^T(\lambda)\, K_0(\lambda) & R_0^{*T}(\lambda) \end{pmatrix}, \\
K_0(\lambda) &:= [R_0^*(\lambda)\, R_0^{*T}(\lambda) + R_0(\lambda)\, R_0^T(\lambda)]^{-1}, \\
V_{N+1}(\lambda) &:= \begin{pmatrix} R_{N+1}^*(\lambda) & R_{N+1}(\lambda) \\ -K_{N+1}(\lambda)\, R_{N+1}(\lambda) & K_{N+1}(\lambda)\, R_{N+1}^*(\lambda) \end{pmatrix}, \\
K_{N+1}(\lambda) &:= [R_{N+1}^*(\lambda)\, R_{N+1}^{*T}(\lambda) + R_{N+1}(\lambda)\, R_{N+1}^T(\lambda)]^{-1}.
\end{aligned}
\right\}
\tag{6.100}
$$

and

$$
\left.
\begin{aligned}
\hat{V}_{-1}(\lambda) &:= \begin{pmatrix} \hat{R}_0^{*T}(\lambda)\, \hat{K}_0(\lambda) & -\hat{R}_0^T(\lambda) \\ \hat{R}_0^T(\lambda)\, \hat{K}_0(\lambda) & \hat{R}_0^{*T}(\lambda) \end{pmatrix}, \\
\hat{K}_0(\lambda) &:= [\hat{R}_0^*(\lambda)\, \hat{R}_0^{*T}(\lambda) + \hat{R}_0(\lambda)\, \hat{R}_0^T(\lambda)]^{-1}, \\
\hat{V}_{N+1}(\lambda) &:= \begin{pmatrix} \hat{R}_{N+1}^*(\lambda) & \hat{R}_{N+1}(\lambda) \\ -\hat{K}_{N+1}(\lambda)\, \hat{R}_{N+1}(\lambda) & \hat{K}_{N+1}(\lambda)\, \hat{R}_{N+1}^*(\lambda) \end{pmatrix}, \\
\hat{K}_{N+1}(\lambda) &:= [\hat{R}_{N+1}^*(\lambda)\, \hat{R}_{N+1}^{*T}(\lambda) + \hat{R}_{N+1}(\lambda)\, \hat{R}_{N+1}^T(\lambda)]^{-1}.
\end{aligned}
\right\}
\tag{6.101}
$$

**Definition 6.21** Let $\hat{Z}^{[R]}(\lambda)$ and $Z^{[L]}(\lambda)$ for $\lambda \in \mathbb{R}$ be symplectic fundamental matrices of (6.1) and (5.1), respectively, satisfying

$$
\hat{Z}_{N+1}^{[R]}(\lambda) = \hat{V}_{N+1}^{-1}(\lambda), \quad Z_0^{[L]}(\lambda) = V_{-1}(\lambda),
\tag{6.102}
$$

where the matrices $\hat{V}_{N+1}(\lambda)$ and $V_{-1}(\lambda)$ are determined by (6.100) and (6.101). Consider conjoined bases $Y_k^{[L]}(\lambda) = Z_k^{[L]}(\lambda)\,(0\ I)^T$ and $\hat{Y}_k^{[R]}(\lambda) = \hat{Z}_k^{[R]}(\lambda)\,(0\ I)^T$. Then the relative oscillation numbers for problems (6.98) with $a, b \in \mathbb{R}$ are defined by the formulas

$$
\left.
\begin{aligned}
\#_k(\hat{Z}^{[R]}(b), Z^{[L]}(a)) &:= \mu\big(\langle \mathcal{G}_k(a,b)\rangle, \langle \mathcal{G}_{k+1}(a,b)\rangle\big) - \mu\big(\langle \hat{\mathcal{S}}_k(b)\rangle, \langle \mathcal{S}_k(a)\rangle\big), \\
\mathcal{G}_k(a,b) &:= \hat{Z}_k^{[R]\,-1}(b) Z_k^{[L]}(a), \quad k \in [0, N]_{\mathbb{Z}}, \\
\#_{-1}(\hat{Z}^{[R]}(b), Z^{[L]}(a)) &:= \mu^*\big(\hat{V}_{-1}^{-1}(b)\hat{Y}_0^{[R]}(b),\ \hat{V}_{-1}^{-1}(b)V_{-1}(a)\,(0\ I)^T\big) \\
&\quad - \mu\big(\hat{V}_{-1}(b)\,(0\ I)^T,\ V_{-1}(a)\,(0\ I)^T\big), \\
\#_{N+1}(\hat{Z}^{[R]}(b), Z^{[L]}(a)) &= \mu\big(\hat{V}_{N+1}(b)Y_{N+1}^{[L]}(a),\ \hat{V}_{N+1}(b)V_{N+1}^{-1}(a)\,(0\ I)^T\big) \\
&\quad - \mu^*\big(\hat{V}_{N+1}^{-1}(b)\,(0\ I)^T,\ V_{N+1}^{-1}(a)\,(0\ I)^T\big),
\end{aligned}
\right\}
\tag{6.103}
$$

Moreover, we define the relative oscillation number on the interval $[-1, N+1]_{\mathbb{Z}}$ for the pair of eigenvalue problems (6.98) by the formula

$$
\left.
\begin{aligned}
\#(\hat{Z}^{[R]}(b), Z^{[L]}(a), -1, N+1) &:= \sum_{k=-1}^{N+1} \#_k(\hat{Z}^{[R]}(b), Z^{[L]}(a)) \\
&= \#(\hat{Z}^{[R]}(b), Z^{[L]}(a), 0, N) + \#_{-1}(\hat{Z}^{[R]}(b), Z^{[L]}(a)) \\
&\qquad + \#_{N+1}(\hat{Z}^{[R]}(b), Z^{[L]}(a)).
\end{aligned}
\right\}
\tag{6.104}
$$

Recall that problems (6.98) are equivalent to extended problems with the Dirichlet boundary conditions (see Lemma 5.40). Then applying Theorems 6.2 and 6.4 to these extended problems, we derive the corresponding results for the separated endpoints.

**Theorem 6.22 (Relative Oscillation Theorem for Separated Endpoints)**
*Assume that for problems (6.98) conditions (5.3), (6.5), (6.99), and (5.156)–(5.158) hold for the boundary conditions in (6.98). Let $Z^{[L]}(\lambda)$ and $\hat{Z}^{[R]}(\lambda)$ be the symplectic fundamental matrices of these systems defined by (6.102) associated with the conjoined bases $Y^{[L]}(\lambda) = Z^{[L]}(\lambda)\,(0\ I)^T$ and $\hat{Y}^{[R]}(\lambda) = \hat{Z}^{[R]}(\lambda)\,(0\ I)^T$. For arbitrary $a, b \in \mathbb{R}$ define the relative oscillation numbers according to (6.103) and (6.104). Then there exists a constant $\mathfrak{P} \in [-(N+2)\,n, (N+2)\,n]_{\mathbb{Z}}$ such that for any $a, b \in \mathbb{R}$ we have for the spectra $\sigma$ and $\hat{\sigma}$ of problems (6.98)*

$$
\begin{aligned}
\#\{v \in \hat{\sigma} \mid v \le b\} &- \#\{v \in \sigma \mid v \le a\} \\
&= \#(\hat{Z}^{[R]}(b), Z^{[L]}(a), -1, N+1) - \mathfrak{P},
\end{aligned}
\tag{6.105}
$$

*where for $\lambda_0$ defined by (6.20) we have*

$$
\mathfrak{P} = \hat{\ell} - \ell = \#(\hat{Z}^{[R]}(\lambda), Z^{[L]}(\lambda), -1, N+1), \quad \lambda < \lambda_0
\tag{6.106}
$$

*with $\ell$ given by (5.215) and similarly defined $\hat{\ell}$.*

*Proof* For the proof we apply Theorem 6.2 to extended problems with the Dirichlet boundary conditions. As in the proof of Theorem 5.50, one can show that the matrices $Q_0(\lambda)$, $Q_{N+1}(\lambda)$, $\hat{Q}_0(\lambda)$, $\hat{Q}_{N+1}(\lambda)$ used in Theorem 5.41 for the construction of the extended problems can be omitted (here we use the notation $\hat{Q}_k$ for $k \in \{0, N+1\}$ for the matrices (5.168) associated with the second problem in (6.98)). Indeed, a direct application of Theorem 6.2 to the extended problems constructed via Theorem 5.41 leads to the relative oscillation numbers

$$
\left.
\begin{aligned}
\#_k(\hat{Z}^{[N+2]}(b), & Z^{[-1]}(a)) \\
&:= \mu\big(\langle \tilde{\mathcal{G}}_k(a, b)\rangle, \langle \tilde{\mathcal{G}}_{k+1}(a, b)\rangle\big) - \mu\big(\langle \hat{\mathcal{S}}_k(b)\rangle, \langle \mathcal{S}_k(a)\rangle\big)
\end{aligned}
\right\}
\tag{6.107}
$$

for $k \in [-1, N+1]_{\mathbb{Z}}$, where

$$\tilde{\mathcal{G}}_k(a,b) := \mathrm{diag}\{\hat{Q}_{N+1}^T(b),\ \hat{Q}_{N+1}^{-1}(b)\}\,[\hat{Z}_k^{[R]}(b)]^{-1} Z_k^{[L]}(a)\,\mathrm{diag}\{Q_0^{T-1}(a),\ Q_0(a)\}$$

for $k \in [0, N]_{\mathbb{Z}}$, and according to Corollary 3.40(i) the symplectic block diagonal matrices (which are a particular case of low triangular matrices) can be omitted, i.e., we have the relative oscillation numbers given by (6.103) for $k \in [0, N]_{\mathbb{Z}}$. For $k = -1$ we have according to Remark 6.3 (see (6.29), where we replace the point 0 by $-1$)

$$\#_{-1}(\hat{Z}^{[N+2]}(b), Z^{[-1]}(a)) = \mu^*\big(\hat{S}_{-1}^{-1}(b)\hat{Y}_0^{[N+2]}(b)\,(0\ I)^T,\ \hat{S}_{-1}^{-1}(b)S_{-1}(a)\,(0\ I)^T\big)$$
$$-\mu\big(\hat{S}_{-1}(b)\,(0\ I)^T,\ S_{-1}(a)\,(0\ I)^T\big).$$

Using $\hat{Y}_{N+1}^{[N+2]}(b) = \hat{S}_{N+1}^{-1}(b)\,(0\ I)^T$ (which implies $\hat{Y}_0^{[N+2]}(b) = \hat{Y}_0^{[R]}(b)\hat{Q}_{N+1}(b)$) and the connection between $\hat{S}_k(\lambda)$, $\mathcal{S}_k(\lambda)$ and $\hat{V}_k(\lambda)$, $V_k(\lambda)$ for $k \in \{-1, N+1\}$ by Theorem 5.41, we derive the representation for $\#_{-1}(\hat{Z}^{[R]}(b), Z^{[L]}(a))$ according to (6.103) omitting all block diagonal matrices with $Q_0(\lambda)$, $Q_{N+1}(\lambda)$, $\hat{Q}_0(\lambda)$, $\hat{Q}_{N+1}(\lambda)$ according to Theorem 3.5(i)–(ii). The proof for the point $N + 1$, where we use (6.30) replacing $N$ by $N + 1$, is similar.

So we have shown that the relative oscillation numbers for the extended problems obey Definition 6.21, and then applying Theorem 6.2 and incorporating (5.215), we complete the proof.                                                                                       □

For the case when $\mathcal{S}_k(\lambda) \equiv \hat{S}_k(\lambda)$ for $k \in [0, N]_{\mathbb{Z}}$ and $R_k(\lambda) \equiv \hat{R}_k(\lambda)$ and $R_k^*(\lambda) \equiv \hat{R}_k^*(\lambda)$ for $k \in \{-1, N+1\}$, we derive a generalization of Theorem 6.4 for separated boundary conditions.

**Theorem 6.23 (Renormalized Oscillation Theorem for Separated Endpoints)**
*Under the assumptions of Theorem 6.22, consider only one problem in (6.98). Let $Z^{[L]}(\lambda)$ and $Z^{[R]}(\lambda)$ be symplectic fundamental matrices of system (5.1) defined by (6.102) (where we put $\hat{V}_{N+1}(\lambda) := V_{N+1}(\lambda)$) associated with the conjoined bases $Y^{[L]}(\lambda) = Z^{[L]}(\lambda)\,(0\ I)^T$ and $Y^{[R]}(\lambda) = Z^{[R]}(\lambda)\,(0\ I)^T$. Then for any $a, b \in \mathbb{R}$ with $a < b$, the number of finite eigenvalues of problem (5.1) with (5.155) in the interval $(a, b]$ is given by the formula*

$$\begin{aligned}
\#\{\nu \in \sigma \mid a < \nu \le b\} = \sum_{k=0}^{N} &\mu\big(\langle \mathcal{G}_k(a,b)\rangle,\ \langle \mathcal{G}_{k+1}(a,b)\rangle\big)\\
&+\mu^*\big(V_{-1}^{-1}(b)\,Y_0^{[R]}(b),\ V_{-1}^{-1}(b)\,V_{-1}(a)\,(0\ I)^T\big)\\
&+\mu\big(V_{N+1}(b)\,Y_{N+1}^{[L]}(a),\ V_{N+1}(b)\,V_{N+1}^{-1}(a)\,(0\ I)^T\big),\\
\mathcal{G}_k(a,b) := &[Z_k^{[R]}(b)]^{-1} Z_k^{[L]}(a).
\end{aligned}$$
$$\hspace{11cm}(6.108)$$

*In particular, for $a := \lambda$, $\lambda < \lambda_1 = \min \sigma$, formula (6.108) presents the number of finite eigenvalues of (5.1), (5.155), which are less than or equal to b.*

*Proof* As in the proof of Theorem 6.4, we have for $S_k(\lambda) \equiv \hat{S}_k(\lambda)$, $k \in [0, N]_{\mathbb{Z}}$, $R_k(\lambda) \equiv \hat{R}_k(\lambda)$, and for $R_k^*(\lambda) \equiv \hat{R}_k^*(\lambda)$, $k \in \{-1, N+1\}$, that

$$\mu\big(\langle S_k(b)\rangle, \langle S_k(a)\rangle\big) = 0, \quad k \in [-1, N+1]_{\mathbb{Z}}, \tag{6.109}$$

where we use Lemma 5.105 and additionally, for the cases $k = -1$ and $k = N+1$, we apply Theorem 5.41 and assumption (6.99). We note that condition (6.109) for $k = -1$ and $k = N+1$ implies

$$\mu\big(S_{-1}(b)\,(0\ I)^T, S_{-1}(a)\,(0\ I)^T\big) = \mu\big(V_{-1}(b)\,(0\ I)^T, V_{-1}(a)\,(0\ I)^T\big) = 0,$$
$$\mu^*\big(S_{N+1}^{-1}(b)\,(0\ I)^T, S_{N+1}^{-1}(a)\,(0\ I)^T\big) = \mu^*\big(V_{N+1}^{-1}(b)\,(0\ I)^T, V_{N+1}^{-1}(a)\,(0\ I)^T\big) = 0,$$

where we use Lemma 3.21(v) and Theorem 3.5(i)–(ii). Then formulas for the relative oscillation numbers in (6.108) follow from (6.109) and Definition 6.21. □

*Remark 6.24* Assume $R_k(\lambda) \equiv \hat{R}_k(\lambda) \equiv R_k$ and $R_k^*(\lambda) \equiv \hat{R}_k^*(\lambda) \equiv R_k^*$ for $k \in \{-1, N+1\}$, i.e., the boundary conditions of problems (6.98) are the same and do not depend on $\lambda$. Then the relative oscillation numbers for $k = -1$ and $k = N+1$ in Theorems 6.22 and 6.23 are equal to zero by Definition 6.21 and by the definition of the comparative index. In particular, for the Dirichlet boundary conditions in (6.17), we have by Definition 6.21 that the relative oscillation numbers for $k \in \{-1, N+1\}$ are equal to zero, and in this case Theorems 6.22 and 6.23 reduce into Theorems 6.2 and 6.4, respectively.

Another important special case of (6.98) is connected with the situation when $S_k(\lambda) \equiv \hat{S}_k(\lambda)$, i.e., when the problems differ in the boundary conditions only. As a corollary to Theorem 6.22, we have the following result.

**Corollary 6.25** *Under all assumptions of Theorem 6.22, suppose additionally that $S_k(\lambda) \equiv \hat{S}_k(\lambda)$ and $a = b$. Then the difference of the numbers of finite eigenvalues of (6.98) is given by the formula*

$$\left.\begin{aligned}
\#\{\nu \in \hat{\sigma} \mid \nu \leq b\} - \#\{\nu \in \sigma \mid \nu \leq b\} \\
= \#_{-1}(\hat{Z}^{[R]}(b), Z^{[L]}(b)) + \#_{N+1}(\hat{Z}^{[R]}(b), Z^{[L]}(b)) - \mathfrak{P},
\end{aligned}\right\} \tag{6.110}$$

*with the constant $\mathfrak{P}$ defined by*

$$\left.\begin{aligned}
\mathfrak{P} := \#_{-1}(\hat{Z}^{[R]}(\lambda), Z^{[L]}(\lambda)) + \#_{N+1}(\hat{Z}^{[R]}(\lambda), Z^{[L]}(\lambda)) = \hat{l} - l, \\
\lambda \leq \lambda_0, \quad \lambda_0 := \min \sigma \cup \hat{\sigma},
\end{aligned}\right\} \tag{6.111}$$

*where the relative oscillation numbers for $k = -1$ and $N + 1$ are given by (6.103) and the constants $l$ and $\hat{l}$ are determined in Theorem 6.23.*

**Remark 6.26** Under the assumptions of Corollary 6.25, suppose that the boundary conditions in (6.98) do not depend on $\lambda$, i.e., $R_k(\lambda) \equiv R_k$, $\hat{R}_k(\lambda) \equiv \hat{R}_k$, $R_k^*(\lambda) \equiv R_k^*$, $\hat{R}_k^*(\lambda) \equiv \hat{R}_k^*$ for $k \in \{-1, N + 1\}$. Then the second addends in the definitions of $\#_{-1}(\hat{Z}^{[R]}(\lambda), Z^{[L]}(\lambda))$ and $\#_{N+1}(\hat{Z}^{[R]}(\lambda), Z^{[L]}(\lambda))$ do not depend on $\lambda$ as well and formula (6.110) can be rewritten in the form

$$\left.\begin{aligned}
\#\{v \in \hat{\sigma} \mid v \leq b\} &- \#\{v \in \sigma \mid v \leq b\} \\
&= \mu^*\big(\hat{V}_{-1}^{-1}\hat{Y}_0^{[R]}(b), \hat{V}_{-1}^{-1}V_{-1}(0\ I)^T\big) \\
&\quad + \mu\big(\hat{V}_{N+1}Y_{N+1}^{[L]}(b), \hat{V}_{N+1}V_{N+1}^{-1}(0\ I)^T\big) - \tilde{\mathfrak{P}},
\end{aligned}\right\} \qquad (6.112)$$

where the new constant $\tilde{\mathfrak{P}}$ is connected with $\mathfrak{P}$ given by (6.111) as follows

$$\left.\begin{aligned}
\tilde{\mathfrak{P}} &= \mathfrak{P} + \mu\big(\hat{V}_{-1}(0\ I)^T_{,}\ V_{-1}(0\ I)^T\big) + \mu^*\big(\hat{V}_{N+1}^{-1}(0\ I)^T_{,}\ V_{N+1}^{-1}(0\ I)^T\big) \\
&= P_1 + P_2, \\
P_1 &:= \mu^*\big(\hat{V}_{-1}^{-1}\hat{Y}_0^{[R]}(\lambda), \hat{V}_{-1}^{-1}V_{-1}(0\ I)^T\big), \\
P_2 &:= \mu\big(\hat{V}_{N+1}Y_{N+1}^{[L]}(\lambda), \hat{V}_{N+1}V_{N+1}^{-1}(0\ I)^T\big), \quad \lambda \leq \lambda_0 := \min \sigma \cup \hat{\sigma}.
\end{aligned}\right\}$$
$$(6.113)$$

### 6.1.7  General Boundary Conditions

In this subsection we consider the general case of problems (5.1), (5.151) and (6.1), (6.3) under the assumptions of Theorem 5.55. In particular, we impose the conditions

$$\left.\begin{aligned}
\operatorname{rank} \mathcal{B}_k(\lambda) \text{ is constant for } \lambda \in \mathbb{R} \text{ for all } k \in [0, N]_{\mathbb{Z}}, \\
\operatorname{rank} R_2(\lambda) \text{ is constant for } \lambda \in \mathbb{R}, \\
\operatorname{rank} \hat{\mathcal{B}}_k(\lambda) \text{ is constant for } \lambda \in \mathbb{R} \text{ for all } k \in [0, N]_{\mathbb{Z}}, \\
\operatorname{rank} \hat{R}_2(\lambda) \text{ is constant for } \lambda \in \mathbb{R}.
\end{aligned}\right\} \qquad (6.114)$$

As for the case of the separated boundary conditions, the main role in the consideration plays Theorem 5.44, which gives the possibility to consider problems (5.1), (5.151) and (6.1), (6.3) as a special case of problems with separated boundary conditions, and then all results of Sect. 6.1.6 can be applied. Applying Theorem 5.44, consider the pair of the augmented problems (5.178) associated with (5.1), (5.151) and (6.1), (6.3)

$$
\left.\begin{aligned}
&\tilde{y}_{k+1}(\lambda) = \{\mathcal{S}_k(\lambda)\}\,\tilde{y}_k(\lambda), \quad k \in [0, N]_{\mathbb{Z}}, \\
&\begin{pmatrix} 0 & 0 \\ -I & I \end{pmatrix}\tilde{x}_0(\lambda) + \begin{pmatrix} -I & -I \\ 0 & 0 \end{pmatrix}\tilde{u}_0(\lambda) = 0, \\
&R_1(\lambda)\,\tilde{x}_{N+1}(\lambda) + R_2(\lambda)\,\tilde{u}_{N+1}(\lambda) = 0
\end{aligned}\right\}
$$

$$
\left.\begin{aligned}
&\bar{y}_{k+1}(\lambda) = \{\hat{\mathcal{S}}_k(\lambda)\}\,\bar{y}_k(\lambda), \quad k \in [0, N]_{\mathbb{Z}}, \\
&\begin{pmatrix} 0 & 0 \\ -I & I \end{pmatrix}\bar{x}_0(\lambda) + \begin{pmatrix} -I & -I \\ 0 & 0 \end{pmatrix}\bar{u}_0(\lambda) = 0, \\
&\hat{R}_1(\lambda)\,\bar{x}_{N+1}(\lambda) + \hat{R}_2(\lambda)\,\bar{u}_{N+1}(\lambda) = 0.
\end{aligned}\right\}
$$

(6.115)

According to Definition 6.21, we introduce the $4n \times 4n$ matrices associated with $R_k(\lambda)$ and $\hat{R}_k(\lambda)$ for $k \in \{1, 2\}$ (see (6.100), (6.101))

$$
\left.\begin{aligned}
&\mathcal{V}_{N+1}(\lambda) := \begin{pmatrix} R_1(\lambda) & R_2(\lambda) \\ -\mathcal{K}(\lambda)\,R_2(\lambda) & \mathcal{K}(\lambda)\,R_1(\lambda) \end{pmatrix}, \\
&\mathcal{K}(\lambda) := [R_1(\lambda)\,R_1^T(\lambda) + R_2(\lambda)\,R_2^T(\lambda)]^{-1}, \\
&\hat{\mathcal{V}}_{N+1}(\lambda) := \begin{pmatrix} \hat{R}_1(\lambda) & \hat{R}_2(\lambda) \\ -\hat{\mathcal{K}}(\lambda)\,\hat{R}_2(\lambda) & \hat{\mathcal{K}}(\lambda)\,\hat{R}_1(\lambda) \end{pmatrix}, \\
&\hat{\mathcal{K}}(\lambda) := [\hat{R}_1(\lambda)\,\hat{R}_1^T(\lambda) + \hat{R}_2(\lambda)\,\hat{R}_2^T(\lambda)]^{-1}.
\end{aligned}\right\}
$$

(6.116)

The boundary conditions at the left endpoint in problems (6.115) are the same and do not depend on $\lambda$, and then one can put $\mathcal{V}_{-1}(\lambda) \equiv \hat{\mathcal{V}}_{-1}(\lambda) \equiv S_0$, where the matrix $S_0$ is given in (3.102). Applying the additional properties of the comparative index for symplectic matrices (see Sect. 3.3.5), we derive the following representation of the relative oscillation numbers for problems (6.115) using the notation introduced in Sect. 3.3.5.

**Lemma 6.27** *Let $\hat{Z}^{[N+1]}(\lambda)$ and $Z^{[0]}(\lambda)$ be the symplectic fundamental matrices of systems (6.1) and (5.1) satisfying for all $\lambda \in \mathbb{R}$ the initial conditions $\hat{Z}_{N+1}^{[N+1]}(\lambda) = I = Z_0^{[0]}(\lambda)$ and*

$$
\left.\begin{aligned}
Z_k^{[L]}(\lambda) &= \{Z_k^{[0]}(\lambda)\}\,S_0, \\
\hat{Z}_k^{[R]}(\lambda) &= \{\hat{Z}_k^{[N+1]}(\lambda)\}\,\hat{\mathcal{V}}_{N+1}^{-1}(\lambda),
\end{aligned}\right\} \quad k \in [0, N+1]_{\mathbb{Z}}, \tag{6.117}
$$

*where $S_0$ and $\hat{\mathcal{V}}_{N+1}(\lambda)$ are defined by (3.102) and (6.116). Then, concerning the relative oscillation numbers $\#_k(\hat{Z}^{[R]}(b), Z^{[L]}(a))$ defined by (6.103) for* *problems* (6.115), *we have the identities*

$$\#_k\big(\hat{\mathcal{Z}}^{[R]}(b),\,\mathcal{Z}^{[L]}(a)\big) = \mu\big(\hat{\mathcal{V}}_{N+1}(b)\langle\mathcal{G}_k(a,b)\rangle,\,\hat{\mathcal{V}}_{N+1}(b)\langle\mathcal{G}_{k+1}(a,b)\rangle\big) \left.\right\}$$
$$-\mu\big(\langle\hat{\mathcal{S}}_k(b)\rangle,\,\langle\mathcal{S}_k(a)\rangle\big),\quad k\in[0,N]_{\mathbb{Z}}, \left.\right\}$$

$$(6.118)$$

*where*

$$\mathcal{G}_k(a,b) = [\hat{Z}_k^{[N+1]}(b)]^{-1} Z_k^{[0]}(a),$$

*and*

$$\#_{-1}\big(\hat{\mathcal{Z}}^{[R]}(b),\,\mathcal{Z}^{[L]}(a)\big) = 0, \left.\right\}$$
$$\#_{N+1}\big(\hat{\mathcal{Z}}^{[R]}(b),\,\mathcal{Z}^{[L]}(a)\big) = \mu\big(\hat{\mathcal{V}}_{N+1}(b)\langle Z_{N+1}^{[0]}(a)\rangle,\,\hat{\mathcal{V}}_{N+1}(b)\mathcal{V}_{N+1}^{-1}(a)\,(0\ I)^T\big) \left.\right\}$$
$$-\mu^*\big(\hat{\mathcal{V}}_{N+1}^{-1}(b)\,(0\ I)^T,\,\mathcal{V}_{N+1}^{-1}(a)\,(0\ I)^T\big). \left.\right\}$$

$$(6.119)$$

*Proof* By (6.103) and (6.117), we have for problems (6.115)

$$\#_k\big(\hat{\mathcal{Z}}^{[R]}(b),\,\mathcal{Z}^{[L}(a)\big)$$
$$= \mu\Big(\langle\hat{\mathcal{V}}_{N+1}(b)\,W_k S_0\rangle,\,\langle\hat{\mathcal{V}}_{N+1}(b)\,W_{k+1} S_0\rangle\Big) - \mu\big(\langle\{\hat{\mathcal{S}}_k(b)\}\rangle,\,\langle\{\mathcal{S}_k(a)\}\rangle\big),$$

where $k\in[0,N]_{\mathbb{Z}}$ and

$$W_k := \big\{[\hat{Z}_k^{[N+1]}(b)]^{-1} Z_k^{[0]}(a)\big\}.$$

According to Corollary 3.40(v), by (3.122) with $\mathcal{R}:=\hat{\mathcal{V}}_{N+1}^{-1}$, we obtain that the first addend on the right-hand side of the previous formula is the same as the first summand on the right-hand side of (6.118). Next, using Proposition 3.37, by (3.109), we obtain

$$\mu\big(\langle\{\hat{\mathcal{S}}_k(b)\}\rangle,\,\langle\{\mathcal{S}_k(a)\}\rangle\big) = \mu\big(\langle\hat{\mathcal{S}}_k(b)\rangle,\,\langle\mathcal{S}_k(a)\rangle\big).$$

Then, formula (6.118) is proved. For the proof of (6.119), we note that the relative oscillation number at $k=-1$ equals to zero by Remark 6.24. The representation of $\#_{N+1}(\hat{\mathcal{Z}}^{[R]}(b),\,\mathcal{Z}^{[L]}(a))$ follows from the definition (see (6.103)), where we incorporate that

$$\mathcal{Y}_{N+1}^{[L]}(a) = Z_{N+1}^{[L]}(a)\,(0\ I)^T = \{Z_{N+1}^{[0]}(a)\}\,S_0(0\ I)^T = \langle Z_{N+1}^{[0]}(a)\rangle,$$

see (6.117) and (3.104). The proof is completed.                                         □

Define the relative oscillation numbers for (6.115)

$$
\begin{aligned}
\#(\hat{\mathcal{Z}}^{[R]}(b), \mathcal{Z}^{[L]}(a), 0, N+1) \\
= \sum_{k=0}^{N+1} \#_k(\hat{\mathcal{Z}}^{[R]}(b), \mathcal{Z}^{[L]}(a)) \\
= \#(\hat{\mathcal{Z}}^{[R]}(b), \mathcal{Z}^{[L]}(a), 0, N) + \#_{N+1}(\hat{\mathcal{Z}}^{[R]}(b), \mathcal{Z}^{[L]}(a)).
\end{aligned}
\tag{6.120}
$$

Based on Lemma 6.27, we present the relative and renormalized oscillation theorems for joint endpoints as corollaries to Theorems 6.22 and 6.23.

**Theorem 6.28 (Relative Oscillation Theorem for Joint Endpoints)** *For problems* (5.1), (5.151) *and* (6.1), (6.3) *assume that* (5.3), (5.152), (5.153), (6.4)–(6.6), *and* (6.114) *hold. Define the relative oscillation numbers according to Lemma 6.27 and* (6.120). *Then there exists a constant* $\mathfrak{P} \in [-(N+2)\,n, (N+2)\,n]_{\mathbb{Z}}$ *such that for any* $a, b \in \mathbb{R}$, *we have for the spectra* $\sigma$ *and* $\hat{\sigma}$ *of problems* (5.1), (5.151) *and* (6.1), (6.3) *the equality*

$$
\begin{aligned}
\#\{\nu \in \hat{\sigma} \mid \nu \le b\} - \#\{\nu \in \sigma \mid \nu \le a\} \\
= \#(\hat{\mathcal{Z}}^{[R]}(b), \mathcal{Z}^{[L]}(a), 0, N+1) - \mathfrak{P},
\end{aligned}
\tag{6.121}
$$

*where for* $\lambda_0$ *in* (6.20) *and* $\ell$ *given by* (5.231) *(and similarly defined* $\hat{\ell}$*), we have*

$$
\mathfrak{P} = \hat{\ell} - \ell = \#(\hat{Z}^{[R]}(\lambda), Z^{[L]}(\lambda), 0, N+1), \quad \lambda < \lambda_0. \tag{6.122}
$$

*Proof* The result follows from the relative oscillation theorem (Theorem 6.22) applied to the two augmented problems (6.115), which have the separated endpoints.

□

**Theorem 6.29 (Renormalized Oscillation Theorem for Joint Endpoints)** *Under the assumptions of Theorem 6.28, consider only one problem* (5.1), (5.151). *Assume that the relative oscillation numbers are defined according to Lemma 6.27, where* $\hat{Z}^{[N+1]}(\lambda) \equiv Z^{[N+1]}(\lambda)$ *and* $\hat{V}_{N+1}(\lambda) \equiv V_{N+1}(\lambda)$. *Then for any* $a, b \in \mathbb{R}$ *with* $a < b$, *the number of finite eigenvalues of problem* (5.1) *with* (5.151) *in the interval* $(a, b]$ *is given by the formula*

$$
\begin{aligned}
\#\{\nu \in \sigma \mid a < \nu \le b\} \\
= \sum_{k=0}^{N} \mu\big(V_{N+1}(b)\,\langle \mathcal{G}_k(a,b)\rangle, V_{N+1}(b)\,\langle \mathcal{G}_{k+1}(a,b)\rangle\big) \\
+ \mu\big(V_{N+1}(b)\,\langle Z_{N+1}^{[0]}(a)\rangle, V_{N+1}(b)V_{N+1}^{-1}(a)\,(0\ I)^T\big), \\
\mathcal{G}_k(a,b) := [Z_k^{[N+1]}(b)]^{-1} Z_k^{[0]}(a).
\end{aligned}
\tag{6.123}
$$

*In particular, for* $a := \lambda$ *and* $\lambda < \lambda_1 = \min \sigma$, *formula* (6.123) *presents the number of finite eigenvalues of* (5.1), (5.151), *which are less than or equal to* $b$.

*Proof* We use that problem (5.1), (5.151) is equivalent to the first problem in (6.115) with separated endpoints. The result then follows from Theorem 6.23.            □

*Remark 6.30*

(i)  Using Lemmas 4.47 and 3.39, it is possible to show (see [121, Remark 4.5.10]) that in case of separated boundary conditions, the identity

$$\#(\hat{\mathcal{Z}}^{[R]}(b), \mathcal{Z}^{[L]}(a), 0, N+1) = \#(\hat{Z}^{[R]}(b), Z^{[L]}(a), -1, N+1)$$

holds, where the number on the left-hand side is given by (6.118), (6.119), (6.120) and on the right-hand side in accordance with Definition 6.21. It should be emphasized that "componentwise" identities

$$\#_{N+1}(\hat{\mathcal{Z}}^{[R]}(b), \mathcal{Z}^{[L]}(a)) = \#_{N+1}(\hat{Z}^{[R]}(b), Z^{[L]}(a)) + \#_{-1}(\hat{Z}^{[R]}(b), Z^{[L]}(a))$$

and

$$\#(\hat{\mathcal{Z}}^{[R]}(b), \mathcal{Z}^{[L]}(a), 0, N) = \#(\hat{Z}^{[R]}(b), Z^{[L]}(a), 0, N)$$

do not necessarily hold in general, but they hold in the case when $\hat{S}_k(\lambda) \equiv S_k(\lambda)$ for all $k \in [0, N]_{\mathbb{Z}}$.

(ii)  For the case $R_1(\lambda) \equiv \hat{R}_1(\lambda) \equiv R_1$ and $R_2(\lambda) \equiv \hat{R}_2(\lambda) \equiv R_2$, i.e., when the boundary conditions are the same and are independent on $\lambda$, we have by (6.119) that $\#_{N+1}(\hat{\mathcal{Z}}^{[R]}(b), \mathcal{Z}^{[L]}(a)) = 0$. In particular, for the Dirichlet boundary conditions, we put $\mathcal{V}_{N+1}(\lambda) \equiv \hat{\mathcal{V}}_{N+1}(\lambda) = I_{4n}$, and then Theorem 6.28 reduces to Theorem 6.2.

By analogy with the case of separated endpoints, consider the case when problems differ in the boundary conditions only.

**Corollary 6.31** *Under all assumptions of Theorem 6.28, suppose additionally that* $S_k(\lambda) \equiv \hat{S}_k(\lambda)$ *for all* $k \in [0, N]_{\mathbb{Z}}$ *and* $a = b$. *Then the difference of the numbers of finite eigenvalues of problems* (5.1), (5.151) *and* (6.1), (6.3) *is given by the formula*

$$\#\{\nu \in \hat{\sigma} \mid \nu \leq b\} - \#\{\nu \in \sigma \mid \nu \leq b\} = \#_{N+1}(\hat{\mathcal{Z}}^{[R]}(b), \mathcal{Z}^{[L]}(b)) - \mathfrak{P}, \qquad (6.124)$$

*with the constant* $\mathfrak{P}$ *defined by*

$$\mathfrak{P} := \#_{N+1}(\hat{\mathcal{Z}}^{[R]}(\lambda), \mathcal{Z}^{[L]}(\lambda)) = \hat{\ell} - \ell, \quad \lambda \leq \lambda_0 := \min \sigma \cup \hat{\sigma}, \qquad (6.125)$$

*where the relative oscillation number for* $k = N + 1$, $a = b$ *is given by* (6.119) *and the constants* $\ell$ *and* $\hat{\ell}$ *are determined in Theorem 6.28.*

*Remark 6.32* As for the case of the separated boundary conditions, under the assumptions of Corollary 6.31, suppose that the boundary conditions in (5.151) , (6.3) do not depend on $\lambda$, i.e., $R_k(\lambda) \equiv R_k$ and $\hat{R}_k(\lambda) \equiv \hat{R}_k$ for $k \in \{1, 2\}$. Then the second addend in the definition of $\#_{N+1}(\hat{Z}^{[R]}(\lambda), Z^{[L]}(\lambda))$ does not depend on $\lambda$ as well and formula (6.124) takes the form

$$
\left.
\begin{aligned}
&\#\{v \in \hat{\sigma} \mid v \le b\} - \#\{v \in \sigma \mid v \le b\} \\
&\qquad = \mu\big(\hat{V}_{N+1}\langle Z^{[0]}_{N+1}(b)\rangle, \hat{V}_{N+1}V^{-1}_{N+1}(0 \; I)^T\big) - \tilde{\mathfrak{P}},
\end{aligned}
\right\}
\tag{6.126}
$$

where the new constant $\tilde{\mathfrak{P}}$ is connected with $\mathfrak{P}$ given by (6.125) as follows

$$
\left.
\begin{aligned}
\tilde{\mathfrak{P}} &= \mathfrak{P} + \mu^*\big(\hat{V}^{-1}_{N+1}(0 \; I)^T_{\!}, V^{-1}_{N+1}(0 \; I)^T\big) \\
&= \mu\big(\hat{V}_{N+1}\langle Z^{[0]}_{N+1}(\lambda)\rangle, \hat{V}_{N+1}V^{-1}_{N+1}(0 \; I)^T\big), \quad \lambda \le \lambda_0 := \min \sigma \cup \hat{\sigma}.
\end{aligned}
\right\}
\tag{6.127}
$$

## 6.2 Inequalities for Finite Eigenvalues

In this section we derive inequalities for eigenvalues of problems (6.17), resp., for (6.98), resp., for (5.1), (5.151) and (6.1), (6.3), by using inequalities for their spectral functions. As in Sect. 6.1, we consider these spectral problems under the assumptions, which guarantee that the finite spectra are bounded from below. We order the finite eigenvalues $\lambda_k \in \sigma$ of (5.1), (5.151), resp., of $\hat{\lambda}_k \in \hat{\sigma}$ of (6.1), (6.3), into nondecreasing sequences (see (6.19)) as

$$
-\infty < \lambda_1 \le \lambda_2 \le \cdots \le \lambda_m \le \ldots, \quad -\infty < \hat{\lambda}_1 \le \hat{\lambda}_2 \le \cdots \le \hat{\lambda}_l \le \ldots.
$$

We also put $\lambda_1 := \infty$ ($\hat{\lambda}_1 := \infty$) if $\sigma_N = \emptyset$ ($\hat{\sigma}_N = \emptyset$). According to (6.20) we define $\lambda_0 := \min\{\lambda_1, \hat{\lambda}_1\}$. For the proofs of the subsequent results, we need the following.

**Proposition 6.33** *Assume that for some $K \in \mathbb{N} \cup \{0\}$ the inequality*

$$
\#\{v \in \hat{\sigma} \mid v \le b\} - \#\{v \in \sigma \mid v \le b\} \ge -K,
\tag{6.128}
$$

*holds and that there exists a finite eigenvalue $\lambda_{j+K} \in \sigma$, $j \ge 1$, of problem (5.1), (5.151). Then there exists the finite eigenvalue $\hat{\lambda}_j \in \hat{\sigma}$ of (6.1), (6.3), and it satisfies*

$$
\hat{\lambda}_j \le \lambda_{j+K}.
\tag{6.129}
$$

*If, instead of* (6.128), *we have*

$$\#\{v \in \hat{\sigma} \mid v < b\} - \#\{v \in \sigma \mid v \le b\} \ge -K, \tag{6.130}$$

*then the inequality in* (6.129) *is strict, i.e.,*

$$\hat{\lambda}_j < \lambda_{j+K}. \tag{6.131}$$

*Proof* If there exists $\lambda_{j+K} \in \sigma$ (with $j \ge 1$), then $\#\{v \in \sigma \mid v \le \lambda_{j+K}\} \ge j + K$ (note that $\#\{v \in \sigma \mid v \le \lambda_{j+K}\} = j + K$ if $\lambda_{j+K}$ is simple) and, according to assumption (6.128),

$$\#\{v \in \hat{\sigma} \mid v \le \lambda_{j+K}\} \ge \#\{v \in \sigma \mid v \le \lambda_{j+K}\} - K \ge j.$$

Consequently, there exist $\hat{\lambda}_j \in \hat{\sigma}$ and (6.129) holds. If (6.130) holds, then repeating the arguments from the proof of the inequality $\hat{\lambda}_j \le \lambda_{j+K}$, we obtain $\#\{v \in \hat{\sigma} \mid v < \lambda_{j+K}\} \ge j$. Consequently, the strict inequality (6.131) holds.                     □

In Sect. 6.2.1 we compare the finite eigenvalues of (5.1), (5.151) and (6.1), (6.3) under some majorant assumptions for their symplectic coefficient matrices and for the matrices in the boundary conditions. There, the consideration is based on Theorems 6.22 and 6.28, whose assumptions guarantee that $\sigma$ and $\hat{\sigma}$ are also bounded from above. In Sect. 6.2.2 we consider problems (5.1), (5.151) and (6.1), (6.3), which differ only in the boundary conditions. For this case we admit the oscillations of the block $\mathcal{B}_k(\lambda)$ of $\mathcal{S}_k(\lambda)$ at $\infty$, which implies that the spectra $\sigma$ and/or $\hat{\sigma}$ may be unbounded from above. We derive the interlacing properties of finite eigenvalues based on the pair of the inequalities

$$\hat{\lambda}_j \le \lambda_{j+K}, \quad \lambda_p \le \hat{\lambda}_{p+\hat{K}}, \quad p \ge 1, \quad j \ge 1, \quad K \ge 0, \quad \hat{K} \ge 0, \tag{6.132}$$

for the finite eigenvalues of problems (5.1), (5.151) and (6.1), (6.3) with joint and separated endpoints.

## 6.2.1   Comparison of Finite Eigenvalues

In this subsection we prove inequality (6.129) under suitable majorant conditions. In particular, we investigate the case when (6.129) holds with $K = 0$ and the finite eigenvalues $\hat{\lambda}_j \in \hat{\sigma}$ for (6.1), (6.3) lie below the finite eigenvalues $\lambda_j \in \sigma$ for (5.1), (5.151), i.e.,

$$\hat{\lambda}_j \le \lambda_j. \tag{6.133}$$

This fact is also proven for problems (6.17) with the Dirichlet boundary conditions, as well as for (6.98) with the separated endpoints. All these results can be derived as corollaries to Theorems 6.2, 6.22, and 6.28.

The following theorem formulates sufficient conditions for inequalities (6.129) and (6.133), when the boundary conditions are separated.

**Theorem 6.34** *Under the assumptions of Theorem 6.22, suppose that for the symplectic coefficient matrices $\mathcal{S}_k(\lambda)$ and $\hat{\mathcal{S}}_k(\lambda)$, we have the majorant condition*

$$\mu\big(\langle\hat{\mathcal{S}}_k(\lambda)\rangle, \langle\mathcal{S}_k(\lambda)\rangle\big) = 0, \quad \lambda \in \mathbb{R}, \quad k \in [0, N]_{\mathbb{Z}}, \tag{6.134}$$

*and the matrix coefficients in the boundary conditions (6.98) obey the assumption*

$$\mu\left(\begin{pmatrix} -\hat{R}_0^T(\lambda) \\ \hat{R}_0^{*T}(\lambda) \end{pmatrix}, \begin{pmatrix} -R_0^T(\lambda) \\ R_0^{*T}(\lambda) \end{pmatrix}\right) = \mu^*\left(\begin{pmatrix} -\hat{R}_{N+1}^T(\lambda) \\ \hat{R}_{N+1}^{*T}(\lambda) \end{pmatrix}, \begin{pmatrix} -R_{N+1}^T(\lambda) \\ R_{N+1}^{*T}(\lambda) \end{pmatrix}\right) = 0, \quad \lambda \in \mathbb{R}. \tag{6.135}$$

*Then for the constant $\mathfrak{P}$ given by (6.106), we have that inequality (6.129) holds with $K := \mathfrak{P} \geq 0$, provided the finite eigenvalue $\lambda_{j+K} \in \sigma$, $j \geq 1$, exists. In particular, inequality (6.133) holds when*

$$\mathfrak{P} = 0. \tag{6.136}$$

*Proof* Note that condition (6.135) can be rewritten in the form

$$\left.\begin{aligned}\mu\big(\hat{V}_{-1}(\lambda)(0\ I)^T, V_{-1}(\lambda)(0\ I)^T\big)\\ = \mu^*\big(\hat{V}_{N+1}^{-1}(\lambda)(0\ I)^T, V_{N+1}^{-1}(\lambda)(0\ I)^T\big) = 0, \quad \lambda \in \mathbb{R},\end{aligned}\right\} \tag{6.137}$$

for the matrices $V_k(\lambda)$ and $\hat{V}_k(\lambda)$ for $k \in \{-1, N+1\}$ defined by (6.100) and (6.101). Then, under assumptions (6.134) and (6.137), the relative oscillation numbers given by (6.103) with $a = b := \lambda$ are nonnegative, i.e.,

$$\#_k\big(\hat{Z}^{[N+2]}(\lambda), Z^{[-1]}(\lambda)\big) \geq 0, \quad k \in [-1, N+1]_{\mathbb{Z}}, \quad \lambda \in \mathbb{R}. \tag{6.138}$$

Next, by (6.138) we derive according to (6.105) that

$$\#\{v \in \hat{\sigma} \mid v \leq \lambda_j\} - \#\{v \in \sigma \mid v \leq \lambda_j\} \geq -\mathfrak{P}, \tag{6.139}$$

where $\mathfrak{P} \geq 0$, because of (6.138) and (6.106). The result then follows from Proposition 6.33. $\qquad\square$

*Remark 6.35*

(i) The majorant condition in (6.134), including the main important special cases of (SDS), is discussed in Remark 4.51. Note also that condition (5.321) from Theorem 5.89 is sufficient for (6.134), because of Lemma 3.25 and formula (3.74); see Remark 4.51(i).

(ii) Note that condition

$$\hat{\ell} = 0 \tag{6.140}$$

with the constant $\hat{\ell}$ given by (5.215) for the second problem in (6.98) (see Theorem 6.22) is sufficient for (6.136). Indeed, according to (6.106), under assumptions (6.134) and (6.137), we have by (6.138) and (6.140) that

$$0 \leq \mathfrak{P} = \#_k\big(\hat{Z}^{[N+2]}(\lambda),\, Z^{[-1]}(\lambda)\big) = -\ell \leq 0, \quad k \in [-1, N+1]_{\mathbb{z}},$$

and then $\mathfrak{P} = -\ell = 0$ as well. Finally, condition (6.140) implies (6.136).

(iii) Recall that condition (6.140) is equivalent to the existence of $\lambda_{00} < 0$ such that the quadratic functional $\hat{\mathcal{G}}(y, \lambda)$ in (5.188) associated with the second problem in (6.98) obeys the condition $\hat{\mathcal{G}}(\cdot, \lambda_{00}) > 0$. For problems (6.98) with the special linear dependence on $\lambda$ (see (5.239) and (5.245)), sufficient conditions for $\hat{\mathcal{G}}(\cdot, \lambda_{00}) > 0$ are given in Propositions 5.66 and 5.67 (see also Remark 5.68). We recall that equivalent conditions to $\hat{\mathcal{G}}(\cdot, \lambda_{00}) > 0$ are presented in Theorem 2.50.

For the problems (6.17) with the Dirichlet boundary conditions, assumption (6.137) is automatically satisfied. Then we have the following corollary to Theorem 6.34.

**Corollary 6.36** *Under the assumptions of Theorem 6.2, suppose that (6.134) holds. Then inequality (6.129) holds for the finite eigenvalues of problems (6.17) with $K := \mathfrak{P} \geq 0$, where the constant $\mathfrak{P}$ is given by (6.28). In particular, for the case (6.136), we have inequality (6.133).*

*Remark 6.37*

(i) It follows from Remark 6.35(ii) that the condition

$$\hat{m} = 0 \tag{6.141}$$

with $\hat{m}$ defined by (5.47) for the second problem in (6.17) (see also Theorem 6.2) is sufficient for (6.136). Moreover, condition (6.141) is equivalent to the existence of $\lambda_{00} < 0$ such that the quadratic functional $\hat{\mathcal{F}}(y, \lambda)$ in (5.138) associated with the second problem in (6.17) obeys the condition $\hat{\mathcal{F}}(\cdot, \lambda_{00}) > 0$ (see Proposition 5.30).

(ii) For the case when problems (6.17) obey assumption (6.35), condition (6.134) can be replaced by

$$[X_{k+1}^{[0]}(\lambda)]^T [C_k(\lambda) \hat{\mathcal{D}}_k^T(\lambda) - \mathcal{D}_k(\lambda) \hat{C}_k^T(\lambda)] X_{k+1}^{[0]}(\lambda) \geq 0, \qquad (6.142)$$

where $C_k(\lambda)$, $\mathcal{D}_k(\lambda)$ and $\hat{C}_k(\lambda)$, $\hat{\mathcal{D}}_k(\lambda)$ are the blocks of $\mathcal{S}_k(\lambda)$ and $\hat{\mathcal{S}}_k(\lambda)$, respectively (see (5.2) and (6.2)). Here $X_{k+1}^{[0]}(\lambda)$ is the upper block of the principal solution $Y^{[0]}(\lambda)$ of the first problem in (6.17) evaluated at $k+1$. In particular, condition (6.142) holds under the assumption

$$C_k(\lambda) \hat{\mathcal{D}}_k^T(\lambda) - \mathcal{D}_k(\lambda) \hat{C}_k^T(\lambda) \geq 0. \qquad (6.143)$$

The proof follows from Theorem 6.8 and from formula (6.52), where we use that (6.142) is equivalent to $\tilde{\mathfrak{B}}_k(a, b) \geq 0$ with $a = b := \lambda$.

(iii) For the pair of eigenvalue problems (6.17) with the special linear dependence on $\lambda$ in (6.42), condition (6.142) takes the form

$$[X_{k+1}^{[0]}(\lambda)]^T [\lambda (\hat{\mathcal{W}}_k - \mathcal{W}_k) + C_k \hat{\mathcal{D}}_k^T - \mathcal{D}_k \hat{C}_k^T] X_{k+1}^{[0]}(\lambda) \geq 0, \qquad (6.144)$$

see (6.43). Note that under the additional assumption $\hat{\mathcal{W}}_k > 0$ for $k \in [0, N]_{\mathbb{Z}}$ condition (6.141) is satisfied due to Proposition 5.69, and then inequality (6.133) holds. In particular, for Sturm-Liouville eigenvalue problems (6.9), (6.10), condition (6.144) takes the form $[x_{k+1}^{[0]}(\lambda)]^2 (p_k - \hat{p}_k) \geq 0$ for $k \in [0, N-1]_{\mathbb{Z}}$, and it certainly implies (6.133), because for this case $\hat{\mathcal{W}}_k = 1$.

(iv) For the discrete matrix Sturm-Liouville spectral problems in Sect. 6.1.4 (see (6.63) and (6.64)), condition (6.134) can be replaced by

$$\left. \begin{array}{l} [U_k^{[0]}(\lambda)]^T [\hat{R}_k^{-1}(\lambda) - R_k^{-1}(\lambda)] U_k^{[0]}(\lambda) \geq 0, \\ [X_{k+1}^{[0]}(\lambda)]^T [Q_k(\lambda) - \hat{Q}_k(\lambda)] X_{k+1}^{[0]}(\lambda) \geq 0, \end{array} \right\} \qquad (6.145)$$

where $X_k^{[0]}(\lambda)$ and $U_k^{[0]}(\lambda)$ are the blocks of the principal solution $Y^{[0]}(\lambda)$ of (6.63), i.e., $X_0^{[0]}(\lambda) = 0$ and $U_0^{[0]}(\lambda) = R_0(\lambda) \Delta X_0^{[0]}(\lambda) = I$. By analogy with the proof of Theorem 6.34, we use Remark 6.18 to show that condition (6.145) implies (6.129) with $K := \tilde{\mathfrak{P}}$ given by (6.91). Moreover, under the assumption $\tilde{\mathfrak{P}} = 0$, we have inequality (6.133) for the finite eigenvalues of (6.63) and (6.64). Note that the conditions in (6.145) follow from the classical majorant assumptions $R_k(\lambda) \geq \hat{R}_k(\lambda) > 0$ and $Q_k(\lambda) \geq \hat{Q}_k(\lambda)$ (see Proposition 6.15 (ii) and Remark 3.35).

The last result of this subsection is devoted to problems (5.1), (5.151) and (6.1), (6.3) with joint endpoints.

**Theorem 6.38** *Under the assumptions of Theorem 6.28, suppose that for the symplectic coefficient matrices $S_k(\lambda)$ and $\hat{S}_k(\lambda)$, we have the majorant condition (6.134), while the matrices in the boundary conditions (5.151) and (6.3) obey*

$$\mu^*\left(\begin{pmatrix} -\hat{R}_2^T(\lambda) \\ \hat{R}_1^T(\lambda) \end{pmatrix}, \begin{pmatrix} -R_2^T(\lambda) \\ R_1^T(\lambda) \end{pmatrix}\right) = 0, \quad \lambda \in \mathbb{R}. \tag{6.146}$$

*Then inequality (6.129) holds for the finite eigenvalues of (5.1), (5.151) and (6.1), (6.3) with the constant $K := \mathfrak{P} \geq 0$ and $\mathfrak{P}$ given by (6.122), provided the finite eigenvalue $\lambda_{j+K}$ exists. In particular, inequality (6.133) holds for the case of (6.136).*

*Proof* Condition (6.146) can be rewritten as

$$\mu^*\left(\hat{\mathcal{V}}_{N+1}^{-1}(\lambda)\,(0\ I)^T, \mathcal{V}_{N+1}^{-1}(\lambda)\,(0\ I)^T\right) = 0, \quad \lambda \in \mathbb{R}, \tag{6.147}$$

where the matrices $\mathcal{V}_{N+1}(\lambda)$ and $\hat{\mathcal{V}}_{N+1}(\lambda)$ are defined by (6.116). It follows that under assumptions (6.134) and (6.146), the relative oscillation numbers given by (6.118) and (6.119) with $a = b := \lambda$ are nonnegative, and then the proof repeats the proof of Theorem 6.34. In particular, we derive inequality (6.139), which holds with $\mathfrak{P}$ given by (6.122).                                                                    □

*Remark 6.39* Note that, by analogy with Remark 6.35(ii), the condition (6.140) with the constant $\hat{\ell}$ defined by (5.436) is sufficient for (6.136) with the constant $\mathfrak{P}$ given by (6.122). In turn, one can apply the statement of Theorem 5.57 saying that $\hat{\ell}$ defined by (5.436) equals to zero if and only if there exists $\lambda_{00} < 0$ such that the quadratic functional $\hat{\mathcal{G}}(y, \lambda_{00})$ in (5.186) over (5.187) associated with (6.1), (6.3) is positive definite. We recall that this positivity condition can be tested by the properties of the principal solution $Y^{[0]}(\lambda_{00})$ in Theorem 2.53.

## 6.2.2  Interlacing of Eigenvalues for Joint Endpoints

In this subsection we consider problems (5.1), (5.151) and (6.1), (6.3) with the same coefficient matrices $S_k(\lambda) \equiv \hat{S}_k(\lambda)$ and with the boundary conditions which are independent on $\lambda$, i.e.,

$$\left.\begin{aligned} y_{k+1}(\lambda) &= S_k(\lambda)\, y_k(\lambda), \quad k \in [0, N]_{\mathbb{Z}}, \\ R_1 \begin{pmatrix} x_0(\lambda) \\ x_{N+1}(\lambda) \end{pmatrix} &+ R_2 \begin{pmatrix} -u_0(\lambda) \\ u_{N+1}(\lambda) \end{pmatrix} = 0, \\ \hat{y}_{k+1}(\lambda) &= S_k(\lambda)\, \hat{y}_k(\lambda), \quad k \in [0, N]_{\mathbb{Z}}, \\ \hat{R}_1 \begin{pmatrix} \hat{x}_0(\lambda) \\ \hat{x}_{N+1}(\lambda) \end{pmatrix} &+ \hat{R}_2 \begin{pmatrix} -\hat{u}_0(\lambda) \\ \hat{u}_{N+1}(\lambda) \end{pmatrix} = 0. \end{aligned}\right\} \tag{6.148}$$

We assume that the symplectic matrix $\mathcal{S}_k(\lambda)$ obeys the monotonicity condition (5.3) and condition (5.333) holds. Then by Corollary 5.123 the spectra $\sigma$ and $\hat{\sigma}$ of (6.148) are bounded from below.

We investigate the case when the finite eigenvalues of the first problem in (6.148) or in (6.181) interlace the finite eigenvalues of the second one, i.e., we derive the inequalities (6.132):

$$\hat{\lambda}_j \leq \lambda_{j+K}, \quad \lambda_p \leq \hat{\lambda}_{p+\hat{K}}, \quad p \geq 1, \ j \geq 1, \quad K \geq 0, \ \hat{K} \geq 0,$$

provided the finite eigenvalues $\lambda_{j+K}$ and $\lambda_{p+\hat{K}}$ exist. Now we formulate conditions, when inequalities (6.132) hold for the finite eigenvalues of (6.148). Introduce the notation

$$R_w := \text{rank} \left[ (R_1 \ R_2) \, \mathcal{J} (\hat{R}_1 \ \hat{R}_2)^T \right], \tag{6.149}$$

for the rank of the Wronskians of boundary conditions in (6.148), and the notation (see (5.221), (5.220), and (3.52))

$$\left. \begin{array}{ll} R_{\mathcal{I}} := \max_{\lambda \in \mathcal{I}} \text{rank} \, \mathcal{L}(\lambda), & \mathcal{L}(\lambda) := (R_1 \ R_2) \langle Z_{N+1}^{[0]}(\lambda) \rangle, \\ \hat{R}_{\mathcal{I}} := \max_{\lambda \in \mathcal{I}} \text{rank} \, \hat{\mathcal{L}}(\lambda), & \hat{\mathcal{L}}(\lambda) := (\hat{R}_1 \ \hat{R}_2) \langle Z_{N+1}^{[0]}(\lambda) \rangle, \end{array} \right\} \tag{6.150}$$

where $Z^{[0]}(\lambda)$ is the fundamental matrix of the symplectic system in (6.148) such that $Z_0^{[0]}(\lambda) = I$ for $\lambda \in \mathbb{R}$ and the interval $\mathcal{I} \subseteq \mathbb{R}$ can be of the form $\mathcal{I} = [a, b]$ or $\mathcal{I} = (\infty, b]$.

*Remark 6.40*

(i) Recall that by the monotonicity assumption for the matrix $\mathcal{S}_k(\lambda)$ (see (5.3)), the functions $\text{rank} \, \mathcal{L}(\lambda)$ and $\text{rank} \, \hat{\mathcal{L}}(\lambda)$ are piecewise constant with respect to $\lambda \in [a, b]$ and then the maxima in (6.150) are attained for some $\lambda \in [a, b]$. Moreover, for the case $\mathcal{I} := (\infty, b]$ we use (5.333) which guaranties that $\text{rank} \, \mathcal{L}(\lambda)$ and $\text{rank} \, \hat{\mathcal{L}}(\lambda)$ are constant for all $\lambda < \lambda_0 = \min\{\lambda_1, \hat{\lambda}_1\}$, and then for this case we evaluate $R_{\mathcal{I}}$ and $\hat{R}_{\mathcal{I}}$ according to

$$R_{(-\infty, b]} = R_{[\lambda, b]}, \quad \hat{R}_{(-\infty, b]} = R_{[\lambda, b]}, \quad \lambda < \lambda_0. \tag{6.151}$$

(ii) For the case when the spectra are unbounded from above (recall that in this subsection we assume only condition (5.333)), we introduce the notation

$$R := \sup_{\lambda \in \mathbb{R}} \text{rank} \, \mathcal{L}(\lambda) \leq 2n, \quad \hat{R} := \sup_{\lambda \in \mathbb{R}} \text{rank} \, \hat{\mathcal{L}}(\lambda) \leq 2n. \tag{6.152}$$

**Theorem 6.41** *Consider the pair of eigenvalue problems* (6.148) *under assumptions* (5.2), (5.3), (5.152), (6.4), *and* (5.333). *Then for* $a, b \in \mathbb{R}$, $a < b$, *we have*

$$
\left.
\begin{aligned}
&\left| \# \{ \nu \in \hat{\sigma} \mid a < \nu \le b \} - \# \{ \nu \in \sigma \mid a < \nu \le b \} \right| \\
&\qquad \le \min\{ r_{a,b}, \hat{r}_{a,b}, R_w \} \le \min\{ R_\mathcal{I}, \hat{R}_\mathcal{I}, R_w \} \le 2n, \quad \mathcal{I} = [a, b], \\
&r_{a,b} := \max\{ \operatorname{rank} \mathcal{L}(a), \operatorname{rank} \mathcal{L}(b) \}, \\
&\hat{r}_{a,b} := \max\{ \operatorname{rank} \hat{\mathcal{L}}(a), \operatorname{rank} \hat{\mathcal{L}}(b) \}.
\end{aligned}
\right\}
$$

$$(6.153)$$

*Proof* The proof is based on Corollary 6.31 and Remark 6.32. The proof of (6.153) is derived by analogy with the proof of Corollary 4.25 in Sect. 4.2.2. According to (6.126) we have

$$
\begin{aligned}
&\#\{ \nu \in \hat{\sigma} \mid a < \nu \le b \} - \#\{ \nu \in \sigma \mid a < \nu \le b \} \\
&\quad = \#\{ \nu \in \hat{\sigma} \mid \nu \le b \} - \#\{ \nu \in \hat{\sigma} \mid \nu \le a \} - \#\{ \nu \in \sigma \mid \nu \le b \} + \#\{ \nu \in \sigma \mid \nu \le a \} \\
&\quad = \mu\big( \hat{\mathcal{V}}_{N+1} \langle Z^{[0]}_{N+1}(\lambda) \rangle, \hat{\mathcal{V}}_{N+1} \mathcal{V}^{-1}_{N+1} (0 \ I)^T \big) \big|_a^b .
\end{aligned}
$$

Then it follows that

$$
\begin{aligned}
&\left| \mu\big( \hat{\mathcal{V}}_{N+1} \langle Z^{[0]}_{N+1}(\lambda) \rangle, \hat{\mathcal{V}}_{N+1} \mathcal{V}^{-1}_{N+1} (0 \ I)^T \big) \big|_a^b \right| \\
&\qquad \le \max \left\{ \mu\big( \hat{\mathcal{V}}_{N+1} \langle Z^{[0]}_{N+1}(\lambda) \rangle, \hat{\mathcal{V}}_{N+1} \mathcal{V}^{-1}_{N+1} (0 \ I)^T \big), \ \lambda \in \{ a, b \} \right\}
\end{aligned}
$$

and by Theorem 3.5(vii) the comparative indices on the right-hand side of the above inequality satisfy the estimate

$$
\mu\big( \hat{\mathcal{V}}_{N+1} \langle Z^{[0]}_{N+1}(\lambda) \rangle, \hat{\mathcal{V}}_{N+1} \mathcal{V}^{-1}_{N+1} (0 \ I)^T \big) \le \min\{ \operatorname{rank} \mathcal{L}(\lambda), R_w \}. \tag{6.154}
$$

Then we have that

$$
\left| \#\{ \nu \in \hat{\sigma} \mid a < \nu \le b \} - \#\{ \nu \in \sigma \mid a < \nu \le b \} \right| \le \min\{ R_w, r_{a,b} \} \le \min\{ R_w, R_\mathcal{I} \},
$$

where $r_{a,b}$ is defined by (6.153). Replacing the roles of the problems (6.148) (and then interchanging $\mathcal{V}_{N+1}$ and $\hat{\mathcal{V}}_{N+1}$ in (6.154)), we have a similar estimate

$$
\left| \#\{ \nu \in \hat{\sigma} \mid a < \nu \le b \} - \#\{ \nu \in \sigma \mid a < \nu \le b \} \right| \le \min\{ R_w, \hat{r}_{a,b} \} \le \min\{ R_w, \hat{R}_\mathcal{I} \}
$$

with $\hat{r}_{a,b}$ defined by (6.153) for the second problem in (6.148). Finally, the combination of the estimates above proves (6.153).                                    $\square$

Using (5.333), which guaranties that the spectra of (6.148) are bounded from below (see Corollary 5.123), one can formulate the following corollary to Theorem 6.41.

**Corollary 6.42** *Under the assumptions of Theorem 6.41, we have for any $b \in \mathbb{R}$*

$$\left|\#\{v \in \hat{\sigma} \mid v \le b\} - \#\{v \in \sigma \mid v \le b\}\right| \le \tilde{K} = \min\{R, \hat{R}, R_w\} \le 2n \qquad (6.155)$$

*with the constants $R$, $\hat{R}$, and $R_w$ given by (6.152) and (6.149). In particular, inequalities (6.132) hold with $K = \hat{K} := \tilde{K}$, where $\tilde{K}$ is defined by (6.155).*

*Proof* The proof of (6.155) follows from (6.153), where one can put $a := \lambda$ with $\lambda < \lambda_0 = \min\{\lambda_1, \hat{\lambda}_1\}$, and then by Proposition 6.33, we have that (6.132) hold with $K = \hat{K} := \tilde{K}$. $\qquad \square$

Now we specify the results of Corollary 6.42 by imposing more restrictions on the behavior of $\mathcal{L}(\lambda)$ and $\hat{\mathcal{L}}(\lambda)$ with respect to $\lambda$. In particular, the restrictions below hold for problems (6.148) with the matrix $S_k(\lambda)$, which is a polynomial in $\lambda$. For this case $\mathcal{L}(\lambda)$ and $\hat{\mathcal{L}}(\lambda)$ are also polynomials. Note that the restrictions are also satisfied for the special linear dependence in $\lambda$; see Sect. 5.3.

**Theorem 6.43** *Under the assumptions of Theorem 6.41, suppose additionally that for $R$ and $\hat{R}$ given by (6.152) we have*

$$\left.\begin{aligned} R &:= \sup_{\lambda \in \mathbb{R}} \operatorname{rank} \mathcal{L}(\lambda) = \operatorname{rank} \mathcal{L}(\lambda), \\ \hat{R} &:= \sup_{\lambda \in \mathbb{R}} \operatorname{rank} \hat{\mathcal{L}}(\lambda) = \operatorname{rank} \hat{\mathcal{L}}(\lambda), \end{aligned}\right\} \quad \lambda < \lambda_0 = \min\{\lambda_1, \hat{\lambda}_1\}. \qquad (6.156)$$

*Then*

$$\left.\begin{aligned} -\hat{K} \le \#\{v \in \sigma \mid v \le b\} - \#\{v \in \hat{\sigma} \mid v \le b\} \le K, \\ K := \min\{\tilde{\mathfrak{P}}, \ \hat{R} - R_w + \tilde{\mathfrak{P}}\} \ge 0, \quad \hat{K} := \min\{R_w - \tilde{\mathfrak{P}}, \ R - \tilde{\mathfrak{P}}\} \ge 0, \\ \hat{K} + K \le \min\{R_w, R, \hat{R}\} \le 2n, \end{aligned}\right\}$$
$$(6.157)$$

*where the constant $\tilde{\mathfrak{P}}$ is defined by (6.127). In particular, inequalities (6.132) for the finite eigenvalues of (6.148) hold with $K$ and $\hat{K}$ given by (6.157).*

*Proof* Under assumptions (6.156) we have for $b \in \mathbb{R}$

$$\operatorname{rank} \mathcal{L}(\tau)\Big|_\lambda^b \le 0, \quad \operatorname{rank} \hat{\mathcal{L}}(\tau)\Big|_\lambda^b \le 0, \quad \lambda < \lambda_0. \qquad (6.158)$$

Next we rewrite (6.126) in the form

$$\left.\begin{aligned} \#\{v \in \sigma \mid v \le b\} - \#\{v \in \hat{\sigma} \mid v \le b\} \\ = \tilde{\mathfrak{P}} - \mu\left(\hat{\mathcal{V}}_{N+1}\langle Z_{N+1}^{[0]}(b)\rangle, \ \hat{\mathcal{V}}_{N+1} \mathcal{V}_{N+1}^{-1}(0\ I)^T\right). \end{aligned}\right\} \qquad (6.159)$$

Then we obtain

$$\#\{v \in \sigma \mid v \le b\} - \#\{v \in \hat\sigma \mid v \le b\} \le \tilde{\mathfrak{P}}. \tag{6.160}$$

The difference in (6.159) can be also estimated using the relation

$$\left.\begin{aligned} &\mu\left(\hat{\mathcal{V}}_{N+1}\langle Z_{N+1}^{[0]}(\lambda)\rangle, \hat{\mathcal{V}}_{N+1}\mathcal{V}_{N+1}^{-1}(0\ I)^T\right) \\ &= R_w - \operatorname{rank} \hat{\mathcal{L}}(\lambda) + \mu^*\left(\hat{\mathcal{V}}_{N+1}\mathcal{V}_{N+1}^{-1}(0\ I)^T, \hat{\mathcal{V}}_{N+1}\langle Z_{N+1}^{[0]}(\lambda)\rangle\right), \end{aligned}\right\} \tag{6.161}$$

where we apply Theorem 3.5(vi) and use the notation (6.149) and (6.150). Applying (6.161) to both addends in (6.159), we obtain

$$\left.\begin{aligned} &\tilde{\mathfrak{P}} - \mu\left(\hat{\mathcal{V}}_{N+1}\langle Z_{N+1}^{[0]}(b)\rangle, \hat{\mathcal{V}}_{N+1}\mathcal{V}_{N+1}^{-1}(0\ I)^T\right) \\ &= \operatorname{rank} \hat{\mathcal{L}}(\tau)|_\lambda^b + \mu^*\left(\hat{\mathcal{V}}_{N+1}\mathcal{V}_{N+1}^{-1}(0\ I)^T, \hat{\mathcal{V}}_{N+1}\langle Z_{N+1}^{[0]}(b)\rangle\right)|_b^\lambda \\ &\le \mu^*\left(\hat{\mathcal{V}}_{N+1}\mathcal{V}_{N+1}^{-1}(0\ I)^T, \hat{\mathcal{V}}_{N+1}\langle Z_{N+1}^{[0]}(\lambda)\rangle\right) = \hat{R} - R_w + \tilde{\mathfrak{P}}, \\ &\lambda < \lambda_0, \end{aligned}\right\} \tag{6.162}$$

where $\lambda_0 = \min\{\lambda_1, \hat\lambda_1\}$ and $\hat{R}$ is given by (6.156). Then we also have

$$\#\{v \in \sigma \mid v \le b\} - \#\{v \in \hat\sigma \mid v \le b\} \le \hat{R} - R_w + \tilde{\mathfrak{P}}. \tag{6.163}$$

Combining (6.159) and (6.163), we derive the right estimate in (6.157). To prove the left estimate, we replace the roles of the problems in the proof given above. Then we have

$$\#\{v \in \hat\sigma \mid v \le b\} - \#\{v \in \sigma \mid v \le b\} \le \hat{K} := \min\{\hat{\mathfrak{P}}, R - R_w + \tilde{\mathfrak{P}}\}, \tag{6.164}$$

where by Theorem 3.5(ix)

$$\begin{aligned} \hat{\mathfrak{P}} &:= \mu\left(\mathcal{V}_{N+1}\langle Z_{N+1}^{[0]}(\lambda)\rangle, \mathcal{V}_{N+1}\hat{\mathcal{V}}_{N+1}^{-1}(0\ I)^T\right) \\ &= R_w - \mu\left(\hat{\mathcal{V}}_{N+1}\langle Z_{N+1}^{[0]}(\lambda)\rangle, \hat{\mathcal{V}}_{N+1}\mathcal{V}_{N+1}^{-1}(0\ I)^T\right) = R_w - \tilde{\mathfrak{P}}, \quad \lambda < \lambda_0; \end{aligned}$$

compare with (6.127). Using the last relation, we see that $\hat{K}$ defined in (6.164) is the same as in (6.157). Finally, the estimate for the sum $\hat{K} + K$ can be proved by direct computations. Indeed, the values of $K$ and $\hat{K}$ depend on the signs of $R_w - \hat{R}$ and $R_w - R$, respectively. So we have

$$\hat{K} + K = \begin{cases} R + \hat{R} - R_w, & R_w > \max\{R, \hat{R}\}, \\ R, & R < R_w \le \hat{R}, \\ \hat{R}, & R \ge R_w > \hat{R}, \\ R_w, & R_w \le \min\{R, \hat{R}\}. \end{cases}$$

From the formula above, we derive the last estimate in (6.157). By Proposition 6.33 we then have that (6.132) hold with $K$ and $\hat{K}$ given by (6.157). The proof is completed. $\qquad\square$

The last result for the case of joint endpoints concerns the strict inequalities in (6.132).

**Theorem 6.44** *Under the assumptions of Theorem 6.43, suppose additionally*

$$\hat{R} \leq R_w. \tag{6.165}$$

*Then the first inequality in (6.132) is strict, i.e., $\hat{\lambda}_j < \lambda_{j+K}$ for $j \geq 1$. Similarly, the condition*

$$R \leq R_w \tag{6.166}$$

*implies that the second inequality in (6.132) is strict, i.e., $\lambda_p < \hat{\lambda}_{p+\hat{K}}$ for $p \geq 1$.*

*Proof* From (6.165) it follows that for $K$ given by (6.157), we have for $\lambda < \lambda_0$

$$K = \hat{R} - R_w + \tilde{\mathfrak{P}} = \mu^*\big(\hat{V}_{N+1} V_{N+1}^{-1}(0\ I)^T, \hat{V}_{N+1}\langle Z_{N+1}^{[0]}(\lambda)\rangle\big).$$

On the other hand, according to (6.159) and (6.162), we have for $\lambda < \lambda_0$

$$\#\{\nu \in \sigma \mid \nu \leq b\} - \#\{\nu \in \hat{\sigma} \mid \nu \leq b\}$$

$$= \operatorname{rank} \hat{\mathcal{L}}(\tau)\big|_\lambda^b + \mu^*\big(\hat{V}_{N+1} V_{N+1}^{-1}(0\ I)^T, \hat{V}_{N+1}\langle Z_{N+1}^{[0]}(b)\rangle\big)\big|_b^\lambda$$

$$= -\hat{\zeta}(b) + \operatorname{rank} \hat{\mathcal{L}}(\tau)\big|_\lambda^{b^-} + \mu^*\big(\hat{V}_{N+1} V_{N+1}^{-1}(0\ I)^T, \hat{V}_{N+1}\langle Z_{N+1}^{[0]}(b)\rangle\big)\big|_b^\lambda$$

$$\leq -\hat{\zeta}(b) + K,$$

where $\hat{\zeta}(b)$ is the multiplicity of $\hat{\lambda} = b \in \hat{\sigma}$, or $\hat{\zeta}(b) = 0$ if $b \notin \hat{\sigma}$, according to Definition 5.54. That is,

$$\hat{\zeta}(\lambda) = \operatorname{rank} \hat{\mathcal{L}}(\lambda^-) - \operatorname{rank} \hat{\mathcal{L}}(\lambda). \tag{6.167}$$

From the last estimate, we derive

$$\#\{\nu \in \sigma \mid \nu \leq b\} - \#\{\nu \in \hat{\sigma} \mid \nu < b\} \leq K,$$

or equivalently

$$\#\{\nu \in \hat{\sigma} \mid \nu < b\} - \#\{\nu \in \sigma \mid \nu \leq b\} \geq -K.$$

From the last inequality and Proposition 6.33, we obtain that $\hat{\lambda}_j < \lambda_{j+K}$ for $j \geq 1$. The proof of the second strict inequality is similar. $\qquad\square$

*Remark 6.45*

(i)  It follows from the results of Corollary 6.42 that between the finite eigenvalues
$\lambda_{j-\tilde{K}}, \lambda_{j+\tilde{K}} \in \sigma$ $(j > \tilde{K} \geq 0)$, there exists the finite eigenvalue $\hat{\lambda}_j \in \hat{\sigma}$.
Similarly, between the finite eigenvalues $\hat{\lambda}_{p-\tilde{K}}, \hat{\lambda}_{p+\tilde{K}} \in \hat{\sigma}$ $(p > \tilde{K} \geq 0)$,
there exists the finite eigenvalue $\lambda_p \in \sigma$. By estimate (6.155) we have
$\tilde{K} \leq 2n$, and then the distance between the indices of the finite eigenvalues
$\lambda_{j+\tilde{K}}, \lambda_{j-\tilde{K}} \in \sigma$ (or $\hat{\lambda}_{p+\tilde{K}}, \hat{\lambda}_{p-\tilde{K}} \in \hat{\sigma}$) is equal to $2\tilde{K} \leq 4n$. At the same time,
by Theorem 6.43, similar arguments applied to $\lambda_{j-\hat{K}}, \lambda_{j+K} \in \sigma$ $(j > \hat{K} \geq 0)$
imply that the distance between the indices of $\lambda_{j+K}, \lambda_{j-\hat{K}} \in \sigma$ $(j > \hat{K} \geq 0)$
is much smaller than $4n$, i.e., $K + \hat{K} \leq 2n$ (see (6.157)). A similar assertion
holds for $\hat{\lambda}_{p-K}, \hat{\lambda}_{p+\hat{K}} \in \hat{\sigma}$ $(p > K \geq 0)$.

(ii)  It follows from the formulas for the number $\hat{K}$ in (6.157) that $\hat{K} = 0$ if and
only if the comparative index

$$\mu\big(\hat{\mathcal{V}}_{N+1}\big(Z^{[0]}_{N+1}(\lambda)\big), \hat{\mathcal{V}}_{N+1}\mathcal{V}^{-1}_{N+1}(0\ I)^T\big) = \tilde{\mathfrak{P}}, \quad \lambda < \lambda_0,$$

attains one of its extremal values $R_w$ or $R$ (see property (vii) of Theorem 3.5
and (6.154)). For the case $\tilde{\mathfrak{P}} = R$, we have $\tilde{\mathfrak{P}} = R_w - R \geq 0$ (see (6.164)), and
then the inequality $\lambda_p < \hat{\lambda}_p$ $(p \geq 1)$ is strict according to the second assertion
in Theorem 6.44. A similar situation occurs for the constant $K$ when changing
the roles of the problems in (6.148). In this case we have to demand $\tilde{\mathfrak{P}} = 0$ or
$\tilde{\mathfrak{P}} = R_w - \hat{R} \geq 0$. Note that for the last case the inequality $\hat{\lambda}_j < \lambda_j$ $(j \geq 1)$ is
strict (see Theorem 6.44).

(iii)  The simplest sufficient condition for $K = 0$ in (6.157) is represented by the
majorant condition (6.146) (see Theorem 6.38) coupled with the condition $\hat{\ell} =
0$, where the constant $\hat{\ell}$ is defined in Theorem 6.28 (see (6.122)). Indeed, by
Remark 6.39 and (6.127), these conditions are sufficient for $\tilde{\mathfrak{P}} = 0$ and then
also for $K = 0$.

(iv)  One can construct an example of problems (6.148) with a piecewise continuously differentiable matrix $\mathcal{S}_k(\lambda)$ such that conditions (6.156) do not hold (see
Example 5.103).

Next we present an example showing that estimates (6.157) are exact.

*Example 6.46* Consider the eigenvalue problem for $n = 2$ and $N = 4$ with the
periodic boundary conditions

$$\hat{y}_{k+1} = \mathcal{S}_k(\lambda)\,\hat{y}_k, \quad \hat{y}_0 = \hat{y}_{N+1}, \tag{6.168}$$

with the matrix $\mathcal{S}_k$ given by

$$\mathcal{S}_k(\lambda) = \begin{pmatrix} 1 & 0 & 0 & 0 \\ 0 & 1 & 0 & 0 \\ 0 & 0 & 1 & 0 \\ 0 & -\lambda & 0 & 1 \end{pmatrix} \begin{pmatrix} 1 & 0 & 0 & 0 \\ 0 & 0 & 0 & 1 \\ 0 & 0 & 1 & 0 \\ 0 & -1 & 0 & 0 \end{pmatrix} = \begin{pmatrix} 1 & 0 & 0 & 0 \\ 0 & 0 & 0 & 1 \\ 0 & 0 & 1 & 0 \\ 0 & -1 & 0 & -\lambda \end{pmatrix}. \tag{6.169}$$

This eigenvalue problem is equivalent to the problem (see Remark 5.58)

$$\widehat{y}_{k+1} = \mathcal{S}_k(\lambda)\,\widehat{y}_k, \quad \begin{pmatrix} 0 & 0 \\ -I & I \end{pmatrix}\begin{pmatrix} \widehat{x}_0 \\ \widehat{x}_{N+1} \end{pmatrix} + \begin{pmatrix} -I & -I \\ 0 & 0 \end{pmatrix}\begin{pmatrix} -\widehat{u}_0 \\ \widehat{u}_{N+1} \end{pmatrix} = 0, \tag{6.170}$$

and consequently to the eigenvalue problem (5.180) with the separated endpoints

$$\left.\begin{aligned} \tilde{y}_{k+1} &= \{\mathcal{S}_k(\lambda)\}\,\tilde{y}_k, \\[4pt] \begin{pmatrix} 0 & 0 \\ -I & I \end{pmatrix}\tilde{x}_0 + \begin{pmatrix} -I & -I \\ 0 & 0 \end{pmatrix}\tilde{u}_0 &= 0, \\[4pt] \begin{pmatrix} 0 & 0 \\ -I & I \end{pmatrix}\tilde{x}_{N+1} + \begin{pmatrix} -I & -I \\ 0 & 0 \end{pmatrix}\tilde{u}_{N+1} &= 0. \end{aligned}\right\} \tag{6.171}$$

The matrix $\mathcal{Y}_k = \langle Z_k \rangle$ with $Z_0 = I$ is a conjoined basis of this system satisfying the required boundary condition at $k = 0$. Hence, the matrix $\hat{\mathcal{L}}(\lambda)$ (see (5.221) and Remark 5.58) for this problem is of the form

$$\hat{\mathcal{L}}(\lambda) = \begin{pmatrix} 0 & 0 & -I & -I \\ -I & I & 0 & 0 \end{pmatrix}\langle Z_{N+1}\rangle = -\mathcal{J}[Z_{N+1}(\lambda) - I], \tag{6.172}$$

and the symplectic fundamental matrix of (6.168) is of the form

$$Z_0(\lambda) = I, \quad Z_1(\lambda) = \begin{pmatrix} 1 & 0 & 0 & 0 \\ 0 & 0 & 0 & 1 \\ 0 & 0 & 1 & 0 \\ 0 & -1 & 0 & -\lambda \end{pmatrix},$$

$$Z_2(\lambda) = \begin{pmatrix} 1 & 0 & 0 & 0 \\ 0 & -1 & 0 & -\lambda \\ 0 & 0 & 1 & 0 \\ 0 & \lambda & 0 & -1+\lambda^2 \end{pmatrix}, \quad Z_3(\lambda) = \begin{pmatrix} 1 & 0 & 0 & 0 \\ 0 & \lambda & 0 & -1+\lambda^2 \\ 0 & 0 & 1 & 0 \\ 0 & 1-\lambda^2 & 0 & 2\lambda & \lambda^3 \end{pmatrix},$$

$$Z_4(\lambda) = \begin{pmatrix} 1 & 0 & 0 & 0 \\ 0 & 1-\lambda^2 & 0 & 2\lambda-\lambda^3 \\ 0 & 0 & 1 & 0 \\ 0 & -2\lambda+\lambda^3 & 0 & 1-3\lambda^2+\lambda^4 \end{pmatrix},$$

$$Z_5(\lambda) = \begin{pmatrix} 1 & 0 & 0 & 0 \\ 0 & -2\lambda+\lambda^3 & 0 & 1-3\lambda^2+\lambda^4 \\ 0 & 0 & 1 & 0 \\ 0 & -1+3\lambda^2-\lambda^4 & 0 & -3\lambda+4\lambda^3-\lambda^5 \end{pmatrix}.$$

Then

$$\text{rank}\,[Z_5(\lambda) - I] = \text{rank}\,T(\lambda),$$

$$T(\lambda) = \begin{pmatrix} -2\lambda + \lambda^3 - 1 & 1 - 3\lambda^2 + \lambda^4 \\ -1 + 3\lambda^2 - \lambda^4 & -3\lambda + 4\lambda^3 - \lambda^5 - 1 \end{pmatrix},\quad \det T(\lambda) = (\lambda + 2)(\lambda^2 - \lambda - 1)^2.$$

The spectrum $\hat{\sigma}$ is given by the equation $\det T(\lambda) = 0$, so that

$$\hat{\sigma} = \left\{ \hat{\lambda}_1 = -2,\ \hat{\lambda}_2 = \hat{\lambda}_3 = \frac{1}{2} - \frac{\sqrt{5}}{2},\ \hat{\lambda}_4 = \hat{\lambda}_5 = \frac{1}{2} + \frac{\sqrt{5}}{2} \right\}.$$

We compare the spectrum $\hat{\sigma}$ of this problem with the spectrum $\sigma$ of the following eigenvalue problem

$$y_{k+1} = \mathcal{S}_k(\lambda)\, y_k, \quad u_0 = x_{N+1},\ x_0 = -u_{N+1}. \tag{6.173}$$

This eigenvalue problem we can rewrite into the form

$$\left. \begin{aligned} y_{k+1} &= \mathcal{S}_k(\lambda)\, y_k, \\ \begin{pmatrix} 0 & -I \\ -I & 0 \end{pmatrix} \begin{pmatrix} x_0 \\ x_{N+1} \end{pmatrix} &+ \begin{pmatrix} -I & 0 \\ 0 & -I \end{pmatrix} \begin{pmatrix} -u_0 \\ u_{N+1} \end{pmatrix} = 0, \end{aligned} \right\} \tag{6.174}$$

which is equivalent to the eigenvalue problem with separated boundary conditions

$$\left. \begin{aligned} \bar{y}_{k+1} &= \{\mathcal{S}_k(\lambda)\}\, \bar{y}_k, \\ \begin{pmatrix} 0 & 0 \\ -I & I \end{pmatrix} \bar{x}_0 &+ \begin{pmatrix} -I & -I \\ 0 & 0 \end{pmatrix} \bar{u}_0 = 0, \\ \begin{pmatrix} 0 & -I \\ -I & 0 \end{pmatrix} \bar{x}_{N+1} &+ \begin{pmatrix} -I & 0 \\ 0 & -I \end{pmatrix} \bar{u}_{N+1} = 0. \end{aligned} \right\} \tag{6.175}$$

The matrix $\mathcal{L}(\lambda)$ (see (5.221)), which we use to determine the finite eigenvalues, has the form

$$\mathcal{L}(\lambda) = \mathcal{J} - Z_{N+1}(\lambda). \tag{6.176}$$

Consequently, for $N = 4$ we have

$$\text{rank}\,[Z_5(\lambda) - \mathcal{J}] = \text{rank} \begin{pmatrix} 1 & 0 & -1 & 0 \\ 0 & -2\lambda + \lambda^3 & 0 & -3\lambda^2 + \lambda^4 \\ 1 & 0 & 1 & 0 \\ 0 & 3\lambda^2 - \lambda^4 & 0 & -3\lambda + 4\lambda^3 - \lambda^5 \end{pmatrix}$$

$$= 2 + \text{rank} \begin{pmatrix} -2\lambda + \lambda^3 & -3\lambda^2 + \lambda^4 \\ 3\lambda^2 - \lambda^4 & -3\lambda + 4\lambda^3 - \lambda^5 \end{pmatrix},$$

with

$$\det \begin{pmatrix} -2\lambda + \lambda^3 & -3\lambda^2 + \lambda^4 \\ 3\lambda^2 - \lambda^4 & -3\lambda + 4\lambda^3 - \lambda^5 \end{pmatrix} = -2\lambda^2(\lambda^2 - 3).$$

Hence, the spectrum is

$$\sigma = \left\{ \lambda_1 = -\sqrt{3}, \ \lambda_2 = 0, \ \lambda_3 = 0, \ \lambda_4 = \sqrt{3} \right\}.$$

We have computed that the finite eigenvalues of (6.168), (6.173) satisfy

$$\lambda_{p-2} < \hat{\lambda}_p, \ p \in \{3, 4, 5\}, \quad \hat{\lambda}_j < \lambda_j, \ j \in \{1, 2, 3, 4\}, \tag{6.177}$$

where we have additionally that

$$\hat{\lambda}_2 = \hat{\lambda}_3 < \hat{\lambda}_4 = \hat{\lambda}_5 \tag{6.178}$$

and

$$\lambda_2 = \lambda_3. \tag{6.179}$$

Let us verify what we obtain as a result of the application of Theorems 6.41, 6.43, and 6.44. The matrices $\mathcal{V}_{N+1}$ and $\hat{\mathcal{V}}_{N+1}$ of the boundary conditions have the form

$$\hat{\mathcal{V}}_{N+1}^{-1} = \begin{pmatrix} 0 & -0.5I & I & 0 \\ 0 & 0.5I & I & 0 \\ -0.5I & 0 & 0 & -I \\ -0.5I & 0 & 0 & I \end{pmatrix}, \quad \mathcal{V}_{N+1}^{-1} = \begin{pmatrix} 0 & -0.5I & I & 0 \\ -0.5I & 0 & 0 & I \\ -0.5I & 0 & 0 & -I \\ 0 & -0.5I & -I & 0 \end{pmatrix}.$$

$$\tag{6.180}$$

and $R_w = 1$, $\hat{R} = 2$, $R = 4$. By Theorem 6.41 (see (6.153)), in any half-open interval $(a, b]$, the number of finite eigenvalues of the investigated problems differs at most by two. Since $\hat{R} < R_w$, we have $K = \hat{R} - R_w + \hat{\mathfrak{P}}$, and from $R = R_w$ we obtain $\hat{K} = R_w - \hat{\mathfrak{P}}$. Consequently, $\hat{K} + K = \hat{R} = 2$. We show that $K = 0$. To this end, note that

$$\mu^*\left( \hat{\mathcal{V}}_{N+1}^{-1}(0\ I)^T, \mathcal{V}_{N+1}^{-1}(0\ I)^T \right) = R_w - \mu^*\left( \mathcal{V}_{N+1}^{-1}(0\ I)^T, \hat{\mathcal{V}}_{N+1}^{-1}(0\ I)^T \right)$$

$$= 4 - \mu_2^*\left( \mathcal{V}_{N+1}^{-1}(0\ I)^T, \hat{\mathcal{V}}_{N+1}^{-1}(0\ I)^T \right)$$

$$= 4 - \mathrm{ind} \left[ \begin{pmatrix} I & 0 \\ I & 0 \end{pmatrix}^T \begin{pmatrix} 0 & -I \\ -I & 0 \end{pmatrix} \begin{pmatrix} I & 0 \\ I & 0 \end{pmatrix} \right] = 4 - 2 = 2.$$

Hence, the majorant condition (6.146) (see Remark 6.45(iii)) is not satisfied. Let us compute the constant $K$. For $\lambda < \lambda_0$ we have

$$K = \hat{R} - R_w + \tilde{\mathfrak{P}} = \mu^*\left(\hat{\mathcal{V}}_{N+1}\mathcal{V}_{N+1}^{-1}(0\ I)^T, \hat{\mathcal{V}}_{N+1}\langle Z_5(\lambda)\rangle\right)$$

$$= \mu^*\left(\begin{pmatrix} I & I \\ -I & I \\ 0.5I & 0.5I \\ -0.5I & 0.5I \end{pmatrix}, \begin{pmatrix} \mathcal{J}[I - Z_5(\lambda)] \\ 0.5[I + Z_5(\lambda)] \end{pmatrix}\right)$$

$$= \mu^*\left(\begin{pmatrix} I & 0 \\ 0 & I \\ 0.5I & 0 \\ 0 & 0.5I \end{pmatrix} C, \begin{pmatrix} \mathcal{J}(I - Z_5(\lambda)) \\ 0.5(I + Z_5(\lambda)) \end{pmatrix}\right),$$

where the matrix $C = \begin{pmatrix} I & I \\ -I & I \end{pmatrix}$ is invertible and it can be neglected in view of the property (i) of Theorem 3.5. Consequently, for $\lambda < \lambda_0$ we have

$$K = \text{ind}\left\{[I - Z_5(\lambda)]^T[I - Z_5(\lambda)] - [I - Z_5(\lambda)]^T\mathcal{J}^T[I + Z_5(\lambda)]\right\}$$

$$= \text{ind}\left\{[I - Z_5(\lambda_0)]^T[I - Z_5(\lambda)] + Z_5(\lambda)^T\mathcal{J}^T + \mathcal{J}Z_5(\lambda)\right\} = \text{ind}\,T_1(\lambda),$$

where we have used that $Z_k(\lambda)$ is a symplectic matrix. The last index leads to the computation of the index of the matrix

$$T_1(\lambda) = A^T A + B + B^T, \quad A := \begin{pmatrix} 2\lambda - \lambda^3 + 1 & -1 + 3\lambda^2 - \lambda^4 \\ 1 - 3\lambda^2 + \lambda^4 & 3\lambda - 4\lambda^3 + \lambda^5 + 1 \end{pmatrix},$$

$$B := \begin{pmatrix} -1 + 3\lambda^2 - \lambda^4 & -3\lambda + 4\lambda^3 - \lambda^5 \\ 2\lambda - \lambda^3 & -1 + 3\lambda^2 - \lambda^4 \end{pmatrix}.$$

We have that $\text{Tr}\,T_1(\lambda) \sim \lambda^{10}$, $\det T_1(\lambda) \sim -4\lambda^9$ as $\lambda \to \pm\infty$. Hence, the eigenvalues of $T_1(\lambda)$ with $\lambda$ sufficiently close to $-\infty$ are positive. Then $K = 0$, $\hat{K} = 2$, and by Theorem 6.44 we derive the same estimate (6.177). Finally, note that Theorems 6.41, 6.43, and 6.44 do not provide the information in (6.178) and (6.179). But if we know (6.178), then (6.177) gives the possibility to order the finite eigenvalues of these problems as

$$\hat{\lambda}_1 < \lambda_1 < \hat{\lambda}_2 = \hat{\lambda}_3 < \lambda_2 \leq \lambda_3 < \hat{\lambda}_4 = \hat{\lambda}_5 < \lambda_4.$$

So we see that this order is correct except of the fact that we cannot conclude whether $\lambda_2 = \lambda_3$ or $\lambda_2 < \lambda_3$.

### 6.2.3 Interlacing of Eigenvalues for Separated Endpoints

In this subsection we investigate the special case of (6.148) considering the problems with the separated boundary conditions, whose coefficient matrices do not depend on $\lambda$, i.e.,

$$
\left.
\begin{aligned}
y_{k+1}(\lambda) &= \mathcal{S}_k(\lambda)\, y_k(\lambda), \quad k \in [0, N]_{\mathbb{Z}}, \\
R_0^*\, x_0(\lambda) &+ R_0\, u_0(\lambda) = 0, \\
R_{N+1}^*\, x_{N+1}(\lambda) &+ R_{N+1}\, u_{N+1}(\lambda) = 0, \\
\hat{y}_{k+1}(\lambda) &= \mathcal{S}_k(\lambda)\, \hat{y}_k(\lambda), \quad k \in [0, N]_{\mathbb{Z}}, \\
\hat{R}_0^*\, \hat{x}_0(\lambda) &+ \hat{R}_0\, \hat{u}_0(\lambda) = 0, \\
\hat{R}_{N+1}^*\, \hat{x}_{N+1}(\lambda) &+ \hat{R}_{N+1}\, \hat{u}_{N+1}(\lambda) = 0.
\end{aligned}
\right\}
\tag{6.181}
$$

As in Sect. 6.2.2, we assume that the symplectic matrix $\mathcal{S}_k(\lambda)$ obeys the monotonicity condition (5.3) and condition (5.333) holds. Then by Corollary 5.123, the spectra $\sigma$ and $\hat{\sigma}$ of (6.148) are bounded from below.

Together with (6.181) we also consider two intermediate problems

$$
\left.
\begin{aligned}
\tilde{y}_{k+1}(\lambda) &= \mathcal{S}_k(\lambda)\, \tilde{y}_k(\lambda), \quad k \in [0, N]_{\mathbb{Z}}, \\
R_0^*\, \tilde{x}_0(\lambda) &+ R_0\, \tilde{u}_0(\lambda) = 0, \\
\hat{R}_{N+1}^*\, \tilde{x}_{N+1}(\lambda) &+ \hat{R}_{N+1}\, \tilde{u}_{N+1}(\lambda) = 0, \\
\bar{y}_{k+1}(\lambda) &= \mathcal{S}_k(\lambda)\, \bar{y}_k(\lambda), \quad k \in [0, N]_{\mathbb{Z}}, \\
\hat{R}_0^*\, \bar{x}_0(\lambda) &+ \hat{R}_0\, \bar{u}_0(\lambda) = 0, \\
R_{N+1}^*\, \bar{x}_{N+1}(\lambda) &+ R_{N+1}\, \bar{u}_{N+1}(\lambda) = 0,
\end{aligned}
\right\}
\tag{6.182}
$$

under the same assumptions for $\mathcal{S}_k(\lambda)$, so that the finite spectra $\tilde{\sigma}$, $\bar{\sigma}$ of (6.182) are also bounded from below. We introduce the notation for the ranks of the Wronskians of the boundary conditions in (6.181) by

$$
\left.
\begin{aligned}
r_L &:= \operatorname{rank}\left[ (R_0\; R_0^*)\, \mathcal{J}(\hat{R}_0\; \hat{R}_0^*)^T \right], \\
r_R &:= \operatorname{rank}\left[ (R_{N+1}\; R_{N+1}^*)\, \mathcal{J}(\hat{R}_{N+1}\; \hat{R}_{N+1}^*)^T \right],
\end{aligned}
\right\}
\tag{6.183}
$$

and the notation (see (5.205), (5.204), and Sect. 6.1.6)

$$
\left.
\begin{aligned}
r_{\mathcal{I}} &:= \max_{\lambda \in \mathcal{I}} \operatorname{rank} \Lambda(\lambda), \quad \Lambda(\lambda) = (R_{N+1}^*\; R_{N+1})\, Z_{N+1}^{[L]}(\lambda)\, (0\; I)^T, \\
\hat{r}_{\mathcal{I}} &:= \max_{\lambda \in \mathcal{I}} \operatorname{rank} \hat{\Lambda}(\lambda), \quad \hat{\Lambda}(\lambda) = (\hat{R}_{N+1}^*\; \hat{R}_{N+1})\, \hat{Z}_{N+1}^{[L]}(\lambda)\, (0\; I)^T,
\end{aligned}
\right\}
\tag{6.184}
$$

for $Z_k^{[L]}(\lambda)$ given by (6.102) and for similarly defined $\hat{Z}_k^{[L]}(\lambda)$ with $\hat{Z}_0^{[L]}(\lambda) = \hat{V}_{-1}$ with the constant matrix $\hat{V}_{-1}(\lambda) \equiv \hat{V}_{-1}$ in (6.101). Finally, in estimates for the finite eigenvalues of problems with the separated boundary conditions, we also use maximal rank conditions for (6.182) with the spectra $\tilde{\sigma}$ and $\bar{\sigma}$ in the form

$$
\left.
\begin{aligned}
\tilde{r}_{\mathcal{I}} &:= \max_{\lambda \in \mathcal{I}} \operatorname{rank} \tilde{\Lambda}(\lambda), \quad \tilde{\Lambda}(\lambda) = (\hat{R}_{N+1}^*\ \hat{R}_{N+1})\, Z_{N+1}^{[L]}(\lambda)\, (0\ I)^T, \\
\bar{r}_{\mathcal{I}} &:= \max_{\lambda \in \mathcal{I}} \operatorname{rank} \bar{\Lambda}(\lambda), \quad \bar{\Lambda}(\lambda) = (R_{N+1}^*\ R_{N+1})\, \hat{Z}_{N+1}^{[L]}(\lambda)\, (0\ I)^T.
\end{aligned}
\right\}
\tag{6.185}
$$

We also use the notation (see Remark 6.40(ii))

$$
\left.
\begin{aligned}
r &:= \sup_{\lambda \in \mathbb{R}} \operatorname{rank} \Lambda(\lambda) \le n, \quad \hat{r} := \sup_{\lambda \in \mathbb{R}} \operatorname{rank} \hat{\Lambda}(\lambda) \le n, \\
\tilde{r} &:= \sup_{\lambda \in \mathbb{R}} \operatorname{rank} \tilde{\Lambda}(\lambda) \le n, \quad \bar{r} := \sup_{\lambda \in \mathbb{R}} \operatorname{rank} \bar{\Lambda}(\lambda) \le n.
\end{aligned}
\right\}
\tag{6.186}
$$

Let $\lambda_0$ be given by the condition

$$
\lambda_0 < \min\{\lambda_1, \hat{\lambda}_1, \tilde{\lambda}_1\},
\tag{6.187}
$$

where $\tilde{\lambda}_1$ is the minimal finite eigenvalue of the first problem in (6.182) (we put $\tilde{\lambda}_1 := \infty$ if $\tilde{\sigma} = \emptyset$). Now we present a modification of Theorem 6.41 for the separated endpoints.

**Theorem 6.47** *Consider the pair of eigenvalue problems* (6.181) *under assumptions* (5.2), (5.3), (5.156) *for all the matrices in the boundary conditions, and* (5.333). *Then for any* $a, b \in \mathbb{R}$ *with* $a < b$, *we have*

$$
\left.
\begin{aligned}
\big|\#\{v \in \hat{\sigma} \mid a < v \le b\} &- \#\{v \in \sigma \mid a < v \le b\}\big| \\
&\le \min\{r_L, \hat{r}_{a,b}, \tilde{r}_{a,b}\} + \min\{r_R, r_{a,b}, \tilde{r}_{a,b}\} \\
&\le \min\{\tilde{r}_{\mathcal{I}}, \hat{r}_{\mathcal{I}}, r_L\} + \min\{\tilde{r}_{\mathcal{I}}, r_{\mathcal{I}}, r_R\} \le 2n, \\
\mathcal{I} &:= [a, b], \\
r_{a,b} &:= \max\{\operatorname{rank} \Lambda(a), \operatorname{rank} \Lambda(b)\}, \\
\hat{r}_{a,b} &:= \max\{\operatorname{rank} \hat{\Lambda}(a), \operatorname{rank} \hat{\Lambda}(b)\}, \\
\tilde{r}_{a,b} &:= \max\{\operatorname{rank} \tilde{\Lambda}(a), \operatorname{rank} \tilde{\Lambda}(b)\},
\end{aligned}
\right\}
\tag{6.188}
$$

*where* $\tilde{\Lambda}(\lambda)$ *given by* (6.185) *is associated with the first problem in* (6.182).

*Proof* For the proof we use Remark 6.26. By (6.112) and similar arguments as in the proof of Theorem 6.41, we derive

$$
\left.
\begin{aligned}
\#\{v \in \hat{\sigma} \mid a < v \le b\} &- \#\{v \in \sigma \mid a < v \le b\} \\
&= \mu^*\big(\hat{V}_{-1}^{-1} \hat{Y}_0^{[R]}(\lambda),\ \hat{V}_{-1}^{-1} V_{-1}(0\ I)^T\big)\big|_a^b \\
&\quad + \mu\big(\hat{V}_{N+1} Y_{N+1}^{[L]}(\lambda),\ \hat{V}_{N+1} V_{N+1}^{-1}(0\ I)^T\big)\big|_a^b.
\end{aligned}
\right\}
\tag{6.189}
$$

Note that the rank of the Wronskian for the first addend is

$$\text{rank } w(\hat{V}_{-1}^{-1}\hat{Y}_0^{[R]}(\lambda), \hat{V}_{-1}^{-1}V_{-1}(0\ I)^T)$$

$$= \text{rank } w((0\ I)^T, [\hat{Z}_0^{[R]}(\lambda)]^{-1}V_{-1}(0\ I)^T)$$

$$= \text{rank } w((0\ I)^T, \hat{V}_{N+1}Z_{N+1}^{[L]}(\lambda)(0\ I)^T) = \text{rank } \tilde{\Lambda}(\lambda) = \tilde{r}(\lambda).$$

Then we estimate

$$\left| \mu^* \big( \hat{V}_{-1}^{-1}\hat{Y}_0^{[R]}(\lambda), \hat{V}_{-1}^{-1}V_{-1}(0\ I)^T \big) \big|_a^b \right|$$

$$\leq \max\{\min\{r_L, \tilde{r}(a)\}, \min\{r_L, \tilde{r}(b)\}\} = \min\{r_L, \tilde{r}_{a,b}\},$$

and using the relation

$$\mu^* \big( \hat{V}_{-1}^{-1}\hat{Y}_0^{[R]}(\lambda), \hat{V}_{-1}^{-1}V_{-1}(0\ I)^T \big) \big|_a^b = \mu^* \big( V_{-1}^{-1}\hat{Y}_0^{[R]}(\lambda), V_{-1}^{-1}\hat{V}_{-1}(0\ I)^T \big) \big|_b^a,$$
$$(6.190)$$

where we apply Theorem 3.5(ix), we derive a similar estimate

$$\left| \mu^* \big( \hat{V}_{-1}^{-1}\hat{Y}_0^{[R]}(\lambda), \hat{V}_{-1}^{-1}V_{-1}(0\ I)^T \big) \big|_a^b \right| = \left| \mu^* \big( V_{-1}^{-1}\hat{Y}_0^{[R]}(\lambda), V_{-1}^{-1}\hat{V}_{-1}(0\ I)^T \big) \big|_b^a \right|$$

$$\leq \max\{\min\{r_L, \hat{r}(a)\}, \min\{r_L, \hat{r}(b)\}\} = \min\{r_L, \hat{r}_{a,b}\}.$$

Combining the estimates derived above, we obtain the estimate for the first addend in (6.189) in the form

$$\left| \mu^* \big( \hat{V}_{-1}^{-1}\hat{Y}_0^{[R]}(\lambda), \hat{V}_{-1}^{-1}V_{-1}(0\ I)^T \big) \big|_a^b \right| \leq \min\{\tilde{r}_{a,b}, \hat{r}_{a,b}, r_L\}. \qquad (6.191)$$

In a similar way, we derive the estimate for the second addend in (6.189), so that finally we prove (6.188). □

By analogy with the consideration for the case of the joint endpoints, we recall that (5.333) guaranties that the spectra of (6.181) and (6.182) are bounded from below (see Corollary 5.118). Then one can formulate the following corollary to Theorem 6.47.

**Corollary 6.48** *Under the assumptions of Theorem 6.47, we have for any $b \in \mathbb{R}$*

$$\left. \begin{array}{l} \big|\#\{v \in \hat{\sigma} \mid v \leq b\} - \#\{v \in \sigma \mid v \leq b\}\big| \leq \tilde{K}, \\ \tilde{K} := \min\{\hat{r}, \tilde{r}, r_L\} + \min\{r, \tilde{r}, r_R\} \leq 2n, \end{array} \right\} \qquad (6.192)$$

*with the constants* $r$, $\hat{r}$, $\tilde{r}$, $r_L$, $r_R$ *given by* (6.186) *and* (6.183). *In particular, inequalities* (6.132) *hold with* $K = \hat{K} := \tilde{K}$, *where* $\tilde{K}$ *is defined by* (6.192).

Now we specify the results of Corollary 6.48 by imposing more restrictions on the behavior of $\Lambda(\lambda)$, $\hat{\Lambda}(\lambda)$, $\tilde{\Lambda}(\lambda)$ with respect to $\lambda$. For example, the restrictions below hold for problems (6.181) with the matrix $\mathcal{S}_k(\lambda)$ being a polynomial in $\lambda$, in particular, for problems with the special linear dependence in $\lambda$ (see Sect. 5.3).

**Theorem 6.49** *Under the assumptions of Theorem 6.47, suppose additionally that for* $r$, $\hat{r}$, $\tilde{r}$ *given by* (6.186) *we have*

$$\left.\begin{aligned} r &:= \sup_{\lambda \in \mathbb{R}} \operatorname{rank} \Lambda(\lambda) = \operatorname{rank} \Lambda(\lambda), \\ \hat{r} &:= \sup_{\lambda \in \mathbb{R}} \operatorname{rank} \hat{\Lambda}(\lambda) = \operatorname{rank} \hat{\Lambda}(\lambda), \\ \tilde{r} &:= \sup_{\lambda \in \mathbb{R}} \operatorname{rank} \tilde{\Lambda}(\lambda) = \operatorname{rank} \tilde{\Lambda}(\lambda), \\ \lambda &< \lambda_0 = \min\{\lambda_1, \hat{\lambda}_1, \tilde{\lambda}_1\}. \end{aligned}\right\} \tag{6.193}$$

*Then*

$$\left.\begin{aligned} \hat{K} &\leq \#\{\nu \in \sigma \mid \nu \leq b\} - \#\{\nu \in \hat{\sigma} \mid \nu \leq b\} \leq K, \\ K = K_1 + K_2 &:= \min\{P_1, \hat{r} - r_L + P_1\} + \min\{P_2, \tilde{r} - r_R + P_2\} \geq 0, \\ \hat{K} = \hat{K}_1 + \hat{K}_2 &:= \min\{r_L - P_1, \tilde{r} - P_1\} + \min\{r_R - P_2, r - P_2\} \geq 0, \\ \hat{K} + K &\leq \min\{r_L, \hat{r}, \tilde{r}\} + \min\{r_R, r, \tilde{r}\} \leq 2n, \end{aligned}\right\} \tag{6.194}$$

*where the constants* $P_1$ *and* $P_2$ *are given by* (6.113). *In particular, inequalities* (6.132) *for the finite eigenvalues of* (6.181) *hold with* $K$ *and* $\hat{K}$ *given by* (6.194).

*Proof* In the proof we follow the approach developed in the proof of Theorem 6.43. Under assumptions (6.193) we have for $b \in \mathbb{R}$

$$\operatorname{rank} \Lambda(\tau)\big|_\lambda^b \leq 0, \quad \operatorname{rank} \hat{\Lambda}(\tau)\big|_\lambda^b \leq 0, \quad \operatorname{rank} \tilde{\Lambda}(\tau)\big|_\lambda^b \leq 0, \quad \lambda < \lambda_0. \tag{6.195}$$

Next we rewrite (6.112) and (6.113) in the form

$$\left.\begin{aligned} &\#\{\nu \in \sigma \mid \nu \leq b\} - \#\{\nu \in \hat{\sigma} \mid \nu \leq b\} \\ &= \mu^*\big(\hat{V}_{-1}^{-1}\hat{Y}_0^{[R]}(\tau), \, \hat{V}_{-1}^{-1}V_{-1}(0\ I)^T\big)\big|_b^\lambda \\ &\quad + \mu\big(\hat{V}_{N+1}Y_{N+1}^{[L]}(\tau), \, \hat{V}_{N+1}V_{N+1}^{-1}(0\ I)^T\big)\big|_b^\lambda, \quad \lambda < \lambda_0. \end{aligned}\right\} \tag{6.196}$$

Then we have for the addends in (6.196) (compare with (6.160))

$$\left.\begin{aligned} \mu^*\big(\hat{V}_{-1}^{-1}\hat{Y}_0^{[R]}(\tau), \, \hat{V}_{-1}^{-1}V_{-1}(0\ I)^T\big)\big|_b^\lambda &\leq P_1, \\ \mu\big(\hat{V}_{N+1}Y_{N+1}^{[L]}(\tau), \, \hat{V}_{N+1}V_{N+1}^{-1}(0\ I)^T\big)\big|_b^\lambda &\leq P_2. \end{aligned}\right\} \tag{6.197}$$

On the other hand, the comparative indices above can be rewritten in the form (compare with (6.161))

$$
\left.
\begin{aligned}
&\mu^* \left( \hat{V}_{-1}^{-1} \hat{Y}_0^{[R]}(\lambda), \ \hat{V}_{-1}^{-1} V_{-1}(0 \ I)^T \right) \\
&\quad = r_L - \operatorname{rank} \hat{\Lambda}(\lambda) + \mu \left( \hat{V}_{-1}^{-1} V_{-1}(0 \ I)^T, \ \hat{V}_{-1}^{-1} \hat{Y}_0^{[R]}(\lambda) \right), \\
&\mu \left( \hat{V}_{N+1} Y_{N+1}^{[L]}(\lambda), \ \hat{V}_{N+1} V_{N+1}^{-1}(0 \ I)^T \right) \\
&\quad = r_R - \operatorname{rank} \tilde{\Lambda}(\lambda) + \mu^* \left( \hat{V}_{N+1} V_{N+1}^{-1}(0 \ I)^T, \ \hat{V}_{N+1} Y_{N+1}^{[L]}(\lambda) \right),
\end{aligned}
\right\}
\tag{6.198}
$$

where we apply Theorem 3.5(vi). Using (6.198) and (6.195), we have for the addends in (6.196) for $\lambda < \lambda_0$ that

$$
\begin{aligned}
&\mu^* \left( \hat{V}_{-1}^{-1} \hat{Y}_0^{[R]}(\tau), \ \hat{V}_{-1}^{-1} V_{-1}(0 \ I)^T \right) \Big|_b^\lambda \\
&\quad = \operatorname{rank} \hat{\Lambda}(\tau) \Big|_\lambda^b + \mu \left( \hat{V}_{-1}^{-1} V_{-1}(0 \ I)^T, \ \hat{V}_{-1}^{-1} \hat{Y}_0^{[R]}(\tau) \right) \Big|_b^\lambda \\
&\quad \leq \mu \left( \hat{V}_{-1}^{-1} V_{-1}(0 \ I)^T, \ \hat{V}_{-1}^{-1} \hat{Y}_0^{[R]}(\lambda) \right) = \hat{r} - r_L + P_1.
\end{aligned}
$$

So we have proved

$$
\mu^* \left( \hat{V}_{-1}^{-1} \hat{Y}_0^{[R]}(\tau), \ \hat{V}_{-1}^{-1} V_{-1}(0 \ I)^T \right) \Big|_b^\lambda \leq \hat{r} - r_L + P_1, \quad \lambda < \lambda_0.
\tag{6.199}
$$

In a similar way we prove the estimate for the second addend in (6.196)

$$
\mu \left( \hat{V}_{N+1} Y_{N+1}^{[L]}(\tau), \ \hat{V}_{N+1} V_{N+1}^{-1}(0 \ I)^T \right) \Big|_b^\lambda \leq \tilde{r} - r_R + P_2, \quad \lambda < \lambda_0.
\tag{6.200}
$$

Combining (6.197), (6.199), and (6.200), we derive the upper estimate in (6.194) with the constant $K$. To prove the lower estimate, we rewrite (6.112) in the equivalent form

$$
\left.
\begin{aligned}
&\#\{\nu \in \hat{\sigma} \mid \nu \leq b\} - \#\{\nu \in \sigma \mid \nu \leq b\} \\
&\quad = \mu^* \left( V_{-1}^{-1} \hat{Y}_0^{[R]}(\tau), \ V_{-1}^{-1} \hat{V}_{-1}(0 \ I)^T \right) \Big|_b^\lambda \\
&\qquad + \mu \left( V_{N+1} Y_{N+1}^{[L]}(\tau), \ V_{N+1} \hat{V}_{N+1}^{-1}(0 \ I)^T \right) \Big|_b^\lambda, \quad \lambda < \lambda_0,
\end{aligned}
\right\}
\tag{6.201}
$$

which is derived using (6.190) for the first addend in (6.112) and a similar relation

$$
\left.
\begin{aligned}
&\mu \left( \hat{V}_{N+1} Y_{N+1}^{[L]}(\tau), \ \hat{V}_{N+1} V_{N+1}^{-1}(0 \ I)^T \right) \Big|_a^b \\
&\quad = \mu \left( V_{N+1} Y_{N+1}^{[L]}(\tau), \ V_{N+1} \hat{V}_{N+1}^{-1}(0 \ I)^T \right) \Big|_b^a
\end{aligned}
\right\}
\tag{6.202}
$$

for the second one (recall that we apply Theorem 3.5(ix)). Repeating the same arguments as in the proof of the upper estimate and using formula (6.201) instead of (6.196), we derive (compare with (6.164))

$$
\left.
\begin{aligned}
\# \, \{v \in \hat{\sigma} \mid v \le b\} & - \# \{v \in \sigma \mid v \le b\} \\
& \le \hat{K} = \min\{\hat{P}_1, \tilde{r} - r_L + \hat{P}_1\} + \min\{\hat{P}_2, r - r_R + \hat{P}_2\}, \quad \lambda < \lambda_0,
\end{aligned}
\right\}
\tag{6.203}
$$

where $\hat{P}_1 = r_L - P_1$ and $\hat{P}_2 = r_R - P_2$ by Theorem 3.5(ix). Substituting the last representations into (6.203), we see that the constant $\hat{K}$ in (6.203) is the same as in the lower estimate in (6.194). The upper bound of the sum $K + \hat{K}$ is derived by analogy with the proof of the similar estimate in Theorem 6.43 (see (6.157)). By Proposition 6.33 we have that (6.132) hold with $K$ and $\hat{K}$ given by (6.194). The proof is completed.                                                                   □

It follows from Theorem 6.49 that the finite eigenvalues of the intermediate problem in (6.182) with the spectrum $\tilde{\sigma}$ separates the finite eigenvalues of (6.181). So we have the following corollary.

**Corollary 6.50** *Under the assumptions of Theorem 6.49, suppose that there exists the finite eigenvalue $\lambda_{j+K} \in \sigma$ ($j \ge 1$), where $K \ge 0$ is defined in (6.194). Then there exist the finite eigenvalues $\tilde{\lambda}_{j+K_1} \in \tilde{\sigma}$ and $\hat{\lambda}_j \in \hat{\sigma}$ and the inequality*

$$
\hat{\lambda}_j \le \tilde{\lambda}_{j+K_1} \le \lambda_{j+K},
\tag{6.204}
$$

*holds with $K_1 = \min\{P_1, \hat{r} - r_L + P_1\}$, where $K_1$ is the first addend in the sum $K = K_1 + K_2$ given by (6.194). Similarly, if $\hat{\lambda}_{p+\hat{K}}$ ($p \ge 1$) exists, then there exist $\tilde{\lambda}_{p+\hat{K}_2} \in \tilde{\sigma}$ and $\lambda_p \in \sigma$, and we have the inequality*

$$
\lambda_p \le \tilde{\lambda}_{p+\hat{K}_2} \le \hat{\lambda}_{p+\hat{K}},
\tag{6.205}
$$

*where $\hat{K}_2$ is the second addend in the sum $\hat{K} = \hat{K}_1 + \hat{K}_2$ given by (6.194).*

*Proof* Replacing $\sigma$ by $\tilde{\sigma}$ in (6.196), we get

$$
\left.
\begin{aligned}
\#\{v \in \tilde{\sigma} \mid v \le b\} & - \#\{v \in \hat{\sigma} \mid v \le b\} \\
& = \mu^* \big( \hat{V}_{-1}^{-1} \hat{Y}_0^{[R]}(\tau), \hat{V}_{-1}^{-1} V_{-1}(0 \ I)^T \big) \big|_b^\lambda, \quad \lambda \le \lambda_0.
\end{aligned}
\right\}
\tag{6.206}
$$

i.e., we cancel the second addend in (6.196). For the first addend in (6.196), we have derived the estimates (6.197) and (6.199). That is, we have the inequality

$$
\#\{v \in \tilde{\sigma} \mid v \le b\} - \#\{v \in \hat{\sigma} \mid v \le b\} \le K_1,
$$

or equivalently the inequality

$$
\#\{v \in \hat{\sigma} \mid v \le b\} - \#\{v \in \tilde{\sigma} \mid v \le b\} \ge -K_1 \ge 0.
$$

Then, by Proposition 6.33 we have $\hat{\lambda}_j \leq \tilde{\lambda}_{j+K_1}$ provided $\tilde{\lambda}_{j+K_1} \in \tilde{\sigma}$ exists. Next we replace $\hat{\sigma}$ by $\tilde{\sigma}$ in (6.196) to get

$$
\left.\begin{aligned}
&\#\{v \in \sigma \mid v \leq b\} - \#\{v \in \tilde{\sigma} \mid v \leq b\} \\
&\quad = \mu\big(\hat{V}_{N+1} Y_{N+1}^{[L]}(\tau),\, \hat{V}_{N+1} V_{N+1}^{-1}(0\ I)^T\big)\big|_b^\lambda, \quad \lambda \leq \lambda_0.
\end{aligned}\right\}
\tag{6.207}
$$

i.e., we cancel the first addend in (6.196). Then, arguing as above we derive the inequality $\tilde{\lambda}_p \leq \lambda_{p+K_2}$, where $K_2$ is the second addend in the sum for $K = K_1 + K_2$ (see (6.194)). Applying the last inequality to the index $p := j + K_1$, we derive $\hat{\lambda}_j \leq \tilde{\lambda}_{j+K_1} \leq \lambda_{j+K}$ or (6.204). For the proof of (6.205), we use (6.201). For the first step, we replace $\sigma$ by $\tilde{\sigma}$ deriving $\tilde{\lambda}_j \leq \hat{\lambda}_{j+\hat{K}_1}$. For the second step, we replace $\hat{\sigma}$ by $\tilde{\sigma}$ and derive $\lambda_p \leq \tilde{\lambda}_{p+\hat{K}_2}$. By putting $j := p + \hat{K}_2$ we derive (6.205). The proof is completed. $\qquad\square$

Next we formulate sufficient conditions for the strict inequalities in (6.132).

**Theorem 6.51** *Under the assumptions of Theorem 6.49, suppose that one of the conditions*

$$
\hat{r} \leq r_L \quad or \quad \tilde{r} \leq r_R
\tag{6.208}
$$

*holds. Then we have instead of* (6.204) *the inequalities*

$$
\hat{\lambda}_j < \tilde{\lambda}_{j+K_1} \leq \lambda_{j+K} \quad or \quad \hat{\lambda}_j \leq \tilde{\lambda}_{j+K_1} < \lambda_{j+K},
\tag{6.209}
$$

*and hence the strict inequality* $\hat{\lambda}_j < \lambda_{j+K}$ *in* (6.132) *holds. Similarly, if one of the conditions*

$$
\tilde{r} \leq r_L \quad or \quad r \leq r_R
\tag{6.210}
$$

*holds, then*

$$
\lambda_p \leq \tilde{\lambda}_{p+\hat{K}_2} < \hat{\lambda}_{p+\hat{K}} \quad or \quad \lambda_p < \tilde{\lambda}_{p+\hat{K}_2} \leq \hat{\lambda}_{p+\hat{K}},
\tag{6.211}
$$

*and we have the strict inequality* $\lambda_p < \hat{\lambda}_{p+\hat{K}}$ *in* (6.132).

*Proof* If $\hat{r} \leq r_L$ holds, then using (6.206) and the same arguments as in case of general boundary conditions (see the proof of Theorem 6.44), we show that $\hat{\lambda}_j < \tilde{\lambda}_{j+K_1}$. Then by combining this estimate with (6.204), we prove the first inequality in (6.209). Similarly, using $\tilde{r} \leq r_R$ and (6.207), by analogy with the proof of Theorem 6.44, we derive $\tilde{\lambda}_p < \lambda_{p+K_2}$. Then, by putting $p := j + K_1$ and combining this inequality with (6.204), we derive the second inequality in (6.209). Inequalities (6.211) and $\lambda_p < \hat{\lambda}_{p+\hat{K}}$, which are connected to at least one of the conditions in (6.210), can be proved analogously by using (6.201). $\qquad\square$

*Remark 6.52*

(i) Observe that the inequalities of Theorem 6.43 for the joint endpoints are improved by formulas of Theorem 6.49 for the separated boundary conditions. Indeed, we have the estimate

$$\min\{P_1, \hat{r} - r_L + P_1\} + \min\{P_2, \tilde{r} - r_R + P_2\}$$
$$\leq \min\{P_1 + P_2, \hat{r} + n - (r_L + r_R) + P_1 + P_2\}$$
$$= \min\{\tilde{\mathfrak{P}}, \hat{R} - R_w + \tilde{\mathfrak{P}}\},$$

where the constants $\tilde{\mathfrak{P}}$, $\hat{R}$, $R_w$, determined by formulas (6.127), (6.152), (6.149), are evaluated for the case, when the boundary conditions are separated. For the proof we use $R_w = r_L + r_R$ (see (5.154)) and

$$\operatorname{rank} \hat{\mathcal{L}}(\lambda) = n + \operatorname{rank} \hat{\Lambda}(\lambda), \tag{6.212}$$

where $\hat{\mathcal{L}}(\lambda)$ in (5.221) is computed for the case of the separated boundary conditions and $\hat{\Lambda}(\lambda)$ is defined by (5.205). Indeed, for this case we have by (5.439) and (3.105)

$$\hat{\mathcal{V}}_{N+1} \langle Z_{N+1}^{[0]}(\lambda) \rangle = \langle \hat{V}_{N+1} Z_{N+1}^{[0]}(\lambda) \, \hat{V}_{-1} \rangle \hat{V}_{-1}^{-1},$$

and then using the relation $\operatorname{rank} \hat{\mathcal{L}}(\lambda) = \operatorname{rank}(I \ 0) \langle \hat{V}_{N+1} Z_{N+1}^{[0]}(\lambda) \, \hat{V}_{-1} \rangle$, we derive (6.212). Moreover, by using Lemma 3.39 (see (3.115)), one can show that the number $\tilde{\mathfrak{P}}$ given by (6.127) for the case of the separated boundary conditions takes the form $\tilde{\mathfrak{P}} = P_1 + P_2$ with the constants $P_1$ and $P_2$ given by (6.113). Similarly,

$$\min\{r_L - P_1, \tilde{r} - P_1\} + \min\{r_R - P_2, r - P_2\}$$
$$\leq \min\{r_L + r_R - P_1 - P_2, n + r - P_1 - P_2\}$$
$$= \min\{R_w - \tilde{\mathfrak{P}}, R - \tilde{\mathfrak{P}}\}.$$

Consequently, inequalities of Theorem 6.43 are improved by formulas of Theorem 6.49 for separated boundary conditions.

(ii) Observe that the consideration above was associated with the first intermediate eigenvalue problem in (6.182) with the spectrum $\tilde{\sigma}$. But for the case of $\bar{r} < \tilde{r}$, the estimates in Theorem 6.49 can be improved by using the second problem in (6.182) with the spectrum $\bar{\sigma}$. For the derivation of the estimates associated with this problem, it is sufficient to replace the roles of problems in (6.181), i.e., replace $\sigma$ by $\hat{\sigma}$ and vice versa.

(iii) As it was already used in the proof of Corollary 6.50, the results in Theorems 6.47, 6.49, and 6.51 and in Corollaries 6.48 and 6.50 can be applied to the case when the problems in (6.181) differ in only one boundary condition. For example, if we consider the first problem in (6.181) with the spectrum $\sigma$ and the first problem in (6.182) with the spectrum $\tilde{\sigma}$ (instead of the second problem in (6.181)), then in all estimates of Theorems 6.47, 6.49, and 6.51 and Corollaries 6.48 and 6.50, we have $r_L = 0$, $P_1 = 0$, $K_1 = 0$, and then the first addend is zero. A similar situation will arise if we consider the second problem in (6.181) with the spectrum $\hat{\sigma}$ and replace the first problem with the spectrum $\sigma$ by the problem with the spectrum $\tilde{\sigma}$. In this case $r_R = 0$, $P_2 = 0$, $K_2 = 0$, and then in all estimates the second addend is zero.

For the sake of convenience (see also Remark 6.52(iii)), we reformulate Theorems 6.49 and 6.51 for a pair of eigenvalue problems (6.181) with the same boundary condition at $k = 0$ but different boundary conditions at $k = N + 1$, i.e.,

$$
\left.
\begin{aligned}
y_{k+1}(\lambda) &= \mathcal{S}_k(\lambda)\, y_k(\lambda), \quad k \in [0, N]_{\mathbb{Z}}, \\
R_0^* x_0(\lambda) &+ R_0\, u_0(\lambda) = 0, \\
R_{N+1}^* x_{N+1}(\lambda) &+ R_{N+1}\, u_{N+1}(\lambda) = 0, \\
\tilde{y}_{k+1}(\lambda) &= \mathcal{S}_k(\lambda)\, \tilde{y}_k(\lambda), \quad k \in [0, N]_{\mathbb{Z}}, \\
R_0^* \tilde{x}_0(\lambda) &+ R_0\, \tilde{u}_0(\lambda) = 0, \\
\hat{R}_{N+1}^* \tilde{x}_{N+1}(\lambda) &+ \hat{R}_{N+1}\, \tilde{u}_{N+1}(\lambda) = 0.
\end{aligned}
\right\}
\tag{6.213}
$$

**Theorem 6.53** *Let the parameters $r_R$, $r$, $\tilde{r}$ for eigenvalue problems (6.213) are determined by (6.183)–(6.186) and for $\lambda_0$ satisfying (6.187) (where we put $\hat{\lambda}_1 = \tilde{\lambda}_1$) assumption (6.193) for $r$ and $\tilde{r}$ holds. Then we have the estimates*

$$
\left.
\begin{aligned}
\hat{K}_2 \leq \#\{\nu \in \sigma \mid \nu \leq b\} &- \#\{\nu \in \tilde{\sigma} \mid \nu \leq b\} \leq K_2, \\
K_2 &:= \min\{P_2, \tilde{r} - r_R + P_2\}, \\
\hat{K}_2 &:= \min\{r_R - P_2, r - P_2\}, \\
\hat{K}_2 + K_2 &\leq \min\{r_R, r, \tilde{r}\} \leq n,
\end{aligned}
\right\}
\tag{6.214}
$$

*where $P_2$ is determined by (6.113) for $\hat{\sigma} := \tilde{\sigma}$. In particular, if there exists the finite eigenvalue $\lambda_{j+K_2} \in \sigma$ ($j \geq 1$) of the first problem in (6.213), then there exists the finite eigenvalue $\tilde{\lambda}_j \in \tilde{\sigma}$ of the second one with $\tilde{\lambda}_j \leq \lambda_{j+K_2}$. If*

$$
\tilde{r} \leq r_R \tag{6.215}
$$

*holds, then we have the strict inequality $\tilde{\lambda}_j < \lambda_{j+K_2}$. Similarly, the existence of $\tilde{\lambda}_{p+\hat{K}_2} \in \tilde{\sigma}$ implies that of $\lambda_p \in \sigma$ with the inequality $\lambda_p \leq \tilde{\lambda}_{p+\hat{K}_2}$. Moreover, if*

$$
r \leq r_R \tag{6.216}
$$

*holds, we have the strict inequality $\lambda_p < \tilde{\lambda}_{p+\hat{K}_2}$.*

*Remark 6.54* We note that Theorem 6.53 can be applied to derive the interlacing properties of the finite eigenvalues of problem (E) considered on the intervals $[0, N+1]_{\mathbb{Z}}$ and $[0, N+2]_{\mathbb{Z}}$. Since the theory developed in this subsection is devoted to the case, when the matrices in the boundary conditions do not depend on $\lambda$, we impose the restriction (6.35) for the blocks $\mathcal{A}_k(\lambda)$ and $\mathcal{B}_k(\lambda)$ (see Sect. 6.1.3), i.e., $\mathcal{A}_k(\lambda) \equiv \mathcal{A}_k$ and $\mathcal{B}_k(\lambda) \equiv \mathcal{B}_k$. In this case problem (E) considered on the subsequent intervals $[0, N+1]_{\mathbb{Z}}$ and $[0, N+2]_{\mathbb{Z}}$ is equivalent to the special case of (6.213) on the interval $[0, N+1]_{\mathbb{Z}}$ with the matrices

$$R_0 = R_{N+1} = 0, \quad R_0^* = R_{N+1}^* = I, \quad \hat{R}_{N+1}^* := \mathcal{A}_{N+1}, \quad \hat{R}_{N+1} := \mathcal{B}_{N+1}.$$

For this case we have $r_R = \text{rank}\, \mathcal{B}_{N+1}$ and $V_{N+1} = I$. By (6.113) the constant $P_2$ in Theorem 6.53 takes the form $P_2 = m_{N+1}(\lambda)$ for $\lambda < \lambda_0$, where $m_{N+1}(\lambda)$ is the multiplicity of a forward focal point of the principal solution $Y^{[0]}(\lambda)$ of the symplectic system in (6.213). In particular, for the nonsingular block $\mathcal{B}_{N+1}$, inequalities (6.215) and (6.216) are necessary satisfied, and then, by Theorem 6.53, we have strict interlacing properties for the finite eigenvalues of problem (E) considered on the subsequent intervals $[0, N+1]_{\mathbb{Z}}$ and $[0, N+2]_{\mathbb{Z}}$ (see Example 6.55).

*Example 6.55* Consider the Sturm-Liouville problem (6.9) on $[0, N+1]_{\mathbb{Z}}$ and the same problem on $[0, N+2]_{\mathbb{Z}}$, i.e., with the Dirichlet boundary conditions $x_0 = 0 = x_{N+2}$. Then, according to Remark 6.54, we have $r_R = \text{rank}\, \mathcal{B}_{N+1} = 1$, $P_2 = 0$, $\tilde{r} = r = 1$, and $K_2 = 0$, $\hat{K}_2 = 1$. Conditions (6.215) and (6.216) are satisfied and imply the strict inequalities

$$\lambda_j^{N+2} < \lambda_j^{N+1} < \lambda_{j+1}^{N+2}.$$

These inequalities are in full agreement with the classical results (see, e.g., [28, Chapter 4]) concerning the separation of roots of the polynomials $\det X_{N+1}^{[0]}(\lambda)$ and $\det X_{N+2}^{[0]}(\lambda)$. In a similar way, consider the Dirichlet eigenvalue problems on $[0, N+1]_{\mathbb{Z}}$ and $[0, N+2]_{\mathbb{Z}}$ for the matrix Sturm-Liouville equation

$$- \Delta(R_k \Delta x_k) + P_k x_{k+1} = \lambda W_k x_{k+1}, \quad x_0 = 0 = x_{N+1}, \tag{6.217}$$

with $\det R_k \neq 0$ and $W_k > 0$. Then, by Remark 6.54 and Corollary 5.69, we have $P_2 = 0$, $r_R = \text{rank}\, \mathcal{B}_{N+1} = n$, and $\tilde{r} = r = n$. Conditions (6.215) and (6.216) (sufficient for the strict inequalities) are then satisfied, $K_2 = 0$, $\hat{K}_2 = n$, and

$$\lambda_j^{N+2} < \lambda_j^{N+1} < \lambda_{j+n}^{N+2}.$$

Next we show an example of spectra of two eigenvalue problems with separated boundary conditions, for which estimates (6.194) are exact.

*Example 6.56* Consider the two problems with the matrix $\mathcal{S}_k(\lambda)$ defined in (6.169) and with the separated boundary conditions on $[0, N+1]_{\mathbb{Z}}$ in the form

$$u_0 = 0 = x_{N+1}, \tag{6.218}$$

$$x_0 = 0 = u_{N+1}. \tag{6.219}$$

Then for problem (6.169), (6.218) with the spectrum $\sigma$, we have $V_{-1} = \mathcal{J}$ and $V_{N+1} = I$ and $\Lambda(\lambda) = (I\ 0)\, Z_5(\lambda)\, (I\ 0)^T$. According to Example 6.46 with $N = 4$ (see the formula for the fundamental matrix $Z_5(\lambda)$), we have

$$\operatorname{rank} \Lambda(\lambda) = 1 + \operatorname{rank}(-2\lambda + \lambda^3),$$

and then

$$\sigma = \{\lambda_1 = -\sqrt{2},\ \lambda_2 = 0,\ \lambda_3 = \sqrt{2}\}.$$

Similarly, for problem (6.169), (6.219) with the spectrum $\hat{\sigma}$, we have $\hat{V}_{-1} = I$ and $\hat{V}_{N+1} = \mathcal{J}$ and $\hat{\Lambda}(\lambda) = (0\ I)\, Z_5(\lambda)\, (0\ I)^T$. According to Example 6.46 with $N = 4$,

$$\operatorname{rank} \hat{\Lambda}(\lambda) = 1 + \operatorname{rank}(-3\lambda + 4\lambda^3 - \lambda^5),$$

and then

$$\hat{\sigma} = \{\hat{\lambda}_1 = -\sqrt{3},\ \hat{\lambda}_2 = -1,\ \hat{\lambda}_3 = 0,\ \hat{\lambda}_4 = 1,\ \hat{\lambda}_5 = \sqrt{3}\}.$$

Let us compare the spectra $\sigma$ and $\hat{\sigma}$. It is easy to see that

$$\lambda_{j-2} < \hat{\lambda}_j\ (j \in \{3, 4, 5\}),\quad \hat{\lambda}_p < \lambda_p\ (p \in \{1, 2, 3\}). \tag{6.220}$$

Now we apply the general theory developed for separated boundary conditions. For the intermediate eigenvalue problems defined by (6.182), we have the boundary conditions

$$u_0 = 0 = u_{N+1}, \tag{6.221}$$

$$x_0 = 0 = x_{N+1}. \tag{6.222}$$

According to formula for $Z_5(\lambda)$ in Example 6.46, the spectra of the intermediate problems (6.169), (6.221) (with the spectrum $\tilde{\sigma}$) and (6.169), (6.222) (with the spectrum $\bar{\sigma}$) can be found by using the functions $\tilde{\Lambda}(\lambda) = (0\ I)\, Z_5(\lambda)\, (I\ 0)^T$ and $\bar{\Lambda}(\lambda) = (I\ 0)\, Z_5(\lambda)\, (0\ I)$, respectively, which have the same maximal ranks $\tilde{r} = \bar{r} = 1$ (and the same spectra $\tilde{\sigma} = \bar{\sigma}$). Finally, we have $r_L = r_R = 2$, $r = \hat{r} = 2$, $\tilde{r} = 1$, and by Theorem 6.49 then $K = P_1 + \tilde{r} - r_R + P_2 = P_1 + P_2 - 1$, where the constants $P_1$ and $P_2$ can be computed by using (6.113). Namely, we have

$$P_1 = \mu^*\left(Z_5^{-1}(\lambda)\,(I\ 0)^T,\,(I\ 0)^T\right) = \text{ind}\,[(1 - 3\lambda + \lambda^4)\,(-3\lambda + 4\lambda^3 - \lambda^5)]$$
$$= 0 \quad \text{for } \lambda \to -\infty,$$

and

$$P_2 = \mu\left(\mathcal{J}Z_5(\lambda)\,(I\ 0)^T,\,(I\ 0)^T\right) = 1 + \text{ind}\,[(-1 + 3\lambda^2 - \lambda^4)\,(-2\lambda + \lambda^3)]$$
$$= 1 \quad \text{for} \quad \lambda \to -\infty.$$

Then $K = 0$, and hence, by Theorem 6.51, the inequality $\hat{\lambda}_p < \lambda_p$ ($p \in \{1, 2, 3\}$) holds. Next, by (6.194) we have $\hat{K} = 2$, and then $\lambda_{j-2} < \hat{\lambda}_j$ ($j \in \{3, 4, 5\}$). Thus, we have proved that (6.220) holds by Theorems 6.49 and 6.51. Observe also that the sum of the comparative indices in (6.113) is

$$\mu\left(\hat{V}_{-1}(0\ I)^T,\ V_{-1}(0\ I)^T\right) + \mu^*\left(\hat{V}_{N+1}^{-1}(0\ I)^T,\ V_{N+1}^{-1}(0\ I)^T\right) = 2,$$

so that the majorant condition (6.137) used in Sect. 6.2.1 is not satisfied.

## 6.3 Symplectic Systems Without Controllability

In this section we will consider symplectic difference system (SDS) on a discrete interval of the form $[0, \infty)_{\mathbb{Z}}$, which is bounded from below and unbounded from above and for which we do not impose any controllability assumption. This assumption was used in Sect. 2.5.1 in order to guarantee the existence of the recessive solution of (SDS) at infinity. We will see in this section that it is possible to develop the theory of recessive and dominant solutions at infinity without the controllability assumption. The organization of this section is the following. In Sect. 6.3.1 we discuss the order of abnormality of system (SDS). In Sect. 6.3.2 we introduce basic notation and derive main properties of conjoined bases of nonoscillatory system (SDS) at $\infty$. The content of Sects. 6.3.3–6.3.5 can be considered as technical results, which are needed for the construction and classification of recessive and dominant solutions of (SDS) at $\infty$. These three subsections can be skipped for the first reading when one aims to see the main results first. The two central objects of this section, i.e., the recessive and dominant solutions of (SDS) at $\infty$, are introduced in Sects. 6.3.6 and 6.3.7, respectively. In Sect. 6.3.8 we introduce the notion of a genus of conjoined bases of (SDS) and study the existence of recessive and dominant solutions of (SDS) at $\infty$ in different genera of conjoined bases. This provides a basis for the investigation of limit properties of the recessive and dominant solutions of (SDS) at $\infty$ in Sect. 6.3.9. In Sect. 6.3.10 we present a special construction of the (unique) minimal recessive solution of (SDS) at $\infty$, which can be potentially useful for applications in the oscillation and spectral theory of (SDS).

### 6.3.1 Order of Abnormality

For $N \in [0, \infty)_{\mathbb{Z}}$ we denote by $\Lambda[N, \infty)_{\mathbb{Z}}$ the linear space of $n$-vector sequences $u = \{u_k\}_{k=N}^{\infty}$ such that $u_{k+1} = \mathcal{D}_k u_k$ and $\mathcal{B}_k u_k = 0$ on $[N, \infty)_{\mathbb{Z}}$. This means that $u \in \Lambda[N, \infty)_{\mathbb{Z}}$ if and only if the pair $(x \equiv 0, u)$ solves (SDS) on $[N, \infty)_{\mathbb{Z}}$. Moreover, we denote by $\Lambda_0[N, \infty)_{\mathbb{Z}}$ the subspace of $\mathbb{R}^n$ consisting of the initial values $u_N$ of the elements $u \in \Lambda[N, \infty)_{\mathbb{Z}}$. Then the number

$$d[N, \infty)_{\mathbb{Z}} := \dim \Lambda[N, \infty)_{\mathbb{Z}} = \dim \Lambda_0[N, \infty)_{\mathbb{Z}}, \qquad 0 \le d[N, \infty)_{\mathbb{Z}} \le n,$$
$$(6.223)$$

is called the *order of abnormality* of (SDS) on the interval $[N, \infty)_{\mathbb{Z}}$. It is obvious that if $(x \equiv 0, u)$ solves (SDS) on $[N_1, \infty)_{\mathbb{Z}}$, then it solves (SDS) also on $[N_2, \infty)_{\mathbb{Z}}$ for every $N_2 \in [N_1, \infty)_{\mathbb{Z}}$. This shows that $\Lambda[N_1, \infty)_{\mathbb{Z}} \subseteq \Lambda[N_2, \infty)_{\mathbb{Z}}$ for all $N_1 \le N_2$, i.e., the function $d[k, \infty)_{\mathbb{Z}}$ is nondecreasing in $k$ on $[0, \infty)_{\mathbb{Z}}$. Hence, the limit

$$d_\infty := \lim_{k \to \infty} d[k, \infty)_{\mathbb{Z}} = \max \left\{ d[k, \infty)_{\mathbb{Z}}, \ k \in [0, \infty)_{\mathbb{Z}} \right\}, \qquad 0 \le d_\infty \le n,$$
$$(6.224)$$

exists and is called the *maximal order of abnormality* of (SDS). In particular, the subspace $\Lambda[N, \infty)_{\mathbb{Z}}$ with $d[N, \infty)_{\mathbb{Z}} = d_\infty$ satisfies

$$\Lambda[N, \infty)_{\mathbb{Z}} = \Lambda[K, \infty)_{\mathbb{Z}} \quad \text{for every } K \in [N, \infty)_{\mathbb{Z}}. \tag{6.225}$$

Note that for an eventually controllable system (SDS), we have $d_\infty = 0$. The number $d_\infty$ is one of the important parameters of system (SDS). In a similar way to (6.223), we define the number $d[N, L]_{\mathbb{Z}}$ as the order of abnormality of system (SDS) on the interval $[N, L]_{\mathbb{Z}}$ with the associated subspaces $\Lambda[N, L]_{\mathbb{Z}}$ and $\Lambda_0[N, L]_{\mathbb{Z}}$. Also, for convenience we set $\Lambda[N, N]_{\mathbb{Z}} = \mathbb{R}^n = \Lambda_0[N, N]_{\mathbb{Z}}$ for the case of $L = N$.

The following result provides a basic connection between the subspaces $\Lambda_0[N, k]$ for $k \in [N, \infty)_{\mathbb{Z}}$ and the principal solution $Y^{[N]}$ of (SDS). We recall that the solution $Y^{[N]}$ is defined by the initial conditions $Y_N^{[N]} = \begin{pmatrix} 0 \\ I \end{pmatrix}$.

**Theorem 6.57** *For any $N \in [0, \infty)_{\mathbb{Z}}$, we have*

$$\Lambda_0[N, k]_{\mathbb{Z}} = \bigcap_{j \in [N, k]} \operatorname{Ker} X_j^{[N]} \quad \text{for every } k \in [N, \infty)_{\mathbb{Z}}, \tag{6.226}$$

$$d[N, \infty)_{\mathbb{Z}} = \dim \bigcap_{k \in [N, \infty)_{\mathbb{Z}}} \operatorname{Ker} X_k^{[N]}. \tag{6.227}$$

*Proof* For $k = N$ the equality in (6.226) holds trivially. Let $k \geq N + 1$. If $c \in \Lambda_0[N, k]$, then $y = (x \equiv 0, u)$ is a solution of (SDS) on $[N, k - 1]_{\mathbb{Z}}$ for some $u \in \Lambda[N, k]$ with $u_N = c$. By the uniqueness of solutions of system (SDS), it follows that $y = Y^{[N]}c$ on $[N, k]$. Hence, $X_j^{[N]}c = 0$ for all $j \in [N, k]$. The opposite direction is trivial. Equality (6.227) then follows from (6.226) and (6.223).

$\square$

**Remark 6.58** We note that when system (SDS) is eventually controllable (in the sense of Sect. 2.5.1) and nonoscillatory at $\infty$, then for every $N \in [0, \infty)_{\mathbb{Z}}$ the principal solution $Y^{[N]}$ has the matrix $X_k^{[N]}$ invertible for large $k$. Therefore, Theorem 6.57 implies that $\Lambda_0[N, k]_{\mathbb{Z}} = \{0\}$ for large $k$ and consequently $d_\infty = 0$.

### 6.3.2  Nonoscillatory Symplectic System

In this section we will use special symplectic fundamental matrices corresponding, for a given index $j \in [0, \infty)_{\mathbb{Z}}$, to the principal solution $Y^{[j]}$ of (SDS). More precisely, we denote

$$\Phi_k^{[j]} = \left( Y_k^{[j]} \ \bar{Y}_k^{[j]} \right), \quad k \in [0, \infty)_{\mathbb{Z}}, \quad w(Y^{[j]}, \bar{Y}^{[j]}) = I, \tag{6.228}$$

where $\bar{Y}^{[j]}$ is the conjoined basis of (SDS) such that $w(Y^{[j]}, \bar{Y}^{[j]}) = I$, that is, $\bar{Y}_j^{[j]} = -\mathcal{J}(0 \ I)^T = (-I \ 0)^T$. This means that the fundamental matrix in (6.228) satisfies the normalization condition $\Phi_j^{[j]} = -\mathcal{J}$. Note that the fundamental matrix $\Phi_k^{[j]}$ corresponds to the symplectic matrix $\tilde{Z}$ in Lemma 1.58(vi). We shall represent conjoined bases of (SDS) in terms of the matrix $\Phi_k^{[j]}$.

**Lemma 6.59** *Let $j \in [0, \infty)_{\mathbb{Z}}$ be fixed, and let the symplectic fundamental matrix $\Phi_k^{[j]}$ of (SDS) be defined by (6.228). Then for every conjoined basis $Y$ of (SDS), there exists a unique constant $2n \times n$ matrix $D_j$ such that*

$$Y_k = \Phi_k^{[j]} D_j, \quad k \in [0, \infty)_{\mathbb{Z}}, \quad \mathcal{J} D_j = \begin{pmatrix} w(Y^{[j]}, Y) \\ w(\bar{Y}^{[j]}, Y) \end{pmatrix}. \tag{6.229}$$

*Proof* The form of the matrix $D_j$ in (6.229) follows directly from the expression $D_j = -\mathcal{J}(\Phi_k^{[j]})^T \mathcal{J} Y_k$ by using the inversion formula $\mathcal{S}^{-1} = -\mathcal{J} \mathcal{S}^T \mathcal{J}$ for a symplectic matrix (see Lemma 1.58(ii)) and from the constancy of the Wronskian of two solutions of (SDS).

$\square$

Later in Sects. 6.3.11, 6.4.2, and 6.4.3, we will extend these considerations to symplectic fundamental matrices of (SDS) involving the (minimal) recessive solutions of (SDS) at $\pm\infty$, i.e., we will use the notation in (6.228) and the result of Lemma 6.59 with $j = \pm\infty$.

By using Lemma 6.59 and the comparative index, we can derive the following additional formulas for the multiplicities of focal points.

*Remark 6.60* Let $Y$ be a conjoined basis of (SDS), and let $Y^{[j]}$ and $Y^{[j+1]}$ be the principal solutions of (SDS) at $j$ and $j+1$. Then we have

$$m(j) = \mu\big(\mathcal{J}D_{j+1}, \mathcal{J}D^{[j]}_{j+1}\big), \quad m^*(j) = \mu^*\big(\mathcal{J}D_j, \mathcal{J}D^{[j+1]}_j\big), \tag{6.230}$$

where $D_{j+1}$ and $D^{[j]}_{j+1}$ are the representation matrices in (6.229) of $Y$ and $Y^{[j]}$ in terms of the symplectic fundamental matrix $\Phi^{[j+1]}$ of (SDS), while $D_j$ and $D^{[j+1]}_j$ are the representation matrices in (6.229) of $Y$ and $Y^{[j+1]}$ in terms of the symplectic fundamental matrix $\Phi^{[j]}$. Indeed, since $\Phi^{[j+1]}_{j+1} = -\mathcal{J} = \Phi^{[j]}_j$, we have

$$\mathcal{J}D_{j+1} = (\Phi^{[j+1]}_k)^T \mathcal{J}\, Y_k = -Y_{j+1}, \quad \mathcal{J}D^{[j]}_{j+1} = (\Phi^{[j+1]}_k)^T \mathcal{J}\, Y^{[j]}_k = -Y^{[j]}_{j+1} \tag{6.231}$$

(when evaluated at $k = j+1$), and similarly

$$\mathcal{J}D_j = (\Phi^{[j]}_k)^T \mathcal{J}\, Y_k = -Y_j, \quad \mathcal{J}D^{[j+1]}_j = (\Phi^{[j]}_k)^T \mathcal{J}\, Y^{[j+1]}_k = -Y^{[j+1]}_j \tag{6.232}$$

(when evaluated at $k = j$). Therefore,

$$\mu\big(\mathcal{J}D_{j+1}, \mathcal{J}D^{[j]}_{j+1}\big) \overset{(6.231)}{=} \mu(-Y_{j+1}, -Y^{[j]}_{j+1}) = \mu(Y_{j+1}, Y^{[j]}_{j+1}) \overset{(4.14)}{=} m(j),$$

$$\mu^*\big(\mathcal{J}D_j, \mathcal{J}D^{[j+1]}_j\big) \overset{(6.232)}{=} \mu^*(-Y_j, -Y^{[j+1]}_j) = \mu^*(Y_j, Y^{[j+1]}_j) \overset{(4.16)}{=} m^*(j).$$

In Sect. 6.4.1 we will derive analogs of (6.230) for unbounded intervals.

Fundamental results from the Sturmian theory of discrete symplectic systems (SDS) are established in Chap. 4. In particular, Theorem 4.24 and its Corollaries 4.25 and 4.26 state that for a given index $N \in [0, \infty)_{\mathbb{Z}}$ the numbers of forward (or left) focal points in the interval $(0, N+1]$ of any two conjoined bases of (SDS) differ by at most $n$. Therefore, based on Corollary 4.25, we formulate the following result, which will be useful for our future reference.

**Proposition 6.61** *The following statements are equivalent.*

(i)   *System (SDS) is (non)oscillatory at $\infty$ (resp., at $-\infty$).*
(ii)  *Every conjoined basis of (SDS) is (non)oscillatory at $\infty$ (resp., at $-\infty$).*
(iii) *There exists a conjoined basis of (SDS), which is (non)oscillatory at $\infty$ (resp., at $-\infty$).*

Therefore, equivalently to Definition 4.37, we say that system (SDS) is *nonoscillatory* at $\infty$ (resp., at $-\infty$) if every conjoined basis of (SDS) has finitely many forward focal points in $(0, \infty)$ (resp., finitely many backward focal points in $(-\infty, 0)$). In the opposite case, we say that system (SDS) is *oscillatory* at $\infty$ (resp., at $-\infty$). We note that by formula (4.13) in Corollary 4.6, it is equivalent to use the forward or the backward focal points to define the (non)oscillation of system (SDS) at $\infty$ (resp., at $-\infty$).

In Example 4.73 we showed that the symplectic system corresponding to the Fibonacci difference equation $x_{k+2} = x_{k+1} + x_k$ for $k \in [0, \infty)_{\mathbb{Z}}$ is oscillatory at $\infty$. Also, the symplectic systems in Examples 4.74 and 4.75 are nonoscillatory at $\infty$.

*Example 6.62* Consider system (SDS) with $\mathcal{S}_k \equiv \mathcal{J}$ on $[0, \infty)_{\mathbb{Z}}$. Then $d[0, \infty)_{\mathbb{Z}} = d_\infty = 0$ and all solutions of (SDS) are periodic with period four. Consider the conjoined basis $Y = \{Y_k\}_{k=0}^\infty$ with $Y_{2j} = (0, \, (-1)^j I)^T$ and $Y_{2j+1} = ((-1)^j I, \, 0)^T$ for $j \in [0, \infty)_{\mathbb{Z}}$. Then for each $k \in [0, \infty)_{\mathbb{Z}}$ either $X_k = 0$ or $X_{k+1} = 0$. Then for every $N \in [0, \infty)_{\mathbb{Z}}$, the kernel of $X_k$ is not constant on $[N, \infty)_{\mathbb{Z}}$, and $Y$ has infinitely many focal points in $(N, \infty)$. These focal points of multiplicity $n$ arise from the first term in (4.3) by violating the kernel condition in (2.37). The focal points are located in the intervals $(k, k + 1] = (2j, 2j + 1]$ (more precisely, at $k + 1 = 2j + 1$) for every $j \in [0, \infty)_{\mathbb{Z}}$ in the terminology of Sect. 4.1.1. Thus, this system (SDS) is oscillatory at $\infty$.

*Example 6.63* Consider system (SDS) with $\mathcal{S}_k \equiv I_{2n}$ on $[0, \infty)_{\mathbb{Z}}$. Then we have $d[0, \infty)_{\mathbb{Z}} = d_\infty = n$, and all solutions of (SDS) are constant on $[0, \infty)_{\mathbb{Z}}$. Therefore, this system is nonoscillatory at $\infty$.

Further examples will be discussed in Sects. 6.3.6 and 6.3.7 in the relation with the recessive and dominant solutions of (SDS) at $\infty$.

*Remark 6.64* If $V$ is a linear subspace in $\mathbb{R}^n$, then we denote by $P_V$ the corresponding $n \times n$ orthogonal projector onto $V$. It follows that the matrix $P_V$ is symmetric, idempotent, and nonnegative definite.

Given a conjoined basis $Y$ of (SDS), we denote by $P_k$ the orthogonal projector onto $\operatorname{Im} X_k^T$ and by $R_k$ the orthogonal projector onto $\operatorname{Im} X_k$. That is, according to Remarks 6.64 and 1.62, we have

$$P_k := P_{\operatorname{Im} X_k^T} = X_k^\dagger X_k, \quad R_k := P_{\operatorname{Im} X_k} = X_k X_k^\dagger. \tag{6.233}$$

Note that $\operatorname{rank} P_k = \operatorname{rank} X_k^T = \operatorname{rank} X_k = \operatorname{rank} R_k$. When $\operatorname{Ker} X_k = \operatorname{Ker} X_{k+1}$, i.e., when $\operatorname{Im} X_k^T = \operatorname{Im} X_{k+1}^T$, we have $P_k = P_{k+1}$. Therefore, the projector

$$P := P_k \tag{6.234}$$

is constant on an interval $[N, \infty)_{\mathbb{Z}}$ where $\operatorname{Ker} X_k$ is constant. For convenience but with slight abuse in terminology, we say that $Y$ *has constant kernel on* $[N, \infty)_{\mathbb{Z}}$ when the kernel of $X_k$ is constant on this interval. Similarly, we say that $Y$ *has rank r on* $[N, \infty)_{\mathbb{Z}}$ when the rank of $X_k$ is equal to $r$ on $[N, \infty)_{\mathbb{Z}}$.

Next we derive important properties of conjoined bases of a nonoscillatory system (SDS) at $\infty$, in particular of conjoined bases with constant kernel on a given interval $[N, \infty)_{\mathbb{Z}}$ and with no forward focal points in $(N, \infty)$. The latter properties are equivalent with the two conditions in (2.37), i.e., with

$$\text{Ker } X_{k+1} = \text{Ker } X_k, \quad X_k X_{k+1}^\dagger \mathcal{B}_k \geq 0, \quad k \in [N, \infty)_{\mathbb{Z}}. \qquad (6.235)$$

Later in this section, we will provide a construction of such conjoined bases from a given conjoined basis with the same properties via the relation "being contained" (see Sect. 6.3.3).

For every conjoined basis $Y$ of (SDS), we define its corresponding $S$-matrix

$$S_k := \sum_{j=N}^{k-1} X_{j+1}^\dagger \, \mathcal{B}_j \, X_j^{\dagger \, T}. \qquad (6.236)$$

We start with two monotonicity properties of the matrix $S_k$ and the set $\text{Im } S_k$, which leads to a definition of the $T$-matrix associated with $Y$.

**Theorem 6.65** *Assume that system* (SDS) *is nonoscillatory at* $\infty$. *Then for every conjoined basis* $Y$ *of* (SDS), *there exists* $N \in [0, \infty)_{\mathbb{Z}}$ *such that the matrix* $X_{k+1}^\dagger \mathcal{B}_k X_k^{\dagger \, T}$ *is symmetric and nonnegative definite on* $[N, \infty)_{\mathbb{Z}}$, *the associated matrix* $S_k$ *in* (6.236) *is symmetric, nonnegative definite, and nondecreasing on* $[N, \infty)_{\mathbb{Z}}$ *with* $S_N = 0$. *Moreover, the matrix* $S_k^\dagger$ *is eventually nonincreasing, the limit*

$$T := \lim_{k \to \infty} S_k^\dagger \qquad (6.237)$$

*exists, and the matrix* $T$ *is symmetric and nonnegative definite.*

*Proof* The result in Proposition 6.61 implies that for a conjoined basis $Y$ of (SDS), there exists $N \in [0, \infty)_{\mathbb{Z}}$ such that the kernel of $X_k$ is constant on $[N, \infty)_{\mathbb{Z}}$ and $X_k X_{k+1}^\dagger \mathcal{B}_k$ is symmetric and nonnegative definite on $[N, \infty)_{\mathbb{Z}}$. This implies that the matrix

$$X_{k+1}^\dagger \mathcal{B}_k X_k^{\dagger \, T} = P_{k+1} X_{k+1}^\dagger \mathcal{B}_k X_k^{\dagger \, T} = P_k X_{k+1}^\dagger \mathcal{B}_k X_k^{\dagger \, T} = X_k^\dagger X_k X_{k+1}^\dagger \mathcal{B}_k X_k^{\dagger \, T}$$

is symmetric and nonnegative definite on $[N, \infty)_{\mathbb{Z}}$ as well. In turn, the matrix $S_k$ in (6.236) is symmetric, nonnegative definite, nondecreasing on $[N, \infty)_{\mathbb{Z}}$, and by convention $S_N = 0$. These properties imply that the matrix $S_k^\dagger$ is also symmetric and nonnegative definite on $[N, \infty)_{\mathbb{Z}}$, and nonincreasing for large $k$. More precisely, $S_k^\dagger$ is nonincreasing on intervals, where $S_k$ has constant image, by Remark 1.60(vi). Hence, the limit in (6.237) exists and the matrix $T$ is symmetric and nonnegative definite. $\qquad \square$

**Theorem 6.66** *Let $Y$ be a conjoined basis of* (SDS) *with constant kernel on* $[N, \infty)_{\mathbb{Z}}$, *and let the matrices $P$, $R_k$, and $S_k$ be defined in* (6.234), (6.233), *and* (6.236). *Then*

*(i) $\operatorname{Im}[U_k(I - P)] = \operatorname{Ker} R_k$, and hence $R_k U_k = R_k U_k P$, for all $k \in [N, \infty)_{\mathbb{Z}}$,*
*(ii) $\mathcal{B}_k = R_{k+1} \mathcal{B}_k$ and $\mathcal{B}_k = \mathcal{B}_k R_k$ for all $k \in [N, \infty)_{\mathbb{Z}}$,*
*(iii) $P S_k = S_k$, i.e., $\operatorname{Im} S_k \subseteq \operatorname{Im} P$, for all $k \in [N, \infty)_{\mathbb{Z}}$.*

*If in addition $Y$ has no forward focal points in $(N, \infty)$, then*

*(iv) the set $\operatorname{Im} S_k$ is nondecreasing on $[N, \infty)_{\mathbb{Z}}$, hence it is eventually constant.*

*Proof* (i) For a vector $v \in \operatorname{Im} U_k(I - P)$, we have $v = U_k(I - P) c$ for some $c \in \mathbb{R}^n$. This yields that $R_k v = X_k^{\dagger T} X_k^T U_k(I - P) c = X_k^{\dagger T} U_k^T X_k(I - P) c = 0$, since $X_k(I - P) = 0$. Therefore, $\operatorname{Im} U_k(I - P) \subseteq \operatorname{Ker} R_k$ holds, which yields the inequality $\operatorname{rank} U_k(I - P) \le \operatorname{def} R_k$. We will show the opposite inequality. Choose $w \in \operatorname{Ker} U_k(I - P)$, i.e., $U_k(I - P) w = 0$. Since also $X_k(I - P) w = 0$ and $\operatorname{rank}(X_k^T, U_k^T) = n$ hold, we have $(I - P) w = 0$, i.e., $w \in \operatorname{Im} P$. Thus, $\operatorname{Ker} U_k(I - P) \subseteq \operatorname{Im} P$ and $\operatorname{def} U_k(I - P) \le \operatorname{rank} P$. By using $\operatorname{rank} R_k = \operatorname{rank} P$, we then obtain the needed estimate

$$\operatorname{def} R_k = n - \operatorname{rank} R_k = n - \operatorname{rank} P \le n - \operatorname{def} U_k(I - P) = \operatorname{rank} U_k(I - P).$$

Consequently, $\operatorname{def} R_k = \operatorname{rank} U_k(I - P)$. But since we have already proved that $\operatorname{Im} U_k(I - P)$ is a subspace of $\operatorname{Ker} R_k$, the statement in part (i) follows. The first identity in part (ii) is a reformulation of the fact that $X_{k+1} X_{k+1}^{\dagger} \mathcal{B}_k = \mathcal{B}_k$ when $\operatorname{Ker} X_{k+1} = \operatorname{Ker} X_k$; see Lemma 2.15 in Sect. 2.2.1. For the second identity in (ii), we note that $X_{k+1} = X_{k+1} P_{k+1} = X_{k+1} P_k$ and that $X_k X_{k+1}^{\dagger} \mathcal{B}_k$ and $R_k$ are symmetric. Therefore, we have

$$\mathcal{B}_k R_k = R_{k+1} \mathcal{B}_k R_k = X_{k+1} P_k X_{k+1}^{\dagger} \mathcal{B}_k R_k = X_{k+1} X_k^{\dagger} \mathcal{B}_k^T X_{k+1}^{\dagger T} X_k^T R_k$$

$$= X_{k+1} X_k^{\dagger} \mathcal{B}_k^T X_{k+1}^{\dagger T} X_k^T = X_{k+1} X_k^{\dagger} X_k X_{k+1}^{\dagger} \mathcal{B}_k = X_{k+1} P_k X_{k+1}^{\dagger} \mathcal{B}_k$$

$$= X_{k+1} X_{k+1}^{\dagger} \mathcal{B}_k = \mathcal{B}_k.$$

For part (iii) we use that the projector $P_j$ is constant on $[N, \infty)_{\mathbb{Z}}$, so that we have

$$P S_k = \sum_{j=N}^{k-1} P P_{j+1} X_{j+1}^{\dagger} \mathcal{B}_j X_j^{\dagger T} = \sum_{j=N}^{k-1} P_{j+1} X_{j+1}^{\dagger} \mathcal{B}_j X_j^{\dagger T} = S_k$$

on $[N, \infty)_{\mathbb{Z}}$. This shows that $\operatorname{Im} S_k = \operatorname{Im} P S_k \subseteq \operatorname{Im} P$ on $[N, \infty)_{\mathbb{Z}}$. If in addition $Y$ has no forward focal points in $(N, \infty)$, then $X_k X_{k+1}^{\dagger} \mathcal{B}_k \ge 0$ on $[N, \infty)_{\mathbb{Z}}$ and so $S_k$ is nonnegative definite and nondecreasing on $[N, \infty)_{\mathbb{Z}}$; compare with Theorem 6.65. And since $S_k$ is symmetric, it follows that $\operatorname{Im} S_k$ is nondecreasing on $[N, \infty)_{\mathbb{Z}}$ and hence eventually constant. $\qquad\square$

Based on Theorem 6.66(iv), we define for a conjoined basis $Y$ of (SDS) with constant kernel on $[N, \infty)_{\mathbb{Z}}$ and no forward focal points in $(N, \infty)$ the orthogonal projectors

$$P_{\mathcal{S}k} := P_{\operatorname{Im} S_k} = S_k^\dagger S_k = S_k S_k^\dagger, \qquad P_{\mathcal{S}\infty} := \lim_{k \to \infty} P_{\mathcal{S}k}, \tag{6.238}$$

where $P_{\mathcal{S}\infty}$ is the constant projector onto the maximal set $\operatorname{Im} S_k$ according to Theorem 6.66(iv). We then have the inclusions

$$\operatorname{Im} S_k \subseteq \operatorname{Im} P_{\mathcal{S}\infty} \subseteq \operatorname{Im} P, \quad k \in [N, \infty)_{\mathbb{Z}}, \quad \operatorname{Im} T \subseteq \operatorname{Im} P_{\mathcal{S}\infty} \subseteq \operatorname{Im} P, \tag{6.239}$$

where we used that $\operatorname{Im} S_k^\dagger = \operatorname{Im} S_k^T = \operatorname{Im} S_k$ by Remark 1.60(i).

In the following result, we show a construction and properties of a special conjoined basis $\bar{Y}$ of (SDS), which completes a given conjoined basis $Y$ with constant kernel on some interval $[N, \infty)_{\mathbb{Z}}$ to a normalized pair (see Sect. 2.1.1). This leads to a special symplectic fundamental matrix $\Phi_k = (Y_k, \bar{Y}_k)$ of (SDS), which will be utilized in further development of the theory.

**Proposition 6.67** *Let $Y$ be a conjoined basis of* (SDS) *with constant kernel on $[N, \infty)_{\mathbb{Z}}$ with the associated matrices $P$, $R_k$, $S_k$, and $P_{\mathcal{S}k}$ defined in (6.234), (6.233), (6.236), and (6.238). Then there exists a conjoined basis $\bar{Y}$ of* (SDS) *such that*

*(i)* $w(Y, \bar{Y}) = I$,
*(ii)* $X_N^\dagger \bar{X}_N = 0$.

*Moreover, such a conjoined basis $\bar{Y}$ then satisfies*

*(iii)* $X_k^\dagger \bar{X}_k P = S_k$ *for $k \in [N, \infty)_{\mathbb{Z}}$,*
*(iv)* $\bar{X}_k P = X_k S_k$ *and $\bar{U}_k P = U_k S_k + X_k^{\dagger T} + U_k (I - P) \bar{X}_k^T X_k^{\dagger T}$ for every $k \in [N, \infty)_{\mathbb{Z}}$ (in particular $\bar{X}_N P = 0$), and the solution $\bar{Y} P$ of* (SDS) *is uniquely determined by $Y$ on $[0, \infty)_{\mathbb{Z}}$,*
*(v)* *the matrices $\bar{X}_k$ for $k \in [N, \infty)_{\mathbb{Z}}$ are uniquely determined by $Y$,*
*(vi)* $\operatorname{Ker} X_k = \operatorname{Im}(P - P_{\mathcal{S}k}) = \operatorname{Im} P \cap \operatorname{Ker} S_k$ *for $k \in [N, \infty)_{\mathbb{Z}}$,*
*(vii)* $\bar{P}_k = I - P + P_{\mathcal{S}k}$ *for $k \in [N, \infty)_{\mathbb{Z}}$, where $\bar{P}_k := \bar{X}_k^\dagger \bar{X}_k$,*
*(viii)* $S_k^\dagger = \bar{X}_k^\dagger X_k P_{\mathcal{S}k} = \bar{X}_k^\dagger X_k \bar{P}_k$ *for $k \in [N, \infty)_{\mathbb{Z}}$,*
*(ix)* $\operatorname{Im} \bar{X}_N = \operatorname{Im}(I - R_N)$ *and $\operatorname{Im} \bar{X}_N^T = \operatorname{Im}(I - P)$,*
*(x)* *the matrix $X_N - \bar{X}_N$ is invertible with $(X_N - \bar{X}_N)^{-1} = X_N^\dagger - \bar{X}_N^\dagger$,*
*(xi)* $\bar{X}_N^\dagger = -(I - P) U_N^T$.

*If in addition the conjoined basis $Y$ has no forward focal points in $(N, \infty)$, then*

*(xii)* $X_k \bar{X}_k^T \geq 0$ *for $k \in [N, \infty)_{\mathbb{Z}}$.*

*Proof* We will prove (i) and (ii). Let $\tilde{Y}$ be a conjoined basis of (SDS) such that $Y$ and $\tilde{Y}$ are normalized, i.e., $w(Y, \tilde{Y}) = I$ (see Sect. 2.1.1). Then according to

Lemma 1.58 applied to the symplectic matrix $S := \Phi_k = (Y_k, \ \bar{Y}_k)$, we have

$$X_k \tilde{U}_k^T - \tilde{X}_k U_k^T = I, \quad X_k \tilde{X}_k^T \text{ and } U_k \tilde{U}_k^T \text{ symmetric}, \quad k \in [0, \infty)_{\mathbb{Z}}. \tag{6.240}$$

Since the projector $P$ is constant on $[N, \infty)_{\mathbb{Z}}$, we define the constant matrix

$$D := X_N^\dagger \tilde{X}_N (P - I) - \tilde{X}_N^T X_N^{\dagger T}.$$

Then $D$ is symmetric, because $X_N^\dagger \tilde{X}_N P = X_N^\dagger \tilde{X}_N X_N^T X_N^{\dagger T}$ is symmetric by (6.240). Furthermore, $PD = -X_N^\dagger \tilde{X}_N$. Define the solution $\bar{Y}$ of (SDS) by $\bar{Y}_k := \tilde{Y}_k + Y_k D$ on $[0, \infty)_{\mathbb{Z}}$. Then $\bar{Y}$ is a conjoined basis of (SDS) satisfying $w(Y, \bar{Y}) = I$ and $X_N^\dagger \bar{X}_N = X_N^\dagger \tilde{X}_N + PD = 0$. This completes the proof of parts (i) and (ii). For part (iii), let $k \in [N, \infty)_{\mathbb{Z}}$ be fixed. Then we have

$$\begin{aligned}
\Delta(X_k^\dagger \bar{X}_k P) &= X_{k+1}^\dagger \bar{X}_{k+1} P - X_k^\dagger \bar{X}_k P \\
&= X_{k+1}^\dagger (\bar{X}_{k+1} - A_k \bar{X}_k) P + (X_{k+1}^\dagger A_k - X_k^\dagger) \bar{X}_k P \\
&= X_{k+1}^\dagger B_k \bar{U}_k X_k^\dagger X_k + X_{k+1}^\dagger A_k X_k \bar{X}_k^T X_k^{\dagger T} - X_k^\dagger X_k \bar{X}_k^T X_k^{\dagger T} \\
&= X_{k+1}^\dagger B_k (\bar{U}_k X_k^T - U_k \bar{X}_k^T) X_k^{\dagger T} + (X_{k+1}^\dagger X_{k+1} - P) \bar{X}_k^T X_k^{\dagger T} \\
&= X_{k+1}^\dagger B_k X_k^{\dagger T}.
\end{aligned}$$

Since we already proved that $X_N^\dagger \bar{X}_N = 0$, we obtain by the summation of the above equality from $j = N$ to $j = k - 1$ that $X_k^\dagger \bar{X}_k P = S_k$, as we claim in (iii). The formulas for $\bar{X}_k P$ and $\bar{U}_k P$ on $[N, \infty)_{\mathbb{Z}}$ in part (iv) follow from the above construction of $\bar{Y}$. Then for $k = N$ we obtain $\bar{X}_N P = X_N S_N = 0$, as $S_N = 0$. Moreover, in view of part (v) proven below, these formulas depend only on the values of $Y_k$ on $[N, \infty)_{\mathbb{Z}}$, so that part (iv) is proven. Regarding part (v), we assume that there exists another conjoined basis $\bar{\bar{Y}}$ of (SDS) satisfying (i)–(iii). Then by [205, Corollary 3.3.9], there exists a symmetric matrix $\bar{D}$ with $\bar{\bar{Y}}_k = \bar{Y}_k + Y_k \bar{D}$ on $[0, \infty)_{\mathbb{Z}}$. From $X_N^\dagger \bar{X}_N = 0 = X_N^\dagger \bar{\bar{X}}_N$, we then obtain $P\bar{D} = X_N^\dagger X_N \bar{D} = X_N^\dagger (\bar{\bar{X}}_N - \bar{X}_N) = 0$, so that $\operatorname{Im} \bar{D} \subseteq \operatorname{Ker} P = \operatorname{Ker} X_k$ on $[N, \infty)_{\mathbb{Z}}$. Therefore, $X_k \bar{D} = 0$ and the equality $\bar{\bar{X}}_k = \bar{X}_k$ on $[N, \infty)_{\mathbb{Z}}$ follows. Therefore, part (v) holds. For part (vi) we note that from the identity $X_k^T \bar{U}_k - U_k^T \bar{X}_k = I$ on $[N, \infty)_{\mathbb{Z}}$ (i.e., from $w(Y, \bar{Y}) = I$), it follows that $\operatorname{Ker} \bar{X}_k \subseteq \operatorname{Im} X_k^T = \operatorname{Im} P$ for every $k \in [N, \infty)_{\mathbb{Z}}$. Moreover, by (iii) and (6.239), we get

$$\operatorname{Ker} \bar{X}_k \subseteq \operatorname{Im} P \cap \operatorname{Ker} S_k = \operatorname{Im} P \cap \operatorname{Ker} P_{Sk} \stackrel{(6.239)}{=} \operatorname{Im} (P - P_{Sk}), \quad k \in [N, \infty)_{\mathbb{Z}}.$$

Conversely, fix an index $k \in [N, \infty)_{\mathbb{Z}}$, and assume that $v \in \operatorname{Im} (P - P_{Sk}) = \operatorname{Im} P \cap \operatorname{Ker} S_k$. The first identity in (iv) then yields $\bar{X}_k v = \bar{X}_k P v = X_k S_k v = 0$, and hence

$v \in \operatorname{Ker} \bar{X}_k$. Therefore, the opposite inclusion $\operatorname{Im}(P - P_{Sk}) \subseteq \operatorname{Ker} \bar{X}_k$ also holds, showing part (vi). In addition, the latter result is equivalent with the fact that the matrix $I - (P - P_{Sk})$ is the orthogonal projector onto the space $(\operatorname{Ker} \bar{X}_k)^{\perp} = \operatorname{Im} \bar{X}_k^T$ for every $k \in [N, \infty)_{\mathbb{Z}}$. More precisely, we have the formula

$$\bar{P} = \bar{X}_k^{\dagger} \bar{X}_k = I - P + P_{Sk}, \quad k \in [N, \infty)_{\mathbb{Z}}, \tag{6.241}$$

which proves part (vii). Next, upon combining the identities $P_{Sk} = S_k S_k^{\dagger}$ and $X_k S_k = \bar{X}_k P$ from (iv) and $P_{Sk} P = P_{Sk}$ on $[N, \infty)_{\mathbb{Z}}$ with (6.241), we obtain

$$\bar{X}_k^{\dagger} X_k P_{Sk} = \bar{X}_k^{\dagger} X_k S_k S_k^{\dagger} \overset{\text{(iv)}}{=} \bar{X}_k^{\dagger} \bar{X}_k P S_k^{\dagger} \overset{(6.241)}{=} (I - P + P_{Sk}) P S_k^{\dagger}$$
$$= P_{Sk} P S_k^{\dagger} = S_k^{\dagger}$$

for every $k \in [N, \infty)_{\mathbb{Z}}$, i.e., part (viii) holds. For part (ix) we first observe that $P_{SN} = S_N^{\dagger} S_N = 0$. Then by (vi) we obtain $\operatorname{Ker} \bar{X}_N = \operatorname{Im} P$, i.e., by taking the orthogonal complements $\operatorname{Im} \bar{X}_N^T = \operatorname{Im}(I - P)$. Moreover, from (ii) we have $\operatorname{Im} \bar{X}_N \subseteq \operatorname{Ker} X_N^{\dagger} = \operatorname{Im}(I - R_N)$. But since $\operatorname{rank} \bar{X}_N = \operatorname{rank} \bar{X}_N^T = \operatorname{rank}(I - P) = \operatorname{rank}(I - R_N)$ (as $\operatorname{rank} P = \operatorname{rank} R_k$), it follows that $\operatorname{Im} \bar{X}_N = \operatorname{Im}(I - R_N)$. For part (x) we note from (ix) and (vii) that

$$\bar{X}_N \bar{X}_N^{\dagger} = I - R_N, \quad \bar{P}_N = \bar{X}_N^{\dagger} \bar{X}_N = I - P. \tag{6.242}$$

Then $X_N \bar{X}_N^{\dagger} = X_N P(I - P) \bar{X}_N^{\dagger} = 0$ and $\bar{X}_N X_N^{\dagger} = \bar{X}_N(I - P) P X_N^{\dagger} = 0$. Thus,

$$(X_N - \bar{X}_N)(X_N^{\dagger} - \bar{X}_N^{\dagger}) = X_N X_N^{\dagger} - X_N \bar{X}_N^{\dagger} - \bar{X}_N X_N^{\dagger} - \bar{X}_N \bar{X}_N^{\dagger}$$
$$\overset{(6.242)}{=} R_N + (I - R_N) = I.$$

This shows that the matrices $X_N - \bar{X}_N$ and $X_N^{\dagger} - \bar{X}_N^{\dagger}$ are invertible and they are inverses of each other. Part (xi) is a direct application of the definition of the Moore-Penrose pseudoinverse (see Sect. 1.6.2). If we set $A := \bar{X}_N$ and $B := -(I - P) U_N^T$, then we verify the four properties in (1.157) to conclude that $B = A^{\dagger}$. In these calculations we use (6.242), property (i) in the form of equations $\bar{X}_N U_N^T = X_N \bar{U}_N^T - I$ and $U_N^T \bar{X}_N = X_N^T \bar{U}_N - I$, and the facts that $(I - R_N) X_N = 0$ and $(I - P) X_N^T = 0$. Finally, property (xii) follows from (iii) by showing that $X_k \bar{X}_k^T = \bar{X}_k S_k \bar{X}_k^T \geq 0$ for $k \in [N, \infty)_{\mathbb{Z}}$, since $S_k \geq 0$ for $k \in [N, \infty)_{\mathbb{Z}}$ under the stated additional assumption on $Y$. The proof of Proposition 6.67 is complete. $\qquad\square$

*Remark 6.68* From the proof of Proposition 6.67(iv), it follows that if $\tilde{Y}$ is any conjoined basis of (SDS) such that $w(Y, \tilde{Y}) = I$, then $S_k = X_k^{\dagger} \tilde{X}_k P - X_N^{\dagger} \tilde{X}_N P$ for all $k \in [N, \infty)_{\mathbb{Z}}$.

In the following result, we describe a mutual representation of conjoined bases of (SDS) with constant kernel on $[N, \infty)_{\mathbb{Z}}$.

**Theorem 6.69** *Let $Y^{(1)}$ and $Y^{(2)}$ be conjoined bases of (SDS) with constant kernels on $[N, \infty)_{\mathbb{Z}}$, and let $P^{(1)}$ and $P^{(2)}$ be the constant projectors defined by (6.234) through the functions $X_k^{(1)}$ and $X_k^{(2)}$, respectively. Let the conjoined basis $Y^{(2)}$ be expressed in terms of $Y^{(1)}$ via the matrices $M^{(1)}$, $N^{(1)}$, and let conjoined basis $Y^{(1)}$ be expressed in terms of $Y^{(2)}$ via the matrices $M^{(2)}$, $N^{(2)}$, that is,*

$$Y_k^{(2)} = \left(Y_k^{(1)} \ \bar{Y}_k^{(1)}\right) \begin{pmatrix} M^{(1)} \\ N^{(1)} \end{pmatrix}, \quad Y_k^{(1)} = \left(Y_k^{(2)} \ \bar{Y}_k^{(2)}\right) \begin{pmatrix} M^{(2)} \\ N^{(2)} \end{pmatrix} \qquad (6.243)$$

*for all $k \in [N, \infty)_{\mathbb{Z}}$, where $\bar{Y}^{(1)}$ and $\bar{Y}^{(2)}$ are conjoined bases of (SDS) satisfying the conclusion in Proposition 6.67 with regard to $Y^{(1)}$ and $Y^{(2)}$, respectively. If $\operatorname{Im} X_N^{(1)} = \operatorname{Im} X_N^{(2)}$ holds, then*

(i) *the matrices $(M^{(1)})^T N^{(1)}$ and $(M^{(2)})^T N^{(2)}$ are symmetric,*
(ii) *the matrices $N^{(1)}$ and $N^{(2)}$ satisfy $N^{(1)} + (N^{(2)})^T = 0$,*
(iii) *the matrices $M^{(1)}$ and $M^{(2)}$ are invertible and $M^{(1)} M^{(2)} = I = M^{(2)} M^{(1)}$,*
(iv) *$\operatorname{Im} N^{(1)} \subseteq \operatorname{Im} P^{(1)}$ and $\operatorname{Im} N^{(2)} \subseteq \operatorname{Im} P^{(2)}$.*

*Moreover, the matrices $M^{(1)}$, $N^{(1)}$ do not depend on the choice of $\bar{Y}^{(1)}$, and the matrices $M^{(2)}$, $N^{(2)}$ do not depend on the choice of $\bar{Y}^{(2)}$. In fact,*

$$\left. \begin{aligned} M^{(1)} &= -w(\bar{Y}^{(1)}, Y^{(2)}), \quad N^{(1)} = w(Y^{(1)}, Y^{(2)}), \\ M^{(2)} &= -w(\bar{Y}^{(2)}, Y^{(1)}), \quad N^{(2)} = w(Y^{(2)}, Y^{(1)}). \end{aligned} \right\} \qquad (6.244)$$

*Proof* Since $Y^{(i)}$ and $\bar{Y}^{(i)}$ for $i \in \{1, 2\}$ are normalized conjoined bases of (SDS), the $2n \times 2n$ fundamental matrices in (6.243) are symplectic. By the formula for the inverse of a symplectic matrix (Lemma 1.58(ii)), we obtain (6.244), i.e.,

$$\begin{pmatrix} M^{(1)} \\ N^{(1)} \end{pmatrix} = \begin{pmatrix} (\bar{U}_N^{(1)})^T X_N^{(2)} - (\bar{X}_N^{(1)})^T U_N^{(2)} \\ (X_N^{(1)})^T U_N^{(2)} - (U_N^{(1)})^T X_N^{(2)} \end{pmatrix}, \qquad (6.245)$$

$$\begin{pmatrix} M^{(2)} \\ N^{(2)} \end{pmatrix} = \begin{pmatrix} (\bar{U}_N^{(2)})^T X_N^{(1)} - (\bar{X}_N^{(2)})^T U_N^{(1)} \\ (X_N^{(2)})^T U_N^{(1)} - (U_N^{(2)})^T X_N^{(1)} \end{pmatrix}. \qquad (6.246)$$

Assertion (i) is then a direct consequence of the above expressions and of formula (6.240) for the pairs $Y^{(1)}$, $\bar{Y}^{(1)}$ and $Y^{(2)}$, $\bar{Y}^{(2)}$. Condition (ii) follows from (6.244), since $w(Y^{(2)}, Y^{(1)}) = -[w(Y^{(1)}, Y^{(2)})]^T$. Regarding item (iii), we define the matrices

$$L^{(1)} := (X_N^{(1)})^\dagger X_N^{(2)}, \quad L^{(2)} := (X_N^{(2)})^\dagger X_N^{(1)}. \qquad (6.247)$$

The assumption $\text{Im } X_N^{(1)} = \text{Im } X_N^{(2)}$ means that the projector $R_N^{(1)}$ onto $\text{Im } X_N^{(1)}$ and the projector $R_N^{(2)}$ onto $\text{Im } X_N^{(2)}$, which are defined in (6.233), satisfy $R_N^{(1)} = R_N^{(2)}$. Consequently, it follows that

$$X_N^{(1)} L^{(1)} = R_N^{(1)} X_N^{(2)} = R_N^{(2)} X_N^{(2)} = X_N^{(2)}, \tag{6.248}$$

$$L^{(1)} L^{(2)} = (X_N^{(1)})^\dagger R_N^{(2)} X_N^{(1)} = (X_N^{(1)})^\dagger R_N^{(1)} X_N^{(1)} = (X_N^{(1)})^\dagger X_N^{(1)} = P^{(1)}. \tag{6.249}$$

Similarly it can be shown that $X_N^{(2)} L^{(2)} = X_N^{(1)}$ and $L^{(2)} L^{(1)} = P^{(2)}$. In addition,

$$L^{(2)} = (L^{(1)})^\dagger, \tag{6.250}$$

which follows by an easy verification of the four equalities in (1.157) and by $\text{Im } L^{(i)} = \text{Im } P^{(i)}$ for $i \in \{1, 2\}$. If we now insert the formula $X_N^{(1)} L^{(1)} = X_N^{(2)}$ from (6.248) into the expression for $M^{(1)}$ in (6.245) and similarly insert the formula $X_N^{(2)} L^{(2)} = X_N^{(1)}$ into the expression for $M^{(2)}$ in (6.246), then we get via $w(Y^{(i)}, \bar{Y}^{(i)}) = I$ for $i \in \{1, 2\}$ that

$$M^{(1)} = (\bar{U}_N^{(1)})^T X_N^{(1)} L^{(1)} - (\bar{X}_N^{(1)})^T U_N^{(2)} = [I + (\bar{X}_N^{(1)})^T U_N^{(1)}] L^{(1)} - (\bar{X}_N^{(1)})^T U_N^{(2)}$$

$$= L^{(1)} + (\bar{X}_N^{(1)})^T [U_N^{(1)} L^{(1)} - U_N^{(2)}], \tag{6.251}$$

$$M^{(2)} = (\bar{U}_N^{(2)})^T X_N^{(2)} L^{(2)} - (\bar{X}_N^{(2)})^T U_N^{(1)} = [I + (\bar{X}_N^{(2)})^T U_N^{(2)}] L^{(2)} - (\bar{X}_N^{(2)})^T U_N^{(1)}$$

$$= L^{(2)} + (\bar{X}_N^{(2)})^T [U_N^{(2)} L^{(2)} - U_N^{(1)}]. \tag{6.252}$$

The product $M^{(1)} M^{(2)}$ is then simplified to the identity matrix by using the identities $L^{(1)} L^{(2)} = P^{(1)}$, Proposition 6.67(iv), (6.240), $R_N^{(1)} = R_N^{(2)}$. This technical and rather lengthy calculation is omitted. For part (iv), the matrices $N^{(1)}$ and $N^{(2)}$ in (6.245) and (6.246) satisfy

$$N^{(1)} = (X_N^{(1)})^T U_N^{(2)} - (U_N^{(1)})^T X_N^{(1)} L^{(1)} = (X_N^{(1)})^T [U_N^{(2)} - U_N^{(1)} L^{(1)}], \tag{6.253}$$

$$N^{(2)} = (X_N^{(2)})^T U_N^{(1)} - (U_N^{(2)})^T X_N^{(2)} L^{(2)} = (X_N^{(2)})^T [U_N^{(1)} - U_N^{(2)} L^{(2)}]. \tag{6.254}$$

Thus, $\text{Im } N^{(1)} \subseteq \text{Im} (X_N^{(1)})^T = \text{Im } P^{(1)}$ and $\text{Im } N^{(2)} \subseteq \text{Im} (X_N^{(2)})^T = \text{Im } P^{(2)}$. In addition, the matrices $M^{(1)}$, $N^{(1)}$ and $M^{(2)}$, $N^{(2)}$ in (6.243) do not depend on the choice of $\bar{Y}^{(1)}$ and $\bar{Y}^{(2)}$, because only the conjoined bases $Y^{(1)}$, $Y^{(2)}$ and the matrices $\bar{X}_N^{(1)}$, $\bar{X}_N^{(2)}$, which are unique by Proposition 6.67(v), are used in expressions (6.251)–(6.254) for $M^{(1)}$, $M^{(2)}$, $N^{(1)}$, $N^{(2)}$.                                       $\square$

*Remark 6.70*

(i) The equality in Proposition 6.67(iv) implies that for $k = N$ in (6.243), the conjoined bases $Y^{(1)}$ and $Y^{(2)}$ satisfy

$$X_N^{(2)} = X_N^{(1)} M^{(1)}, \quad U_N^{(2)} = U_N^{(1)} M^{(1)} + (X_N^{(1)})^{\dagger T} N^{(1)}, \tag{6.255}$$

$$X_N^{(1)} = X_N^{(2)} M^{(2)}, \quad U_N^{(1)} = U_N^{(2)} M^{(2)} + (X_N^{(2)})^{\dagger T} N^{(2)}. \tag{6.256}$$

(ii) Equations (6.249) and (6.250) are summarized as

$$L^{(1)}(L^{(1)})^{\dagger} = P^{(1)}, \quad L^{(2)}(L^{(2)})^{\dagger} = P^{(2)}. \tag{6.257}$$

This means that $P^{(1)}$ is the orthogonal projector onto $\operatorname{Im} L^{(1)}$ and similarly $P^{(2)}$ is the orthogonal projector onto $\operatorname{Im} L^{(2)}$. Moreover, expressions (6.251) and (6.252) for matrices $M^{(1)}$ and $M^{(2)}$ provide a connection between $M^{(1)}$ and $L^{(1)}$ and between $M^{(2)}$ and $L^{(2)}$. In particular,

$$L^{(1)} = P^{(1)} M^{(1)}, \quad L^{(2)} = P^{(2)} M^{(2)}. \tag{6.258}$$

These equalities follow from equations (6.251) and (6.257), since $\bar{X}_N^{(1)} P^{(1)} = 0$ and $\bar{X}_N^{(2)} P^{(2)} = 0$, by Proposition 6.67(iv). Moreover, condition (iv) in Theorem 6.69 means that $N^{(1)} = P^{(1)} N^{(1)}$ and $N^{(2)} = P^{(2)} N^{(2)}$, which shows that the representations in (6.243) in fact contain the uniquely determined solutions $\bar{Y}^{(1)} P^{(1)}$ and $\bar{Y}^{(2)} P^{(2)}$. These facts allows one to rewrite expressions (6.243) into a form, which is more convenient for the further analysis of the associated $S$-matrices. More precisely, from Proposition 6.67(iv), we get

$$\left. \begin{array}{l} X_k^{(2)} = X_k^{(1)} (L^{(1)} + S_k^{(1)} N^{(1)}), \\ X_k^{(1)} = X_k^{(2)} (L^{(2)} + S_k^{(2)} N^{(2)}), \end{array} \right\} \quad k \in [N, \infty)_{\mathbb{Z}}. \tag{6.259}$$

(iii) A more detailed analysis of the statements in parts (i) and (ii) of this remark shows that if only $Y^{(1)}$ has constant kernel on $[N, \infty)_{\mathbb{Z}}$ and (6.255) holds (so that $\operatorname{Im} X_N^{(2)} \subseteq \operatorname{Im} X_N^{(1)}$), then the first equality in (6.259) is satisfied. Similarly, if $Y^{(2)}$ has constant kernel on $[N, \infty)_{\mathbb{Z}}$ and (6.256) holds (so that $\operatorname{Im} X_N^{(1)} \subseteq \operatorname{Im} X_N^{(2)}$), then the second equality in (6.259) is satisfied.

*Remark 6.71* From the previous remark and Theorem 6.69, it also follows that the matrices $(L^{(1)})^T N^{(1)}$ and $(L^{(2)})^T N^{(2)}$ are symmetric. This can be seen from the calculation $(L^{(1)})^T N^{(1)} = (M^{(1)})^T P^{(1)} N^{(1)} = (M^{(1)})^T N^{(1)}$ and similarly for $(L^{(2)})^T N^{(2)} = (M^{(2)})^T N^{(2)}$.

*Remark 6.72* As a completion of Theorem 6.69, we show that the matrices $M^{(1)} + S_k^{(1)} N^{(1)}$ and $M^{(2)} + S_k^{(2)} N^{(2)}$ are invertible for $k \in [N, \infty)_{\mathbb{Z}}$. Indeed, from conditions (i) and (iv) in Theorem 6.69, we obtain that $\operatorname{Im}(N^{(1)})^T \subseteq \operatorname{Im} P^{(2)}$, so that $\operatorname{Ker} X_k^{(2)} = \operatorname{Ker} P^{(2)} \subseteq \operatorname{Ker} N^{(1)}$ on $[N, \infty)_{\mathbb{Z}}$. If for some vector $v \in \mathbb{R}^n$ we have $(M^{(1)} + S_k^{(1)} N^{(1)}) v = 0$, then $v \in \operatorname{Ker} X_k^{(2)}$ by (6.259) and by $L^{(1)} = P^{(1)} M^{(1)}$, $P^{(1)} S_k^{(1)} = S_k^{(1)}$, and $X_k^{(1)} P^{(1)} = X_k^{(1)}$. In turn, $v \in \operatorname{Ker} N^{(1)}$, and so $v \in \operatorname{Ker} M^{(1)}$. But since $M^{(1)}$ is invertible by (iii) of Theorem 6.69, it follows that $v = 0$. Similarly, one has that $M^{(2)} + S_k^{(2)} N^{(2)}$ is invertible on $[N, \infty)_{\mathbb{Z}}$.

The next statement is essentially a continuation of Theorem 6.69 and its proof. It provides a relation of the $S$-matrices corresponding to the conjoined bases $Y^{(1)}$ and $Y^{(2)}$. It will become one of the crucial tools for the development of the theory of recessive and dominant solutions of (SDS) at $\infty$ in Sects. 6.3.6 and 6.3.7.

**Theorem 6.73** *With the assumptions and notation of Theorem 6.69 and Remark 6.70, if the condition* $\operatorname{Im} X_N^{(1)} = \operatorname{Im} X_N^{(2)}$ *holds, then for all* $k \in [N, \infty)_{\mathbb{Z}}$ *we have* $\operatorname{Im} X_k^{(1)} = \operatorname{Im} X_k^{(2)}$ *and*

$$(L^{(1)} + S_k^{(1)} N^{(1)})^{\dagger} = L^{(2)} + S_k^{(2)} N^{(2)}, \tag{6.260}$$

$$\operatorname{Im} (L^{(1)} + S_k^{(1)} N^{(1)}) = \operatorname{Im} P^{(1)}, \quad \operatorname{Im} (L^{(2)} + S_k^{(2)} N^{(2)}) = \operatorname{Im} P^{(2)}, \tag{6.261}$$

$$S_k^{(2)} = (L^{(1)} + S_k^{(1)} N^{(1)})^{\dagger} S_k^{(1)} L_1^{\dagger T}. \tag{6.262}$$

*Proof* Fix $k \in [N, \infty)_{\mathbb{Z}}$. The equality $\operatorname{Im} X_k^{(1)} = \operatorname{Im} X_k^{(2)}$ is a direct consequence of the identities in (6.259). This means that $R_k^{(1)} = R_k^{(2)}$. Using Theorem 6.66(iii), (6.259), $\operatorname{Im} L^{(1)} = \operatorname{Im} P^{(1)}$, and $\operatorname{Im} L^{(2)} = \operatorname{Im} P^{(2)}$, we have

$$L^{(1)} + S_k^{(1)} N^{(1)} = P^{(1)} (L^{(1)} + S_k^{(1)} N^{(1)})$$
$$= (X_k^{(1)})^{\dagger} X_k^{(1)} (L^{(1)} + S_k^{(1)} N^{(1)}) = (X_k^{(1)})^{\dagger} X_k^{(2)}, \tag{6.263}$$

$$L^{(2)} + S_k^{(2)} N^{(2)} = P^{(2)} (L^{(2)} + S_k^{(2)} N^{(2)})$$
$$= (X_k^{(2)})^{\dagger} X_k^{(2)} (L^{(2)} + S_k^{(2)} N^{(2)}) = (X_k^{(2)})^{\dagger} X_k^{(1)}. \tag{6.264}$$

Expressions (6.263) and (6.264) then imply formula (6.260) by the verification of the four equalities in (1.157). In particular, the third and fourth identity in (1.157) read as

$$(L^{(1)} + S_k^{(1)} N^{(1)}) (L^{(2)} + S_k^{(2)} N^{(2)}) = (X_k^{(1)})^{\dagger} R_k^{(2)} X_k^{(1)}$$
$$= (X_k^{(1)})^{\dagger} R_k^{(1)} X_k^{(1)} = P^{(1)}, \tag{6.265}$$

$$(L^{(2)} + S_k^{(2)} N^{(2)}) (L^{(1)} + S_k^{(1)} N^{(1)}) = (X_k^{(2)})^{\dagger} R_k^{(1)} X_k^{(2)}$$
$$= (X_k^{(2)})^{\dagger} R_k^{(2)} X_k^{(2)} = P^{(2)}, \tag{6.266}$$

which imply the relations in (6.261). The proof of formula (6.262) is slightly more complicated. Since by (6.249) we have $L^{(1)}L^{(2)} = P^{(1)}$, from (6.265) we get

$$(L^{(1)} + S_k^{(1)} N^{(1)}) S_k^{(2)} N^{(2)} = -S_k^{(1)} N^{(1)} L^{(2)}.$$

Using (6.266), the fact $\operatorname{Im} S_k^{(2)} \subseteq \operatorname{Im} P^{(2)}$, and (6.260) we then obtain

$$S_k^{(2)} N^{(2)} = -(L^{(1)} + S_k^{(1)} N^{(1)})^\dagger S_k^{(1)} N^{(1)} L^{(2)}.$$

By Theorem 6.69(ii) and (6.250), we have $N^{(2)} = -(N^{(1)})^T$ and $L^{(2)} = (L^{(1)})^\dagger$, while from Remark 6.71 we know that $(N^{(2)})^T L^{(2)}$ is symmetric. This implies $N^{(1)} L^{(2)} = -(L^{(1)})^{\dagger T} N^{(2)}$, so that

$$S_k^{(2)} (N^{(1)})^T = (L^{(1)} + S_k^{(1)} N^{(1)})^\dagger S_k^{(1)} (L^{(1)})^{\dagger T} (N^{(1)})^T. \tag{6.267}$$

We will show that the matrix $(N^{(1)})^T$ in (6.267) can be canceled. Indeed, if the equality $\operatorname{Im}(N^{(1)})^T = \operatorname{Im} N^{(2)} = \operatorname{Im} P^{(2)}$ holds, then it suffices to multiply (6.267) from the right by $(N^{(1)})^{\dagger T}$, because $(N^{(1)})^T (N^{(1)})^{\dagger T} = (N^{(1)})^\dagger N^{(1)} = P^{(2)}$, $\operatorname{Im} S_k^{(2)} \subseteq \operatorname{Im} P^{(2)}$, and $\operatorname{Im}(L^{(1)})^\dagger = \operatorname{Im} L^{(2)} \subseteq \operatorname{Im} P^{(2)}$. But in general we only have $\operatorname{Im}(N^{(1)})^T = \operatorname{Im} N^{(2)} \subseteq \operatorname{Im} P^{(2)}$, which shows that more analysis is required in order to cancel $(N^{(1)})^T$ in (6.267). Let us denote $G := (L^{(1)})^T N^{(1)}$. The matrix $G$ is symmetric, $\operatorname{Im} G \subseteq \operatorname{Im} P^{(2)}$, and $N^{(1)} = (L^{(1)})^{\dagger T} G$. According to Lemma 1.89, there exists a sequence $\{G^{\{\nu\}}\}_{\nu=1}^\infty$ of symmetric matrices with $\operatorname{Im} G^{\{\nu\}} = \operatorname{Im} P^{(2)}$ for all $\nu \in \mathbb{N}$ such that $G^{\{\nu\}} \to G$ for $\nu \to \infty$. Furthermore, with $N^{\{\nu\}} := (L^{(1)})^{\dagger T} G^{\{\nu\}}$, we have $N^{\{\nu\}} \to (L^{(1)})^{\dagger T} G = N^{(1)}$ for $\nu \to \infty$, and in addition $\operatorname{Im} N^{\{\nu\}} = \operatorname{Im} P^{(1)}$ and $\operatorname{Im}(N^{\{\nu\}})^T = \operatorname{Im} P^{(2)}$ for all $\nu \in \mathbb{N}$. By verifying the identities in (1.157), it follows that $N^{\{\nu\}\dagger} = G^{\{\nu\}\dagger} (L^{(1)})^T$, and in particular $N^{\{\nu\}} N^{\{\nu\}\dagger} = P^{(1)}$ and $N^{\{\nu\}\dagger} N^{\{\nu\}} = P^{(2)}$ hold. Observe that the matrices $(M^{(1)})^T N^{\{\nu\}}$ and $(L^{(1)})^T N^{\{\nu\}}$ are symmetric, since

$$(M^{(1)})^T N^{\{\nu\}} = (M^{(1)})^T (L^{(1)})^{\dagger T} G^{\{\nu\}} = (M^{(1)})^T (L^{(2)})^T G^{\{\nu\}}$$

$$= (M^{(1)})^T (M^{(2)})^T P^{(2)} G^{\{\nu\}} = G^{\{\nu\}},$$

$$(L^{(1)})^T N^{\{\nu\}} = (M^{(1)})^T P^{(1)} N^{\{\nu\}} = (M^{(1)})^T N^{\{\nu\}},$$

where we used Remark 6.70(ii). For each $\nu \in \mathbb{N}$, we now define the solution $Y^{\{\nu\}}$ of system (SDS) by

$$Y_k^{\{\nu\}} := \left( Y_k^{(1)} \; \bar{Y}_k^{(1)} \right) \begin{pmatrix} M^{(1)} \\ N^{\{\nu\}} \end{pmatrix}, \quad k \in [N, \infty)_{\mathbb{Z}}. \tag{6.268}$$

Since $M_1^T N^{\{\nu\}}$ is symmetric and rank $((M^{(1)})^T, (N^{\{\nu\}})^T) = n$ (as $M^{(1)}$ is invertible by Theorem 6.69), it follows that $Y^{\{\nu\}}$ is a conjoined basis of (SDS). Moreover, the sequence $\{Y^{\{v\}}\}_{v=1}^{\infty}$ converges on $[N, \infty)_{\mathbb{Z}}$ to the conjoined basis $Y^{(2)}$, which follows from (6.243) and from the convergence of $\{(N^{(1)})^{\{v\}}\}_{v=1}^{\infty}$ to $N^{(1)}$. Since $\operatorname{Im} N^{\{v\}} = \operatorname{Im} P^{(1)}$, the function $X_k^{\{v\}}$ in (6.268) will have the form as in (6.259). That is, we have $X_k^{\{v\}} = X_k^{(1)}(L^{(1)} + S_k^{(1)} N^{\{v\}})$ on $[N, \infty)_{\mathbb{Z}}$ for every $v \in \mathbb{N}$. Thus, for each $k \in [N, \infty)_{\mathbb{Z}}$, we have the inclusions $\operatorname{Im} X_k^{\{v\}} \subseteq \operatorname{Im} X_k^{(1)} = \operatorname{Im} X_k^{(2)}$ and $\operatorname{Im}(X_k^{\{v\}})^T \subseteq \operatorname{Im}[(L^{(1)})^T + (N^{\{v\}})^T S_k^{(1)}] \subseteq \operatorname{Im} P^{(2)} = \operatorname{Im}(X_k^{(2)})^T$ for all $v \in \mathbb{N}$. Fix an index $K \in [N, \infty)_{\mathbb{Z}}$. Then by Lemma 1.61 (with $j := v$ and using that the interval $[N, K]_{\mathbb{Z}}$ is a finite set), there exists $v_0 \in \mathbb{N}$ such that for all $v \geq v_0$

$$\operatorname{Im} X_k^{\{v\}} = \operatorname{Im} X_k^{(2)}, \quad \operatorname{Im}(X_k^{\{v\}})^T = \operatorname{Im}(X_k^{(2)})^T \quad \text{for all } k \in [N, K]_{\mathbb{Z}}, \qquad (6.269)$$

$$\operatorname{rank} X_k^{\{v\}} = \operatorname{rank} X_k^{(2)} \quad \text{for all } k \in [N, K]_{\mathbb{Z}}. \qquad (6.270)$$

Therefore, by (6.270) and the limit property of the Moore-Penrose pseudoinverse in Remark 1.60(v), we obtain

$$(X_k^{\{v\}})^{\dagger} \to (X_k^{(2)})^{\dagger} \quad \text{for } v \to \infty \text{ and every } k \in [N, K]_{\mathbb{Z}}. \qquad (6.271)$$

Fix an index $v \geq v_0$. Then from the second equality in (6.269), it follows that $\operatorname{Ker} X_k^{\{v\}} = \operatorname{Ker} X_k^{(2)} = \operatorname{Ker} P^{(2)}$ on $[N, K]_{\mathbb{Z}}$, so that $Y^{\{v\}}$ is a conjoined basis of (SDS) with constant kernel on $[N, K]_{\mathbb{Z}}$ and $\operatorname{Im} X_N^{\{v\}} = \operatorname{Im} X_N^{(2)} = \operatorname{Im} X_N^{(1)}$. Hence, Theorem 6.69 (on the interval $[N, K]_{\mathbb{Z}}$) can be applied, and the first formula in (6.261) proven above holds for the pair $Y^{(1)}$, $Y^{\{v\}}$, i.e.,

$$\operatorname{Im}(L^{(1)} + S_k^{(1)} N^{\{v\}}) = \operatorname{Im} P^{(1)}, \quad k \in [N, K]_{\mathbb{Z}}. \qquad (6.272)$$

Let $S_k^{\{v\}}$ be the $S$-matrix corresponding to the conjoined basis $Y^{\{v\}}$. Then

$$S_k^{\{v\}}(N^{\{v\}})^T = (L^{(1)} + S_k^{(1)} N^{\{v\}})^{\dagger} S_k^{(1)}(L^{(1)})^{\dagger T}(N^{\{v\}})^T, \quad k \in [N, K]_{\mathbb{Z}},$$

according to (6.267). Since $\operatorname{Im}(N^{\{v\}})^T = \operatorname{Im} P^{(2)}$, the matrix $(N^{\{v\}})^T$ can be canceled as we showed above, and then we obtain

$$S_k^{\{v\}} = (L^{(1)} + S_k^{(1)} N^{\{v\}})^{\dagger} S_k^{(1)}(L^{(1)})^{T\dagger}, \quad k \in [N, K]_{\mathbb{Z}}. \qquad (6.273)$$

Assertion (6.271) yields that $S_k^{\{v\}} \to S_k^{(2)}$ as $v \to \infty$ for every $k \in [N, K]_{\mathbb{Z}}$. Since $L^{(1)} + S_k^{(1)} N^{\{v\}} \to L^{(1)} + S_k^{(1)} N^{(1)}$ as $v \to \infty$ for each $k \in [N, K]_{\mathbb{Z}}$ and (6.272) holds, it follows from Remark 1.60(v) that

$$(L^{(1)} + S_k^{(1)} N^{\{v\}})^{\dagger} \to (L^{(1)} + S_k^{(1)} N^{(1)})^{\dagger} \quad \text{as } v \to \infty \text{ for each } k \in [N, K]_{\mathbb{Z}}.$$

In turn, equation (6.273) implies that $S_k^{\{\nu\}} \to S_k^{(2)}$ as $\nu \to \infty$ for each $k \in [N, K]_{\mathbb{Z}}$ and that formula (6.262) holds for all $k \in [N, K]_{\mathbb{Z}}$. But since the index $K \in [N, \infty)_{\mathbb{Z}}$ was arbitrary, it follows that (6.262) holds for all $k \in [N, \infty)_{\mathbb{Z}}$. The proof of Theorem 6.73 is complete.                                                                $\square$

*Remark 6.74* Under the assumptions of Theorem 6.73, we take $k = N$ in (6.260) and obtain by using (6.258) and (6.250) the equality

$$(P^{(1)} M^{(1)})^\dagger = (L^{(1)})^\dagger = L^{(2)} = P^{(2)} M^{(2)} = P^{(2)} (M^{(1)})^{-1}. \tag{6.274}$$

The following result provides a connection between a conjoined basis of (SDS) with constant kernel on $[N, \infty)_{\mathbb{Z}}$ and no forward focal points in $(N, \infty)$ with the principal solution $Y^{[N]}$ of (SDS) and with the order of abnormality. This result, and in particular inequality (6.279), will serve as another crucial tool for the analysis of recessive and dominant solutions of (SDS) at $\infty$.

**Theorem 6.75** *Let $Y$ be a conjoined basis of (SDS) with constant kernel on $[N, \infty)_{\mathbb{Z}}$ and no forward focal points in $(N, \infty)$. Let the matrices $P$, $S_k$, $P_{S\infty}$, and $T$ be defined by (6.234), (6.236), (6.233), (6.238), and (6.237). Then for all $k \in [N, \infty)_{\mathbb{Z}}$*

$$X_k^{[N]} = X_k S_k X_N^T, \quad \operatorname{rank} S_k = \operatorname{rank} X_k^{[N]} = n - d[N, k]_{\mathbb{Z}}, \tag{6.275}$$

$$\operatorname{rank} P_{S\infty} = n - d[N, \infty)_{\mathbb{Z}}, \tag{6.276}$$

$$0 \leq \operatorname{rank} T \leq n - d[N, \infty)_{\mathbb{Z}}, \tag{6.277}$$

$$\Lambda_0[N, \infty)_{\mathbb{Z}} = \operatorname{Im}[X_N^{\dagger T}(I - P_{S\infty})] \oplus \operatorname{Im}[U_N(I - P)], \tag{6.278}$$

$$n - d[N, \infty)_{\mathbb{Z}} \leq \operatorname{rank} X_k \leq n. \tag{6.279}$$

*Proof* As a consequence of estimate (4.62) (with $Y^{[0]} := Y^{[N]}$) or a consequence of Theorems 2.39 and 2.36(iv), we know that the principal solution $Y^{[N]}$ has no forward focal point in $[N, \infty)_{\mathbb{Z}}$. This means that the kernel of $X_k^{[N]}$ is nonincreasing on $[N, \infty)_{\mathbb{Z}}$. Using (6.226) in Theorem 6.57 obtain

$$\Lambda_0[N, k]_{\mathbb{Z}} = \operatorname{Ker} X_k^{[N]} \quad \text{for all } k \in [N, \infty)_{\mathbb{Z}}. \tag{6.280}$$

Let $Y^{[N]}$ be expressed in terms of $Y$ via the matrices $M^{[N]}$, $N^{[N]}$, i.e.,

$$Y_k^{[N]} = (Y_k, \ \bar{Y}_k) \begin{pmatrix} M^{[N]} \\ N^{[N]} \end{pmatrix}, \quad k \in [N, \infty)_{\mathbb{Z}}, \tag{6.281}$$

where $\bar{Y}$ is a conjoined basis of (SDS) from Proposition 6.67. By (6.244) in Theorem 6.69, we have from (6.281) at $k = N$

$$M^{[N]} = -w(\bar{Y}, Y^{[N]}) = -\bar{X}_N^T, \quad N^{[N]} = w(Y, Y^{[N]}) = X_N^T. \tag{6.282}$$

Since $\bar{X}_N$ is uniquely determined by Proposition 6.67(v), the matrices $M^{[N]}$, $N^{[N]}$ do not depend on the choice of $\bar{Y}$. Furthermore, inserting expressions (6.282) into the representation of $X^{[N]}$ in (6.281) and using the definition of $P$ and Proposition 6.67(iv) yield for all $k \in [N, \infty)_{\mathbb{Z}}$

$$X_k^{[N]} = -X_k \bar{X}_N^T + \bar{X}_k X_N^T = -X_k P \bar{X}_N^T + \bar{X}_k P X_N^T$$

$$= -X_k (X_N S_N)^T + (X_k S_k) X_N^T = X_k S_k X_N^T. \qquad (6.283)$$

Since by Theorem 6.66(iii) the equalities $P S_k = S_k = S_k P$ hold, (6.283) gives

$$X_k^\dagger X_k^{[N]} X_N^{\dagger T} \stackrel{(6.283)}{=} X_k^\dagger X_k S_k X_N^T X_N^{\dagger T} = P S_k P = S_k, \quad k \in [N, \infty)_{\mathbb{Z}}. \qquad (6.284)$$

The expressions in (6.283) and (6.284) show that rank $X_k^{[N]}$ = rank $S_k$, while (6.280) implies that rank $X_k^{[N]}$ = $n - d[N, k]_{\mathbb{Z}}$. Therefore, the equalities in (6.275) are established. The constancy of Im $S_k$ for large $k$ and the identity rank $S_k$ = rank $P S_k$ then yield (6.276). Next, by Remark 1.60(i) we know that rank $S_k^\dagger$ = rank $S_k = n - d[N, \infty)_{\mathbb{Z}}$ for large $k$, so that the definition of $T$ in (6.237) yields (6.277). In order to prove (6.278), fix $k \in [N, \infty)_{\mathbb{Z}}$, and let $v \in \Lambda_0[N, k]_{\mathbb{Z}}$. By (6.280) and (6.283), we have $X_k S_k X_N^T v = 0$, and using Theorem 6.66(iii), we get

$$P S_k X_N^T v = S_k^\dagger P S_k X_N^T v = S_k^\dagger X_k^\dagger X_k S_k X_N^T v = 0.$$

Hence, $X_N^T v = (I - P S_k) v_*$ holds for some $v_* \in \mathbb{R}^n$. Now we use that the vector $v$ can be uniquely decomposed as the sum $v = v_1 + v_2$, where $v_1 \in \text{Im } X_N = \text{Im } X_N^{\dagger T}$ and $v_2 \in (\text{Im } X_N)^\perp = \text{Ker } X_N^T = \text{Ker } R_N = \text{Im}[U_N(I - P)]$, by (6.233) and Theorem 6.66(i). Consequently, $X_N^T v_1 = (I - P S_k) v_*$ which in turn implies that $v_1 \in \text{Im}[X_N^{\dagger T}(I - P S_k)]$. Thus, $v \in \text{Im}[X_N^{\dagger T}(I - P S_k)] \oplus \text{Im}[U_N(I - P)]$. Conversely, every vector $w \in \text{Im}[X_N^{\dagger T}(I - P S_k)] \oplus \text{Im}[U_N(I - P)]$ has the form

$$w = X_N^{\dagger T}(I - P S_k) w_1 + U_N(I - P) w_2 \quad \text{for some } w_1, w_2 \in \mathbb{R}^n.$$

Then by the aid of (6.283), $S_k P = S_k$, and $S_k P S_k = S_k$, we get

$$X_k^{[N]} w = X_k S_k X_N^T X_N^{\dagger T}(I - P S_k) w_1 + X_k S_k X_N^T U_N(I - P) w_2$$

$$= X_k S_k P(I - P S_k) w_1 + X_k S_k U_N^T X_N(I - P) w_2$$

$$= X_k S_k(I - P S_k) w_1 + X_k S_k U_N^T X_N P(I - P) w_2 = 0.$$

Thus $w \in \mathrm{Ker}\, X_k^{[N]} = \Lambda_0[N, k]_{\mathbb{Z}}$ by (6.280), and

$$\Lambda_0[N, k]_{\mathbb{Z}} = \mathrm{Im}\,[X_N^{\dagger T}(I - P_{Sk})] \oplus \mathrm{Im}\,[U_N(I - P)], \quad k \in [N, \infty)_{\mathbb{Z}}.$$

In view of (6.238), we conclude, by taking $k \to \infty$, that (6.278) holds. For (6.279) we note that $n - d[N, k]_{\mathbb{Z}} = \mathrm{rank}\, X_k^{[N]} \le \mathrm{rank}\, X_k$ holds on $[N, \infty)_{\mathbb{Z}}$, by (6.275). Moreover, the equality $\mathrm{rank}\, S_k = \mathrm{rank}\, P_{Sk}$ and (6.276) yield that the number $n - d[N, \infty)_{\mathbb{Z}}$ is the maximum of $\mathrm{rank}\, X_k^{[N]}$ on $[N, \infty)_{\mathbb{Z}}$. Therefore, $\mathrm{rank}\, X_k$ lies necessarily between $n - d[N, \infty)_{\mathbb{Z}}$ and $n$, as we claim in (6.279).              □

*Remark 6.76* Formula (6.275) shows that for a conjoined basis $Y$ of (SDS) with constant kernel on $[N, \infty)_{\mathbb{Z}}$ and no forward focal points in $(N, \infty)$, the rank of its corresponding $S$-matrix depends only on the rank of $X^{[N]}$ and hence, on the abnormality of (SDS) on $[N, \infty)_{\mathbb{Z}}$, but not on the choice of $Y$ itself. This means that the changes in $\mathrm{Im}\,(X_k^{[N]})^T$ and $\mathrm{Im}\, S_k$ occur at the same points, i.e., according to Theorem 6.66(iv), there exist a finite partition $N_0 := N < N_1 < N_2 < \cdots < N_m < \infty$ of $[N, \infty)_{\mathbb{Z}}$, which does not depend on the matrix function $S$, such that $\mathrm{Im}\, S_k$ is constant on each subinterval $[N_j, N_{j+1} - 1]_{\mathbb{Z}}$, $j \in \{0, \dots, m - 1\}$ and

$$\mathrm{Im}\, S_k \equiv \mathrm{Im}\, S_{N_j} \subsetneqq \mathrm{Im}\, S_{N_{j+1}}, \quad k \in [N_j, N_{j+1} - 1]_{\mathbb{Z}}, \quad j \in \{0, 1, \dots, m - 1\}$$

$$\mathrm{Im}\, S_k \equiv \mathrm{Im}\, S_{N_m}, \quad \mathrm{rank}\, S_k = n - d[N, \infty)_{\mathbb{Z}}, \quad k \in [N_m, \infty)_{\mathbb{Z}}.$$

The estimate in (6.279) is extremely important in the theory of conjoined bases $Y$ of (SDS). It gives in particular the lower bound for the rank of $X_k$. We shall see that all integer values between $n - d[N, \infty)_{\mathbb{Z}}$ and $n$ are indeed attained by the rank of $X_k$. Moreover, the second condition in (6.275) implies that for all conjoined bases $Y$ of (SDS) with constant kernel on $[N, \infty)_{\mathbb{Z}}$ and no forward focal points in $(N, \infty)$, the rank of the matrices $S_k$ changes at the same points in $[N, \infty)_{\mathbb{Z}}$, i.e., the points where $\mathrm{rank}\, S_k$ changes (increases) do not depend on the particular choice of the conjoined basis $Y$.

At the end of this section, we present three extensions of the previous results. The first one is a generalization of Theorem 6.73 in a sense that the considered initial condition can be taken at any index $L \in [N, \infty)_{\mathbb{Z}}$. This statement then leads to the definition of a genus of conjoined bases of (SDS) in Sect. 6.3.8.

**Theorem 6.77** *Let $Y^{(1)}$ and $Y^{(2)}$ be conjoined bases of* (SDS) *with constant kernel on $[N, \infty)_{\mathbb{Z}}$, and let the equality $\mathrm{Im}\, X_L^{(1)} = \mathrm{Im}\, X_L^{(2)}$ be satisfied for some index $L \in [N, \infty)_{\mathbb{Z}}$. Then $\mathrm{Im}\, X_k^{(1)} = \mathrm{Im}\, X_k^{(2)}$ holds for all $k \in [N, \infty)_{\mathbb{Z}}$.*

*Proof* If $k \in [L, \infty)_{\mathbb{Z}}$, then the result follows from Theorem 6.73 (with $N := L$), i.e., by (6.259) from the equations

$$\left.\begin{aligned} X_k^{(3-i)} &= X_k^{(i)}(M_{(L)}^{(i)} + S_k^{(i)(L)} N^{(i)}), \quad k \in [L, \infty)_{\mathbb{Z}}, \\ P^{(i)} M_{(L)}^{(i)} &= (X_L^{(i)})^{\dagger} X_L^{(3-i)}, \end{aligned}\right\} \quad i \in \{1, 2\}, \quad (6.285)$$

where $S_k^{(i)(L)} := S_k^{(i)} - S_L^{(i)}$ and where $M_{(L)}^{(i)}$ is the invertible matrix associated with $\bar{Y}^{(i)}$ in Proposition 6.67 and Theorem 6.69 (with $N := L$). We note that the matrix $N^{(i)}$ does not depend on $L \in [N, \infty)_{\mathbb{Z}}$ by (6.244) in Theorem 6.69.

For $k \in [N, L]_{\mathbb{Z}}$ we use the backward system (2.10). Define the Riccati quotients

$$Q_k^{(i)} := U_k^{(i)}(X_k^{(i)})^\dagger + (X_k^{(i)})^{\dagger\,T}(U_k^{(i)})^T(I - R_k^{(i)}), \quad k \in [N, \infty)_{\mathbb{Z}}, \quad i \in \{1, 2\}, \tag{6.286}$$

where $R_k^{(i)}$ is the orthogonal projector onto $\operatorname{Im} X_k^{(i)}$ as defined in (6.233). By (2.10), (6.286), and Theorem 6.66, we obtain

$$\left.\begin{array}{l} X_k^{(i)} = (\mathcal{D}_k^T - \mathcal{B}_k^T Q_{k+1}^{(i)})\, X_{k+1}^{(i)}, \\ U_k^{(i)}(X_k^{(i)})^\dagger = Q_k^{(i)} R_k^{(i)}, \end{array}\right\} \quad k \in [N, \infty)_{\mathbb{Z}}, \quad i \in \{1, 2\}. \tag{6.287}$$

The Wronskian $N^{(1)} = w(Y^{(1)}, Y^{(2)})$ is constant, which implies by (6.287) that

$$(X_k^{(1)})^{\dagger\,T} N^{(1)} = R_k^{(1)} U_k^{(2)} - R_k^{(1)} Q_k^{(1)} X_k^{(2)}, \quad k \in [N, \infty)_{\mathbb{Z}}. \tag{6.288}$$

We now apply Theorem 6.66(ii) to $Y^{(1)}$ and use (2.10) and (6.288) to get

$$\begin{aligned} X_k^{(2)} &= \mathcal{D}_k^T X_{k+1}^{(2)} - \mathcal{B}_k^T R_{k+1}^{(1)} U_{k+1}^{(2)} \\ &\overset{(6.288)}{=} (\mathcal{D}_k^T - \mathcal{B}_k^T Q_{k+1}^{(1)})\, X_{k+1}^{(2)} - \mathcal{B}_k^T (X_{k+1}^{(1)})^{\dagger\,T} N^{(1)} \end{aligned} \tag{6.289}$$

for all $k \in [N, \infty)_{\mathbb{Z}}$. We will show that $\operatorname{Im} X_k^{(2)} \subseteq \operatorname{Im} X_k^{(1)}$ on $[N, L]_{\mathbb{Z}}$. Define

$$Z_k := X_k^{(1)}(M_{(L)}^{(1)} - F_k N^{(1)}), \quad F_k := \sum_{j=k}^{L-1}(X_{j+1}^{(1)})^\dagger B_j (X_j^{(1)})^{\dagger\,T}, \quad k \in [N, L]_{\mathbb{Z}},$$

with $F_L := 0$. It follows by the symmetry of $X_k^{(1)}(X_{k+1}^{(1)})^\dagger B_k$ and by the identity $P^{(1)}(X_{k+1}^{(1)})^\dagger = (X_{k+1}^{(1)})^\dagger$, as $X^{(1)}$ has constant kernel on $[N, \infty)_{\mathbb{Z}}$, that

$$\begin{aligned} &Z_k - (\mathcal{D}_k^T - \mathcal{B}_k^T Q_{k+1}^{(1)})\, Z_{k+1} \\ &= X_k^{(1)}(M_{(L)}^{(1)} - F_k N^{(1)}) - (\mathcal{D}_k^T - \mathcal{B}_k^T Q_{k+1}^{(1)})\, X_{k+1}^{(1)}(M_{(L)}^{(1)} - F_{k+1} N^{(1)}) \\ &\overset{(6.287)}{=} X_k^{(1)}(\Delta F_k)\, N^{(1)} = -R_k^{(1)} X_k^{(1)}(X_{k+1}^{(1)})^\dagger B_k (X_k^{(1)})^{\dagger\,T} N^{(1)} \\ &= -R_k^{(1)} \mathcal{B}_k^T (X_{k+1}^{(1)})^{\dagger\,T} P^{(1)} N^{(1)} = -\mathcal{B}_k^T (X_{k+1}^{(1)})^{\dagger\,T} N^{(1)} \end{aligned}$$

for $k \in [N, L - 1]_{\mathbb{Z}}$. Therefore, the sequence $Z_k$ satisfies the nonhomogeneous equation (6.289) on $[N, L - 1]_{\mathbb{Z}}$. Moreover, since $R_L^{(1)} = R_L^{(2)}$ by the assumption $\operatorname{Im} X_L^{(1)} = \operatorname{Im} X_L^{(2)}$, we have

$$Z_L = X_L^{(1)} P^{(1)} M_{(L)}^{(1)} \overset{(6.285)}{=} R_L^{(1)} X_L^{(2)} = R_L^{(2)} X_L^{(2)} = X_L^{(2)}.$$

Thus, the uniqueness of backward solutions of equation (6.289) yields

$$X_k^{(2)} = Z_k = X_k^{(1)} (M_{(L)}^{(1)} - F_k N^{(1)}), \quad k \in [N, L]_{\mathbb{Z}}.$$

This implies that $\operatorname{Im} X_k^{(2)} \subseteq \operatorname{Im} X_k^{(1)}$ on $[N, L]_{\mathbb{Z}}$. Upon interchanging the roles of the conjoined bases $Y^{(1)}$ and $Y^{(2)}$, we derive that $\operatorname{Im} X_k^{(1)} \subseteq \operatorname{Im} X_k^{(2)}$ on $[N, L]_{\mathbb{Z}}$. Therefore, $\operatorname{Im} X_k^{(1)} = \operatorname{Im} X_k^{(2)}$ holds for $k \in [N, L]_{\mathbb{Z}}$, which completes the proof. $\qquad \square$

*Remark 6.78* The proof of Theorem 6.77 shows that the representation formulas in (6.285) are in fact satisfied for every index $k \in [N, \infty)_{\mathbb{Z}}$.

In the next result, we extend Theorem 6.69 by using a notion of representable conjoined bases of (SDS).

**Definition 6.79** Let $Y$ be a conjoined basis of (SDS) with constant kernel on $[N, \infty)_{\mathbb{Z}}$. We say that a solution $\tilde{Y}$ is *representable by* $Y$ on the interval $[N, \infty)_{\mathbb{Z}}$ if in the relation

$$\tilde{Y}_k = \Phi_k \tilde{D}, \quad k \in [0, \infty)_{\mathbb{Z}}, \quad \mathcal{J}\tilde{D} = \begin{pmatrix} w(Y, \tilde{Y}) \\ w(\bar{Y}, \tilde{Y}) \end{pmatrix}, \tag{6.290}$$

where $\Phi_k = (Y_k, \bar{Y}_k)$ is a symplectic fundamental matrix of (SDS) such that $w(Y, \bar{Y}) = I$, the matrix $\tilde{D}$ does not depend on the conjoined basis $\bar{Y}$.

**Theorem 6.80** *Assume that $Y$ is a conjoined basis of* (SDS) *with constant kernel on $[N, \infty)_{\mathbb{Z}}$ with $P$ defined by (6.234), and let $\tilde{Y}$ be a solution of* (SDS). *Then the following conditions are equivalent.*

(i) *The solution $\tilde{Y}$ is representable by $Y$ on $[N, \infty)_{\mathbb{Z}}$.*
(ii) *The (constant) Wronskian $w(Y, \tilde{Y})$ of $Y$ and $\tilde{Y}$ satisfies*

$$\operatorname{Im} w(Y, \tilde{Y}) \subseteq \operatorname{Im} P. \tag{6.291}$$

(iii) *The inclusion $\operatorname{Im} \tilde{X}_k \subseteq \operatorname{Im} X_k$ holds for all $k \in [N, \infty)_{\mathbb{Z}}$.*
(iv) *The inclusion $\operatorname{Im} \tilde{X}_K \subseteq \operatorname{Im} X_K$ holds for some $K \in [N, \infty)_{\mathbb{Z}}$.*

*Proof* Let $\tilde{D}$ be the matrix in (6.290), where $\bar{Y}$ is the conjoined basis in Proposition 6.67 associated with $Y$. First we assume condition (i). Since $w(Y, \bar{Y}) = I$, $w(Y, Y) = 0$, and $X_N^\dagger X_N = P$ hold, it is straightforward to verify that the solution

$\bar{Y}^* := \bar{Y} + Y(I - P)$ of (SDS) also satisfies $w(Y, \bar{Y}^*) = I$ and $X_N^\dagger \bar{X}_N^* = 0$. Denote by $\tilde{D}^*$ the matrix in (6.290) representing $\tilde{Y}$ in terms of the symplectic fundamental matrix $\Phi_k^* := (Y_k, \bar{Y}_k^*)$, i.e.,

$$\tilde{Y}_k = \Phi_k^* \tilde{D}^*, \quad k \in [0, \infty)_\mathbb{Z}, \quad \mathcal{J}\tilde{D}^* = \begin{pmatrix} w(Y, \tilde{Y}) \\ w(\bar{Y}^*, \tilde{Y}) \end{pmatrix}. \tag{6.292}$$

Observe that $w(\bar{Y}^*, \bar{Y}) = I - P$. Then we have

$$\tilde{D}^* \overset{(6.292)}{=} (\Phi_k^*)^{-1} \tilde{Y}_k \overset{(6.290)}{=} -\mathcal{J}(\Phi_k^*)^T \mathcal{J} \, \Phi_k \tilde{D}$$

$$= -\mathcal{J} \begin{pmatrix} w(Y, Y) & w(Y, \bar{Y}) \\ w(\bar{Y}^*, Y) & w(\bar{Y}^*, \bar{Y}) \end{pmatrix} \tilde{D} = \begin{pmatrix} I & P - I \\ 0 & I \end{pmatrix} \tilde{D}.$$

Using the block notation $\tilde{D} = \begin{pmatrix} F \\ G \end{pmatrix}$ and $\tilde{D}^* = \begin{pmatrix} F^* \\ G^* \end{pmatrix}$, we then obtain the equalities $F^* = F - (I - P)G$ and $G^* = w(Y, \tilde{Y}) = G$. Since we now assume that $\tilde{Y}$ is representable by $Y$, it follows that $\tilde{D}^* = \tilde{D}$, and hence $F^* = F$. This yields that $(I - P)G = 0$, i.e., $\operatorname{Im} G \subseteq \operatorname{Ker}(I - P) = \operatorname{Im} P$. Thus, condition (6.291) is proven. Assume now (ii) and let $G := w(Y, \tilde{Y})$. Combining the identities $PG = G$ and $PX_k^T = X_k^T$ on $[N, \infty)_\mathbb{Z}$ yields that $(I - P)U_k^T \tilde{X}_k = 0$ on $[N, \infty)_\mathbb{Z}$. By using the property $\operatorname{Im}[U_k(I - P)] = \operatorname{Ker} R_k$ on $[N, \infty)_\mathbb{Z}$ from Theorem 6.66(i), we then get

$$\operatorname{Im} \tilde{X}_k \subseteq \operatorname{Ker}[(I - P)U_k^T] = \left\{ \operatorname{Im}[U_k(I - P)] \right\}^\perp = (\operatorname{Ker} R_k)^\perp = \operatorname{Im} R_k = \operatorname{Im} X_k$$

on $[N, \infty)_\mathbb{Z}$, which shows (iii). Next, part (iii) implies (iv) trivially. Finally, assume that (iv) is satisfied. Since $\operatorname{Im} X_K = \operatorname{Im} R_K$, it follows that $R_K \tilde{X}_K = \tilde{X}_K$. Hence, $\tilde{X}_K = X_K X_K^\dagger \tilde{X}_K$ and consequently the matrix $F := -w(\bar{Y}, \tilde{Y})$ satisfies

$$F - -(\bar{X}_K^T \tilde{U}_K \quad \bar{U}_K^T \tilde{X}_K) = \bar{U}_K^T X_K X_K^\dagger \tilde{X}_K - \bar{X}_K^T \tilde{U}_K$$

$$= (I + \bar{X}_K^T U_K) X_K^\dagger \tilde{X}_K - \bar{X}_K^T \tilde{U}_K.$$

This shows that the matrix $F$ is determined by $Y$, $\tilde{Y}$, and $\tilde{X}_K$. But since $\tilde{X}_k$ is uniquely determined by $Y$ on $[N, \infty)_\mathbb{Z}$ by Proposition 6.67(v), we get that $F$ is determined by $Y$ and $\tilde{Y}$ only. This implies that the matrix $D = \begin{pmatrix} F \\ G \end{pmatrix}$ does not depend on the choice of $\bar{Y}$, so that the solution $\tilde{Y}$ is representable by $Y$ on $[N, \infty)_\mathbb{Z}$. The proof is complete.                                                                                            $\square$

### 6.3.3  Conjoined Bases with Given Rank

In this subsection we present a construction of conjoined bases $Y$ with a given rank satisfying (6.279). This construction is based on the notion of an equivalence of two solutions of (SDS) and the relation being contained for two conjoined bases of (SDS). More precisely, we say that two solutions $Y$ and $\tilde{Y}$ are *equivalent* on the interval $[N, \infty)_{\mathbb{Z}}$ if

$$X_k = \tilde{X}_k \quad \text{for all } k \in [N, \infty)_{\mathbb{Z}}. \tag{6.293}$$

This means that

$$X_N = \tilde{X}_N \quad \text{and} \quad \operatorname{Im}(U_N - \tilde{U}_N) \subseteq \Lambda_0[N, \infty)_{\mathbb{Z}}. \tag{6.294}$$

By using the representation of the space $\Lambda_0[N, \infty)_{\mathbb{Z}}$ in (6.278) we obtain the following more precise statement.

**Proposition 6.81** *Let $Y$ be a conjoined basis of* (SDS) *with constant kernel on $[N, \infty)_{\mathbb{Z}}$ and no forward focal points in $(N, \infty)$. Let the matrices $P$ and $P_{S\infty}$ be defined by (6.234) and (6.238). Then two solutions $Y^{(1)}$ and $Y^{(2)}$ of* (SDS) *are equivalent on $[N, \infty)_{\mathbb{Z}}$ if and only if there exist unique $n \times n$ matrices $G$ and $H$ such that*

$$X_N^{(1)} = X_N^{(2)}, \quad U_N^{(1)} - U_N^{(2)} = X_N^{\dagger T} G + U_N H, \tag{6.295}$$

$$\operatorname{Im} G \subseteq \operatorname{Im}(P - P_{S\infty}), \quad \operatorname{Im} H \subseteq \operatorname{Im}(I - P). \tag{6.296}$$

*Proof* The existence of matrices $G$ and $H$ satisfying (6.295) follows from (6.294) and from the representation of $\Lambda_0[N, \infty)_{\mathbb{Z}}$ in (6.278) in Theorem 6.75. The latter reference also gives the second property in (6.296) and the inclusion $\operatorname{Im} G \subseteq \operatorname{Im}(I - P_{S\infty})$. However, the matrix $G$ can be chosen so that $\operatorname{Im} G \subseteq \operatorname{Im} P$, because in (6.295) we may take $X_N^{\dagger T} X^T X^{\dagger T} G = X^{\dagger T} P G$ instead of $X^{\dagger T} G$, by the properties of the Moore-Penrose pseudoinverse. Equation (6.296) then also implies the uniqueness of $G$ and $H$, because $Y$ is a conjoined basis. Conversely, the conditions in (6.295) and (6.296) imply that (6.294) holds, and thus $Y^{(1)}$ and $Y^{(2)}$ are equivalent. $\qquad\square$

The equivalence of solutions on $[N, \infty)_{\mathbb{Z}}$ leads to an ordering in the set of all conjoined bases of (SDS). This ordering is phrased in terms of the relation "being contained" defined below.

**Definition 6.82** Let $Y$ be a conjoined basis of (SDS) with constant kernel on $[N, \infty)_{\mathbb{Z}}$ and no forward focal points in $(N, \infty)$. Let $P$ and $P_{S\infty}$ be the associated orthogonal projectors in (6.234) and (6.238). Consider an orthogonal projector $P^*$ satisfying

$$\operatorname{Im} P_{S\infty} \subseteq \operatorname{Im} P^* \subseteq \operatorname{Im} P. \tag{6.297}$$

We say that a conjoined basis $Y^*$ is *contained* in $Y$ on $[N, \infty)_{\mathbb{Z}}$ with respect to $P^*$, or that $Y$ *contains* $Y^*$ on $[N, \infty)_{\mathbb{Z}}$ with respect to $P^*$, if the solutions $Y^*$ and $Y P^*$ are equivalent on $[N, \infty)_{\mathbb{Z}}$, i.e., if $X_k^* = X_k P^*$ for all $k \in [N, \infty)_{\mathbb{Z}}$.

*Remark 6.83* The conjoined basis $Y^*$, which is contained in $Y$ on $[N, \infty)_{\mathbb{Z}}$ with respect to the projector $P^*$, has also a constant kernel on $[N, \infty)_{\mathbb{Z}}$ and no forward focal points in $(N, \infty)$. The first property can be seen from $X_k^* = X_k P^*$ and $\operatorname{Ker} X_k = \operatorname{Ker} P \subseteq \operatorname{Ker} P^*$, which imply that $\operatorname{Ker} X_k^* = \operatorname{Ker} P^*$ on $[N, \infty)_{\mathbb{Z}}$. This also shows that $P^* = P_{\operatorname{Im}(X_k^*)^T}$ is the projector associated with $Y^*$ through (6.233) and (6.234). The fact that $Y^*$ has no forward focal points in $(N, \infty)$ is proven as follows. From the identity $P^* = P P^*$, we get

$$(X_k^*)^\dagger = P^*(X_k^*)^\dagger = P P^*(X_k^*)^\dagger = X_k^\dagger X_k P^*(X_k^*)^\dagger$$

$$= X_k^\dagger X_k^*(X_k^*)^\dagger = X_k^\dagger R_k^*. \tag{6.298}$$

Therefore, by Theorem 6.66(ii), we obtain on $[N, \infty)_{\mathbb{Z}}$

$$X_k^*(X_{k+1}^*)^\dagger \mathcal{B}_k = X_k P^*(X_{k+1}^*)^\dagger \mathcal{B}_k = X_k(X_{k+1}^*)^\dagger \mathcal{B}_k$$

$$\stackrel{(6.298)}{=} X_k X_{k+1}^\dagger R_{k+1}^* \mathcal{B}_k = X_k X_{k+1}^\dagger \mathcal{B}_k. \tag{6.299}$$

The last term is nonnegative definite, since $Y$ has no forward focal points in $(N, \infty)$.

We now utilize the matrices $G$ and $H$ in (6.295)–(6.296) in order to describe the conjoined bases of (SDS), which are contained in a given conjoined basis (Theorem 6.84) or which contain a given conjoined basis (Theorem 6.87). We now introduce the set $\mathcal{M}(P^{**}, P^*, P)$ of pairs of matrices $(G, H)$ associated with orthogonal projectors $P^{**}, P^*, P$ satisfying the inclusions $\operatorname{Im} P^{**} \subseteq \operatorname{Im} P^* \subseteq \operatorname{Im} P$. We define

$$\mathcal{M}(P^{**}, P^*, P) = \left\{ (G, H) \in \mathbb{R}^{n \times n} \times \mathbb{R}^{n \times n}, \ \operatorname{rank}(G^T, H^T, P^*) = n, \\ P^{**} G = 0, \ P G = G, \ P^* G = G^T P^*, \ P H = 0 \right\}, \tag{6.300}$$

Note that the set $\mathcal{M}(P^{**}, P^*, P)$ is always nonempty, because the pair $(G, H)$ with $G := P - P^*$ and $H := I - P$ belongs to $\mathcal{M}(P^{**}, P^*, P)$. Note also that if $P = I$, then $H = 0$. The following result describes all conjoined bases $Y^*$ of (SDS), which are contained in a given conjoined basis $Y$ with respect to a fixed projector $P^*$.

**Theorem 6.84** *Let $Y$ be a conjoined basis of (SDS) with constant kernel on $[N, \infty)_{\mathbb{Z}}$ and no forward focal points in $(N, \infty)$. Let $P$ and $P_{S\infty}$ be the associated orthogonal projectors in (6.234) and (6.238). Consider an orthogonal projector $P^*$ satisfying (6.297). Then a conjoined basis $Y^*$ of (SDS) is contained in $Y$ on $[N, \infty)_{\mathbb{Z}}$ with respect to $P^*$ if and only if for some $(G, H) \in \mathcal{M}(P_{S\infty}, P^*, P)$*

$$X_N^* = X_N P^*, \quad U_N^* = U_N P^* + X_N^{\dagger T} G + U_N H. \tag{6.301}$$

*Proof* Let $Y^*$ be a conjoined basis of (SDS) which is contained in $Y$ on $[N, \infty)_{\mathbb{Z}}$ with respect to $P^*$. Then $Y$ and $Y P^*$ are equivalent on $[N, \infty)_{\mathbb{Z}}$ by Definition 6.82. From Proposition 6.81 (with $Y^{(1)} := Y P^*$ and $Y^{(2)} := Y^*$), it follows that $Y^*$ satisfies the initial conditions in (6.301) with the matrices $G$ and $H$ such that $P_{S\infty} G = 0$, $PG = G$, and $PH = 0$. We will show that $(G, H) \in \mathcal{M}(P_{S\infty}, P^*, P)$. Multiplying (6.301) by $(X_N^*)^T$, we get

$$(X_N^*)^T U_N^* = P^* X_N^T U_N P^* + P^* X_N^T X_N^{T\dagger} G + P^* X_N^T U_N H. \tag{6.302}$$

The symmetry of $X_N^T U_N$ and the identities $X_N P = X_N$ and $PH = 0$ imply that

$$P^* X_N^T U_N H = P^* U_N^T X_N H = P^* U_N^T X_N P H = 0. \tag{6.303}$$

Upon inserting (6.303) into (6.302) and using $X_N^T X_N^{T\dagger} = P$ and $PG = G$, we obtain

$$P^* G = P^* P G = P^* X_N^T X_N^{T\dagger} G = (X_N^*)^T U_N^* - P^* X_N^T U_N P^*.$$

This shows that the matrix $P^* G$ is symmetric, i.e., $P^* G = G^T P^*$. Furthermore, if $v \in \mathbb{R}^n$ is a vector such that $v \in \operatorname{Ker} G \cap \operatorname{Ker} H \cap \operatorname{Ker} P^*$, then (6.301) implies that $v \in \operatorname{Ker} X_N^* \cap \operatorname{Ker} U_N^* = \{0\}$, because $Y^*$ is a conjoined basis. Therefore, $\operatorname{Ker} G \cap \operatorname{Ker} H \cap \operatorname{Ker} P^* = \{0\}$, which is equivalent with $\operatorname{rank}(G^T, H^T, P^*) = n$. The above properties of $G$ and $H$ imply that $(G, H) \in \mathcal{M}(P_{S\infty}, P^*, P)$. Conversely, it is easy to see that for any pair $(G, H) \in \mathcal{M}(P_{S\infty}, P^*, P)$ the solution $Y^*$ of (SDS) satisfying the initial conditions in (6.301) is a conjoined basis, which is contained in $Y$ on $[N, \infty)_{\mathbb{Z}}$ with respect to $P^*$.                                                                 □

It follows from Proposition 6.81 that the pair $(G, H) \in \mathcal{M}(P_{S\infty}, P^*, P)$, which determines the conjoined basis $Y^*$ in Theorem 6.84, is unique. For this reason we also say that $Y^*$ is contained in $Y$ on $[N, \infty)_{\mathbb{Z}}$ *through the pair* $(G, H)$. Moreover, the matrix $G$ in (6.301) satisfies $G = w(Y, Y^*)$, which can be verified by the calculation of the Wronskian at $k = N$.

*Remark 6.85* For every orthogonal projector $P^*$ satisfying (6.297), there always exists a conjoined basis $Y^*$ which is contained in $Y$ with respect to $P^*$ on $[N, \infty)_{\mathbb{Z}}$, for example, it is the conjoined basis $Y^*$ of (SDS) given by the initial conditions (6.301) with $G := P - P^*$ and $H := I - P$. This follows from Theorem 6.84 and from the fact that the above choice of $(G, H)$ belongs to $\mathcal{M}(P_{S\infty}, P^*, P)$.

In the following result, we derive an additional property of the conjoined bases $Y^*$ of (SDS) which are contained in $Y$ on $[N, \infty)_{\mathbb{Z}}$.

**Proposition 6.86** *Let $Y$ be a conjoined basis of (SDS) with constant kernel on $[N, \infty)_{\mathbb{Z}}$ and no forward focal points in $(N, \infty)$. Let $S_k$ in (6.236) be its corresponding S-matrix. If $Y^*$ is any conjoined basis of (SDS) which is contained* *in* $Y$ *on* $[N, \infty)_{\mathbb{Z}}$ *and if* $S_k^*$ *is its corresponding S-matrix, then* $S_k^* = S_k$ *for all* $k \in [N, \infty)_{\mathbb{Z}}.$

*Proof* Let $P^*$ be an orthogonal projector satisfying (6.297) such that $Y^*$ is contained in $Y$ with respect to $P^*$ on $[N, \infty)_{\mathbb{Z}}.$, i.e., $X_k^* = X_k P^*$ on $[N, \infty)_{\mathbb{Z}}.$ Then by Remark 6.83, the conjoined basis $Y^*$ has constant kernel on $[N, \infty)_{\mathbb{Z}}$ and no forward focal points in $(N, \infty)$. In turn, by Theorem 6.66(ii) and identity (6.298), we have

$$
S_k \overset{(6.236)}{=} \sum_{j=N}^{k-1} X_{j+1}^{\dagger} \mathcal{B}_j X_j^{\dagger T} = \sum_{j=N}^{k-1} X_{j+1}^{\dagger} R_{j+1}^* \mathcal{B}_j R_j^* X_j^{\dagger T}
$$

$$
= \sum_{j=N}^{k-1} (X_{j+1}^*)^{\dagger} \mathcal{B}_j (X_j^*)^{\dagger T} \overset{(6.236)}{=} S_k^* \tag{6.304}
$$

for all $k \in [N, \infty)_{\mathbb{Z}}$, which completes the proof. $\qquad\qquad\square$

  Next we present a supplement to Theorem 6.84 in the sense that we construct conjoined bases $\tilde{Y}$ of (SDS) with constant kernel on $[N, \infty)_{\mathbb{Z}}$, which contain a given conjoined basis $Y^*$ on $[N, \infty)_{\mathbb{Z}}$ according to Definition 6.82. This construction is based on a suitable choice of the initial conditions of $\tilde{Y}$ at $k = N$. We shall see that this choice is closely related with the set in (6.300)

**Theorem 6.87** *Let* $Y^*$ *be a conjoined basis of* (SDS) *with constant kernel on* $[N, \infty)_{\mathbb{Z}}$ *and no forward focal points in* $(N, \infty)$. *Let* $P^*$, $R_k^*$, *and* $P_{S^*\infty}$ *be the associated projectors defined in* (6.234), (6.233), *and* (6.238) *through* $X^*$. *Let* $P$ *and* $R$ *be orthogonal projectors satisfying*

$$
\mathrm{Im}\, P^* \subseteq \mathrm{Im}\, P, \quad \mathrm{Im}\, R_N^* \subseteq \mathrm{Im}\, R, \quad \mathrm{rank}\, P = \mathrm{rank}\, R, \tag{6.305}
$$

*and let* $(G, H) \in \mathcal{M}(P_{S^*\infty}, P^*, P)$. *If the* $n \times n$ *matrices* $X$ *and* $U$ *solve the system*

$$
XX^{\dagger} = R, \quad X^{\dagger}X = P, \tag{6.306}
$$

$$
XP^* = X_N^*, \quad X^T U = U^T X, \quad X^{\dagger T} G + U(P^* + H) = U_N^*, \tag{6.307}
$$

*we denote by* $\tilde{Y}$ *the solution of* (SDS) *given by the initial conditions* $\tilde{X}_N = X$ *and* $\tilde{U}_N = U$. *Then* $\tilde{Y}$ *has the following properties.*

(i) *The solution* $\tilde{Y}$ *is a conjoined basis of* (SDS).
(ii) *The conjoined basis* $\tilde{Y}$ *has constant kernel on* $[N, \infty)_{\mathbb{Z}}$ *and no forward focal points in* $(N, \infty)$. *In addition, the associated projectors* $\tilde{P}$ *and* $\tilde{R}_k$ *defined in* (6.234) *and* (6.233) *through* $\tilde{X}$ *satisfy* $\tilde{P} = P$ *and* $\tilde{R}_N = R$.
(iii) *The conjoined basis* $Y^*$ *is contained in* $\tilde{Y}$ *on* $[N, \infty)_{\mathbb{Z}}$ *through the pair* $(G, H)$ *or equivalently with respect to the projector* $P^*$.

*Proof* Let the conjoined basis $Y^*$ and the projectors $P^*$, $R^*$, $P_{S^*\infty}$ and $P$, $R$ be as in the above statement and let $(G, H) \in \mathcal{M}(P_{S^*\infty}, P^*, P)$. Let $X$ and $U$ solve system (6.306)–(6.307), and let $\tilde{Y}$ be the solution of (SDS) given by the initial conditions $\tilde{X}_N = X$ and $\tilde{U}_N = U$. Let $\tilde{P}_k := \tilde{X}_k^\dagger \tilde{X}_k$ and $\tilde{R}_k := \tilde{X}_k \tilde{X}_k^\dagger$ be the associated orthogonal projectors in (6.233). Then (6.306) implies that $\tilde{P}_N = P$ and $\tilde{R}_N = R$. For part (i) we observe that it is enough to check the two defining properties of a conjoined basis at $k = N$. The symmetry of $\tilde{X}_N^T \tilde{U}_N$ follows from the second equation in (6.307). Suppose now that $\tilde{X}_N v = 0$ and $\tilde{U}_N v = 0$ for some $v \in \mathbb{R}^n$. Then $Pv = 0$ and $P^*v = 0$, by the first inclusion in (6.305). From [285, Lemmas 5.3(iv) and 5.4], we then obtain that $v \in \text{Im } P^*$. This means that $v \in \text{Ker } P^* \cap \text{Im } P^* = \{0\}$, i.e., $v = 0$. Thus, $\text{rank}(\tilde{X}_N^T, \tilde{U}_N^T) = n$ and $\tilde{Y}$ is a conjoined basis. For part (ii) we first observe that $G = w(\tilde{Y}, Y^*)$, because by the definition of $\mathcal{M}(P_{S^*\infty}, P^*, P)$ in (6.300), we have $PG = G$ and $PH = 0$ and

$$\tilde{X}_N^T U_N^* - \tilde{U}_N^T X_N^* \overset{(6.307)}{=} X^T [X^{\dagger T} G + U(P^* + H)] - U^T X P^* \overset{(6.306)}{=} PG + U^T X P H = G.$$

We define on $[N, \infty)_\mathbb{Z}$ the symmetric matrix $Q_k^*$ with the following property

$$Q_k^* := U_k^* (X_k^*)^\dagger + [U_k^* (X_k^*)^\dagger]^T (I - R_k^*), \quad Q_k^* X_k^* = U_k^* (X_k^*)^\dagger X_k^* = U_k^* P^*;$$
$$\tag{6.308}$$

see [257, Section 2.1.1]. By (6.308) and Theorem 6.66(i)–(ii) applied to $Y^*$, we get

$$\left.\begin{array}{c} R_k^* Q_k^* \tilde{X}_k - R_k^* \tilde{U}_k = (X_k^*)^{\dagger T} G^T, \\ \mathcal{B}_k R_k^* = \mathcal{B}_k = R_{k+1}^* \mathcal{B}_k, \quad R_k^* U_k^* = R_k^* U_k^* P^*, \end{array}\right\} \tag{6.309}$$

where the equalities in (6.309) are satisfied for $k \in [N, \infty)_\mathbb{Z}$. It now follows that

$$X_{k+1}^* = \mathcal{A}_k X_k^* + \mathcal{B}_k U_k^* = \mathcal{A}_k X_k^* + \mathcal{B}_k R_k^* U_k^* P^* = (\mathcal{A}_k + \mathcal{B}_k Q_k^*) X_k^* \tag{6.310}$$

on $[N, \infty)_\mathbb{Z}$. From (6.309) it follows that the function $\tilde{X}$ solves on $[N, \infty)_\mathbb{Z}$ the nonhomogeneous first-order linear difference equation

$$\tilde{X}_{k+1} = \mathcal{A}_k \tilde{X}_k + \mathcal{B}_k R_k^* \tilde{U}_k = (\mathcal{A}_k + \mathcal{B}_k Q_k^*) \tilde{X}_k - \mathcal{B}_k (X_k^*)^{\dagger T} G^T. \tag{6.311}$$

Let $\Phi_k$ be the solution of the associated homogeneous equation

$$\Phi_{k+1} = (\mathcal{A}_k + \mathcal{B}_k Q_k^*) \Phi_k, \quad k \in [N, \infty)_\mathbb{Z}, \quad \Phi_N = X. \tag{6.312}$$

Note that $\Phi_k$ is correctly defined in forward time on $[N, \infty)_\mathbb{Z}$ and that the matrices $\mathcal{A}_k + \mathcal{B}_k Q_k^*$ and hence $\Phi_k$ are not necessarily invertible. From (6.310) and the fact $\Phi_N P^* = X_N^*$ (see the first condition in (6.307)), the uniqueness of solutions of (6.312) yields the equality $X_k^* = \Phi_k P^*$ on $[N, \infty)_{\mathbb{Z}}$. Moreover, from the variation of constants principle for equations (6.311) and (6.312), we get

$$\tilde{X}_k = \Phi_k (P - S_k^* G^T), \quad k \in [N, \infty)_{\mathbb{Z}}. \tag{6.313}$$

Indeed, for $k = N$ we have $\Phi_N (P - S_N^* G^T) = XP = X = \tilde{X}_N$, because $S_N^* = 0$, and by (6.313), (6.312), $P^*(X_{k+1}^*)^\dagger = (X_{k+1}^*)^\dagger$, and $XP^* = X_N^*$ from (6.307), we get

$$\tilde{X}_{k+1} - (\mathcal{A}_k + \mathcal{B}_k \mathcal{Q}_k^*) \tilde{X}_k = -\Phi_{k+1} (\Delta S_k^*) G^T = -\Phi_{k+1} (X_{k+1}^*)^\dagger \mathcal{B}_k (X_k^*)^{\dagger T} G^T$$

$$= -\Phi_{k+1} P^* (X_{k+1}^*)^\dagger \mathcal{B}_k (X_k^*)^{\dagger T} G^T$$

$$= -X_{k+1}^* (X_{k+1}^*)^\dagger \mathcal{B}_k (X_k^*)^{\dagger T} G^T$$

$$= -R_{k+1}^* \mathcal{B}_k (X_k^*)^{\dagger T} G^T \overset{(6.309)}{=} -\mathcal{B}_k (X_k^*)^{\dagger T} G^T.$$

Equality (6.313) and identities $PP^* = P^*$, $G^T P = G^T$, $G^T P^* = P^* G$, $P_{S^* \infty} G = 0$, and $P^* S_k^* = S_k^* = S_k^* P^*$, $S_k^* = S_k^* P_{S^* \infty}$, from (6.305), (6.300), and (6.239) imply that

$$\tilde{X}_k P = \tilde{X}_k, \quad \tilde{X}_k P^* = X_k^* \quad \text{on } [N, \infty)_{\mathbb{Z}}. \tag{6.314}$$

This shows that $\operatorname{Im} \tilde{X}_k^T \subseteq \operatorname{Im} P$ and $\operatorname{Im} X_k^* \subseteq \operatorname{Im} \tilde{X}_k$ on $[N, \infty)_{\mathbb{Z}}$. Next we will prove that for all $k \in [N, \infty)_{\mathbb{Z}}$, we have

$$\operatorname{Im} (P - S_k^* G^T) = \operatorname{Im} P = \operatorname{Im} (P - S_k^* G^T)^T = \operatorname{Im} (P - S_k^* G^T)^\dagger. \tag{6.315}$$

Fix an index $k \in [N, \infty)_{\mathbb{Z}}$. Since $\operatorname{Im} S_k^* \subseteq \operatorname{Im} P^* \subseteq \operatorname{Im} P$ by Theorem 6.66(iv) (with $Y := Y^*$) and assumption (6.305), the equality $S_k^* = PS_k^*$ holds. This in turn implies that $P - S_k^* G^T = P - PS_k^* G^T$ and hence $\operatorname{Im} (P - S_k^* G^T) \subseteq \operatorname{Im} P$. Next we consider a vector $v \in \operatorname{Ker} (P - S_k^* G^T)$, so that by $G^T = G^T P$ we get $Pv = S_k^* G^T v = S_k^* G^T Pv$. Then the vector $w := Pv$ satisfies $w = S_k^* G^T w$, i.e., $w \in \operatorname{Im} S_k^* \subseteq \operatorname{Im} P_{S^* \infty}$. Thus, $w = P_{S^* \infty} w$, and consequently with the aid of $G^T P_{S^* \infty} = 0$, we obtain $w = S_k^* G^T w = S_k^* G^T P_{S^* \infty} w = 0$. This shows that $Pv = w = 0$, so that $v \in \operatorname{Ker} P$. Therefore, we proved that $\operatorname{Ker} (P - S_k^* G^T) \subseteq \operatorname{Ker} P$, which is equivalent with $\operatorname{Im} P \subseteq \operatorname{Im} (P - S_k^* G^T)^T$. Altogether, we showed that

$$\operatorname{Im} (P - S_k^* G^T) \subseteq \operatorname{Im} P \subseteq \operatorname{Im} (P - S_k^* G^T)^T. \tag{6.316}$$

Since the dimensions of the subspaces on the left-hand and right-hand sides above are equal (as $\operatorname{rank} A = \operatorname{rank} A^T$), it follows that in (6.316) we actually have the equalities. This proves the first two equalities in (6.315). Finally, the last equality in (6.315) follows from Remark 1.60(i). We now consider the time-reversed symplectic system (2.47) for $\tilde{Y}$, i.e.,

$$\tilde{X}_k = \mathcal{D}_k^T \tilde{X}_{k+1} - \mathcal{B}_k^T \tilde{U}_{k+1}, \quad \tilde{U}_k = -\mathcal{C}_k^T \tilde{X}_{k+1} + \mathcal{A}_k^T \tilde{U}_{k+1}, \quad k \in [0, \infty)_{\mathbb{Z}}.$$

Its first equation in combination with $\mathcal{B}_k = R_{k+1}^* \mathcal{B}_k$ and with (6.308) and (6.309) at $k+1$ implies that $\tilde{X}_k$ satisfies on $[N, \infty)_{\mathbb{Z}}$ the nonhomogeneous first-order linear difference equation

$$\tilde{X}_k = \mathcal{D}_k^T \tilde{X}_{k+1} - \mathcal{B}_k^T \tilde{U}_{k+1} = (\mathcal{D}_k^T - \mathcal{B}_k^T Q_{k+1}^*) \tilde{X}_{k+1} + \mathcal{B}_k^T (X_{k+1}^*)^{\dagger T} G^T. \quad (6.317)$$

Fix an index $M \in [N, \infty)_{\mathbb{Z}}$, and consider the solution $\Psi_k$ of the associated homogeneous equation

$$\Psi_k = (\mathcal{D}_k^T - \mathcal{B}_k^T Q_{k+1}^*) \Psi_{k+1}, \quad k \in [N, M-1]_{\mathbb{Z}}, \quad \Psi_M = \tilde{X}_M (P - S_M^* G^T)^{\dagger}. \tag{6.318}$$

From (6.318), (6.315), (6.313), and $P P^* = P^*$, we now obtain the equalities

$$\Psi_k P = \Psi_k, \quad \Psi_k P^* = X_k^*, \quad k \in [N, M]_{\mathbb{Z}}, \tag{6.319}$$

since the sequences $\Psi_k P$, $\Psi_k$, $\Psi_k P^*$, and $X_k^*$ satisfy the same difference equation in (6.318) and

$$\Psi_M P = \tilde{X}_M (P - S_M^* G^T)^{\dagger} P \overset{(6.315)}{=} \tilde{X}_M (P - S_M^* G^T)^{\dagger} = \Psi_M,$$

$$\Psi_M P^* = \tilde{X}_M (P - S_M^* G^T)^{\dagger} P^* \overset{(6.313)}{=} \Phi_M (P - S_M^* G^T) (P - S_M^* G^T)^{\dagger} P^*$$

$$\overset{(6.315)}{=} \Phi_M P P^* = \Phi_M P^* = X_M^*.$$

We shall prove that

$$\tilde{X}_k = \Psi_k (P - S_k^* G^T), \quad k \in [N, M]_{\mathbb{Z}}. \tag{6.320}$$

Indeed, the sequence $V_k = \Psi_k (P - S_k^* G^T)$ satisfies on $[N, M-1]_{\mathbb{Z}}$ the equation

$$V_k - (\mathcal{D}_k^T - \mathcal{B}_k^T Q_{k+1}^*) V_{k+1} \overset{(6.318)}{=} \Psi_k (\Delta S_k^*) G^T = \Psi_k (\Delta S_k^*)^T G^T$$

$$= \Psi_k P^* (X_k^*)^{\dagger} \mathcal{B}_k^T (X_{k+1}^*)^{\dagger T} G^T$$

$$\overset{(6.319)}{=} R_k^* \mathcal{B}_k^T (X_{k+1}^*)^{\dagger T} G^T = \mathcal{B}_k^T (X_{k+1}^*)^{\dagger T} G^T,$$

which is the same equation as (6.317), and

$$V_M = \Psi_M (P - S_M^* G^T) \overset{(6.318)}{=} \tilde{X}_M (P - S_M^* G^T)^{\dagger} (P - S_M^* G^T) \overset{(6.315)}{=} \tilde{X}_M P \overset{(6.314)}{=} \tilde{X}_M.$$

Therefore, the uniqueness of solutions of (6.317) in backward time yields that equality (6.320) holds. Our next claim is to prove the equalities

$$\text{Ker } \Psi_k = \text{Ker } P, \quad \Psi_k^\dagger \Psi_k = P, \quad \text{Ker } \tilde{X}_k = \text{Ker } P, \quad k \in [N, M]_{\mathbb{Z}}. \quad (6.321)$$

We start with the first condition in (6.321). By (6.318), the kernel of $\Psi_k$ is nonincreasing on $[N, M]_{\mathbb{Z}}$, which yields through (6.320) that

$$\text{Ker } \Psi_k \subseteq \text{Ker } \Psi_N = \text{Ker } \tilde{X}_N = \text{Ker } X = \text{Ker } P, \quad k \in [N, M]_{\mathbb{Z}}.$$

On the other hand, the first equality in (6.319) implies that $\text{Ker } P \subseteq \text{Ker } \Psi_k$ on $[N, M]_{\mathbb{Z}}$. Thus, $\text{Ker } \Psi_k = \text{Ker } P$, which also yields the second condition in (6.321). Finally, by Remark 1.62(iv) and $S_k^* = P S_k^*$, we have for $k \in [N, M]_{\mathbb{Z}}$

$$\text{Ker } \tilde{X}_k \overset{(6.320)}{=} \text{Ker } \Psi_k \, (P - S_k^* G^T) = \text{Ker } \Psi_k^\dagger \Psi_k \, (P - S_k^* G^T) = \text{Ker } P(P - S_k^* G^T)$$

$$= \text{Ker } (P - S_k^* G^T) = [\,\text{Im } (P - S_k^* G^T)^T\,]^\perp \overset{(6.315)}{=} (\text{Im } P)^\perp = \text{Ker } P,$$

which completes the proof of (6.321). Since the index $M \in [N, \infty)_{\mathbb{Z}}$ was arbitrary, it follows from the third condition in (6.321) that $\text{Ker } \tilde{X}_k = \text{Ker } P$ on $[N, \infty)_{\mathbb{Z}}$, i.e., $\tilde{Y}$ has constant kernel on $[N, \infty)_{\mathbb{Z}}$ and $\tilde{P}_k = \tilde{P} = P$ on $[N, \infty)_{\mathbb{Z}}$. In addition, by the same calculations as in (6.298) and (6.299) with $Y := \tilde{Y}$, we can show that $\tilde{X}_k^\dagger R_k^* = (X_k^*)^\dagger$ and $\tilde{X}_k \tilde{X}_{k+1}^\dagger B_k = X_k^* (X_{k+1}^*)^\dagger B_k \geq 0$ on $[N, \infty)_{\mathbb{Z}}$. This means that the conjoined basis $\tilde{Y}$ has no forward focal points in $(N, \infty)$ and the proof of part (ii) is complete. For part (iii), we first observe that Theorem 6.66(ii) and the same calculation as in (6.304) imply $\tilde{S}_k = S_k^*$ on $[N, \infty)_{\mathbb{Z}}$. Therefore, $P_{\tilde{S}_\infty} = P_{S^*_\infty}$ and consequently, $\text{Im } P_{\tilde{S}_\infty} = \text{Im } P_{S^*_\infty} \subseteq \text{Im } P^* \subseteq \text{Im } P$. Hence, $Y^*$ is contained in $\tilde{Y}$ by Definition 6.82. Moreover, since we now have $(G, H) \in \mathcal{M}(P_{\tilde{S}_\infty}, P^*, P)$, it follows from (6.307) and Theorem 6.84 that $Y^*$ is contained in $\tilde{Y}$ through the pair $(G, H)$. The proof of Theorem 6.87 is complete.                                    $\square$

*Remark 6.88*  Note that in the discrete case we utilize both forward and backward (time-reversed) systems (6.312) and (6.318) and that their solutions $\Phi_k$ and $\Psi_k$ are in general singular. In the continuous case, both systems (6.312) and (6.318) coincide, and the argument therein is more straightforward; see the proof of [285, Theorem 5.6]. Moreover, the results in Theorem 6.87(ii) and Remark 6.83 show that the relation being contained preserves not only the constant kernel on $[N, \infty)_{\mathbb{Z}}$ of the involved conjoined bases but also the property of having no forward focal points in $(N, \infty)$. This points to a fundamental difference in the discrete theory compared with the continuous case.

In the next result, we provide a converse to Theorem 6.87. Namely, we prove that the initial conditions of the conjoined bases $\tilde{Y}$ of (SDS), which contain a given $Y^*$ on $[N, \infty)_{\mathbb{Z}}$, satisfy system (6.306)–(6.307).

**Theorem 6.89** *Let $\tilde{Y}$ and $Y^*$ be conjoined bases of* (SDS) *such that $\tilde{Y}$ has constant kernel on $[N, \infty)_{\mathbb{Z}}$ and no forward focal points in $(N, \infty)$. If $\tilde{Y}$ contains $Y^*$ on $[N, \infty)_{\mathbb{Z}}$ with respect to the projector $P^*$ or equivalently through a pair $(G, H) \in \mathcal{M}(P_{S^*\infty}, P^*, \tilde{P})$, then the matrices $\tilde{X}_N$ and $\tilde{U}_N$ solve system* (6.306)–(6.307) *with $P := \tilde{P}$ and $R := \tilde{R}_N$.*

*Proof* By Theorem 6.84 with $Y := \tilde{Y}$, condition (6.301) holds with a (unique) pair $(G, H) \in \mathcal{M}(P_{\tilde{S}\infty}, P^*, \tilde{P})$. Since by Proposition 6.86 the corresponding $S$-matrices satisfy $\tilde{S}_k = S_k^*$ on $[N, \infty)_{\mathbb{Z}}$, it follows that $P_{\tilde{S}\infty} = P_{S^*\infty}$, and consequently $(G, H) \in \mathcal{M}(P_{S^*\infty}, P^*, \tilde{P})$. Set $X := \tilde{X}_N$ and $U := \tilde{U}_N$. Then the above choice of $P$ and $R$ yields that (6.305) as well as (6.306)–(6.307) are satisfied. $\square$

The results in Theorems 6.87 and 6.89 show that the construction of conjoined bases $\tilde{Y}$ of (SDS) with constant kernel on $[N, \infty)_{\mathbb{Z}}$ and no forward focal points in $(N, \infty)$, which contain a given conjoined basis $Y^*$ with the same properties, is completely characterized by the solutions of the algebraic system (6.306)–(6.307). We note that this system is the same as in the continuous case; see [285, Section 5]. There it is known that system (6.306)–(6.307) is solvable with a suitable choice of the matrices $G$ and $H$; see [285, Theorem 5.7 and formula (5.25)] for more details. At the same time, Theorem 6.84 allows to construct all conjoined bases of (SDS) with the same properties as above, which are contained in the given $Y^*$. Combining these two results with estimate (6.279) yields a construction of conjoined bases $Y$ of (SDS) with constant kernel on $[N, \infty)_{\mathbb{Z}}$ and no forward focal points in $(N, \infty)$, which have the rank of $X_k$ equal to any value between $n - d[N, \infty)_{\mathbb{Z}}$ and $n$. This construction is based on the choice of the projectors $P$ and $P^*$ in (6.297) and (6.305).

**Theorem 6.90** *Assume that there exists a conjoined basis of* (SDS) *with constant kernel on $[N, \infty)_{\mathbb{Z}}$ and no forward focal points in $(N, \infty)$. Then for any integer value $r$ between $n - d[N, \infty)_{\mathbb{Z}}$ and $n$, there exists a conjoined basis $Y$ of* (SDS) *with constant kernel on $[N, \infty)_{\mathbb{Z}}$ and no forward focal points in $(N, \infty)$ such that* rank $X_k = r$ *on $[N, \infty)_{\mathbb{Z}}$.*

*Proof* Let $Y^*$ be the conjoined basis of (SDS) from the assumption of the theorem, and let $P^*$, $R_k^*$, and $P_{S^*\infty}$ be its associated projectors from (6.234), (6.233), and (6.238). Let $r$ be an integer between $n - d[N, \infty)_{\mathbb{Z}}$ and $n$. If $r \leq$ rank $P^*$, then we choose an orthogonal projector $P^{**}$ such that rank $P^{**} = r$ and Im $P_{S^*\infty} \subseteq$ Im $P^{**} \subseteq$ Im $P^*$ holds; compare with (6.297). It follows by Theorem 6.84 that for this projector $P^{**}$, there exists a conjoined basis $Y^{**}$ of (SDS) with constant kernel on $[N, \infty)_{\mathbb{Z}}$ and no forward focal points in $(N, \infty)$ such that Ker $X_k^{**} =$ Ker $P^{**}$ on $[N, \infty)_{\mathbb{Z}}$, i.e., rank $X_k^{**} =$ rank $P^{**} = r$ on $[N, \infty)_{\mathbb{Z}}$. Similarly, if $r \geq$ rank $P^*$, then we choose orthogonal projectors $P$ and $R$ such that (6.305) holds and rank $P = r$. Then the conjoined basis $Y := \tilde{Y}$ from Theorem 6.87 has the required properties and rank $X_k =$ rank $P = r$ on $[N, \infty)_{\mathbb{Z}}$. The proof is complete. $\square$

In the next lemma, we derive a property of the Wronskian of two equivalent conjoined bases of (SDS) with constant kernel on $[N, \infty)_{\mathbb{Z}}$.

**Lemma 6.91** *Let $Y^{(1)}$ and $Y^{(2)}$ be conjoined bases of (SDS) with constant kernel on $[N, \infty)_{\mathbb{Z}}$, and let $P^{(1)}$, $P^{(2)}$ and $P_{S^{(1)}\infty}$, $P_{S^{(2)}\infty}$ be the corresponding orthogonal projectors defined in (6.234) and (6.238) through $X_k^{(1)}$ and $X_k^{(2)}$, respectively. If $Y^{(1)}$ is equivalent with $Y^{(2)}$ on $[N, \infty)_{\mathbb{Z}}$, then*

$$\left.\begin{aligned}
\operatorname{Im} w(Y^{(1)}, Y^{(2)}) &\subseteq \operatorname{Im}(P^{(1)} - P_{S^{(1)}\infty}), \\
\operatorname{Im}[w(Y^{(1)}, Y^{(2)})]^T &\subseteq \operatorname{Im}(P^{(2)} - P_{S^{(2)}\infty}).
\end{aligned}\right\} \tag{6.322}$$

*Proof* Let $Y^{(1)}$ be equivalent with $Y^{(2)}$ on $[N, \infty)_{\mathbb{Z}}$. By Proposition 6.81 (with $Y := Y^{(1)}$ and $\tilde{Y} := Y^{(2)}$), there exist matrices $G, H \in \mathbb{R}^{n \times n}$ such that

$$X_N^{(2)} = X_N^{(1)}, \quad U_N^{(2)} - U_N^{(1)} = (X_N^{(1)})^{\dagger T} G + U_N^{(1)} H, \tag{6.323}$$

and $\operatorname{Im} G \subseteq \operatorname{Im}(P^{(1)} - P_{S^{(1)}\infty})$ and $\operatorname{Im} H \subseteq \operatorname{Im}(I - P^{(1)})$. By (6.323) and the symmetry of $(X^{(1)})^T U^{(1)}$, we obtain for the Wronskian $N^{(1)} := w(Y^{(1)}, Y^{(2)})$ at $k = N$

$$N^{(1)} = (X_N^{(1)})^T (U_N^{(2)} - U_N^{(1)}) = (X_N^{(1)})^T (X_N^{(1)})^{\dagger T} G + (X_N^{(1)})^T U_N^{(1)} H. \tag{6.324}$$

From the equalities $(X_N^{(1)})^T (X_N^{(1)})^{\dagger T} = P^{(1)}$ and $X_N^{(1)} P^{(1)} = X_N^{(1)}$, we obtain $(X_N^{(1)})^T (X_N^{(1)})^{\dagger T} G = P^{(1)} G = G$ and

$$(X_N^{(1)})^T U_N^{(1)} H = (U_N^{(1)})^T X_N^{(1)} H = (U_N^{(1)})^T X_N^{(1)} P^{(1)} H = 0.$$

Therefore, (6.324) gives $N^{(1)} = G$ and so $\operatorname{Im} N^{(1)} = \operatorname{Im} G \subseteq \operatorname{Im}(P^{(1)} - P_{S^{(1)}\infty})$. The second inclusion in (6.322) follows from the fact that $-[N^{(1)}]^T$ is the Wronskian of the solutions $Y^{(2)}$ and $Y^{(1)}$. □

In the next two results, we provide additional information about the relation being contained, which will be utilized in the construction of dominant solutions of (SDS) at $\infty$ in Sect. 6.3.7. First we show that the relation being contained for conjoined bases of (SDS) with constant kernel on $[N, \infty)_{\mathbb{Z}}$ and no forward focal points in $(N, \infty)$ is invariant under suitable change of the interval $[N, \infty)_{\mathbb{Z}}$. Namely, the point $N$ can always be moved forward and under some additional conditions also backward. We recall from Sect. 6.3.1 that the order of abnormality $d[k, \infty)_{\mathbb{Z}}$ of (SDS) is nondecreasing in $k$ on $[0, \infty)_{\mathbb{Z}}$.

**Proposition 6.92** *Let $Y$ and $Y^*$ be two conjoined bases of (SDS) with constant kernel on $[N, \infty)_{\mathbb{Z}}$ and no forward focal points in $(N, \infty)$. Then the following hold.*

*(i) If $Y$ contains $Y^*$ on the interval $[N, \infty)_\mathbb{Z}$, then $Y$ contains $Y^*$ also on the interval $[L, \infty)_\mathbb{Z}$ for all $L \in [N, \infty)_\mathbb{Z}$.*

*(ii) Assume that $d[N, \infty)_\mathbb{Z} = d_\infty$. If $Y$ contains $Y^*$ on the interval $[L, \infty)_\mathbb{Z}$ for some index $L \in [N, \infty)_\mathbb{Z}$, then $Y$ contains $Y^*$ also on the interval $[N, \infty)_\mathbb{Z}$, and hence on $[K, \infty)_\mathbb{Z}$ for every $K \in [N, \infty)_\mathbb{Z}$.*

*Proof* Fix $L \in [N, \infty)_\mathbb{Z}$. We denote by $S_k$, $S_k^*$, resp., $S_k^{(L)}$, $S_k^{*(L)}$, the $S$-matrices corresponding to $Y$ and $Y^*$ on the interval $[N, \infty)_\mathbb{Z}$, resp., on the interval $[L, \infty)_\mathbb{Z}$. Then $S_k^{(L)} = S_k - S_L$ and $S_k^{*(L)} = S_k^* - S_L^*$ on $[L, \infty)_\mathbb{Z}$. Let $P$ and $P^*$ be the projectors in (6.234) defined by the functions $X_k$ and $X_k^*$. Moreover, let $P_{S\infty}$, $P_{S^*\infty}$ and $P_{S^{(L)}\infty}$, $P_{S^{*(L)}\infty}$ be the projectors associated with the matrices $S_k$, $S_k^*$ and $S_k^{(L)}$, $S_k^{*(L)}$ through (6.238). The inequalities $0 \leq S_k^{(L)} \leq S_k$ and $0 \leq S_k^{*(L)} \leq S_k^*$ for $k \geq L$ or the inclusions in (6.239) then imply

$$\operatorname{Im} P_{S^{(L)}\infty} \subseteq \operatorname{Im} P_{S\infty}, \quad \operatorname{Im} P_{S^{*(L)}\infty} \subseteq \operatorname{Im} P_{S^*\infty}. \tag{6.325}$$

For part (i) we suppose that $Y$ contains $Y^*$ on $[N, \infty)_\mathbb{Z}$, that is, the inclusions in (6.297) hold and $Y^*$ is equivalent with $YP^*$ on $[N, \infty)_\mathbb{Z}$ by Definition 6.82. Then $\operatorname{Im} P_{S^{(L)}\infty} \subseteq \operatorname{Im} P^* \subseteq \operatorname{Im} P$ as well, by the first inclusion in (6.325), and $Y^*$ is equivalent with $YP^*$ on $[L, \infty)_\mathbb{Z}$. Therefore, $Y$ contains $Y^*$ also on $[L, \infty)_\mathbb{Z}$, by Definition 6.82. For the proof of part (ii), we assume that $d[N, \infty)_\mathbb{Z} = d_\infty$. Then $d[N, \infty)_\mathbb{Z} = d[L, \infty)_\mathbb{Z}$ and by (6.276) and (6.325),

$$P_{S^{(L)}\infty} = P_{S\infty}, \quad P_{S^{*(L)}\infty} = P_{S^*\infty}. \tag{6.326}$$

Now suppose that $Y$ contains $Y^*$ on $[L, \infty)_\mathbb{Z}$. Moreover, let $Y^{**}$ be a conjoined basis of (SDS) with constant kernel on $[N, \infty)_\mathbb{Z}$ such that $Y$ contains $Y^{**}$ on $[N, \infty)_\mathbb{Z}$ with respect to the projector $P^*$. Such a conjoined basis always exists, by Remark 6.85. According to part (i) of this theorem, $Y$ contains $Y^{**}$ also on $[L, \infty)_\mathbb{Z}$ with respect to $P^*$, and hence, $Y^*$ and $Y^{**}$ are equivalent on $[L, \infty)_\mathbb{Z}$. This means that $X_k^* = X_k^{**}$ on $[L, \infty)_\mathbb{Z}$. We will show that the assumption $d[N, \infty)_\mathbb{Z} = d_\infty$ allows to extend the latter equality to the whole interval $[N, \infty)_\mathbb{Z}$. For this we define $N^* := w(Y^*, Y^{**})$. By Lemma 6.91 (with the initial index $N := L$, $Y^{(1)} := Y^*$, $Y^{(2)} := Y^{**}$, $P^{(1)} := P^*$, $P_{S^{(1)}\infty} := P_{S^{*(L)}\infty}$, and the Wronskian $w(Y^*, Y^{**})$), it follows that $\operatorname{Im} N^* \subseteq \operatorname{Im}(P^* - P_{S^{*(L)}\infty})$. Consequently, $\operatorname{Im} N^* \subseteq \operatorname{Im}(P^* - P_{S^*\infty})$, by the second equality in (6.326). On the other hand, the equality $X_k^* = X_k^{**}$ on $[L, \infty)_\mathbb{Z}$ implies that $\operatorname{Im} X_k^* = \operatorname{Im} X_k^{**}$ on $[N, \infty)_\mathbb{Z}$, by Theorem 6.77. Therefore, the conjoined bases $Y^*$ and $Y^{**}$ are mutually representable on $[N, \infty)_\mathbb{Z}$ in the sense of Theorem 6.69. In particular, we have

$$Y_k^{**} = Y_k^* M^* + \bar{Y}_k^* N^*, \quad k \in [N, \infty)_\mathbb{Z}, \tag{6.327}$$

where $\bar{Y}^*$ is a conjoined basis from Proposition 6.67 associated with $Y^*$ and where $M^* \in \mathbb{R}^{n \times n}$ is a constant invertible matrix. By using (6.327), $P^* N^* = N^*$, and

Proposition 6.67(iv) (with $Y := Y^*$ and $\bar{Y} := \bar{Y}^*$), we obtain on $[N, \infty)_{\mathbb{Z}}$

$$X_k^{**} = X_k^* M^* + \bar{X}_k^* P^* N^* = X_k^* M^* + X_k^* S_k^* N^* = X_k^* (M^* + S_k^* N^*). \tag{6.328}$$

But $S_k^* N^* = S_k^* P_{\mathcal{S}^*\infty} N^* = 0$ on $[N, \infty)_{\mathbb{Z}}$. Therefore, (6.328) becomes $X_k^{**} = X_k^* M^*$ on $[N, \infty)_{\mathbb{Z}}$. At the same time, we have $X_k^{**} = X_k^*$ on $[L, \infty)_{\mathbb{Z}}$, which gives the formula $X_k^* M^* = X_k^*$ on $[L, \infty)_{\mathbb{Z}}$. Multiplying the latter equation by $(X_k^*)^{\dagger}$ from the left and using the identity $(X_k^*)^{\dagger} X^* = P^*$ on $[L, \infty)_{\mathbb{Z}}$, we get $P^* M^* = P^*$. Hence, $X_k^{**} = X_k^* M^* = X_k^* P^* M^* = X_k^*$ on $[N, \infty)_{\mathbb{Z}}$. This shows that $Y^*$ and $Y^{**}$ are equivalent on $[N, \infty)_{\mathbb{Z}}$, so that $Y$ contains $Y^*$ also on $[N, \infty)_{\mathbb{Z}}$ with respect to $P^*$. Finally, the fact that $Y$ contains $Y^*$ on $[K, \infty)_{\mathbb{Z}}$ for every $K \geq N$ now follows from part (i) of this theorem. □

**Proposition 6.93** *Let $Y^*$ be a conjoined basis of* (SDS) *with constant kernel on* $[N, \infty)_{\mathbb{Z}}$ *and no forward focal points in* $(N, \infty)$ *with* $d[N, \infty)_{\mathbb{Z}} = d_{\infty}$. *Then the following statements hold for every index* $L \in [N, \infty)_{\mathbb{Z}}$.

(i) *If $Y^{**}$ is a conjoined basis of* (SDS) *with constant kernel on $[L, \infty)_{\mathbb{Z}}$ and no forward focal points in $(L, \infty)$ and it is contained in $Y^*$ on $[L, \infty)_{\mathbb{Z}}$, then $Y^{**}$ has constant kernel also on $[N, \infty)_{\mathbb{Z}}$ and no forward focal points in $(N, \infty)$.*

(ii) *If $Y$ is a conjoined basis of* (SDS) *with constant kernel on $[L, \infty)_{\mathbb{Z}}$ and no forward focal points in $(L, \infty)$ and it contains $Y^*$ on $[L, \infty)_{\mathbb{Z}}$, then $Y$ has constant kernel also on $[N, \infty)_{\mathbb{Z}}$ and no forward focal points in $(N, \infty)$.*

*Proof* (i) Fix $L \in [N, \infty)_{\mathbb{Z}}$ and let $Y^*$ and $Y^{**}$ be as in the theorem. Furthermore, let $P_{\mathcal{S}^*\infty}$ and $P_{\mathcal{S}^{*(L)}\infty}$ be the orthogonal projectors in (6.238) associated with $Y^*$ on the intervals $[N, \infty)_{\mathbb{Z}}$ and $[L, \infty)_{\mathbb{Z}}$, respectively, and let $P^{**}$ be the orthogonal projector in (6.234) associated with $Y^{**}$ on $[L, \infty)_{\mathbb{Z}}$. Then $X_k^{**} = X_k^* P^{**}$ for every $k \in [L, \infty)_{\mathbb{Z}}$. As in the proof of Proposition 6.92, the projector $P_{\mathcal{S}^{*(L)}\infty}$ is defined by the matrix $S_k^{*(L)} = S_k^* - S_L^*$. Then $0 \leq S_k^{*(L)} \leq S_k^*$ for $k \in [L, \infty)_{\mathbb{Z}}$ and hence $\operatorname{Im} P_{\mathcal{S}^{*(L)}\infty} \subseteq \operatorname{Im} P_{\mathcal{S}^*\infty}$. The assumption $d[N, \infty)_{\mathbb{Z}} = d_{\infty}$ then implies the identity $\Lambda[N, \infty)_{\mathbb{Z}} = \Lambda[I_\cdot, \infty)_{\mathbb{Z}}$ by (6.225), and the equality $P_{\mathcal{S}^{*(L)}\infty} = P_{\mathcal{S}^*\infty}$ by rank $P_{\mathcal{S}^{*(L)}\infty} = n - d[L, \infty)_{\mathbb{Z}} = n - d[N, \infty)_{\mathbb{Z}} = \operatorname{rank} P_{\mathcal{S}^*\infty}$ in (6.276). From the former equality, it then follows that $X_k^{**} = X_k^* P^{**}$ holds for every $k \in [N, \infty)_{\mathbb{Z}}$, which shows that $Y^{**}$ is contained in $Y^*$ also on $[N, \infty)_{\mathbb{Z}}$. Therefore, $Y^{**}$ has constant kernel on $[N, \infty)_{\mathbb{Z}}$ and no forward focal points in $(N, \infty)$, by Remark 6.83 (with $Y := Y^*$ and $Y^* := Y^{**}$). For the proof of part (ii), assume that $Y$ is a conjoined basis of (SDS) with constant kernel on $[L, \infty)_{\mathbb{Z}}$ and no forward focal points in $(L, \infty)$ such that $Y^*$ is contained in $Y$ on $[L, \infty)_{\mathbb{Z}}$. Using similar arguments as above, it then follows that $X_k^* = X_k P^*$ for every $k \in [N, \infty)_{\mathbb{Z}}$, where $P^*$ is the constant orthogonal projector in (6.234) which corresponds to $Y^*$. Let $S_k^*$ and $Q_k^*$ be the matrices in (6.236) and (6.286) associated with $Y^*$ on $[N, \infty)_{\mathbb{Z}}$, and let $W := w(Y, Y^*)$ be the (constant) Wronskian of $Y$ and $Y^*$. Then

$$P_{\mathcal{S}^*\infty} W = P_{\mathcal{S}^{*(L)}\infty} W = 0, \quad PW = W, \quad P^* W = W^T P^*, \tag{6.329}$$

where $P$ is the constant orthogonal projectors in (6.234) which corresponds to $Y$ on $[L, \infty)_{\mathbb{Z}}$. Similarly as in the proof of Theorem 6.87, we then obtain that

$$\text{Im}\,(P - S_k^* W^T) = \text{Im}\,P = \text{Im}\,(P - S_k^* W^T)^T = \text{Im}\,(P - S_k^* W^T)^\dagger, \quad (6.330)$$

$$X_{k+1} = (\mathcal{A}_k + \mathcal{B}_k Q_k^*)\, X_k - \mathcal{B}_k (X_k^*)^{\dagger\,T} W^T, \quad (6.331)$$

$$X_k = (\mathcal{D}_k^T - \mathcal{B}_k^T Q_{k+1}^*)\, X_{k+1} + \mathcal{B}_k^T (X_{k+1}^*)^{\dagger\,T} W^T, \quad (6.332)$$

for all $k \in [N, \infty)_{\mathbb{Z}}$. Furthermore, let $\Phi_k$ and $\Psi_k$ be, respectively, the solutions of the associated homogeneous equations

$$\Phi_{k+1} = (\mathcal{A}_k + \mathcal{B}_k Q_k^*)\, \Phi_k, \quad k \in [N, \infty)_{\mathbb{Z}}, \quad \Phi_N = X_N, \quad (6.333)$$

$$\Psi_k = (\mathcal{D}_k^T - \mathcal{B}_k^T Q_{k+1}^*)\, \Psi_{k+1}, \quad k \in [N, L-1]_{\mathbb{Z}}, \quad \Psi_L = X_L (P - S_L^* W^T)^\dagger. \quad (6.334)$$

The initial condition in (6.334) together with (6.330) and $X_L^\dagger X_L = P$ imply

$$\text{Ker}\,\Psi_L = \text{Ker}\,[X_L^\dagger X_L (P - S_L^* W^T)^\dagger] = \text{Ker}\,[P(P - S_L^* W^T)^\dagger] = \text{Ker}\,P. \quad (6.335)$$

The equalities $X_L P^* = X_L^*$ and $P^* = P P^*$ and the initial condition (6.334) give

$$X_L^* = X_L P P^* \overset{(6.330)}{=} X_L (P - S_L^* W^T)^\dagger (P - S_L^* W^T)\, P^* \overset{(6.329)}{=} \Psi_L\,(P P^* - S_L^* P^* W)$$

$$= \Psi_L\,(P^* - S_L^* W) = \Psi_L\,(P^* - S_L^* P_{S^*\infty} W) = \Psi_L\, P^*.$$

On the other hand, the matrix $\Phi_N$ satisfies $\Phi_N P^* = X_N P^* = X_N^*$ by (6.333). By using similar arguments as in the proof of Theorem 6.87, we get

$$X_k = \Psi_k\,(P - S_k^*\, W^T), \quad k \in [N, L]_{\mathbb{Z}}, \quad (6.336)$$

$$X_k = \Phi_k\,(P - S_k^*\, W^T), \quad k \in [N, \infty)_{\mathbb{Z}}, \quad (6.337)$$

Moreover, the kernel of $\Psi_k$ is nonincreasing on $[N, L]_{\mathbb{Z}}$, while the kernel of $\Phi_k$ is nondecreasing on $[N, \infty)_{\mathbb{Z}}$ by (6.333) and (6.334). In turn, by (6.335) we obtain the identity $\Psi_k P = \Psi_k$ for all $k \in [N, L]_{\mathbb{Z}}$. Furthermore, equations (6.337) and (6.330) yield the equality $X_k P = X_k$ on $[N, \infty)_{\mathbb{Z}}$. In particular, for $k = N$ we have $\Phi_N P = X_N P = X_N = \Phi_N$, which implies through (6.333) that $\Phi_k P = \Phi_k$ for all $k \in [N, \infty)_{\mathbb{Z}}$. From (6.330), (6.336), and (6.337), it then follows that

$$\Psi_k = \Psi_k P = \Psi_k\,(P - S_k^* W^T)\,(P - S_k^* W^T)^\dagger = \Phi_k\,(P - S_k^* W^T)\,(P - S_k^* W^T)^\dagger$$

$$= \Phi_k P = \Phi_k$$

for all $k \in [N, L]_{\mathbb{Z}}$. Therefore, the matrix $\Psi_k = \Phi_k$ has constant kernel on $[N, L]_{\mathbb{Z}}$ equal to $\operatorname{Ker} P$ by (6.335), and consequently $\operatorname{Ker} X_k = \operatorname{Ker} P$ for all $k \in [N, L]_{\mathbb{Z}}$ by (6.336) and (6.330). Thus, the conjoined basis $Y$ has constant kernel on $[N, \infty)_{\mathbb{Z}}$. Finally, according to (6.299) we have $X_k X_{k+1}^{\dagger} \mathcal{B}_k = X_k^*(X_{k+1}^*)^{\dagger} \mathcal{B}_k \geq 0$ for every $k \in [N, \infty)_{\mathbb{Z}}$. This means that $Y$ has no forward focal points in $(N, \infty)$ and the proof is complete. $\qquad \square$

## 6.3.4 Minimal Conjoined Bases

Inequality (6.279) motivates the following notions of a minimal conjoined basis of (SDS) and a maximal conjoined basis of (SDS).

**Definition 6.94** A conjoined basis $Y$ of on $[N, \infty)_{\mathbb{Z}}$ if $\operatorname{rank} X_k = n - d[N, \infty)_{\mathbb{Z}}$ for all $k \in [N, \infty)_{\mathbb{Z}}$. Similarly, $Y$ is called a *maximal conjoined basis* on $[N, \infty)_{\mathbb{Z}}$ if $\operatorname{rank} X_k = n$ for all $k \in [N, \infty)_{\mathbb{Z}}$.

*Remark 6.95* The terminology in Definition 6.94 follows the estimate in (6.279), as the minimal conjoined bases $Y$ of (SDS) attain the smallest possible rank of $X_k$ on $[N, \infty)_{\mathbb{Z}}$, while the maximal conjoined bases $Y$ of (SDS) attain the largest possible rank of $X_k$ on $[N, \infty)_{\mathbb{Z}}$ (i.e., $X_k$ is invertible on $[N, \infty)_{\mathbb{Z}}$ in this case). Sometimes we will call conjoined bases $Y$ of (SDS), which satisfy the rank condition $n - d[N, \infty)_{\mathbb{Z}} < \operatorname{rank} X_k < n$ on $[N, \infty)_{\mathbb{Z}}$, as *intermediate* conjoined bases on the interval $[N, \infty)_{\mathbb{Z}}$.

Minimal conjoined bases of (SDS) constitute an important tool in the investigation of recessive solutions of (SDS). For example, given a conjoined basis $Y$ of (SDS) with constant kernel on $[N, \infty)_{\mathbb{Z}}$ and no forward focal points in $(N, \infty)$ with the projectors $P$ and $P_{S\infty}$ in (6.234) and (6.238), then any conjoined basis $Y^*$ of (SDS) which is contained in $Y$ with respect to the projector $P^* := P_{S\infty}$ is a minimal conjoined basis of (SDS) on $[N, \infty)_{\mathbb{Z}}$. In fact, the property $P = P_{S\infty}$ can be shown to be characterizing the minimal conjoined bases $Y$ of (SDS) on $[N, \infty)_{\mathbb{Z}}$.

*Remark 6.96* If $Y$ is a minimal conjoined basis of (SDS) on $[N, \infty)_{\mathbb{Z}}$, then the abnormality of (SDS) on the interval $[N, \infty)_{\mathbb{Z}}$ is necessarily maximal, i.e., $d[N, \infty)_{\mathbb{Z}} = d_\infty$. This follows from estimate (6.279) in Theorem 6.75, which yields

$$n - d[N, \infty)_{\mathbb{Z}} \leq \operatorname{rank} X_k = n - d_\infty, \quad k \in [N, \infty)_{\mathbb{Z}}, \quad \text{i.e.,} \quad d[N, \infty)_{\mathbb{Z}} \geq d_\infty.$$

The opposite inequality $d[N, \infty)_{\mathbb{Z}} \leq d_\infty$ holds by the definition of $d_\infty$ in (6.224), so that $d[N, \infty)_{\mathbb{Z}} = d_\infty$ follows.

In the next result, we show further basic properties of minimal conjoined bases of (SDS) on $[N, \infty)_{\mathbb{Z}}$.

**Theorem 6.97** *The following properties of minimal conjoined bases hold.*

(i) *Let $Y$ be a minimal conjoined basis of* (SDS) *on $[N, \infty)_{\mathbb{Z}}$ with the associated projector $P$ in (6.234). Then*

$$\Lambda_0[N, \infty)_{\mathbb{Z}} = \mathrm{Im}\,[U_N(I - P)], \quad \mathrm{Im}\,X_N = \left(\Lambda_0[N, \infty)_{\mathbb{Z}}\right)^{\perp}. \qquad (6.338)$$

*Consequently, the initial subspace $\mathrm{Im}\,X_N$ is the same for all minimal conjoined bases $Y$ of* (SDS) *on $[N, \infty)_{\mathbb{Z}}$.*

(ii) *Let $Y^{(1)}$ and $Y^{(2)}$ be two minimal conjoined bases of* (SDS) *on $[N, \infty)_{\mathbb{Z}}$ with the corresponding projectors $P^{(1)}$ and $P^{(2)}$ defined in (6.234) through $X^{(1)}$ and $X^{(2)}$. Then $Y^{(1)}$ and $Y^{(2)}$ are equivalent on $[N, \infty)_{\mathbb{Z}}$ if and only if*

$$P^{(2)} = P^{(1)}, \quad Y_k^{(2)} = Y_k^{(1)}M, \quad k \in [N, \infty)_{\mathbb{Z}}, \qquad (6.339)$$

*where $M$ is a constant nonsingular matrix satisfying $P^{(1)}M = P^{(1)}$.*

(iii) *Let $Y^{(1)}$ and $Y^{(2)}$ be two minimal conjoined bases of* (SDS) *on $[N, \infty)_{\mathbb{Z}}$ with their associated S-matrices $S_k^{(1)}$ and $S_k^{(2)}$ defined in (6.236). If $K \in [N, \infty)_{\mathbb{Z}}$ is an index such that*

$$\mathrm{rank}\,S_k^{(1)} = n - d[N, \infty)_{\mathbb{Z}} = \mathrm{rank}\,S_k^{(2)} \quad \text{for all } k \in [K, \infty)_{\mathbb{Z}},$$

*then for $k \in [K, \infty)_{\mathbb{Z}}$, we have the equality*

$$(S_k^{(3-i)})^{\dagger} = (L^{(i)})^T (S_k^{(i)})^{\dagger} L^{(i)} + (L^{(i)})^T N^{(i)}, \quad i \in \{1, 2\}, \qquad (6.340)$$

*where the matrices $L^{(i)}$ and $N^{(i)}$ are from Theorem 6.69 and its proof, resp., from Remarks 6.70 and 6.71.*

*Proof* The first conclusion in (6.338) follows from equation (6.278) in Theorem 6.75, since for a minimal conjoined basis $Y$ on $[N, \infty)_{\mathbb{Z}}$, we have $P = P_{\mathcal{S}\infty}$. The second conclusion in (6.338) then follows from Theorem 6.66(i), since

$$\mathrm{Im}\,X_N = \mathrm{Im}\,R_N = \left(\mathrm{Im}\,[U_N(I - P)]\right)^{\perp} = \left(\Lambda_0[N, \infty)_{\mathbb{Z}}\right)^{\perp}.$$

This shows that the set $\mathrm{Im}\,X_N$ does not depend on a choice of $Y$. Part (ii) is a consequence of Proposition 6.81 with $M := I + H$. More precisely, by using Proposition 6.81, the equivalence of the minimal conjoined bases $Y^{(1)}$ and $Y^{(2)}$ on $[N, \infty)_{\mathbb{Z}}$ means that $X_N^{(2)} = X_N^{(1)}$ and $\mathrm{Im}\,(U_N^{(2)} - U_N^{(1)}) \subseteq \Lambda_0[N, \infty)_{\mathbb{Z}}$, while from (6.338) we get $\Lambda_0[N, \infty)_{\mathbb{Z}} = \mathrm{Im}\,[U_N^{(1)}(I - P^{(1)})]$. Therefore, the projectors $P^{(1)}$ and $P^{(2)}$ satisfy $P^{(2)} = P^{(1)}$ and $U_N^{(2)} - U_N^{(1)} = U_N^{(1)}H$ with a unique matrix $H$ satisfying $P^{(1)}H = 0$. Consequently, for $M := I + H$, we have $X_N^{(1)}M = X_N^{(2)}$ and $U_N^{(1)}M = U_N^{(2)}$. This completes the formulas in (6.339) by the uniqueness of solutions of (SDS). In addition, the constant matrix $M$ is nonsingular, because rank $Y_k^{(2)} = n$. Finally, $P^{(1)}M = P^{(1)}$ follows by the definition of $M$. Conversely, if the minimal conjoined bases $Y^{(1)}$ and $Y^{(2)}$ satisfy the equalities in (6.339) with a nonsingular matrix $M$ such that $P^{(1)}M = P^{(1)}$, then they are equivalent on $[N, \infty)_{\mathbb{Z}}$. This follows from $X_N^{(1)} = X_N^{(1)}P^{(1)}M = X_N^{(1)}$ and $U_N^{(2)} - U_N^{(1)} = U_N^{(1)}(M - I)$, where $\text{Im}\,(M - I) \subseteq \text{Im}\,(I - P^{(1)})$. For part (iii) we first note that equality (6.262) holds on $[N, \infty)_{\mathbb{Z}}$, since $\text{Im}\,X_N^{(1)} = \text{Im}\,X_N^{(2)}$ by part (i) of this theorem. We multiply (6.262) by $(L^{(1)})^T$ from the right and use the symmetry of $L^{(1)}(L^{(1)})^{\dagger} = P^{(1)}$ and the identity $S_k^{(1)}P^{(1)} = S_k^{(1)}$ on $[N, \infty)_{\mathbb{Z}}$ from Theorem 6.66(iii) (with $Y := Y^{(1)}$) to get

$$S_k^{(2)}(L^{(1)})^T = (L^{(1)} + S_k^{(1)}N^{(1)})^{\dagger}\,S_k^{(1)} \quad \text{for all } k \in [N, \infty)_{\mathbb{Z}}. \tag{6.341}$$

Fix $k \in [K, \infty)_{\mathbb{Z}}$, where the index $K$ satisfies $K \geq N_m$ by Remark 6.76. Then by the same remark, we have $\text{Im}\,S_k^{(1)} = \text{Im}\,P^{(1)}$ and $\text{Im}\,S_k^{(2)} = \text{Im}\,P^{(2)}$, since $P^{(1)} = P_{S^{(1)}\infty}$ and $P^{(2)} = P_{S^{(2)}\infty}$ as $Y^{(1)}$ and $Y^{(2)}$ are minimal conjoined bases on $[N, \infty)_{\mathbb{Z}}$. Moreover, with the aid of Remark 1.60(i) and (6.250), we get

$$\text{Im}\,(L^{(1)})^T = \text{Im}\,(L^{(1)})^{\dagger} = \text{Im}\,L^{(2)} = \text{Im}\,P^{(2)}, \tag{6.342}$$

and from (6.261) and Theorem 6.69(iv), we obtain

$$(L^{(1)} + S_k^{(1)}N^{(1)})\,(L^{(1)} + S_k^{(1)}N^{(1)})^{\dagger} = P^{(1)}, \quad P^{(1)}N^{(1)} = N^{(1)}. \tag{6.343}$$

By using Remark 1.62 for the pseudoinverse of a product of two matrices and the equalities $S_k^{(1)}(S_k^{(1)})^{\dagger} = P_{S^{(1)}\infty} = P^{(1)}$ and $(S_k^{(2)})^{\dagger}S_k^{(2)} = P_{S^{(2)}\infty} = P^{(2)}$, and $(L^{(1)})^{\dagger}L^{(1)} = P^{(2)}$ by (6.342), the Moore-Penrose pseudoinverse of the left-hand side of (6.341) is equal to

$$[S_k^{(2)}(L^{(1)})^T]^{\dagger} = [P^{(2)}(L^{(1)})^T]^{\dagger}\,[S_k^{(2)}P^{(2)}]^{\dagger} \overset{(6.342)}{=} (L^{(1)})^{\dagger T}(S_k^{(2)})^{\dagger}, \tag{6.344}$$

while the Moore-Penrose pseudoinverse of the right-hand side of (6.341) is

$$[(L^{(1)} + S_k^{(1)}N^{(1)})^{\dagger}\,S_k^{(1)}]^{\dagger} = (P^{(1)}S_k^{(1)})^{\dagger}\,[(L^{(1)} + S_k^{(1)}N^{(1)})^{\dagger}\,P^{(1)}]^{\dagger}$$

$$\overset{(6.343)}{=} (S_k^{(1)})^{\dagger}(L^{(1)} + S_k^{(1)}N^{(1)}) = (S_k^{(1)})^{\dagger}L^{(1)} + P^{(1)}N^{(1)}$$

$$= (S_k^{(1)})^{\dagger}L^{(1)} + N^{(1)}. \tag{6.345}$$

Using (6.344) and (6.345) in the pseudoinverse of both sides of (6.341) then yields

$$(L^{(1)})^{\dagger T}(S_k^{(2)})^{\dagger} = (S_k^{(1)})^{\dagger}L^{(1)} + N^{(1)}.$$

If we now multiply the latter equation by $(L^{(1)})^T$ from the left and use the symmetry of $(L^{(1)})^\dagger L^{(1)} = P^{(2)}$, then formula (6.340) with $i = 1$ follows. The same formula with $i = 2$ is then obtained by interchanging the roles of $Y^{(1)}$ and $Y^{(2)}$.     □

*Remark 6.98* The result in Theorem 6.97 implies that that any two minimal conjoined bases $Y^{*(1)}$ and $Y^{*(2)}$ on $[N, \infty)_{\mathbb{Z}}$ are always mutually representable in the sense of Theorem 6.69, since by (6.338) they satisfy

$$\operatorname{Im} X_N^{*(1)} = (\Lambda_0[N, \infty)_{\mathbb{Z}})^\perp = \operatorname{Im} X_N^{*(2)},$$

where $\Lambda_0[N, \infty)_{\mathbb{Z}}$ is the subspace of the initial conditions $u_N$ of the elements $u \in \Lambda[N, \infty)_{\mathbb{Z}}$. Hence, there exist matrices $M^{*(1)}$, $N^{*(1)}$, $M^{*(2)}$, $N^{*(2)}$ as in Theorem 6.69 such that

$$X_N^{*(2)} = X_N^{*(1)} M^{*(1)}, \quad U_N^{*(2)} = U_N^{*(1)} M^{*(1)} + (X_N^{*(1)})^{\dagger T} N^{*(1)}, \tag{6.346}$$

$$X_N^{*(1)} = X_N^{*(2)} M^{*(2)}, \quad U_N^{*(1)} = U_N^{*(2)} M^{*(2)} + (X_N^{*(2)})^{\dagger T} N^{*(2)}. \tag{6.347}$$

Moreover, by taking the limit as $k \to \infty$ in (6.340), we obtain that the corresponding matrices $T^{*(1)}$ and $T^{*(2)}$ in (6.237) associated with the minimal conjoined bases $Y^{*(1)}$ and $Y^{*(2)}$ on $[N, \infty)_{\mathbb{Z}}$ satisfy the equality

$$T^{*(2)} = (M^{*(1)})^T T^{*(1)} M^{*(1)} + (M^{*(1)})^T N^{*(1)}. \tag{6.348}$$

In the next result, we study the situation when the minimal conjoined bases $Y^{*(1)}$ and $Y^{*(2)}$ on $[N, \infty)_{\mathbb{Z}}$ are constructed from already mutually representable conjoined bases $Y^{(1)}$ and $Y^{(2)}$ with matrices $M^{(1)}$, $N^{(1)}$, $M^{(2)}$, $N^{(2)}$ from Theorem 6.69.

**Proposition 6.99** *Let $Y^{(1)}$ and $Y^{(2)}$ be conjoined bases of (SDS) with constant kernel on $[N, \infty)_{\mathbb{Z}}$ and no forward focal points in $(N, \infty)$, and let $P^{(1)}$, $P^{(2)}$ and $P_{S^{(1)}\infty}$, $P_{S^{(2)}\infty}$ be the corresponding orthogonal projectors in (6.234) and (6.238). Moreover, let $Y^{*(1)}$ be a minimal conjoined basis of (SDS), which is contained in $Y^{(1)}$ on $[N, \infty)_{\mathbb{Z}}$ with respect to $P_{S^{(1)}\infty}$, and similarly, let $Y^{*(2)}$ be a minimal conjoined basis of (SDS) which is contained in $Y^{(2)}$ on $[N, \infty)_{\mathbb{Z}}$ with respect to $P_{S^{(2)}\infty}$. Suppose that $Y^{(1)}$ and $Y^{(2)}$ are mutually representable on $[N, \infty)_{\mathbb{Z}}$ through the matrices $M^{(1)}$, $N^{(1)}$, $M^{(2)}$, $N^{(2)}$ as in Theorem 6.69. If $M^{*(1)}$, $N^{*(1)}$ and $M^{*(2)}$, $N^{*(2)}$ are the matrices corresponding to $Y^{*(1)}$ and $Y^{*(2)}$ in (6.346)–(6.347), then*

$$\left.\begin{aligned} P^{(i)} M^{(i)} P_{S^{(3-i)}\infty} &= P_{S^{(i)}\infty} M^{*(i)}, \\ N^{*(i)} (M^{*(i)})^{-1} &= P_{S^{(i)}\infty} N^{(i)} (M^{(i)})^{-1} P_{S^{(i)}\infty}, \end{aligned}\right\} \quad i \in \{1, 2\}. \tag{6.349}$$

*Proof* With the notation from the proposition, we have by Theorem 6.84

$$X_N^{*(1)} = X_N^{(1)} P_{\mathcal{S}^{(1)}\infty}, \quad U_N^{*(1)} = U_N^{(1)} P_{\mathcal{S}^{(1)}\infty} + (X_N^{(1)})^{\dagger T} G_1 + U_N^{(1)} H_1, \tag{6.350}$$

$$X_N^{*(2)} = X_N^{(2)} P_{\mathcal{S}^{(2)}\infty}, \quad U_N^{*(2)} = U_N^{(2)} P_{\mathcal{S}^{(2)}\infty} + (X_N^{(2)})^{\dagger T} G_2 + U_N^{(2)} H_2 \tag{6.351}$$

with $(G_i, H_i) \in \mathcal{M}(P_{\mathcal{S}_i\infty}, P_{\mathcal{S}_i\infty}, P_i)$ for $i = 1, 2$. We consider the case $i = 1$, since the other case $i = 2$ is obtained by interchanging the roles of the involved conjoined bases. Inserting the first equality from (6.350) and from (6.351) into the first formula in (6.346) gives $X_N^{(2)} P_{\mathcal{S}^{(2)}\infty} = X_N^{(1)} P_{\mathcal{S}^{(1)}\infty} M^{*(1)}$, from which we obtain by (6.255) that $X_N^{(1)} M^{(1)} P_{\mathcal{S}^{(2)}\infty} = X_N^{(1)} P_{\mathcal{S}^{(1)}\infty} M^{*(1)}$. Consequently, multiplying the latter equality by $(X_N^{(1)})^{\dagger}$ from the left and using the identities $(X_N^{(1)})^{\dagger} X_N^{(1)} = P^{(1)}$, $P^{(1)} P_{\mathcal{S}^{(1)}\infty} = P_{\mathcal{S}^{(1)}\infty}$, we get the first formula in (6.349). For the proof of the second formula in (6.349), we recall from Theorem 6.69 that $N^{(1)} = w(Y^{(1)}, Y^{(2)})$ and $N^{*(1)} = w(Y^{*(1)}, Y^{*(2)})$. In particular, at the point $k = N$, we have

$$\left. \begin{array}{l} N^{(1)} = (X_N^{(1)})^T U_N^{(2)} - (U_N^{(1)})^T X_N^{(2)}, \\ N^{*(1)} = (X_N^{*(1)})^T U_N^{*(2)} - (U_N^{*(1)})^T X_N^{*(2)}. \end{array} \right\} \tag{6.352}$$

Combining (6.352) with (6.350)–(6.351) leads to the expression

$$N^{*(1)} = P_{\mathcal{S}^{(1)}\infty} N^{(1)} P_{\mathcal{S}^{(2)}\infty}$$

$$+ P_{\mathcal{S}^{(1)}\infty} (X_N^{(1)})^T (X_N^{(2)})^{\dagger T} G_2 + P_{\mathcal{S}^{(1)}\infty} (X_N^{(1)})^T U_N^{(2)} H_2$$

$$- G_1^T (X_N^{(1)})^{\dagger} X_N^{(2)} P_{\mathcal{S}^{(2)}\infty} - H_1^T (U_N^{(1)})^T X_N^{(2)} P_{\mathcal{S}^{(2)}\infty}. \tag{6.353}$$

We now calculate the last four terms on the right-hand side of (6.353) separately. By the first equality in (6.256) and the already proven first formula in (6.349), by the symmetry of $(X^{(2)})^T U^{(2)}$ and the identities $X_N^{(2)} H_2 = 0$, $(X_N^{(2)})^T (X_N^{(2)})^{\dagger T} = P^{(2)}$, and $P_{\mathcal{S}^{(2)}\infty} G_2 = 0$, we have

$$P_{\mathcal{S}^{(1)}\infty} (X_N^{(1)})^T (X_N^{(2)})^{\dagger T} G_2 = P_{\mathcal{S}^{(1)}\infty} (M^{(2)})^T (X_N^{(2)})^T (X_N^{(2)})^{\dagger T} G_2$$

$$= P_{\mathcal{S}^{(1)}\infty} (M^{(2)})^T P^{(2)} G_2 = (M^{*(2)})^T P_{\mathcal{S}^{(2)}\infty} G_2 = 0,$$

$$P_{\mathcal{S}^{(1)}\infty} (X_N^{(1)})^T U_N^{(2)} H_2 = P_{\mathcal{S}^{(1)}\infty} (M^{(2)})^T (X_N^{(2)})^T U_N^{(2)} H_2$$

$$= P_{\mathcal{S}^{(1)}\infty} (M^{(2)})^T (U_N^{(2)})^T X_N^{(2)} H_2 = 0.$$

Similarly, it follows by the first equality in (6.255) and the first part of (6.349), by the symmetry of $(X^{(1)})^T U^{(1)}$ and the identities $X_N^{(1)} H_1 = 0$, $(X_N^{(1)})^{\dagger} X_N^{(1)} = P^{(1)}$, and $P_{\mathcal{S}^{(1)}\infty} G_1 = 0$ that the last two terms in (6.353) are equal to zero. Therefore, we have $N^{*(1)} = P_{\mathcal{S}^{(1)}\infty} N^{(1)} P_{\mathcal{S}^{(2)}\infty}$. Now we use the properties $M^{*(2)} = (M^{*(1)})^{-1}$, $M^{(2)} = (M^{(1)})^{-1}$, $N^{(1)} P^{(2)} = N^{(1)}$ from Theorem 6.69 to get

$$N^{*(1)}(M^{*(1)})^{-1} = P_{\mathcal{S}^{(1)}\infty} N^{(1)} P_{\mathcal{S}^{(2)}\infty} M^{*(2)} = P_{\mathcal{S}^{(1)}\infty} N^{(1)} P^{(2)} M^{(2)} P_{\mathcal{S}^{(1)}\infty}$$

$$= P_{\mathcal{S}^{(1)}\infty} N^{(1)}(M^{(1)})^{-1} P_{\mathcal{S}^{(1)}\infty}.$$

This shows the second equality in (6.349).                                                       □

In the last result of this subsection, we prove a relationship between conjoined bases $\bar{Y}^{(1)}$ and $\bar{Y}^{(2)}$ in Proposition 6.67 associated with two minimal conjoined bases $Y^{(1)}$ and $Y^{(2)}$.

**Lemma 6.100** *Let $Y^{(1)}$ and $Y^{(2)}$ be minimal conjoined bases of* (SDS) *on the interval $[N, \infty)_{\mathbb{Z}}$, and let $\bar{Y}^{(1)}$ and $\bar{Y}^{(2)}$ be their associated conjoined bases from Proposition 6.67, respectively. Then there exists a constant invertible $n \times n$ matrix $G$ such that*

$$\bar{X}_k^{(2)} = \bar{X}_k^{(1)} G, \quad k \in [N, \infty)_{\mathbb{Z}}. \tag{6.354}$$

*Proof* Let $P^{(1)}$, $R_k^{(1)}$ and $P^{(2)}$, $R_k^{(2)}$ be the orthogonal projectors in (6.234), (6.233) associated with $X^{(1)}$ and $X^{(2)}$, respectively. First we note that $\operatorname{Im} X_N^{(1)} = \operatorname{Im} X_N^{(2)}$ by Remark 6.98, so that $R_N^{(1)} = R_N^{(2)}$. Next we represent $Y^{(1)}$ in terms of $Y^{(2)}$ and $\bar{Y}^{(2)}$, and both $Y^{(2)}$ and $\bar{Y}^{(2)}$ in terms of $Y^{(1)}$ and $\bar{Y}^{(1)}$. Thus, for $i \in \{1, 2\}$ and any $k \in [N, \infty)_{\mathbb{Z}}$, we have

$$Y_k^{(3-i)} = Y_k^{(i)} M^{(i)} + \bar{Y}_k^{(i)} N^{(i)}, \quad \bar{Y}_k^{(2)} = Y_k^{(1)} \bar{M}^{(1)} + \bar{Y}_k^{(1)} \bar{N}^{(1)}, \tag{6.355}$$

where according to Theorem 6.69, the matrices $M^{(1)}$ and $M^{(2)}$ are invertible with $M^{(2)} = (M^{(1)})^{-1}$, $N^{(2)} = -(N^{(1)})^T$, and

$$\bar{M}^{(1)} = -w(\bar{Y}^{(1)}, \bar{Y}^{(2)}) = (\bar{U}_N^{(1)})^T \bar{X}_N^{(2)} - (\bar{X}_N^{(1)})^T \bar{U}_N^{(2)}, \tag{6.356}$$

$$\bar{N}^{(1)} = w(Y^{(1)}, \bar{Y}^{(2)}) = (X_N^{(1)})^T \bar{U}_N^{(2)} - (U_N^{(1)})^T \bar{X}_N^{(2)}$$

$$= (M^{(2)})^T = (M^{(1)})^{T-1}. \tag{6.357}$$

Therefore, the matrix $\bar{N}^{(1)}$ is invertible. Since by Proposition 6.67(iv) at $k = N$, we have $P^{(1)}(\bar{U}_N^{(1)})^T = (X_N^{(1)})^\dagger$ and $P^{(1)}(\bar{X}_N^{(1)})^T = S_N^{(1)}(X_N^{(1)})^T = 0$, as well as $R_N^{(2)} \bar{X}_N^{(2)} = X_N^{(2)}(X_N^{(2)})^\dagger \bar{X}_N^{(2)} = 0$, it follows from (6.356) that

$$P^{(1)} \bar{M}^{(1)} = (X_N^{(1)})^\dagger \bar{X}_N^{(2)} = (X_N^{(1)})^\dagger R_N^{(1)} \bar{X}_N^{(2)} = (X_N^{(1)})^\dagger R_N^{(2)} \bar{X}_N^{(2)} = 0.$$

Therefore, again by (6.355) we get for $k \in [N, \infty)_{\mathbb{Z}}$

$$\bar{X}_k^{(2)} = X_k^{(1)} \bar{M}^{(1)} + \bar{X}_k^{(1)} \bar{N}^{(1)} = X_k^{(1)} P^{(1)} \bar{M}^{(1)} + \bar{X}_k^{(1)} \bar{N}^{(1)} = \bar{X}_k^{(1)} \bar{N}^{(1)}.$$

This shows that (6.354) holds with the matrix $G := \bar{N}^{(1)}$, which is by equality (6.357) invertible. □

## 6.3.5 Asymptotics of S-Matrices

In this subsection we derive some additional properties of the $S$-matrices for conjoined bases of (SDS) with constant kernel on $[N, \infty)_{\mathbb{Z}}$ and no forward focal points in $(N, \infty)$. First of all, by a diagonalization argument for symmetric matrices, we can write the matrices $S_k$, $S_k^{\dagger}$, $T$ in (6.236) and (6.237) in Theorem 6.65 as

$$\left.\begin{array}{c} S_k = V \begin{pmatrix} W_k & 0 \\ 0 & 0_{n-r_k} \end{pmatrix} V^T, \quad S_k^{\dagger} = V \begin{pmatrix} W_k^{-1} & 0 \\ 0 & 0_{n-r_k} \end{pmatrix} V^T, \\ T = V \begin{pmatrix} T_\star & 0 \\ 0 & 0_{n-r_\infty} \end{pmatrix} V^T, \quad k \in [N, \infty)_{\mathbb{Z}}. \end{array}\right\}$$
(6.358)

Here $V$ is a constant orthogonal $n \times n$ matrix, $W_k$ are symmetric and positive definite $r_k \times r_k$ matrices for $k \in [N, \infty)_{\mathbb{Z}}$, and $T_\star := \lim_{k\to\infty} W_k^{-1}$ is symmetric and nonnegative definite $r_\infty \times r_\infty$ matrix. Note that by (6.239) the matrix $T$ satisfies $\operatorname{Im} T \subseteq \operatorname{Im} P_{S\infty}$. The dimension $r_k = \operatorname{rank} S_k = \operatorname{rank} \hat{X}_k^{[N]}$ of the matrices $W_k$ in (6.358) is calculated in the second condition in (6.275). This implies that the rank of $S_k$ changes (i.e., increases in view of Theorem 6.66(iv)) independently of the conjoined basis $Y$ from which $S_k$ is constructed; compare with Remark 6.76. Moreover, by using the formulas in (6.358), we can write the orthogonal projectors $P_{Sk}$ for $k \in [N, \infty)_{\mathbb{Z}}$ and $P_{S\infty}$ defined in (6.238) as

$$P_{Sk} = V \begin{pmatrix} I_{r_k} & 0 \\ 0 & 0_{n-r_k} \end{pmatrix} V^T, \quad P_{S\infty} = V \begin{pmatrix} I_{r_\infty} & 0 \\ 0 & 0_{n-r_\infty} \end{pmatrix} V^T.$$
(6.359)

In the following results, we analyze the properties of the $S$-matrices for such conjoined bases of (SDS). In particular, we will see that the asymptotic properties of the $S$-matrices are affected by the condition

$$d[N, \infty)_{\mathbb{Z}} = d[k, \infty)_{\mathbb{Z}} = d_\infty \quad \text{for all } k \in [N, \infty)_{\mathbb{Z}}$$
(6.360)

on the maximal order of abnormality of (SDS) on $[N, \infty)_{\mathbb{Z}}$.

**Proposition 6.101** *Let $Y$ be a conjoined basis of* (SDS) *with constant kernel on* $[N, \infty)_{\mathbb{Z}}$ *and no forward focal points in* $(N, \infty)$*, let $S_k$ be its corresponding S-matrix in* (6.236)*, and let $T$ be the limit in* (6.237)*. If* (6.360) *holds, then there exists an index $K \in [N, \infty)_{\mathbb{Z}}$ such that*

$$S_k^\dagger \geq T \geq 0, \quad \operatorname{rank}(S_k^\dagger - T) = n - d_\infty, \quad k \in [K, \infty)_{\mathbb{Z}}. \tag{6.361}$$

*Proof* Without loss of generality, we may assume that $Y$ is a minimal conjoined basis of (SDS) on $[N, \infty)_{\mathbb{Z}}$, since from a given $Y$ we can always construct by Theorem 6.84 a minimal conjoined basis, which is contained in $Y$ and which by Proposition 6.86 has the same matrix $S_k$ (and hence $T$). The assumptions then imply that the associated projectors $P$ and $P_{S_\infty}$ satisfy rank $P_{S_\infty} = n - d_\infty$ and $P_{S_\infty} = P$. This implies that $P = S_k^\dagger S_k = S_k S_k^\dagger$ for $k \in [N_m, \infty)_{\mathbb{Z}}$, where $N_m \in [N, \infty)_{\mathbb{Z}}$ is the index from Remark 6.76, i.e., $\operatorname{Im} S_k = \operatorname{Im} P$ is maximal on $[N_m, \infty)_{\mathbb{Z}}$. Consider the auxiliary conjoined basis $\hat{Y} := Y - \bar{Y}T$, where $\bar{Y}$ is the conjoined basis from Proposition 6.67. By Proposition 6.61, the conjoined basis $\hat{Y}$ has constant kernel on $[K, \infty)_{\mathbb{Z}}$ and no forward focal points in $(K, \infty)$ for some index $K \in [N_m, \infty)_{\mathbb{Z}}$. By (6.239) the inclusion $\operatorname{Im} T \subseteq \operatorname{Im} P$ holds, i.e., $T = PT$. Moreover, Proposition 6.67(iv), $X_k^\dagger X_k = P$, and $P S_k = S_k$ imply that for $k \in [K, \infty)_{\mathbb{Z}}$ we have

$$\hat{X}_k = X_k - \bar{X}_k T = X_k P - \bar{X}_k PT = X_k(P - S_k T), \tag{6.362}$$

$$P - S_k T = P(P - S_k T) = X_k^\dagger X_k(P - S_k T) \overset{(6.362)}{=} X_k^\dagger \hat{X}_k. \tag{6.363}$$

Similarly, from $S_k^\dagger S_k = P = S_k S_k^\dagger$, we obtain for $k \in [K, \infty)_{\mathbb{Z}}$

$$S_k^\dagger - T = S_k^\dagger - PT = S_k^\dagger(P - S_k T), \quad P - S_k T = S_k(S_k^\dagger - T). \tag{6.364}$$

Equations (6.362)–(6.364) imply that

$$\operatorname{Ker}(P - S_k T) = \operatorname{Ker} \hat{X}_k = \operatorname{Ker}(S_k^\dagger - T), \quad k \in [K, \infty)_{\mathbb{Z}}. \tag{6.365}$$

Let $\hat{P}$ be the orthogonal projector defined in (6.234) through the conjoined basis $\hat{Y}$ on the interval $[K, \infty)_{\mathbb{Z}}$. We will show that $\operatorname{Im} \hat{P} = \operatorname{Im} P$, i.e., $\hat{P} = P$ by the uniqueness of orthogonal projectors. One inclusion follows from (6.365), since $\operatorname{Im} \hat{P} = \operatorname{Im} \hat{X}_k^T = \operatorname{Im}(S_k^\dagger - T) \subseteq \operatorname{Im} P$ for $k \in [K, \infty)_{\mathbb{Z}}$. On the other hand, by assumption (6.360) and inequality (6.279) in Theorem 6.75 (with $Y := \hat{Y}$ and with $N := K$), we have

$$\operatorname{rank} P = n - d[N, \infty)_{\mathbb{Z}} \overset{(6.360)}{=} n - d[K, \infty)_{\mathbb{Z}} \overset{(6.279)}{\leq} \operatorname{rank} \hat{P}.$$

Thus, $\operatorname{Im} P = \operatorname{Im} \hat{P}$ follows. Since $\hat{P}$ is the orthogonal projector onto $\operatorname{Im} \hat{X}_k^T$ on $[K, \infty)_{\mathbb{Z}}$ and since $\hat{P} = P$ as we just proved, the second equality in (6.365) reads as $\operatorname{Ker}(S_k^\dagger - T) = \operatorname{Ker} P$ for all $k \in [K, \infty)_{\mathbb{Z}}$. Consequently,

$$\operatorname{rank}(S_k^\dagger - T) = \operatorname{rank} P = n - d[K, \infty)_{\mathbb{Z}} = n - d_\infty, \quad k \in [K, \infty)_{\mathbb{Z}}, \quad (6.366)$$

as we claim in (6.361). Finally, the first condition in (6.361) follows from the fact that $T$ is the limit of $S_k^\dagger$ for $k \to \infty$ and from the monotonicity of $S_k^\dagger$ on $[K, \infty)_{\mathbb{Z}}$, which is guaranteed on this interval by Theorem 6.65. □

*Remark 6.102* The proof of Proposition 6.101 shows that the statement in (6.361) can be extended to the whole interval $[N, \infty)_{\mathbb{Z}}$ (instead only for large $k$), when the maximal orthogonal projector $P_{S\infty}$ is replaced by the projector $P_{Sk}$. Then

$$S_k^\dagger - P_{Sk} T P_{Sk} \geq 0, \qquad \operatorname{Ker}(S_k^\dagger - P_{Sk} T P_{Sk}) = \operatorname{Ker} P_{Sk}, \quad k \in [N, \infty)_{\mathbb{Z}}. \tag{6.367}$$

The result in Proposition 6.101 implies that $\operatorname{rank}(S_k^\dagger - T)$ does not depend on the choice of the matrices $S_k$ and $T$, i.e., it does not depend on the choice of the conjoined basis $Y$ with constant kernel on $[N, \infty)_{\mathbb{Z}}$ and no forward focal points in $(N, \infty)$. In fact, we prove for such a conjoined basis $Y$ of (SDS) the following equivalence.

**Theorem 6.103** *Let $Y$ be a conjoined basis of* (SDS) *with constant kernel on $[N, \infty)_{\mathbb{Z}}$ and no forward focal points in $(N, \infty)$, and let $S_k$, $T$, $P_{S\infty}$ be its corresponding matrix in* (6.236), (6.237), (6.238). *Then the following two conditions are equivalent:*

$$\operatorname{rank}(P_{S\infty} - S_k T) = n - d[N, \infty)_{\mathbb{Z}}, \quad k \in [N, \infty)_{\mathbb{Z}}, \tag{6.368}$$

$$\operatorname{rank}(S_k^\dagger - T) = n - d[N, \infty)_{\mathbb{Z}}, \quad k \in [M, \infty)_{\mathbb{Z}} \text{ for some } M \in [N, \infty)_{\mathbb{Z}}. \tag{6.369}$$

*In this case we have the equalities*

$$\operatorname{Im}(P_{S\infty} - S_k T) = \operatorname{Im} P_{S\infty} = \operatorname{Im}(P_{S\infty} - S_k T)^T, \quad k \in [N, \infty)_{\mathbb{Z}}. \tag{6.370}$$

*Proof* In the proof of Proposition 6.101, we showed that (6.365) and (6.366) hold, i.e., in the setting of the present theorem, we have

$$\operatorname{Ker}(P_{S\infty} - S_k T) = \operatorname{Ker} P_{S\infty} = \operatorname{Ker}(S_k^\dagger - T), \quad k \in [K, \infty)_{\mathbb{Z}}, \tag{6.371}$$

where $K \in [N_m, \infty)_{\mathbb{Z}}$ and $N_m$ is the index from Remark 6.76. That is, the space $\operatorname{Im} S_k = \operatorname{Im} P_{S\infty}$ is maximal on $[K, \infty)_{\mathbb{Z}}$. The implication $(6.368) \Rightarrow (6.369)$ (with $M := K$) then follows from (6.371) and from (6.276) in Theorem 6.75. We will prove the converse implication. Assume (6.369) and set $L := \max\{M, K\}$.

Then (6.371) yields

$$\operatorname{Im}(P_{S\infty} - S_k T) = \operatorname{Im} P_{S\infty}, \quad k \in [L, \infty)_{\mathbb{Z}}. \tag{6.372}$$

Multiplying (6.372) by $T$ from the left and using that $T P_{S\infty} = T$, we get

$$\operatorname{Im}(T - T S_k T)] = \operatorname{Im}[T(P_{S\infty} - S_k T)] = \operatorname{Im} T, \quad k \in [L, \infty)_{\mathbb{Z}}. \tag{6.373}$$

We will show that this equality is satisfied even for all $k \in [N, \infty)_{\mathbb{Z}}$. The matrix function $T S_k T$ is symmetric, nondecreasing, and nonnegative definite on $[N, \infty)_{\mathbb{Z}}$, and $S_k^{\dagger} \to T$ for $k \to \infty$ by (6.237). Therefore, the limit theorem for monotone matrix-valued functions (Theorem 1.88 in Sect. 1.6.5) implies that $T S_k T \to T$ for $k \to \infty$, where the convergence is monotone (nondecreasing). This implies that the symmetric matrix function $G_k := T - T S_k T = T(P_{S\infty} - S_k T)$ is nonincreasing and nonnegative definite on $[N, \infty)_{\mathbb{Z}}$ and hence,

$$\operatorname{Im}[T(P_{S\infty} - S_k T)] = \operatorname{Im}(T - T S_k T)] \subseteq \operatorname{Im} T, \quad k \in [N, \infty)_{\mathbb{Z}}. \tag{6.374}$$

The combination of (6.373) and (6.374) and the monotonicity of $G_k$ implies that

$$\operatorname{Im}[T(P_{S\infty} - S_k T)] = \operatorname{Im} T, \quad k \in [N, \infty)_{\mathbb{Z}}, \tag{6.375}$$

or equivalently by taking the orthogonal complements

$$\operatorname{Ker}[T(P_{S\infty} - S_k T)] = \operatorname{Ker} T, \quad k \in [N, \infty)_{\mathbb{Z}}. \tag{6.376}$$

This then implies that

$$\operatorname{Ker}(P_{S\infty} - S_k T) \subseteq \operatorname{Ker} T, \quad k \in [N, \infty)_{\mathbb{Z}}, \tag{6.377}$$

since if $(P_{S\infty} - S_k T) d = 0$ for some vector $d \in \mathbb{R}^n$, then $T(P_{S\infty} - S_k T) d = 0$ and so $d \in \operatorname{Ker} T$ by (6.376). Moreover, in this case $P_{S\infty} d = (P_{S\infty} - S_k T) d = 0$ as well, so that we also have the inclusion

$$\operatorname{Ker}(P_{S\infty} - S_k T) \subseteq \operatorname{Ker} P_{S\infty}, \quad k \in [N, \infty)_{\mathbb{Z}}. \tag{6.378}$$

On the other hand, we know by $T = T P_{\hat{S}\infty}$ (even without assumption (6.369)) that

$$\operatorname{Ker}(P_{S\infty} - S_k T) = \operatorname{Ker}(P_{S\infty} - S_k T P_{S\infty}) \supseteq \operatorname{Ker} P_{S\infty}, \quad k \in [N, \infty)_{\mathbb{Z}}. \tag{6.379}$$

Combining (6.378) and (6.379) yields that

$$\operatorname{Ker}(P_{S\infty} - S_k T) = \operatorname{Ker} P_{S\infty}, \quad k \in [N, \infty)_{\mathbb{Z}}, \tag{6.380}$$

which implies through (6.276) the desired equality (6.368). Finally, the equality $S_k = P_{S\infty} S_k$ implies the inclusion $\text{Im}\,(P_{S\infty} - S_k T) \subseteq \text{Im}\,P_{S\infty}$ on $[N, \infty)_{\mathbb{Z}}$, while under (6.368) or (6.369) the orthogonal complement of (6.380) yields the inclusion $\text{Im}\,P_{S\infty} \subseteq \text{Im}\,(P_{S\infty} - S_k T)^T$ on $[N, \infty)_{\mathbb{Z}}$. Therefore, the rank condition (6.368) guarantees the equality of the subspaces in (6.370). The proof is complete.  $\square$

*Remark 6.104* We note that the inequality $G_k = T - T S_k T \geq 0$ for large $k$, derived in the proof of Theorem 6.103 through the limit theorem for monotone matrix-valued functions (Theorem 1.88), also follows from the properties of the Moore-Penrose pseudoinverse. Indeed, since we know that $0 \leq T \leq S_k^\dagger$ for large $k$ and $S_k^\dagger \to T$ monotonically for $k \to \infty$, it follows from Remark 1.60(vi) (with $A := T$ and $B := S_k^\dagger$) that $0 \leq T S_k T \leq T$ for large $k$, i.e., $G_k \geq 0$ for large $k$. In this context the value of the limit $\lim_{k \to \infty} T S_k T = T$ is an additional property, which is in fact not needed for the proof of Theorem 6.103.

The next result shows that condition (6.368) turns out to be a characterization of the interval, where the abnormality of system (SDS) is maximal.

**Proposition 6.105** *Equality* (6.368) *holds for some (and hence for any) S-matrix $S_k$ associated with a conjoined basis $Y$ of* (SDS) *with constant kernel on $[N, \infty)_{\mathbb{Z}}$ and no forward focal points in $(N, \infty)$ if and only if condition* (6.360) *holds.*

*Proof* We have already proven in Proposition 6.101 and Theorem 6.103 that condition (6.360) implies (6.368) for any $S$-matrix $S_k$ associated with a conjoined basis $Y$ of (SDS) on $[N, \infty)_{\mathbb{Z}}$. Thus, we suppose that (6.368) holds for some such a matrix $S_k$, which corresponds to a conjoined basis $Y$ of (SDS) with constant kernel on $[N, \infty)_{\mathbb{Z}}$ and no forward focal points in $(N, \infty)$. Without loss of generality (by Theorem 6.84), we may assume that $Y$ is a minimal conjoined basis on $[N, \infty)_{\mathbb{Z}}$. Fix any index $M \in [N, \infty)_{\mathbb{Z}}$. We will show that $d[M, \infty)_{\mathbb{Z}} = d[N, \infty)_{\mathbb{Z}}$ by using Theorem 6.75 and Remark 6.76. Consider the matrix function $S_k^{(M)}$ defined by $S_k^{(M)} := S_k - S_M$ for all $k \in [M, \infty)_{\mathbb{Z}}$. Since the kernel of $Y$ is constant on $[N, \infty)_{\mathbb{Z}}$, and hence on $[M, \infty)_{\mathbb{Z}}$, it is straightforward to see from the definition of $S_k$ in (6.236) that $S_k^{(M)}$ is the $S$-matrix corresponding to $Y$ on $[M, \infty)_{\mathbb{Z}}$. The following analysis shows that $\text{Im}\,S_k^{(M)} = \text{Im}\,S_k$ for large $k$. By (6.239) and the definition of $P_{S\infty}$ in (6.238), we have $\text{Im}\,S_M \subseteq \text{Im}\,P_{S\infty}$ and $\text{Im}\,S_k = \text{Im}\,P_{S\infty}$ for large $k$. This implies that $S_M = P_{S\infty} S_M = S_M P_{S\infty}$ and $P_{S\infty} S_k = S_k$ with $P_{S\infty} = S_k^\dagger S_k = S_k S_k^\dagger$ for large $k$. Consequently,

$$S_k^{(M)} = P_{S\infty} S_k - S_M S_k^\dagger S_k = (P_{S\infty} - S_M S_k^\dagger)\, S_k \quad \text{for large } k. \tag{6.381}$$

If we now let $k \to \infty$, then $P_{S\infty} - S_M S_k^\dagger \to P_{S\infty} - S_M T$. Moreover, this limiting matrix satisfies $\text{Im}\,(P_{S\infty} - S_M T) \subseteq \text{Im}\,P_{S\infty}$ and also $\text{Im}\,(P_{S\infty} - S_M T)^T \subseteq \text{Im}\,P_{S\infty}$, because $\text{Im}\,T \subseteq \text{Im}\,P_{S\infty}$. By using assumption (6.368) and Theorem 6.103, we get

$$\text{Im}\,(P_{S\infty} - S_M T) = \text{Im}\,P_{S\infty} = \text{Im}\,(P_{S\infty} - S_M T)^T.$$

In a similar way, we have $\mathrm{Im}\,(P_{\mathcal{S}\infty} - S_M S_k^\dagger) \subseteq \mathrm{Im}\,P_{\mathcal{S}\infty} = \mathrm{Im}\,(P_{\mathcal{S}\infty} - S_M T)$ and $\mathrm{Im}\,(P_{\mathcal{S}\infty} - S_M S_k^\dagger)^T \subseteq \mathrm{Im}\,P_{\mathcal{S}\infty} = \mathrm{Im}\,(P_{\mathcal{S}\infty} - S_M T)^T$ for all $k \in [N, \infty)_{\mathbb{Z}}$. Therefore,

$$\mathrm{Im}\,(P_{\mathcal{S}\infty} - S_M S_k^\dagger) = \mathrm{Im}\,(P_{\mathcal{S}\infty} - S_M S_k^\dagger)^T = \mathrm{Im}\,P_{\mathcal{S}\infty} \quad \text{for large } k, \tag{6.382}$$

and by Lemma 1.61 (with $A_j := P_{\mathcal{S}\infty} - S_M S_k^\dagger$ and $A := P_{\mathcal{S}\infty} - S_M T$), we obtain

$$(P_{\mathcal{S}\infty} - S_M S_k^\dagger)^\dagger \to (P_{\mathcal{S}\infty} - S_M T)^\dagger \quad \text{for } k \to \infty. \tag{6.383}$$

By Remark 1.62 and the equalities in (6.381) and (6.382), we now calculate

$$(S_k^{(M)})^\dagger = (P_{\mathcal{S}\infty} S_k)^\dagger\,[(P_{\mathcal{S}\infty} - S_M S_k^\dagger)\,P_{\mathcal{S}\infty}]^\dagger = S_k^\dagger (P_{\mathcal{S}\infty} - S_M S_k^\dagger)^\dagger \tag{6.384}$$

for large $k$. By using Remark 1.62, the matrix $S_k^{(M)} (S_k^{(M)})^\dagger$ is the orthogonal projector onto $\mathrm{Im}\,S_k^{(M)}$. Thus, by (6.381) and (6.384), we have for large $k$ that

$$\begin{aligned} S_k^{(M)} (S_k^{(M)})^\dagger &= (P_{\mathcal{S}\infty} - S_M S_k^\dagger)\,S_k S_k^\dagger\,(P_{\mathcal{S}\infty} - S_M S_k^\dagger)^\dagger \\ &= (P_{\mathcal{S}\infty} - S_M S_k^\dagger)\,(P_{\mathcal{S}\infty} - S_M S_k^\dagger)^\dagger, \end{aligned} \tag{6.385}$$

where we used the identities $S_k S_k^\dagger = P_{\mathcal{S}\infty}$ and $S_k^\dagger P_{\mathcal{S}\infty} = S_k^\dagger$ for large $k$. But since by Remark 1.62 the matrix in (6.385) is the orthogonal projector onto the subspace $\mathrm{Im}\,(P_{\mathcal{S}\infty} - S_M S_k^\dagger)$, we conclude from (6.382) and (6.385) that $S_k^{(M)} (S_k^{(M)})^\dagger = P_{\mathcal{S}\infty}$ for large $t$. This means that the two projectors onto $\mathrm{Im}\,S_k^{(M)}$ and $\mathrm{Im}\,S_k$ are the same for large $k$ (and they are equal to $P_{\mathcal{S}\infty}$), so that $\mathrm{Im}\,S_k^{(M)} = \mathrm{Im}\,S_k$ for large $k$. This implies through Remark 6.76 that

$$n - d[M, \infty)_{\mathbb{Z}} = \mathrm{rank}\,S_k^{(M)} = \mathrm{rank}\,S_k = n - d[N, \infty)_{\mathbb{Z}} \quad \text{for large } k.$$

This shows that $d[N, \infty)_{\mathbb{Z}} = d[M, \infty)_{\mathbb{Z}}$. Since the index $M \in [N, \infty)_{\mathbb{Z}}$ was arbitrary, condition (6.368) holds, and the proof is complete.  $\square$

In our next result, we use the knowledge of the asymptotic properties of the $S$-matrices and the $T$-matrices to provide a complete classification of all minimal conjoined bases of (SDS) on the given interval $[N, \infty)_{\mathbb{Z}}$, where the order of abnormality $d[N, \infty)_{\mathbb{Z}}$ is maximal. This turns out to be one of the crucial results of this subsection, as it will be utilized in the characterization of the matrices $T$ in Theorem 6.107 below, as well as in the construction of minimal dominant solutions of (SDS) at $\infty$ in Sect. 6.3.7.

Below the letter $N$ denotes also the matrix $N$ from the representation formula in Theorem 6.69. No confusion should arise regarding the notation for this matrix $N$ and for the index $N$ used throughout this section for the interval $[N, \infty)_{\mathbb{Z}}$.

**Theorem 6.106** *Let Y be a minimal conjoined basis of* (SDS) *on* $[N, \infty)_{\mathbb{Z}}$ *with the matrices* $P_{S\infty}$ *and* $T$ *defined in* (6.238) *and* (6.237), *and assume* $d[N, \infty)_{\mathbb{Z}} = d_\infty$. *Then a solution* $\tilde{Y}$ *of* (SDS) *is a minimal conjoined basis on* $[N, \infty)_{\mathbb{Z}}$ *if and only if there exist matrices* $M, N \in \mathbb{R}^{n \times n}$ *such that*

$$\tilde{X}_N = X_N M, \quad \tilde{U}_N = U_N M + X_N^{\dagger T} N, \tag{6.386}$$

$$M \text{ is nonsingular}, \quad M^T N = N^T M, \quad \text{Im } N \subseteq \text{Im } P_{S\infty}, \tag{6.387}$$

$$N M^{-1} + T \geq 0. \tag{6.388}$$

*In this case the matrix* $\tilde{T}$ *in* (6.237), *which corresponds to* $\tilde{Y}$, *satisfies*

$$\text{rank } \tilde{T} = \text{rank}\,(N M^{-1} + T). \tag{6.389}$$

*Proof* Let $Y$ and the index $N \in [0, \infty)_{\mathbb{Z}}$ be as in the theorem. Then the orthogonal projector $P$ defined in (6.234) satisfies $P = P_{S\infty}$. If $\tilde{Y}$ is also a minimal conjoined basis on $[N, \infty)_{\mathbb{Z}}$, then $\text{Im } \tilde{X}_N = \text{Im } X_N$ by Theorem 6.97(i). Therefore, by Theorem 6.69 (with $Y^{(1)} := Y$ and $Y^{(2)} := \tilde{Y}$), there exist matrices $M, N \in \mathbb{R}^{n \times n}$ such that (6.386) and (6.387) hold. Moreover, let $T$ and $\tilde{T}$ be the $T$-matrices defined in (6.237) through the functions $S_k$ and $\tilde{S}_k$ in (6.236), which are associated with $Y$ and $\tilde{Y}$, respectively. By using formula (6.348) (with $T^{*(1)} := T$, $T^{*(2)} := \tilde{T}$, $M^{*(1)} := M$, and $N^{*(1)} := N$), we have

$$\tilde{T} = M^T T M + M^T N, \quad \text{i.e.,} \quad N M^{-1} + T = M^{T-1} \tilde{T} M^{-1} \geq 0, \tag{6.390}$$

since $\tilde{T} \geq 0$. This shows condition (6.388). Conversely, let $\tilde{Y}$ be a solution of (SDS) satisfying (6.386)–(6.388). Then the conditions in (6.387) together with the identity $X_N^T X_N^{\dagger T} = P = P_{S\infty}$ and the fact that $Y$ is a conjoined basis imply that $\tilde{Y}$ is also a conjoined basis of (SDS). Let $S_k$ be the $S$-matrix in (6.236) corresponding to $Y$ on $[N, \infty)_{\mathbb{Z}}$. By Remark 6.70(iii), condition (6.386) then yields

$$\tilde{X}_k = X_k(P_{S\infty} M + S_k N) \quad k \in [N, \infty)_{\mathbb{Z}}. \tag{6.391}$$

We will show that $\tilde{Y}$ has constant kernel on $[N, \infty)_{\mathbb{Z}}$ and that $\text{Ker } \tilde{X}_k = \text{Ker } P_{S\infty} M$ on $[N, \infty)_{\mathbb{Z}}$. First we note that by the symmetry of $M^T N$ and $P_{S\infty} N = N$,

$$N M^{-1} P_{S\infty} = M^{T-1} N^T P_{S\infty} = M^{T-1} N^T = N M^{-1}$$

holds. Hence, by (6.391), we have for any $k \in [N, \infty)_{\mathbb{Z}}$

$$\tilde{X}_k = X_k(P_{S\infty} M + S_k N M^{-1} M) = X_k(I + S_k N M^{-1}) P_{S\infty} M. \tag{6.392}$$

Therefore, $\text{Ker } P_{S\infty} M \subseteq \text{Ker } \tilde{X}_k$ on $[N, \infty)_{\mathbb{Z}}$. Fix now $k \in [N, \infty)_{\mathbb{Z}}$, $v \in \text{Ker } \tilde{X}_k$, and set $w := P_{S\infty} M v$. Then $X_k(w + S_k N M^{-1} w) = 0$ by (6.392). Multiplying the latter equality by $X_k^\dagger$ from the left and using the identities $X_k^\dagger X_k = P_{S\infty}$, $P_{S\infty} S_k = S_k$, and $w = P_{S\infty} w$, we get $w = -S_k N M^{-1} w$. This implies by using (6.238) and (6.239) that $w \in \operatorname{Im} S_k = \operatorname{Im} P_{Sk}$ and consequently,

$$w^T S_k^\dagger w = -w^T S_k^\dagger S_k N M^{-1} w = -w^T P_{Sk} N M^{-1} P_{Sk} w. \qquad (6.393)$$

Equality (6.393) and condition (6.388) then yield $w^T S_k^\dagger w \leq w^T P_{Sk} T P_{Sk} w$, or equivalently $w^T (S_k^\dagger - P_{Sk} T P_{Sk}) w \leq 0$. But $S_k^\dagger - P_{Sk} T P_{Sk} \geq 0$ according to Remark 6.102 and thus, $w \in \operatorname{Ker}(S_k^\dagger - P_{Sk} T P_{Sk}) = \operatorname{Ker} P_{Sk}$, by the second formula in (6.367). Hence we obtain that $w \in \operatorname{Ker} P_{Sk} \cap \operatorname{Im} P_{Sk} = \{0\}$. This shows that $w = 0$, and then $v \in \operatorname{Ker} P_{S\infty} M$, i.e., $\operatorname{Ker} \tilde{X}_k \subseteq \operatorname{Ker} P_{S\infty} M$. Finally, (6.386) and the invertibility of $M$ imply that rank $\tilde{X}_k = \operatorname{rank} \tilde{X}_N = \operatorname{rank} X_N = n - d_\infty$ on $[N, \infty)_\mathbb{Z}$. This shows that $\tilde{Y}$ is a minimal conjoined basis of (SDS) on $[N, \infty)_\mathbb{Z}$. The proof is complete.                                                        $\square$

In the last result of this subsection, we present a criterion for the classification of all $T$-matrices, which correspond to (minimal) conjoined bases of (SDS) on an interval $[N, \infty)_\mathbb{Z}$ with the maximal order of abnormality.

**Theorem 6.107** *Assume that system* (SDS) *is nonoscillatory. Then* $D \in \mathbb{R}^{n \times n}$ *is a $T$-matrix of some minimal conjoined basis* $Y$ *of* (SDS) *on* $[N, \infty)_\mathbb{Z}$ *with* $d[N, \infty)_\mathbb{Z} = d_\infty$ *if and only if*

$$D \text{ is symmetric}, \quad D \geq 0, \quad \operatorname{rank} D \leq n - d_\infty. \qquad (6.394)$$

*Proof* Let $D$ be a $T$-matrix associated with a minimal conjoined basis $Y$ a given interval $[N, \infty)_\mathbb{Z} \subseteq [0, \infty)_\mathbb{Z}$. Let $S_k$ and $T$ be defined in (6.236) and (6.237), so that $D = T$. By Remark 6.96 we have $d[N, \infty)_\mathbb{Z} = d_\infty$. From Theorem 6.65 and (6.239), we obtain that $D$ is symmetric, nonnegative definite and $\operatorname{Im} D \subseteq \operatorname{Im} P_{S\infty}$ with $P_{S\infty}$ defined in (6.238). But since rank $P_{S\infty} = n - d[N, \infty)_\mathbb{Z} = n - d_\infty$ by (6.276), the condition rank $D \leq n - d_\infty$ follows. Conversely, assume that $D \in \mathbb{R}^{n \times n}$ satisfies (6.394). From the third condition in (6.394), we have that there exists an orthogonal projector $Q$ such that $\operatorname{Im} D \subseteq \operatorname{Im} Q$ and rank $Q = n - d_\infty$. Furthermore, the nonoscillation of (SDS) and Theorem 6.90 (with $r := n - d_\infty$) imply that there exists a minimal conjoined basis $Y$ of (SDS) on an interval $[N, \infty)_\mathbb{Z} \subseteq [0, \infty)_\mathbb{Z}$. Let $S_k$, $P_{S\infty}$, and $T$ be the matrices associated with $Y$ in (6.236), (6.238), and (6.237). Since $d[N, \infty)_\mathbb{Z} = d_\infty$, we have rank $P_{S\infty} = n - d_\infty = \operatorname{rank} Q$, and hence there exists an invertible matrix $E$ satisfying $\operatorname{Im} E P_{S\infty} = \operatorname{Im} Q$. The matrix $E$ can be obtained, e.g., from the diagonalization of $P_{S\infty}$ and $Q$ or from Theorem 1.91 in Sect. 1.6.6 (with $P_* := 0$). In particular, we then have $\operatorname{Im} E^{-1} Q = \operatorname{Im} P_{S\infty}$, i.e., $P_{S\infty} E^{-1} Q = E^{-1} Q$. Define now the matrices $M, N \in \mathbb{R}^{n \times n}$ by

$$M := E^T, \quad N := E^{-1} D - T E^T. \qquad (6.395)$$

We show that these matrices satisfy conditions (6.387) and (6.388) in Theorem 6.106. The matrix $M$ is invertible by its definition. The symmetry of $D$ and $T$ implies that $M^T N = D - E T E^T$ is also symmetric. Moreover, the equalities $QD = D$, $P_{S_\infty} E^{-1} Q = E^{-1} Q$, and $P_{S_\infty} T = T$ yield

$$P_{S_\infty} N = P_{S_\infty} E^{-1} QD - T E^T = E^{-1} QD - T E^T = E^{-1} D - T E^T = N.$$

This means that $\operatorname{Im} N \subseteq \operatorname{Im} P_{S_\infty}$. Finally, the inequality $D \geq 0$ implies (6.388), since $N M^{-1} + T = (E^{-1} D - T E^T) E^{T-1} + T = E^{-1} D E^{T-1} \geq 0$. Therefore, we proved that for a given $D$ satisfying (6.394) and for any minimal conjoined basis $Y$ of (SDS) on $[N, \infty)_{\mathbb{Z}}$, the matrices $M$ and $N$ in (6.395) satisfy the conditions in (6.387) and (6.388). Consider now the solution $\tilde{Y}$ of (SDS) given by the initial conditions (6.386). By Theorem 6.106 it follows that $\tilde{Y}$ is a minimal conjoined basis on $[N, \infty)_{\mathbb{Z}}$. Moreover, if $\tilde{T}$ is the matrix in (6.237) associated with $\tilde{Y}$, then by (6.390) satisfies $\tilde{T} = M^T T M + M^T N$. By using (6.395) we then obtain that $\tilde{T} = D$. Therefore, the matrix $D$ is a $T$-matrix associated with the minimal conjoined basis $\tilde{Y}$ of (SDS) on $[N, \infty)_{\mathbb{Z}}$. □

*Remark 6.108* We note that with the aid of Proposition 6.86 and Theorem 6.90, the statement in Theorem 6.107 extends directly to any conjoined basis $Y$ of (SDS) with constant kernel on $[N, \infty)_{\mathbb{Z}}$ and no forward focal points in $(N, \infty)$.

### 6.3.6 Recessive Solutions at Infinity

In this subsection we present the concept of a recessive solution of (SDS) at $\infty$. It is defined by the property $T = 0$ in (6.237). It is a generalization of the corresponding notion in Definition 2.63 in Sect. 2.5.

**Definition 6.109 (Recessive Solution at $\infty$)** A conjoined basis $\hat{Y}$ of (SDS) is said to be a *recessive solution* of (SDS) at $\infty$ if there exists $N \in [0, \infty)_{\mathbb{Z}}$ such that $\hat{Y}$ has constant kernel on $[N, \infty)_{\mathbb{Z}}$ and no forward focal points in $(N, \infty)$ and satisfying

$$\lim_{k \to \infty} \hat{S}_k^\dagger = 0, \qquad \hat{S}_k := \sum_{j=N}^{k-1} \hat{X}_{j+1}^\dagger B_j \hat{X}_j^{T\,\dagger}, \tag{6.396}$$

i.e., the matrix $\hat{T}$ in (6.237) associated with $\hat{Y}$ satisfies $\hat{T} = 0$.

If $\hat{Y}$ is a recessive solution of (SDS) at $\infty$ such that $\hat{Y}$ has constant kernel on $[N, \infty)_{\mathbb{Z}}$ and no forward focal points in $(N, \infty)$, then $\hat{X}_k$ satisfies the rank condition in (6.279). Therefore, we introduce the following terminology regarding the classification of recessive solutions of (SDS) according to their rank.

*Remark 6.110* Let $\hat{Y}$ be a recessive solution of (SDS) at $\infty$, and let $r$ be its rank on $[N, \infty)_{\mathbb{Z}}$. If $r = n - d_\infty$, then $\hat{Y}$ is called a *minimal recessive solution* of (SDS) at $\infty$, while if $r = n$, then $\hat{Y}$ is called a *maximal recessive solution* of (SDS) at $\infty$. This terminology corresponds to the two extreme cases in (6.279) or in Theorem 6.115 below. We will use the special notation $\hat{Y}^{\min} = Y^{[\infty]}$ and $\hat{Y}^{\max}$ for the recessive solutions of (SDS) at $\infty$, which are minimal and maximal, respectively, according to the above definition. If $n - d_\infty < r < n$, then the recessive solution $\hat{Y}$ will be called *intermediate* (of the rank $r$). We note that in the eventually controllable case, we have $d_\infty = 0$ (see Remark 6.58), so that the minimal and maximal recessive solutions of (SDS) at $\infty$ coincide. Therefore, in this case all recessive solutions $\hat{Y}$ of (SDS) at $\infty$ have eventually $\hat{X}_k$ invertible, as it is known in Sect. 2.5, or in [16, Section 3.11] and [81].

The next statement shows that moving the initial point of the interval $[N, \infty)_{\mathbb{Z}}$, with respect to which a recessive solution at $\infty$ is considered, to the right does not change the property of being a recessive solution of (SDS).

**Proposition 6.111** *Let $\hat{Y}$ be a recessive solution of (SDS) at $\infty$ on $[N, \infty)_{\mathbb{Z}}$. Then condition (6.360) holds and $\hat{Y}$ is a recessive solution of (SDS) at $\infty$ on $[M, \infty)_{\mathbb{Z}}$ for every $M \in [N, \infty)_{\mathbb{Z}}$.*

*Proof* Fix an index $M \in [N, \infty)_{\mathbb{Z}}$. Let $\hat{S}_k$ defined in (6.396) be the $S$-matrix corresponding to $\hat{Y}$ on $[N, \infty)_{\mathbb{Z}}$ with $\hat{T} := \lim_{k \to \infty} \hat{S}_k^\dagger = 0$. Since by (6.276) we have rank $P_{\hat{S}_\infty} = n - d[N, \infty)_{\mathbb{Z}}$, it follows that condition (6.368) in Theorem 6.103 holds with $T := \hat{T} = 0$. This yields via Proposition 6.105 that (6.360) is satisfied, in particular $d[N, \infty)_{\mathbb{Z}} = d[M, \infty)_{\mathbb{Z}}$. Now we consider the $S$-matrix

$$\hat{S}_k^{(M)} := \sum_{j=M}^{k-1} \hat{X}_{j+1}^\dagger \mathcal{B}_j \hat{X}_j^{\dagger T} = \hat{S}_k - \hat{S}_M, \quad k \in [M, \infty)_{\mathbb{Z}}, \tag{6.397}$$

for $\hat{Y}$ on $[M, \infty)_{\mathbb{Z}}$. Then as in the proof of Proposition 6.105, we have on $[M, \infty)_{\mathbb{Z}}$

$$(\hat{S}_k^{(M)})^\dagger = (P_{\hat{S}_\infty} \hat{S}_k)^\dagger [(P_{\hat{S}_\infty} - \hat{S}_M \hat{S}_k^\dagger) P_{\hat{S}_\infty}]^\dagger = \hat{S}_k^\dagger (P_{\hat{S}_\infty} - \hat{S}_M \hat{S}_k^\dagger)^\dagger. \tag{6.398}$$

Upon taking the limit as $k \to \infty$ in (6.398), we obtain that

$$\hat{T}^{(M)} := \lim_{k \to \infty} (\hat{S}_k^{(M)})^\dagger = \hat{T} (P_{\hat{S}_\infty} - \hat{S}_M \hat{T})^\dagger. \tag{6.399}$$

Note that the limit of $(P_{\hat{S}_\infty} - \hat{S}_M \hat{S}_k^\dagger)^\dagger$ for $k \to \infty$ indeed exists and is equal to $(P_{\hat{S}_\infty} - \hat{S}_M \hat{T})^\dagger$, because the matrices $P_{\hat{S}_\infty} - \hat{S}_M \hat{S}_k^\dagger$ have constant rank for large $k$, which is equal to the rank of the limit matrix $P_{\hat{S}_\infty} - \hat{S}_M \hat{T}$; see (6.368) and Lemma 1.61 or Remark 1.60(v). Since $\hat{T} = 0$, equality (6.399) yields that $\hat{T}^{(M)} = 0$ and $\hat{Y}$ is a recessive solution of (SDS) at $\infty$ on the interval $[M, \infty)_{\mathbb{Z}}$, by Definition 6.109.                                                                                    $\square$

The following result shows that the relation "being contained" preserves the property of being a recessive solution of (SDS).

**Proposition 6.112** *Let $\hat{Y}$ be a recessive solution of (SDS) at $\infty$ on $[N, \infty)_{\mathbb{Z}}$. Then every conjoined basis of (SDS) with constant kernel on $[N, \infty)_{\mathbb{Z}}$ and no forward focal points in $(N, \infty)$, which is either contained in $\hat{Y}$ on $[N, \infty)_{\mathbb{Z}}$ or which contains $\hat{Y}$ on $[N, \infty)_{\mathbb{Z}}$, is also a recessive solution of (SDS) at $\infty$ on the interval $[N, \infty)_{\mathbb{Z}}$.*

*Proof* The result follows from Proposition 6.86 and from the proof of Theorem 6.87(iii), since the relation "being contained" for conjoined bases of (SDS) with constant kernel on $[N, \infty)_{\mathbb{Z}}$ and no forward focal points in $(N, \infty)$ preserves the corresponding $S$-matrices. Specifically, the conjoined bases $Y^*$ which are contained in $\hat{Y}$ on $[N, \infty)_{\mathbb{Z}}$ satisfy $S_k^* = \hat{S}_k$ on $[N, \infty)_{\mathbb{Z}}$ by Proposition 6.86, while the conjoined bases $\bar{Y}$ of (SDS) which contain $\hat{Y}$ on $[N, \infty)_{\mathbb{Z}}$ satisfy $\tilde{S}_k = \hat{S}_k$ on $[N, \infty)_{\mathbb{Z}}$ by the proof of Theorem 6.87(iii). □

In the next results, we discuss the existence and uniqueness of the minimal recessive solution at $\infty$ (Theorem 6.113) and the relationship of the nonoscillation of (SDS) at $\infty$ with the existence of the recessive solutions at $\infty$ (Theorem 6.115). The first statement is a direct generalization of the results in Theorem 2.66 and Remark 2.69, as we now drop the eventual controllability assumption.

**Theorem 6.113** *System (SDS) is nonoscillatory at $\infty$ if and only if there exists a minimal recessive solution of (SDS) at $\infty$. In this case the minimal recessive solution is unique up to a right nonsingular multiple, that is, if $\hat{Y}$ is a minimal recessive solution of (SDS) at $\infty$, then a solution $\hat{Y}^{(0)}$ of (SDS) is a minimal recessive solution at $\infty$ if and only if $\hat{Y}_k^{(0)} = \hat{Y}_k \hat{M}$ on $[0, \infty)_{\mathbb{Z}}$ for some invertible matrix $\hat{M}$.*

*Proof* Assume that system (SDS) is nonoscillatory. Let $Y$ be any fixed conjoined basis of (SDS). Then there exists a sufficiently large $N \in [0, \infty)_{\mathbb{Z}}$ such that $Y$ has constant kernel on $[N, \infty)_{\mathbb{Z}}$ and no forward focal points in $(0, \infty)$ and such that condition (6.360) holds. Let $S_k$ and $T$ be given in (6.236) and (6.237), and let $P$ and $P_{S\infty}$ be the associated orthogonal projectors in (6.234) and (6.238). Without loss of generality (by Theorem 6.84 and Remark 6.85 with $P^* := P_{S\infty}$), we may assume that $Y$ is a minimal conjoined basis on $[N, \infty)_{\mathbb{Z}}$, i.e., $P_{S\infty} = P$. Consider the conjoined basis $\hat{Y} := Y - \bar{Y} T$, where $\bar{Y}$ is given in Proposition 6.67. Then

$$\hat{X}_k = X_k (P - S_k T), \quad \text{Ker } \hat{X}_k = \text{Ker} (P - S_k T) = \text{Ker } P, \quad k \in [N, \infty)_{\mathbb{Z}},$$
(6.400)

where the second condition above follows from (6.370). Thus, $\hat{Y}$ has constant kernel on $[N, \infty)_{\mathbb{Z}}$ and the corresponding orthogonal projector $\hat{P}$ satisfies $\hat{P} = P$. We shall prove that $\hat{Y}$ has no forward focal points in $(N, \infty)$. From (6.400) and (6.236), it follows that

$$\hat{X}_k^\dagger = (P - S_k T)^\dagger X_k^\dagger, \quad k \in [N, \infty)_{\mathbb{Z}}, \tag{6.401}$$

$$P - S_k T = P - S_{k+1} T + (\Delta S_k) T$$

$$= (P - S_{k+1} T) + X_{k+1}^\dagger \mathcal{B}_k X_k^{\dagger T} T, \quad k \in [N, \infty)_{\mathbb{Z}}. \tag{6.402}$$

Note that by (6.370) we have $(P - S_{k+1} T)(P - S_{k+1} T)^\dagger = P$. Therefore, by (6.400)–(6.402) we obtain

$$\hat{X}_k \hat{X}_{k+1}^\dagger \mathcal{B}_k = X_k (P - S_k T)(P - S_{k+1} T)^\dagger X_{k+1}^\dagger \mathcal{B}_k$$

$$= X_k [(P - S_{k+1} T) + X_{k+1}^\dagger \mathcal{B}_k X_k^{\dagger T} T](P - S_{k+1} T)^\dagger X_{k+1}^\dagger \mathcal{B}_k$$

$$= X_k P X_{k+1}^\dagger \mathcal{B}_k + X_k X_{k+1}^\dagger \mathcal{B}_k X_k^{\dagger T} T^{[k+1]} X_{k+1}^\dagger \mathcal{B}_k$$

$$= X_k X_{k+1}^\dagger \mathcal{B}_k + \mathcal{B}_k^T X_{k+1}^{\dagger T} T^{[k+1]} X_{k+1}^\dagger \mathcal{B}_k, \tag{6.403}$$

where, similarly to (6.399), the matrix $T^{[k+1]} = T(P - S_{k+1} T)^\dagger$ and where $X_{k+1}^{\dagger T} P = X_{k+1}^{\dagger T}$, as the projector $P = P_{S\infty}$ is constant on $[N, \infty)_{\mathbb{Z}}$. Each term in the sum in (6.403) is nonnegative definite, because $Y$ has no forward focal points in $(0, \infty)$ and $T^{[k+1]} \geq 0$. This shows that $\hat{X}_k \hat{X}_{k+1}^\dagger \mathcal{B}_k \geq 0$ on $[N, \infty)_{\mathbb{Z}}$, i.e., $\hat{Y}$ has no forward focal points in $(0, \infty)$ as well. Thus, $\hat{Y}$ is a minimal conjoined basis on $[N, \infty)_{\mathbb{Z}}$. Let $\hat{S}_k$ be its corresponding $S$-matrix. From Proposition 6.97(iii) (with $Y^{(1)} := Y$, $Y^{(2)} := \hat{Y}$, $L^{(1)} := X_N^\dagger \hat{X}_N = X_N^\dagger X_N = P$, and $N^{(1)} := -T$), we obtain the equality $\hat{S}_k^\dagger = S_k^\dagger - T$ for all $k$ large enough. Formula (6.237) now implies that $\hat{S}_k^\dagger \to 0$ for $k \to \infty$, i.e., $\hat{Y}$ is a minimal recessive solution of (SDS) at $\infty$. Conversely, the existence of a minimal recessive solution of (SDS) at $\infty$, which is a nonoscillatory conjoined basis of (SDS) at $\infty$, implies the nonoscillation of (SDS) at $\infty$ by Proposition 6.61.

Now we prove the uniqueness. Let $\hat{Y}$ be a minimal recessive solution of (SDS) at $\infty$ with respect to the interval $[N, \infty)_{\mathbb{Z}}$ and let $\hat{Y}^{(0)}$ be a minimal recessive solution of (SDS) at $\infty$ with respect to $[N_0, \infty)_{\mathbb{Z}}$. Without loss of generality, we may assume that $N_0 = N$, since shifting the initial point to the right preserves the property of being a recessive solution of (SDS) at $\infty$, by Proposition 6.111. Let $\hat{P}$ and $\hat{P}^{(0)}$ be the corresponding orthogonal projectors in (6.233) defined through $\hat{X}$ and $\hat{X}^{[0]}$. By Proposition 6.97(i), we know that $\operatorname{Im} \hat{X}_N = \operatorname{Im} \hat{X}_N^{[0]}$, which in turn implies by Theorem 6.69 (with $Y^{(1)} := \hat{Y}^{(0)}$ and $Y^{(2)} := \hat{Y}$) that $\hat{Y}_k^{(0)} = \hat{Y}_k \hat{M} + \bar{Y}_k^{(2)} \hat{N}$ on $[N, \infty)_{\mathbb{Z}}$, where the matrix $\hat{M}$ is constant and nonsingular and the matrix $\hat{N}$ satisfies $\operatorname{Im} \hat{N} \subseteq \operatorname{Im} \hat{P}$. If the matrices $\hat{S}_k$, $\hat{T}$ and $\hat{S}^{(0)}$, $\hat{T}^{(0)}$ are now defined in (6.236), (6.237) through $\hat{X}$ and $\hat{X}^{(0)}$, respectively, then formula (6.340) in Proposition 6.97(iii) with $i = 2$ has the form

$$(\hat{S}_k^{(0)})^\dagger = \hat{L}^T \hat{S}_k^\dagger \hat{L} + \hat{L}^T \hat{N} \quad \text{for large } k, \tag{6.404}$$

with $\hat{L} = \hat{P}\hat{M}$ by Remark 6.70(ii). Upon taking the limit as $k \to \infty$ in (6.404) and using that $\hat{Y}$ and $\hat{Y}^{(0)}$ are recessive solutions of (SDS) at $\infty$, i.e., $\hat{T} = 0 = \hat{T}^{(0)}$, we obtain from (6.404) the equality $0 = \hat{T}^{(0)} = \hat{L}^T \hat{T} \hat{L} + \hat{L}^T \hat{N} = \hat{L}^T \hat{N}$. Multiplying this equation by $\hat{L}^{\dagger T}$ from the left and using $\hat{L}^{\dagger T} \hat{L}^T = \hat{P}$ and $\hat{P}\hat{N} = \hat{N}$ (see Remark 6.70(ii) again), we obtain $\hat{N} = 0$. This means that $\hat{Y}_k^{(0)} = \hat{Y}_k \hat{M}$ on $[N, \infty)_{\mathbb{Z}}$ with $\hat{M}$ invertible. Conversely, if $\hat{Y}$ is a minimal recessive solution of (SDS) at $\infty$ with respect to the interval $[N, \infty)_{\mathbb{Z}}$, then the solution $\hat{Y}_k^{(0)} := \hat{Y}_k \hat{M}$ with an invertible $\hat{M}$ obviously has constant kernel on $[N, \infty)_{\mathbb{Z}}$. Moreover, $\hat{Y}^{(0)}$ has also no forward focal points in $(N, \infty)$, because the matrix $\hat{P} = \hat{X}_k^{\dagger} \hat{X}_k = \hat{X}_{k+1}^{\dagger} \hat{X}_{k+1}$ is constant on $[N, \infty)_{\mathbb{Z}}$, the matrix

$$(\hat{X}_{k+1}^{(0)})^{\dagger} = (\hat{X}_{k+1}\hat{M})^{\dagger} = (\hat{X}_{k+1}^{\dagger}\hat{X}_{k+1}\hat{M})^{\dagger}(\hat{X}_{k+1}\hat{M}\hat{M}^{-1})^{\dagger} = (\hat{P}\hat{M})^{\dagger} \hat{X}_{k+1}^{\dagger}$$
$$(6.405)$$

by Remark 1.62, and for $k \in [N, \infty)_{\mathbb{Z}}$

$$\hat{X}_k^{(0)}(\hat{X}_{k+1}^{(0)})^{\dagger}\mathcal{B}_k \overset{(6.405)}{=} \hat{X}_k \hat{M}(\hat{P}\hat{M})^{\dagger} \hat{X}_{k+1}^{\dagger}\mathcal{B}_k = \hat{X}_k \hat{P}\hat{M}(\hat{P}\hat{M})^{\dagger} \hat{P}\hat{X}_{k+1}^{\dagger}\mathcal{B}_k$$
$$= \hat{X}_k(\hat{P}\hat{M})(\hat{P}\hat{M})^{\dagger}(\hat{P}\hat{M})\hat{M}^{-1}\hat{X}_{k+1}^{\dagger}\mathcal{B}_k$$
$$= \hat{X}_k(\hat{P}\hat{M})\hat{M}^{-1}\hat{X}_{k+1}^{\dagger}\mathcal{B}_k = \hat{X}_k \hat{X}_{k+1}^{\dagger}\mathcal{B}_k \geq 0.$$

Hence, $\hat{Y}^{(0)}$ is a minimal conjoined basis of (SDS) on $[N, \infty)_{\mathbb{Z}}$. Moreover, we have the equality $\operatorname{Im} \hat{X}_N^{(0)} = \operatorname{Im} \hat{X}_N$ and, by Theorem 6.69 (with the same notation as above), $\hat{N} := N^{(2)} = 0$. As in (6.404) we then obtain $(\hat{S}_k^{(0)})^{\dagger} = \hat{L}^T \hat{S}_k^{\dagger} \hat{L}$ for large $k$. Since $\hat{S}_k^{\dagger} \to \hat{T} = 0$ as $k \to \infty$, it follows that $(\hat{S}_k^{(0)})^{\dagger} \to \hat{T}^{(0)} = 0$ for $k \to \infty$ as well. Therefore, $\hat{Y}^{(0)}$ is a minimal recessive solution of (SDS) at $\infty$ and the proof is complete. □

In Proposition 6.151 and Theorem 6.153 in Sect. 6.3.10 below, we present further characterizations of the minimal recessive solution of (SDS) at $\infty$.

*Remark 6.114* In the last part of Theorem 6.113, we showed that if $\hat{Y}$ is a recessive solution of (SDS) at $\infty$ on $[N, \infty)_{\mathbb{Z}}$, then $\hat{Y}\hat{M}$ is also a recessive solution of (SDS) at $\infty$ on $[N, \infty)_{\mathbb{Z}}$ with the same rank for any constant invertible matrix $\hat{M}$. This statement extends [16, Theorem 3.43(i)] to the general concept of recessive solutions at $\infty$ in this paper.

We note that Theorem 6.115 below is an extension of the existence part of Theorem 6.113. In particular, we shall see that the existence of the minimal recessive solution of (SDS) from Theorem 6.113 is used in the proof of the implication (i) $\Rightarrow$ (ii) in Theorem 6.115. Also, the uniqueness (or some other classification) of the recessive solutions of (SDS) at $\infty$ with rank $r > n - d_{\infty}$ is not guaranteed; see, e.g., Remark 6.116.

**Theorem 6.115** *The following statements are equivalent.*

*(i)  System* (SDS) *is nonoscillatory at* $\infty$.
*(ii)  There exists a recessive solution* $\hat{Y}$ *of* (SDS) *at* $\infty$.
*(iii)  For any integer value r between* $n - d_\infty$ *and n, there exists a recessive solution* $\hat{Y}$ *of* (SDS) *at* $\infty$ *with the rank of* $\hat{X}_k$ *equal to r for large k.*

*Proof* (i)$\Rightarrow$(ii) This follows from Theorem 6.113, as the nonoscillation of (SDS) at $\infty$ implies the existence of the minimal recessive solution of (SDS) at $\infty$.

(ii)$\Rightarrow$(iii) Let $\hat{Y}$ be a recessive solution of (SDS) at $\infty$. By Definition 6.109, there exists an index $N \in [0, \infty)_{\mathbb{Z}}$ such that $\hat{Y}$ is a conjoined basis of (SDS) with constant kernel on $[N, \infty)_{\mathbb{Z}}$, no forward focal points in $(N, \infty)$, and its associated matrix $\hat{S}_k$ satisfies (6.396). From Proposition 6.111 we then have $d[N, \infty)_{\mathbb{Z}} = d_\infty$. By Theorem 6.90, for any integer $r$ between $n - d_\infty = n - d[N, \infty)_{\mathbb{Z}}$ and $n$, there exists a conjoined basis $Y$ of (SDS) with constant kernel on $[N, \infty)_{\mathbb{Z}}$ and no forward focal points in $(N, \infty)$ such that rank $X_k = r$ on $[N, \infty)_{\mathbb{Z}}$. Moreover, the conjoined basis $Y$ is in Theorem 6.90 constructed in such a way that it is either contained in or contains the recessive solution $\hat{Y}$ at $\infty$ on $[N, \infty)_{\mathbb{Z}}$. In turn, Proposition 6.112 yields that $Y$ is also a recessive solution of (SDS) at $\infty$ on $[N, \infty)_{\mathbb{Z}}$.

(iii)$\Rightarrow$(i) The choice $r := n - d_\infty$ yields the existence of the minimal recessive solution of (SDS) at $\infty$, which through Theorem 6.113 implies the nonoscillation of system (SDS) at $\infty$. Alternatively, we may use Proposition 6.61 to obtain the same conclusion.                                                                                                         $\square$

In Theorem 6.113 we guarantee the uniqueness of the minimal recessive solution of (SDS) at $\infty$. In the following remark, we show that the minimal recessive solution is the only one for which this property is satisfied. This remark also shows that nonunique recessive solutions of (SDS) will always exist as long as $d_\infty \geq 1$.

*Remark 6.116* Let $\hat{Y}$ be a recessive solution of (SDS) at $\infty$ with rank $r$ satisfying $n - d_\infty \leq r \leq n$. Then $\hat{Y}$ is unique up to a right nonsingular multiple if and only if $r = n - d_\infty$, that is, $\hat{Y}$ is a minimal recessive solution of (SDS). We shall prove by construction the implication "$\Rightarrow$", as the opposite direction "$\Leftarrow$" is contained in Theorem 6.113. Let $\hat{Y}$ be a recessive solution of (SDS) at $\infty$ on the interval $[N, \infty)_{\mathbb{Z}}$ with the projectors $\hat{P}$ and $P_{\hat{S}_\infty}$ in (6.234) and (6.238). Set $\hat{M} := 2I - \hat{P}$ and $\hat{N} := \hat{P} - P_{\hat{S}_\infty}$, and define the solution $\hat{Y}_k^{(0)} := \hat{Y}_k \hat{M} + \bar{Y}_k \hat{N}$ on $[0, \infty)_{\mathbb{Z}}$, where $\bar{Y}$ is the conjoined basis of (SDS) associated with $\hat{Y}$ in Proposition 6.67. Since the equality $\hat{P} P_{\hat{S}_\infty} = P_{\hat{S}_\infty}$ holds by (6.239), it follows that $\hat{M}^T \hat{N} = \hat{P} - P_{\hat{S}_\infty}$ is symmetric. The invertibility of $\hat{M}$ then yields that $\hat{Y}^{(0)}$ is a conjoined basis of (SDS) and $\hat{N} = w(\hat{Y}, \hat{Y}^{(0)})$. Moreover, $\bar{X}_k \hat{P} = \hat{X}_k \hat{S}_k$ by Proposition 6.67(iv) and $\hat{S}_k (I - P_{\hat{S}_\infty}) = 0$ by the first inclusion in (6.239) for all $k \in [N, \infty)_{\mathbb{Z}}$. Then

$$X_k^{[0]} = \hat{X}_k \hat{M} + \bar{X}_k \hat{N} = \hat{X}_k \hat{P}(2I - \hat{P}) + \bar{X}_k \hat{P}(I - P_{\hat{S}_\infty})$$

$$= \hat{X}_k \hat{P} + \hat{X}_k \hat{S}_k (I - P_{\hat{S}_\infty}) = \hat{X}_k$$

on $[N, \infty)_{\mathbb{Z}}$, which shows that the solutions $\hat{Y}$ and $\hat{Y}^{(0)}$ are equivalent on $[N, \infty)_{\mathbb{Z}}$. Therefore, $\hat{Y}^{(0)}$ is also a recessive solution of (SDS) at $\infty$ on $[N, \infty)_{\mathbb{Z}}$ with the same rank $r$. Now if $\hat{Y}$ is unique up to a right nonsingular multiple, then necessarily $\hat{N} = 0$. This means that $\hat{P} = P_{\hat{S}\infty}$, $r = n - d_{\infty}$, and $\hat{Y}$ is a minimal recessive solution of (SDS) at $\infty$.

*Remark 6.117* In the literature one can find an alternative definition of a recessive solution of (SDS) in terms of a limit. More precisely, by [16, pg. 115] a conjoined basis $\hat{Y}$ of (SDS) is a recessive solution at $\infty$ if for any other linearly independent conjoined basis $Y$ of (SDS) with $X_k$ invertible for large $k$ we have

$$\lim_{k \to \infty} X_k^{-1} \hat{X}_k = 0. \qquad (6.406)$$

This property is known as the limit characterization (or the limit definition) of the recessive solution at $\infty$ and it goes back to the historical papers by Olver and Sookne [238] and Gautschi [151]. When system (SDS) is eventually controllable, then both concepts in Definition 2.63 and (6.406) coincide; see Theorem 2.67 and [51, pg. 965] or [104, pg. 211]. However, the above definition (6.406) allows also an eventually noncontrollable system (SDS) and its recessive solutions $\hat{Y}$ at $\infty$ with noninvertible $\hat{X}_k$ on $[N, \infty)_{\mathbb{Z}}$; see [16, pp. 116–117]. This poses a question on the exact relationship between the limit property in (6.406) and the summation property in (6.396) of a recessive solution of (SDS) at $\infty$. This question is also related with the so-called *dominant solutions* of (SDS) at $\infty$ considered in Sect. 2.5 or [16, Theorems 3.35 and 3.43]. In the next subsections, we will complete this study and show that the recessive solutions of (SDS) at $\infty$ in Definition 6.109 are indeed the smallest solutions at $\infty$ when they are compared with suitable dominant solutions of (SDS) at $\infty$.

As a final result in this subsection, we present a block diagonal construction of certain recessive solutions at $\infty$ of symplectic systems arising in higher dimension. With system (SDS) we consider another symplectic system (with possibly different dimension $\underline{n} \in \mathbb{N}$)

$$\underline{y}_{k+1} = \underline{S}_k \underline{y}_k, \quad k \in [0, \infty)_{\mathbb{Z}}, \quad \underline{y}_k = \begin{pmatrix} \underline{x}_k \\ \underline{u}_k \end{pmatrix}, \quad \underline{S}_k = \begin{pmatrix} \underline{A}_k & \underline{B}_k \\ \underline{C}_k & \underline{D}_k \end{pmatrix}, \qquad \text{(SDS)}$$

where $\underline{S}_k$ and $\underline{\mathcal{J}} := \begin{pmatrix} 0 & I \\ -I & 0 \end{pmatrix}$ are $2\underline{n} \times 2\underline{n}$ matrices such that $\underline{S}_k$ is real symplectic, i.e., $\underline{S}_k^T \underline{\mathcal{J}} \underline{S}_k = \underline{\mathcal{J}}$ for all $k \in [0, \infty)_{\mathbb{Z}}$. Define the block diagonal matrices

$$A_k^* := \begin{pmatrix} A_k & 0 \\ 0 & \underline{A}_k \end{pmatrix}, \ B_k^* := \begin{pmatrix} B_k & 0 \\ 0 & \underline{B}_k \end{pmatrix}, \ C_k^* := \begin{pmatrix} C_k & 0 \\ 0 & \underline{C}_k \end{pmatrix}, \ D_k^* := \begin{pmatrix} D_k & 0 \\ 0 & \underline{D}_k \end{pmatrix}$$

in the dimension $n^* := n + \underline{n}$, and consider the "augmented" system

$$y_k^* = S_k^* y_k^*, \quad k \in [0, \infty)_{\mathbb{Z}}, \quad y_k^* = \begin{pmatrix} x_k^* \\ u_k^* \end{pmatrix}, \quad S_k^* = \begin{pmatrix} A_k^* & B_k^* \\ C_k^* & D_k^* \end{pmatrix}, \quad \text{(SDS*)}$$

where the dimension of $S_k^*$ and $\mathcal{J}^* := \begin{pmatrix} 0 & I \\ -I & 0 \end{pmatrix}$ is $2n^*$. Since the matrices $S_k$ and $\underline{S}_k$ are symplectic, it follows that $(S_k^*)^T \mathcal{J}^* S_k^* = \mathcal{J}^*$, i.e., the augmented system (SDS*) is a symplectic system. The following result shows that certain recessive solutions of system (SDS*) at $\infty$ can be constructed from the recessive solutions of (SDS) and (SDS) at $\infty$.

**Theorem 6.118** *Assume that the systems* (SDS) *and* (SDS) *are nonoscillatory at* $\infty$. *Let* $\hat{Y}$ *and* $\underline{\hat{Y}}$ *be recessive solutions of* (SDS) *and* (SDS) *at* $\infty$ *with rank equal to* $r$ *and* $\underline{r}$, *respectively. Then the sequence* $\hat{Y}_k^*$ *defined by*

$$\hat{X}_k^* := \begin{pmatrix} \hat{X}_k & 0 \\ 0 & \underline{\hat{X}}_k \end{pmatrix}, \quad \hat{U}_k^* := \begin{pmatrix} \hat{U}_k & 0 \\ 0 & \underline{\hat{U}}_k \end{pmatrix}, \quad k \in [0, \infty)_{\mathbb{Z}}, \quad (6.407)$$

*is a recessive solution of system* (SDS*) *at* $\infty$ *with rank equal to* $r^* := r + \underline{r}$. *Moreover, the recessive solution* $\hat{Y}^*$ *at* $\infty$ *constructed in* (6.407) *is minimal (maximal) if and only if the recessive solutions* $\hat{Y}$ *and* $\underline{\hat{Y}}$ *at* $\infty$ *are minimal (maximal).*

*Proof* By Proposition 6.111, there exists a common index $N \in [0, \infty)_{\mathbb{Z}}$ such that $\hat{Y}$ and $\underline{\hat{Y}}$ are recessive solutions of (SDS) and (SDS) at $\infty$ with respect to the interval $[N, \infty)_{\mathbb{Z}}$. It is obvious that the rank of $\hat{Y}^*$ is equal to $r^* = r + \underline{r}$ and that the orders of abnormality $d[N, \infty)_{\mathbb{Z}}, \underline{d}[N, \infty)_{\mathbb{Z}}, d^*[N, \infty)_{\mathbb{Z}}$ of systems (SDS), (SDS), (SDS*) satisfy $d^*[N, \infty)_{\mathbb{Z}} = d[N, \infty)_{\mathbb{Z}} + \underline{d}[N, \infty)_{\mathbb{Z}}$. This follows from the structure of the coefficients in $\underline{S}_k$. Moreover, if $\hat{M}_k, \underline{M}_k, M_k^*$ and $\hat{T}_k, \underline{T}_k, T_k^*$ and $\hat{P}_k, \underline{P}_k, P_k^*$ and $\hat{S}_k$, $\underline{\hat{S}}_k, \hat{S}_k^*$ are the matrices in (4.1) and (4.2) and (6.396) associated with the conjoined bases $\hat{Y}, \underline{\hat{Y}}, \hat{Y}^*$, then

$$M_k^* = \text{diag}\{\hat{M}_k, \underline{M}_k\}, \quad T_k^* = \text{diag}\{\hat{T}_k, \underline{T}_k\}, \quad P_k^* = \text{diag}\{\hat{P}_k, \underline{P}_k\},$$

$$\hat{S}_k^* = \text{diag}\{\hat{S}_k, \underline{\hat{S}}_k\}, \quad (\hat{S}_k^*)^\dagger = \text{diag}\{\hat{S}_k^\dagger, \underline{\hat{S}}_k^\dagger\}.$$

This means that $\hat{Y}^*$ has constant kernel on $[N, \infty)_{\mathbb{Z}}$, no forward focal points in $(N, \infty)$, and $(\hat{S}_k^*)^\dagger \to 0$ as $k \to \infty$. Therefore, $\hat{Y}^*$ is a recessive solution of the augmented system (SDS*) at $\infty$ according to Definition 6.109.                                        $\square$

We conclude this subsection with three examples, which illustrate the presented theory of recessive solutions of (SDS) at $\infty$.

*Example 6.119* Let $n = 1$ and consider a scalar system (SDS) with $S_k \equiv \begin{pmatrix} 1 & 1 \\ 0 & 1 \end{pmatrix}$. This system is nonoscillatory at $\infty$, because it corresponds to the nonoscillatory second-order Sturm-Liouville difference equation $\Delta^2 y_k = 0$ on $[0, \infty)_{\mathbb{Z}}$, and

$d[0, \infty)_\mathbb{Z} = d_\infty = 0$. Therefore, by Theorem 6.115 (or Theorem 6.113 or Theorem 2.66), the conjoined basis $\hat{Y}_k \equiv \left(\begin{smallmatrix} 1 \\ 0 \end{smallmatrix}\right)$ of (SDS) is the (unique) recessive solution of (SDS) at $\infty$. The second linearly independent solution $\tilde{Y}_k = \left(\begin{smallmatrix} k \\ 1 \end{smallmatrix}\right)$ of (SDS) is not in this case a recessive solution of (SDS) at $\infty$ according to Definition 6.109.

*Example 6.120* Consider system (SDS) with $\mathcal{S}_k \equiv I_{2n}$. This system is nonoscillatory at $\infty$, its solutions are constant on $[0, \infty)_\mathbb{Z}$, and its order of abnormality is $d[0, \infty)_\mathbb{Z} = d_\infty = n$. Every conjoined basis of (SDS) is then a (constant) recessive solution at $\infty$ on the interval $[0, \infty)_\mathbb{Z}$. The recessive solutions of (SDS) at $\infty$ with a given rank $r$ between 0 and $n$ in Theorem 6.115 can be constructed by a suitable choice of the orthogonal projector $\hat{P} \in \mathbb{R}^{n \times n}$ with rank $\hat{P} = r$. More precisely, the solution $\hat{Y}$ with $\hat{X} = \hat{P}$ and $\hat{U} = I - \hat{P}$ is a recessive solution of (SDS) at $\infty$ with rank equal to rank $\hat{P}$. If $\hat{P} = 0$, then $\hat{Y} = \hat{Y}^{\min} = Y^{[0]} = \left(\begin{smallmatrix} 0 \\ I \end{smallmatrix}\right)$ is the (unique) minimal recessive solution of (SDS) at $\infty$. In fact, it is this minimal recessive solution at $\infty$, which is derived in [16, Example 3.39]. On the other hand, if $\hat{P} = I$, then $\hat{Y} = \hat{Y}^{\max} = \left(\begin{smallmatrix} I \\ 0 \end{smallmatrix}\right)$ is an example of a maximal recessive solution of (SDS) at $\infty$. Note that $\hat{Y}^{(0)} = \left(\begin{smallmatrix} I \\ I \end{smallmatrix}\right)$ is also a maximal recessive solution of (SDS) at $\infty$, which is not a constant multiple of the maximal recessive solution $\hat{Y}^{\max} = \left(\begin{smallmatrix} I \\ 0 \end{smallmatrix}\right)$ at $\infty$, demonstrating the nonuniqueness of the recessive solutions at $\infty$ with rank $r \geq 1$ in Remark 6.116.

Finally, we illustrate the construction of recessive solutions of (SDS) at $\infty$ by using Theorem 6.118.

*Example 6.121* Consider symplectic system (SDS) with the coefficients $\mathcal{A}_k \equiv I_3 = \mathrm{diag}\{1, 1, 1\}$, $\mathcal{B}_k \equiv \mathrm{diag}\{1, 0, 0\}$, $\mathcal{C}_k \equiv 0_3 = \mathrm{diag}\{0, 0, 0\}$, and $\mathcal{D}_k \equiv I_3 = \mathrm{diag}\{1, 1, 1\}$ on $[0, \infty)_\mathbb{Z}$. This system corresponds to the partitioned system (SDS*), which arises from the symplectic systems in Example 6.119 and Example 6.120 (with dimension two). System (SDS) is nonoscillatory at $\infty$ and $d[0, \infty)_\mathbb{Z} = d_\infty = 2$. By Theorem 6.118, $\hat{Y}^{\min} = (\mathrm{diag}\{1, 0, 0\}, \mathrm{diag}\{0, 1, 1\})^T$ with rank $r = n - d_\infty = 1$ is the minimal recessive solution of (SDS) at $\infty$, while $\hat{Y}^{\max} = (I_3, 0_3)^T$ with rank $r = n = 3$ is one of the maximal recessive solutions of (SDS) at $\infty$. Note that for both recessive solutions at $\infty$, we have $\hat{S}_k = \mathrm{diag}\{k, 0, 0\}$ on $[0, \infty)_\mathbb{Z}$, so that $\hat{S}_k^\dagger = \mathrm{diag}\{1/k, 0, 0\} \to 0$ for $k \to \infty$, as required in (6.396).

### 6.3.7  Dominant Solutions at Infinity

In this subsection we develop the notion of a dominant solution of (SDS) at $\infty$ without any eventual controllability assumption. As a main result, we prove (Theorem 6.128) an analog of Proposition 6.115 for dominant solutions at $\infty$.

**Definition 6.122 (Dominant Solution at $\infty$)**  A conjoined basis $Y$ of (SDS) is said to be a *dominant solution* of (SDS) at $\infty$ if there exists an index $N \in [0, \infty)_\mathbb{Z}$ with

$d[N, \infty)_{\mathbb{Z}} = d_\infty$ such that $Y$ has constant kernel on $[N, \infty)_{\mathbb{Z}}$ and no forward focal points in $(N, \infty)$ and the corresponding matrix $T$ defined in (6.237) satisfies

$$\operatorname{rank} T = n - d_\infty. \tag{6.408}$$

From Theorem 6.107 and Remark 6.108, it follows that the dominant solutions of (SDS) at $\infty$ are defined by the maximal possible rank of the associated matrix $T$, while the recessive solutions of (SDS) at $\infty$ are defined by the minimal rank of $T$. In this respect the dominant solution of (SDS) at $\infty$ can also be called as an *antirecessive solution* of (SDS) at $\infty$. This alternative terminology then complies with the continuous time notions of principal and antiprincipal solutions at $\infty$ of linear Hamiltonian differential systems (1.103); see [285, Definition 7.1] and [286, Definition 5.1]. We also note that with respect to (6.239) and (6.276) the matrix $T$ associated with a dominant solution $Y$ of (SDS) at $\infty$ has the property

$$\operatorname{Im} T = \operatorname{Im} P_{S\infty}. \tag{6.409}$$

In the following we introduce the notation for dominant solutions at $\infty$, which is analogous to the terminology *minimal recessive solution* $\hat{Y}^{\min}$ and *maximal recessive solution* $\hat{Y}^{\max}$ of (SDS) at $\infty$ used in Remark 6.110.

*Remark 6.123* Let $Y$ be a dominant solution of (SDS) at $\infty$ with $r = \operatorname{rank} X_k$ on $[N, \infty)_{\mathbb{Z}}$. If $r = n - d_\infty$, then $Y$ is called a *minimal dominant solution* at $\infty$, while if $r = n$, then $Y$ is called a *maximal dominant solution* at $\infty$. This terminology corresponds to the two extreme cases in formula (6.279). As before, we will use the notation $Y^{\min}$ and $Y^{\max}$ for the minimal and maximal dominant solutions of (SDS) at $\infty$, respectively. Moreover, if $n - d_\infty < r < n$, then the dominant solution $Y$ at $\infty$ is called *intermediate*.

When the system (SDS) is eventually controllable, then $d_\infty = 0$ and hence the matrix $T$ is positive definite. In this case Definition 6.122 reduces to Definition 2.63.

Our first result shows that the dominant solutions of (SDS) at $\infty$ can be characterized by the limit of $S_k$ alone instead of the limit of $S_k^\dagger$ as $k \to \infty$. In some situations this condition may be easier to verify in comparison with Definition 6.122.

**Theorem 6.124** *Let $Y$ be a conjoined basis of* (SDS) *with constant kernel on* $[N, \infty)_{\mathbb{Z}}$ *and no forward focal points in* $(N, \infty)$, *and assume that $d[N, \infty)_{\mathbb{Z}} = d_\infty$. Let $S_k$ and $T$ be defined in (6.236) and (6.237) through $Y$. Then the following statements are equivalent.*

  (i) *The conjoined basis $Y$ is a dominant solution of* (SDS) *at $\infty$.*
 (ii) *The limit of $S_k$ exists as $k \to \infty$.*
(iii) *The matrices $S_k$ and $T$ satisfy the condition*

$$\lim_{k \to \infty} S_k = T^\dagger. \tag{6.410}$$

*Proof* From (6.238) and (6.276) and from the assumption $d[N, \infty)_{\mathbb{Z}} = d_\infty$, we know that the equalities rank $S_k^\dagger = $ rank $S_k = $ rank $P_{S\infty} = n - d_\infty$ hold for large $k$. And since rank $T = n - d_\infty$ is the defining property for $Y$ being a dominant solution at $\infty$, it follows that rank $S_k^\dagger = n - d_\infty = $ rank $T$ for large $k$. This is equivalent by Remark 1.60(v) (with $A_j := S_j^\dagger$ and $A := T$) with the existence of the limit of $(S_k^\dagger)^\dagger = S_k$ as $k \to \infty$, i.e., with condition (6.410). $\qquad \square$

Next we discuss the dependence of Definition 6.122 on the initial index $N$ in the interval $[0, \infty)_{\mathbb{Z}}$. In this respect we obtain a similar statement to Proposition 6.111.

**Proposition 6.125** *Let $Y$ be a dominant solution of* (SDS) *at $\infty$ with respect to the interval $[N, \infty)_{\mathbb{Z}}$. Then $Y$ is a dominant solution at $\infty$ also with respect to the interval $[M, \infty)_{\mathbb{Z}}$ for every $M \in [N, \infty)_{\mathbb{Z}}$.*

*Proof* Fix an index $M \in [N, \infty)_{\mathbb{Z}}$, and let $S_k$ and $S_k^{(M)}$ be the matrices in (6.236) and (6.397) corresponding to $Y$ on the intervals $[N, \infty)_{\mathbb{Z}}$ and $[M, \infty)_{\mathbb{Z}}$, respectively. Furthermore, let $P_{S\infty}$ and $T$ be the associated matrices in (6.238) and (6.237). From Definition 6.122 we have the conditions rank $T = n - d_\infty$ and $d[N, \infty)_{\mathbb{Z}} = d_\infty$. The latter equation implies that $d[M, \infty)_{\mathbb{Z}} = d_\infty$, by (6.224). By Proposition 6.101 and Theorem 6.103, we have that (6.370) holds, while from the proof of Proposition 6.111, we conclude that (6.398) is satisfied. Summarizing, the conjoined basis $Y$ satisfies

$$\text{Im}\,(P_{S\infty} - S_k T) = \text{Im}\,P_{S\infty} = \text{Im}\,(P_{S\infty} - S_k T)^T, \quad k \in [N, \infty)_{\mathbb{Z}}, \qquad (6.411)$$

$$(S_k^{(M)})^\dagger = S_k^\dagger\,(P_{S\infty} - S_M\,S_k^\dagger)^\dagger, \quad k \in [M, \infty)_{\mathbb{Z}}. \qquad (6.412)$$

The matrices $G_k := P_{S\infty} - S_M\,S_k^\dagger$ satisfy $\text{Im}\,G_k \subseteq \text{Im}\,P_{S\infty}$, which implies that rank $G_k \leq $ rank $P_{S\infty}$ for all $k \in [M, \infty)_{\mathbb{Z}}$. Moreover, by (6.411) the limit matrix $G := \lim_{k \to \infty} G_k = P_{S\infty} - S_M T$ satisfies rank $G = $ rank $P_{S\infty}$. On the other hand, the inequality rank $G_k \geq $ rank $G$ for large $k$ always holds for a limit of a sequence of matrices. Therefore, rank $G_k = $ rank $G = $ rank $P_{S\infty}$ for large $k$. Then the properties of Moore-Penrose pseudoinverse in Remark 1.60(v) imply that by taking the limit as $k \to \infty$ in (6.412) the matrix $T^{(M)}$, defined in (6.237) and corresponding to the $S$-matrix $S_k^{(M)}$, satisfies

$$T^{(M)} = \lim_{k \to \infty} S_k^\dagger G_k^\dagger = T G^\dagger = T(P_{S\infty} - S_M T)^\dagger. \qquad (6.413)$$

We will show that $\text{Im}\,T^{(M)} = \text{Im}\,T$. Indeed, identity (6.413) yields the inclusion $\text{Im}\,T^{(M)} \subseteq \text{Im}\,T$. On the other hand, by (6.413) and (6.411), we obtain

$$T^{(M)}\,(P_{S\infty} - S_M T) \overset{(6.413)}{=} T(P_{S\infty} - S_M T)^\dagger (P_{S\infty} - S_M T) \overset{(6.411)}{=} T P_{S\infty} = T,$$

since $\operatorname{Im} T \subseteq \operatorname{Im} P_{S\infty}$. This shows the opposite inclusion $\operatorname{Im} T \subseteq \operatorname{Im} T^{(M)}$. Finally, the equalities $\operatorname{rank} T^{(M)} = \operatorname{rank} T = n - d_\infty$ then imply that $Y$ is a dominant solution of (SDS) at $\infty$ with respect to the interval $[M, \infty)_{\mathbb{Z}}$, by Definition 6.122.

$\square$

*Remark 6.126* In the proof of Proposition 6.125 we show that the set $\operatorname{Im} T^{(M)}$ is preserved within the interval $[N, \infty)_{\mathbb{Z}}$, i.e., $\operatorname{Im} T^{(M)} = \operatorname{Im} T$ for every $M \in [N, \infty)_{\mathbb{Z}}$. Moreover, this statement obviously holds for any conjoined basis $Y$ with constant kernel on $[N, \infty)_{\mathbb{Z}}$ and no forward focal points in $(N, \infty)$ such that $d[N, \infty)_{\mathbb{Z}} = d_\infty$.

In the following statement, we show that the property of $Y$ being a dominant solution of (SDS) at $\infty$ is preserved under the relation being contained from Sect. 6.3.3.

**Proposition 6.127** *Let $Y$ be a dominant solution of (SDS) at $\infty$ with respect to the interval $[N, \infty)_{\mathbb{Z}}$. Then every conjoined basis of (SDS) with constant kernel on $[N, \infty)_{\mathbb{Z}}$ and no forward focal points in $(N, \infty)$, which is either contained in $Y$ on $[N, \infty)_{\mathbb{Z}}$ or which contains $Y$ on $[N, \infty)_{\mathbb{Z}}$, is also a dominant solution of (SDS) at $\infty$ with respect to the interval $[N, \infty)_{\mathbb{Z}}$.*

*Proof* The result follows directly from Proposition 6.86 and Definition 6.122.   $\square$

Next we characterize the nonoscillation of system (SDS) in terms of the existence of dominant solutions of (SDS) at $\infty$. It is an analog of Theorem 6.115.

**Theorem 6.128** *The following statements are equivalent.*

(i) *System (SDS) is nonoscillatory.*
(ii) *There exists a dominant solution of (SDS) at $\infty$.*
(iii) *For any integer $r$ satisfying $n - d_\infty \le r \le n$, there exists a dominant solution $Y$ of (SDS) at $\infty$ with the rank of $X_k$ equal to $r$ for large $k$.*

*Proof* If (SDS) is nonoscillatory, then by Theorem 6.107 for any symmetric and nonnegative definite matrix $D$ with $\operatorname{rank} D = n - d_\infty$, there exists $N \in [0, \infty)_{\mathbb{Z}}$ and a minimal conjoined basis $Y$ of (SDS) on $[N, \infty)_{\mathbb{Z}}$ such that $d[N, \infty)_{\mathbb{Z}} = d_\infty$ and the corresponding matrix $T$ in (6.237) satisfies $T = D$, i.e., $\operatorname{rank} T = n - d_\infty$. From Definition 6.122 and Remark 6.123, it then follows that $Y$ is a minimal dominant solution of (SDS) at $\infty$. Suppose now that (ii) holds, and let $Y$ be a dominant solution of (SDS) at $\infty$, i.e., there exists $N \in [0, \infty)_{\mathbb{Z}}$ with $d[N, \infty)_{\mathbb{Z}} = d_\infty$ such that $Y$ is a conjoined basis of (SDS) with constant kernel on $[N, \infty)_{\mathbb{Z}}$ and no forward focal points in $(N, \infty)$, and its associated matrix $T$ satisfies $\operatorname{rank} T = n - d_\infty$. By Theorem 6.90, for any integer $r$ between $n - d_\infty$ and $n$, there exists a conjoined basis $\tilde{Y}$ of (SDS) with constant kernel on $[N, \infty)_{\mathbb{Z}}$ and no forward focal points in $(N, \infty)$ and with rank $\tilde{X}_k = r$ on $[N, \infty)_{\mathbb{Z}}$ such that $\tilde{Y}$ is either contained in $Y$ or $\tilde{Y}$ contains $Y$ on $[N, \infty)_{\mathbb{Z}}$. Therefore, $\tilde{Y}$ is also a dominant solution of (SDS) at $\infty$, by Proposition 6.127, showing part (iii). Finally, if (iii) is satisfied, then $Y$ is a conjoined basis of (SDS) with finitely many focal points in $(0, \infty)$. Therefore, system (SDS) is nonoscillatory at $\infty$ by Proposition 6.61. The proof is complete. □

For an eventually controllable system (SDS), we obtain from Theorem 6.128 the following counterpart of Theorem 2.66 and Remark 2.69 in Sect. 2.5.1.

**Corollary 6.129** *Assume that* (SDS) *is eventually controllable. System* (SDS) *is nonoscillatory at* $\infty$ *if and only if there exists a dominant solution $Y$ of* (SDS) *at* $\infty$ *with rank equal to n, i.e., with $X_k$ eventually invertible. In this case the corresponding matrix $T$ in* (6.237) *is positive definite.*

In the last result of this section, we present a construction of all recessive and dominant solutions of (SDS) at $\infty$ from the minimal recessive and dominant solutions at $\infty$, respectively. In this construction we utilize two main ingredients: (i) the properties in Theorem 6.90 applied to the minimal recessive and dominant solutions at $\infty$ and (ii) the uniqueness of the minimal recessive solution $\hat{Y}^{\min}$ of (SDS) at $\infty$ in Theorem 6.113.

*Remark 6.130* If $\hat{Y}^{\min}$ is the (unique) minimal recessive solution of (SDS) at $\infty$, then we define the point $\hat{K}_{\min} \in [0, \infty)_{\mathbb{Z}}$ as the smallest index $N \in [0, \infty)_{\mathbb{Z}}$ such that $\hat{Y}^{\min}$ has constant kernel on $[N, \infty)_{\mathbb{Z}}$ and no forward focal points in $(N, \infty)$. In particular, we have the identities

$$d[\hat{K}_{\min}, \infty)_{\mathbb{Z}} = d[N, \infty)_{\mathbb{Z}} = d_\infty \quad \text{for all } N \in [\hat{K}_{\min}, \infty)_{\mathbb{Z}}. \tag{6.414}$$

These equalities follow from Theorem 6.90, the definition of $d_\infty$ in (6.224), and from the fact that rank $\hat{X}_k^{\min} = n - d_\infty$ on $[\hat{K}_{\min}, \infty)_{\mathbb{Z}}$. Moreover, by Proposition 6.111 the conjoined basis $\hat{Y}^{\min}$ is a minimal recessive solution of (SDS) at $\infty$ with respect to the interval $[N, \infty)_{\mathbb{Z}}$ for every $N \in [\hat{K}_{\min}, \infty)_{\mathbb{Z}}$.

**Theorem 6.131** *Assume that system* (SDS) *is nonoscillatory at* $\infty$ *with the index $\hat{K}_{\min}$ defined in Remark 6.130. Then the following statements hold.*

(i) *A solution $Y$ of* (SDS) *is a recessive solution at* $\infty$ *if and only if $Y$ is a conjoined basis of* (SDS) *with constant kernel on $[\hat{K}_{\min}, \infty)_{\mathbb{Z}}$ and no forward focal points in $(\hat{K}_{\min}, \infty)$, which contains some minimal recessive solution of* (SDS) *on $[N, \infty)_{\mathbb{Z}}$ for some (and hence every) $N \in [\hat{K}_{\min}, \infty)_{\mathbb{Z}}$.*

(ii) *A solution $Y$ of* (SDS) *is a dominant solution at* $\infty$ *if and only if $Y$ is a conjoined basis of* (SDS)*, which contains some minimal dominant solution of* (SDS) *at* $\infty$ *on $[N, \infty)_{\mathbb{Z}}$ for some $N \in [\hat{K}_{\min}, \infty)_{\mathbb{Z}}$.*

*Proof*

(i) Let $Y$ be a recessive solution of (SDS) at $\infty$. Then there exists an index $N \in [0, \infty)_{\mathbb{Z}}$ such that $Y$ has constant kernel on $[N, \infty)_{\mathbb{Z}}$ and no forward focal points in $(N, \infty)$, and the corresponding matrix $S_k$ in (6.236) satisfies $S_k^\dagger \to 0$ for $k \to \infty$. From Proposition 6.111, we know that this property of $S_k$ is preserved under shifting the index $N$ to the right. Therefore, we may assume that $N \in [\hat{K}_{\min}, \infty)_{\mathbb{Z}}$ and hence, we have $d[N, \infty)_{\mathbb{Z}} = d_\infty$. Consequently, by Theorem 6.90 there exists a conjoined basis $Y^*$ of (SDS) with constant kernel on $[N, \infty)_{\mathbb{Z}}$ and no forward focal points in $(N, \infty)$ and with rank $X_k^* = n - d_\infty$ on $[N, \infty)_{\mathbb{Z}}$ such that $Y$ contains $Y^*$ on $[N, \infty)_{\mathbb{Z}}$. In turn, Proposition 6.112 implies that $Y^*$ is a minimal recessive solution of (SDS) at $\infty$ with respect to the interval $[N, \infty)_{\mathbb{Z}}$. From Proposition 6.92(i), we then obtain that $Y$ contains $Y^*$ also on $[L, \infty)_{\mathbb{Z}}$ for all $L \in [N, \infty)_{\mathbb{Z}}$. It remains to show that $Y$ contains $Y^*$ on $[L, \infty)_{\mathbb{Z}}$ for all $L \in [\hat{K}_{\min}, N - 1]_{\mathbb{Z}}$. Let us fix such an index $L$. By Remark 6.130 we know that $d[L, \infty)_{\mathbb{Z}} = d_\infty$ and that $Y^*$ has constant kernel on $[L, \infty)_{\mathbb{Z}}$ and no forward focal points in $(L, \infty)$. Consequently, $Y$ has also constant kernel on $[L, \infty)_{\mathbb{Z}}$ and no forward focal points in $(L, \infty)$ according to Proposition 6.93(ii). On the other hand, Proposition 6.92(ii) implies that $Y$ contains $Y^*$ also on $[L, \infty)_{\mathbb{Z}}$. This completes the proof of the first implication. Conversely, suppose that $Y$ is a conjoined basis of (SDS) with constant kernel on $[\hat{K}_{\min}, \infty)_{\mathbb{Z}}$ and no forward focal points in $(\hat{K}_{\min}, \infty)$. Let $\hat{Y}^{\min}$ be a minimal recessive solution of (SDS) at $\infty$, which is contained in $Y$ on $[L, \infty)_{\mathbb{Z}}$ for some $L \in [\hat{K}_{\min}, \infty)_{\mathbb{Z}}$. By Remark 6.130, $\hat{Y}^{\min}$ has constant kernel on $[\hat{K}_{\min}, \infty)_{\mathbb{Z}}$ and no forward focal points in $(\hat{K}_{\min}, \infty)$, and it is a minimal recessive solution at $\infty$ with respect to $[N, \infty)_{\mathbb{Z}}$ for every $N \in [\hat{K}_{\min}, \infty)_{\mathbb{Z}}$. Consequently, Proposition 6.92(i) implies that $\hat{Y}^{\min}$ is contained in $Y$ on $[N, \infty)_{\mathbb{Z}}$ for every $N \in [\hat{K}_{\min}, \infty)_{\mathbb{Z}}$. Finally, from Proposition 6.111 it then follows that $Y$ is a recessive solution of (SDS) at $\infty$.

(ii) Let $Y$ be a dominant solution of (SDS) at $\infty$ with respect to an interval $[N, \infty)_{\mathbb{Z}}$. By Proposition 6.125, we may assume that $N \in [\hat{K}_{\min}, \infty)_{\mathbb{Z}}$. From Theorem 6.90 we know that there exists a conjoined basis $Y^*$ of (SDS) with constant kernel on $[N, \infty)_{\mathbb{Z}}$ and no forward focal points in $(N, \infty)$ such that $Y$ contains $Y^*$ on $[N, \infty)_{\mathbb{Z}}$ and rank $X_k^* = n - d[N, \infty)_{\mathbb{Z}} = n - d_\infty$ on $[N, \infty)_{\mathbb{Z}}$. In turn, Proposition 6.127 and Remark 6.123 then imply that $Y^*$ is a minimal dominant solution of (SDS) at $\infty$ with respect to $[N, \infty)_{\mathbb{Z}}$. Conversely, let $Y$ be a conjoined basis of (SDS) with constant kernel on $[N, \infty)_{\mathbb{Z}} \subseteq [\hat{K}_{\min}, \infty)_{\mathbb{Z}}$ and no forward focal points in $(N, \infty)$ such that $Y$ contains some minimal dominant solution of (SDS) on $[N, \infty)_{\mathbb{Z}}$. Then $d[N, \infty)_{\mathbb{Z}} = d_\infty$ by Definition 6.122, and $Y$ is also a dominant solution of (SDS) at $\infty$, by Proposition 6.127.

$\square$

*Remark 6.132* From Theorem 6.131(i), it follows that every recessive solution $\hat{Y}$ of (SDS) at $\infty$ is a recessive solution at $\infty$ with respect to the interval $[N, \infty)_{\mathbb{Z}}$ for every $N \in [\hat{K}_{\min}, \infty)_{\mathbb{Z}}$. In addition, the orthogonal projector $P_{\hat{\mathcal{S}}_\infty}$ in (6.238) associated with $\hat{Y}$ through the matrix $\hat{S}_k$ in (6.236) is the same for all initial indices $N \in [\hat{K}_{\min}, \infty)_{\mathbb{Z}}$, since in this case $d[N, \infty)_{\mathbb{Z}} = d_\infty$ for every $N \in [\hat{K}_{\min}, \infty)_{\mathbb{Z}}$, by Remark 6.130. Similarly, if $Y$ is a dominant solution at $\infty$ with respect to an interval $[N, \infty)_{\mathbb{Z}} \subseteq [\hat{K}_{\min}, \infty)_{\mathbb{Z}}$, then the corresponding orthogonal projector $P_{\mathcal{S}_\infty}$ in (6.238) is the same for all initial indices $L \in [N, \infty)_{\mathbb{Z}}$.

In Theorem 6.118 we showed that recessive solutions of (SDS) at $\infty$ can be constructed in higher dimensions from the recessive solutions of the corresponding symplectic systems in lower dimensions by a block diagonal procedure. We note that exactly the same statement holds also for dominant solutions of (SDS) at $\infty$.

**Theorem 6.133** *Assume that the systems (SDS) and (SDS) in Theorem 6.118 are nonoscillatory at $\infty$. Let $Y$ and $\underline{Y}$ be dominant solutions of (SDS) and (SDS) at $\infty$ with rank equal to $r$ and $\underline{r}$, respectively. Then the sequence $Y_k^*$ defined by*

$$X_k^* := \begin{pmatrix} X_k & 0 \\ 0 & \underline{X}_k \end{pmatrix}, \quad U_k^* := \begin{pmatrix} U_k & 0 \\ 0 & \underline{U}_k \end{pmatrix}, \quad k \in [0, \infty)_{\mathbb{Z}}, \tag{6.415}$$

*is a dominant solution of system (SDS\*) at $\infty$ with rank equal to $r^* := r + \underline{r}$. Moreover, the dominant solution $Y^*$ at $\infty$ constructed in (6.415) is minimal (maximal) if and only if the dominant solutions $Y$ and $\underline{Y}$ at $\infty$ are minimal (maximal).*

*Proof* The statement follows by the same arguments as in the proof of Theorem 6.118 by noting that the maximal order of abnormality of the augmented system (SDS\*) is $d_\infty^* = d_\infty + \underline{d}_\infty$, so that the matrix $T^*$ in (6.237) associated with $Y^*$ in (6.415) satisfies (using that $n^* = n + \underline{n}$)

$$\operatorname{rank} T^* = \operatorname{rank} \operatorname{diag}\{T, \underline{T}\} = \operatorname{rank} T + \operatorname{rank} \underline{T}$$

$$= n - d_\infty + \underline{n} - \underline{d}_\infty = n^* - d_\infty^*.$$

The result then follows from Definition 6.122 applied system (SDS\*).          □

## 6.3.8  Genus Conjoined Bases

In this section we introduce the concept of a genus of conjoined bases of (SDS) and study its properties. As our main results, we prove the existence (Theorem 6.138) and classification (Theorems 6.139 and 6.141) of recessive and dominant solutions of (SDS) in every such a genus.

**Definition 6.134 (Genus of Conjoined Bases)** Let $Y^{(1)}$ and $Y^{(2)}$ be two conjoined bases of (SDS). We say that $Y^{(1)}$ and $Y^{(2)}$ *have the same genus* (or they *belong to the same genus*) if there exists an index $N \in [0, \infty)_{\mathbb{Z}}$ such that the equality $\operatorname{Im} X_k^{(1)} = \operatorname{Im} X_k^{(2)}$ holds for all $k \in [N, \infty)_{\mathbb{Z}}$.

The relation "having (or belonging to) the same genus" is an equivalence relation on the set of all conjoined bases of (SDS). Therefore, there exists a partition of this set into disjoint classes of conjoined bases of (SDS), which belong to the same genus. This allows to interpret each such an equivalence class $\mathcal{G}$ as a genus itself. In particular, when system (SDS) is nonoscillatory, we have the following property of conjoined bases of (SDS) in one genus $\mathcal{G}$.

**Proposition 6.135** *Assume that system* (SDS) *is nonoscillatory at* $\infty$. *Let* $Y^{(1)}$ *and* $Y^{(2)}$ *be conjoined bases of* (SDS). *Then the following are equivalent.*

(i) *The conjoined bases* $Y^{(1)}$ *and* $Y^{(2)}$ *belong to the same genus* $\mathcal{G}$.
(ii) *The equality* $\operatorname{Im} X_k^{(1)} = \operatorname{Im} X_k^{(2)}$ *holds on some subinterval* $[N, \infty)_{\mathbb{Z}}$, *where* $Y^{(1)}$ *and* $Y^{(2)}$ *have constant kernel.*
(iii) *The equality* $\operatorname{Im} X_k^{(1)} = \operatorname{Im} X_k^{(2)}$ *holds on every subinterval* $[N, \infty)_{\mathbb{Z}}$, *where* $Y^{(1)}$ *and* $Y^{(2)}$ *have constant kernel.*

*Proof* The statement follows from Definition 6.134 and Theorem 6.77.                □

*Remark 6.136* For a nonoscillatory system (SDS) at $\infty$, there is only one genus of conjoined bases (denoted by $\mathcal{G}_{\min}$) containing all conjoined bases $Y$ of (SDS) with the minimal eventual rank of $X_k$ in (6.279), i.e., with rank $X_k = n - d_\infty$ for large $k$. This is a direct consequence of the fact that any two conjoined bases $Y^{(1)}$ and $Y^{(2)}$ of (SDS) with the eventual rank of $X_k^{(1)}$ and $X_k^{(2)}$ equal to $n - d_\infty$ necessarily satisfy $\operatorname{Im} X_k^{(1)} = \operatorname{Im} X_k^{(2)}$ for large $k$. Indeed, if $Y^{(1)}$ and $Y^{(2)}$ have constant kernel on $[N, \infty)_{\mathbb{Z}}$ and no forward focal points in $(N, \infty)$, then the equality $d[N, \infty)_{\mathbb{Z}} = d_\infty$ holds by Remark 6.96. Therefore, $Y^{(1)}$ and $Y^{(2)}$ are minimal conjoined bases on $[N, \infty)_{\mathbb{Z}}$, and by Remark 6.98 they satisfy $\operatorname{Im} X_N^{(1)} = (\Lambda_0[N, \infty)_{\mathbb{Z}})^\perp = \operatorname{Im} X_N^{(2)}$. In turn, Theorem 6.77 implies that $\operatorname{Im} X_k^{(1)} = \operatorname{Im} X_k^{(2)}$ on $[N, \infty)_{\mathbb{Z}}$. In particular, all minimal recessive and dominant solutions of (SDS) at $\infty$ belong to the minimal genus $\mathcal{G}_{\min}$.

*Remark 6.137* Similarly to Remark 6.136, there is only one genus of conjoined bases (denoted by $\mathcal{G}_{\max}$) containing all conjoined bases $Y$ of (SDS) satisfying $\operatorname{Im} X_k = \mathbb{R}^n$ for large $k$, i.e., with $X_k$ eventually invertible. All maximal recessive and dominant solutions of (SDS) at $\infty$ then belong to the maximal genus $\mathcal{G}_{\max}$.

In the following result, we show that there exists both recessive and dominant solutions of (SDS) at $\infty$ in every genus $\mathcal{G}$.

**Theorem 6.138** *Assume that system* (SDS) *is nonoscillatory at* $\infty$. *Let* $\mathcal{G}$ *be a genus of conjoined bases of* (SDS). *Then there exists a recessive solution and a dominant solution of* (SDS) *at* $\infty$ *in the genus* $\mathcal{G}$.

*Proof* First we focus on the case of the recessive solutions at $\infty$. Let $\hat{Y}^{\min}$ be the minimal recessive solution of (SDS) at $\infty$ with $\hat{K}_{\min}$ defined in Remark 6.130. Furthermore, let $Y$ be a conjoined basis of (SDS), which belongs to the genus $\mathcal{G}$. Then there exists an index $N \in [\hat{K}_{\min}, \infty)_{\mathbb{Z}}$ such that $Y$ has constant kernel on $[N, \infty)_{\mathbb{Z}}$ and no forward focal points in $(N, \infty)$. By Remark 6.130 we know that the equality $d[N, \infty)_{\mathbb{Z}} = d_\infty$ holds and at the same time $\hat{Y}^{\min}$ is a minimal recessive solution with respect to $[N, \infty)_{\mathbb{Z}}$. Moreover, by Theorem 6.90 there exists a conjoined basis $Y^*$ of (SDS) with constant kernel on $[N, \infty)_{\mathbb{Z}}$ and no forward focal points in $(N, \infty)$ such that rank $X_k^* = n - d_\infty$ on $[N, \infty)_{\mathbb{Z}}$ and such that $Y$ contains $Y^*$ on $[N, \infty)_{\mathbb{Z}}$. Therefore, $\hat{Y}^{\min}$ and $Y^*$ are minimal conjoined bases of (SDS) on $[N, \infty)_{\mathbb{Z}}$, so that $\operatorname{Im} \hat{X}_N^{\min} = \operatorname{Im} X_N^*$ by Remark 6.98. Denote by $\hat{R}_k^{\min}$, $R_k^*$, $R_k$ the orthogonal projectors in (6.233) defined by $\hat{X}_k^{\min}$, $X_k^*$, $X_k$, respectively. Then $\hat{R}_N^{\min} = R_N^*$ and $\operatorname{Im} \hat{R}_N^{\min} = \operatorname{Im} R_N^* \subseteq \operatorname{Im} R_N$, by Remark 6.83. From Theorem 6.87 we know that there exists a conjoined basis $\hat{Y}$ of (SDS) with constant kernel on $[N, \infty)_{\mathbb{Z}}$ and no forward focal points in $(N, \infty)$ such that $\hat{Y}$ contains $\hat{Y}^{\min}$ on $[N, \infty)_{\mathbb{Z}}$ and the equality $\operatorname{Im} \hat{X}_N = \operatorname{Im} R_N$ holds. Consequently, by Proposition 6.86 the conjoined basis $\hat{Y}$ is a recessive solution of (SDS) at $\infty$ with respect to $[N, \infty)_{\mathbb{Z}}$. And since $\operatorname{Im} \hat{X}_N = \operatorname{Im} R_N = \operatorname{Im} X_N$, we have that $\operatorname{Im} \hat{X}_k = \operatorname{Im} X_k$ on $[N, \infty)_{\mathbb{Z}}$ by Theorem 6.77. This shows that the recessive solution $\hat{Y}$ belongs to the genus $\mathcal{G}$, according to Definition 6.134. The proof for the dominant solution at $\infty$ in $\mathcal{G}$ can be carried out by exactly the same arguments, by considering a minimal dominant solution $Y^{\min}$ at $\infty$ from Theorem 6.128 instead of the minimal recessive solution $\hat{Y}^{\min}$ at $\infty$ in the above proof. $\qquad\square$

In the next result, we provide a complete classification of all recessive solutions of (SDS) at $\infty$ in a given genus $\mathcal{G}$.

**Theorem 6.139** *Assume that system* (SDS) *is nonoscillatory at $\infty$ with $\hat{K}_{\min}$ defined in Remark 6.130. Let $\hat{Y}$ be a recessive solution of* (SDS) *at $\infty$, which belongs to a genus $\mathcal{G}$, and let $\hat{P}$ and $P_{\hat{S}\infty}$ be the constant orthogonal projectors defined in* (6.234), (6.238), *and Remark 6.132 on $[\hat{K}_{\min}, \infty)_{\mathbb{Z}}$ through the matrix $\hat{X}_k$. Then a solution $Y$ of* (SDS) *is a recessive solution at $\infty$ belonging to the genus $\mathcal{G}$ if and only if for some (and hence for every) index $N \in [\hat{K}_{\min}, \infty)_{\mathbb{Z}}$ there exist matrices $\hat{M}, \hat{N} \in \mathbb{R}^{n \times n}$ such that*

$$X_N = \hat{X}_N \hat{M}, \quad U_N = \hat{U}_N \hat{M} + \hat{X}_N^{\dagger T} \hat{N}, \tag{6.416}$$

$$\hat{M} \text{ is nonsingular}, \quad \hat{M}^T \hat{N} = \hat{N}^T \hat{M}, \quad \operatorname{Im} \hat{N} \subseteq \operatorname{Im} \hat{P}, \tag{6.417}$$

$$P_{\hat{S}\infty} \hat{N} \hat{M}^{-1} P_{\hat{S}\infty} = 0. \tag{6.418}$$

*Proof* Let $Y$ be a recessive solution of (SDS), which belongs to the genus $\mathcal{G}$. From Theorem 6.131 we know that $\hat{Y}$ and $Y$ have constant kernel on $[\hat{K}_{\min}, \infty)_{\mathbb{Z}}$ and no forward focal points in $(\hat{K}_{\min}, \infty)$, and consequently, according to Proposition 6.135 they satisfy $\operatorname{Im} \hat{X}_k = \operatorname{Im} X_k$ on $[\hat{K}_{\min}, \infty)_{\mathbb{Z}}$. Therefore, by Theorem 6.69 and Remark 6.70 (with $Y^{(1)} := \hat{Y}$ and $Y^{(2)} := Y$) for every $N \in [\hat{K}_{\min}, \infty)_{\mathbb{Z}}$, there exist $\hat{M}, \hat{N} \in \mathbb{R}^{n \times n}$ such that (6.416) and (6.417) hold. We will prove (6.418). Fix $N \in [\hat{K}_{\min}, \infty)_{\mathbb{Z}}$. Denote by $\hat{Y}^{\min}$ and $Y^*$ the minimal recessive solutions of (SDS) at $\infty$ from Theorem 6.131, which are contained in $\hat{Y}$ and $Y$ on $[N, \infty)_{\mathbb{Z}}$, respectively. By Theorem 6.113 (or Theorem 6.69), we know that $Y_k^* = \hat{Y}_k^{\min} \hat{M}^{\min}$ on $[0, \infty)_{\mathbb{Z}}$ for some nonsingular matrix $\hat{M}^{\min}$. This means that the Wronskian $\hat{W}^{\min} := w(\hat{Y}^{\min}, Y^*) = 0$. On the other hand, the conjoined bases $\hat{Y}^{\min}$ and $Y^*$ are minimal on $[N, \infty)_{\mathbb{Z}}$, and therefore, the formulas in (6.346) hold (with $Y^{*(1)} := \hat{Y}^{\min}$, $Y^{*(2)} := Y^*$, $M^{*(1)} := \hat{M}^{\min}$, and $N^{*(1)} := \hat{W}^{\min} = 0$). Consequently, by

Proposition 6.99 (with $P_{\mathcal{S}^{(1)}\infty} := P_{\hat{\mathcal{S}}\infty}$), we obtain that $P_{\hat{\mathcal{S}}\infty} \hat{N}\hat{M}^{-1}P_{\hat{\mathcal{S}}\infty} = 0$ is satisfied.

Conversely, fix $N \in [\hat{K}_{\min}, \infty)_{\mathbb{Z}}$, and suppose that a solution $Y$ of (SDS) satisfies (6.416)–(6.418). From Remark 6.130 we have $d[N, \infty)_{\mathbb{Z}} = d_\infty$. The third condition in (6.417) yields that $\hat{P}\hat{N} = \hat{N}$, so that $X_N^T U_N = \hat{M}^T \hat{X}_N^T \hat{U}_N \hat{M} + \hat{M}^T \hat{N}$ by (6.416). Therefore, $X_N^T U_N$ is symmetric. If $Y_N d = 0$ for some $d \in \mathbb{R}^n$, then $\hat{X}_N \hat{M}d = 0$ and $\hat{U}_N \hat{M}d = -\hat{X}_N^{\dagger T} \hat{N}d$ again by (6.416). Multiplying the last equality by $\hat{X}_N^T$, we obtain that $0 = \hat{U}_N^T \hat{X}_N \hat{M}d = -\hat{P}\hat{N}d = -\hat{N}d$. Therefore, $\hat{U}_N \hat{M}d = 0$ as well. But since $\hat{Y}$ is a conjoined basis (rank $\hat{Y}_N = n$), it follows that $\hat{M}d = 0$. The invertibility of $\hat{M}$ then implies that $d = 0$, i.e., $Y$ is a conjoined basis of (SDS).

Let $\hat{S}_k$ be the $S$-matrix in (6.236) corresponding to $\hat{Y}$ on $[N, \infty)_{\mathbb{Z}}$, and let $\bar{Y}$ be a conjoined basis of (SDS) from Proposition 6.67 (with $Y := \hat{Y}$). With the aid of the identities $\hat{P}\hat{N} = \hat{N}$, $\hat{X}_N^\dagger \bar{X}_N = 0$, and $\bar{X}_N \hat{P} = \hat{X}_N \hat{S}_N = 0$ from Proposition 6.67 (with $Y := \hat{Y}$ at the index $k = N$), we can rewrite (6.416) as $Y_N = \hat{Y}_N \hat{M} + \bar{Y}_N \hat{N}$. Hence, $Y_k = \hat{Y}_k \hat{M} + \bar{Y}_k \hat{N}$ for all $k \in [N, \infty)_{\mathbb{Z}}$, by the uniqueness of solutions of (SDS). In particular, by Proposition 6.67 again and $\hat{X}_k \hat{P} = \hat{X}_k$ on $[N, \infty)_{\mathbb{Z}}$, we get

$$X_k = \hat{X}_k \hat{M} + \hat{X}_k \hat{S}_k \hat{N} = \hat{X}_k (\hat{P}\hat{M} + \hat{S}_k \hat{N}), \quad k \in [N, \infty)_{\mathbb{Z}}. \tag{6.419}$$

We will show that $Y$ has constant kernel on $[N, \infty)_{\mathbb{Z}}$. Namely, we will prove that $\operatorname{Ker} X_k = \operatorname{Ker} \hat{P}\hat{M}$ on $[N, \infty)_{\mathbb{Z}}$. First we note that the symmetry of $\hat{M}^T \hat{N}$ implies the symmetry of $\hat{N}\hat{M}^{-1}$ and that $\hat{N}\hat{M}^{-1}\hat{P} = \hat{N}\hat{M}^{-1}$ holds. Hence, by (6.419) we obtain

$$X_k = \hat{X}_k (\hat{P}\hat{M} + \hat{S}_k \hat{N}\hat{M}^{-1}\hat{M}) = \hat{X}_k (I + \hat{S}_k \hat{N}\hat{M}^{-1}) \hat{P}\hat{M}, \quad k \in [N, \infty)_{\mathbb{Z}}.$$

This implies that $\operatorname{Ker} \hat{P}\hat{M} \subseteq \operatorname{Ker} X_k$ for all $k \in [N, \infty)_{\mathbb{Z}}$. Fix now $k \in [N, \infty)_{\mathbb{Z}}$, $v \in \operatorname{Ker} X_k$, and put $w := \hat{P}\hat{M}v$. Then $\hat{X}_k (w + \hat{S}_k \hat{N}\hat{M}^{-1}w) = 0$. Multiplying the latter equality by $\hat{X}_k^\dagger$ from the left and using the identities $\hat{P}\hat{S}_k = \hat{S}_k$ and $w = \hat{P}w$, we get $w = -\hat{S}_k \hat{N}\hat{M}^{-1}w$. Therefore, $w \in \operatorname{Im} \hat{S}_k \subseteq \operatorname{Im} P_{\hat{\mathcal{S}}\infty}$ by (6.239), and thus, $w = -\hat{S}_k P_{\hat{\mathcal{S}}\infty} \hat{N}\hat{M}^{-1}P_{\hat{\mathcal{S}}\infty}w = 0$ by (6.418). This shows that $v \in \operatorname{Ker} \hat{P}\hat{M}$, i.e., $\operatorname{Ker} X_k \subseteq \operatorname{Ker} \hat{P}\hat{M}$. Therefore, $\operatorname{Ker} X_k = \operatorname{Ker} \hat{P}\hat{M}$ on $[N, \infty)_{\mathbb{Z}}$, i.e., the conjoined basis $Y$ has constant kernel on $[N, \infty)_{\mathbb{Z}}$.

Note that the first equation in (6.416) and the invertibility of $\hat{M}$ yield that $\operatorname{Im} X_N = \operatorname{Im} \hat{X}_N$. Then it follows from Theorem 6.77 that $\operatorname{Im} X_k = \operatorname{Im} \hat{X}_k$ for all $k \in [N, \infty)_{\mathbb{Z}}$. In particular, $Y$ belongs to the genus $\mathcal{G}$.

We shall prove that $Y$ has no forward focal points in $(N, \infty)$. Since the conjoined basis $Y$ has constant kernel on $[N, \infty)_{\mathbb{Z}}$, we denote by $P$ the associated projector in (6.234). The fact that $\operatorname{Im} X_k = \operatorname{Im} \hat{X}_k$ on $[N, \infty)_{\mathbb{Z}}$ implies that $Y$ and $\hat{Y}$ are mutually representable on $[N, \infty)_{\mathbb{Z}}$ in the spirit of Theorem 6.69 and the subsequent remarks. This means that by this theorem (with the notation $Y^{(1)} := \hat{Y}$, $Y^{(2)} := Y$ and $S_k^{(1)} := \hat{S}_k$, $S_k^{(2)} := S_k$, $P^{(1)} := \hat{P}$, $P^{(2)} := P$, and $M^{(1)} := \hat{M}$, $N^{(1)} := \hat{N}$), we have

$$
\left.\begin{array}{l}
\mathrm{Im}\,(\hat{P}\hat{M} + \hat{S}_k\hat{N}) = \mathrm{Im}\,\hat{P}, \quad \mathrm{Im}\,\hat{M}S_k = \mathrm{Im}\,\hat{S}_k, \\[4pt]
(\hat{P}\hat{M} + \hat{S}_k\hat{N})^\dagger = P\hat{M}^{-1} - S_k\hat{N}^T,
\end{array}\right\} \quad k \in [N, \infty)_\mathbb{Z}. \tag{6.420}
$$

In turn, equalities (6.419), (6.420), and (6.236) yield on $[N, \infty)_\mathbb{Z}$ the formulas

$$
X_k^\dagger = (\hat{P}\hat{M} + \hat{S}_k\hat{N})^\dagger \hat{X}_k^\dagger, \tag{6.421}
$$

and

$$
\begin{aligned}
\hat{P}\hat{M} + \hat{S}_k\hat{N} &= \hat{P}\hat{M} + \hat{S}_{k+1}\hat{N} - (\Delta\hat{S}_k)\,\hat{N} \\
&= (\hat{P}\hat{M} + \hat{S}_{k+1}\hat{N}) - \hat{X}_{k+1}^\dagger\,\mathcal{B}_k\,\hat{X}_k^{\dagger\,T}\hat{N}.
\end{aligned} \tag{6.422}
$$

Note that by the first identity in (6.420), we have $(\hat{P}\hat{M} + \hat{S}_k\hat{N})(\hat{P}\hat{M} + \hat{S}_k\hat{N})^\dagger = \hat{P}$ on $[N, \infty)_\mathbb{Z}$. Therefore, equations (6.419)–(6.422) and the symmetry of $\hat{X}_k\hat{X}_{k+1}^\dagger\,\mathcal{B}_k$ on $[N, \infty)_\mathbb{Z}$ imply that

$$
\begin{aligned}
X_k X_{k+1}^\dagger\,\mathcal{B}_k &\overset{(6.419),\,(6.421)}{=} \hat{X}_k(\hat{P}\hat{M} + \hat{S}_k\hat{N})(\hat{P}\hat{M} + \hat{S}_{k+1}\hat{N})^\dagger\hat{X}_{k+1}^\dagger\mathcal{B}_k \\
&\overset{(6.422)}{=} \hat{X}_k[(\hat{P}\hat{M} + \hat{S}_{k+1}\hat{N}) - \hat{X}_{k+1}^\dagger\mathcal{B}_k\hat{X}_k^{\dagger\,T}\hat{N}](\hat{P}\hat{M} + \hat{S}_{k+1}\hat{N})^\dagger\hat{X}_{k+1}^\dagger\mathcal{B}_k \\
&= \hat{X}_k\hat{P}\hat{X}_{k+1}^\dagger\mathcal{B}_k - (\hat{X}_k\hat{X}_{k+1}^\dagger\mathcal{B}_k)\,\hat{X}_k^{\dagger\,T}\hat{N}(\hat{P}\hat{M} + \hat{S}_{k+1}\hat{N})^\dagger\hat{X}_{k+1}^\dagger\mathcal{B}_k \\
&= \hat{X}_k\hat{X}_{k+1}^\dagger\mathcal{B}_k - \mathcal{B}_k^T\hat{X}_{k+1}^{\dagger\,T}\hat{P}\hat{N}(\hat{P}\hat{M} + \hat{S}_{k+1}\hat{N})^\dagger\hat{X}_{k+1}^\dagger\mathcal{B}_k \\
&\overset{(6.420)}{=} \hat{X}_k\hat{X}_{k+1}^\dagger\mathcal{B}_k - \mathcal{L}_k, \quad \mathcal{L}_k := \mathcal{B}_k^T\hat{X}_{k+1}^{\dagger\,T}\hat{N}(P\hat{M}^{-1} - S_{k+1}\hat{N}^T)\,\hat{X}_{k+1}^\dagger\mathcal{B}_k.
\end{aligned} \tag{6.423}
$$

Our claim is to show that the matrix $\mathcal{L}_k$ in (6.423) is zero. Let $\hat{Y}^{\min}$ be the minimal recessive solution of (SDS) from Theorem 6.131(i), which is contained in $\hat{Y}$ on $[N, \infty)_\mathbb{Z}$, and let $\hat{R}_k^{\min}$ be the corresponding orthogonal projector in (6.234). Then $\hat{Y}^{\min}$ is a minimal conjoined basis of (SDS) on $[N, \infty)_\mathbb{Z}$, so that $P_{\hat{S}_\infty}(\hat{X}_k^{\min})^\dagger = (\hat{X}_k^{\min})^\dagger$ on $[N, \infty)_\mathbb{Z}$, by Proposition 6.86. Moreover, the equality $(\hat{X}_k^{\min})^\dagger = \hat{X}_k^\dagger\hat{R}_k^{\min}$ holds for every $k \in [N, \infty)_\mathbb{Z}$ by (6.298) (with $Y^* := \hat{Y}^{\min}$). Then by $\hat{N}P = \hat{N}$, the second identity in (6.420), and Theorem 6.66(ii), we obtain

$$
\begin{aligned}
\mathcal{L}_k &= \mathcal{B}_k^T\hat{R}_k^{\min}\hat{X}_{k+1}^{\dagger\,T}\hat{N}(P\hat{M}^{-1} - S_{k+1}\,\hat{N}^T)\,\hat{X}_{k+1}^\dagger\hat{R}_k^{\min}\mathcal{B}_k \\
&= \mathcal{B}_k^T(\hat{X}_{k+1}^{\min})^{\dagger\,T}\hat{N}(P\hat{M}^{-1} - \hat{M}^{-1}\hat{M}S_{k+1}\hat{N}^T)(\hat{X}_{k+1}^{\min})^\dagger\mathcal{B}_k \\
&\overset{(6.420)}{=} \mathcal{B}_k^T(\hat{X}_{k+1}^{\min})^{\dagger\,T}P_{\hat{S}_\infty}\hat{N}(P\hat{M}^{-1} - \hat{M}^{-1}P_{\hat{S}_\infty}\hat{M}S_{k+1}\hat{N}^T)P_{\hat{S}_\infty}(\hat{X}_{k+1}^{\min})^\dagger\mathcal{B}_k
\end{aligned}
$$

$$= \mathcal{B}_k^T (\hat{X}_{k+1}^{\min})^{\dagger T} P_{\hat{S}_\infty} (\hat{N}\hat{M}^{-1} P_{\hat{S}_\infty} - \hat{N}\hat{M}^{-1} P_{\hat{S}_\infty} \hat{M} S_{k+1} \hat{N}^T P_{\hat{S}_\infty}) (\hat{X}_{k+1}^{\min})^{\dagger} \mathcal{B}_k$$

$$= \mathcal{B}_k^T (\hat{X}_{k+1}^{\min})^{\dagger T} P_{\hat{S}_\infty} \hat{N}\hat{M}^{-1} P_{\hat{S}_\infty} (I - \hat{M} S_{k+1} \hat{N}^T P_{\hat{S}_\infty}) (\hat{X}_{k+1}^{\min})^{\dagger} \mathcal{B}_k \stackrel{(6.418)}{=} 0.$$

Hence, from equality (6.423) we get $X_k X_{k+1}^{\dagger} \mathcal{B}_k = \hat{X}_k \hat{X}_{k+1}^{\dagger} \mathcal{B}_k$ on $[N, \infty)_{\mathbb{Z}}$. Since $\hat{Y}$ has no forward focal points in $(N, \infty)$, i.e., $\hat{X}_k \hat{X}_{k+1}^{\dagger} \mathcal{B}_k \geq 0$ on $[N, \infty)_{\mathbb{Z}}$, we conclude that $Y$ has no forward focal points in $(N, \infty)$ as well.

In the final step of the proof, we will show that $Y$ is a recessive solution of (SDS) at $\infty$. By Theorem 6.90, there exists a minimal conjoined basis $Y^*$ of (SDS) on $[N, \infty)_{\mathbb{Z}}$, which is contained in $Y$ on $[N, \infty)_{\mathbb{Z}}$. In particular, the equality rank $X_k^* = n - d_\infty$ holds on $[N, \infty)_{\mathbb{Z}}$. Then according to Remark 6.98, there exist matrices $\hat{M}^{\min}, \hat{N}^{\min} \in \mathbb{R}^{n \times n}$ such that (6.346) is satisfied (with $Y^{*(1)} := \hat{Y}^{\min}$, $Y^{*(2)} := Y^*$, $M^{*(1)} := \hat{M}^{\min}$, and $N^{*(1)} := \hat{N}^{\min}$). In particular, $\hat{M}^{\min}$ is invertible and $\hat{N}^{\min} = w(\hat{Y}^{\min}, Y^*)$ by (6.244). Consequently, by Proposition 6.99 (with $Y^{(1)} := \hat{Y}$, $Y^{(2)} := Y$, $M^{(1)} := \hat{M}$, $N^{(1)} := \hat{N}$, and $P_{S^{(1)}\infty} := P_{\hat{S}_\infty}$), we derive that $\hat{N}^{\min}(\hat{M}^{\min})^{-1} = P_{\hat{S}_\infty} \hat{N}\hat{M}^{-1} P_{\hat{S}_\infty}$. In view of (6.418), we get $\hat{N}^{\min}(\hat{M}^{\min})^{-1} = 0$, and hence $\hat{N}^{\min} = 0$. This implies by (6.346) that $Y_N^* = \hat{Y}_N^{\min} \hat{M}^{\min}$, so that $Y_k^* = \hat{Y}_k^{\min} \hat{M}^{\min}$ on $[0, \infty)_{\mathbb{Z}}$ by the uniqueness of solutions of (SDS). According to Theorem 6.113, $Y^*$ is a minimal recessive solution of (SDS) at $\infty$ with respect to the interval $[N, \infty)_{\mathbb{Z}}$. Thus, we proved that $Y^*$, being a minimal recessive solution of (SDS) at $\infty$, is contained in $Y$ on $[N, \infty)_{\mathbb{Z}}$. Therefore, $Y$ is also a recessive solution of (SDS) at $\infty$ by Proposition 6.112. The proof is complete. $\qquad\square$

As a special case of Theorem 6.139, we obtain a classification of the recessive solutions of (SDS) at $\infty$ in the maximal genus $\mathcal{G}_{\max}$.

**Corollary 6.140** *Assume that system (SDS) is nonoscillatory at $\infty$ with $\hat{K}_{\min}$ defined in Remark 6.130. Let $\hat{Y}$ be a maximal recessive solution of (SDS) at $\infty$. Moreover, let $P_{\hat{S}_\infty}$ be the orthogonal projector defined through the invertible matrix $\hat{X}_k$ on $[\hat{K}_{\min}, \infty)_{\mathbb{Z}}$ in (6.238) and Remark 6.132. Then a solution $Y$ of (SDS) is a maximal recessive solution at $\infty$ if and only if for some (and hence for every) index $N \in [\hat{K}_{\min}, \infty)_{\mathbb{Z}}$, there exist matrices $\hat{M}, \hat{N} \in \mathbb{R}^{n \times n}$ such that*

$$X_N = \hat{X}_N \hat{M}, \quad U_N = \hat{U}_N \hat{M} + \hat{X}_k^{T-1} \hat{N},$$

$$\hat{M} \text{ is nonsingular}, \quad \hat{M}^T \hat{N} = \hat{N}^T \hat{M}, \quad P_{\hat{S}_\infty} \hat{N}\hat{M}^{-1} P_{\hat{S}_\infty} = 0.$$

*Proof* The result follows from Theorem 6.139 with $\hat{P} = I$. $\qquad\square$

In remaining part of this section, we will derive several classifications of the dominant solutions of (SDS) at $\infty$ in a given genus $\mathcal{G}$ in terms of recessive solutions of (SDS) at $\infty$ from this genus.

**Theorem 6.141** *Assume that system (SDS) is nonoscillatory at $\infty$. Let $\mathcal{G}$ be a genus of conjoined bases of (SDS). Let $\hat{Y}$ be a recessive solution of (SDS) at*

$\infty$ *belonging to* $\mathcal{G}$, *and let* $Y$ *be a conjoined basis from* $\mathcal{G}$. *Denote by* $P_{\hat{\mathcal{S}}_\infty}$ *and* $P_{\mathcal{S}_\infty}$ *their associated orthogonal projectors in* (6.238) *and Remark* 6.132. *Then* $Y$ *is a dominant solution of* (SDS) *at* $\infty$ *if and only if*

$$\operatorname{rank} P_{\hat{\mathcal{S}}_\infty} w(\hat{Y}, Y) P_{\mathcal{S}_\infty} = n - d_\infty. \tag{6.424}$$

*Proof* Let $\hat{Y}$ be a recessive solution of (SDS) at $\infty$ belonging to $\mathcal{G}$, and let $Y$ be a conjoined basis from $\mathcal{G}$. We denote $\hat{N} := w(\hat{Y}, Y)$. Then there exists an index $N \in [\hat{K}_{\min}, \infty)_{\mathbb{Z}}$ such that $\hat{Y}$ and $Y$ have constant kernel on $[N, \infty)_{\mathbb{Z}}$ and no forward focal points in $(N, \infty)$. By Remark 6.130 we have $d[N, \infty)_{\mathbb{Z}} = d_\infty$. Since $\hat{Y}$ and $Y$ belong to the same genus $\mathcal{G}$, we may assume without loss of generality that $\operatorname{Im} \hat{X}_k = \operatorname{Im} X_k$ on $[N, \infty)_{\mathbb{Z}}$. From Theorem 6.69 and Remark 6.70, it then follows that $\hat{Y}$ and $Y$ are mutually representable on $[N, \infty)_{\mathbb{Z}}$ and the matrix $\hat{N}$ satisfies $\hat{P}\hat{N} = \hat{N} = \hat{N}P$, where $\hat{P}$ and $P$ are the corresponding orthogonal projectors in (6.234). Let $\hat{Y}^*$ and $Y^*$ be minimal conjoined bases of (SDS) on $[N, \infty)_{\mathbb{Z}}$, which are contained in $\hat{Y}$ and $Y$ on $[N, \infty)_{\mathbb{Z}}$, respectively, and let $\hat{N}^* := w(\hat{Y}^*, Y^*)$. In particular, $\hat{Y}^*$ is a minimal recessive solution of (SDS) by Proposition 6.112, i.e., the associated matrix $\hat{T}^*$ in (6.237) satisfies $\hat{T}^* = 0$. We apply the representations of $\hat{Y}$, $Y$ and of $\hat{Y}^*$, $Y^*$ in Theorem 6.69 and Remark 6.98, and in formula (6.349) in Proposition 6.99 (with $Y^{(1)} := \hat{Y}$, $Y^{(2)} := Y$, $Y^{*(1)} := \hat{Y}^*$, $Y^{*(2)} := Y^*$, $P^{(1)} := \hat{P}$, $P^{(2)} := P$, $P_{\mathcal{S}^{(1)}\infty} := P_{\hat{\mathcal{S}}_\infty}$, $P_{\mathcal{S}^{(2)}\infty} := P_{\mathcal{S}_\infty}$, and $N^{(1)} := \hat{N}$). Then there exist invertible matrices $\hat{M}$ and $\hat{M}^*$ such that

$$\left.\begin{array}{ll} \hat{P}\hat{M}P_{\mathcal{S}_\infty} = P_{\hat{\mathcal{S}}_\infty}\hat{M}^*, & P\hat{M}^{-1}P_{\hat{\mathcal{S}}_\infty} = P_{\mathcal{S}_\infty}(\hat{M}^*)^{-1}, \\ \hat{N}^*(\hat{M}^*)^{-1} = P_{\hat{\mathcal{S}}_\infty}\hat{N}\hat{M}^{-1}P_{\hat{\mathcal{S}}_\infty}. & \end{array}\right\} \tag{6.425}$$

By using (6.425) and the equality $\hat{N}P = \hat{N}$, we then obtain

$$\hat{N}^*(\hat{M}^*)^{-1} = P_{\hat{\mathcal{S}}_\infty}\hat{N}P\hat{M}^{-1}P_{\hat{\mathcal{S}}_\infty} = P_{\hat{\mathcal{S}}_\infty}\hat{N}P_{\mathcal{S}_\infty}(\hat{M}^*)^{-1}. \tag{6.426}$$

Let $Y$ be a dominant solution of (SDS) at $\infty$. Then also $Y^*$ is a dominant solution at $\infty$, by Proposition 6.127. From (6.389) we know that the matrix $T^*$ in (6.237) associated with $Y^*$ satisfies $\operatorname{rank} T^* = \operatorname{rank}[\hat{N}^*(\hat{M}^*)^{-1} + \hat{T}^*] = \operatorname{rank} \hat{N}^*(\hat{M}^*)^{-1}$. Since $\operatorname{rank} T^* = n - d_\infty$ by Definition 6.122, we get from (6.426) that

$$\operatorname{rank} P_{\hat{\mathcal{S}}_\infty}\hat{N}P_{\mathcal{S}_\infty} = \operatorname{rank} P_{\hat{\mathcal{S}}_\infty}\hat{N}P_{\mathcal{S}_\infty}(\hat{M}^*)^{-1} = \operatorname{rank}\hat{N}^*(\hat{M}^*)^{-1} = n - d_\infty,$$

i.e., formula (6.424) holds. Conversely, if (6.424) is satisfied, then from (6.426) we obtain that $\operatorname{rank}\hat{N}^*(\hat{M}^*)^{-1} = n - d_\infty$. Therefore, $\operatorname{rank} T^* = n - d_\infty$ by (6.389), and hence $Y^*$ is a dominant solution of (SDS) at $\infty$ by Definition 6.122. Finally, Proposition 6.127 implies that $Y$ is a dominant solution at $\infty$ as well. □

When we specialize Theorem 6.141 to the minimal genus $\mathcal{G}_{\min}$, we obtain a classification of all minimal dominant solutions of (SDS) at $\infty$.

**Corollary 6.142** *Assume that system* (SDS) *is nonoscillatory at* $\infty$*. Let* $\hat{Y}^{\min}$ *be the minimal recessive solution of* (SDS) *at* $\infty$*, and let* $Y$ *be a minimal conjoined basis of* (SDS)*. Then* $Y$ *is a minimal dominant solution of* (SDS) *at* $\infty$ *if and only if* rank $w(\hat{Y}^{\min}, Y) = n - d_\infty$.

*Proof* By Theorem 6.141 and its proof with $\hat{P} := P_{\hat{S}\infty}$ and $P := P_{S\infty}$, we have that $P_{\hat{S}\infty} \hat{N} = \hat{N}$ and $\hat{N} P_{S\infty} = \hat{N}$, where $\hat{N} := w(\hat{Y}^{\min}, Y)$. Therefore, the equality $P_{\hat{S}\infty} \hat{N} P_{S\infty} = \hat{N}$ holds, and the statement follows from (6.424).  □

In the following result, we provide important examples of minimal dominant solutions of (SDS) at $\infty$. We recall that the *principal solution* $Y^{[N]}$ of (SDS) at the index $N \in [0, \infty)_{\mathbb{Z}}$ is defined as the solution of (SDS) given by the initial conditions $X_N^{[N]} = 0$ and $U_N^{[N]} = I$.

**Theorem 6.143** *Assume that system* (SDS) *is nonoscillatory at* $\infty$ *with* $\hat{K}_{\min}$ *defined in Remark 6.130. Then for every* $N \in [\hat{K}_{\min}, \infty)_{\mathbb{Z}}$*, the principal solution* $Y^{[N]}$ *at the index* $N$ *is a minimal dominant solution of* (SDS) *at* $\infty$.

*Proof* Let $\hat{Y}^{\min}$ be the minimal recessive solution of (SDS) at $\infty$ from Theorem 6.113, and let $N \in [\hat{K}_{\min}, \infty)_{\mathbb{Z}}$ be fixed. Then $\hat{Y}^{\min}$ has constant kernel on $[N, \infty)_{\mathbb{Z}}$ and no forward focal points in $(N, \infty)$, by Remark 6.130. In particular, the condition $d[N, \infty)_{\mathbb{Z}} = d_\infty$ holds, by (6.414). In order to simplify the notation, we put $\hat{Y} := \hat{Y}^{\min}$ and $Y := Y^{[N]}$. Let $\hat{P}$, $\hat{R}_k$, $\hat{S}_k$, and $P_{\hat{S}\infty}$ be the matrices in (6.234), (6.236), (6.238), and Remark 6.132 associated with $\hat{Y}$. Then $\hat{P} = P_{\hat{S}\infty}$ and by Theorem 6.75 the conjoined basis $Y$ satisfies

$$X_k = \hat{X}_k \hat{S}_k \hat{X}_N^T, \quad \text{rank } \hat{S}_k = \text{rank } X_k, \quad k \in [N, \infty)_{\mathbb{Z}}. \tag{6.427}$$

Now fix an index $L \in [N, \infty)_{\mathbb{Z}}$ such that $Y$ has constant kernel on $[L, \infty)_{\mathbb{Z}}$ and no forward focal points in $(L, \infty)$. In particular, the rank of $X_k$ is constant on $[L, \infty)_{\mathbb{Z}}$. The second identity in (6.427) then implies that also the rank of $\hat{S}_k$ is constant on $[L, \infty)_{\mathbb{Z}}$, which by Theorem 6.66(iv) and property (6.239) yields the equalities rank $\hat{S}_k = n - d[N, \infty)_{\mathbb{Z}} = n - d_\infty$ on $[L, \infty)_{\mathbb{Z}}$. Therefore, rank $X_k = n - d_\infty$ on $[L, \infty)_{\mathbb{Z}}$ by (6.427), i.e., $Y$ is a minimal conjoined basis of (SDS) on $[L, \infty)_{\mathbb{Z}}$. We will show that

$$X_k^\dagger = \hat{X}_N^{\dagger T} \hat{S}_k^\dagger \hat{X}_k^\dagger, \quad k \in [L, \infty)_{\mathbb{Z}}. \tag{6.428}$$

Setting $A := \hat{X}_k \hat{S}_k \hat{X}_N^T$ and $B := \hat{X}_N^{\dagger T} \hat{S}_k^\dagger \hat{X}_k^\dagger$ for a fixed index $k \in [L, \infty)_{\mathbb{Z}}$, we verify that the matrices $A$ and $B$ satisfy the four defining equations for the Moore-Penrose pseudoinverse in (1.157). Namely, the identities $\hat{X}_k^\dagger \hat{X}_k = \hat{P}$, $\hat{X}_k \hat{X}_k^\dagger = \hat{R}_k$, and $\hat{S}_k^\dagger \hat{S}_k = P_{\hat{S}\infty}$ imply that $BA = \hat{R}_N$ and $AB = \hat{R}_k$ are symmetric. Moreover,

$$BAB = (BA)\,B = \hat{R}_N \hat{X}_N^{\dagger T} \hat{S}_k^\dagger \hat{X}_k^\dagger = \hat{X}_N^{\dagger T} \hat{S}_k^\dagger \hat{X}_k^\dagger = B,$$

$$ABA = (AB)\,A = \hat{R}_k \hat{X}_k \hat{S}_k \hat{X}_N^T = \hat{X}_k \hat{S}_k \hat{X}_N^T = A.$$

Therefore, we get $A^\dagger = B$, showing equality (6.428). Let $S_k^{(L)}$ be the $S$-matrix in (6.236), which corresponds to $Y$ on $[L, \infty)_{\mathbb{Z}}$, that is,

$$S_k^{(L)} := \sum_{j=L}^{k-1} X_{j+1}^\dagger \, \mathcal{B}_j \, X_j^{\dagger\,T}, \quad k \in [L, \infty)_{\mathbb{Z}}. \tag{6.429}$$

By inserting (6.428) into (6.429) and using the identity $\Delta \hat{S}_k = \hat{X}_{k+1}^\dagger \mathcal{B}_k \hat{X}_k^{\dagger\,T}$ together with $\hat{S}_k \hat{S}_k^\dagger = \hat{S}_k^\dagger \hat{S}_k = P_{\hat{S}_\infty}$ on $[L, \infty)_{\mathbb{Z}}$, we obtain for every $k \in [L, \infty)_{\mathbb{Z}}$ that

$$S_k^{(L)} \overset{(6.428)}{=} \sum_{j=L}^{k-1} \hat{X}_N^{\dagger\,T} \hat{S}_{j+1}^\dagger \hat{X}_{j+1}^\dagger \mathcal{B}_j \hat{X}_j^{\dagger\,T} \hat{S}_j^\dagger \hat{X}_N = \hat{X}_N^{\dagger\,T} \left( \sum_{j=L}^{k-1} \hat{S}_{j+1}^\dagger (\Delta \hat{S}_j) \, \hat{S}_j^\dagger \right) \hat{X}_N$$

$$= \hat{X}_N^{\dagger\,T} \left( \sum_{j=L}^{k-1} (\hat{S}_{j+1}^\dagger \hat{S}_{j+1} \hat{S}_j^\dagger - \hat{S}_{j+1}^\dagger \hat{S}_j \hat{S}_j^\dagger) \right) \hat{X}_N$$

$$= \hat{X}_N^{\dagger\,T} \left( \sum_{j=L}^{k-1} (P_{\hat{S}_\infty} \hat{S}_j^\dagger - \hat{S}_{j+1}^\dagger P_{\hat{S}_\infty}) \right) \hat{X}_N$$

$$= -\hat{X}_N^{\dagger\,T} \left( \sum_{j=L}^{k-1} \Delta \hat{S}_j^\dagger \right) \hat{X}_N = \hat{X}_N^{\dagger\,T} (\hat{S}_L^\dagger - \hat{S}_k^\dagger) \, \hat{X}_N. \tag{6.430}$$

Since $\hat{S}_k^\dagger \to 0$ as $k \to \infty$, equality (6.430) then implies that the limit of $S_k^{(L)}$ as $k \to \infty$ exists and it is equal to $\hat{X}_N^{\dagger\,T} \hat{S}_L^\dagger \hat{X}_N^\dagger$. Therefore, by Theorem 6.124 we conclude that $Y$ is a (minimal) dominant solution of (SDS) at $\infty$. We note that by (6.410) the matrix $T^{(L)}$ in (6.237) associated with $Y$ on $[L, \infty)_{\mathbb{Z}}$ satisfies the equality $T^{(L)} = \left( \hat{X}_N^{\dagger\,T} \hat{S}_L^\dagger \hat{X}_N^\dagger \right)^\dagger = \hat{X}_N \hat{S}_L \hat{X}_N^T$ by (6.428) and (6.427) at $k = N$. This additional information is however not needed in this proof. $\qquad \square$

*Remark 6.144* The result of Theorem 6.143 shows that a minimal dominant solution of (SDS) at $\infty$ is not uniquely determined (up to a constant invertible multiple), in contrast with the minimal recessive solution of (SDS) at $\infty$. Moreover, it implies that one cannot expect to have a unifying classification of all minimal dominant solutions of (SDS) at $\infty$ in the spirit of Theorem 6.139. In addition, the nonuniqueness of minimal dominant solutions of (SDS) at $\infty$ (in the minimal genus $\mathcal{G}_{\min}$) yields through Theorem 6.131(ii) the same property for all dominant solutions at $\infty$ in every other genus $\mathcal{G}$.

In the case of an eventually controllable system (SDS), we obtain from Corollary 6.142 and Theorem 6.143 an interesting characterization of the dominant solutions of (SDS) at $\infty$.

**Corollary 6.145** *Assume that system* (SDS) *is nonoscillatory at* $\infty$ *and eventually controllable. Let* $\hat{Y}$ *be the recessive solution of* (SDS) *at* $\infty$. *Then a conjoined basis* $Y$ *of* (SDS) *is a dominant solution at* $\infty$ *if and only if the Wronskian* $w(\hat{Y}, Y)$ *is invertible. In particular, for every index* $N \in [\hat{K}_{\min}, \infty)_{\mathbb{Z}}$ *the principal solution* $Y^{[N]}$ *of* (SDS) *at the index* $N$ *is a dominant solution at* $\infty$. *More generally, for any index* $N \in [0, \infty)_{\mathbb{Z}}$, *the principal solution* $Y^{[N]}$ *at the index* $N$ *is a dominant solution at* $\infty$ *if and only if the matrix* $\hat{X}_N$ *is invertible.*

*Proof* If (SDS) is eventually controllable, then $d_\infty = 0$, and for every conjoined basis $Y$ of (SDS), the matrix $X_k$ is eventually invertible, by (6.279). Therefore, according to Remarks 6.136 and 6.137, there is only one genus $\mathcal{G} = \mathcal{G}_{\min} = \mathcal{G}_{\max}$ of conjoined bases of (SDS). Let $\hat{Y}$ be the recessive solution of (SDS) at $\infty$ and $Y$ be a conjoined basis of (SDS). The first and second statements follow directly from Corollary 6.142 and Theorem 6.143, respectively. Finally, for $Y := Y^{[N]}$ we have $w(\hat{Y}, Y) = \hat{X}_N^T$, so that by the first part $Y^{[N]}$ is a dominant solution of (SDS) at $\infty$ if and only if the matrix $\hat{X}_N$ is invertible.                                                    $\square$

## 6.3.9   Limit Properties of Recessive and Dominant Solutions

In this subsection we derive characterizations of recessive solutions of (SDS) at $\infty$ in terms of a limit involving dominant solutions of (SDS) at $\infty$. This is a generalization of the limit property (2.134) in Theorem 2.67 to the abnormal case, in particular to an arbitrary genus $\mathcal{G}$; compare with Remark 6.117.

**Theorem 6.146** *Assume that system* (SDS) *is nonoscillatory at* $\infty$ *with* $\hat{K}_{\min}$ *defined in Remark 6.130. Let* $\hat{Y}$ *and* $Y$ *be two conjoined bases of* (SDS) *from a given genus* $\mathcal{G}$, *and let* $N \in [\hat{K}_{\min}, \infty)_{\mathbb{Z}}$ *be such that* $\hat{Y}$ *and* $Y$ *have constant kernel on* $[N, \infty)_{\mathbb{Z}}$ *and no forward focal points in* $(N, \infty)$. *Denote by* $\hat{P}$, $P_{\hat{S}\infty}$, *and* $P_{S\infty}$ *their associated projectors in* (6.234), (6.238), *and Remark 6.132. Then* $\hat{Y}$ *is a recessive solution of* (SDS) *at* $\infty$ *and* $\operatorname{rank} P_{\hat{S}\infty} w(\hat{Y}, Y) P_{S\infty} = n - d_\infty$ *if and only if*

$$\lim_{k \to \infty} X_k^\dagger \hat{X}_k = V \quad \text{with} \quad \operatorname{Im} V^T = \operatorname{Im}(\hat{P} - P_{\hat{S}\infty}). \tag{6.431}$$

*In this case* $Y$ *is a dominant solution of* (SDS) *at* $\infty$.

*Proof* Let $\hat{S}_k$ and $S_k$ be the $S$-matrices in (6.236) which are associated with $\hat{Y}$ and $Y$ on $[N, \infty)_{\mathbb{Z}}$. Denote $\hat{N} := w(\hat{Y}, Y)$. By (6.259) and (6.258) in Remark 6.70 (with $Y^{(1)} := \hat{Y}$ and $Y^{(2)} := Y$) and by $w(Y, \hat{Y}) = -[w(\hat{Y}, Y)]^T$, we have

$$X_k = \hat{X}_k\,(\hat{P}\hat{M} + \hat{S}_k\hat{N}), \quad \hat{X}_k = X_k\,(P\hat{M}^{-1} - S_k\hat{N}^T), \quad k \in [N, \infty)_{\mathbb{Z}}, \tag{6.432}$$

where the matrix $\hat{M}$ is invertible. By using the identities $X_k^\dagger X_k = P$ and $P S_k = S_k$ on $[N, \infty)_{\mathbb{Z}}$, it follows from (6.432) that

$$X_k^\dagger \hat{X}_k = P\hat{M}^{-1} - S_k \hat{N}^T, \quad k \in [N, \infty)_{\mathbb{Z}}. \tag{6.433}$$

Let $\hat{T}^*$ and $T^*$ be the matrices in (6.237) defined through the minimal conjoined bases $\hat{Y}^*$ and $Y^*$, which are contained in $\hat{Y}$ and $Y$ on $[N, \infty)_{\mathbb{Z}}$, respectively, as in the proof of Theorem 6.141. It follows by (6.348) that

$$T^* = (\hat{M}^*)^T \hat{T}^* \hat{M}^* + (\hat{M}^*)^T \hat{N}^*, \tag{6.434}$$

where $\hat{M}^*$ is invertible and $\hat{N}^* = w(\hat{Y}^*, Y^*)$.

Suppose that $\hat{Y}$ is a recessive solution of (SDS) at $\infty$ and (6.424) holds. Then $\hat{T}^* = 0$, and by (6.434), (6.426), and (6.424), we get

$$\text{rank } T^* \overset{(6.434)}{=} \text{rank } \hat{N}^* \overset{(6.426)}{=} \text{rank } P_{\hat{S}\infty} \hat{N} P_{S\infty} \overset{(6.424)}{=} n - d_\infty. \tag{6.435}$$

Therefore, $Y$ is a dominant solution of (SDS) at $\infty$ by Definition 6.122. This means that $\text{Im } T^* = \text{Im } P_{S\infty}$ since $S_k^* = S_k$ on $[N, \infty)_{\mathbb{Z}}$, $\text{Im } T^* \subseteq \text{Im } P_{S^*\infty} = \text{Im } P_{S\infty}$ by (6.239), and rank $T^* = n - d_\infty$ by (6.435). From (6.434) and $\hat{T}^* = 0$, we know that $T^* = (\hat{M}^*)^T \hat{N}^* = (\hat{N}^*)^T \hat{M}^*$. Multiplying this equality by $(T^*)^\dagger$ from the left and by $(\hat{M}^*)^{-1}$ from the right and using the identity $(T^*)^\dagger T^* = P_{S\infty}$ yield

$$P_{S\infty} (\hat{M}^*)^{-1} = (T^*)^\dagger (\hat{N}^*)^T. \tag{6.436}$$

Furthermore, by (6.433) and Theorem 6.124, we obtain

$$\lim_{k\to\infty} X_k^\dagger \hat{X}_k = \lim_{k\to\infty} (P\hat{M}^{-1} - S_k \hat{N}^T) = V := P\hat{M}^{-1} - (T^*)^\dagger \hat{N}^T. \tag{6.437}$$

We will show that $\text{Im } V^T = \text{Im}(\hat{P} - P_{\hat{S}\infty}) = \text{Im } \hat{P} \cap \text{Ker } P_{\hat{S}\infty}$. By using (6.437) and the identities $P\hat{M}^{-1}\hat{P} = P\hat{M}^{-1}$, $\hat{N}^T \hat{P} = \hat{N}^T$, we get $V\hat{P} = V$, which yields that $\text{Im } V^T \subseteq \text{Im } \hat{P}$. Moreover, the equality $(T^*)^\dagger P_{S\infty} = (T^*)^\dagger$ and formulas (6.425), (6.426), and (6.436) imply that

$$V P_{\hat{S}\infty} \overset{(6.437)}{=} P\hat{M}^{-1} P_{\hat{S}\infty} - (T^*)^\dagger P_{S\infty} \hat{N}^T P_{\hat{S}\infty}$$

$$= P_{S\infty} (\hat{M}^*)^{-1} - (T^*)^\dagger (\hat{N}^*)^T \overset{(6.436)}{=} 0.$$

Thus, $\text{Im } V^T \subseteq \text{Ker } P_{\hat{S}\infty}$. Hence, we proved that $\text{Im } V^T \subseteq \text{Im } \hat{P} \cap \text{Ker } P_{\hat{S}\infty}$. Now we show the opposite inclusion $\text{Im } \hat{P} \cap \text{Ker } P_{\hat{S}\infty} \subseteq \text{Im } V^T$, or equivalently the inclusion $\text{Ker } V \subseteq \text{Ker } \hat{P} \oplus \text{Im } P_{\hat{S}\infty}$. Let $v \in \text{Ker } V$. Then $v$ can be uniquely decomposed as $v = v_1 + v_2$ with $v_1 \in \text{Ker } \hat{P}$ and $v_2 \in \text{Im } \hat{P}$. The identity $V\hat{P} = V$

then implies that $V v_2 = V \hat{P} v = V v = 0$, so that by the definition of $V$ in (6.437) we have $[P \hat{M}^{-1} - (T^*)^\dagger \hat{N}^T] v_2 = V v_2 = 0$. Hence, $P \hat{M}^{-1} v_2 = (T^*)^\dagger \hat{N}^T v_2$. The vector $w := P \hat{M}^{-1} v_2$ then satisfies $w \in \mathrm{Im}\,(T^*)^\dagger = \mathrm{Im}\, P_{S\infty}$. By the equalities $\hat{P} M P \hat{M}^{-1} = \hat{P}$, $\hat{P} v_2 = v_2$, $P_{S\infty} w = w$, and the first formula in (6.425), we get

$$v_2 = \hat{P} \hat{M} P \hat{M}^{-1} v_2 = \hat{P} \hat{M} w = \hat{P} \hat{M} P_{S\infty} w = P_{\hat{S}\infty} \hat{M}^* w, \quad \text{i.e.,} \quad v_2 \in \mathrm{Im}\, P_{\hat{S}\infty}.$$

This shows that $v = v_1 + v_2 \in \mathrm{Ker}\, \hat{P} \oplus \mathrm{Im}\, P_{\hat{S}\infty}$, which completes the proof of this direction.

Conversely, assume that (6.431) is satisfied. Denote by $V_0 := P \hat{M}^{-1} - V$, where $V$ is given in (6.431). Then by (6.433) we get $S_k \hat{N}^T \to V_0$ as $k \to \infty$. The equality $S_k = S_k P_{S\infty}$ implies the inclusion $\mathrm{Ker}\, P_{S\infty} \hat{N}^T \subseteq \mathrm{Ker}\, V_0$, and similarly $S_k = P_{S\infty} S_k$ implies that $\mathrm{Im}\, V_0 \subseteq \mathrm{Im}\, P_{S\infty}$. In particular, $\mathrm{rank}\, V_0 \leq \mathrm{rank}\, P_{S\infty} \hat{N}^T$. Moreover, by using the identities $P \hat{M}^{-1} P_{\hat{S}\infty} = P_{S\infty} (\hat{M}^*)^{-1}$ and $V P_{\hat{S}\infty} = 0$, we get $V_0 P_{\hat{S}\infty} = P_{S\infty} (\hat{M}^*)^{-1}$, which implies that $\mathrm{Im}\, P_{S\infty} \subseteq \mathrm{Im}\, V_0$. Hence, we proved the equality $\mathrm{Im}\, V_0 = \mathrm{Im}\, P_{S\infty}$, so that $\mathrm{rank}\, V_0 = \mathrm{rank}\, P_{S\infty} = n - d_\infty$. In turn, the inequality $n - d_\infty = \mathrm{rank}\, V_0 \leq \mathrm{rank}\, P_{S\infty} \hat{N}^T$ holds. On the other hand, we always have $\mathrm{rank}\, P_{S\infty} \hat{N}^T \leq \mathrm{rank}\, P_{S\infty} = n - d_\infty$. Thus, we conclude that $\mathrm{rank}\, P_{S\infty} \hat{N}^T = n - d_\infty$. The definition of $T^*$ in (6.237) now yields

$$P_{S\infty} \hat{N}^T = \lim_{k\to\infty} S_k^\dagger S_k \hat{N}^T = \lim_{k\to\infty} (S_k^*)^\dagger (S_k \hat{N}^T) = T^* V_0. \tag{6.438}$$

From (6.438) we then obtain the inequality

$$n - d_\infty = \mathrm{rank}\, P_{S\infty} \hat{N}^T = \mathrm{rank}\, T^* V_0 \leq \mathrm{rank}\, T^*.$$

This yields by the third condition in (6.394) that $\mathrm{rank}\, T^* = n - d_\infty$. Therefore, $Y$ is a dominant solution of (SDS) at $\infty$, by Definition 6.122. Moreover, by using (6.426), (6.438), the symmetry of $\hat{N}^* (\hat{M}^*)^{-1}$, and the equalities $V_0 P_{\hat{S}\infty} = P_{S\infty} (\hat{M}^*)^{-1}$ and $T^* P_{S\infty} = T^*$, we get

$$\hat{N}^* (\hat{M}^*)^{-1} \overset{(6.426)}{=} (\hat{M}^*)^{T-1} P_{S\infty} \hat{N}^T P_{\hat{S}\infty} \overset{(6.438)}{=} (\hat{M}^*)^{T-1} T^* V_0 P_{\hat{S}\infty}$$

$$= (\hat{M}^*)^{T-1} T^* P_{S\infty} (\hat{M}^*)^{-1} = (\hat{M}^*)^{T-1} T^* (\hat{M}^*)^{-1}.$$

This implies that $T^* = (\hat{M}^*)^T \hat{N}^*$. From (6.434) we then obtain the equality $(\hat{M}^*)^T \hat{T}^* \hat{M}^* = 0$, i.e., $\hat{T}^* = 0$, as the matrix $\hat{M}^*$ is invertible. Therefore, $\hat{Y}$ is a recessive solution of (SDS) at $\infty$. Finally, Theorem 6.141 yields that (6.424) holds, which completes the proof. $\square$

Motivated by Theorem 2.67, we can now determine when is the limit $V$ in (6.431) equal to zero. It follows from condition (6.431) that this happens if and only if $\hat{P} = P_{\hat{S}\infty}$, i.e., if and only if Theorem 6.146 is applied to the minimal genus $\mathcal{G}_{\min}$.

We note that if the system (SDS) is eventually controllable, then Corollary 6.147 below reduces to Theorem 2.67.

**Corollary 6.147** *Assume that system* (SDS) *is nonoscillatory at* $\infty$. *Let* $\hat{Y}$ *and* $Y$ *be two conjoined bases of* (SDS) *from the minimal genus* $\mathcal{G}_{\min}$. *Then* $\hat{Y}$ *is a minimal recessive solution of* (SDS) *at* $\infty$ *and* rank $w(\hat{Y}, Y) = n - d_\infty$ *if and only if*

$$\lim_{k\to\infty} X_k^\dagger \hat{X}_k = 0. \tag{6.439}$$

*In this case* $Y$ *is a minimal dominant solution of* (SDS) *at* $\infty$.

*Proof* Let $\hat{Y}$ and $Y$ be as in the corollary, and let $K \in [\hat{K}_{\min}, \infty)_\mathbb{Z}$ be such that $\hat{Y}$ and $Y$ have constant kernel on $[N, \infty)_\mathbb{Z}$ and no forward focal points in $(N, \infty)$. Then $\hat{Y}$ and $Y$ are minimal conjoined bases on $[N, \infty)_\mathbb{Z}$. Moreover, let $\hat{P}$, $P$, $P_{\hat{S}\infty}$, and $P_{S\infty}$ be the corresponding orthogonal projectors in (6.234), (6.238), and Remark 6.132. Then $\hat{P} = P_{\hat{S}\infty}$, $P = P_{S\infty}$, and $P_{\hat{S}\infty} \hat{N} P_{S\infty} = \hat{N}$, where $\hat{N} := w(\hat{Y}, Y)$. The statement now follows from Theorem 6.146 (with $\mathcal{G} := \mathcal{G}_{\min}$). $\qquad\square$

Similarly, the limit in (6.431) involves an invertible $X_k$ when Theorem 6.146 is applied to the maximal genus $\mathcal{G}_{\max}$.

**Corollary 6.148** *Assume that system* (SDS) *is nonoscillatory at* $\infty$ *with* $\hat{K}_{\min}$ *defined in Remark 6.130. Let* $\hat{Y}$ *and* $Y$ *be two maximal conjoined bases of* (SDS), *and let* $N \in [\hat{K}_{\min}, \infty)_\mathbb{Z}$ *be such that* $\hat{Y}$ *and* $Y$ *have invertible* $\hat{X}_k$ *and* $X_k$ *on* $[N, \infty)_\mathbb{Z}$ *and no forward focal points in* $(N, \infty)$. *Let* $P_{\hat{S}\infty}$ *and* $P_{S\infty}$ *be their associated projectors in* (6.238) *and Remark 6.132. Then* $\hat{Y}$ *is a maximal recessive solution of* (SDS) *at* $\infty$ *and* rank $P_{\hat{S}\infty} w(\hat{Y}, Y) P_{S\infty} = n - d_\infty$ *if and only if*

$$\lim_{k\to\infty} X_k^{-1} \hat{X}_k = V \quad with \quad \operatorname{Im} V^T = \operatorname{Ker} P_{\hat{S}\infty}.$$

*In this case* $Y$ *is a maximal dominant solution of* (SDS) *at* $\infty$.

*Proof* The statement follows directly from Theorem 6.146 by using the fact that the orthogonal projector $\hat{P}$ in (6.234) associated with the maximal conjoined basis $\hat{Y}$ satisfies $\hat{P} = I$. $\qquad\square$

In the last result of this subsection, we present an extension of Theorem 6.146. More precisely, we show that the limit in (6.431) always exists when $Y$ is a dominant solution of (SDS) at $\infty$. In this case we also obtain an additional information about the structure of the space $\operatorname{Im} V^T$ in (6.431).

**Theorem 6.149** *Assume that system* (SDS) *is nonoscillatory at* $\infty$ *with* $\hat{K}_{\min}$ *defined in Remark 6.130. Let* $\tilde{Y}$ *and* $Y$ *be two conjoined bases from a given genus* $\mathcal{G}$, *such that* $Y$ *is a dominant solution of* (SDS) *at* $\infty$ *and such that* $\tilde{Y}$ *and* $Y$ *have constant kernel on* $[N, \infty)_\mathbb{Z}$ *and no forward focal points in* $(N, \infty)$ *for some index* $N \in [\hat{K}_{\min}, \infty)_\mathbb{Z}$. *Let* $\tilde{P}$, $P_{\tilde{S}\infty}$, $\tilde{T}$ *be the matrices in* (6.234), (6.238), (6.237) *defined* *through $\tilde{X}_k$. Then the limit of $X_k^\dagger \tilde{X}_k$ as $k \to \infty$ exists and satisfies*

$$\lim_{k\to\infty} X_k^\dagger \tilde{X}_k = V \quad \text{with} \quad \text{Im } V^T = \text{Im } \tilde{T} \oplus \text{Im} (\tilde{P} - P_{\tilde{S}\infty}). \tag{6.440}$$

*Proof* We proceed similarly as in the proof of Theorem 6.146 (with $\hat{Y} := \tilde{Y}$), since some of those arguments did not depend on the fact that $\hat{Y}$ was a recessive solution of (SDS) at $\infty$. Let $\tilde{N} := w(\tilde{Y}, Y)$. Then, as in (6.432) and (6.433),

$$X_k = \tilde{X}_k \,(\tilde{P}\tilde{M} + \tilde{S}_k \tilde{N}), \quad \tilde{X}_k = X_k \,(P\tilde{M}^{-1} - S_k \tilde{N}^T), \quad X_k^\dagger \tilde{X}_k = P\tilde{M}^{-1} - S_k \tilde{N}^T$$

on $[N, \infty)_{\mathbb{Z}}$, where $\tilde{M}$ is invertible. Let $\tilde{T}^*$ and $T^*$ be the matrices in (6.237) defined through the minimal conjoined bases $\tilde{Y}^*$ and $Y^*$, which are contained in $\tilde{Y}$ and $Y$ on $[N, \infty)_{\mathbb{Z}}$, respectively. Then by (6.348) we have

$$T^* = (\tilde{M}^*)^T \,\tilde{T}^* \tilde{M}^* + (\tilde{M}^*)^T \,\tilde{N}^*, \tag{6.441}$$

as in (6.434). Since $Y$ is a dominant solution of (SDS) at $\infty$, we get by (6.409) the equality $\text{Im } T^* = \text{Im } T = \text{Im } P_{S\infty}$ with $T$ given in (6.237). Moreover, as in (6.437),

$$\lim_{k\to\infty} X_k^\dagger \tilde{X}_k = \lim_{k\to\infty} (P\tilde{M}^{-1} - S_k \tilde{N}^T) = V := P\tilde{M}^{-1} - (T^*)^\dagger \tilde{N}^T \tag{6.442}$$

with $V\tilde{P} = V$ and $\text{Ker } V \subseteq \text{Ker } \tilde{P} \oplus \text{Im } P_{\tilde{S}\infty}$. This shows that every vector $v \in \text{Ker } V$ can be uniquely decomposed as $v = v_1 + v_2$, where $v_1 \in \text{Ker } \tilde{P}$ and $v_2 \in \text{Im } P_{\tilde{S}\infty}$. Then for the vector $w := P\tilde{M}^{-1} v_2$, we have the equality $w = P\tilde{M}^{-1} P_{\tilde{S}\infty} v_2 = P_{S\infty} (\tilde{M}^*)^{-1} v_2$ by (6.349). Therefore, $w \in \text{Im } P_{S\infty}$, and $w = (T^*)^\dagger \tilde{N}^T v_2$ by (6.442). Moreover, since $(P_{\tilde{S}\infty} \tilde{M}^*)^\dagger = P_{S\infty} (\tilde{M}^*)^{-1}$ holds by (6.274), we then obtain

$$v_2 = P_{\tilde{S}\infty} v_2 = (P_{\tilde{S}\infty} \tilde{M}^*) \,(P_{\tilde{S}\infty} \tilde{M}^*)^\dagger v_2$$
$$= P_{\tilde{S}\infty} \tilde{M}^* P_{S\infty} (\tilde{M}^*)^{-1} v_2 = P_{\tilde{S}\infty} \tilde{M}^* w. \tag{6.443}$$

With the identities $(T^*)^\dagger P_{S\infty} = (T^*)^\dagger$ and $P_{S\infty} \tilde{N}^T P_{\tilde{S}\infty} = (\tilde{N}^*)^T$, we then obtain

$$w = (T^*)^\dagger \tilde{N}^T v_2 \overset{(6.443)}{=} (T^*)^\dagger P_{S\infty} \tilde{N}^T P_{\tilde{S}\infty} \tilde{M}^* w = (T^*)^\dagger (\tilde{N}^*)^T \tilde{M}^* w. \tag{6.444}$$

This expression for the vector $w$ will be utilized in the subsequent proof. Next we will derive the formula

$$\text{Ker } V = \left( \text{Ker } \tilde{T}^* \cap \text{Im } P_{\tilde{S}\infty} \right) \oplus \text{Ker } \tilde{P}. \tag{6.445}$$

Let $v \in \operatorname{Ker} V$ and let $v_1$, $v_2$, and $w$ be its associated vectors defined above. If we multiply (6.444) by $T^*$ from the left and use the identities $T^*(T^*)^\dagger = P_{S_\infty}$ and $P_{S_\infty}(\tilde{N}^*)^T = (\tilde{N}^*)^T$, then we get $T^*w = (\tilde{N}^*)^T \tilde{M}^*w = (\tilde{M}^*)^T \tilde{N}^*w$. By using (6.441) in the last equality, it follows that $(\tilde{M}^*)^T \tilde{T}^* \tilde{M}^*w = 0$. The invertibility of $\tilde{M}^*$ and the equality $\tilde{T}^* = \tilde{T}^* P_{\tilde{S}_\infty}$ then imply that $\tilde{T}^* P_{\tilde{S}_\infty} \tilde{M}^*w = 0$. Therefore, the vector $v_2 = P_{\tilde{S}_\infty} \tilde{M}^*w$ satisfies $v_2 \in \operatorname{Ker} \tilde{T}^* \cap \operatorname{Im} P_{\tilde{S}_\infty}$. Hence, the inclusion $\subseteq$ in (6.445) holds. Conversely, assume that $v \in \left(\operatorname{Ker} \tilde{T}^* \cap \operatorname{Im} P_{\tilde{S}_\infty}\right) \oplus \operatorname{Ker} \tilde{P}$. Then we can write $v = v_1 + v_2$ with $v_1 \in \operatorname{Ker} \tilde{T}^* \cap \operatorname{Im} P_{\tilde{S}_\infty}$ and $v_2 \in \operatorname{Ker} \tilde{P}$. Since $V\tilde{P} = V$, it follows from (6.442) that $Vv = Vv_1 = [P\tilde{M}^{-1} - (T^*)^\dagger \tilde{N}^T]v_1$. In turn, applying the identities $v_1 = P_{\tilde{S}_\infty} v_1$, $P\tilde{M}^{-1} P_{\tilde{S}_\infty} = P_{S_\infty}(\tilde{M}^*)^{-1}$ from (6.349), $(T^*)^\dagger P_{S_\infty} = (T^*)^\dagger$, $P_{S_\infty} \tilde{N}^T P_{\tilde{S}_\infty} = (\tilde{N}^*)^T$, and $(T^*)^\dagger T^* = P_{S_\infty}$ then yields

$$Vv = [P\tilde{M}^{-1} - (T^*)^\dagger \tilde{N}^T] P_{\tilde{S}_\infty} v_1 = [P_{S_\infty}(\tilde{M}^*)^{-1} - (T^*)^\dagger (\tilde{N}^*)^T]v_1$$

$$= (T^*)^\dagger [T^*(\tilde{M}^*)^{-1} - (\tilde{N}^*)^T]v_1 = (T^*)^\dagger (\tilde{M}^*)^T \tilde{T}^*v_1, \qquad (6.446)$$

because $T^*(\tilde{M}^*)^{-1} - (\tilde{N}^*)^T = (\tilde{M}^*)^T \tilde{T}^*$ by (6.441), the invertibility of $\tilde{M}^*$, and the symmetry of $(\tilde{M}^*)^T \tilde{N}^*$. Since $v_1 \in \operatorname{Ker} \tilde{T}^*$, formula (6.446) yields that $Vv = 0$, i.e., $v \in \operatorname{Ker} V$. Thus, the inclusion $\supseteq$ in (6.445) is satisfied as well. According to Proposition 6.86, we have $\tilde{T} = \tilde{T}^*$, and hence, the matrix $\tilde{T}^*$ in (6.445) can be replaced by $\tilde{T}$. Finally, by (6.239) we have the inclusions $\operatorname{Im} \tilde{T} \subseteq \operatorname{Im} \tilde{P}$ and $\operatorname{Im} \tilde{T} \cap \operatorname{Ker} P_{\tilde{S}_\infty} \subseteq \operatorname{Im} P_{\tilde{S}_\infty} \cap \operatorname{Ker} P_{\tilde{S}_\infty} = \{0\}$, which implies that

$$\operatorname{Im} V^T = (\operatorname{Ker} V)^\perp = \left(\operatorname{Im} \tilde{T} \oplus \operatorname{Ker} P_{\tilde{S}_\infty}\right) \cap \operatorname{Im} \tilde{P} = \operatorname{Im} \tilde{T} \oplus \left(\operatorname{Ker} P_{\tilde{S}_\infty} \cap \operatorname{Im} \tilde{P}\right).$$

This shows the second condition in (6.440) and the proof is complete. $\qquad\square$

The rank of the limiting matrix $V$ in (6.440) can be connected with the notions of rank and defect of the genus $\mathcal{G}$, which we introduce in the following remark.

*Remark 6.150* Let (SDS) be a nonoscillatory symplectic system at $\infty$. For every genus $\mathcal{G}$, we introduce its rank and defect as follows. The number $\operatorname{rank} \mathcal{G}$ is defined as the rank of $\tilde{Y}$, where $\tilde{Y}$ is any conjoined basis from $\mathcal{G}$. This quantity is well defined, since any two conjoined bases from $\mathcal{G}$ have eventually the same image of their first components. Then $n - d_\infty \leq \operatorname{rank} \mathcal{G} \leq n$. Also, we define the quantity $\operatorname{def} \mathcal{G} := n - \operatorname{rank} \mathcal{G}$, for which we have the inequalities $0 \leq \operatorname{def} \mathcal{G} \leq d_\infty$. From Theorem 6.149 it then follows that the matrix $V$ in (6.440) satisfies

$$\operatorname{rank} V = \operatorname{rank} \tilde{T} + d_\infty - \operatorname{def} \mathcal{G}, \qquad (6.447)$$

since in this case $\operatorname{rank} V = \operatorname{rank} \tilde{T} + \operatorname{rank} \tilde{P} - \operatorname{rank} P_{\tilde{S}_\infty}$, while $\operatorname{rank} \tilde{P} = \operatorname{rank} \mathcal{G}$ and $\operatorname{rank} P_{\tilde{S}_\infty} = n - d_\infty$. Formula (6.447) shows that the actual value of the rank of $V$ depends primarily on the rank of $\tilde{T}$, because the quantities $d_\infty$ and $\operatorname{def} \mathcal{G}$ are fixed within one system (SDS) and its genus $\mathcal{G}$. In particular, the rank of $V$ is minimal (equal to $d_\infty - \operatorname{def} \mathcal{G}$) if and only if the conjoined basis $\tilde{Y}$ is a recessive solution of (SDS) at $\infty$. This property is well-known in the controllable case, for which $d_\infty = 0 = \operatorname{def} \mathcal{G}$ and hence, rank $V = \operatorname{rank} \tilde{T} = 0$ as in Theorem 2.67.

## 6.3.10   Reid's Construction of Minimal Recessive Solution

In this subsection we present the so-called Reid construction of the (minimal) recessive solution of (SDS) at $\infty$ by means of a pointwise limit of certain specifically chosen solutions of (SDS). First we derive a representation of the minimal recessive solution of (SDS) at $\infty$ in terms of any minimal conjoined basis $Y$ of (SDS). This result is based on the properties of the associated conjoined basis $\bar{Y}$ from Proposition 6.67 and Theorem 6.69.

**Proposition 6.151** *Assume that system* (SDS) *is nonoscillatory at* $\infty$. *Suppose that* $N \in [0, \infty)_{\mathbb{Z}}$ *is an index such that* $d[N, \infty)_{\mathbb{Z}} = d_\infty$ *and there exists a conjoined basis of* (SDS) *with constant kernel on* $[N, \infty)_{\mathbb{Z}}$ *and no forward focal points in* $(N, \infty)$. *Then a solution* $\hat{Y}$ *of* (SDS) *is a minimal recessive solution at* $\infty$ *with respect to the interval* $[N, \infty)_{\mathbb{Z}}$ *if and only if*

$$\hat{Y}_k = Y_k - \bar{Y}_k T, \quad k \in [N, \infty)_{\mathbb{Z}}, \tag{6.448}$$

*for some minimal conjoined basis* $Y$ *of* (SDS) *on* $[N, \infty)_{\mathbb{Z}}$. *Here* $\bar{Y}$ *is the conjoined basis of* (SDS) *from Proposition 6.67 associated with* $Y$, *and the matrix* $T$ *is defined in* (6.237).

*Proof* Let $N$ be as in the proposition. If $\hat{Y}$ is a minimal recessive solution at $\infty$ with respect to $[N, \infty)_{\mathbb{Z}}$, then it is a minimal conjoined basis on $[N, \infty)_{\mathbb{Z}}$, and the associated matrix $\hat{T}$ in (6.237) satisfies $\hat{T} = 0$. Formula (6.448) then holds trivially with $Y := \hat{Y}$. The opposite implication follows from the proof of Theorem 6.113, where it is shown that $Y - \bar{Y}T$ is a minimal recessive solution of (SDS) at $\infty$.   □

*Remark 6.152* In the proof of Theorem 6.113, we showed that the minimal recessive solution $\hat{Y}$ of (SDS) at $\infty$ in Proposition 6.151 has constant kernel on $[N, \infty)_{\mathbb{Z}}$ and no forward focal points in $(N, \infty)$. The uniqueness of the minimal recessive solution at $\infty$ in Theorem 6.113 and the definition of the index $\hat{K}_{\min}$ in Remark 6.130 then imply that the index $N$ appearing in Proposition 6.151 satisfies $N \in [\hat{K}_{\min}, \infty)_{\mathbb{Z}}$.

Next we state and prove the main result of this subsection. We recall that by $\hat{Y}^{\min}$ we denote the minimal recessive solution of (SDS) at $\infty$.

**Theorem 6.153** *Let* $Y$ *be a minimal conjoined basis of* (SDS) *on* $[N, \infty)_{\mathbb{Z}}$ *with* $d[N, \infty)_{\mathbb{Z}} = d_\infty$, *and let* $\bar{Y}$ *be the associated conjoined basis of* (SDS) *from Proposition 6.67. Then there exists an index* $L \in [N, \infty)_{\mathbb{Z}}$ *such that* $\bar{X}_k$ *is invertible* *for all $k \in [L, \infty)_{\mathbb{Z}}$ and the solutions $Y^{(j)}$ of (SDS) given by the initial conditions*

$$X_j^{(j)} = 0, \quad U_j^{(j)} = -\bar{X}_j^{T-1}, \quad j \in [L, \infty)_{\mathbb{Z}}, \tag{6.449}$$

*are conjoined bases satisfying*

$$\hat{Y}_k^{\min} = \lim_{j \to \infty} Y_k^{(j)}, \quad k \in [0, \infty)_{\mathbb{Z}}. \tag{6.450}$$

*Proof* Let $Y$ and $\bar{Y}$ be as in the theorem, and let $P$, $S_k$, $P_{Sk}$, and $P_{S\infty}$ be the matrices in (6.234), (6.236), and (6.238), which correspond to $Y$. Since $Y$ is a minimal conjoined basis on $[N, \infty)_{\mathbb{Z}}$, we have the equality $P = P_{S\infty}$. Moreover, with

$$L := \min\{k \in [N, \infty)_{\mathbb{Z}}, \text{ rank } S_k = n - d[N, \infty)_{\mathbb{Z}} = n - d_\infty\} \tag{6.451}$$

it follows that $P_{Sk} = P_{S\infty}$ for every $k \in [L, \infty)_{\mathbb{Z}}$. The equality $\text{Ker}\, \bar{X}_k = \text{Im}\,(P - P_{Sk})$ on $[N, \infty)_{\mathbb{Z}}$ in Proposition 6.67(vi) then implies that $\text{Ker}\, \bar{X}_k = \{0\}$ for all $k \in [L, \infty)_{\mathbb{Z}}$, i.e., the matrix $\bar{X}_k$ is invertible on $[L, \infty)_{\mathbb{Z}}$. Consequently, Proposition 6.67(viii) then reads as

$$S_k^\dagger = \bar{X}_k^{-1} X_k P_{S\infty} = \bar{X}_k^{-1} X_k, \quad k \in [L, \infty)_{\mathbb{Z}}. \tag{6.452}$$

Fix now an index $j \in [L, \infty)_{\mathbb{Z}}$, and let $Y^{(j)}$ be the solution of (SDS) given by the initial conditions (6.449). It is easy to check that $Y^{(j)}$ is a conjoined basis of (SDS). Moreover, consider the constant matrices $M^{(j)}$ and $N^{(j)}$ such that

$$Y_k^{(j)} = Y_k M^{(j)} + \bar{Y}_k N^{(j)}, \quad k \in [0, \infty)_{\mathbb{Z}}. \tag{6.453}$$

Since the conjoined bases $Y$ and $\bar{Y}$ are normalized by Proposition 6.67(i), the fundamental matrix $(Y_k, \bar{Y}_k)$ in (6.453) is symplectic with the formula for its inverse $(Y_k, \bar{Y}_k)^{-1} = -\mathcal{J}\,(Y_k, \bar{Y}_k)^T \mathcal{J}$, by Lemma 1.58(ii). Thus, $M^{(j)} = -w(\bar{Y}, Y^{(j)})$ and $N^{(j)} = w(Y, Y^{(j)})$, i.e.,

$$M^{(j)} = \bar{U}_k^T X_k^{(j)} - \bar{X}_k^T U_k^{(j)}, \quad N^{(j)} = X_k^T U_k^{(j)} - U_k^T X_k^{(j)}, \quad k \in [0, \infty)_{\mathbb{Z}}. \tag{6.454}$$

In particular, by evaluating (6.454) at $k = j$ and using (6.449), we obtain that

$$M^{(j)} = I, \quad N^{(j)} = -X_j^T \bar{X}_j^{T-1} \overset{(6.452)}{=} -(S_j^\dagger)^T = -S_j^\dagger. \tag{6.455}$$

It follows from (6.455) that $M^{(j)} \to I$ and $N^{(j)} \to -T$ as $j \to \infty$, where $T$ is the matrix in (6.237) associated with $Y$. Finally, the representation in (6.453) and Proposition 6.151 then imply that for each $k \in [0, \infty)_{\mathbb{Z}}$ the limit of $Y_k^{(j)}$ as $j \to \infty$ exists and it is equal to $\hat{Y}_k^{\min}$. Therefore, formula (6.450) holds and the proof is complete. $\qquad\qquad\square$

*Remark 6.154* The initial conditions in (6.449) mean that for each $j \in [N, \infty)_{\mathbb{Z}}$, the solution $Y^{(j)}$ is a conjoined basis, which is a constant nonsingular multiple of the principal solution $Y^{[j]}$, namely, $Y_k^{(j)} = Y_k^{[j]} M_j$ for all $k \in [0, \infty)_{\mathbb{Z}}$, where $M_j := -\bar{X}_j^{T-1}$. Then it follows from Remark 6.152 and Theorem 6.143 that for every index $j \in [L, \infty)_{\mathbb{Z}}$, the conjoined basis $Y_k^{(j)}$ used in the Reid construction of $\hat{Y}^{\min}$ in Theorem 6.153 is a (minimal) dominant solution of (SDS) at $\infty$.

In Theorem 6.143 we presented examples of minimal dominant solutions of (SDS) at $\infty$. Based on formula (6.452), we can now prove that the conjoined basis $\bar{Y}$ considered in Theorem 6.153 is an example of a maximal dominant solution of (SDS) at $\infty$.

**Proposition 6.155** *Let $Y$ be a minimal conjoined basis of* (SDS) *on an interval $[N, \infty)_{\mathbb{Z}}$ satisfying $d[N, \infty)_{\mathbb{Z}} = d_\infty$. Then the associated conjoined basis $\bar{Y}$ from Proposition 6.67 is a maximal dominant solution of* (SDS) *at $\infty$.*

*Proof* Let $R_k$, $S_k$, $T$, $P_{S\infty}$ be the matrices in (6.233), (6.236), (6.237), (6.238) corresponding to $Y$. From the proof of Theorem 6.153, we know that the matrix $\bar{X}_k$ is invertible on the interval $[L, \infty)_{\mathbb{Z}}$, where $L$ is defined in (6.451). Let $\bar{S}_k^{(L)}$ be the matrix in (6.236) defined through $\bar{X}_k$ on $[L, \infty)_{\mathbb{Z}}$, i.e.,

$$\bar{S}_k^{(L)} := \sum_{j=L}^{k-1} \bar{X}_{j+1}^{-1} \mathcal{B}_j \bar{X}_j^{T-1}, \quad k \in [L, \infty)_{\mathbb{Z}}. \tag{6.456}$$

We show that $\bar{S}_k^{(L)}$ has a limit as $k \to \infty$. First we note that equality (6.452) yields $S_k^\dagger X_k^\dagger = \bar{X}_k^{-1} X_k X_k^\dagger = \bar{X}_k^{-1} R_k$ on $[L, \infty)_{\mathbb{Z}}$. By using (6.456), Theorem 6.66(ii), and the identities $\Delta S_k = X_{k+1}^\dagger \mathcal{B}_k X_k^{\dagger T}$ and $S_k S_k^\dagger = S_k^\dagger S_k = P_{S\infty}$ on $[L, \infty)_{\mathbb{Z}}$, we get for every $k \in [L, \infty)_{\mathbb{Z}}$

$$\bar{S}_k^{(L)} = \sum_{j=L}^{k-1} \bar{X}_{j+1}^{-1} R_{j+1} \mathcal{B}_j R_j \bar{X}_j^{T-1} = \sum_{j=L}^{k-1} S_{j+1}^\dagger X_{j+1}^\dagger \mathcal{B}_j X_j^{\dagger T} S_j^\dagger$$

$$= \sum_{j=L}^{k-1} S_{j+1}^\dagger (\Delta S_j) S_j^\dagger = \sum_{j=L}^{k-1} (S_{j+1}^\dagger S_{j+1} S_j^\dagger - S_{j+1}^\dagger S_j S_j^\dagger)$$

$$= \sum_{j=L}^{k-1} (P_{S\infty} S_j^\dagger - S_{j+1}^\dagger P_{S\infty}) = -\sum_{j=L}^{k-1} (\Delta S_j^\dagger) = S_L^\dagger - S_k^\dagger.$$

Therefore, the limit of $\bar{S}_k^{(L)}$ as $k \to \infty$ exists and is equal to $S_L^\dagger - T$. By Theorem 6.124 we conclude that $\bar{Y}$ is a (maximal) dominant solution of (SDS) at $\infty$. $\qquad\square$

*Remark 6.156* Assume, as in Proposition 6.155, that $[N, \infty)_{\mathbb{Z}}$ is an interval such that there exists a minimal conjoined basis of (SDS) on $[N, \infty)_{\mathbb{Z}}$. From Theorem 6.75 it then follows that the index $L$ in (6.451) also satisfies

$$L = \min\{k \in [N, \infty)_{\mathbb{Z}}, \ Y^{[N]} \text{ has constant kernel on } [k, \infty)_{\mathbb{Z}}\}, \qquad (6.457)$$

where $Y^{[N]}$ is the principal solution of (SDS) at the index $N$.

Next we shall comment the existence and uniqueness of the limit in (6.450).

*Remark 6.157* The proof of Theorem 6.153 solves also the problem, when the limit in (6.450) exists and what is its value depending on the chosen initial conditions in (6.449). More precisely, let $Y$ be a minimal conjoined basis of (SDS) on $[N, \infty)_{\mathbb{Z}}$ with $d[N, \infty)_{\mathbb{Z}} = d_\infty$, and let $\bar{Y}$ be the associated conjoined basis from Proposition 6.67. Assume that $Y^{(j)}$ is the solution of (SDS) given by the initial conditions $X_j^{(j)} = 0$ and $U_j^{(j)} = W_j$ with an invertible matrix $W_j$ for all $j \in [\tilde{L}, \infty)_{\mathbb{Z}}$ for some $\tilde{L} \in [L, \infty)_{\mathbb{Z}}$, where $L$ is defined in (6.457). Then for $k \in [0, \infty)_{\mathbb{Z}}$, the limit of $Y_k^{(j)}$ as $j \to \infty$ exists if and only if $W_j = -\bar{X}_j^{T-1} E_j$ for $j \in [L, \infty)_{\mathbb{Z}}$, where $E_j$ are invertible matrices such that the limit $E := \lim_{j \to \infty} E_j$ exists. In this case

$$\lim_{j \to \infty} Y_k^{(j)} = \hat{Y}_k^{\min} E, \quad k \in [0, \infty)_{\mathbb{Z}},$$

where $\hat{Y}^{\min}$ is the minimal recessive solution of (SDS) at $\infty$ from Remark 6.130. This statement follows from the proof of Theorem 6.153, in which the matrices $M^{(j)}$ and $N^{(j)}$ in (6.454) are given by $M^{(j)} = E_j$ and $N^{(j)} = -S_j^\dagger E_j$.

In view of Proposition 6.155, we can deduce that the choice of initial conditions (6.449) with $\bar{Y}$ being a maximal dominant solution of (SDS) at $\infty$ is natural and the only possible in order to guarantee the existence of the limit in (6.450).

*Remark 6.158* Let $N$, $L$, $Y$, and $\bar{Y}$ be as in Remark 6.157. Let $\tilde{Y}$ be a conjoined basis of (SDS) belonging to the maximal genus $\mathcal{G}_{\max}$, i.e., the matrix $\tilde{X}_k$ is invertible for every $k \in [\tilde{L}, \infty)_{\mathbb{Z}}$ for some $\tilde{L} \in [L, \infty)_{\mathbb{Z}}$. Without loss of generality, assume that $\bar{Y}$ and $\tilde{Y}$ have no forward focal points in $(\tilde{L}, \infty)$. For $j \in [\tilde{L}, \infty)_{\mathbb{Z}}$, let $Y^{(j)}$ be the solution of (SDS) given by the initial conditions $X_j^{(j)} = 0$ and $U_j^{(j)} = \tilde{X}_j^{T-1}$. By applying Remark 6.157 with $W_j := \tilde{X}_j^{T-1}$ and $E_j := -\bar{X}_j^T \tilde{X}_j^{T-1}$, we will show that

$$\left.\begin{array}{l} \lim_{j \to \infty} Y^{(j)} \text{ exists if and only if} \\[4pt] \tilde{Y} \text{ is a (maximal) dominant solution at } \infty. \end{array}\right\} \qquad (6.458)$$

In this case the limit in (6.458) is the minimal recessive solution of (SDS) at $\infty$. Assume first that $\tilde{Y}$ is a (maximal) dominant solution of (SDS) at $\infty$. From Theorem 6.153 we know that the conjoined basis $\tilde{Y}$ belongs to the maximal genus

$\mathcal{G}_{\max}$. By Theorem 6.149 and Remark 6.150 (with $\mathcal{G} := \mathcal{G}_{\max}$, $N := \tilde{L}$, $Y := \tilde{Y}$, and $\tilde{Y} := \bar{Y}$), we then obtain

$$\lim_{j \to \infty} (-E_j^T) = \lim_{j \to \infty} \tilde{X}_j^{-1} \bar{X}_j = V, \quad \operatorname{rank} V = \operatorname{rank} \bar{T} + d_\infty, \tag{6.459}$$

where $\bar{T}$ is the matrix in (6.237), which corresponds to $\bar{Y}$ on $[\tilde{L}, \infty)_{\mathbb{Z}}$. Moreover, from Proposition 6.155 it follows that $\bar{Y}$ is a (maximal) dominant solution of (SDS) at $\infty$, so that rank $\bar{T} = n - d_\infty$ by Definition 6.122. Hence, the matrix $V$ in (6.459) is invertible and the sequence $E_j \to E := -V^T$ as $j \to \infty$ with $E$ being invertible. Therefore, the limit in (6.458) exists, and it is equal to the minimal recessive solution of (SDS) at $\infty$, by Remark 6.157.

Conversely, assume that the limit in (6.458) exists. This means that

$$E := \lim_{j \to \infty} E_j = -\lim_{j \to \infty} \bar{X}_j^T \tilde{X}_j^{T-1} \tag{6.460}$$

exists, by Remark 6.157. By using the fact that $\bar{Y}$ is a maximal dominant solution at $\infty$, we obtain (through Theorem 6.149 and Remark 6.150) that

$$F := \lim_{j \to \infty} \bar{X}_j^{-1} \tilde{X}_j = -\lim_{j \to \infty} E_j^{T-1}, \quad \operatorname{rank} F = \operatorname{rank} \tilde{T} + d_\infty \tag{6.461}$$

also exists, where $\tilde{T}$ is the matrix in (6.237) associated with $\tilde{Y}$ on $[\tilde{L}, \infty)_{\mathbb{Z}}$. Identities (6.460) and (6.461) then imply that both the matrices $E$ and $F$ are invertible with $E = -F^{T-1}$. In turn, the limit in (6.458) is the minimal recessive solution of (SDS) at $\infty$, again by Remark 6.157. Finally, from the second equality in (6.461), it follows that rank $\tilde{T} = \operatorname{rank} F - d_\infty = n - d_\infty$. Thus, $\tilde{Y}$ is a (maximal) dominant solution of (SDS) at $\infty$ by Definition 6.122.

In the last part of this section, we will discuss the dependence of the construction of the minimal recessive solution $\hat{Y}^{\min}$ at $\infty$ in Theorem 6.153 with respect to the used initial data, in particular with respect to the choice of the conjoined bases $\bar{Y}$, $Y$, and the index $N$. First we observe that the representation of $Y^{(j)}$ in (6.453) with $M^{(j)} = I$ and $N^{(j)} = -S_j^\dagger$ from (6.455) does not depend on the choice of $\bar{Y}$, since in this case $N^{(j)} = PN^{(j)}$ and the solution $\bar{Y}P$ is uniquely determined by $Y$ on $[N, \infty)_{\mathbb{Z}}$ by Proposition 6.67(iv). Therefore, the construction of $\hat{Y}^{\min}$ in (6.450) does not depend on the choice of $\bar{Y}$ either. Moreover, in the following remark, we will show the independence of the construction in Theorem 6.153 also on the minimal conjoined basis $Y$.

*Remark 6.159* Assume now that in addition to $Y$ in Theorem 6.153, we start with another minimal conjoined basis $Y^*$ of (SDS) on $[N, \infty)_{\mathbb{Z}}$ and let $\bar{Y}^*$ be its associated conjoined basis from Proposition 6.67. By Lemma 6.100 (with $Y^{(1)} := Y$ and $Y^{(2)} := Y^*$), we then have $\bar{X}_k^* = \bar{X}_k G$ for all $k \in [N, \infty)_{\mathbb{Z}}$ with a constant invertible matrix $G$. Following (6.449) we construct for $j \in [L, \infty)_{\mathbb{Z}}$ the conjoined bases $Y^{*(j)}$ of (SDS) satisfying $X_j^{*(j)} = 0$ and $U_j^{*(j)} = -(\bar{X}_j^*)^{T-1} = -\bar{X}_j^{T-1}G^{T-1}$. Hence, we obtain by the uniqueness of solutions of (SDS) that $Y_k^{*(j)} = Y_k^{(j)}M$ on $[0, \infty)_{\mathbb{Z}}$ with $M := G^{T-1}$. This implies that

$$\lim_{j \to \infty} Y_k^{*(j)} = \lim_{j \to \infty} Y_k^{(j)}M = \hat{Y}_k^{\min}M, \quad k \in [0, \infty)_{\mathbb{Z}}. \tag{6.462}$$

Thus, using a different minimal conjoined basis $Y^*$ of (SDS) in Theorem 6.153 leads through (6.462) to a constant right nonsingular multiple of the minimal recessive solution $\hat{Y}^{\min}$. But since $\hat{Y}^{\min}$ is essentially unique by Theorem 6.113, it follows that the construction in Theorem 6.153 does not depend on the choice of the minimal conjoined basis $Y$.

*Remark 6.160* Finally, we note that the result in Theorem 6.153 also does not depend on the index $N$, i.e., on the interval $[N, \infty)_{\mathbb{Z}}$ on which $Y$ is a minimal conjoined basis of (SDS). This follows immediately from the uniqueness of the minimal recessive solution of (SDS) at $\infty$ in Theorem 6.113. Hence, moving the index $N$ to the right yields a constant right nonsingular multiple in the representation (6.450), similarly to formula (6.462) in the previous remark.

## 6.3.11 Additional Properties of Minimal Recessive Solution

In this subsection we derive some additional properties of the minimal recessive solution $\hat{Y}^{\min}$ at $\infty$. These properties will be utilized in Sect. 6.4 in the singular Sturmian theory for system (SDS). Our first result represents a variant of Corollary 6.147, but here we consider the conjoined bases $\hat{Y}^{\min} \in \mathcal{G}_{\min}$ and $Y \in \mathcal{G}$ to be from different genera of conjoined bases of (SDS).

**Theorem 6.161** *Assume that (SDS) is nonoscillatory at $\infty$, and let $\hat{K}_{\min}$ be the index defined in Remark 6.130 for $\hat{Y}^{\min}$. Then for any $N \in [\hat{N}^{\min}, \infty)_{\mathbb{Z}}$, the conjoined basis $\bar{Y}^{[\infty]}$ of (SDS), which is associated with $\hat{Y}^{\min}$ on $[N, \infty)_{\mathbb{Z}}$ in Proposition 6.67, is a maximal dominant solution of (SDS) at $\infty$, and it satisfies*

$$\lim_{k \to \infty} (\bar{X}_k^{[\infty]})^{-1}\hat{X}_k^{\min} = 0. \tag{6.463}$$

*Proof* Fix $N \in [\hat{K}_{\min}, \infty)_{\mathbb{Z}}$, and let $\bar{Y}^{[\infty]}$ be the conjoined basis of (SDS) from Proposition 6.67, which is associated with $\hat{Y}^{\min}$ on $[N, \infty)_{\mathbb{Z}}$. Then (6.414) holds, and the result in Proposition 6.155 (with $Y := \hat{Y}^{\min}$) implies that $\bar{Y}^{[\infty]}$ is a maximal dominant solution of (SDS) at $\infty$. It remains to prove (6.463). For simplicity of the notation, we set $\hat{Y} := \hat{Y}^{\min}$ and $Y := \bar{Y}^{[\infty]}$. First we consider the maximal recessive solution $\hat{Y}^{\max}$ of (SDS) at $\infty$, which contains $\hat{Y}$ on $[N, \infty)_{\mathbb{Z}}$ with respect to the orthogonal projector $P_{\hat{S}_\infty}$ defined in (6.238) through $\hat{Y}$. Such a maximal recessive solution of (SDS) at $\infty$ exists by Theorem 6.115, and it can be chosen so that it contains $\hat{Y}$ on $[N, \infty)_{\mathbb{Z}}$ by Theorem 6.87 (with $P = I = R$ and $P^* = P_{\hat{S}\infty}$). Then we have

$$\hat{X}_k = \hat{X}_k^{\max} P_{\hat{S}\infty}, \quad k \in [N, \infty)_{\mathbb{Z}}. \tag{6.464}$$

By Theorem 6.141 for the maximal genus $\mathcal{G} := \mathcal{G}_{\max}$ (with the given $Y$ and with $\hat{Y} := \hat{Y}^{\max}$), we obtain that $\text{rank}\,[P_{\hat{S}\infty}\, w(\hat{Y}^{\max}, Y)\, P_{S\infty}] = n - d_\infty$. In turn, by Corollary 6.148 (with $\hat{Y} := \hat{Y}^{\max}$), we get

$$\lim_{k\to\infty} X_k^{-1} \hat{X}_k^{\max} = V \quad \text{with} \quad \text{Im}\, V^T = \text{Ker}\, P_{\hat{S}\infty}. \tag{6.465}$$

Therefore, we conclude that

$$\lim_{k\to\infty} X_k^{-1} \hat{X}_k \overset{(6.464)}{=} \lim_{k\to\infty} X_k^{-1} \hat{X}_k^{\max} P_{\hat{S}\infty} = V P_{\hat{S}\infty} \overset{(6.465)}{=} 0,$$

since $\text{Im}\, P_{\hat{S}\infty} = \text{Ker}\, V$ by (6.465). This shows that (6.463) holds. $\qquad\square$

In the next statement, we apply Theorem 6.161 in order to derive a new property of the Wronskians of the minimal recessive solution $\hat{Y}^{\min}$ of (SDS) at $\infty$ with a conjoined basis $Y$ and its associated conjoined basis $\bar{Y}$. Here we use a symplectic fundamental matrix $\Phi_k^{[\infty]}$ of (SDS), which is associated with the minimal recessive solution $\hat{Y}^{\min}$, i.e., following (6.228) we introduce the matrix

$$\Phi_k^{[\infty]} := \left( \hat{Y}_k^{\min}\ \bar{Y}_k^{[\infty]} \right), \quad k \in [N, \infty)_{\mathbb{Z}}, \quad w(\hat{Y}^{\min}, \bar{Y}^{[\infty]}) = I. \tag{6.466}$$

Then, as in Lemma 6.59, every conjoined basis $Y$ of (SDS) can be uniquely represented by a constant $2n \times n$ matrix $D_\infty$ such that

$$Y_k = \Phi_k^{[\infty]} D_\infty, \quad k \in [0, \infty)_{\mathbb{Z}}, \quad \mathcal{J} D_\infty = \begin{pmatrix} w(Y^{[\infty]}, Y) \\ w(\bar{Y}^{[\infty]}, Y) \end{pmatrix}. \tag{6.467}$$

**Theorem 6.162** *Assume that (SDS) is nonoscillatory at $\infty$. Let $Y$ be a conjoined basis of (SDS) with constant kernel on $[N, \infty)_{\mathbb{Z}}$ and no forward focal points in $(N, \infty)$ for some $N \in [0, \infty)_{\mathbb{Z}}$, and let $\bar{Y}$ be the conjoined basis of (SDS) from Proposition 6.67 associated with $Y$ on $[N, \infty)_{\mathbb{Z}}$. Then*

$$w(\hat{Y}^{\min}, Y)\, [w(\hat{Y}^{\min}, \bar{Y})]^T \geq 0.$$

*Proof* First we observe that by Lemma 2.4 (with the choice $\tilde{Y} := \hat{Y}^{\min}$) the matrix $w(\hat{Y}^{\min}, Y)\, [w(\hat{Y}^{\min}, \bar{Y})]^T$ is symmetric. Choose an index $K \in [N, \infty)_{\mathbb{Z}}$ so that $d[K, \infty)_{\mathbb{Z}} = d_\infty$ and the conjoined bases $Y$, $\bar{Y}$, and $\hat{Y}^{\min}$ have constant kernel on $[K, \infty)_{\mathbb{Z}}$ and no forward focal points in $(K, \infty)$. Let $\bar{Y}^{[\infty]}$ be the conjoined basis of (SDS) from Proposition 6.67 associated with $\hat{Y}^{\min}$ on the interval $[K, \infty)_{\mathbb{Z}}$, i.e., $(\hat{X}_K^{\min})^\dagger \bar{X}_K^{[\infty]} = 0$. Let $D_\infty$ be the representing matrix of $Y$ in terms of the symplectic fundamental matrix $\Phi_k^{[\infty]}$, i.e., (6.228) with $j = \infty$ and (6.467) hold. As in the proof of Theorem 6.80, we split $D_\infty = \left(\begin{smallmatrix} F \\ G \end{smallmatrix}\right)$ with $F = -w(\bar{Y}^{[\infty]}, Y)$ and $G = w(\hat{Y}^{\min}, Y)$. Then

$$X_k = \hat{X}_k^{\min} F + \bar{X}_k^{[\infty]} G, \quad k \in [0, \infty)_{\mathbb{Z}}. \tag{6.468}$$

From Theorem 6.161 we know that $\bar{Y}^{[\infty]}$ is a maximal dominant solution of (SDS) at $\infty$, so that the matrix $\bar{X}_k^{[\infty]}$ is invertible for all $k$ large enough. By applying Theorem 6.161, we then obtain

$$\lim_{k\to\infty} (\bar{X}_k^{[\infty]})^{-1} X_k \overset{(6.468)}{=} \lim_{k\to\infty} (\bar{X}_k^{[\infty]})^{-1} \hat{X}_k^{\min} F + G \overset{(6.463)}{=} G = w(\hat{Y}^{\min}, Y).$$
$$\tag{6.469}$$

By repeating the above argument with the conjoined basis $\bar{Y}$ instead of $Y$, we conclude similarly as in (6.469) that

$$\lim_{k\to\infty} (\bar{X}_k^{[\infty]})^{-1} \bar{X}_k = w(\hat{Y}^{\min}, \bar{Y}). \tag{6.470}$$

Therefore, upon combining (6.469) and (6.470), we get

$$w(\hat{Y}^{\min}, Y)\,[w(\hat{Y}^{\min}, \bar{Y})]^T = \lim_{k\to\infty} (\bar{X}_k^{[\infty]})^{-1} X_k \bar{X}_k^T (\bar{X}_k^{[\infty]})^{T-1} \geq 0,$$

where the last inequality follows from Proposition 6.67(xii). □

## 6.3.12 Further Examples

In this subsection we present three examples, in which we illustrate the theory of dominant solutions at $\infty$—their limit comparison with recessive solutions at $\infty$ and the Reid construction of the (minimal) recessive solution at $\infty$ in terms of (maximal) dominant solutions at $\infty$. For this purpose we will utilize Examples 6.119–6.121 from Sect. 6.3.6. In agreement with the notation in Remarks 6.110 and 6.123, recessive solutions at $\infty$ will be denoted by $\hat{Y}$ and in the special case of minimal and maximal recessive solutions at $\infty$ by $\hat{Y}^{\min}$ and $\hat{Y}^{\max}$. Similarly, dominant solutions at $\infty$ will be denoted by $Y$ and in the special case of minimal and maximal dominant solutions at $\infty$ by $Y^{\min}$ and $Y^{\max}$.

*Example 6.163* Let us continue the considerations initiated in Example 6.119. Consider a nonoscillatory scalar system (SDS) with $n = 1$ and $S_k \equiv \left(\begin{smallmatrix} 1 & 1 \\ 0 & 1 \end{smallmatrix}\right)$ on $[0, \infty)_{\mathbb{Z}}$. From Example 6.119 we know that $d_\infty = 0$ and that the conjoined basis

$\hat{Y}_k \equiv \left(\begin{smallmatrix}1\\0\end{smallmatrix}\right)$ of (SDS) is the unique recessive solution at $\infty$. In this case we have $\hat{K}_{\min} = 0$. On the other hand, according to Corollary 6.145, the conjoined basis $Y = Y^{[0]} = \left(\begin{smallmatrix}k\\1\end{smallmatrix}\right)$ of (SDS) is a dominant solution at $\infty$, being at the same time the principal solution of (SDS) at the index $k = 0$. Moreover, the solutions $\hat{Y}$ and $Y$ satisfy $X_k^{-1}\hat{X}_k = 1/k \to 0$ as $k \to \infty$, as we claim in formula (6.439) of Corollary 6.147. Finally, the Reid construction of the recessive solution $\hat{Y}$ in Theorem 6.153 is the following. With $\bar{Y} := Y$ we have $N = 1$, and for any index $j \in [1, \infty)_{\mathbb{Z}}$, the solution $Y^{(j)}$ of (SDS) from (6.449) satisfying the initial conditions $X_j^{(j)} = 0$ and $U_j^{(j)} = -\bar{X}_j^{T-1} = -1/j$ is

$$X_k^{(j)} = (j - k)/j, \quad U_k^{(j)} = -1/j, \quad k \in [0, \infty)_{\mathbb{Z}}.$$

Therefore, $Y_k^{(j)} \to \left(\begin{smallmatrix}1\\0\end{smallmatrix}\right) = \hat{Y}_k$ as $j \to \infty$ for every $k \in [0, \infty)_{\mathbb{Z}}$, as we claim in formula (6.450).

*Example 6.164* We continue with Example 6.120. Consider a nonoscillatory system (SDS) with $n \in \mathbb{N}$ and $\mathcal{S}_k \equiv I_{2n}$ on $[0, \infty)_{\mathbb{Z}}$. Then $d[0, \infty)_{\mathbb{Z}} = d_\infty = n$ and every conjoined basis of (SDS) is a (constant) recessive and also dominant solution at $\infty$ with respect to the interval $[0, \infty)_{\mathbb{Z}}$. Therefore, we have $\hat{K}_{\min} = 0$. Every genus $\mathcal{G}$ of conjoined bases is associated with a unique orthogonal projector $P \in \mathbb{R}^{n \times n}$ such that the conjoined basis $\hat{Y} = Y = \left(\begin{smallmatrix}P\\I-P\end{smallmatrix}\right)$ is a recessive and dominant solution at $\infty$ belonging to $\mathcal{G}$. In addition, we have $X_k^\dagger \hat{X}_k = P^\dagger P = P$ for all $k \to \infty$, so that $V = P$, $\hat{P} = P$, and $P_{\hat{\mathcal{S}}_\infty} = 0$ in formula (6.431). The special choice of $P = 0$ then yields the solutions $\hat{Y}^{\min} = Y^{\min} = \left(\begin{smallmatrix}0\\I\end{smallmatrix}\right)$, while the choice of $P = I$ yields the solutions $\hat{Y}^{\max} = Y^{\max} = \left(\begin{smallmatrix}I\\0\end{smallmatrix}\right)$. Note that $Y^* = \left(\begin{smallmatrix}I\\I\end{smallmatrix}\right)$ is another maximal recessive and dominant solution at $\infty$, which illustrates the nonuniqueness of these solutions in Remark 6.144.

*Example 6.165* We continue with Example 6.121. Consider a nonoscillatory system (SDS) with $\mathcal{A}_k = \mathcal{D}_k \equiv I_3$, $\mathcal{B}_k \equiv \text{diag}\{1, 0, 0\}$, and $\mathcal{C}_k \equiv 0_3$ on $[0, \infty)_{\mathbb{Z}}$. This system arises from the scalar system in Example 6.163 and from the system in Example 6.164 with dimension two by a block diagonal construction. In this case $d[0, \infty)_{\mathbb{Z}} = d_\infty = 2$ and $\hat{K}_{\min} = 0$. From Theorems 6.118 and 6.133, we know that some recessive and dominant solutions of (SDS) at $\infty$ can be constructed from the recessive and dominant solutions at $\infty$ in Examples 6.163 and 6.164 by the same block diagonal procedure.

(a) First we analyze the minimal genus $\mathcal{G}_{\min}$ with rank $r = n - d_\infty = 1$. Then

$$\hat{Y}_k^{\min} = \left(\text{diag}\{1, 0, 0\}, \text{diag}\{0, 1, 1\}\right)^T,$$
$$Y_k^{\min} = \left(\text{diag}\{k, 0, 0\}, \text{diag}\{1, 1, 1\}\right)^T.$$

In this case $(X_k^{\min})^\dagger \hat{X}_k^{\min} = \mathrm{diag}\{1/k, 0, 0\} \to 0_3$ as $k \to \infty$, as we claim in formula (6.439) of Corollary 6.147.

(b) Further, we examine the maximal genus $\mathcal{G}_{\max}$, whose rank is $r = n = 3$. Then we have

$$\hat{Y}_k^{\max} = \big(\, \mathrm{diag}\{1, 1, 1\}, \ \mathrm{diag}\{0, 0, 0\} \,\big)^T,$$

$$Y_k^{\max} = \big(\, \mathrm{diag}\{k, 1, 1\}, \ \mathrm{diag}\{1, 0, 0\} \,\big)^T,$$

so that $(X_k^{\max})^\dagger \hat{X}_k^{\max} = \mathrm{diag}\{1/k, 1, 1\} \to V := \mathrm{diag}\{0, 1, 1\}$ as $k \to \infty$. In this case we have $P_{\hat{\mathcal{S}}\infty} = \mathrm{diag}\{1, 0, 0\}$ with $\mathrm{Im}\, V^T = \{0\} \times \mathbb{R}^2 = \mathrm{Ker}\, P_{\hat{\mathcal{S}}\infty}$, as we state in Corollary 6.148.

(c) Next we discuss three different genera with rank equal to $r = 2$. We note that only two of them arise from the diagonal construction mentioned above. Let $\mathcal{G}_1$ be the genus with rank $r = 2$, which contains the recessive and dominant solutions at $\infty$

$$\hat{Y}_k^{(1)} = \big(\, \mathrm{diag}\{1, 1, 0\}, \ \mathrm{diag}\{0, 0, 1\} \,\big)^T,$$

$$Y_k^{(1)} = \big(\, \mathrm{diag}\{k, 1, 0\}, \ \mathrm{diag}\{1, 0, 1\} \,\big)^T.$$

In Theorem 6.146 we then have $(X_k^{(1)})^\dagger \hat{X}_k^{(1)} = \mathrm{diag}\{1/k, 1, 0\} \to V := \mathrm{diag}\{0, 1, 0\}$ as $k \to \infty$, and $\hat{P} = \mathrm{diag}\{1, 1, 0\}$ and $P_{\hat{\mathcal{S}}\infty} = \mathrm{diag}\{1, 0, 0\}$. Let $\mathcal{G}_2$ be the genus with rank $r = 2$ given by the recessive and dominant solutions at $\infty$

$$\hat{Y}_k^{(2)} = \big(\, \mathrm{diag}\{1, 0, 1\}, \ \mathrm{diag}\{0, 1, 0\} \,\big)^T,$$

$$Y_k^{(2)} = \big(\, \mathrm{diag}\{k, 0, 1\}, \ \mathrm{diag}\{1, 1, 0\} \,\big)^T.$$

In this case we have $(X_k^{(2)})^\dagger \hat{X}_k^{(2)} = \mathrm{diag}\{1/k, 0, 1\} \to V := \mathrm{diag}\{0, 0, 1\}$ as $k \to \infty$, and $\hat{P} = \mathrm{diag}\{1, 0, 1\}$ and $P_{\hat{\mathcal{S}}\infty} = \mathrm{diag}\{1, 0, 0\}$. Now we consider the nondiagonal genus $\mathcal{G}_3$ with rank $r = 2$ defined by the recessive and dominant solutions at $\infty$

$$\hat{Y}_k^{(3)} = \left( \begin{pmatrix} 1 & 0 & 0 \\ 0 & 1 & 0 \\ 0 & 1 & 0 \end{pmatrix}, \ \begin{pmatrix} 0 & 0 & 0 \\ 0 & 1 & 1 \\ 0 & -1 & -1 \end{pmatrix} \right)^T,$$

$$Y_k^{(3)} = \left( \begin{pmatrix} k & 0 & 0 \\ 0 & 1 & -1 \\ 0 & 1 & -1 \end{pmatrix}, \ \begin{pmatrix} 1 & 0 & 0 \\ 0 & 3 & -2 \\ 0 & 1 & -2 \end{pmatrix} \right)^T.$$

In Theorem 6.146 we then have

$$(X_k^{(3)})^\dagger \hat{X}_k^{(3)} = \begin{pmatrix} 1/k & 0 & 0 \\ 0 & 1/2 & 0 \\ 0 & -1/2 & 0 \end{pmatrix} \to V := \begin{pmatrix} 0 & 0 & 0 \\ 0 & 1/2 & 0 \\ 0 & -1/2 & 0 \end{pmatrix} \quad \text{as } k \to \infty.$$

The orthogonal projectors $\hat{P}$ and $P_{\hat{S}_\infty}$ in (6.431) are given by $\hat{P} = \mathrm{diag}\{1, 1, 0\}$, $P_{\hat{S}_\infty} = \mathrm{diag}\{1, 0, 0\}$, and $\hat{P} - P_{\hat{S}_\infty} = \mathrm{diag}\{0, 1, 0\}$. Hence, in this case we indeed have $\mathrm{Im}\, V^T = \mathrm{Im}\,(\hat{P} - P_{\hat{S}_\infty})$, although $V^T \neq \hat{P} - P_{\hat{S}_\infty}$.

(d) Finally, we present the Reid construction of the minimal recessive solution $\hat{Y}^{\min}$ in Theorem 6.153. With $Y := Y^{\min}$ given above in part (a) and with

$$\bar{Y}_k = \big(\,\mathrm{diag}\{k - 1, -1, -1\},\ \mathrm{diag}\{1, 0, 0\}\,\big)^T$$

we have $N = 1$ and $L = 2$. For $j \in [2, \infty)_{\mathbb{Z}}$ the solution in (6.449) is

$$Y_k^{(j)} = \big(\,\mathrm{diag}\{(j - k)/(j - 1), 0, 0\},\ \mathrm{diag}\{-1/(j - 1), 1, 1\}\,\big)^T, \quad k \in [0, \infty)_{\mathbb{Z}}.$$

Then we have $Y_k^{(j)} \to (\mathrm{diag}\{1, 0, 0\},\ \mathrm{diag}\{0, 1, 1\})^T = \hat{Y}_k^{\min}$ as $j \to \infty$ for every $k \in [0, \infty)_{\mathbb{Z}}$, as we claim in formula (6.450) of Theorem 6.153.

## 6.4  Singular Sturmian Separation Theorems

In this section we establish singular Sturmian separation theorems for conjoined bases of (SDS) on unbounded intervals. We will consider discrete intervals, which are unbounded from above, unbounded from below, or unbounded at both endpoints. By using the theory of recessive and dominant solutions of (SDS) at $\infty$, we essentially improve the results on singular Sturmian theory in Theorem 4.39 and Remark 4.40 and at the same time provide singular versions of the separation theorems presented in Sects. 4.2.2 and 4.2.3.

For convenience and easier reference, we will use the notation

$$Y_k = \begin{pmatrix} X_k \\ U_k \end{pmatrix}, \quad \tilde{Y}_k = \begin{pmatrix} \tilde{X}_k \\ \tilde{U}_k \end{pmatrix}, \quad Y_k^{[j]} = \begin{pmatrix} X_k^{[j]} \\ U_k^{[j]} \end{pmatrix}, \quad Y_k^{[\infty]} := \hat{Y}_k^{\min}, \quad E := \begin{pmatrix} 0 \\ I \end{pmatrix},$$

$$\tag{6.471}$$

for generic conjoined bases $Y$ and $\tilde{Y}$ of (SDS), for the principal solution $Y^{[j]}$ of (SDS) at the index $j$ (satisfying $Y_j^{[j]} = E$), and for the minimal recessive solution $\hat{Y}^{\min}$ at $\infty$. For a conjoined basis $Y$ of (SDS), the multiplicity of forward (or left) focal point in $(k, k + 1]$ and the multiplicity of the backward (or right) focal point in $[k, k + 1)$ will be denoted by

$$m_L(k, k + 1] := m(k) = \mu(Y_{k+1}, S_k E) = \mu(Y_{k+1}, Y_{k+1}^{[k]}), \qquad (6.472)$$

$$m_R[k, k + 1) := m^*(k) = \mu^*(Y_k, S_k^{-1} E) = \mu^*(Y_k, Y_k^{[k+1]}), \qquad (6.473)$$

where $m(k)$ and $m^*(k)$ are defined by (4.3) and (4.6) and where we use the results of Lemmas 4.7 and 4.8 to connect these multiplicities with the comparative index. Moreover, using (4.10) and (4.11), we denote by

$$\left. \begin{aligned} m_L(M, N] &:= l(Y, M, N) = \sum_{k=M}^{N-1} m_L(k, k + 1], \\ m_R[M, N) &:= l^*(Y, M, N) = \sum_{k=M}^{N-1} m_R[k, k + 1), \end{aligned} \right\} \qquad (6.474)$$

the numbers of forward and backward focal points of $Y$, including their multiplicities, in the intervals $(M, N]$ and $[M, N)$. We will use similar notation $\tilde{m}_L(M, N]$ and $\tilde{m}_R[M, N)$ for the numbers of focal points of another conjoined basis $\tilde{Y}$ of (SDS) in these intervals. We summarize the regular Sturmian separation theorems from Sects. 4.1 and 4.2 as follows (see Theorems 4.23 and 4.24 and Corollary 4.6).

**Proposition 6.166** *For any conjoined bases $Y$ and $\tilde{Y}$ of (SDS), we have*

$$m_L(M, N] - \tilde{m}_L(M, N] = \mu(Y_N, \tilde{Y}_N) - \mu(Y_M, \tilde{Y}_M), \qquad (6.475)$$

$$m_R[M, N) - \tilde{m}_R[M, N) = \mu^*(Y_M, \tilde{Y}_M) - \mu^*(Y_N, \tilde{Y}_N), \qquad (6.476)$$

$$m_L(M, N] + \operatorname{rank} X_N = m_R[M, N) + \operatorname{rank} X_M. \qquad (6.477)$$

Similarly as above, for the principal solution $Y^{[j]}$ of (SDS), we denote by $m_L^{[j]}(M, N]$ and $m_R^{[j]}[M, N)$ the numbers of its forward and backward focal points in the indicated intervals. Then we summarize the corresponding regular Sturmian separation theorems as follows (see Theorems 4.34 and 4.35, formulas (4.62) and (4.63), and Remark 4.28(iii)).

**Proposition 6.167** *For the principal solutions $Y^{[M]}$ and $Y^{[N]}$ of (SDS), we have the identities*

$$m_L^{[M]}(M, N] = m_R^{[N]}[M, N), \quad m_R^{[M]}[M, N) = m_L^{[N]}(M, N]. \qquad (6.478)$$

*For any conjoined basis $Y$ of (SDS), we have the estimates*

$$m_L^{[M]}(M, N] \leq m_L(M, N] \leq m_L^{[N]}(M, N], \qquad (6.479)$$

$$m_R^{[N]}[M, N) \leq m_R[M, N) \leq m_R^{[M]}[M, N). \qquad (6.480)$$

*For any conjoined bases $Y$ and $\tilde{Y}$ of (SDS), we have the estimates*

$$\left| m_L(M, N) - \tilde{m}_L(M, N) \right| \leq \operatorname{rank} X_N^{[M]} \leq n, \tag{6.481}$$

$$\left| m_R[M, N) - \tilde{m}_R[M, N) \right| \leq \operatorname{rank} X_M^{[N]} \leq n, \tag{6.482}$$

$$\left| m_L(M, N] - \tilde{m}_R[M, N) \right| \leq \operatorname{rank} X_M^{[N]} \leq n. \tag{6.483}$$

In the following subsections, we will extend the results in Propositions 6.166 and 6.167 to unbounded intervals.

### 6.4.1  Multiplicity of Focal Point at Infinity

In this subsection we assume that system (SDS) is defined on the interval $[0, \infty)_{\mathbb{Z}}$ and that it is nonoscillatory at $\infty$. As a new notion, we define for a conjoined basis $Y$ of (SDS) the multiplicity of its focal point at $\infty$. In this definition we employ the matrix $T$, which is associated with $Y$ through (6.237). We then prove a representation formula for the multiplicity at $\infty$ in terms of the Wronskian of $Y$ with the minimal recessive solution $Y^{[\infty]}$ of (SDS) at $\infty$.

**Definition 6.168**  For a conjoined basis $Y$ of (SDS), we define the quantity

$$m_L(\infty) := n - d_\infty - \operatorname{rank} T = \operatorname{def} T - d_\infty, \tag{6.484}$$

where $T$ is the matrix in (6.237) associated with $Y$ on an interval $[N, \infty)_{\mathbb{Z}}$ satisfying $d[N, \infty)_{\mathbb{Z}} = d_\infty$. Moreover, we say that $Y$ has a *focal point at* $\infty$ if $m_L(\infty) \geq 1$, and then $m_L(\infty)$ is called its *multiplicity*.

*Remark 6.169*

(i) Estimate (6.277) shows that the quantity $m_L(\infty)$ in (6.484) is correctly defined and that $0 \leq m_L(\infty) \leq n - d_\infty$. Moreover, by Remark 6.126 the number $m_L(\infty)$ does not depend on the index $N$ satisfying $d[N, \infty)_{\mathbb{Z}} = d_\infty$.

(ii) It follows from Definition 6.168 that a conjoined basis $Y$ is a recessive solution of (SDS) at $\infty$ if and only if $m_L(\infty) = n - d_\infty$ (i.e., $m_L(\infty)$ is maximal). In particular, for the minimal recessive solution $Y^{[\infty]}$ of (SDS) at $\infty$, we have $m_L^{[\infty]}(\infty) = n - d_\infty$. Similarly, a conjoined basis $Y$ is a dominant solution of (SDS) at $\infty$ if and only if $Y$ has no focal point at $\infty$ (i.e., $m_L(\infty) = 0$ is minimal).

(iii) The quantity $m_L(\infty)$ is preserved under the relation being contained, since this relation preserves the corresponding matrices $S_k$ and hence $T$ (see Definition 6.82 and Proposition 6.86).

In the following result, we provide an alternative formula for the multiplicity $m_L(\infty)$ in Definition 6.168 in terms of the Wronskian of $Y$ with the minimal recessive solution $Y^{[\infty]}$ at $\infty$. We recall the quantity $\operatorname{rank} \mathcal{G}$ for a genus $\mathcal{G}$ of conjoined bases of (SDS) defined in Remark 6.150.

**Theorem 6.170** *Assume that system* (SDS) *is nonoscillatory at* $\infty$, *and let* $Y$, *belonging to a genus* $\mathcal{G}$, *be a conjoined basis of* (SDS) *with constant kernel on* $[N, \infty)_\mathbb{Z}$ *and no forward focal points in* $(N, \infty)$, *where the index* $N \in [0, \infty)_\mathbb{Z}$ *is such that* $d[N, \infty)_\mathbb{Z} = d_\infty$. *Then*

$$\operatorname{Im} [w(Y^{[\infty]}, Y)]^T = \operatorname{Im} T \oplus \operatorname{Im} (P - P_{S\infty}), \tag{6.485}$$

$$\operatorname{rank} T = \operatorname{rank} w(Y^{[\infty]}, Y) - \operatorname{rank} \mathcal{G} + n - d_\infty, \tag{6.486}$$

$$m_L(\infty) = \operatorname{rank} \mathcal{G} - \operatorname{rank} w(Y^{[\infty]}, Y), \tag{6.487}$$

*where* $P$, $P_{S\infty}$, $T$ *are the matrices in* (6.234), (6.238), (6.237) *associated with* $Y$.

*Proof* Let the index $N$ be as in the theorem. By Remark 6.126 the space $\operatorname{Im} T$ is preserved, when $N$ is replaced by any larger index. Therefore, without loss of generality, we may assume that the index $N \in [0, \infty)_\mathbb{Z}$ is such that $d[N, \infty)_\mathbb{Z} = d_\infty$ holds and both conjoined bases $Y$ and $Y^{[\infty]}$ have constant kernel on $[N, \infty)_\mathbb{Z}$ and no forward focal points in $(N, \infty)$, i.e., $N \geq \hat{K}_{\min}$ according to Remark 6.130. Let $\bar{Y}^{[\infty]}$ be the conjoined basis from Proposition 6.67 associated with $Y^{[\infty]}$ on $[N, \infty)_\mathbb{Z}$. Then by Theorem 6.161, we know that $\bar{Y}^{[\infty]}$ is a maximal dominant solution of (SDS) at $\infty$. This yields that there exists an index $M > N$ such that $\bar{Y}^{[\infty]}$ has no forward focal points in $(M, \infty)$ and $\bar{X}_k^{[\infty]}$ is invertible on $[M, \infty)_\mathbb{Z}$. Moreover, by (6.469) in the proof of Theorem 6.162, we have

$$\lim_{k \to \infty} (\bar{X}_k^{[\infty]})^{-1} X_k = w(Y^{[\infty]}, Y). \tag{6.488}$$

Let $R_k$ for $k \in [N, \infty)_\mathbb{Z}$ be the orthogonal projector onto $\operatorname{Im} X_k$ defined in (6.233). If $\bar{Y}^{[\infty]*} \in \mathcal{G}$ is a conjoined basis of (SDS), which is contained in $\bar{Y}^{[\infty]}$ on $[M, \infty)_\mathbb{Z}$ according to Definition 6.82 and which belongs to the same genus $\mathcal{G}$ as $Y$, then by (6.298) in Remark 6.83, we have the formula

$$(\bar{X}_k^{[\infty]*})^\dagger = (\bar{X}_k^{[\infty]})^{-1} R_k, \quad k \in [M, \infty)_\mathbb{Z}.$$

Moreover, $\bar{Y}^{[\infty]*}$ is also a dominant solution of (SDS) at $\infty$ by Proposition 6.127. Therefore, upon applying Theorem 6.149 (with $Y := \bar{Y}^{[\infty]*}$ and $\tilde{Y} := Y$), we obtain

$$\lim_{k \to \infty} (\bar{X}_k^{[\infty]*})^\dagger X_k = V \quad \text{with} \quad \operatorname{Im} V^T = \operatorname{Im} T \oplus \operatorname{Im} (P - P_{S\infty}). \tag{6.489}$$

Consequently, by combining (6.488)–(6.489), we get

$$V = \lim_{k \to \infty} (\bar{X}_k^{[\infty]})^{-1} R_k X_k = \lim_{k \to \infty} (\bar{X}_k^{[\infty]})^{-1} X_k \overset{(6.488)}{=} w(Y^{[\infty]}, Y).$$

This implies through the second part of (6.489) that the equality in (6.485) holds. By evaluating the ranks of the subspaces in (6.485), we then get

$$\text{rank } w(Y^{[\infty]}, Y) = \text{rank } T + \text{rank } P - \text{rank } P_{S\infty} = \text{rank } T + \text{rank } \mathcal{G} - (n - d_\infty),$$

which shows (6.486). Finally, formula (6.487) follows from (6.486) by using the definition of the multiplicity $m_L(\infty)$ in (6.484). The proof is complete.    □

*Remark 6.171* If system (SDS) is nonoscillatory at $\infty$ and eventually controllable near $\infty$, then rank $\mathcal{G}_{\min} = n$, i.e., $\mathcal{G} = \mathcal{G}_{\min} = \mathcal{G}_{\max}$ holds for every genus $\mathcal{G}$ (see Remark 6.58). Hence, in this case (6.484) and (6.487) yield that for every conjoined basis $Y$ of (SDS)

$$m_L(\infty) = \text{def } T = n - \text{rank } w(Y^{[\infty]}, Y) = \text{def } w(Y^{[\infty]}, Y). \qquad (6.490)$$

In order to count the numbers of focal points in unbounded intervals, we adopt as in (6.474) for any $M \in [0, \infty)_{\mathbb{Z}}$ the notation

$$\left.\begin{aligned}
m_L(M, \infty] &:= m_L(M, \infty) + m_L(\infty), \\
m_L(M, \infty) &:= l(Y, M, \infty) = \sum_{k=M}^{\infty} m_L(k, k+1], \\
m_R[M, \infty) &:= l^*(Y, M, \infty) = \sum_{k=M}^{\infty} m_R[k, k+1),
\end{aligned}\right\} \qquad (6.491)$$

where we used the definitions of $l(Y, M, \infty)$ and $l^*(Y, M, \infty)$ in (4.87).

The following result is an analog of the formulas in (6.472), (6.473), and (6.230) for the unbounded interval $[N, \infty)_{\mathbb{Z}}$. We will use the representation of conjoined bases of (SDS) in terms of the symplectic fundamental matrix $\Phi_k^{[\infty]}$ in (6.466) and (6.467).

**Theorem 6.172** *If $Y$ is a conjoined basis of (SDS) with constant kernel on $[N, \infty)_{\mathbb{Z}}$ and no forward focal points in $(N, \infty)$, then*

$$m_R[N, \infty) = 0 = \mu^*(Y_N, Y_N^{[\infty]}), \qquad (6.492)$$

$$m_L(N, \infty] = m_L(\infty) = \mu(\mathcal{J}D_\infty, \mathcal{J}D_\infty^{[N]}), \qquad (6.493)$$

*where $D_\infty$ and $D_\infty^{[N]}$ are the constant matrices in (6.467) corresponding to $Y$ and to the principal solution $Y^{[N]}$.*

*Proof* Since $Y^{[\infty]}$ is a minimal conjoined basis near $\infty$ and $Y$ has constant kernel on $[N, \infty)_{\mathbb{Z}}$ and no forward focal points in $(N, \infty)$, there exists $K \in [N, \infty)_{\mathbb{Z}}$ such that both $Y^{[\infty]}$ and $Y$ have constant kernel on $[K, \infty)_{\mathbb{Z}}$ and no forward focal points in $(K, \infty)$, and hence Im $X_k^{[\infty]} \subseteq$ Im $X_k$ on $[K, \infty)_{\mathbb{Z}}$. This means by Theorem 6.80 (with $\tilde{Y} := Y^{[\infty]}$) that $Y^{[\infty]}$ is representable by $Y$ on $[K, \infty)_{\mathbb{Z}}$, i.e., the inclusion

$\operatorname{Im} w(Y, Y^{[\infty]}) \subseteq \operatorname{Im} P$ or equivalently the equality $P w(Y, Y^{[\infty]}) = w(Y, Y^{[\infty]})$ holds. We first prove (6.492). Let $\bar{Y}$ be the conjoined basis in Proposition 6.67 associated with $Y$ on $[N, \infty)_{\mathbb{Z}}$ (i.e., $X_N^{\dagger} \bar{X}_N = 0$ holds). Similarly, let $\bar{Y}^{[\infty]}$ be the conjoined basis in Proposition 6.67 associated with $Y^{[\infty]}$ on $[K, \infty)_{\mathbb{Z}}$. Following (6.466), we denote by $\Phi_k := (Y, \bar{Y})$ and $\Phi_k^{[\infty]} := (Y_k^{[\infty]}, \bar{Y}_k^{[\infty]})$ the corresponding symplectic fundamental matrices of (SDS), so that $Y_k = \Phi_k \mathcal{J}E$ and $Y_k^{[\infty]} = \Phi_k^{[\infty]} \mathcal{J}E$ for all $k \in [0, \infty)_{\mathbb{Z}}$. Then by Theorem 3.5(iii) (with $Z := -\mathcal{J}\Phi_N^{-1}$ and $\hat{Z} := -\mathcal{J}\Phi_N^{-1}\Phi_N^{[\infty]}\mathcal{J}$) we obtain

$$\mu^*(Y_N, Y_N^{[\infty]}) = \mu^*(Z^{-1}E, Z^{-1}\hat{Z}E) = \mu(ZE, \hat{Z}E)$$

$$= \mu\bigl(-\Phi_N^T \mathcal{J}E, -\Phi_N^T \mathcal{J}\Phi_N^{[\infty]}\mathcal{J}E\bigr)$$

$$= \mu\left(\begin{pmatrix} X_N^T \\ \bar{X}_N^T \end{pmatrix}, \begin{pmatrix} w(Y, Y^{[\infty]}) \\ w(\bar{Y}, Y^{[\infty]}) \end{pmatrix}\right), \tag{6.494}$$

where in the last equality we also applied the property $\mu(-Y, -\tilde{Y}) = \mu(Y, \tilde{Y})$ (Theorem 3.5(i) with $C_1 = C_2 := -I$). We now calculate the comparative index in (6.494) by Definition 3.1. Since $\bar{X}_N P = 0$ (see Proposition 6.67(iv)), we have

$$w = X_N w(\bar{Y}, Y^{[\infty]}) - \bar{X}_N P w(Y, Y^{[\infty]}) = X_N w(\bar{Y}, Y^{[\infty]}),$$

$$\mathcal{M} = (I - R_N) w = (I - R_N) X_N w(\bar{Y}, Y^{[\infty]}) = 0, \quad \mathcal{T} = I,$$

$$\mathcal{P} = [w(\bar{Y}, Y^{[\infty]})]^T P w(Y, Y^{[\infty]}) = w(Y^{[\infty]}, \bar{Y}) [w(Y^{[\infty]}, Y)]^T.$$

Then by Theorem 6.162, we know that $\mathcal{P} \geq 0$, and hence the comparative index in (6.494) is equal to rank $\mathcal{M} + \operatorname{ind} \mathcal{P} = 0$. Therefore, we proved $\mu^*(Y_N, Y_N^{[\infty]}) = 0$. But since rank $X_k$ is constant on $[N, \infty)_{\mathbb{Z}}$ and $Y$ has no forward focal points in $(N, \infty)$, the calculation (using Lemma 4.38)

$$m_R[N, \infty) \stackrel{(4.88)}{=} m_L(N, \infty) + \lim_{k \to \infty} \operatorname{rank} X_k - \operatorname{rank} X_N = 0$$

then completes the proof of (6.492). To prove (6.493), we start with the facts that

$$D_\infty = (\Phi_N^{[\infty]})^{-1} Y_N = (\Phi_N^{[\infty]})^{-1} \Phi_N \mathcal{J}E, \quad D_\infty^{[N]} = (\Phi_N^{[\infty]})^{-1} E,$$

$$w(\mathcal{J}D_\infty, \mathcal{J}D_\infty^{[N]}) = Y_N^T \mathcal{J}E = X_N^T.$$

Therefore, we obtain by Theorem 3.5(v) (with $Y := \mathcal{J}D_\infty$ and $\tilde{Y} := \mathcal{J}D_\infty^{[N]}$)

$$\mu(\mathcal{J}D_\infty, \mathcal{J}D_\infty^{[N]}) = \operatorname{rank} X_N - \mu(\mathcal{J}D_\infty^{[N]}, \mathcal{J}D_\infty). \tag{6.495}$$

Now rank $X_N = \operatorname{rank} \mathcal{G}$ and the last term in (6.495) we calculate by Theorem 3.5(iii) (with $Z := \mathcal{J}(\Phi_N^{[\infty]})^{-1}$ and $\hat{Z} := \mathcal{J}(\Phi_N^{[\infty]})^{-1}\Phi_N \mathcal{J}$). Then

$$
\mu(\mathcal{J}D_\infty, \mathcal{J}D_\infty^{[N]}) = \operatorname{rank}\mathcal{G} - \mu\big(\mathcal{J}(\Phi_N^{[\infty]})^{-1}E, \mathcal{J}(\Phi_N^{[\infty]})^{-1}\Phi_N \mathcal{J}E\big)
$$

$$
\overset{(\mathrm{iii})}{=} \operatorname{rank}\mathcal{G} - \mu^*\big(-\Phi_N^{[\infty]}\mathcal{J}E,\ \Phi_N \mathcal{J}E\big)
$$

$$
\overset{(\mathrm{i})}{=} \operatorname{rank}\mathcal{G} - \mu^*(Y_N^{[\infty]}, Y_N)
$$

$$
\overset{(\mathrm{v})}{=} \operatorname{rank}\mathcal{G} - \operatorname{rank} w(Y_N^{[\infty]}, Y_N) + \mu^*(Y_N, Y_N^{[\infty]})
$$

$$
\overset{(6.492)}{=} \operatorname{rank}\mathcal{G} - \operatorname{rank} w(Y_N, Y_N^{[\infty]}) \overset{(6.487)}{=} m_L(\infty).
$$

where above the equality signs we indicate the application of the corresponding properties (iii) and (i) (with $C_1 = -C_2 := -I$) and (v) of Theorem 3.5. Observe that in the last equality we applied Theorem 6.170. Finally, since $Y$ has no forward focal points in $(N, \infty)$, it follows from (6.491) that $m_L(N, \infty] = m_L(\infty)$, which completes the proof of (6.493). ☐

## 6.4.2   Singular Separation Theorems I

In this subsection we derive proper extensions of Propositions 6.166 and 6.167 to unbounded intervals with singular right endpoint. Our results show that, instead of considering the principal solution $Y^{[N]}$ at the right endpoint $k = N$, we have to use the minimal recessive solution $Y^{[\infty]}$ at $\infty$ in the singular case. For convenience we will utilize the notation in (6.491) for the numbers of forward and backward focal points of conjoined bases $Y$, $\tilde{Y}$, $Y^{[M]}$, $Y^{[\infty]}$ in the corresponding unbounded intervals. Our first result is a singular version of Proposition 6.166.

**Theorem 6.173 (Singular Sturmian Separation Theorem)**   *Assume that system* (SDS) *is nonoscillatory at* $\infty$. *Then for any conjoined bases $Y$ and $\tilde{Y}$ of* (SDS), *we have the equalities*

$$
m_L(M, \infty] - \tilde{m}_L(M, \infty] = \mu(\mathcal{J}D_\infty, \mathcal{J}\tilde{D}_\infty) - \mu(Y_M, \tilde{Y}_M) \tag{6.496}
$$

$$
m_R[M, \infty) - \tilde{m}_R[M, \infty) = \mu^*(Y_M, \tilde{Y}_M) - \mu^*(\mathcal{J}D_\infty, \mathcal{J}\tilde{D}_\infty), \tag{6.497}
$$

$$
m_L(M, \infty] + \operatorname{rank} w(Y^{[\infty]}, Y) = m_R[M, \infty) + \operatorname{rank} X_M, \tag{6.498}
$$

*where $D_\infty$ and $\tilde{D}_\infty$ are the constant matrices in (6.467) corresponding to $Y$ and $\tilde{Y}$, respectively.*

*Proof* Since system (SDS) is nonoscillatory at $\infty$, we have $m_L(M, \infty] < \infty$ and $\tilde{m}_L(M, \infty] < \infty$. Then we can choose $N \in [M, \infty)_{\mathbb{Z}}$ such that both conjoined bases $Y$ and $\tilde{Y}$ have constant kernel on $[N, \infty)_{\mathbb{Z}}$ and no forward focal points in $(N, \infty)$. First we will prove that

$$m_L(N, \infty] - \tilde{m}_L(N, \infty] = \mu(\mathcal{J}D_\infty, \mathcal{J}\tilde{D}_\infty) - \mu(Y_N, \tilde{Y}_N) \tag{6.499}$$

$$m_R[N, \infty) - \tilde{m}_R[N, \infty) = \mu^*(Y_N, \tilde{Y}_N) - \mu^*(\mathcal{J}D_\infty, \mathcal{J}\tilde{D}_\infty). \tag{6.500}$$

Let $\Phi_k^{[\infty]}$ be the symplectic fundamental matrix of (SDS) in (6.467) with the associated matrices $D_\infty$, $\tilde{D}_\infty$, $D_\infty^{[N]}$, and $D_\infty^{[\infty]} = \mathcal{J}E$ for $Y$, $\tilde{Y}$, $Y^{[N]}$, and $Y^{[\infty]}$, respectively. Further, consider the symplectic fundamental matrices $\Phi_k$ and $\tilde{\Phi}_k$ of (SDS) such that $\Phi_k E = Y_k$ and $\tilde{\Phi}_k E = \tilde{Y}_k$ on $[0, \infty)_{\mathbb{Z}}$, i.e., $\Phi_k = (*, Y_k)$ and $\tilde{\Phi}_k = (*, \tilde{Y}_k)$ according to the definition of $E$ in (6.471). Define the symplectic matrix $\mathcal{R} := -\mathcal{J}(\Phi_N^{[\infty]})^{-1}$. Then (6.467) yields

$$\mathcal{J}D_\infty = -\mathcal{R}\,\Phi_N E, \quad \mathcal{J}\tilde{D}_\infty = -\mathcal{R}\,\tilde{\Phi}_N E, \quad \mathcal{J}D_\infty^{[N]} = -\mathcal{R}E. \tag{6.501}$$

By Theorem 6.172 and by the transformation formula (3.16) in Theorem 3.6 for the comparative index (with $W := -\mathcal{R}$, $Z := \Phi_N$, $\hat{Z} := \tilde{\Phi}_N$), we get

$$m_L(N, \infty] - \tilde{m}_L(N, \infty] \overset{(6.493)}{=} \mu(\mathcal{J}D_\infty, \mathcal{J}D_\infty^{[N]}) - \mu(\mathcal{J}\tilde{D}_\infty, \mathcal{J}D_\infty^{[N]})$$

$$\overset{(6.501)}{=} \mu(-\mathcal{R}\,\Phi_N E, -\mathcal{R}E) - \mu(-\mathcal{R}\,\tilde{\Phi}_N E, -\mathcal{R}E)$$

$$\overset{(3.16)}{=} \mu(-\mathcal{R}\,\Phi_N E, -\mathcal{R}\,\tilde{\Phi}_N E) - \mu(\Phi_N E, \tilde{\Phi}_N E)$$

$$\overset{(6.501)}{=} \mu(\mathcal{J}D_\infty, \mathcal{J}\tilde{D}_\infty) - \mu(Y_N, \tilde{Y}_N),$$

showing (6.499). Next, we have

$$Y_N^{[\infty]} = \mathcal{R}^{-1}E, \quad Y_N = \Phi_N E, \quad \tilde{Y}_N = \tilde{\Phi}_N E. \tag{6.502}$$

By Theorem 6.172 and by the transformation formula (3.27) for the dual comparative index (with $W := \mathcal{R}^{-1}$, $Z := \mathcal{R}\,\Phi_N$, $\hat{Z} := \mathcal{R}\,\tilde{\Phi}_N$), we get

$$m_R[N, \infty) - \tilde{m}_R[N, \infty) \overset{(6.492)}{=} \mu^*(Y_N, Y_N^{[\infty]}) - \mu^*(\tilde{Y}_N, Y_N^{[\infty]})$$

$$\overset{(6.502)}{=} \mu^*(\mathcal{R}^{-1}\mathcal{R}\,\Phi_N E, \mathcal{R}^{-1}E) - \mu^*(\mathcal{R}^{-1}\mathcal{R}\,\tilde{\Phi}_N E, \mathcal{R}^{-1}E)$$

$$\overset{(3.27)}{=} \mu^*(\mathcal{R}^{-1}\mathcal{R}\,\Phi_N E, \mathcal{R}^{-1}\mathcal{R}\,\tilde{\Phi}_N E) - \mu^*(\mathcal{R}\,\Phi_N E, \mathcal{R}\,\tilde{\Phi}_N E)$$

$$\overset{(6.501),\,(6.502)}{=} \mu^*(Y_N, \tilde{Y}_N) - \mu^*(-\mathcal{J}D_\infty, -\mathcal{J}\tilde{D}_\infty)$$

$$= \mu^*(Y_N, \tilde{Y}_N) - \mu^*(\mathcal{J}D_\infty, \mathcal{J}\tilde{D}_\infty),$$

showing (6.500), where in the last step we used that $\mu^*(-Y, -\tilde{Y}) = \mu^*(Y, \tilde{Y})$ (see Theorem 3.5(i) with $C_1 = C_2 = -I$). Next we combine the above formulas (6.499) and (6.500) with the formulas on the bounded interval in Proposition 6.166 to get

$$m_L(M, \infty] - \tilde{m}_L(M, \infty] = m_L(M, N] + m_L(N, \infty] - \tilde{m}_L(M, N] - \tilde{m}_L(N, \infty]$$

$$\overset{(6.475),\,(6.499)}{=} \mu(\mathcal{J}D_\infty, \mathcal{J}\tilde{D}_\infty) - \mu(Y_M, \tilde{Y}_M),$$

$$m_R[M, \infty) - \tilde{m}_R[M, \infty) = m_R[M, N) + m_R[N, \infty) - \tilde{m}_R[M, N) - \tilde{m}_R[N, \infty)$$

$$\overset{(6.476),\,(6.500)}{=} \mu^*(Y_M, \tilde{Y}_M) - \mu^*(\mathcal{J}D_\infty, \mathcal{J}\tilde{D}_\infty).$$

Therefore, we proved (6.496) and (6.497). Finally, since the index $N$ is chosen so that $m_L(N, \infty) = 0$ and $m_R[N, \infty) = 0$, it follows that $m_L(M, N] = m_L(M, \infty)$ and $m_R[M, N) = m_R[M, \infty)$ and rank $X_N = \text{rank}\,\mathcal{G}$, where $\mathcal{G}$ is the genus of $Y$ near $\infty$. By (6.477) in Proposition 6.166, we then get

$$m_L(M, \infty) + \text{rank}\,\mathcal{G} = m_R[M, \infty) + \text{rank}\,X_M,$$

which upon substituting for rank $\mathcal{G} = m_L(\infty) + \text{rank}\,w(Y^{[\infty]}, Y)$ from formula (6.487) in Theorem 6.170 yields the required equation (6.498).                    □

The result in Theorem 4.39, or equivalently the limiting case of (6.475) and (6.476) as $N \to \infty$, implies the equalities

$$m_L(M, \infty) - \tilde{m}_L(M, \infty) = \mu_\infty(Y, \tilde{Y}) - \mu(Y_M, \tilde{Y}_M), \qquad (6.503)$$

$$m_R[M, \infty) - \tilde{m}_R[M, \infty) = \mu^*(Y_M, \tilde{Y}_M) - \mu^*_\infty(Y, \tilde{Y}), \qquad (6.504)$$

where the numbers $\mu_\infty(Y, \tilde{Y})$ and $\mu^*_\infty(Y, \tilde{Y})$ are defined, respectively, as the limits

$$\mu_\infty(Y, \tilde{Y}) := \lim_{k \to \infty} \mu(Y_k, \tilde{Y}_k), \quad \mu^*_\infty(Y, \tilde{Y}) := \lim_{k \to \infty} \mu^*(Y_k, \tilde{Y}_k); \qquad (6.505)$$

see also (4.91) and (4.93). In other words, $\mu_\infty(Y, \tilde{Y})$ and $\mu^*_\infty(Y, \tilde{Y})$ are defined by equations (6.503) and (6.504), since the quantities $m_L(M, \infty)$, $\tilde{m}_L(M, \infty)$ and $m_R[M, \infty)$, $\tilde{m}_R[M, \infty)$ are finite for a nonoscillatory system (SDS) at $\infty$. The results in Theorem 6.173 then allow to calculate these numbers explicitly as values of the comparative index, which was not possible by the methods of Sect. 4.2.4.

**Corollary 6.174** *Assume that system (SDS) is nonoscillatory at $\infty$. Then for any conjoined bases $Y$ and $\tilde{Y}$ of (SDS), the limits in (6.505) satisfy*

$$\mu_\infty(Y, \tilde{Y}) = \mu(\mathcal{J}D_\infty, \mathcal{J}\tilde{D}_\infty) - m_L(\infty) + \tilde{m}_L(\infty), \qquad (6.506)$$

$$\mu^*_\infty(Y, \tilde{Y}) = \mu^*(\mathcal{J}D_\infty, \mathcal{J}\tilde{D}_\infty), \qquad (6.507)$$

where $D_\infty$ and $\tilde{D}_\infty$ are the constant matrices in (6.467) corresponding to $Y$ and $\tilde{Y}$, respectively.

*Proof* We use identities (6.503) and (6.504) to derive

$$\mu_\infty(Y, \tilde{Y}) \overset{(6.503)}{=} m_L(M, \infty) - \tilde{m}_L(M, \infty) + \mu(Y_M, \tilde{Y}_M)$$

$$= m_L(M, \infty] - m_L(\infty) - \tilde{m}_L(M, \infty] + \tilde{m}_L(\infty) + \mu(Y_M, \tilde{Y}_M)$$

$$\overset{(6.496)}{=} \mu(\mathcal{J}D_\infty, \mathcal{J}\tilde{D}_\infty) - m_L(\infty) + \tilde{m}_L(\infty),$$

$$\mu_\infty^*(Y, \tilde{Y}) \overset{(6.504)}{=} \mu^*(Y_M, \tilde{Y}_M) - m_R[M, \infty) + \tilde{m}_R[M, \infty) \overset{(6.497)}{=} \mu^*(\mathcal{J}D_\infty, \mathcal{J}\tilde{D}_\infty),$$

which shows the statement.                                                                     □

We also note that the formulas in (6.506) and (6.507) are new even for a controllable system (SDS) near $\infty$.

*Remark 6.175* Given a conjoined basis $Y$ of (SDS), equation (6.506) allows to connect the limit $\mu_\infty(Y, Y^{[\infty]})$ with the multiplicity $m_L(\infty)$ and with the rank of the associated matrix $T$. Namely, by taking $\tilde{Y} := Y^{[\infty]}$, we get $\tilde{D}_\infty = D_\infty^{[\infty]} = \mathcal{J}E$ (see the proof of Theorem 6.173), and then the property $\mu(Y, -E) = 0$ implies that $\mu(\mathcal{J}D_\infty, \mathcal{J}D_\infty^{[\infty]}) = \mu(\mathcal{J}D_\infty, -E) = 0$. Consequently, equation (6.506) and Remark 6.169 yield the formula

$$\mu_\infty(Y, Y^{[\infty]}) \overset{(6.506)}{=} n - d_\infty - m_L(\infty) \overset{(6.484)}{=} \text{rank } T. \tag{6.508}$$

The results in Theorem 6.173 allow to present exact formulas for the numbers of focal points of a given conjoined basis $Y$ of (SDS) in terms of the corresponding numbers of focal points of the the the minimal recessive solution $Y^{[\infty]}$ of (SDS) at $\infty$ and of the principal solution $Y^{[M]}$.

**Corollary 6.176** *Assume that system (SDS) is nonoscillatory at $\infty$. Then for any conjoined basis $Y$ of (SDS), we have the equalities*

$$m_L(M, \infty] = m_L^{[M]}(M, \infty] + \mu(\mathcal{J}D_\infty, \mathcal{J}D_\infty^{[M]}), \tag{6.509}$$

$$m_L(M, \infty] = m_L^{[\infty]}(M, \infty] - \mu(Y_M, Y_M^{[\infty]}), \tag{6.510}$$

$$m_R[M, \infty) = m_R^{[M]}[M, \infty) - \mu^*(\mathcal{J}D_\infty, \mathcal{J}D_\infty^{[M]}), \tag{6.511}$$

$$m_R[M, \infty) = m_R^{[\infty]}[M, \infty) + \mu^*(Y_M, Y_M^{[\infty]}), \tag{6.512}$$

*where $D_\infty$ and $D_\infty^{[M]}$ are the constant matrices in (6.467) corresponding to $Y$ and $Y^{[M]}$, respectively.*

*Proof* For (6.509) and (6.511), we apply (6.496) and (6.497) with $\tilde{Y} := Y^{[M]}$. In this case $\tilde{D}_\infty = D_\infty^{[M]}$ and $\mu(Y_M, E) = 0 = \mu^*(Y_M, E)$. For (6.510) and (6.512), we apply (6.496) and (6.497) with $\tilde{Y} := Y^{[\infty]}$ and $\tilde{D}_\infty = D_\infty^{[\infty]} = \mathcal{J}E$. In this case $\mu(\mathcal{J}D_\infty, -E) = 0 = \mu^*(\mathcal{J}D_\infty, -E)$. □

In the following result, we relate the numbers of forward and backward focal points of the minimal recessive solution $Y^{[\infty]}$ at $\infty$, resp., of the principal solution $Y^{[M]}$, according to equalities (6.498), (6.509), and (6.512).

**Corollary 6.177** *Assume that system* (SDS) *is nonoscillatory at* $\infty$. *Then*

$$m_L^{[\infty]}(M, \infty] = m_R^{[\infty]}[M, \infty) + \operatorname{rank} X_M^{[\infty]}, \tag{6.513}$$

$$m_R^{[M]}[M, \infty) = m_L^{[M]}(M, \infty] + \operatorname{rank} X_M^{[\infty]}, \tag{6.514}$$

$$m_L^{[\infty]}(M, \infty] = m_L^{[M]}(M, \infty] + \operatorname{rank} X_M^{[\infty]}, \tag{6.515}$$

$$m_R^{[M]}[M, \infty) = m_R^{[\infty]}[M, \infty) + \operatorname{rank} X_M^{[\infty]}. \tag{6.516}$$

*Proof* For (6.513) we apply (6.498) with $Y := Y^{[\infty]}$, while for (6.514) we apply (6.498) with $Y := Y^{[M]}$, where we utilize the fact that $w(Y^{[\infty]}, Y^{[M]}) = (X_M^{[\infty]})^T$. Next, for (6.515) we use (6.509) with $Y := Y^{[\infty]}$ and $D_\infty = D_\infty^{[\infty]} = \mathcal{J}E$. Since by (6.467) the upper block of $\mathcal{J}D_\infty^{[M]}$ is equal to $w(Y^{[\infty]}, Y^{[M]}) = (X_M^{[\infty]})^T$, it follows by $\mu(E, Y) = \operatorname{rank} X$ (see Remark 3.4(iii)) that $\mu(\mathcal{J}D_\infty, \mathcal{J}D_\infty^{[M]}) = \mu(-E, \mathcal{J}D_\infty^{[M]}) = \operatorname{rank} X_M^{[\infty]}$ and hence, (6.509) implies (6.515). Finally, for (6.516) we use (6.512) with $Y := Y^{[M]}$, where $\mu^*(E, Y_M^{[\infty]}) = \operatorname{rank} X_M^{[\infty]}$ according to the property $\mu^*(E, Y) = \operatorname{rank} X$. □

In the next result, we present a complete singular version of Proposition 6.167.

**Theorem 6.178 (Singular Sturmian Separation Theorem)** *Assume that system* (SDS) *is nonoscillatory at* $\infty$. *Then for the minimal recessive solution* $Y^{[\infty]}$ *at* $\infty$ *and for the principal solution* $Y^{[M]}$, *we have the identities*

$$m_L^{[M]}(M, \infty] = m_R^{[\infty]}[M, \infty), \quad m_R^{[M]}[M, \infty) = m_L^{[\infty]}(M, \infty]. \tag{6.517}$$

*For any conjoined basis Y of* (SDS), *we have the estimates*

$$m_L^{[M]}(M, \infty] \leq m_L(M, \infty] \leq m_L^{[\infty]}(M, \infty], \tag{6.518}$$

$$m_R^{[\infty]}[M, \infty) \leq m_R[M, \infty) \leq m_R^{[M]}[M, \infty). \tag{6.519}$$

*For any conjoined bases Y and $\tilde{Y}$ of* (SDS), *we have the estimates*

$$\left| m_L(M, \infty] - \tilde{m}_L(M, \infty] \right| \leq \operatorname{rank} X_M^{[\infty]} \leq n, \tag{6.520}$$

$$\left| m_R[M, \infty) - \tilde{m}_R[M, \infty) \right| \leq \operatorname{rank} X_M^{[\infty]} \leq n, \tag{6.521}$$

$$\left| m_L(M, \infty] - \tilde{m}_R[M, \infty) \right| \leq \operatorname{rank} X_M^{[\infty]} \leq n. \tag{6.522}$$

*Proof* The equalities in (6.517) follow by subtracting (6.514) and (6.516), respectively, by subtracting (6.514) and (6.515). Next, the comparative index and the dual comparative index are nonnegative, so that estimate (6.518) is a consequence of (6.509) and (6.510). Similarly, estimate (6.519) follows from (6.512) and (6.511). Finally, the lower and upper bounds in (6.518) and (6.519) differ by the same number rank $X_M^{[\infty]}$ (by Corollary 6.177). Therefore, the inequalities in (6.520)–(6.522) follow from the estimates in (6.518) and (6.519). □

We note that the lower and upper bounds in (6.518)–(6.522) are optimal in a sense that they cannot be improved by better bounds, which would be independent on the arbitrarily chosen conjoined bases $Y$ and $\tilde{Y}$. In this respect the two quantities in (6.517) together with the number rank $X_M^{[\infty]}$ represent important parameters or characteristics of the symplectic system (SDS) on the given unbounded interval $[M, \infty)_{\mathbb{Z}}$.

In the remaining part of this subsection, we will study the multiplicities of focal points of conjoined bases of (SDS) in the open interval $(M, \infty)$. The above problem is closely related with limiting the inequalities in Proposition 6.167 for $N \to \infty$. In particular, we will see that by taking $N \to \infty$ in Proposition 6.167, we do not obtain the statements in Theorem 6.178. First we consider the upper bound in (6.479). We have

$$\lim_{N \to \infty} m_L^{[N]}(M, N] \stackrel{(6.478)}{=} \lim_{N \to \infty} m_R^{[M]}[M, N) \stackrel{(6.491)}{=} m_R^{[M]}[M, \infty)$$

$$\stackrel{(6.517)}{=} m_L^{[\infty]}(M, \infty], \tag{6.523}$$

which is the correct upper bound obtained in (6.518). Another expression of this limit can be obtained as

$$\lim_{N \to \infty} m_L^{[N]}(M, N] = \lim_{N \to \infty} \left\{ m_L^{[M]}(M, N] + \operatorname{rank} X_N^{[M]} \right\}$$

$$\stackrel{(6.491)}{=} m_L^{[M]}(M, \infty) + \operatorname{rank} \mathcal{G}^{[M]}, \tag{6.524}$$

where the first equality follows from (6.475) with $\tilde{Y} := Y^{[M]}$ and where $\mathcal{G}^{[M]}$ is the genus of $Y^{[M]}$ near $\infty$. On the other hand, by considering the lower bound in (6.480), we get

$$\lim_{N \to \infty} m_R^{[N]}[M, N) \stackrel{(6.478)}{=} \lim_{N \to \infty} m_L^{[M]}(M, N] \stackrel{(6.491)}{=} m_L^{[M]}(M, \infty)$$

$$= m_L^{[M]}(M, \infty] - m_L^{[M]}(\infty) \stackrel{(6.517)}{=} m_R^{[\infty]}[M, \infty) - m_L^{[M]}(\infty),$$

which is in general smaller than the optimal lower bound in (6.519). Moreover, it is equal to the optimal lower bound $m_R^{[\infty]}[M, \infty)$ if and only if $m_L^{[M]}(\infty) = 0$, i.e., if and only if the principal solution $Y^{[M]}$ is a dominant solution of (SDS) at infinity (by Remark 6.169(ii)). Altogether, limiting the inequalities in (6.480) for $N \to \infty$, we obtain the estimate

$$m_R^{[\infty]}[M, \infty) - m_L^{[M]}(\infty) \le m_R[M, \infty) \le m_R^{[M]}[M, \infty),$$

which contains a nonoptimal lower bound for the number of backward focal points of $Y$ in $[M, \infty)$, while limiting the inequalities in (6.479) for $N \to \infty$, we obtain a result counting the forward focal points of $Y$ in the open interval $(M, \infty)$.

**Corollary 6.179** *Assume that system* (SDS) *is nonoscillatory at* $\infty$. *Then for any conjoined basis* $Y$ *of* (SDS), *we have the estimates*

$$m_L^{[M]}(M, \infty) \le m_L(M, \infty) \le m_L^{[\infty]}(M, \infty] = m_L^{[M]}(M, \infty) + \operatorname{rank} \mathcal{G}^{[M]},$$
$$(6.525)$$

*where* $\mathcal{G}^{[M]}$ *is the genus of the principal solution* $Y^{[M]}$ *of* (SDS) *near* $\infty$.

*Proof* The statement follows from (6.479) and from the calculations in (6.523) and (6.524).                                                                       □

Next we derive an exact relationship between the numbers of forward and backward focal points of $Y^{[\infty]}$ and $Y^{[M]}$ in the open interval $(M, \infty)$.

**Corollary 6.180** *Assume that system* (SDS) *is nonoscillatory at* $\infty$. *Then*

$$m_L^{[M]}(M, \infty) + \operatorname{rank} \mathcal{G}^{[M]} = m_L^{[\infty]}(M, \infty) + n - d_\infty, \qquad (6.526)$$

$$m_L^{[M]}(M, \infty) = m_L^{[\infty]}(M, \infty) \iff \operatorname{rank} \mathcal{G}^{[M]} = n - d_\infty \iff Y^{[M]} \in \mathcal{G}_{\min},$$
$$(6.527)$$

*where* $\mathcal{G}^{[M]}$ *and* $\mathcal{G}_{\min}$ *are the genera of conjoined bases of* (SDS) *near* $\infty$ *corresponding to* $Y^{[M]}$ *and* $Y^{[\infty]}$. *If in addition system* (SDS) *is controllable near* $\infty$, *then we have the equality*

$$m_L^{[M]}(M, \infty) = m_L^{[\infty]}(M, \infty). \qquad (6.528)$$

*Proof* Equation (6.526) is a reformulation of (6.515), since $m_L^{[\infty]}(\infty) = n - d_\infty$ by Remark 6.169(ii) and $m_L^{[M]}(\infty) = \operatorname{rank} \mathcal{G}^{[M]} - \operatorname{rank} w(Y^{[\infty]}, Y^{[M]})$, where $w(Y^{[\infty]}, Y^{[M]}) = (X_M^{[\infty]})^T$. The equivalences in (6.527) then follow from equation (6.526). If in addition the system (SDS) is controllable near $\infty$, then $d_\infty = 0$ and $\operatorname{rank} \mathcal{G}_{\min} = n = \operatorname{rank} \mathcal{G}^{[M]}$. In this case we obtain from (6.527) that (6.528) is indeed satisfied.                                                                       □

As the last result in this subsection, we present limit properties of the comparative index and the dual comparative index involving a given conjoined basis $Y$ and the principal solution $Y^{[k]}$ for $k \to \infty$. Namely, we show that these limits are related to the multiplicity $m_L(\infty)$ and to the maximal order of abnormality $d_\infty$.

**Theorem 6.181** *Assume that system* (SDS) *is nonoscillatory at* $\infty$. *Then for any conjoined basis* $Y$ *of* (SDS), *the comparative indices* $\mu(\mathcal{J}D_\infty, \mathcal{J}D_\infty^{[k]})$ *and* $\mu^*(\mathcal{J}D_\infty, \mathcal{J}D_\infty^{[k]})$ *have limits for* $k \to \infty$, *which satisfy*

$$\lim_{k\to\infty} \mu(\mathcal{J}D_\infty, \mathcal{J}D_\infty^{[k]}) = m_L(\infty), \tag{6.529}$$

$$\lim_{k\to\infty} \mu^*(\mathcal{J}D_\infty, \mathcal{J}D_\infty^{[k]}) = n - d_\infty, \tag{6.530}$$

*where* $D_\infty$ *and* $D_\infty^{[k]}$ *are the constant matrices in* (6.467) *corresponding to* $Y$ *and* $Y^{[k]}$, *respectively.*

*Proof* For any $k \in [0, \infty)_\mathbb{Z}$, we have by Corollary 6.176 and Theorem 6.178 on the interval $[k, \infty)_\mathbb{Z}$ that

$$\mu(\mathcal{J}D_\infty, \mathcal{J}D_\infty^{[k]}) \overset{(6.509)}{=} m_L(k, \infty] - m_L^{[k]}(k, \infty]$$

$$\overset{(6.517)}{=} m_L(k, \infty) + m_L(\infty) - m_R^{[\infty]}[k, \infty). \tag{6.531}$$

Since system (SDS) is nonoscillatory at $\infty$, the conjoined bases $Y$ and $Y^{[\infty]}$ have for large $k$ no forward and backward focal points in the intervals $(k, \infty)$ and $[k, \infty)$, i.e., $m_L(k, \infty) = 0 = m_R^{[\infty]}[k, \infty)$ for large $k$. Therefore, equality (6.529) follows from (6.531). Similarly, we have

$$\mu^*(\mathcal{J}D_\infty, \mathcal{J}D_\infty^{[k]}) \overset{(6.511)}{=} m_R^{[k]}[k, \infty) - m_R[k, \infty)$$

$$\overset{(6.517)}{=} m_L^{[\infty]}(k, \infty) + m_L^{[\infty]}(\infty) - m_R[k, \infty). \tag{6.532}$$

And since $m_L^{[\infty]}(k, \infty) = 0 = m_R[k, \infty)$ for large $k$ and $m_L^{[\infty]}(\infty) = n - d_\infty$, we obtain equality (6.530) from taking the limit $k \to \infty$ in (6.532). $\qquad\square$

It is interesting to realize that the value of the limit in (6.530) does not depend on the chosen conjoined basis $Y$.

### 6.4.3  Singular Separation Theorems II

In this subsection we present singular separation theorems for system (SDS) on unbounded intervals involving $-\infty$, i.e., with the left singular endpoint. These intervals will be either bounded from above or unbounded from above. We will present results, which are in some sense analogous to those in Sect. 6.4.3.

The maximal order of abnormality of (SDS) near $-\infty$ is defined by

$$d_{-\infty} := \lim_{k\to-\infty} d(-\infty, k]_\mathbb{Z}, \quad 0 \le d_{-\infty} \le n,$$

where $d(-\infty, N]_{\mathbb{Z}}$ is the order of abnormality of system (SDS) on the interval $(-\infty, N]_{\mathbb{Z}}$. The nonoscillation and eventual controllability of (SDS) near $-\infty$ then imply that $d_{-\infty} = 0$ and that every conjoined basis $Y$ of (SDS) has the matrix $X_k$ invertible for all negative $k$ large enough.

If (SDS) is nonoscillatory at $-\infty$, then we denote by $Y^{[-\infty]}$ the (unique) minimal recessive solution of (SDS) at $-\infty$ (by an analog with Theorem 6.115 and the notation in (6.471)). For completeness we note that a conjoined basis $Y$ of (SDS) is a *recessive solution* of (SDS) at $-\infty$ if there exists $N \in (-\infty, -1]_{\mathbb{Z}}$ such that $Y$ has constant kernel on $(-\infty, N + 1]_{\mathbb{Z}}$ and no backward focal points in $(-\infty, N + 1)$ and the corresponding negative semidefinite matrix $T_{-\infty}$ defined by

$$T_{-\infty} := \lim_{k \to -\infty} (S_k^{-\infty})^\dagger, \quad S_k^{-\infty} := -\sum_{j=k+1}^{N} X_j^\dagger B_j^T X_{j+1}^{\dagger T}, \quad k \in (-\infty, N]_{\mathbb{Z}},$$

(6.533)

with $S_N^{-\infty} = 0$ satisfies $T_{-\infty} = 0$; compare with [300, Definition 4.2] in the controllable case near $-\infty$. Similarly, a conjoined basis $Y$ of (SDS) is a *dominant solution* of (SDS) at $-\infty$ if there exists $N \in (-\infty, -1]_{\mathbb{Z}}$ such that

$$d(-\infty, N + 1]_{\mathbb{Z}} = d_{-\infty}$$

(6.534)

holds, the conjoined basis $Y$ has constant kernel on $(-\infty, N + 1]_{\mathbb{Z}}$ and no backward focal points in $(-\infty, N + 1)$, and the corresponding matrix $T_{-\infty}$ defined in (6.237) satisfies rank $T_{-\infty} = n - d_{-\infty}$ (i.e., the rank of $T_{-\infty}$ is maximal).

*Remark 6.182* In order to distinguish the notation for possibly different genera of conjoined bases of (SDS) near $\infty$ and $-\infty$, we will denote the genera near $\infty$ with the upper or lower index $+$ and the genera near $-\infty$ with upper or lower index $-$. That is, $\mathcal{G}_+$ will denote a genus near $\infty$ (with $\mathcal{G}_{\min}^+$ and $\mathcal{G}_{\max}^+$ being the minimal and maximal genus near $\infty$), while $\mathcal{G}_-$ will denote a genus near $-\infty$ (with $\mathcal{G}_{\min}^-$ and $\mathcal{G}_{\max}^-$ being the minimal and maximal genus near $-\infty$).

In view of Lemma 6.59 with $j = -\infty$, every conjoined basis $Y$ of (SDS) can be uniquely represented by a constant $2n \times n$ matrix $D_{-\infty}$ such that

$$Y_k = \Phi_k^{[-\infty]} D_{-\infty}, \quad k \in \mathcal{I}_{\mathbb{Z}}, \quad \mathcal{J} D_{-\infty} = \begin{pmatrix} w(Y^{[-\infty]}, Y) \\ w(\bar{Y}^{[-\infty]}, Y) \end{pmatrix},$$

(6.535)

where $\Phi_k^{[-\infty]} = (Y_k^{[-\infty]}, \bar{Y}_k^{[-\infty]})$ is the symplectic fundamental matrix of (SDS) defined by (6.228) with $j = -\infty$ and where $\mathcal{I}_{\mathbb{Z}}$ is the unbounded interval $(-\infty, 0]_{\mathbb{Z}}$ or $(-\infty, \infty)_{\mathbb{Z}} = \mathbb{Z}$.

In the rest of this subsection, we assume that system (SDS) is nonoscillatory at $-\infty$. In analogous way to Definition 6.168, we introduce the multiplicity $m_R(-\infty)$ as follows.

**Definition 6.183** For a conjoined basis $Y$ of (SDS), we define the quantity

$$m_R(-\infty) := n - d_{-\infty} - \operatorname{rank} T_{-\infty} = \operatorname{def} T_{-\infty} - d_{-\infty}, \qquad (6.536)$$

where $T_{-\infty}$ is the matrix in (6.533) associated with $Y$ on $(-\infty, N]_{\mathbb{Z}}$, which satisfies (6.534). Moreover, we say that $Y$ has a *focal point at* $-\infty$ if $m_R(-\infty) \geq 1$ and then $m_R(-\infty)$ is called its *multiplicity*.

Following (6.491) and (4.89), we adopt for any conjoined basis $Y$ of (SDS) and any $N \in (-\infty, 0]_{\mathbb{Z}}$ the notation

$$\left. \begin{aligned}
m_R[-\infty, N) &:= m_R(-\infty, N) + m_R(-\infty), \\
m_R(-\infty, N) &:= \sum_{k=-\infty}^{N-1} m_R[k, k+1), \\
m_L(-\infty, N] &:= l(Y, -\infty, N) = \sum_{k=-\infty}^{N-1} m_L(k, k+1].
\end{aligned} \right\} \qquad (6.537)$$

For intervals with two singular endpoints, we denote in a similar way

$$\left. \begin{aligned}
m_L(-\infty, \infty] &:= m_L(-\infty, \infty) + m_L(\infty), \\
m_L(-\infty, \infty) &:= \sum_{k=-\infty}^{\infty} m_L(k, k+1], \\
m_R[-\infty, \infty) &:= m_R(-\infty, \infty) + m_R(-\infty), \\
m_R(-\infty, \infty) &:= \sum_{k=-\infty}^{\infty} m_R[k, k+1).
\end{aligned} \right\} \qquad (6.538)$$

Similar notation will be used for another conjoined basis $\tilde{Y}$, for the principal solution $Y^{[N]}$, and for the minimal recessive solution $Y^{[-\infty]}$ at $-\infty$. Then we have the following results.

**Theorem 6.184** *Assume that system (SDS) is nonoscillatory at* $-\infty$, *and let* $Y$, *belonging to a genus* $\mathcal{G}_-$, *be a conjoined basis of (SDS) with constant kernel on* $(-\infty, N+1]_{\mathbb{Z}}$ *and no backward focal points in* $(-\infty, N+1)$, *where the index* $N \in (-\infty, -1]_{\mathbb{Z}}$ *is such that (6.534) holds. Then*

$$\operatorname{Im}[w(Y^{[-\infty]}, Y)]^T = \operatorname{Im} T_{-\infty} \oplus \operatorname{Im}(P - P_{\mathcal{S}-\infty}), \qquad (6.539)$$

$$\operatorname{rank} T_{-\infty} = \operatorname{rank} w(Y^{[-\infty]}, Y) - \operatorname{rank} \mathcal{G}_- + n - d_{-\infty}, \qquad (6.540)$$

$$m_R(-\infty) = \operatorname{rank} \mathcal{G}_- - \operatorname{rank} w(Y^{[-\infty]}, Y). \qquad (6.541)$$

*Proof* The results are proven by analogy with Theorem 6.170. □

**Theorem 6.185 (Singular Sturmian Separation Theorem)** *If* (SDS) *is nonoscillatory at* $-\infty$,, *then for any conjoined bases* $Y$ *and* $\tilde{Y}$ *of* (SDS), *we have*

$$m_L(-\infty, N] - \tilde{m}_L(-\infty, N] = \mu(Y_N, \tilde{Y}_N) - \mu(\mathcal{J}D_{-\infty}, \mathcal{J}\tilde{D}_{-\infty}), \qquad (6.542)$$

$$m_R[-\infty, N) - \tilde{m}_R[-\infty, N) = \mu^*(\mathcal{J}D_{-\infty}, \mathcal{J}\tilde{D}_{-\infty}) - \mu^*(Y_N, \tilde{Y}_N), \qquad (6.543)$$

$$m_L(-\infty, N] + \operatorname{rank} X_N = m_R[-\infty, N) + \operatorname{rank} w(Y^{[-\infty]}, Y), \qquad (6.544)$$

*where* $D_{-\infty}$ *and* $\tilde{D}_{-\infty}$ *are the constant matrices in* (6.535) *corresponding to* $Y$ *and* $\tilde{Y}$, *respectively. If* (SDS) *is nonoscillatory at* $\pm\infty$, *then for any conjoined bases* $Y$ *and* $\tilde{Y}$ *of* (SDS), *we have the equalities*

$$m_L(-\infty, \infty] - \tilde{m}_L(-\infty, \infty] = \mu(\mathcal{J}D_\infty, \mathcal{J}\tilde{D}_\infty) - \mu(\mathcal{J}D_{-\infty}, \mathcal{J}\tilde{D}_{-\infty}), \qquad (6.545)$$

$$m_R[-\infty, \infty) - \tilde{m}_R[-\infty, \infty) = \mu^*(\mathcal{J}D_{-\infty}, \mathcal{J}\tilde{D}_{-\infty}) - \mu^*(\mathcal{J}D_\infty, \mathcal{J}\tilde{D}_\infty), \qquad (6.546)$$

$$m_L(-\infty, \infty] + \operatorname{rank} w(Y^{[\infty]}, Y) = m_R[-\infty, \infty) + \operatorname{rank} w(Y^{[-\infty]}, Y), \qquad (6.547)$$

*where* $D_{\pm\infty}$ *and* $\tilde{D}_{\pm\infty}$ *are the constant matrices in* (6.467) *and* (6.535) *corresponding to* $Y$ *and* $\tilde{Y}$, *respectively.*

*Proof* Identities (6.542)–(6.544) are proven by analogy with Theorem 6.173. Identities (6.545)–(6.547) follow by adding the corresponding identities in (6.496)–(6.498) and (6.542)–(6.544). □

Following (6.506)–(6.507), Theorem 4.39, and Theorems 4.59 and 4.61 (for one system (SDS)), we consider the equalities

$$m_L(-\infty, N] - \tilde{m}_L(-\infty, N] = \mu(Y_N, \tilde{Y}_N) - \mu_{-\infty}(Y, \tilde{Y}), \qquad (6.548)$$

$$m_R(-\infty, N) - \tilde{m}_R(-\infty, N) = \mu^*_{-\infty}(Y, \tilde{Y}) - \mu^*(Y_N, \tilde{Y}_N), \qquad (6.549)$$

as the limiting case of (6.475)–(6.476) when the left endpoint approaches $-\infty$, where the numbers $\mu_{-\infty}(Y, \tilde{Y})$ and $\mu^*_{-\infty}(Y, \tilde{Y})$ are defined as the limits

$$\mu_{-\infty}(Y, \tilde{Y}) := \lim_{k \to -\infty} \mu(Y_k, \tilde{Y}_k), \quad \mu^*_{-\infty}(Y, \tilde{Y}) := \lim_{k \to -\infty} \mu^*(Y_k, \tilde{Y}_k). \qquad (6.550)$$

When considering system (SDS) with two singular endpoints, we also have

$$m_L(-\infty, \infty) - \tilde{m}_L(-\infty, \infty) = \mu_\infty(Y, \tilde{Y}) - \mu_{-\infty}(Y, \tilde{Y}),$$

$$m_R(-\infty, \infty) - \tilde{m}_R(-\infty, \infty) = \mu^*_{-\infty}(Y, \tilde{Y}) - \mu^*_\infty(Y, \tilde{Y}),$$

as a sum of (6.503) and (6.548), respectively, as a sum of (6.504) and (6.549). Then by Theorem 6.185, we can calculate the limits in (6.550) explicitly as values of the comparative index, which could not be done by the methods in Sects. 4.2.4 and 4.3.3.

**Corollary 6.186** *Assume that system* (SDS) *is nonoscillatory at* $-\infty$. *Then for any conjoined bases* $Y$ *and* $\tilde{Y}$ *of* (SDS), *the limits in* (6.550) *satisfy*

$$\mu_{-\infty}(Y, \tilde{Y}) = \mu(\mathcal{J}D_{-\infty}, \mathcal{J}\tilde{D}_{-\infty}), \tag{6.551}$$

$$\mu^*_{-\infty}(Y, \tilde{Y}) = \mu^*(\mathcal{J}D_{-\infty}, \mathcal{J}\tilde{D}_{-\infty}) - m_R(-\infty) + \tilde{m}_R(-\infty), \tag{6.552}$$

*where* $D_{-\infty}$ *and* $\tilde{D}_{-\infty}$ *are the constant matrices in* (6.535) *corresponding to* $Y$ *and* $\tilde{Y}$, *respectively. Moreover,*

$$\mu^*_{-\infty}(Y, Y^{[-\infty]}) = \text{rank } T_{-\infty}. \tag{6.553}$$

*Proof* The statements in (6.551) and (6.552) follow from Theorem 6.185; compare also with the proof of Corollary 6.174. Equality (6.553) then follows from (6.552) with $\tilde{Y} := Y^{[-\infty]}$; compare with formula (6.508) in Remark 6.175. $\qquad\square$

Based on Theorem 6.185, we derive exact formulas for the numbers of focal points of a given conjoined basis $Y$ of (SDS) in terms of $Y^{[-\infty]}$ and $Y^{[N]}$, respectively, in terms of $Y^{[-\infty]}$ and $Y^{[\infty]}$.

**Corollary 6.187** *If* (SDS) *is nonoscillatory at* $-\infty$, *then for any conjoined basis* $Y$ *of* (SDS), *we have*

$$m_L(-\infty, N] = m_L^{[N]}(-\infty, N] - \mu\big(\mathcal{J}D_{-\infty}, \mathcal{J}D^{[N]}_{-\infty}\big), \tag{6.554}$$

$$m_L(-\infty, N] = m_L^{[-\infty]}(-\infty, N] + \mu(Y_N, Y_N^{[-\infty]}), \tag{6.555}$$

$$m_R[-\infty, N) = m_R^{[N]}[-\infty, N) + \mu^*\big(\mathcal{J}D_{-\infty}, \mathcal{J}D^{[N]}_{-\infty}\big), \tag{6.556}$$

$$m_R[-\infty, N) = m_R^{[-\infty]}[-\infty, N) - \mu^*(Y_N, Y_N^{[-\infty]}), \tag{6.557}$$

*where* $D_{-\infty}$ *and* $D^{[N]}_{-\infty}$ *are the constant matrices in* (6.535) *corresponding to* $Y$ *and* $Y^{[N]}$, *respectively. If* (SDS) *is nonoscillatory at* $\pm\infty$, *then for any conjoined basis* $Y$ *of* (SDS), *we have*

$$m_L(-\infty, \infty] = m_L^{[-\infty]}(-\infty, \infty] + \mu\big(\mathcal{J}D_{\infty}, \mathcal{J}D^{[-\infty]}_{\infty}\big), \tag{6.558}$$

$$m_L(-\infty, \infty] = m_L^{[\infty]}(-\infty, \infty] - \mu\big(\mathcal{J}D_{-\infty}, \mathcal{J}D^{[\infty]}_{-\infty}\big), \tag{6.559}$$

$$m_R[-\infty, \infty) = m_R^{[-\infty]}[-\infty, \infty) - \mu^*\big(\mathcal{J}D_{\infty}, \mathcal{J}D^{[-\infty]}_{\infty}\big), \tag{6.560}$$

$$m_R[-\infty, \infty) = m_R^{[\infty]}[-\infty, \infty) + \mu^*\big(\mathcal{J}D_{-\infty}, \mathcal{J}D^{[\infty]}_{-\infty}\big), \tag{6.561}$$

*where $D_\infty$ and $D_\infty^{[-\infty]}$ are the constant matrices in (6.467) corresponding to $Y$ and $Y^{[-\infty]}$, respectively, and $D_{-\infty}$ and $D_{-\infty}^{[\infty]}$ are the constant matrices in (6.535) corresponding to $Y$ and $Y^{[\infty]}$, respectively.*

*Proof* Equalities (6.554) and (6.556) follow from (6.542) and (6.543) with the choice $\tilde{Y} := Y^{[N]}$, while equalities (6.555) and (6.557) follow from (6.542) and (6.543) with the choice $\tilde{Y} := Y^{[-\infty]}$. Equalities (6.558) and (6.560) follow from (6.545) and (6.546) with the choice $\tilde{Y} := Y^{[-\infty]}$, while equalities (6.559) and (6.561) follow from (6.545) and (6.546) with $\tilde{Y} := Y^{[\infty]}$.                                 □

Next we present the relationship between the numbers of forward and backward focal points of $Y^{[N]}$, $Y^{[-\infty]}$, and $Y^{[\infty]}$.

**Corollary 6.188** *If* (SDS) *is nonoscillatory at* $-\infty$, *then*

$$m_R^{[-\infty]}[-\infty, N) = m_L^{[-\infty]}(-\infty, N] + \operatorname{rank} X_N^{[-\infty]}, \tag{6.562}$$

$$m_L^{[N]}(-\infty, N] = m_R^{[N]}[-\infty, N) + \operatorname{rank} X_N^{[-\infty]}, \tag{6.563}$$

$$m_L^{[N]}(-\infty, N] = m_L^{[-\infty]}(-\infty, N] + \operatorname{rank} X_N^{[-\infty]}, \tag{6.564}$$

$$m_R^{[-\infty]}[-\infty, N) = m_R^{[N]}[-\infty, N) + \operatorname{rank} X_N^{[-\infty]}. \tag{6.565}$$

*If* (SDS) *is nonoscillatory at* $\pm\infty$, *then*

$$m_R^{[-\infty]}[-\infty, \infty) = m_L^{[-\infty]}(-\infty, \infty] + \operatorname{rank} w(Y^{[\infty]}, Y^{[-\infty]}), \tag{6.566}$$

$$m_L^{[\infty]}(-\infty, \infty] = m_R^{[\infty]}[-\infty, \infty) + \operatorname{rank} w(Y^{[-\infty]}, Y^{[\infty]}), \tag{6.567}$$

$$m_L^{[\infty]}(-\infty, \infty] = m_L^{[-\infty]}(-\infty, \infty] + \operatorname{rank} w(Y^{[\infty]}, Y^{[-\infty]}), \tag{6.568}$$

$$m_R^{[-\infty]}[-\infty, \infty) = m_R^{[\infty]}[-\infty, \infty) + \operatorname{rank} w(Y^{[-\infty]}, Y^{[\infty]}). \tag{6.569}$$

*Proof* Equations (6.562) and (6.566) follow from (6.544) and (6.547) with the choice $Y := Y^{[-\infty]}$. Equation (6.563) follows from (6.544) with $Y := Y^{[N]}$, while (6.567) follows from (6.547) with $Y := Y^{[\infty]}$. Next, equations (6.564) and (6.565) follow from (6.555) and (6.557) with the choice $Y := Y^{[N]}$ or (6.554) and (6.556) with the choice $Y := Y^{[-\infty]}$, while equations (6.568) and (6.569) follow from (6.558) and (6.560) with $Y := Y^{[\infty]}$ or from (6.559) and (6.561) with the choice $Y := Y^{[-\infty]}$.                                 □

The following two results are analogs of Theorem 6.178 for the singular endpoint of (SDS) at $-\infty$ or for two singular endpoints of (SDS).

**Theorem 6.189 (Singular Sturmian Separation Theorem)** *If* (SDS) *is nonoscillatory at* $-\infty$, *then*

$$m_L^{[-\infty]}(-\infty, N] = m_R^{[N]}[-\infty, N), \quad m_R^{[-\infty]}[-\infty, N) = m_L^{[N]}(-\infty, N]. \tag{6.570}$$

*Moreover, for any conjoined basis Y of* (SDS), *we have the estimates*

$$m_L^{[-\infty]}(-\infty, N] \le m_L(-\infty, N] \le m_L^{[N]}(-\infty, N], \tag{6.571}$$

$$m_R^{[N]}[-\infty, N) \le m_R[-\infty, N) \le m_R^{[-\infty]}[-\infty, N), \tag{6.572}$$

*and for any conjoined bases Y and $\tilde{Y}$ of* (SDS), *we have the estimates*

$$\left| m_L(-\infty, N] - \tilde{m}_L(-\infty, N] \right| \le \operatorname{rank} X_N^{[-\infty]} \le n, \tag{6.573}$$

$$\left| m_R[-\infty, N) - \tilde{m}_R[-\infty, N) \right| \le \operatorname{rank} X_N^{[-\infty]} \le n, \tag{6.574}$$

$$\left| m_L(-\infty, N] - \tilde{m}_R[-\infty, N) \right| \le \operatorname{rank} X_N^{[-\infty]} \le n. \tag{6.575}$$

*Proof* The equalities in (6.570) follow from (6.562) in combination with (6.565) and (6.564). The estimates in (6.571) follow from (6.555) and (6.554), while the estimates in (6.572) follow from (6.556) and (6.557). Finally, the estimates in (6.573)–(6.575) follow from (6.571)–(6.572) by using (6.570). $\quad\square$

**Theorem 6.190 (Singular Sturmian Separation Theorem)** *If* (SDS) *is nonoscillatory at $\pm\infty$, then*

$$m_L^{[-\infty]}(-\infty, \infty] = m_R^{[\infty]}[-\infty, \infty), \quad m_R^{[-\infty]}[-\infty, \infty) = m_L^{[\infty]}(-\infty, \infty]. \tag{6.576}$$

*Moreover, for any conjoined basis Y of* (SDS), *we have the estimates*

$$m_L^{[-\infty]}(-\infty, \infty] \le m_L(-\infty, \infty] \le m_L^{[\infty]}(-\infty, \infty], \tag{6.577}$$

$$m_R^{[\infty]}[-\infty, \infty) \le m_R[-\infty, \infty) \le m_R^{[-\infty]}[-\infty, \infty), \tag{6.578}$$

*and for any conjoined bases Y and $\tilde{Y}$ of* (SDS), *we have the estimates*

$$\left| m_L(-\infty, \infty] - \tilde{m}_L(-\infty, \infty] \right| \le \operatorname{rank} w(Y^{[\infty]}, Y^{[-\infty]}) \le n, \tag{6.579}$$

$$\left| m_R[-\infty, \infty) - \tilde{m}_R[-\infty, \infty) \right| \le \operatorname{rank} w(Y^{[\infty]}, Y^{[-\infty]}) \le n, \tag{6.580}$$

$$\left| m_L(-\infty, \infty] - \tilde{m}_R[-\infty, \infty) \right| \le \operatorname{rank} w(Y^{[\infty]}, Y^{[-\infty]}) \le n. \tag{6.581}$$

*Proof* The equalities in (6.576) follow from (6.566) in combination with (6.569) and (6.568). The estimates in (6.577) follow from (6.558) and (6.559), while the estimates in (6.578) follow from (6.561) and (6.560). Finally, the estimates in (6.579)–(6.581) follow from (6.577)–(6.578) by using (6.576). $\quad\square$

We note that the lower and upper bounds in Theorems 6.189 and 6.190 are optimal in a sense that they cannot be improved by better bounds, which would be independent on the arbitrarily chosen conjoined bases $Y$ and $\tilde{Y}$.

In the remaining part of this section, we will present the results regarding the multiplicities of focal points of conjoined bases of (SDS) in the open intervals

$(-\infty, N)$ or $(-\infty, \infty)$. Similarly to Corollaries 6.179 and 6.180, we obtain the following.

**Corollary 6.191** *If* (SDS) *is nonoscillatory at* $-\infty$, *then for any conjoined basis* $Y$ *of* (SDS), *we have the estimates*

$$\left.\begin{aligned} m_R^{[N]}(-\infty, N) \le m_R(-\infty, N) &\le m_R^{[-\infty]}[-\infty, N) \\ &= m_R^{[N]}(-\infty, N) + \operatorname{rank} \mathcal{G}_-^{[N]}, \end{aligned}\right\} \tag{6.582}$$

*where* $\mathcal{G}_-^{[N]}$ *is the genus of the principal solution* $Y^{[N]}$ *of* (SDS) *near* $-\infty$. *Moreover, we have*

$$m_R^{[-\infty]}(-\infty, N) + n - d_{-\infty} = m_R^{[N]}(-\infty, N) + \operatorname{rank} \mathcal{G}_-^{[N]}, \tag{6.583}$$

$$\left.\begin{aligned} m_R^{[-\infty]}(-\infty, N) = m_R^{[N]}(-\infty, N) &\Leftrightarrow \operatorname{rank} \mathcal{G}_-^{[N]} = n - d_{-\infty} \\ &\Leftrightarrow Y^{[N]} \in \mathcal{G}_{\min}^-, \end{aligned}\right\} \tag{6.584}$$

*where* $\mathcal{G}_{\min}^-$ *is the genus of conjoined bases of* (SDS) *near* $-\infty$ *corresponding to* $Y^{[-\infty]}$. *If in addition system* (SDS) *is controllable near* $-\infty$, *then*

$$m_R^{[-\infty]}(-\infty, N) = m_R^{[N]}(-\infty, N). \tag{6.585}$$

**Corollary 6.192** *If* (SDS) *is nonoscillatory at* $\pm\infty$, *then for any conjoined basis* $Y$ *of* (SDS), *we have the estimates*

$$\left.\begin{aligned} m_L^{[-\infty]}(-\infty, \infty) \le m_L(-\infty, \infty) &\le m_L^{[\infty]}(-\infty, \infty] \\ &= m_L^{[-\infty]}(-\infty, \infty) + \operatorname{rank} \mathcal{G}_+^{[-\infty]}, \end{aligned}\right\} \tag{6.586}$$

$$\left.\begin{aligned} m_R^{[\infty]}(-\infty, \infty) \le m_R(-\infty, \infty) &\le m_R^{[-\infty]}[-\infty, \infty) \\ &= m_R^{[\infty]}(-\infty, \infty) + \operatorname{rank} \mathcal{G}_-^{[\infty]}, \end{aligned}\right\} \tag{6.587}$$

*where* $\mathcal{G}_+^{[-\infty]}$ *is the genus of* $Y^{[-\infty]}$ *near* $\infty$ *and* $\mathcal{G}_-^{[\infty]}$ *is the genus of* $Y^{[\infty]}$ *near* $-\infty$. *Moreover, we have*

$$m_L^{[-\infty]}(-\infty, \infty) + \operatorname{rank} \mathcal{G}_+^{[-\infty]} = m_L^{[\infty]}(-\infty, \infty) + n - d_\infty, \tag{6.588}$$

$$m_R^{[-\infty]}(-\infty, \infty) + n - d_{-\infty} = m_R^{[\infty]}(-\infty, \infty) + \operatorname{rank} \mathcal{G}_-^{[\infty]}, \tag{6.589}$$

$$\left.\begin{aligned} m_L^{[-\infty]}(-\infty, \infty) = m_L^{[\infty]}(-\infty, \infty) &\Leftrightarrow \operatorname{rank} \mathcal{G}_+^{[-\infty]} = n - d_\infty \\ &\Leftrightarrow Y^{[-\infty]} \in \mathcal{G}_{\min}^+, \end{aligned}\right\} \tag{6.590}$$

$$\left.\begin{aligned} m_R^{[-\infty]}(-\infty, \infty) = m_R^{[\infty]}(-\infty, \infty) &\Leftrightarrow \operatorname{rank} \mathcal{G}_-^{[\infty]} = n - d_{-\infty} \\ &\Leftrightarrow Y^{[\infty]} \in \mathcal{G}_{\min}^-, \end{aligned}\right\} \tag{6.591}$$

*and consequently*

$$m_L^{[-\infty]}(-\infty, \infty) + \text{rank}\,\mathcal{G}_+^{[-\infty]} = m_R^{[\infty]}(-\infty, \infty) + \text{rank}\,\mathcal{G}_-^{[\infty]}, \qquad (6.592)$$

$$m_R^{[-\infty]}(-\infty, \infty) - d_{-\infty} = m_L^{[\infty]}(-\infty, \infty) - d_\infty, \qquad (6.593)$$

$$m_L^{[-\infty]}(-\infty, \infty) = m_R^{[\infty]}(-\infty, \infty) \;\Leftrightarrow\; \text{rank}\,\mathcal{G}_+^{[-\infty]} = \text{rank}\,\mathcal{G}_-^{[\infty]}, \qquad (6.594)$$

$$m_R^{[-\infty]}(-\infty, \infty) = m_L^{[\infty]}(-\infty, \infty) \;\Leftrightarrow\; d_{-\infty} = d_\infty. \qquad (6.595)$$

*If in addition system* (SDS) *is controllable near* $\pm\infty$, *then*

$$m_L^{[-\infty]}(-\infty, \infty) = m_R^{[-\infty]}(-\infty, \infty) = m_L^{[\infty]}(-\infty, \infty) = m_R^{[\infty]}(-\infty, \infty). \qquad (6.596)$$

In the final result of this subsection, we present the limit properties of the comparative index at $-\infty$ and its relation with the multiplicity $m_R(-\infty)$ for a conjoined basis $Y$ and with the maximal order of abnormality $d_{-\infty}$. It represents an analog of Theorem 6.181.

**Theorem 6.193** *Assume that system* (SDS) *is nonoscillatory at* $-\infty$. *Then for any conjoined basis $Y$ of* (SDS), *the comparative indices* $\mu(\mathcal{J}D_{-\infty}, \mathcal{J}D_{-\infty}^{[k]})$ *and* $\mu^*(\mathcal{J}D_{-\infty}, \mathcal{J}D_{-\infty}^{[k]})$ *have limits for $k \to -\infty$, which satisfy*

$$\lim_{k \to -\infty} \mu(\mathcal{J}D_{-\infty}, \mathcal{J}D_{-\infty}^{[k]}) = n - d_{-\infty}, \qquad (6.597)$$

$$\lim_{k \to -\infty} \mu^*(\mathcal{J}D_{-\infty}, \mathcal{J}D_{-\infty}^{[k]}) = m_R(-\infty), \qquad (6.598)$$

*where $D_{-\infty}$ and $D_{-\infty}^{[k]}$ are the constant matrices in* (6.535) *corresponding to $Y$ and $Y^{[k]}$, respectively.*

*Proof* The proof is analogous to the proof of Theorem 6.181 by using formulas (6.554), (6.556), and (6.570). The details are therefore omitted. $\qquad\square$

## 6.5 Notes and References

The results in Sect. 6.1.1 are based on [22, 314] and [315, Chapter 4]. In the paper [21] the authors considered the case when equations (6.9), (6.10) differ also in $r_k \neq \hat{r}_k$. Note that all results in [21, 22, 315] are related to the case of the special linear dependence on $\lambda$ (see Sect. 5.3). The relative oscillation theorems for two Sturm-Liouville equations with general nonlinear dependence on $\lambda$ (see Sect. 1.2.5) can be derived from [124, Theorems 3.8 and 3.9] (see Sect. 6.1.4). Moreover, in [124, Subsection 3.2], we discussed connections between the results in [21] and [124]. For differential Sturm-Liouville eigenvalue problems, the relative oscillation theory is developed in [153, 213] (see also the references in these papers). The first results concerning the relative oscillation theory for discrete symplectic systems are presented in [117, Section 3] and [118, 120, 121, 123, 124].

The results in Sect. 6.1.2 (see Theorems 6.2 and 6.4) are direct consequence of the comparison results for focal points of the principal solutions of two symplectic systems from [117, Corollary 2.4] (see Sect. 4.3.2). The consideration in Sect. 6.1.3 is based on algebraic properties of the comparative index for two symplectic matrices under assumption (6.35) (see [117, Lemma 2.2] and Sect. 4.3.1, in particular Lemma 4.43). The first applications of the comparison results for this special case to the spectral theory of symplectic systems with the special linear dependence on $\lambda$ (see Sect. 5.3) can be found in [117, Section 3] and [118]. In particular, the results in Lemma 6.6 are from [118, Lemma 1] (note that the proofs for the linear and nonlinear cases are the same). Formula (6.60) in Remark 6.10(ii) concerning the number of eigenvalues in $(-\infty, b)$ (instead of $(-\infty, b]$) was derived in [118, Theorem 3, formula (3.7)]. In this connection we remark that the renormalized version of (6.60), formula (6.62) (concerning the number of eigenvalues of problem (E) in $[a, b)$, compare with Theorem 6.9), is proved in [118, Theorem 4, formula (3.11)]. Observe that the comments in Remark 6.10(ii) (as well as [118, Theorem 4, formula (3.11)]) refer to the special linear dependence of on $\lambda$. A generalization of (6.60) and (6.62) to the case of general nonlinear dependence on $\lambda$ demands introducing the notion of the so-called *right* finite eigenvalues, because condition (5.274) is not in general satisfied.

Majority of the results in Sects. 6.1.4 and 6.1.5 are from the paper [124]. Note that from the results of Sect. 6.1.4, one can easily derive the results for the scalar case $n = 1$ concerning a pair of Sturm-Liouville eigenvalue problems with general nonlinear dependence on $\lambda$. The main results in Sects. 6.1.6 and 6.1.7 are from [121, Chapter 4], where the consideration is restricted to the special linear dependence on $\lambda$ (see Sect. 5.3). Note, however, that all algebraic properties of the comparative index used in the results of Sects. 6.1.6 and 6.1.7 remain the same for the general nonlinear dependence on $\lambda$. Observe also, that the renormalized oscillation theorems for separated and joint endpoints (see Theorems 6.23 and 6.29), as well as Corollaries 6.25 and 6.31 and Remarks 6.26 and 6.32, for the case when the boundary conditions do not depend on $\lambda$ can be found in [120, Theorems 1 and 2] (for separated boundary conditions) and in [123, Theorems 5 and 6] (for general boundary conditions).

The results of Sect. 6.2 belong to the classical research topic strongly developed for differential Sturm-Liouville eigenvalue problems with some generalizations to vector case, i.e., for differential matrix Sturm-Liouville eigenvalue problems and for boundary value problems for linear Hamiltonian differential systems. Comparison theorems for eigenvalues under some majorant conditions for scalar and vector differential Sturm-Liouville equations can be found in [280, Chapters 1 and 3] and in [205, Section 7.5]. For the discrete case, comparison results for eigenvalues of scalar and vector Sturm-Liouville eigenvalue problems can be found in [262, Theorem 3.21], [268, Theorem 3.7], and [267, Theorem 5.5]. See also the references therein.

First comparison theorems for eigenvalues of discrete symplectic eigenvalue problems with nonlinear dependence on $\lambda$ were proven in [301, Theorems 1–3]. Similar results were proven in [121, Chapter 4], for the case of the special linear dependence on $\lambda$ (see Sect. 5.3). In Sect. 6.2.1 we present a modification of the results in [301] and [121, Chapter 4] for the general nonlinear dependence on $\lambda$ formulating majorant conditions for the symplectic matrices $S_k(\lambda)$ and $\hat{S}_k(\lambda)$ and for the matrices of the boundary conditions in terms of the comparative index and using the relative oscillation theorems from Sect. 6.1 as the main tool for the proofs.

In Sects. 6.2.2 and 6.2.3, we present the comparative index theory as a new tool for the proof of the interlacing properties of finite eigenvalues of symplectic eigenvalue problems with nonlinear dependence on $\lambda$. For the scalar differential Sturm-Liouville eigenvalue problems, the classical interlacing properties of eigenvalues are proven in [38] and [69, Chapter 8]; see also [37] and the references in the latter paper. For the matrix case, these properties are proven in [28, Theorem 10.5.1], [254, Theorem 2] (for matrix Sturm-Liouville eigenvalue problems), in [28, Theorem 10.9.1], [156, Chapter 4] (for linear Hamiltonian differential systems with general linear dependence on $\lambda$), and in [205, Section 7.5]. For geometrical interpretation of interlacing properties, we refer to [26]. For the discrete case, results concerning interlacing properties of zeros for orthogonal polynomials for scalar and matrix case can be found in [28, Chapters 4 and 6] (see Theorems 4.3.2, 4.3.3, and 6.7.7 therein). Different results concerning the interlacing properties of eigenvalues for discrete Sturm-Liouville problems can be found in [149, 322] and the references therein.

For the discrete symplectic systems with special linear dependence on $\lambda$ (see Sect. 5.3), some interlacing properties of finite eigenvalues are proved in [120] (for separated boundary conditions) and in [123] (for joint endpoints). These results were generalized in [121, Chapter 4], but the consideration was restricted to the case of the special linear dependence on $\lambda$.

Majority of the results in Sect. 6.3 about recessive and dominant solutions of (SDS) at $\infty$ in the possibly uncontrollable case are derived from [284] and [290]. More precisely, most of Sects. 6.3.1–6.3.6 are from [284]. The properties of the important auxiliary conjoined basis $\bar{Y}$ in Proposition 6.67 are collected from several sources: parts (i)–(v) from [284, Proposition 4.3 and Remark 4.5]; parts (vi)–(viii) from [290, Proposition 2.6]; and parts (ix)–(xi) are proven as a discrete version of [281, Theorem 2.2.11]. The proof of Lemma 6.100 is a discrete version of the proof of [288, Lemma 1]. The proofs of Theorems 6.69, 6.73, 6.75, 6.97(ii)–(iii) and of Propositions 6.81 and 6.105 were not discussed in details in [284]. In the presented form, they are discrete analogs of the proofs of [283, Theorems 4.6, 4.10, 5.2, 5.7, 5.15, 5.17, 6.9]. The proof of Theorem 6.84 is a discrete analog of the proof of [285, Theorem 4.13]. Similarly, the proofs of Propositions 6.92 and 6.99 and Theorems 6.106 and 6.107 were not discussed in details in [290]. Their presented versions are discrete analogs of the proofs of [285, Theorem 6.7 and Lemma 6.9] and [286, Theorems 4.4 and 4.9]. Moreover, Lemma 6.91 is the discrete version of [285, Lemma 6.6]. The concept of representable conjoined bases in Definition 6.79 and Theorem 6.80 is a discrete version of [281, Theorem 2.3.3].

The statement in Theorem 6.153 extends the Reid construction of the recessive solution of (SDS) at $\infty$ known for the controllable case in [81, Section 4(iii)]. The corresponding result for controllable linear Hamiltonian differential systems (1.103) can be found in [70, pg. 44] and [248, Theorem VII.3.4], while for general abnormal linear Hamiltonian systems, we refer to [288, Theorem 1]. Singular Sturmian separation theorems on unbounded intervals presented in Sect. 6.4, including the necessary tools in Sect. 6.3.11, are derived from [292]. The results in Corollary 6.177 (singular Sturmian separation theorem) are correct singular versions of [94, Eq. (4.14) and (4.29)]. The results in Corollaries 6.180 and 6.191 represent generalizations of [94, Corollary 4.2, Eqs. (4.14a), (4.16a)] to a possibly uncontrollable system (SDS) near $\infty$ or $-\infty$. A similar statement was obtained in [91, Theorem 2] and [104, Theorem 1] for an eventually controllable system (SDS) on the interval $(-\infty, \infty)$; compare with Corollaries 6.191 and 6.192. The equalities in (6.596) are known in [104, Theorem 1] and in [91, Theorem 2]. The equivalences in (6.594) and (6.595) then generalize the result in (6.596) to possibly uncontrollable system (SDS).

Regarding the basic notions in Sect. 6.4.3, it is shown in [300, Section 4] how the notion of a recessive/dominant solution of (SDS) at $-\infty$ can be transformed into the corresponding notion at $\infty$ and vice versa. Several examples, which illustrate the concepts and results in Sects. 6.4.2 and 6.4.3, are presented in [292, Section 7].

We note that the development of the presented discrete-time theory was influenced by the corresponding theory of linear Hamiltonian differential systems (1.103) in [283, 285–289, 293].

We recommend the following additional related references for further reading about the topics presented in this chapter: [5, 52, 76] for Sturmian theory of symplectic systems and their recessive and dominant solutions.

# References

1. A.A. Abramov, A modification of one method for solving nonlinear self-adjoint eigenvalue problems for Hamiltonian systems of ordinary differential equations, Comput. Math. Math. Phys. **51**(1), 35–39 (2011)
2. R.P. Agarwal, *Difference Equations and Inequalities. Theory, methods, and applications*, 2nd edn. Monographs and Textbooks in Pure and Applied Mathematics, vol. 228 (Marcel Dekker, New York, 2000)
3. R.P. Agarwal, C.D. Ahlbrandt, M. Bohner, A. Peterson, Discrete linear Hamiltonian systems: A survey. Dynam. Syst. Appl. **8**(3–4), 307–333 (1999)
4. R.P. Agarwal, M. Bohner, S.R. Grace, D. O'Regan, *Discrete Oscillation Theory* (Hindawi Publishing, New York, NY, 2005)
5. D. Aharonov, M. Bohner, U. Elias, Discrete Sturm comparison theorems on finite and infinite intervals. J. Differ. Equ. Appl. **18**, 1763–1771 (2012)
6. C.D. Ahlbrandt, Principal and antiprincipal solutions of self-adjoint differential systems and their reciprocals. Rocky Mt. J. Math. **2**, 169–182 (1972)
7. C.D. Ahlbrandt, Linear independence of the principal solutions at $\infty$ and $-\infty$ for formally self-adjoint differential systems. J. Differ. Equ. **29**(1), 15–27 (1978)
8. C.D. Ahlbrandt, Equivalence of discrete Euler equations and discrete Hamiltonian systems. J. Math. Anal. Appl. **180**, 498–517 (1993)
9. C.D. Ahlbrandt, Continued fractions representation of maximal and minimal solutions of a discrete matrix Riccati equation. SIAM J. Math. Anal. **24**, 1597–1621 (1993)
10. C.D. Ahlbrandt, Geometric, analytic, and arithmetic aspects of symplectic continued fractions, in *Analysis, Geometry and Groups: A Riemann Legacy Volume*, Hadronic Press Collect. Orig. Artic. (Hadronic Press, Palm Harbor, FL, 1993), pp. 1–26
11. C.D. Ahlbrandt, Dominant and recessive solutions of symmetric three term recurrences. J. Differ. Equ. **107**, 238–258 (1994)
12. C.D. Ahlbrandt, M. Bohner, J. Ridenhour, Hamiltonian systems on time scales. J. Math. Anal. Appl. **250**(2), 561–578 (2000)
13. C.D. Ahlbrandt, C. Chicone, S.L. Clark, W.T. Patula, D. Steiger, Approximate first integrals for discrete Hamiltonian systems. Dynam. Contin. Discrete Impuls. Syst. **2**, 237–264 (1996)
14. C.D. Ahlbrandt, M. Heifetz, Discrete Riccati equations of filtering and control, in *Proceedings of the First International Conference on Difference Equations* (San Antonio, TX, 1994), ed. by S. Elaydi, J. Graef, G. Ladas, A. Peterson (Gordon and Breach, Newark, NJ, 1996), pp. 1–16

© Springer Nature Switzerland AG 2019

O. Došlý et al., *Symplectic Difference Systems: Oscillation and Spectral Theory*, Pathways in Mathematics, https://doi.org/10.1007/978-3-030-19373-7

15. C.D. Ahlbrandt, J.W. Hooker, Recessive solutions of symmetric three term recurrence relations, in *Oscillations, Bifurcation and Chaos* (Toronto, 1986), Canadian Math. Soc. Conference Proceedings, Vol. 8, ed. by F. Atkinson, W. Langford, A. Mingarelli (Amer. Math. Soc., Providence, RI, 1987), pp. 3–42

16. C.D. Ahlbrandt, A.C. Peterson, *Discrete Hamiltonian Systems. Difference Equations, Continued Fractions, and Riccati Equations.* Kluwer Texts in the Mathematical Sciences, Vol. 16 (Kluwer Academic Publishers, Dordrecht, 1996)

17. C.D. Ahlbrandt, A.C. Peterson, A general reduction of order theorem for discrete linear symplectic systems, in *Dynamical Systems and Differential Equations*, Vol. I (Springfield, MO, 1996). Discrete Contin. Dynam. Syst. **1998**, Added Volume I, 7–18 (1998)

18. L.D. Akulenko, S.V. Nesterov, A frequency-parametric analysis of natural oscillations of non-uniform rods. J. Appl. Math. Mech. **67**, 525–537 (2003)

19. L.D. Akulenko, S.V. Nesterov, *High-Precision Methods in Eigenvalue Problems and Their Applications* (Chapman and Hall/Crc, Boca Raton, FL, 2005)

20. L.D. Akulenko, S.V. Nesterov, Flexural vibrations of a moving rod. J. Appl. Math. Mech. **72**, 550–560 (2008)

21. K. Ammann, Relative oscillation theory for Jacobi matrices extended. Oper. Matrices **1**, 99–115 (2014)

22. K. Ammann, G. Teschl, Relative oscillation theory for Jacobi matrices, in *Proceedings of the 14th International Conference on Difference Equations and Applications* (Istanbul, 2008), ed. by M. Bohner, Z. Došlá, G. Ladas, M. Ünal, A. Zafer (Uğur-Bahçeşehir University Publishing Company, Istanbul, 2009), pp. 105–115

23. W.O. Amrein, A.M. Hinz, D.B. Pearson (eds.), *Sturm–Liouville Theory. Past and Present.* Including papers from the International Colloquium held at the University of Geneva (Geneva, 2003) (Birkhäuser Verlag, Basel, 2005)

24. D.R. Anderson, Discrete trigonometric matrix functions. PanAmer. Math. J. **7**(1), 39–54 (1997)

25. D.R. Anderson, Normalized prepared bases for discrete symplectic matrix systems. Dynam. Syst. Appl. **8**(3–4), 335–344 (1999)

26. V.I. Arnold, Sturm theorems and symplectic geometry. Funct. Anal. Appl. **19**(4), 251–259 (1985)

27. V.I. Arnold, *Mathematical Methods of Classical Mechanics*, 2nd edn. (Springer, Berlin, 1989)

28. F.V. Atkinson, *Discrete and Continuous Boundary Value Problems* (Academic Press, New York, NY, 1964)

29. T. Auckenthalera, V. Blumb, H.J. Bungartza, et al., Parallel solution of partial symmetric eigenvalue problems from electronic structure calculations. Parallel Comput. **37**, 783–794 (2011)

30. P.B. Bailey, W.N. Everitt, A. Zettl, The SLEIGN2 Sturm–Liouville Code. ACM Trans. Math. Software **21**, 143–192 (2001)

31. B. Bandyrskii, I. Gavrilyuk, I. Lazurchak, V. Makarov, Functional-discrete method (fd-method) for matrix Sturm–Liouville problems. Comput. Methods Appl. Math. **5**, 362–386 (2005)

32. J.H. Barrett, A Prüfer transformation for matrix differential systems. Proc. Am. Math. Soc. **8**, 510–518 (1957)

33. H. Behncke, Spectral theory of Hamiltonian difference systems with almost constant coefficients. J. Differ. Equ. Appl. **19**(1), 1–12 (2013)

34. A. Ben-Israel, T.N.E. Greville, *Generalized Inverses: Theory and Applications* (Wiley, New York, 1974)

35. P. Benner, Symplectic balancing of Hamiltonian matrices. SIAM J. Sci. Comput. **22**, 1885–1904 (2000)

36. D.S. Bernstein, *Matrix Mathematics. Theory, Facts, and Formulas with Application to Linear Systems Theory* (Princeton University Press, Princeton, 2005)

37. P.A. Binding, H. Volkmer, Interlacing and oscillation for Sturm–Liouville problems with separated and coupled boundary conditions. J. Comput. Appl. Math. **194**(1), 75–93 (2006)

38. G.D. Birkhoff, Existence and oscillation theorem for a certain boundary value problem. Trans. Am. Math. Soc. **10**(2), 259–270 (1909)
39. M. Bohner, Linear Hamiltonian difference systems: disconjugacy and Jacobi-type conditions. J. Math. Anal. Appl. **199**(3), 804–826 (1996)
40. M. Bohner, Riccati matrix difference equations and linear Hamiltonian difference systems. Dynam. Contin. Discrete Impuls. Syst. **2**(2), 147–159 (1996)
41. M. Bohner, On disconjugacy for Sturm–Liouville difference equations, in *Difference Equations: Theory and Applications* (San Francisco, CA, 1995). J. Differ. Equ. Appl. **2**(2), 227–237 (1996)
42. M. Bohner, Symplectic systems and related discrete quadratic functionals. Facta Univ. Ser. Math. Inform. **12**, 143–156 (1997)
43. M. Bohner, Discrete Sturmian theory. Math. Inequal. Appl. **1**(3), 375–383 (1998)
44. M. Bohner, Discrete linear Hamiltonian eigenvalue problems. Comput. Math. Appl. **36**(10–12), 179–192 (1998)
45. M. Bohner, O. Došlý, Disconjugacy and transformations for symplectic systems. Rocky Mt. J. Math. **27**, 707–743 (1997)
46. M. Bohner, O. Došlý, Trigonometric transformations of symplectic difference systems. J. Differ. Equ. **163**, 113–129 (2000)
47. M. Bohner, O. Došlý, The discrete Prüfer transformation. Proc. Am. Math. Soc. **129**, 2715–2726 (2001)
48. M. Bohner, O. Došlý, Trigonometric systems in oscillation theory of difference equations, in *Dynamic Systems and Applications, Proceedings of the Third International Conference on Dynamic Systems and Applications* (Atlanta, GA, 1999), Vol. 3 (Dynamic, Atlanta, GA, 2001), pp. 99–104
49. M. Bohner, O. Došlý, R. Hilscher, Linear Hamiltonian dynamic systems on time scales: Sturmian property of the principal solution, in *Proceedings of the Third World Congress of Nonlinear Analysts* (Catania, 2000). Nonlinear Anal. **47**, 849–860 (2001)
50. M. Bohner, O. Došlý, R. Hilscher, W. Kratz, Diagonalization approach to discrete quadratic functionals. Arch. Inequal. Appl. **1**(2), 261–274 (2003)
51. M. Bohner, O. Došlý, W. Kratz, Inequalities and asymptotics for Riccati matrix difference operators. J. Math. Anal. Appl. **221**, 262–286 (1998)
52. M. Bohner, O. Došlý, W. Kratz, A Sturmian theorem for recessive solutions of linear Hamiltonian difference systems. Appl. Math. Lett. **12**, 101–106 (1999)
53. M. Bohner, O. Došlý, W. Kratz, Discrete Reid roundabout theorems. Dynam. Syst. Appl. **8**(3–4), 345–352 (1999)
54. M. Bohner, O. Došlý, W. Kratz, Positive semidefiniteness of discrete quadratic functionals. Proc. Edinb. Math. Soc. (2) **46**(3), 627–636 (2003)
55. M. Bohner, O. Došlý, W. Kratz, An oscillation theorem for discrete eigenvalue problems. Rocky Mt. J. Math. **33**(4), 1233–1260 (2003)
56. M. Bohner, O. Došlý, W. Kratz, Sturmian and spectral theory for discrete symplectic systems. Trans. Am. Math. Soc. **361**, 3019–3123 (2009)
57. M. Bohner, W. Kratz, R. Šimon Hilscher, Oscillation and spectral theory for linear Hamiltonian systems with nonlinear dependence on the spectral parameter. Math. Nachr. **285**(11–12), 1343–1356 (2012)
58. M. Bohner, A. Peterson, *Dynamic Equations on Time Scales. An Introduction with Applications* (Birkhäuser, Boston, 2001)
59. M. Bohner, A. Peterson (eds.), *Advances in Dynamic Equations on Time Scales* (Birkhäuser, Boston, 2003)
60. M. Bohner, S. Sun, Weyl–Titchmarsh theory for symplectic difference systems. Appl. Math. Comput. **216**(10), 2855–2864 (2010)
61. D.L. Boley, P. Van Dooren, Placing zeroes and the Kronecker canonical form. Circuits Systems Signal Process. **13**(6), 783–802 (1994)
62. V.G. Boltyanskii, *Optimal Control of Discrete Systems* (Wiley, New York, NY, 1978)

63. A. Bunse-Gerstner, Matrix factorizations for symplectic QR-like methods. Linear Algebra Appl. **83**, 49–77 (1986)
64. S.L. Campbell, C.D. Meyer, *Generalized Inverses of Linear Transformations*. Reprint of the 1991 corrected reprint of the 1979 original, Classics in Applied Mathematics, Vol. 56 (Society for Industrial and Applied Mathematics (SIAM), Philadelphia, PA, 2009)
65. P.J. Channell, C. Scovel, Symplectic integration of Hamiltonian systems. Nonlinearity **3**(2), 231–259 (1990)
66. S. Clark, F. Gesztesy, On Weyl–Titchmarsh theory for singular finite difference Hamiltonian systems. J. Comput. Appl. Math. **171**(1–2), 151–184 (2004)
67. S. Clark, P. Zemánek, On a Weyl–Titchmarsh theory for discrete symplectic systems on a half line. Appl. Math. Comput. **217**(7), 2952–2976 (2010)
68. S. Clark, P. Zemánek, On discrete symplectic systems: Associated maximal and minimal linear relations and nonhomogeneous problems. J. Math. Anal. Appl. **421**(1), 779–805 (2014)
69. E.A. Coddington, N. Levinson, *Theory of Ordinary Differential Equations* (McGraw–Hill, New York, 1955)
70. W.A. Coppel, *Disconjugacy*. Lecture Notes in Mathematics, Vol. 220 (Springer, Berlin, 1971)
71. J.W. Demmel, *Applied Numerical Linear Algebra* (Society for Industrial and Applied Mathematics (SIAM), Philadelphia, 1997)
72. J.W. Demmel, I.S. Dhillon, H. Ren, On the correctness of some bisection-like parallel algorithms in floating point arithmetic. Electron. Trans. Numer. Anal. **3**, 116–140 (1995)
73. J. Demmel, A. McKenney, A test matrix generation suite: LAPACK Working Note 9, technical report, Department of Computer Science, Courant Institute, New York, 1989
74. F.M. Dopico, C.R. Johnson, Complementary bases in symplectic matrices and a proof that their determinant is one. Linear Algebra Appl. **419**, 772–778 (2006)
75. F.M. Dopico, C.R. Johnson, Parametrization of matrix symplectic group and applications. SIAM J. Matrix Anal. **419**, 1–24 (2009)
76. Z. Došlá, D. Škrabáková, Phases of second order linear difference equations and symplectic systems. Math. Bohem. **128**(3), 293–308 (2003)
77. O. Došlý, On transformations of self-adjoint linear differential systems and their reciprocals. Ann. Polon. Math. **50**, 223–234 (1990)
78. O. Došlý, Principal solutions and transformations of linear Hamiltonian systems. Arch. Math. (Brno) **28**, 113–120 (1992)
79. O. Došlý, Oscillation criteria for self-adjoint linear differential equations. Math. Nachr. **166**, 141–153 (1994)
80. O. Došlý, Oscillation criteria for higher order Sturm–Liouville difference equations. J. Differ. Equ. Appl. **4**(5), 425–450 (1998)
81. O. Došlý, Principal and nonprincipal solutions of symplectic dynamic systems on time scales, in *Proceedings of the Sixth Colloquium on the Qualitative Theory of Differential Equations* (Szeged, Hungary, 1999), No. 5, 14 pp. (electronic). Electron. J. Qual. Theory Differ. Equ., Szeged, 2000
82. O. Došlý, Oscillation theory of linear difference equations, in *CDDE Proceedings* (Brno, 2000). Arch. Math. (Brno) **36**, 329–342 (2000)
83. O. Došlý, Trigonometric transformation and oscillatory properties of second order difference equations, in *Communications in Difference Equations* (Poznan, 1998) (Gordon and Breach, Amsterdam, 2000), pp. 125–133
84. O. Došlý, Discrete quadratic functionals and symplectic difference systems. Funct. Differ. Equ. **11**(1–2), 49–58 (2004)
85. O. Došlý, Symplectic difference systems: oscillation theory and hyperbolic Prüfer transformation. Abstr. Appl. Anal. **2004**(4), 285–294 (2004)
86. O. Došlý, The Bohl transformation and its applications, in *2004–Dynamical Systems and Applications, Proceedings of the International Conference* (Antalya, 2004) (GBS Publishers & Distributors, Delhi, 2005), pp. 371–385
87. O. Došlý, Oscillation theory of symplectic difference systems, in *Advances in Discrete Dynamical Systems, Proceedings of the Eleventh International Conference on Difference*

*Equations and Applications* (Kyoto, 2006), ed. by S. Elaydi, K. Nishimura, M. Shishikura, N. Tose. Adv. Stud. Pure Math., Vol. 53 (Mathematical Society of Japan, Tokyo, 2009), pp. 41–50

88. O. Došlý, Oscillation and conjugacy criteria for two-dimensional symplectic difference systems. Comput. Math. Appl. **64**(7), 2202–2208 (2012)

89. O. Došlý, Symplectic difference systems with periodic coefficients: Krein's traffic rules for multipliers. Adv. Differ. Equ. **2013**(85), 13 pp. (2013)

90. O. Došlý, Symplectic difference systems: natural dependence on a parameter, in *Proceedings of the International Conference on Differential Equations, Difference Equations, and Special Functions* (Patras, 2012). Adv. Dyn. Syst. Appl. **8**(2), 193–201 (2013)

91. O. Došlý, Focal points and recessive solutions of discrete symplectic systems. Appl. Math. Comput. **243**, 963–968 (2014)

92. O. Došlý, Relative oscillation of linear Hamiltonian differential systems. Math. Nachr. **290**(14–15), 2234–2246 (2017)

93. O. Došlý, On some aspects of the Bohl transformation for Hamiltonian and symplectic systems. J. Math. Anal. Appl. **448**(1), 281–292 (2017)

94. O. Došlý, J. Elyseeva, Singular comparison theorems for discrete symplectic systems. J. Differ. Equ. Appl. **20**(8), 1268–1288 (2014)

95. O. Došlý, J. Elyseeva, Discrete oscillation theorems and weighted focal points for Hamiltonian difference systems with nonlinear dependence on a spectral parameter. Appl. Math. Lett. **43**, 114–119 (2015)

96. O. Došlý, J. Elyseeva, An oscillation criterion for discrete trigonometric systems. J. Differ. Equ. Appl. **21**(12), 1256–1276 (2015)

97. O. Došlý, S. Hilger, R. Hilscher, Symplectic dynamic systems, in *Advances in Dynamic Equations on Time Scales*, ed. by M. Bohner, A. Peterson (Birkhäuser, Boston, 2003), pp. 293–334

98. O. Došlý, R. Hilscher, Linear Hamiltonian difference systems: transformations, recessive solutions, generalized reciprocity. Dynam. Syst. Appl. **8**(3–4), 401–420 (1999)

99. O. Došlý, R. Hilscher, Disconjugacy, transformations and quadratic functionals for symplectic dynamic systems on time scales. J. Differ. Equ. Appl. **7**, 265–295 (2001)

100. O. Došlý, R. Hilscher, V. Zeidan, Nonnegativity of discrete quadratic functionals corresponding to symplectic difference systems. Linear Algebra Appl. **375**, 21–44 (2003)

101. O. Došlý, W. Kratz, A Sturmian separation theorem for symplectic difference systems. J. Math. Anal. Appl. **325**, 333–341 (2007)

102. O. Došlý, W. Kratz, Oscillation theorems for symplectic difference systems. J. Differ. Equ. Appl. **13**, 585–60 (2007)

103. O. Došlý, W. Kratz, Oscilation and spectral theory for symplectic difference systems with separated boundeary conditions. J. Differ. Equ. Appl. **16**, 831–846 (2010)

104. O. Došlý, W. Kratz, A remark on focal points of recessive solutions of discrete symplectic systems. J. Math. Anal. Appl. **363**, 209–213 (2010)

105. O. Došlý, W. Kratz, Singular Sturmian theory for linear Hamiltonian differential systems. Appl. Math. Lett. **26**, 1187–1191 (2013)

106. O. Došlý, Š. Pechancová, Trigonometric recurrence relations and tridiagonal trigonometric matrices. Int. J. Differ. Equ. **1**(1), 19–29 (2006)

107. O. Došlý, Š. Pechancová, Generalized zeros of $2 \times 2$ symplectic difference system and of its reciprocal system. Adv. Differ. Equ. **2011**(Article ID 571935), 23 pp. (2011)

108. O. Došlý, Z. Pospíšil, Hyperbolic transformation and hyperbolic difference systems. Fasc. Math. **32**, 26–48 (2001)

109. H.I. Dwyer, A. Zettl, Computing eigenvalues of regular Sturm–Liouville problems. Electron. J. Differ. Equ. **1994**(6), 10 pp. (1994)

110. S. Elaydi, *An Introduction to Difference Equations*, 3rd edn. Undergraduate Texts in Mathematics (Springer, New York, NY, 2005)

111. Yu.V. Eliseeva, An algorithm for solving the matrix difference Riccati equation. Comput. Math. Math. Phys. **39**(2), 177–184 (1999)

112. J.V. Elyseeva, A transformation for symplectic systems and the definition of a focal point. Comput. Math. Appl. **47**, 123–134 (2004)

113. J.V. Elyseeva, Symplectic factorizations and the definition of a focal point, in *Proceedings of the Eighth International Conference on Difference Equations and Applications* (Brno, 2003), ed. by S. Elaydi, G. Ladas, B. Aulbach, O. Došlý (Chapman & Hall/CRC, Boca Raton, FL, 2005), pp. 127–135

114. J.V. Elyseeva, The comparative index for conjoined bases of symplectic difference systems, in *Difference Equations, Special Functions, and Orthogonal Polynomials, Proceedings of the International Conference* (Munich, 2005), ed. by S. Elaydi, J. Cushing, R. Lasser, A. Ruffing, V. Papageorgiou, W. Van Assche (World Scientific, London, 2007), pp. 168–177

115. Yu.V. Eliseeva, Comparative index for solutions of symplectic difference systems. Differential Equations **45**(3), 445–459 (2009)

116. J.V. Elyseeva, Transformations and the number of focal points for conjoined bases of symplectic difference systems. J. Differ. Equ. Appl. **15**(11–12), 1055–1066 (2009)

117. Yu.V. Eliseeva, Comparison theorems for symplectic systems of difference equations. Differential Equations **46**(9), 1339–1352 (2010)

118. J.V. Elyseeva, On relative oscillation theory for symplectic eigenvalue problems. Appl. Math. Lett. **23**, 1231–1237 (2010)

119. J.V. Elyseeva, The comparative index and the number of focal points for conjoined bases of symplectic difference systems, in *Discrete Dynamics and Difference Equations, Proceedings of the Twelfth International Conference on Difference Equations and Applications* (Lisbon, 2007), ed. by S. Elaydi, H. Oliveira, J.M. Ferreira, J.F. Alves (World Scientific, London, 2010), pp. 231–238

120. Yu.V. Eliseeva, Spectra of discrete symplectic eigenvalue problems with separated boundary conditions. Russ. Math. (Iz. VUZ) **55**(11), 71–75 (2011)

121. J.V. Elyseeva, *Comparative Index in Mathematical Modelling of Oscillations of Discrete Symplectic Systems*, (in Russian) (Moscow State University of Technology "Stankin", Moscow, 2011)

122. Yu.V. Eliseeva, An approach for computing eigenvalues of discrete symplectic boundary-value problems. Russ. Math. (Iz. VUZ) **56**(7), 47–51 (2012)

123. J.V. Elyseeva, A note on relative oscillation theory for symplectic difference systems with general boundary conditions. Appl. Math. Lett. **25**(11), 1809–1814 (2012)

124. J.V. Elyseeva, Relative oscillation theory for matrix Sturm-Liouville difference equations extended. Adv. Differ. Equ. **2013**(328), 25 pp. (2013)

125. J.V. Elyseeva, Generalized oscillation theorems for symplectic difference systems with nonlinear dependence on spectral parameter. Appl. Math. Comput. **251**, 92–107 (2015)

126. J.V. Elyseeva, Generalized reciprocity principle for discrete symplectic systems. Electron. J. Qual. Theory Differ. Equ. **2015**(95), 12 pp. (2015) (electronic)

127. J.V. Elyseeva, Comparison theorems for conjoined bases of linear Hamiltonian differential systems and the comparative index. J. Math. Anal. Appl. **444**(2), 1260–1273 (2016)

128. J.V. Elyseeva, Comparison theorems for weighted focal points of conjoined bases of Hamiltonian difference systems, in *Differential and Difference Equations with Applications, Proceedings of the International Conference on Differential & Difference Equations and Applications* (Amadora, 2015), ed. by S. Pinelas, Z. Došlá, O. Došlý, P.E. Kloeden. Springer Proceedings in Mathematics & Statistics, Vol. 164 (Springer, Berlin, 2016), pp. 225–233

129. J.V. Elyseeva, On symplectic transformations of linear Hamiltonian differential systems without normality. Appl. Math. Lett. **68**, 33–39 (2017)

130. J.V. Elyseeva, The comparative index and transformations of linear Hamiltonian differential systems. Appl. Math. Comput. **330**, 185–200 (2018)

131. J.V. Elyseeva, Oscillation theorems for linear Hamiltonian systems with nonlinear dependence on the spectral parameter and the comparative index. Appl. Math. Lett. **90**, 15–22 (2019)

132. J.V. Elyseeva, A.A. Bondarenko, The Schur complement in an algorithm for calculation of focal points of conjoined bases of symplectic difference systems. Int. J. Pure Appl. Math. **67**(4), 455–474 (2011)

133. J.V. Elyseeva, R. Šimon Hilscher, Discrete oscillation theorems for symplectic eigenvalue problems with general boundary conditions depending nonlinearly on spectral parameter. Linear Algebra Appl. **558**, 108–145 (2018)

134. L. Erbe, P. Yan, Disconjugacy for linear Hamiltonian difference systems. J. Math. Anal. Appl. **167**, 355–367 (1992)

135. L. Erbe, P. Yan, Qualitative properties of Hamiltonian differece systems. J. Math. Anal. Appl. **171**, 334–345 (1992)

136. L. Erbe, P. Yan, Oscillation criteria for Hamiltonian matrix difference systems. Proc. Am. Math. Soc. **119**, 525–533 (1993)

137. L. Erbe, P. Yan, On the discrete Riccati equation and its application to discrete Hamiltonian systems. Rocky Mt. J. Math. **25**, 167–178 (1995)

138. R. Fabbri, R. Johnson, S. Novo, C. Núñez, Some remarks concerning weakly disconjugate linear Hamiltonian systems. J. Math. Anal. Appl. **380**(2), 853–864 (2011)

139. H. Fassbender, *Symplectic Methods for the Symplectic Eigenproblem* (Kluwer, New York. NY, 2000)

140. K. Feng, The Hamiltonian way for computing Hamiltonian dynamics, in *Applied and Industrial Mathematics* (Venice, 1989). Math. Appl., Vol. 56 (Kluwer, Dordrecht, 1991), pp. 17–35

141. K. Feng, M. Qin, *Symplectic Geometric Algorithms for Hamiltonian Systems.* Translated and revised from the Chinese original. With a foreword by Feng Duan, Zhejiang Science and Technology Publishing House, Hangzhou, Springer, Heidelberg, 2010

142. K. Feng, D. Wang, A note on conservation laws of symplectic difference schemes for Hamiltonian systems. J. Comput. Math. **9**(3), 229–237 (1991)

143. K. Feng, H. Wu, M. Qin, Symplectic difference schemes for linear Hamiltonian canonical systems. J. Comput. Math. **8**(4), 371–380 (1990)

144. A. Fischer, C. Remling, The absolutely continuous spectrum of discrete canonical systems. Trans. Am. Math. Soc. **361**(2), 793–818 (2009)

145. S. Flach, A.V. Gorbach, Discrete breathers – advances in theory and applications. Phys. Rep. **467**, 1–116 (2008)

146. T.E. Fortmann, A matrix inversion identity. IEEE Trans. Automat. Control **15**, 599–599 (1970)

147. F.R. Gantmacher, *Lectures in Analytical Mechanics* (MIR Publischers, Moscow, 1975)

148. F.R. Gantmacher, *Theory of Matrices* (AMS Chelsea Publishing, Providence, RI, 1998)

149. C. Gao, R. Ma, Eigenvalues of discrete linear second-order periodic and antiperiodic eigenvalue problems with sign-changing weight. Linear Algebra Appl. **467**, 40–56 (2015)

150. C. Gao, R. Ma, Eigenvalues of discrete Sturm–Liouville problems with eigenparameter dependent boundary conditions. Linear Algebra Appl. **503**, 100–119 (2016)

151. W. Gautschi, Computational aspects of three term recurrence relations. SIAM Rev. **9**, 24–82 (1967)

152. I.M. Gelfand, M.M. Kapranov, A.V. Zelevinsky, *Discriminants, Resultants, and Multidimensional Determinants* (Birkhäuser, Boston, 1994)

153. F. Gesztesy, B. Simon, G. Teschl, Zeros of the Wronskian and renormalized oscillation theory. Am. J. Math. **118**(3), 571–594 (1996)

154. S.K. Godunov, Verification of boundedness for the powers of symplectic matrices with the help of averaging. Sib. Math. J. **33**(6), 939–949 (1992)

155. S.K. Godunov, M. Sadkane, Numerical determination of a canonical form of a symplectic matrix. Sib. Math. J. **42**(4), 629–647 (2001)

156. I.C. Gohberg, M.G. Krein, *Theory and Applications of Volterra Operators in Hilbert Space.* Translations of Mathematical Monographs, Vol. 24 (American Mathematical Society, 1970)

157. G.H. Golub, C.F. Van Loan, *Matrix Computations.* John Hopkins Series in Mathematical Sciences, 2nd edn. (John Hopkins University Press, Baltimore, MD, 1989)

158. L. Greenberg, An oscillation method for fourth order self-adjoint two-point boundary value problems with nonlinear eigenvalues. SIAM J. Math. Anal. **22**, 1021–1042 (1991)

159. L. Greenberg, A Prüfer method for calculating eigenvalues of self-adjoint systems of ordinary differential equations, technical report Tr91-24, University of Maryland, College Park, MD, 1991

160. L. Greenberg, M. Marletta, Algorithm 775: The code SLEUTH for solving fourth-order Sturm–Liouville problems. ACM Trans. Math. Software **23**, 453–493 (1997)

161. L. Greenberg, M. Marletta, Numerical methods for higher order Sturm–Liouville problems. J. Comput. Appl. Math. **125**, 367–383 (2000)

162. D. Greenspan, *Discrete Models* (Addison-Wesley, London, 1973)

163. E. Hairer, S.P. Norsett, G. Wanner, *Ordinary Differential Equations I: Nonstiff Problems* (Springer, New York, NY, 2008)

164. E. Hairer, G. Wanner, *Solving Ordinary Differential Equations II: Stiff and Differential-Algebraic Problems* (Springer, New York, NY, 2010)

165. A. Halanay, V. Răsvan, Stability and boundary value problems for discrete-time linear Hamiltonian systems. Dynam. Syst. Appl. **8**(3–4), 439–459 (1999)

166. B.J. Harmsen, A. Li, Discrete Sturm–Liouville problems with parameter in the boundary conditions. J. Differ. Equ. Appl. **8**(11), 969–981 (2002)

167. B.J. Harmsen, A. Li, Discrete Sturm–Liouville problems with nonlinear parameter in the boundary conditions. J. Differ. Equ. Appl. **13**(7), 639–653 (2007)

168. P. Hartman, *Ordinary Differential Equations* (Wiley, New York, NY, 1964)

169. P. Hartman, Difference equations: disconjugacy, principal solutions, Green's function, complete monotonicity. Trans. Am. Math. Soc. **246**, 1–30 (1978)

170. R.E. Hartwig, A note on the partial ordering of positive semi-definite matrices. Linear Multilinear Algebra **6**(3), 223–226 (1978/79)

171. R. Hilscher, Disconjugacy of symplectic systems and positivity of block tridiagonal matrices. Rocky Mt. J. Math. **29**(4), 1301–1319 (1999)

172. R. Hilscher, Reid roundabout theorem for symplectic dynamic systems on time scales. Appl. Math. Optim. **43**(2), 129–146 (2001)

173. R. Hilscher, V. Růžičková, Implicit Riccati equations and discrete symplectic systems. Int. J. Differ. Equ. **1**, 135–154 (2006)

174. R. Hilscher, V. Růžičková, Riccati inequality and other results for discrete symplectic systems. J. Math. Anal. Appl. **322**(2), 1083–1098 (2006)

175. R. Hilscher, V. Růžičková, Perturbation of time scale quadratic functionals with variable endpoints. Adv. Dyn. Syst. Appl. **2**(2), 207–224 (2007)

176. R. Hilscher, P. Řehák, Riccati inequality, disconjugacy, and reciprocity principle for linear Hamiltonian dynamic systems. Dynam. Syst. Appl. **12**(1–2), 171–189 (2003)

177. R. Hilscher, V. Zeidan, Discrete optimal control: the accessory problem and necessary optimality conditions. J. Math. Anal. Appl. **243**(2), 429–452 (2000)

178. R. Hilscher, V. Zeidan, Second order sufficiency criteria for a discrete optimal control problem. J. Differ. Equ. Appl. **8**(6), 573–602 (2002)

179. R. Hilscher, V. Zeidan, Discrete optimal control: second order optimality conditions. J. Differ. Equ. Appl. **8**(10), 875–896 (2002)

180. R. Hilscher, V. Zeidan, Coupled intervals in the discrete calculus of variations: necessity and sufficiency. J. Math. Anal. Appl. **276**(1), 396–421 (2002)

181. R. Hilscher, V. Zeidan, Symplectic difference systems: variable stepsize discretization and discrete quadratic functionals. Linear Algebra Appl. **367**, 67–104 (2003)

182. R. Hilscher, V. Zeidan, Nonnegativity of a discrete quadratic functional in terms of the (strengthened) Legendre and Jacobi conditions. Comput. Math. Appl. **45**(6–9), 1369–1383 (2003)

183. R. Hilscher, V. Zeidan, A remark on discrete quadratic functionals with separable endpoints. Rocky Mt. J. Math. **33**(4), 1337–1351 (2003)

184. R. Hilscher, V. Zeidan, Coupled intervals in the discrete optimal control. J. Differ. Equ. Appl. **10**(2), 151–186 (2004)

185. R. Hilscher, V. Zeidan, Discrete quadratic functionals with jointly varying endpoints via separable endpoints, in *New Progress in Difference Equations, Proceedings of the Sixth International Conference on Difference Equations* (Augsburg, 2001), ed. by B. Aulbach, S. Elaydi, G. Ladas (Chapman & Hall/CRC, Boca Raton, FL, 2004), pp. 461–470

186. R. Hilscher, V. Zeidan, Nonnegativity and positivity of a quadratic functional in the discrete calculus of variations: A survey. J. Differ. Equ. Appl. **11**(9), 857–875 (2005)

187. R. Hilscher, V. Zeidan, Time scale symplectic systems without normality. J. Differ. Equ. **230**(1), 140–173 (2006)

188. R. Hilscher, V. Zeidan, Coupled intervals for discrete symplectic systems. Linear Algebra Appl. **419**(2–3), 750–764 (2006)

189. R. Hilscher, V. Zeidan, Extension of discrete LQR–problem to symplectic systems. Int. J. Differ. Equ. **2**(2), 197–208 (2007)

190. R. Hilscher, V. Zeidan, Applications of time scale symplectic systems without normality. J. Math. Anal. Appl. **340**(1), 451–465 (2008)

191. R. Hilscher, V. Zeidan, Riccati equations for abnormal time scale quadratic functionals. J. Differ. Equ. **244**(6), 1410–1447 (2008)

192. R. Hilscher, V. Zeidan, Weak maximum principle and accessory problem for control problems on time scales. Nonlinear Anal. **70**(9), 3209–3226 (2009)

193. R. Hilscher, V. Zeidan, Multiplicities of focal points for discrete symplectic systems: revisited. J. Differ. Equ. Appl. **15**(10), 1001–1010 (2009)

194. R. Hilscher, P. Zemánek, Trigonometric and hyperbolic systems on time scales. Dynam. Syst. Appl. **18**(3–4), 483–506 (2009)

195. R.A. Horn, C.R. Johnson, *Topics in Matrix Analysis* (Cambridge University Press, Cambridge, 1991)

196. P. Howard, S. Jung, B. Kwon, The Maslov index and spectral counts for linear Hamiltonian systems on [0, 1]. J. Dynam. Differ. Equ. **30**(4), 1703–1729 (2018)

197. X. Huang, A. Jiang, Z. Zhang, H. Hua, Design and optimization of periodic structure mechanical filter in suppression of foundation resonances. J. Sound Vib. **330**, 4689–4712 (2011)

198. Y. Huanga, Z. Denga, L. Yaoa, An improved symplectic precise integration method for analysis of the rotating rigid-flexible coupled system. J. Sound Vib. **299**, 229–246 (2007)

199. J. Ji, B. Yang, Eigenvalue comparison for boundary value problems for second order difference equations. J. Math. Anal. Appl. **320**(2), 964–972 (2006)

200. R. Johnson, S. Novo, C. Núñez, R. Obaya, Uniform weak disconjugacy and principal solutions for linear Hamiltonian systems, in *Recent Advances in Delay Differential and Difference Equations* (Balatonfuered, Hungary, 2013). Springer Proceedings in Mathematics & Statistics, Vol. 94 (Springer, Berlin, 2014), pp. 131–159

201. R. Johnson, S. Novo, C. Núñez, R. Obaya, Nonautonomous linear-quadratic dissipative control processes without uniform null controllability. J. Dynam. Differ. Equ. **29**(2), 355–383 (2017)

202. R. Johnson, C. Núñez, R. Obaya, Dynamical methods for linear Hamiltonian systems with applications to control processes. J. Dynam. Differ. Equ. **25**(3), 679–713 (2013)

203. R. Johnson, R. Obaya, S. Novo, C. Núñez, R. Fabbri, *Nonautonomous Linear Hamiltonian Systems: Oscillation, Spectral Theory and Control*. Developments in Mathematics, Vol. 36 (Springer, Cham, 2016)

204. W.G. Kelley, A. Peterson, *Difference Equations: An Introduction with Applications* (Academic Press, San Diego, CA, 1991)

205. W. Kratz, *Quadratic Functionals in Variational Analysis and Control Theory*. Mathematical Topics, Vol. 6 (Akademie Verlag, Berlin, 1995)

206. W. Kratz, Banded matrices and difference equations. Linear Algebra Appl. **337**(1–3), 1–20 (2001)

207. W. Kratz, Definitnes of quadratic functionals. Analysis (Munich) **23**, 163–183 (2003)

208. W. Kratz, Discrete oscillation. J. Differ. Equ. Appl. **9**, 127–135 (2003)

209. W. Kratz, R. Šimon Hilscher, Rayleigh principle for linear Hamiltonian systems without controllability. ESAIM Control Optim. Calc. Var. **18**(2), 501–519 (2012)

210. W. Kratz, R. Šimon Hilscher, A generalized index theorem for monotone matrix-valued functions with applications to discrete oscillation theory. SIAM J. Matrix Anal. Appl. **34**(1), 228–243 (2013)

211. W. Kratz, R. Šimon Hilscher, V. Zeidan, Eigenvalue and oscillation theorems for time scale symplectic systems. Int. J. Dyn. Syst. Differ. Equ. **3**(1–2), 84–131 (2011)

212. W. Kratz, M. Tentler, Recursion formulae for the characteristic polynomial of symmetric banded matrices. Linear Algebra Appl. **428**, 2482–2500 (2008)

213. H. Krüger, G. Teschl, Relative oscillation theory, weighted zeros of the Wronskian, and spectral shift function. Comm. Math. Phys. **287**, 613–640 (2009)

214. G. Ladas, E.A. Grove, M.R.S. Kulenovic, Progress report on rational difference equations. J. Differ. Equ. Appl. **10**, 1313–1327 (2004)

215. A.J. Laub, *Matrix Analysis for Scientists and Engineers* (SIAM, Philadelphia, PA, 2005)

216. P.D. Lax, *Linear Algebra*, Pure and Applied Mathematics (New York), A Wiley-Interscience Publication (Wiley, New York, 1997)

217. V. Ledoux, M. Van Daele, G. Vanden Berghe, Efficient computation of high index Sturm–Liouville eigenvalues for problems in physics. Comput. Phys. Comm. **180**, 241–250 (2009)

218. J.V. Lill, T.G. Schmalz, J.C. Light, Imbedded matrix Green's functions in atomic and molecular scattering theory. J. Chem. Phys. **78**, 4456–4463 (1983)

219. W.W. Lin, V. Mehrmann, H. Xu, Canonical forms for Hamiltonian and symplectic matrices and pencils. Linear Algebra Appl. **302–303**, 469–533 (1999)

220. X.-S. Liu, Y.-Y. Qi, J.-F. He, P.-Z. Ding, Recent progress in symplectic algorithms for use in quantum systems. Commun. Comput. Phys. **2**(1), 1–53 (2007)

221. Y. Liu, Y. Shi, Regular approximations of spectra of singular discrete linear Hamiltonian systems with one singular endpoint. Linear Algebra Appl. **451**, 94–130 (2018)

222. V. Loan, A Symplectic method for approximating all the eigenvalues of a Hamiltonian matrix. Linear Algebra Appl. **61**, 233–251 (1984)

223. D.G. Luenberger, *Linear and Nonlinear Programming*, 2nd edn. (Addison-Wesley, Reading, MA, 1984)

224. D.S. Mackey, N. Mackey, F. Tisseur, Structured tools for structured matrices. Electron. J. Linear Algebra **10**, 106–145 (2003)

225. B. Marinković, Optimality conditions in discrete optimal control problems with state constraints. Numer. Funct. Anal. Optim. **28**(7–8), 945–955 (2007)

226. B. Marinković, Optimality conditions for discrete optimal control problems with equality and inequality type of constraints. Positivity **12**(3), 535–545 (2008)

227. B. Marinković, Second order optimality conditions in a discrete optimal control problem. Optimization **57**(4), 539–548 (2008)

228. C.R. Maple, M. Marletta, Solving Hamiltonian systems arising from ODE eigenproblems. Numer. Algorithms **22**(3), 263–284 (1999)

229. M. Marletta, Automatic solution of regular and singular vector Sturm–Liouville problems. Numer. Algorithms **4**, 65–99 (1993)

230. M. Marletta, Numerical solution of eigenvalue problems for Hamiltonian systems. Adv. Comput. Math. **2**(2), 155–184 (1994)

231. G. Marsaglia, G.P.H. Styan, Equalities and inequalities for ranks of matrices. Linear Multilinear Algebra **2**, 269–292 (1974)

232. J. Marsden, M. West, Discrete mechanics and variational integrators. Acta Numer. **10**, 357–514 (2001)

233. J. McMahona, S. Grayb, G. Schatza, A discrete action principle for electrodynamics and the construction of explicit symplectic integrators for linear, non-dispersive media. J. Comput. Phys. **228**, 3421–3432 (2009)

234. V. Mehrmann, A symplectic orthogonal method for single input or single output discrete time optimal quadratic control problems. SIAM J. Matrix Anal. Appl. **9**, 221–247 (1988)

235. S.J. Monaquel, K.M. Schmidt, On $M$-functions and operator theory for non-self-adjoint discrete Hamiltonian systems. J. Comput. Appl. Math. **208**(1), 82–101 (2007)
236. M. Morse, *Variational Analysis: Critical Extremals and Sturmian Extensions* (Willey, New York, NY, 1973)
237. P. Nelson, On the effectiveness of the inverse Riccati transformation in the matrix case. J. Math. Anal. Appl. **67**, 201–210 (1978)
238. F.W.J. Olver, D.J. Sookne, Note on backward recurrence algorithms. Math. Comp. **26**, 941–947 (1972)
239. C. Paige, C. Van Loan, A Schur decomposition for Hamiltonian matrices. Linear Algebra Appl. **41**, 11–32 (1981)
240. H.J. Peng, Q. Gao, Z.G. Wu, W.X. Zhong, Symplectic adaptive algorithm for solving nonlinear two-point boundary value problems in astrodynamics. Celestial Mech. Dynam. Astronom. **110**(4), 319–342 (2011)
241. C.H. Rasmussen, Oscillation and asymptotic behaviour of systems of ordinary linear differential equations. Trans. Am. Math. Soc. **256**, 1–49 (1979)
242. V. Răsvan, Stability zones for discrete time Hamiltonian systems. Arch. Math. (Brno) **36**(5), 563–573 (2000) (electronic)
243. V. Răsvan, Stability zones and parametric resonance for discrete-time Hamiltonian systems. *Dynamic Systems and Applications*, Vol. 4 (Dynamic, Atlanta, GA, 2004), pp. 367–373
244. V. Răsvan, On stability zones for discrete-time periodic linear Hamiltonian systems. Adv. Differ. Equ. **2006**(Art. 80757), 1–13 (2006)
245. W.T. Reid, A Prüfer transformation for differential systems. Pacific J. Math. **8**, 575–584 (1958)
246. W.T. Reid, Riccati matrix differential equations and non-oscillation criteria for associated linear differential systems. Pacific J. Math. **13**, 665–685 (1963)
247. W.T. Reid, Generalized polar coordinate transformations for differential systems. Rocky Mountain J. Math. **1**(2), 383–406 (1971)
248. W.T. Reid, *Ordinary Differential Equations* (Wiley, New York, NY, 1971)
249. W.T. Reid, *Riccati Differential Equations* (Academic Press, New York, NY, 1972)
250. W.T. Reid, *Sturmian Theory for Ordinary Differential Equations* (Springer, New York, NY, 1980)
251. G. Ren, On the density of the minimal subspaces generated by discrete linear Hamiltonian systems. Appl. Math. Lett. **27**, 1–5 (2014)
252. G. Ren, Y. Shi, Defect indices and definiteness conditions for a class of discrete linear Hamiltonian systems. Appl. Math. Comput. **218**(7), 3414–3429 (2011)
253. G. Ren, Y. Shi, Self-adjoint extensions for discrete linear Hamiltonian systems. Linear Algebra Appl. **454**, 1–48 (2014)
254. F.S. Rofe-Beketov, A.M. Kholkin, On the connection between spectral and oscillation properties of the Sturm–Liouville matrix problem. Math. USSR-Sb. **31**(3), 365–378 (1977)
255. F.S. Rofe-Beketov, A.M. Kholkin, *Spectral Analysis of Differential Operators*. World Scientific Monograph Series in Mathematics, Vol. 7 (World Scientific, Hackensack, NJ, 2005)
256. R.D. Ruth, A canonical integration technique. IEEE Trans. Nuclear Sci. **30**, 2669–2671 (1983)
257. V. Růžičková, *Discrete Symplectic Systems and Definiteness of Quadratic Functionals*, PhD dissertation, Masaryk University, Brno, 2006. Available at https://is.muni.cz/th/p9iz7/?lang=en
258. V. Růžičková, Perturbation of discrete quadratic functionals. Tatra Mountains Math. Publ. **38**(1), 229–241 (2007)
259. P. Řehák, Oscillatory properties of second order half-linear difference equations. Czechoslovak Math. J. **51(126)**(2), 303–321 (2001)
260. J.M. Sanz-Serna, A. Portillo, Classical numerical integrators for wave-packet dynamics. J. Chem. Phys. **104**(6). 2349–2355 (1996)
261. H. Schulz-Baldes, Sturm intersection theory for periodic Jacobi matrices and linear Hamiltonian systems. Linear Algebra Appl. **436**(3), 498–515 (2012)

262. G. Shi, H. Wu, Spectral theory of Sturm–Liouville difference operators. Linear Algebra Appl. **430**, 830–846 (2009)
263. Y. Shi, Symplectic structure of discrete Hamiltonian systems. J. Math. Anal. Appl. **266**(2), 472–478 (2002)
264. Y. Shi, Spectral theory of discrete linear Hamiltonian systems. J. Math. Anal. Appl. **289**(2), 554–570 (2004)
265. Y. Shi, Weyl–Titchmarsh theory for a class of discrete linear Hamiltonian systems. Linear Algebra Appl. **416**, 452–519 (2006)
266. Y. Shi, Transformations for complex discrete linear Hamiltonian and symplectic systems. Bull. Aust. Math. Soc. **75**(2), 179–191 (2007)
267. Y. Shi, S. Chen, Spectral theory of second-order vector difference equations. J. Math. Anal. Appl. **239**(2), 195–212 (1999)
268. Y. Shi, S. Chen, Spectral theory of higher-order discrete vector Sturm—Liouville problems. Linear Algebra Appl. **323**, 7–36 (2001)
269. Y. Shi, C. Shao, G. Ren, Spectral properties of self-adjoint subspaces. Linear Algebra Appl. **438**(1), 191–218 (2013)
270. G.W. Stewart, J. Sun, *Matrix Perturbation Theory* (Academic Press, Boston, MA, 1990)
271. H. Sun, Q. Kong, Y. Shi, Essential spectrum of singular discrete linear Hamiltonian systems. Math. Nachr. **289**(2–3), 343–359 (2016)
272. H. Sun, Y. Shi, Strong limit point criteria for a class of singular discrete linear Hamiltonian systems. J. Math. Anal. Appl. **336**(1), 224–242 (2007)
273. H. Sun, Y. Shi, Self-adjoint extensions for linear Hamiltonian systems with two singular endpoints. J. Funct. Anal. **259**(8), 2003–2027 (2010)
274. H. Sun, Y. Shi, Self-adjoint extensions for singular linear Hamiltonian systems. Math. Nachr. **284**(5–6), 797–814 (2011)
275. H. Sun, Y. Shi, Spectral properties of singular discrete linear Hamiltonian systems. J. Differ. Equ. Appl. **20**(3), 379–405 (2014)
276. H. Sun, Y. Shi, On essential spectra of singular linear Hamiltonian systems. Linear Algebra Appl. **469**, 204–229 (2015)
277. S. Sun, Y. Shi, S. Chen, The Glazman–Krein–Naimark theory for a class of discrete Hamiltonian systems. J. Math. Anal. Appl. **327**(2), 1360–1380 (2007)
278. N.G. Stephen, Transfer matrix analysis of the elastostatics of one-dimensional repetitive structures. Proc. R. Soc. Lond. Ser. A Math. Phys. Eng. Sci. **462**(2072), 2245–2270 (2006)
279. J.C.F. Sturm, Memoire sur une classe d'équations différénces partielles. J. Math. Pures Appl. **1**, 373–444 (1836)
280. C.A. Swanson, *Comparison and Oscillation Theory of Linear Differential Equations* (Academic Press, New York, NY, 1968)
281. P. Šepitka, *Theory of Principal Solutions at Infinity for Linear Hamiltonian Systems*, PhD dissertation, Masaryk University, Brno, 2014. Available at https://is.muni.cz/th/vqad7/?lang=en
282. P. Šepitka, Riccati equations for linear Hamiltonian systems without controllability condition. Discrete Contin. Dyn. Syst. **39**(4), 1685–1730 (2019)
283. P. Šepitka, R. Šimon Hilscher, Minimal principal solution at infinity of nonoscillatory linear Hamiltonian systems. J. Dynam. Differ. Equ. **26**(1), 57–91 (2014)
284. P. Šepitka, R. Šimon Hilscher, Recessive solutions for nonoscillatory discrete symplectic systems. Linear Algebra Appl. **469**, 243–275 (2015)
285. P. Šepitka, R. Šimon Hilscher, Principal solutions at infinity of given ranks for nonoscillatory linear Hamiltonian systems. J. Dynam. Differ. Equ. **27**(1), 137–175 (2015)
286. P. Šepitka, R. Šimon Hilscher, Principal and antiprincipal solutions at infinity of linear Hamiltonian systems. J. Differ. Equ. **259**(9), 4651–4682 (2015)
287. P. Šepitka, R. Šimon Hilscher, Genera of conjoined bases of linear Hamiltonian systems and limit characterization of principal solutions at infinity. J. Differ. Equ. **260**(8), 6581–6603 (2016)

288. P. Šepitka, R. Šimon Hilscher, Reid's construction of minimal principal solution at infinity for linear Hamiltonian systems, in *Differential and Difference Equations with Applications, Proceedings of the International Conference on Differential & Difference Equations and Applications* (Amadora, 2015), S. Pinelas, Z. Došlá, O. Došlý, P.E. Kloeden, (eds.), Springer Proceedings in Mathematics & Statistics, Vol. 164 (Springer, Berlin, 2016), pp. 359–369

289. P. Šepitka, R. Šimon Hilscher, Comparative index and Sturmian theory for linear Hamiltonian systems. J. Differ. Equ. **262**(2), 914–944 (2017)

290. P. Šepitka, R. Šimon Hilscher, Dominant and recessive solutions at infinity and genera of conjoined bases for discrete symplectic systems. J. Differ. Equ. Appl. **23**(4), 657–698 (2017)

291. P. Šepitka, R. Šimon Hilscher, Focal points and principal solutions of linear Hamiltonian systems revisited. J. Differ. Equ. **264**(9), 5541–5576 (2018)

292. P. Šepitka, R. Šimon Hilscher, Singular Sturmian separation theorems for nonoscillatory symplectic difference systems. J. Differ. Equ. Appl. **24**(12), 1894–1934 (2018)

293. P. Šepitka, R. Šimon Hilscher, Singular Sturmian separation theorems on unbounded intervals for linear Hamiltonian systems. J. Differ. Equ. **266**(11), 7481–7524 (2019)

294. R. Šimon Hilscher, A note on the time scale calculus of variations problems, in *Ulmer Seminare über Funktionalanalysis und Differentialgleichungen*, Vol. 14 (University of Ulm, Ulm, 2009), pp. 223–230

295. R. Šimon Hilscher, Sturmian theory for linear Hamiltonian systems without controllability. Math. Nachr. **284**(7), 831–843 (2011)

296. R. Šimon Hilscher, On general Sturmian theory for abnormal linear Hamiltonian systems, in: *Dynamical Systems, Differential Equations and Applications, Proceedings of the 8th AIMS Conference on Dynamical Systems, Differential Equations and Applications* (Dresden, 2010), W. Feng, Z. Feng, M. Grasselli, A. Ibragimov, X. Lu, S. Siegmund, J. Voigt (eds.), *Discrete Contin. Dynam. Systems, Suppl. 2011* (American Institute of Mathematical Sciences (AIMS), Springfield, MO, 2011), pp. 684–691

297. R. Šimon Hilscher, Oscillation theorems for discrete symplectic systems with nonlinear dependence in spectral parameter. Linear Algebra Appl. **437**(12), 2922–2960 (2012)

298. R. Šimon Hilscher, Spectral and oscillation theory for general second order Sturm–Liouville difference equations. Adv. Differ. Equ. **2012**(82), 19 pp. (2012)

299. R. Šimon Hilscher, Eigenvalue theory for time scale symplectic systems depending nonlinearly on spectral parameter. Appl. Math. Comput. **219**(6), 2839–2860 (2012)

300. R. Šimon Hilscher, Asymptotic properties of solutions of Riccati matrix equations and inequalities for discrete symplectic systems. Electron. J. Qual. Theory Differ. Equ. **2015**(54), 16 pp. (2015) (electronic)

301. R. Šimon Hilscher, Eigenvalue comparison for discrete symplectic systems, in *Difference Equations, Discrete Dynamical Systems and Applications, Proceedings of the 20th International Conference on Difference Equations and Applications* (Wuhan, 2014), ed. by M. Bohner, Y. Ding, O. Došlý (Springer, Berlin, 2015), pp. 95–107

302. R. Šimon Hilscher, V. Zeidan, Picone type identities and definiteness of quadratic functionals on time scales. Appl. Math. Comput. **215**(7), 2425–2437 (2009)

303. R. Šimon Hilscher, V. Zeidan, Symplectic structure of Jacobi systems on time scales. Int. J. Differ. Equ. **5**(1), 55–81 (2010)

304. R. Šimon Hilscher, V. Zeidan, Symmetric three-term recurrence equations and their symplectic structure. Adv. Differ. Equ. **2010**(Article ID 626942), 17 pp. (2010)

305. R. Šimon Hilscher, V. Zeidan, Oscillation theorems and Rayleigh principle for linear Hamiltonian and symplectic systems with general boundary conditions. Appl. Math. Comput. **218**(17), 8309–8328 (2012)

306. R. Šimon Hilscher, P. Zemánek, Definiteness of quadratic functionals for Hamiltonian and symplectic systems: A survey. Int. J. Differ. Equ. **4**(1), 49–67 (2009)

307. R. Šimon Hilscher, P. Zemánek, New results for time reversed symplectic dynamic systems and quadratic functionals, in *Proceedings of the 9th Colloquium on Qualitative Theory of Differential Equations*, Vol. 9, No. 15, Electron. J. Qual. Theory Differ. Equ. (electronic) (Szeged, 2012), 11 pp.

308. R. Šimon Hilscher, P. Zemánek, Weyl disks and square summable solutions for discrete symplectic systems with jointly varying endpoints. Adv. Differ. Equ. **2013**(232), 18 pp. (2013)

309. R. Šimon Hilscher, P. Zemánek, Weyl–Titchmarsh theory for discrete symplectic systems with general linear dependence on spectral parameter. J. Differ. Equ. Appl. **20**(1), 84–117 (2014)

310. R. Šimon Hilscher, P. Zemánek, Limit point and limit circle classification for symplectic systems on time scales. Appl. Math. Comput. **233**, 623–646 (2014)

311. R. Šimon Hilscher, P. Zemánek, Generalized Lagrange identity for discrete symplectic systems and applications in Weyl–Titchmarsh theory, in *Theory and Applications of Difference Equations and Discrete Dynamical Systems, Proceedings of the 19th International Conference on Difference Equations and Applications* (Muscat, 2013), ed. by Z. AlSharawi, J. Cushing, S. Elaydi. Springer Proceedings in Mathematics & Statistics, Vol. 102 (Springer, Berlin, 2014), pp. 187–202

312. R. Šimon Hilscher, P. Zemánek, Time scale symplectic systems with analytic dependence on spectral parameter. J. Differ. Equ. Appl. **21**(3), 209–239 (2015)

313. R. Šimon Hilscher, P. Zemánek, Limit circle invariance for two differential systems on time scales. Math. Nachr. **288**(5–6), 696–709 (2015)

314. G. Teschl, Oscillation theory and renormalized oscillation theory for Jacobi operators. J. Differ. Equ. **129**, 532–558 (1996)

315. G. Teschl, *Jacobi Operators and Completely Integrable Nonlinear Lattices* (AMS Mathematical Surveys and Monographs, Providence, RI, 1999)

316. Y. Tian, Equalities and inequalities for inertias of Hermitian matrices with applications. Linear Algebra Appl. **433**, 263–296 (2010)

317. Y. Tian, Rank and inertia of submatrices of the Moore–Penrose inverse of a Hermitian matrix. Electron. J. Linear Algebra **20**, 226–240 (2010)

318. P. Van Dooren, The computation of Kronecker's canonical form of a singular pencil. Linear Algebra Appl. **27**, 103–140 (1979)

319. P. Van Dooren, P. Dewilde, The eigenstructure of an arbitrary polynomial matrix. Computational aspects. Linear Algebra Appl. **50**, 545–579 (1983)

320. G. Verghese, P. Van Dooren, T. Kailath, Properties of the system matrix of a generalized state-space system. Int. J. Control **30**(2), 235–243 (1979)

321. M. Wahrheit, Eigenvalue problems and oscillation of linear Hamiltonian systems. Int. J. Differ. Equ. **2**, 221–244 (2007)

322. Y. Wang, Y. Shi, Eigenvalues of second-order difference equations with periodic and antiperiodic boundary conditions. J. Math. Anal. Appl. **309**(1), 56–69 (2005)

323. Y. Wang, Y. Shi, G. Ren, Transformations for complex discrete linear Hamiltonian and symplectic systems. Bull. Aust. Math. Soc. **75**(2), 179–191 (2007)

324. Y. Wu, Symplectic transformation and symplectic difference schemes. Chin. J. Numer. Math. Appl. **12**(1), 23–31 (1990)

325. L. Xue-Shen, Q. Yue-Ying, H. Jian-Feng, D. Pei-Zhu, Recent progress in symplectic algorithms for use in quantum systems. Commun. Comput. Phys. **2**(1), 1–53 (2007)

326. V.A. Yakubovich, Arguments on the group of symplectic matrices. Mat. Sb. **55**, 255–280 (1961)

327. V.A. Yakubovich, Oscillatory properties of solutions of canonical equations. Mat. Sb. **56**, 3–42 (1962)

328. V.A. Yakubovich, V.M. Starzhinskii, *Linear Differential Equations with Periodic Coefficients*, 2 volumes (Wiley, New York, 1975)

329. Y. Yalçin, L. Gören Sümer, S. Kurtulan, Discrete-time modeling of Hamiltonian systems. Turk. J. Electr. Eng. Comput. Sci. **23**, 149–170 (2015)

330. V. Zeidan, Continuous versus discrete nonlinear optimal control problems, in *Proceedings of the 14th International Conference on Difference Equations and Applications* (Istanbul, 2008), ed. by M. Bohner, Z. Došlá, G. Ladas, M. Ünal, A. Zafer (Uğur-Bahçeşehir University Publishing Company, Istanbul, 2009), pp. 73–93

331. V. Zeidan, Constrained linear-quadratic control problems on time scales and weak normality. Dynam. Syst. Appl. **26**(3–4), 627–662 (2017)

332. M.I. Zelikin, *Control Theory and Optimization. I. Homogeneous Spaces and the Riccati Equation in the Calculus of Variations*. Encyclopaedia of Mathematical Sciences, Vol. 86 (Springer, Berlin, 2000)

333. P. Zemánek, Discrete trigonometric and hyperbolic systems: An overview, in *Ulmer Seminare über Funktionalanalysis und Differentialgleichungen*, Vol. 14 (University of Ulm, Ulm, 2009), pp. 345–359

334. P. Zemánek, Rofe-Beketov formula for symplectic systems. Adv. Differ. Equ. **2012**(104), 9 pp. (2012)

335. P. Zemánek, S. Clark, Characterization of self-adjoint extensions for discrete symplectic systems. J. Math. Anal. Appl. **440**(1), 323–350 (2016)

336. A. Zettl, *Sturm–Liouville Theory* (AMS Mathematical Surveys and Monographs, Providence, RI, 2005)

337. Z. Zheng, Invariance of deficiency indices under perturbation for discrete Hamiltonian systems. J. Differ. Equ. Appl. **19**(8), 1243–1250 (2013)

# Index

© Springer Nature Switzerland AG 2019

O. Došlý et al., *Symplectic Difference Systems: Oscillation and Spectral Theory*,
Pathways in Mathematics, https://doi.org/10.1007/978-3-030-19373-7

Printed in the United States
By Bookmasters